HANDBOOK
OF
PROOF THEORY

STUDIES IN LOGIC

AND

THE FOUNDATIONS OF MATHEMATICS

VOLUME 137

ELSEVIER

AMSTERDAM • LAUSANNE • NEW YORK • OXFORD • SHANNON • SINGAPORE • TOKYO

HANDBOOK OF PROOF THEORY

Edited by

SAMUEL R. BUSS

University of California, San Diego

1998

ELSEVIER

AMSTERDAM • LAUSANNE • NEW YORK • OXFORD • SHANNON • SINGAPORE • TOKYO

ELSEVIER SCIENCE B.V.
Sara Burgerhartstraat 25
P.O. Box 211, 1000 AE Amsterdam
The Netherlands

Library of Congress Cataloging-in-Publication Data

Handbook of proof theory / edited by Samuel R. Buss.
p. cm. -- (Studies in logic and the foundations of mathematics ; v. 137)
Includes bibliographic references and indexes.
ISBN 0-444-89840-9 (alk. paper)
1. Proof theory. I. Buss, Samuel R. II. Series.
QA9.54.H35 1998
511.3--dc21 98-18922
CIP

ISBN: 0-444-89840-9

⊗ The paper used in this publication meets the requirements of ANSI/NISO Z39.48-1992 (Permanence of Paper).

Transferred to digital printing 2005

Preface

Proof theory is the study of proofs as formal objects and is concerned with a broad range of related topics. It is one of the central topics of mathematical logic and has applications in many areas of mathematics, philosophy, and computer science. Historically, proof theory was developed by mathematicians and philosophers as a formalization for mathematical reasoning; however, proof theory has gradually become increasingly important for computer science, and nowadays proof theory and theoretical computer science are recognized as being very closely connected.

This volume contains articles covering a broad spectrum of proof theory, with an emphasis on its mathematical aspects. The articles should not only be interesting to specialists in proof theory, but should also be accessible to a diverse audience, including logicians, mathematicians, computer scientists and philosophers. We have attempted to include many of the central topics in proof theory; but have opted to have self-contained expository articles, rather than to have encyclopedic coverage. Thus, a number of important topics have been largely omitted, but with the advantage that the included material is covered in more detail and at greater depth.

The chapters are arranged so that the two introductory articles come first; these are then followed by articles from the core classical areas of proof theory; finally the handbook ends with articles that deal with topics closely related to computer science.

This handbook was initiated at the suggestion of the publisher, as a partial successor to the very successful *Handbook of Mathematical Logic*, edited by J. Barwise. Only one quarter of the 1977 *Handbook of Mathematical Logic* was devoted to proof theory, and since then there has been considerable progress in this area; as a result, there is remarkably little overlap between the contents of the *Handbook of Mathematical Logic* and the present volume.

Sam Buss
La Jolla, California
November 1997

List of Contributors

J. Avigad, *Carnegie Mellon University* (Ch. V)

S.R. Buss, *University of California, San Diego* (Ch. I and II)

R.L. Constable, *Cornell University* (Ch. X)

M. Fairtlough, *University of Sheffield* (Ch. III)

S. Feferman, *Stanford University* (Ch. V)

G. Jäger, *Universität Bern* (Ch. IX)

G. Japaridze, *University of Pennsylvania* (Ch. VII)

D. de Jongh, *University of Amsterdam* (Ch. VII)

P. Pudlák, *Academy of Sciences of the Czech Republic* (Ch. VIII)

W. Pohlers, *Westfälische Wilhelms-Universität* (Ch. IV)

R.F. Stärk, *Universität Freiburg* (Ch. IX)

A.S. Troelstra, *University of Amsterdam* (Ch. VI)

S.S. Wainer, *University of Leeds* (Ch. III)

Table of Contents

CHAPTER I

An Introduction to Proof Theory

Samuel R. Buss
Departments of Mathematics and Computer Science, University of California, San Diego
La Jolla, California 92093-0112, USA

Contents

HANDBOOK OF PROOF THEORY
Edited by S. R. Buss

Proof Theory is the area of mathematics which studies the concepts of mathematical proof and mathematical provability. Since the notion of "proof" plays a central role in mathematics as the means by which the truth or falsity of mathematical propositions is established; Proof Theory is, in principle at least, the study of the foundations of all of mathematics. Of course, the use of Proof Theory as a foundation for mathematics is of necessity somewhat circular, since Proof Theory is itself a subfield of mathematics.

There are two distinct viewpoints of what a mathematical proof is. The first view is that proofs are social conventions by which mathematicians convince one another of the truth of theorems. That is to say, a proof is expressed in natural language plus possibly symbols and figures, and is sufficient to convince an expert of the correctness of a theorem. Examples of social proofs include the kinds of proofs that are presented in conversations or published in articles. Of course, it is impossible to precisely define what constitutes a valid proof in this social sense; and, the standards for valid proofs may vary with the audience and over time. The second view of proofs is more narrow in scope: in this view, a proof consists of a string of symbols which satisfy some precisely stated set of rules and which prove a theorem, which itself must also be expressed as a string of symbols. According to this view, mathematics can be regarded as a 'game' played with strings of symbols according to some precisely defined rules. Proofs of the latter kind are called "formal" proofs to distinguish them from "social" proofs.

In practice, social proofs and formal proofs are very closely related. Firstly, a formal proof can serve as a social proof (although it may be very tedious and unintuitive) provided it is formalized in a proof system whose validity is trusted. Secondly, the standards for social proofs are sufficiently high that, in order for a proof to be socially accepted, it should be possible (in principle!) to generate a formal proof corresponding to the social proof. Indeed, this offers an explanation for the fact that there are generally accepted standards for social proofs; namely, the implicit requirement that proofs can be expressed, in principle, in a formal proof system enforces and determines the generally accepted standards for social proofs.

Proof Theory is concerned almost exclusively with the study of formal proofs: this is justified, in part, by the close connection between social and formal proofs, and it is necessitated by the fact that only formal proofs are subject to mathematical analysis. The principal tasks of Proof Theory can be summarized as follows. First, to formulate systems of logic and sets of axioms which are appropriate for formalizing mathematical proofs and to characterize what results of mathematics follow from certain axioms; or, in other words, to investigate the proof-theoretic strength of particular formal systems. Second, to study the structure of formal proofs; for instance, to find normal forms for proofs and to establish syntactic facts about proofs. This is the study of proofs as objects of independent interest. Third, to study what kind of additional information can be extracted from proofs beyond the truth of the theorem being proved. In certain cases, proofs may contain computational or constructive information. Fourth, to study how best to construct formal proofs; e.g., what kinds of proofs can be efficiently generated by computers?

The study of Proof Theory is traditionally motivated by the problem of formalizing mathematical proofs; the original formulation of first-order logic by Frege [1879] was the first successful step in this direction. Increasingly, there have been attempts to extend Mathematical Logic to be applicable to other domains; for example, intuitionistic logic deals with the formalization of constructive proofs, and logic programming is a widely used tool for artificial intelligence. In these and other domains, Proof Theory is of central importance because of the possibility of computer generation and manipulation of formal proofs.

This handbook covers the central areas of Proof Theory, especially the mathematical aspects of Proof Theory, but largely omits the philosophical aspects of proof theory. This first chapter is intended to be an overview and introduction to mathematical proof theory. It concentrates on the proof theory of *classical* logic, especially propositional logic and first-order logic. This is for two reasons: firstly, classical first-order logic is by far the most widely used framework for mathematical reasoning, and secondly, many results and techniques of classical first-order logic frequently carryover with relatively minor modifications to other logics.

This introductory chapter will deal primarily with the sequent calculus, and resolution, and to lesser extent, the Hilbert-style proof systems and the natural deduction proof system. We first examine proof systems for propositional logic, then proof systems for first-order logic. Next we consider some applications of cut elimination, which is arguably the central theorem of proof theory. Finally, we review the proof theory of some non-classical logics, including intuitionistic logic and linear logic.

1. Proof theory of propositional logic

Classical propositional logic, also called sentential logic, deals with sentences and propositions as abstract units which take on distinct True/False values. The basic syntactic units of propositional logic are *variables* which represent atomic propositions which may have value either *True* or *False.* Propositional variables are combined with Boolean functions (also called connectives): a *k-ary Boolean function* is a mapping from $\{T, F\}^k$ to $\{T, F\}$ where we use T and F to represent *True* and *False.* The most frequently used examples of Boolean functions are the connectives $\top$ and $\bot$ which are the 0-ary functions with values T and F, respectively; the binary connectives $\land$, $\lor$, $\supset$, $\leftrightarrow$ and $\oplus$ for "and", "or", "if-then", "if-and-only-if" and "parity"; and the unary connective $\neg$ for negation. Note that $\lor$ is the inclusive-or and $\oplus$ is the exclusive-or.

We shall henceforth let the set of propositional variables be $V = \{p_1, p_2, p_3, \ldots\}$; however, our theorems below hold also for uncountable sets of propositional variables. The set of formulas is inductively defined by stating that every propositional variable is a formula, and that if A and B are formulas, then $(\neg A)$, $(A \land B)$, $(A \lor B)$, $(A \supset B)$, etc., are formulas. A *truth assignment* consists of an assignment of True/False values to the propositional variables, i.e., a truth assignment is a mapping $\tau : V \to \{T, F\}$.

A truth assignment can be extended to have domain the set of all formulas in the obvious way, according to Table 1; we write $\overline{\tau}(A)$ for the truth value of the formula A induced by the truth assignment τ.

Table 1
Values of a truth assignment $\overline{\tau}$

A	B	$(\neg A)$	$(A \land B)$	$(A \lor B)$	$(A \supset B)$	$(A \leftrightarrow B)$	$(A \oplus B)$
T	T	F	T	T	T	T	F
T	F	F	F	T	F	F	T
F	T	T	F	T	T	F	T
F	F	T	F	F	T	T	F

A formula A involving only variables among $p_1, \ldots, p_k$ defines a k-ary Boolean function f_A, by letting $f_A(x_1, ..., x_k)$ equal the truth value $\overline{\tau}(A)$ where $\tau(p_i) = x_i$ for all i. A *language* is a set of connectives which may be used in the formation of L*-formulas.* A language L is *complete* if and only if every Boolean function can be defined by an L-formula. Propositional logic can be formulated with any complete (usually finite) language L — for the time being, we shall use the language $\neg$, $\land$, $\lor$ and $\supset$.

A propositional formula A is said to be a *tautology* or to be *(classically) valid* if A is assigned the value T by every truth assignment. We write $\models A$ to denote that A is a tautology. The formula A is *satisfiable* if there is some truth assignment that gives it value T. If Γ is a set of propositional formulas, then Γ is *satisfiable* if there is some truth assignment that simultaneously satisfies all members of Γ. We say Γ *tautologically implies* A, or $\Gamma \models A$, if every truth assignment which satisfies Γ also satisfies A.

One of the central problems of propositional logic is to find useful methods for recognizing tautologies; since A is a tautology if and only if $\neg A$ is not satisfiable, this is essentially the same as the problem of finding methods for recognizing satisfiable formulas. Of course, the set of tautologies is decidable, since to verify that a formula A with n distinct propositional variables is a tautology, one need merely check that the 2^n distinct truth assignments to these variables all give A the value T. This brute-force 'method of truth-tables' is not entirely satisfactory; firstly, because it can involve an exorbitant amount of computation, and secondly, because it provides no intuition as to *why* the formula is, or is not, a tautology.

For these reasons, it is often advantageous to *prove* that A is a tautology instead of using the method of truth-tables. The next three sections discuss three commonly used propositional proof systems. The so-called Frege proof systems are perhaps the most widely used and are based on modus ponens. The sequent calculus systems provide an elegant proof system which combines both the possibility of elegant proofs and the advantage of an extremely useful normal form for proofs. The resolution refutation proof systems are designed to allow for efficient computerized search for proofs. Later, we will extend these three systems to first-order logic.

1.1. Frege proof systems

The mostly commonly used propositional proof systems are based on the use of *modus ponens* as the sole rule of inference. Modus ponens is the inference rule, which allows, for arbitrary A and B, the formula B to be inferred from the two hypotheses $A \supset B$ and A; this is pictorially represented as

$$\frac{A \qquad A \supset B}{B}$$

In addition to this rule of inference, we need *logical axioms* that allow the inference of 'self-evident' tautologies from no hypotheses. There are many possible choices for sets of axioms: obviously, we wish to have a sufficiently strong set of axioms so that every tautology can be derived from the axioms by use of modus ponens. In addition, we wish to specify the axioms by a finite set of schemes.

1.1.1. Definition. A *substitution* σ is a mapping from the set of propositional variables to the set of propositional formulas. If A is a propositional formula, then the result of applying σ to A is denoted $A\sigma$ and is equal to the formula obtained by simultaneously replacing each variable appearing in A by its image under σ.

An example of a set of axiom schemes over the language $\neg$, $\wedge$, $\vee$ and $\supset$ is given in the next definition. We adopt conventions for omitting parentheses from descriptions of formulas by specifying that unary operators have the highest precedence, the connectives $\wedge$ and $\vee$ have second highest precedence, and that $\supset$ and $\leftrightarrow$ have lowest precedence. All connectives of the same precedence are to be associated from right to left; for example, $A \supset \neg B \supset C$ is a shorthand representation for the formula $(A \supset ((\neg B) \supset C))$.

1.1.2. Definition. Consider the following set of axiom schemes:

$$\begin{array}{ll}
p_1 \supset (p_2 \supset p_1) & (p_1 \supset p_2) \supset (p_1 \supset \neg p_2) \supset \neg p_1 \\
(p_1 \supset p_2) \supset (p_1 \supset (p_2 \supset p_3)) \supset (p_1 \supset p_3) & (\neg\neg p_1) \supset p_1 \\
p_1 \supset p_1 \vee p_2 & p_1 \wedge p_2 \supset p_1 \\
p_2 \supset p_1 \vee p_2 & p_1 \wedge p_2 \supset p_2 \\
(p_1 \supset p_3) \supset (p_2 \supset p_3) \supset (p_1 \vee p_2 \supset p_3) & p_1 \supset p_2 \supset p_1 \wedge p_2
\end{array}$$

The propositional proof system $\mathcal{F}$ is defined to have as its axioms every substitution instance of the above formulas and to have modus ponens as its only rule. An $\mathcal{F}$-proof of a formula A is a sequence of formulas, each of which is either an $\mathcal{F}$-axiom or is inferred by modus ponens from two earlier formulas in the proof, such that the final formula in the proof is A.

We write $\vdash_{\mathcal{F}} A$, or just $\vdash A$, to say that A has an $\mathcal{F}$-proof. We write $\Gamma \vdash_{\mathcal{F}} A$, or just $\Gamma \vdash A$, to say that A has a proof in which each formula either is deduced according the axioms or inference rule of $\mathcal{F}$ or is in Γ. In this case, we say that A is proved from the extra-logical hypotheses Γ; note that Γ may contain formulas which are not tautologies.

1.1.3. Soundness and completeness of $\mathcal{F}$. It is easy to prove that every $\mathcal{F}$-provable formula is a tautology, by noting that all axioms of $\mathcal{F}$ are valid and that modus ponens preserves the property of being valid. Similarly, whenever $\Gamma \vdash_{\mathcal{F}} A$, then Γ tautologically implies A. In other words, $\mathcal{F}$ is *(implicationally) sound*; which means that all provable formulas are valid (or, are consequences of the extra-logical hypotheses Γ).

Of course, any useful proof system ought to be sound, since the purpose of creating proofs is to establish the validity of a sentence. Remarkably, the system $\mathcal{F}$ is also *complete* in that it can prove any valid formula. Thus the semantic notion of validity and the syntactic notion of provability coincide, and a formula is valid if and only if it is provable in $\mathcal{F}$.

Theorem. *The propositional proof system $\mathcal{F}$ is complete and is implicationally complete; namely,*
(1) *If A is a tautology, then $\vdash_{\mathcal{F}} A$.*
(2) *If $\Gamma \vDash A$, then $\Gamma \vdash_{\mathcal{F}} A$.*

The philosophical significance of the completeness theorem is that a finite set of (schematic) axioms and rules of inference are sufficient to establish the validity of any tautology. In hindsight, it is not surprising that this holds, since the method of truth-tables already provides an algorithmic way of recognizing tautologies. Indeed, the proof of the completeness theorem given below, can be viewed as showing that the method of truth tables can be formalized within the system $\mathcal{F}$.

1.1.4. Proof. We first observe that part (2) of the completeness theorem can be reduced to part (1) by a two step process. Firstly, note that the compactness theorem for propositional logic states that if $\Gamma \vDash A$ then there is a finite subset Γ_0 of Γ which also tautologically implies A. A topological proof of the compactness theorem for propositional logic is sketched in 1.1.5 below. Thus Γ may, without loss of generality, be assumed to be a finite set of formulas, say $\Gamma = \{B_1, \ldots, B_k\}$. Secondly, note that $\Gamma \vDash A$ implies that $B_1 \supset B_2 \supset \cdots \supset A$ is a tautology. So, by part (1), the latter formula has an $\mathcal{F}$-proof, and by k additional modus ponens inferences, $\Gamma \vdash A$. (To simplify notation, we write $\vdash$ instead of $\vdash_{\mathcal{F}}$.)

It remains to prove part (1). We begin by establishing a series of special cases, (a)–(k), of the completeness theorem, in order to "bootstrap" the propositional system $\mathcal{F}$. We use symbols ϕ, ψ, χ for arbitrary formulas and Π to represent any set of formulas.

(a) $\vdash \phi \supset \phi$.
Proof: Combine the three axioms $(\phi \supset \phi \supset \phi) \supset (\phi \supset (\phi \supset \phi) \supset \phi) \supset (\phi \supset \phi)$, $\phi \supset (\phi \supset \phi)$ and $\phi \supset (\phi \supset \phi) \supset \phi$ with two uses of modus ponens.

(b) Deduction Theorem: $\Gamma, \phi \vdash \psi$ if and only if $\Gamma \vdash \phi \supset \psi$.
Proof: The reverse implication is trivial. To prove the forward implication, suppose $C_1, C_2, \ldots, C_k$ is an $\mathcal{F}$-proof of ψ from Γ, ϕ. This means that C_k is ψ and that each

C_i is ϕ, is in Γ, is an axiom, or is inferred by modus ponens. It is straightforward to prove, by induction on i, that $\Gamma \vdash \phi \supset C_i$ for each C_i.

(c) $\phi \supset \psi \vdash \neg\psi \supset \neg\phi$.
Proof: By the deduction theorem, it suffices to prove that $\phi \supset \psi, \neg\psi \vdash \neg\phi$. To prove this, use the two axioms $\neg\psi \supset (\phi \supset \neg\psi)$ and $(\phi \supset \psi) \supset (\phi \supset \neg\psi) \supset \neg\phi$ and three uses of modus ponens.

(d) $\phi, \neg\phi \vdash \psi$.
Proof: From the axiom $\phi \supset (\neg\psi \supset \phi)$, we have $\phi \vdash \neg\psi \supset \phi$. Thus, by (c) we get $\phi \vdash \neg\phi \supset \neg\neg\psi$, and by (b), $\phi, \neg\phi \vdash \neg\neg\psi$. Finally modus ponens with the axiom $\neg\neg\psi \supset \psi$ gives the desired result.

(e) $\neg\phi \vdash \phi \supset \psi$ and $\psi \vdash \phi \supset \psi$.
Proof: This former follows from (d) and the deduction theorem, and the latter follows from the axiom $\psi \supset (\phi \supset \psi)$.

(f) $\phi, \neg\psi \vdash \neg(\phi \supset \psi)$.
Proof: It suffices to prove $\phi \vdash \neg\psi \supset \neg(\phi \supset \psi)$. Thus, by (c) and the deduction theorem, it suffices to prove $\phi, \phi \supset \psi \vdash \psi$. The latter assertion is immediate from modus ponens.

(g) $\phi, \psi \vdash \phi \wedge \psi$.
Proof: Two uses of modus ponens with the axiom $\phi \supset \psi \supset (\phi \wedge \psi)$.

(h) $\neg\phi \vdash \neg(\phi \wedge \psi)$ and $\neg\psi \vdash \neg(\phi \wedge \psi)$.
Proof: For the first part, it suffices to show $\vdash \neg\phi \supset \neg(\phi \wedge \psi)$, and thus, by (c), it suffices to show $\vdash (\phi \wedge \psi) \supset \phi$, which is an axiom. The proof that $\neg\psi \vdash \neg(\phi \wedge \psi)$ is similar.

(i) $\phi \vdash \phi \vee \psi$ and $\psi \vdash \phi \vee \psi$.
Proof: $\phi \supset (\phi \vee \psi)$ and $\psi \supset (\phi \vee \psi)$ are axioms.

(j) $\neg\phi, \neg\psi \vdash \neg(\phi \vee \psi)$.
Proof: It suffices to prove $\neg\phi \vdash \neg\psi \supset \neg(\phi \vee \psi)$, and this, by (c), follows from $\neg\phi \vdash (\phi \vee \psi) \supset \psi$. For this, we combine

(*i*) $\neg\phi \vdash \phi \supset \psi$, by (e),
(*ii*) $\vdash \psi \supset \psi$, by (a),
(*iii*) $\vdash (\phi \supset \psi) \supset (\psi \supset \psi) \supset ((\phi \vee \psi) \supset \psi)$, an axiom,

with two uses of modus ponens.

(k) $\phi \vdash \neg\neg\phi$.
Proof: By (d), $\phi, \neg\phi \vdash \neg\neg\phi$, and obviously, $\phi, \neg\neg\phi \vdash \neg\neg\phi$. So $\phi \vdash \neg\neg\phi$ follows from the next lemma.

1.1.4.1. Lemma. *If* $\Gamma, \phi \vdash \psi$ *and* $\Gamma, \neg\phi \vdash \psi$, *then* $\Gamma \vdash \psi$.

Proof. By (b) and (c), the two hypotheses imply that $\Gamma \vdash \neg\psi \supset \neg\phi$ and $\Gamma \vdash \neg\psi \supset \neg\neg\phi$. These plus the two axioms $(\neg\psi \supset \neg\phi) \supset (\neg\psi \supset \neg\neg\phi) \supset \neg\neg\psi$ and $\neg\neg\psi \supset \psi$ give $\Gamma \vdash \psi$. □

1.1.4.2. Lemma. *Let the formula A involve only the propositional variables among $p_1, \ldots, p_n$. For $1 \le i \le n$, suppose that B_i is either p_i or $\neg p_i$. Then, either*

$$B_1, \ldots, B_n \vdash A \qquad \text{or} \qquad B_1, \ldots, B_n \vdash \neg A.$$

Proof. Define τ to be a truth assignment that makes each B_i true. By the soundness theorem, A (respectively, $\neg A$), can be proved from the hypotheses $B_1, \ldots, B_n$ only if $\overline{\tau}(A) = T$ (respectively $\overline{\tau}(A) = F$). Lemma 1.1.4.2 asserts that the converse holds too.

The lemma is proved by induction on the complexity of A. In the base case, A is just p_i: this case is trivial to prove since B_i is either p_i or $\neg p_i$. Now suppose A is a formula $A_1 \vee A_2$. If $\sigma(A) = T$, then we must have $\tau(A_i) = T$ for some $i \in \{1, 2\}$; the induction hypothesis implies that $B_1, \ldots, B_n \vdash A_i$ and thus, by (i) above, $B_1, \ldots, B_n \vdash A$. On the other hand, if $\tau(A) = F$, then $\tau(A_1) = \tau(A_2) = F$, so the induction hypothesis implies that $B_1, \ldots, B_n \vdash \neg A_i$ for both $i = 1$ and $i = 2$. From this, (j) implies that $B_1, \ldots, B_n \vdash \neg A$. The cases where A has outermost connective $\wedge$, $\supset$ or $\neg$ are proved similarly. □.

We are now ready to complete the proof of the Completeness Theorem 1.1.3. Suppose A is a tautology. We claim that Lemma 1.1.4.2 can be strengthened to have

$$B_1, \ldots, B_k \vdash A$$

where, as before each B_i is either p_i or $\neg p_i$, but now $0 \le k \le n$ is permitted. We prove this by induction on $k = n, n-1, \ldots, 1, 0$. For $k = n$, this is just Lemma 1.1.4.2. For the induction step, note that $B_1, \ldots, B_k \vdash A$ follows from $B_1, \ldots, B_k, p_{k+1} \vdash A$ and $B_1, \ldots, B_k, \neg p_{k+1} \vdash A$ by Lemma 1.1.4.1. When $k = 0$, we have that $\vdash A$, which proves the Completeness Theorem.
Q.E.D. Theorem 1.1.3

1.1.5. It still remains to prove the compactness theorem for propositional logic. This theorem states:

Compactness Theorem. *Let Γ be a set of propositional formulas.*
(1) *Γ is satisfiable if and only if every finite subset of Γ is satisfiable.*
(2) *$\Gamma \vDash A$ if and only if there is a finite subset Γ_0 of Γ such that $\Gamma_0 \vDash A$.*

Since $\Gamma \vDash A$ is equivalent to $\Gamma \cup \{\neg A\}$ being unsatisfiable, (2) is implied by (1). It is fairly easy to prove the compactness theorem directly, and most introductory books in mathematical logic present such a proof. Here, we shall instead, give a proof based on the Tychonoff theorem; obviously this connection to topology is the reason for the name 'compactness theorem.'

Proof. Let V be the set of propositional variables used in Γ; the sets Γ and V need not necessarily be countable. Let 2^V denote the set of truth assignments on V and endow 2^V with the product topology by viewing it as the product of $|V|$ copies of the two element space with the discrete topology. That is to say, the subbasis elements of 2^V are the sets $B_{p,i} = \{\tau : \tau(p) = i\}$ for $p \in V$ and $i \in \{T, F\}$. Note that these subbasis elements are both open and closed. Recall that the Tychonoff theorem states that an arbitrary product of compact spaces is compact; in particular, 2^V is compact. (See Munkres [1975] for background material on topology.)

For $\phi \in \Gamma$, define $D_\phi = \{\tau \in 2^V : \tau \models \phi\}$. Since ϕ only involves finitely many variables, each D_ϕ is both open and closed. Now Γ is satisfiable if and only if $\cap_{\phi\in\Gamma} D_\phi$ is non-empty. By the compactness of 2^V, the latter condition is equivalent to the sets $\cap_{\phi\in\Gamma_0} D_\phi$ being non-empty for all finite $\Gamma_0 \subset \Gamma$. This, in turn is equivalent to each finite subset Γ_0 of Γ being satisfiable. □

The compactness theorem for first-order logic is more difficult; a purely model-theoretic proof can be given with ultrafilters (see, e.g., Eklof [1977]). We include a proof-theoretic proof of the compactness theorem for first-order logic for countable languages in section 2.3.7 below.

1.1.6. Remarks. There are of course a large number of possible ways to give sound and complete proof systems for propositional logic. The particular proof system $\mathcal{F}$ used above is adapted from Kleene [1952]. A more detailed proof of the completeness theorem for $\mathcal{F}$ and for related systems can be found in the textbook of Mendelson [1987]. The system $\mathcal{F}$ is an example of a class of proof systems called *Frege proof systems*: a Frege proof system is any proof system in which all axioms and rules are schematic and which is implicationally sound and implicationally complete. Most of the commonly used proof systems similar to $\mathcal{F}$ are based on modus ponens as the only rule of inference; however, some (non-Frege) systems also incorporate a version of the deduction theorem as a rule of inference. In these systems, if B has been inferred from A, then the formula $A \supset B$ may also be inferred. An example of such a system is the propositional fragment of the natural deduction proof system described in section 2.4.8 below.

Other rules of inference that are commonly allowed in propositional proof systems include the *substitution rule* which allows any instance of ϕ to be inferred from ϕ, and the *extension rule* which permits the introduction of abbreviations for long formulas. These two systems *appear* to be more powerful than Frege systems in that they seem to allow substantially shorter proofs of certain tautologies. However, whether they actually are significantly more powerful than Frege systems is an open problem. This issues are discussed more fully by Pudlák in Chapter VIII.

There are several currently active areas of research in the proof theory of propositional logic. Of course, the central open problem is the P versus NP question of whether there exists a polynomial time method of recognizing tautologies. Research on the proof theory of propositional logic can be, roughly speaking, separated into three problem areas. Firstly, the problem of "proof-search" is the question of what are the best algorithmic methods for searching for propositional proofs. The proof-search problem is important for artificial intelligence, for automated theorem proving and for logic programming. The most common propositional proof systems used for proof-search algorithms are variations of the resolution system discussed in 1.3 below. A second, related research area is the question of proof lengths. In this area, the central questions concern the minimum lengths of proofs needed for tautologies in particular proof systems. This topic is treated in more depth in Chapter VIII in this volume.

A third research area concerns the investigation of fragments of the propositional proof system $\mathcal{F}$. For example, propositional intuitionist logic is the logic which is axiomatized by the system $\mathcal{F}$ without the axiom scheme $\neg\neg A \supset A$. Another important example is linear logic. Brief discussions of these two logics can be found in section 3.

1.2. The propositional sequent calculus

The sequent calculus, first introduced by Gentzen [1935] as an extension of his earlier natural deduction proof systems, is arguably the most elegant and flexible system for writing proofs. In this section, the propositional sequent calculus for classical logic is developed; the extension to first-order logic is treated in 2.3 below.

1.2.1. Sequents and Cedents. In the Hilbert-style systems, each line in a proof is a formula; however, in sequent calculus proofs, each line in a proof is a *sequent*: a sequent is written in the form

$$A_1, \ldots, A_k \rightarrow B_1, \ldots, B_\ell$$

where the symbol $\rightarrow$ is a new symbol called the sequent arrow (not to be confused with the implication symbol $\supset$) and where each A_i and B_j is a formula. The intuitive meaning of the sequent is that the conjunction of the A_i's implies the disjunction of the B_j's. Thus, a sequent is equivalent in meaning to the formula

$$\bigwedge_{i=1}^{k} A_i \supset \bigvee_{j=1}^{\ell} B_j.$$

The symbols $\bigwedge$ and $\bigvee$ represent conjunctions and disjunctions, respectively, of multiple formulas. We adopt the convention that an empty conjunction (say, when $k = 0$ above) has value "True", and that an empty disjunction (say, when $\ell = 0$ above) has value "False". Thus the sequent $\rightarrow A$ has the same meaning as the formula A, and the *empty sequent* $\rightarrow$ is false. A sequent is defined to be valid or a tautology if and only if its corresponding formula is.

The sequence of formulas $A_1, \ldots, A_k$ is called the *antecedent* of the sequent displayed above; $B_1, \ldots, B_\ell$ is called its *succedent*. They are both referred to as *cedents*.

1.2.2. Inferences and proofs. We now define the propositional sequent calculus proof system PK. A sequent calculus proof consists of a rooted tree (or sometimes a directed acyclic graph) in which the nodes are sequents. The root of the tree, written at the bottom, is called the *endsequent* and is the sequent proved by the proof. The leaves, at the top of the tree, are called *initial sequents* or *axioms*. Usually, the only initial sequents allowed are the logical axioms of the form $A \rightarrow A$, where we further require that A be atomic.

Other than the initial sequents, each sequent in a PK-proof must be inferred by one of the rules of inference given below. A rule of inference is denoted by a figure $\frac{S_1}{S}$ or $\frac{S_1 \quad S_2}{S}$ indicating that the sequent S may be inferred from S_1 or from the pair S_1 and S_2. The conclusion, S, is called the *lower sequent* of the inference; each hypotheses is an *upper sequent* of the inference. The valid rules of inference for PK are as follows; they are essentially schematic, in that A and B denote arbitrary formulas and Γ, Δ, etc. denote arbitrary cedents.

Weak Structural Rules

Exchange:left $$\frac{\Gamma, A, B, \Pi \rightarrow \Delta}{\Gamma, B, A, \Pi \rightarrow \Delta}$$

Exchange:right $$\frac{\Gamma \rightarrow \Delta, A, B, \Lambda}{\Gamma \rightarrow \Delta, B, A, \Lambda}$$

Contraction:left $$\frac{A, A, \Gamma \rightarrow \Delta}{A, \Gamma \rightarrow \Delta}$$

Contraction:right $$\frac{\Gamma \rightarrow \Delta, A, A}{\Gamma \rightarrow \Delta, A}$$

Weakening:left $$\frac{\Gamma \rightarrow \Delta}{A, \Gamma \rightarrow \Delta}$$

Weakening:right $$\frac{\Gamma \rightarrow \Delta}{\Gamma \rightarrow \Delta, A}$$

The weak structural rules are also referred to as just *weak* inference rules. The rest of the rules are called *strong* inference rules. The *structural* rules consist of the weak structural rules and the cut rule.

The Cut Rule

$$\frac{\Gamma \rightarrow \Delta, A \qquad A, \Gamma \rightarrow \Delta}{\Gamma \rightarrow \Delta}$$

The Propositional Rules[1]

¬:left $$\frac{\Gamma \rightarrow \Delta, A}{\neg A, \Gamma \rightarrow \Delta}$$

¬:right $$\frac{A, \Gamma \rightarrow \Delta}{\Gamma \rightarrow \Delta, \neg A}$$

∧:left $$\frac{A, B, \Gamma \rightarrow \Delta}{A \land B, \Gamma \rightarrow \Delta}$$

∧:right $$\frac{\Gamma \rightarrow \Delta, A \qquad \Gamma \rightarrow \Delta, B}{\Gamma \rightarrow \Delta, A \land B}$$

∨:left $$\frac{A, \Gamma \rightarrow \Delta \qquad B, \Gamma \rightarrow \Delta}{A \lor B, \Gamma \rightarrow \Delta}$$

∨:right $$\frac{\Gamma \rightarrow \Delta, A, B}{\Gamma \rightarrow \Delta, A \lor B}$$

⊃:left $$\frac{\Gamma \rightarrow \Delta, A \qquad B, \Gamma \rightarrow \Delta}{A \supset B, \Gamma \rightarrow \Delta}$$

⊃:right $$\frac{A, \Gamma \rightarrow \Delta, B}{\Gamma \rightarrow \Delta, A \supset B}$$

[1] We have stated the ∧*:left* and the ∨*:right* rules differently than the traditional form. The traditional definitions use the following two ∨*:right* rules of inference

The above completes the definition of *PK*. We write $PK \vdash \Gamma \rightarrow \Delta$ to denote that the sequent $\Gamma \rightarrow \Delta$ has a *PK*-proof. When A is a formula, we write $PK \vdash A$ to mean that $PK \vdash \; \rightarrow A$.

The cut rule plays a special role in the sequent calculus, since, as is shown in section 1.2.8, the system *PK* is complete even without the cut rule; however, the use of the cut rule can significantly shorten proofs. A proof is said to be *cut-free* if does not contain any cut inferences.

1.2.3. Ancestors, descendents and the subformula property. All of the inferences of *PK*, with the exception of the cut rule, have a *principal formula* which is, by definition, the formula occurring in the lower sequent of the inference which is not in the cedents Γ or Δ (or Π or Λ). The exchange inferences have two principal formulas. Every inference, except weakenings, has one or more *auxiliary formulas* which are the formulas A and B, occurring in the upper sequent(s) of the inference. The formulas which occur in the cedents Γ, Δ, Π or Λ are called *side formulas* of the inference. The two auxiliary formulas of a cut inference are called the *cut formulas.*

We now define the notions of descendents and ancestors of formulas occurring in a sequent calculus proof. First we define *immediate descendents* as follows: If C is a side formula in an upper sequent of an inference, say C is the i-th subformula of a cedent Γ, Π, Δ or Λ, then C's only immediate descendent is the corresponding occurrence of the same formula in the same position in the same cedent in the lower sequent of the inference. If C is an auxiliary formula of any inference except an exchange or cut inference, then the principal formula of the inference is the immediate descendent of C. For an exchange inference, the immediate descendent of the A or B in the upper sequent is the A or B, respectively, in the lower sequent. The cut formulas of a cut inference do not have immediate descendents. We say that C is an *immediate ancestor* of D if and only if D is an immediate descendent of C. Note that the only formulas in a proof that do not have immediate ancestors are the formulas in initial sequents and the principal formulas of weakening inferences.

The *ancestor* relation is defined to be the reflexive, transitive closure of the immediate ancestor relation; thus, C is an ancestor of D if and only if there is a chain of zero or more immediate ancestors from D to C. A *direct ancestor* of D is an ancestor C of D such that C is the same formula as D. The concepts of *descendent* and *direct descendent* are defined similarly as the converses of the ancestor and direct ancestor relations.

A simple, but important, observation is that if C is an ancestor of D, then C is a subformula of D. This immediately gives the following subformula property:

$$\frac{\Gamma \rightarrow \Delta, A}{\Gamma \rightarrow \Delta, A \vee B} \quad \text{and} \quad \frac{\Gamma \rightarrow \Delta, A}{\Gamma \rightarrow \Delta, B \vee A}$$

and two dual rules of inference for $\wedge$*:left*. Our method has the advantage of reducing the number of rules of inference, and also simplifying somewhat the upper bounds on cut-free proof length we obtain below.

1.2.4. Proposition. *(The Subformula Property) If P is a cut-free PK-proof, then every formula occurring in P is a subformula of a formula in the endsequent of P.*

1.2.5. Lengths of proofs. There are a number of ways to measure the length of a sequent calculus proof P; most notably, one can measure either the number of symbols or the number of sequents occurring in P. Furthermore, one can require P to be tree-like or to be dag-like; in the case of dag-like proofs no sequent needs to be derived, or counted, twice. ('Dag' abbreviates 'directed acyclic graph', another name for such proofs is 'sequence-like'.)

For this chapter, we adopt the following conventions for measuring lengths of sequent calculus proofs: proofs are always presumed to be tree-like, unless we explicitly state otherwise, and we let $||P||$ denote the number of **strong** inferences in a tree-like proof P. The value $||P||$ is polynomially related to the number of sequents in P. If P has n sequents, then, of course, $||P|| < n$. On the other hand, it is not hard to prove that for any tree-like proof P of a sequent $\Gamma \longrightarrow \Delta$, there is a (still tree-like) proof of an endsequent $\Gamma' \longrightarrow \Delta'$ with at most $||P||^2$ sequents and with $\Gamma' \subseteq \Gamma$ and $\Delta' \subseteq \Delta$. The reason we use $||P||$ instead of merely counting the actual number of sequents in P, is that using $||P||$ often makes bounds on proof size significantly more elegant to state and prove.

Occasionally, we use $||P||_{dag}$ to denote the number of strong inferences in a dag-like proof P.

1.2.6. Soundness Theorem. *The propositional sequent calculus PK is sound. That is to say, any PK-provable sequent or formula is a tautology.*

The soundness theorem is proved by observing that the rules of inference of *PK* preserve the property of sequents being tautologies.

The implicational form of the soundness theorem also holds. If $\mathfrak{S}$ is a set of sequents, let an $\mathfrak{S}$-proof be any sequent calculus proof in which sequents from $\mathfrak{S}$ are permitted as initial sequents (in addition to the logical axioms). The implicational soundness theorem states that if a sequent $\Gamma \longrightarrow \Delta$ has an $\mathfrak{S}$-proof, then $\Gamma \longrightarrow \Delta$ is made true by every truth assignment which satisfies $\mathfrak{S}$.

1.2.7. The inversion theorem. The inversion theorem is a kind of inverse to the implicational soundness theorem, since it says that, for any inference except weakening inferences, if the conclusion of the inference is valid, then so are all of its hypotheses.

Theorem. *Let I be a propositional inference, a cut inference, an exchange inference or a contraction inference. If I's lower sequent is valid, then so are are all of I's upper sequents. Likewise, if I's lower sequent is true under a truth assignment τ, then so are are all of I's upper sequents.*

The inversion theorem is easily proved by checking the eight propositional inference rules; it is obvious for exchange and contraction inferences.

Note that the inversion theorem can fail for weakening inferences. Most authors define the $\wedge$*:left* and $\vee$*:right* rules of inference differently than we defined them for *PK*, and the inversion theorem can fail for these alternative formulations (see the footnote on page 11).

1.2.8. The completeness theorem. The completeness theorem for *PK* states that every valid sequent (tautology) can be proved in the propositional sequent calculus. This, together with the soundness theorem, shows that the *PK*-provable sequents are precisely the valid sequents.

Theorem. *If* $\Gamma \longrightarrow \Delta$ *is a tautology, then it has a PK-proof in which no cuts appear.*

1.2.9. In order to prove Theorem 1.2.8 we prove the following stronger lemma which includes bounds on the size of the *PK*-proof.

Lemma. *Let* $\Gamma \longrightarrow \Delta$ *be a valid sequent in which there are* m *occurrences of logical connectives. Then there is a tree-like, cut free PK-proof* P *of* $\Gamma \longrightarrow \Delta$ *containing fewer than* 2^m *strong inferences.*

Proof. The proof is by induction on m. In the base case, $m = 0$, the sequent $\Gamma \longrightarrow \Delta$ contains no logical connectives and thus every formula in the sequent is a propositional variable. Since the sequent is valid, there must be some variable, p, which occurs both in Γ and in Δ. Thus $\Gamma \longrightarrow \Delta$ can be proved with zero strong inferences from the initial sequent $p \longrightarrow p$.

The induction step, $m > 0$, is handled by cases according to which connectives are used as outermost connectives of formulas in the cedents Γ and Δ. First suppose there is a formula of the form $(\neg A)$ in Γ. Letting Γ' be the cedent obtained from Γ by removing occurrences of $\neg A$, we can infer $\Gamma \longrightarrow \Delta$ by:

$$\dfrac{\dfrac{\Gamma' \longrightarrow \Delta, A}{\neg A, \Gamma' \longrightarrow \Delta}}{\Gamma \longrightarrow \Delta}$$

where the double line indicates a series of weak inferences. By the inversion theorem, $\Gamma' \longrightarrow \Delta, A$ is valid, and hence, since it has at most $m-1$ logical connectives, the induction hypothesis implies that it has a cut-free proof with fewer than 2^{m-1} strong inferences. This gives $\Gamma \longrightarrow \Delta$ a cut-free proof with fewer than $2^{m-1} + 1 \leq 2^m$ strong inferences. The case where a formula $(\neg A)$ occurs in Δ is handled similarly.

Second, consider the case where a formula of the form $A \wedge B$ appears in Γ. Then, letting Γ' be the cedent Γ minus the formula $A \wedge B$, we can infer $\Gamma \longrightarrow \Delta$ by:

$$\dfrac{\dfrac{A, B, \Gamma' \longrightarrow \Delta}{A \wedge B, \Gamma' \longrightarrow \Delta}}{\Gamma \longrightarrow \Delta}$$

By the inversion theorem and the induction hypothesis, $A, B, \Gamma' \longrightarrow \Delta$ has a cut-free proof with fewer than 2^{m-1} strong inferences. Thus $\Gamma \longrightarrow \Delta$ has a cut-free proof with fewer than 2^m strong inferences. Third, suppose there is a formula of the $A \wedge B$

appearing in the succedent Δ. Letting Δ' be the the succedent Δ minus the formula $A \land B$, we can infer

$$\dfrac{\dfrac{\Gamma \rightarrow \Delta', A \qquad \Gamma \rightarrow \Delta', B}{\Gamma \rightarrow \Delta', A \land B}}{\Gamma \rightarrow \Delta}$$

By the inversion theorem, both of upper sequents above are valid. Furthermore, they each have fewer than m logical connectives, so by the induction hypothesis, they have cut-free proofs with fewer than 2^{m-1} strong inferences. This gives the sequent $\Gamma \rightarrow \Delta$ a cut-free proof with fewer than 2^m strong inferences.

The remaining cases are when a formula in the sequent $\Gamma \rightarrow \Delta$ has outermost connective $\lor$ or $\supset$. These are handled with the inversion theorem and the induction hypothesis similarly to the above cases. □

1.2.10. The bounds on the proof size in Lemma 1.2.9 can be improved somewhat by counting only the occurrences of distinct subformulas in $\Gamma \rightarrow \Delta$. To make this precise, we need to define the concepts of positively and negatively occurring subformulas. Given a formula A, an occurrence of a subformula B of A, and a occurrence of a logical connective α in A, we say that B is negatively bound by α if either (1) α is a negation sign, $\neg$, and B is in its scope, or (2) α is an implication sign, $\supset$, and B is a subformula of its first argument. Then, B is said to occur *negatively* (respectively, *positively*) in A if B is negatively bound by an odd (respectively, even) number of connectives in A. A subformula occurring in a sequent $\Gamma \rightarrow \Delta$ is said to be positively occurring if it occurs positively in Δ or negatively in Γ; otherwise, it occurs negatively in the sequent.

Lemma. *Let $\Gamma \rightarrow \Delta$ be a valid sequent. Let m' equal the number of distinct subformulas occurring positively in the sequent and m'' equal the number of distinct subformulas occurring negatively in the sequent. Let $m = m' + m''$. Then there is a tree-like, cut free PK-proof P containing fewer than 2^m strong inferences.*

Proof. (Sketch) Recall that the proof of Lemma 1.2.9 built a proof from the bottom-up, by choosing a formula in the endsequent to eliminate (i.e., to be inferred) and thereby reducing the total number of logical connectives and then appealing to the induction hypothesis. The construction for the proof of the present lemma is exactly the same, except that now care must be taken to reduce the total number of distinct positively or negatively occurring subformulas, instead of just reducing the total number of connectives. This is easily accomplished by always choosing a formula from the endsequent which contains a maximal number of connectives and which is therefore not a proper subformula of any other subformula in the endsequent. □

1.2.11. The cut elimination theorem states that if a sequent has a *PK*-proof, then it has a cut-free proof. This is an immediate consequence of the soundness and completeness theorems, since any *PK*-provable sequent must be valid, by the soundness theorem, and hence has a cut-free proof, by the completeness theorem.

This is a rather slick method of proving the cut elimination theorem, but unfortunately, does not shed any light on how a given *PK*-proof can be constructively transformed into a cut-free proof. In section 2.3.7 below, we shall give a step-by-step procedure for converting first-order sequent calculus proofs into cut-free proofs; the same methods work also for propositional sequent calculus proofs. We shall not, however, describe this constructive proof transformation procedure here; instead, we will only state, without proof, the following upper bound on the increase in proof length which can occur when a proof is transformed into a cut-free proof. (A proof can be given using the methods of Sections 2.4.2 and 2.4.3.)

Cut-Elimination Theorem. *Suppose P be a (possibly dag-like) PK-proof of $\Gamma \rightarrow \Delta$. Then $\Gamma \rightarrow \Delta$ has a cut-free, tree-like PK-proof with less than or equal to $2^{||P||_{dag}}$ strong inferences.*

1.2.12. Free-cut elimination. Let $\mathfrak{S}$ be a set of sequents and, as above, define an $\mathfrak{S}$-proof to be a sequent calculus proof which may contain sequents from $\mathfrak{S}$ as initial sequents, in addition to the logical sequents. If I is a cut inference occurring in an $\mathfrak{S}$-proof P, then we say I's cut formulas are *directly descended from* $\mathfrak{S}$ if they have at least one direct ancestor which occurs as a formula in an initial sequent which is in $\mathfrak{S}$. A cut I is said to be *free* if neither of I's auxiliary formulas is directly descended from $\mathfrak{S}$. A proof is *free-cut free* if and only if it contains no free cuts. (See Definition 2.4.4.1 for a better definition of free cuts.) The cut elimination theorem can be generalized to show that the free-cut free fragment of the sequent calculus is implicationally complete:

Free-cut Elimination Theorem. *Let S be a sequent and $\mathfrak{S}$ a set of sequents. If $\mathfrak{S} \models \mathfrak{S}$, then there is a free-cut free $\mathfrak{S}$-proof of S.*

We shall not prove this theorem here; instead, we prove the generalization of this for first-order logic in 2.4.4 below. This theorem is essentially due to Takeuti [1987] based on the cut elimination method of Gentzen [1935].

1.2.13. Some remarks. We have developed the sequent calculus only for classical propositional logic; however, one of the advantages of the sequent calculus is its flexibility in being adapted for non-classical logics. For instance, propositional intuitionistic logic can be formalized by a sequent calculus PJ which is defined exactly like *PK* except that succedents in the lower sequents of *strong* inferences are restricted to contain at most one formula. As another example, minimal logic is formalized like PJ, except with the restriction that every succedent contain exactly one formula. Linear logic, relevant logic, modal logics and others can also be formulated elegantly with the sequent calculus.

1.2.14. The Tait calculus. Tait [1968] gave a proof system similar in spirit to the sequent calculus. Tait's proof system incorporates a number of simplifications with regard to the sequent calculus; namely, it uses sets of formulas in place of sequents, it allows only propositional formulas to be negated, and there are no weak structural rules at all.

Since the Tait calculus is often used for analyzing fragments of arithmetic, especially in the framework of infinitary logic, we briefly describe it here. Formulas are built up from atomic formulas p, from negated atomic formulas $\neg p$, and with the connectives $\wedge$ and $\vee$. A negated atomic formula is usually denoted $\overline{p}$; and the negation operator is extended to all formulas by defining $\overline{\overline{p}}$ to be just p, and inductively defining $\overline{A \vee B}$ and $\overline{A \wedge B}$ to be $\overline{A} \wedge \overline{B}$ and $\overline{A} \vee \overline{B}$, respectively.

Frequently, the Tait calculus is used for infinitary logic. In this case, formulas are defined so that whenever Γ is a set of formulas, then so are $\bigvee \Gamma$ and $\bigwedge \Gamma$. The intended meaning of these formulas is the disjunction, or conjunction, respectively, of all the formulas in Γ.

Each line in a Tait calculus proof is a set Γ of formulas with the intended meaning of Γ being the disjunction of the formulas in Γ. A Tait calculus proof can be tree-like or dag-like. The initial sets, or logical axioms, of a proof are sets of the form $\Gamma \cup \{p, \overline{p}\}$. In the infinitary setting, there are three rules of inference; namely,

$$\frac{\Gamma \cup \{A_j\}}{\Gamma \cup \{\bigvee_{i \in I} A_i\}} \quad \text{where } j \in I,$$

$$\frac{\Gamma \cup \{A_j : j \in I\}}{\Gamma \cup \{\bigwedge_{j \in I} A_j\}} \quad \text{(there are } |I| \text{ many hypotheses), and}$$

$$\frac{\Gamma \cup \{A\} \qquad \Gamma \cup \{\overline{A}\}}{\Gamma} \quad \text{the cut rule.}$$

In the finitary setting, the same rules of inference may also be used. It is evident that the Tait calculus is practically isomorphic to the sequent calculus. This is because a sequent $\Gamma \rightarrow \Delta$ may be transformed into the equivalent set of formulas containing the formulas from Δ and the negations of the formulas from Γ. The exchange and contraction rules are superfluous once one works with sets of formulas, the weakening rule of the sequent calculus is replaced by allowing axioms to contain extra side formulas (this, in essence, means that weakenings are pushed up to the initial sequents of the proof). The strong rules of inference for the sequent calculus translate, by this means, to the rules of the Tait calculus.

Recall that we adopted the convention that the length of a sequent calculus proof is equal to the number of strong inferences in the proof. When we work with tree-like proofs, this corresponds exactly to the number of inferences in the corresponding Tait-style proof.

The cut elimination theorem for the (finitary) sequent calculus immediately implies the cut elimination theorem for the Tait calculus for finitary logic; this is commonly called the *normalization theorem* for Tait-style systems. For general infinitary logics, the cut elimination/normalization theorems may not hold; however, Lopez-Escobar [1965] has shown that the cut elimination theorem does hold for infinitary logic with formulas of countably infinite length. Also, Chapters III and IV of this volume discuss cut elimination in some infinitary logics corresponding to theories of arithmetic.

1.3. Propositional resolution refutations

The Hilbert-style and sequent calculus proof systems described earlier are quite powerful; however, they have the disadvantage that it has so far proved to be very difficult to implement computerized procedures to search for propositional Hilbert-style or sequent calculus proofs. Typically, a computerized procedure for proof search will start with a formula A for which a proof is desired, and will then construct possible proofs of A by working backwards from the conclusion A towards initial axioms. When cut-free proofs are being constructed this is fairly straightforward, but cut-free proofs may be much longer than necessary and may even be too long to be feasibly constructed by a computer. General, non-cut-free, proofs may be quite short; however, the difficulty with proof search arises from the need to determine what formulas make suitable cut formulas. For example, when trying to construct a proof of $\Gamma \longrightarrow \Delta$ that ends with a cut inference; one has to consider *all* formulas C and try to construct proofs of the sequents $\Gamma \longrightarrow \Delta, C$ and $C, \Gamma \longrightarrow \Delta$. In practice, it has been impossible to choose cut formulas C effectively enough to effectively generate general proofs. A similar difficulty arises in trying to construct Hilbert-style proofs which must end with a modus ponens inference.

Thus to have a propositional proof system which would be amenable to computerized proof search, it is desirable to have a proof system in which (1) proof search is efficient and does not require too many 'arbitrary' choices, and (2) proof lengths are not excessively long. Of course, the latter requirement is intended to reflect the amount of available computer memory and time; thus proofs of many millions of steps might well be acceptable. Indeed, for computerized proof search, having an easy-to-find proof which is millions of steps long may well be preferable to having a hard-to-find proof which has only hundreds of steps.

The principal propositional proof system which meets the above requirements is based on resolution. As we shall see, the expressive power and implicational power of resolution is weaker than that of the full propositional logic; in particular, resolution is, in essence, restricted to formulas in conjunctive normal form. However, resolution has the advantage of being amenable to efficient proof search.

Propositional and first-order resolution were introduced in the influential work of Robinson [1965b] and Davis and Putnam [1960]. Propositional resolution proof systems are discussed immediately below. A large part of the importance of propositional resolution lies in the fact that it leads to efficient proof methods in first-order logic: first-order resolution is discussed in section 2.6 below.

1.3.1. Definition. A *literal* is defined to be either a propositional variable p_i or the negation of a propositional variable $\neg p_i$. The literal $\neg p_i$ is also denoted $\overline{p_i}$; and if x is the literal $\overline{p_i}$, then $\overline{x}$ denotes the unnegated literal p_i. The literal $\overline{x}$ is called the *complement* of x. A *positive* literal is one which is an unnegated variable; a *negative* literal is one which is a negated variable.

A *clause* C is a finite set of literals. The intended meaning of C is the disjunction of its members; thus, for σ a truth assignment, $\sigma(C)$ equals *True* if and only if $\sigma(x) = True$ for some $x \in C$. Note that the empty clause, $\emptyset$, always has value *False.* Since clauses that contain both x and $\overline{x}$ are always true, it is often assumed w.l.o.g. that no clause contains both x and $\overline{x}$. A clause is defined to be *positive* (respectively, *negative*) if it contains only positive (resp., only negative) literals. The empty clause is the only clause which is both positive and negative. A clause which is neither positive nor negative is said to be *mixed.*

A non-empty set Γ of clauses is used to represent the conjunction of its members. Thus $\sigma(\Gamma)$ is *True* if and only if $\sigma(C)$ is *True* for all $C \in \Gamma$. Obviously, the meaning of Γ is the same as the conjunctive normal form formula consisting of the conjunction of the disjunctions of the clauses in Γ. A set of clauses is said to be *satisfiable* if there is at least one truth assignment that makes it true.

Resolution proofs are used to prove that a set Γ of clauses is unsatisfiable: this is done by using the resolution rule (defined below) to derive the empty clause from Γ. Since the empty clause is unsatisfiable this will be sufficient to show that Γ is unsatisfiable.

1.3.2. Definition. Suppose that C and D are clauses and that $x \in C$ and $\overline{x} \in D$ are literals. The *resolution rule* applied to C and D is the inference

$$\frac{C \qquad D}{(C \setminus \{x\}) \cup (D \setminus \{\overline{x}\})}$$

The conclusion $(C \setminus \{x\}) \cup (D \setminus \{\overline{x}\})$ is called the *resolvent* of C and D (with respect to x).[2]

Since the resolvent of C and D is satisfied by any truth assignment that satisfies both C and D, the resolution rule is *sound* in the following sense: if Γ is satisfiable and B is the resolvent of two clauses in Γ, then $\Gamma \cup \{B\}$ is satisfiable. Since the empty clause is not satisfiable, this yields the following definition of the resolution proof system.

Definition. A *resolution refutation* of Γ is a sequence $C_1, C_2, \ldots, C_k$ of clauses such that each C_i is either in Γ or is inferred from earlier member of the sequence by the resolution rule, and such that C_k is the empty clause.

1.3.3. Resolution is defined to be a refutation procedure which refutes the satisfiability of a set of clauses, but it also functions as a proof procedure for proving the validity of propositional formulas; namely, to prove a formula A, one forms a set Γ_A of clauses such that A is a tautology if and only if Γ_A is unsatisfiable. Then

[2]Note that x is uniquely determined by C and D if we adopt the (optional) convention that clauses never contain any literal and its complement.

a resolution proof of A is, by definition, a resolution refutation of Γ_A. There are two principal means of forming Γ_A from A. The first method is to express the negation of A in conjunctive normal form and let Γ_A be the set of clauses which express that conjunctive normal form formula. The principal drawback of this definition of Γ_A is that the conjunctive normal form of $\neg A$ may be exponentially longer than A, and Γ_A may have exponentially many clauses.

The second method is the method of *extension*, introduced by Tsejtin [1968], which involves introducing new propositional variables in addition to the variables occurring in the formula A. The new propositional variables are called *extension variables* and allow each distinct subformula B in A to be assigned a literal x_B by the following definition: (1) for propositional variables appearing in A, x_p is p; (2) for negated subformulas $\neg B$, $x_{\neg B}$ is $\overline{x_B}$; (3) for any other subformula B, x_B is a new propositional variable. We then define Γ_A to contain the following clauses: (1) the singleton clause $\{\overline{x_A}\}$, (2) for each subformula of the form $B \land C$ in A, the clauses

$$\{\overline{x_{B\land C}}, x_B\}, \quad \{\overline{x_{B\land C}}, x_C\} \quad \text{and} \quad \{\overline{x_B}, \overline{x_C}, x_{B\land C}\};$$

and (3) for each subformula of the form $B \lor C$ in A, the clauses

$$\{x_{B\lor C}, \overline{x_B}\}, \quad \{x_{B\lor C}, \overline{x_C}\} \quad \text{and} \quad \{x_B, x_C, \overline{x_{B\lor C}}\}.$$

If there are additional logical connectives in A then similar sets of clauses can be used. It is easy to check that the set of three clauses for $B \land C$ (respectively, $B \lor C$) is satisfied by exactly those truth assignments that assign $x_{B\land C}$ (respectively, $x_{B\lor C}$) the same truth value as $x_B \land x_C$ (resp., $x_B \lor x_C$). Thus, a truth assignment satisfies Γ_A only if it falsifies A and, conversely, if a truth assignment falsifies A then there is a unique assignment of truth values to the extension variables of A which lets it satisfy Γ_A. The primary advantage of forming Γ_A by the method of extension is that Γ_A is linear-size in the size of A. The disadvantage, of course, is that the introduction of additional variables changes the meaning and structure of A.

1.3.4. Completeness Theorem for Propositional Resolution. *If Γ is an unsatisfiable set of clauses, then there is a resolution refutation of Γ.*

Proof. We shall briefly sketch two proofs of this theorem. The first, and usual, proof is based on the Davis-Putnam procedure. First note, that by the Compactness Theorem we may assume w.l.o.g. that Γ is finite. Therefore, we may use induction on the number of distinct variables appearing in Γ. The base case, where no variables appear in Γ is trivial, since Γ must contain just the empty clause. For the induction step, let p be a fixed variable in Γ and define Γ' to be the set of clauses containing the following clauses:

a. For all clauses C and D in Γ such that $p \in C$ and $\overline{p} \in D$, the resolvent of C and D w.r.t. p is in Γ', and

b. Every clause C in Γ which contains neither p nor $\overline{p}$ is in Γ'.

Assuming, without loss of generality, that no clause in Γ contained both p and $\overline{p}$, it is clear that the variable p does not occur in Γ'. Now, it is not hard to show that Γ' is satisfiable if and only if Γ is, from whence the theorem follows by the induction hypothesis.

The second proof reduces resolution to the free-cut free sequent calculus. For this, if C is a clause, let Δ_C be the cedent containing the variables which occur positively in C and Π_C be the variables which occur negatively in C. Then the sequent $\Pi_C \rightarrow \Delta_C$ is a sequent with no non-logical symbols which is identical in meaning to C. For example, if $C = \{p_1, \overline{p_2}, p_3\}$, then the associated sequent is $p_2 \rightarrow p_1, p_3$. Clearly, if C and D are clauses with a resolvent E, then the sequent $\Pi_E \rightarrow \Delta_E$ is obtained from $\Pi_C \rightarrow \Delta_C$ and $\Pi_D \rightarrow \Delta_D$ with a single cut on the resolution variable. Now suppose Γ is unsatisfiable. By the completeness theorem for the free-cut free sequent calculus, there is a free-cut free proof of the empty sequent from the sequents $\Pi_C \rightarrow \Delta_C$ with $C \in \Gamma$. Since the proof is free-cut free and there are no non-logical symbols appearing in any initial sequents, every cut formula in the proof must be atomic. Therefore no non-logical symbol appear anywhere in the proof and, by identifying the sequents in the free-cut free proof with clauses and replacing each cut inference with the corresponding resolution inference, a resolution refutation of the empty clause is obtained. □

1.3.5. Restricted forms of resolution. One of the principal advantages of resolution is that it is easier for computers to search for resolution refutations than to search for arbitrary Hilbert-style or sequent calculus proofs. The reason for this is that resolution proofs are less powerful and more restricted than Hilbert-style and sequent calculus proofs and, in particular, there are fewer options on how to form resolution proofs. This explains the paradoxical situation that a less-powerful proof system can be preferable to a more powerful system. Thus it makes sense to consider further restrictions on resolution which may reduce the proof search space even more. Of course there is a tradeoff involved in using more restricted forms of resolution, since one may find that although restricted proofs are easier to search for, they are a lot less plentiful. Often, however, the ease of proof search is more important than the existence of short proofs; in fact, it is sometimes even preferable to use a proof system which is not complete, provided its proofs are easy to find.

Although we do not discuss this until section 2.6, the second main advantage of resolution is that propositional refutations can be 'lifted' to first-order refutations of first-order formulas. It is important that the restricted forms of resolution discussed next also apply to first-order resolution refutations.

One example of a restricted form of resolution is implicit in the first proof of the Completeness Theorem 1.3.4 based on the Davis-Putnam procedure; namely, for any ordering of the variables $p_1, \ldots, p_m$, it can be required that a resolution refutation has first resolutions with respect to p_1, then resolutions with respect to p_2, etc., concluding with resolutions with respect to p_m. This particular strategy is not particularly useful since it does not reduce the search space sufficiently. We consider next several strategies that have been somewhat more useful.

1.3.5.1. Subsumption. A clause C is said to *subsume* a clause D if and only if $C \subseteq D$. The subsumption principle states that if two clauses C and D have been derived such that C subsumes D, then D should be discarded and not used further in the refutation. The subsumption principle is supported by the following theorem:

Theorem. *If Γ is unsatisfiable and if $C \subset D$, then $\Gamma' = (\Gamma \setminus \{D\}) \cup \{C\}$ is also unsatisfiable. Furthermore, Γ' has a resolution refutation which is no longer than the shortest refutation of Γ.*

1.3.5.2. Positive resolution and hyperresolution. Robinson [1965a] introduced positive resolution and hyperresolution. A *positive* resolution inference is one in which one of the hypotheses is a positive clause. The completeness of positive resolution is shown by:

Theorem. (Robinson [1965a]) *If Γ is unsatisfiable, then Γ has a refutation containing only positive resolution inferences.*

Proof. (Sketch) It will suffice to show that it is impossible for an unsatisfiable Γ to be closed under positive resolution inferences and not contain the empty clause. Let A be the set of positive clauses in Γ; A must be non-empty since Γ is unsatisfiable. Pick a truth assignment τ that satisfies all clauses in A and assigns the minimum possible number of "true" values. Pick a clause L in $\Gamma \setminus A$ which is falsified by τ and has the minimum number of negative literals; and let $\overline{p}$ be one of the negative literals in L. Note that L exists since Γ is unsatisfiable and that $\tau(p) = True$. Pick a clause $J \in A$ that contains p and has the rest of its members assigned false by τ; such a clause J exists by the choice of τ. Considering the resolvent of J and L, we obtain a contradiction. □

Positive resolution is nice in that it restricts the kinds of resolution refutations that need to be attempted; however, it is particularly important as the basis for *hyperresolution.* The basic idea behind hyperresolution is that multiple positive resolution inferences can be combined into a single inference with a positive conclusion. To justify hyperresolution, note that if R is a positive resolution refutation then the inferences in R can be uniquely partitioned into subproofs of the form

$$
\cfrac{A_n \qquad \cfrac{\vdots}{B_n}}{A_{n+1}}
\qquad\qquad
\cfrac{A_3 \qquad \cfrac{A_2 \qquad \cfrac{A_1 \qquad B_1}{B_2}}{B_3}}{B_4}
$$

where each of the clauses $A_1, \ldots, A_{n+1}$ are positive (and hence the clauses $B_1, \ldots, B_n$ are not positive). These $n+1$ positive resolution inferences are combined into the single *hyperresolution* inference

$$\frac{A_1 \quad A_2 \quad A_3 \quad \cdots \quad A_n \quad B_1}{A_{n+1}}$$

(This construction is the definition of hyperresolution inferences.)

It follows immediately from the above theorem that hyperresolution is complete. The importance of hyperresolution lies in the fact that one can search for refutations containing only positive resolutions and that as clauses are derived, only the positive clauses need to be saved for possible future use as hypotheses.

Negative resolution is defined similarly to positive resolution and is likewise complete.

1.3.5.3. Semantic resolution. Semantic resolution, independently introduced by Slagle [1967] and Luckham [1970], can be viewed as a generalization of positive resolution. For semantic resolution, one uses a fixed truth assignment (interpretation) τ to restrict the permissible resolution inferences. A resolution inference is said to be τ-supported if one of its hypotheses is given value *False* by τ. Note that at most one hypothesis can have value *False*, since the hypotheses contain complementary occurrences of the resolvent variable.

A resolution refutation is said to be τ-supported if each of its resolution inferences are τ-supported. If τ_F is the truth assignment which assigns every variable the value *False*, then a τ_F-supported resolution refutation is definitionally the same as a positive resolution refutation. Conversely, if Γ is a set of clauses and if τ is any truth assignment, then one can form a set Γ' by complementing every variable in Γ which has τ-value *True*: clearly, a τ-supported resolution refutation of Γ is isomorphic to a positive resolution refutation of Γ'. Thus, Theorem 1.3.5.2 is equivalent to the following Completeness Theorem for semantic resolution:

Theorem. *For any τ and Γ, Γ is unsatisfiable if and only if Γ has a τ-supported resolution refutation.*

It is possible to define semantic-hyperresolution in terms of semantic resolution, just as hyperresolution was defined in terms of positive resolution.

1.3.5.4. Set-of-support resolution. Wos, Robinson and Carson [1965] introduced *set-of-support* resolution as another principle for guiding a search for resolution refutations. Formally, set of support is defined as follows: if Γ is a set of clauses and if $\Pi \subset \Gamma$ and $\Gamma \setminus \Pi$ is satisfiable, then Π is a set of support for Γ; a refutation R of Γ is said to be *supported by* Π if every inference in R is derived (possibly indirectly) from at least one clause in Π. (An alternative, almost equivalent, definition would be to require that no two members of $\Gamma \setminus \Pi$ are resolved together.) The intuitive idea behind set of support resolution is that when trying to refute Γ, one should concentrate on trying to derive a contradiction from the part Π of Γ which is not known to be consistent. For example, $\Gamma \setminus \Pi$ might be a database of facts which is presumed to be consistent, and Π a clause which we are trying to refute.

Theorem. *If* Γ *is unsatisfiable and* Π *is a set of support for* Γ*, then* Γ *has a refutation supported by* Π.

This theorem is immediate from Theorem 1.3.5.3. Let τ be any truth assignment which satisfies $\Gamma \setminus \Pi$, then a τ-supported refutation is also supported by Π.

The main advantage of set of support resolution over semantic resolution is that it does not require knowing or using a satisfying assignment for $\Gamma \setminus \Pi$.

1.3.5.5. Unit and input resolution. A *unit* clause is defined to be a clause containing a single literal; a *unit resolution inference* is an inference in at least one of the hypotheses is a unit clause. As a general rule, it is desirable to perform unit resolutions whenever possible. If Γ contains a unit clause $\{x\}$, then by combining unit resolutions with the subsumption principle, one can remove from Γ every clause which contains x and also every occurrence of $\overline{x}$ from the rest of the clauses in Γ. (The situation is a little more difficult when working in first-order logic, however.) This completely eliminates the literal x and reduces the number of and sizes of clauses to consider.

A unit resolution refutation is a refutation which contains only unit resolutions. Unfortunately, unit resolution is not complete: for example, an unsatisfiable set Γ with no unit clauses cannot have a unit resolution refutation.

An *input* resolution refutation of Γ is defined to be a refutation of Γ in which every resolution inference has at least one of its hypotheses in Γ. Obviously, a minimal length input refutation will be tree-like. Input resolution is also not complete; in fact, it can refute exactly the same sets as unit resolution:

Theorem. (Chang [1970]) *A set of clauses has a unit refutation if and only if it has a input refutation.*

1.3.5.6. Linear resolution. Linear resolution is a generalization of input resolution which has the advantage of being complete: a *linear resolution refutation* of Γ is a refutation $A_1, A_2, \ldots, A_{n-1}, A_n = \emptyset$ such that each A_i is either in Γ or is obtained by resolution from A_{i-1} and A_j for some $j < i - 1$. Thus a linear refutation has the same linear structure as an input resolution, but is allowed to reuse intermediate clauses which are not in Γ.

Theorem. (Loveland [1970] and Luckham [1970]) *If* Γ *is unsatisfiable, then* Γ *has a linear resolution refutation.*

Linear and input resolution both lend themselves well to depth-first proof search strategies. Linear resolution is still complete when used in conjunction with set-of-support resolution.

Further reading. We have only covered some of the basic strategies for proposition resolution proof search. The original paper of Robinson [1965b] still provides an excellent introduction to resolution; this and many other foundational papers on this topic have been reprinted in Siekmann and Wrightson [1983]. In addition, the textbooks by Chang and Lee [1973], Loveland [1978], and Wos et al. [1992] give a more complete description of various forms of resolution than we have given above.

Horn clauses. A *Horn clause* is a clause which contains at most one positive literal. Thus a Horn clause must be of the form $\{p, \overline{q_1}, \ldots, \overline{q_n}\}$ or of the form $\{\overline{q_1}, \ldots, \overline{q_n}\}$ with $n \geq 0$. If a Horn clause is rewritten as sequents of atomic variables, it will have at most one variable in the antecedent; typically, Horn clauses are written in reverse-sequent format so, for example, the two Horn clauses above would be written as implications

$$p \Leftarrow q_1, \ldots, q_n$$

and

$$\Leftarrow q_1, \ldots, q_n.$$

In this reverse-sequent notation, the antecedent is written after the $\Leftarrow$, and the commas are interpreted as conjunctions ($\wedge$'s). Horn clauses are of particular interest both because they are expressive enough to handle many situations and because deciding the satisfiability of sets of Horn clauses is more feasible than deciding the satisfiability of arbitrary sets of clauses. For these reasons, many logic programming environments such as PROLOG are based partly on Horn clause logic.

In propositional logic, it is an easy matter to decide the satisfiability of a set of Horn clauses; the most straightforward method is to restrict oneself to positive unit resolution. A *positive unit* inference is a resolution inference in which one of the hypotheses is a unit clause containing a positive literal only. A positive unit refutation is a refutation containing only positive unit resolution inferences.

Theorem. *A set of Horn clauses is unsatisfiable if and only if it has a positive unit resolution refutation.*

Proof. Let Γ be an unsatisfiable set of Horn clauses. Γ must contain at least one positive unit clause $\{p\}$, since otherwise the truth assignment that assigned *False* to all variables would satisfy Γ. By resolving $\{p\}$ against all clauses containing $\overline{p}$, and then discarding all clauses which contain p or $\overline{p}$, one obtains a smaller unsatisfiable set of Horn clauses. Iterating this yields the desired positive unit refutation. □

Positive unit resolutions are quite adequate in propositional logic, however, they do not lift well to applications in first-order logic and logic programming. For this, a more useful method of search for refutations is based on combining semantic resolution, linear resolution and set-of-support resolution:

1.3.5.7. Theorem. Henschen and Wos [1974]. *Suppose Γ is an unsatisfiable set of Horn clauses with $\Pi \subseteq \Gamma$ a set of support for Γ, and suppose that every clause in $\Gamma \setminus \Pi$ contains a positive literal. Then Γ has a refutation which is simultaneously a negative resolution refutation and a linear refutation and which is supported by Π.*

Note that the condition that every clause in $\Gamma \setminus \Pi$ contains a positive literal means that the truth assignment τ that assigns *True* to every variable satisfies $\Gamma \setminus \Pi$. Thus a negative resolution refutation is the same as a τ-supported refutation and hence is supported by Π.

The theorem is fairly straightforward to prove, and we leave the details to the reader. However, note that since every clause in $\Gamma \setminus \Pi$ is presumed to contain a positive literal, it is impossible to get rid of all positive literals only by resolving against clauses in $\Gamma \setminus \Pi$. Therefore, Π must contain a negative clause C such that there is a linear derivation that begins with C, always resolves against clauses in $\Gamma \setminus \Pi$ yielding negative clauses only, and ending with the empty clause. The resolution refutations of Theorem 1.3.5.7, or rather the lifting of these to Horn clauses described in section 2.6.5, can be combined with restrictions on the order in which literals are resolved to give what is commonly called SLD-resolution.

2. Proof theory of first-order logic

2.1. Syntax and semantics

2.1.1. Syntax of first-order logic. First-order logic is a substantial extension of propositional logic, and allows reasoning about individuals using functions and predicates that act on individuals. The symbols allowed in first-order formulas include the propositional connectives, quantifiers, variables, function symbols, constant symbols and relation symbols. We take $\neg$, $\land$, $\lor$ and $\supset$ as the allowed propositional connectives. There is an infinite set of variable symbols; we use $x, y, z, \ldots$ and $a, b, c, \ldots$ as metasymbols for variables. The quantifiers are the existential quantifiers, $(\exists x)$, and the universal quantifiers, $(\forall x)$, which mean "there exists x" and "for all x". A given first-order language contains a set of function symbols of specified arities, denoted by metasymbols $f, g, h, \ldots$ and a set of relation symbols of specified arities, denoted by metasymbols $P, Q, R, \ldots$. Function symbols of arity zero are called *constant symbols.* Sometimes the first-order language contains a distinguished two-place relation symbol $=$ for equality.

The formulas of first-order logic are defined as follows. Firstly, *terms* are built up from function symbols, constant symbols and variables. Thus, any variable x is a term, and if $t_1, \ldots, t_k$ are terms and f is k-ary, then $f(t_1, \ldots, t_k)$ is a term. Second, *atomic formulas* are defined to be of the form $P(t_1, \ldots, t_k)$ for P a k-ary relation symbol. Finally, *formulas* are inductively to be built up from atomic formulas and logical connectives; namely, any atomic formula is a formula, and if A and B are formulas, then so are $(\neg A)$, $(A \land B)$, $(A \lor B)$, $(A \supset B)$, $(\forall x)A$ and $(\exists x)A$. To avoid writing too many parentheses, we adopt the conventions on omitting parentheses in propositional logic with the additional convention that quantifiers bind more tightly than binary propositional connectives. In addition, binary predicate symbols, such as $=$ and $<$, are frequently written in infix notation.

Consider the formula $(x = 0 \lor (\forall x)(x \neq f(x)))$. (We use $x \neq f(x)$ to abbreviate $(\neg x = f(x))$.) This formula uses the variable x in two different ways: on one hand, it asserts something about an object x; and on the other hand, it (re)uses the variable x to state a general property about all objects. Obviously it would be less confusing to write instead $(x = 0 \vee (\forall y)(y \neq f(y)))$; however, there is nothing wrong, formally speaking, with using x in two ways in the same formula. To keep track of this, we need to define free and bound occurrences of variables. An *occurrence* of a variable x in a formula A is defined to be any place that the symbol x occurs in A, except in quantifer symbols (Qx). (We write (Qx) to mean either $(\forall x)$ or $(\exists x)$.) If $(Qx)(\cdots)$ is a subformula of A, then the *scope* of this occurrence of (Qx) is defined to be the subformula denoted $(\cdots)$. An occurrence of x in A is said to be *bound* if and only if it is in the scope of a quantifier (Qx); otherwise the occurrence of x is called *free*. If x is a bound occurrence, it is *bound by* the rightmost quantifier (Qx) which it is in the scope of. A formula in which no variables appear freely is called a *sentence*.

The intuitive idea of free occurrences of variables in A is that A says something about the free variables. If t is a term, we define the *substitution of t for x in A*, denoted $A(t/x)$ to be the formula obtained from A by replacing each free occurrence of x in A by the term t. To avoid unwanted effects, we generally want t to be *freely substitutable* for x, which means that no free variable in t becomes bound in A as a result of this substitution; formally defined, this means that no free occurrence of x in A occurs in the scope of a quantifier (Qy) with y a variable occurring in t. The *simultaneous substitution of $t_1, \ldots, t_k$ for $x_1, \ldots, x_k$ in A*, denoted $A(t_1/x_1, \ldots, t_k/x_k)$, is defined similarly in the obvious way.

To simplify notation, we adopt some conventions for denoting substitution. Firstly, if we write $A(x)$ and $A(t)$ in the same context, this indicates that $A = A(x)$ is a formula, and that $A(t)$ is $A(t/x)$. Secondly, if we write $A(s)$ and $A(t)$ in the same context, this is to mean that A is formula, x is some variable, and that $A(s)$ is $A(s/x)$ and $A(t)$ is $A(t/x)$.

The sequent calculus, discussed in 2.3, has different conventions on variables than the Hilbert-style systems discussed in 2.2; most notably, it has distinct classes of symbols for free variables and bound variables. See section 2.3.1 for the discussion of the usage of variables in the sequent calculus.

2.1.2. Semantics of first-order logic. In this section, we define the semantics, or 'meaning', of first-order formulas. Since this is really model theory instead of proof theory, and since the semantics of first-order logic is well-covered in any introductory textbook on mathematical logic, we give only a very concise description of the notation and conventions used in this chapter.

In order to ascribe a truth value to a formula, it is necessary to give an interpretation of the non-logical symbols appearing in it; namely, we must specify a *domain* or *universe* of objects, and we must assign meanings to each variable occurring freely and to each function symbol and relation symbol appearing in the formula. A *structure* $\mathcal{M}$ (also called an *interpretation*), for a given language L, consists of the following:

(1) A non-empty universe M of objects, intended to be the universe of objects over which variables and terms range;

(2) For each k-ary function f of the language, an interpretation $f^{\mathcal{M}} : M^k \mapsto M$; and

(3) For each k-ary relation symbol P of the language, an interpretation $P^{\mathcal{M}} \subseteq M^k$ containing all k-tuples for which P is intended to hold. If the first-order language contains the symbol for equality, then $=^{\mathcal{M}}$ must be the true equality predicate on M.

We shall next define the true/false value of a sentence in a structure. This is possible since the structure specifies the meaning of all the function symbols and relation symbols in the formula, and the quantifiers and propositional connectives take their usual meanings. For A a sentence and $\mathcal{M}$ a structure, we write $\mathcal{M} \vDash \mathcal{A}$ to denote A being true in the structure $\mathcal{M}$. In this case, we say that $\mathcal{M}$ is a *model* of A, or that A is *satisfied* by $\mathcal{M}$. Often, we wish to talk about the meaning of a formula A in which free variables occur. For this, we need not only a structure, but also an *object assignment*, which is a mapping σ from the set of variables (at least the ones free in A) to the universe M. The object assignment σ gives meanings to the freely occurring variables, and it is straightforward to define the property of A being true in a given structure $\mathcal{M}$ under a given object assignment σ, denoted $\mathcal{M} \vDash A[\sigma]$.

To give the formal definition of $\mathcal{M} \vDash A[\sigma]$, we first need to define the interpretation of terms, i.e, we need to formally define the manner in which arbitrary terms represent objects in the universe M. To this end, we define $t^{\mathcal{M}}[\sigma]$ by induction on the complexity of t. For x a variable, $x^{\mathcal{M}}[\sigma]$ is just $\sigma(x)$. For t a term of the form $f(t_1, \ldots, t_k)$, we define $t^{\mathcal{M}}[\sigma]$ to equal $f^{\mathcal{M}}(t_1^{\mathcal{M}}[\sigma], \ldots, t_k^{\mathcal{M}}[\sigma])$. If t is a *closed term*, i.e., contains no variables, then $t^{\mathcal{M}}[\sigma]$ is independent of σ and is denoted by just $t^{\mathcal{M}}$.

If σ is an object assignment and $m \in M$, then $\sigma(m/x)$ is the object assignment which is identical to σ except that it maps x to m. We are now ready to define the truth of A in $\mathcal{M}$ with respect to σ, by induction on the complexity of A. For A an atomic formula $P(t_1, \ldots, t_k)$, then $\mathcal{M} \vDash A[\sigma]$ holds if and only if the k-tuple $(t_1^{\mathcal{M}}[\sigma], \ldots, t_k^{\mathcal{M}}[\sigma])$ is in $P^{\mathcal{M}}$. Also, $\mathcal{M} \vDash \neg A[\sigma]$ holds if and only $\mathcal{M} \nvDash A[\sigma]$, where $\nvDash$ is the negation of $\vDash$. Likewise, the value of $\mathcal{M} \vDash A \odot B[\sigma]$ with $\odot$ one of the binary propositional connectives depends only on the truth values of $\mathcal{M} \vDash A[\sigma]$ and $\mathcal{M} \vDash B[\sigma]$, in the obvious way according to Table 1 above. If A is $(Qx)B$ with Q denoting either $\exists$ or $\forall$, then $\mathcal{M} \vDash A[\sigma]$ holds if and only if the property $\mathcal{M} \vDash B[\sigma(m/x)]$ holds for some (respectively, for all) $m \in M$.

We say that a formula A is *valid in* $\mathcal{M}$, if $\mathcal{M} \vDash A[\sigma]$ holds for all appropriate object assignments. A formula is defined to be *valid* if and only if it is valid in all structures. If Γ is a set of formulas, we say Γ is valid in $\mathcal{M}$, written $\mathcal{M} \vDash \Gamma$, if and only if every formula in Γ is valid in M. When $\mathcal{M} \vDash \Gamma$, we say that Γ is *satisfied* by $\mathcal{M}$. The set Γ is *satisfiable* if and only if it is satisfied by some model.

For Γ a set of formulas and A a formula, we define Γ *logically implies* A, $\Gamma \vDash A$, to hold if A is valid in every structure in which Γ is valid. If Γ is the empty set, we write just $\vDash A$, which means of course that A is valid.[3]

[3]This definition of $\Gamma \vDash A$ has the slightly unexpected result that free variables appearing in Γ are treated as if they are universally quantified. For example, according to our definition, we have

2.1.3. For languages which contain the equality symbol, $=$, the interpretation of $=^{\mathcal{M}}$ in any structure $\mathcal{M}$ must be true equality. Of course, this restriction influences the definition of $\Gamma \vDash A$ in languages which contain equality. Let $\vDash'$ indicate logical implication defined with respect to all structures, including structures in which equality is not interpreted by true equality. It is an elementary, but important, fact that $\vDash$ can be defined in terms of $\vDash'$. Namely, let EQ be the set of equality axioms defined in section 2.2.1 below; then $\Gamma \vDash A$ holds if and only if $\Gamma, EQ \vDash' A$ holds.

2.1.4. Definition. A first-order *theory* is a set T of sentences closed under logical implication; i.e., if $T \vDash A$ then $A \in T$. An *axiomatization* of T is a set Γ of sentences such that T is precisely the set of sentences logically implied by Γ.

2.2. Hilbert-style proof systems

2.2.1. A proof system. We give here an example of a proof system $\mathcal{F}_{FO}$ for first-order logic, which is an extension of $\mathcal{F}$ to first-order logic. In addition to the axioms of $\mathcal{F}$ and the rule modus ponens, there are two axiom schemes:

$$A(t) \supset (\exists x)A(x) \qquad \text{and} \qquad (\forall x)A(x) \supset A(t)$$

where A may be any formula and t any term. There are also two quantifier rules of inference:

$$\frac{C \supset A(x)}{C \supset (\forall x)A(x)} \qquad \text{and} \qquad \frac{A(x) \supset C}{(\exists x)A(x) \supset C}$$

where, in both inferences, x may not appear freely in C.

If the first-order language under consideration contains the distinguished equality sign $(=)$, then the *equality axioms* must also be included; namely,

$$(\forall x)(x = x)$$
$$(\forall \vec{x})(\forall \vec{y})(x_1 = y_1 \land \cdots \land x_k = y_k \supset f(\vec{x}) = f(\vec{y}))$$
$$(\forall \vec{x})(\forall \vec{y})(x_1 = y_1 \land \cdots \land x_k = y_k \land P(\vec{x}) \supset P(\vec{y}))$$

where f and P are arbitrary function and predicate symbols and k is their arity.

We leave it to the reader to check that the symmetry and transitivity of equality follow from these axioms, since in the third equality axiom, P may be the equality sign. We write $\mathcal{F}_{\rm FO} \vdash A$ to denote A having an $\mathcal{F}_{\rm FO}$-proof.

$A(x) \vDash (\forall x)A(x)$, and thus $A(x) \vDash A(y)$. An alternative definition of logical implication is often used (but not in this chapter!) in which free variables occurring in Γ are treated as syntactically as constants; under this definition $A(x) \vDash (\forall x)A(x)$ does not hold in general.

The advantage our choice for the definition of $\vDash$, is that it yields a simpler and more elegant proof-theory. This choice does not involve any loss of expressive power since, instead of free variables in Γ, one can use new constant symbols, and thus obtain the same effect as the alternative definition of $\vDash$. In any event, the two definitions of $\vDash$ coincide when Γ is a set of sentences.

2.2.2. The Soundness Theorem.

(1) *If* $\mathcal{F}_{\mathrm{FO}} \vdash A$*, then* $\vDash A$.

(2) *Let* Γ *be a set of sentences. If there is an* $\mathcal{F}_{\mathrm{FO}}$*-proof of* A *using sentences from* Γ *as additional axioms, then* $\Gamma \vDash A$.

The Soundness Theorem states that $\mathcal{F}_{\mathrm{FO}}$ proves only valid formulas. Both parts of the soundness theorem are readily proved by induction on the number of lines in a proof.

2.2.3. The Completeness Theorem.

(1) *If* $\vDash A$*, then* $\mathcal{F}_{\mathrm{FO}} \vdash A$.

(2) *Let* Γ *be a set of formulas. If* $\Gamma \vDash A$*, then there is an* $\mathcal{F}_{\mathrm{FO}}$*-proof of* A *using sentences from* Γ *as additional axioms.*

The completeness theorem and soundness theorem together show that $\mathcal{F}_{\mathrm{FO}}$ is an adequate system for formalizing first-order logic. The completeness theorem was proved originally by Gödel [1930]; the now-standard textbook proof is due to Henkin [1949]. We shall not give a proof for the completeness of $\mathcal{F}_{\mathrm{FO}}$ here; instead, we shall prove the completeness of the cut-free fragment of LK in 2.3.7 below. It is straightforward to see that $\mathcal{F}_{\mathrm{FO}}$ can simulate the LK proof system, and this gives an indirect proof of the completeness of $\mathcal{F}_{\mathrm{FO}}$.

2.2.4. A set Γ of sentences is *consistent* if and only if there is no sentence A such that both A and $\neg A$ are provable from Γ and, equivalently, if and only if there is some formula A such that there is no proof of A from Γ. The soundness and completeness theorems immediately imply that Γ is consistent if and only if Γ is satisfiable.

2.2.5. Historical remarks. Frege [1879] gave the first full formulation of first-order logic. Frege used a pictorial representation of propositional connectives and quantifiers; one remnant of his notation that is still in use is "$\vdash A$", which was Frege's notation for "A is asserted to be true". Frege used the pictorial notation —— A to express a proposition A; whereas he used the notation $\vdash$—— A to express the assertion that the proposition A is true (or has been proved). Negation, implication, and universal quantification were represented by Frege with pictures such as

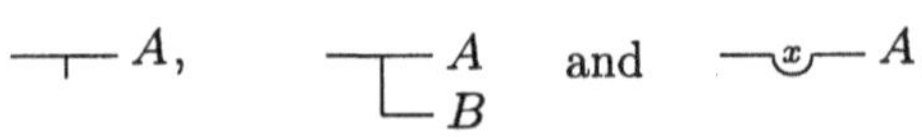

which represent the propositions $\neg A$, $B \supset A$, and $(\forall x)A$, respectively. These constructions can be iterated to form arbitrary first-order expression; that is to say, in the pictures above, A and B may be replaced by arbitrary pictorial propositions. The addition of a vertical line to the left end of a proposition indicates that the proposition has been established to be true (e.g., proved). Thus, for instance

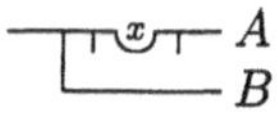

and

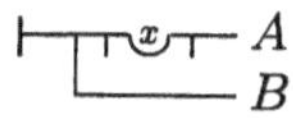

denote the proposition $B \supset (\exists x)A$ and the assertion that this proposition has been established, respectively.

Frege also developed extensions to his first-order logic to include functionals and predicates; however, these extensions were later discovered by Russell to introduce set-theoretic paradoxes.

Peano [1889] introduced, independently of Frege, a fragment of first-order logic for reasoning about integers. Based on the work of Frege and Peano, Whitehead and Russell [1910] gave a consistent framework for formalizing mathematics based on first-order logic, using universal and existential quantifiers denoted (x) and $(\exists x)$.

The 'Hilbert-style' system given above is apparently so-named because it is closely related to a system used by Hilbert and Ackermann [1928], which is based on earlier lectures of Hilbert. The later work of Hilbert and Bernays [1934-39], however, used the ϵ-calculus instead of a 'Hilbert-style' system. The ϵ-calculus contains no quantifiers, but in their place uses symbols $\epsilon_x(A(x))$ which are intended to denote an object x such that $A(x)$ holds, if there is such an object. Note that other free variables and other uses of ϵ symbols may appear in $A(x)$. The ϵ-symbol can be used to express quantifiers by the informal equivalences $(\exists x)A(x) \Leftrightarrow A(\epsilon_x(A(x)))$ and $(\forall x)A(x) \Leftrightarrow A(\epsilon_x(\neg A(x)))$.

The Hilbert-style system we used above is essentially that of Kleene [1952].

2.3. The first-order sequent calculus

In this section, the propositional sequent calculus, introduced in section 1.2, is enlarged to a proof system for first-order logic.

2.3.1. Free and bound variables. The first-order sequent calculus has two classes of variables, called *free variables* and *bound variables.* There are infinitely many variables of each type; free variables are denoted by the metavariables $a, b, c, \ldots$, and bound variables by the metavariables $z, y, x, \ldots$. The essential idea is that free variables may not be quantified, while bound variables may not occur freely in formulas. The syntactic distinction between free and bound variables necessitates a change to the definitions of terms and formulas. Firstly, *terms* are now defined as being built up from free variables and function symbols; whereas, the *semiterms* are defined as being built up from free and bound variables and function symbols. Secondly, only bound variables may ever be quantified. The set of *formulas* is now redefined with the additional requirement that only free variables may occur freely in formulas. *Semiformulas* are like formulas, except that bound variables may occur freely in semiformulas. We henceforth use $r, s, t, \ldots$ as metavariables for terms, and $A, B, C, \ldots$ as metavariables for formulas.

Note that in general, a subformula of a formula will be a semiformula instead of a formula.

One advantage of these conventions on free and bound variables, is that it avoids some of the difficulties involved in defining $A(t)$, since it is always the case that the term t will be freely substitutable for b in a formula $A(b)$. Also, without these conventions, the cut elimination theorem proved below would have to be reformulated slightly. For example, it is not hard to see that $P(x,y) \rightarrow (\exists y)(\exists x)P(y,x)$ would not have a cut-free proof (this example is from Feferman [1968]).

2.3.2. Definition. The first-order sequent calculus LK is defined as an extension of the propositional system PK. LK contains all the rules of inference of PK plus the following additional rules of inference:

The Quantifier Rules

$$\forall\text{:}left\ \frac{A(t), \Gamma \rightarrow \Delta}{(\forall x)A(x), \Gamma \rightarrow \Delta} \qquad \forall\text{:}right\ \frac{\Gamma \rightarrow \Delta, A(b)}{\Gamma \rightarrow \Delta, (\forall x)A(x)}$$

$$\exists\text{:}left\ \frac{A(b), \Gamma \rightarrow \Delta}{(\exists x)A(x), \Gamma \rightarrow \Delta} \qquad \exists\text{:}right\ \frac{\Gamma \rightarrow \Delta, A(t)}{\Gamma \rightarrow \Delta, (\exists x)A(x)}$$

In quantifier rules, A may be an arbitrary formula, t an arbitrary term, and the free variable b of the $\forall$*:right* and $\exists$*:left* inferences is called the *eigenvariable* of the inference and must not appear in Γ, Δ. The propositional rules and the quantifier rules are collectively called *logical rules.*

Most of the syntactic definitions of PK carry over to LK. For example, the notions of (direct) descendents and ancestors are identically defined; and proof lengths are still measured in terms of the number of strong inferences in the proof. The quantifier inferences are, of course, considered strong rules of inferences.

If S is a sequent $\Gamma \rightarrow \Delta$, then we let A_S be the formula $(\bigwedge \Gamma) \supset (\bigvee \Delta)$. Taking S to have the same meaning as A_S, all the definitions of 'validity' and 'logical implication' of section 2.1.2 apply also to sequents. Let the free variables of A_S be $\vec{b}$, so $A_S = A_S(\vec{b})$. We let $\forall S$ denote the *universal closure*, $(\forall \vec{x})A_S(\vec{x})$, of the formula A_S.

2.3.3. LK is defined to allow only initial sequents of the form $A \rightarrow A$ with A atomic. However, it is often convenient to allow other initial sequents; so if $\mathfrak{S}$ is a set of sequents, we define $LK_{\mathfrak{S}}$ to be the proof system defined like LK, but allowing initial sequents to be from $\mathfrak{S}$ too.

An important example is the theory LK_e for first-order logic with equality. LK_e is LK with the addition of the following initial sequents for equality:

$$\rightarrow s = s$$

$$s_1 = t_1, \ldots, s_k = t_k \rightarrow f(\vec{s}) = f(\vec{t})$$

$$s_1 = t_1, \ldots, s_k = t_k, P(\vec{s}) \rightarrow P(\vec{t})$$

We say that a set $\mathfrak{S}$ is *closed under substitution*, if whenever $\Gamma(a) \rightarrow \Delta(a)$ is in $\mathfrak{S}$ and t is a term, then $\Gamma(t) \rightarrow \Delta(t)$ is also in $\mathfrak{S}$.

2.3.4. When P is a proof, we write $P(a)$ and $P(t)$ to indicate that $P(t)$ is the result of replacing every free occurrence of a in formulas of P with t.

Theorem. *Let $\mathfrak{S}$ be a set of sequents which is closed under substitution. If $P(b)$ is an $LK_{\mathfrak{S}}$-proof, and if neither b nor any variable in t is used as an eigenvariable in $P(b)$, then $P(t)$ is a valid $LK_{\mathfrak{S}}$-proof.*

2.3.5. Definition. A free variable in the endsequent of a proof is called a *parameter variable* of the proof. A proof P is said to be *in free variable normal form* provided that (1) no parameter variable is used as an eigenvariable, and (2) every other free variable appearing in P is used exactly once as an eigenvariable and appears in P only in sequents above the inference for which it is used as an eigenvariable.

In this chapter, we consider only tree-like proofs and thus any proof may be put in free variable normal form by merely renaming variables.

2.3.6. Soundness Theorem. *Let $\Gamma \rightarrow \Delta$ be an arbitrary sequent.*

(1) *If $\Gamma \rightarrow \Delta$ has an LK-proof, then $\Gamma \rightarrow \Delta$ is valid.*

(2) *Let $\mathfrak{S}$ be a set of sequents. If $\Gamma \rightarrow \Delta$ has an $LK_{\mathfrak{S}}$-proof, then $\mathfrak{S} \models \Gamma \rightarrow \Delta$.*

The soundness theorem is readily proved by induction on the number of inferences in a proof.

2.3.7. Completeness of cut-free LK. We next prove the completeness of the cut-free fragment of LK and, more generally, the completeness of $LK_{\mathfrak{S}}$. Since our proofs of completeness also give a proof of the (countable) compactness theorem, we include the compactness theorem as part (2) of the completeness theorem.

Cut-free Completeness Theorem. *Let $\Gamma \rightarrow \Delta$ be a sequent in a first-order language L which does not contain equality.*

(1) *If $\Gamma \rightarrow \Delta$ is valid, then it has a cut-free LK-proof.*

(2) *Let Π be a set of sentences. If Π logically implies $\Gamma \rightarrow \Delta$, then there are $C_1, \ldots, C_k \in \Pi$ such that $C_1, \ldots, C_k, \Gamma \rightarrow \Delta$ has a cut-free LK-proof.*

Corollary. *Let $\mathfrak{S}$ be a set of sequents. If $\mathfrak{S}$ logically implies $\Gamma \rightarrow \Delta$, then $\Gamma \rightarrow \Delta$ has an $LK_{\mathfrak{S}}$-proof.*

Note that in the corollary, the $LK_{\mathfrak{S}}$-proof may not be cut-free. An important special case of the corollary is that LK_e is complete. Although, in general, the cut-free fragment of $LK_{\mathfrak{S}}$ may not be complete; we shall see later that it is always possible to get $LK_{\mathfrak{S}}$ proofs which contain no 'free cuts'.

The corollary is an immediate consequence of the cut-free completeness theorem. This is because, if $\mathfrak{S} \vDash \Gamma \rightarrow \Delta$, then part (2) of the theorem implies that there are $S_1, \ldots, S_k \in \mathfrak{S}$ so that $\forall S_1, \ldots, \forall S_k, \Gamma \rightarrow \Delta$ has an LK-proof. But clearly each $\rightarrow \forall S_i$ has an $LK_{\mathfrak{S}}$-proof, so with k further cut inferences, $\Gamma \rightarrow \Delta$ also has an $LK_{\mathfrak{S}}$-proof.

Proof. Part (2) of the cut-free completeness theorem clearly implies part (1); we shall prove (2) only for the case where Π is countable (and hence the language L is, w.l.o.g., countable). Assume Π logically implies $\Gamma \rightarrow \Delta$. The general idea of the argument is to try to build an LK-proof of $\Gamma \rightarrow \Delta$ from the bottom up, working backwards from $\Gamma \rightarrow \Delta$ to initial sequents. This is, of course, analogous to the procedure used to prove Theorem 1.2.8; but because of the presence of quantifiers, the process of searching for a proof of $\Gamma \rightarrow \Delta$ is no longer finite. Thus, we shall have to show the proof-search process eventually terminates — this will be done by showing that if the proof-search does not yield a proof, then $\Gamma \rightarrow \Delta$ is not valid.

Since the language is countable, we may enumerate all L-formulas as $A_1, A_2, A_3, \ldots$ so that every L-formula occurs infinitely often in the enumeration. Likewise, we enumerate all L-terms as $t_1, t_2, t_3, \ldots$ with each term repeated infinitely often. We shall attempt to construct a cut-free proof P of $\Gamma \rightarrow \Delta$. The construction of P will proceed in stages: initially, P consists of just the sequent $\Gamma \rightarrow \Delta$; at each stage, P will be modified (we keep the same name P for the new partially constructed P). A sequent in P is said to be *active* provided it is a leaf sequent of P and there is no formula that occurs in both its antecedent and its succedent. Note that a non-active leaf sequent is derivable with a cut-free proof.

We enumerate all pairs $\langle A_i, t_j \rangle$ by a diagonal enumeration; each stage of the construction of P considers one such pair. Initially, P is the single sequent $\Gamma \rightarrow \Delta$. At each stage we do the following:

Loop: Let $\langle A_i, t_j \rangle$ be the next pair in the enumeration.

Step (1): If A_i is in Π, then replace every sequent $\Gamma' \rightarrow \Delta'$ in P with the sequent $\Gamma', A_i \rightarrow \Delta'$.

Step (2): If A_i is atomic, do nothing and proceed to the next stage. Otherwise, we will modify P at the active sequents which contain A_i by doing one of the following:

Case (2a): If A_i is $\neg B$, then every active sequent in P which contains A_i, say of form $\Gamma', \neg B, \Gamma'' \rightarrow \Delta'$, is replaced by the derivation

$$\frac{\Gamma', \neg B, \Gamma'' \rightarrow \Delta', B}{\Gamma', \neg B, \Gamma'' \rightarrow \Delta'}$$

and similarly, every active sequent in P of the form $\Gamma' \rightarrow \Delta', \neg B, \Delta''$ is replaced by the derivation

$$\frac{B, \Gamma' \rightarrow \Delta', \neg B, \Delta''}{\Gamma' \rightarrow \Delta', \neg B, \Delta''}$$

Case (2b): If A_i is of the form $B \vee C$, then every active sequent in P of the form $\Gamma', B \vee C, \Gamma'' \rightarrow \Delta'$, is replaced by the derivation

$$\frac{B, \Gamma', B \vee C, \Gamma'' \rightarrow \Delta' \qquad C, \Gamma', B \vee C, \Gamma'' \rightarrow \Delta'}{\Gamma', B \vee C, \Gamma'' \rightarrow \Delta'}$$

And, every active sequent in P of the form $\Gamma' \rightarrow \Delta', B \vee C, \Delta''$ is replaced by the derivation

$$\frac{\Gamma' \rightarrow \Delta', B \vee C, \Delta'', B, C}{\Gamma' \rightarrow \Delta', B \vee C, \Delta''}$$

Cases (2c)-(2d): The cases where A_i has outermost connective $\supset$ or $\wedge$ are similar (dual) to case (2b), and are omitted here.

Case (2e): If A_i is of the form $(\exists x)B(x)$, then every active sequent in P of the form $\Gamma', (\exists x)B(x), \Gamma'' \rightarrow \Delta'$ is replaced by the derivation

$$\frac{B(c), \Gamma', (\exists x)B(x), \Gamma'' \rightarrow \Delta'}{\Gamma', (\exists x)B(x), \Gamma'' \rightarrow \Delta'}$$

where c is a new free variable, not used in P yet. In addition, any active sequent of the form $\Gamma' \rightarrow \Delta', (\exists x)B(x), \Delta''$ is replaced by the derivation

$$\frac{\Gamma' \rightarrow \Delta', (\exists x)B(x), \Delta'', B(t_j)}{\Gamma' \rightarrow \Delta', (\exists x)B(x), \Delta''}$$

Note that this, and the dual $\forall$*:left* case, are the only cases where t_j is used. These two cases are also the only two cases where it is really necessary to keep the formula A_i in the new active sequent; we have however, always kept A_i since it makes our arguments a little simpler.

Case (2f): The case where A_i begins with a universal quantifier is dual to case (2e).

Step (3): If there are no active sequents remaining in P, exit from the loop; otherwise, continue with the next loop iteration.

End loop.

If the algorithm constructing P ever halts, then P gives a cut-free proof of $C_1, \ldots, C_k, \Gamma \rightarrow \Delta$ for some $C_1, \ldots, C_k \in \Pi$. This is since each (non-active) initial sequent has a formula A which appears in both its antecedent and succedent and since, by induction on the complexity of A, every sequent $A \rightarrow A$ is derivable with a cut-free proof.

It remains to show that if the above construction of P never halts, then the sequent $\Gamma \rightarrow \Delta$ is not logically implied by Π. So suppose the above construction of P never halts and consider the result of applying the entire infinite construction process. From the details of the construction of P, P will be an infinite tree (except in the exceptional case where $\Gamma \rightarrow \Delta$ contains only atomic formulas and Π is empty, in which case P is a single sequent). If Π is empty, then each node in the infinite tree P will be a sequent; however, in the general case, each node in the infinite tree will be a generalized sequent of the form $\Gamma', \Pi \longrightarrow \Delta'$ with an infinite number of formulas in its antecedent. (At each stage of the construction of P, the sequents contain only finitely many formulas, but in the limit the antecedents contain every formula from Π since these are introduced by step (1).) P is a finitely branching tree, so by König's lemma, there is at least one infinite branch π in P starting at the roots and proceeding up through the tree (except, in the exceptional case, π is to contain just the endsequent). We use π to construct a structure $\mathcal{M}$ and object assignment σ, for which $\Gamma \longrightarrow \Delta$ is not true. The universe of $\mathcal{M}$ is equal to the set of L-terms. The object assignment σ just maps a variable a to itself. The interpretation of a function symbol is defined so that $f^{\mathcal{M}}(r_1, \ldots, r_k)$ is the term $f(r_1, \ldots, r_k)$. Finally, the interpretation of a predicate symbol P is defined by letting $P^{\mathcal{M}}(r_1, \ldots, r_k)$ hold if and only the formula $P(r_1, \ldots, r_k)$ appears in an antecedent of a sequent contained in the branch π.

To finish the proof of the theorem, it suffices to show that every formula A occurring in an antecedent (respectively, a succedent) along π is true (respectively, false) in $\mathcal{M}$ with respect to σ; since this implies that $\Gamma \longrightarrow \Delta$ is not valid. This claim is proved by induction on the complexity of A. For A atomic, it is true by definition. Consider the case where A is of the form $(\exists x)B(x)$. If A appears in an antecedent, then so does a formula of the form $B(c)$; by the induction hypothesis, $B(c)$ is true in $\mathcal{M}$ w.r.t. σ, so hence A is. If A appears in an succedent, then, for every term t, $B(t)$ eventually appears in an succedent; hence every $B(t)$ is false in $\mathcal{M}$ w.r.t. σ, which implies A is also false. Note that A cannot appear in both an antecedent and a succedent along π, since the nodes in π were never active initial sequents. The rest of cases, for different outermost connectives of A, are similar and we omit their proofs.

2.3.8. There are number of *semantic tableau* proof systems, independently due to Beth [1956], Hintikka [1955], Kanger [1957] and Schütte [1965], which are very similar to the first-order cut-free sequent calculus. The proof above, of the completeness of the cut-free sequent calculus, is based on the proofs of the completeness of semantic tableau systems given by these four authors. The original proof, due to Gentzen, was based on the completeness of LK with cut and on a process of eliminating cuts from a proof similar to the construction of the next section.

2.4. Cut elimination. The cut-free completeness theorem, as established above, has a couple of drawbacks. Firstly, we have proved cut-free completeness only for pure LK and it is desirable to also establish a version of cut-free completeness (called 'free-cut free' completeness) for LK_e and for more general systems $LK_{\mathfrak{S}}$. Secondly, the proof is completely non-constructive and gives no bounds on the sizes of cut-free proofs. Of course, the undecidability of validity in first-order logic implies that the size of proofs (cut-free or otherwise) cannot be recursively bounded in terms of the formula being proved; instead, we wish to give an upper bound on the size of a cut-free LK-proof in terms of the size of a general LK-proof.

Accordingly, we shall give a constructive proof of the cut elimination theorem — this proof will give an effective procedure for converting a general LK-proof into a cut-free LK-proof. As part of our analysis, we compute an upper bound on the size of the cut-free proof constructed by this procedure. With some modification, the procedure can also be used to construct free-cut free proofs in LK_e or $LK_{\mathfrak{S}}$; this will be discussed in section 2.4.4.

It is worth noting that the cut elimination theorem proved next, together with the Completeness Theorem 2.2.3 for the Hilbert-style calculus, implies the Cut-free Completeness Theorem 2.3.7. This is because LK can easily simulate Hilbert-style proofs and is thus complete; and then, since any valid sequent has an LK-proof, it also has a cut-free LK-proof.

2.4.1. Definition. The *depth*, $dp(A)$, of a formula A is defined to equal the height of the tree representation of the formula; that is to say:

- $dp(A) = 0$, for A atomic,
- $dp(A \land B) = dp(A \lor B) = dp(A \supset B) = 1 + \max\{dp(A), dp(B)\}$,
- $dp(\neg A) = dp((\exists x)A) = dp((\forall x)A) = 1 + dp(A)$.

The *depth* of a cut inference is defined to equal the depth of its cut formula.

Definition. The *superexponentiation* function 2^x_i, for $i, x \geq 0$, is defined inductively by $2^x_0 = x$ and $2^x_{i+1} = 2^{2^x_i}$. Thus 2^x_i is the number which is expressed in exponential notation as a stack of i many 2's with an x at the top.

2.4.2. Cut-Elimination Theorem. *Let P be an LK-proof and suppose every cut formula in P has depth less than or equal to d. Then there is a cut-free LK-proof P^* with the same endsequent as P, with size*

$$||P^*|| < 2^{||P||}_{2d+2}.$$

Most proofs of this theorem are based on the original proof of Gentzen [1935] which involved making local changes to a proof to reduce the depth of cuts, the number of cuts, or the so-called rank of a cut. We present here a somewhat differently structured proof in which the depth or number of cuts is reduced by making *global* changes to a proof. We feel that our approach has the advantage of making the overall cut elimination process clearer and more intuitive.

The main step in proving the cut elimination theorem will be to establish the following lemma:

2.4.2.1. Lemma. *Let P be an LK-proof with final inference a cut of depth d such that every other cut in P has depth strictly less than d. Then there is an LK-proof P^* with the same endsequent as P with all cuts in P^* of depth less than d and with $||P^*|| < ||P||^2$.*

Proof. The proof P ends with a cut inference

$$\dfrac{\dfrac{\vdots\, Q}{\Gamma \longrightarrow \Delta, A} \qquad \dfrac{\vdots\, R}{A, \Gamma \longrightarrow \Delta}}{\Gamma \longrightarrow \Delta}$$

where the depth of the cut formula A equals d and where all cuts in the subproofs Q and R have depth strictly less than d. The lemma is proved by cases, based on the outermost logical connective of the cut formula A. We can assume w.l.o.g. that both Q and R contain at least one strong inference; since otherwise, we must have A in Γ or in Δ, or have a formula which occurs in both Γ and Δ, and in the former case, the sequent $\Gamma \longrightarrow \Delta$ is obtainable by weak inferences from one of the upper sequents and the cut can therefore be eliminated, and in the second case, $\Gamma \longrightarrow \Delta$ can be inferred with no cut inference at all. The proof P is also assumed to be in free variable normal form.

Case (a): Suppose A is a formula of the form $\neg B$. We shall form new proofs Q^* and R^* of the sequents $B, \Gamma \longrightarrow \Delta$ and $\Gamma \longrightarrow \Delta, B$, which can then be combined with a cut inference of depth $d-1$ to give the proof P^* of $\Gamma \longrightarrow \Delta$. To form Q^*, first form Q' by replacing every sequent $\Pi \longrightarrow \Lambda$ in Q with the sequent $\Pi, B \longrightarrow \Lambda^-$, where Λ^- is obtained from Λ by removing all direct ancestors of the cut formula A. Of course, Q' is not a valid proof; for example, a $\neg$*:right* inference in Q of the form

$$\frac{B, \Pi \longrightarrow \Lambda}{\Pi \longrightarrow \Lambda, \neg B}$$

could become in Q'

$$\frac{B, \Pi, B \longrightarrow \Lambda^-}{\Pi, B \longrightarrow \Lambda^-}$$

This is not, strictly speaking, a valid inference; but of course, it can be modified to be valid by inserting some exchanges and a contraction. In this fashion, it is straightforward to modify Q' so that it becomes a valid proof Q^* by removing some $\neg$*:left* inferences and inserting some weak inferences. We leave it to the reader to check that Q^* can be validly formed in this way: it should be noted that we are using the assumption that initial sequents $C \longrightarrow C$ must have C atomic. The proof, R^*, of $\Gamma \longrightarrow \Delta, B$ is formed in a similar fashion from R. Obviously, no new cuts are introduced by this process and, since we do not count weak inferences, $||Q^*|| \le ||Q||$ and $||R^*|| \le ||R||$; thus P^* has only cuts of depth $< d$ and has $||P^*|| \le ||P||$.

Case (b): Now suppose the cut formula A is of the form $B \lor C$. We define Q' as a tree of sequents, with root labeled $\Gamma \longrightarrow \Delta, B, C$, by replacing every sequent $\Pi \longrightarrow \Lambda$ in Q with the sequent $\Pi \longrightarrow \Lambda^-, B, C$, where Λ^- is Λ minus all occurrences of direct ancestors of the cut formula. By removing some formerly $\lor$*:right* inferences from Q' and by adding some weak inferences, Q' can be transformed into a valid proof Q^*. Now construct R_B from R by replacing every occurrence in R of $B \lor C$ as a direct ancestor of the cut formula with just the formula B. One way that R_B can fail to be a valid proof is that an $\lor$*:left* inference

$$\frac{B, \Pi \rightarrow \Lambda \qquad C, \Pi \rightarrow \Lambda}{B \vee C, \Pi \rightarrow \Lambda}$$

may become just

$$\frac{B, \Pi \rightarrow \Lambda \qquad C, \Pi \rightarrow \Lambda}{B, \Pi \rightarrow \Lambda}$$

in R_B. This is no longer a valid inference, but it can be fixed up by discarding the inference and its upper right hypothesis, including discarding the entire subproof of the upper right hypothesis. The only other changes needed to make R_B valid are the addition of weak inferences, and in this way, a valid proof R_B of $B, \Gamma \rightarrow \Delta$ is formed. A similar process forms a valid proof R_C of $C, \Gamma \rightarrow \Delta$. The proof P^* can now be defined to be

$$\dfrac{\dfrac{\dfrac{\vdots\, Q^*}{\Gamma \rightarrow \Delta, B, C} \qquad \dfrac{\vdots\, R_C}{C, \Gamma \rightarrow \Delta}}{\Gamma \rightarrow \Delta, B} \qquad \dfrac{\vdots\, R_B}{B, \Gamma \rightarrow \Delta}}{\Gamma \rightarrow \Delta}$$

The processes of forming Q^*, R_B, and R_C did not introduce any new cuts or any new strong inferences. Thus we clearly have that every cut in P^* has depth $< d$, and that $||P^*|| \le ||Q|| + 2||R|| + 2$. Since $||P|| = ||Q|| + ||R|| + 1$ and $||Q||, ||R|| \ge 1$, this suffices to prove the lemma for this case.

Cases (c),(d): The cases where A has outermost connective $\wedge$ or $\supset$ are very similar to the previous case, and are omitted.

Case (e): Now suppose A is of the form $(\exists x)B(x)$. First consider how the formula $(\exists x)B(x)$ can be introduced in the subproof Q. Since it is not atomic, it cannot be introduced in an initial sequent; thus, it can only be introduced by weakenings and by $\exists$*:right* inferences. Suppose that there are $k \ge 0$ many $\exists$*:right* inferences in Q which have their principal formula a direct ancestor of the cut formula. These can be enumerated as

$$\frac{\Pi_i \rightarrow \Lambda_i, B(t_i)}{\Pi_i \rightarrow \Lambda_i, (\exists x)B(x)}$$

with $1 \le i \le k$. Similarly, we locate all the $\exists$*:left* inferences in R which have principal formula a direct ancestor of the cut formula and enumerate these as

$$\frac{B(a_i), \Pi'_i \rightarrow \Lambda'_i}{(\exists x)B(x), \Pi'_i \rightarrow \Lambda'_i}$$

for $1 \le i \le \ell$.

For each $i \le k$, we form a proof R_i of the sequent $B(t_i), \Gamma \rightarrow \Delta$ by replacing all ℓ of the variables a_j with the term t_i everywhere in R, replacing every direct ancestor of the cut formula $(\exists x)B(x)$ in R with $B(t_i)$, and then removing the ℓ $\exists$*:left* inferences. It is easy to see that this yields a valid proof; note that the fact that P is in free variable normal form ensures that replacing the a_j's with t_i will not impact the eigenvariable conditions for inferences in R.

Now, form Q' from Q by replacing each sequent $\Pi \rightarrow \Lambda$ in Q with the sequent $\Pi, \Gamma \rightarrow \Delta, \Lambda^-$, where Λ^- is Λ minus all direct ancestors of the cut formula A.

Clearly, Q' ends with the sequent $\Gamma, \Gamma \longrightarrow \Delta, \Delta$; however, it is not a valid proof. To fix it up to be a valid proof, we need to do the following. First, an initial sequent in Q' will be of the form $A, \Gamma \longrightarrow \Delta, A$; this can be validly derived by using the initial sequent $A \longrightarrow A$ followed by weakenings and exchanges. Second, for $1 \le i \le k$, the i-th $\exists$*:right* inference enumerated above, will be of the form

$$\frac{\Pi_i, \Gamma \longrightarrow \Delta, \Lambda_i, B(t_i)}{\Pi_i, \Gamma \longrightarrow \Delta, \Lambda_i}$$

in Q. This can be replaced by the following inferences

$$\dfrac{\Pi_i, \Gamma \longrightarrow \Delta, \Lambda_i, B(t_i) \qquad \overset{\ddots \vdots \cdot^{\cdot} R_i}{B(t_i), \Gamma \longrightarrow \Delta}}{\overline{\Pi_i, \Gamma \longrightarrow \Delta, \Lambda_i}}$$

Note that this has replaced the $\exists$*:right* inference of Q with a cut of depth $d-1$ and some weak inferences.

No further changes are needed to Q' to make it a valid proof. In particular, the eigenvariable conditions still hold since no free variable in $\Gamma \longrightarrow \Delta$ is used as an eigenvariable in Q. By adding some exchanges and contractions to the end of this proof, the desired proof P^* of $\Gamma \longrightarrow \Delta$ is obtained. It is clear that every cut in P^* has depth $< d$. From inspection of the construction of P^*, the size of P^* can be bounded by

$$||P^*|| \le ||Q|| \cdot (||R|| + 1) < ||P||^2.$$

Case (f): The case where A is of the form $(\forall x)B(x)$ is completely dual to case (e), and is omitted here.

Case (g): Finally, consider the case where A is atomic. Form R' from R by replacing every sequent $\Pi \longrightarrow \Lambda$ in R with the sequent $\Pi^-, \Gamma \longrightarrow \Delta, \Lambda$, where Π^- is Π minus all occurrences of direct ancestors of A. R' will end with the sequent $\Gamma, \Gamma \longrightarrow \Delta, \Delta$ and will be valid as a proof, except for its initial sequents. The initial sequents $B \longrightarrow B$ in R, with B not a direct ancestor of the cut formula A, become $B, \Gamma \longrightarrow \Delta, B$ in R'; these are readily inferred from the initial sequent $B \longrightarrow B$ with only weak inferences. On the other hand, the other initial sequents $A \longrightarrow A$ in R become $\Gamma \longrightarrow \Delta, A$ which is just the endsequent of Q. The desired proof P^* of $\Gamma \longrightarrow \Delta$ is thus formed from R' by adding some weak inferences and adding some copies of the subproof Q to the leaves of R', and by adding some exchanges and contractions to the end of R'.

Since Q and R have only cuts of degree $< d$ (i.e., have no cuts, since $d = 0$), P^* likewise has only cuts of degree $< d$. Also, since the number of initial sequents in R' is bounded by $||R|| + 1$, the size of P^* can be bounded by

$$||P^*|| \le ||R|| + ||Q|| \cdot (||R|| + 1) < (||Q|| + 1)(||R|| + 1) < ||P||^2$$

That completes the proof of Lemma 2.4.2.1.

Lemma 2.4.2.1 shows how to replace a single cut by lower depth cut inferences. By iterating this construction, it is possible to remove all cuts of the maximum depth d in a proof. This is stated as Lemma 2.4.2.2: the Cut-Elimination Theorem is an immediate consequence of this lemma.

2.4.2.2. Lemma. *If P is an LK-proof with all cuts of depth at most d, there is an LK-proof P^* with the same endsequent which has all cuts of depth strictly less than d and with size $||P^*|| < 2^{2^{||P||}}$.*

Proof. Lemma 2.4.2.2 will be proved by induction on the number of depth d cuts in P. The base case, where there are no depth d cuts is trivial, of course, since $||P|| < 2^{2^{||P||}}$. For the induction step, it suffices to prove the lemma in the case where P ends with a cut inference

$$\dfrac{\begin{array}{c}\vdots\, Q \\ \Gamma \rightarrow \Delta, A\end{array} \qquad \begin{array}{c}\vdots\, R \\ A, \Gamma \rightarrow \Delta\end{array}}{\Gamma \rightarrow \Delta}$$

where the cut formula A is of depth d.

First suppose that one of the subproofs, say R, does not have any strong inferences; i.e., $||R|| = 0$. Therefore, R must either contain the axiom $A \rightarrow A$, or must have direct ancestors of the cut formula A introduced only by weakenings. In the former case, A must appear in Δ, and the desired proof P^* can be obtained from Q by adding some exchanges and a contraction to the end of Q. In the second case, P^* can be obtained from R by removing all the *Weakening:left* inferences that introduce direct ancestors of the cut formula A (and possibly removing some exchanges and contractions involving these A's). A similar argument works for the case $||Q|| = 0$. In both cases, $||P^*|| < ||P|| < 2^{2^{||P||}}$.

Second, suppose that $||Q||$ and $||P||$ are both nonzero. By the induction hypothesis, there are proofs Q^* and R^* of the same endsequents, with all cuts of depth less than d, and with

$$||Q^*|| < 2^{2^{||Q||}} \qquad \text{and} \qquad ||R^*|| < 2^{2^{||R||}}$$

Applying Lemma 2.4.2.1 to the proof

$$\dfrac{\begin{array}{c}\vdots\, Q^* \\ \Gamma \rightarrow \Delta, A\end{array} \qquad \begin{array}{c}\vdots\, R^* \\ A, \Gamma \rightarrow \Delta\end{array}}{\Gamma \rightarrow \Delta}$$

gives a proof P^* of $\Gamma \rightarrow \Delta$ with all cuts of depth $< d$, so that

$$||P^*|| \;<\; (||Q^*|| + ||R^*|| + 1)^2 \;\leq\; \left(2^{2^{||Q||}} + 2^{2^{||R||}} - 1\right)^2 \;<\; 2^{2^{||Q||+||R||+1}} \;=\; 2^{2^{||P||}}.$$

The final inequality holds since $||Q||, ||R|| \geq 1$.
Q.E.D. Lemma 2.4.2.2 and the Cut Elimination Theorem.

2.4.3. A general bound on cut elimination. The upper bound $2_{2d+2}^{||P||}$ in the Cut Elimination Theorem is based not only on the size of P, but also on the maximum depth of the cut formulas in P. However, there is a general method that allows the bound to be expressed in terms of just $||P||$. This is based on the following:

Proposition. *Suppose P is an LK-proof of the sequent $\Gamma \longrightarrow \Delta$. Then there is a cut-free proof P^* of the same sequent with size $||P^*|| < 2_{2||P||}^{||P||}$.*

Proof. The proposition is obviously true if P has no cuts, so suppose it contains at least one. We need some definitions: Two semiformulas are said to be *term variants* if they have identical logical structure and thus one can be transformed into the other by changing only its semiterms. Obviously the 'term variant' relation is an equivalence relation. For any equivalence class of term variants, there is a formula skeleton $R(\tau_1, \ldots, \tau_k)$ such that the equivalence class contains exactly the semiformulas of the form $R(t_1, \ldots, t_k)$ where $t_1, \ldots, t_k$ are semiterms.

Let $c_1, c_2, c_3, \ldots$ be an infinite sequence of new free variables. For $R(\cdots)$ a formula skeleton for a term variant class, we pick a new predicate symbol P_R. Given a member $R(t_1, \ldots, t_k)$ of the term variant class, we can form the atomic semiformula $P_R(t_1, \ldots, t_k)$ and its bound variables correspond to the bound variables which are freely occurring in $R(\vec{t})$.

A formula A is defined to be *active* in P if some strong inference in P has a term variant of A as a principal formula. Suppose A is non-atomic and not active in P; let $R(\cdots)$ give the term variant class of A. Consider the following transformation of P: for each term variant $R(t_1, \ldots, t_k)$ which appears as a subformula of a formula in P, replace it with $P_R(t_1, \ldots, t_k)$. It can be checked that this transformation yields a valid proof, which contains exactly the same kinds of inferences as P.

By repeatedly applying this transformation, a valid proof P' is obtained in which every subformula either is atomic or is a term variant of a principal formula of a strong inference in P'. Since there are at most $||P||$ many strong inferences in P', including at least one cut inference, and since initial sequents contain only atomic formulas, this implies that every formula in P' has depth less than $||P||$. (To prove the last assertion, note that every formula in P' either is atomic or has an atomic ancestor.) Therefore, by the cut elimination theorem, there is a cut-free proof P'' with the same endsequent of size $||P''|| < 2_{2||P||}^{||P||}$.

This proof P'' has the desired size, but it may no longer be a proof of $\Gamma \longrightarrow \Delta$, since it may contain subformulas of the form $P_R(t_1, \ldots, t_k)$. However, we may (iteratively) replace all such subformulas in P'' with the formula $R(t_1, \ldots, t_k)$. This yields the desired proof of $\Gamma \longrightarrow \Delta$. □

2.4.4. Free-cut elimination. We next investigate the possibility of eliminating cuts in $LK_{\mathfrak{S}}$-proofs; that is to say, in proofs in which initial sequents may come from $\mathfrak{S}$. The set $\mathfrak{S}$ is presumed to be a fixed set of sequents closed under substitution. An important example of such an $\mathfrak{S}$ is the set of equality axioms of LK_e; however, $\mathfrak{S}$ can also be the axioms of any first-order theory.

The Cut Elimination Theorem 2.4.2 applied only to LK-proofs; on the other hand, the proof of Corollary 2.3.7 gave a partial cut elimination theorem for $LK_{\mathfrak{S}}$-proofs. These results can be significantly improved by introducing a notion of 'free' cuts and giving a construction eliminating free cuts from $LK_{\mathfrak{S}}$-proofs by a method similar to the proof of Theorem 2.4.2.

2.4.4.1. Definition. Let P be an $LK_{\mathfrak{S}}$-proof. A formula occurring in P is *anchored* (by an $\mathfrak{S}$-sequent) if it is a direct descendent of a formula occurring in an initial sequent in $\mathfrak{S}$. A cut inference in P is *anchored* if either:

(i) the cut formula is not atomic and at least one of the two occurrences of the cut formula in the upper sequents is anchored, or

(ii) the cut formula is atomic and both of the occurrences of the cut formula in the upper sequents are anchored.

A cut inference which is not anchored is said to be *free*. The proof P is *free-cut free* if it contains no free cuts.

2.4.4.2. An occurrence of a formula in a proof P is said to be *only weakly introduced* in P if it does not have a direct ancestor which appears in an initial sequent or which is a principal formula of a strong inference. It is often convenient to assume that a proof P satisfies the condition that no cut formula is only weakly introduced; that is to say, that every cut inference satisfies the condition that neither occurrence of its cut formula is only weakly introduced in P.

Note that if P is an $LK_{\mathfrak{S}}$-proof, then P can be assumed w.l.o.g. to satisfy this extra condition without an increase in the size of the proof or in the depth of cuts in the proof. However, conforming to this convention might increase the number and depth of free cuts in the proof P. To see this suppose that P contains the cut inference with one of its cut formulas weakly introduced; for instance, suppose that an inference

$$\frac{\Gamma \longrightarrow \Delta, A \qquad A, \Gamma \longrightarrow \Delta}{\Gamma \longrightarrow \Delta}$$

has the A occurring in the right upper sequent $A, \Gamma \longrightarrow \Delta$ only weakly introduced. Now, it is possible that the subproof of $\Gamma \longrightarrow \Delta, A$ causes a formula B in Γ or Δ to be anchored, and that the corresponding B is not anchored in the subproof of $A, \Gamma \longrightarrow \Delta$. The obvious way to eliminate this cut is to remove all direct ancestors of the A in the right upper sequent, thereby getting a proof of $\Gamma \longrightarrow \Delta$, and then discard the left upper sequent and its proof. This, of course, causes B to be only weakly introduced in the new subproof of $\Gamma \longrightarrow \Delta$. In other words, the occurrence of B in the sequent $\Gamma \longrightarrow \Delta$ becomes unanchored. Then, if a direct descendent of B is later used as a cut formula, the elimination of the cut on A could have changed the cut from an anchored cut into a free cut.

Our proof of the free-cut elimination theorem will use induction on the maximum depth of free cuts appearing in a proof. For this induction to work, it is important that anchored cuts do not change into free cuts, introducing new free cuts of higher depth. To avoid this, we will assume that no cut formulas in the proof are only weakly introduced.

2.4.5. Free-cut Elimination Theorem. *Let $\mathfrak{S}$ be a set of sequents closed under substitution.*
(1) *If $LK_{\mathfrak{S}} \vdash \Gamma \longrightarrow \Delta$, then there is a free-cut free $LK_{\mathfrak{S}}$-proof of $\Gamma \longrightarrow \Delta$.*
(2) *Let P be an $LK_{\mathfrak{S}}$-proof satisfying condition 2.4.4.2 and suppose every anchored cut formula in P has depth less than or equal to d. Then there is a free-cut free $LK_{\mathfrak{S}}$-proof P^* with the same endsequent as P, with size*

$$||P^*|| < 2^{||P||}_{2d+2}.$$

The Free-cut Elimination Theorem is essentially due to Gentzen; Takeuti [1987] states a related construction for use with induction rules, which we describe in section 2.4.6.

Proof. Obviously, it suffices to prove part (2) of Theorem 2.4.5. For this proof, we shall give a procedure (Lemma 2.4.5.1) for removing one maximum depth free cut. It is not possible to use the procedure from the proof of Lemma 2.4.2.1 without modification, because it may remove $\mathfrak{S}$-sequents from the proof and thereby change some anchored cuts into free cuts (and thereby increase the maximum depth of free cuts). This is unacceptable since our proof uses induction on the maximum depth of free cuts in P. This undesirable situation can happen in case (b) of the proof of Lemma 2.4.2.1 where the outermost connective of the cut formula A is $\lor$, since at various points, subproofs ending with $C, \Pi \longrightarrow \Lambda$ or with $B, \Pi \longrightarrow \Lambda$ are just discarded in R'_B and in R'_C. Subformulas in Π and Λ may become unanchored by this process. This can happen also in the similar cases (c) and (d). In addition, it could also happen in cases (e), (f) and (g) if the cut formula occurring in the right upper sequent is only weakly introduced, since in these cases the subproof on the left is completely discarded; however, since condition 2.4.4.2 holds, this never occurs. Therefore, we need the following analogue of Lemma 2.4.2.1:

2.4.5.1. Lemma. *Let P be an $LK_{\mathfrak{S}}$-proof with final inference a free cut of depth d such that every other free cut in P has depth strictly less than d. Then there is an $LK_{\mathfrak{S}}$-proof P^* with the same endsequent as P with all free cuts in P^* of depth less than d and with $||P^*|| < ||P||^2$. Furthermore, every formula occurring in the endsequent of P which was anchored by an $\mathfrak{S}$-sequent is still anchored in the proof P^*, and every formula in the endsequent of P^* which is only weakly introduced in P^* was already only weakly introduced in P.*

Proof. We indicate how to modify the proof of Lemma 2.4.2.1. Assume that the proof P ends with a free cut inference

$$\dfrac{\begin{matrix} \vdots\, Q \\ \Gamma \longrightarrow \Delta, A \end{matrix} \qquad \begin{matrix} \vdots\, R \\ A, \Gamma \longrightarrow \Delta \end{matrix}}{\Gamma \longrightarrow \Delta}$$

If the cut formula A is nonatomic, then Q and R both contain at least one strong inference with principal formula equal to A. As before the proof splits into cases depending on the outermost connective of A. When A has outermost connective $\neg$, $\exists$ or $\forall$ then the argument used in cases (a), (e) and (f) of Lemma 2.4.2.1 still works: we leave it to the reader to check that, in these cases, any formulas in the sequent $\Gamma \longrightarrow \Delta$ which were anchored or were not only weakly introduced in P still have these properties in P^*.

For the case where A is of the form $B \vee C$, a different construction is needed. In this case, form a proof Q^* with endsequent $\Gamma \longrightarrow \Delta, B, C$ by the construction used in case (b) of the proof of Lemma 2.4.2.1. Also form R' from R be replacing every sequent $\Pi \longrightarrow \Lambda$ in R with the sequent $\Pi^-, \Gamma \longrightarrow \Delta, \Lambda$ where Π^- is Π minus all direct ancestors of the cut formula $B \vee C$. An $\vee$*:left* inference in R with principal formula $B \vee C$ will no longer be a valid inference since, in R' it will become

$$\frac{B, \Pi^-, \Gamma \longrightarrow \Delta, \Lambda \qquad C, \Pi^-, \Gamma \longrightarrow \Delta, \Lambda}{\Pi^-, \Gamma \longrightarrow \Delta, \Lambda}$$

This can be transformed into a valid proof by replacing it with

$$\dfrac{\dfrac{\dfrac{\vdots\, Q}{\Gamma \longrightarrow \Delta, B, C} \qquad \dfrac{\vdots}{C, \Pi^-, \Gamma \longrightarrow \Delta, \Lambda}}{\Pi^-, \Gamma \longrightarrow \Delta, \Lambda, B} \qquad \dfrac{\vdots}{B, \Pi^-, \Gamma \longrightarrow \Delta, \Lambda}}{\Pi^-, \Gamma \longrightarrow \Delta, \Lambda}$$

Fixing up R' in this way, plus adding some weak inferences yields a valid proof R^* of $\Gamma, \Gamma \longrightarrow \Delta, \Delta$. Appending some exchange and contraction inferences yields the desired proof P^*. It is readily checked that $||P^*|| \leq ||R|| \cdot (||Q|| + 1) < ||P||^2$.

The final case to consider is the case where A is atomic. Since the cut is not anchored, the cut formula will either not be anchored in Q or not be anchored in R. In the latter case, since the cut formula is not only weakly introduced, it must have a direct ancestor in an initial sequent $A \longrightarrow A$ of R. Therefore, the argument from case (g) of the proof of Lemma 2.4.2.1 still works. In the former case, where the cut formula is not anchored in Q, the dual argument works.

That completes the proof of Lemma 2.4.5.1. The Free-cut Elimination Theorem follows immediately from this lemma in the same way that the Cut-Elimination Theorem followed from Lemma 2.4.2.1.
Q.E.D. Free-cut Elimination Theorem.

2.4.6. Free-cut elimination with induction rules. A very useful application of free-cut elimination is for analyzing subtheories of arithmetic in which induction is restricted to certain classes of formulas. For these fragments of arithmetic, the free-cut elimination theorem becomes easier to use if induction rules are used in place of induction axioms. The most common formulation of induction rules is as inferences of the form:

$$\frac{A(b), \Gamma \longrightarrow \Delta, A(b+1)}{A(0), \Gamma \longrightarrow \Delta, A(t)}$$

where b is an eigenvariable and may appear only as indicated, and t is an arbitrary term. It is easily checked that, because of the presence of the side formulas Γ and Δ, the induction rule for A is equivalent to the induction axiom for A:

$$A(0) \wedge (\forall x)(A(x) \supset A(x+1)) \supset (\forall x)A(x).$$

For Φ a set of formulas, Φ-IND indicates the theory with the above induction rules for all formulas $A \in \Phi$. We shall always assume that such a set Φ is closed under (term) substitution.

The classic examples of arithmetic theories axiomatized by induction are the theories $I\Sigma_n$. These are axiomatized by the six axioms of Robinson's theory Q plus induction for Σ_n formulas (see Chapter II). When $I\Sigma_n$ is formalized in the sequent calculus, it has the non-logical initial sequents expressing the axioms of Q plus the induction rule for Σ_n formulas. The initial sequents of $I\Sigma_n$ proofs are always purely existential. Along the same lines, the fragments T_2^n and S_2^n of bounded arithmetic are commonly formalized in the sequent calculus with quantifier-free non-logical initial sequents from BASIC, plus induction on Σ_n^b-formulas (see Chapter II or Buss [1986]). For T_2^n, the induction rules are as shown above; for S_2^n the following *PIND* rules are used:

$$\frac{A(\lfloor \frac{1}{2}b \rfloor), \Gamma \longrightarrow \Delta, A(b)}{A(0), \Gamma \longrightarrow \Delta, A(t)}$$

Definition. Let P be a sequent calculus proof in an arithmetic theory $\mathfrak{S} + \Phi$-IND (or $\mathfrak{S} + \Phi$-PIND). The *principal formulas* of an induction inference are the formulas denoted by $A(0)$ and $A(t)$ in the lower sequent. An occurrence of a formula in P is defined to be *anchored* if it is a direct descendent of a formula occurring in an initial sequent from $\mathfrak{S}$ or it is a direct descendent of a principal formula of an induction inference.

Using this definition of anchored formulas, the notions of *anchored* cut inference and *free* cut inference are defined identically to Definition 2.4.4.1.

Free-cut Elimination Theorem. *Let $T = \mathfrak{S} + \Phi$-IND be a theory of arithmetic, with $\mathfrak{S}$ and Φ closed under term substitution. Let $\Gamma \longrightarrow \Delta$ be a logical consequence of T. Then there is a free-cut free T-proof of $\Gamma \longrightarrow \Delta$. Furthermore, the size bounds of Theorem 2.4.5(2) apply to T-proofs.*

This theorem is proved by an argument identical to the proof of Theorem 2.4.5. The theorem also applies without modification to theories axiomatized with PIND in place of IND. It is also straightforward to modify the free cut elimination to work with sequent calculus formulations of inference rules other than induction. For example, the collection (replacement) axioms for fragments of Peano arithmetic can be equivalently formulated as sequent calculus rules. The obvious straightforward modification of the Free-cut Elimination theorem applies to theories with collection rules.

A particularly useful corollary of the Free-Cut Elimination Theorem is:

2.4.7. Corollary. *Let $\mathfrak{S}$ be a set of sequents and Φ be a set of formulas closed under term substitution and under subformulas, such that every sequent in $\mathfrak{S}$ contains only formulas from Φ. Let T be the theory $\mathfrak{S} + \Phi$-IND (or $\mathfrak{S} + \Phi$-PIND). Suppose that $\Gamma \longrightarrow \Delta$ is a consequence of T and that every formula in $\Gamma \longrightarrow \Delta$ is in Φ. Then there is a T-proof P of $\Gamma \longrightarrow \Delta$ such that every formula appearing in P is in Φ.*

To prove this corollary, just note that it holds for every free-cut free T-proof of $\Gamma \longrightarrow \Delta$; this is because every formula in the proof must be an ancestor of either a cut formula or a formula in the endsequent.

2.4.8. Natural deduction. Natural deduction proof systems were introduced by Gentzen [1935] in the same paper which introduced the first-order sequent calculus. Gentzen's motivation in defining natural deduction was (in his words) "to set up a formula system which comes as close as possible to actual reasoning."[4] The importance of natural deduction was established by the classic result of Prawitz [1965] that a version of the cut elimination holds also for natural deduction.

It is fair to say that the process of constructing natural deduction proofs is indeed 'natural' in that it corresponds closely to the human reasoning process. On the other hand, a fully constructed natural deduction proof can be very confusing to read; in particular, because of the non-local nature of natural deduction proofs, it is difficult to quickly ascertain which formulas depends on which hypotheses.

Natural deduction proof systems are particularly elegant for intuitionistic logic, especially with respect to the Curry-Howard formulas-as-types interpretation (see section 3.1.5). A good modern treatment of applications of natural deduction is given by Girard [1989].

We give a short definition of the natural deduction proof system here; however, the reader should refer to Prawitz [1965] for a full treatment and for statements of the normalization theorems. The definitions of terms and formulas and the conventions on free and bound variables are the same for natural deduction as for the sequent calculus, except that negation, $\neg$, is not a basic symbol; instead, a new atomic formula $\perp$ is used to denote absurdity (the constant *False*), and $\neg A$ abbreviates $A \supset \perp$. A natural deduction proof is a tree of formulas; any formula may appear at a leaf, as a hypothesis. Various inferences may close or *discharge* the hypotheses; in a completed proof all hypotheses must be discharged and the proof is a proof of the formula appearing at the root node at the bottom of the tree. It is best to picture the construction of a natural deduction proof as an ongoing process; at any point in this process some hypotheses may already be discharged, whereas others remain open. The valid rules of inference are given below; they are classified as *introduction* rules

[4]Gentzen [1969], p. 68

or *elimination* rules. Hypotheses discharged by an inference are shown in square brackets.

$\wedge$-intro $\dfrac{A \quad B}{A \wedge B}$ $\wedge$-elim $\dfrac{A \wedge B}{A}$ $\dfrac{A \wedge B}{B}$

$\vee$-intro $\dfrac{A}{A \vee B}$ $\dfrac{B}{A \vee B}$ $\vee$-elim $\dfrac{A \vee B \quad \begin{matrix}[A]\\ C\end{matrix} \quad \begin{matrix}[B]\\ C\end{matrix}}{C}$

$\supset$-intro $\dfrac{\begin{matrix}[A]\\ B\end{matrix}}{A \supset B}$ $\supset$-elim $\dfrac{A \quad A \supset B}{B}$

$\forall$-intro $\dfrac{A(b)}{(\forall x)A(x)}$ $\forall$-elim $\dfrac{(\forall x)A(x)}{A(t)}$

$\exists$-intro $\dfrac{A(t)}{(\exists x)A(x)}$ $\exists$-elim $\dfrac{(\exists x)A(x) \quad \begin{matrix}[A(b)]\\ B\end{matrix}}{B}$

$\bot_I$ $\dfrac{\bot}{A}$ $\bot_C$ $\dfrac{\begin{matrix}[A \supset \bot]\\ \bot\end{matrix}}{A}$

In the $\forall$-intro and $\exists$-elim rules, the free variable b is an eigenvariable: this means it must not appear in any non-discharged hypothesis above the $\forall$-intro inference or in any non-discharged hypotheses other than $A(b)$ above the hypothesis B of the $\exists$-elim inference. The $\bot_I$ rule is used for both intuitionistic and classical logic; the $\bot_C$ rule is included for classical logic. If both rules for $\bot$ are omitted, then *minimal* logic is obtained.

2.5. Herbrand's theorem, interpolation and definability theorems

2.5.1. Herbrand's theorem. One of the fundamental theorems of mathematical logic is the theorem of Herbrand [1930] which allows a certain type of reduction of first-order logic to propositional logic. In its basic form it states:

Herbrand's Theorem. *Let T be a theory axiomatized by purely universal formulas. Suppose that $T \models (\forall \vec{x})(\exists y_1, \ldots, y_k)B(\vec{x}, \vec{y})$ with $B(\vec{x}, \vec{y})$ a quantifier-free formula. Then there is a finite sequence of terms $t_{i,j} = t_{i,j}(\vec{x})$, with $1 \le i \le r$ and $1 \le j \le k$ so that*

$$T \vdash (\forall \vec{x}) \left(\bigvee_{i=1}^{r} B(\vec{x}, t_{i,1}, \ldots, t_{i,k}) \right).$$

Proof. Since T is axiomatized by purely universal formulas, it may, without loss of generality, be axiomatized by quantifier-free formulas (obtained by removing the universal quantifiers). Let $\mathfrak{T}$ be the set of sequents $\rightarrow A$ with A a (quantifier-free) axiom of T. Since $T \models (\forall \vec{x})(\exists \vec{y})B(\vec{x}, \vec{y})$, there is a $LK_{\mathfrak{T}}$-proof of the sequent

$\rightarrow (\exists\vec{y})B(\vec{a},\vec{y})$. By the free-cut elimination theorem, there is a free-cut free proof P of this sequent.

Since the $\mathfrak{T}$-sequents contain only quantifier-free formulas, all cut formulas in P are quantifier-free. Thus, any non-quantifier-free formula in P must be of the form $(\exists y_j)\cdots(\exists y_k)B(\vec{a},t_1,\ldots,t_{j-1},y_j,\ldots,y_k)$ with $1 \le j < k$. We claim that P can be modified to be a valid proof of a sequent of the form

$$\rightarrow B(\vec{a},t_{1,1},\ldots,t_{1,k}),\ldots,B(\vec{a},t_{r,1},\ldots,t_{r,k}).$$

The general idea is to remove all $\exists$*:right* inferences in P and remove all existential quantifiers, replacing the bound variables by appropriate terms. Since there may have been contractions on existential formulas that are no longer identical after terms are substituted for variables it will also be necessary to remove contractions and add additional formulas to the sequents. To do this more formally, we know that any sequent in P is of the form $\Gamma \rightarrow \Delta, \Delta'$ (up to order of the formulas in the sequent), where each formula in Γ and Δ is quantifier-free and where each formula in Δ' is not quantifier-free but is purely existential. We can then prove by induction on the number of lines in the free-cut free proof of $\Gamma \rightarrow \Delta, \Delta'$ that there is an $r \ge 0$ and a cedent Δ'' of the form

$$B(\vec{a},t_{1,1},\ldots,t_{1,k}),\ldots,B(\vec{a},t_{r,1},\ldots,t_{r,k})$$

such that $\Gamma \rightarrow \Delta, \Delta''$ is provable. We leave the rest of the details to the reader. □

We momentarily define an *instance* of a universal formula $(\forall\vec{x})A(\vec{x})$ to be any quantifier-free formula $A(\vec{t})$. It is not hard to see using cut elimination, that if a quantifier-free formula C is a consequence of a universal theory T, then it is a tautological consequence of some finite set of instances of axioms of T and of equality axioms. In the special case where T is the null theory, we have that C is a consequence of instances of equality axioms (and C is therefore called a *quasitautology*). If, in addition, C does not involve equality, C will be tautologically valid. Thus, Herbrand's theorem reduces provability in first-order logic to generation of (quasi)tautologies.

2.5.2. As stated above, Herbrand's theorem has limited applicability since it applies only to Π_2-consequences of universal theories: fortunately, however, there are several ways to extend Herbrand's theorem to more general situations. In 2.5.3 below, we explain one such generalization; but first, in this paragraph, we give a simpler method of widening the applicability of Herbrand's theorem, based on the introduction of new function symbols, which we call *Herbrand* and *Skolem* functions, that allow quantifier alternations to be reduced.

For notational convenience, we will consider only formulas in prenex form in this paragraph; however, the definitions and proposition below can be readily generalized to arbitrary formulas.

Definition. Let $(\exists x)A(x,\vec{c})$ be a formula with $\vec{c}$ all of its free variables. The *Skolem function* for $(\exists x)A$ is represented by a function symbol $f_{\exists xA}$ and has the defining axiom:

$$\textit{Sk-def}(f_{\exists xA}): \quad (\forall\vec{y})(\forall x)\,(A(x,\vec{y}) \supset A(f_{\exists xA}(\vec{y}),\vec{y}))\,.$$

Note that $\textit{Sk-def}(f_{\exists xA})$ implies $(\forall\vec{y})\,((\exists x)A(x,\vec{y}) \leftrightarrow A(f_{\exists xA}(\vec{y}),\vec{y}))$.

Definition. Let $A(\vec{c})$ be a formula in prenex form. The *Skolemization*, $A^S(\vec{c})$, of A is the formula defined inductively by:

(1) If $A(\vec{c})$ is quantifier-free, then $A^S(\vec{c})$ is $A(\vec{c})$.

(2) If $A(\vec{c})$ is $(\forall y)B(\vec{c},y)$, then $A^S(\vec{c})$ is the formula $(\forall y)B^S(\vec{c},y)$.

(3) If $A(\vec{c})$ is $(\exists y)B(\vec{c},y)$, then $A^S(\vec{c})$ is $B^S(\vec{c},f_A(\vec{c}))$, where f_A is the Skolem function for A.

It is a simple, but important fact that $A^S \vDash A$.

The *Skolemization* of a theory T is the theory $T^S = \{A^S : A \in T\}$. Note that T^S is a purely universal theory. Incidentally, the set of all *Sk-def* axioms of the Skolem functions is equivalent to a set of universal formulas; however, they are not included in theory T^S. From model-theoretic considerations, it is not difficult to see that T^S contains and is conservative over T.

We next define the concept of 'Herbrandization' which is completely dual to the notion of Skolemization:

Definition. Let $(\forall x)A(x,\vec{c})$ be a formula with $\vec{c}$ all of its free variables. The *Herbrand function* for $(\forall x)A$ is represented by a function symbol $h_{\forall xA}$ and has the defining axiom:

$$(\forall\vec{y})(\forall x)\,(\neg A(x,\vec{y}) \supset \neg A(h_{\forall xA}(\vec{y}),\vec{y}))\,.$$

Note that this implies $(\forall\vec{y})\,((\forall x)A(x,\vec{y}) \leftrightarrow A(h_{\forall xA}(\vec{y}),\vec{y}))$. The Herbrand function can also be thought of as a 'counterexample function'; in that $(\forall x)A(x)$ is false if and only if $h_{\forall xA}$ provides a value x which is a counterexample to the truth of $(\forall x)A$.

Definition. Let $A(\vec{c})$ be a formula in prenex form. The *Herbrandization*, $A^H(\vec{c})$, of A is the formula defined inductively by:

(1) If $A(\vec{c})$ is quantifier-free, then $A^H(\vec{c})$ is $A(\vec{c})$.

(2) If $A(\vec{c})$ is $(\exists y)B(\vec{c},y)$, then $A^H(\vec{c})$ is the formula $(\exists y)B^H(\vec{c},y)$.

(3) If $A(\vec{c})$ is $(\forall y)B(\vec{c},y)$, then $A^H(\vec{c})$ is $B^H(\vec{c},h_A(\vec{c}))$, where h_A is the Herbrand function for A.

It is not hard to see that $A \vDash A^H$. Note that A^H is purely existential.

Proposition. *Let T be a set of prenex formulas and A any prenex formula. Then the following are equivalent:*

(1) $T \vDash A$,

(2) $T^S \vDash A$,

(3) $T \vDash A^H$,

(4) $T^S \vDash A^H$,

This proposition is easily proved from the above definitions and remarks. The importance of the proposition lies in the fact that T^S is a universal theory and that A^H is an existential formula, and that therefore Herbrand's theorem applies to $T^S \vDash A^H$. Therefore, Herbrand's theorem can be applied to an arbitrary logical implication $T \vDash A$, at the cost of converting formulas to prenex form and introducing Herbrand and Skolem functions.

2.5.3. A generalized Herbrand's theorem. Herbrand actually proved a more general version of Theorem 2.5.1 which applied directly whenever $\vDash A$, for A a general formula, not necessarily existential. His result avoids the use of Skolem/Herbrand functions but is somewhat more difficult to state and comprehend. The theorem we state next is a generalization of Herbrand's theorem which is quite similar in spirit and power to the theorem as stated originally by Herbrand [1930].

In this section, we shall consider a first-order formula A such that $\vDash A$. Without loss of generality, we shall suppose that the propositional connectives in A are restricted to be $\land$, $\lor$ and $\neg$, and that the $\neg$ connective appears only in front of atomic subformulas of A. (The only reason for this convention is that it avoids having to keep track of whether quantifiers appear positively and negatively in A.)

Definition. Let A satisfy the above convention on negations. An $\lor$*-expansion* of A is any formula that can be obtained from A by a finite number of applications of the following operation:

(α) If B is a subformula of an $\lor$-expansion A' of A, replacing B in A' with $B \lor B$ produces another $\lor$-expansion of A.

A *strong* $\lor$*-expansion* of A is defined similarly, except that now the formula B is restricted to be a subformula with outermost connective an existential quantifier.

Definition. Let A be a formula. A *prenexification* of A is a formula obtained from A by first renaming bound variables in A so that no variable is quantified more than once in A and then using prenex operations to put the formula in prenex normal form.

Note that there will generally be more than one prenexification of A since prenex operations may be applied in different orders resulting in a different order of the quantifiers in the prenex normal form formula.

Definition. Let A be a valid first-order formula in prenex normal form, with no variable quantified twice in A. If A has $r \geq 0$ existential quantifiers, then A is of the following form with B quantifier-free:

$$(\forall x_1 \cdots x_{n_1})(\exists y_1)(\forall x_{n_1+1} \cdots x_{n_2})(\exists y_2) \cdots (\exists y_r)(\forall x_{n_r+1} \cdots x_{n_{r+1}})B(\vec{x}, \vec{y})$$

with $0 \leq n_1 \leq n_2 \leq \cdots \leq n_{r+1}$. A *witnessing substitution* for A is a sequence of terms (actually, semiterms) $t_1, \ldots t_r$ such that (1) each t_i contains arbitrary free variables but only bound variables from $x_1, \ldots, x_{n_i}$ and (2) the formula $B(\vec{x}, t_1, \ldots, t_r)$ is a quasitautology (i.e., a tautological consequence of instances of equality axioms only). In the case where B does not contain the equality sign, then (2) is equivalent to B being a tautology.

Let T be a first-order theory. A sequence of terms is said to *witness A over T* if the above conditions hold except with condition (2) replaced by the weaker condition that $T \vDash (\forall \vec{x})B(x, \vec{t})$.

Definition. A *Herbrand proof* of a first-order formula A consists of a prenexification A^* of a strong $\vee$-expansion of A plus a witnessing substitution σ for A^*.

A *Herbrand T-proof* of A consists of a prenexification A^* of a strong $\vee$-expansion of A plus a substitution which witnesses A over T.

We are now in a position to state the general form of Herbrand's theorem:

Theorem. *A first-order formula A is valid if and only if A has a Herbrand proof. More generally, if T is a universal theory, then $T \vDash A$ if and only if A has a Herbrand T-proof.*

Proof. We shall sketch a proof of only the first part of the theorem since the proof of the second part is almost identical. Of course it is immediate from the definitions that if A has a Herbrand proof, then A is valid. So suppose A is valid, and therefore has a cut-free LK-proof P. We shall modify P in stages so as to extract a Herbrand proof of P.

The first stage will involve restricting the formulas which can be combined by a contraction inference. One problem that arises in this regard is that inferences with two hypothesis, such as $\vee$*:left* and $\wedge$*:right*, contain implicit contractions on side formulas. To avoid dealing with this complication, we modify the rules of inference so that no implicit contractions occur; e.g., the $\vee$*:left* inference rule is replaced by the rule

$$\frac{A, \Gamma \rightarrow \Delta \qquad B, \Gamma' \rightarrow \Delta'}{A \vee B, \Gamma, \Gamma' \rightarrow \Delta, \Delta'}$$

and the $\wedge$*:right* and $\supset$*:left* are modified analogously. It is easy to see that changing the rules of inference in this way, makes no difference to what strong inferences are needed in a proof: instead, it merely makes contractions explicit. In particular, the cut elimination theorems still hold with the new inference rules.

A contraction inference is said to be a *propositional contraction* (resp., an $\exists$-*contraction*) provided that the principal formula of the contraction is quantifier-free (resp., its outermost connective is an existential quantifier). The first step in modifying P is to form a cut-free proof P_1, also with endsequent $\rightarrow A$ such that all contraction inferences in P_1 are propositional or $\exists$-contractions. The construction of P_1 from P is formally done by an induction process analogous to the proof of the Cut-elimination Theorem 2.4.2. Namely, define the *E-depth* of a formula similarly to the definition of *depth* in 2.4.1, except define the E-depth of a quantifier-free formula or of a formula with outermost connective an existential quantifier to be zero. Then prove, by double induction on the maximum E-depth d of contraction formulas and the number of contractions of formulas of this maximum E-depth, that P_1 can be transformed into a proof in which all contractions are on formulas of E-depth zero. The induction step consists of removing a topmost contraction inference of the maximum E-depth d. For example, suppose that the following inference is a topmost contraction with principal formula of E-depth d;

$$\dfrac{\begin{matrix}\ddots\vdots\;\cdot^{\cdot^{\cdot}} R\\ \Gamma \rightarrow \Delta, (\forall x)B, (\forall x)B\end{matrix}}{\Gamma \rightarrow \Delta, (\forall x)B}$$

Since P_1 is w.l.o.g. in free variable normal form and since this is a topmost contraction of E-depth d, we can modify the subproof R of P_1 by removing at most two $\forall$*:right* inferences and/or changing some *Weakening:right* inferences to get a proof of $\Gamma \rightarrow \Delta, B(a), B(a')$, where a and a' are free variables not appearing in the endsequent of R. Further replacing a' everywhere by a gives a proof of $\Gamma \rightarrow \Delta, B(a), B(a)$: we use this get to a proof ending:

$$\dfrac{\dfrac{\begin{matrix}\ddots\vdots\;\cdot^{\cdot^{\cdot}}\\ \Gamma \rightarrow \Delta, B(a), B(a)\end{matrix}}{\Gamma \rightarrow \Delta, B(a)}}{\Gamma \rightarrow \Delta, (\forall x)B}$$

Thus we have reduced the E-depth of the contraction inference. A similar procedure works for contractions of E-depth d with outermost connective a propositional connective—we leave the details to the reader. Note that the construction of P_1 depends on the fact that propositional inferences and $\forall$*:right* inferences can be pushed downward in the proof. It is not generally possible to push $\exists$*:right* inferences downward in a proof without violating eigenvariable conditions.

The second step in modifying P is to convert P_1 into a cut-free proof P_2 of some strong $\vee$-expansion A' of A such that every contraction in P_2 is propositional. This is done by the simple expedient of replacing every $\exists$-contraction in P_1 with an $\vee$*:right* inference, and then making the induced changes to all descendents of the principal formula of the inference. More precisely, starting with a lowermost $\exists$-contraction in P_1, say

$$\dfrac{\Gamma \rightarrow \Delta, (\exists x)B, (\exists x)B}{\Gamma \rightarrow \Delta, (\exists x)B}$$

replace this with an $\vee$*:left* inference

$$\frac{\Gamma \longrightarrow \Delta, (\exists x)B, (\exists x)B}{\Gamma \longrightarrow \Delta, (\exists x)B \vee (\exists x)B}$$

and then, in order to get a syntactically correct proof, replace, as necessary, subformulas $(\exists x)B'$ of formulas in P with $(\exists x)B' \vee (\exists x)B'$ (we use the notation B' since terms in B may be different in the descendents). Iterating this process yields the desired proof P_2 of a strong $\vee$-expansion A' of A. By renaming bound variables in P_2 we can assume w.l.o.g. that no variable is quantified twice in any single sequent in P_2.

Thirdly, from P_2 we can construct a prenexification A^* of A' together with a witnessing substitution, thereby obtaining a Herbrand proof of A. To do this, we iterate the following procedure for pulling quantifiers to the front of the proved formula. Find any lowest quantifier inference in P_2 which has not already been handled: this quantifier inference corresponds to a unique quantifier, (Qx), appearing in the endsequent of the proof (and conversely, each quantifier in the endsequent of the proof corresponds to a unique quantifier inference, since all contraction formulas are quantifier-free). Use prenex operations to pull (Qx) as far to the front of the endsequent formula as possible (but not past the quantifiers that have already been moved to the front of the endsequent formula). Also, push the quantifier inference downward in the proof until it reaches the group of quantifier inferences that have already been pushed downward in the proof. It is straightforward to check that this procedure preserves the property of having a syntactically valid proof. When we are done iterating this procedure, we obtain a proof P_3 of a prenexification $\longrightarrow A^*$ of A. It remains to define a witnessing substitution for A^*, which is now easy: for each existential quantifier $(\exists y_i)$ in A^*, find the corresponding $\exists$*:right* inference

$$\frac{\Gamma \longrightarrow \Delta, B(t_i)}{\Gamma \longrightarrow \Delta, (\exists y_i)B(y_i)}$$

and let the term t_i be from this inference. That this is a witnessing substitution for A^* is easily proved by noting that by removing the $\exists$*:right* inference from P_3, a proof of $A^*_M(\vec{x}, \vec{t})$ is obtained where A^*_M is the quantifier-free portion of A^*. □

The above theorem can be used to obtain the following 'no-counterexample interpretation' which has been very useful recently in the study of bounded arithmetic (see Krajíček, Pudlák and Takeuti [1991], or section 3.3.2 of Chapter II).[5]

Corollary. *Let T be a universal theory and suppose $T \vDash (\exists x)(\forall y)A(x, y, \vec{c})$ with A a quantifier-free formula. Then there is a $k > 0$ and terms $t_1(\vec{c})$, $t_2(\vec{c}, y_1)$, $t_3(\vec{c}, y_1, y_2), \ldots, t_k(\vec{c}, y_1, \ldots y_{k-1})$ such that*

$$T \vDash (\forall y_1)[A(t_1(\vec{c}), y_1, \vec{c}) \vee (\forall y_2)[A(t_2(\vec{c}, y_1), y_2, \vec{c}) \vee (\forall y_3)[A(t_3(\vec{c}, y_1, y_2), y_3, \vec{c}) \vee \cdots \vee (\forall y_k)[A(t_k(\vec{c}, y_1, \ldots, y_{k-1}), y_k, \vec{c}))] \cdots]]]$$

[5]This corollary is named after the more sophisticated no-counterexample interpretations of Kreisel [1951,1952].

To prove the corollary, note that the only strong $\vee$-expansions of A are formulas of the form $\bigvee(\exists x)(\forall y)A(x,y,\vec{c})$ and apply the previous theorem.

2.5.4. No recursive bounds on number of terms. It is interesting to ask whether it is possible to bound the value of r in Herbrand's Theorem 2.5.1. For this, consider the special case where the theory T is empty, so that we have an LK-proof P of $(\exists x_1,\ldots,x_k)B(\vec{a},\vec{x})$ where B is quantifier-free. There are two ways in which one might wish to bound the number r needed for Herbrand's theorem: as a function of the size of P, or alternatively, as a function of the size of the formula $(\exists\vec{x})B$. For the first approach, it follows immediately from Theorem 2.4.3 and the proof of Herbrand's theorem, that $r \le 2^{||P||}_{2||P||}$. For the second approach, we shall sketch a proof below that r can not be recursively bounded as a function of the formula $(\exists\vec{x})B$.

To show that r cannot be recursively bounded as a function of $(\exists\vec{x})B$, we shall prove that having a recursive bound on r would give a decision procedure for determining if a given existential formula is valid. Since it is well known that validity of existential first-order formulas is undecidable, this implies that r cannot be recursively bounded in terms of the formula size.

What we shall show is that, given a formula B as in Theorem 2.5.1 and given an $r > 0$, it is decidable whether there are terms $t_{1,1},\ldots,t_{r,k}$ which make the formula

$$\bigvee_{i=1}^{r} B(\vec{a},t_{i,1},\ldots,t_{i,k}) \tag{1}$$

a tautology.[6] This will suffice to show that r cannot be recursively bounded. The quantifier-free formula B is expressible as a Boolean combination $C(D_1,\ldots,D_\ell)$ where each D_i is an atomic formula and $C(\cdots)$ is a propositional formula. If the formula (1) is a tautology, it is by virtue of certain formulas $D_i(\vec{a},t_{i,1},\ldots,t_{i,k})$ being identical. That is to say there is a finite set X of equalities of the form

$$D_i(\vec{a},t_{i,1},\ldots,t_{i,k}) \;=\; D_{i'}(\vec{a},t_{i',1},\ldots,t_{i',k})$$

such that, any set of terms $t_{1,1},\ldots,t_{r,k}$ which makes all the equalities in X true will make (1) a tautology.

But now the question of whether there exist terms $t_{1,1},\ldots,t_{r,k}$ which satisfy such a finite set X of equations is easily seen to be a first-order unification problem, as described in section 2.6.1 below. This means that there is an algorithm which can either determine that no choice of terms will satisfy all the equations in X or will find a most general unifier which specifies all possible ways to satisfy the equations of X.

Since, for a fixed $r > 0$, there are only finitely many possible sets X of equalities, we have the following algorithm for determining if there are terms which make (1) a tautology: for each possible set X of equalities, check if it has a solution (i.e., a

[6] This was first proved by Herbrand [1930] by the same argument that we sketch here.

most general unifier), and if so, check if the equalities are sufficient to make (1) a tautology. □

2.5.5. Interpolation theorem

Suppose we are given two formulas A and B such that $A \supset B$ is valid. An *interpolant* for A and B is a formula C such that $A \supset C$ and $C \supset B$ are both valid. It is a surprising, and fundamental, fact that it is always possible to find an interpolant C such that C contains only non-logical symbols which occur in both A and B.

We shall assume for this section that first-order logic has been augmented to include the logical symbols $\top$ and $\bot$. For this, the sequent calculus has two new initial sequents $\rightarrow\top$ and $\bot\rightarrow$. We write $L(A)$ to denote the set of non-logical symbols occurring in A plus all free variables occurring in A, i.e., the constant, symbols, function symbols, predicate symbols and free variables used in A. For Π a cedent, $L(\Pi)$ is defined similarly.

Craig's Interpolation Theorem. Craig [1957a].

(a) *Let A and B be first-order formulas such that $\models A \supset B$. Then there is a formula C such that $L(C) \subseteq L(A) \cap L(B)$ and such that $\models A \supset C$ and $\models C \supset B$.*

(b) *Suppose $\Gamma_1, \Gamma_2 \rightarrow \Delta_1, \Delta_2$ is a valid sequent. Then there is a formula C such that $L(C)$ is a subset of $L(\Gamma_1, \Delta_1) \cap L(\Gamma_2, \Delta_2)$ and such that $\Gamma_1 \rightarrow \Delta_1, C$ and $C, \Gamma_2 \rightarrow \Delta_2$ are both valid.*

Craig's interpolation can be proved straightforwardly from the cut elimination theorem. We shall outline some of the key points of the proof, but leave a full proof to the reader. First it is easy to see that part (a) is just a special case of (b), so it suffices to prove (b). To prove (b), we first use the cut elimination theorem to obtain a cut-free proof P of $\Gamma_1, \Gamma_2 \rightarrow \Delta_1, \Delta_2$. We then prove by induction on the number of strong inferences in P that there is a formula C with only the desired non-logical symbols and there are proofs P_1 and P_2 of $\Gamma_1 \rightarrow \Delta_1, C$ and $C, \Gamma_2 \rightarrow \Delta_2$. In fact, the proofs P_1 and P_2 are also cut-free and have lengths linearly bounded by the length of P. For an example of how the proof by induction goes, let's make the simplifying assumption that there are no function symbols in our languages, and then assume that the final strong inference of P is an $\exists$*:right* inference with principal formula in Δ_2. That is to say, suppose P ends with the inference

$$\dfrac{\dfrac{\vdots}{\Gamma_1, \Gamma_2 \rightarrow \Delta_1, \Delta_2', A(t)}}{\Gamma_1, \Gamma_2 \rightarrow \Delta_1, \Delta_2', (\exists x)A(x)}$$

with Δ_2 is the cedent $\Delta_2', (\exists x)A(x)$. Since we are assuming there are no function symbols, t is just a free variable or a constant symbol. The induction hypothesis states that there is an interpolant $C(t)$ with an appropriate first-order language such that $\Gamma_1 \rightarrow \Delta_1, C(t)$ and $C(t), \Gamma_2 \rightarrow \Delta_2, A(t)$ are LK-provable. The interpolant C^* for the endsequent of P is defined as follows: if the symbol t does not appear in the sequent $\Gamma_2 \rightarrow \Delta_2$, then C^* is $(\exists y)C(y)$; otherwise, if the symbol t does not appear in the sequent $\Gamma_1 \rightarrow \Delta_1$, then C^* is $(\forall y)C(y)$; and if t appears in both sequents, C^* is just C. It can be checked that in all three cases, the sequents $\Gamma_1 \rightarrow \Delta_1, C^*$ and $C^*, \Gamma_2 \rightarrow \Delta_2, (\exists x)A(x)$ are LK-provable. Therefore, C^* is an interpolant for the endsequent of P; also, it is obvious that the language $L(C)$ of C is still appropriate for the endsequent.

Secondly, suppose P ends with the inference

$$\dfrac{\dfrac{\ddots\vdots\cdot^{\cdot}}{\Gamma_1, \Gamma_2 \rightarrow \Delta_1, \Delta_2', A(b)}}{\Gamma_1, \Gamma_2 \rightarrow \Delta_1, \Delta_2', (\forall x)A(x)}$$

with the principal formula still presumed to be in Δ_2. The induction hypothesis states that there is an interpolant C with an appropriate first-order language such that $\Gamma_1 \rightarrow \Delta_1, C(b)$ and $C(b), \Gamma_2 \rightarrow \Delta_2, A(b)$. Since, by the eigenvariable condition, b does not occur except as indicated; we get immediately LK-proofs of the sequents $\Gamma_1 \rightarrow \Delta_1, (\forall y)C(y)$ and $(\forall y)C(y), \Gamma_2 \rightarrow \Delta_2, (\forall x)A(x)$. Therefore, $(\forall y)C(y)$ serves as an interpolant for the endsequent of P.

There are a number of other cases that must be considered, depending on the type of the final strong inference in P and on whether its principal formula is (an ancestor of) a formula in Δ_1, Δ_2, Γ_1 or Γ_2. These cases are given in full detail in textbooks such as Takeuti [1987] or Girard [1987b].

It remains to consider the case where there are function symbols in the language. The usual method of handling this case is to just reduce it to the case where there are no function symbols by removing function symbols in favor of predicate symbols which define the graphs of the functions. Alternatively, one can carry out directly a proof on induction on the number of strong inferences in P even when function symbols are present. This involves a more careful analysis of the 'flow' of terms in the proofs, but still gives cut-free proofs P_1 and P_2 with size linear in the size of P. We leave the details of the function-symbol case to the reader.

Other interpolation theorems. A useful strengthening of the Craig interpolation theorem is due to Lyndon [1959]. This theorem states that Craig's interpolation theorem may be strengthened by further requiring that every predicate symbol which occurs positively (resp., negatively) in C also occurs positively (resp., negatively) in both A and B. The proof of Lyndon's theorem is identical to the proof sketched above, except that now one keeps track of positive and negative occurrences of predicate symbols.

Craig [1957b] gives a generalization of the Craig interpolation theorem which applies to interpolants of cedents.

Lopez-Escobar [1965] proved that the interpolation theorem holds for some infinitary logics. Barwise [1975,§III.6] proved that the interpolation theorem holds for a wider class of infinitary logics. Lopez-Escobar's proof was proof-theoretic, based on a sequent calculus formalization of infinitary logic. Barwise's proof was model-theoretic; Feferman [1968] gives a proof-theoretic treatment of these general interpolation theorems, based on the sequent calculus.

2.5.6. Beth's definability theorem

Definition. Let P and P' be predicate symbols with the same arity. Let $\Gamma(P)$ be an arbitrary set of first-order sentences not involving P', and let $\Gamma(P')$ be the same set of sentences with every occurrence of P replaced with P'.

The set $\Gamma(P)$ is said to *explicitly define* the predicate P if there is a formula $A(\vec{c})$ such that

$$\Gamma(P) \vdash (\forall \vec{x})(A(\vec{x}) \leftrightarrow P(\vec{x})).$$

The set $\Gamma(P)$ is said to *implicitly define* the predicate P if

$$\Gamma(P) \cup \Gamma(P') \vDash (\forall \vec{x})(P(\vec{x}) \leftrightarrow P'(\vec{x})).$$

The Definability Theorem of Beth [1953] states the fundamental fact that the notions of explicit and implicit definability coincide. One way to understand the importance of this is to consider *implicit* definability of P as equivalent to being able to uniquely characterize P. Thus, Beth's theorem states, loosely speaking, that if a predicate can be uniquely characterized, then it can be explicitly defined by a formula not involving P.

One common, elementary mistake is to confuse implicit definability by a set of sentences $\Gamma(P)$ with implicit definability in a particular model. For example, consider the theory T of sentences which are true in the standard model $(\mathbf{N}, 0, S, +)$ of natural numbers with zero, successor and addition. One might attempt to implicitly define multiplication in terms of zero and addition letting $\Gamma(M)$ be the theory

$$T \cup \{(\forall x)(M(x,0) = 0), (\forall x)(\forall y)(M(x,S(y)) = M(x,y) + x)\}$$

It is true that this uniquely characterizes the multiplication function $M(x,y)$ in the sense that there is only one way to expand $(\mathbf{N}, 0, S, +)$ to a model of $\Gamma(M)$; however, this is not an implicit definition of M since there are nonstandard models of T which have more than one expansion to a model of $\Gamma(M)$.

Beth's Definability Theorem. *$\Gamma(P)$ implicitly defines P if and only if it explicitly defines P.*

Proof. Beth's theorem is readily proved from the Craig interpolation theorem as follows. First note that if P is explicitly definable, then it is clearly implicitly definable. For the converse, assume that P is implicitly definable. By compactness, we may assume without loss of generality that $\Gamma(P)$ is a single sentence. Then we have that

$$\Gamma(P) \wedge P(\vec{c}) \vDash \Gamma(P') \supset P'(\vec{c}).$$

By the Craig Interpolation Theorem, there is a interpolant $A(\vec{c})$ for $\Gamma(P) \wedge P(\vec{c})$ and $\Gamma(P') \supset P'(\vec{c})$. This interpolant is the desired formula explicitly defining P. □

It is also possible to prove the Craig Interpolation Theorem from the Beth Definability Theorem. In addition, both theorems are equivalent to the model-theoretic Joint Consistency Theorem of Robinson [1956].

2.6. First-order logic and resolution refutations

The importance of the resolution proof method for propositional logic (described in section 1.3) lies in large part in the fact that it also serves as a foundation for theorem-proving in first-order logic. Recall that by introducing Herbrand and Skolem functions, theorem-proving in first-order logic can be reduced to proving Π_2-formulas of the form $(\forall\vec{x})(\exists\vec{y})A(\vec{x},\vec{y})$ with A quantifier-free (see §2.5.2). Also, by Herbrand's Theorem 2.5.1, the problem of proving $(\forall\vec{x})(\exists\vec{y})A(\vec{x},\vec{y})$ is reducible to the problem of finding terms $\vec{r}_1, \vec{r}_2, \ldots, \vec{r}_k$ so that $\bigvee_i A(\vec{x},\vec{r}_i)$ is tautologically valid. Now, given the terms $\vec{r}_1, \ldots, \vec{r}_k$, determining tautological validity is 'merely' a problem in propositional logic; and hence is amenable to theorem proving methods such as propositional resolution. Thus one hopes that if one had a good scheme for choosing terms $\vec{r}_1, \ldots, \vec{r}_k$, then one could have a reasonable method of first-order theorem proving.

This latter point is exactly the problem that was solved by Robinson [1965b]; namely, he introduced the resolution proof method and showed that by using a unification algorithm to select terms, the entire problem of which terms to use could be solved efficiently by using the "most general" possible terms. In essence, this reduces the problem of first-order theorem proving to propositional theorem proving. (Of course, this last statement is not entirely true for two reasons: firstly, there may be a very large number (not recursively bounded) of terms that are needed, and secondly, it is entirely possible that foreknowledge of what terms are sufficient, might help guide the search for a propositional proof.)

2.6.1. Unification. We shall now describe the unification algorithm for finding most general unifiers. We shall let t denote a term containing function symbols, constant symbols and variables. A *substitution*, σ, is a partial map from variables to terms; we write $x\sigma$ to denote $\sigma(x)$, and when x is not in the domain of σ, we let $x\sigma$ be x. If σ is a substitution, then $t\sigma$ denotes the result of simultaneously replacing every variable x in t with $x\sigma$. We extend σ to atomic relations by letting $R(t_1, \ldots, t_k)\sigma$ denote $R(t_1\sigma, \ldots, t_k\sigma)$. We use concatenation $\sigma\tau$ to denote the substitution which is equivalent to an application of σ followed by an application of τ.

Definition. Let $A_1, \ldots, A_k$ be atomic formulas. A *unifier* for the set $\{A_1, \ldots, A_k\}$ is a substitution σ such that $A_1\sigma = A_2\sigma = \cdots = A_k\sigma$, where $=$ represents the property of being the identical formula.

A substitution is said to be a *variable renaming* substitution, if the substitution maps variables only to terms which are variables.

A unifier σ is said to be a *most general unifier* if, for every unifier τ for the same set, there is a unifier ρ such that $\tau = \sigma\rho$. Note that up to renaming of variables, a most general unifier must be unique.

Unification Theorem. *If* $\{A_1, \ldots, A_k\}$ *has a unifier then it has a most general unifier.*

Proof. We shall prove the theorem by outlining an efficient algorithm for determining whether a unifier exists and, if so, finding a most general unifier. The algorithm is described as an iterative procedure which, at stage s has a set E_s of equations and a substitution σ_s. The equations in E_s are of the form $\alpha \doteq \beta$ where α and β may be formulas or terms. The meaning of this equation is that the sought-for most general unifier must be a substitution which makes α and β identical. Initially, E_0 is the set of $k-1$ equations $A_j \doteq A_{j+1}$ and σ_0 is the identity. Given E_s and σ_s, the algorithm does any one of the following operations to choose E_{s+1} and σ_{s+1}:

(1) If E_s is empty, we are done and σ_s is a most general unifier.

(2) If E_s contains an equation of the form

$$F(t_1, \ldots, t_i) \doteq F(t'_1, \ldots, t'_i),$$

then $\sigma_{s+1} = \sigma_s$ and E_{s+1} is obtained from E_s by removing this equation and adding the i equations $t_i \doteq t'_i$. Here F is permitted to be a function symbol, a constant symbol or a predicate symbol.

(3) Suppose E_s contains an equation of the form

$$x \doteq t \qquad \text{or} \qquad t \doteq x,$$

with x a variable and t a term. Firstly, if t is equal to x, then this equation is discarded, so E_{s+1} is E_s minus this equation and $\sigma_{s+1} = \sigma_s$. Secondly, if t is a non-trivial term in which x occurs, then the algorithm halts outputting that no unifier exists.[7] Thirdly, if x does not occur in t, then let $[x/t]$ denote the substitution that maps x to t and define E_{s+1} to be the set of equations $s[x/t] \doteq s'[x/t]$ such that $s \doteq s'$ is in E_s, and define $\sigma_{s+1} = \sigma_s[x/t]$.

We leave it to the reader to prove that this algorithm always halts with a most general unifier if a unifier exists. The fact that the algorithm does halt can be proved by noting that each iteration of the algorithm either reduces the number of distinct variables in the equations, or reduces the sum of the lengths of the terms occurring in the equations. □

The algorithm as given above is relatively efficient; however, the sizes of the terms involved may grow exponentially large. Indeed, there are unification problems where the size of the most general unifier is exponential in the size of the unification problem; for example, the most general unifier for $\{f(x_1, x_2, \ldots x_k), f(g(x_2, x_2), \ldots, g(x_{k+1}, x_{k+1}))\}$ maps x_1 to a term of height k with 2^k atoms.

[7]This failure condition is known as the *occurs check*.

If one is willing to use a dag (directed acyclic graph) representation for terms, then this exponential growth rate does not occur. Paterson and Wegman [1978] have given an efficient linear-time unification algorithm based on representing terms as dags.

2.6.2. Resolution and factoring inferences. We now describe the resolution inference used by Robinson [1965b] for first-order logic. The starting point is Herbrand's theorem: we assume that we are attempting to prove a first-order sentence, which without loss of generality is of the form $(\exists\vec{x})A(\vec{y})$. (This may be assumed without loss of generality since, if necessary, Herbrand functions may be introduced.) We assume, in addition, that A is in disjunctive normal form. Instead of proving this sentence, we shall instead attempt to refute the sentence $(\forall\vec{x})(\neg A(x))$. Since A is in disjunctive normal form, we may view $\neg A$ as a set Γ_A of clauses with the literals in the clauses being atomic formulas or negated atomic formulas. We extend the definition of substitutions to act on clauses in the obvious way so that $\{C_1, \ldots, C_k\}\sigma$ is defined to be $\{C_1\sigma, \ldots, C_k\sigma\}$.

When we refute $(\forall\vec{x})(\neg A)$, we are showing that there is no structure M which satisfies $(\forall\vec{x})(\neg A)$. Consider a clause C in Γ_A. The clause C is a set of atomic and negated atomic formulas, and a structure M is said to *satisfy* C provided it satisfies $(\forall\vec{x})(\bigvee_{\phi\in C}\phi(x))$. Thus, it is immediate that if M satisfies C, and if σ is a substitution, then M also satisfies $C\sigma$. From this, we see that the following version of resolution is sound in that it preserves the property of being satisfied by a model M: If B and C are clauses, if ϕ is an atomic formula and if σ and τ are substitutions, let D be the clause $(B\sigma \setminus \{\phi\}) \cup (C\tau \setminus \{\overline{\phi}\})$. It is easy to verify that if B and C are satisfied in M, then so is D.

Following Robinson [1965b], we use a restricted form of this inference principle as the sole rule of inference for first-order resolution refutations:

Definition. Let B and C be clauses and suppose $P(\vec{s}_1), P(\vec{s}_2), \ldots P(\vec{s}_k)$ are atomic formulas in B and that $\neg P(\vec{t}_1), \neg P(\vec{t}_2), \ldots \neg P(\vec{t}_\ell)$ are negated atomic formulas in C. Choose a variable renaming substitution τ so that $C\tau$ has no variables in common with B. Also suppose that the $k+\ell$ formulas $P(\vec{s}_i)$ and $P(\vec{t}_i)\tau$ have a most general unifier σ. Then the clause D defined by

$$(B\sigma \setminus \{P(\vec{s}_1)\sigma\}) \cup (C\tau\sigma \setminus \{\neg P(\vec{s}_1)\tau\sigma\})$$

is defined to be an *R-resolvent* of B and C.

The reason for using the renaming substitution τ is that the variables in the clauses B and C are implicitly universally quantified; thus if the same variable occurs in both B and C we allow that variable to be instantiated in B by a different term than in C when we perform the unification. Applying τ before the unification allows this to happen automatically.

One often views R-resolution as the amalgamation of two distinct operations: first, the *factoring* operation finds a most general unifier of a subset of clause, and second, the unitary resolution operation which resolves two clauses with respect to a single literal. Thus, R-resolution consists of (a) choosing subsets of the clauses B and C and factoring them, and then (b) applying resolution w.r.t. to the literal obtained by the factoring.

Completeness of R-resolution. *A set Γ of first-order clauses is unsatisfiable if and only if the empty clause can be derived from Γ by R-resolution.*

This theorem is proved by the discussion in the next paragraph.

2.6.3. Lifting ground resolution to first-order resolution. A *ground literal* is defined to be a literal in which no variables occur; a *ground clause* is a set of ground literals. We assume, with no loss of generality, that our first-order language contains a constant symbol and that therefore ground literals exist. Ground literals may independently be assigned truth values[8] and therefore play the same role that literals played in propositional logic. A *ground* resolution refutation is a propositional-style refutation involving ground clauses only, with ground literals in place of propositional literals. By the Completeness Theorem 1.3.4 for propositional resolution, a set of ground clauses is unsatisfiable if and only if it has a ground resolution refutation.

For sets of ground clauses, R-resolution is identical to propositional-style resolution. Suppose, however, that Γ is an unsatisfiable set of first-order (not necessarily ground) clauses. Since Γ is unsatisfiable there is, by Herbrand's theorem, a set of substitutions $\sigma_1, \ldots, \sigma_r$ so that each $\Gamma\sigma_r$ is a set of ground clauses and so that the set $\Pi = \bigcup_i \Gamma\sigma_i$ of clauses is propositionally unsatisfiable. Therefore there is a ground resolution refutation of Π.

To justify the completeness of R-resolution, we shall show that any ground resolution refutation of Π can be 'lifted' to an R-resolution refutation of Γ. In fact, we shall prove the following: if $C_1, C_2, \ldots, C_n = \emptyset$ is a resolution refutation of Π, then there are clauses $D_1, D_2, \ldots, D_m = \emptyset$ which form an R-resolution refutation of Γ and there are substitutions $\sigma_1, \sigma_2, \ldots, \sigma_m$ so that $D_i\sigma_i = C_i$. We define D_i and σ_i by induction on i as follows. Firstly, if $C_i \in \Pi$, then it must be equal to $D_i\sigma_i$ for some $D_i \in \Gamma$ and some substitution σ_i by the definition of Π. Secondly, if C_i is inferred from C_j and C_k, with $j, k < i$ by resolution w.r.t. the literal $P(\vec{r})$, then define E_j to be the subset of D_j which is mapped to $P(\vec{r})$ by σ_j, and define E_k similarly. Now, form the R-resolution inference which factors the subsets E_j and E_k of D_j and D_k and forms the resolvent. This resolvent is D_i and it is straightforward to show that the desired σ_i exists.

That finishes the proof of the Completeness Theorem for R-resolution. It should be noted that the method of proof shows that R-resolution refutations are the shortest possible refutations, even if arbitrary substitutions are allowed for factoring inferences. Even more importantly, the method by which ground resolution refutations were 'lifted' to R-resolution refutations preserves many of the search strategies that

[8]We are assuming that the equality sign ($=$) is not present.

were discussed in section 1.3.5. This means that these search strategies can be used for first-order theorem proving.[9]

2.6.4. Paramodulation. The above discussion of R-resolution assumed that equality was not present in the language. In the case where equality is in the language, one must either add additional initial clauses as axioms that express the equality axioms or one must add additional inference rules. For the first approach, one could add clauses which express the equality axioms from section 2.2.1; for instance the third equality axiom can be expressed with the clause

$$\{x_1 \neq y_1, ..., x_k \neq y_k, \neg P(\vec{x}), P(\vec{y})\},$$

and the other equality axioms can similarly be expressed as clauses. More computational efficiency can be obtained with equality clauses of the form

$$\{x \neq y, \overline{A(x)}, A(y)\}$$

where $A(x)$ indicates an arbitrary literal.

For the second approach, the *paramodulation* inference is used instead of equality clauses; this inference is a little complicated to define, but goes as follows: Suppose B and C are clauses with no free variables in common and that $r = s$ is a literal in C; let t be a term appearing somewhere in B and let σ be a most general unifier of r and t (or of s and t); let B' be the clause which is obtained from $B\sigma$ by replacing occurrences of $t\sigma$ with $s\sigma$ (or with $r\sigma$, respectively) and let C' be $(C \setminus \{r = s\})\sigma$. Under these circumstances, paramodulation allows $B' \cup C'$ to be inferred from B and C. Paramodulation was introduced and shown to be complete by Robinson and Wos [1969] and Wos and Robinson [1973]: for completeness, paramodulation must be combined with R-resolution, with factoring and with application of variable renaming substitutions.

2.6.5. Horn clauses. An important special case of first-order resolution is when the clauses are restricted to be Horn clauses. The propositional refutation search strategies described in section 1.3.5.6 still apply; and, in particular, an unsatisfiable set Γ of Horn clauses always has a linear refutation supported by a negative clause in Γ. In addition, the factoring portion of R-resolution is not necessary in refutations of Γ.

A typical use of Horn clause refutations is as follows: a set Δ of Horn clauses is assumed as a 'database' of knowledge, such that every clause in Δ contains a positive literal. A query, which is an atomic formula $P(s_1, ..., s_k)$, is chosen; the object is to determine if there is an instance of $P(\vec{s})$ which is a logical consequence of Δ. In other words, the object is to determine if $(\exists\vec{x})P(\vec{s})$ is a consequence of Δ where $\vec{x}$ is the vector of variables in $P(\vec{s})$. To solve this problem, one forms the clause $\{\overline{P(\vec{s})}\}$

[9]Historically, it was the desire to find strategies for first-order theorem proving and the ability to lift results from propositional theorem proving, that motivated the research into search strategies for propositional resolution.

and lets Γ be the set $\Delta \cup \{\overline{P(\vec{s})}\}$; one then searches for a linear refutation of Γ which is supported by $\overline{P(\vec{s})}$. If successful, such a linear refutation R also yields a substitution σ, such that $\Delta \vdash P(\vec{s})\sigma$; and indeed, σ is the most general substitution such that R gives a refutation of $\Delta \cup \{\overline{P(\vec{s})\sigma}\}$. From this, what one actually has is a proof of $(\forall \vec{y})(P(\vec{s})\sigma)$ where $\vec{y}$ is the vector of free variables in the terms $P(\vec{s})\sigma$. Note that there may be more than one refutation, and that different refutations can give different substitutions σ, so there is not necessarily a unique most general unifier.

What we have described is essentially a pure form of PROLOG, which is a logic programming language based on searching for refutations of Horn clauses, usually in a depth-first search. PROLOG also contains conventions for restricting the order of the proof search procedure.

For further reading. There is an extensive literature on logic programming, automated reasoning and automated theorem proving which we cannot survey here. The paper of Robinson [1965b] still provides a good introduction to the foundations of logic programming; the textbooks of Chang and Lee [1973] and Loveland [1978] provide a more detailed treatment of the subject matter above, and the textbooks of Kowalski [1979] and Clocksin and Mellish [1981] provide good detailed introductions to logic programming and PROLOG. Chapter IX, by G. Jäger and R. Stärk, discusses the proof-theory and semantics of extensions of logic programming to non-Horn clauses.

3. Proof theory for other logics

In the final section of this chapter, we shall briefly discuss two important non-classical logics, intuitionistic logic and linear logic.

3.1. Intuitionistic logic

Intuitionistic logic is a subsystem of classical logic which historically arose out of various attempts to formulate a more constructive foundation for mathematics. For example, in intuitionistic logic, the law of the excluded middle, $A \vee \neg A$, does not hold in general; furthermore, it is not possible to intuitionistically prove $A \vee B$ unless already at least one of A or B is already intuitionistically provable. We shall discuss below primarily mathematical aspects of the intuitionistic logic, and shall omit philosophical or foundational issues: the books of Troelstra and van Dalen [1988] provide a good introduction to the latter aspects of intuitionistic logic.

The intuitionistic sequent calculus, LJ, is defined similarly to the classical sequent calculus LK, except with the following modifications:

(1) To simplify the exposition, we adopt the convention that negation ($\neg$) is not a propositional symbol. In its place, we include the absurdity symbol $\bot$ in the language; $\bot$ is a nullary propositional symbol which is intended to always have value *False*. The two $\neg$ rules of LK are replaced with the single $\bot$*:left* initial sequent, namely $\bot \longrightarrow$. Henceforth, $\neg A$ is used as an abbreviation for $A \supset \bot$.

(2) In LJ, the $\lor$*:right* rule used in the definition of LK in section 1.2.2 is replaced by the two rules

$$\frac{\Gamma \rightarrow \Delta, A}{\Gamma \rightarrow \Delta, A \lor B} \quad \text{and} \quad \frac{\Gamma \rightarrow \Delta, A}{\Gamma \rightarrow \Delta, B \lor A}$$

(3) Otherwise, LJ is defined like LK, except with the important proviso that at most one formula may appear in the antecedent of any sequent. In particular, this means that rules which have the principal formula to the right of the sequent arrow may not have any side formulas to the right of the sequent arrow.

3.1.1. Cut elimination. An important property of intuitionistic logic is that the cut elimination and free-cut elimination theorems still apply:

Theorem.
(1) *Let* $\Gamma \rightarrow A$ *be* LJ*-provable. Then there is a cut-free* LJ*-proof of* $\Gamma \rightarrow A$.
(2) *Let* $\mathfrak{S}$ *be a set of sequents closed under substitution such that each sequent in* $\mathfrak{S}$ *has at most one formula in its antecedent. Let* $LJ_{\mathfrak{S}}$ *be* LJ *augmented with initial sequents from* $\mathfrak{S}$. *If* $LK_{\mathfrak{S}} \vdash \Gamma \rightarrow A$*, then there is a free-cut free* $LK_{\mathfrak{S}}$*-proof of* $\Gamma \rightarrow A$.

The (free) cut elimination theorem for LJ can be proved similarly to the proof-theoretic proofs used above for classical logic.

3.1.2. Constructivism and intuitionism. Intuitionistic logic is intended to provide a 'constructive' subset of classical logic: that is to say, it is designed to not allow non-constructive methods of reasoning. An important example of the constructive aspect of intuitionistic logic is the Brouwer-Heyting-Kolmogorov (BHK) constructive interpretation of logic. In the BHK interpretation, the logical connectives $\lor$, $\exists$ and $\supset$ have non-classical, constructive meanings. In particular, in order to have a proof of (or knowledge of) a statement $(\exists x)A(x)$, it is necessary to have a particular object t and a proof of (or knowledge of) $A(t)$. Likewise, in order to have a proof of $A \lor B$ the BHK interpretation requires one to have a proof either of A or of B, along with a indication of which one it is a proof of. In addition, in order to have a proof of $A \supset B$, one must have a method of converting any proof of A into a proof B. The BHK interpretation of the connectives $\land$ and $\forall$ are similar to the classical interpretation; thus, a proof of $A \land B$ or of $(\forall x)A(x)$, consists of proofs of both A and B or of a method of constructing a proof of $A(t)$ for all objects t.

It is not difficult to see that the intuitionistic logic LJ is sound under the BHK interpretation, in that any LJ-provable formula has a proof in the sense of the BHK interpretation.

The BHK interpretation provides the philosophical and historical underpinnings of intuitionistic logic; however, it has the drawback of characterizing what constitutes a valid proof rather than characterizing the meanings of the formulas directly. It is possible to extend the BHK interpretation to give meanings to formulas directly, e.g., by using realizability; under this approach, an existential quantifier $(\exists x)$ might mean "there is a constructive procedure which produces x, and (possibly) produces evidence that x is correct". This approach has the disadvantage of requiring one to pick a notion of 'constructive procedure' in an ad-hoc fashion; nonetheless it can be very fruitful in applications such a intuitionistic theories of arithmetic where there are natural notions of 'constructive procedure'.

For pure intuitionistic logic, there is a very satisfactory model theory based on Kripke models. Kripke models provide a semantics for intuitionistic formulas which is analogous to model theory for classical logic in many ways, including a model theoretic proof of the cut-free completeness theorem, analogous to the proof of Theorem 2.3.7 given above.

For an accessible account of the BHK interpretation and Kripke model semantics for intuitionistic logic, the reader can refer to Troelstra and van Dalen [1988,vol. 1]; Chapter VI contains an thorough discussion of realizability. In this section we shall give only the following theorem of Harrop, which generalizes the statements that $A \lor B$ is intuitionistically valid if and only if either A or B is, and that $(\exists x)A(x)$ is intuitionistically valid if and only if there is a term t such that $A(t)$ is intuitionistically valid.

3.1.3. Definition. The *Harrop formulas* are inductively defined by:

(1) Every atomic formula is a Harrop formula,

(2) If B is a Harrop formula and A is an arbitrary formula, then $A \supset B$ is a Harrop formula.

(3) If B is a Harrop formula, then $(\forall x)B$ is a Harrop formula.

(4) If A and B are Harrop formulas, then so is $A \land B$.

Intuitively, a Harrop formula is a formula in which all existential quantifiers and all disjunctions are 'hidden' inside the left-hand scope of an implication.

Theorem. (Harrop [1960]) *Let Γ be a cedent containing only Harrop formulas, and let A and B be arbitrary formulas.*

(a) *If $LJ \vdash \Gamma \rightarrow (\exists x)B(x)$, then there exists a term t such that LJ proves $\Gamma \rightarrow B(t)$.*

(b) *If $LJ \vdash \Gamma \rightarrow A \lor B$, then at least one of $\Gamma \rightarrow A$ and $\Gamma \rightarrow B$ is LJ-provable.*

We omit the proof of the theorem, which is readily provable by using induction on the length of cut-free LJ-proofs.

3.1.4. Interpretation into classical logic. The 'negative translation' provides a translation of classical logic into intuitionistic logic; an immediate corollary of the negative translation is a simple, constructive proof of the consistency of classical logic from the consistency of intuitionistic logic. There are a variety of negative translations: the first ones were independently discovered by Kolmogorov, Gödel and Gentzen.

Definition. Let A be a formula. The *negative translation*, A^-, of A is inductively defined by:

(1) If A is atomic, then A^- is $\neg\neg A$,

(2) $(\neg B)^-$ is $\neg(B^-)$.

(3) $(B \wedge C)^-$ is $(B^-) \wedge (C^-)$,

(4) $(B \supset C)^-$ is $(B^-) \supset (C^-)$,

(5) $(B \vee C)^-$ is $\neg(\neg(B^-) \wedge \neg(C^-))$,

(6) $(\forall x B)^-$ is $\forall x(B^-)$,

(7) $(\exists x B)^-$ is $\neg\forall x(\neg(B^-))$,

Theorem. *Let A be a formula. Then $LK \vdash A$ if and only if $LJ \vdash A^-$.*

Clearly A^- is classically equivalent to A, so if A^- is intuitionistically valid, then A is classically valid. For the converse, we use the following two lemmas which immediately imply the theorem.

Lemma. *Let A be a formula. Then LJ proves $\neg\neg A^- \rightarrow A^-$.*

Proof. The proof is by induction on the complexity of A. We will do only one of the more difficult cases. Suppose A is $B \supset C$. We need to show that LJ proves $\neg\neg(B^- \supset C^-) \rightarrow B^- \supset C^-$, for which, it suffices to prove $B^-, \neg\neg(B^- \supset C^-) \rightarrow C^-$. By the induction hypothesis, LJ proves $\neg\neg C^- \rightarrow C^-$, so it will suffice to show that $B^-, \neg\neg(B^- \supset C^-), \neg C^- \rightarrow$ is LJ-provable. To do this it suffices to show that $B^-, B^- \supset C^- \rightarrow C^-$ is LJ-provable, which is easy to prove. □

If Γ is a cedent $B_1, \ldots, B_k$, then Γ^- denotes the cedent $B_1^-, \ldots, B_k^-$ and $\neg\Gamma^-$ denotes the cedent $\neg B_1^-, \ldots, \neg B_k^-$.

Lemma. *Suppose that LK proves $\Gamma \rightarrow \Delta$. Then LJ proves the sequent $\Gamma^-, \neg\Delta^- \rightarrow$.*

Proof. The proof of the lemma is by induction on the number of inferences in an LK-proof, possibly containing cuts. We'll do only one case and leave the rest to the reader. Suppose the LK-proof ends with the inference

$$\frac{B, \Gamma \rightarrow \Delta, C}{\Gamma \rightarrow \Delta, B \supset C}$$

The induction hypothesis is that $B^-, \Gamma^-, \neg\Delta^-, \neg C^- \rightarrow$ is provable in LJ. By the previous lemma, the induction hypothesis implies that $B^-, \Gamma^-, \neg\Delta^- \rightarrow C^-$ is provable, and from this it follows that $\Gamma^-, \neg\Delta^-, \neg(B^- \supset C^-) \rightarrow$ is also LJ-provable, which is what we needed to show. □

One consequence of the proof of the above theorem is a constructive proof that the consistency of intuitionistic logic is equivalent to the consistency of classical logic; thus intuitionistic logic does not provide a better foundations for mathematics from the point of view of sheer consistency. This is, however, of little interest to a constructivist, since the translation of classical logic into intuitionistic logic does not preserve the constructive meaning (under, e.g., the BHK interpretation) of the formula.

It is possible to extend the negative translation to theories of arithmetic and to set theory; see, e.g., Troelstra and van Dalen [1988,vol. 1].

3.1.5. Formulas as types. Section 3.1.2 discussed a connection between intuitionistic logic and constructivism. An important strengthening of this connection is the Curry-Howard formulas-as-types isomorphism, which provides a direct correspondence between intuitionistic proofs and λ-terms representing computable functions, under which intuitionistic proofs can be regarded as computable functions. This section will discuss the formulas-as-types isomorphism for intuitionistic propositional logic; our treatment is based on the development by Howard [1980]. Girard [1989] contains further material, including the formulas-as-types isomorphism for first-order intuitionistic logic.

The $\{\supset\}$-fragment.. We will begin with the fragment of propositional logic in which the only logical connective is $\supset$. We work in the sequent calculus, and the only strong rules of inference are the $\supset$*:left* and $\supset$*:right* rules. Firstly we must define the set of *types*:

Definition. The *types* are defined as follows:

(a) Any propositional variable p_i is a type.

(b) If σ and τ are types, then $(\sigma \to \tau)$ is a type.

If we identify the symbols $\supset$ and $\to$, the types are exactly the same as formulas, since $\supset$ is the only logical connective allowed. We shall henceforth make this identification of types and formulas without comment.

Secondly, we must define the terms of the λ-calculus; each term t has a unique associated type. We write t^τ to mean that t is a term of type τ.

Definition. For each type τ, there is an infinite set of variables, $x_1^\tau, x_2^\tau, x_3^\tau, \ldots$ of type τ. Note that if $\sigma \neq \tau$, then x_i^σ and x_i^τ are distinct variables.

The *terms* are inductively defined by:

(a) Any variable of type τ is a term of type τ.

(b) If $s^{\sigma\to\tau}$ and t^σ are terms of the indicated types, then (st) is a term of type τ. This term may also be denoted $(st)^\tau$ or even $(s^{\sigma\to\tau}t^\sigma)^\tau$.

(c) If t^τ is a term then $\lambda x^\sigma.t$ is a term of type $\sigma \to \tau$. This term can also be denoted $(\lambda x^\sigma.t)^{\sigma\to\tau}$ or $(\lambda x^\sigma.t^\tau)^{\sigma\to\tau}$

Traditionally, a type $\sigma \to \tau$ is viewed as a set of mappings from the objects of type σ into the objects of type τ. For intuitionistic logic, it is often useful to think of a type σ as being the set of proofs of the formula σ. Note that under the BKH-interpretation this is compatible with the traditional view of types as a set of mappings.

The λx connective serves to bind occurrences of x; thus one can define the notions of bound and free variables in terms in the obvious way. A term is *closed* provided it has no free variables. The computational content of a term is defined by letting (st) represent the composition of the terms s and t, and letting $\lambda x^\sigma.t$ represent the mapping that takes objects d of type τ to the object $t(x/d)$. Clearly this gives a constructive computational meaning to terms.

Theorem. *Let B be a formula. $LJ \vdash B$ if and only if there is a closed term of type B.*

An example of this theorem is illustrated by the closed term K defined as $\lambda x.\lambda y.x$ where x and y are of type σ and τ, respectively, and therefore the term K has type $\sigma \to (\tau \to \sigma)$, which is a valid formula. A second important example is the closed term S which is defined by $\lambda x.\lambda y.\lambda z.((xz)(yz))$ where the types of x, y and z are $\sigma \to (\tau \to \mu)$, $\sigma \to \tau$ and σ, respectively, and therefore, S has type $(\sigma \to (\tau \to \mu)) \to ((\sigma \to \tau) \to (\sigma \to \mu))$ which is a valid formula. The terms K and S are examples of combinators.

The import of this theorem is that a closed term of type B corresponds to a proof of B. We shall briefly sketch the proof of this for the sequent calculus; however, the correspondence between closed terms and natural deduction proofs is even stronger, namely, the closed terms are isomorphic to natural deduction proofs in intuitionistic logic (see Girard [1989]). To prove the theorem, we prove the following lemma:

Lemma. *LJ proves the sequent $A_1, \ldots, A_n \longrightarrow B$ if and only if there is a term t^B of the indicated type involving only the free variables $x_1^{A_1}, \ldots, x_n^{A_n}$.*

Proof. It is straightforward to prove, by induction on the complexity of t^B, that the desired LJ-proof exists. For the other direction, use induction on the length of the LJ-proof to prove that the term t^B exists. We will consider only the case where the last inference of the LJ-proof is

$$\frac{\Gamma \longrightarrow A \qquad B, \Gamma \longrightarrow C}{A \supset B, \Gamma \longrightarrow C}$$

The induction hypotheses give terms $r^A(x^\Gamma)$ and $s^C(x_1^B, x^\Gamma)$ where x^Γ actually represents a vector of variables, one for each formula in Γ. The desired term $t^C(x_2^{A\supset B}, x^\Gamma)$ is defined to be $s^C((x_2^{A\supset B} r^A(x^\Gamma)), x^\Gamma)$. $\square$

The $\{\supset, \bot\}$-fragment. The above treatment of the formulas-as-types isomorphism allowed only the connective $\supset$. The formulas-as-types paradigm can be extended to allow also the symbol $\bot$ and thereby negation, by making the following changes.

Firstly, enlarge the definition of types, by adding a clause specifying that $\emptyset$ is a type. Identifying $\emptyset$ with $\bot$ makes types identical to formulas. The type $\emptyset$ corresponds to the empty set; this is consistent with the BHK interpretation since $\bot$ has no proof. Secondly, enlarge the definition of terms by adding to the definition a new clause stating that, for every type σ, there is a term $f^{\emptyset\to\sigma}$ of type $\emptyset \to \sigma$.

Now the Theorem and Lemma of section 3.1.5 still hold verbatim for formulas with connectives $\{\supset, \bot\}$ and with the new definitions of types and terms.

The $\{\supset, \bot, \land\}$-fragment. To further expand the formulas-as-types paradigm to intuitionistic logic with connectives $\supset$, $\bot$ and $\land$, we must make the following modifications to the definitions of terms and types. Firstly, add to the definition of types, a clause stating that if σ and τ are types, then $(\sigma \times \tau)$ is a type. By identifying $\times$ with $\land$, we still have the types identified with the formulas. Consistent with the BHK-interpretation, the type $(\sigma \times \tau)$ may be viewed as the cross-product of its constituent types. Secondly, add to the definition of terms the following two clauses: (*i*) if s^σ and t^τ are terms, then $\langle s, t\rangle$ is a term of type $\sigma \times \tau$, and (*ii*) if $s^{\sigma\times\tau}$ is a term, then $(\pi_1 s)^\sigma$ and $(\pi_2 s)^\tau$ are terms. The former term uses the pairing function; and the latter terms use the projection functions.

Again, the Theorem and Lemma above still holds with these new definitions of types and terms.

The $\{\supset, \bot, \land, \lor\}$-fragment. To incorporate also the connective $\lor$ into the formulas-as-types isomorphism we expand the definitions of types and terms as follows. Firstly, add to the definition of types that if σ and τ are types, then $(\sigma + \tau)$ is a type. To identify formulas with types, the symbol $+$ is identified with $\lor$. The type $\sigma + \tau$ is viewed as the disjoint union of σ and τ, consistent with the BHK-interpretation. Secondly, add to the definitions of terms the following two clauses

- If s^σ and t^τ are terms of the indicated types, then $(\iota_1^{\sigma\to(\sigma+\tau)} s)$ and $(\iota_2^{\tau\to(\sigma+\tau)} t)$ are terms of type $\sigma + \tau$.
- For all types σ, τ and μ, there is a constant symbol $d^{(\sigma\to\mu)\to((\tau\to\mu)\to((\sigma+\tau)\to\mu))}$ of the indicated type. In other words, if $s^{\sigma\to\mu}$ and $t^{\tau\to\mu}$ are terms, then (dst) is a term of type $(\sigma + \tau) \to \mu$.

Once again, the Theorem and Corollary of section 3.1.5 holds with these enlarged definitions for types and terms.

3.2. Linear logic

Linear logic was first introduced by Girard [1987a] as a refinement of classical logic, which uses additional connectives and in which weakening and contractions are not allowed. Linear logic is best understood as a 'resource logic' in that proofs in linear logic must use each formula exactly once. In this point view, assumptions in a proof are viewed as resources; each resource must be used once and only once during the proof. Thus, loosely speaking, an implication $A \vdash B$ can be proved in linear logic only if A is *exactly* what is needed to prove B; thus if the assumption A is either too weak or too strong, it is not possible to give a linear logic proof of 'if A then B'. As an example of this, $A \vdash B$ being provable in linear logic does not generally imply that $A, A \vdash B$ is provable in linear logic; this is because $A \vdash B$ is asserting that one use of the resource A gives B, whereas $A, A \vdash B$ asserts that two uses of the resource A gives B.

We shall introduce only the propositional fragment of linear logic, since already the main features of linear logic are present in its propositional fragment. We are particularly interested in giving a intuitive understanding of linear logic; accordingly, we shall frequently make vague or even not-quite-true assertions, sometimes, but not always, preceded by phrases such as "loosely speaking" or "intuitively", etc.

Linear logic will be formalized as a sequent calculus.[10] The initial logical sequents are just $A \rightarrow A$; the weakening and contraction structural rules are not allowed and the only structural rules are the exchange rule and the cut rule. Since contraction is not permitted, this prompts one to reformulate the rules, such as $\land$*:right* and $\lor$*:left*, which contain implicit contraction of side formulas. To begin with conjunction, linear logic has two different different conjunction symbols, denoted $\otimes$ and &: the *multiplicative* connective $\otimes$ does not allow contractions on side formulas, whereas the *additive* conjunction & does. The rules of inference for the two varieties of conjunction are:

$$\otimes{:}left\ \frac{A, B, \Gamma \rightarrow \Delta}{A \otimes B, \Gamma \rightarrow \Delta} \qquad \otimes{:}right\ \frac{\Gamma_1 \rightarrow \Delta_1, A \qquad \Gamma_2 \rightarrow \Delta_2, B}{\Gamma_1, \Gamma_2 \rightarrow \Delta_1, \Delta_2, A \otimes B}$$

and

$$\&{:}left\ \frac{A, \Gamma \rightarrow \Delta}{A \& B, \Gamma \rightarrow \Delta} \qquad \&{:}right\ \frac{\Gamma \rightarrow \Delta, A \qquad \Gamma \rightarrow \Delta, B}{\Gamma \rightarrow \Delta, A \& B}$$

$$\&{:}left\ \frac{B, \Gamma \rightarrow \Delta}{A \& B, \Gamma \rightarrow \Delta}$$

The obvious question now arises of what the difference is in the meanings of the two different conjunctions $\otimes$ and &. For this, one should think in terms of resource usage. Consider first the $\otimes$*:left* inference: the upper sequent contains both A and B so that resources for both A and B are available. The lower sequent contains $A \otimes B$ in their stead; the obvious intuition is that the resource (for) $A \otimes B$ is equivalent to having resources for both A and B. Thus, loosely speaking, we translate $\otimes$ as meaning "both". Now consider the &*:left* inference; here the lower sequent has the principal formula $A\&B$ in place of either A or B, depending on which form of the inference is used. This means that a resource $A\&B$ is equivalent to having a resource for A or B, whichever is desired. So we can translate & as "whichever".

The translations "both" and "whichever" also make sense for the other inferences for the conjunctions. Consider the $\otimes$*:right* inference and, since we have not yet discussed the disjunctions, assume Δ_1 and Δ_2 are empty. In this case, the upper hypotheses say that from a resource Γ_1 one can obtain A, and from a resource Γ_2

[10] The customary convention for linear logic is to use "$\vdash$" as the symbol for the sequent connection; we shall continue to use the symbol " $\rightarrow$ " however.

one can obtain B. The conclusion then says that from (the disjoint union of) both these resources, one obtains both A and B. Now consider a $\&$*:right* inference

$$\frac{\Gamma \rightarrow A \qquad \Gamma \rightarrow B}{\Gamma \rightarrow A\&B}$$

For the lower sequent, consider an agent who has the responsibility of providing (outputting) a resource $A\&B$ using exactly the resource Γ. To provide the resource $A\&B$, the agent must be prepared to either provide a resource A or a resource B, whichever one is needed; in either case, there is an upper sequent which says the agent can accomplish this.

Linear logic also has two different forms of disjunction: the multiplicative disjunction $\invamp$ and the additive disjunction $\oplus$. The former connective is dual to $\otimes$ and the latter is dual to $\&$. Accordingly the rule of inference for the disjunctions are:

$$\invamp{:}left\,\frac{A,\Gamma_1 \rightarrow \Delta_1 \qquad B,\Gamma_2 \rightarrow \Delta_2}{A\invamp B,\Gamma_1,\Gamma_2 \rightarrow \Delta_1,\Delta_2} \qquad \invamp{:}right\,\frac{\Gamma \rightarrow \Delta, A, B}{\Gamma \rightarrow \Delta, A\invamp B}$$

and

$$\oplus{:}left\,\frac{A,\Gamma \rightarrow \Delta \qquad B,\Gamma \rightarrow \Delta}{A \oplus B,\Gamma \rightarrow \Delta} \qquad \oplus{:}right\,\frac{\Gamma \rightarrow \Delta, A}{\Gamma \rightarrow \Delta, A \oplus B}$$

$$\oplus{:}right\,\frac{\Gamma \rightarrow \Delta, B}{\Gamma \rightarrow \Delta, A \oplus B}$$

By examining these inferences, we can see that the multiplicative disjunction, $\invamp$, is a a kind of 'indeterminate OR' or 'unspecific OR', in that a resource for $A\invamp B$ consists of a either a resource either for A or for B, but without a specification of which one it is a resource for. On the other hand, the additive conjunction, $\oplus$, is a 'determinate OR' or a 'specific OR', so that a resource for $A \oplus B$ is either a resource for A or a resource for B together with a specification of which one it is a resource for. To give a picturesque example, consider a military general who has to counter an invasion which will come either on the western front *or* on the eastern front. If the connective 'or' is the multiplicative or, then the general needs sufficient resources to counter both threats, whereas if the 'or' is an additive disjunction then the general needs only enough resources to counter either threat. In the latter case, the general is told ahead of time where the invasion will occur and has the ability to redeploy resources so as to counter the actual invasion; whereas in the former case, the general must separately deploy resources sufficient to counter the attack from the west and resources sufficient to counter the attack from the east. From the rules of inference for disjunction, it is clear that the commas in the succedent of a sequent intended to be interpreted as the unspecific or, $\invamp$.

The linear implication symbol, $\multimap$, is defined by letting $A \multimap B$ be an abbreviation for $A^{\perp}\invamp B$, where $A^{\perp}$ is defined below.

Linear logic also has two nullary connectives, $\mathbf{1}$ and $\top$, for truth and two nullary connectives, $\mathbf{0}$ and $\perp$, for falsity. Associated to these connectives are the initial sequents $\rightarrow \mathbf{1}$ and $\mathbf{0}, \Gamma \rightarrow \Delta$, and the one rule of inference:

$$\frac{\Gamma \rightarrow \Delta}{\mathbf{1}, \Gamma \rightarrow \Delta}$$

The intuitive idea for the multiplicatives is that **1** has the null resource and that there is no resource for $\bot$. For the additive connectives, $\top$ has a 'arbitrary' resource, which means that anything is a resource for $\top$, whereas any resource for **0** is a 'wild card' resource which may be used for any purpose.

In addition, linear logic has a negation operation $A^\bot$ which is involutive in that $(A^\bot)^\bot$ is the same as formula A. For each propositional variable p, $p^\bot$ is the negation of p. The operation of the negation is inductively defined by letting $(A \otimes B)^\bot$ be $A^\bot \parr B^\bot$, $(A \& B)^\bot$ be $A^\bot \oplus B$, $\mathbf{1}^\bot$ be $\bot$, and $\top^\bot$ be $\mathbf{0}$. This plus the involution property defines the operation $A^\bot$ for all formulas A. The two rules of inference associated with negation are

$$^\bot{:}right\,\frac{A, \Gamma \to \Delta}{\Gamma \to \Delta, A^\bot} \qquad \text{and} \qquad ^\bot{:}left\,\frac{\Gamma \to \Delta, A}{A^\bot, \Gamma \to \Delta}$$

Definition. The multiplicative/additive fragment of propositional linear logic is denoted **MALL** and contains all the logical connectives, initial sequents and rules of inference discussed above.

The absence of weakening and contraction rules, makes the linear logic sequent calculus particularly well behaved; in particular, the cut elimination theorem for **MALL** is quite easy to prove. In fact, if P is a **MALL** proof, then there is a *shorter* **MALL** proof P^* with the same endsequent which contains no cut inferences. In addition, the number of strong inferences in P^* is bounded by the number of connectives in the endsequent. Bellin [1990] contains an in-depth discussion of **MALL** and related systems.

As an exercise, consider the distributive laws $A \otimes (B \oplus C) \to (A \otimes B) \oplus (A \otimes B)$ and $(A \otimes B) \oplus (A \otimes B) \to A \otimes (B \oplus C)$. The reader should check that these are valid under the intuitive interpretation of the connectives given above; and, as expected, they can be proved in **MALL**. Similar reasoning shows that & is distributive over $\parr$. On the other hand, $\oplus$ is not distributive over $\parr$; this can be seen intuitively by considering the meanings of the connectives, or formally by using the cut elimination theorem.

MALL is not the full propositional linear logic. Linear logic (**LL**) has, in addition, two modalities ! and ? which allow contraction and weakening to be used in certain situations. The negation operator is extended to **LL** by defining $(!A)^\bot$ to be $?(A^\bot)$. The four rules of inference for ! are:

$$!{:}weakening_0\,\frac{\Gamma \to \Delta}{!A, \Gamma \to \Delta} \qquad !{:}weakening_1\,\frac{A, \Gamma \to \Delta}{!A, \Gamma \to \Delta}$$

$$!{:}contraction\,\frac{!A, !A, \Gamma \to \Delta}{!A, \Gamma \to \Delta} \qquad \frac{!\Gamma \to ?\Delta, A}{!\Gamma \to ?\Delta, !A}$$

where in the last inference, $!\Gamma$ (respectively, $?\Delta$) represents a cedent containing only formulas beginning with the ! symbol (respectively, the ? symbol).

The dual rules for ? are obtained from these using the negation operation. One should intuitively think of a resource for $!A$ as consisting of zero or more resources for A. The nature of full linear logic with the modalities is quite different from that of either **MALL** or propositional classical logic; in particular, Lincoln et al. [1992] show that **LL** is undecidable.

The above development of the intuitive meanings of the connectives in linear logic has not been as rigorous as one would desire; in particular, it would be nice have a completeness theorem for linear logic based on these intuitive meanings. This has been achieved for some fragments of linear logic by Blass [1992] and Abramsky and Jagadeesan [1994], but has not yet been attained for the full propositional linear logic. In addition to the references above, more information on linear logic may be found in Troelstra [1992], although his notation is different from the standard notation we have used.

Acknowledgements. We are grateful to S. Cook, E. Gertz, C. Gutiérrez, A. Jonasson, C. Lautemann, A. Maciel, R. Parikh C. Pollett, R. Stärk, J. Torán and S. Wainer for reading and suggesting corrections to preliminary versions of this chapter. Preparation of this article was partially supported by NSF grant DMS-9503247 and by cooperative research grant INT-9600919/ME-103 of the NSF and the Czech Republic Ministry of Education.

References

S. Abramsky and R. Jagadeesan
[1994] Games and full completeness for multiplicative linear logic, *Journal of Symbolic Logic*, 59, pp. 543–574.

J. Barwise
[1975] *Admissable Sets and Structures: An Approach to Definability Theory*, Springer-Verlag, Berlin.
[1977] *Handbook of Mathematical Logic*, North-Holland, Amsterdam.

G. Bellin
[1990] *Mechanizing Proof Theory: Resource-Aware Logics and Proof-Transformations to Extract Explicit Information*, PhD thesis, Stanford University.

E. W. Beth
[1953] On Padoa's method in the theory of definition, *Indagationes Mathematicae*, 15, pp. 330–339.
[1956] Semantic entailment and formal derivability, *Indagationes Mathematicae*, 19, pp. 357–388.

A. Blass
[1992] A game semantics for linear logic, *Annals of Pure and Applied Logic*, 56, pp. 183–220.

S. R. Buss
[1986] *Bounded Arithmetic*, Bibliopolis, Napoli. Revision of 1985 Princeton University Ph.D. thesis.

C.-L. Chang
[1970] The unit proof and the input proof in theorem proving, *J. Assoc. Comput. Mach.*, 17, pp. 698–707. Reprinted in: Siekmann and Wrightson [1983,vol 2].

C.-L. Chang and R. C.-T. Lee
[1973] *Symbolic Logic and Mechanical Theorem Proving*, Academic Press, New York.

W. F. Clocksin and C. S. Mellish

[1981] *Programming in Prolog*, North-Holland, Amsterdam, 4th ed.

W. Craig

[1957a] Linear reasoning. A new form of the Herbrand-Gentzen theorem, *Journal of Symbolic Logic*, 22, pp. 250–268.

[1957b] Three uses of the Herbrand-Gentzen theorem in relating model theory and proof theory, *Journal of Symbolic Logic*, 22, pp. 269–285.

M. Davis and H. Putnam

[1960] A computing procedure for quantification theory, *J. Assoc. Comput. Mach.*, 7, pp. 201–215. Reprinted in: Siekmann and Wrightson [1983,vol 1].

P. C. Eklof

[1977] Ultraproducts for algebraists, in: Barwise [1977], pp. 105–137.

S. Feferman

[1968] Lectures on proof theory, in: *Lectures on proof theory, Proceedings of the Summer School in Logic, Leeds, 1967*, M. H. Löb, ed., Lecture Notes in Mathematics #70, Springer-Verlag, Berlin, pp. 1–107.

G. Frege

[1879] *Begriffsschrift, eine der arithmetischen nachgebildete Formelsprache des reinen Denkens*, Halle. English translation: in van Heijenoort [1967], pp. 1-82.

G. Gentzen

[1935] Untersuchungen über das logische Schliessen, *Mathematische Zeitschrift*, 39, pp. 176–210, 405–431. English translation in: Gentzen [1969], pp. 68-131.

[1969] *Collected Papers of Gerhard Gentzen*, North-Holland, Amsterdam. Edited by M. E. Szabo.

J.-Y. Girard

[1987a] Linear logic, *Theoretical Computer Science*, 50, pp. 1–102.

[1987b] *Proof Theory and Logical Complexity*, vol. I, Bibliopolis, Napoli.

[1989] *Proofs and Types*, Cambridge tracts in theoretical computer science #7, Cambridge University Press. Translation and appendices by P. Taylor and Y. Lafont.

K. Gödel

[1930] Die Vollständigkeit der Axiome des logischen Funktionenkalküls, *Monatshefte für Mathematik und Physik*, 37, pp. 349–360.

R. Harrop

[1960] Concerning formulas of the types $A \to B \vee C$, $A \to (Ex)B(x)$ in intuitionistic formal systems, *Journal of Symbolic Logic*, 25, pp. 27–32.

J. van Heijenoort

[1967] *From Frege to Gödel: A sourcebook in mathematical logic, 1879-1931*, Harvard University Press.

L. Henkin

[1949] The completeness of the first-order functional calculus, *Journal of Symbolic Logic*, 14, pp. 159–166.

L. Henschen and L. Wos

[1974] Unit refutations and Horn sets, *J. Assoc. Comput. Mach.*, 21, pp. 590–605.

J. Herbrand

[1930] *Recherches sur la théorie de la démonstration*, PhD thesis, University of Paris. English translation in Herbrand [1971] and translation of chapter 5 in van Heijenoort [1967], pp. 525-581.

[1971] *Logical Writings*, D. Reidel, Dordrecht, Holland. ed. by W. Goldfarb.

D. Hilbert and W. Ackermann
[1928] *Grundzüge der theoretischen Logik*, Springer-Verlag, Berlin.

D. Hilbert and P. Bernays
[1934-39] *Grundlagen der Mathematik, I & II*, Springer, Berlin.

J. Hintikka
[1955] Form and content in quantification theory, two papers on symbolic logic, *Acta Philosophica Fennica*, 8, pp. 7–55.

W. A. Howard
[1980] The formulas-as-types notion of construction, in: *To H. B. Curry: Essays in Combinatory Logic, Lambda Calculus and Formalism*, J. P. Seldin and J. R. Hindley, eds., Academic Press, New York, pp. 479–491.

S. Kanger
[1957] *Provability in Logic*, Almqvist & Wiksell, Stockholm.

S. C. Kleene
[1952] *Introduction to Metamathematics*, Wolters-Noordhoff, Groningen and North-Holland, Amsterdam.

R. Kowalski
[1979] *Logic for Problem Solving*, North-Holland, Amsterdam.

J. Krajíček, P. Pudlák, and G. Takeuti
[1991] Bounded arithmetic and the polynomial hierarchy, *Annals of Pure and Applied Logic*, 52, pp. 143–153.

G. Kreisel
[1951] On the interpretation of non-finitist proofs–part I, *Journal of Symbolic Logic*, 16, pp. 241–267.
[1952] On the interpretation of non-finitist proofs, part II. interpretation of number theory, applications, *Journal of Symbolic Logic*, 17, pp. 43–58.

P. Lincoln, J. C. Mitchell, A. Scedrov, and N. Shankar
[1992] Decision problems for linear logic, *Annals of Pure and Applied Logic*, 56, pp. 239–311.

E. G. K. Lopez-Escobar
[1965] An interpolation theorem for denumerably long formulas, *Fundamenta Mathematicae*, 57, pp. 253–272.

D. W. Loveland
[1970] A linear format for resolution, in: *Symp. on Automatic Demonstration*, Lecture Notes in Mathematics #125, Springer-Verlag, Berlin, pp. 147–162.
[1978] *Automated Theorem Proving: A Logical Basis*, North-Holland, Amsterdam.

D. Luckham
[1970] Refinement theorems in resolution theory, in: *Symp. on Automatic Demonstration*, Lecture Notes in Mathematics #125, Springer-Verlag, Berlin, pp. 163–190.

R. C. Lyndon
[1959] An interpolation theorem in the predicate calculus, *Pacific Journal of Mathematics*, 9, pp. 129–142.

E. Mendelson
[1987] *Introduction to Mathematical Logic*, Wadsworth & Brooks/Cole, Monterey.

J. R. Munkres
[1975] *Topology: A First Course*, Prentice-Hall, Englewood Cliffs, New Jersey.

M. S. Paterson and M. N. Wegman
[1978] Linear unification, *J. Comput. System Sci.*, 16, pp. 158–167.

G. Peano
[1889] *Arithmetices Principia, noca methodo exposito*, Turin. English translation in: van Heijenoort [1967], pp. 83-97.

D. Prawitz
[1965] *Natural Deduction: A Proof-Theoretical Study*, Almqvist & Wiksell, Stockholm.

A. Robinson
[1956] A result on consistency and it application to the theory of definability, *Indagationes Mathematicae*, 18, pp. 47–58.

G. Robinson and L. Wos
[1969] Paramodulation and theorem-proving in first-order theories with equality, in: *Machine Intelligence 4*, pp. 135–150.

J. A. Robinson
[1965a] Automatic deduction with hyper-resolution, *International Journal of Computer Mathematics*, 1, pp. 227–234. Reprinted in: Siekmann and Wrightson [1983,vol 1].
[1965b] A machine-oriented logic based on the resolution principle, *J. Assoc. Comput. Mach.*, 12, pp. 23–41. Reprinted in: Siekmann and Wrightson [1983,vol 1].

K. Schütte
[1965] Ein System des verknüpfenden Schliessens, *Archiv für Mathematische Logik und Grundlagenforschung*, 2, pp. 55–67.

J. Siekmann and G. Wrightson
[1983] *Automation of Reasoning*, vol. 1&2, Springer-Verlag, Berlin.

J. R. Slagle
[1967] Automatic theorem proving with renamable and semantic resolution, *J. Assoc. Comput. Mach.*, 14, pp. 687–697. Reprinted in: Siekmann and Wrightson [1983,vol 1].

W. W. Tait
[1968] Normal derivability in classical logic, in: *The Syntax and Semantics of Infinitary Languages, Lecture Notes in Mathematics #72*, J. Barwise, ed., Springer-Verlag, Berlin, pp. 204–236.

G. Takeuti
[1987] *Proof Theory*, North-Holland, Amsterdam, 2nd ed.

A. S. Troelstra
[1992] *Lectures on Linear Logic*, Center for the Study of Logic and Information, Stanford.

A. S. Troelstra and D. van Dalen
[1988] *Constructivism in Mathematics: An Introduction*, vol. I&II, North-Holland, Amsterdam.

G. S. Tsejtin
[1968] On the complexity of derivation in propositional logic, *Studies in Constructive Mathematics and Mathematical Logic*, 2, pp. 115–125. Reprinted in: Siekmann and Wrightson [1983,vol 2].

A. N. Whitehead and B. Russell
[1910] *Principia Mathematica*, vol. 1, Cambridge University Press.

L. Wos, R. Overbeek, E. Lusk, and J. Boyle
[1992] *Automated Reasoning: Introduction and Applications*, McGraw-Hill, New York, 2nd ed.

L. Wos and G. Robinson

[1973] Maximal models and refutation completeness: Semidecision procedures in automatic theorem proving, in: *Word Problems: Decision Problems and the Burnside Problem in Group Theory*, W. W. Boone, F. B. Cannonito, and R. C. Lyndon, eds., North-Holland, Amsterdam, pp. 609–639.

L. Wos, G. Robinson, and D. F. Carson

[1965] Efficiency and completeness of the set of support stategy in theorem proving, *J. Assoc. Comput. Mach.*, 12, pp. 201–215. Reprinted in: Siekmann and Wrightson [1983,vol 1].

CHAPTER II

First-Order Proof Theory of Arithmetic

Samuel R. Buss

Departments of Mathematics and Computer Science, University of California, San Diego, La Jolla, CA 92093-0112, USA

Contents

HANDBOOK OF PROOF THEORY
Edited by S. R. Buss

This chapter discusses the proof-theoretic foundations of the first-order theory of the non-negative integers. This first-order theory of numbers, also called 'first-order arithmetic', consists of the first-order sentences which are true about the integers. The study of first-order arithmetic is important for several reasons. Firstly, in the study of the foundation of mathematics, arithmetic and set theory are two of the most important first-order theories; indeed, the usual foundational development of mathematical structures begins with the integers as fundamental and from these constructs mathematical constructions such as the rationals and the reals. Secondly, the proof theory for arithmetic is highly developed and serves as a basis for proof-theoretic investigations of many stronger theories. Thirdly, there are intimate connections between subtheories of arithmetic and computational complexity; these connections go back to Gödel's discovery that the numeralwise representable functions of arithmetic theories are exactly the recursive functions and are recently of great interest because some weak theories of arithmetic have very close connection of feasible computational classes.

Because of Gödel's second incompleteness theorem that the theory of numbers is not recursive, there is no good proof theory for the complete theory of numbers; therefore, proof-theorists consider axiomatizable subtheories (called fragments) of first-order arithmetic. These fragments range in strength from the very weak theories R and Q up to the very strong theory of Peano arithmetic (PA).

The outline of this chapter is as follows. Firstly, we shall introduce the most important fragments of arithmetic and discuss their relative strengths and the bootstrapping process. Secondly, we give an overview of the incompleteness theorems. Thirdly, section 3 discusses the topics of what functions are provably total in various fragments of arithmetic and of the relative strengths of different fragments of arithmetic. Finally, we conclude with a proof of a theorem of J. Paris and A. Wilkie which improves Gödel's incompleteness theorem by showing that $I\Delta_0 + \exp$ cannot prove the consistency of Q. The main prerequisite for reading this chapter is knowledge of the sequent calculus and cut-elimination, as contained in Chapter I of this volume. The proof theory of arithmetic is a major subfield of logic and this chapter necessarily omits many important and central topics in the proof theory of arithmetic; the most notable omission is theories stronger than Peano arithmetic. Our emphasis has instead been on weak fragments of arithmetic and on finitary proof theory, especially on applications of the cut-elimination theorem. The articles of Fairtlough-Wainer, Pohlers, Troelstra and Avigad-Feferman in this volume also discuss the proof theory of arithmetic.

There are a number of book length treatments of the proof theory and model theory of arithmetic. Takeuti [1987], Girard [1987] and Schütte [1977] discuss the classical proof theory of arithmetic, Buss [1986] discusses the proof of the bounded arithmetic, and Kaye [1991] and Hájek and Pudlák [1993] treat the model theory of arithmetic. The last reference gives an in-depth and modern treatment both of classical fragments of Peano arithmetic and of bounded arithmetic.

1. Fragments of arithmetic

This section introduces the most commonly used axiomatizations for fragments of arithmetic. These axiomatizations are organized into the categories of 'strong fragments', 'weak fragments' and 'very weak fragments'. The line between strong and weak fragments is somewhat arbitrarily drawn between those theories which can prove the arithmetized version of the cut-elimination theorem and those which cannot; in practice, this is equivalent to whether the theory can prove that the superexponential function $i \mapsto 2^1_i$ is total. The very weak theories are theories which do not admit any induction axioms.

Non-logical symbols for arithmetic. We will be working exclusively with first-order theories of arithmetic: these have all the usual first-order symbols, including propositional connectives and quantifiers and the equality symbol ($=$). In addition, they have *non-logical* symbols specific to arithmetic. These will always include the constant symbol 0, the unary successor function S, the binary functions symbols $+$ and $\cdot$ for addition and multiplication, and the binary predicate symbol $\leq$ for 'less than or equal to'.[1] Very often, terms are abbreviated by omitting parentheses around the arguments of the successor function, and we write St instead of $S(t)$. In addition, for $n \geq 0$ an integer, we write $S^n t$ to denote the term with n applications of S to t.

For weak theories of arithmetic, especially for bounded arithmetic, it is common to include further non-logical symbols. These include a unary function $\lfloor \frac{1}{2}x \rfloor$ for division by two, a unary function $|x|$ which is defined by

$$|n| \;=\; \lceil \log_2(n+1) \rceil,$$

and Nelson's binary function # (pronounced 'smash') which we define by

$$m\#n \;=\; 2^{|m|\cdot|n|}.$$

It is easy to check that $|n|$ is equal to the number of bits in the binary representation of n.

An alternative to the # function is the unary function ω_1, which is defined by $\omega_1(n) = n^{\lfloor \log_2 n \rfloor}$ and has growth rate similar to #. The importance of the use of the ω_1 function and the # function lies mainly in their growth rate. In this regard, they are essentially equivalent since $\omega_1(n) \approx n\#n$ and $m\#n = O(\omega_1(\max\{m,n\}))$. Both of these functions are generalizable to faster growing functions by defining $\omega_n(x) = x^{\omega_{n-1}(\lfloor \log_2 x \rfloor)}$ and $x\#_{n+1}y = 2^{|x|\#_n|y|}$ where $\#_2$ is #. It is easy to check that the growth rates of ω_n and $\#_{n+1}$ are equivalent in the sense that any term involving one of the function symbols can be bounded by a term involving the other function symbol.

[1] Many authors use $<$ instead of $\leq$; however, we prefer the use of $\leq$ since this sometimes makes axioms and theorems more elegant to state.

For strong theories of arithmetic, it is sometimes convenient to enlarge the set of non-logical symbols to include function symbols for all primitive recursive functions. The usual way to do this is to inductively define the primitive recursive functions as the smallest class of functions which contains the constant function 0 and the successor function S, is closed under a general form of composition, and is closed under primitive recursion. The closure under primitive recursion means that if g and h are primitive recursive functions of arities n and $n+2$, then the $(n+1)$-ary function f defined by

$$\begin{aligned} f(\vec{x},0) &= g(\vec{x}) \\ f(\vec{x},m+1) &= h(\vec{x},m,f(\vec{x},m)) \end{aligned}$$

is also primitive recursive. These equations are called the *defining equations* of f.

A *bounded* quantifier is of the form $(\forall x \le t)(\cdots)$ or $(\exists x \le t)(\cdots)$ where t is a term not involving x. These may be used as abbreviations for $(\forall x)(x \le t \supset \cdots)$ and $(\exists x)(x \le t \wedge \cdots)$, respectively; or, alternatively, the syntax of first-order logic may be expanded to incorporate bounded quantifiers directly. In the latter case, the sequent calculus is enlarged with additional inference rules, shown in section 1.4. A usual quantifier is called an *unbounded* quantifier; when $|x|$ is in the language, a bounded quantifier of the form $(Qx \le |t|)$ is called a *sharply bounded* quantifier.

A theory is said to be *bounded* if it is axiomatizable with a set of bounded formulas. Since free variables in axioms are implicitly universally quantified, this is equivalent to being axiomatized with a set of Π_1-sentences (which are defined in section 1.2.1).

1.1. Very weak fragments of arithmetic

The most commonly used induction-free fragment of arithmetic is Robinson's theory Q, introduced by Tarski, Mostowski and Robinson [1953]. The theory Q has non-logical symbols 0, S, $+$ and $\cdot$ and is axiomatized by the following six axioms:

$$\begin{aligned} &(\forall x)(\neg Sx \neq 0) \\ &(\forall x)(\forall y)(Sx = Sy \supset x = y) \\ &(\forall x)(x \neq 0 \supset (\exists y)(Sy = x)) \\ &(\forall x)(x + 0 = x) \\ &(\forall x)(\forall y)(x + Sy = S(x+y)) \\ &(\forall x)(x \cdot 0 = 0) \\ &(\forall x)(x \cdot Sy = x \cdot y + x) \end{aligned}$$

Unlike most of the theories of arithmetic we consider, the language of Q does not contain the inequality symbol; however, we can conservatively extend Q to include $\le$ by giving it the defining axiom:

$$x \le y \leftrightarrow (\exists z)(x + z = y).$$

This conservative extension of Q is denoted $Q_{\leq}$.

A yet weaker theory is the theory R, also introduced by Tarski, Mostowski and Robinson [1953]. This has the same language as Q and is axiomatized by the following infinite set of axioms, where we let $s \leq t$ abbreviate $(\exists z)(s + z = t)$.

$$\begin{array}{ll} S^m 0 \neq S^n 0 & \text{for all } 0 \leq m < n, \\ S^m 0 + S^n 0 = S^{m+n} 0 & \text{for all } m, n \geq 0, \\ S^m 0 \cdot S^n 0 = S^{m \cdot n} 0 & \text{for all } m, n \geq 0, \\ (\forall x)(x \leq S^m 0 \vee S^m 0 \leq x) & \text{for all } m \geq 0, \text{ and} \\ (\forall x)(x \leq S^m 0 \leftrightarrow x = 0 \vee x = S0 \vee x = S^2 0 \vee \cdots \vee x = S^m 0) & \\ & \text{for all } m \geq 0. \end{array}$$

We leave it to the reader to prove that $Q \models R$.

1.2. Strong fragments of arithmetic

This section presents the definitions and the basic capabilities of some strong fragments of arithmetic. These fragments are defined by using induction axioms, minimization axioms or collection axioms; these axioms do not always apply to all first-order formulas, but rather apply to formulas that satisfy certain restrictions on quantifier alternation. For this purpose, we make the following definitions:

1.2.1. Definition. A formula is called a *bounded formula* if it contains only bounded quantifiers. The set of bounded formulas is denoted Δ_0. For $n \geq 0$, the classes Σ_n and Π_n of first-order formulas are inductively defined by:

(1) $\Sigma_0 = \Pi_0 = \Delta_0$,

(2) Σ_{n+1} is the set of formulas of the form $(\exists \vec{x})A$ where $A \in \Pi_n$ and $\vec{x}$ is a possibly empty vector of variables.

(3) Π_{n+1} is the set of formulas of the form $(\exists \vec{x})A$ where $A \in \Sigma_n$ and $\vec{x}$ is a possibly empty vector of variables.

These classes Σ_n, Π_n form the *arithmetic hierarchy.*

1.2.2. Definition. The *induction axioms* are specified as an axiom scheme; that is, if Φ is a set of formulas then the Φ-IND axiom are the formulas

$$A(0) \wedge (\forall x)(A(x) \supset A(Sx)) \supset (\forall x)A(x),$$

for all formulas $A \in \Phi$. Note that $A(x)$ is permitted to have other free variables in addition to x. Similarly, the *least number principle* axioms or *minimization* axioms for ϕ are denoted Φ-MIN and consist of all formulas

$$(\exists x)A(x) \supset (\exists x)(A(x) \wedge \neg(\exists y)(y < x \wedge A(y))),$$

for all $A \in \Phi$. Likewise, the *collection* or *replacement* axioms for Φ are denoted Φ-REPL and consist of the formulas

$$(\forall x \le t)(\exists y)A(x,y) \supset (\exists z)(\forall x \le t)(\exists y \le z)A(x,y),$$

for all $A \in \Phi$.

1.2.3. Definition. The above axioms form the basis for a hierarchy of strong fragments of arithmetic over the language containing the non-logical symbols 0, S, $+$, $\cdot$ and $\le$. The theory $I\Sigma_n$ is defined to be the theory axiomatized by the eight axioms of $Q_\le$ plus the Σ_n-IND axioms. Of particular importance is the special case of the theory $I\Delta_0$ which defined as $Q_\le$ plus the Δ_0-IND axioms. The theory $L\Sigma_n$ is defined to be the theory $I\Delta_0$ plus the Σ_n-MIN axioms. Similarly, $B\Sigma_n$ is the theory consisting of $I\Delta_0$ plus the Σ_n-REPL axioms. Other theories, especially $I\Pi_n$, $L\Pi_n$ and $B\Pi_n$, can be defined similarly.

The theory of *Peano arithmetic*, *PA*, is defined to be the theory Q plus induction for all first-order formulas. The figure below shows that *PA* also admits the minimization and replacement axioms for all formulas.

1.2.4. The figure below shows the containments between the various strong fragments of arithmetic, where $S \Rightarrow T$ indicates that the theory S logically implies the theory T. The two arrows $I\Sigma_{n+1} \Rightarrow B\Sigma_{n+1}$ and $B\Sigma_{n+1} \Rightarrow I\Sigma_n$ do not reverse, i.e., the containments are proper. These facts are due to Parsons [1970] and Paris and Kirby [1978]. (The figure is taken from the latter reference.)

$$\begin{array}{l} I\Sigma_{n+1} \\ \Downarrow \\ B\Sigma_{n+1} \Longleftrightarrow B\Pi_n \\ \Downarrow \\ I\Sigma_n \Longleftrightarrow I\Pi_n \Longleftrightarrow L\Sigma_n \Longleftrightarrow L\Pi_n \end{array}$$

Most of these containments are proved in section 1.2.9. The fact that $B\Sigma_{n+1}$ is a subtheory of $I\Sigma_{n+1}$ is proved as Theorem 1.2.9 below. The fact that it is a *proper* subtheory of $I\Sigma_{n+1}$ is proved as Theorem 3.4.2.

1.2.5. Σ_n^+ and Π_n^+ formulas. Some authors use a different definition of the arithmetic hierarchy than definition 1.2.1. These alternative classes, which we denote Σ_n^+ and Π_n^+, are inductively defined by

(1) $\Sigma_0^+ = \Pi_0^+ = \Delta_0$,

(2) Σ_{n+1}^+ is the set of formulas obtained by prepending an arbitrary block of existential quantifiers and bounded universal quantifiers to Π_n^+-formulas.

(3) Π_{n+1}^+ is the set of formulas obtained by prepending an arbitrary block of universal quantifiers and bounded existential quantifiers to Σ_n^+-formulas.

Thus Σ_n^+ and Π_n^+ are defined analogously to Σ_n and Π_n, except arbitrary bounded quantifiers may be inserted without adding to the quantifier complexity.

It is straightforward to prove that Σ_n-REPL proves that every Σ_n^+-formula is equivalent to a Σ_n-formula, using induction on the number of unbounded quantifiers which are in the scope of a bounded quantifier, with a sub-induction on the number of bounded quantifiers which have the outermost unbounded quantifier in their scope. Therefore, $I\Sigma_n^+$ is equivalent to $I\Sigma_n$ and $B\Sigma_n^+$ is equivalent to $B\Sigma_n$.

1.2.6. Bootstrapping $I\Delta_0$, Phase 1

The axioms of Q are very simplistic and, by themselves, do not imply many elementary facts about addition and multiplication, such as commutativity and associativity. When combined with induction axioms, however, the axioms of Q imply many basic facts about the integers. The process of establishing such basic facts as commutativity and associativity of addition and multiplication, the transitivity of $\leq$, the totality of subtraction, etc. is called *bootstrapping*, named after the expression "to lift oneself by one's bootstraps". That is to say, in order to use the full power of a set of axioms, it is necessary to do some relatively tedious work establishing that the axioms of Q are sufficient as a base theory.

This section will give a sketch of the bootstrapping process for $I\Delta_0$; to keep things brief, only an outline will be given, with most of the proofs left to the reader. Because $I\Delta_0$ is a subtheory of all the strong fragments defined above, this bootstrapping applies equally well to all of them.

To begin the bootstrapping process, show that the following formulas are $I\Delta_0$ provable.

(a) *Addition is commutative:* $(\forall x, y)(x + y = y + x)$. In order to prove this, first prove the formulas (a.1) $(\forall x)(0 + x = x)$ and (a.2) $(\forall x, y)(Sx + y = S(x + y))$. In order to prove (a.1), use induction on the formula $0 + x = x$, and to prove (a.2), use induction on $Sx + y = S(x + y)$ with respect to the variable y. Note that the variable x is being used as a parameter in the latter induction. From these two, one can use induction on $x + y = y + x$ and prove the commutativity of addition.

(b) *Addition is associative:* $(\forall x, y, z)((x + y) + z = x + (y + z))$. Use induction on $(x + y) + z = x + (y + z)$.

(c) *Multiplication is commutative:* $(\forall x, y)(x \cdot y = y \cdot x)$. Analogously to (a), first prove (c.1) $0 \cdot x = 0$ and (c.2) $(Sx) \cdot y = x \cdot y + y$ by induction with respect to x and y, respectively.

(d) *Distributive law:* $(\forall x, y, z)((x + y) \cdot z = x \cdot z + y \cdot z)$. Use induction.

(e) *Multiplication is associative:* $(\forall x, y, z)((x \cdot y) \cdot z = x \cdot (y \cdot z))$. Use induction.

(f) *Cancellation laws for addition:* $(\forall x, y, z)(x + z = y + z \leftrightarrow x = y)$ and $(\forall x, y, z)(x+z \leq y+z \leftrightarrow x \leq y)$. Use induction w.r.t. z for the forward implications.

(g) *Discreteness of* $\leq$*:* $(\forall x, y)(x \leq Sy \supset x \leq y \vee x = Sy)$. This can be proved from Q without any induction: if $x \leq Sy$, then $x + z = Sy$ for some z; either $z = 0$ so $x = Sy$, or there is a u such that $u = Sz$ and so $x + Su = Sy$, in which case $x + u = y$ and thus $x \leq y$.

(h) *Transitivity of* $\leq$: $(\forall x, y, z)(x \leq y \wedge y \leq z \supset x \leq y)$. Follows from Q and the associativity of addition.

(i) *Anti-idempotency laws:* $(\forall x, y)(x + y = 0 \supset x = 0 \wedge y = 0)$ and $(\forall x, y)(x \cdot y = 0 \supset x = 0 \vee y = 0)$. These both follow from Q without any induction. Use the fact that if $y \neq 0$, then $y = Sz$ for some z.

(j) *Reflexivity, trichotomy and antisymmetry of the* $\leq$ *ordering:* $(\forall x)(x \leq x)$, $(\forall x, y)(x \leq y \vee y \leq x)$ and $(\forall x, y)(x \leq y \wedge y \leq x \supset x = y)$. To prove trichotomy, use induction on y; the argument splits into two cases, depending on whether $x \leq y$ or $y + Sz = x$ for some z. To prove antisymmetry, reason as follows: if $x + u = y$ and $y + v = x$, then $x + u + v = x$, so by the cancellation law for addition, $u + v = 0$ and by anti-idempotency $u = v = 0$ and thus $x = y$.

(k) *Cancellation laws for multiplication:* $(\forall x, y, z)(z \neq 0 \wedge x \cdot z = y \cdot z \supset x = y)$ and $(\forall x, y, z)(z \neq 0 \wedge x \cdot z \leq y \cdot z \supset x \leq y)$. These can be proved using (j) to have $x = y$ or $x + Sv = y$ or $y + Sv = x$ for some v. Then use the distributive law, the anti-idempotency of multiplication, and the cancellation laws for addition.

(l) *Strict inequality:* $s < t$ is an abbreviation for $S(s) \leq t$. Thus, we can use bounded existential quantifiers $(\forall x < t)(\cdots)$ to mean $(\forall x \leq t)(x < t \supset \cdots)$, and similarly for bounded universal quantifiers.

Theorem. $I\Delta \vdash \Delta_0$-*MIN*.

Proof. The minimization axiom for $A(x)$ is easily seen to be equivalent to complete induction on $\neg A(x)$, namely to

$$(\forall y)(((\forall z < y)\neg A(z)) \supset \neg A(y)) \supset (\forall x)\neg A(x).$$

This is equivalent to induction on the bounded formula $(\forall y \leq x)(\neg A(y))$, and therefore is provable in $I\Delta_0$.

1.2.7. Extending the language of arithmetic.

We now introduce two useful means of conservatively extending the language of arithmetic with definitions of new predicate symbols and function symbols. It will be of particular importance that we can use the new function and predicate symbols in induction formulas.

Definition. A predicate symbol $R(\vec{x})$ is Δ_0-*defined* if it has a defining axiom

$$R(\vec{x}) \leftrightarrow \phi(\vec{x})$$

with ϕ a Δ_0-formula with all free variables as indicated.

The predicate R is Δ_1-*defined by a theory* T if there are Σ_1 formulas $\phi(\vec{x})$ and $\psi(\vec{x})$ such that R has the defining axiom above and $T \vdash (\forall \vec{x})(\phi \leftrightarrow \neg\psi)$.

Definition. Let T be a theory of arithmetic. A function symbol $f(\vec{x})$ is Σ_1*-defined by* T if it has a defining axiom

$$y = f(\vec{x}) \leftrightarrow \phi(\vec{x}, y),$$

where ϕ is a Σ_1 formula with all free variables indicated, such that T proves $(\forall\vec{x})(\exists!y)\phi(\vec{x}, y)$.[2]

The Σ_1-definable functions of a theory are sometimes referred to as the *provably recursive* or *provably total* functions of the theory. To see that this a reasonable definition for "provably recursive", let M be a Turing machine which computes a function $y = M(x)$. Also choose some scheme for encoding computations of M and let $T_M(x, w, y)$ be the predicate expressing "w encodes a computation of M on input x which outputs y." From the bootstrapping below, it can be seen that the predicate T_M can be a bounded formula. Therefore, the function computed by M can be Σ_1-defined by the (true) formula $(\forall x)(\exists!y)(\exists w)(T_M(x, w, y))$. Conversely, for any true sentence $(\forall\vec{x})(\exists!y)\phi(\vec{x}, y)$ with ϕ a Σ_1-formula, the function mapping $\vec{x}$ to y can be computed by a Turing machine that, given input values for $\vec{x}$, searches for a value for y and for values of the existential quantified variables in ϕ which witness the truth of $(\exists y)\phi(\vec{x}, y)$.

In the case of Σ_1-definability of functions in $I\Delta_0$, it is possible to give a stronger equivalent condition; this is based on the following theorem of Parikh [1971]:

1.2.7.1. Parikh's Theorem. *Let* $A(\vec{x}, y)$ *be a bounded formula and* T *be a bounded theory. Suppose* $T \vdash (\forall\vec{x})(\exists y)A(\vec{x}, y)$*. Then there is a term* t *such that* T *also proves* $(\forall\vec{x})(\exists y \leq t)A(\vec{x}, y)$*.*

The above theorem is stated with y a single variable, but it also holds for a vector of existentially quantified variables. A proof-theoretic proof of a generalization of this theorem is sketched in section 1.4.3 below.

It is easily seen that $I\Delta_0$ is a bounded theory, since the defining axiom for $\leq$ may be replaced by the ($I\Delta_0$-provable) formula

$$x \leq y \leftrightarrow (\exists z \leq y)(x + z = y),$$

and the induction axioms may be replaced by the equivalent axioms

$$(\forall z)(A(0) \land (\forall x \leq z)(A(x) \supset A(Sx)) \supset A(z)).$$

Thus applying Parikh's theorem gives the following theorem. Its proof is straightforward and we leave it to the reader.

[2]The notation $(\exists!y)$ means "there exists a unique y such that $\cdots$". This is not part of the syntax of first-order logic; but is rather an abbreviation for a more complicated first-order formula.

1.2.7.2. Theorem. *A function symbol $f(\vec{x})$ is Σ_1-*defined by $I\Delta_0$ *if and only if it has a defining axiom*

$$y = f(\vec{x}) \leftrightarrow \phi(\vec{x}, y),$$

and it has a bounding term $t(\vec{x})$ *such that ϕ is a Δ_0 formula with all free variables indicated and $I\Delta_0$ proves $(\forall\vec{x})(\exists! y \le t)\phi(\vec{x}, y)$.*[3]

A predicate symbol R is Δ_1-defined by $I\Delta_0$ if and only if it is Δ_0-defined by $I\Delta_0$ (and furthermore, $I\Delta_0$ can prove the equivalence of the two definitions).

The next theorem states the crucial fact about Σ_1-definable functions that they may be used freely without increasing the quantifier complexity of formulas, even when reexpressed in the original language of arithmetic.

1.2.7.3. Theorem.
(Gaifman and Dimitracopoulos [1982,Prop 2.3], see also Buss [1986,Thm 2.2].)

(a) Let $T \supseteq B\Sigma_1$ be a theory of arithmetic. Let T be extended to a theory T^+ in an enlarged language L^+ by adding Δ_1-defined predicate symbols, Σ_1-defined function symbols and their defining equations. Then T^+ is conservative over T. Also, if $i \ge 1$ and if A is a Σ_i- (respectively, Π_i-) formula in the enlarged language L^+, then there is a formula A^- in the language of T such that A is also in Σ_i (respectively, Π_i) and such that

$$T^+ \vdash (A \leftrightarrow A^-).$$

(b) The same results holds for $T \supseteq Q$ a bounded theory in the language $0, S, +, \cdot$ and $i = 0$, i.e., for the class Δ_0.

Proof. We first sketch the proof for part (b). The proof shows that the new symbols can be eliminated from a formula A by induction on the number of occurrences of the new Δ_0-defined predicate symbols and Σ_1-defined function symbols in A. Firstly, any atomic formula involving a Δ_0-defined R may just be replaced by the defining equation for R. Secondly, eliminate Σ_1-defined function symbols from terms in quantifier bounds, by replacing each bounded quantifier $(\forall x \le t)(\cdots)$ by $(\forall x \le t^*)(x \le t \supset \cdots)$ where t^* is obtained from t by replacing every new Σ_1-defined function symbol with its bounding term; and by similarly replacing bounds on existential bounded quantifiers. Thirdly, repeatedly replace any atomic formula $P(f(\vec{s}))$ where s does not involve any new function symbols by either of the formulas

$$(\exists z \le t(\vec{s}))(A_f(\vec{s}, z) \wedge P(z))$$

or

$$(\forall z \le t(\vec{s}))(A_f(\vec{s}, z) \supset P(z)),$$

where A_f is the Δ_0-formula which Σ_1-defined f, and t is the bounding term of f.

It is easy to see that each step removes new function and predicate symbols from A and preserves equivalence to A and this proves (b).

[3]The notation $(\exists! y \le t)(\cdots)$ means "there is a unique y such that $\cdots$, and this y is $\le t$".

The proof of part (a) is similar, but needs a little modification. Most notably, there is no bounding term t, so the two formulas which can replace $P(f(\vec{s}))$ use an unbounded quantification of z and are thus in Σ_1 and Π_1, respectively. Since A_f is a Σ_i-formula, it is necessary to pick the correct one of the two formulas for replacing $P(f(\vec{s}))$: since the first is Σ_1 and the second is Π_1 there is always an appropriate choice that does not increase the alternation of unbounded quantifiers. Also, in order to remove new function symbols from the terms in bounded quantifiers, it is necessary to use the Σ_1-replacement axioms. We leave the details to the reader. □

As an immediate corollary to the previous theorem, we get the following important bootstrapping fact.

Corollary. *Let T be $I\Delta_0$, $I\Sigma_n$ or $B\Sigma_n$ for $n \geq 1$. Then in the conservative extension of T with Σ_1-defined function symbols and Δ_1-defined function symbols, the new function and predicate symbols may be used freely in induction, minimization and replacement axioms.*

1.2.8. Bootstrapping $I\Delta_0$, Phase 2

To begin the second phase of the bootstrapping for $I\Delta_0$, several elementary functions and relations are shown to be Σ_1- and Δ_0-definable in $I\Delta_0$.

(a) *Restricted subtraction.* The function $x \dot{-} y$ which equals $\max\{0, x-y\}$ can be Σ_1-defined by $I\Delta_0$ by the formula

$$M(x,y,z) \Leftrightarrow (y+z=x) \vee (x \leq y \wedge z = 0).$$

The existence of z follows immediately from the trichotomy of $\leq$; thus $I\Delta_0$ can prove $(\forall x,y)(\exists z \leq x)M(x,y,z)$. Furthermore, $I\Delta_0$ can prove the uniqueness of z satisfying $M(x,y,z)$ using the cancellation law for addition. This then is a Σ_1-definition of the restricted subtraction function.

(b) *Predecessor.* The predecessor function is easily Σ_1-defined by $y = x \dot{-} 1$.

(c) *Division.* The division function $(x,y) \mapsto \lfloor x/y \rfloor$ can be Σ_1-defined by Δ_0 using the formula

$$M(x,y,z) \Leftrightarrow (y \cdot z \leq x \wedge x < y(z+1)) \vee (y = 0 \wedge z = 0).$$

Note that in order to make the division function total, we have arbitrarily defined $x/0$ to equal 0. The existence of z is proved using induction on the formula $(\exists z \leq x)M(x,y,z)$. The uniqueness of z is provable as follows, arguing inside $I\Delta_0$: suppose $M(x,y,z)$ and $M(x,y,z')$, w.l.o.g. $z \leq z'$; thus, using restricted subtraction and the distributive law, $(z' \dot{-} z)(y+1) < y+1$; and from this, $z' = z$ follows easily.

Two particularly useful special cases of division are when the divisor is two or four, i.e., $\lfloor \frac{1}{2}x \rfloor$ and $\lfloor \frac{1}{4}x \rfloor$.

(d) *Remainder.* The remainder function is Σ_1-definable in $I\Delta_0$ since $x \bmod y = x \dot{-} y \cdot \lfloor x/y \rfloor$. The *divides* relation, $x|y$, is defined by $y \bmod x = 0$.

(e) *Square root.* The square root function $x \mapsto \lfloor \sqrt{x} \rfloor$ is Σ_1-definable $I\Delta_0$ with the formula

$$M(x, y) \Leftrightarrow y \cdot y \leq x \wedge x < (y+1)(y+1).$$

(f) *Primes.* The set of primes is Δ_0-definable by the formula

$$(\forall y \leq x)(y|x \supset y = x \vee y = 1) \wedge 1 < x.$$

$I\Delta_0$ can prove many useful facts about primes and remainders. In particular, it proves that if x is prime and $x|ab$ then $x|a$ or $x|b$. The sequence coding tools developed below will enable $I\Delta_0$ to prove the unique factorization theorem; however, more bootstrapping is needed before we can even express the unique factorization theorem in first-order logic.

(g) *Prime powers.* The predicate "x is prime and y is a power of x" is Δ_0-definable by

$$x \text{ is prime and } (\forall z \leq y)(1 < z \wedge z|y \supset x|z).$$

$I\Delta_0$ can prove simple properties about prime powers, such as the fact that if y is a power of the prime x then $x \cdot y$ is the least power of x greater than y. This fact can be proved by using Δ_0-minimization with respect to y.

The bootstrapping is not yet sufficiently developed for us to give Δ_0 definitions of powers of composite numbers; however, we shall next define powers of prime powers.

(h) *Powers of two, of four, and of prime powers.* We already have shown how to define powers of the prime two. For powers of fours, we can give two equivalent definitions:

$$y \text{ is a power of four} \Leftrightarrow y \text{ is a power of two and } y \bmod 3 = 1,$$

and

$$y \text{ is a power of four} \Leftrightarrow y \text{ is a power of two and } y = (\lfloor \sqrt{y} \rfloor)^2.$$

The equivalence of these definitions can be proved using Δ_0-induction with respect to y. $I\Delta_0$ can also prove that when $y < y'$ are a powers of four, then $y|y'$ and that when y is a power of four, then $4y$ is the least power of four greater then y.

More generally, the predicate "x is a prime power and y is a power of x" can be Δ_0-defined by

$$(\exists p \leq x)(p \text{ is prime, } x \text{ and } y \text{ are powers of } p, \text{ and } y \bmod (x-1) = 1).$$

(i) The *LenBit* function is defined so that $LenBit(2^i, x)$ is equal to the i-th bit in the binary expansion of x, where the least significant bit is by convention the zeroth bit. This is Σ_1-definable by $I\Delta_0$ since $LenBit(y, x) = \lfloor x/y \rfloor \bmod 2$. Although it is defined for all values y, we shall use $LenBit(y, x)$ only when y is a power of two.

The next theorem states that $I\Delta_0$ can prove that the binary representation of a number uniquely determines the number. This theorem also introduces a new notation; namely, we will write quantifiers of the form $(\forall 2^i)$ and $(\exists 2^i)$ to mean, "for all powers of two" and "there exists a power of two." It is important to note that although we use this notation for quantifying over powers of two, we have not yet shown how to Δ_0-define i in terms of 2^i.

Theorem. *$I\Delta_0$ proves* $(\forall x)(\forall y < x)(\exists 2^i \le x)(LenBit(2^i, x) > LenBit(2^i, y))$.

To prove the theorem, $I\Delta_0$ uses strong induction with respect to x and argues that if 2^i is the greatest power of two less than x, then $LenBit(2^i, x)$ equals one, and then, when $LenBit(2^i, y)$ also equals one, applies the induction hypothesis to $x \dot{-} 2^i$ and $y \dot{-} 2^i$. □

(j) We next show how to Δ_0-define the relation $x = 2^y$ as a predicate of x and y. As a preliminary step, we consider numbers of the form

$$m_p = \sum_{i=0}^{p} 2^{2^p}$$

for $p \ge 0$ and show that these numbers are Δ_0-definable. In fact, the set $\{m_p\}_p$ is definable by the formula

$$\begin{aligned} &LenBit(1, x) = 0 \wedge LenBit(2, x) = 1 \wedge \\ &\quad \wedge (\forall 2^i \le x)(2 < 2^i \supset \\ &\qquad [LenBit(2^i, x) = 1 \leftrightarrow (2^i \text{ is a power of } 4 \wedge LenBit(\lfloor \sqrt{2^i} \rfloor, x) = 1)]) \end{aligned}$$

As an immediate corollary we get a Δ_0-formula defining the the numbers of the form $x = 2^{2^p}$; namely, they are the powers of two for which $LenBit(x, m_p) = 1$ holds for some $m_p < 2x$.

Now the general idea of defining 2^y is to express y in binary notation as $y = 2^{p_1} + 2^{p_2} + \cdots + 2^{p_k}$ with distinct values p_j, and thus define $x = \prod_{j=1}^{k} 2^{2^{p_j}}$. To carry this out, we define an extraction function $Ext(u, x)$ which will be applied when u is of a number of the form 2^{2^p}. Formally we define

$$Ext(u, x) = \lfloor x/u \rfloor \bmod u.$$

Note that when $u = 2^{2^p}$, then $Ext(u, x)$ returns the number with binary expansion equal to the 2^p-th bit through the $(2^{p+1} - 1)$-th bit of x. We will think of x coding the sequence of numbers $Ext(2^{2^p}, x)$ for $p = 0, 1, 2, \ldots$. We also define $Ext'(u, x)$ as $Ext(u^2, x)$; this is of course the number which succeeds $Ext(u, x)$ in the sequence of numbers coded by x.

We are now ready to Δ_0-define $x = 2^i$. We define it with a formula $\phi(x, i)$ which states there are numbers $a, b, c, d \le x^2$ such that the following hold:

(1) a is of the form m_p and $a > x$.
(2) $Ext(2, b) = 1$, $Ext(2, c) = 0$ and $Ext(2, d) = 1$.
(3) For all u of the form 2^{2^p} such that $a > u^2$, $Ext'(u, b) = 2 \cdot Ext(u, b)$,
(4) For all u of the form 2^{2^p} such that $a > u^2$, either
 (a) $Ext'(u, c) = Ext(u, c)$ and $Ext'(u, d) = Ext(u, d)$, or
 (b) $Ext'(u, c) = Ext(u, c) + Ext(u, b)$ and $Ext'(u, d) = Ext(u, d) \cdot Ext(u, a)$.
(5) There is a $u < a$ of the form 2^{2^p} such that $Ext(u, c) = i$ and $Ext(u, d) = x$.

Obviously this is a Δ_0-formula; we leave it to the reader the nontrivial task of checking that $I\Delta_0$ can prove simple facts about this definition of 2^i, including (1) the fact that if $\phi(x,i)$ and $\phi(x,j)$ both hold, then $i=j$, (2) the fact that $2^i \cdot 2^j = 2^{i+j}$, (3) the fact that if x is a power of two, then $x = 2^i$ for some $i < x$, and (4) that if $\phi(x,i)$ and $\phi(y,i)$ then $x=y$.

(k) *Length function.* The length function is $|x| = \lceil \log_2(x+1) \rceil$ and can be Δ_0-defined in $I\Delta_0$ as the value i such that $y = 2^i$ is the least power of two greater than x. Note that $|0| = 0$ and for $x > 0$, $|x|$ is the number of bits in the binary representation of x.

The reader should check that $I\Delta_0$ can prove elementary facts about the $|x|$ function, including that $|x| \le x$ and that $36 < x \supset |x|^2 < x$.

(l) The $Bit(i,x)$ function is definable as $LenBit(2^i, x)$. This is equivalent to a Δ_0-definition, since when $2^i > x$, $Bit(i,x) = 0$. $Bit(i,x)$ is the i-th bit of the binary representation of x; by convention, the lowest order bit is bit number 0.

(m) *Sequence coding.* Sequences will be coded in the base 4 representation used by Buss [1986]; many prior works have used similar encodings. A number x is viewed as a bit string in which pairs of bits code one of the three symbols comma, "0" or "1". The i-th symbol from the end is coded by the two bits $Bit(2i+1,x)$ and $Bit(2i,x)$. This is best illustrated by an example: consider the sequence $\langle 3,0,4 \rangle$. Firstly, a comma is prepended to the the sequence and the entries are written in base two, preserving the commas, as the string: ",11,,100"; leading zeros are optionally dropped in this process. Secondly, each symbol in the string is replaced by a two bit encoding by replacing each "1" with "11", each "0" with "10", and each comma with "01". This yields "0111110101111010" in our example. Thirdly, the result is interpreted as a binary representation of a number; in our example it is the integer 32122. This then is a Gödel number of the sequence $\langle 3,0,4 \rangle$.

This scheme for encoding sequences has the advantage of being relatively efficient from an information theoretic point of view and of making it easy to manipulate sequences. It does have the minor drawbacks that not every number is the Gödel number of a sequence and that Gödel numbers of sequences are non-unique since it is allowable that elements of the sequence be coded with excess leading zeros.

Towards arithmetizing Gödel numbers, we define predicates $Comma(i,x)$ and $Digit(i,x)$ as

$$Comma(i,x) \Leftrightarrow Bit(2i+1,x) = 0 \land Bit(2i,x) = 1$$

$$Digit(i,x) = 2 \cdot (1 \dot{-} Bit(2i+1,x)) + Bit(2i,x).$$

Note *Digit* equals zero or one for encodings of "0" and "1" and equals 2 or 3 otherwise.

It is now fairly simple to recognize and extract values from a sequence's Gödel number. We define $IsEntry(i,j,x)$ as

$$(i = 0 \lor Comma(i \dot{-} 1, x)) \land Comma(j,x) \land (\forall k < j)(k \ge i \supset Digit(j,x) \le 1)$$

which states that the i-th through $(j-1)$-st symbols coded by x code an entry in the sequence. And we define $Entry(i, j, x) = y$ by

$$|y| \leq j \dot{-} i \wedge (\forall k < j \dot{-} i)(Bit(k, y) = Digit(i + k, x)).$$

When $IsEntry(i, j, x)$ is true, then $Entry(i, j, x)$ equals the value of that entry in the sequence coded by x. Checking that *Entry* is Δ_0-definable by $I\Delta_0$ is left to the reader; note that the quantifier $(\forall k < j \dot{-} i)$ may be replaced by a sharply bounded quantifier since, w.l.o.g., $j \leq |x|$.

(n) *Length-bounded counting* and *Numones.* Although we have defined *Entry* already, we are not quite done with arithmetizing sequence coding; in particular, we would like to define the Gödel beta function, $\beta(i, x)$, which equals the i-th entry of the sequence coded by x. One way to do this would be by encoding a sequence of numbers $\langle a_n, a_{n-1}, \ldots, a_1 \rangle$ as the sequence $\langle b_n, \ldots, b_1 \rangle$ where $b_i = \langle i, a_i \rangle$. The drawback of this approach is that when the values a_i are small, the length of the Gödel number encoding the sequence $\langle \vec{b} \rangle$ is longer than the length of the Gödel number encoding the sequence $\langle \vec{a} \rangle$; in fact, it is longer by a logarithmic factor and thus the function $\langle \vec{a} \rangle \mapsto \langle \vec{b} \rangle$ cannot be Δ_0-defined by $I\Delta_0$ by virtue of the function's superlinear growth rate.

Upon reflection, one sees that the basic difficulty in defining the β function is the difficulty of counting the number of commas encoded in a Gödel number of a sequence. This basically the same as the problem as counting of ones in the binary representation of a number x. Supposing x has binary representation $(x_n x_{n-1} \cdots x_1 x_0)_2$, we would like to be able to let $a_0 = x_0$ and $a_i = a_{i-1} + x_i$ and then let $b_i = \langle i, a_i \rangle$ and finally let y be the Gödel number of the sequence $\langle b_n, \ldots, b_1 \rangle$. Now, $I\Delta_0$ can prove that, if y exists, it is unique, and from y the number of 1's in the binary representation of x is easily computed. The catch is that, as above, $\langle \vec{b} \rangle$ will in general not be bounded by a term involving x since its length is not necessarily $O(|x|)$. However, the length of the Gödel number of $\langle \vec{b} \rangle$ is $O(|x|^2)$ so this this method does work when x is small; in particular, it works if $x = |y|$ for some y. Thus, $I\Delta_0$ can Σ_1-define the function *LenNumones* defined so that

$$LenNumones(y) = \text{the number of 1's in the binary representation of } |y|.$$

To define a *Numones* function that works for all numbers, we use a trick that allows more efficient encoding of successive numbers. The basic idea is that a sequence $a_1, a_2, a_3, \ldots, a_k$ of numbers can be encoded with fewer bits if, when writing the number a_{i+1}, one only writes the bits of a_{i+1} which are different from the corresponding bits in a_i. This works particularly well when we have $a_i \leq a_{i+1} \leq a_i + 1$ for all i; in this case we formally define the succinct encoding as follows: for $i > 0$, define i^* to be the greatest power of 2 which divides i; and define 0^* to equal 0. Now define $a'_0 = a_0$ and define a'_i to be a^*_i if $a_i \neq a_{i-1}$ or to be 0 otherwise. (Example: if $a = 24$, then $a' = 8$.) Then the sequence $\vec{a}$ can be more succinctly described by $\vec{a'}$.

It is now important to see that $I\Delta_0$ can extract the sequence $\langle \vec{a} \rangle$ from the sequence $\langle \vec{a'} \rangle$, at least in a certain limited sense. In particular, we have that if $x = \langle a'_k, \ldots, a'_1, a'_0 \rangle$ and if $IsEntry(i, j, x)$ and if this entry is the entry for a'_ℓ, then the value a_ℓ can be Σ_1-defined in terms of i, j, x. To describe this Σ_1-definition, note that the k-th bit of the binary representation of a_ℓ is computed by finding the maximum values $i_0 < j_0 \le i$ such that $IsEntry(i_0, j_0, x)$ and such that $|Entry(i_0, j_0, x)| > k$ and letting the k-th bit of a_ℓ equal the k-th bit of $Entry(i_0, j_0, x)$.

The whole point of using $\langle \vec{a'} \rangle$ is to give a sufficiently succinct encoding of $\langle \vec{a} \rangle$. Of course, the fact that the encoding is sufficiently succinct also needs to be provable in $I\Delta_0$. It is easily checked that the Gödel number of the sequence $\langle 0^*, 1^*, \ldots, x^* \rangle$ uses exactly $6x - 2Numones(x) + 2$ many bits; this is proved by first showing that there are $2x - Numones(x)$ bits in the numbers in the sequence, i.e., $\sum_{i=0}^{x} |i^*| = 2x - Numones(x)$, and second noting that there are $x+1$ commas, and noting that each bit and comma is encoded by two bits in the Gödel number. Furthermore, when $x = |y|$ for some y, $I\Delta_0$ can prove this fact, using *LenNumones* in place of *Numones*.

We are now able to Σ_1-define the function $Numones(x)$ equal to the number of 1's in the binary representation of x. This is done by defining the sequence

$$u = \langle \langle k^*, a'_k \rangle, \langle (k-1)^*, a'_{k-1} \rangle, \ldots, \langle 0, a'_0 \rangle \rangle$$

such that $k = |x|$, $a_0 = 0$, and each a_{i+1} is equal to $a_i + Bit(i, x)$. By the considerations in the previous paragraph, $I\Delta_0$ can prove that this sequence is bounded by a term involving only x; also, $I\Delta_0$ can compute the values of $0, \ldots, k$ from $0^*, \ldots, k^*$ and therefore can compute the values of a_i as a function of i and u.

(o) *Sequence coding.* Once we have the *Numones* function, it is an easy matter to define the Gödel β function by counting commas. The β function is defined so that $\beta(m, x) = a_m$ provided x is the Gödel number of a sequence $\langle a_1, \ldots, a_k \rangle$ with $m \le k$. It is also useful to define the length function $Len(x)$ which equals k when x is as above. These are defined easily in terms of the *Numones* function: the value $\beta(m, x)$ equals $Entry(i, j, x)$ where there are $m-1$ commas encoded in x to the left of bit i; and $Len(x)$ equals the number commas coded by x.

Once sequence encoding has been achieved, the rest of the bootstrapping process is fairly straightforward. The next stage in bootstrapping is to arithmetize metamathematics, and this is postponed until section 2 below. Stronger theories, such as $I\Sigma_1$, can define all primitive recursive functions: this is discussed in section 1.2.10.

1.2.9. Relationships amongst the axioms of *PA*

We are now ready to sketch the proofs of the relationships between the various fragments of Peano arithmetic pictured in paragraph 1.2.4 above.

Theorem. *Let* $n \ge 0$.
(a) $B\Pi_n \vDash B\Sigma_{n+1}$.
(b) $I\Sigma_{n+1} \vDash B\Sigma_{n+1}$.
(c) *If* $A(x, \vec{w}) \in \Sigma_n$ *and* t *is a term, then* $B\Sigma_n$ *can prove that* $(\forall x \le t)A(x, \vec{w})$ *is equivalent to a* Σ_n*-formula.*

Proof. To prove (a), suppose $A(x,y)$ is a formula in Σ_{n+1}. We want to show that

$$(\forall x \le u)(\exists y)A(x,y) \supset (\exists v)(\forall x \le u)(\exists y \le v)A(x,y).$$

is a consequence of $B\Pi_n$. This is proved by the following trick: take all leading existential quantifiers $(\exists\vec{z})$ from the beginning of A and replace these quantifiers and the existential quantifier $(\exists y)$ by a single existential quantifier $(\exists w)$ which is intended to range over Gödel numbers of sequences coding values for all of the variables y and $\vec{z}$, say by letting $\beta(1,w) = y$ and letting $\beta(i+1,w)$ be a value for the i-th variable in $\vec{z}$. Since $y = \beta(1,w) < w$, it follows that the collection axiom for this new formula implies the collection axiom for A.

Part (c) is proved by induction on n. Note that (c) is obvious when $n = 0$. For $n > 0$, (c) is proved by noting that, by using a sequence to code multiple values, we may assume without loss of generality that there is only one (unbounded) existential quantifier at the front of A, so A is $(\exists y)B$ with $B \in \Pi_{n-1}$. Then $(\forall x \le t)A$ is equivalent to $(\exists u)(\forall x \le t)(\exists y \le u)B$; and finally by using the induction hypothesis that (c) holds for $n-1$, we have that $(\exists y \le u)B$ is equivalent to a Π_{n-1}-formula. From this $(\exists u)(\forall x \le t)A$ is equivalent to a Σ_n-formula.

We prove (b) by induction on n. Suppose $A(x,y)$ is a Σ_{n+1}-formula, possibly containing other free variables. We need to show that $I\Sigma_{n+1}$ proves the formula displayed above, and by part (a) we may assume that A is a Π_n-formula. We argue informally inside $I\Sigma_{n+1}$, assuming that $(\forall x \le u)(\exists y)A(x,y)$ holds. Let $\phi(a)$ be the formula

$$(\exists v)(\forall x \le a)(\exists y \le v)A(x,y).$$

It follows from our assumption that $\phi(0)$ and that $\phi(a) \supset \phi(a+1)$ for all $a < u$. The induction hypothesis that $I\Sigma_n \models B\Sigma_n$ together with part (c) implies that the formula $(\forall x \le u)(\exists y \le v)A$ is equivalent to a Π_{n-1}-formula and thus ϕ is equivalent to a Σ_n-formula. Therefore, by induction on ϕ, $\phi(u)$ holds; this is what we needed to show. □

With the aid of the above theorem, the other relationships between fragments of Peano arithmetic are relatively easy to prove. To prove that $I\Sigma_n$ implies Π_n-IND, let $A(x)$ be a Π_n formula and argue informally inside $I\Sigma_n$ assuming $A(0)$ and $(\forall x)(A(x) \supset A(x+1))$. Letting a be arbitrary, and letting $B(x)$ be the formula $\neg A(a \dot{-} x)$, one has $\neg B(a)$ and $B(x) \supset B(x+1)$. Thus, by induction, $\neg B(0)$, and this is equivalent to $A(a)$. Since a was arbitrary, $(\forall x)A(x)$ follows. A similar argument shows that $I\Pi_n$ implies Σ_n-IND.

To show that the Σ_n-MIN axioms are consequences of $I\Sigma_n$, note that by the argument given at the end of section 1.2.6 above, the minimization axiom for $A(x)$ follows from induction on the formula $(\forall x \le y)\neg A(x)$ with respect to the variable y. If $A \in \Sigma_n$, then from part (c) of the above theorem, the formula $(\forall x \le y)(\neg A)$ is equivalent to a Π_n-formula, so the minimization axiom for A is a consequence of $I\Pi_n = I\Sigma_n$.

It is easy to derive the induction axiom for A from the minimization axiom for A, so $L\Sigma_n = L\Pi_n = I\Sigma_n = I\Pi_n$.

Finally, the theorem of Clote [1985] that the strong Σ_n-replacement axioms are consequences of $I\Sigma_n$ can be proved as follows. Assume $n \geq 1$ and $A \in \Sigma_n$ and consider the strong replacement axiom

$$(\exists w)(\forall x \leq a)[(\exists y)A(x,y) \leftrightarrow A(x, \beta(x+1, w))].$$

for A. (Note A may have free variables other than x, y.) Let $Num_A(u)$ be a Σ_n-formula which expresses the property that there exists a w for which $A(x, \beta(x+1, w))$ holds for at least u many values of $x \leq a$. Clearly $Num_A(0)$ holds and $Num_A(a+2)$ fails. So, by Σ_n-maximization (which follows easily from Σ_n-minimization), there is a maximum value $u_0 \leq a+1$ for which $Num_A(u_0)$ holds. A value w that works for this u_0 satisfies the strong replacement axioms for A.

1.2.10. Definable functions of $I\Sigma_n$.

When bootstrapping theories stronger than $I\Delta_0$, such as $I\Sigma_n$ for $n > 0$, the main theorem of section 1.2.7 still applies, and Δ_1-definable predicates and Σ_1-definable functions may be introduced into the language of arithmetic and used freely in induction axioms, without increasing the strength of the theory. Of particular importance is the fact that the primitive recursive functions can be Σ_1-defined in (any theory containing) $I\Sigma_1$.

Definition. The *primitive recursive* functions are functions on N and are inductively defined as follows:

(1) The constant function with value 0 is primitive recursive. We can view this a nullary function.

(2) The unary successor function $S(x) = x + 1$ is primitive recursive.

(3) For each $1 \leq k \leq n$, then n-ary projection function $\pi_k^n(x_1, \ldots, x_n) = x_k$ is primitive recursive.

(4) If g is an n-ary primitive recursive function and $h_1, \ldots, h_n$ are m-ary primitive recursive functions, then the m-ary function f defined by $f(\vec{x}) = g(h_1(\vec{x}), \ldots h_n(\vec{x}))$ is primitive recursive.

(5) If $n \geq 1$ and g is an $(n-1)$-ary primitive recursive function and h is an $(n+1)$-ary primitive recursive function, then the n-ary function f defined by:

$$\begin{aligned} f(0, \vec{x}) &= g(\vec{x}) \\ f(m+1, \vec{x}) &= h(m, f(m, \vec{x}), \vec{x}) \end{aligned}$$

is primitive recursive.

The only use of the projection functions is as a technical tool to allow generalized substitutions with case (4) above.

A predicate is *primitive recursive* if its characteristic function is primitive recursive.

Theorem. *$I\Sigma_1$ can Σ_1-define the primitive recursive functions.*

The converse to this theorem holds as well; namely, $I\Sigma_1$ can Σ_1-define exactly the primitive recursive functions. This converse is proved later as Theorem 3.1.1.

Proof. It is obvious that the base functions, zero, successor and projection, are Σ_1-definable in $I\Sigma_1$. It is easy to check that set of functions Σ_1-definable by $I\Sigma_1$ is closed under composition. Finally suppose that g and h are $I\Sigma_1$-definable in $I\Sigma_1$. Then, the function f defined from g and h by primitive recursion can be Σ_1-defined with the following formula expressing $f(m, \vec{x}) = y$:

$$(\exists w)[Len(w) = m + 1 \wedge y = \beta(m + 1, w) \wedge \beta(1, w) = g(\vec{x}) \wedge \\ (\forall i < m)(\beta(i + 2, w) = h(i, \beta(i + 1, w), \vec{x}))].$$

This formula expresses the condition that there is a sequence, coded by w, containing all the values $f(0, x), \ldots, f(m, \vec{x})$, such that each value in the sequence is correctly computed from the preceding value and such that the final value is y. The theorem of section 1.2.7 shows that the above formula defining f is (equivalent to) a Σ_1-formula. We leave it to the reader to check that $I\Sigma_1$ can prove the requisite existence and uniqueness conditions for this definition of f. □

As an easy consequence, we have

Corollary. *Every primitive recursive predicate is Δ_1-definable by $I\Sigma_1$.*

In the theories $I\Sigma_n$ with $n > 1$, even more functions are Σ_1-definable. A characterization of the functions Σ_1-definable in $I\Sigma_n$ is given in Chapter III of this volume. Other proof-theoretic characterizations of these functions can be found in Takeuti [1987], Buss [1994] and in references cited therein.

1.3. Fragments of bounded arithmetic

A subtheory of Peano arithmetic is called a bounded theory of arithmetic, or a theory of *bounded arithmetic*, if it is axiomatized by Π_1-formulas. The potential strength of such theories depends partly on the growth rates of the function symbols in the language, and usually bounded arithmetic theories have only functions of subexponential growth rate, including addition, multiplication and possibly polynomial growth rate functions such as ω_1 or $\#$. These theories are typically weaker than the strong theories considered in section 1.2, but stronger than the theories Q and R discussed in section 1.1.

There are two principal approaches to bounded arithmetic. The original approach involved theories such as $I\Delta_0$ and $I\Delta_0 + \Omega_1$; more recently, bounded theories such as S_2^i and T_2^i have been extensively studied. One of the main motivations for studying bounded arithmetics is their close connection to low-level computational complexity, especially regarding questions relating expressibility and provability in bounded arithmetics to questions about the linear time hierarchy and the polynomial time hierarchy.

1.3.1. $I\Delta_0$ and Ω_n

We have already defined $I\Delta_0$ and described its bootstrapping process in fairly complete detail in section 1.2. One corollary of the bootstrapping process is that the graph of exponentiation is Δ_0-definable in $I\Delta_0$; that is to say, there is a bounded formula $exp(x,y,z)$ which expresses the condition $x^y = z$ and such that $I\Delta_0$ can prove facts like $exp(x,0,1)$, $exp(x,1,x)$,

$$exp(x,y,z) \land exp(x,y',z') \Rightarrow exp(x,y+y',z\cdot z')$$

and that for any x and y, there is at most one z such that $exp(x,y,z)$. The underlying idea of the Δ_0-definition of $exp(x,y,z)$ is to define the sequence $\langle x^{\lfloor y/2^i \rfloor}\rangle_i$ where i ranges from $|y|$ down to 0; however, we leave it to the reader to supply the details behind this Δ_0-definition. The fact that exponentiation is Δ_0-definable is essentially due to Bennett [1962] and was first (and independently) proved in the setting of $I\Delta_0$ by Gaifman and Dimitracopoulos [1982].

Once the graph of exponentiation has been shown to be Δ_0-definable, one can formulate the axioms Ω_k. Firstly, when working in bounded arithmetic, we define $\log x$ to equal the greatest y such that $2^y \le x$. Then the function $\omega_1(x,y)$ is defined to equal $x^{\log y}$. Since $|\omega_1(x,y)| = \Theta(|x|\cdot|y|)$, it is evident $\omega_1(x,y)$ cannot be bounded by a polynomial of x and y. Therefore, by Parikh's Theorem 1.2.7.1, the function ω_1 is not Σ_1-definable in $I\Delta_0$. As we shall see later, it is often very desirable to have the ω_1 function be total; therefore it is common to extend $I\Delta_0$ to a stronger theory containing the axiom

$$\Omega_1: \qquad (\forall x)(\forall y)(\exists z)(z = \omega_1(x,y)).$$

This stronger theory is called $I\Delta_0 + \Omega_1$.

The function ω_1 has what is called *polynomial growth rate*, i.e., for any term $t(\vec{a})$ constructed with the functions S, $+$, $\cdot$ and ω_1 there is a polynomial p_t such that for all $\vec{a}$, $|t(\vec{a})| \le p_t(|a_1|,\ldots,|a_n|)$. There is also a hierarchy of functions ω_n, $n \ge 1$, which have subexponential growth rates, defined by $\omega_{n+1}(x,y) = 2^{\omega_n(\log x, \log y)}$. The axioms Ω_n are Π_2-axioms which say that the function ω_n is total. By using Parikh's Theorem 1.2.7.1, it is immediate that $I\Delta_0 + \Omega_n \nvdash \Omega_{n+1}$.

Although the ω_n functions, for $n \ge 2$, are superpolynomial, they are much more similar in nature to polynomial growth rate functions than to exponential growth rate functions. Using a technique due to Solovay [1976], it can be shown that, for each n, $I\Delta_0$ can define an inductive cut on which the ω_n function is provably total; for an explanation of this construction, see Pudlák [1983], Nelson [1986] or Chapter VIII of this volume. However, Paris and Dimitracopoulos [1982] showed that it is not possible to define an inductive cut on which the exponentiation function is provably total. For this reason, we view the ω_n functions as being more akin to feasible polynomial growth rate functions than to the infeasible exponential function (see Nelson [1986] for a strong expression of this viewpoint).

1.3.2. Δ_0-formulas and the linear-time hierarchy

There is a very close connection between Δ_0-expressibility and computational complexity. Recall that the linear time hierarchy consists of those predicates which can be recognized by some Turing machine which runs in linear time and which makes a bounded number of alternations between existential and universal states. Lipton [1978,sect. 4], building on work of Smullyan, Bennett and Wrathall, proved that the Δ_0 definable predicates on $\mathbf{N}$ are precisely the subsets which are in the linear time hierarchy.

The original motivation for the definition of the theory $I\Delta_0$ by Parikh [1971] was to give a proof theory that would be appropriate to linear bounded automata, i.e., to predicates computable by linear space bounded Turing machines. It is still an open problem whether the linear time hierarchy equals linear space; although it is commonly conjectured that they are not equal. It is known that the linear time hierarchy contains log space, and also contains the predicates which can be computed by a Turing machine which simultaneously polynomial time and $n^{1-\epsilon}$ space for a constant $\epsilon > 0$ (see Bennett [1962] and Nepomnjaščii [1970]).

1.3.3. The theories S_2^i and T_2^i of bounded arithmetic

The second approach to theories of bounded arithmetic is due to Buss [1986] and gives a (conjectured) hierarchy of fragments of $I\Delta_0 + \Omega_1$, which are very closely related to the computational complexity classes of the polynomial time hierarchy. These fragments, S_2^i and T_2^i and others, use the language 0, S, $+$, $\cdot$, $\#$, $|x|$, $\lfloor \frac{1}{2}x \rfloor$, and $\leq$; where the $\#$ function (pronounced 'smash') is defined so that $x\#y = 2^{|x|\cdot|y|}$. The $\#$ function was first introduced by Nelson [1986], and it is evident that the $\#$ function has essentially the same growth rate as the ω_1-function.

The second difference between the S_2^i and the T_2^i theories and the $I\Delta_0 + \Omega_1$ approach is that the former theories are based on restricting the power of induction; firstly by further restricting the formulas for which induction holds, and secondly by using (apparently) weaker forms of induction. It is for this reason that the functions $|x|$ and $\lfloor \frac{1}{2}x \rfloor$ are included in the non-logical language, since they are needed to elegantly state the axioms of the theories S_2^i and T_2^i.

Before defining the theories S_2^i and T_2^i, we define the classes Σ_i^b and Π_i^b of formulas, which are defined by counting alternations of bounded quantifiers, ignoring sharply bounded quantifiers. (Bounded and sharply bounded quantifiers are defined in section 1 above.)

Definition. The set $\Delta_0^b = \Sigma_0^b = \Pi_0^b$ is equal to the set of formulas in which all quantifiers are sharply bounded. For $i \geq 1$, the sets Σ_i^b and Π_i^b are inductively defined by the following conditions:

(a) If A and B are Σ_i^b-formulas, then so are $A \vee B$ and $A \wedge B$. If A is a Π_i^b formula and B is a Σ_i^b-formula, then $A \supset B$ and $\neg A$ are Σ_i^b-formulas.
(b) If A is a Π_{i-1}^b-formula, then A is a Σ_i^b-formula.
(c) If A is a Σ_i^b-formula and t is a term, then $(\forall x \le |t|)A$ is a Σ_i^b-formula.
(d) If A is a Σ_i^b-formula and t is a term, then $(\exists x \le t)A$ is a Σ_i^b-formula. Note this quantifier may be sharply bounded.

The four inductive conditions defining Π_i^b are dual to (a)-(d) with the roles of existential and universal quantifiers and the roles of Π_i^b and Σ_i^b reversed.

This is a good place to justify the presence, in bounded arithmetic, of the # function or the equivalently growing ω_1. There are essentially three reasons why it is natural to include # or ω_1. Firstly, it gives a natural bound to the Gödel number of a formula $A(t)$ in terms of the Gödel numbers of A and t; namely, the number of symbols in $A(t)$ is bounded by the product of the numbers of symbols in A and in t. This allows a smooth arithmetization of metamathematics. Secondly, it arises naturally from consideration of bounded versus sharply bounded quantifiers, since it has exactly the growth rate necessary to make the following *quantifier exchange property* hold:

$$(\forall x \le |a|)(\exists y \le b)A(x,y)$$
$$\leftrightarrow \quad (\exists w \le SqBd(b,a))(\forall x \le |a|)(A(x,\beta(x+1,y)) \wedge \beta(x+1,y) \le b)$$

where $SqBd$ is a term involving #. In fact, the size of w can be bounded in terms of a and b, by noting that w must encode $|a|+1$ many numbers of at most $|b|$ bits each; therefore, $w \le 2^{c\cdot|a|\cdot|b|}$ for some constant c, and $SqBd$ can easily be expressed using #. The quantifier exchange property allows sharply bounded quantifiers to be pushed inside non-sharply bounded quantifiers (at least when the β function is available). Thirdly, the use of # function means that any term $t(x)$ can be bounded by $2^{|x|^c}$ for some constant c, and conversely, any $2^{|x|^c}$ can be bounded by a term $t(x)$ in the language of bounded arithmetic. In other words, the terms define functions of *polynomial growth rate.* This leads to the principal importance of the classes Σ_i^b and Π_i^b of formulas, which is that they express precisely the corresponding classes of the polynomial time hierarchy. This fact is discussed in more depth in section 1.3.6 below, but in brief, a set of natural numbers is definable by a Σ_i^b-formula (respectively, a Π_i^b-formula) if and only if the set is recognizable by a predicate in the class Σ_i^p (respectively, Π_i^p) from the polynomial time hierarchy. This is essentially due to Wrathall [1976] and Stockmeyer [1976] and was first proved in this exact form by Kent and Hodgson [1982]. Thus we have that NP, the set of nondeterministic polynomial time predicates, consists of precisely the predicates expressible by Σ_1^b-formulas, etc.

1.3.3.1. The theory T_2^i will be defined by restricting induction to Σ_i^b-formulas, where by induction we mean the usual 'IND' flavor of induction. For S_2^i, we need some additional varieties of induction:

Definition. Let Φ be a set of formulas. The Φ-PIND axioms are the formulas

$$A(0) \land (\forall x)(A(\lfloor \tfrac{1}{2}x \rfloor) \supset A(x)) \supset (\forall x)A(x)$$

for all formulas $A \in \Phi$. As usual, A may have other free variables in addition to x that serve as parameters. The length-induction Φ-LIND axioms are the formulas

$$A(0) \land (\forall x)(A(x) \supset A(Sx)) \supset (\forall x)A(|x|)$$

for all $A \in \Phi$. The length-minimization axioms, Φ-LMIN, are the formulas

$$(\exists x)A(x) \supset (\exists x)(A(x) \land (\forall y)(|y| < |x| \supset \neg A(y)))$$

for all $A \in \Phi$.

In addition to induction and minimization axioms, there are replacement axioms that will be defined below after the Gödel β function has been introduced. All of these axiom schemes are used in conjunction with a set of purely universal axioms called the BASIC axioms. The set of BASIC axioms consists of:

$a \le b \supset a \le Sb$
$a \neq Sa$
$0 \le a$
$a \le b \land a \neq b \leftrightarrow Sa \le b$
$a \neq 0 \supset 2 \cdot a \neq 0$
$a \le b \lor b \le a$
$a \le b \land b \le a \supset a = b$
$a \le b \land b \le c \supset a \le c$
$|0| = 0$
$|S0| = S0$
$a \neq 0 \supset |2 \cdot a| = S(|a|) \land |S(2 \cdot a)| = S(|a|)$
$a \le b \supset |a| \le |b|$
$|a\#b| = S(|a| \cdot |b|)$
$0\#a = S0$
$a \neq 0 \supset 1\#(2 \cdot a) = 2 \cdot (1\#a)$
$\quad \land 1\#(S(2 \cdot a)) = 2 \cdot (1\#a)$
$a\#b = b\#a$
$|a| = |b| \supset a\#c = b\#c$
$|a| = |b| + |c| \supset a\#d = (b\#d) \cdot (c\#d)$
$a \le a + b$
$a \le b \land a \neq b \supset$
$\quad S(2 \cdot a) \le 2 \cdot b \land S(2 \cdot a) \neq 2 \cdot b$
$a + b = b + a$
$a + 0 = a$
$a + Sb = S(a + b)$
$(a + b) + c = a + (b + c)$
$a + b \le a + c \leftrightarrow b \le c$
$a \cdot 0 = 0$
$a \cdot (Sb) = (a \cdot b) + a$
$a \cdot b = b \cdot a$
$a \cdot (b + c) = (a \cdot b) + (a \cdot c)$
$S0 \le a \supset (a \cdot b \le a \cdot c \leftrightarrow b \le c)$
$a \neq 0 \supset |a| = S(|\lfloor \tfrac{1}{2}a \rfloor|)$
$a = \lfloor \tfrac{1}{2}b \rfloor \leftrightarrow 2 \cdot a = b \lor S(2 \cdot a) = b$

These BASIC axioms serve the same role for S_2^i and T_2^i that the axioms of Q served for the fragments $I\Sigma_n$ of Peano arithmetic. There is a certain amount of flexibility in the choice of BASIC axioms; essentially any finite set of purely universal axioms which both are sufficiently strong to carry out the bootstrapping of S_2^1 and are contained in the theory S_2^1 would serve as well for the BASIC axioms.[4]

[4]We have given the BASIC axioms as defined by Buss [1986]. This choice is not entirely optimal, since, for instance, the second axiom $a \le S(a)$, follows from the first, fourth and sixth axioms. An alternative, and weaker, set of BASIC axioms are given by Cook and Urquhart [1993]; see Buss [1992] for a discussion of their BASIC axioms. Buss and Ignjatović [1995] propose that $|a| \le a$ should be added to the BASIC axioms.

Definition. Let $i \geq 0$. S_2^i is the theory axiomatized by the BASIC axioms plus Σ_i^b-PIND. T_2^i is the theory axiomatized by BASIC plus Σ_i^b-IND. The theories $S_2^{(-1)}$ and $T_2^{(-1)}$ are equal to just BASIC.

S_2 is $\cup_{i\geq 0} S_2^i$ and T_2 is $\cup_{i\geq 0} T_2^i$. Section 1.3.5 shows that S_2 and T_2 are the same theory.

1.3.3.2. Bootstrapping and Σ_1^b-definable functions. The bootstrapping of S_2^1 and T_2^1 is analogous to the bootstrapping of $I\Delta_0$ as described in sections 1.2.6-1.2.8 above. There is now the additional difficulty that the induction axioms are more severely restricted; but on the other hand, the language of S_2^i and T_2^i is richer since it contains the function symbol $|x|$ and its BASIC axioms and this makes the definition of the graph of $y = 2^x$ essentially trivial, and thereby helps with defining Gödel numbering of sequences. The most outstanding difference between the bootstrapping of S_2^1 and T_2^1 and the above bootstrapping of $I\Delta_0$ is that quantifiers are more carefully counted; namely, whereas $I\Delta_0$ could use Δ_0-defined predicates and Σ_1-defined functions, the theories S_2^1 and T_2^1 can introduce Δ_1^b-defined predicates and Σ_1^b-defined functions. Accordingly, we make the following important definitions:

Definition. A predicate symbol $R(\vec{x})$ is *Δ_i^b-defined by a theory T* if there is a Σ_i^b-formula $\phi(\vec{x})$ and a Π_i^b-formula $\psi(\vec{x})$ such that R has *defining axiom*

$$R(\vec{x}) \leftrightarrow \phi(\vec{x})$$

and such that $T \vdash (\forall \vec{x})(\phi \leftrightarrow \psi)$.

Definition. Let T be a theory of arithmetic. A function symbol $f(\vec{x})$ is *Σ_i^b-defined by T* if it has a defining axiom

$$y = f(\vec{x}) \quad \leftrightarrow \quad \phi(\vec{x}, y),$$

where ϕ is a Σ_i^b formula with all free variables indicated such that T proves $(\forall \vec{x})(\exists! y)\phi(\vec{x}, y)$.

By Parikh's theorem 1.2.7.1, when f is Σ_i^b-defined then $T \vdash (\forall \vec{x})(\exists y \leq t(\vec{x}))\phi(\vec{x}, y)$ for some term t.

The analogue of Theorem 1.2.7.3 for fragments of bounded arithmetic is the following theorem.

1.3.3.3. Theorem. (Buss [1986,Thm 2.2]) *Let $T \supseteq BASIC$ be a theory of arithmetic. Let T be extended to a theory T^+ in an enlarged language L^+ by adding Δ_1^b-defined predicate symbols, Σ_1^b-defined function symbols and their defining equations. Then T^+ is conservative over T. Also, if A is a Σ_i^b (respectively, a Π_i^b) formula in the enlarged language L^+, then there is a formula A^- in the language of T such that A^- is also in Σ_i^b (respectively, Π_i^b) and such that*

$$T^+ \vdash (A \leftrightarrow A^-).$$

An immediate corollary to this theorem is that, for $i \geq 1$, theories such as S_2^i and T_2^i can introduce Σ_1^b-defined function symbols and Δ_1^b-predicate symbols and use them freely in induction axioms.

With the aid of Theorem 1.3.3.3, the bootstrapping for S_2^1 and T_2^1 is analogous to the bootstrapping for $I\Delta_0$ in section 1.2.8; indeed, every single function and predicate symbol which was claimed to be Σ_1-definable or Δ_0-definable (respectively) in $I\Delta_0$ in section 1.2.8 is likewise Σ_1^b-definable or Δ_1^b-definable in each of the six theories S_2^1, T_2^1, $BASIC + \Pi_1^b$-*PIND*, $BASIC + \Sigma_1^b$-*LIND*, $BASIC + \Pi_1^b$-*LIND* and $BASIC + \Pi_1^b$-*IND*. We shall omit the details of this bootstrapping here; they can be found in Buss [1986,1992] and Buss and Ignjatović [1995].

One consequence of the bootstrapping process is that some of the other forms of induction follow from Σ-*PIND* and Π-*IND*:

1.3.3.4. Theorem. (Buss [1986]) *Let $i \geq 1$.*

(1) *T_2^i proves Π_i^b-IND and $T_2^i \vDash S_2^i$.*

(2) *S_2^i proves Σ_i^b-LIND, Π_i^b-PIND and Π_i^b-LIND.*

1.3.4. Polynomial time computable functions in S_2^1

The last section discussed the fact that Σ_1^b-definable functions and Δ_1^b-defined predicates can be introduced into theories of bounded arithmetic and used freely in induction axioms. Of particular importance is the fact that these include all polynomial time computable functions and predicates.

A function or predicate is said to be *polynomial time* computable provided there exists a Turing machine M and a polynomial $p(n)$, such that M computes the function or recognizes the predicate, and such that M runs in time $\leq p(n)$ for all inputs of length n. The inputs and outputs for M are integers coded in binary notation, thus the length of an input is proportional to the total length of its binary representation.

For our purposes, it is convenient to use an alternative definition of the polynomial time computable functions; the equivalence of this definition is due to Cobham [1965].

Definition. The *polynomial time* functions on $\mathbb{N}$ are inductively defined by

(1) The following functions are polynomial time:
   - The nullary constant function 0.
   - The successor function $x \mapsto S(x)$.
   - The doubling function $x \mapsto 2x$.
   - The conditional function $Cond(x, y, z) = \begin{cases} y & \text{if } x = 0 \\ z & \text{otherwise.} \end{cases}$

(2) The projection functions are polynomial time functions and the composition of polynomial time functions is a polynomial time function.

(3) If g is a $(n-1)$-ary polynomial time function and h is a $(n+1)$-ary polynomial time function and p is a polynomial, then the following function f, defined by *limited iteration on notation from g and h*, is also polynomial time:

$$\begin{aligned} f(0,\vec{x}) &= g(\vec{x}) \\ f(z,\vec{x}) &= h(z,\vec{x},f(\lfloor \tfrac{1}{2}z \rfloor,\vec{x})) \quad \text{for } z \neq 0 \end{aligned}$$

provided $|f(z,\vec{x})| \le p(|z|,|\vec{x}|)$ for all $z, \vec{x}$.

A predicate is *polynomial time* computable provided its characteristic function is polynomial time. The class of polynomial time functions is denoted $\square_1^p$, and the class of polynomial time predicates is denoted Δ_1^p.

1.3.4.1. Theorem. (Buss [1986])
(a) *Every polynomial time function is Σ_1^b-definable in S_2^1.*
(b) *Every polynomial time predicate is Δ_1^b-definable in S_2^1.*

Once one has bootstrapped S_2^1 sufficiently to intensionally introduce sequence coding functions, it is fairly straightforward to prove this theorem using Cobhams's inductive definition of polynomial time computability. The main case in the proof by induction is the case where f is defined from g and h by limited iteration on notation: in this case the predicate $f(z,\vec{x}) = y$ is defined similarly to the way $f(m,\vec{x}) = y$ was defined in the proof of Theorem 1.2.10; the main difference now is that Σ_1^b-PIND is used to prove w exists, and for this it is necessary to bound w with a term. Fortunately, the bounding condition $|f(z,\vec{x})| \le p(|z|,|\vec{x}|)$ makes it possible to bound the elements of w, and hence w, with a term. We leave the details to the reader. □

A second way to approach defining the polynomial time function in S_2^1 is to directly formalize polynomial time computability using Turing machine computations, instead of using Cobham's definition. This can also be formalized in S_2^1; furthermore S_2^1 can prove the equivalence of the two approaches. See Buss [1986] for more details.

For $i \ge 1$, $S_2^i \supseteq S_2^1$ and also, by Theorem 1.3.5 below, $T_2^i \supseteq S_2^1$. Therefore, the above theorem, combined with Theorem 1.3.3.3 gives:

1.3.4.2. Theorem. (Buss [1986]) *Let $i \ge 1$. The theories S_2^i and T_2^i can introduce symbols for polynomial time computable functions and predicates and use them freely in induction axioms.*

We shall show later (Theorem 3.2) that the converse to Theorem 1.3.4.1 also holds and that S_2^1 can Σ_1^b-define and Δ_1^b-define precisely the polynomial time computable functions and predicates, respectively.

1.3.5. Relating S_2^i and T_2^i

It is clear that $S_2^i \supseteq S_2^1$ and $T_2^i \supset T_2^1$, for $i \ge 1$. In addition we have the following relationships among these theories:

Theorem. (Buss [1986]) *Let* $i \geq 1$.
(1) $T_2^i \supseteq S_2^i$.
(2) $S_2^i \supseteq T_2^{i-1}$.

It is however open whether the theories

$$S_2^1 \subseteq T_2^1 \subseteq S_2^2 \subseteq T_2^2 \subseteq S_2^3 \subseteq \cdots$$

are distinct.

Proof. A proof of (1) can be found in Buss [1986,sect 2.6]: this proof mostly involves bootstrapping of T_2^i, and we shall not present it here.

The proof of (2) uses a divide-and-conquer method. Fix $i \geq 1$ and fix a Σ_{i-1}^b-formula $A(x)$; we must prove that S_2^i proves the IND axiom for A. We argue informally inside S_2^i, assuming $(\forall x)(A(x) \supset A(x+1))$. Let $B(x,z)$ be the formula

$$(\forall w \leq x)(\forall y \leq z+1)(A(w \dot{-} y) \supset A(w)).$$

Clearly B is equivalent to a Π_i^b-formula. By the definition of B, it follows that

$$(\forall x)(\forall z)(B(x, \lfloor \tfrac{1}{2}z \rfloor) \supset B(x,z)),$$

and hence by Π_i^b-PIND on $B(x,z)$ with respect to z,

$$(\forall x)(B(x,0) \supset B(x,x)).$$

Now, $(\forall x)B(x,0)$ holds as it is equivalent to the assumption $(\forall x)(A(x) \supset A(x+1))$, and therefore $(\forall x)B(x,x)$ holds. Finally, $(\forall x)B(x,x)$ immediately implies $(\forall x)(A(0) \supset A(x))$: this completes the proof of the IND axiom for A.

The theorem immediately implies the following corollary:

Corollary. (Buss [1986]) $S_2 = T_2$.

In the proof of the above theorem, it would suffice for $A(x)$ to be Δ_i^b with respect to S_2^i. Therefore, S_2^i proves Δ_i^b-IND.

1.3.6. Polynomial hierarchy functions in bounded arithmetic

The polynomial time hierarchy is a hierarchy of bounded alternation polynomial time computability; the base classes are the class $P = \Delta_1^p$ of polynomial time recognizable predicates, the class $FP = \square_1^p$ of polynomial time computable functions, the class $NP = \Sigma_1^p$ of predicates computable in nondeterministic polynomial time, the class $coNP = \Pi_1^p$ of complements of NP predicates, etc. More generally, Δ_i^p, $\square_i^p$, Σ_i^p and Π_i^p are defined as follows:

Definition. The classes Δ_1^p and $\square_1^p$ have already been defined. Further define, by induction on i,

(1) Σ_i^p is the class of predicates $R(\vec{x})$ definable by

$$R(\vec{x}) \Leftrightarrow (\exists y \leq s(\vec{x}))(Q(\vec{x}, y))$$

for some term s in the language of bounded arithmetic, and some Δ_i^p predicate Q.

(2) Π_i^p is the class of complements of predicates in Σ_i^p.

(3) $\square_{i+1}^p$ is class of predicates computable on a Turing in polynomial time using an oracle from Σ_i^p.[5]

(4) Δ_{i+1}^p is the class of predicates which have characteristic function in $\square_{i+1}^p$.

The connection between the syntactically defined classes of formulas Σ_i^b defined by counting alternations of quantifiers and the computationally defined classes Σ_i^p is given by the next theorem.

Theorem. (Wrathall [1976], Stockmeyer [1976], Kent and Hodgson [1982])
A predicate is Σ_i^p if and only if there is a Σ_i^b-formula which defines it.

Proof. The easier part of the proof is that every Σ_i^b-formula defines a Σ_i^p-predicate. For this, start by noting that a sharply bounded formula defines a polynomial time predicate, even when the β function and pairing functions are present. Then, given a Σ_i^b-formula, one can use the quantifier exchange property to push sharply bounded quantifiers inward and can use pairing functions to combine adjacent like quantifiers; this transforms the formula into an equivalent formula which explicitly defines a Σ_i^p property according to the above definition.

For the reverse inclusion, use induction on i. To start the induction, note that Theorem 1.3.4.1 already implies that every Δ_1^p predicate is defined by both a Σ_1^b- and a Π_1^b-formula. For the first part of the induction step, assume that every Δ_i^p predicate is definable by both a Σ_1^b and a Π_i^b-formula. Then it is immediate that every Σ_i^p predicate is definable by a Σ_i^b-formula. For the second part of the induction step, we must prove that every Δ_{i+1}^p-predicate is definable by both a Σ_{i+1}^p- and a Π_{i+1}^p-formula. For this, note that it suffices to prove that every $\square_{i+1}^p$-function has its graph defined by a Σ_{i+1}^b-formula. To prove this last fact, use induction on the definition of the functions in $\square_{i+1}^p$: it is necessary to show that this condition is preserved by definition using composition as well as by definition by limited iteration on notation. The proofs of these facts are fairly straightforward and can even be formalized by S_2^i, which gives the following theorem:

[5]An oracle from Σ_i^p is just a predicate from Σ_i^p. For our purposes, the most convenient way to define the class of functions *polynomial time relative to an oracle* R is as the smallest class of functions containing all polynomial time functions and the characteristic function of R and closed under composition and limited iteration on notation.

Theorem. (Buss [1986])) *Let* $i \geq 1$.
(a) *Every* $\square_i^p$ *function is* Σ_i^b*-definable in* S_2^i.
(b) *Every* Δ_i^p *predicate is* Δ_i^b*-definable in* S_2^i.

Proof. The proof proceeds by induction on i. The base case has already been done as Theorem 1.3.4.1. Part (b) is implied by (a), so it suffices to prove (a). To prove the inductive step, we must show the following three things (and show they are provable in S_2^i):
(1) If $f(\vec{x}, y)$ is a $\square_{i-1}^p$-function, then the characteristic function $\chi(\vec{x})$ of $(\exists y \leq t(\vec{x}))(f(\vec{x}, y) = 0)$ is Σ_i^b-definable. To prove this, we have by the induction hypothesis that $f(\vec{x}, y) = z$ is equivalent to a Σ_{i-1}^b formula $A(\vec{x}, y, z)$. The Σ_i^b-definition of $\chi(\vec{x})$ is thus[6]

$$\chi(\vec{x}) = z \Leftrightarrow (z = 0 \wedge (\exists y \leq t) A(\vec{x}, y, 0)) \vee (z = 1 \wedge \neg(\exists y \leq t) A(\vec{x}, y, 0))$$

which is clearly equivalent to a Σ_i^b-formula by prenex operations.
(2) If functions g and $\vec{h}$ have graph definable by Σ_i^b-formulas, then so does their composition. As an example of how to prove this, suppose $f(\vec{x}) = g(\vec{x}, h(\vec{x}))$; then the graph of f can be defined by

$$f(\vec{x}) = y \Leftrightarrow (\exists z \leq t_h(\vec{x}))(h(\vec{x}) = z \wedge g(\vec{x}, z) = y),$$

where t_h is a term bounding the function h.
(3) If f is defined by limited iteration from g and h with bounding polynomial p, and g and h have Σ_i^b-definable graphs, then so does f. To prove this, show that $f(z, \vec{x}) = y$ is expressed by the formula

$$(\exists w \leq SqBd(2^{p(|z|,|\vec{x}|)}, z))[\beta(|z|+1, \vec{x}) = y \wedge \beta(1, w) = g(\vec{x}) \wedge$$
$$\wedge (\forall i < |z|)(\beta(i+2, w) = \min\{h(\lfloor \tfrac{z}{2^{|z|-i-1}} \rfloor, \vec{x}, \beta(i+1, w)), 2^{p(|i+1|,|\vec{x}|)}\})].$$

Here the term $SqBd(\cdots)$ has been chosen sufficiently large to bound the size of the sequence w encoding the steps in the computation of $f(z, \vec{x})$. The formula is clearly in Σ_i^b, and the theory S_2^i can prove the existence and uniqueness of w by PIND induction up to z. □

A more complicated proof can establish the stronger result that T_2^{i-1} can also Σ_i^b-define the $\square_i^p$-functions:

Theorem. (Buss [1990]) *Let* $i > 1$.
(a) *Every* $\square_i^p$ *function is* Σ_i^b*-definable in* T_2^{i-1}.
(b) *Every* Δ_i^p *predicate is* Δ_i^b*-definable in* T_2^{i-1}.

[6]We use the convention that a characteristic function of a predicate equals zero when the predicate is true.

It is a very interesting question whether the possible collapse of the polynomial-time hierarchy is related to the possible collapse of the hierarchy of bounded arithmetic theories. So far what is known is that if S_2 is finitely axiomatized (more precisely, if $T_2^i = S_2^{i+1}$ for some $i \geq 1$), then the polynomial time hierarchy collapses provably in T_2 (see Krajíček, Pudlák and Takeuti [1991], Buss [1995], Zambella [1996], and section 3.3.2). This means that the hierarchy of theories of bounded arithmetic collapses if and only if the polynomial time hierarchy collapses S_2-provably.

1.3.7. The theories PV_i

Since T_2^{i-1} and S_2^i can Σ_i^b-define the $\square_i^p$ functions, it is often convenient to conservatively extend the language of bounded arithmetic with symbols for these functions. Accordingly, we define $T_2^{i-1}(\square_i^p)$ and $S_2^i(\square_i^p)$ to be the (conservative) extensions of T_2^{i-1} and S_2^i to the language containing symbols for the $\square_i^p$-functions with their Σ_i^b-defining equations as new axioms. For $i = 1$, the theory $T_2^0(\square_1^p)$ has to be defined slightly differently, since T_2^0 does not have sufficient bootstrapping power to Σ_1^b-define the polynomial time functions. Instead, $T_2^0(\square_1^p)$ is defined to have first-order language consisting of symbols for all polynomial time functions and predicates, and to have as axioms (1) the *BASIC* axioms, (2) axioms that define the non-logical symbols in the spirit of Cobham's definition of the polynomial time and (3) IND for all sharply bounded (equivalently, all atomic) formulas.[7]

One must be careful when working with $T_2^{i-1}(\square_i^p)$ and $S_2^i(\square_i^p)$ since, for $i > 1$, the functions symbols for $\square_i^p$ cannot be used freely in induction axioms (modulo some open questions).

Since the notation $T_2^{i-1}(\square_i^p)$ is so atrocious, it is sometimes denoted PV_i instead. Krajíček, Pudlák and Takeuti [1991] prove that PV_i can be axiomatized by purely universal axioms: to see the main idea of the universal axiomatization, note that if A is Δ_i^b, then PV_i proves A is equivalent to a quantifier-free formula via Skolemization and thus induction on $A(x, \vec{c})$, can be obtained from the universal formula

$$(\forall \vec{c})(\forall t)[A(0, \vec{c}) \land \neg A(t, \vec{c})) \supset A(f_A(t, \vec{c}) \dot{-} 1, \vec{c}) \land \neg A(f_A(t, \vec{c}), \vec{c})]$$

where f_A is computed by a binary search procedure which asks Δ_i^p queries to find a value b for which $A(b-1, \vec{c})$ is true and $A(b, \vec{c})$ is false. Of course, this f is a $\square_i^p$-function and therefore is a symbol in the language of PV_i.

1.3.8. More axiomatizations of bounded arithmetic

For any theory T in which the Gödel β function is present or is Σ_1^b-definable, in particular, for any theory $T \supseteq S_2^1$, there are two further possible axiomatizations that are useful for bounded arithmetic:

---

[7]The original definition of a theory of this type was the definition of equational theory PV of polynomial time functions by Cook [1975]. $T_2^0(\square_1^p)$ can also be defined as the conservative extension of PV to first-order logic.

Definition. Let Φ be a set of formulas. The *Φ-replacement* axioms are the formulas

$$(\forall x \leq |s|)(\exists y \leq t)A(x,y) \supset (\exists w)(\forall x \leq |s|)(A(x, \beta(x+1,w)) \wedge \beta(x+1,w) \leq t)$$

for all formulas $A \in \Phi$ and all appropriate (semi)terms s and t. As usual A may have other free variables in addition to x that serve as parameters.

The *strong Φ-replacement* axioms are similarly defined to be the formulas

$$(\exists w)(\forall x \leq |s|)[(\exists y \leq t)A(x,y) \leftrightarrow A(x, \beta(x+1,w)) \wedge \beta(x+1,w) \leq t].$$

The replacement and strong replacement axioms contain an apparently unbounded quantifier $(\exists w)$; however, S^1_2 can always bound w by a term $SqBd(t,s)$ which is large enough to bound a sequence of $|s|+1$ values $\leq t$. For example, setting $SqBd(t,s)$ equal to $(2t+1)\#(2(2s+1)^2)$ will work for the sequence encoding given in section 1.2.8.

It is known that the Σ^b_i-replacement axioms are consequences of the Σ^b_i-PIND axioms, and that the strong Σ^b_i-replacement axioms are equivalent to the Σ^b_i-PIND axioms (for $i \geq 1$, and over the base theory S^1_2). Figure 1 shows these and other relationships among the axiomatizations of bounded arithmetic.

1.4. Sequent calculus formulations of arithmetic

This section discusses the proof theory of theories of arithmetic in the setting of the sequent calculus: this will be an essential tool for our analysis of the proof-theoretic strengths of fragments of arithmetic and of their interrelationships. The sequent calculus used for arithmetic is based on the system LK_e described in Chapter I of this volume; LK_e will be enlarged with additional rules of inference for induction, minimization, etc., and for theories of bounded arithmetic, LK_e is enlarged to include inference rules for bounded quantifiers.

1.4.1. Definition. *LKB* (or LKB_e) is the sequent calculus *LK* (respectively, LK_e) extended as follows: First, the language of first-order arithmetic is expanded to allow bounded quantifiers as a basic part of the syntax. Second, the following new rules of inference are allowed:

Bounded quantifier rules

$$\forall\leq\text{:left}\ \frac{A(t), \Gamma \rightarrow \Delta}{t \leq s, (\forall x \leq s)A(x), \Gamma \rightarrow \Delta} \qquad \forall\leq\text{:right}\ \frac{b \leq s, \Gamma \rightarrow \Delta, A(b)}{\Gamma \rightarrow \Delta, (\forall x \leq s)A(x)}$$

$$\exists\leq\text{:left}\ \frac{b \leq s, A(b), \Gamma \rightarrow \Delta}{(\exists x \leq s)A(x), \Gamma \rightarrow \Delta} \qquad \exists\leq\text{:right}\ \frac{\Gamma \rightarrow \Delta, A(t)}{t \leq s, \Gamma \rightarrow \Delta, (\exists x \leq s)A(x)}$$

where the variable b is an eigenvariable and may not occur in in s or in Γ, Δ.

The Cut Elimination and Free-cut Elimination Theorems still hold for *LKB* and LKB_e, in the exact same form as they were proved to hold for *LK* and LK_e

$$\Sigma_i^b\text{-IND} \iff \Pi_i^b\text{-IND} \iff \Sigma_i^b\text{-MIN} \iff \Delta_{i+1}^b\text{-IND}$$

$$\Downarrow$$

$$\Sigma_i^b\text{-PIND} \iff \Pi_i^b\text{-PIND} \iff \Sigma_i^b\text{-LIND} \iff \Pi_i^b\text{-LIND}$$

$$\Updownarrow$$

$$\Sigma_i^b\text{-LMIN} \iff (\Sigma_{i+1}^b \cap \Pi_{i+1}^b)\text{-PIND}$$

$$\Downarrow$$

$$\Sigma_{i-1}^b\text{-IND}$$

$$\Sigma_{i+1}^b\text{-MIN} \iff \Pi_i^b\text{-MIN}$$

$$S_2^i \underset{\Sigma_i^b}{\succ} T_2^{i-1}$$

$$S_2^i \underset{\mathcal{B}(\Sigma_i^b)}{\succ} T_2^{i-1} + \Sigma_i^b\text{-replacement}$$

$$\Sigma_1^b\text{-}PIND + \Sigma_{i+1}^b\text{-replacement} \Longrightarrow \Sigma_i^b\text{-PIND} \Longrightarrow \Sigma_i^b\text{-replacement}$$

$$\Sigma_i^b\text{-PIND} \iff \Sigma_1^b\text{-}PIND + \text{strong } \Sigma_i^b\text{-replacement}$$

Figure 1

Relationships among axiomatizations for Bounded Arithmetic relative to the base theory $BASIC$ with $i \geq 1$; T_2^0 should be interpreted as PV_1. See Buss [1986,1990], Buss and Ignjatović [1995] for proofs.

in Chapter I. The principal formulas of the bounded quantifier inferences are the formulas $t \leq s$ and $(Qx \leq s)A$ introduced in the lower sequent; as usual, a cut on a direct descendent of a principal formula is anchored.

1.4.2. Rule forms of induction. We next introduce inference rules which are equivalent to induction axioms; the reason for using rules of inference for induction in place of induction axioms is that the use of free-cut free proofs provides a powerful proof-theoretic tool for the analysis of fragments of arithmetic.

Definition. Let Ψ be a class of formulas. Then Ψ-IND induction rules are the inferences of the form

$$\frac{A(b), \Gamma \longrightarrow \Delta, A(b+1)}{A(0), \Gamma \longrightarrow \Delta, A(t)}$$

where $A \in \Psi$ and where the *eigenvariable* b does not occur except as indicated.

The Ψ-PIND induction rules are the inferences of the form

$$\frac{A(\lfloor\frac{1}{2}b\rfloor), \Gamma \longrightarrow \Delta, A(b)}{A(0), \Gamma \longrightarrow \Delta, A(t)}$$

where again $A \in \Psi$ and b occurs only as indicated.

We leave it to the reader to check that the induction rules Ψ-IND and Ψ-PIND are equivalent to the induction axioms Ψ-IND and Ψ-PIND (respectively); in fact, this is true for *any* class Ψ of formulas. The fact that the induction rules are equivalent to the induction axioms depends crucially on the presence of the the side formulas Γ and Δ in the inference; when side formulas are not allowed, the inference rules are often slightly weaker than the induction axioms; see, e.g., Parsons [1972] and Sieg [1985]. It follows that theories such as $I\Delta_0$, $I\Sigma_n$, $I\Pi_n$, S_2^i and T_2^i can be equivalently formulated using induction rules instead of induction axioms. For the rest of this chapter, we will presume that these theories are formulated with the induction rules.

As was discussed in Chapter I, the free-cut elimination theorem holds for theories such as $I\Sigma_n$, $I\Pi_n$, S_2^i and T_2^i. In particular, we have the following corollary to the free-cut elimination theorem, which generalizes the subformula property to fragments of arithmetic. For this theorem, Ψ must be a class of formulas which is closed under the operations of taking subformulas and freely substituting terms for variables. Strictly speaking, classes such as Σ_i are not closed under subformulas, since a Σ_i-formula may contain a (negated) Π_i-subformula; however, one may instead use the class of Σ_i-formulas in which all negation signs are in front of atomic subformulas. This can be done without loss of generality and then this class of formulas is closed both under subformulas and under term substitution.

Theorem. *Let Ψ be a class of formulas closed under subformulas and under term substitution and containing the atomic formulas. Let R be a fragment of arithmetic axiomatized by Ψ-IND (or by Ψ-PIND) plus initial sequents containing only formulas from Ψ. Also suppose that the sequent $\Gamma \longrightarrow \Delta$ contains only formulas from Ψ and that $R \vdash \Gamma \longrightarrow \Delta$. Then there is an R-proof of $\Gamma \longrightarrow \Delta$ such that every formula appearing in the proof is a Ψ-formula.*

The proof of this theorem is of course based on the fact that every formula appearing in an R-proof either is a direct descendent of a formula in an initial sequent or is an ancestor (and hence subformula in the wide sense) of either a cut formula or a formula in the endsequent. Furthermore, in a free-cut free R-proof, all cut-formulas are in Ψ and by free-cut elimination, $\Gamma \longrightarrow \Delta$ has a free-cut free proof.

The above theorem turns out be an extremely powerful tool for the proof-theoretic analysis of fragments of arithmetic.

1.4.3. We now state and prove a generalization of Theorem 1.2.7.1 which applies to very general bounded theories R, possibly including induction inferences for bounded formulas. Assume that R contains $\leq$ in its language and that R proves that $\leq$ is reflexive and transitive. Also suppose that for all terms r and s, there is a term t so that R proves $r \leq t$ and $s \leq t$. Further suppose that for all terms $t(\vec{a}, b)$ and $r(\vec{a})$, there is a term s so that R proves that $b \leq r(\vec{a}) \supset t(\vec{a}, b) \leq s(\vec{a})$.

Parikh's Theorem. *Let R be a bounded theory satisfying the above conditions and $A(\vec{x}, y)$ a bounded formula. Suppose $R \vdash (\forall \vec{x})(\exists y)A(\vec{x}, y)$. Then there is a term t such that R also proves $(\forall \vec{x})(\exists y \leq t)A(\vec{x}, y)$.*

1.4.4. Proof. (Sketch). By the free-cut elimination theorem, there is a free-cut free R-proof P of $(\exists y)A(\vec{b}, y)$, where the b's are new free variables. By the subformula property, every sequent $\Gamma \longrightarrow \Delta$ in the proof P contains only bounded formulas in its antecedent Γ and its antecedent Δ contains only bounded formulas plus possibly occurrences of the formula $(\exists y)A(\vec{b}, y)$. Given such a Δ and given a term t, let $\Delta^{\leq t}$ denote the result of removing all occurrences of $(\exists y)A(\vec{b}, y)$ from Δ and adding the formula $(\exists y \leq t)A(\vec{b}, y)$. It is straightforward to prove by induction on the number of inferences in P that, for each sequent $\Gamma \longrightarrow \Delta$ in P, there is a term t such that R proves $\Gamma \longrightarrow \Delta^{\leq t}$. □

1.4.5. Inference rules for collection. Just as it possible to replace induction axioms with induction inferences, it is also possible to formulate the collection axioms of $B\Sigma_i$ as rules of inference. The Σ_i-collection inferences, Σ_i-REPL, are

$$\frac{\Gamma \longrightarrow \Delta, (\forall x \leq t)(\exists y)A(x, y)}{\Gamma \longrightarrow \Delta, (\exists z)(\forall x \leq t)(\exists y \leq z)A(x, y)}$$

It is not difficult to check that the inference rule for collection (replacement) is equivalent to the axiom form of collection. Furthermore, the free-cut elimination theorem holds as before; however, the notion of 'free-cut' is changed by also declaring every direct descendent of the principal formula of a collection inference to be anchored.

One easy consequence of free-cut elimination for collection inferences is that Parikh's Theorem 1.4.3 holds also for theories R that contain Σ_1-REPL; compare this to Theorem 3.4.1 about the conservativity of $B\Sigma_{i+1}$ over $I\Sigma_i$.

2. Gödel incompleteness

Gödel's incompleteness theorems, on the impossibility of giving an adequate and complete axiomatization for mathematics, were of great philosophical and foundational importance to mathematics. They are arguably the most important results in mathematical logic since the development of first-order logic. Loosely speaking, the incompleteness theorems state that any sufficiently expressive, consistent theory with a decidable axiomatization is not complete; and furthermore, for any such theory, the second incompleteness theorem gives an explicit, non self-referential true statement which is not a consequence of the theory. More generally, the set of Π_1-sentences true about the integers[8] is not recursively enumerable, so there is no way to generalize or replace first-order logic with any other kind of formal system which both admits a decidable notion of 'provability' and is complete in the sense of 'proving' every true Π_1-sentence of the integers.

2.1. Arithmetization of metamathematics

The usual methods of proving Gödel's incompleteness theorems involve coding metamathematical concepts (i.e., coding the syntax of first-order logic) with integers and then using a self-referential or diagonal construction to obtain non-provable true statements;[9] this process of coding syntactic aspects of logic with integers is called 'arithmetization'. There are essentially two different approaches to the arithmetization of syntax. The first approach uses numeralwise representability as a means of representing computable functions: a numeralwise representation of a function gives a characterization of the function's values for particular choice of inputs to the function. To be precise, a formula $A(\vec{x}, y)$ *numeralwise represents* a function $f(\vec{x}) = y$ in a theory T if and only if, for every particular integers $n_1, \ldots, n_k$ with $f(\vec{n}) = m$, the theory T proves

$$(\forall y)(A(S^{n_1}0, \ldots, S^{n_k}, y) \leftrightarrow y = S^m 0).$$

It turns out that every recursive function is numeralwise representable even in very weak theories such as R and Q; and conversely, only recursive functions are numeralwise representable in *any* axiomatizable theory, no matter how strong.

However, numeralwise representation of f in the theory T only implies that T can 'represent' all particular, fixed values of f; this in no way implies that T can prove general properties of the function f. This in is in contrast to the second approach to the arithmetization of syntax which involves giving intensional definitions of certain (but not all) recursive functions. In the intensional approach to arithmetization of metamathematics, one gives formulas which define concepts such as "formula", "term", "substitution", "proof", "theorem", etc; these definitions are said to be *intensional* provided the theory T can prove simple properties of these concepts. For instance, one wants the theory T to be able to define the notion of substituting a term into a formula and prove the result is a formula; similarly, T should be able to prove that the set of theorems is closed under modus ponens; etc. (See Feferman [1960] for a comprehensive discussion of intensionality.)

It is significantly more work to carry out the details an intensional arithmetization of syntax than a numeralwise representation of recursive functions; indeed, an

[8] By Matijacevič's theorem, the same holds for the true, purely universal sentences (in the language of PA).

[9] There are approaches to the first incompleteness theorem that avoid this arithmetization of metamathematics; for instance, one can directly prove that the true Π_1 sentences of arithmetic do not form a recursively enumerable set, say by encoding Turing machine computations with integers. For somewhat different approaches based on Berry's paradox, see Chaitin [1974] and Boolos [1989].

intensional definition of a function is typically a numeralwise representation of the same function. Furthermore, the intensional representation requires the additional verification that the underlying theory can prove simple facts about the function. Nonetheless, the intensional definition has significant advantages, most notably, in allowing a smoother treatment of the Gödel incompleteness theorem, especially the second incompleteness theorem.

Since many textbooks discuss the first approach based on numeralwise representability and since we prefer the intensional approach, this article will deal only with the intensional approach. The reader who wants to see the numeralwise representability approach can consult Smorynski [1977] and any number of textbooks such as Mendelson [1987]. The intensional approach is due to Feferman [1960]. An effective unification of the two approaches can be given using the fact (independently due to Wilkie and to Nelson [1986]) that $I\Delta_0+\Omega_1$ and S^1_2 are interpretable in Q; since both $I\Delta_0+\Omega_1$ and S^1_2 admit a relatively straightforward intensional arithmetization of metamathematics (see Wilkie and Paris [1987] and Buss [1986]), this allows strong forms of incompleteness obtained via the intensional approach to apply also to the theory Q; paragraph 2.1.4 bclow sketches how the interpretation of S^1_2 in Q can be used to give an intensional arithmetization in Q. The book of Smullyan [1992] gives a modern, in-depth treatment of Gödel's incompleteness theorems.

2.1.1. Overview of an intensional arithmetization of metamathematics. We now sketch some of the details of an arithmetization of metamathematics; this arithmetization can be carried out intensionally in $I\Delta_0 + \Omega_1$ and in S^1_2. Detailed explanations of similar arithmetizations in these theories can be found in Wilkie and Paris [1987] and in Buss [1986]. We shall always work in the (apparently) weaker theory S^1_2.

To arithmetize metamathematics, we need to assign Gödel numbers to syntactic objects such as 'terms', 'formulas', 'proofs', etc. Each such syntactic object is viewed as an expression consisting of string of symbols from a finite alphabet. This finite alphabet contains logical connective symbols "$\land$", "$\lor$", "$\neg$", "$\supset$", "$\exists$", "$\forall$", etc., and the comma symbol and parentheses; it also contains non-logical connective symbols for the function and relation symbols of arithmetic. The alphabet also needs symbols for variables: for this there is a variable symbol "x" (and possibly "a" for free variables) and there are symbols "0" and "1" used to write the values of subscripts of variables in binary notation. In addition, when first-order proofs are formalized in the sequent calculus, it will contain a symbol "$\longrightarrow$" for the sequent connective and the semicolon symbol for separating sequents. At times, it will be convenient to enlarge the finite alphabet with other symbols that can be used for describing the skeleton of a proof.

Since the alphabet is finite, we can identity the alphabet with some finite set $\{0, \ldots, s\}$ of integers, thereby giving each alphabet symbol σ a *Gödel number* which is denoted $\ulcorner\sigma\urcorner$. Then, given an expression α which consists of the symbols $a_1 a_2 \cdots a_m$, let $n_1, \ldots, n_m$ be their Gödel numbers, then the least Gödel number of the sequence $\langle n_1, \ldots, n_m\rangle$ is, by definition, the *Gödel number* of the expression α. The Gödel number of an expression α is denoted $\ulcorner\alpha\urcorner$. There is a subtle difference between the Godel number of symbol σ and the Gödel number of the expression containing just σ; these are both denoted $\ulcorner\sigma\urcorner$ and it should always be clear from context which one is intended.[10]

Since we have already discussed that intensional definitions of Gödel numbers of sequences can be given; it is straightforward to further give intensional definitions of (Gödel numbers of) syntactic objects; in particular, S^1_2 can Δ^b_1-define the following predicates:

$FreeVar(w)$ - w codes a free variable

$BdVar(w)$ - w codes a bound variable

$Term(w)$ - w codes a term

$Fmla(w)$ - w codes a formula

$Sequent(w)$ - w codes a sequent

For example, the formula $FreeVar(w)$ asserts that either $w = \langle\ulcorner a\urcorner, \ulcorner 0\urcorner\rangle$ (which codes the free variable "a_0") or w is of the form $\langle\ulcorner a\urcorner, \ulcorner 1\urcorner, w_1, \ldots, w_k\rangle$ with $k \geq 0$ and with each w_i equal to $\ulcorner 0\urcorner$ or $\ulcorner 1\urcorner$ (this encodes "a_i" where $i > 0$ has binary representation $1w_1 \cdots w_k$). The Δ^b_1 definitions of *Term*, *Fmla* and *Sequent* are somewhat more complicated: they depend crucially on the fact that length-bounded counting is Σ^b_1-definable and that therefore terms and formulas may be parsed by means of counting parentheses. Counting commas also allows notions such as the i-th formula of a sequent to be Σ^b_1-defined.

In keeping with the convention that all syntactic objects are coded by expressions, we let the Gödel number of a proof be defined to the Gödel number of an expression consisting of a sequence of sequents separated by semicolons. A proof is intended to be valid provided that each sequent in the proof can be inferred by a valid rule of inference from sequents appearing earlier in the proof. Of course the notion of *valid inference* depends on the formal proof system in which the proof is being carried out. Accordingly, for an appropriate fixed formal system T, we wish S^1_2 to be able to Δ^b_1-define the predicate to define $Proof_T(w)$ which states that w codes a valid T-proof. For this to be possible, there must be a polynomial time procedure which determines whether w codes a valid T-proof; in general, this will be based on a polynomial time procedure which checks whether a given inference is valid for T.

The theories T that we will consider, such as S^i_2, T^i_2, $I\Sigma_i$, $B\Sigma_i$, etc., will have only axioms and unary and binary inference rules, and these will be specified by a finite set of schemes. For such schematic theories, S^1_2 can Δ^b_1-define the relation[11]

$ValidInference_T(u, v, w)$ - w can be inferred with a single T-inference from zero, one or both of u and v.

[10]We have defined $\ulcorner A\urcorner$ in a nonconventional manner: the usual definition is to let $\ulcorner A\urcorner$ represent a closed term whose value is equal to the Gödel number of A. We shall represent this alternative concept with the notation $\underline{\ulcorner A\urcorner}$. The definition for $\ulcorner A\urcorner$ that we are using is better for our *intensional* development.

[11]We discuss the situation for non-schematic theories below in section 2.1.3.

With this, it is easy for S^1_2 to Δ^b_1-define $Proof_T(w)$. In addition, S^1_2 can Δ^b_1-define the predicate $Prf_T(w,u)$ which states that $Proof_T(w)$ and $Fmla(u)$ and that w is a proof of the sequent $\longrightarrow A$ where $u = \ulcorner A \urcorner$ (i.e., that w is a T-proof of the formula A). Finally, the set of theorems of T can be defined by

$$Thm_T(u) \Leftrightarrow (\exists w) Prf_T(w,u).$$

However, Thm_T is not generally Δ^b_1-definable, since it not generally even decidable (a consequence of Gödel's incompleteness theorems, see below).[12]

A particularly important syntactic operation is the substitution of a term into a formula. Let A be a formula and let t be a term; because of the sequent calculus' conventions on free and bound variables, one can always form the formula $A(t/a_0)$, which is A with t substituted in for the free variable a_0, merely by replacing each occurrence of a_0 as a subexpression of A with the expression t. This is clearly Σ^b_1-definable in S^1_2 and $Sub(u,v)$ denotes the function such that $Sub(\ulcorner A \urcorner, \ulcorner t \urcorner) = \ulcorner A(t/a_0) \urcorner$ for all formulas A and terms t.

One final simple, but important formalization, is the definition of closed *canonical terms* $\underline{n}$ which represent an integer n. The term $\underline{0}$ is just the constant symbol 0. And inductively, the term $\underline{2m}$ is $(SS0) \cdot \underline{m}$ and the term $\underline{2m+1}$ is $\underline{2m} + S0$. Note that the number of symbols in $\underline{n}$ is $O(|n|)$; also, S^1_2 can Σ^b_1-define the map $n \mapsto \ulcorner \underline{n} \urcorner$.

2.1.2. Intensionality of the arithmetization. In order for the above-sketched arithmetization to be considered *intensional*, it is necessary that S^1_2 can prove basic facts about the arithmetization. To simplify notation, we shall use abbreviations such as $(\forall \ulcorner A \urcorner)(\cdots \ulcorner A \urcorner \cdots)$, which abbreviates the formula $(\forall u)(Fmla(u) \supset (\cdots u \cdots))$. Some examples of what S^1_2 can prove include:

1. $(\forall u, v, w)(Fmla(u) \land Term(v) \supset Fmla(Sub(u,v)))$.
2. $(\forall \ulcorner A \urcorner)(\forall \ulcorner B \urcorner)(Thm_T(\ulcorner A \urcorner) \land Thm_T(\ulcorner A \supset B \urcorner) \supset Thm_T(\ulcorner B \urcorner))$.
3. $(\forall u)(Proof_T(u) \supset Thm_{S^1_2}(\ulcorner Proof_T(\underline{u}) \urcorner))$.
4. $(\forall \ulcorner A \urcorner)(\forall u)(Prf_T(u, \ulcorner A \urcorner) \supset Thm_{S^1_2}(\ulcorner Prf(\underline{u}, \underline{\ulcorner A \urcorner}) \urcorner))$.
5. $(\forall \ulcorner A \urcorner)(Thm_T(\ulcorner A \urcorner) \supset Thm_{S^1_2}(\ulcorner Thm_T(\underline{\ulcorner A \urcorner}) \urcorner))$.

These five formulas require some explanation. The first just states that when a term is substituted into a formula, a formula is obtained. The second codes the fact that the consequences of the sequent calculus are closed under modus ponens. S^1_2 proves this by the simple argument that if there are sequent calculus proofs of $\longrightarrow A$ and $\longrightarrow A \supset B$, then these can be combined with the simple sequent calculus proof of $A, A \supset B \longrightarrow B$ using two cuts to obtain a sequent calculus proof of B. The third formula states that any u which encodes a T-proof can in fact be proved to be a T-proof. The intuitive idea behind the fact that S^1_2 can prove this formula is

[12] Even for decidable theories T, if $T \nvDash (\forall x)(\forall y)(x = y)$, then the predicate $Thm_T(u)$ is PSPACE-hard.

that, given u encoding a proof as a string of symbols, it is possible to construct a S^1_2-proof of the statement $Proof_T(\underline{u})$ that u codes a T-proof. In fact, the S^1_2-proof of $Proof_T(\underline{u})$, proceeds by verifying that u, viewed as a string of symbols, satisfies all the properties of being a valid T-proof. The provability of the fourth and fifth formulas in S^1_2 is similar to the provability of the third.

The fact that S^1_2 can prove the third formula is a special case of a more general fact:

Theorem. (Buss [1986,Thm 7.4]) *Let $A(b)$ be a Σ^b_1-formula with only the variable b free. Then, S^1_2 can prove*

$$(\forall u)(A(u) \supset Thm_{S^1_2}(\ulcorner A(\underline{u})\urcorner)).$$

Hilbert-Bernays-Löb derivability conditions. The following three derivability conditions, introduced by Hilbert and Bernays [1934-39] and Löb [1955], give sufficient conditions on an arithmetization for the second incompleteness theorem to hold for a theory T, with respect to a given formalization Thm_T of provability:

HBL1: For all A and B, $T \vdash Thm_T(\underline{\ulcorner A\urcorner}) \wedge Thm_T(\underline{\ulcorner A \supset B\urcorner}) \supset Thm_T(\underline{\ulcorner B\urcorner})$.
HBL2: For all A, if $T \vdash A$, then $T \vdash Thm_T(\underline{\ulcorner A\urcorner})$.
HBL3: For all A, $T \vdash Thm_T(\underline{\ulcorner A\urcorner}) \supset Thm_T(\underline{Thm_T(\underline{\ulcorner A\urcorner})})$.

Assuming $T \supseteq S^1_2$, the first and third conditions follow from the fact that formulas 2 and 5 above are provable in S^1_2. The second condition is the fact that formula 5 is true; which of course is an immediate consequence of fact that formula 5 above is provable in S^1_2 and hence is true.

2.1.3. Arithmetization of syntax for non-schematic theories. So far, we have considered only schematically axiomatized theories. This is not unreasonable, since many of the theories we are interested in, such as Peano arithmetic are schematically axiomatized. Many other theories such as S^i_2, $I\Delta_0$, $I\Sigma_1$, etc. are not schematic but are at least nearly schematic in that they are axiomatized by a finite set of schemes with substitution restricted to certain formula classes. The metamathematics for these latter theories can be arithmetized with only a slight modification of the above methods.

A theory is said to be *axiomatizable* provided that it has a recursive (i.e., decidable) set of axioms. By a theorem of Craig's this is equivalent to having a recursively enumerable set of axioms. In general, there are many axiomatizable theories which are not schematic; nonetheless, the arithmetization of metamathematics can be modified to apply to any axiomatizable theory as follows.

Let T be an axiomatizable theory. Since the predicate $ValidInference_T$ may no longer be polynomial-time, it may not be Δ^b_1-definable in S^1_2. However, it is possible to express $ValidInference_T(u, v, w)$ in equivalent form as

$$(\exists a)\, ValidInfEvidence(a, u, v, w),$$

where *ValidInfEvidence* is Δ_1^b w.r.t. S_2^1.[13] With this, a T-proof is then coded as a sequence containing the lines of the proof, plus any necessary evidence values, a, justifying the steps in the proof. In this way, the predicates $Proof_T(w)$ and $Prf_T(w,u)$ are Δ_1^b-definable in S_2^1; likewise, Thm_T is definable from Prf_T as before.

It is easy to check that formulas 1-5 are still provable in S_2^1. Also, if $T \supseteq S_2^1$, the Hilbert derivability conditions hold as well.

2.1.4. Arithmetization in theories which contain only Q. All our proofs of incompleteness theorems will assume that the theory T under consideration contains S_2^1. However, the results all hold as well for theories which only contain Q. The intensional arithmetization in S_2^1 can be extended to an intensional arithmetization in Q based on the fact that S_2^1 is interpretable in Q. The interpretation of S_2^1 in Q is a very special kind based on an inductive cut J; namely, there is a formula $J(a)$ such that Q proves J is closed downwards and is closed under 0, S, $+$, $\cdot$ and $\#$. In addition, for ϕ any sentence, the sentence ϕ^J, *ϕ relativized by J*, is obtained from ϕ by replacing every quantifier (Qx) with $(Qx.J(x))$. Then, we have that $Q \vdash \phi^J$ for all theorems ϕ of S_2^1.

The first use of inductive cuts for interpretations was by Solovay. The fact that $I\Delta_0$ can be interpreted in Q was first discovered by Wilkie; a local interpretation of $I\Delta_0$ in Q was independently discovered by E. Nelson. For more details of inductive cuts and this interpretation of S_2^1 in Q see Theorem 4.3.3 below, or Pudlák [1983], Nelson [1986] or Chapter VIII of this volume. Pudlák [1983] gives a very general form of the interpretation of $I\Delta_0$ in Q.

An intensional arithmetization of metamathematics can be given in Q by replacing predicates such as $Proof(w)$ with their relativizations $J(w) \wedge (Proof_T(w))^J$. The reader can check that the proofs given in the next sections all still work with these relativized predicates.

2.2. The Gödel incompleteness theorems

In this section, we discuss the diagonal, or fixpoint, lemma, the first and second incompleteness theorems, and Löb's theorem.

2.2.1. The Gödel diagonal lemma. The Gödel diagonal, or fixpoint, lemma is a crucial ingredient in the proof of the incompleteness theorems. This lemma states that, for any first-order property A, there is a formula B that states that the property A holds of the Gödel number of B. Thus, since we know that provability is a first-order property, it will be possible to construct a formula which asserts "I am not provable".

[13] Our formulation works for any decidable set of axioms and rules of inference; we do require always that all the usual logical axioms and rules of inference are present. A similar construction will work for inference rules with any finite number of hypotheses.

Gödel's Diagonal Lemma. *Let $A(a_0)$ be a formula. Then there is a formula B such that S^1_2 proves*

$$B \leftrightarrow A(\underline{\ulcorner B \urcorner}).$$

Furthermore, if A is a Σ^b_i, Π^b_i, Σ_i or Π_i formula (respectively), then so is B; and if A involves free variables other than a_0, then so does B.

Proof. This proof quite simple but rather tricky and difficult to conceptualize. We first define a diagonalization function f which satisfies

$$f(\ulcorner C \urcorner) = \ulcorner C(\underline{\ulcorner C \urcorner}) \urcorner$$

for all formulas C, where $C(\underline{\ulcorner C \urcorner})$ means $C(\underline{\ulcorner C \urcorner}/a_0)$. To define f, recall that the function $n \mapsto Num(n) = \ulcorner \underline{n} \urcorner$ is Σ^b_1-definable in S^1_2. Then the function f is Σ^b_1-definable by S^1_2 since

$$f(x) = Sub(x, Num(x)).$$

Next, we wish to let the formula $C(a_0)$ be $A(f(a_0))$; however, since f is not a function symbol in the language of S^1_2, we must be more careful in defining C. Let $f(a) = b$ be Σ^b_1-defined with the formula $G_f(a, b)$ which defines the graph of f and let t_f be a term such that S^1_2 proves $f(a) \le t_f$. Now the formula C can be taken to be either

$$(\exists x \le t_f)(G_f(a_0, x) \land A(x)) \qquad \text{or} \qquad (\forall x \le t_f)(G_f(a_0, x) \supset A(x)).$$

(With a little more care, we can choose C to be in the same quantifier complexity class as A.) Finally, define B to be the formula $C(\underline{\ulcorner C \urcorner})$.

We claim that S^1_2 proves $B \leftrightarrow A(\underline{\ulcorner B \urcorner})$. The proof of this claim is almost immediate. First, by the definitions of f and B, we have $\ulcorner B \urcorner$ is equal to $f(\ulcorner C \urcorner)$; of course S^1_2 proves this fact. Second, by the definition of C, B is S^1_2-provably equivalent to $A(f(\underline{\ulcorner C \urcorner}))$. Therefore, B is S^1_2-provably equivalent to $A(\underline{\ulcorner B \urcorner})$. Q.E.D.

2.2.2. The first incompleteness theorem. Gödel's first incompleteness theorem states that there is no complete, axiomatizable, consistent theory T extending Q. We shall prove several variants of this in this section.

Definition. Let T be an axiomatizable theory. Con_T is the $\forall\Delta^b_1$-formula $\neg Thm_T(0 \neq 0)$ which expresses the condition "*T is consistent.*"

T is said to be ω-consistent if there does not exist a formula $B(a)$ such that $T \vdash (\exists x)B(x)$ and such that $T \vdash \neg B(\underline{n})$ for all $n \ge 0$. $T \supseteq S^1_2$ is *weakly ω-consistent* provided there is no such formula B which is Δ^b_1 w.r.t. S^1_2. Since every true Δ^b_1-sentence is provable in S^1_2, T is weakly ω-consistent if and only if T is consistent and proves only true $\exists\Delta^b_1$-sentences.

Gödel's First Incompleteness Theorem. *Let T be an consistent, axiomatizable theory containing Q. Then there is a true sentence ϕ such that $T \nvdash \phi$. Further, if T is weakly ω-consistent, then $T \nvdash \neg\phi$.*

The formula ϕ is explicitly constructible from the axiomatization of T.

Proof. We'll prove this theorem under the assumption that $T \supseteq S_2^1$. Choose ϕ to a formula such that

$$S_2^1 \vdash \phi \leftrightarrow \neg Thm_T(\underline{\ulcorner\phi\urcorner}).$$

Intuitively, the formula ϕ is asserting "I am not provable in T"; the Diagonal Lemma 2.2.1 guarantees that ϕ exists.

First, let's show that ϕ is true. Suppose ϕ were false. Then, by the choice of ϕ and since S_2^1 is a true theory, $T \vdash \phi$. Therefore, $S_2^1 \vdash Thm_T(\underline{\ulcorner\phi\urcorner})$; and again, by the choice of ϕ, $S_2^1 \vdash \neg\phi$. Since $T \supseteq S_2^1$, we also have $T \vdash \neg\phi$, which contradicts the consistency of T. So ϕ cannot be false.

Second, we show that $T \nvdash \phi$. Suppose, for sake of a contradiction, $T \vdash \phi$. Then $S_2^1 \vdash Thm_T(\underline{\ulcorner\phi\urcorner})$. By choice of ϕ, $S_2^1 \vdash \neg\phi$. So also $T \vdash \neg\phi$, which again contradicts the consistency of T.

Third, we assume T is weakly ω-consistent and prove that $T \nvdash \neg\phi$. Suppose T does prove $\neg\phi$. Then since $T \supseteq S_2^1$ and by choice of ϕ, $T \vdash Thm_T(\underline{\ulcorner\phi\urcorner})$. But $Thm_T(\underline{\ulcorner\phi\urcorner})$ is a false $\exists\Delta_1^b$-sentence, which contradicts the weak ω-consistency of T. Q.E.D.

The obvious question at this point is whether the hypothesis of weak ω-consistency can be removed from the First Incompleteness Theorem; i.e., whether there is a consistent, axiomatizable, complete theory extending Q. It turns out that this hypothesis can be removed:

Rosser's Theorem. (Rosser [1936]) *There is no consistent, axiomatizable, complete theory $T \supseteq Q$.*

The proof of this theorem will give a constructive method of obtaining a formula ϕ from the axiomatization of a consistent theory T such that ϕ is independent of T.

Proof. As before, we give the proof assuming $T \supseteq S_2^1$. We need to define a modified notion of provability called "Rosser provability". Let Neg be the unary function, Σ_1^b-definable in S_2^1, such that $Neg(\ulcorner A\urcorner) = \ulcorner\neg A\urcorner$ for all formulas A. Then we define the predicate $R\text{-}Prf_T(w,a)$ as

$$R\text{-}Prf_T(w,a) \;\Leftrightarrow\; Prf_T(w,a) \wedge (\forall v \le w)(\neg Prf_T(v, Neg(a))).$$

Intuitively, A has a Rosser proof if and only if A has a (ordinary) proof such that its negation, $\neg A$, has no smaller proof. Note that, since T is consistent, $R\text{-}Prf_T(n, \ulcorner A\urcorner)$

is true exactly when $Prf_T(n, \ulcorner A \urcorner)$ is; however, this fact is not provable in S^1_2.[14] The predicate $R\text{-}Thm_T(u)$ is defined to be the $\exists\Pi^b_1$-formula $(\exists w)(R\text{-}Prf_T(w,u))$.

Now use the Diagonal Lemma to choose a sentence ϕ so that

$$S^1_2 \vdash \phi \leftrightarrow \neg R\text{-}Thm(\underline{\ulcorner \phi \urcorner}).$$

As before, we have ϕ is true. In fact, the same proof works as in the proof of the First Incompleteness Theorem, since Thm_T and $R\text{-}Thm_T$ are extensionally equivalent. Secondly, $T \nvdash \phi$ again by the same argument as in the previous theorem; in brief: if $T \vdash \phi$, then ϕ is false by choice of ϕ, but we just claimed ϕ is true.

Thirdly, we want to show $T \nvdash \neg\phi$. Suppose T does prove $\neg\phi$; let n be a Gödel number of a T-proof of $\neg\phi$. Since T is consistent, we have that there is no T-proof of ϕ; thus S^1_2 proves the true Δ_0-sentence $(\forall v \le n)\neg Prf_T(v, \underline{\ulcorner \phi \urcorner})$. And since $Prf_T(n, Neg(\phi))$ is true, S^1_2 also proves $(\forall v)(v > n \supset \neg R\text{-}Prf_T(v, \underline{\ulcorner \phi \urcorner}))$. Hence, S^1_2 proves $\neg R\text{-}Thm_T(\underline{\ulcorner \phi \urcorner})$. By the choice of ϕ and since $T \supseteq S^1_2$, this implies $T \vdash \phi$, which contradicts the consistency of T.
Q.E.D.

2.2.3. The second incompleteness theorem. Gödel's second incompleteness theorem improves on his first incompleteness theorem by giving an example of a true formula with an intuitive meaning which is not provable by a decidable, consistent theory T. This formula is the formula Con_T which expresses the consistency of T. Note, however, that unlike the the formula ϕ in the first incompleteness theorem, Con_T is not necessarily independent of T since there are consistent theories that prove their own inconsistency. An example of such a theory is $T + \neg Con_T$.

Gödel's Second Incompleteness Theorem. *Let T be a decidable, consistent theory and suppose $T \supseteq Q$. Then $T \nvdash Con_T$.*

Proof. As usual, we assume $T \supseteq S^1_2$. Let ϕ be the formula from the proof of Gödel's First Incompleteness Theorem which is S^1_2-provably equivalent to $\neg Thm_T(\underline{\ulcorner \phi \urcorner})$. Recall that we proved $T \nvdash \phi$. We shall prove that S^1_2 proves $Con_T \supset \phi$; which will suffice to show that $T \nvdash Con_T$, since $T \supseteq S^1_2$.

By choice of ϕ, S^1_2 proves $\neg\phi \supset Thm_T(\underline{\ulcorner \phi \urcorner})$. Also, by the formula 5 in section 2.1.2, S^1_2 proves $Thm_T(\underline{\ulcorner \phi \urcorner}) \supset Thm_T(\underline{\ulcorner Thm_T(\underline{\ulcorner \phi \urcorner}) \urcorner})$. Also, by choice of ϕ and by formula 1 of section 2.1.2, S^1_2 proves $Thm_T(\underline{\ulcorner Thm_T(\underline{\ulcorner \phi \urcorner}) \urcorner}) \supset Thm_T(\underline{\ulcorner \neg\phi \urcorner})$. Putting these together shows that S^1_2 proves that

$$\neg\phi \supset [Thm_T(\underline{\ulcorner \phi \urcorner}) \wedge Thm_T(\underline{\ulcorner \neg\phi \urcorner})].$$

From whence $\neg\phi \supset \neg Con_T$ is easily proved. Therefore, S^1_2 proves $Con_T \supset \phi$. □

[14]This is a good example (see Feferman [1960]) of an extensional definition which is not an intensional definition. For consistent theories T, $R\text{-}Prf_T$ and $R\text{-}Thm_T$ provide an extensionally correct definition of provability, since $R\text{-}Prf_T(n, \ulcorner A \urcorner)$ has the correct truth value for all particular n and $\ulcorner A \urcorner$. However, they are not intensionally correct; since, in general, T cannot prove that $R\text{-}Thm_T(\ulcorner A \urcorner)$ and $R\text{-}Thm_T(\ulcorner A \supset B \urcorner)$ implies $R\text{-}Thm_T(B)$.

The formula ϕ not only implies Con_T, but is actually S^1_2-equivalent to Con_T. For this, note that since ϕ implies $\neg Thm_T(\ulcorner\phi\urcorner)$, it can be proved in S^1_2 that ϕ implies $\neg Thm_T(\ulcorner 0=1\urcorner)$. (Since if a contradiction is provable, then every formula is provable.)

2.2.4. Löb's theorem. The self-referential formula constructed for the proof of the First and Second Incompleteness Theorems asserted "I am not provable". A related problem would be to consider formulas which assert "I *am* provable". As the next theorem shows, such formulas are necessarily provable. In fact, if a formula is implied by its provability, then the formula is already provable. This gives a strengthening of the Second Incompleteness Theorem, which implies that, in order to prove a formula A, one is not substantially helped by the assuming that A is provable. More precisely, the assumption $Thm_T(\ulcorner A\urcorner)$ will not significantly aid a theory T in proving A.

Löb's Theorem. *Let $T \supseteq Q$ be an axiomatizable theory and A be any sentence. If T proves $Thm_T(\ulcorner A\urcorner) \supset A$, then T proves A.*

Proof. As usual, we assume $T \supseteq S^1_2$. Let T' be the axiomatizable theory $T \cup \{\neg A\}$. The proof of Löb's Theorem uses the fact that T' is consistent if and only if $T \nvdash A$; and furthermore, that S^1_2 proves $Con(T')$ is equivalent to $\neg Thm_T(\ulcorner A\urcorner)$. From these considerations, the proof is almost immediate from the second incompleteness theorem. Namely, since T proves $\neg A \supset \neg Thm_T(\ulcorner A\urcorner)$ by choice of A, T also proves $\neg A \supset Con(T')$. Therefore, by the Deduction Theorem, $T' \vdash Con(T')$ so by Gödel's Second Incompleteness Theorem, T' is inconsistent, i.e., $T \vdash A$.

2.2.5. Further reading. The above material gives only an introduction to the incompleteness theorems. Other significant aspects of incompleteness include: (1) the strength of reflection principles which state that the provability of a formula implies the truth of the formula, see, e.g., Smorynski [1977]; (2) provability and interpretability logics, for which see Boolos [1993], Lindström [1997], and Chapter VII of this handbook; and (3) concrete, combinatorial examples of independence statements, such as the Ramsey theorems shown by Paris and Harrington [1977] to be independent of Peano arithmetic.

3. On the strengths of fragments of arithmetic

3.1. Witnessing theorems

In section 1.2.10, it was shown that every primitive recursive function is Σ_1-definable by the theory $I\Sigma_1$. We shall next establish the converse which implies that the Σ_1-definable functions of $I\Sigma_1$ are precisely the primitive recursive functions. The principal method of proof is the 'witnessing theorem method': $I\Sigma_1$ provides the simplest and most natural application of the witnessing method.

3.1.1. Theorem. (Parsons [1970], Mints [1973] and Takeuti [1987]). *Every Σ_1-definable function of $I\Sigma_1$ is primitive recursive.*

Parsons' proof of this theorem was based on the Gödel Dialectica theorem and a similar proof is given by Avigad and Feferman in Chapter V in this volume. Takeuti's proof was based on a Gentzen-style assignment of ordinals to proofs. Mints's proof was essentially the same as the witness function proof presented next; except his proof was presented with a functional language.

3.1.2. The Witness predicate for Σ_1-formulas. For each Σ_1-formula $A(\vec{b})$, we define a Δ_0-formula $Witness_A(w, \vec{b})$ which states that w is a witness for the truth of A.

Definition. Let $A(\vec{b})$ be a formula of the form $(\exists x_1, \ldots, x_k)B(x_1, \ldots, x_k, \vec{b})$, where B is a Δ_0-formula. Then the formula $Witness_A(w, \vec{b})$ is is defined to be the formula

$$B(\beta(1, w), \ldots, \beta(k, w), \vec{b}).$$

If $\Delta = \Delta', A$ is a succedent, then $Witness_{\bigvee\Delta}(w, \vec{c})$ is defined to be

$$Witness_A(\beta(1, w), \vec{c}) \vee Witness_{\bigvee\Delta'}(\beta(2, w), \vec{c}).$$

Dually, if $\Gamma = A, \Gamma'$ is an antecedent, then $Witness_{\bigwedge\Gamma}$ is defined similarly as

$$Witness_A(\beta(1, w), \vec{c}) \wedge Witness_{\bigwedge\Gamma'}(\beta(2, w), \vec{c}).$$

Note the different conventions on ordering disjunctions and conjunctions; these are not intrinsically important, but merely reflect the conventions for the sequent calculus are that active formulas of strong inferences are at the beginning of an antecedent and at the end of a succedent.

It is, of course, obvious that $Witness_A$ is a Δ_0-formula, and that $I\Delta_0$ can prove

$$A(b) \leftrightarrow (\exists w)\, Witness_A(w, \vec{b}).$$

3.1.3. Proof. (Sketch of the proof of Theorem 3.1.1.) Suppose $I\Sigma_1$ proves $(\forall x)(\exists y)A(x, y)$ where $A \in \Sigma_1$. Then there is a sequent calculus proof P in the theory $I\Sigma_1$ of the sequent $(\exists y)A(c, y)$. We must prove that there is a primitive recursive function f such that $A(n, f(n))$ is true, in the standard integers, for all $n \geq 0$. In fact, we shall prove more than this: we will prove that there is a primitive recursive function f, with a Σ_1-definition in $I\Sigma_1$, such that $I\Sigma_1$ proves $(\forall x)A(x, f(x))$. This will be a corollary to the next lemma.

Witnessing Lemma for $I\Sigma_1$. *Let Γ and Δ be cedents of Σ_1-formulas and suppose $I\Sigma_1$ proves the sequent $\Gamma \longrightarrow \Delta$. Then there is a function h such that the following hold:*

(1) *h is* Σ_1*-defined by* $I\Sigma_1$ *and is primitive recursive, and*
(2) $I\Sigma_1$ *proves*

$$(\forall\vec{c})(\forall w)[\mathit{Witness}_{\bigwedge\Gamma}(w,\vec{c}) \supset \mathit{Witness}_{\bigvee\Delta}(h(w,\vec{c}),\vec{c})].$$

Note that Theorem 3.1.1 is an immediate corollary to the lemma, since we may take Γ to be the empty sequent, Δ to be the sequent containing just $(\exists y)A(c,y)$, and let $f(x) = \beta(1,\beta(1,h(x)))$ where h is the function guaranteed to exist by the lemma. This is because $h(x)$ will be a sequence of length one witnessing the cedent $(\exists y)A$, so its first and only element is a witness for the formula $(\exists y)A$, and the first element of that is a value for y that makes A true.

It remains to prove the Witnessing Lemma. For this, we know by the Cut Elimination Theorem 1.4.2, that there is a free-cut free proof P of the sequent $\Gamma \longrightarrow \Delta$ in the theory $I\Sigma_1$; in this proof, every formula in every sequent can be assumed to be a Σ_1-formula. Therefore, we may prove the Witnessing Lemma by induction on the number of steps in the proof P.

The base case is where there are zero inferences in the proof P and so $\Gamma \longrightarrow \Delta$ is an initial sequent. Since the initial sequents allowed in an $I\Sigma_1$ proof contain only atomic formulas, the Witnessing Lemma is trivial for this case.

For the induction step, the argument splits into cases, depending on the final inference of the proof. There are a large number of cases, one for each inference rule of the sequent calculus; for brevity, we present only three cases below and leave the rest for the reader.

For the first case, suppose the final inference of the proof P is an $\exists$*:right* inference, namely,

$$\frac{\begin{array}{c}\vdots\\ \Gamma \longrightarrow \Delta, A(t)\end{array}}{\Gamma \longrightarrow \Delta, (\exists x)A(x)}$$

Let c be the free variables in the upper sequent. The induction hypothesis gives a Σ_1-defined, primitive recursive function $g(w,\vec{c})$ such that $I\Sigma_1$ proves

$$\mathit{Witness}_{\bigwedge\Gamma}(w,\vec{c}) \longrightarrow \mathit{Witness}_{\bigvee\{\Delta,A(t)\}}(g(w,\vec{c}),\vec{c}).$$

In order for $\mathit{Witness}_{\bigvee\{\Delta,A(t)\}}(g(w,\vec{c}),\vec{c})$ to hold, either $\beta(2,g(w,\vec{c}))$ witnesses $\bigvee\Delta$ or $\beta(1,g(w,\vec{c}))$ witnesses $A(t)$. So letting $h(w,\vec{c})$ be Σ_1-defined by

$$h(w,\vec{c}) = \langle\langle t(\vec{c})\rangle * \beta(1,g(w,\vec{c})), \beta(2,g(w,\vec{c}))\rangle,$$

where $*$ denotes sequence concatenation. It is immediate from the definition of *Witness* that

$$\mathit{Witness}_{\bigwedge\Gamma}(w,\vec{c}) \longrightarrow \mathit{Witness}_{\bigvee\{\Delta,(\exists x)A(x)\}}(h(w,\vec{c}),\vec{c}).$$

For the second case, suppose the final inference of the proof P is an $\exists$*:left* inference, namely,

$$\frac{\begin{array}{c}\because\vdots\because\\ A(b),\Gamma \longrightarrow \Delta\end{array}}{(\exists x)A(x),\Gamma \longrightarrow \Delta}$$

where b is an eigenvariable which occurs only as indicated. The induction hypothesis gives us a Σ_1-defined, primitive recursive function $g(w,\vec{c},b)$ such that $I\Sigma_1$ proves

$$Witness_{\bigwedge\{A(b),\Gamma\}}(w,\vec{c}) \longrightarrow Witness_{\bigvee\Delta}(g(w,\vec{c},b),\vec{c}).$$

Let $tail(w)$ be the Σ_1-defined function so that $tail(\langle w_0,w_1,\ldots,w_n\rangle) = \langle w_1,\ldots,w_n\rangle$. Letting $h(w,\vec{c})$ be the function $g(\langle tail(\beta(1,w)),\beta(2,w)\rangle,\vec{c},\beta(1,\beta(1,w)))$, it is easy to check that h satisfies the desired conditions of the Witnessing Lemma.

For the third case, suppose the final inference of P is a Σ_1-IND inference:

$$\frac{\begin{array}{c}\because\vdots\because\\ A(b),\Gamma \longrightarrow \Delta, A(Sb)\end{array}}{A(0),\Gamma \longrightarrow \Delta, A(t)}$$

where b is the eigenvariable and does not occur in the lower sequent. The induction hypothesis gives a Σ_1-defined, primitive recursive function $g(w,\vec{c},b)$ such that $I\Sigma_1$ proves

$$Witness_{\bigwedge\{A(b),\Gamma\}}(w,\vec{c},b) \longrightarrow Witness_{\bigvee\{\Delta,A(Sb)\}}(g(w,\vec{c},b),\vec{c},b).$$

Let $k(\vec{c},v,w)$ be defined as

$$k(\vec{c},v,w) = \begin{cases} v & \text{if } Witness_{\bigvee\{\Delta\}}(v,\vec{c}) \\ w & \text{otherwise} \end{cases}$$

Since *Witness* is a Δ_0-predicate, k is Σ_1-defined by $I\Sigma_1$. Now define the primitive recursive function $f(w,\vec{c},b)$ by

$$\begin{aligned} f(w,\vec{c},0) &= \langle \beta(1,w),0\rangle \\ f(w,\vec{c},b+1) &= \langle \beta(1,g(\langle\beta(1,f(w,\vec{c},b)),\beta(2,w)\rangle,\vec{c},b)), \\ &\qquad k(\vec{c},\beta(2,f(w,\vec{c},b)),\beta(2,g(\langle\beta(1,f(w,\vec{c},b)),\beta(2,w)\rangle,\vec{c},b)))\rangle \end{aligned}$$

By Theorem 1.2.10, f is Σ_1 definable by $I\Sigma_1$, and since f may be used in induction formulas, Σ_1 can prove

$$Witness_{\bigwedge\{A(0),\Gamma\}}(w,\vec{c}) \longrightarrow Witness_{\bigvee\{\Delta,A(b)\}}(f(w,\vec{c},b),\vec{c},b).$$

using Σ_1-IND with respect to b. Setting $h(w,\vec{c}) = f(w,\vec{c},t)$ establishes the desired conditions of the Witnessing Lemma.
Q.E.D. Witnessing Lemma and Theorem 3.1.1.

3.1.4. Corollary. *The Δ_1-definable predicates of $I\Sigma_1$ are precisely the primitive recursive predicates.*

Proof. Corollary 1.2.10 already established that every primitive recursive predicate is Δ_1-definable by $I\Sigma_1$. For the converse, suppose $A(c)$ and $B(c)$ are Σ_1-formulas such that $I\Sigma_1$ proves $(\forall x)(A(x) \leftrightarrow \neg B(x))$. Then the characteristic function of the predicate $A(c)$ is Σ_1-definable in $I\Sigma_1$ since $I\Sigma_1$ can prove

$$(\forall x)(\exists! y)[(A(x) \land y = 0) \lor (B(x) \land y = 1)].$$

By Theorem 3.1.1, this characteristic function is primitive recursive, hence so is the predicate $A(c)$.

3.1.5. Total functions of $I\Sigma_n$. Theorem 1.2.1 provided a characterization of the Σ_1-definable functions of $I\Sigma_1$ as being precisely the primitive recursive functions. It is also possible to characterize the Σ_1-definable functions of $I\Sigma_n$ for $n > 1$ in terms of computational complexity; however, the $n > 1$ situation is substantially more complicated. This problem of characterizing the provably total functions of fragments of Peano arithmetic is classically one of the central problems of proof theory; and a number of important and elegant methods are available to solve it. Space prohibits us from explaining these methods, so we instead mention only a few references.

The first method of analyzing the strength of fragments of Peano is based on Gentzen's assignment of ordinals to proofs; Gentzen [1936,1938] used Cantor normal form to represent ordinals less than ϵ_0 and gave a constructive method of assigning ordinals to proofs in such a way that allowed cuts and inductions to be removed from PA-proofs of sentences. This can then be used to characterize the primitive recursive functions of fragments of Peano arithmetic in terms of recursion on ordinals less than ϵ_0. The textbooks of Takeuti [1987] and Girard [1987] contain descriptions of this approach. A second version of this method is based on the infinitary proof systems of Tait: Chapter III of this volume describes this for Peano arithmetic, and Chapter IV describes extensions of this ordinal assignment method to much stronger second-order theories of arithmetic. The books of Schütte [1977] and Pohlers [1980] also describe ordinal assignments and infinitary proofs for strong theories of arithmetic. A further use of ordinal notations is to characterize natural theories of arithmetic in terms of transfinite induction.

A second approach to analyzing the computational strength of theories of arithmetic is based on model-theoretic constructions; see Paris and Harrington [1977], Ketonen and Solovay [1981], Sommer [1990], and Avigad and Sommer [1997].

A third method is based on the Dialectica interpretation of Gödel [1958] and on Howard's [1970] assignment of ordinals to terms that arise in the Dialectica interpretation. Chapter V of this volume discusses the Dialectica interpretation.

A fourth method, due to Ackermann [1941] uses an ordinal analysis of ϵ-calculus proofs.

More recently, Buss [1994] has given a characterization of the provably total functions of the theories $I\Sigma_n$ based on an extension of the witness function method used above.

3.2. Witnessing theorem for S^i_2

Theorem 1.3.4.1 stated that every polynomial time function and every polynomial time predicate is Σ^b_1-definable or Δ^b_1-definable (respectively) by S^1_2. More generally, Theorem 1.3.6 stated that every $\square^p_i$-function and every Δ^p_i-predicate is Σ^b_i-definable or Δ^b_i-definable by S^i_2. The next theorem states the converse; this gives a precise characterization of the Σ^b_1-definable functions of S^1_2 and of the Σ^b_i-definable functions of S^i_2 in terms of their complexity in the polynomial hierarchy. The most interesting case is probably the base case $i = 1$, where S^1_2 is seen to have proof-theoretic strength that corresponds precisely to polynomial time.

Theorem. (Buss [1986])
(1) *Every Σ^b_1-definable function of S^1_2 is polynomial time computable.*
(2) *Let $i \geq 1$. Every Σ^b_i-definable function of S^i_2 is in the i-th level, $\square^p_i$, of the polynomial hierarchy.*

Corollary. (Buss [1986])
(1) *Every Δ^b_1-definable predicate of S^1_2 is polynomial time.*
(2) *Let $i \geq 1$. Every Δ^b_i-definable predicate of S^i_2 is in the i-th level, Δ^p_i, of the polynomial hierarchy.*

The corollary follows from the theorem by exactly the same argument as was used to prove Corollary 3.1.4 from Theorem 3.1.1. To prove the theorem, we shall use a witnessing argument analogous to the one use for $I\Sigma_1$ above. First, we need a revised form of the *Witness* predicate; unlike the usual definition of the *Witness* predicate for bounded arithmetic formulas, we define the *Witness* predicate only for prenex formulas, since this provides some substantial simplification. This simplification is obtained without loss of generality since every Σ^b_i-formula is logically equivalent to a Σ^b_i-formula in prenex form.

3.2.1. Definition. Fix $i \geq 1$. Let $A(\vec{c})$ be a Σ^b_i-formula which is in prenex form. Then $Witness^i_A(w, \vec{c})$ is defined by induction on the complexity of A as follows:

(1) If A is a Π^b_{i-1}-formula, then $Witness^i_A(w, \vec{c})$ is just the formula $A(\vec{c})$,

(2) If $A(\vec{c})$ is not in Π^b_{i-1} and is of the form $(\exists x \leq t)B(\vec{c}, x)$, then $Witness^i_A(w, \vec{c})$ is the formula

$$Witness^i_{B(\vec{c},b)}(\beta(2, w), \vec{c}, \beta(1, w)) \wedge \beta(1, w) \leq t.$$

Intuitively, a witness for $(\exists x \leq t)B(x)$ is a pair w, the first element of the pair giving a value for x and the second element witnessing the truth of $B(x)$ for that value of x.

(3) If $A(\vec{c})$ is not in Π^b_{i-1} and is of the form $(\forall x \leq |t|)B(\vec{c}, x)$, then $Witness^i_A(w, \vec{c})$ is the formula

$$(\forall x \leq |t|)\, Witness^i_{B(\vec{c},d)}(\beta(x+1, w), \vec{c}, x).$$

Intuitively, a witness for $(\forall x \leq |t|)B(x)$ is a sequence w of length $|t|+1$, $w = \langle w_0, w_1, \ldots, w_{|t|}\rangle$ such that each w_x witnesses the truth of $B(x)$, for $0 \leq x \leq |t|$.

Lemma. *Let $i \geq 1$ and $A \in \Sigma_i^b$. Then*
(a) *$Witness_A^i$ is Δ_i^b with respect to S_2^1. If $i > 1$, then $Witness_A^i$ is a Π_{i-1}^b-formula.*
(b) *$Witness_A^i$ defines a Δ_i^p-predicate.*
(c) *S_2^i proves*

$$A(\vec{c}) \leftrightarrow (\exists w)\, Witness_A^i(w, \vec{c}).$$

(d) *There is a term t_A and a polynomial time, Σ_1^b-definable function g_A such that S_2^1 proves*

$$Witness_A^i(w, \vec{c}) \supset g_A(w) < t_A(\vec{c}) \wedge Witness_A^i(g_A(w), \vec{c}).$$

The lemma is easily proved by induction on the complexity of A. For part (d), the function $g_A(w)$ merely computes a succinct Gödel number of w; this just involves removing unnecessary leading zeros and removing unnecessary elements from the witness.

We extend the witness predicate to cedents of prenex form formulas as follows. If $\Delta = \Delta', A$ is a succedent, then $Witness_{\bigvee\Delta}^i(w, \vec{c})$ is defined to be

$$Witness_A^i(\beta(1, w), \vec{c}) \vee Witness_{\bigvee\Delta'}^i(\beta(2, w), \vec{c}).$$

Dually, if $\Gamma = A, \Gamma'$ is an antecedent, then $Witness_{\bigwedge\Gamma}^i$ is defined similarly as

$$Witness_A^i(\beta(1, w), \vec{c}) \wedge Witness_{\bigwedge\Gamma'}^i(\beta(2, w), \vec{c}).$$

3.2.2. Proof. (Proof sketch for Theorem 3.2.) We shall prove Theorem 3.2 by proving a slightly more general witnessing lemma that applies to sequents of Σ_i^b-formulas. Although the lemma holds for sequents of general Σ_i^b-formulas, we state it only for formulas in prenex form, since this simplifies the *Witness* predicate and the proof.

Witnessing Lemma for S_2^i. *Let $i \geq 1$. Let $\Gamma \longrightarrow \Delta$ be a sequent of formulas in Σ_i^b in prenex form, and suppose S_2^i proves $\Gamma \longrightarrow \Delta$. Let $\vec{c}$ include all free variables in the sequent. Then there is a $\square_i^p$-function $h(w, \vec{c})$ which is Σ_i^b-defined in S_2^i such that S_2^i proves*

$$Witness_{\bigwedge\Gamma}^i(w, \vec{c}) \longrightarrow Witness_{\bigvee\Delta}^i(h(w, \vec{c}), \vec{c}).$$

The proof of the Witnessing Lemma is by induction on the number of sequents in a free-cut free proof P of $\Gamma \longrightarrow \Delta$. Since every Σ_i^b-formula is equivalent to a Σ_i^b-formula in prenex form, we may assume w.l.o.g. that every induction formula in the free-cut free proof P is a prenex form Σ_i^b-formula. Then, by the subformula property, every formula appearing anywhere in the proof is also a Σ_i^b-formula in prenex form. The base case of the induction proof is when $\Gamma \longrightarrow \Delta$ is an initial sequent; in this case, every formula in the sequent is atomic, so the Witnessing Lemma trivially holds. The induction step splits into cases depending on the final inference of the proof. The structural inferences and the propositional inferences are essentially trivial, the latter because of our assumption that all formulas are in prenex form. So it remains to consider the quantifier inferences and the induction inferences. The cases where the final inference of P is an $\exists\le$*:right* inference or an $\exists\le$*:left* inference are similar to the $\exists$*:right* and $\exists$*:left* cases of the proof of the Witnessing Lemma 3.1.3 for $I\Sigma_i$, so we omit these cases too.

Now suppose the final inference of P is a Σ_i^b-PIND inference:

$$\frac{A(\lfloor\frac{1}{2}b\rfloor),\Gamma\longrightarrow\Delta,A(b)}{A(0),\Gamma\longrightarrow\Delta,A(t)}$$

where $A\in\Sigma_i^b\setminus\Pi_{i-1}^b$ and where b is the eigenvariable and does not occur in the lower sequent. The induction hypothesis gives a Σ_i^b-defined, $\square_i^p$-function $g(w,\vec{c},b)$ such that S_2^i proves

$$Witness^i_{\bigwedge\{A(\lfloor\frac{1}{2}b\rfloor),\Gamma\}}(w,\vec{c},b)\longrightarrow Witness^i_{\bigvee\{\Delta,A(b)\}}(g(w,\vec{c},b),\vec{c},b).$$

Let $k(\vec{c},v,w)$ be defined as

$$k(\vec{c},v,w) = \begin{cases} v & \text{if } Witness^i_{\bigvee\Delta}(v,\vec{c}) \\ w & \text{otherwise}\end{cases}$$

Since $Witness^i$ is a Δ_i^b-predicate, k is Σ_i^b-defined by S_2^i; and since $Witness^i$ is in Δ_i^p, k is in $\square_i^p$. Define the $\square_i^p$-function $f(w,\vec{c},b)$ by

$$\begin{aligned} f(w,\vec{c},0) &= \langle\beta(1,w),0\rangle \\ f(w,\vec{c},b) &= \langle\beta(1,g(\langle\beta(1,f(w,\vec{c},\lfloor\tfrac{1}{2}b\rfloor)),\beta(2,w)\rangle,\vec{c},b)), \\ &\qquad k(\vec{c},\beta(2,f(w,\vec{c},\lfloor\tfrac{1}{2}b\rfloor)),\beta(2,g(\langle\beta(1,f(w,\vec{c},\lfloor\tfrac{1}{2}b\rfloor)),\beta(2,w)\rangle,\vec{c},b)))\rangle \\ &\qquad \text{for } b>0. \end{aligned}$$

Since f is defined by limited recursion on notation from g, and since g is Σ_i^b-defined by S_2^i, f is also Σ_i^b-defined by S_2^i. Therefore, f may be used in induction formulas and S_2^i can prove

$$Witness^i_{\bigwedge\{A(0),\Gamma\}}(w,\vec{c})\longrightarrow Witness^i_{\bigvee\{\Delta,A(b)\}}(f(w,\vec{c},b),\vec{c},b).$$

using Σ_i^b-PIND with respect to b. Setting $h(w,\vec{c})=f(w,\vec{c},t)$ establishes the desired conditions of the Witnessing Lemma.

Finally, we consider the inferences involving bounded universal quantifiers. The cases where the principal formula of the inference is a Π_{i-1}^b-formula are essentially trivial, since such formulas do not require a witness value, i.e., they are their own witnesses. This includes any inference where the principal connective is a non-sharply bounded universal quantifier. A $\forall\leq$*:left* where the principal formula is in Σ_i^b must have a sharply bounded universal quantifier as its principal connective; this case of the Witnessing Lemma is fairly simple and we leave it to the reader. Finally, we consider the case where the last inference of P is a $\forall\leq$*:right* inference

$$\frac{\begin{array}{c}\ddots\vdots\cdot^{\cdot^{\cdot}}\\ b\leq|t|,\Gamma\longrightarrow\Delta,A(b)\end{array}}{\Gamma\longrightarrow\Delta,(\forall x\leq|t|)A(t)}$$

where $A\in\Sigma_i^b\setminus\Pi_{i-1}^b$ and where the eigenvariable b occurs only as indicated. The induction hypothesis gives a Σ_i^b-defined, $\square_i^p$-function g such that S_2^i proves

$$\begin{aligned}&b\leq|t|\wedge Witness^i_{\bigwedge\Gamma}(\beta(2,w),\vec{c})\longrightarrow\\&\qquad Witness^i_{A(b)}(\beta(1,g(w,\vec{c},b)),\vec{c},b)\vee Witness^i_{\bigvee\Delta}(\beta(2,g(w,\vec{c},b)),\vec{c}).\end{aligned}$$

Let $f_1(w,\vec{c})$ be defined to equal $\beta(2,g(w,\vec{c},b))$ for the least value $b\leq|t|$ such this value witnesses $\bigvee\Delta$, or if there is no such value of $b\leq t$, let $f_1(w,\vec{c})=0$. Since the predicate $Witness^i_{\bigvee\Delta}(\beta(2,g(w,\vec{c},b)),\vec{c})$ is Δ_i^p and is Δ_i^b-defined by S_2^i, the function f_1 is in $\square_i^p$ and is Σ_i^b-defined by S_2^i. Also, let $f_2(w,\vec{c})$ equal

$$\langle\beta(1,g(w,\vec{c},0)),\beta(1,g(w,\vec{c},1)),\beta(1,g(w,\vec{c},2)),\ldots,\beta(1,g(w,\vec{c},|t|))\rangle\,.$$

It is easy to verify that f_2 also is in $\square_i^p$ and is Σ_i^b-definable by S_2^i. Now let $h(w,\vec{c})$ equal $\langle f_2(\langle 0,w\rangle,\vec{c}),f_1(\langle 0,w\rangle,\vec{c})\rangle$. It is easy to check that h satisfies the desired conditions of the Witnessing Lemma.
Q.E.D. Witnessing Lemma and Theorem 3.2.

3.3. Witnessing theorems and conservation results for T_2^i

This section takes up the question of the definable functions of T_2^i. For these theories, there are three witnessing theorems, one for each of the Σ_i^b-definable, the Σ_{i+1}^b-definable and the Σ_{i+2}^b-definable functions. In addition, there is a close connection between S_2^{i+1} and T_2^i; namely, the former theory is conservative over the latter. We'll present most of these results without proof, leaving the reader to look up the original sources.

The results stated in this section will apply to T_2^i for $i\geq 0$; however, for $i=0$, the bootstrapping process does not allow us to introduce many simple functions. Therefore, when $i=0$, instead of T_2^0, we must use the theory $PV_1=T_2^0(\square_1^p)$ as defined in section 1.3.7. To avoid continually treating $i=0$ as a special case, we let T_2^0 denote PV_1 for the rest of this section.

3.3.1. The Σ_{i+1}^b-definable functions of T_2^i

Theorem. (Buss [1990]) *Let* $i\geq 0$.
(1) T_2^i *can* Σ_{i+1}^b*-define every* $\square_{i+1}^p$*-function.*

(2) *Conversely, every Σ_{i+1}^b-definable function of T_2^i is a $\square_{i+1}^p$-function.*
(3) *S_2^{i+1} is Σ_{i+1}^b-conservative over T_2^i.*
(4) *S_2^{i+1} is conservative over $T_2^i + \Sigma_{i+1}^b$-replacement with respect to Boolean combinations of Σ_{i+1}^b formulas.*

We'll just state some of the main ideas of the proof of this theorem, and let the reader refer to Buss [1990] or Krajíček [1995] for the details. Part (1) with $i = 0$ is trivial because of the temporary convention that T_2^0 denotes PV_1. To prove part (1) for $i > 0$, one shows that T_2^i can "Q_i-define" every $\square_{i+1}^p$ formula, where Q_i-definability is a strong form of Σ_{i+1}^b-definability. The general idea of a Q_i-definable function is that it is computed by a Turing machine with a Σ_i^p-oracle such that every "yes" answer of the oracle must be supported by a witness. In the correct computation, the sequence of "yes/no" answers is maximum in a lexicographical ordering, and thus T_2^i can prove that the correct computation exists using Σ_i^b-MAX axioms (which can be derived similarly to minimization axioms). This proof of (1) is reminiscent of the theorem of Krentel [1988] that MINSAT is complete for $\square_1^p$.

Part (2) of the theorem is immediate from Theorem 3.2 and the fact that $T_2^i \subseteq S_2^{i+1}$. Part (3) is based on the following strengthening of the Witnessing Theorem 3.2.2 for S_2^{i+1}:

Witnessing Lemma for S_2^{i+1}. *Let $i \geq 1$. Let $\Gamma \longrightarrow \Delta$ be a sequent of formulas in Σ_{i+1}^b in prenex form, and suppose S_2^{i+1} proves $\Gamma \longrightarrow \Delta$; let $\vec{c}$ include all free variables in the sequent. Then there is a $\square_{i+1}^p$-function $h(w, \vec{c})$ which is Q_i-defined in T_2^i such that T_2^i proves*

$$Witness_{\bigwedge\Gamma}^{i+1}(w, \vec{c}) \longrightarrow Witness_{\bigvee\Delta}^{i+1}(h(w, \vec{c}), \vec{c}).$$

The proof of this Witnessing Lemma is almost exactly the same as the proof of the Witnessing Lemma in section 3.2.2; the only difference is that the witnessing functions are now proved to be Q_i-definable in T_2^i. In fact, (1) implies the necessary functions are Q_i-defined by T_2^i since we already know they are Σ_{i+1}^b-defined by S_2^{i+1}. So the main new aspect is showing that T_2^i can prove that the witnessing functions work.

Part (3) of the theorem is an immediate consequence of the Witnessing Lemma. Part (4) can also be obtained from the Witnessing Lemma using the fact that $T_2^i + \Sigma_{i+1}^b$-replacement can prove that $A(\vec{c})$ is equivalent to $(\exists w)\, Witness_A^{i+1}(w, \vec{c})$ for any $A \in \Sigma_{i+1}^b$.

3.3.2. The Σ_{i+2}^b-definable functions of T_2^i

The Σ_{i+2}^b-definable functions of T_2^i can be characterized by the following theorem, due to Krajíček, Pudlák and Takeuti [1991].

KPT Witnessing Theorem. *Let $i \geq 0$. Suppose T_2^i proves*

$$(\forall x)(\exists y)(\forall z \leq t(x))A(y, x, z)$$

where $A \in \Pi_i^b$. *Then there is a* $k > 0$ *and there are* Σ_{i+1}^b*-definable function symbols* $f_1(x)$, $f_2(x, z_1), \ldots, f_k(x, z_1, \ldots z_{k-1})$ *such that* T_2^i *proves*

$$(\forall x)(\forall z_1 \leq t)[A(f_1(x), x, z_1) \vee (\forall z_2 \leq t)[A(f_2(x, z_1), x, z_2)$$
$$\vee(\forall z_3 \leq t)[A(f_3(x, z_1, z_2), x, z_3)$$
$$\vee \cdots \vee (\forall z_k \leq t)[A(f_k(x, z_1, \ldots, z_{k-1}), x, z_k)] \cdots]]]$$

Conversely, whenever the above formula is provable, then T_2^i *can also prove* $(\forall x)(\exists y)(\forall z \leq t)A(y, x, z)$.

The variables x, y and z could just as well have been vectors of variables, since the replacement axioms and sequence coding can be used to combine adjacent like quantifiers. Also, the first half of the theorem holds even if t involves both x and y. The proof of the KPT Witnessing Theorem is now quite simple: by the discussion in section 1.3.7, we can replace each T_2^i by its conservative, universally axiomatized extension PV_{i+1}, and now the theorem is an immediate corollary of the corollary to the generalized Herbrand's theorem in section 2.5.3 of Chapter I.

3.3.2.1. Applications to the polynomial hierarchy. The above theorem has had a very important application in showing an equivalence between the collapse of the hierarchy of theories of bounded arithmetic and the (provable) collapse of the polynomial time hierarchy. This equivalence was first proved by Krajíček, Pudlák and Takeuti [1991]; we state two improvements to their results. (We continue the convention that T_1^0 denotes PV_1.)

Theorem. (Buss [1995], Zambella [1996]) *Let* $i \geq 0$. *If* $T_2^i \vDash S_2^{i+1}$, *then (1)* $T_2^i = S_2$ *and therefore* S_2 *is finitely axiomatized, and (2)* T_2^i *proves the polynomial hierarchy collapses, and in fact, (2.a)* T_2^i *proves that every* Σ_{i+3}^b*-formula is equivalent to a Boolean combination of* Σ_{i+2}^b*-formulas and (2.b)* T_2^i *proves the polynomial time hierarchy collapses to* $\Sigma_{i+1}^p/poly$.

Corollary. S_2 *is finitely axiomatized if and only if* S_2 *proves the polynomial hierarchy collapses.*

Let $g(x)$ be a Σ_1^b-definable function of T_2^i such that for each $n > 0$ there is an $m > 0$ so that $T_2^i \vdash (\forall x)(x > n \supset g(x) > m)$ (for example, $g(n) = |n|$ or $g(n) = ||n||$, etc.) Let $g\Sigma_i^b$-IND denote the axioms

$$A(0) \wedge (\forall x)(A(x) \supset A(x+1)) \supset (\forall z \leq g(x))A(z).$$

Let $\forall\Pi_i^b(\mathbf{N})$ denote the set of all $\forall\Pi_i^b$ sentences (in the language of S_2) true about the standard integers.

3.3.2.2. Theorem. (essentially Krajíček, Pudlák and Takeuti [1991])
If $T_2^i + \forall\Pi_i^b(\mathbb{N}) \models g\Sigma_{i+1}^b$-IND, then the polynomial time hierarchy collapses to $\Delta_{i+1}^p/poly$.

Note that second theorem differs from the first in that there is no mention of the provability of the collapse of the polynomial hierarchy; on the other hand, the second theorem states a stronger collapse. Krajíček, Pudlák and Takeuti [1991] prove the second theorem with $g(n) = |n|$ and without the presence of $\forall\Pi_i^b(\mathbb{N})$: their proof gives the stronger form stated here with only minor modifications.

3.3.3. The Σ_1^b-definable functions of T_2^1

Buss and Krajíček [1994] characterize the Σ_1^b-definable functions of T_2^1 as being precisely the functions which are projections of PLS functions.

Polynomial Local Search. Johnson, Papadimitriou and Yannakakis [1988] defined a *Polynomial Local Search* problem (PLS-problem) L to be a maximization problem satisfying the following conditions: (we have made some inessential simplifications to their definition)

(1) For every instance $x \in \{0,1\}^*$, there is a set $F_L(x)$ of *solutions,* an integer valued cost function $c_L(s,x)$ and a neighborhood function $N_L(s,x)$,

(2) The binary predicate $s \in F_L(x)$ and the functions $c_L(s,x)$ and $N_L(s,x)$ are polynomial time computable. There is a polynomial p_L so that for all $s \in F_L(x)$, $|s| \le p_L(|x|)$. Also, $0 \in F_L(x)$.

(3) For all $s \in \{0,1\}^*$, $N_L(s,x) \in F_L(x)$.

(4) For all $s \in F_L(x)$, if $N_L(s,x) \neq s$ then $c_L(s,x) < c_L(N_L(s,x),x)$.

(5) The problem is solved by finding a locally optimal $s \in F_L(x)$, i.e., an s such that $N_L(s,x) = s$.

It follows from these conditions that all $s \in F_L(x)$ are polynomial size.

A PLS-problem L can be expressed as a Π_1^b-sentence saying that the conditions above hold; if these are provable in T_2^1 then we say L is a *PLS-problem in T_2^1*. The formula $Opt_L(x,s)$ is the Δ_1^b-formula $N_L(s,x) = s$. A multivalued function g such that for all x, $N_L(g(x),x) = g(x)$, is called a *PLS function*; g must be total, but may be multivalued, since there may exist more than one optimal cost solution. The next theorem states, loosely speaking, that the (multivalued) Σ_1^b-definable functions of T_2^1 are precisely the functions f which can be expressed in the form $f = \pi \circ g$, where g is a PLS function and where π is a polynomial time function (in fact, $\pi(y) = \beta(1,y)$ can always be used).

Theorem. (Buss and Krajíček [1994])
(1) *For every PLS problem L, T_2^1 can prove $(\forall x)(\exists y)Opt_L(x,y)$.*

(2) *If* $A \in \Sigma_1^b$ *and if* T_2^1 *proves* $(\forall\vec{x})(\exists y)A(\vec{x},y)$, *then there is a polynomial time (projection) function* $\pi(y)$ *and a PLS problem* L *such that* T_2^1 *proves*

$$(\forall\vec{x})(\forall y)(Opt_L(\vec{x},y) \supset A(\vec{x},\pi(y))).$$

In other words, if g *is a PLS function solving* L, *then* $A(\vec{x},\pi\circ g(\vec{x}))$ *holds for all* $\vec{x}$ *and all values of* $g(\vec{x})$.

Natural Proofs. The above theorem characterizing the Σ_1^b consequences of T_2^1 in terms of PLS functions was used in an important way to establish the independence of some computational complexity conjectures from $S_2^2(\alpha)$. Razborov and Rudich [1994] introduced a notion of "P-natural proofs" of $P \neq NP$; which intuitively are proofs which provide a polynomial time method of separating out truth tables of Boolean functions that do not have polynomial size circuits. They then showed that under a certain strong pseudo-random number generator conjecture (henceforth: the SPRNG conjecture) that there cannot be P-natural proofs of $P \neq NP$. Razborov [1995] then showed that $S_2^2(\alpha)$ cannot prove superpolynomial lower bounds on the size of circuits for predicates in the polynomial hierarchy unless there are P-natural proofs that $P \neq NP$. This latter condition of course implies the SPRNG conjecture is false; however, most researchers in cryptography apparently do believe the SPRNG conjecture. Thus commonly believed cryptographic conjectures imply that $S_2^2(\alpha)$ cannot prove superpolynomial lower bounds for NP predicates. A further observation of Widgerson is that S_2^2 cannot prove the SPRNG conjecture. Razborov's proof used the conservativity of S_2^2 over T_2^1, and the above characterization of the Σ_1^b-consequences of T_2^1; he then combined this with a communication complexity result (analogous to Craig interpolation) to extract a P-natural proof from the resulting PLS function.

Razborov [1994] has subsequently given a simpler proof of the above-discussed theorem which uses the translations from bounded arithmetic into propositional logic (see Chapter VIII of this volume) plus interpolation theorems for propositional logic. A complete account of this simpler proof can be found in our survey article, Buss [1997].

3.4. Relationships between $B\Sigma_n$ and $I\Sigma_n$

Recall from section 1.2.9, that $B\Sigma_{n+1} \vdash I\Sigma_n \vdash B\Sigma_n$. We show in the next paragraphs that these three theories are distinct and that $B\Sigma_{n+1}$ is conservative over $I\Sigma_n$.

3.4.1. Conservation of $B\Sigma_{n+1}$ over $I\Sigma_n$. In this section we outline a proof of the well-known theorem that the $B\Sigma_{n+1}$ is Π_{n+2}-conservative over $I\Sigma_n$; this was first proved by Parsons [1970]. A model-theoretic proof was later given by Paris and Kirby [1978], and we sketch below a proof-theoretic proof from Buss [1994].

Theorem. *$B\Sigma_{n+1}$ is Π_{n+2}-conservative over $I\Sigma_n$.*

Recall that $B\Sigma_{n+1}$ is equivalent to the theory $B\Pi_n$, which has Π_n-REPL axioms of the form

$$(\forall x \le t)(\exists y)A(x,y) \longrightarrow (\exists z)(\forall x \le t)(\exists y \le z)A(x,y)$$

where $A \in \Pi_n$. In the above sequent, there are unbounded quantifiers in the scope of bounded quantifiers, so the formula in the antecedent is a Σ^+_{n+1}-formula, not a Σ_{n+1}-formula.

Definition. Fix n and suppose $A \in \Sigma^+_{n+1}$.
(1) If $A \in \Pi^+_n$, then $A^{\leqslant s}$ is defined to be A.
(2) If A is $(\exists x)B$ and $A \notin \Pi^+_n$, then $A^{\leqslant s}$ is defined to be $(\exists x \le s)B$.
(3) If A is $(Qx \le t)B$ then $A^{\leqslant s}$ is defined to be $(Qx \le t)(B^{\leqslant s})$.
Let $\Gamma \longrightarrow \Delta$ be a sequent $A_1, \ldots, A_k \longrightarrow B_1, \ldots, B_\ell$ of Σ^+_{n+1}-formulas. Then $\Gamma^{\leqslant s}$ is the formula $\bigwedge_{i=1}^k A_i^{\leqslant s}$ and $\Delta^{\leqslant s}$ is the formula $\bigvee_{j=1}^\ell B_j^{\leqslant s}$. This notation should cause no confusion since antecedents and succedents are always clearly distinguished.

If $\vec{c} = c_1, \ldots c_s$ is a vector of free variables, then $\vec{c} \le u$ abbreviates the formula $c_1 \le u \wedge \cdots \wedge c_s \le u$. $(\forall \vec{c} \le u)$ and $(\exists \vec{c} \le u)$ abbreviate the corresponding vectors of bounded quantifiers.

Lemma. *Let $n \ge 1$. Suppose $\Gamma \longrightarrow \Delta$ is a sequent of Σ^+_{n+1}-formulas that is provable in $B\Sigma_{n+1}$. Let $\vec{c}$ include all the free variables occurring in $\Gamma \longrightarrow \Delta$. Then*

$$I\Sigma_n \vdash (\forall u)(\exists v)(\forall \vec{c} \le u)\Big(\Gamma^{\leqslant u} \supset \Delta^{\leqslant v}\Big).$$

Intuitively, this theorem is saying that given a bound u on the sizes of the free variables and on the sizes of the witnesses for the formulas in Γ, there is a bound v for the values of a witness for a formula in Δ. The conservation theorem above is an immediate corollary of the lemma, so it remains to prove the lemma.

Proof. We give only a short sketch of the proof of the lemma here; a more detailed version is given in Buss [1994] although the definitions are slightly different there.

Firstly, formulate $B\Pi_n$-proofs in the sequent calculus, using the inference rule form of the Π_n-REPL axioms described in section 1.4.5. Secondly, since $B\Pi_n = B\Sigma_{n+1}$, we may assume there is a $B\Pi_n$-proof P of $\Gamma \longrightarrow \Delta$, and by the Free-cut Elimination Theorem, we may assume that every formula appearing in P is a Σ^+_{n+1}-formula. Thirdly, we shall use induction on the number of sequents in P to prove that the Lemma holds for every sequent in P. The induction step involves a number of cases; we shall do only the two cases where the final inference of P is a replacement inference and where the final inference of P is a $\forall$*:right* inference. The latter case is the hardest of all the cases; the rest of the cases are left to the reader.

Suppose the final inference of P is a replacement inference:

$$\frac{\begin{array}{c}\vdots\\ \Gamma \rightarrow \Delta, (\forall x \le t)(\exists y)A(x,y)\end{array}}{\Gamma \rightarrow \Delta, (\exists z)(\forall x \le t)(\exists y \le z)A(x,y)}$$

The induction hypothesis is that $I\Sigma_n$ proves

$$(\forall u)(\exists v)(\forall \vec{c} \le u)[\Gamma^{\le u} \supset \Delta^{\le v} \vee (\forall x \le t)(\exists y \le v)A(x,y)].$$

From this, the desired result that

$$(\forall u)(\exists v)(\forall \vec{c} \le u)[\Gamma^{\le u} \supset \Delta^{\le v} \vee (\exists z \le v)(\forall x \le t)(\exists y \le z)A(x,y)].$$

follows immediately.

Now suppose that P ends with a $\forall$*:right* inference:

$$\frac{\begin{array}{c}\vdots\\ \Gamma \rightarrow \Delta, B(\vec{c},d)\end{array}}{\Gamma \rightarrow \Delta, (\forall x)B(\vec{c},x)}$$

Note that $B \in \Pi_n^+$ since $(\forall x)B$ is a Σ_{n+1}^+-formula. We reason inside $I\Sigma_n$. Let u be arbitrary. By strong Σ_n-replacement (see the end of section 1.2.9) there is a $u' \ge u$ such that

$$(\forall \vec{c} \le u)\Big((\forall x)B(\vec{c},x) \leftrightarrow (\forall x \le u')B(\vec{c},u')\Big).$$

Let $v \ge u'$ be given by the induction hypothesis so that

$$(\forall \vec{c}, d \le u')\Big(\Gamma^{\leqslant u'} \supset \Delta^{\leqslant v} \vee B(\vec{c},d)\Big).$$

Now let $\vec{c} \le u$ be arbitrary such that $\Gamma^{\leqslant u}$. We need to show $\Delta^{\leqslant v} \vee (\forall x)B(\vec{c},x)$. Suppose that $(\forall x)B(\vec{c},x)$ is false: then there is a $d \le u'$ such that $\neg B(\vec{c},d)$, and by the induction hypothesis, $\Delta^{\leqslant v}$ holds. Thus $\Delta^{\leqslant v} \vee (\forall x)B(\vec{c},x)$ holds.

3.4.2. $I\Sigma_{n+1}$ properly contains $B\Sigma_{n+1}$

Theorem. (Parsons [1970]) *Let $n \ge 1$. $I\Sigma_{n+1}$ is not equal to $B\Sigma_{n+1}$.*

Proof. We'll give only a quick sketch of a proof-theoretic proof based on Gödel's second incompleteness theorem; see Paris and Kirby [1978] for a model-theoretic proof. The two main steps in our proof are:

(1) $I\Sigma_1 \vdash Con(I\Sigma_n) \supset Con(B\Sigma_{n+1})$. This is proved by formalizing, in $I\Sigma_1$, the proof of Theorem 3.4.1 sketched above. That proof was quite constructive: any $B\Sigma_{n+1}$ proof of a Σ_n-formula can be transformed into a free-cut free proof by a primitive recursive process. Then the transformation of the free-cut free $B\Sigma_{n+1}$-proof into a $I\Sigma_n$-proof, as in the proof of Lemma 3.4.1, is primitive recursive (in fact it is polynomial time).
Therefore, $I\Sigma_1$ proves that if $B\Sigma_{n+1}$ proves a contradiction, $0 = 1$, then so does $I\Sigma_n$; i.e., $I\Sigma_1$ proves that if $B\Sigma_{n+1}$ is inconsistent, then so is $I\Sigma_n$.

(2) $I\Sigma_{n+1} \vdash Con(I\Sigma_n)$. To prove this, first note that, since $I\Sigma_1$ can prove the free-cut elimination theorem, it is sufficient to prove that $I\Sigma_{n+1}$ can prove that there is no free-cut free $I\Sigma_n$-proof of $0 = 1$; in particular, it suffices to show that $I\Sigma_{n+1}$ can prove that there is no $I\Sigma_n$ sequent calculus proof of $0 = 1$ in which every formula is a Σ_n-formula. Second, $I\Sigma_{n+1}$ has a truth definition for Σ_n-formulas; i.e., there is a formula $Tr(x, y)$ such that when x is the Gödel number of a Σ_n-formula and y is a sequence encoding values for the free variables of the formula, then $Tr(x, y)$ defines the truth of the formula for those values. In addition, $I\Sigma_{n+1}$ can prove that the truth definition satisfies the usual properties of truth, in that it obeys the meanings of the logical connectives. Chapter VIII discusses truth definitions in depth, and the reader should refer to that for more details. Third, using the truth definition for Σ_n-formulas, $I\Sigma_{n+1}$ can prove, by induction on the number of lines in the free-cut free $I\Sigma_n$-proof, that every sequent in the proof is valid. Therefore, it cannot be a proof of $0 = 1$, since that is not valid. So by this means, $I\Sigma_{n+1}$ proves the consistency of $I\Sigma_n$.

(1) and (2) immediately that $I\Sigma_{n+1}$ proves the consistency of $B\Sigma_{n+1}$; therefore, by Gödel's incompleteness theorem, $I\Sigma_{n+1}$ is not equal to $B\Sigma_{n+1}$.

3.4.3. $B\Sigma_{n+1}$ properly contains $I\Sigma_n$. The fact that $I\Sigma_n$ does not prove the Σ_{n+1}-replacement axioms was first established independently by Lessan [1978] and Paris and Kirby [1978]. Their proofs were model-theoretic; Kaye [1993] gave a proof-theoretic proof based on an argument analogous to the proof of Theorem 3.3.2.1 using a Herbrand-style nocounterexample interpretation.

4. Strong incompleteness theorems for $I\Delta_0 + \exp$

4.1. Gödel's Second Incompleteness Theorem states that a sufficiently strong, axiomatized, consistent theory T cannot prove its own consistency. One way to strengthen this incompleteness theorem is by working with two theories, S and T, such that S is a subtheory of T: in some cases, one can establish that T cannot prove the consistency of its subtheory S.

There are many cases in which this strengthening of the second incompleteness theorem can be achieved. One important situation is where T is conservative over S; for example, $B\Sigma_{n+1}$ cannot prove the consistency of $I\Sigma_n$, since $B\Sigma_{n+1}$ is conservative over $I\Sigma_n$ and the latter theory cannot prove its own consistency. A second important example is where S is interpretable in T and thus $Con(S) \supset Con(T)$; for example, it is known that S_2 is interpretable in Q (see Wilkie and Paris [1987] and Nelson [1986]) and therefore S_2 cannot prove $Con(Q)$.

A third example, and the one that is the main subject of this section, is the theorem of Wilkie and Paris [1987] that $I\Delta_0 + \exp$ cannot prove the consistency of Q. This example does not fall into either of the above examples, since $I\Delta_0 + \exp$ is not interpretable in Q.

4.2. Before beginning a discussion of the proof that $I\Delta_0 + \exp$ does not prove *Con(Q)*, we discuss a few other extensions of the second incompleteness theorem. The first extension is that the second incompleteness theorem applies also to restricted notions of provability, such as "bounded consistency" and "Σ_k-consistency".

Definition. Let S be a theory, formalized in the sequent calculus. We say that S is *bounded consistent* if there is no S-proof of the empty sequent $\longrightarrow$ in which only bounded formulas appear. For $k \geq 0$, S is Σ_k-consistent provided there is no S-proof of the the empty sequent in which only Σ_k-formulas appear. S is *free-cut free consistent* if there is no formula A such that S has free-cut free proofs of both $\longrightarrow A$ and $A \longrightarrow$.

The formulas *BdCon*(S), $Con_{\Sigma_k}(S)$ and *FCFCon*(S) are $\forall\Pi_1^b$-formulas which express the bounded consistency, the Σ_k-consistency and the free-cut free consistency of S, respectively.

Of course, the cut-elimination theorem implies that a bounded theory S satisfies these three notions of consistency if and only if S is consistent in the usual sense; however, since the cut-elimination theorem is not provable in weak theories where the superexponentiation function is not provably total, these three new notions of consistency will not generally be provably equivalent to each other or to *Con*(S).

Definition. We say that a proof is *bounded* provided every formula in the proof is Δ_0. Similarly a proof is Σ_k, if every formula in the proof is in Σ_k. We write $S \vdash_{\Delta_0} A$ and $S \vdash_{\Sigma_k} A$ to denote the condition that A has a sequent calculus S-proof in which every formula is in Δ_0 or Σ_k, respectively.

Buss [1986] proved that if S is a bounded theory (such as $I\Delta_0$, S_2^i, T_2^i, S_2, etc) then S cannot prove its own free-cut free consistency; i.e., S cannot prove *FCFCon*(S) and hence $S \nvdash$ *BdCon*(S) and $S \nvdash \Sigma_k$-*Con* for all $k > 0$. This was later strengthened to show that S_2 does not prove *BdCon*$(BASIC)$, the bounded consistency of its induction-free base theory: this result first appeared in Takeuti [1990] and Buss and Ignjatović [1995] building on the earlier work of Pudlák [1990]. A related result, proved by Buss and Ignjatović [1995], is that the theory PV (and hence S_2^1) does not prove the consistency of a finitely axiomatized, induction-free fragment PV^- of PV.

Like the theorem of Wilkie and Paris [1987] that we discuss below, these independence results provide situations where a stronger theory cannot prove a consistency statement about a weaker theory. These results are interesting in their own right of course; but in addition, they are motivated by a yet-unfulfilled hope that independence results of these kinds could lead to a proof that $P \neq NP$. This wistful hope is based on the intuitive idea that $P \neq NP$ is analogous to a finitary incompleteness theorem.

4.2.1. More strengthenings of Gödel's second incompleteness theorem can be found in Chapters VII and VIII of this volume.

4.3. A theorem of Wilkie and Paris

This section outlines a proof of the theorem that $I\Delta_0 + \exp$ cannot prove the consistency of $I\Delta_0$. Earlier, we used the notation $exp(x, y, z)$ as a Δ_0-predicate expressing the condition that $x^y = z$. We also define "exp" to be the sentence

$$(\forall x)(\forall y)(\exists z)exp(x, y, z)$$

stating that the exponentiation function is total.

Theorem. (Wilkie and Paris [1987]) *$I\Delta_0 + \exp$ cannot prove $Con(I\Delta_0)$. Therefore, $I\Delta_0 + \exp$ cannot prove $Con(Q)$.*

It is worth noting some theories that are sufficiently strong to prove the consistency of $I\Delta_0$. First, if one considers bounded consistency, we have that

$$I\Delta_0 + \exp \vdash BdCon(I\Delta).$$

To prove this fact, one shows that $I\Delta_0 + \exp$ can define a truth definition for bounded formulas which is sufficient to allow $I\Delta_0 + \exp$ to prove the validity of every sequent which has a bounded Δ_0-proof.

Second, let 2^x_k be the superexponentiation function defined by $2^x_0 = x$ and $2^x_{i+1} = 2^{2^x_i}$. By the bootstrapping techniques used earlier, there is a Δ_0-formula $superexp(i, x, z)$ which intensionally expresses $2^x_i = z$. Similarly, $superexp(i, x, z)$ is Δ^b_1-definable with respect to S^1_2. We let "superexp" be the axiom stating

$$(\forall x)(\forall y)(\exists z)superexp(x, y, z).$$

Since $I\Delta_0 + \text{superexp}$ can prove the free-cut elimination theorem, it can also prove that $BdCon(I\Delta_0)$ implies $Con(I\Delta_0)$. Therefore, $I\Delta_0 + \text{superexp} \vdash Con(I\Delta_0)$.

4.3.1. We are now ready to outline the proof of Theorem 4.3. It is more convenient to work with S_2 instead of $I\Delta_0$ and so we shall prove that $S_2 + \exp \nvdash Con(S_2)$. Note that we still have $S_2 + \exp \vdash BdCon(S_2)$. Also, S^1_2 (and $I\Delta_0 + \Omega_1$) proves $Con(Q) \supset Con(S_2)$, so $Con(S_2)$ and $Con(I\Delta_0)$ are equivalent.[15] We shall prove Theorem 4.3, by proving a series of lemmas, theorems and corollaries, namely, Lemma 4.3.2 through Theorem 4.3.10. The proof is a modified version of the original proof of Wilkie and Paris [1987].

4.3.2. Lemma. *Let $\phi(x)$ be a Σ_1-formula, and suppose $S_2 + \exp \vdash (\forall x)\phi(x)$. Then there is a constant $k > 0$ such that*

$$S_2 \vdash (\forall x)(2^x_k \text{ exists} \supset \phi(x)).$$

To improve readability, we shall often write, as above, a shorthand notation such as "2^x_k exists" as an abbreviation for $(\exists y)superexp(k, x, y)$.

[15] The proof below that $S_2 + \exp \nvdash Con(S_2)$ can be understood without knowing how to prove in S^1_2 that $I\Delta_0$ and S_2 are equiconsistent.

Proof. Without loss of generality, $\phi(x)$ is of the form $(\exists u)\phi_M(x,u)$ with $\phi_M \in \Delta_0$. The hypothesis implies that S_2 proves $(\forall y)(\exists z)(2^y = z) \supset (\forall x)\phi(x)$, which can be put in prenex form as

$$(\forall x)(\exists u)(\exists y)(\forall z)[2^y \neq z \vee \phi_M(x,u)].$$

Now, momentarily enlarge S_2 to have Skolem functions for all Δ_0 formulas, thereby making S_2 axiomatized by purely universal formulas. The strong form of Herbrand's theorem (section 2.5.3 of Chapter I) implies that there is an $\ell > 0$ and there are terms $s_1(u)$, $t_1(x)$, $s_2(x,u_1,y_1)$, $t_2(x,u_1,y_1), \ldots,$ $s_\ell(x,u_1,\ldots,u_{\ell-1},y_1,\ldots,y_{\ell-1})$, $t_\ell(x,u_1,\ldots,u_{\ell-1},y_1,\ldots,y_{\ell-1})$ so that S_2 proves

$$(\forall x)[(\forall z_1)[2^{t_1(x)} \neq z_1 \vee \phi_M(x,s_1(x)) \vee (\forall z_2)[2^{t_2(x,z_1)} \neq z_2 \vee \phi_M(x,s_2(x,z_1)) \vee \\ \cdots \vee (\forall z_\ell)[2^{t_2(x,z_1,\ldots,z_{\ell-1})} \neq z_\ell \vee \phi_M(x,s_2(x,z_1,\ldots,z_{\ell-1}))]\cdots]]].$$

Since $\neg\phi \supset \neg\phi_M(x,s_i(x,\vec{z}))$, we immediately have that S_2 also proves

$$(\forall x)[\phi(x) \vee (\forall z_1)(2^{t_1(x)} \neq z_1 \vee (\forall z_2)(2^{t_2(x,z_1)} \neq z_2 \vee \\ \cdots \vee (\forall z_\ell)(2^{t_\ell(x,z_1,\ldots,z_{\ell-1})} \neq z_\ell)\cdots))]$$

where each t_i is a function with polynomial growth rate with graph defined by a Δ_0-formula. Thus, S_2 proves that $\phi(x)$ holds, provided there exists $z_1 = 2^{t_1(x)}$, $z_2 = 2^{t_2(x,z_1)}, \ldots,$ $z_\ell = 2^{t_\ell(x,z_1,\ldots,z_{\ell-1})}$. Since each t_i has polynomial growth rate, the values of the z_i's are bounded by $2^x_{\ell+1}$ for sufficiently large $x \in \mathbb{N}$; therefore, S_2 proves that if $2^x_{\ell+1}$ exists, then $\phi(x)$ holds. Taking $k = \ell+1$, Lemma 4.3.2 is proved.

4.3.3. Theorem. (Solovay [1976]) *For each $n, k \geq 0$, there is a S^1_2-proof P of $(\exists x)(superexp(\underline{k},\underline{n},x))$ with size polynomially bounded in terms of $|n|$ and k. In addition, P is a Σ_{2k+1}-proof.*

Proof. The proof is based on using formulas that define inductive cuts. The particular ones we need are formulas $J_i(x)$ and $K_i(x)$ defined as:

$$\begin{aligned} J_0(x) &\Leftrightarrow 0=0 \quad \text{(always true)} \\ K_0(x) &\Leftrightarrow (\exists y)(2^x = y) \\ J_{i+1}(x) &\Leftrightarrow (\forall z)(K_i(z) \supset K_i(z+x)) \\ K_{i+1}(x) &\Leftrightarrow (\exists y)(2^x = y \wedge J_{i+1}(y)) \end{aligned}$$

Lemma.
(a) $S^1_2 \vdash J_k(0)$
(b) $S^1_2 \vdash J_k(x) \supset J_k(x+1)$

(c) $S_2^1 \vdash J_k(x) \wedge u < x \supset J_k(u)$

(d) $S_2^1 \vdash J_k(x) \supset J_k(x + x)$

(e) $S_2^1 \vdash K_k(0)$

(f) $S_2^1 \vdash K_k(x) \wedge u < x \supset K_k(u)$

(g) $S_2^1 \vdash K_k(x) \supset K_k(x + 1)$

(h) $S_2^1 \vdash K_k(x) \supset (\exists z) superexp(\underline{k+1}, x, z)$.

Parts (a)-(g) are proved simultaneously by induction on k. Part (h) is likewise proved using induction on k. Moreover, it is easy to verify that the S_2-proofs of formulas (a)-(g) are polynomial size in k, and involve only Σ_{2k+1}-formulas.

By using (d) and (c) of the lemma, it is straightforward now to give find an S_2^1-proof of $J_k(\underline{n})$ of size polynomial in $|n|$ and k; from this, (e) and (h) give the desired proof of P of $(\exists z) superexp(\underline{k}, \underline{n}, z)$.

4.3.4. Lemma. *Suppose $\phi(x) \in \Sigma_1$ and $S_2 + \exp \vdash (\forall x)\phi(x)$. Then, there is a $k \geq 0$ such that*

$$S_2^1 \vdash (\forall x)\left(S_2 \vdash_{\overline{\Sigma_k}} \phi(\underline{x})\right).$$

Lemma 4.3.4 is proved from Lemma 4.3.2 by formalizing the argument of Lemma 4.3.3 in S_2^1.

4.3.5. Lemma. *Let $\phi(x)$ be a $\forall\Pi_1^b$-formula, which is without loss of generality of the form $(\forall y)\phi_M(x, y)$ where $\phi_M \in \Pi_1^b$. Then there is a term t such that*

$$S_2^1 \vdash \neg\phi(x) \longrightarrow \left(S_2 \vdash_{\overline{\Delta_0}} (\exists y \leq t(x))\neg\phi_M(\underline{x}, y)\right).$$

This lemma is a special case of Theorem 2.1.2.

4.3.6. Lemma. *Let ϕ be a $\forall\Pi_1^b$-sentence such that $S_2 + \exp \vdash \phi$. Then there is a $k \geq 0$ such that*

$$S_2^1 \vdash \neg\phi \longrightarrow \neg Con_{\Sigma_k}(S_2).$$

Proof. Without loss of generality, ϕ is of the form $(\forall x)\phi_M(x)$ with ϕ_M a Π_1^b-formula. By Lemma 4.3.4, S_2 proves $(\forall x)(S_2 \vdash_{\overline{\Sigma_k}} \phi_M(\underline{x}))$. On the other hand, Lemma 4.3.5 implies that S_2 proves

$$\neg\phi_M(x) \supset \left(S_2 \vdash_{\overline{\Delta_0}} \neg\phi_M(\underline{x})\right).$$

These two facts suffice to prove Lemma 4.3.6.

4.3.7. Lemma. *Let $k > 0$. Then $S_2 + \exp$ proves $Con_{\Sigma_k}(S_2)$.*

Proof. (Sketch). The proof of this has two main steps:

(1) Firstly, one shows that $S_2 + \exp$ proves $BdCon(S_2) \supset Con_{\Sigma_k}(S_2)$. This is done, by formalizing the following argument: (a) Assume that P is a Σ_k-proof of $0 = 1$ in the theory S_2. (b) By using sequence encoding to collapse adjacent like quantifiers, we may assume w.l.o.g. that each formula in P has at most $k+1$ unbounded quantifiers. (c) By applying the process used to prove the Cut-Elimination Theorem 2.4.2 of Chapter I, there is a bounded S_2-proof of $0 = 1$ of size at most $2_{2k+4}^{||P||}$. Since only finitely many iterations of exponentiation are needed, the last step can be formalized in $S_2 + \exp$.

(2) Secondly, one shows that $S_2 + \exp$ can prove the bounded consistency of S_2. The general idea is that if there is a bounded S_2-proof P of $0 = 1$, then there is a a fixed value ℓ so that all variables appearing in P can be implicitly bounded by $L = 2_\ell^{size(P)}$ where $size(P)$ is the number of symbols in P. (In fact, $\ell = 3$ works.) Once all variables are bounded by L, a truth definition can be given based on the fact that $2^{L^{size(P)}}$ exists. With this truth definition, $S_2 + \exp$ can prove that every sequent in the S_2-proof is valid.

4.3.8. Corollary. *The theory $S_2 + \exp$ is conservative over the theory $S_2 \cup \{Con_{\Sigma_k}(S_2) : k \geq 0\}$ with respect to $\forall\Pi_1^b$-consequences.*

Proof. The fact that the first theory includes the second theory is immediate from Theorem 4.3.7. The conservativity is immediate from Lemma 4.3.6.

Incidentally, since S_2 is globally interpretable in Q, we also have that the theories $S_2 + \{Con_{\Sigma_k}(S_2) : k \geq 0\}$ and $S_2 + \{Con_{\Sigma_k}(Q) : k \geq 0\}$ are equivalent.

4.3.9. Theorem. $S_2 \cup \{Con_{\Sigma_k}(S_2) : k \geq 0\} \nvdash Con(S_2)$.

It is an immediate consequence of Theorem 4.3.9 and Corollary 4.3.8 that $S_2 + \exp \nvdash Con(S_2)$, which is the main result we are trying to establish. So it remains to prove Theorem 4.3.9:

Proof. Let $k > 0$ be fixed. Use Gödel's Diagonal Lemma to choose an $\exists\Sigma_1^b$-sentence ϕ_k such that

$$S_2^1 \vdash \phi_k \leftrightarrow (S_2 + Con_{\Sigma_k}(S_2) \vdash_{\overline{\Delta_0}} \neg\phi_k).$$

Now ϕ_k is certainly false, since otherwise $S_2 + Con_{\Sigma_k}(S_2)$ proves $\neg\phi_k$, which would be false if ϕ_k were true. Furthermore, $S_2 + \exp$ can formalize the previous sentence, since as sketched above in the part (2) of the proof of Theorem 4.3.7, $S_2 + \exp$ can prove the validity of every formula appearing in a bounded proof in the theory $S_2 + Con_{\Sigma_k}(S_2)$. Therefore, $S_2 + \exp$ proves $\neg\phi_k$.

Since $\neg\phi_k$ is a $\forall\Pi_1^b$-sentence, Corollary 4.3.8 implies that there is some $m > 0$ such that $S_2 + Con_{\Sigma_m}(S_2) \vdash \neg\phi_k$. It is evident that $S_2 + Con_{\Sigma_k}(S_2)$ cannot prove $Con_{\Sigma_m}(S_2)$ since this would contradict the fact that ϕ_k is false, which implies that

$S_2 + Con_{\Sigma_k}(S_2)$ does not prove $\neg\phi_k$. Therefore $S_2 + Con_{\Sigma_k}(S_2)$ also cannot prove $Con(S_2)$.

Since this argument works for arbitrary k, the Compactness Theorem implies that $S_2 + \{Con_{\Sigma_k}(Q) : k \geq 0\}$ cannot prove $Con(S_2)$.

The proofs above actually proved something slightly stronger than Theorem 4.3.9; namely,

4.3.10. Theorem. *There is an $m = O(k)$ such that $S_2 + Con_{\Sigma_k}(S_2) \nvdash Con_{\Sigma_m}(S_2)$.*

We conjecture that $m = k + 1$ also works, but do not have a proof at hand.

4.3.11. A related result, which was stated as an open problem by Wilkie and Paris [1987] and was later proved by Hájek and Pudlák [1993,Coro. 5.34], is the fact that there is a $\forall\Pi_1^b$-sentence ϕ, such that $S_2^1 + \exp \vdash \phi$ but such that $S_k^1 \nvdash \phi$ for all $k \geq 0$.

Acknowledgements. We are grateful to J. Avigad, C. Pollett, and J. Krajíček for suggesting corrections to preliminary versions of this chapter. Preparation of this article was partially supported by NSF grant DMS-9503247 and by cooperative research grant INT-9600919/ME-103 of the NSF and the Czech Republic Ministry of Education.

References

W. Ackermann

[1941] Zur Widerspruchsfreiheit der Zahlentheorie, *Mathematische Annallen*, 117, pp. 162–194.

J. Avigad and R. Sommer

[1997] A model-theoretic approach to ordinal analysis, *Bulletin of Symbolic Logic*, 3, pp. 17–52.

J. Barwise

[1977] *Handbook of Mathematical Logic*, North-Holland, Amsterdam.

J. H. Bennett

[1962] *On Spectra*, PhD thesis, Princeton University.

G. Boolos

[1989] A new proof of the Gödel incompleteness theorem, *Notices of the American Mathematical Society*, 36, pp. 388–390.

[1993] *The Logic of Provability*, Cambridge University Press.

S. R. Buss

[1986] *Bounded Arithmetic*, Bibliopolis, Napoli. Revision of 1985 Princeton University Ph.D. thesis.

[1990] Axiomatizations and conservation results for fragments of bounded arithmetic, in: *Logic and Computation, proceedings of a Workshop held Carnegie-Mellon University, 1987*, W. Sieg, ed., vol. 106 of Contemporary Mathematics, American Mathematical Society, Providence, Rhode Island, pp. 57–84.

[1992] A note on bootstrapping intuitionistic bounded arithmetic, in: *Proof Theory: A selection of papers from the Leeds Proof Theory Programme 1990*, P. H. G. Aczel, H. Simmons, and S. S. Wainer, eds., Cambridge University Press, pp. 149–169.

[1994] The witness function method and fragments of Peano arithmetic, in: *Proceedings of the Ninth International Congress on Logic, Methodology and Philosophy of Science, Uppsala, Sweden, August 7-14, 1991*, D. Prawitz, B. Skyrms, and D. Westerståhl, eds., Elsevier, North-Holland, Amsterdam, pp. 29–68.

[1995] Relating the bounded arithmetic and polynomial-time hierarchies, *Annals of Pure and Applied Logic*, 75, pp. 67–77.

[1997] Bounded arithmetic and propositional proof complexity, in: *Logic of Computation*, H. Schwichtenberg, ed., Springer-Verlag, Berlin, pp. 67–121.

S. R. Buss and A. Ignjatović

[1995] Unprovability of consistency statements in fragments of bounded arithmetic, *Annals of Pure and Applied Logic*, 74, pp. 221–244.

S. R. Buss and J. Krajíček

[1994] An application of Boolean complexity to separation problems in bounded arithmetic, *Proceedings of the London Mathematical Society*, 69, pp. 1–21.

G. J. Chaitin

[1974] Information-theoretic limitations of formal systems, *J. Assoc. Comput. Mach.*, 21, pp. 403–424.

P. Clote

[1985] Partition relations in arithmetic, in: *Methods in Mathematical Logic*, C. A. Di Prisco, ed., Lecture Notes in Computer Science #1130, Springer-Verlag, Berlin, pp. 32–68.

A. Cobham

[1965] The intrinsic computational difficulty of functions, in: *Logic, Methodology and Philosophy of Science, proceedings of the second International Congress, held in Jerusalem, 1964*, Y. Bar-Hillel, ed., North-Holland, Amsterdam.

S. A. Cook

[1975] Feasibly constructive proofs and the propositional calculus, in: *Proceedings of the Seventh Annual ACM Symposium on Theory of Computing*, Association for Computing Machinery, New York, pp. 83–97.

S. A. Cook and A. Urquhart

[1993] Functional interpretations of feasibly constructive arithmetic, *Annals of Pure and Applied Logic*, 63, pp. 103–200.

S. Feferman

[1960] Arithmetization of metamathematics in a general setting, *Fundamenta Mathematicae*, 49, pp. 35–92.

H. Gaifman and C. Dimitracopoulos

[1982] Fragments of Peano's arithmetic and the MRDP theorem, in: *Logic and Algorithmic: An International Symposium held in honour of Ernst Specker*, Monographie #30 de L'Enseignement Mathématique, pp. 187–206.

G. Gentzen

[1936] Die Widerspruchsfreiheit der reinen Zahlentheorie, *Mathematische Annalen*, 112, pp. 493–565. English translation in: Gentzen [1969], pp. 132-213.

[1938] Neue Fassung des Widerspruchsfreiheitbeweis für der reinen Zahlentheorie, *Forschungen zur Logik end zur Grundlegung der exacten Wissenschaften, New Series*, 4, pp. 19–44. English translation in: Gentzen [1969], pp. 252-286.

[1969] *Collected Papers of Gerhard Gentzen*, North-Holland, Amsterdam. Edited by M. E. Szabo.

J.-Y. Girard

[1987] *Proof Theory and Logical Complexity*, vol. I, Bibliopolis, Napoli.

K. GÖDEL

[1958] Über eine bisher noch nicht benützte Erweiterung des finiten Standpunktes, *Dialectica*, 12, pp. 280–287.

P. HÁJEK AND P. PUDLÁK

[1993] *Metamathematics of First-order Arithmetic*, Perspectives in Mathematical Logic, Springer-Verlag, Berlin.

D. HILBERT AND P. BERNAYS

[1934-39] *Grundlagen der Mathematik, I & II*, Springer, Berlin.

W. A. HOWARD

[1970] Assignment of ordinals to terms for primitive recursive functionals of finite type, in: *Intuitionism and Proof Theory: Proceedings of the Summer Conference at Buffalo N.Y. 1968*, A. Kino, J. Myhill, and R. E. Vesley, eds., North-Holland, Amsterdam, pp. 443–458.

D. S. JOHNSON, C. H. PAPADIMITRIOU, AND M. YANNAKAKIS

[1988] How easy is local search?, *Journal of Computer and System Science*, 37, pp. 79–100.

R. W. KAYE

[1991] *Models of Peano arithmetic*, Oxford Logic Guides #15, Oxford University Press.

[1993] Using Herbrand-type theorems to separate strong fragments of arithmetic, in: *Arithmetic, Proof Theory and Computational Complexity*, P. Clote and J. Krajíček, eds., Clarendon Press (Oxford University Press), Oxford.

C. F. KENT AND B. R. HODGSON

[1982] An arithmetic characterization of NP, *Theoretical Computer Science*, 21, pp. 255–267.

J. KETONEN AND R. M. SOLOVAY

[1981] Rapidly growing Ramsey functions, *Annals of Mathematics*, 113, pp. 267–314.

J. KRAJÍČEK

[1995] *Bounded Arithmetic, Propositional Calculus and Complexity Theory*, Cambridge University Press.

J. KRAJÍČEK, P. PUDLÁK, AND G. TAKEUTI

[1991] Bounded arithmetic and the polynomial hierarchy, *Annals of Pure and Applied Logic*, 52, pp. 143–153.

M. W. KRENTEL

[1988] The complexity of optimization problems, *Journal of Computer and System Sciences*, 36, pp. 490–509.

H. LESSAN

[1978] *Models of Arithmetic*, PhD thesis, Manchester University.

P. LINDSTRÖM

[1997] *Aspects of Incompleteness*, Lecture Notes in Logic #10, Springer-Verlag, Berlin.

R. J. LIPTON

[1978] Model theoretic aspects of computational complexity, in: *Proceedings of the 19th Annual Symposium on Foundations of Computer Science*, IEEE Computer Society, Piscataway, New Jersey, pp. 193–200.

M. H. LÖB

[1955] Solution of a problem of Leon Henkin, *Journal of Symbolic Logic*, 20, pp. 115–118.

E. MENDELSON

[1987] *Introduction to Mathematical Logic*, Wadsworth & Brooks/Cole, Monterey.

G. E. MINTS

[1973] Quantifier-free and one-quantifier systems, *Journal of Soviet Mathematics*, 1, pp. 71–84.

E. NELSON

[1986] *Predicative Arithmetic*, Princeton University Press.

V. A. NEPOMNJAŠČII

[1970] Rudimentary predicates and Turing calculations, *Kibernetika*, 6, pp. 29–35. English translation in *Cybernetics* 8 (1972) 43-50.

R. PARIKH

[1971] Existence and feasibility in arithmetic, *Journal of Symbolic Logic*, 36, pp. 494–508.

J. B. PARIS AND C. DIMITRACOPOULOS

[1982] Truth definitions for Δ_0 formulae, in: *Logic and Algorithmic, Monographie no 30 de L'Enseignement Mathematique*, University of Geneva, pp. 317–329.

J. B. PARIS AND L. HARRINGTON

[1977] A mathematical incompleteness in Peano arithmetic, in: *Handbook of Mathematical Logic*, North-Holland, Amsterdam, pp. 1133–1142.

J. B. PARIS AND L. A. S. KIRBY

[1978] Σ_n-collection schemes in arithmetic, in: *Logic Colloquium '77*, North-Holland, Amsterdam, pp. 199–210.

C. PARSONS

[1970] On a number-theoretic choice schema and its relation to induction, in: *Intuitionism and Proof Theory: Proceedings of the Summer Conference at Buffalo N.Y. 1968*, A. Kino, J. Myhill, and R. E. Vesley, eds., North-Holland, Amsterdam, pp. 459–473.

[1972] On n-quantifier induction, *Journal of Symbolic Logic*, 37, pp. 466–482.

W. POHLERS

[1980] *Proof Theory: An Introduction*, Lecture Notes in Mathematics #1407, Springer-Verlag, Berlin.

P. PUDLÁK

[1983] Some prime elements in the lattice of interpretability types, *Transactions of the American Mathematical Society*, 280, pp. 255–275.

[1990] A note on bounded arithmetic, *Fundamenta Mathematicae*, 136, pp. 85–89.

A. A. RAZBOROV

[1994] *On provably disjoint NP-pairs*, Tech. Rep. RS-94-36, Basic Research in Computer Science Center, Aarhus, Denmark, November. http://www.brics.dk/index.html.

[1995] Unprovability of lower bounds on the circuit size in certain fragments of bounded arithmetic, *Izvestiya of the RAN*, 59, pp. 201–224.

A. A. RAZBOROV AND S. RUDICH

[1994] Natural proofs, in: *Proceedings of the Twenty-Sixth Annual ACM Symposium on Theory of Computing*, Association for Computing Machinery, New York, pp. 204–213.

J. B. ROSSER

[1936] Extensions of some theorems of Gödel and Church, *Journal of Symbolic Logic*, 1, pp. 87–91.

K. SCHÜTTE

[1977] *Proof Theory*, Grundlehren der mathematischen Wissenschaften #225, Springer-Verlag, Berlin.

W. SIEG

[1985] Fragments of arithmetic, *Annals of Pure and Applied Logic*, 28, pp. 33–71.

C. SMORYNSKI
[1977] The incompleteness theorems, in: Barwise [1977], pp. 821–865.

R. M. SMULLYAN
[1992] *Gödel's Incompleteness Theorems*, Oxford Logic Guides #19, Oxford University Press.

R. M. SOLOVAY
[1976] *Letter to P. Hájek* Unpublished.

R. SOMMER
[1990] *Transfinite Induction and Hierarchies Generated by Transfinite Recursion within Peano Arithmetic*, PhD thesis, U.C. Berkeley.

L. J. STOCKMEYER
[1976] The polynomial-time hierarchy, *Theoretical Computer Science*, 3, pp. 1–22.

G. TAKEUTI
[1987] *Proof Theory*, North-Holland, Amsterdam, 2nd ed.
[1990] Some relations among systems for bounded arithmetic, in: *Mathematical Logic, Proceedings of the Heyting 1988 Summer School*, P. P. Petkov, ed., Plenum Press, New York, pp. 139–154.

A. TARSKI, A. MOSTOWSKI, AND R. M. ROBINSON
[1953] *Undecidable Theories*, North-Holland, Amsterdam.

A. J. WILKIE AND J. B. PARIS
[1987] On the scheme of induction for bounded arithmetic formulas, *Annals of Pure and Applied Logic*, 35, pp. 261–302.

C. WRATHALL
[1976] Complete sets and the polynomial-time hierarchy, *Theoretical Computer Science*, 3, pp. 23–33.

D. ZAMBELLA
[1996] Notes on polynomially bounded arithmetic, *Journal of Symbolic Logic*, 61, pp. 942–966.

CHAPTER III

Hierarchies of Provably Recursive Functions

Matt Fairtlough
Department of Computer Science, University of Sheffield, Sheffield S1 4DP, England

Stanley S. Wainer[1]
Department of Pure Mathematics, University of Leeds, Leeds LS2 9JT, England

Contents

[1] *The second author thanks the Department of Philosophy at Carnegie Mellon University for generous hospitality and the opportunity to teach some of this material, during his year as a Fulbright Scholar 1992-93.*

HANDBOOK OF PROOF THEORY
Edited by S. R. Buss

1. Introduction

Since the recursive functions are of fundamental importance in logic and computer science, it is a natural pure-mathematical exercise to attempt to classify them in some way according to their logical and computational complexity. We hope to convince the reader that this is also an interesting and a useful thing to do: interesting because it brings to bear, in a clear and simple context, some of the most basic techniques of proof theory such as cut-elimination and ordinal assignments; and useful because it brings out deep theoretical connections with program-verification, program complexity and finite combinatorics. One might wonder why this branch of recursive function theory should most appropriately be viewed in a proof-theoretic light, but this is simply because the underlying concerns are of an intensional character, to do with computations or derivations of functions according to given programs rather than merely their definitions in extenso as sets of ordered pairs. The proof-theoretic connection is immediately observable by considering the most basic recursive operation of all, namely composition: given functions f and g define $h := f \circ g$ by the rule

$$(g(x) = y) \rightarrow (f(y) = z) \rightarrow (h(x) = z).$$

Then the usual quantifier rules of logic yield

$$\forall x . \exists y . (g(x) = y) \rightarrow \forall y . \exists z . (f(y) = z) \rightarrow \forall x . \exists z . (h(x) = z)$$

and so the totality/termination of h follows from that of g and f respectively by means of two applications of Cut. As we shall see, cut-elimination then yields a "direct" proof from which the complexity of h can be read off.

It is the relationship between computational complexity on the one hand, and logical complexity (of termination proofs) on the other, which forms our principal theme here. Put simply, a program satisfies a specification

$$\forall \mathit{input} . \exists \mathit{output} . \mathrm{Spec}(\mathit{input}, \mathit{output})$$

if for each input x it computes an output y such that $\mathrm{Spec}(x, y)$ holds. Mere knowledge that the specification is true tells us only that *there exists* a while-program satisfying it. But to gain information about the possible structure and complexity of such a program we need to know *why* the specification is true, in other words we need to be given a proof. Thus our primary interest will be with those (recursively enumerable) classes of functions which are "verifiably computable" in given subsystems of arithmetic and analysis whose proof-theoretic strength is well-understood. This is not to say that the problem of classifying all recursive functions "in one go" is uninteresting—far from it. The known general results of Feferman [1962] in that direction—on completeness and incompleteness of hierarchies generated along paths in Kleene's $\mathcal{O}$—raise further deep questions which remain unanswered, e.g. "what is a natural well-ordering?"

Our aim then is to find uniform scales against which we can measure the computational complexity of functions verifiably computable in "known" theories. By "complexity" we mean "complexity in the large", as measured by the rates of growth of resource-bounding functions irrespective of whether they be polynomial, exponential or much worse. We do not wish to place prior restrictions on their size, but rather to have the means of comparing one with another. How might this be achieved? What form should a "subrecursive scale" take? To answer this we need first to ask what kind of features of recursive definitions we are actually trying to measure and compare.

Suppose given a number-theoretic program of some kind, together with an operational semantics determining for each number n a space $C(n)$ consisting of all computations and sub-computations of the program, starting on inputs $\leq n$. The sub-computation relation induces a tree structure on $C(n)$ and we will assume further that it has been linearly ordered by a suitable Kleene-Brouwer ordering $\prec_n$. Then if $n_1 < n_2$ there will be an order-preserving embedding $C(n_1, n_2) : (C(n_1), \prec_{n_1}) \to (C(n_2), \prec_{n_2})$ such that $C(n_2, n_3) \circ C(n_1, n_2) = C(n_1, n_3)$ whenever $n_1 < n_2 < n_3$. Thus C is a functor from the category $\mathbf{N} = \{0 < 1 < 2 < \cdots\}$ into the category of linear orderings with order-preserving maps, and the idea is that C abstracts the uniform computational structure from the given program.

Now C has a direct limit. This limit will be a countable linear ordering $\gamma = (X, \prec)$ together with maps $g(n) : C(n) \to X$ satisfying $n_1 < n_2 \to g(n_1) = g(n_2) \circ C(n_1, n_2)$, and it will be initial among all such objects. Therefore γ can be represented as the union of a nested, increasing sequence of sub-orderings $\gamma[n] := (\mathit{Image}\, g(n), \prec)$.

Let us now make the further assumption that the given program always terminates. Then computations are finite and the sub-computation relation is well-founded. So each $C(n)$ and hence each $\gamma[n]$ will be finite and γ will be a *well-ordering*. In other words, the computational structure of terminating programs can be described abstractly in terms of countable well-orderings γ, presented as the unions of ascending chains of finite sub-orderings $\gamma[0] \subset \gamma[1] \subset \gamma[2] \subset \cdots$. This is our basis for a uniform theory of "ordinal assignments" and as we shall see, it applies equally well to proofs as to computations, thus providing a link between the two.

There is still more useful structure to be extracted from the above presentation of a well-ordering $\gamma = (X, \prec)$: given an arbitrary point $\alpha \in X \cup \{\gamma\}$ and any number n, define $P_n(\alpha)$, the "n-predecessor" of α, to be the topmost element of $\gamma[n] \cap \{\beta : \beta \prec \alpha\}$ if this is non-empty, and 0 (the least element) otherwise. Then these predecessor functions provide a uniform method of generating hierarchies of number-theoretic functions and functionals, by transfinite recursion:

Let $J : \mathbf{N}^{\mathbf{N}} \to \mathbf{N}^{\mathbf{N}}$ be any given operator taking unary functions to unary functions. Then for each $\alpha \preceq \gamma$ define the α-th iterate of J,

$$J^{\alpha} : \mathbf{N}^{\mathbf{N}} \to \mathbf{N}^{\mathbf{N}}$$

as follows:

$$J^\alpha(g) := \begin{cases} g & \text{if } \alpha = 0 \\ \lambda n \,.\, J(J^{P_n(\alpha)}(g))(n) & \text{if } \alpha \neq 0 \end{cases}$$

Two simple examples spring immediately to mind, but the reader should not be deceived by their apparent simplicity since they turn out to be fundamental tools of the sub-recursive hierarchy trade. Both are obtained as iterates of the composition operator $J(g,h) := g \circ h$, the first with g fixed and h varying, the second with h fixed and g varying.

I Fix $g \in \mathbf{N}^\mathbf{N}$ and define $J_g(h) = g \circ h$. Then for every $h \in \mathbf{N}^\mathbf{N}$ and every $n \in \mathbf{N}$,

$$J_g^\alpha(h)(n) = g^k(h(n))$$

where k is the length of the descending chain

$$P_n(\alpha) \succ P_n(P_n(\alpha)) \succ P_n(P_n(P_n(\alpha))) \succ \cdots \succ 0.$$

In particular with g the successor function and h constantly zero we obtain for each $\alpha \preceq \gamma$ the functions

$$G_\alpha(n) := \text{least } k \,.\, (P_n^k(\alpha) = 0)$$

constituting the so-called *Slow-Growing Hierarchy*. Notice that $G_\gamma(n)$ measures the size of $\gamma[n]$.

II Fix $h \in \mathbf{N}^\mathbf{N}$ and define $J_h(g) = g \circ h$. Then for every $g \in \mathbf{N}^\mathbf{N}$ and every $n \in \mathbf{N}$,

$$J_h^\alpha(g)(n) = g(h^k(n))$$

where k is the length of the descending chain

$$P_n(\alpha) \succ P_{h(n)}(P_n(\alpha)) \succ P_{h^2(n)}(P_{h(n)}(P_n(\alpha))) \succ \cdots \succ 0.$$

In particular with h the successor and g the identity function we obtain for every $\alpha \preceq \gamma$ the functions

$$H_\alpha(n) := \text{least } m \,.\, (P_{m-1}P_{m-2}\cdots P_{n+1}P_n(\alpha) = 0)$$

constituting the *Hardy Hierarchy* (so called because Hardy [1904] was the first to make use of them, in "exhibiting" a set of reals with cardinality $\aleph_1$). As we shall see however, this hierarchy also contains, embedded within it, the *Fast-Growing Hierarchies* B_α and F_α which provide crucial links between proof theory on the one hand and recursive function theory on the other.

This paper is about these hierarchies, their interconnections, and their relationships with proof theory as displayed in "subrecursive classification theorems" of the following form

Given an arithmetical theory T with proof-theoretic ordinal $\|T\|$, then the provably recursive functions of T are exactly those functions computable within complexity-bounds H_α for $\alpha \prec \|T\|$.

We shall concentrate on the cases where T is Peano Arithmetic (PA), a fragment of it or an extension of it by some axiom of transfinite induction. Pohlers, in this volume, provides an ordinal analysis for many richer theories, but once this is done for a theory T its subrecursive classification may then be reduced to that of a corresponding theory of transfinite induction over order types $\prec \|T\|$. Thus the two papers together supply the methods one needs to classify the provably recursive functions of a wide spectrum of theories.

Throughout this paper we shall only consider theories based on classical logic. It is known by e.g. Friedman [1978] that in general, the provably recursive functions will be the same whether the underlying logic is intuitionistic or classical.

We view proof theory as an attempt to analyse the truth definition and to classify truths. Consequently we here use proof theory to measure the complexity of recursive functions according to the logical complexity of the judgements that they terminate. We begin by developing in section 2 a general inductive class Ω^S of ordinal presentations, together with their arithmetic and associated function hierarchies. Section 3 presents arithmetic truth as a cut-free infinitary calculus in the style of Tait [1968], with an assignment of ordinal bounds based on Buchholz [1987] but modified and simplified here following Fairtlough [1991]. In this presentation the slow-growing hierarchy arises naturally as the class of bounding functions for the truth of existential statements. Addition of the Cut rule then yields an infinitary proof theory for which the fast-growing hierarchies now supply the bounding functions. In section 4 we read off classification theorems for PA and its fragments by embedding them into the infinitary framework. These results are subsequently extended in section 7 to theories of transfinite induction. Section 5 gives an application to a well-known mathematical independence result for PA due originally to Kirby and Paris [1982]. However the proof given here is due to Cichon [1983] and displays a clear connection with the Hardy hierarchy. Section 6 reduces fast- to slow-growing by computing a map $\alpha \mapsto \alpha^+$ such that $B_\alpha = G_{\alpha^+}$. This amounts to a complete cut-elimination since proofs of Π^0_2 sentences are now reduced to their cut-free truth-definitions. The result is a new and more subtle ordinal assignment to theories T which was first discovered by Girard [1981]. Only the initial stages of this reduction are needed here (those appropriate for PA) but they already serve to illustrate the subtlety and power of ordinal assignments in providing uniform measurements of proof-theoretic complexity.

2. Structured ordinals and associated hierarchies

A "structured" countable ordinal is one for which an arbitrary but fixed "fundamental sequence" has been assigned to each limit below it. The most convenient way to introduce such ordinals is in two stages. First the set Ω of abstract "tree-ordinals" is defined inductively, in a manner reminiscent of Kleene's $\mathcal{O}$ but without any conditions (effectivity or otherwise) being placed on sequences. Then the subset Ω^S of "structured tree-ordinals" is obtained by requiring that fundamental sequences mesh together in an appropriate way, as in Schmidt [1976] and Ketonen and Solovay [1981].

It will be intuitively clear from their uniform construction that the particular tree-ordinals named and used in what follows are all *recursive* ones.

2.1. Definition. The set Ω of countable *tree-ordinals* $\alpha, \beta, \gamma, \ldots, \lambda, \ldots$ is generated inductively according to the rules:

- $0 \in \Omega$
- $\alpha \in \Omega \Rightarrow \alpha + 1 := \alpha \cup \{\alpha\} \in \Omega$
- $\forall x \in \mathbf{N}\,(\alpha_x \in \Omega) \Rightarrow \alpha := \langle \alpha_x \rangle_{x \in \mathbf{N}} \in \Omega$.

2.2. Note. λ will always denote a "limit" $\lambda = \langle \lambda_x \rangle_{x \in \mathbf{N}}$. We usually write, more suggestively, $\lambda = \sup \lambda_x$.

2.3. Definition. The *sub-tree ordering* $\prec$ is the transitive closure of the rules

- $\alpha \prec \alpha + 1$ for all $\alpha \in \Omega$
- $\alpha_m \prec \langle \alpha_x \rangle$ for all $\alpha \in \Omega$ and $m \in \mathbf{N}$.

2.4. Remark. One of the immediate consequences of this intensional approach to ordinals is that $\prec$ is only a *partial* ordering on Ω. For example, if we identify $x \in \mathbf{N}$ with $0 + 1 + \cdots + 1 (x \text{ times}) \in \Omega$ then we can define ω_0 and $\omega \in \Omega$ by $\omega_0 := \langle x \rangle$ and $\omega := \langle 1 + x \rangle$. Clearly ω_0 and ω are incomparable under $\prec$.

2.5. Ω-induction Principle. If Q is a property of tree-ordinals such that

- $Q(0)$
- $Q(\alpha) \Rightarrow Q(\alpha + 1)$
- $\forall x \in \mathbf{N} . Q(\lambda_x) \Rightarrow Q(\lambda)$.

then $Q(\alpha)$ holds for every $\alpha \in \Omega$.

For each $\alpha \in \Omega$, the ordering $\prec_\alpha$ is well-founded, where $\prec_\alpha$ is the restriction of $\prec$ below α. Henceforward, $|\alpha|$ denotes the set-theoretic ordinal height of $\prec_\alpha$.

2.6. Definitions. For each $n \in \mathbf{N}$ and each $\alpha \in \Omega$ the finite set $\alpha[n]$ of *n-predecessors* of α is defined recursively by

- $0[n] := \varnothing$
- $(\alpha + 1)[n] := \alpha[n] \cup \{\alpha\}$
- $\lambda[n] := \lambda_n[n]$.

The *immediate n-predecessor* of α is $P_n(\alpha) :=$ the $\prec$-maximum element of $\alpha[n]$ if $\alpha[n] \neq \varnothing$, and 0 otherwise.

2.7. Definition. The set Ω^S of *structured tree-ordinals* consists of those $\alpha \in \Omega$ such that

$$\forall \lambda \preceq \alpha . \forall x \in \mathbf{N} . (\lambda_x \in \lambda[x + 1]).$$

Note that if $\alpha \in \Omega^S$ and $\beta \prec \alpha$ then $\beta \in \Omega^S$.

2.8. Theorem. *For every $\alpha \in \Omega^S$ we have*

1. $\alpha[0] \subseteq \alpha[1] \subseteq \cdots \subseteq \alpha[n] \subseteq \alpha[n+1] \subseteq \cdots$
2. $\beta \prec \alpha$ *if and only if* $\beta \in \alpha[n]$ *for some* n.

Proof. We proceed by Ω-inductions on α:

1. If $\alpha = 0$ then $\alpha[n] = \varnothing = \alpha[n+1]$. For the successor case assume $\alpha[n] \subseteq \alpha[n+1]$. Then $(\alpha+1)[n] = \alpha[n] \cup \{\alpha\} \subseteq \alpha[n+1] \cup \{\alpha\} = (\alpha+1)[n+1]$. For the limit case $\alpha = \langle \alpha_x \rangle$, assume that $\alpha_n[n] \subseteq \alpha_n[n+1]$. The structuredness condition in 2.7 gives $\alpha_n \in \alpha[n+1]$ and it is easy to see that from definition 2.6 that whenever $\gamma \in \alpha[x]$ then $\gamma[x] \subset \alpha[x]$. Therefore $\alpha_n[n+1] \subset \alpha[n+1]$ and hence $\alpha[n] = \alpha_n[n] \subseteq \alpha_n[n+1] \subset \alpha[n+1]$ as required.
2. The implication from right to left is immediate. The converse is vacuously true when $\alpha = 0$, and when α is a successor, say $\alpha' + 1$ then $\beta \prec \alpha$ implies $\beta \preceq \alpha'$ so the induction hypothesis gives an $n \in \mathbf{N}$ such that $\beta \in \alpha'[n] \cup \{\alpha'\} = \alpha[n]$. Finally if $\beta \prec \alpha$ where $\alpha = \langle \alpha_x \rangle \in \Omega^S$ then $\beta \preceq \alpha_m \prec \alpha_{m+1}$ for some m, so by the induction hypothesis $\beta \in \alpha_{m+1}[n]$ for some n. By part 1 we may take $n > m+1$ so that by the structuredness of α, $\alpha_{m+1} \in \alpha[m+2] \subseteq \alpha[n]$. Then since $\gamma \in \alpha[n]$ implies $\gamma[n] \subset \alpha[n]$ we obtain $\beta \in \alpha_{m+1}[n] \subset \alpha[n]$ as desired.

2.9. Corollary. *For each non-zero $\alpha \in \Omega^S$ the set $\{\beta : \beta \prec \alpha\}$ is linearly and hence well-ordered by $\prec$, with least element 0 and such that $\beta \prec \alpha$ implies $\beta + 1 \preceq \alpha$. This well-ordering is the direct union of its finite sub-orderings $\alpha[n]$ for $n \in \mathbf{N}$.*

Proof. By 2.8 if $\beta_0, \beta_1 \prec \alpha$ then for some n we have $\beta_0, \beta_1 \in \alpha[n]$. But it is immediate from definition 2.6 that $\alpha[n]$ is linearly ordered by $\prec$, so $\beta_0 \prec \beta_1$ or $\beta_0 = \beta_1$ or $\beta_1 \prec \beta_0$. A simple induction on $\alpha \in \Omega^S$ shows that $\beta + 1 \preceq \alpha$ whenever $\beta \prec \alpha$. The final sentence is just a restatement of theorem 2.8.1.

2.10. Note. There are continuum-many $\prec$-incomparable $\alpha \in \Omega^S$ each with $|\alpha| = \omega$ (just take $\alpha_f := \langle f(x) \rangle$ for each strictly increasing function $f : \mathbf{N} \to \mathbf{N}$; then $G_{\alpha_f} = f$ and hence every such function already crops up at the first limit level in Ω). Some of these are more "natural" than others however, the most natural being ω_0 and ω as in 2.4. We will be concerned principally with those tree-ordinals which can be "constructed" from ω_0 or ω.

2.11. Arithmetic on Ω. Any primitive recursive definition over $\mathbf{N}$, say

$$\begin{aligned} f(a,0) &= f_0(a) \\ f(a,b+1) &= f_1(a,b,f(a,b)) \end{aligned}$$

can be extended straightforwardly to Ω by the addition of a third "continuity" clause. Thus, using Greek letters instead, we have the corresponding definition over Ω:

$$\begin{aligned} \varphi(\alpha,0) &= \varphi_0(\alpha) \\ \varphi(\alpha,\beta+1) &= \varphi_1(\alpha,\beta,\varphi(\alpha,\beta)) \end{aligned}$$

$$\varphi(\alpha, \lambda) = \sup_x \varphi(\alpha, \lambda_x).$$

The connection between f and φ can be expressed simply in terms of the *slow-growing* operator G. Recall that $G_\alpha(n) :=$ card $\alpha[n]$, so from definition 2.6 we immediately have $G_0(n) = 0$, $G_{\alpha+1}(n) = G_\alpha(n) + 1$ and $G_\lambda(n) = G_{\lambda_n}(n)$. Since $G_\alpha(n)$ is defined pointwise-at-n, we shall sometimes find it notationally convenient to write $G_n(\alpha)$ instead of $G_\alpha(n)$ so that, for each fixed n, $G_n : \Omega \to \mathbf{N}$.

Now suppose we know already that for fixed n, and all α, β and $\gamma \in \Omega$,

$$G_n(\varphi_0(\alpha)) = f_0(G_n(\alpha))$$
$$G_n(\varphi_1(\alpha, \beta, \gamma)) = f_1(G_n(\alpha), G_n(\beta), G_n(\gamma)).$$

Then by a simple Ω-induction on β we have for all α, $\beta \in \Omega$,

$$G_n(\varphi(\alpha, \beta)) = f(G_n(\alpha), G_n(\beta)).$$

Thus G_n is a homomorphism, collapsing the arithmetic of Ω onto the arithmetic of $\mathbf{N}$.

For example, defining addition, multiplication and exponentiation on Ω in the obvious way:

Addition. $\alpha + 0 := \alpha$; $\alpha + (\beta + 1) := (\alpha + \beta) + 1$; $\alpha + \lambda := \sup(\alpha + \lambda_x)$
Multiplication. $\alpha \cdot 0 := 0$; $\alpha \cdot (\beta + 1) := (\alpha \cdot \beta) + \alpha$; $\alpha \cdot \lambda := \sup(\alpha \cdot \lambda_x)$
Exponentiation. $\alpha^0 := 1$; $\alpha^{(\beta+1)} := \alpha^\beta \cdot \alpha$; $\alpha^\lambda := \sup(\alpha^{\lambda_x})$

we have for each fixed $n \in \mathbf{N}$,

1. $G_n(\alpha + \beta) = G_n(\alpha) + G_n(\beta)$
2. $G_n(\alpha \cdot \beta) = G_n(\alpha) \cdot G_n(\beta)$
3. $G_n(\alpha^\beta) = G_n(\alpha)^{G_n(\beta)}$.

What we need to know is that these operations are well-defined on *structured* tree-ordinals.

2.12. Lemma. *For all α, β and $\gamma \in \Omega$ we have*

1. $\gamma \in \beta[n] \Longrightarrow \alpha + \gamma \in (\alpha + \beta)[n]$.
2. $\gamma \in \beta[n] \Longrightarrow \alpha \cdot \gamma \in (\alpha \cdot \beta)[n]$ *if* $0 \in \alpha[n]$.
3. $\gamma \in \beta[n] \Longrightarrow \alpha^\gamma \in \alpha^\beta[n]$ *if* $1 \in \alpha[n]$.

Proof. We proceed by Ω-induction on β. All three are similar so we just prove (3) assuming (2). The case $\beta = 0$ is trivial since $0[n] = \varnothing$.

For the successor step from β to $\beta+1$, suppose $\gamma \in (\beta+1)[n] = \beta[n] \cup \{\beta\}$. Then by the induction hypothesis, $\alpha^\gamma \in \alpha^\beta[n] \cup \{\alpha^\beta\}$. Since $1 \in \alpha[n]$ we also have $0 \in \alpha^\beta[n]$ by a simple induction on β using (2), and hence $\alpha^\beta = \alpha^\beta \cdot 1 \in (\alpha^\beta \cdot \alpha)[n] = \alpha^{\beta+1}[n]$ again using (2).

For the limit case suppose $\beta = \sup \beta_x$ and $\gamma \in \beta[n] = \beta_n[n]$, so that, by the induction hypothesis, $\alpha^\gamma \in \alpha^{\beta_n}[n]$. But $\alpha^\beta = \sup(\alpha^{\beta_x})$, so $\alpha^\beta[n] = \alpha^{\beta_n}[n]$ and $\alpha^\gamma \in \alpha^\beta[n]$.

2.13. Theorem.

1. $\alpha, \beta \in \Omega^S \Longrightarrow \alpha + \beta \in \Omega^S$
2. $\alpha, \beta \in \Omega^S \Longrightarrow \alpha \cdot \beta \in \Omega^S$ *provided* $0 \in \alpha[1]$
3. $\alpha, \beta \in \Omega^S \Longrightarrow \alpha^\beta \in \Omega^S$ *provided* $1 \in \alpha[1]$

Proof. We proceed by Ω-induction on β. Again, all three are similar so we prove (3) assuming (2). If $\beta = 0$ then $\alpha^\beta = 1$ and $1 \in \Omega^S$.

For the successor step from β to $\beta + 1$ assume $\alpha^\beta \in \Omega^S$. Then $\alpha^{\beta+1} = \alpha^\beta \cdot \alpha$ and this belongs to Ω^S by (2), because the proviso $1 \in \alpha[1]$ implies $0 \in \alpha^\beta[1]$ from the proof of Lemma 2.12.

For the limit case $\beta = \sup \beta_x \in \Omega^S$ we must check the structuredness condition $\lambda_n \in \lambda[n+1]$ for every limit $\lambda \preceq \alpha^\beta$. This follows immediately from the induction hypotheses if $\lambda \prec \alpha^\beta$ because in that case we have $\lambda \preceq \alpha^{\beta_x}$ for some x. It therefore remains to verify it in the case $\lambda = \alpha^\beta = \sup(\alpha^{\beta_x})$. But then for any fixed n we have $\beta_n \in \beta[n+1]$ since $\beta \in \Omega^S$, and $1 \in \alpha[1] \subset \alpha[n+1]$ by the proviso. Therefore $\lambda_n = \alpha^{\beta_n} \in \alpha^\beta[n+1] = \lambda[n+1]$ immediately from part (3) of Lemma 2.12. This completes the proof.

2.14. Notation. In what follows we shall often drop round brackets from $(\alpha+\beta)[n]$ or $(\alpha \cdot \beta)[n]$ and write $\alpha + \beta[n]$, $\alpha \cdot \beta[n]$ instead.

2.15. Examples.

1. $\omega_0 \in \Omega^S$ and $\omega = 1 + \omega_0 = \sup(1 + x) \in \Omega^S$.
2. $\alpha_1, \ldots, \alpha_k \in \Omega^S \Longrightarrow \omega^{\alpha_1} \cdot n_1 + \cdots + \omega^{\alpha_k} \cdot n_k \in \Omega^S$.
3. $\omega^\omega[n] = \omega^{n+1}[n] = \omega^n \cdot n + \omega^{n-1} \cdot n + \cdots + \omega \cdot n + n + 1[n]$. Thus the elements of $\omega^\omega[n]$ are exactly those ordinals of the form $\omega^n \cdot m_0 + \cdots + \omega \cdot m_{n-1} + m_n$ where $m_i \leq n$ for every $i \leq n$.
4. More generally, by induction on α, the finite set $\omega^\alpha[n]$ consists of all tree-ordinals of the form $\omega^{\beta_0} \cdot m_0 + \omega^{\beta_1} \cdot m_1 + \cdots + \omega^{\beta_k} \cdot m_k$ where $\beta_k \prec \cdots \prec \beta_0$ all lie in $\alpha[n]$ and $m_i \leq n$ for every $i \leq k$.
5. $\varepsilon_0 = \sup(1, \omega, \omega^\omega, \omega^{\omega^\omega}, \ldots) \in \Omega^S$. Thus $\alpha \prec (\varepsilon_0)_n$ if and only if α is expressible in Cantor normal form $\alpha = \omega^{\beta_0} \cdot m_0 + \omega^{\beta_1} \cdot m_1 + \cdots + \omega^{\beta_k} \cdot m_k$ where $\beta_k \prec \cdots \prec \beta_0 \prec (\varepsilon_0)_{n-1}$.
6. We define $\exp_\alpha(\beta)$ to be α^β and then the n-th iterate $\exp_\alpha^n(\beta) := \left.\alpha^{\cdot^{\cdot^{\alpha^\beta}}}\right\} n\ \alpha\text{'s}$ is also structured provided $\alpha, \beta \in \Omega^S$ and $1 \in \alpha[1]$.
7. For each $\alpha \in \Omega^S$, $\varepsilon(\alpha) := \sup_n(\exp_\omega^n(1 + \alpha + 1)) \in \Omega^S$ is a natural tree-ordinal notation for the next ε-number above $|\alpha|$.

2.16. The Function-Hierarchies.

1. The *Slow-Growing* functions $G_\alpha : \mathbf{N} \to \mathbf{N}$ are defined by the recursion

$$\begin{aligned} G_0(n) &:= 0, \\ G_{\alpha+1}(n) &:= G_\alpha(n) + 1, \\ G_\lambda(n) &:= G_{\lambda_n}(n). \end{aligned}$$

2. The *Hardy* functions $H_\alpha : \mathbf{N} \to \mathbf{N}$ are defined by the recursion

$$\begin{aligned} H_0(n) &:= n, \\ H_{\alpha+1}(n) &:= H_\alpha(n+1), \\ H_\lambda(n) &:= H_{\lambda_n}(n). \end{aligned}$$

3. The *Fast-Growing* functions $B_\alpha : \mathbf{N} \to \mathbf{N}$ and $F_\alpha : \mathbf{N} \to \mathbf{N}$ are defined by the recursions

$$\begin{aligned} B_0(n) &:= n+1, \\ B_{\alpha+1}(n) &:= B_\alpha(B_\alpha(n)), \\ B_\lambda(n) &:= B_{\lambda_n}(n) \end{aligned}$$

$$\begin{aligned} F_0(n) &:= n+1, \\ F_{\alpha+1}(n) &:= F_\alpha^{n+1}(n), \\ F_\lambda(n) &:= F_{\lambda_n}(n). \end{aligned}$$

2.17. Lemma. *For all α, $\beta \in \Omega$,*

$$H_{\alpha+\beta} = H_\alpha \circ H_\beta$$

Proof. We proceed by Ω-induction on β. The case $\beta = 0$ is obvious because H_0 is the identity. For the successor case we have, by the induction hypothesis and definition 2.16 (2),

$$\begin{aligned} H_{\alpha+(\beta+1)}(n) &= H_{(\alpha+\beta)+1}(n) \\ &= H_{\alpha+\beta}(n+1) \\ &= H_\alpha(H_\beta(n+1)) \\ &= H_\alpha(H_{\beta+1}(n)). \end{aligned}$$

For the limit case $\beta = \sup \beta_x$ we have

$$\begin{aligned} H_{\alpha+\beta}(n) &= H_{\alpha+\beta_n}(n) \\ &= H_\alpha(H_{\beta_n}(n)) \\ &= H_\alpha(H_\beta(n)). \end{aligned}$$

2.18. Theorem. *For all $\alpha \in \Omega$,*

$$B_\alpha = H_{2^\alpha} \quad \textit{and} \quad F_\alpha = H_{\omega^\alpha}$$

Proof. We proceed by Ω-inductions on α. Both are similar and we only do the second. If $\alpha = 0$ then $\omega^\alpha = 1$ and it is clear that both F_0 and H_1 are the successor function. For the successor case α to $\alpha+1$ we have, using the induction hypothesis, 2.17 and the fact that $\omega^{\alpha+1} = \omega^\alpha \cdot \omega = \sup \omega^\alpha \cdot (x+1)$,

$$\begin{aligned} H_{\omega^{\alpha+1}}(n) &= H_{\omega^\alpha \cdot (n+1)}(n) \\ &= H_{\omega^\alpha}^{n+1}(n) \\ &= F_\alpha^{n+1}(n) \\ &= F_{\alpha+1}(n). \end{aligned}$$

For the limit case $\alpha = \sup \alpha_x$, we have $\omega^\alpha = \sup \omega^{\alpha_x}$ and so by the induction hypothesis,

$$\begin{aligned} H_{\omega^\alpha}(n) &= H_{\omega^{\alpha_n}}(n) \\ &= F_{\alpha_n}(n) \\ &= F_\alpha(n). \end{aligned}$$

2.19. Theorem. *For each $\alpha \in \Omega$, define its* descent functional D_α *by*

$$D_\alpha(f)(n) = \textit{least}\ k.(P_{f^{k-1}(n)} \cdots P_{f^2(n)} P_{f(n)} P_n(\alpha) = 0).$$

Then

$$G_\alpha(n) = D_\alpha(\textit{identity})(n)$$

and

$$H_\alpha(n) = D_\alpha(\textit{successor})(n) + n.$$

Proof. Note that D_α satisfies the recursion:

$$\begin{aligned} D_0(f)(n) &= 0 \\ D_{\alpha+1}(f)(n) &= D_\alpha(f)(f(n)) + 1 \\ D_\lambda(f)(n) &= D_{\lambda_n}(f)(n) \end{aligned}$$

because $P_n(\alpha+1) = \alpha$ and $P_n(\lambda) = P_n(\lambda_n)$. Then with $f = \textit{identity}$ we get exactly the definition of G_α, and with $f = \textit{successor}$ we get exactly the definition of $H_\alpha(n) - n$.

2.20. Remark. We are going to need various "majorization properties" of the above hierarchies, but these cannot be expected to hold for arbitrary tree-ordinals $\alpha \in \Omega$. However for structured $\alpha \in \Omega^S$ they do.

2.21. Theorem. *For all $\alpha \in \Omega^S$,*

1. *G_α is increasing (strictly if α is infinite) and $\beta \in \alpha[n] \Longrightarrow G_\beta(n) < G_\alpha(n)$*
2. *H_α is strictly increasing and if $\beta \in \alpha[n]$ then $H_\beta(n) < H_\alpha(n)$*
3. *Same for B_α and for F_α provided $n \neq 0$.*

Proof. (1) is fairly obvious since $G_\alpha(n)$ is the size of $\alpha[n]$ and $\beta \in \alpha[n]$ implies $\beta[n] \subset \alpha[n]$.

(2) is proved by Ω-induction. The case $\alpha = 0$ is trivial. For the successor case α to $\alpha + 1$ we have by the induction hypothesis

$$H_{\alpha+1}(n) = H_\alpha(n+1) < H_\alpha(n+2) = H_{\alpha+1}(n+1)$$

and if $\beta \in \alpha + 1[n] = \alpha[n] \cup \{\alpha\}$ then

$$H_\beta(n) \leq H_\alpha(n) < H_\alpha(n+1) = H_{\alpha+1}(n).$$

For the limit case $\alpha = \sup \alpha_x \in \Omega^S$ we have $\alpha_n \in \alpha[n+1] = \alpha_{n+1}[n+1]$ so by the induction hypothesis,

$$H_\alpha(n) = H_{\alpha_n}(n) < H_{\alpha_n}(n+1) < H_{\alpha_{n+1}}(n+1) = H_\alpha(n+1)$$

and if $\beta \in \alpha[n] = \alpha_n[n]$,

$$H_\beta(n) < H_{\alpha_n}(n) = H_\alpha(n).$$

(3) is immediate from (2) since $B_\alpha = H_{2^\alpha}$, $F_\alpha = H_{\omega^\alpha}$ and $\beta \in \alpha[n]$ implies $2^\beta \in 2^\alpha[n]$ and $\omega^\beta \in \omega^\alpha[n]$, except that the latter only holds when $n \neq 0$. In fact $F_\alpha(0) = 1$.

2.22. Corollary. *Each of the hierarchies G_α, H_α, B_α and F_α forms a majorization hierarchy on Ω^S, in the sense that the functions are strictly increasing (except for the constant functions $G_\alpha(n) = \alpha$ when α is finite) and each function at level α eventually dominates the corresponding function at level β whenever $\beta \prec \alpha$.*

Proof. Simply note that if $\beta \prec \alpha \in \Omega^S$ then by 2.8, $\beta \in \alpha[n]$ for all but finitely many n, and then by 2.21, $G_\beta(n) < G_\alpha(n)$, $H_\beta(n) < H_\alpha(n)$ etc.

2.23. Comparisons Lemma. *For all non-zero $\alpha \in \Omega^S$ and $n > 1$,*

$$G_\alpha(n) < H_\alpha(n) < B_\alpha(n) < F_\alpha(n) < B_{\omega\cdot\alpha}(n).$$

Proof. We proceed by simple Ω-inductions on α. For example, the least straightforward inequality is the final one:

$$\begin{aligned} F_1(n) &= F_0^{n+1}(n) = 2n+1 < n + 2^{n+1} = B_\omega(n) \\ F_{\alpha+1}(n) &= F_\alpha^{n+1}(n) < B_{\omega\cdot\alpha}^{n+1}(n) < B_{\omega\cdot\alpha}^{2^{n+1}}(n) = B_{\omega\cdot\alpha+\omega}(n) \\ F_\lambda(n) &= F_{\lambda_n}(n) < B_{\omega\cdot\lambda_n}(n) = B_{\omega\cdot\lambda}(n). \end{aligned}$$

2.24. The computational honesty of G, H, B and F

We call a computable function "honest" if its complexity is bounded by some iterate of itself. Since our concern here is with "large" functions, bigger than exponential, it does not matter whether we measure computational complexity in terms of space or time, but it will be convenient to think in terms of space-complexity.

Clearly, the computability and complexity of the functions G_α, H_α, B_α and F_α, for α's ranging below some fixed tree ordinal γ depend on the computability and complexity of the operation

$$(\alpha, x) \longmapsto \begin{cases} (\alpha - 1, x) & \text{if } \alpha \text{ is a successor} \\ (\alpha_x, x) & \text{if } \alpha \text{ is limit.} \end{cases}$$

2.25. Definition. Let q be a strictly increasing number-theoretic function. Then a tree-ordinal γ is *q-space-representable* if there is a uniform method of representing each $\alpha \prec \gamma$ as a word $\ulcorner\alpha\urcorner$ on a Turing Machine tape (or more generally as a term in some finite language), and a standard representation of numbers x by words $\ulcorner x\urcorner$, such that the transition

$$(\ulcorner\alpha\urcorner, \ulcorner x\urcorner) \longmapsto \begin{cases} (\ulcorner\alpha - 1\urcorner, \ulcorner x\urcorner) & \text{if } \alpha \text{ is a successor} \\ (\ulcorner\alpha_x\urcorner, \ulcorner x\urcorner) & \text{if } \alpha \text{ is limit} \end{cases}$$

is computable within space less than $q(l(\ulcorner\alpha\urcorner) + l(\ulcorner x\urcorner))$, where $l(\ulcorner\alpha\urcorner)$ denotes the length of the word $\ulcorner\alpha\urcorner$.

2.26. Lemma. *Suppose that $\gamma = \sup \gamma_x \in \Omega^S$ is q-space-representable and let $q_1(n) = q(n)\cdot 2$ and $q_2(n) = q(n)\cdot n$. Then for $\alpha \prec \gamma$ and $x \in \mathrm{N}$ the space-complexities of the computations of $G_\alpha(x)$, $H_\alpha(x)$, $B_\alpha(x)$ and $F_\alpha(x)$ are bounded as follows:*

$$\begin{aligned} S_G(\alpha, x) &\le q^{G_\alpha(x+1)}(l(\ulcorner\alpha\urcorner) + l(\ulcorner x\urcorner)) \\ S_H(\alpha, x) &\le q^{H^2_\alpha(x)}(l(\ulcorner\alpha\urcorner) + l(\ulcorner x\urcorner)) \\ S_B(\alpha, x) &\le q_1^{B^2_\alpha(x)}(l(\ulcorner\alpha\urcorner) + l(\ulcorner x\urcorner)) \\ S_F(\alpha, x) &\le q_2^{F^2_\alpha(x)}(l(\ulcorner\alpha\urcorner) + l(\ulcorner x\urcorner)) \end{aligned}$$

Proof. We prove the result for B, and then indicate briefly what modifications are needed for G, H and F. For B and F we cannot prove the desired inequality directly; instead we need to prove something stronger, and then we can extract what we want easily. First some notation: let $\beta = \beta_0, \beta_1, \ldots, \beta_{k-1}$ denote any finite sequence (possibly empty) of tree-ordinals $\prec \gamma$, and let $\ulcorner\beta\urcorner = \ulcorner\beta_1\urcorner, \ulcorner\beta_2\urcorner, \ldots, \ulcorner\beta_{k-1}\urcorner$ be the tape-representation of β. Define

$$B(\beta, x) = B_{\beta_0} \circ B_{\beta_1} \circ \cdots \circ B_{\beta_{k-1}}(x).$$

Then $B(\beta, x)$ will be computed according to the following tape-transitions where $\ulcorner\beta\urcorner, \ulcorner\alpha\urcorner$ denotes the tape-word $\ulcorner\beta_0\urcorner, \ulcorner\beta_1\urcorner, \ldots, \ulcorner\beta_{k-1}\urcorner, \ulcorner\alpha\urcorner$ with length $l(\ulcorner\beta\urcorner, \ulcorner\alpha\urcorner) = l(\ulcorner\beta_0\urcorner) + l(\ulcorner\beta_1\urcorner) + \cdots + l(\ulcorner\beta_{k-1}\urcorner) + l(\ulcorner\alpha\urcorner) + k$:

$$\begin{aligned} (\langle\rangle, \ulcorner x\urcorner) &\longmapsto \ulcorner x\urcorner \text{ (and halt)} \\ (\ulcorner\beta\urcorner, \ulcorner 0\urcorner, \ulcorner x\urcorner) &\longmapsto (\ulcorner\beta\urcorner, \ulcorner x+1\urcorner) \\ (\ulcorner\beta\urcorner, \ulcorner\alpha+1\urcorner, \ulcorner x\urcorner) &\longmapsto (\ulcorner\beta\urcorner, \ulcorner\alpha\urcorner, \ulcorner\alpha\urcorner, \ulcorner x\urcorner) \\ (\ulcorner\beta\urcorner, \ulcorner\lambda\urcorner, \ulcorner x\urcorner) &\longmapsto (\ulcorner\beta\urcorner, \ulcorner\lambda_x\urcorner, \ulcorner x\urcorner) \end{aligned}$$

The space used in the computation of $B(\beta, x)$ will be denoted $S_B(\beta, x)$.

Now let $L_\alpha(x)$ be the step-counting function associated with the recursive definition of $B_\alpha(x)$, thus $L_0(x) = x + 1$; $L_{\alpha+1}(x) = L_\alpha(L_\alpha(x)) + 1$; $L_\lambda(x) = L_{\lambda_x}(x) + 1$. As above, define $L(\beta, x) = L_{\beta_0} \circ L_{\beta_1} \circ \cdots \circ L_{\beta_{k-1}}(x)$. Then what we need to prove is the following:

Claim. *For every $\alpha \prec \gamma$ and every sequence β, if for every x*

$$S_B(\beta, x) \leq q_1^{L(\beta,x)}(l(\ulcorner\beta\urcorner) + l(\ulcorner x\urcorner))$$

then for every x

$$S_B(\beta, \alpha, x) \leq q_1^{L(\beta,\alpha,x)}(l(\ulcorner\beta\urcorner, \ulcorner\alpha\urcorner) + l(\ulcorner x\urcorner)).$$

The desired result follows almost immediately from this because with $\beta = \langle\rangle$ we obtain

$$S_B(\alpha, x) \leq q_1^{L(\alpha,x)}(l(\ulcorner\alpha\urcorner) + l(\ulcorner x\urcorner))$$

and it is an easy matter to check that

$$L_\alpha(x) < B_{\alpha+1}(x) = B_\alpha^2(x).$$

We prove the claim by induction on $\alpha \prec \gamma$, assuming in each case that β is an arbitrary finite sequence such that for every x,

$$S_B(\beta, x) \leq q_1^{L(\beta,x)}(l(\ulcorner\beta\urcorner) + l(\ulcorner x\urcorner)).$$

If $\alpha = 0$ then by the second transition above,

$$\begin{aligned} S_B(\beta, 0, x) &= S_B(\beta, x+1) \\ &\leq q_1^{L(\beta,x+1)}(l(\ulcorner\beta\urcorner) + l(\ulcorner x+1\urcorner)) \\ &\leq q_1^{L(\beta,0,x)}(l(\ulcorner\beta\urcorner, \ulcorner 0\urcorner) + l(\ulcorner x\urcorner)) \end{aligned}$$

since $L(\beta, 0, x) = L(\beta, x+1)$ and $l(\ulcorner x+1\urcorner) \leq l(\ulcorner 0\urcorner) + l(\ulcorner x\urcorner)$.

For the successor step from α to $\alpha + 1$, assume inductively that the claim holds for α. By the third transition, and using the q-space representability of γ, we have

$$S_B(\beta, \alpha+1, x) \leq \max\{l(\ulcorner\beta\urcorner) + q(l(\ulcorner\alpha+1\urcorner) + l(\ulcorner x\urcorner)) \cdot 2, S_B(\beta, \alpha, \alpha, x)\}$$

since we first have to pass from $(\ulcorner\alpha+1\urcorner, \ulcorner x\urcorner)$ to $(\ulcorner\alpha\urcorner, \ulcorner x\urcorner)$ in $q(l(\ulcorner\alpha+1\urcorner)+l(\ulcorner x\urcorner))$ space, and then we have to copy $\ulcorner\alpha\urcorner$ again which would take no more than another $q(l(\ulcorner\alpha+1\urcorner)+l(\ulcorner x\urcorner))$ space units, before beginning the subsequent computation from $(\ulcorner\beta\urcorner, \ulcorner\alpha\urcorner, \ulcorner\alpha\urcorner, x)$. Now by two successive applications of the induction hypothesis for α, first with sequence β, and then with β replaced by β, α, we obtain

$$\begin{aligned} S_B(\beta, \alpha, \alpha, x) &\leq q_1^{L(\beta,\alpha,\alpha,x)}(l(\ulcorner\beta\urcorner, \ulcorner\alpha\urcorner, \ulcorner\alpha\urcorner) + l(\ulcorner x\urcorner)) \\ &\leq q_1^{L(\beta,\alpha,\alpha,x)}(l(\ulcorner\beta\urcorner) + q(l(\ulcorner\alpha+1\urcorner) + l(\ulcorner x\urcorner)) \cdot 2). \end{aligned}$$

Hence

$$\begin{aligned} S_B(\beta, \alpha+1, x) &\leq q_1^{L(\beta,\alpha,\alpha,x)}(q(l(\ulcorner\beta\urcorner, \ulcorner\alpha+1\urcorner) + l(\ulcorner x\urcorner)) \cdot 2) \\ &= q_1^{L(\beta,\alpha,\alpha,x)+1}(l(\ulcorner\beta\urcorner, \ulcorner\alpha+1\urcorner) + l(\ulcorner x\urcorner)) \\ &\leq q_1^{L(\beta,\alpha+1,x)}(l(\ulcorner\beta\urcorner, \ulcorner\alpha+1\urcorner) + l(\ulcorner x\urcorner)) \end{aligned}$$

since $L(\beta, \alpha, \alpha, x) + 1 \leq L(\beta, L_\alpha^2(x) + 1) = L(\beta, \alpha + 1, x)$.

For the limit case $\alpha = \sup \alpha_x$ use the fourth transition, the q-space representability of γ and the induction hypothesis for α_x to obtain in a similar fashion to the above,

$$\begin{aligned} S_B(\beta, \alpha, x) &= \max\{l(\ulcorner\beta\urcorner) + q(l(\ulcorner\alpha\urcorner) + l(\ulcorner x\urcorner)), S_B(\beta, \alpha_x, x)\} \\ &\leq q_1^{L(\beta,\alpha_x,x)}(l(\ulcorner\beta\urcorner, \ulcorner\alpha\urcorner) + l(\ulcorner x\urcorner)) \\ &\leq q_1^{L(\beta,\alpha_x,x)+1}(l(\ulcorner\beta\urcorner, \ulcorner\alpha\urcorner) + l(\ulcorner x\urcorner)) \\ &\leq q_1^{L(\beta,\alpha,x)}(l(\ulcorner\beta\urcorner, \ulcorner\alpha\urcorner) + l(\ulcorner x\urcorner)). \end{aligned}$$

This now completes the proof for B. The proof for F is almost identical except that the successor-transition is then

$$(\ulcorner\beta\urcorner, \ulcorner\alpha+1\urcorner, \ulcorner x\urcorner) \longmapsto (\ulcorner\beta\urcorner, \ulcorner\alpha\urcorner, \ulcorner\alpha\urcorner, \ldots, \ulcorner\alpha\urcorner, \ulcorner x\urcorner)$$

with $x+1$ copies of α rather than just 2, and so $q_1(n) = q(n) \cdot 2$ is replaced by $q_2(n) = q(n) \cdot n$. The results for G and H can be proved directly without the more complicated claim. The step-counting function for $G_\alpha(x)$ is easily seen to be dominated by $G_\alpha(x+1)$, and the step-counting function for $H_\alpha(x)$ is $L_0(x) = x$, $L_{\alpha+1}(x) = L_\alpha(x+1) + 1$, $L_\lambda(x) = L_{\lambda_x}(x) + 1$, which gives $L_\alpha(x) \leq H_\alpha^2(x)$ by an easy induction on α.

The reader may have noticed that we have made one or two natural assumptions about the nature of the encoding $\ulcorner\_\urcorner$ in the proof of this lemma, which should strictly speaking form part of the definition of q-space-representability.

2.27. Honesty Theorem. *Suppose $\gamma \in \Omega^S$ is q-space-representable, and suppose $\beta \in \Omega^S$ is such that for all n and m,*

$$q_1^n(m) \leq B_\beta(\max(n, m)).$$

Then for every $\alpha \prec \gamma$ we have

$$B_\alpha \in \textsc{Space}(B_\beta \circ B_\alpha \circ B_\alpha),$$

and so B_α is "honest" when $\beta \preceq \alpha$. Similar results holds for G, H and F.

Proof. From the above Lemma we have

$$S_B(\alpha, x) \leq B_\beta(B_\alpha^2(l(\ulcorner\alpha\urcorner) + l(\ulcorner x\urcorner))),$$

and similarly for G, H and F.

2.28. Definition. A function f is *elementary* (in a function g) if f is definable explicitly from 0, 1, +, $\dot{-}$ (and g), using bounded sums and products. $E(g)$ denotes the class of all such functions f. A relation is elementary if its characteristic function is.

2.29. Fact. If g is honest and $g(n) \geq 2^n$ then $E(g)$ consists exactly of those functions which are computable within time or space bounded by some fixed iterate of g. See e.g. Cutland [1981].

2.30. Corollary. *Suppose $\omega \prec \gamma = \sup \gamma_x \in \Omega^S$ is q-space-representable, and suppose there is $\beta \prec \gamma$ such that $q_1^n(m) \leq B_\beta(\max(n,m))$ for all n, m. Then*

$$\bigcup_{\alpha \prec \gamma} \text{SPACE}(B_\alpha) = \bigcup_{\alpha \prec \gamma} E(B_\alpha).$$

Similar results hold for G, H and F.

3. Complete ω-arithmetic

In this section we formalize the truth-definition for first-order arithmetic as an infinitary proof-system; infinitary because of the "ω-rule" which allows us to derive $\forall x A(x)$ from the infinite sequence of premises $A(n)$, $n \in \mathbf{N}$. A careful assignment of tree-ordinal bounds to derivations, due originally to Buchholz [1987], but modified and somewhat simplified here following Fairtlough [1991], then provides a direct link with the fast-growing B_α's, which act as bounding functions for existential quantifiers.

Schütte [1977] was the first to develop the ω-rule as a systematic basis for proof-theoretic ordinal analysis, and many others have significantly extended these methods since. See Buchholz et al. [1981], Girard [1981,1987], Howard [1970], Pohlers' article in this volume, Schwichtenberg [1977] and Tait [1968]. Our development here follows most closely the ideas of Buchholz and Tait, see also Buchholz and Wainer [1987].

3.1. The language and rules of ω-arithmetic

The language of arithmetic consists of those formulas A, B, C, ... built up from atoms using only the symbols $\wedge$, $\vee$, $\forall$ and $\exists$. We do not need to be very specific about the atomic formulas, except to insist that they occur in complementary pairs $E_i(t_1, \ldots, t_n)$, $\overline{E_i}(t_1, \ldots, t_n)$ and that in the standard model

$$\langle \mathbf{N}, 0, succ, \ldots, E_i^{\mathbf{N}}, \ldots \rangle$$

they are interpreted as certain *elementary* relations, in the sense of 2.28. In particular E_0 and $\overline{E_0}$ will stand for $=$ and $\neq$, E_1 and $\overline{E_1}$ for $<$ and $\geq$ etc. Note that negation is not included as a logical symbol, instead it is *defined* by de Morgan's laws thus: $\neg E \equiv \overline{E}$, $\neg \overline{E} \equiv E$, $\neg(A \wedge B) \equiv \neg A \vee \neg B$, $\neg \forall x A \equiv \exists x \neg A$, etc. Hence $\neg\neg A$ *is* A.

In addition there is to be one further special atomic formula $x : \mathsf{N}$ meaning "x is a natural number". It will occur only in this atomic state and never negated. Its purpose is to keep track of the computations of witnesses for $\exists x$.

In what follows we shall not bother to distinguish between a number m and its numeral $\overline{m}$ as the context will make it clear which is intended. For formulas A with free-variables $x_1, \ldots x_k$ we will simply write $A(m_1, \ldots, m_k)$ for the instantiation $A[\overline{m_1}/x_1, \ldots, \overline{m_k}/x_k]$.

Our infinitary proof-system will be formalized in the style of Tait [1968] so as to give an inductive definition of a set of judgements

$$n : \mathsf{N} \vdash^{\alpha} \mathbf{A}$$

where $n : \mathsf{N}$ declares a fixed natural number parameter n, $\alpha \in \Omega^S$ bounds the height of the derivation, and $\mathbf{A} = \{A_1, \ldots, A_k\}$ is a *finite set* of *closed* formulas, possibly including some of the form $m : \mathsf{N}$. The intended meaning of $\mathbf{A}$ is the disjunction A_1 or ... or A_k and we write $\mathbf{A}, B$ for $\mathbf{A} \cup \{B\}$, $\mathbf{A}, \mathbf{B}$ for $\mathbf{A} \cup \mathbf{B}$, etc. Thus $\neg\mathbf{A}, B$ represents $A_1 \to (\cdots \to (A_k \to B))$.

The rules are given in Figure 1 and fall into two categories: N-rules to do with computation, and L-rules to do with the logic. In the L-Cut rule the right-hand formula C is called the *cut-formula* of that rule, or sometimes the L-Cut-formula. It must be an ordinary logical formula and cannot be of the special atomic form $n : \mathsf{N}$.

3.2. Definition. If $n : \mathsf{N} \vdash^{\alpha} \mathbf{A}$ is derivable without use of either of the cut-rules N-Cut or L-Cut we shall write instead $n : \mathsf{N} \models^{\alpha} \mathbf{A}$ or simply $\models^{\alpha} \mathbf{A}$ if $n = 0$.

The reason for this notation is clear, for in this case the definition is nothing other than the truth definition of arithmetic enhanced with two parameters n and α which together provide a uniform control over the heights of sub-derivations as follows:

- if $\models^{\alpha} \forall x.A(x)$ then for each n, $n : \mathsf{N} \models^{P_n(\alpha)} A(n)$,
- if $n : \mathsf{N} \models^{\beta} \exists y.B(y)$ for quantifier-free B, then the derivation will be finite with height bounded by the cardinality of $\beta[n]$ and furthermore there will be an existential witness bounded by $n + \mathrm{card}(\beta[n])$.

Thus in particular, since the cardinality of $\beta[n]$ is $G_\beta(n)$, if $\models^{\alpha} \forall x.\exists y.B(x,y)$ with B quantifier-free then $G_{\omega+\alpha}(n) = n + 1 + G_\alpha(n)$ will bound the size and complexity of the recursive function $f(n) = \text{least } m.B(n,m)$.

The proof of the following result underlines our claim that $\models^{\alpha}$ is little more than a dressed-up truth definition.

3.3. Completeness Theorem. *If A is true (in the standard model) then $\models^{\alpha} A$ for some $\alpha \in \Omega^S$ with $|\,\alpha\,| < \omega^{\omega}$.*

Proof. We proceed by induction on the complexity of A. If A is a true atomic formula the result is immediate by L-Ax.

If $A \equiv \exists x B(x)$ is true then $B(m)$ holds for some m and hence by the induction hypothesis and by 3.5 below, $\models^{m+\beta} B(m)$ for some β. Also $\models^{m} m : \mathsf{N}$ by N-Ax, N-Succ. Thus $\models^{\alpha} A$ by the $\exists$-rule with $\alpha = m + \beta + 1$.

(N-Ax)	$n : \mathsf{N} \vdash^{\alpha} \mathbf{A}, m : \mathsf{N}$	if $m \leq n+1$.
(N-Succ)	$\dfrac{n : \mathsf{N} \vdash^{\beta} \mathbf{A}, m : \mathsf{N}}{n : \mathsf{N} \vdash^{\alpha} \mathbf{A}, m+1 : \mathsf{N}}$	if $\beta \in \alpha[n]$.
(N-Cut)	$\dfrac{n : \mathsf{N} \vdash^{\beta_0} m : \mathsf{N} \quad m : \mathsf{N} \vdash^{\beta_1} \mathbf{A}}{n : \mathsf{N} \vdash^{\alpha} \mathbf{A}}$	if $\beta_0, \beta_1 \in \alpha[n]$.
(L-Ax)	$n : \mathsf{N} \vdash^{\alpha} \mathbf{A}$	if $\mathbf{A}$ contains a true atom.
($\vee$)	$\dfrac{n : \mathsf{N} \vdash^{\beta} \mathbf{A}, B_i}{n : \mathsf{N} \vdash^{\alpha} \mathbf{A}, (B_0 \vee B_1)}$	if $\beta \in \alpha[n]$ and $i = 0$ or $i = 1$.
($\wedge$)	$\dfrac{n : \mathsf{N} \vdash^{\beta_0} \mathbf{A}, B_0 \quad n : \mathsf{N} \vdash^{\beta_1} \mathbf{A}, B_1}{n : \mathsf{N} \vdash^{\alpha} \mathbf{A}, (B_0 \wedge B_1)}$	if $\beta_0, \beta_1 \in \alpha[n]$.
($\exists$)	$\dfrac{n : \mathsf{N} \vdash^{\beta_0} m : \mathsf{N} \quad n : \mathsf{N} \vdash^{\beta_1} \mathbf{A}, B(m)}{n : \mathsf{N} \vdash^{\alpha} \mathbf{A}, \exists x B(x)}$	if $\beta_0, \beta_1 \in \alpha[n]$.
($\forall$)	$\dfrac{\{\max(n,m) : \mathsf{N} \vdash^{\beta_m} \mathbf{A}, B(m)\}_{m \in \mathsf{N}}}{n : \mathsf{N} \vdash^{\alpha} \mathbf{A}, \forall x B(x)}$	if $\beta_m \in \alpha[\max(n,m)]$.
(L-Cut)	$\dfrac{n : \mathsf{N} \vdash^{\beta_0} \mathbf{A}, \neg C \quad n : \mathsf{N} \vdash^{\beta_1} \mathbf{A}, C}{n : \mathsf{N} \vdash^{\alpha} \mathbf{A}}$	if $\beta_0, \beta_1 \in \alpha[n]$.

Figure 1: Inductive Definition of $n : \mathsf{N} \vdash^{\alpha} \mathbf{A}$

If $A \equiv \forall x B(x)$ is true the induction hypothesis gives for each m, $\models^{\beta_m} B(m)$. Let $\alpha = \sup \alpha_x \in \Omega^S$ where $\alpha_x = \alpha_{x-1} + \beta_x + 1$ and $\alpha_{-1} = 0$. Then by 3.5 below, $\models^{\alpha_{m-1}+\beta_m} B(m)$ and $\alpha_{m-1} + \beta_m \in \alpha[m]$ for each m. So $\models^{\alpha} A$ by the $\forall$-rule.

The remaining cases $A \equiv B_0 \vee B_1$ and $A \equiv B_0 \wedge B_1$ are easy.

Finally notice that in each case the α so generated has rank $|\alpha| \leq \omega^k$ where k measures the height of A.

3.4. Corollary. *If $\vdash^{\alpha} A$ then A is true and hence derivable without any cuts of either kind. Thus we have complete cut-elimination for the system $\vdash^{\alpha}$.*

An important question which arises immediately is "how do we measure the increase in ordinal complexity in going from a proof $n : \mathsf{N} \vdash^{\alpha} \mathbf{A}$ with cuts to a cut-free proof $n : \mathsf{N} \models^{\alpha'} \mathbf{A}$?" This is not at all a trivial question and as the above proof is non-constructive it gives us no information about the increase. To answer the question we must turn to a syntactic treatment of cut-elimination, part of which stems from Gentzen's elimination of logical cuts and part from the application of hierarchy theory to the subsequent elimination of N-Cuts. The latter is covered in section 6. The rest of this section is devoted to L-Cut-elimination and the corresponding bounding results for the full system $n : \mathsf{N} \vdash^{\alpha} \mathbf{A}$ on which the results in section 4 are based.

3.5. Weakening Lemma. *If* $n : \mathsf{N} \vdash^{\alpha} \mathbf{A}$, $n \leq n'$, $\mathbf{A} \subseteq \mathbf{A}'$, $\delta \in \Omega^S$ *and* $\alpha[m] \subseteq \alpha'[m]$ *for every* $m \geq n$, *then* $n' : \mathsf{N} \vdash^{\delta+\alpha'} \mathbf{A}'$.

Proof. We proceed by induction over the derivation of $n : \mathsf{N} \vdash^{\alpha} \mathbf{A}$ according to the nine rules. For example, we consider two of the cases:

1. Assume $n : \mathsf{N} \vdash^{\beta_0} m : \mathsf{N}$, $m : \mathsf{N} \vdash^{\beta_1} \mathbf{A}$ where $\beta_0, \beta_1 \in \alpha[n]$. Applying the induction hypothesis to each of these premises we get $n' : \mathsf{N} \vdash^{\delta+\beta_0} m : \mathsf{N}$ and $m : \mathsf{N} \vdash^{\delta+\beta_1} \mathbf{A}'$. By 2.12(1), $\delta + \beta_0, \delta + \beta_1 \in \delta + \alpha'[n] \subseteq \delta + \alpha'[n']$ and so by re-applying N-Cut, $n' : \mathsf{N} \vdash^{\delta+\alpha'} \mathbf{A}'$.
2. Assume $\max(n, m) : \mathsf{N} \vdash^{\delta+\beta_m} \mathbf{A}, B(m)$ for every m, where $\beta_m \in \alpha[\max(n, m)]$. For each m, the induction hypothesis gives

$$\max(n', m) : \mathsf{N} \vdash^{\delta+\beta_m} \mathbf{A}', B(m)$$

with $\delta+\beta_m \in \delta+\alpha'[\max(n, m)] \subseteq \delta+\alpha'[\max(n', m)]$. Therefore by re-applying the $\forall$-rule, $n' : \mathsf{N} \vdash^{\delta+\alpha'} \mathbf{A}', \forall x B(x)$, as required.

3.6. Inversion Lemma.

$\wedge$. *If* $n : \mathsf{N} \vdash^{\alpha} \mathbf{A}, (B_0 \wedge B_1)$ *then* $n : \mathsf{N} \vdash^{\alpha} \mathbf{A}, B_i$ *for each* $i = 0, 1$.

$\forall$. *If* $n : \mathsf{N} \vdash^{\alpha} \mathbf{A}, \forall x B(x)$ *then* $\max(n, m) : \mathsf{N} \vdash^{\alpha} \mathbf{A}, B(m)$ *for each* m.

Proof. For example we prove $\forall$-inversion, the proof of $\wedge$-inversion being similar. Proceed by induction on the derivation of $n : \mathsf{N} \vdash^{\alpha} \mathbf{A}, \forall x B(x)$.

1. Suppose the final rule applied is a $\forall$-rule with $\forall x B(x)$ the "main formula" derived. Then the m-th premise is

$$\max(n, m) : \mathsf{N} \vdash^{\beta_m} \mathbf{A}', B(m)$$

where $\mathbf{A}' = \mathbf{A}, \forall x B(x)$ or $\mathbf{A}' = \mathbf{A}$ and where $\beta_m \in \alpha[\max(n, m)]$ so $\beta_m[k] \subset \alpha[k]$ for every $k \geq \max(n, m)$. In the first case we need to apply the induction hypothesis to obtain

$$\max(n, m) : \mathsf{N} \vdash^{\beta_m} \mathbf{A}, B(m).$$

The desired result then follows immediately by Weakening.

2. Otherwise, whatever final rule is applied $\forall x B(x)$ remains as a "side formula" in the premises. So we can apply the induction hypothesis to each one, thus replacing $n : \mathsf{N}$ by $\max(n, m) : \mathsf{N}$ and $\forall x B(x)$ by $B(m)$. The desired result is then obtained by re-applying the same final rule, since $\beta \in \alpha[k]$ implies $\beta \in \alpha[\max(k, m)]$. Note that if $n : \mathsf{N} \vdash^{\alpha} \mathbf{A}, \forall x B(x)$ is an axiom then $n : \mathsf{N} \vdash^{\alpha} \mathbf{A}$ is also an axiom, and hence so is $\max(n, m) : \mathsf{N} \vdash^{\alpha} \mathbf{A}, B(m)$.

3.7. Definition. The height $|A|$ of a formula A is defined as follows:

1. $|n : \mathsf{N}| = 0$
2. $|E(\ldots)| = |\overline{E}(\ldots)| = 1$
3. $|A_0 \wedge A_1| = |A_0 \vee A_1| = \max(|A_0|, |A_1|) + 1$
4. $|\forall x A(x)| = |\exists x A(x)| = |A| + 1$

The *cut-rank* of a derivation is then defined as the supremum of all the values of $|C|$ where C is a cut-formula appearing in it. Thus if a derivation has cut-rank 0 then there are no L-Cuts used in it, though there may still be many N-Cuts. We need only be concerned here with derivations of finite cut-rank and we denote the fact that a derivation of $n : \mathsf{N} \vdash^{\alpha} \mathbf{A}$ has cut-rank $\leq r$ by writing $n : \mathsf{N} \vdash^{\alpha}_{r} \mathbf{A}$.

3.8. Note. Weakening and inversion do not affect cut-rank.

3.9. Lemma. *Suppose $n : N \vdash^{\alpha} \mathbf{A}, E$ where E is a false atom. Then $n : N \vdash^{\alpha} \mathbf{A}$.*

This lemma is required for the following result (the proof is an easy induction on the derivation of $n : N \vdash^{\alpha} \mathbf{A}, E$).

3.10. Cut-Reduction Lemma. *Suppose $n : \mathsf{N} \vdash^{\alpha}_{r} \mathbf{A}, C$ and $n : \mathsf{N} \vdash^{\gamma}_{r} \mathbf{A}', \neg C$ where C is a formula of the form $E(\ldots)$ or $B_0 \vee B_1$ or $\exists x B(x)$, and $|C| = r + 1$. Suppose also that $\alpha[n'] \subseteq \gamma[n']$ for every $n' \geq n$. Then*

$$n : \mathsf{N} \vdash^{\gamma+\alpha}_{r} \mathbf{A}', \mathbf{A}$$

Proof. We proceed by induction over the derivation of $n : \mathsf{N} \vdash^{\alpha}_{r} \mathbf{A}, C$.

1. If C is an inactive side-formula in the final rule applied then either $n : \mathsf{N} \vdash^{\alpha}_{r} \mathbf{A}$ is an axiom in which case so is $n : \mathsf{N} \vdash^{\gamma+\alpha}_{r} \mathbf{A}', \mathbf{A}$; or else C remains in the premises of that final rule thus:

$$k : \mathsf{N} \vdash^{\beta}_{r} \mathbf{A}'', C$$

where we may assume, by weakening if necessary, that $k \geq n$ and $\beta \in \alpha[k]$, so $\beta[n'] \subseteq \gamma[n']$ for every $n' \geq k$. Therefore by weakening $n : \mathsf{N} \vdash^{\gamma}_{r} \mathbf{A}', \neg C$ (if necessary) to $k : \mathsf{N} \vdash^{\gamma}_{r} \mathbf{A}', \neg C$ and applying the induction hypothesis, we obtain

$$k : \mathsf{N} \vdash^{\gamma+\beta}_{r} \mathbf{A}', \mathbf{A}''$$

and hence the final rule can be re-applied to obtain $n : \mathsf{N} \vdash^{\gamma+\alpha}_{r} \mathbf{A}', \mathbf{A}$ as required.

2. If $C \equiv E(\ldots)$ is the "main" formula of $n : \mathsf{N} \vdash_r^\alpha \mathbf{A}, C$ then this means $E(\ldots)$ is true and we have an axiom. Consequently $\neg C \equiv \overline{E}(\ldots)$ is false. Now $n : \mathsf{N} \vdash_r^\gamma \mathbf{A}'$ follows from $n : \mathsf{N} \vdash_r^\gamma \mathbf{A}', \neg C$ by Lemma 3.9 and Weakening then gives $n : \mathsf{N} \vdash_r^{\gamma+\alpha} \mathbf{A}', \mathbf{A}$.
3. If $C \equiv (B_0 \vee B_1)$ is the main formula the proof is similar to the next case.
4. If $C \equiv \exists x B(x)$ is the main formula then the last rule applied in obtaining $n : \mathsf{N} \vdash_r^\alpha \mathbf{A}, C$ is an $\exists$-rule and we may assume that the two premises are, for some m,

$$n : \mathsf{N} \vdash_r^{\beta_0} m : \mathsf{N} \quad \text{and} \quad n : \mathsf{N} \vdash_r^{\beta_1} \mathbf{A}, B(m), C$$

where β_0, $\beta_1 \in \alpha[n] \subseteq \gamma[n]$.
Now apply the induction hypothesis to the right-hand premise so as to obtain

$$n : \mathsf{N} \vdash_r^{\gamma+\beta_1} \mathbf{A}', \mathbf{A}, B(m)$$

and apply $\forall$-inversion and Weakening to the assumption $n : \mathsf{N} \vdash_r^\gamma \mathbf{A}', \forall x \neg B(x)$ to obtain

$$\max(n, m) : \mathsf{N} \vdash_r^\gamma \mathbf{A}', \mathbf{A}, \neg B(m) \qquad (*)$$

Then one N-Cut with $n : \mathsf{N} \vdash_r^{\beta_0} \max(n, m) : \mathsf{N}$ yields

$$n : \mathsf{N} \vdash_r^{\gamma+\beta_1} \mathbf{A}', \mathbf{A}, \neg B(m)$$

provided $\beta_1[n] \neq \varnothing$. For if $\delta \in \beta_1[n]$ we may weaken the ordinal bound in $(*)$ from γ to $\gamma+\delta$. Then the N-Cut applies since both β_0 and $\gamma+\delta$ lie in $\gamma+\beta_1[n]$. Thus if $\beta_1[n] \neq \varnothing$ we can apply an L-Cut with cut-formula $B(m)$ of height r to obtain $n : \mathsf{N} \vdash_r^{\gamma+\alpha} \mathbf{A}', \mathbf{A}$ as required.
On the other hand if $\beta_1[n] = \varnothing$ then $n : \mathsf{N} \vdash^{\beta_1} \mathbf{A}, B(m), C$ is an axiom and $n : \mathsf{N} \vdash^{\beta_1} \mathbf{A}', \mathbf{A}, B(m)$ is too. Thus either $n : \mathsf{N} \vdash_r^\gamma \mathbf{A}', \mathbf{A}$ is already an axiom, or else $\neg B(m)$ is a false atom in which case we may use 3.9 and $(*)$ to obtain $\max(n, m) : \mathsf{N} \vdash_r^\gamma \mathbf{A}', \mathbf{A}$, from which we may deduce $n : \mathsf{N} \vdash_r^{\gamma+\alpha} \mathbf{A}', \mathbf{A}$ by an N-Cut as before.

3.11. Cut-Elimination Theorem. (Gentzen [1936] and Schütte [1977])
If $n : \mathsf{N} \vdash_{r+1}^\alpha \mathbf{A}$ *then* $n : \mathsf{N} \vdash_r^{2^\alpha} \mathbf{A}$.

Hence $n : \mathsf{N} \vdash_r^\alpha \mathbf{A}$ *implies* $n : \mathsf{N} \vdash_0^{\alpha^*} \mathbf{A}$ *where* $\alpha^* = \exp_2^r(\alpha) = 2^{\cdot^{\cdot^{2^\alpha}}} \Big\} r\ 2\text{'s}$.

Proof. We proceed by induction on the derivation of $n : \mathsf{N} \vdash_{r+1}^\alpha \mathbf{A}$. If the final rule applied is anything other than an L-Cut of rank $r+1$, apply the induction hypothesis to the premises and then re-apply the same rule using the fact that if $\beta \in \alpha[n]$ then $2^\beta \in 2^\alpha[n]$. If on the other hand, the final rule is an L-Cut with premises

$$n : \mathsf{N} \vdash_{r+1}^{\beta_0} \mathbf{A}, \neg C \quad \text{and} \quad n : \mathsf{N} \vdash_{r+1}^{\beta_1} \mathbf{A}, C$$

where $|C| = r + 1$, then the induction hypothesis and possibly a Weakening gives

$$n : \mathsf{N} \vdash_r^{2^\beta} \mathbf{A}, \neg C \quad \text{and} \quad n : \mathsf{N} \vdash_r^{2^\beta} \mathbf{A}, C$$

where β is the greatest of β_0, β_1 in $\alpha[n]$. Now the Cut-Reduction lemma 3.10 applies immediately, with $\gamma = \alpha = 2^\beta$, $\mathbf{A}' = \mathbf{A}$ and one of C, $\neg C$ of the required form. Hence $n : \mathsf{N} \vdash_r^{2^\beta + 2^\beta} \mathbf{A}$ and since $2^\beta + 2^\beta[n'] \subseteq 2^\alpha[n']$ when $n' \geq n$, the desired result follows by a Weakening.

3.12. Note. It may later be convenient to replace the exponential 2^α (in 3.11) by ω^α. The above proof still works in that case, but with the proviso that the declared parameter n should be nonzero since only then do we have $\omega^\beta + \omega^\beta[n'] \subseteq \omega^\alpha[n']$ for all $n' \geq n$, whenever $\beta \in \alpha[n]$.

3.13. Definitions. A Σ_1^0-formula is one of the form

$$\exists z_1 \cdots \exists z_k B(z_1, \ldots, z_k)$$

where B is "bounded" in the sense that all quantifiers occurring in it are bounded quantifiers $\exists y(y < t \wedge \ldots)$, $\forall y(y \not< t \vee \ldots)$ with t either a variable or a *closed* term. Note that this restriction is not a severe one, for if the quantifier bound t were an *open* term then we could rewrite $\exists y(y < t \wedge \ldots)$ as $\exists z(z = t \wedge \exists y(y < z \wedge \ldots))$ and $\forall y(y \not< t \vee \ldots)$ as $\exists z(z = t \wedge \forall y(y < z \wedge \ldots))$.

A closed Σ_1^0-formula as above is said to be *true at* $m \in \mathrm{N}$ if there are $m_1, \ldots, m_k$, all less than m, such that $B(m_1, \ldots, m_k)$ is true (in the standard model).

A set $\mathbf{A}$ of closed Σ_1^0-formulas is true at m if at least one $A_i \in \mathbf{A}$ is true at m.

3.14. Note. If a set $\mathbf{A}$ of closed Σ_1^0-formulas is true at m then it is automatically true at any greater m'. This property is called Σ_1^0-persistence. Also if $\mathbf{A}$ contains a true formula all of whose quantifiers are bounded then $\mathbf{A}$ is true at any m.

3.15. Bounding Lemma for $\vdash$.

1. $n : \mathsf{N} \vdash_0^\alpha m : \mathsf{N}$ *if and only if* $m \leq B_\alpha(n)$.
2. *Suppose* $\mathbf{A}$ *is a set of closed* Σ_1^0*-formulas and* $n : \mathsf{N} \vdash^\alpha \mathbf{A}$ *by a derivation in which all the* L*-Cut-formulas are* Σ_1^0. *Suppose also that* n *bounds the value of every closed term occurring in* $\mathbf{A}$ *or in any cut-formula of the derivation. Then* $\mathbf{A}$ *is true at* $B_\alpha(n)$.
3. *If* $A \equiv \exists z_1 \cdots \exists z_k B(z_1, \ldots, z_k)$ *is a closed* Σ_1^0*-formula true at* $B_\alpha(n)$ *and* n *bounds the value of every closed term in* A *then* $n : \mathsf{N} \vdash_0^{\alpha + |A|} A$.

Proof.

1. Suppose $n : \mathsf{N} \vdash_0^\alpha m : \mathsf{N}$. Since no L-cuts are involved only the rules N-Ax, N-Succ, N-Cut can have been applied in the derivation. If it arises by N-Ax then $m \leq n + 1 \leq B_\alpha(n)$. If it arises by N-Succ from $n : \mathsf{N} \vdash_0^\beta m - 1 : \mathsf{N}$ then inductively we have $m-1 \leq B_\beta(n)$ and since $\beta \in \alpha[n]$, $m \leq B_\beta(n)+1 \leq B_\alpha(n)$. If it arises by N-Cut from the premises $n : \mathsf{N} \vdash_0^{\beta_0} k : \mathsf{N}$ and $k : \mathsf{N} \vdash_0^{\beta_1} m : \mathsf{N}$ then by the induction hypothesis $k \leq B_{\beta_0}(n)$ and $m \leq B_{\beta_1}(k)$ so as β_0, $\beta_1 \in \alpha[n]$ we have $m \leq B_{\beta_1}(B_{\beta_0}(n)) \leq B_\alpha(n)$.

Conversely suppose $m \leq B_\alpha(n)$. We show $n : \mathsf{N} \vdash_0^\alpha m : \mathsf{N}$ by induction on α. If $\alpha = 0$ then $m \leq n+1$ so $n : \mathsf{N} \vdash_0^\alpha m : \mathsf{N}$ by N-Ax. If $\alpha = \beta + 1$ then $m \leq B_\beta(k)$ where $k = B_\beta(n)$ so by the induction hypothesis we have $n : \mathsf{N} \vdash_0^\beta k : \mathsf{N}$ and $k : \mathsf{N} \vdash_0^\beta m : \mathsf{N}$ and hence $n : \mathsf{N} \vdash_0^\alpha m : \mathsf{N}$ by N-Cut. Finally if $\alpha = \sup \alpha_x$ then $m \leq B_\alpha(n) = B_{\alpha_n}(n)$ so $n : \mathsf{N} \vdash_0^{\alpha_n} m : \mathsf{N}$ by the induction hypothesis. But α is structured so $\alpha_n[n'] \subseteq \alpha[n']$ for every $n' \geq n$. Therefore $n : \mathsf{N} \vdash_0^\alpha m : \mathsf{N}$ by Weakening.

2. Suppose $\mathbf{A}$ is a set of closed Σ_1^0-formulas and $n : \mathsf{N} \vdash^\alpha \mathbf{A}$ by a derivation involving only Σ_1^0 cut-formulas. Suppose also that n is large enough to bound the values of all closed terms in $\mathbf{A}$ and all closed terms occurring in the cut-formulas used (note that in general no such bound need exist).
   We prove that $\mathbf{A}$ is true at $B_\alpha(n)$ by induction on α, with cases according to the last rule applied in deriving $n : \mathsf{N} \vdash^\alpha \mathbf{A}$.

L-Ax. Then $\mathbf{A}$ contains a true atom so $\mathbf{A}$ is automatically true at $B_\alpha(n)$.

$\wedge, \vee$. These cases are also immediate since we only need take disjunctions or conjunctions of "bounded" formulas in this context, and so it is only necessary that the rules are sound.

$\exists$. Suppose $\mathbf{A} = \mathbf{A}', \exists x D(x)$ and $n : \mathsf{N} \vdash^{\beta_0} m : \mathsf{N}$ and $n : \mathsf{N} \vdash^{\beta_1} \mathbf{A}', D(m)$ where $\beta_0, \beta_1 \in \alpha[n]$. Then by the induction hypothesis $\mathbf{A}', D(m)$ is true at $B_{\beta_1}(\max(n,m))$, and by part (1) above,

$$m < B_{\beta_1}(\max(n,m)) \leq B_{\beta_1}(B_{\beta_0}(n)) \leq B_\alpha(n).$$

Hence by persistence, either $\mathbf{A}'$ is true at $B_\alpha(n)$ or $\exists x D(x)$ is true at $B_\alpha(n)$.

$\forall$. The only way the $\forall$-rule can be applied is in a "bounded" context $\mathbf{A} \equiv \mathbf{A}', \forall x(x \not< t \vee D(x))$ where t is a closed term and D only contains bounded quantifiers. The premises of this rule will then be

$$\max(n,m) : \mathsf{N} \vdash^{\beta_m} \mathbf{A}',\, m \not< t \vee D(m)$$

for every m, with $\beta_m \in \alpha[\max(n,m)]$. Suppose $\forall x(x \not< t \vee D(x))$ is false and let k be the value of the closed term t. Then by the induction hypothesis, for some $m < k$ we have $\mathbf{A}'$ true at $B_{\beta_m}(\max(n,m))$. But by the standing assumption on n we know $m < k \leq n$, so $\beta_m \in \alpha[n]$ and by persistence $\mathbf{A}'$ is therefore true at $B_\alpha(n)$. Hence $\mathbf{A}$ is true at $B_\alpha(n)$ as required.

L-Cut. Finally suppose the last rule used in deriving $n : \mathsf{N} \vdash^\alpha \mathbf{A}$ was an L-Cut with cut-formula $C \equiv \exists \vec{z}. D(\vec{z})$. The premises are therefore of the forms

$$n : \mathsf{N} \vdash^{\beta_0} \mathbf{A}, \exists \vec{z}. D(\vec{z}) \quad \text{and} \quad n : \mathsf{N} \vdash^{\beta_1} \mathbf{A}, \forall \vec{z}. \neg D(\vec{z})$$

where $\beta_0, \beta_1 \in \alpha[n]$, so both $B_{\beta_0}(n)$ and $B_{\beta_1}(n)$ are less than $B_\alpha(n)$. The standing assumption on n means that n bounds the values of closed terms in C. Now if $\mathbf{A}$ is true at $B_{\beta_0}(n)$ then it's true at $B_\alpha(n)$ by persistence, and we're done. Otherwise by the induction hypothesis, $\exists\vec{z}D(\vec{z})$ is true at $B_{\beta_0}(n)$, so there are $\vec{m} < B_{\beta_0}(n)$ such that the bounded formula $D(\vec{m})$ is true. But $\forall$-inversion applied to the second premise of the cut gives $\max(n,\vec{m}) : \mathsf{N} \vdash^{\beta_1} \mathbf{A}, \neg D(\vec{m})$ and we can now apply the induction hypothesis again to obtain $\mathbf{A}$ true at $B_{\beta_1}(\max(n,\vec{m}))$, because $\neg D(\vec{m})$ is false. Since $B_{\beta_1}(\max(n,\vec{m})) < B_{\beta_1}(B_{\beta_0}(n)) \leq B_\alpha(n)$ we again conclude $\mathbf{A}$ true at $B_\alpha(n)$ by persistence, and this completes part 2.

3. Suppose $A \equiv \exists z_1 \cdots \exists z_k D(z_1, \ldots, z_k)$ is a closed Σ^0_1-formula true at $B_\alpha(n)$. Then we have $\vec{m} < B_\alpha(n)$ such that the bounded formula $D(\vec{m})$ is true. It is an easy exercise to check that $n : \mathsf{N} \vdash_0^{\alpha+b} D(\vec{m})$ where $b = |D|$, as long as n bounds the value of any closed term in A. Then since $n : \mathsf{N} \vdash^\alpha m_i : \mathsf{N}$ for each i by part 1, we obtain $n : \mathsf{N} \vdash_0^{\alpha+|A|} A$ by k applications of the $\exists$-rule.

3.16. Bounding Lemma for $\models$.

1. $n : \mathsf{N} \models^\alpha m : \mathsf{N}$ *if and only if* $m \leq G_{\omega+\alpha}(n)$.
2. *If* $\mathbf{A}$ *is a set of closed* Σ^0_1*-formulas and* $n : \mathsf{N} \models^\alpha \mathbf{A}$ *where* n *bounds the value of all closed terms in* $\mathbf{A}$ *then* $\mathbf{A}$ *is true at* $G_{\omega+\alpha}(n)$.
3. *If* A *is a closed* Σ^0_1*-formula true at* $G_{\omega+\alpha}(n)$ *where* n *bounds the value of all closed terms in* A *then* $n : \mathsf{N} \models^{\alpha+|A|} A$.

Proof. As for 3.15 but without L-cuts or N-Cuts.

3.17. Definitions. Recall that a number-theoretic function f is partial recursive iff its graph is Σ^0_1-definable. i.e. there is a bounded formula $C_f(\vec{x}, y, z)$ such that for all $\vec{m} = m_1, \ldots, m_k$ and $n \in \mathbb{N}$,

$$f(\vec{m}) \simeq n \iff \exists z \, . \, C_f(\vec{m}, n, z) \quad \text{is true in } \mathbb{N}\ .$$

Also f is recursive iff it is partial recursive and totally-defined, i.e. if the Π^0_2-formula

$$\forall \vec{x}\, \exists y\, \exists z \, . \, C_f(\vec{x}, y, z)$$

is true (in the standard model $\mathbb{N}$). Roughly speaking, the "computation formula" $C_f(\vec{x}, y, z)$ expresses that z encodes a terminating computation of $f(\vec{x})$ with output value y. There are canonical choices of such computation formulas, depending on the underlying model of computation. With this in mind, we make the following definitions.

1. Call f *γ-recursive* iff for some computation formula C_f and some $\alpha \prec \gamma$,

$$\vdash_0^\alpha \forall \vec{x}\, \exists y\, \exists z \, . \, C_f(\vec{x}, y, z)$$

and let $\mathrm{REC}(\gamma)$ denote the class of all γ-recursive functions.

2. Call f *γ-definable* iff for some computation formula C_f and some $\alpha \prec \gamma$,

$$\models^\alpha \forall \vec{x}\, \exists y\, \exists z \, . \, C_f(\vec{x}, y, z)$$

and let $\Sigma^0_1\text{-}\mathrm{DEF}(\gamma)$ denote the class of all γ-definable functions.

3. Call f *provably recursive* in a given arithmetical theory T iff for some computation formula C_f
$$T \vdash \forall \vec{x}\, \exists y\, \exists z\,.\, C_f(\vec{x}, y, z)$$
and let $\text{PROVREC}(T)$ denote the class of all functions provably recursive in T.

3.18. Remark. Normally definitions of provable recursiveness require not only a proof of the existence of the output value but in addition a proof of its uniqueness. However, for standard choices of computation formula, such as those we shall construct later, uniqueness will be a corollary of existence. Alternatively, in the presence of Σ_1^0 induction, the least number principle would allow any computation formula to be modified to satisfy the stronger definition.

3.19. Hierarchy Theorem.

1. *Suppose $\gamma = \sup \gamma_x \in \Omega^S$ is q-space-representable with $\omega \prec \gamma$, and suppose there is $\beta \prec \gamma$ such that $q_1^n(m) \leq B_\beta(\max(n, m))$ for all n, m. Then*
$$\text{REC}(\gamma) = \bigcup_{\alpha \prec \gamma} E(B_\alpha) = \bigcup_{\alpha \prec \gamma} \text{SPACE}(B_\alpha).$$

2. *Suppose $\gamma = \sup \gamma_x \in \Omega^S$ is q-space-representable with $2^\omega \prec \gamma$, and suppose there is $\beta \prec \gamma$ such that $q^n(m) \leq G_\beta(\max(n, m))$ for all n, m. Also suppose that $\forall \alpha_0, \alpha_1 \prec \gamma\, \exists \alpha \prec \gamma\ (G_{\alpha_0} \circ G_{\alpha_1} \leq G_\alpha)$. Then*
$$\Sigma_1^0\text{-DEF}(\gamma) = \bigcup_{\alpha \prec \gamma} E(G_\alpha) = \bigcup_{\alpha \prec \gamma} \text{SPACE}(G_\alpha).$$

Proof.

1. Suppose $f : \mathbf{N}^k \to \mathbf{N}$ is γ-recursive. Then the graph of f is defined by a Σ_1^0-formula $\exists z\,.\, C_f(\vec{x}, y, z)$ and for some $\alpha \prec \gamma$,
$$\vdash_0^\alpha \forall \vec{x}\, \exists y\, \exists z\,.\, C_f(\vec{x}, y, z).$$
For any $\vec{m} \in \mathbf{N}^k$, $\forall$-inversion gives
$$\max \vec{m} : \mathsf{N} \vdash_0^\alpha \exists y\, \exists z\,.\, C_f(\vec{m}, y, z).$$
The Bounding Lemma 3.15 (2) then shows that the Σ_1^0-formula $\exists y\, \exists z\,.\, C_f(\vec{m}, y, z)$ is true at $B_\alpha(\max \vec{m})$ provided $\max \vec{m}$ is at least as big as any number-parameter occurring in C_f (there are no cut-formulas since the proof above has cut-rank 0). Thus for all but finitely-many inputs $\vec{m}$ there are numbers n_0 $(= f(\vec{m}))$ and $\vec{n} = n_1, \ldots, n_l$, all less than $B_\alpha(\max \vec{m})$, such that $C_f(\vec{m}, n_0, \vec{n})$ holds.
Now let $\langle n_0, \ldots, n_l \rangle$ be any standard polynomial coding of $l+1$-tuples into $\mathbf{N}$, with inverses $(n)_i$ so that if $n = \langle n_0, \ldots, n_l \rangle$ then $(n)_i = n_i$ for each $i \leq l$.

Furthermore let $b(\vec{m}) := \langle B_\alpha(\max\vec{m}), \ldots, B_\alpha(\max\vec{m})\rangle$. Then we can compute f as follows:

$$f(\vec{m}) = (\text{least } n \le b(\vec{m}) \,.\, C_f(\vec{m}, (n)_0, (n)_1, \ldots, (n)_l)_0$$

But C_f is built up from elementary relations using only propositional connectives and bounded quantifiers, so C_f itself defines an elementary relation. Also the coding $\langle\cdots\rangle$ and decoding $()_i$ functions are elementary and elementary functions are closed under bounded minimisation. So f is elementary in B_α.
Conversely suppose $f \in E(B_\alpha)$ for some $\alpha \prec \gamma$. Since $E(B_\alpha)$ expands as α increases, and since $\omega \prec \gamma$, we may as well suppose also that $\omega \preceq \alpha$. So $B_\alpha(m) \ge B_\omega(m) = m + 2^{m+1}$ and by 2.29, f is computable within time or space bounded by some fixed iterate of B_α, say $B_\alpha^{2^i} = B_{\alpha+i}$. But for any reasonable model of computation there is a bounded "computation formula" $C_f(\vec{x}, y, z)$ which expresses the fact that $f(\vec{x})$ is computable within resource z and with output-value y. Since we know that our f is computable within resource bounded by $B_{\alpha+i}$, it follows that the formula $\exists y \exists z \,.\, C_f(\vec{m}, y, z)$ is true at $B_{\alpha+i}(\max\vec{m})$, for all inputs $\vec{m}$. Therefore by Bounding Lemma 3.15 (3), we have for $j = |\,C_f\,| + 2$,

$$\max\vec{m} : \mathsf{N} \vdash_0^{\alpha+i+j} \exists y \exists z \,.\, C_f(\vec{m}, y, z)$$

and then by k applications of the $\forall$-rule,

$$\vdash_0^{\alpha+i+j+k} \forall\vec{x}\, \exists y \exists z \,.\, C_f(\vec{m}, y, z).$$

Since γ is a limit and $\alpha \prec \gamma$ we also have $\alpha + i + j + k \prec \gamma$ and hence f is γ-recursive.
The second equality was previously established at the end of section 2.

2. This follows exactly the same lines as (1) but using Bounding Lemma 3.16 for $\models^\alpha$. The closure conditions imposed on γ ensure that if $f \in E(G_\alpha)$ for some $\alpha \prec \gamma$ we may assume $2^\omega \preceq \alpha$ so that $G_\alpha(m) \ge 2^{m+1}$, and then find $\alpha' \prec \gamma$ such that $G_{\alpha'}$ bounds a fixed iterate of G_α. Hence f will be computable within resource bounded by $G_{\alpha'}$ and for each input $\vec{m}$, the Σ_1^0-formula $\exists y \exists z \,.\, C_f(\vec{m}, y, z)$ will be true at $G_{\alpha'}(\max\vec{m})$, and so $\max\vec{m} \models^{\alpha'} \exists y \exists z \,.\, C_f(\vec{m}, y, z)$ and f is therefore γ-definable.

3.20. Example. First notice that for each integer $k \ge 0$, $B_{\alpha+k}$ is just B_α iterated 2^k times, so $E(B_\alpha) = E(B_{\alpha+k})$. However $B_{\alpha+\omega}(m) = B_\alpha^{2^{m+1}}(m)$ and so $B_{\alpha+\omega}$ eventually dominates every function in $E(B_\alpha)$ if $\omega \preceq \alpha$. Thus $B_{\alpha+\omega} \notin E(B_\alpha)$. In particular then, the functions $B_{\omega\cdot k}$ with $k = 1, 2, \ldots$ play the role of the Ackermann/Grzegorczyk functions in such a way that $E(B_{\omega\cdot k}) =$ Grzegorczyk's $\mathcal{E}^{k+2}$. Therefore for each $k \ge 2$ we have by 3.19 (1), since $\omega \cdot k$ is q-space representable for some polynomial q,

$$\textsc{Rec}(\omega \cdot k) = \mathcal{E}^{k+1}$$

and hence

$$\text{REC}(\omega^2) = \text{PRIMITIVE RECURSIVE}.$$

Note that on the other hand,

$$\Sigma_1^0\text{-DEF}(\varepsilon_0) = \mathcal{E}^3.$$

3.21. Example. It should be fairly clear that for limits γ satisfying mild representability and closure conditions, that the γ-recursive functions are exactly those defined by nested recursions over initial segments of γ. For any such recursion f, a computation formula C_f would be definable and an informal termination proof by transfinite induction could be translated into a formal proof within the system $\vdash^\gamma$. Conversely, the Hierarchy Theorem would show that any γ-recursive function is elementary in some B_α for $\alpha \prec \gamma$, and so definable by nested recursions because the B_α functions are.

3.22. Note. From the example on primitive recursion, we see that part 1 of the Hierarchy Theorem applies to any γ which is q-space representable for some primitive recursive q, and such that $\omega^2 \preceq \gamma$.

Similarly part 2 of the Hierarchy Theorem applies to any γ which is q-space representable for some primitive recursive q, provided γ is at least as big as the first primitive recursively closed ordinal and also satisfies the stated compositionality requirement.

These conditions will indeed be satisfied by any γ to which we later apply the Hierarchy Theorem, but we leave the reader to convince him or herself of this fact. In particular, it is quite easy to see that ε_0 is polynomial-space representable. See e.g. Sommer [1992].

4. Provably recursive functions of PA

Here we characterize the provably recursive functions of Peano Arithmetic (PA) and its fragments:

$$\begin{aligned} \text{PROVREC(PA)} &= \text{REC}(\varepsilon_0) \\ \text{PROVREC}(\Sigma_n^0\text{-IND}) &= \text{REC}((\varepsilon_0)_n) \quad \text{if } n \geq 2 \\ \text{PROVREC}(\Sigma_1^0\text{-IND}) &= \text{REC}(\omega^2) \end{aligned}$$

These results go back to Kreisel [1952] for PA, Parsons [1966] for the fragments, and to Wainer [1970] and Schwichtenberg [1971] for their corresponding subrecursive hierarchy classifications.

4.1. Peano Arithmetic (PA). Our version of PA is formalized classically in a Tait-style calculus PA $\vdash$ **A** where as before, **A** is a finite set of formulas built out of atoms $E(\ldots)$, $\overline{E}(\ldots)$ using $\vee$, $\wedge$, $\exists$, $\forall$; but now the formulas may contain free variables. It is sometimes convenient to display the free variables occurring in **A** by

L-Ax	$\vdash \mathbf{A}, E(t_1, \ldots, t_k), \overline{E}(t_1, \ldots, t_k)$		
$\vee$	$\dfrac{\vdash \mathbf{A}, B_i}{\vdash \mathbf{A}, (B_0 \vee B_1)}$		$i = 0$ or $i = 1$
$\wedge$	$\dfrac{\vdash \mathbf{A}, B_0 \quad \vdash \mathbf{A}, B_1}{\vdash \mathbf{A}, (B_0 \wedge B_1)}$		
$\exists$	$\dfrac{\vdash \mathbf{A}, B(t)}{\vdash \mathbf{A}, \exists x B(x)}$		
$\forall$	$\dfrac{\vdash \mathbf{A}, B(y)}{\vdash \mathbf{A}, \forall x B(x)}$		y not free in $\mathbf{A}$
L-Cut	$\dfrac{\vdash \mathbf{A}, \neg C \quad \vdash \mathbf{A}, C}{\vdash \mathbf{A}}$		

Figure 2: Logical axioms and rules of PA

writing $\mathbf{A}(x_1, \ldots, x_k)$ etc. The special atoms $x : \mathsf{N}$ do not occur in the language of PA.

1. The *logical* axioms and rules are as in figure 2.
2. The principle of induction is formulated here as a rule (but see Note 4.3 below):

$$\text{Ind} \quad \frac{\vdash \mathbf{A}, B(0) \quad \vdash \mathbf{A}, \neg B(x), B(x+1)}{\vdash \mathbf{A}, B(t)}$$

   where the variable x is not free in $\mathbf{A}$, and t is any term.

3. To get the theory off the ground, we also need to add certain *arithmetical* axioms defining the atomic relations between basic terms. Among these will be the constant 0, the successor function $+1$ and the equality and inequality relations $E_0(x, y) \equiv (x = y)$ and $\overline{E_0}(x, y) \equiv (x \neq y)$, $E_1(x, y) \equiv (x < y)$ and $\overline{E_1}(x, y) \equiv (x \not< y)$. Their defining axioms will include all substitution instances of the axioms in figure 3. Instead of adding terms for addition and multiplication as in more usual formulations of PA, we choose to add a pairing function p and its inverses u and v satisfying the axioms in figure 4, so that by iterating p we can then encode finite lists directly as

$$\langle x_0, \ldots, x_{k-1} \rangle \equiv p(\cdots p(p(0, x_0), x_1), \ldots, x_{k-1})$$

$$
\begin{array}{ll}
\vdash & \mathbf{A}, x+1 \neq 0 \\
\vdash & \mathbf{A}, x+1 \neq y+1, x = y \\
\vdash & \mathbf{A}, x = x \\
\vdash & \mathbf{A}, x \neq y, y = x \\
\vdash & \mathbf{A}, x \neq y, y \neq z, x = z \\
\vdash & \mathbf{A}, x \not< 0 \\
\vdash & \mathbf{A}, x < x+1 \\
\vdash & \mathbf{A}, x \not< y+1, x < y, x = y \\
\vdash & \mathbf{A}, x \not< x \\
\vdash & \mathbf{A}, x \not< y, y \not< z, x < z \\
\vdash & \mathbf{A}, x < y, y < x, x = y \\
\vdash & \mathbf{A}, x \neq y, \overline{E}(\dots, x, \dots), E(\dots, y, \dots) \\
\vdash & \mathbf{A}, x \neq y, t(\dots, x, \dots) = t(\dots, y, \dots)
\end{array}
$$

Figure 3: Arithmetical axioms of PA

$$
\begin{array}{ll}
\vdash & \mathbf{A}, p(x,y) \neq 0 \\
\vdash & \mathbf{A}, u(p(x,y)) = x \\
\vdash & \mathbf{A}, v(p(x,y)) = y \\
\vdash & \mathbf{A}, x = 0, x = p(u(x), v(x))
\end{array}
$$

Figure 4: Pairing axioms of PA

where 0 denotes the empty list (thus we can think of $p(x, y)$ as representing the list with head y and tail x). Together with this list-constructor we require three further terms $lh(x)$, $in(x, j)$, $(x)_j$ such that if $x = \langle x_0, \dots, x_{k-1}\rangle$ as above, then $lh(x) = k$, $in(x, j) = \langle x_0, \dots, x_{j-1}\rangle$ and $(x)_j = x_j$, provided $j < lh(x)$.

We shall not write down all the axioms needed to define these term-constructors p, u, v, lh, in, $()_j$. It suffices to say that in the standard model $\mathbf{N}$, they can all be interpreted as *elementary* functions, for example take:

$$
\begin{aligned}
p(m,n) &= \frac{1}{2}(m+n)(m+n+1)+m+1 \\
lh(n) &= \text{least } i < n+2.\,(u^i(n) = 0) \\
in(n,j) &= u^{lh(n)-j}(n) \\
(n)_j &= v(in(n, j+1))
\end{aligned}
$$

Thus they can all be introduced in PA by sets of axioms corresponding to their elementary defining equations. Further terms denoting primitive recursive functions could be added at will.

4.2. Definition. The fragments Σ^0_n-IND and Π^0_n-IND of arithmetic are obtained simply by restricting the logical complexity of the induction-formula B in the Ind rule to $B \in \Sigma^0_n$, $B \in \Pi^0_n$ respectively. Recall that a Σ^0_n (respectively Π^0_n) formula is one obtained from a bounded formula by prefixing n alternating quantifier blocks, starting with existentials (respectively universals). Note however that the sequence coding machinery allows each block of like quantifiers to be contracted to a single quantifier and any Σ^0_n or Π^0_n formula is provably equivalent to its contracted form without the use of induction.

4.3. Note. We do not place any restrictions on the logical complexity of the side-formulas $\mathbf{A}$ in the induction rule (but compare Sieg [1991]). Consequently it is very easy to see that in PA or any of the above fragments, the induction rule may equivalently be re-formulated as an *axiom scheme*

$$\vdash \mathbf{A}, \neg B(0), \exists x(B(x) \wedge \neg B(x+1)), B(t).$$

Note also that in the presence of the modified subtraction function $\dot{-}$, the fragments Π^0_n-IND and Σ^0_n-IND are equivalent. For it is easy to see that the above induction axiom can be derived from its dual form

$$\vdash \mathbf{A}, B(t \dot{-} 0), \exists x(\neg B(t \dot{-} x) \wedge B(t \dot{-} (x+1))), \neg B(t \dot{-} s).$$

4.4. Definition. The *rank* of a PA-derivation is the maximum of all the heights $|C|$, $|B|$ of cut-formulas and induction-formulas occurring in it.

4.5. Definition. For each PA-derivation D define its *height* d by induction as follows (so as to incorporate an estimate of the complexity of witnessing terms in addition to the depth of the derivation):

1. Any axiom has height $d = 1$.
2. If D is constructed from sub-derivations D_i of heights d_i ($i \leq 1$) by an application of any of the rules $\vee$, $\wedge$, $\forall$, L-Cut, then $d = \max d_i + 1$.
3. If D is constructed from a sub-derivation D_0 of $\mathbf{A}, B(t)$ by applying the $\exists$-rule then $d = \max(d_0, |t|) + 1$ where $|t|$ is the least number k such that the term t (regarded as an elementary function of its number variables) is bounded by $B_{\omega \cdot k}$ (it follows from 3.20 that such a k exists).
4. If D is a derivation of $\mathbf{A}, B(t)$ by the induction rule from subderivations D_0, D_1, then $d = \max(d_0, d_1, |t|) + 2$.

4.6. Notation. PA $\vdash^d_r \mathbf{A}$ indicates that there is a PA-derivation of $\mathbf{A}$ with height d and rank r.

4.7. Embedding of PA. *If* PA $\vdash_r^d \mathbf{A}(t_1(\vec{x}),\ldots,t_k(\vec{x}))$ *then for all assignments of numbers* $\vec{n}=n_1,\ldots,n_l$ *to the free variables* $\vec{x}=x_1,\ldots,x_l$ *we have*

$$\max \vec{n} : \mathsf{N} \vdash_r^{\omega\cdot d} \mathbf{A}(m_1,\ldots,m_k)$$

where $m_1,\ldots,m_k$ *are the numerical values of the terms* $t_1(\vec{n}),\ldots,t_k(\vec{n})$.

Proof. By induction over the PA-derivation of $\mathbf{A}(t_1,\ldots,t_k)$. Let $n=\max\vec{n}$ where $\vec{n}$ is a fixed assignment and note that $\omega\cdot(d-1)\in\omega\cdot d[n]$.

1. If $\mathbf{A}(t_1,\ldots,t_k)$ is either a logical or arithmetical axiom then on the assignment $\vec{x}:=\vec{n}$, the resulting set of closed formulas $\mathbf{A}(m_1,\ldots,m_k)$ must contain a true atom, so automatically we have $n:\mathsf{N}\vdash^{\omega}\mathbf{A}(m_1,\ldots,m_k)$.
2. The $\vee$, $\wedge$, and L-Cut cases are immediate since the corresponding rules in the infinitary system are virtually the same.
3. Suppose PA $\vdash_r^d \mathbf{A}(t_1,\ldots,t_k),\forall z B(z,t_1,\ldots,t_k)$ follows by an application of the $\forall$-rule from PA $\vdash_r^{d-1}\mathbf{A}(t_1,\ldots,t_k),B(z,t_1,\ldots,t_k)$. Then the induction hypothesis gives for every $n'\in\mathbf{N}$,

$$\max(n,n'):\mathsf{N}\vdash_r^{\omega\cdot(d-1)}\mathbf{A}(\vec{m}),B(n',\vec{m}).$$

   But then the infinitary $\forall$-rule applies immediately to give

$$n:\mathsf{N}\vdash_r^{\omega\cdot d}\mathbf{A}(\vec{m}),\forall z B(z,\vec{m}).$$

4. Suppose PA $\vdash_r^d \mathbf{A}(t_1,\ldots,t_k),\exists z B(z,t_1,\ldots,t_k)$ follows by an application of the $\exists$-rule from PA $\vdash_r^{d_0}\mathbf{A}(t_1,\ldots,t_k),B(t_0,t_1,\ldots,t_k)$. Then the induction hypothesis gives

$$n:\mathsf{N}\vdash_r^{\omega\cdot(d-1)}\mathbf{A}(\vec{m}),B(m_0,\vec{m})$$

   where m_0 is the value of t_0 after assigning $\vec{n}$ to the variables $\vec{x}$ and 0 to any other variables in t_0. Now the derivation height d is defined so that the witnessing term t_0 has value $m_0\le B_{\omega\cdot(d-1)}(n)$ (here B is the fast-growing hierarchy, not the formula). Therefore by the Bounding Lemma 3.15, $n:\mathsf{N}\vdash_0^{\omega\cdot(d-1)} m_0:\mathsf{N}$ and so the $\exists$-rule in the infinitary system immediately gives

$$n:\mathsf{N}\vdash_r^{\omega\cdot d}\mathbf{A}(\vec{m}),\exists z B(z,\vec{m}).$$

5. Suppose PA $\vdash_r^d \mathbf{A}(t_1,\ldots,t_k),B(t_0,t_1,\ldots,t_k)$ follows by the induction rule from

$$\text{PA}\vdash_r^{d_0}\mathbf{A}(t_1,\ldots,t_k),B(0,t_1,\ldots,t_k)$$

   and

$$\text{PA}\vdash_r^{d_1}\mathbf{A}(t_1,\ldots,t_k),\neg B(x,t_1,\ldots,t_k),B(x+1,t_1,\ldots,t_k)$$

   where $|B|\le r$ and x is not free in $\mathbf{A}(t_1,\ldots,t_k)$. Let m_0 be the value of $t_0(\vec{n})$. Then as before $n:\mathsf{N}\vdash_0^{\omega\cdot(d-1)} m_0:\mathsf{N}$. By the induction hypothesis,

$$n:\mathsf{N}\vdash_r^{\omega\cdot d_0}\mathbf{A}(\vec{m}),B(0,\vec{m})$$

and for every $n' \in \mathbf{N}$,

$$\max(n, n') \vdash_r^{\omega \cdot d_1} \mathbf{A}(\vec{m}), \neg B(n', \vec{m}), B(n'+1, \vec{m}).$$

So by m_0 successive applications of L-Cut with cut-formulas $B(0, \vec{m})$, $B(1, \vec{m})$, ..., $B(m_0 - 1, \vec{m})$ we obtain

$$\max(n, m_0) : \mathsf{N} \vdash_r^{\omega \cdot (d-2)+m_0} \mathbf{A}(\vec{m}), B(m_0, \vec{m}).$$

Since $\omega \cdot (d-2) + m_0 \in \omega \cdot (d-1)[\max(n, m_0)]$ we thus obtain by Weakening,

$$\max(n, m_0) : \mathsf{N} \vdash_r^{\omega \cdot (d-1)} \mathbf{A}(\vec{m}), B(m_0, \vec{m}).$$

But $n : \mathsf{N} \vdash_0^{\omega \cdot (d-1)} \max(n, m_0) : \mathsf{N}$ from above, so a final N-Cut gives the desired result:

$$n : \mathsf{N} \vdash_r^{\omega \cdot d} \mathbf{A}(\vec{m}), B(m_0, \vec{m})$$

and this completes the proof.

4.8. Corollary. *Suppose* PA $\vdash_r^d \mathbf{A}(t_1(\vec{x}), \ldots, t_k(\vec{x}))$. *Then for all numerical assignments $\vec{n}$ to the free variables $\vec{x}$ we have, by the Embedding and Cut-Elimination Theorems 4.7, 3.11,*

$$\max \vec{n} : \mathsf{N} \vdash_0^\alpha \mathbf{A}(m_1, \ldots, m_k)$$

where as before m_i is the value of term $t_i(\vec{n})$ and where $\alpha = \exp_2^r(\omega \cdot d)$ or $\exp_\omega^r(\omega \cdot d)$.

4.9. Theorem.

$$\begin{aligned} \textsc{ProvRec}(\text{PA}) &\subseteq \textsc{Rec}(\varepsilon_0) \\ \textsc{ProvRec}(\Sigma_1^0\text{-}\textsc{Ind}) &\subseteq \textsc{Rec}(\omega^2) \\ \textsc{ProvRec}(\Sigma_n^0\text{-}\textsc{Ind}) &\subseteq \textsc{Rec}((\varepsilon_0)_n) \quad \textit{if } n \geq 2 \end{aligned}$$

Proof.

1. Immediate from 4.8, for if f is Σ_1^0-defined by $\exists y \exists z \, . \, C_f(\vec{x}, y, z)$ and PA $\vdash \forall \vec{x} \, \exists y \exists z \, . \, C_f(\vec{x}, y, z)$ then for some $\alpha \prec \varepsilon_0$ we have $\vdash_0^\alpha \forall \vec{x} \, \exists y \exists z \, . \, C_f(\vec{x}, y, z)$.
2. Suppose Σ_1^0-Ind $\vdash \forall \vec{x} \, \exists y \exists z \, . \, C_f(\vec{x}, y, z)$. The cut-reduction method of 3.10 applies to Σ_1^0-Ind in much the same way as for the infinitary system, so that cuts of (disjunctive or) existential form:

$$\frac{\mathbf{A}, \forall x \neg B(x) \qquad \dfrac{\mathbf{A}, B(t)}{\mathbf{A}, \exists x B(x)}}{\mathbf{A}}$$

get reduced (by inverting the left premise) to

$$\frac{\mathbf{A}, \neg B(t) \qquad \mathbf{A}, B(t)}{\mathbf{A}}$$

However if the right premise $\mathbf{A}, \exists x B(x)$ comes about by a Σ_1^0-induction instead of an $\exists$-rule, the method comes unstuck and we can reduce the cuts no further. Nevertheless, this means that any Σ_1^0-IND derivation can be transformed into one in which all cut-formulas are (at worst) Σ_1^0. Then applying the Embedding Theorem 4.7 to the given proof of $\forall \vec{x}\ \exists y \exists z\ .\ C_f(\vec{x}, y, z)$ and inverting the $\forall$ yields an infinitary derivation $\max \vec{n} : \mathsf{N} \vdash^{\omega \cdot d} \exists y \exists z\,.\, C_f(\vec{n}, y, z)$ in which all cut-formulas are Σ_1^0 and in which there is a finite bound k on any additional numerical constants appearing in them. Note that embedding replaces induction by cuts. The Bounding Lemma 3.15 (2) then applies to give $\exists y \exists z\,.\, C_f(\vec{n}, y, z)$ true at $B_{\omega \cdot d}(\max(\vec{n}, k))$, for all $\vec{n}$. As before f is therefore elementary in $B_{\omega \cdot d}$ and by the Hierarchy Theorem 3.19 and 3.20, $f \in \mathrm{REC}(\omega^2)$.

3. Suppose Σ_n^0-IND $\vdash^d \forall \vec{x}\ \exists y \exists z\,.\, C_f(\vec{x}, y, z)$. As above we may assume that the proof is reduced to one in which all cut-formulas are in prenex form with at most n alternating quantifiers. Recall that the sequence-coding apparatus allows us to assume further that blocks of like quantifiers are contracted to one such. Thus by the Embedding Theorem and Inversion we have, for all assignments $\vec{x} := \vec{m}$,

$$\max \vec{m} : \mathsf{N} \vdash^{\omega \cdot d} \exists y \exists z\,.\, C_f(\vec{m}, y, z)$$

by a derivation in which all cut-formulas have at most $n - 1$ alternating quantifiers prefixed to a Σ_1^0-formula or its negation, and in which there is a finite bound k on the numerical constants originating from the given (Σ_n^0-IND)-proof. The Cut-Elimination method then allows us to reduce the quantifier-complexity of cut-formulas one step at a time, to obtain a derivation of

$$\max \vec{m} : \mathsf{N} \vdash^{\alpha} \exists y \exists z\,.\, C_f(\vec{m}, y, z)$$

in which only Σ_1^0-cuts remain. By the Bounding Lemma we then have for every $\vec{m}$,

$$\exists y \exists z\,.\, C_f(\vec{m}, y, z) \text{ true at } B_\alpha(\max(\vec{m}, k))$$

and therefore f is elementary in B_α, as before. The desired result then follows immediately provided we can ensure that $\alpha \prec (\varepsilon_0)_n$.

Unfortunately this is not straightforward. If we were to apply step-by-step cut reduction in the most obvious way, the final bound α would be too large. Instead we need to insert a further "refining" step, which is step 2 below.

We begin with a derivation of

$$\max \vec{m} : \mathsf{N} \vdash^{\omega \cdot d} \exists y \exists z\,.\, C_f(\vec{m}, y, z)$$

in which the cut-formulas have at most $n - 1$ alternating quantifiers prefixed to a Σ_1^0-formula or its negation.

Step 1: Remove the outermost quantifier from all cut-formulas by cut elimination. This increases the bound from $\omega \cdot d$ to $2^{\omega \cdot d}$.

Step 2: Transform the resulting derivation into one with bound $\omega^{d+1} + \omega^d \cdot 3 + 2$. We show how this is done afterwards.

3 to n: Successively remove outermost quantifiers from all cut-formulas, each time increasing the bound by an exponential to the base ω as in Note 3.12. What remains is a derivation of

$$\max \vec{m} : \mathsf{N} \vdash^{\alpha} \exists y \exists z . C_f(\vec{m}, y, z)$$

with only Σ_1^0-cuts, and where

$$\alpha = \exp_\omega^{n-2}(\omega^{d+1} + \omega^d \cdot 3 + 2) \prec \exp_\omega^n(1) = (\varepsilon_0)_n$$

as required.

It remains to justify step 2. Given any tree-ordinal γ of the form

$$\gamma = 2^{\omega \cdot d_1} \cdot c_1 + \cdots + 2^{\omega \cdot d_k} \cdot c_k$$

where $d_1 > \cdots > d_k$ and $c_i > 0$, define γ' by

$$\gamma' = \omega^{d_1} \cdot (3c_1) + \cdots + \omega^{d_k} \cdot (3c_k).$$

Then by induction on $\gamma \prec 2^{\omega \cdot \omega}$ we have

$$\beta \in \gamma[m] \text{ implies } \beta' + 2 \in \gamma'[e(m)] \qquad (\dagger)$$

where $e(m) = 3 \cdot 2^{m+1} - 1$.

The case $\gamma = 0$ holds vacuously. Otherwise we can write $\gamma = \delta + 2^{\omega \cdot d} \cdot c$. If γ is a successor then $d = 0$ and $\beta \in \gamma[m]$ implies $\beta \in \delta + (c-1)[m] \cup \{\delta + (c-1)\}$. By the induction hypothesis we then have

$$\beta' + 2 \in \delta' + 3(c-1)[e(m)] \cup \{\delta' + 3c - 1\} \subset \gamma'[e(m)].$$

Finally if γ is a limit then $d > 0$ and $\beta \in \gamma[m]$ implies $\beta \in \gamma_m[m]$. So by the induction hypothesis $\beta' + 2 \in \gamma'_m[e(m)]$. But $\gamma_m = \delta + 2^{\omega \cdot d} \cdot (c-1) + 2^{\omega \cdot (d-1)} . 2^{m+1}$ and so

$$\gamma'_m = \delta' + \omega^d \cdot (3c-3) + \omega^{d-1} \cdot (e(m)+1).$$

Therefore

$$\beta' + 2 \in \gamma'_m[e(m)] = \delta' + \omega^d \cdot (3c-2)[e(m)] \subset \gamma'[e(m)].$$

This completes the proof of ($\dagger$).

We now use this to prove

$$n : \mathsf{N} \vdash^{\gamma} \mathbf{A} \text{ implies } n : \mathsf{N} \vdash^{\gamma''} \mathbf{A}$$

where if $\gamma = 2^{\omega \cdot d_1} \cdot c_1 + \cdots + 2^{\omega \cdot d_k} \cdot c_k$ as above, then $\gamma'' = \omega^{d_1+1} + \gamma' + 2$. Furthermore the transformation from bound γ to bound γ'' does not alter any of the cut-formulas. The proof is by induction over derivations. The axiom case is immediate and all other cases have the same form. For suppose

$n : \mathsf{N} \vdash^{\gamma} \mathbf{A}$ is derived from premises $m : \mathsf{N} \vdash^{\beta} \mathbf{A}, B$ by any rule, where $\beta \in \gamma[m]$. By the induction hypothesis we have $m : \mathsf{N} \vdash^{\beta''} \mathbf{A}, B$ and by Weakening, $e(m) : \mathsf{N} \vdash^{\gamma''-2} \mathbf{A}, B$ since by (†), $\beta' + 2 \in \gamma'[e(m)]$ and hence $\beta'' \in \gamma'' - 2[e(m)]$. The only reason for the extra ω^{d_1+1} on the front of γ'' is to ensure that $B_{\gamma''-2}(m) \geq B_{\omega+1}(m) \geq e(m)$. Hence $m : \mathsf{N} \vdash^{\gamma''-2} e(m) : \mathsf{N}$ by Bounding Lemma 3.15 (1). Therefore by an N-cut, $m : \mathsf{N} \vdash^{\gamma''-1} \mathbf{A}, B$. We can now re-apply the rule in question to obtain $n : \mathsf{N} \vdash^{\gamma''} \mathbf{A}, B$ as desired.
Step 2 now follows by putting $\gamma = 2^{\omega \cdot d}$ so that $\gamma'' = \omega^{d+1} + \omega^d \cdot 3 + 2$. This completes 4.9.

4.10. Theorem.

$$\text{REC}(\omega^2) \subseteq \text{PROVREC}(\Sigma_1^0\text{-IND})$$

Proof. By 3.20 $B_{\omega \cdot k}$ is primitive recursive, for each $k \in \mathbf{N}$ and so by the Hierarchy Theorem, every function in $\text{REC}(\omega^2)$ is primitive recursive. We therefore only need show that every primitive recursive function is provably recursive in Σ_1^0-IND. This is done by assigning to each primitive recursive definition of a function f, a bounded formula $C_f(\vec{x}, y, z)$ with the intuitive meaning: "z is a sequence code which describes the step-by-step computation of $f(\vec{x})$, ending with output y". The formula $\exists y \exists z . C_f(\vec{x}, y, z)$ then Σ_1^0-defines $f(\vec{x})$ and we merely have to show it to be provable in Σ_1^0-IND.

1. If f is defined by one of the initial schemes:

$$f(\vec{x}) = 0 \quad \text{or} \quad f(\vec{x}) = x_1 + 1 \quad \text{or} \quad f(\vec{x}) = x_i$$

then take $C_f(\vec{x}, y, z)$ to be the conjunction of $z = 0$ (the empty sequence) with

$$y = 0 \quad \text{or} \quad y = x_1 + 1 \quad \text{or} \quad y = x_i$$

respectively. Then in each case we have

$$C_f(\vec{x}, 0, 0) \quad \text{or} \quad C_f(\vec{x}, x_1 + 1, 0) \quad \text{or} \quad C_f(\vec{x}, x_i, 0)$$

provable immediately by identity axioms, hence $\vdash \exists y \exists z . C_f(\vec{x}, y, z)$ by the $\exists$-rule and hence $\vdash \forall \vec{x}\, \exists y \exists z . C_f(\vec{x}, y, z)$ by the $\forall$-rule.

2. Suppose f is defined from g_0, g_1 and g_2 by the substitution scheme:

$$f(\vec{x}) = g_0(g_1(\vec{x}), g_2(\vec{x}))$$

and assume inductively that g_0, g_1 and g_2 have already been assigned "computation formulas" C_0, C_1 and C_2 so that in Σ_1^0-IND

$$\vdash \forall \vec{x}\, \exists y \exists z . C_i(\vec{x}, y, z).$$

Then take $C_f(\vec{x}, y, z)$ to be the formula

$$\begin{aligned} &lh(z) = 3 \wedge (z)_0 \neq 0 \wedge (z)_1 \neq 0 \wedge (z)_2 \neq 0 \wedge y = u((z)_0) \\ &\wedge C_1(\vec{x}, u((z)_1), v((z)_1)) \wedge C_2(\vec{x}, u((z)_2), v((z)_2)) \\ &\wedge C_0(u((z)_1), u((z)_2), u((z)_0), v((z)_0)). \end{aligned}$$

Now from the arithmetical axioms in PA it's easy to see that we can derive

$$\vdash \neg C_1(\vec{x}, y_1, z_1), \neg C_2(\vec{x}, y_2, z_2), \neg C_0(y_1, y_0, z_0),$$
$$C_f(\vec{x}, y_0, \langle p(y_0, z_0), p(y_1, z_1), p(y_2, z_2)\rangle).$$

Then by applying the quantifier rules in the correct order we obtain

$$\vdash \exists\vec{x}\forall y\forall z\,.\,\neg C_1(\vec{x}, y, z), \exists\vec{x}\forall y\forall z\,.\,\neg C_2(\vec{x}, y, z),$$
$$\exists\vec{x}\forall y\forall z\,.\,\neg C_0(\vec{x}, y, z), \forall\vec{x}\,\exists y\exists z\,.\,C_f(\vec{x}, y, z),$$

and from this by three successive cuts,

$$\vdash \forall\vec{x}\,\exists y\exists z\,.\,C_f(\vec{x}, y, z).$$

3. Suppose f is defined from g_0 and g_1 by primitive recursion:

$$f(\vec{x}, 0) = g_0(\vec{x}) \quad , \quad f(\vec{x}, w+1) = g_1(\vec{x}, w, f(\vec{x}, w)).$$

Assume g_0 and g_1 have already been assigned formulas C_0, C_1 such that in Σ_1^0-IND we have

$$\vdash \forall\vec{x}\,\exists y\exists z\,.\,C_0(\vec{x}, y, z)$$
$$\vdash \forall\vec{x}\,\forall w\forall w'\exists y\exists z\,.\,C_1(\vec{x}, w, w', y, z).$$

Then take $C_f(\vec{x}, w, y, z)$ to be the formula

$$lh(z) = w + 1 \land \forall i < w + 1\,.\,((z)_i \neq 0) \land y = u((z)_w)$$
$$\land\, C_0(\vec{x}, u((z)_0), v((z)_0))$$
$$\land\, \forall i < w\,.\,C_1(\vec{x}, i, u((z)_i), u((z)_{i+1}), v((z)_{i+1})).$$

Using the arithmetical axioms we obtain

$$\vdash \neg C_0(\vec{x}, y, z), C_f(\vec{x}, 0, u((t)_0), t)$$

by choosing the term $t = \langle p(y, z)\rangle$, and

$$\vdash \neg C_f(\vec{x}, w, y, z), \neg C_1(\vec{x}, w, y, y', z'), C_f(\vec{x}, w+1, y', t')$$

by choosing $t' = p(z, p(y', z'))$. Thus by appropriate quantifier rules,

$$\vdash \exists\vec{x}\forall y\forall z\,.\,\neg C_0(\vec{x}, y, z), \exists y\exists z\,.\,C_f(\vec{x}, 0, y, z)$$

and

$$\vdash \forall y\forall z\,.\,\neg C_f(\vec{x}, w, y, z),$$
$$\exists\vec{x}\exists w\exists w'\forall y\forall z\,.\,\neg C_1(\vec{x}, w, w', y, z),$$
$$\exists y\exists z\,.\,C_f(\vec{x}, w+1, y, z)$$

and hence by applying cut to each one:

$$\vdash \exists y \exists z . C_f(\vec{x}, 0, y, z) \quad \text{and} \quad \vdash \forall y \forall z . \neg C_f(\vec{x}, w, y, z), \exists y \exists z . C_f(\vec{x}, w+1, y, z).$$

Then by Σ_1^0-IND we obtain

$$\vdash \exists y \exists z . C_f(\vec{x}, w, y, z)$$

and by $\forall$-rules

$$\vdash \forall \vec{x}\, \forall w \exists y \exists z . C_f(\vec{x}, w, y, z).$$

This completes the proof since every primitive recursive function is definable by a sequence of the above schemes.

4.11. Theorem.

$$\text{REC}((\varepsilon_0)_n) \subseteq \text{PROVREC}(\Pi_n^0\text{-IND}) \qquad (n \geq 2)$$

Proof. By the Hierarchy Theorem and the Comparisons Lemma it suffices to show that for every $\alpha \prec (\varepsilon_0)_n$, the fast-growing F_α is provably recursive in Π_n^0-IND. For if $f \in E(B_\alpha)$ then $f \in E(F_\alpha)$ since $B_\alpha \leq F_\alpha$ on all non-zero arguments. Thus f is primitive recursively definable from F_α and so by the proof of 4.10, f will also be provably recursive by some further Σ_1^0-inductions.

We need to construct an appropriate "computation formula" C_F for the F-functions. To do this we must first define a coding of tree-ordinals $\alpha, \beta, \gamma, \ldots, \prec \varepsilon_0$ as numbers $a = \ulcorner \alpha \urcorner$, $b = \ulcorner \beta \urcorner$, $c = \ulcorner \gamma \urcorner$, ... respectively, so that the basic ordinal operations can be simulated by the arithmetical term-structure of PA. This is easy to do because each $\alpha \prec (\varepsilon_0)_n$ can be represented in Cantor normal form thus

$$\alpha = \omega^{\alpha_0} \cdot m_0 + \omega^{\alpha_1} \cdot m_1 + \cdots + \omega^{\alpha_{k-1}} \cdot m_{k-1}$$

where $\alpha_{k-1} \prec \cdots \prec \alpha_1 \prec \alpha_0 \prec \exp_\omega^{n-1}(1)$ and $m_0, \ldots, m_{k-1}$ are positive integers. So if α_i is coded as the number a_i we can take as code for α:

$$a = \langle p(a_0, m_0), p(a_1, m_1), \ldots, p(a_{k-1}, m_{k-1}) \rangle$$

The ordinal 0 will be coded as the empty sequence 0. Succ(a) then stands for the formula $(u(v(a)) = 0 \wedge v(v(a)) \neq 0)$ and Lim(a) stands for the formula $(u(v(a)) \neq 0 \wedge v(v(a)) \neq 0)$.

We can then define elementary number theoretic functions

1. $(a, b) \mapsto a \oplus b$
2. $(b, n) \mapsto \omega^b \cdot (n+1)$
3. $(a, x) \mapsto l(a, x)$

such that, if a encodes α and b encodes β then

1. $a \oplus b$ encodes $\alpha + \beta$ (provided the lowest exponent of ω in α is no smaller than the highest exponent of ω in β)
2. $\omega^b \cdot (n+1)$ encodes $\omega^\beta \cdot (n+1)$

3. $l(a,x)$ encodes α_x if α is a limit.

Since these functions are primitive recursive, they are provably recursive in Σ_1^0-IND by the foregoing theorem and we may therefore use them freely to form new terms, in such a way as to render the equation

$$(a \oplus \omega^b \cdot (u+1)) \oplus \omega^b \quad = \quad a \oplus \omega^b \cdot (u+2)$$

provable in Σ_1^0-IND. We sometimes write ω^a for $\omega^a \cdot 1$.

Now we can easily define a computation formula $C_H(a,x,y,z)$ for the Hardy functions $H_\alpha(x) = y$, by formalizing the statement:

z is a sequence $\langle p(a_0,x_0), \dots, p(a_{k-1},x_{k-1})\rangle$ where $a_0 = 0$, $x_0 = y$, $a_{k-1} = a$, $x_{k-1} = x$ and for each $i < k-1$; if $\mathrm{Succ}(a_{i+1})$ then $a_i \oplus \ulcorner 1 \urcorner = a_{i+1}$ and $x_i = x_{i+1} + 1$, and if $\mathrm{Lim}(a_{i+1})$ then $a_i = l(a_{i+1}, x_{i+1})$ and $x_i = x_{i+1}$.

But since $F_\alpha = H_{\omega^\alpha}$ we can therefore take as the computation formula for F:

$$C_F(a,x,y,z) \equiv C_H(\omega^a, x, y, z).$$

The following are derivable in Σ_1^0-IND for all terms a and b:

1. $\vdash \exists y \exists z . C_H(\ulcorner 1 \urcorner, x, y, z)$ since $\vdash C_H(\ulcorner 1 \urcorner, x, x+1, p(p(0, p(0, x+1)), p(\ulcorner 1 \urcorner, x)))$.
2. $\vdash \neg\forall x \exists y \exists z . C_H(a,x,y,z), \neg\exists y \exists z . C_H(b,x,y,z), \exists y \exists z . C_H(a \oplus b, x, y, z)$.
For if $z' = \langle p(0,y), \dots, p(b,x)\rangle$ codes the computation of $H_\beta(x) = y$ and $z'' = \langle p(0,y'), \dots, p(a,y)\rangle$ codes the computation of $H_\alpha(y)$, then $z = \langle p(0,y'), \dots, p(a,y), \dots, p(a \oplus b, x)\rangle$ codes the computation of $H_{\alpha+\beta}(x) = H_\alpha(H_\beta(x)) = y'$. Clearly z is primitive recursively (and so provably recursively in Σ_1^0-IND) definable from z' and z''. Therefore in Σ_1^0-IND we have
$$\vdash \neg C_H(a,y,y',z''), \neg C_H(b,x,y,z'), \exists z . C_H(a \oplus b, x, y', z)$$
and (2) follows by the quantifier rules.
3. $\vdash \neg\mathrm{Lim}(a), \neg\exists y \exists z . C_H(a_x, x, y, z), \exists y \exists z . C_H(a,x,y,z)$
since $\vdash \neg\mathrm{Lim}(a), \neg C_H(a_x,x,y,z), C_H(a,x,y,p(z,p(a,x)))$.
4. $\vdash \exists y \exists z . C_F(0,x,y,z)$ by (1) since $\omega^0 = 1$.
5. $\vdash \neg\forall x \exists y \exists z . C_F(a,x,y,z), \exists y \exists z . C_F(a \oplus \ulcorner 1 \urcorner, x, y, z)$.
First substitute ω^a for a and $\omega^a \cdot (u+1)$ for b throughout the derivation of (2) to obtain
$$\begin{aligned}\vdash\ & \neg\forall x \exists y \exists z . C_H(\omega^a, x, y, z), \\ & \neg\exists y \exists z . C_H(\omega^a \cdot (u+1), x, y, z), \\ & \exists y \exists z . C_H(\omega^a \cdot (u+2), x, y, z)\end{aligned}$$
using $\vdash \omega^a \oplus \omega^a \cdot (u+1) = \omega^a \cdot (u+2)$. Since the base case
$$\vdash \neg\forall x \exists y \exists z . C_H(\omega^a, x, y, z), \exists y \exists z . C_H(\omega^a, x, y, z)$$

follows by logic, we can now apply Σ_1^0-induction over u to obtain

$$\vdash \neg\forall x\exists y\exists z\,.\,C_H(\omega^a, x, y, z), \exists y\exists z\,.\,C_H(\omega^a\cdot(x+1), x, y, z).$$

Hence by (3) with a replaced by $\omega^{a\oplus\ulcorner 1\urcorner}$,

$$\vdash \neg\forall x\exists y\exists z\,.\,C_H(\omega^a, x, y, z), \exists y\exists z\,.\,C_H(\omega^{a\oplus\ulcorner 1\urcorner}, x, y, z)$$

which is what was required since $C_F(a, x, y, z)$ is $C_H(\omega^a, x, y, z)$.

6. $\vdash \neg\mathrm{Lim}(a), \neg\exists y\exists z\,.\,C_F(a_x, x, y, z), \exists y\exists z\,.\,C_F(a, x, y, z)$
   again by (3) with a replaced by ω^a.

Now in order to prove F_α totally-defined we clearly need the principle of transfinite induction up to α. This can be formalized—and appropriate instances of it proved—in PA, using a neat method of Gentzen [1943]. See also Schütte [1977], Feferman [1968] and Takeuti [1987]. Let

$$TI(a, A) \equiv Prog(A) \to \forall c\,.\,(c \preceq a \to A(c))$$

where $Prog(A)$ meaning A is "progressive", is the formula

$$\forall d(A(0) \wedge (A(d) \to A(d \oplus \ulcorner 1\urcorner)) \wedge (\mathrm{Lim}(d) \wedge \forall x\,.\,A(d_x) \to A(d)))$$

and where $(c \prec a)$ is a Σ_1^0-formula defining the partial ordering $\gamma \prec \alpha$ on tree-ordinals, i.e, "there is a sequence $z = \langle p(c_0, z_0), \ldots, p(c_k, z_k)\rangle$ with $c_0 = c$, $c_k = a$, and such that for each $i < k$, if $\mathrm{Succ}(c_{i+1})$ then $c_i \oplus \ulcorner 1\urcorner = c_{i+1}$, and if $\mathrm{Lim}(c_{i+1})$ then $c_i = l(c_{i+1}, z_{i+1})$".

For any formula $A(c)$ with a distinguished free variable c define $A'(b)$ to be the formula $\forall a(A(a) \to A(a \oplus \omega^b))$, and let $A^{(n+1)} \equiv (A^{(n)})'$. Note that if A is Π_m^0 then A' is Π_{m+1}^0 when brought to prenex form.

We have the following three lemmas:

7. If A is a Π_n^0-formula then

$$\Pi_n^0\text{-}\mathrm{IND} \vdash \neg Prog(A), Prog(A')$$

   We argue informally. First assume $Prog(A)$. The first conjunct of $Prog(A')$ is $A'(0)$ which follows straight from $A(a) \to A(a \oplus \ulcorner 1\urcorner)$. The second conjunct is $A'(d) \to A'(d\oplus\ulcorner 1\urcorner)$. To prove this assume $A'(d)$ and $A(a)$. Then $A(a\oplus\omega^{d\oplus\ulcorner 1\urcorner})$ comes about by a Π_n^0-induction on the formula $A(a \oplus \omega^d\cdot(u+1))$. The base case $A(a \oplus \omega^d)$ comes from $A'(d)$ and $A(a)$; the induction step comes from $A'(d)$ with a instantiated by $a \oplus \omega^d\cdot(u+1)$. Hence $\forall x A(a \oplus \omega^d\cdot(x+1))$ and then $A(a \oplus \omega^{d\oplus\ulcorner 1\urcorner})$ follows from $Prog(A)$. Therefore we have proved $\neg A'(d), A(a) \to A(a\oplus\omega^{d\oplus\ulcorner 1\urcorner})$ and hence by $\forall$-introduction and $\vee$-introduction, $A'(d) \to A'(d\oplus\ulcorner 1\urcorner)$. The third conjunct of $Prog(A')$ is $(\mathrm{Lim}(d)\wedge\forall x\,.\,A'(d_x)) \to A'(d)$. But if $\mathrm{Lim}(d)$ and $\forall x\,.\,A'(d_x)$ then for any a, $A(a) \to \forall x\,.\,A(a \oplus \omega^{d_x})$, so by $Prog(A)$ yet again, $A(a) \to A(a \oplus \omega^d)$. Hence $(\mathrm{Lim}(d) \wedge \forall x\,.\,A'(d_x)) \to A'(d)$. Putting these cases together gives $Prog(A')$.

8. If A is a Π^0_n-formula then

$$\Pi^0_n\text{-IND} \vdash \neg TI(b, A'), TI(\omega^b, A).$$

For, assuming $TI(b, A')$ and $Prog(A)$, we have by (7) $Prog(A')$ and therefore $\forall c \preceq b . \forall a . (A(a) \to A(a \oplus \omega^c))$. Using this we can then prove $\forall d \preceq \omega^b . A(d)$ as follows. Suppose $d \preceq \omega^b$. Recall that for *fixed* d and i, $in(d, i)$ is the initial part of sequence d, with length i, so if $i \leq lh(d)$ then $in(d, i) = \omega^{d_0} \cdot m_0 \oplus \cdots \oplus \omega^{d_{i-1}} \cdot m_{i-1}$ where $d_0, \ldots, d_{i-1} \preceq b$. However, in the formal theory we can actually prove $(i \leq lh(d)) \to in(d, i+1) = in(d, i) \oplus \omega^{d_i} \cdot m_i$. We are going to prove $A(in(d, i))$ by a Π^0_n-induction on i. The base case is just $A(0)$ which follows immediately from $Prog(A)$. The induction step is $A(in(d, i)) \to A(in(d, i+1))$ which we prove by another Π^0_n-induction. For informally, assume $A(in(d, i))$, $i \leq lh(d)$ and $d_i \preceq b$. For any a we have $A(a) \to A(a \oplus \omega^{d_i})$. Therefore putting a equal to $in(d, i)$ gives

$$A(in(d, i) \oplus \omega^{d_i})$$

and putting a equal to $in(d, i) \oplus \omega^{d_i} \cdot (m+1)$ gives

$$A(in(d, i) \oplus \omega^{d_i} \cdot (m+1)) \to A(in(d, i) \oplus \omega^{d_i} \cdot (m+2)).$$

Hence by Π^0_n-induction on m we obtain $A(in(d, i) \oplus \omega^{d_i} \cdot m_i)$. Consequently $A(in(d, i+1))$ and this completes the inductive proof of $A(in(d, i))$. If we now put $i = lh(d)$ and use $d = in(d, lh(d))$ we finally obtain $A(d)$. Hence $TI(b, A')$ implies $TI(\omega^b, A)$.

9. If A is a Π^0_2-formula then for any $k \in \mathbf{N}$,

$$\Pi^0_n\text{-IND} \vdash TI(\ulcorner \exp^{n-1}_\omega(k) \urcorner, A) \qquad \text{for } n \geq 2.$$

This is proved by iterating (8) above. Note that A' is Π^0_3, A'' is Π^0_4, ..., $A^{(n-2)}$ is Π^0_n. For any fixed $k \in \mathbf{N}$ we have in Π^0_n-IND, $\vdash TI(\ulcorner k \urcorner, A^{(n-1)})$, hence $\vdash TI(\omega^{\ulcorner k \urcorner}, A^{(n-2)})$, hence $\vdash TI(\omega^{\omega^{\ulcorner k \urcorner}}, A^{(n-3)})$ etc. until finally $\vdash TI(a, A)$ where $a = \ulcorner \exp^{n-1}_\omega(k) \urcorner$.

Now we can complete the proof of the theorem by choosing A to be the formula $\forall x \exists y \exists z . C_F(a, x, y, z)$. By (4), (5) and (6) this formula is progressive. So by (9),

$$\Pi^0_n\text{-IND} \vdash \forall a \prec \ulcorner \exp^{n-1}_\omega(k) \urcorner . \forall x \exists y \exists z . C_F(a, x, y, z)$$

for every fixed $k \in \mathbf{N}$. Therefore if a is a *fixed* code for a tree-ordinal $\alpha \prec (\varepsilon_0)_n$, the formula $\forall x \exists y \exists z . C_F(a, x, y, z)$ defining F_α is provable in Π^0_n-IND.

Putting 3.19(1), 3.20, 4.9, 4.10 and 4.11 together and recalling 2.23 and 4.3, we obtain our main subrecursive classification for arithmetic.

4.12. Classifications.

1. The primitive recursive functions can be classified as

$$\begin{aligned}\text{ProvRec}(\Sigma_1^0\text{-Ind}) &= \text{Rec}(\omega^2)\\ &= \bigcup_{\alpha\prec\omega^2} E(B_\alpha)\\ &= \bigcup_{\alpha\prec\omega} E(F_\alpha) = \bigcup_{\alpha\prec\omega^\omega} E(H_\alpha).\end{aligned}$$

2. The multiply-recursive functions of Péter [1967] can be classified as

$$\begin{aligned}\text{ProvRec}(\Sigma_2^0\text{-Ind}) &= \text{Rec}(\omega^\omega)\\ &= \bigcup_{\alpha\prec\omega^\omega} E(B_\alpha)\\ &= \bigcup_{\alpha\prec\omega^\omega} E(F_\alpha) = \bigcup_{\alpha\prec\omega^{\omega^\omega}} E(H_\alpha)\end{aligned}$$

3. More generally, for $n \geq 2$, and recalling $(\varepsilon_0)_0 = 1$, $(\varepsilon_0)_{m+1} = \omega^{(\varepsilon_0)_m}$;

$$\begin{aligned}\text{ProvRec}(\Sigma_n^0\text{-Ind}) &= \text{Rec}((\varepsilon_0)_n)\\ &= \bigcup_{\alpha\prec(\varepsilon_0)_n} E(B_\alpha)\\ &= \bigcup_{\alpha\prec(\varepsilon_0)_n} E(F_\alpha) = \bigcup_{\alpha\prec(\varepsilon_0)_{n+1}} E(H_\alpha)\end{aligned}$$

4. And finally,

$$\begin{aligned}\text{ProvRec}(\text{PA}) &= \text{Rec}(\varepsilon_0)\\ &= \bigcup_{\alpha\prec\varepsilon_0} E(B_\alpha)\\ &= \bigcup_{\alpha\prec\varepsilon_0} E(F_\alpha) = \bigcup_{\alpha\prec\varepsilon_0} E(H_\alpha)\end{aligned}$$

4.13. Remarks. Kreisel [1952], using ideas from Ackermann [1940], was first to show that the provably recursive functions of PA are definable by transfinite recursions over order-types $\prec \varepsilon_0$. The refinements of this result, showing the provably recursive functions of Π_n^0-Ind to be those definable by recursion over order-types $\prec (\varepsilon_0)_n$, are due to Parsons [1966]. The result that the provably recursive functions of Σ_1^0-Ind are exactly the primitive recursive functions is due variously to Parsons [1970], Mints [1973] and Takeuti [1987]. The corresponding subrecursive classifications in terms of the fast-growing F_αs follow from Grzegorczyk [1953] for $\alpha \prec \omega$; from Robbin [1965] for $\alpha \prec \omega^\omega$; and from Löb and Wainer [1970], Wainer [1970] and Schwichtenberg [1971] for $\alpha \prec \varepsilon_0$. See also Constable [1971], Buchholz and Wainer [1987] and Rose [1984].

The H_α's first appeared in a subrecursive context in Wainer [1972], though their definition and first use (exhibiting a set of reals with cardinality $\aleph_1$) appears very early in Hardy [1904]. The development of the $\mathrm{REC}(\alpha)$ classes and corresponding B_α-functions stems from Fairtlough [1991]. See also Fairtlough and Wainer [1992]. Many other significant contributions to the theory of provably recursive functions have been (and are still being) made—see for instance Buchholz [1987], Buchholz, Cichon and Weiermann [1994], Buss [1994], Friedman and Sheard [1995], Girard [1981], Leivant [1995], Ratajczyk [1993], Schwichtenberg [1977], Sieg [1985,1991], Tucker and Zucker [1992] and Weiermann [1996]. Furthermore there are non-standard model-theoretic methods running parallel to the proof-theoretic ones employed here—see for example Paris [1980], Hajek and Pudlak [1991], Avigad and Sommer [1997] and Sommer [1995]. The field is now quite broad, and we can only apologise to those many contributors whom we have not mentioned.

5. Independence results for PA

From 4.12(4) we see immediately that B_{ε_0}, F_{ε_0} and H_{ε_0} eventually dominate all the provably recursive functions of PA, and so they cannot themselves be provably recursive in PA. Consequently

$$\mathrm{PA} \nvdash \forall n \forall x \exists y \exists z \,.\, C_H((\varepsilon_0)_n, x, y, z).$$

In other words, the proof of 4.11 shows that although for any fixed n we have

$$\mathrm{PA} \vdash TI((\varepsilon_0)_n, A)$$

there are Π^0_2-instances of A for which

$$\mathrm{PA} \nvdash TI(\varepsilon_0, A).$$

Another way to see this would be to note that, again for an appropriate formula A

$$\mathrm{PA} + TI(\varepsilon_0, A) \vdash \text{“PA is consistent”}.$$

For, by Gentzen, if a false atom were derivable in PA, then by the Embedding and Cut-Elimination theorems, it would be derivable in the infinitary system with cut-rank 0 and ordinal bound $\alpha \prec \varepsilon_0$. By induction up to ε_0, this is impossible since an infinitary rank-0 proof of an atom cannot use logic. Formalisation of this argument would lead to a proof of $TI(\varepsilon_0, A) \vdash \mathrm{Con(PA)}$ in Primitive Recursive Arithmetic. Of course, Gödel's second theorem tells us that there is no proof of Con(PA) inside PA, hence no proof of $TI(\varepsilon_0, A)$.

Thus ε_0 is the least upper bound of the “provable ordinals” of PA. See Gentzen [1943].

These may be termed “logic” independence results. The question then was whether they might be equivalent to other independence results of a more clear mathematical character. This was shown to be the case by Paris and Harrington [1977] who proved the independence (from PA) of a certain finite version of Ramsey's theorem, and by Ketonen and Solovay [1981] who then gave a "rate-of-growth" analysis of it in terms of the fast-growing hierarchy. In this section we treat another simpler independence result of Kirby and Paris [1982] concerning so-called "Goodstein sequences", but the proof we give is due to Cichon [1983] since it is directly related to the H-functions.

5.1. Definition. Given numbers a, $x \geq 1$ with $a \leq \exp_{x+1}^{x}(1)$ the "complete base-$(x+1)$ form" of a is

$$a = (x+1)^{a_1} \cdot m_1 + (x+1)^{a_2} \cdot m_2 + \cdots + (x+1)^{a_k} \cdot m_k$$

wherein $m_1, m_2, \ldots, m_k \leq x$ and all exponents $a_1 > \cdots > a_k$, and their exponents etc, are again written in base-$(x+1)$ form.

Let $g(x, a)$ be the number which results by writing $a-1$ in complete base-$(x+1)$ form, and then increasing the base in this expression from $(x+1)$ to $(x+2)$, leaving all coefficients m_i fixed.

The *Goodstein sequence* on (x, a) is then the sequence of numbers $\{a_i\}_{i \geq x}$ where $a_x = a$ and $a_{x+j+1} = g(x+j, a_{x+j})$.

5.2. Definition. Given a number a written in complete base-$(x+1)$ form, let $\text{ord}_x(a)$ be the tree-ordinal obtained by replacing the base throughout by ω.

5.3. Lemma. $\text{ord}_x(a-1) = P_x(\text{ord}_x(a))$ *for* $a > 0$.

Proof. The proof is by induction on a. If $a = 1$ then $\text{ord}_x(a-1) = 0 = P_x(1)$. Suppose $a > 1$ and that the complete base-$(x+1)$ form of a is

$$a = (x+1)^{a_1} \cdot m_1 + (x+1)^{a_2} \cdot m_2 + \cdots + (x+1)^{a_k} \cdot m_k.$$

If $a_k = 0$ then $\text{ord}_x(a-1) = \text{ord}_x(a) - 1 = P_x(\text{ord}_x(a))$. If $a_k > 0$ then let

$$b = (x+1)^{a_1} \cdot m_1 + (x+1)^{a_2} \cdot m_2 + \cdots + (x+1)^{a_k} \cdot (m_k - 1).$$

Then

$$a - 1 = b + (x+1)^{a_k - 1} \cdot x + (x+1)^{a_k - 2} \cdot x + \cdots + (x+1)^0 \cdot x.$$

Let $\alpha = \text{ord}_x(a)$, $\beta = \text{ord}_x(b)$ and $\alpha_k = \text{ord}_x(a_k)$. Then by the induction hypothesis we have

$$\text{ord}_x(a-1) = \beta + \omega^{P_x(\alpha_k)} \cdot x + \omega^{P_x^2(\alpha_k)} \cdot x + \cdots + x.$$

Therefore by the properties of the function P_x we obtain

$$\text{ord}_x(a-1) = P_x(\beta + \omega^{\alpha_k}) = P_x(\text{ord}_x(a))$$

5.4. Lemma. $g(x, a) = G_{x+1}(P_x(\text{ord}_x(a)))$.

Proof. By the definitions, note that

$$g(x,a) = G_{x+1}(\mathrm{ord}_x(a-1))$$

since G_{x+1} replaces base ω by $(x+2)$, as in 2.11. The result then follows from 5.3.

5.5. Lemma. *Let $a_x, a_{x+1}, a_{x+2}, \ldots$ be the Goodstein sequence on (x,a). Then for each j*

1. $\mathrm{ord}_{x+j}(a_{x+j}) = P_{x+j-1}\, P_{x+j-2} \cdots P_x(\mathrm{ord}_x(a))$.
2. $a_{x+j} = G_{x+j}(\mathrm{ord}_{x+j}(a_{x+j}))$.

Proof.

1. By induction on j. The base case is trivial and for the induction step we have by 5.3,

$$\begin{aligned}\mathrm{ord}_{x+j+1}(g(x+j, a_{x+j})) &= \mathrm{ord}_{x+j}(a_{x+j}-1) \\ &= P_{x+j}(\mathrm{ord}_{x+j}(a_{x+j})).\end{aligned}$$

   Hence $\mathrm{ord}_{x+j+1}(a_{x+j+1}) = P_{x+j}(\mathrm{ord}_{x+j}(a_{x+j}))$ and the result follows immediately from the induction hypothesis.

2. This is immediate by iterating 5.4.

5.6. Theorem. (Cichon [1983]) *Let $\{a_i\}_{i\geq x}$ be the Goodstein sequence on (x,a). Then there is a y such that $a_y = 0$, and the least such y is given by $y = H_{\mathrm{ord}_x(a)}(x)$.*

Proof. By 5.5, $\mathrm{ord}_{x+j+1}(a_{x+j+1}) \prec \mathrm{ord}_{x+j}(a_{x+j})$ if $\mathrm{ord}_{x+j}(a_{x+j}) \neq 0$. By well-foundedness there must be a first stage k at which $\mathrm{ord}_{x+k}(a_{x+k}) = 0$ and hence $a_{x+k} = 0$. By Theorem 2.19 we can express this k as

$$\begin{aligned}k &= \text{least } k\,.\,(P_{x+k-1}\, P_{x+k-2} \cdots P_x(\mathrm{ord}_x(a)) = 0) \\ &= D_{\mathrm{ord}_x(a)}(\mathrm{succ})(x)\end{aligned}$$

and therefore $x + k = H_{\mathrm{ord}_x(a)}(x)$.

5.7. Theorem. (Kirby and Paris [1982]) *Let Good(a,x,y) be a Σ^0_1-formula of arithmetic expressing the fact that the Goodstein sequence on (x,a) terminates at y, i.e. $a_y = 0$. Then $\forall a \forall x \exists y\,.\,\mathrm{Good}(a,x,y)$ is true by 5.6, but not provable in* PA.

Proof. If it were a theorem of PA, the function $h(a,x) = \text{least } y\,.\,\mathrm{Good}(a,x,y)$ would be provably recursive in PA. For each x, set $a(x) = \exp^x_{x+1}(1)$. Then $a(x)$ is primitive recursive and so the function $h(a(x),x)$ would also be provably recursive in PA. However, by 5.6 and since $\mathrm{ord}_x(a(x)) = (\varepsilon_0)_x$, we have $h(a(x),x) = H_{\varepsilon_0}(x)$. This contradicts 4.12(4).

6. The "true" ordinal of PA

Section 4 characterizes the provably recursive functions of PA in terms of ε_0-recursiveness but, recalling definitions 3.17, it still remains to characterize them in terms of γ-definability. We shall now "compute" the appropriate γ by appealing to the Hierarchy Theorems 3.19 and finding an ordinal map $\alpha \mapsto \alpha^+$ such that for $\alpha \preceq \varepsilon_0$, and even much larger α's,

$$B_\alpha = G_{\alpha^+}$$

We then have

$$\text{ProvRec(PA)} = \text{Rec}(\varepsilon_0) = \Sigma^0_1\text{-Def}(\varepsilon_0^+).$$

For related results and an alternative treatment in terms of Gödel's system T of primitive recursive functionals, see Schwichtenberg and Wainer [1995]. ε_0^+ is the proof-theoretic ordinal of the theory of one inductive definition and is usually referred to as the Bachmann-Howard ordinal (see Howard [1970]). Girard [1981] was the first to give a detailed analysis of the relationship between the fast-growing and the slow-growing hierarchies and once the correct result was known, many others gave more direct and simpler analyses. We shall follow the treatment in Cichon and Wainer [1983] and more generally, Wainer [1989]. The main point is that, in order to describe the map $\alpha \mapsto \alpha^+$, one needs to make use of "higher number classes" of uncountable tree-ordinals. However, since we are only concerned here with "small" α's below ε_0, we only need to go to the "next" number class over Ω.

6.1. Definition. Let $\Omega_0 = \mathrm{N}$ and $\Omega_1 = \Omega$. Then the set Ω_2 is generated inductively according to the four rules:

Zero. $0 \in \Omega_2$

Succ. $\alpha \in \Omega_2 \Longrightarrow \alpha + 1 \in \Omega_2$

Lim_0. $\forall x \in \Omega_0(\alpha_x \in \Omega_2) \Longrightarrow \alpha = \langle \alpha_x \rangle \in \Omega_2$

Lim_1. $\forall \xi \in \Omega_1(\alpha_\xi \in \Omega_2) \Longrightarrow \alpha = \langle \alpha_\xi \rangle \in \Omega_2$

Note: we sometimes write $\alpha = \sup \alpha_x$ or $\alpha = \text{SUP}\alpha_\xi$ according to whether $\alpha = \langle \alpha_x \rangle$ or $\alpha = \langle \alpha_\xi \rangle$ in Ω_2.

6.2. Definition. The (well-founded) "subtree" partial ordering $\prec$ on Ω_2 is defined as the transitive closure of the rules

- $\alpha \prec \alpha + 1$
- $\forall x \in \Omega_0(\alpha_x \prec \sup \alpha_x)$
- $\forall \xi \in \Omega_1(\alpha_\xi \prec \text{SUP}\alpha_\xi)$

6.3. Arithmetic on Ω_2. Addition, multiplication and exponentiation of Ω_2 are defined exactly as in 2.11 for Ω_1, but with an extra limit clause in each case, viz.

$$\begin{aligned} \alpha + \text{SUP}\beta_\xi &= \text{SUP}(\alpha + \beta_\xi) \\ \alpha \cdot \text{SUP}\beta_\xi &= \text{SUP}(\alpha \cdot \beta_\xi) \\ \alpha^{\text{SUP}\beta_\xi} &= \text{SUP}(\alpha^{\beta_\xi}). \end{aligned}$$

6.4. Examples.

1. $\omega_0 = \sup x$, $\omega_1 = \mathrm{SUP}(1+\xi)$
2. $\varepsilon_{\omega_1+1} = \sup(1, \omega_1, \omega_1^{\omega_1}, \omega_1^{\omega_1^{\omega_1}}, \ldots)$.

6.5. Definition. The slow growing function $G : \Omega_1 \times \Omega_0 \to \Omega_0$ is now extended to a map $G : \Omega_2 \times \Omega_0 \to \Omega_1$. As in 2.11 it will be notationally convenient to swap the arguments and write, for each fixed $n \in \mathrm{N}$, $G_n(\alpha)$ instead of $G_\alpha(n)$. Thus for each n we define $G_n : \Omega_2 \to \Omega_1$ by the following recursion:

$$\begin{aligned} G_n(0) &= 0 \\ G_n(\alpha+1) &= G_n(\alpha)+1 \\ G_n(\sup \alpha_x) &= G_n(\alpha_n) \\ G_n(\mathrm{SUP}\alpha_\xi) &= \sup G_n(\alpha_x) \end{aligned}$$

Note that we immediately have, for every n,

$$G_n(\omega_1) = 1 + \omega_0 = \omega \in \Omega_1.$$

6.6. Lemma. *For each fixed $n \in \mathrm{N}$ and all α, $\beta \in \Omega_2$,*

$$\begin{aligned} G_n(\alpha+\beta) &= G_n(\alpha)+G_n(\beta) \\ G_n(\alpha\cdot\beta) &= G_n(\alpha)\cdot G_n(\beta) \\ G_n(\alpha^\beta) &= G_n(\alpha)^{G_n(\beta)} \end{aligned}$$

Proof. This is by easy inductions on $\beta \in \Omega_2$.

6.7. Definition. Let $\mathrm{EXP} \subseteq \Omega_2$ be generated inductively according to the rules:

- $\Omega_1 \cup \{\omega_1\} \subseteq \mathrm{EXP}$
- $\alpha, \beta \in \mathrm{EXP} \Longrightarrow \alpha+\beta, \alpha\cdot\beta, \alpha^\beta \in \mathrm{EXP}$.

6.8. Lemma. *Fix $n \in \mathrm{N}$. Then for every $\gamma \in \mathrm{EXP}$ of the form $\gamma = \mathrm{SUP}\gamma_\xi$ we have*

$$\forall \xi \in \Omega_1 (G_n(\gamma_\xi) = G_n(\gamma)_{G_n(\xi)})$$

Proof. This is by induction on the generation of $\gamma \in \mathrm{EXP}$. The base case 1 is easy because the only possibility is $\gamma = \omega_1$, so then $\gamma_\xi = 1+\xi$ and $G_n(\gamma_\xi) = 1 + G_n(\xi) = \omega_{G_n(\xi)} = G_n(\gamma)_{G_n(\xi)}$. In case 2 suppose for example that $\gamma = \alpha\cdot\beta$ where $\alpha \in \mathrm{EXP}$ and $\beta \in \mathrm{EXP}$. Then there are two sub-cases. Either $\beta = \beta'+1$, in which case $\gamma = \alpha\cdot\beta' + \alpha$ and hence by the induction hypothesis applied to α,

$$G_n(\gamma_\xi) = G_n(\alpha\cdot\beta' + \alpha_\xi) = G_n(\alpha\cdot\beta') + G_n(\alpha)_{G_n(\xi)} = G_n(\gamma)_{G_n(\xi)}.$$

Or $\beta = \mathrm{SUP}\beta_\xi$ in which case $\gamma_\xi = \alpha\cdot\beta_\xi$, and hence by induction hypothesis applied to β,

$$G_n(\gamma_\xi) = G_n(\alpha)\cdot G_n(\beta_\xi) = G_n(\alpha)\cdot G_n(\beta)_{G_n(\xi)} = G_n(\gamma)_{G_n(\xi)}.$$

6.9. Definition. The fast-growing hierarchy $B : \Omega_1 \times \Omega_0 \to \Omega_0$ is "lifted" to a hierarchy $\varphi : \Omega_2 \times \Omega_1 \to \Omega_1$ as follows, writing $\varphi_\alpha(\beta)$ instead of $\varphi(\alpha, \beta)$:

$$\begin{aligned} \varphi_0(\beta) &= \beta + 1 \\ \varphi_{\alpha+1}(\beta) &= \varphi_\alpha(\varphi_\alpha(\beta)) \\ \varphi_\alpha(\beta) &= \sup \varphi_{\alpha_x}(\beta) \quad \text{if } \alpha = \sup \alpha_x \\ \varphi_\alpha(\beta) &= \varphi_{\alpha_\beta}(\beta) \quad \text{if } \alpha = \mathrm{SUP}\alpha_\xi \end{aligned}$$

6.10. Collapsing Theorem. *Fix $n \in \mathrm{N}$. Then for every $\alpha \in \mathrm{EXP}$ and all $\beta \in \Omega_1$ we have*

$$G_n(\varphi_\alpha(\beta)) = B_{G_n(\alpha)}(G_n(\beta)).$$

Proof. First note that

- if $\alpha + 1 \in \mathrm{EXP}$ then $\alpha \in \mathrm{EXP}$
- if $\alpha = \sup \alpha_x \in \mathrm{EXP}$ then $\alpha_x \in \mathrm{EXP}$ for every $x \in \mathrm{N}$
- if $\alpha = \mathrm{SUP}\alpha_\xi \in \mathrm{EXP}$ then $\alpha_\xi \in \mathrm{EXP}$ for every $\xi \in \Omega_1$.

The proof of these facts is by an exhaustive case-analysis according to the inductive definition of EXP. We leave it as an exercise. However, (1), (2) and (3) mean that we can proceed by $\prec$-induction to show that for every $\alpha \in \Omega_2$,

$$\alpha \in \mathrm{EXP} \Longrightarrow \forall \beta \in \Omega_1 (G_n(\varphi_\alpha(\beta)) = B_{G_n(\alpha)}(G_n(\beta))).$$

The cases are easy:

If $\alpha = 0$ then $G_n(\alpha) = 0$ and therefore we have

$$G_n(\varphi_0(\beta)) = G_n(\beta + 1) = G_n(\beta) + 1 = B_0(G_n(\beta)).$$

If $\alpha = \alpha' + 1$ then $G_n(\alpha) = G_n(\alpha') + 1$ and $\varphi_\alpha = \varphi_{\alpha'} \circ \varphi_{\alpha'}$, so by the induction hypothesis (twice) we have

$$G_n(\varphi_\alpha(\beta)) = B_{G_n(\alpha')} \circ B_{G_n(\alpha')}(G_n(\beta)) = B_{G_n(\alpha)}(G_n(\beta)).$$

If $\alpha = \sup \alpha_x$ then $G_n(\alpha) = G_n(\alpha_n)$ and $\varphi_\alpha = \sup \varphi_{\alpha_x}$, so by the induction hypothesis applied to α_n we have

$$G_n(\varphi_\alpha(\beta)) = G_n(\varphi_{\alpha_n}(\beta)) = B_{G_n(\alpha_n)}(G_n(\beta)) = B_{G_n(\alpha)}(G_n(\beta)).$$

If $\alpha = \mathrm{SUP}\alpha_\xi$ then $G_n(\alpha) = \sup G_n(\alpha_x)$ and $\varphi_\alpha(\beta) = \varphi_{\alpha_\beta}(\beta)$, so by the induction hypothesis applied to α_β, and since $G_n(\alpha_\beta) = G_n(\alpha)_{G_n(\beta)}$ by Lemma 6.8, we have

$$G_n(\varphi_\alpha(\beta)) = G_n(\varphi_{\alpha_\beta}(\beta)) = B_{G_n(\alpha)_{G_n(\beta)}}(G_n(\beta)) = B_{G_n(\alpha)}(G_n(\beta)).$$

Note that this proof will go through for any $\prec$-closed subset of Ω_2 satisfying 6.8. We did it for EXP since that is all we need here.

6.11. Corollary. *Suppose $\alpha \prec \varepsilon_0$. Then α can be written in exponential "Cantor normal form"*

$$\alpha = \omega^{\alpha_0} \cdot m_0 + \omega^{\alpha_1} \cdot m_1 + \cdots + \omega^{\alpha_k} \cdot m_k$$

where $\alpha_k \prec \alpha_{k-1} \prec \cdots \prec \alpha_1 \prec \alpha_0$ are also of this form, and their exponents, etc.

Let $\hat{\alpha} \in$ EXP be the result of replacing ω by ω_1 throughout this normal form. Now define $\alpha^+ = \varphi_{\hat{\alpha}}(\omega_0) \in \Omega_1$. Then by 6.6 and the facts that $G_n(\omega_1) = \omega$ and $G_n(m) = m$ for $m \in \mathbf{N}$, we have

$$G_n(\hat{\alpha}) = \alpha$$

and therefore by 6.10 we have for every n,

$$G_n(\alpha^+) = B_\alpha(n).$$

Thus, reverting to the old notation for G,

$$B_\alpha = G_{\alpha_+}.$$

6.12. Examples.

1. $(\omega^\omega)^+ = \varphi_{\omega_1^{\omega_1}}(\omega_0)$
2. The obvious and natural extension of $\hat{\alpha}$ to the case $\alpha = \varepsilon_0$ is $\hat{\varepsilon_0} = \varepsilon_{\omega_1+1}$, since for every n, $(\varepsilon_{\omega_1+1})_n = (\hat{\varepsilon_0})_n$. In this case we obtain, for each n,

$$\begin{aligned} G_n(\varepsilon_0^+) &= G_n(\varphi_{\varepsilon_{\omega_1+1}}(\omega_0)) \\ &= G_n(\varphi_{(\varepsilon_{\omega_1+1})_n}(\omega_0)) \\ &= B_{(\varepsilon_0)_n}(n) = B_{\varepsilon_0}(n). \end{aligned}$$

Thus $B_{\varepsilon_0} = G_{\varepsilon_0^+}$ where $\varepsilon_0^+ = \varphi_{\varepsilon_{\omega_1+1}}(\omega_0)$.

6.13. Remark. $\varepsilon_0^+ = \varphi_{\varepsilon_{\omega_1+1}}(\omega_0)$ is our tree-ordinal representation of the Bachmann-Howard ordinal. The reader will note that we have not yet proved that this ordinal is structured. We *will* do so, but not until after the main result below:

6.14. Theorem.

1. $\bigcup_{\alpha \prec \varepsilon_0} E(B_\alpha) = \bigcup_{\alpha \prec \varepsilon_0^+} E(G_\alpha)$ *and hence*
2. $\text{ProvRec(PA)} = \text{Rec}(\varepsilon_0) = \Sigma_1^0\text{-Def}(\varepsilon_0^+)$.

Proof.

1. If $\alpha \prec \varepsilon_0$ then $\alpha \preceq (\varepsilon_0)_n$ for some n, and we then have $E(B_\alpha) \subseteq E(B_{(\varepsilon_0)_n})$. But by 6.11 $B_{(\varepsilon_0)_n} = G_{(\varepsilon_0)_n^+} = G_{(\varepsilon_0^+)_n}$ and hence

$$E(B_\alpha) \subseteq E(G_{(\varepsilon_0^+)_n}).$$

This proves the containment from left to right.
Conversely if $\alpha \prec \varepsilon_0^+$ then $\alpha \preceq (\varepsilon_0^+)_n$ for some n and so, assuming the structuredness of ε_0^+, we have

$$E(G_\alpha) \subseteq E(G_{(\varepsilon_0^+)_n}) = E(B_{(\varepsilon_0)_n})$$

and this proves the containment from right to left.

2. This follows almost immediately from (1) by means of the Hierarchy Theorems 3.19. The only necessity is to check the conditions on the second Hierarchy Theorem in the case where $\alpha = \varepsilon_0^+$. An easy computation gives

$$\omega_0 + 2^\omega = \varphi_\omega(\omega_0) = \varphi_{\omega_1}(\omega_0) \prec \varepsilon_0^+,$$

and if $\alpha_0,\ \alpha_1 \prec \varepsilon_0^+$ then $\alpha_0,\ \alpha_1 \preceq (\varepsilon_0^+)_n$ for some fixed n; so again assuming the structuredness of ε_0^+, both G_{α_0} and G_{α_1} are eventually dominated by $G_{(\varepsilon_0^+)_n} = B_{(\varepsilon_0)_n}$, and therefore $G_{\alpha_0} \circ G_{\alpha_1}$ is eventually dominated by $B_{(\varepsilon_0)_n+1}$ and hence by $B_{(\varepsilon_0)_{n+1}} = G_{(\varepsilon_0^+)_{n+1}}$. We can then ensure that the composition $G_{\alpha_0} \circ G_{\alpha_1}$ is completely dominated by some G_α with $\alpha \prec \varepsilon_0^+$, by choosing α to be $(\varepsilon_0^+)_{n+1} + k$ for some large enough k.

6.15. Note. One can easily read off corresponding results for the fragments of PA. For example, for $n > 1$,

$$\text{PROVREC}(\Pi_n^0\text{-IND}) = \text{REC}((\varepsilon_0)_n) = \Sigma_1^0\text{-DEF}((\varepsilon_0^+)_n).$$

6.16. It now remains to prove that

$$\varepsilon_0^+ = \varphi_{\varepsilon_{\omega_1}+1}(\omega_0)$$

is structured. This has been done by Kadota [1993] and independently (though previously unpublished) by the second author. We first need to develop an appropriate notion of structuredness for Ω_2.

Definition. (cf. 2.6) For each $\alpha \in \Omega_2$ and all $n \in \mathbf{N}$, $\beta \in \Omega_1$, define the finite set $\alpha[n, \beta]$ of $\prec$-predecessors of α by

$$\begin{aligned} 0[n,\beta] &= \varnothing \\ \alpha+1[n,\beta] &= \alpha[n,\beta] \cup \{\alpha\} \\ \alpha[n,\beta] &= \alpha_n[n,\beta] \quad \text{if } \alpha = \sup \alpha_x \\ \alpha[n,\beta] &= \alpha_\beta[n,\beta] \quad \text{if } \alpha = \text{SUP}\alpha_\xi. \end{aligned}$$

6.17. Lemma. *For all $\alpha, \gamma \in \Omega_2$, all $n \in \mathbf{N}$ and all $\beta \in \Omega_1$ we have*

$$\gamma \in \alpha[n,\beta] \Longrightarrow \varphi_\gamma(\beta) \in \varphi_\alpha(\beta)[n].$$

Proof. We proceed by induction over $\alpha \in \Omega_2$. The zero case is trivial and both limit cases follow immediately from the induction hypothesis and the definitions of $\varphi_\alpha(\beta)$ and $\alpha[n,\beta]$. For the successor case suppose $\gamma \in \alpha+1[n,\beta] = \alpha[n,\beta] \cup \{\alpha\}$. Then $\varphi_\gamma(\beta) \in \varphi_\alpha(\beta) + 1[n] \subseteq \varphi_{\alpha+1}(\beta)[n]$ since $\delta + 1[n] = \varphi_0(\delta)[n] \subseteq \varphi_\alpha(\delta)[n]$ for any δ.

6.18. Definition. Let $\prec^S$ be the transitive closure of the rules $\alpha \prec^S \alpha + 1$; $\forall n(\alpha_n \prec^S \sup \alpha_x)$; $\forall \gamma \in \Omega_1^S(\alpha_\gamma \prec^S \text{SUP}\alpha_\xi)$.

6.19. Definition. Call $\alpha \in \Omega_2$ *structured* if for all "small limits" $\lambda \preceq^S \alpha$,

$$\forall \gamma \in \Omega_1^S - \{0\}.\forall n \in \mathbf{N}.(\lambda_n \in \lambda[n+1,\gamma]).$$

Let Ω_2^S denote the set of all structured $\alpha \in \Omega_2$ and note that $\alpha \in \Omega_2^S$ and $\beta \prec^S \alpha$ imply that $\beta \in \Omega_2^S$.

6.20. Lemma. *For every $\alpha \in \Omega_2^S$ we have:*

$$\varphi_\alpha : (\Omega_1^S - \{0\}) \to (\Omega_1^S - \{0\}).$$

Proof. We proceed by induction on $\alpha \in \Omega_2^S$. Let $\beta \in \Omega_1^S - \{0\}$. If $\alpha = 0$ then $\varphi_\alpha(\beta) = \beta + 1 \in \Omega_1^S - \{0\}$. For the successor case $\alpha \to \alpha + 1$ we have $\varphi_{\alpha+1}(\beta) = \varphi_\alpha(\varphi_\alpha(\beta)) \in \Omega_1^S - \{0\}$ by two applications of the induction hypothesis. If $\alpha = \sup \alpha_x$ then $\varphi_\alpha(\beta) = \sup \varphi_{\alpha_x}(\beta)$ and by the induction hypothesis $\varphi_{\alpha_x}(\beta) \in \Omega_1^S - \{0\}$ for each x. Thus we only need check the structuredness condition 6.19 for $\varphi_\alpha(\beta)$ itself. Now $\alpha_x \in \alpha[x+1,\beta]$ for each x, because α is structured, and therefore by Lemma 6.17, $\varphi_{\alpha_x}(\beta) \in \varphi_\alpha(\beta)[x+1]$. If $\alpha = \mathrm{SUP}\alpha_\xi$ then $\varphi_\alpha(\beta) = \varphi_{\alpha_\beta}(\beta)$ and $\alpha_\beta \prec^S \alpha$ and this case is then immediate by the induction hypothesis.

6.21. Lemma. (cf. 2.12) *For all α, β, $\delta \in \Omega_2$, all $\gamma \in \Omega_1$ and all $n \in \mathbf{N}$ we have*

1. $\delta \in \beta[n,\gamma] \Longrightarrow \alpha + \delta \in \alpha + \beta[n,\gamma]$
2. $\delta \in \beta[n,\gamma] \Longrightarrow \alpha \cdot \delta \in \alpha \cdot \beta[n,\gamma]$ *if* $0 \in \alpha[n,\gamma]$
3. $\delta \in \beta[n,\gamma] \Longrightarrow \alpha^\delta \in \alpha^\beta[n,\gamma]$ *if* $1 \in \alpha[n,\gamma]$.

Proof. This is almost identical to that of 2.12 but with, in each case, an additional trivial step corresponding to "big limits" $\beta = \mathrm{SUP}\beta_\xi$.

6.22. Theorem. (cf. 2.13) *If α, $\beta \in \Omega_2^S$ then*

1. $\alpha + \beta \in \Omega_2^S$
2. $\alpha \cdot \beta \in \Omega_2^S$ *provided* $\forall \gamma \in \Omega_1^S - \{0\}.\forall n > 0.(0 \in \alpha[n,\gamma])$
3. $\alpha^\beta \in \Omega_2^S$ *provided* $\forall \gamma \in \Omega_1^S - \{0\}.\forall n > 0.(1 \in \alpha[n,\gamma])$.

Proof. All parts are similar, by induction on $\beta \in \Omega_2^S$. We just do (3) assuming (2). If $\beta = 0$ then $\alpha^0 = 1 \in \Omega_2^S$. For the successor case $\beta \to \beta + 1$ we have $\alpha^{\beta+1} = \alpha^\beta \cdot \alpha \in \Omega_2^S$ by the induction hypothesis and part (2), since the proviso on α ensures that α^β satisfies the proviso in (2). If $\beta = \sup \beta_x$ then $\alpha^\beta = \sup \alpha^{\beta_x}$ and by the induction hypothesis $\alpha^{\beta_x} \in \Omega_2^S$ for each $x \in \mathbf{N}$. Also $\beta_x \in \beta[x+1,\gamma]$ for all $\gamma \in \Omega_1^S - \{0\}$ and all $x \in \mathbf{N}$, so $\alpha^{\beta_x} \in \alpha^\beta[x+1,\gamma]$ by 6.21 (3). Therefore $\alpha^\beta \in \Omega_2^S$. If $\beta = \mathrm{SUP}\beta_\xi$ then $\alpha^\beta = \mathrm{SUP}\alpha^{\beta_\xi}$ and since, by the induction hypothesis, $\alpha^{\beta_\xi} \in \Omega_2^S$ whenever $\xi \in \Omega_1^S$, we automatically have $\alpha^\beta \in \Omega_2^S$.

6.23. Theorem. $\varepsilon_0^+ = \varphi_{\varepsilon_{\omega_1+1}}(\omega_0) \in \Omega_1^S$.

Proof. First note that $\omega_1 = \mathrm{SUP}(1+\xi) \in \Omega_2^S$, since if $\lambda = \sup \lambda_x \preceq^S \omega_1$ then $\lambda \preceq 1+\gamma$ for some $\gamma \in \Omega_1^S$. Therefore $\lambda \in \Omega_1^S$ and so $\lambda_n \in \lambda[n+1]$ for all $n \in \mathbb{N}$. But since λ is countable, $\lambda[n+1] = \lambda[n+1,\xi]$ for all $\xi \in \Omega_1$. Hence we have $\forall \xi \in \Omega_1^S - \{0\} \,.\, \forall n \in \mathbb{N} \,.\, (\lambda_n \in \lambda[n+1,\xi])$. Note also that $1 \in \omega_1[n,\xi]$ whenever $\xi \in \Omega_1^S - \{0\}$ and $n > 0$. Therefore by 6.22 (3), $\omega_1^\beta \in \Omega_2^S$ whenever $\beta \in \Omega_2^S$. Hence $1, \omega_1, \omega_1^{\omega_1}, \omega_1^{\omega_1^{\omega_1}}, \ldots$ all belong to Ω_2^S.

Now let $\lambda = \varepsilon_{\omega_1+1}$. Then $\lambda = \sup \lambda_x$ where $\lambda_0 = 1$ and $\lambda_{n+1} = \omega_1^{\lambda_n}$. Fixing $n \in \mathbb{N}$ and $\gamma \in \Omega_1^S - \{0\}$ we have $1 \in \omega_1[n+1,\gamma]$, and therefore by n successive applications of 6.21 (3), $\lambda_n \in \lambda_{n+1}[n+1,\gamma] = \lambda[n+1,\gamma]$. Hence $\varepsilon_{\omega_1+1} \in \Omega_2^S$. Thus $\varphi_{\varepsilon_{\omega_1+1}}(\omega_0) \in \Omega_1^S$ by Lemma 6.20.

The results of this section suggest that Peano Arithmetic could be reformulated with a weaker "pointwise" induction scheme, sufficient only to prove termination of the slow-growing G functions. But then, in order to capture all the provably recursive functions of PA, these inductions would have to extend over all initial segments of the Howard ordinal. In this way, the Howard ordinal becomes the ordinal of PA with pointwise induction. The realisation of this idea, and appropriate formulation of pointwise induction schemes, is due to Schmerl [1982].

7. Theories with transfinite induction

This final section shows how the foregoing subrecursive classifications for PA can be extended quite easily to theories obtained from it by adding the Principle of Transfinite Induction over given well-orderings of order type $\succ \varepsilon_0$. Since proof-theoretic ordinal analysis seeks to compute for a given theory T, the least upper bound τ of its "provable ordinals" (see Pohlers in this volume), the results here will then immediately give a classification of the provably recursive functions of T viz.

$$\textsc{ProvRec}(T) = \textsc{Rec}(\tau) = \bigcup_{\alpha \prec \tau} E(F_\alpha).$$

Of course, τ must be shown to satisfy the conditions of our Hierarchy Theorem, or something like them, and this often requires some checking! See Buchholz, Cichon and Weiermann [1994] for related work involving similar kinds of conditions. See also Weiermann [1996] for an alternative treatment of PA and transfinite induction in terms of "ordinal majorisation" relations.

We shall assume henceforth that $\gamma = \sup \gamma_x \succ \omega$ is a structured, countable tree-ordinal which is "primitive recursively representable", i.e. q-space representable for some primitive recursive q. The representability of γ ensures that it will be possible to code the well-ordering relation $\alpha \prec \beta$ for $\alpha, \beta \preceq \gamma$, by a Σ_1^0-formula of arithmetic "$a \prec b$" (just as was done for ε_0 in the proof of Theorem 4.12 and for quite general systems of ordinal notations in Sommer [1992]). There will be primitive recursive functions $a \mapsto a \oplus \ulcorner 1 \urcorner$ representing the successor, and $(a,x) \mapsto l(a,x)$ such that if a encodes α then $l(a,x)$ encodes α_x when α is a limit, and $l(a,x) = 0$

otherwise. For simplicity we shall assume that PA is extended to include them as new terms, also denoted $a \oplus \ulcorner 1 \urcorner$ and $l(a,x)$, with appropriate defining axioms for them. We shall also assume that 0 is the least element of the well-ordering and that the value of $a \oplus \ulcorner 1 \urcorner$ is always numerically larger than a. The top element of the well-ordering, representing γ itself, will be denoted by c and we shall write $\mathrm{Lim}(a)$ for "$l(a,1) \neq 0$".

7.1. Definition. PA + TI(γ) is the theory obtained by adding to PA the Principle of Transfinite Induction up to γ, formulated either as an axiom-scheme or as a rule. We choose the rule TI(γ):

$$\frac{\vdash \mathbf{A}, B(0) \quad \vdash \mathbf{A}, \neg B(a), B(a \oplus \ulcorner 1 \urcorner) \quad \vdash \mathbf{A}, \neg \mathrm{Lim}(a), \exists x \neg B(l(a,x)), B(a)}{\vdash \mathbf{A}, \forall a(a \not\prec c \vee B(a))}$$

with the restriction that a is not free in $\mathbf{A}$.

The embedding of PA + TI(γ) into ω-Arithmetic follows the same lines as the embedding of PA in 4.7. Applications of the TI(γ) rule are dealt with as follows:

7.2. Lemma. *Suppose the three premises of the* TI(γ) *rule have been embedded into ω-Arithmetic with a fixed ordinal bound $\delta \in \Omega^S$, so that* $\vdash^\delta \mathbf{A}, B(0)$ *and for every $a \in \mathbf{N}$,*

$$a : \mathsf{N} \vdash^\delta \mathbf{A}, \neg B(a), B(a \oplus \ulcorner 1 \urcorner)$$
$$a : \mathsf{N} \vdash^\delta \mathbf{A}, \neg \mathrm{Lim}(a), \exists x \neg B(l(a,x)), B(a).$$

Furthermore suppose that δ is so chosen that for all a, $n \in \mathbf{N}$,

1. *the term $l(a, n \dot{-} 1)$ has numerical value bounded by $B_\delta(\max(a,n))$,*
2. *the cardinality of $\delta[a]$ is at least $|B|$, the height of the induction formula.*

Then for every $\alpha \prec \gamma$, if a is the number encoding α we have

$$a : \mathsf{N} \vdash^{\delta+5\cdot\alpha+3} \mathbf{A}, B(a).$$

Proof. Proceed by induction on $\alpha \prec \gamma$ (for notational simplicity we shall suppress the side formulas $\mathbf{A}$). The case $\alpha = 0$ is immediate and the successor case from α to $\alpha + 1$ is also straightforward, since from the assumption $a : \mathsf{N} \vdash^\delta \neg B(a), B(a \oplus \ulcorner 1 \urcorner)$ and the induction hypothesis $a : \mathsf{N} \vdash^{\delta+5\cdot\alpha+3} B(a)$, we obtain $a : \mathsf{N} \vdash^{\delta+5\cdot\alpha+4} B(a \oplus \ulcorner 1 \urcorner)$ by Cut and hence $a \oplus \ulcorner 1 \urcorner : \mathsf{N} \vdash^{\delta+5\cdot(\alpha+1)+3} B(a \oplus \ulcorner 1 \urcorner)$ by Weakening.

Now suppose $\alpha = \sup \alpha_x$ and choose any $n \in \mathbf{N}$. Then letting m denote the numerical value of $l(a, n \dot{-} 1)$, we have $m : \mathsf{N} \vdash^{\delta+5\cdot\alpha_{n\dot{-}1}+3} B(l(a, n \dot{-} 1))$ by the induction hypothesis, and $\max(a,n) : \mathsf{N} \vdash^\delta m : \mathsf{N}$ by the Bounding Lemma 3.15 and condition 1. Therefore by an N-Cut we obtain for every n,

$$\max(a,n) : \mathsf{N} \vdash^{\delta+5\cdot\alpha_{n\dot{-}1}+4} B(l(a, n \dot{-} 1)).$$

The structuredness of α gives $\delta + 5 \cdot \alpha_{n \dot{-} 1} + 4 \in \delta + 5 \cdot \alpha[\max(a, n)]$ and so an application of the $\forall$-rule yields

$$a : \mathsf{N} \vdash^{\delta+5\cdot\alpha} \forall x B(l(a, x \dot{-} 1)).$$

The reason for the second condition on δ is that it ensures (we leave the reader to check it),

$$a : \mathsf{N} \vdash^{\delta+5\cdot\alpha} \exists x \neg B(l(a, x \dot{-} 1)), \forall x B(l(a, x)).$$

Hence by Cut we obtain $a : \mathsf{N} \vdash^{\delta+5\cdot\alpha+1} \forall x B(l(a, x))$. Therefore from the assumption

$$a : \mathsf{N} \vdash^{\delta} \neg \mathrm{Lim}(a), \exists x \neg B(l(a, x)), B(a)$$

and since $\vdash^{\delta} \mathrm{Lim}(a)$ is an axiom, we obtain by two further Cuts,

$$a : \mathsf{N} \vdash^{\delta+5\cdot\alpha+3} B(a)$$

and this completes the proof.

Now in order to prove the Embedding Theorem for $\mathrm{PA} + \mathrm{TI}(\gamma)$ there is one further crucial requirement to be placed on γ. Clearly if γ is coded as a number-theoretic well-ordering there must be a "norm" function with the property that whenever $\alpha \prec \gamma$ is encoded by the number a, we have $\alpha \in \gamma[\mathrm{norm}(a)]$. See Buchholz, Cichon and Weiermann [1994]. For "standard" codings of proof-theoretic ordinals this norm function will often be just the identity, but we shall merely require that it be primitive recursive.

7.3. Embedding of $\mathrm{PA} + \mathrm{TI}(\gamma)$. *Suppose γ is primitive recursively representable, with a primitive recursive norm as above. Suppose*

$$\mathrm{PA} + \mathrm{TI}(\gamma) \vdash \mathbf{A}(t_1(\vec{x}), \ldots, t_k(\vec{x})).$$

Then there is a number d, measuring the "size" of this proof, such that for every assignment of numbers $\vec{n}$ to the variables $\vec{x}$, we have

$$\max \vec{n} : \mathsf{N} \vdash^{5\cdot\gamma\cdot d} \mathbf{A}(m_1, \ldots, m_k)$$

where $m_1, \ldots, m_k$ are the numerical values of the terms $t_1(\vec{n}), \ldots, t_k(\vec{n})$. Furthermore this infinitary derivation has finite cut-rank.

Proof. All the cases of PA-rules carry over straightforwardly just as in 4.8, but with ω now replaced by $5 \cdot \gamma$. The only case we need worry about is the application of the $\mathrm{TI}(\gamma)$ rule. Assume inductively that its premises are all embedded in ω-Arithmetic with ordinal bound $\delta = 5 \cdot \gamma \cdot d$. Assume also that d is chosen large enough so that conditions 1 and 2 of the previous Lemma are satisfied by δ. Then what we need to prove is

$$\vdash^{5\cdot\gamma\cdot(d+2)} \mathbf{A}, \forall a(a \not\prec c \vee B(a))$$

(we suppress the parameters occurring in the side-formulas **A** since they play no active part in this case). Now the previous lemma gives, for every $\alpha \prec \gamma$ with code a,

$$a : \mathsf{N} \vdash^{\delta+5\cdot\alpha+3} \mathbf{A}, B(a).$$

Let $M(a,n)$ be a Σ^0_1-formula expressing the relation "norm$(a) = n$", and recall that γ itself is coded by the top element c in its number-theoretic well-ordering. Then for a sufficiently large d we have, for all a, $n \in \mathbf{N}$,

$$\max(n,a) : \mathsf{N} \vdash^{\beta_a} \mathbf{A}, \neg M(a,n) \vee a \not\prec c \vee B(a)$$

where $\beta_a = \delta + 5\cdot\alpha + 4$ if $a \prec c$ and norm$(a) = n$, and $\beta_a = \delta$ otherwise. Hence $\beta_a \in \delta + 5\cdot\gamma[\max(n,a)]$ for every a and so by the $\forall$-rule,

$$n : \mathsf{N} \vdash^{\delta+5\cdot\gamma} \mathbf{A}, \forall a(\neg M(a,n) \vee a \not\prec c \vee B(a))$$

for every n, so by the $\forall$-rule again,

$$\vdash^{\delta+5\cdot\gamma+1} \mathbf{A}, \forall x \forall a(\neg M(a,x) \vee a \not\prec c \vee B(a)).$$

The desired result

$$\vdash^{5\cdot\gamma\cdot(d+2)} \mathbf{A}, \forall a(a \not\prec c \vee B(a))$$

will then follow by a Cut if we can derive

$$\vdash^{\delta+5\cdot\gamma+1} \exists x \exists a(M(a,x) \wedge a \prec c \wedge \neg B(a)), \forall a(a \not\prec c \vee B(a)).$$

This is done as follows: for each $a \in \mathbf{N}$ with norm$(a) = m$ we have (again for a large enough d) $\vdash^{\delta} M(a,m)$ and $\vdash^{\delta} a \prec c, a \not\prec c$ and $\vdash^{\delta} \neg B(a), B(a)$. Hence $\vdash^{\delta+3} (M(a,m) \wedge a \prec c \wedge \neg B(a)), (a \not\prec c \vee B(a))$. But since the norm function is primitive recursive and $\gamma \succeq \omega$, we can assume that the chosen d is large enough to ensure $a : \mathsf{N} \vdash^{\delta} m : \mathsf{N}$ by the Bounding Lemma 3.15. Hence

$$a : \mathsf{N} \vdash^{\delta+5} \exists x \exists a(M(a,x) \wedge a \prec c \wedge \neg B(a)), (a \not\prec c \vee B(a))$$

follows by two applications of the $\exists$-rule, and a final application of the $\forall$-rule yields

$$\vdash^{\delta+5\cdot\gamma+1} \exists x \exists a(M(a,x) \wedge a \prec c \wedge \neg B(a)), \forall a(a \not\prec c \vee B(a)).$$

This completes the proof.

7.4. Classification Theorem. *Suppose $\gamma = \omega^\alpha$ is primitive recursively representable with a primitive recursive norm. Recall $\varepsilon(\alpha) := \sup_x \exp^x_\omega(1+\alpha+1)$. Then*

$$\textsc{ProvRec}(\mathrm{PA} + \mathrm{TI}(\gamma)) = \textsc{Rec}(\varepsilon(\alpha)) = \bigcup_{\beta \prec \varepsilon(\alpha)} E(F_\beta).$$

Proof. For the first containment $\subseteq$ suppose f is provably recursive in $\mathrm{PA} + \mathrm{TI}(\gamma)$. Then for an appropriate computation formula C_f we have

$$\mathrm{PA} + \mathrm{TI}(\gamma) \vdash \forall x \exists y \exists z \,.\, C_f(x, y, z).$$

The Embedding Theorem gives (for appropriate r, $d \in \mathrm{N}$)

$$\vdash_r^{5 \cdot \gamma \cdot d} \forall x \exists y \exists z \,.\, C_f(x, y, z)$$

and, with only minor modifications to the proof, the 5 could be replaced by ω thus:

$$\vdash_r^{\omega \cdot \gamma \cdot d} \forall x \exists y \exists z \,.\, C_f(x, y, z).$$

By the Cut-Elimination Theorem 3.11 we then obtain

$$\vdash_0^{\delta} \forall x \exists y \exists z \,.\, C_f(x, y, z)$$

where $\delta = \exp_\omega^r(\omega \cdot \gamma \cdot d) \prec \exp_\omega^{r+1}(1 + \alpha + 1) \prec \varepsilon(\alpha)$, and hence $f \in \mathrm{REC}(\varepsilon(\alpha))$.

The second containment follows from the Hierarchy Theorem 3.19 since $B_\beta \le F_\beta$.

For the third containment, suppose $f \in E(F_\beta)$ for some $\beta \prec \varepsilon(\alpha)$. Then f will be provably recursive in $\mathrm{PA} + \mathrm{TI}(\gamma)$ if F_β is and for some fixed $m \in \mathrm{N}$ we have $\beta \prec \exp_\omega^m(1 + \alpha + 1)$.

But in $\mathrm{PA} + \mathrm{TI}(\gamma)$ we can prove $\mathrm{TI}(\alpha)$ and hence $\mathrm{TI}(1 + \alpha + 1)$. Then by iterating part 8 of the proof of Theorem 4.11 we can prove transfinite induction up to $\exp_\omega^m(1 + \alpha + 1)$ and hence up to β. As before F_β is then provably recursive in $\mathrm{PA} + \mathrm{TI}(\gamma)$, and this completes the proof.

As a concluding example, let ID_n be the theory of an n-times iterated inductive definition and let τ_n denote its proof-theoretic ordinal (see Buchholz et al. [1981] for detailed analyses of these and other theories). Then the τ_n's have structured tree-ordinal representations and by Wainer [1989] we have $\tau_n^+ = \tau_{n+1}$. Therefore

$$\mathrm{PROVREC}(\mathrm{ID}_n) = \mathrm{REC}(\tau_n) = \Sigma_1^0\text{-}\mathrm{DEF}(\tau_{n+1})$$

and letting $\tau = \sup \tau_x$,

$$\mathrm{PROVREC}(\Pi_1^1 - \mathrm{CA}_0) = \mathrm{REC}(\tau) = \Sigma_1^0\text{-}\mathrm{DEF}(\tau).$$

A more direct analysis of general ID–theories in terms of the slow growing hierarchy is given by Arai [1991].

References

W. ACKERMANN

[1940] Zur Widerspruchsfreiheit der Zahlenentheorie, *Mathematische Annalen*, 117, pp. 162–194.

T. Arai
[1991] A slow growing analogue of Buchholz' proof, *Annals of Pure and Applied Logic*, 54, pp. 101–120.

J. Avigad and R. Sommer
[1997] A model-theoretic approach to ordinal analysis, *Bulletin of Symbolic Logic*, 3, pp. 17–52.

W. Buchholz
[1987] An independence result for (Π_1^1-CA) + (BI), *Annals of Pure and Applied Logic*, 23, pp. 131–155.

W. Buchholz, E. A. Cichon, and A. Weiermann
[1994] A uniform approach to fundamental sequences and subrecursive hierarchies, *Mathematical Logic Quarterly*, 40, pp. 273–286.

W. Buchholz, S. Feferman, W. Pohlers, and W. Sieg
[1981] *Iterated Inductive Definitions and Subsystems of Analysis*, Lecture Notes in Mathematics #897, Springer-Verlag, Berlin.

W. Buchholz and S. S. Wainer
[1987] Provably computable functions and the fast growing hierarchy, in: *Logic and Combinatorics*, S. G. Simpson, ed., vol. 65 of Contemporary Mathematics, American Mathematical Society, Providence, R.I., pp. 179–198.

S. R. Buss
[1994] The witness function method and provably recursive functions of Peano Arithmetic, in: *Proceedings of the 9th. International Congress of Logic, Methodology and Philosophy of Science*, D. Prawitz, B. Skyrms, and D. Westerstahl, eds., North-Holland, Amsterdam.

E. A. Cichon
[1983] A short proof of two recently discovered independence proofs using recursion theoretic methods, *Proceedings of the American Mathematical Society*, 87, pp. 704–706.

E. A. Cichon and S. S. Wainer
[1983] The slow growing and Grzegorczyk hierarchies, *Journal of Symbolic Logic*, 48, pp. 399–408.

R. L. Constable
[1971] Subrecursive programming languages III, the multiple recursive functions, in: *Proceedings of the 21st International Symposium on Computers and Automata*, Brooklyn Polytechnic Institute, NY, pp. 393–410.

N. J. Cutland
[1981] *Computability*, Cambridge University Press.

M. V. H. Fairtlough
[1991] *Ordinal Complexity of Recursive Programs and their Termination Proofs*, PhD thesis, Leeds University Department of Pure Mathematics.

M. V. H. Fairtlough and S. S. Wainer
[1992] Ordinal complexity of recursive definitions, *Information and Computation*, 99, pp. 123–153.

S. Feferman
[1962] Classification of recursive functions by means of hierarchies, *Journal of the American Mathematical Society*, 104, pp. 101–122.
[1968] Systems of predicative analysis II, *Journal of Symbolic Logic*, 33, pp. 193–220.

H. M. Friedman
[1978] Classically and intuitionistically provably recursive functions, in: *Proc. Higher Set Theory, Oberwolfach*, G. H. Müller and D. S. Scott, eds., Lecture Notes in Mathematics #669, Springer-Verlag, Berlin, pp. 21–27.

H. M. FRIEDMAN AND M. SHEARD

[1995] Elementary descent recursion and proof theory, *Annals of Pure and Applied Logic*, 71, pp. 1–45.

G. GENTZEN

[1936] Die Widerspruchsfreiheit der reinen Zahlentheorie, *Mathematische Annalen*, 112, pp. 493–565.

[1943] Beweisbarkeit und Unbeweisbarkeit von Anfangsfällen der Transfiniten Induction in der reiner Zahlentheorie, *Mathematische Annalen*, 119, pp. 140–161.

J.-Y. GIRARD

[1981] Π^1_2-logic part I, *Annals of Mathematical Logic*, 21, pp. 75–219.

[1987] *Proof Theory and Logical Complexity*, Bibliopolis, Naples.

A. GRZEGORCZYK

[1953] Some classes of recursive functions, *Rozprawy Matem. IV, Warsaw.*

P. HAJEK AND P. PUDLAK

[1991] *The Metamathematics of First Order Arithmetic*, Perspectives in Mathematical Logic, Springer-Verlag, Berlin.

G. H. HARDY

[1904] A theorem concerning the infinite cardinal numbers, *Quarterly Journal of Mathematics*, 35, pp. 87–94.

W. A. HOWARD

[1970] A system of abstract constructive ordinals, *Journal of Symbolic Logic*, 37, pp. 355–374.

N. KADOTA

[1993] On Wainer's notation for a minimal subrecursive inaccessible ordinal, *Mathematical Logic Quarterly*, 39, pp. 217–227.

J. KETONEN AND R. M. SOLOVAY

[1981] Rapidly growing Ramsey functions, *Annals of Mathematics*, 113, pp. 267–314.

L. A. S. KIRBY AND J. B. PARIS

[1982] Accessible independence results for Peano Arithmetic, *Bulletin of the American Mathematical Society*, pp. 285–293.

G. KREISEL

[1952] On the interpretation of non-finitist proofs II, *Journal of Symbolic Logic*, 17, pp. 43–58.

D. LEIVANT

[1995] Intrinsic theories and computational complexity, in: *Logic and Computational Complexity, International Workshop LCC'94*, D. Leivant, ed., Lecture Notes in Computer Science #960, Springer-Verlag, Berlin, pp. 177–194.

M. H. LÖB AND S. S. WAINER

[1970] Hierarchies of number theoretic functions I and II, *Arkiv für mathematische Logik und Grundlagenforschung*, 14, pp. 39–51 and 97–113. Correction in vol. 14, pages 198–199.

G. E. MINTS

[1973] Quantifier-free and one quantifier systems, *Journal of Soviet Mathematics*, 1, pp. 71–84.

J. B. PARIS

[1980] A hierarchy of cuts in models of arithmetic, in: *Model Theory of Algebra and Arithmetic*, L. Pacholski and et al., eds., Lecture Notes in Mathematics #834, Springer-Verlag, Berlin, pp. 312–337.

J. B. PARIS AND L. HARRINGTON

[1977] A mathematical incompleteness in Peano Arithmetic, in: *Handbook of Mathematical Logic*, J. Barwise, ed., North-Holland, Amsterdam, pp. 1133–1142.

C. PARSONS

[1966] Ordinal recursion in partial systems of number theory (abstract), *Notices of the American Mathematical Society*, 13, pp. 857–858.

[1970] On a number-theoretic choice schema and its relation to induction, in: *Intuitionism and Proof Theory: proceedings of the summer conference at Buffalo N.Y. 1968*, North-Holland, Amsterdam, pp. 459–473.

R. PÉTER

[1967] *Recursive Functions*, Academic Press, New York, 3rd ed.

Z. RATAJCZYK

[1993] Subsystems of true arithmetic and hierarchies of functions, *Annals of Pure and Applied Logic*, 64, pp. 95–152.

J. ROBBIN

[1965] *Subrecursive Hierarchies*, PhD thesis, Princeton University.

H. E. ROSE

[1984] *Subrecursion: Functions and Hierarchies*, vol. 9 of Oxford Logic Guides, Clarendon Press, Oxford.

U. SCHMERL

[1982] Number theory and the Bachmann-Howard ordinal, in: *Logic Colloquium '81*, J. Stern, ed., North-Holland, Amsterdam, pp. 287–298.

D. SCHMIDT

[1976] Built-up systems of fundamental sequences and hierarchies of number-theoretic functions, *Arkiv für mathematische Logik und Grundlagenforschung*, 18, pp. 47–53.

K. SCHÜTTE

[1977] *Proof Theory*, Springer-Verlag, Berlin. Translation by J. N. Crossley

H. SCHWICHTENBERG

[1971] Eine Klassifikation der ε_0-rekursiven Functionen, *Zeitschrift für mathematische Logik und Grundlagen der Mathematik*, 17, pp. 61–74.

[1977] Proof theory: Some applications of cut-elimination, in: *Handbook of Mathematical Logic*, J. Barwise, ed., North-Holland, Amsterdam, pp. 867–895.

H. SCHWICHTENBERG AND S. S. WAINER

[1995] Ordinal bounds for programs, in: *Feasible Mathematics II*, P. Clote and J. B. Remmel, eds., vol. 13 of Progress in Computer Science and Applied Logic, Birkhäuser, Boston, pp. 387–406.

W. SIEG

[1985] Fragments of arithmetic, *Annals of Pure and Applied Logic*, 28, pp. 33–71.

[1991] Herbrand analyses, *Archive for Mathematical Logic*, 30, pp. 409–441.

R. SOMMER

[1992] Ordinal arithmetic in $I\Delta_0$, in: *Arithmetic, Proof Theory and Computational Complexity*, P. Clote and J. Krajicek, eds., Oxford University Press.

[1995] Transfinite induction within Peano Arithmetic, *Annals of Pure and Applied Logic*, 76, pp. 231–289.

W. W. TAIT

[1968] Normal derivability in classical logic, in: *The Syntax and Semantics of Infinitary Languages*, J. Barwise, ed., Lecture Notes in Mathematics #72, Springer-Verlag, Berlin, pp. 204–236.

G. TAKEUTI

[1987] *Proof Theory*, North-Holland, Amsterdam, 2nd ed.

J. V. TUCKER AND J. I. ZUCKER

[1992] Provable computable selection functions on abstract structures, in: *Proof Theory*, P. H. G. Aczel, H. Simmons, and S. S. Wainer, eds., Cambridge University Press, pp. 277–306.

S. S. WAINER

[1970] A classification of the ordinal recursive functions, *Arkiv für mathematische Logik und Grundlagenforschung*, 13, pp. 136–153.

[1972] Ordinal recursion and a refinement of the extended Grzegorczyk hierarchy, *Journal of Symbolic Logic*, 38, pp. 281–292.

[1989] Slow growing versus fast growing, *Journal of Symbolic Logic*, 54, pp. 608–614.

A. WEIERMANN

[1996] How to characterise provably total functions by local predicativity, *Journal of Symbolic Logic*, 61, pp. 52–69.

CHAPTER IV

Subsystems of Set Theory and Second Order Number Theory

Wolfram Pohlers

Institut für mathematische Logik und Grundlagenforschung
Westfälische Wilhelms-Universität, D-48149 Münster, Germany

Contents

HANDBOOK OF PROOF THEORY
Edited by S. R. Buss

1. Preliminaries

The aim of the following contribution is to present a sample of ordinal analyses of subsystems of Set Theory and Second Order Number Theory.[1] But before we start the presentation of results we think we should enter a general discussion about the type of results we are going to obtain.

We want to keep close to Hilbert's program and try to give consistency proofs for axiom systems as constructively as possible – being well aware of all the obstacles to this enterprise resulting from Gödel's Theorems.

The emphasis will be on impredicative subsystems of Set Theory and Second Order Number Theory. Here we opt for a seemingly unconventional approach. We will try to construct partial models of theories within the constructible sets. The hierarchy $\mathbf{L}$ of constructible sets is determined by the ordinal line. Therefore special care will be given to notations for ordinals. In the Preliminaries we will introduce our concept of ordinal analysis. First we introduce some basic facts on ordinals, then introduce our concept of partial models and finally show how this is connected to the more conventional approach to ordinal analysis.

1.1. Ordinals

Ordinals will play the crucial role for all what follows. Therefore we start with a short introduction to ordinals as we will use them in the following contribution. Ordinals are regarded in their set theoretic sense, i.e., an ordinal is a hereditarily transitive set. Assuming that the membership relation $\in$ is a well-founded relation, this entails that every ordinal is well-founded with respect to the membership relation and every ordinal is the set of its $\in$-predecessors. For the reader who is not so familiar with set theory we give a brief sketch of the theory of ordinals as far as it will be needed for this paper. Besides transfinite recursion (which may be regarded as a generalization of primitive recursion) all we need from Set Theory are the facts *(On1)* – *(On4)* below. They may be viewed as axioms for the theory. To make the article not too long we will not give proofs here. Detailed information how to prove the results of this section from *(On1)* – *(On4)* can be found in Pohlers [1989].

A linear order relation $\prec$ well-orders its field iff it does not contain infinite $\prec$-descending sequences. A class M is transitive iff $a \in M \Rightarrow a \subseteq M$.

(On1) The class **On** *of ordinals is a non void transitive class, which is well-ordered by the membership relation* $\in$. *We define* $\alpha < \beta$ *as* $\alpha \in \mathbf{On} \land \beta \in \mathbf{On} \land \alpha \in \beta$.

In general we use lower case Greek letters as syntactical variables for ordinals. The well-foundedness of $\in$ on the class **On** implies the principle of *transfinite induction*

$$(\forall \xi \in \mathbf{On})[(\forall \eta < \xi)F(\eta) \Rightarrow F(\xi)] \Rightarrow (\forall \xi \in \mathbf{On})F(\xi)$$

[1] I am indebted to Dr. Arnold Beckmann for proofreading. He not only detected a series of errors in the first versions but also made many valuable suggestions.

and transfinite recursion which, for a given function g, allows the definition of a function f satisfying the recursion equation

$$f(\eta) = g(\{f(\xi) \mid \xi < \eta\}).$$

(On2) *The class* $\mathbf{On}$ *of ordinals is unbounded, i.e.,* $(\forall \xi \in \mathbf{On})(\exists \eta \in \mathbf{On}))[\xi < \eta]$. *The cardinality* $|M|$ *of a set* M *is the least ordinal* α *such that* M *can be mapped bijectively onto* α. *An ordinal* α *is a cardinal if* $|\alpha| = \alpha$.

(On3) *If* $M \subseteq \mathbf{On}$ *and* $|M| \in \mathbf{On}$ *then* M *is bounded in* $\mathbf{On}$, *i.e., there is an* $\alpha \in \mathbf{On}$ *such that* $M \subseteq \alpha$.

For every ordinal α we have by *(On1)* and *(On2)* a least ordinal α' which is bigger than α. We call α' the *successor* of α. There are three types of ordinals:

- *the least ordinal* 0,
- *successor ordinals, i.e., ordinals of the form* α',
- *ordinals which are neither* 0 *nor successor ordinals. Such ordinals are called limit ordinals. We denote the class of limit ordinals by* Lim.

Considering these three types of ordinals we reformulate transfinite induction and recursion as follows:

Transfinite induction: If $F(0)$ *and* $(\forall \alpha \in \mathbf{On})[F(\alpha) \Rightarrow F(\alpha')]$ *as well as* $(\forall \xi < \lambda)F(\xi) \Rightarrow F(\lambda)$ *for* $\lambda \in \mathsf{Lim}$ *then* $(\forall \xi \in \mathbf{On})F(\xi)$.

Transfinite recursion: For given $\alpha \in \mathbf{On}$ *and functions* g, h *there is a function* f *satisfying the recursion equations*

$$\begin{aligned} &f(0) = \alpha \\ &f(\xi') = g(f(\xi)) \\ &f(\lambda) = h(\{f(\eta) \mid \eta < \lambda\}) \text{ for } \lambda \in \mathsf{Lim}. \end{aligned}$$

An ordinal κ satisfying

(R1) $\kappa \in \mathsf{Lim}$

(R2) *If* $M \subseteq \kappa$ *and* $|M| < \kappa$ *then* M *is bounded in* κ, *i.e., there is an* $\alpha \in \kappa$ *such that* $M \subseteq \alpha$

is called *regular*. The class of regular ordinals is denoted by $\mathbb{R}$.

(On4) *The class* $\mathbb{R}$ *is unbounded, i.e.,* $(\forall \xi \in \mathbf{On})(\exists \eta \in \mathbb{R})[\xi \leq \eta]$.

We define

$$\sup M := \min \{\xi \in \mathbf{On} \mid (\forall \eta \in M)(\eta \leq \xi)\}$$

as the least upper bound for a set $M \subseteq \mathbf{On}$. In set theoretic terms it is $\sup M = \bigcup M$. It follows that $\sup M$ is either the biggest ordinal in M, i.e., $\sup M = \max M$, or $\sup M \in \mathsf{Lim}$. By ω we denote the least limit ordinal. It exists according to *(On4)* and *(On1)*. The ordinal ω_1 denotes the first uncountable ordinal, i.e., the first ordinal whose cardinality is bigger than that of ω. It exists by *(On3)*.

For every class $M \subseteq \mathbf{On}$ there is a uniquely determined transitive class $\mathsf{otyp}(M) \subseteq \mathbf{On}$ and an order preserving function $\mathrm{en}_M\colon \mathsf{otyp}(M) \overset{\text{onto}}{\longrightarrow} M$. The function en_M enumerates the elements of M in increasing order. Since $\mathsf{otyp}(M)$ is transitive it is either $\mathsf{otyp}(M) = \mathbf{On}$ or $\mathsf{otyp}(M) \in \mathbf{On}$. We call $\mathsf{otyp}(M)$ the *order type* of M. In fact $\mathsf{otyp}(M)$ is the Mostowski collapse of M and en_M the inverse of the collapsing function (usually denoted by π). By *(On3)* we have $\mathsf{otyp}(M) \in \mathbf{On}$ iff M is bounded in $\mathbf{On}$. Unbounded, i.e., proper classes of ordinals have order type $\mathbf{On}$. If M is a set of ordinals then $\mathsf{otyp}(M) \in \mathbf{On}$.

If M is a transitive class and $f\colon M \longrightarrow \mathbf{On}$ an order preserving function then we have $\alpha \le f(\alpha)$ for all $\alpha \in M$.

A class M is *closed* (in a regular ordinal κ) iff $\sup N \in M$ holds for every class $N \subseteq M$ such that $|N| \in \mathbf{On}$ ($|N| < \kappa$). We call M *club* (in κ) iff M is closed and unbounded (in κ).

We call an order preserving function $f\colon M \longrightarrow \mathbf{On}$ (κ-) continuous iff M is (κ-) closed and f preserves suprema, i.e., $\sup\{f(\xi) \mid \xi \in N\} = f(\sup(N))$ for any $N \subseteq M$ such that $|N| \in \mathbf{On}$ ($|N| < \kappa$).

A normal (κ-normal) function is an order-preserving continuous function

$$f\colon \mathbf{On} \longrightarrow \mathbf{On} \text{ or } f\colon \kappa \longrightarrow \kappa \text{ respectively.}$$

For $M \subseteq \mathbf{On}$ ($M \subseteq \kappa$) the enumerating function en_M is a (κ-)normal function iff M is club (in κ).

Extending their primitive recursive definitions continuously into the transfinite we obtain the basic arithmetical functions $+$, $\cdot$ and exponentiation for all ordinals. The ordinal sum, for example, satisfies the recursion equations

$$\begin{aligned}
&\alpha + 0 = \alpha\\
&\alpha + \beta' = (\alpha + \beta)'\\
&\alpha + \lambda = \sup_{\xi < \lambda}(\alpha + \xi) \text{ for } \lambda \in \mathsf{Lim}.
\end{aligned}$$

It is easy to see that the function $\lambda\xi\,.\,\alpha + \xi$ is the enumerating function of the class $\{\xi \in \mathbf{On} \mid \alpha \le \xi\}$ which is club in all regular $\kappa > \alpha$. Hence $\lambda\xi\,.\,\alpha + \xi$ is a κ-normal function for all regular $\kappa > \alpha$. We define

$$\mathbb{H} := \{\alpha \in \mathbf{On} \mid \alpha \neq 0 \wedge (\forall \xi < \alpha)(\forall \eta < \alpha)[\xi + \eta < \alpha]\}$$

and call the ordinals in $\mathbb{H}$ *additively indecomposable.* Then $\mathbb{H}$ is club (in any regular ordinal $> \omega$), $1 := 0' \in \mathbb{H}$, $\omega \in \mathbb{H}$ and $\omega \cap \mathbb{H} = \{1\}$. Hence $\mathrm{en}_{\mathbb{H}}(0) = 1$ and $\mathrm{en}_{\mathbb{H}}(1) = \omega$ which are the first two examples of the fact that

$$(\forall \xi \in \mathbf{On})[\mathrm{en}_{\mathbb{H}}(\xi) = \omega^{\xi}]. \tag{1}$$

Thus $\lambda\xi\,.\,\omega^{\xi}$ is a (κ-)normal function (for all $\kappa \in \mathbb{R}$ bigger than ω). We have

$$\mathbb{H} \subseteq \mathsf{Lim} \cup \{1\}$$

and obtain

$$\alpha \in \mathbb{H} \textit{ iff } (\forall \xi < \alpha)[\xi + \alpha = \alpha].$$

Thus for a finite set $\{\alpha_1, \ldots, \alpha_n\} \subseteq \mathbb{H}$ we get

$$\alpha_1 + \cdots + \alpha_n = \alpha_{k_1} + \cdots + \alpha_{k_m}$$

for $\{k_1, \ldots, k_m\} \subseteq \{1, \ldots, n\}$ such that $k_i < k_{i+1}$ and $\alpha_{k_i} \geq \alpha_{k_{i+1}}$. By induction on α we obtain thus ordinals $\{\alpha_1, \ldots, \alpha_n\} \subseteq \mathbb{H}$ such that for $\alpha \neq 0$ we have

$$\alpha = \alpha_1 + \cdots + \alpha_n \ \textit{and} \ \alpha_1 \geq \cdots \geq \alpha_n. \tag{2}$$

This is obvious for $\alpha \in \mathbb{H}$ and immediate from the induction hypothesis and the above remark if $\alpha = \xi + \eta$ for $\xi, \eta < \alpha$. It follows by induction on n that the ordinals $\alpha_1, \ldots, \alpha_n$ in (2) are uniquely determined. We therefore define an *additive normal form*

$$\alpha =_{\mathsf{NF}} \alpha_1 + \cdots + \alpha_n :\Leftrightarrow \alpha = \alpha_1 + \cdots + \alpha_n, \ \{\alpha_1, \ldots, \alpha_n\} \subseteq \mathbb{H} \ \textit{and} \ \alpha_1 \geq \cdots \geq \alpha_n.$$

We call $\{\alpha_1, \ldots, \alpha_n\}$ the set of *additive components* of α if $\alpha =_{\mathsf{NF}} \alpha_1 + \cdots + \alpha_n$.

We use the additive components to define the *symmetric sum* of ordinals $\alpha =_{\mathsf{NF}} \alpha_1 + \cdots + \alpha_n$ and $\beta =_{\mathsf{NF}} \alpha_{n+1} + \cdots + \alpha_m$ by

$$\alpha \# \beta := \alpha_{\pi(1)} + \cdots + \alpha_{\pi(m)}$$

where π is a permutation of the numbers $\{1, \ldots, m\}$ such that

$$1 \leq i < j \leq m \Rightarrow \alpha_{\pi(i)} \geq \alpha_{\pi(j)}.$$

In contrast to the "ordinary ordinal sum" the symmetric sum does not cancel additive components. By definition we have

$$\alpha \# \beta = \beta \# \alpha.$$

It is easy to check that the symmetric sum is order preserving in its both arguments.

As another consequence of (2) we obtain the *Cantor normal form* for ordinals for the basis ω, which says that for every ordinal $\alpha \neq 0$ there are ordinals $\xi_1, \ldots, \xi_n$ such that

$$\alpha =_{\mathsf{NF}} \omega^{\xi_1} + \cdots + \omega^{\xi_n}.$$

Since $\lambda\xi.\omega^\xi$ is a normal function we have $\alpha \leq \omega^\alpha$ for all ordinals α. We call α an ε-number if $\omega^\alpha = \alpha$ and define

$$\varepsilon_0 := \min\{\alpha \mid \omega^\alpha = \alpha\}.$$

more generally let $\lambda\xi.\varepsilon_\xi$ enumerate the fixed points of $\lambda\xi.\omega^\xi$. If we put

$$exp^0(\alpha, \beta) := \beta \ \textit{and} \ exp^{n+1}(\alpha, \beta) := \alpha^{exp^n(\alpha, \beta)}$$

we obtain

$$\varepsilon_0 := \sup_{n<\omega} exp^n(\omega, 0).$$

For $0 < \alpha < \varepsilon_0$ we have $\alpha < \omega^\alpha$ and obtain by the Cantor Normal Form Theorem uniquely determined ordinals $\alpha_1, \ldots, \alpha_n < \alpha$ such that $\alpha =_{\mathsf{NF}} \omega^{\alpha_1} + \cdots + \omega^{\alpha_n}$.

For a class $M \subseteq \mathbf{On}$ we define its *derivative*

$$M' := \{\xi \in \mathbf{On} |\ \mathrm{en}_M(\xi) = \xi\}.$$

The derivative f' of a function f is defined by $f' := \mathrm{en}_{\mathrm{Fix}(f)}$, where

$$\mathrm{Fix}(f) := \{\xi |\ f(\xi) = \xi\}.$$

Thus f' enumerates the fixed-points of f. If M is club (in some regular κ) then M' is also club (in κ). Thus if f is a normal function f' is a normal function, too.

If $\{M_\iota |\ \iota \in I\}$ is a collections of classes club (in some regular κ) and $|I| \in \mathbf{On}$ ($|I| \in \kappa$) then $\bigcap_{\iota \in I} M_\iota$ is also club (in κ).

These facts give raise to a hierarchy of club classes. We define

$$\begin{aligned} &Cr(0) := \mathbb{H} \\ &Cr(\alpha') := Cr(\alpha)' \\ &Cr(\lambda) := \textstyle\bigcap_{\xi<\lambda} Cr(\xi) \ \ for\ \lambda \in \mathsf{Lim}. \end{aligned}$$

If we put

$$\varphi_\alpha := \mathrm{en}_{Cr(\alpha)},$$

then all φ_α are normal functions and we have by definition

$$\alpha < \beta \Rightarrow \varphi_\alpha(\varphi_\beta\gamma) = \varphi_\beta\gamma. \tag{3}$$

The function φ is commonly called Veblen function. From (3) we obtain immediately

$$\begin{aligned} \varphi_{\alpha_1}\beta_1 \le \varphi_{\alpha_2}\beta_2 \ \ \textit{iff} \quad &\alpha_1 < \alpha_2 \ \ and\ \ \beta_1 \le \varphi_{\alpha_2}\beta_2 \\ \text{or }&\alpha_1 = \alpha_2 \ \ and\ \ \beta_1 \le \beta_2 \\ \text{or }&\alpha_2 < \alpha_1 \ \ and\ \ \varphi_{\alpha_1}\beta_1 \le \beta_2. \end{aligned} \tag{4}$$

We define the Veblen normal form for ordinals $\varphi_\xi\eta$ by

$$\alpha =_{\mathsf{NF}} \varphi_\xi\eta :\Leftrightarrow \ \ \alpha = \varphi_\xi\eta \ \ and\ \ \eta < \alpha.$$

Then $\alpha =_{\mathsf{NF}} \varphi_{\xi_1}\eta_1$ *and* $\alpha =_{\mathsf{NF}} \varphi_{\xi_2}\eta_2 \Rightarrow \xi_1 = \xi_2$ *and* $\eta_1 = \eta_2$. Since $\xi < \alpha$ and $\eta < \beta \in Cr(\alpha)$ implies $\varphi_\xi\eta < \beta$ we call $Cr(\alpha)$ the class of α–critical ordinals.. If α is itself α–critical then $\xi, \eta < \alpha \Rightarrow \varphi_\xi\eta < \alpha$. Therefore we define the class SC of *strongly critical ordinals* by

$$\mathsf{SC} := \{\alpha \in \mathbf{On} |\ \alpha \in Cr(\alpha)\}.$$

The class SC is club (in all regular ordinals $\kappa > \omega$). Its enumerating function is denoted by $\lambda\xi\,.\,\Gamma_\xi$. Regarding that by (4) $\lambda\xi\,.\,\varphi_\xi 0$ is order preserving one easily proves

$$\mathsf{SC} = \{\alpha |\ \varphi_\alpha 0 = \alpha\}.$$

If we define $\gamma_0 := 0$ and $\gamma_{n+1} := \varphi_{\gamma_n} 0$ then we obtain

$$\Gamma_0 = \sup_{n<\omega} \gamma_n.$$

For every $\alpha < \Gamma_0$ there are uniquely determined ordinals $\xi_1, \ldots, \xi_n < \alpha$ and $\eta_1, \ldots, \eta_n < \alpha$ such that

$$\alpha =_{\mathsf{NF}} \varphi_{\xi_1}\eta_1 + \cdots + \varphi_{\xi_n}\eta_n \textit{ and } \eta_i < \varphi_{\xi_i}\eta_i \textit{ for } i \in \{1, \ldots, n\}. \tag{5}$$

This is all we need to know about ordinals for the moment. We will have to come back to the theory later.

1.2. Partial models for axiom systems of set theory

Let $\mathcal{L}(\in)$ denote a language of Set Theory, i.e., a first order language which contains a symbol $\in$ for the membership relation. Later we will add a unary relation symbol Ad.

We will distinguish between restricted quantifiers $(\forall x \in a)$ and $(\exists x \in a)$ and unrestricted quantifiers of the form $(\forall x)$ and $(\exists x)$.

Recall the Levy hierarchy of $\mathcal{L}$-formulas. We say that a formula is Δ_0 if it contains only restricted quantifiers. A formula F of $\mathcal{L}$ is a Π_n-formula, if

$$F \equiv (\forall x_1)(\exists x_2)\cdots(\mathsf{Q}_n x_n)G(x_1, \ldots, x_n),$$

where $G(x_1, \ldots, x_n)$ is a Δ_0- formula and $\forall\exists\cdots(\mathsf{Q}_n)$ an alternating string of unrestricted quantifiers. Dually, a formula is Σ_n if $\neg F$ is logically equivalent to a Π_n-formula.

A formula $F(x_1, \ldots, x_n)$ is a Δ_n-formula of an $\mathcal{L}(\in)$-structure $\mathfrak{A}$ iff there are a Π_n-formula $F_\Pi(x_1, \ldots, x_n)$ and a Σ_n-formula $F_\Sigma(x_1, \ldots, x_n)$ such that

$$\mathfrak{A} \models (\forall \vec{x})(F(\vec{x}) \leftrightarrow F_\Pi(\vec{x})) \wedge (\forall \vec{x})(F(\vec{x}) \leftrightarrow F_\Sigma(\vec{x})). \tag{6}$$

We call F a Δ_n-formula of a theory **Ax** iff **Ax** proves the formula in (6).

The class of Σ–formulas is the smallest class which contains the Δ_0–formulas and is closed under the positive boolean operations $\vee$ and $\wedge$, restricted quantification and unrestricted existential quantification. The class of Π formulas is the dual of the class of Σ–formulas. A formula $F(x_1, \ldots, x_n)$ is Δ in a structure $\mathfrak{A}$ if there is a Σ–formula $F_\Sigma(x_1, \ldots, x_n)$ and a Π–formula $F_\Pi(x_1, \ldots, x_n)$ such that $\mathfrak{A} \models (\forall \vec{x})[F(\vec{x}) \leftrightarrow F_\Sigma(\vec{x})]$ and $\mathfrak{A} \models (\forall \vec{x})[F(\vec{x}) \leftrightarrow F_\Pi(\vec{x})]$. It is Δ for a theory **Ax** iff it is Δ for all models of **Ax**.

Recall that the constructible hierarchy $\mathbf{L}$ is the union of its stages $\mathbf{L}_\alpha$ given by the following definition.

1.2.1. Definition.

$$\mathbf{L}_0 = \emptyset$$
$$\mathbf{L}_{\alpha+1} = \mathsf{Def}(\mathbf{L}_\alpha)$$
$$\mathbf{L}_\lambda = \bigcup\{\mathbf{L}_\xi \mid \xi < \lambda\} \text{ for } \lambda \in \mathsf{Lim},$$

where $a \in \mathsf{Def}(\mathbf{L}_\alpha)$ iff there is an $\mathcal{L}(\in)$-formula $F(x, y_1, \ldots, y_n)$ with no other free variables and a tuple $(b_1, \ldots, b_n)$ of elements of $\mathbf{L}_\alpha$ such that $a = \{s \in \mathbf{L}_\alpha \mid \mathbf{L}_\alpha \models F(s, b_1, \ldots, b_n)\}$.

The formula F^z is obtained from F by restricting all unrestricted quantifiers in F to z, i.e., replacing $(\forall x)$ by $(\forall x \in z)$ and $(\exists x)$ by $(\exists x \in z)$, respectively. The formula F^z is obviously Δ_0.

We say that a formula G is Π_n^κ or Σ_n^κ if

$$G \equiv F^{\mathbf{L}_\kappa}$$

for a Π_n-formula F or Σ_n-formula F, respectively.

Let $\mathbf{Ax}$ be an axiom system formulated in the language $\mathcal{L}$. We define

$$\|\mathbf{Ax}\|_\infty := \min\{\alpha \mid \mathbf{L}_\alpha \models \mathbf{Ax}\} \tag{7}$$

and call $\|\mathbf{Ax}\|_\infty$ the ∞-norm of $\mathbf{Ax}$. By Löwenheim-Skolem downwards and the Condensation Lemma for the constructible hierarchy we know that $\|\mathbf{Ax}\|_\infty$ is a countable ordinal which may be quite big. Much too big for our purpose as we will see in a moment. To obtain smaller ordinals which may be characteristic for $\mathbf{Ax}$ we regard partial models for $\mathbf{Ax}$. For any complexity class $\mathcal{F}$ of $\mathcal{L}$-sentences we define

$$\mathsf{Con}_{\mathcal{F}}(\mathbf{Ax}) := \{F \mid F \in \mathcal{F} \wedge \mathbf{Ax} \vdash F\} \tag{8}$$

and

$$\|\mathbf{Ax}\|_{\mathcal{F}} := \min\{\alpha \mid \mathbf{L}_\alpha \models \mathsf{Con}_{\mathcal{F}}(\mathbf{Ax})\}. \tag{9}$$

The complexity of the sentences in the axiom systems which we are going to study will never exceed Π_3. Therefore we always have

$$\|\mathbf{Ax}\|_{\Pi_3} = \|\mathbf{Ax}\|_\infty.$$

We will only consider axiom systems which comprise the following basis theory $\mathbf{KP}^-$ whose axioms are:

The *ontological axioms* of Extensionality and Foundation

(Ext) $(\forall u)(\forall v)[u = v \leftrightarrow (\forall x \in u)(x \in v) \wedge (\forall x \in v)(x \in u)]$

(Found) $(\forall u)[(\exists x \in u)(x \in u) \rightarrow (\exists x \in u)(\forall z \in x)\neg(z \in u)]$.

The *closure axioms* of Pairing and Union

(Pair) $(\forall u)(\forall v)(\exists w)(\forall x)[x \in w \leftrightarrow x = u \vee x = v]$

(Union) $(\forall u)(\exists w)(\forall x)[x \in w \leftrightarrow (\exists y \in u)(x \in y)]$

The *set existence axiom schemes* of absolute Separation and Collection

(Δ_0-Sep) $(\forall \vec{v})(\forall a)(\exists z)(\forall x)[x \in z \leftrightarrow x \in a \wedge F(x, \vec{v})]$

(Δ_0-Col) $(\forall \vec{v})(\forall u)[(\forall x \in u)(\exists y)F(x, y, \vec{v}) \rightarrow (\exists z)(\forall x \in u)(\exists y \in z)F(x, y, \vec{v})]$.

In both schemes we allow only Δ_0-formulas $F(x, \vec{v})$ or $F(x, y, \vec{v})$ of the language $\mathcal{L}(\in)$, respectively. This requirement guarantees the absoluteness of the sets whose existence is postulated.

We use the abbreviations which are common in Set Theory. E.g. $a = \{x \in u \mid F(x)\}$ stands for $(\forall x \in a)[x \in u \wedge F(x)] \wedge (\forall x \in u)[F(x) \rightarrow x \in a]$;

$\{x \in u \mid F(x)\} \in b$ stands for $(\exists y)[y = \{x \in u \mid F(x)\} \land y \in b]$, etc. Lower case Greek letters are supposed to range over ordinals.

Adding the ontological axiom of *Infinity*

$$\text{(Inf)} \qquad (\exists u)[\emptyset \in u \land (\forall x \in u)(x \cup \{x\} \in u)]$$

we obtain the theory $\mathbf{KP}\omega^-$ and adding the *Foundation Scheme*

$$\text{(FOUND)}\, (\exists x)F(x) \to (\exists x)[F(x) \land (\forall y \in x)\neg F(y)]$$

we obtain the theories **KP** or $\mathbf{KP}\omega$ respectively. If we restrict the formula $F(x)$ in (FOUND) to a complexity class $\mathcal{F}$ we talk about $\mathcal{F}$-Foundation, denoted by ($\mathcal{F}$-FOUND).

An ordinal α is called *admissible* if $\mathbf{L}_\alpha \models \mathbf{KP}$.

The theories **KP** and $\mathbf{KP}\omega$ are profoundly studied in Barwise [1975]. They prove Σ-recursion – a very important theorem – and Σ-reflection, i.e., the scheme

$$F \;\to\; (\exists x)F^x$$

for Σ-formulas F. As a consequence both theories prove the equivalence of any Σ-formula to a Σ_1-formula.

Recall that a partial function $f\colon \mathbf{L}_\kappa \longrightarrow_{\mathrm{p}} \mathbf{L}_\kappa$ is called *κ-partial recursive* if its graph is Σ-definable over $\mathbf{L}_\kappa$ i.e., if we have a Σ–formula $F(x, y, \vec{z})$ without further free variables and a tuple $\vec{c}$ of elements of $\mathbf{L}_\kappa$ such that

$$f(a) \simeq b \text{ iff } \mathbf{L}_\kappa \models F[a, b, \vec{c}]$$

for all $a, b \in \mathbf{L}_\kappa$. Admissible ordinals are important for generalized recursion theory because $\mathbf{L}_\kappa$ is closed under all κ-recursive functions if κ is admissible.

Obviously ω is an admissible ordinal and it is a folklore result that ω_1^{CK}, i.e., the least countable ordinal which cannot be represented by a recursive well-ordering on the natural numbers, is the next admissible ordinal. Therefore

$$\mathbf{L}_{\omega_1^{\mathrm{CK}}} \models \mathbf{KP}\boldsymbol{\omega}$$

and, since there are no further admissibles between ω and ω_1^{CK}, even $\|\mathbf{KP}\boldsymbol{\omega}\|_\infty = \omega_1^{\mathrm{CK}}$. Hence $\|\mathbf{KP}\boldsymbol{\omega}\|_{\Pi_2} \leq \omega_1^{\mathrm{CK}}$ as well as $\|\mathbf{KP}\boldsymbol{\omega}\|_{\Sigma_1} \leq \omega_1^{\mathrm{CK}}$. For most impredicative theories **Ax**, however, we have $\omega_1^{\mathrm{CK}} < \|\mathbf{Ax}\|_{\Sigma_1}$. Therefore we need to introduce the following ordinals.

1.2.2. Definition. Let **Ax** be a theory which contains $\mathbf{KP}^-$. Then we define

$$\|\mathbf{Ax}\|_{\Sigma_1^\kappa} := \min\{\alpha \mid (\forall F)[F \text{ is a } \Sigma_1\text{-sentence} \land \mathbf{Ax} \vdash F^{\mathbf{L}_\kappa} \;\Rightarrow\; \mathbf{L}_\alpha \models F]\}$$

and analogously

$$\|\mathbf{Ax}\|_{\Pi_2^\kappa} := \min\{\alpha \mid (\forall F)[F \text{ is a } \Pi_2\text{-sentence} \land \mathbf{Ax} \vdash F^{\mathbf{L}_\kappa} \;\Rightarrow\; \mathbf{L}_\alpha \models F]\}$$

The notation $\mathbf{Ax} \vdash F^{\mathbf{L}_\kappa}$ has to be read with the necessary care. It anticipates that $\mathbf{L}_\kappa$ can be defined in some way from the axioms in **Ax**. We need Σ-recursion to define the function $\alpha \mapsto \mathbf{L}_\alpha$. To prove the Σ-recursion theorem it suffices to have $\mathbf{KP}^- + \Sigma_1$-FOUND. So for theories extending $\mathbf{KP}^- + \Sigma_1$-FOUND we need only a description of α in order to characterize $\mathbf{L}_\alpha$. We say that $\mathbf{Ax}$ *believes that* κ *is admissible* iff $\mathbf{L}_\kappa$ is definable in $\mathbf{Ax}$ and

$$\mathbf{Ax} \vdash G^{\mathbf{L}_\kappa} \tag{10}$$

holds for all sentences $G \in \mathbf{KP}^-$. We say that $\mathbf{Ax}$ *proves the* $\mathbf{L}$- *or* $\mathbf{L}_\kappa$*-reflection rule* for a complexity class $\mathcal{F}$ iff

$$\mathbf{Ax} \vdash G \;\Rightarrow\; \mathbf{Ax} \vdash (\exists\gamma)(\exists u)[u = \mathbf{L}_\gamma \wedge G^u]$$

or

$$\mathbf{Ax} \vdash G^{\mathbf{L}_\kappa} \;\Rightarrow\; \mathbf{Ax} \vdash ((\exists\gamma)(\exists u)[u = \mathbf{L}_\gamma \wedge G^u])^{\mathbf{L}_\kappa},$$

respectively, for all formulas $G \in \mathcal{F}$. The following lemma is a first easy observation about partial models.

1.2.3. Lemma. *Let* $\kappa \in (\omega, \|\mathbf{Ax}\|_\infty]$ *be an ordinal such that* $\mathbf{Ax}$ *believes that* κ *is admissible and* $\mathbf{Ax}$ *proves the* $\mathbf{L}_\kappa$*-reflection rule for* Σ_1*-formulas. Then* $\|\mathbf{Ax}\|_{\Pi_2^\kappa} = \|\mathbf{Ax}\|_{\Sigma_1^\kappa}$.

Proof. We obviously have $\|\mathbf{Ax}\|_{\Sigma_1^\kappa} \le \|\mathbf{Ax}\|_{\Pi_2^\kappa}$ and need only to show the converse inequality. Thus put $\alpha := \|\mathbf{Ax}\|_{\Sigma_1^\kappa}$, let $(\forall x)(\exists y)F(x,y)$ be a Π_2-sentence such that $\mathbf{Ax} \vdash (\forall x \in \mathbf{L}_\kappa)(\exists y \in \mathbf{L}_\kappa)F(x,y)$ and choose some $a \in \mathbf{L}_\alpha$. We have to show that $\mathbf{L}_\alpha \models (\exists y)F(a,y)$. Since α is obviously a limit ordinal there is a $\beta < \alpha$ such that $a \in \mathbf{L}_\beta$. By definition of α there is a Σ_1-sentence G such that $\mathbf{Ax} \vdash G^{\mathbf{L}_\kappa}$ but $\mathbf{L}_\beta \not\models G$. Since $\mathbf{Ax}$ proves $\mathbf{L}_\kappa$-reflection we obtain $\mathbf{Ax} \vdash (\exists\gamma \in \mathbf{L}_\kappa)(\exists u \in \mathbf{L}_\kappa)(u = \mathbf{L}_\gamma \wedge G^u)$ and since $\mathbf{Ax}$ believes that κ is admissible by Δ_0–collection relativized to $\mathbf{L}_\kappa$ also $\mathbf{Ax} \vdash v \in \mathbf{L}_\kappa \to (\exists z \in \mathbf{L}_\kappa)(\forall x \in v)(\exists y \in z)F(x,y)$. Choosing $v = u$ we thus get

$$\mathbf{Ax} \vdash (\exists\gamma \in \mathbf{L}_\kappa)(\exists u \in \mathbf{L}_\kappa)(\exists z \in \mathbf{L}_\kappa)[u = \mathbf{L}_\gamma \wedge G^u \wedge (\forall x \in u)(\exists y \in z)F(x,y)].$$

Since $\mathbf{Ax}$ believes that κ is admissible this is equivalent to a Σ_1-sentence relativized to $\mathbf{L}_\kappa$. Hence $\mathbf{L}_\alpha \models (\exists\gamma)(\exists u)(\exists z)[u = \mathbf{L}_\gamma \wedge G^u \wedge (\forall x \in u)(\exists y \in z)F(x,y)]$. Because $u = \mathbf{L}_\gamma$ is absolute for $\mathbf{L}_\alpha$ we finally get $G^{\mathbf{L}_\gamma}$ and $(\forall x \in \mathbf{L}_\gamma)(\exists y \in \mathbf{L}_\alpha)F(x,y)$ for some $\gamma < \alpha$. Because of $\mathbf{L}_\beta \not\models G$ we have $\beta < \gamma$. Hence $a \in \mathbf{L}_\beta \subseteq \mathbf{L}_\gamma$ and it follows $\mathbf{L}_\alpha \models (\exists y)F(a,y)$ as desired. □

If $\mathbf{KP}\boldsymbol{\omega}^- + \Sigma_1$–FOUND $\subseteq \mathbf{Ax}$ then $\omega < \|\mathbf{Ax}\|_\infty =: \alpha$ and $\mathbf{Ax}$ proves the $\mathbf{L}$-reflection rule for Σ-formulas. Interpreting the provable sentences of $\mathbf{Ax}$ it makes no difference if we think that every unrestricted quantifier is restricted by $\mathbf{L}_\alpha$. Since $\mathbf{KP}^- \subseteq \mathbf{Ax}$ this has the same effect as if $\mathbf{Ax}$ believes that α is admissible. Therefore we obtain as a corollary of Lemma 1.2.3

1.2.4. Corollary. *If* $\mathbf{KP}\boldsymbol{\omega}^- + \Sigma_1$- FOUND $\subseteq \mathbf{Ax}$ *then* $\|\mathbf{Ax}\|_{\Sigma_1} = \|\mathbf{Ax}\|_{\Pi_2}$.

Another observation is that adding true Π_1^κ-sentences does not increase the Σ_1^κ ordinal of an axiom system.

1.2.5. Theorem. *Let G be a true Π_1^κ sentence. Then $\|\mathbf{Ax}+G\|_{\Sigma_1^\kappa} = \|\mathbf{Ax}\|_{\Sigma_1^\kappa}$.*

Proof. Let $G \equiv H^{\mathbf{L}_\kappa}$ for a Π_1-sentence H. Assume that $\mathbf{Ax}+G \vdash F^{\mathbf{L}_\kappa}$ for a Σ_1 sentence F. Then $\mathbf{Ax} \vdash (H \to F)^{\mathbf{L}_\kappa}$ and $H \to F$ is Σ_1. For $\alpha := \|\mathbf{Ax}\|_{\Sigma_1^\kappa}$ we thus have $\mathbf{L}_\alpha \models H \to F$. From $\alpha \leq \kappa$, $\mathbf{L}_\kappa \models H$ and the downwards persistency of Π_1-sentences we get $\mathbf{L}_\alpha \models H$ which in turn entails $\mathbf{L}_\alpha \models F$. Hence $\|\mathbf{Ax}+G\|_{\Sigma_1^\kappa} \leq \|\mathbf{Ax}\|_{\Sigma_1^\kappa}$. But the converse inequality holds trivially and we have $\|\mathbf{Ax}+G\|_{\Sigma_1^\kappa} = \|\mathbf{Ax}\|_{\Sigma_1^\kappa}$. □

We introduce the following notation.

1.2.6. Definition. $\|\mathbf{Ax}\|_\kappa := \|\mathbf{Ax}\|_{\Sigma_1^\kappa}$

Because of Lemma 1.2.3 we get $\|\mathbf{Ax}\|_\kappa = \|\mathbf{Ax}\|_{\Pi_2^\kappa}$ for theories $\mathbf{Ax}$ satisfying the hypotheses of the lemma.

We call the computation of the ordinal $\|\mathbf{Ax}\|_\kappa$ a κ–ordinal analysis for $\mathbf{Ax}$. It will turn out that $\|\mathbf{Ax}\|_{\omega_1^{CK}}$ will be the most important ordinal. In Section 2.1.4 we will see that there is also something as an ω–ordinal which gives a characterization of the Skolem functions of the provable Π_2^ω–sentences of an axiom system $\mathbf{Ax}$ in terms of a sub-recursive hierarchy.

1.3. Connections to subsystems of second order number theory

Let $\mathcal{L}_{\mathrm{N}}^2$ be the language of Second Order Arithmetic. We assume that $\mathcal{L}_{\mathrm{N}}^2$ contains a constant $\underline{0}$ for 0 and constants for all primitive recursive functions and predicates. We restrict the language to unary predicate variables and talk about set variables. This means no real restriction since we have a primitive recursive coding machinery. We use capital Latin letters as syntactical variables for sets and write $t \in X$ instead of $X(t)$. We assume familiarity with the complexity classes in the arithmetical and analytical hierarchy.

Since all primitive recursive functions and predicates have Δ_0-definitions in $\mathbf{KP}\omega^- + \Sigma_1 - \mathrm{FOUND}$, we may regard $\mathcal{L}_{\mathrm{N}}^2$ as a sublanguage of $\mathcal{L}(\in)$ by restricting all first order quantifiers to ω and replacing all second order quantifiers $(\forall X)$ and $(\exists X)$ by $(\forall X \subseteq \omega)$ and $(\exists X \subseteq \omega)$, respectively. We may therefore transfer the notions of the arithmetical and analytical hierarchy to the language of $\mathcal{L}(\in)$. Whenever we talk of a Π_n^0, Σ_n^0, Π_1^1, …- sentences in the language of Set Theory without further comments we think of a translation of the corresponding $\mathcal{L}_{\mathrm{N}}^2$-sentence.

One of the basic facts for the things to come is the ω-Completeness Theorem for Π_1^1-sentences. We will use the ω-Completeness Theorem to introduce the notion of *truth complexity* for Π_1^1-sentences. The value $t^{\mathbf{N}}$ of a closed term t and the truth value of an atomic sentence in the standard structure $\mathbf{N}$ are primitive recursively computable. Since there are symbols for all primitive recursive functions and predicates we obtain the diagram of $\mathbf{N}$

$$\mathrm{D}(\mathbf{N}) := \{A \mid A \textit{ is an atomic sentence and } \mathbf{N} \models A\}$$

as a recursive set. For arithmetical sentences which are not atomic the truth definition is given inductively by

$$\begin{aligned}
&\mathbb{N} \models A_1 \ \textit{and}\ \mathbb{N} \models A_2 \Rightarrow \mathbb{N} \models A_1 \wedge B_1\\
&\mathbb{N} \models A_i \ \textit{for some}\ i \in \{1,2\} \ \Rightarrow \mathbb{N} \models A_1 \vee A_2\\
&\mathbb{N} \models A(\underline{n}) \ \textit{for all}\ n \in \mathbb{N} \ \Rightarrow \mathbb{N} \models (\forall x)A(x)\\
&\mathbb{N} \models A(\underline{n}) \ \textit{for some}\ n \in \mathbb{N} \ \Rightarrow \mathbb{N} \models (\exists x)A(x).
\end{aligned}$$

To extend this truth definition to Π^1_1-sentences we introduce an infinitary calculus. For technical reasons we opt for a one sided sequent calculus à la Tait. First we fix the language of the Tait calculus.

The non logical symbols for the Tait-language of $\mathcal{L}^2_{\mathbb{N}}$ are:

- *The constant* $\underline{0}$ *as well as constants for all primitive recursive functions and – relations.*

The logical symbols comprise:

- *Bounded number-variables, denoted by* $x,\ y,\ z,\ x_1, \ldots$ *and set variables, denoted by* $X,\ Y,\ Z,\ X_1, \ldots$
- *The logical connectives* $\wedge$, $\vee$ *and the quantifiers* $\forall$, $\exists$.
- *The membership symbol* $\in$ *and its negation* $\notin$.

Terms are built up from $\underline{0}$ and function symbols in the familiar way. We use S as a symbol for the *successor* function. Terms of the shape $(\underbrace{\underline{\mathsf{S}} \cdots \underline{\mathsf{S}}}_{n-\text{times}} 0)$ are *numerals* and will be denoted by $\underline{n}$.

Atomic formulas are $t \in X$, $t \notin X$ and $R(t_1, \ldots, t_n)$, where $t,\ t_1, \ldots, t_n$ are terms, X is a set variable and R is a symbol for an n-ary relation symbol.

From the atomic formulas we obtain the formulas of $\mathcal{L}^2_{\mathbb{N}}$ in the familiar way. Notice that we do not have free number variables in the language.

The negation symbol is not a basic symbol of the Tait-language. We define the negation of a formula by de Morgan's laws.

$$\begin{aligned}
&\neg(t \in X) :\equiv (t \notin X);\ \ \neg(t \notin X) :\equiv (t \in X)\\
&\neg(Rt_1 \ldots t_n) :\equiv (\overline{R}t_1 \ldots t_n) \ \ \textit{where}\ \overline{R}\ \textit{is a symbol for the complement of}\ R\\
&\neg(A \wedge B) :\equiv (\neg A \vee \neg B);\ \ \neg(A \vee B) :\equiv (\neg A \wedge \neg B)\\
&\neg(\forall x)A(x) :\equiv (\exists x)\neg A(x);\ \ \neg(\exists x)A(x) :\equiv (\forall x)\neg A(x).
\end{aligned}$$

It is obvious that we have

$$\neg\neg A \equiv A. \tag{11}$$

The semantics for the Tait-language is straightforward. We easily check

$$\mathbb{N} \models (\neg A)[S_1, \ldots, S_n] \ \textit{iff}\ \mathbb{N} \not\models A[S_1, \ldots, S_n]$$

for any assignment of sets $S_1, \ldots, S_n$ to the set variables occurring in A.

We use capital Greek letters Δ, Γ, Λ, Δ_1, ... as syntactical variables for finite sets of $\mathcal{L}^2_{\mathbb{N}}$-formulas.

1.3.1. Definition. We define $\models^{\alpha} \Delta$ inductively by the following clauses:

(AxM) *If* $\Delta \cap D(\mathbb{N}) \neq \emptyset$ *then* $\models^{\alpha} \Delta$ *for all ordinals* α.

(AxL) *If* $t^{\mathbb{N}} = s^{\mathbb{N}}$ *then* $\models^{\alpha} \Delta, s \notin X, t \in X$ *for all ordinals* α.

($\wedge$) *If* $\models^{\alpha_i} \Delta, A_i$ *and* $\alpha_i < \alpha$ *for* $i = 1, 2$ *then* $\models^{\alpha} \Delta, A_1 \wedge A_2$.

($\vee$) *If* $\models^{\alpha_i} \Delta, A_i$ *and* $\alpha_i < \alpha$ *for some* $i \in \{1, 2\}$ *then* $\models^{\alpha} \Delta, A_1 \vee A_2$.

($\forall$) *If* $\models^{\alpha_i} \Delta, A(\underline{i})$ *and* $\alpha_i < \alpha$ *for all* $i \in \mathbb{N}$ *then* $\models^{\alpha} \Delta, (\forall x)A(x)$.

($\exists$) *If* $\models^{\alpha_0} \Delta, A(\underline{i})$ *and* $\alpha_0 < \alpha$ *for some* $i \in \mathbb{N}$ *then* $\models^{\alpha} \Delta, (\exists x)A(x)$.

The relations $\models^{\alpha} \Delta$ is to be read that there is an infinite proof tree of $\bigvee \Delta$ whose depth is bounded by the ordinal α. It is obvious from the definition that we have

$$\models^{\alpha} \Delta, \qquad \alpha \leq \beta, \qquad \Delta \subseteq \Gamma \qquad \Rightarrow \qquad \models^{\beta} \Gamma \tag{12}$$

and a simple induction on α shows

$$\models^{\alpha} F_1, \ldots, F_n \Rightarrow \mathbb{N} \models (F_1 \vee \cdots \vee F_n)[S_1, \ldots, S_m]$$

for all assignments $S_1, \ldots, S_m$ to the set parameters occurring in $(F_1 \vee \cdots \vee F_n)$. We thus have

$$\models^{\alpha} F(\vec{X}) \;\Rightarrow\; \mathbb{N} \models (\forall \vec{X}) F(\vec{X}) \tag{13}$$

for all Π^1_1-sentences $(\forall \vec{X})F(\vec{X})$. If F is a true arithmetical sentence, i.e., if F does not contain set parameters, we obtain by a simple induction on the complexity $\mathsf{rk}(F)$ of the formula F (which can be taken to be the number of logical symbols occurring in F)

$$\models^{\mathsf{rk}(F)} F\,. \tag{14}$$

This shows that $\models^{\alpha} F$ is indeed an extension of the truth definition for arithmetical sentences. However, to prove that $\models^{\alpha} F$ is also complete for Π^1_1-sentences, i.e., the opposite direction in (13), needs harder work. We follow Schütte's proof via search trees.

Let *Seq* be the set of (codes for) finite sequences of natural numbers. For $s, t \in \mathit{Seq}$ we denote by $s \subseteq t$ that s is an initial segment of t. A *tree* is a set of finite number–sequences which is closed under initial segments. The elements of a tree are called *nodes.* A set of nodes in a tree is a *thread,* if it is linearly ordered by $\subseteq$. A maximal thread in a tree is a *path.*

A binary relation $\prec \subseteq \omega \times \omega$ is *well-founded* iff there are no infinite $\prec$-descending chains. For well-founded $\prec \subseteq \omega \times \omega$ we define

$$\mathsf{otyp}_{\prec}(n) := \begin{cases} \sup \{\mathsf{otyp}_{\prec}(m) + 1 \mid m \prec n\} & \text{if } n \in \mathsf{field}(\prec) \\ \omega_1 & \text{otherwise} \end{cases}$$

and

$$\mathsf{otyp}(\prec) := \sup\{\mathsf{otyp}_{\prec}(n) \mid n \in \mathsf{field}(\prec)\}.$$

For a tree T we define the *tree–relation* $\prec_T$ by

$$s \prec_T t \text{ iff } s \in T \wedge t \in T \wedge t \subsetneqq s.$$

A tree T is well-founded iff $\prec_T$ is a well-founded relation, i.e. iff there is no infinite path through T. For a well-founded tree T we define

$$\mathsf{otyp}_T(s) := \mathsf{otyp}_{\prec_T}(s)$$

and

$$\mathsf{otyp}(T) = \mathsf{otyp}_T(\langle\rangle).$$

Recall that the first non recursive ordinal is defined by

$$\omega_1^{\mathrm{CK}} := \sup\{\mathsf{otyp}(\prec) \mid \prec \quad \text{is a recursive well-ordering}\}.$$

We are going to define search trees for finite sequences of formulas. Such a sequence is called *reducible* if it contains at least one non atomic formula. The left most non atomic formula in a reducible sequence is called *distinguished.* The *reduced sequence* Δ^r of a reducible sequence Δ is obtained by removing the distinguished formula from the sequence.

The search tree for a finite sequence Δ of $\mathcal{L}^2_{\mathrm{N}}$-formulas is a tree S_Δ together with a label function which assigns a finite sequence $\delta(s)$ of $\mathcal{L}^2_{\mathrm{N}}$-formulas to each node $s \in S_\Delta$. It is inductively defined by the following clauses:

($S_{\langle\rangle}$) $\langle\rangle \in S_\Delta$ *and* $\delta(\langle\rangle) = \Delta$.

(S_{Ax}) *If* $s \in S_\Delta$ *and* $\delta(s)$ *is an axiom according to* (AxM) *or* (AxL), *then* $s^\frown\langle i\rangle \notin S_\Delta$ *for all* $i \in \mathbb{N}$. *(I.e. s is a topmost node of S_Δ.)*

For the following clauses assume $s \in S_\Delta$ such that $\delta(s)$ is not an axiom.

(S_{id}) *If* $\delta(s)$ *is not reducible then* $s^\frown\langle 0\rangle \in S_\Delta$ *and* $\delta(s^\frown\langle 0\rangle) = \delta(s)$.

($S_\wedge$) *If* $F_0 \wedge F_1$ *is the distinguished formula in* $\delta(s)$ *then* $s^\frown\langle i\rangle \in S_\Delta$ *for* $i = 0, 1$ *and* $\delta(s^\frown\langle i\rangle) := \delta(s)^r, F_i$.

($S_\vee$) *Let* $F_0 \vee F_1$ *be the distinguished formula in* $\delta(s)$. *Then* $s^\frown\langle i_0\rangle \in S_\Delta$ *and* $\delta(s^\frown\langle i_0\rangle) := \delta(s)^r, F_0, F_1$.

($S_\forall$) *If the distinguished formula in* $\delta(s)$ *is* $(\forall x)F(x)$, *then* $s^\frown\langle i\rangle \in S_\Delta$ *for all* $i \in \mathbb{N}$ *and* $\delta(s^\frown\langle i\rangle) = \delta(s)^r, F(\underline{i})$.

($S_\exists$) *If the distinguished formula in* $\delta(s)$ *is* $(\exists x)F(x)$, *then* $s^\frown\langle 0\rangle \in S_\Delta$ *and* $\delta(s^\frown\langle 0\rangle) = \delta(s)^r, F(\underline{n}), (\exists x)F(x)$, *where n is the least natural number such that* $n \neq t^{\mathbb{N}}$ *for all formulas* $F(t) \in \bigcup_{s_0 \subseteq s} \delta(s_0)$.

Observe that we introduced clause (S_{Ax}) only for better readability. It follows from the other clauses and the fact that S_Δ is inductively defined.

There are two main lemmas.

1.3.2. Syntactical Main Lemma. *If S_Δ is well–founded then* $\mathsf{otyp}(S_\Delta) < \omega_1^{\mathrm{CK}}$ *and* $\overset{\mathsf{otyp}(S_\Delta)}{\Longleftarrow\!\!\models} \Delta$.

Proof. Let S_Δ be well–founded. The tree relation $\prec_{S_\Delta}$ is obviously recursive, hence $\mathsf{otyp}(S_\Delta) < \omega_1^{\mathrm{CK}}$. Every path in S_Δ is finite. Thus $\delta(s)$ has to be an axiom for every topmost node s and the second claim follows easily by induction on $\mathsf{otyp}(S_\Delta)$. □

1.3.3. Semantical Main Lemma. *If the search tree S_Δ is not well–founded then there is an assignment $S_1, \ldots, S_n$ of subsets of* N *to the set variables in Δ such that* $\mathbb{N} \not\models \bigvee \{F \mid F \in \Delta\}[S_1, \ldots, S_n]$.

To sketch the proof let f be an infinite path in S_Δ. We say sloppily that a formula F occurs in $s \in S_\Delta$ if $F \in \delta(s)$. Let $f[n] := \langle f(0), \ldots, f(n-1)\rangle$ and call $f[n]$ the course of values of f below n. We observe:

(1) *If A is an atomic formula occurring in $s \in S_\Delta$ then A occurs in all t such that $s \subseteq t \in S_\Delta$.*

(2) *If a non atomic formula F occurs in some $f[n]$ then there is an $m \geq n$ such that F is distinguished in $f[m]$.*

The proof of (2) is an easy induction on the number of non atomic formulas occurring left of F in $\delta(f[n])$. Using *(2)* the proofs of the following observations are almost immediate from the definition of S_Δ.

(3) *If a formula $A \wedge B$ occurs in $f[n]$ then there is an m such that either A or B occurs in $f[m]$.*

(4) *If a formula $A \vee B$ occurs in $f[n]$ then there are m_A and m_B such that A occurs in $f[m_A]$ and B occurs in $f[m_B]$.*

(5) *If a formula $(\forall x)F(x)$ occurs in $f[n]$ then there are numbers i and m such that $F(\underline{i})$ occurs in $f[m]$.*

(6) *If a formula $(\exists x)F(x)$ occurs in $f[n]$ then for every number i there is a term t and an m_i such that $t^{\mathbb{N}} = i$ and $F(t)$ occurs in $f[m_i]$.*

To prove fact *(6)* we assume that $(\exists x)F(x)$ is distinguished in $\delta(f[n])$. By *(2)* this means no loss of generality. Then $\delta(f[n+1]) = \delta(f[n])^r, F(\underline{j}), (\exists x)F(x)$ and, if $i < j$ we have $F(t)$ in $f[m]$ for some t and $m \leq n$ such that $t^{\mathbb{N}} = i$. If $j \leq i$ then $F(t)$ will occur in $f[m+1]$ for some $m \geq n$ and t with $t^{\mathbb{N}} = i$ as soon as $(\exists x)F(x)$ becomes distinguished in $\delta(f[m])$ and $F(t)$ has occurred for all $t^{\mathbb{N}} < i$.

We define an assignment

$$\Phi(X) := \{t^{\mathbb{N}} \mid (t \notin X) \ \textit{occurs in } f\}.$$

Here F occurs in f means that F occurs in $f[n]$ for some n.

An easy induction on the length of a formula G, using observations *(3) – (6)* and the fact that f must not contain an axiom, shows $\mathbb{N} \not\models G[\Phi]$ for all formulas G occurring in f. Since all formulas of Δ occur in $f[0]$ this yields $\mathbb{N} \not\models \bigvee \{F \mid F \in \Delta\}[\Phi]$.

□

The Syntactical Main Lemma together with the Semantical Main Lemma prove the following theorem.

1.3.4. ω–Completeness Theorem. *Let $(\forall\vec{X})F(\vec{X})$ be a Π^1_1-sentence. Then we have $\mathbf{N} \models (\forall\vec{X})F(\vec{X})$ iff there is an $\alpha < \omega_1^{\mathrm{CK}}$ such that $\models^{\alpha} F(\vec{X})$.*

Proof. The soundness is already stated in (13). For the completeness direction we assume $\not\models^{\alpha} F(\vec{X})$ for all ordinals $\alpha < \omega_1^{\mathrm{CK}}$. Then, by the Syntactical Main-Lemma, the search tree for $F(\vec{X})$ cannot be well-founded. Applying the Semantical Main-Lemma we obtain an assignment Φ over $\mathbf{N}$ such that $\mathbf{N} \not\models F(\vec{X})[\Phi]$. Hence $\mathbf{N} \not\models (\forall\vec{X})F(\vec{X})$. □

In view of Theorem 1.3.4 we call formulas which contain at most free set variables sloppily Π^1_1-sentences. Semantically we treat these pseudo Π^1_1-sentences as their universal closure, i.e.,

$$\mathbf{N} \models F(\vec{X}) \;:\Leftrightarrow\; \mathbf{N} \models (\forall\vec{X})F(\vec{X}).$$

We use Theorem 1.3.4 to define the truth complexity of Π^1_1-sentences.

1.3.5. Definition. (Truth Complexity) For a Π^1_1-sentence $G := (\forall\vec{X})F(\vec{X})$ or $G := F(\vec{X})$ we define

$$\mathsf{tc}(G) := \begin{cases} \omega_1 & \text{if } \mathbf{N} \not\models G \\ \min\{\alpha \mid\ \models^{\alpha} F(\vec{X})\} & \text{otherwise.} \end{cases}$$

Using truth complexities we may restate Theorem 1.3.4 for Π^1_1-sentences F as

$$\mathbf{N} \models F \textit{ iff } \mathsf{tc}(F) < \omega_1^{\mathrm{CK}}. \tag{15}$$

As we have seen in (14) we have

$$\mathsf{tc}(F) < \omega \tag{16}$$

for all true arithmetical sentences F. But in contrast to that we have by Corollary 1.3.9 and Theorem 1.3.10 below

$$\sup\{\mathsf{tc}(F) \mid\ (\forall\vec{X})F(\vec{X}) \text{ is a } \Pi^1_1\text{–sentence and } \mathbf{N} \models F(\vec{X})\} = \omega_1^{\mathrm{CK}}$$

which shows that for "real" Π^1_1-sentences truth complexity is a non-trivial notion.

We call a binary relation $\prec$ *arithmetical* if there is an arithmetical formula $F(x, y)$ such that

$$m \prec n \textit{ iff } \mathbf{N} \models F(\underline{m}, \underline{n}).$$

The following Boundedness Theorem is one of the most important theorems for this contribution and will return in different variations.

1.3.6. Boundedness Theorem. *Let $\prec$ be an arithmetical definable relation and $\mathsf{TI}(\prec, X)$ be the formula*

$$(\forall x \in \mathsf{field}(\prec))[(\forall y)(y \prec x \to y \in X) \to x \in X] \to (\forall x \in \mathsf{field}(\prec))[x \in X]$$

expressing induction along $\prec$. *Then* $\mathsf{otyp}(\prec) \le \mathsf{tc}\big((\forall X)\, \mathit{TI}(\prec, X)\big)$.

To obtain the Boundedness Theorem we prove the more general Boundedness Lemma. We prepare the Boundedness Lemma by a few notions and observations.

An $\mathcal{L}^2_{\mathbb{N}}$-formula is X-positive if it does not contain occurrences of $t \notin X$. An obvious property of X-positive formulas is stated in the next lemma.

1.3.7. Monotonicity Lemma. *Let F be an X-positive formula not containing further set variables and $M \subseteq N \subseteq \mathbb{N}$. Then $\mathbb{N} \models F[M]$ entails $\mathbb{N} \models F[N]$.*

The proof is straightforward by induction on the length of the sentence F. □

By induction on α we obtain the following inversion properties:

$$\models^{\alpha} \Delta, A_1 \vee A_2 \Rightarrow \models^{\alpha} \Delta, A_1, A_2 \ , \tag{17}$$

$$\models^{\alpha} \Delta, A_1 \wedge A_2 \Rightarrow \models^{\alpha} \Delta, A_i \quad \text{for } i = 1, 2 \tag{18}$$

and

$$\models^{\alpha} \Delta, (\forall x)F(x) \Rightarrow \models^{\alpha} F(\underline{i}) \quad \text{for all numbers } i. \tag{19}$$

Let $\mathit{Prog}(\prec, X)$ denote the premise

$$(\forall x \in \mathsf{field}(\prec))[(\forall y)(y \prec x \to y \in X) \to x \in X]$$

of the Π^1_1-sentence $\mathit{TI}(\prec, X)$. For a fixed well-founded binary relation $\prec$ let $\overline{\mathrm{en}}^{\prec}_{\{z_1,\ldots,z_n\}}$ denote the enumerating function of the complement of the set $\{\mathsf{otyp}_{\prec}(z_1), \ldots, \mathsf{otyp}_{\prec}(z_n)\}$, i.e. of $\mathbf{On} \setminus \{\mathsf{otyp}_{\prec}(z_1), \ldots, \mathsf{otyp}_{\prec}(z_n)\}$. Define

$$M^{\prec}_{\{z_1,\ldots,z_n\}}(\alpha) := \{m \mid \mathsf{otyp}_{\prec}(m) \le \overline{\mathrm{en}}^{\prec}_{\{z_1,\ldots,z_n\}}(\alpha)\} \cup \{z_1, \ldots, z_n\}.$$

1.3.8. Boundedness Lemma. *Let $\prec$ be a well-founded transitive binary relation and assume that*

$$\models^{\alpha} \neg \mathit{Prog}(\prec, X), \underline{z}_1 \notin X, \ldots, \underline{z}_n \notin X, \Delta$$

for a set of X-positive Π^1_1-sentences $\Delta = \{F_1, \ldots, F_n\}$ not containing further set-variables. Then

$$\mathbb{N} \models (F_1 \vee \cdots \vee F_n)[M^{\prec}_{\{z_1,\ldots,z_n\}}(\alpha)].$$

The Boundedness Theorem follows from the Boundedness Lemma. Assume $\mathsf{tc}\big(\mathit{TI}(\prec, X)\big) \le \alpha$. Then there is for every $n \in \mathsf{field}(\prec)$ an $\alpha_0 < \alpha$ such that $\models^{\alpha_0} \neg \mathit{Prog}(\prec, X), \underline{n} \notin \mathsf{field}(\prec), \underline{n} \in X$. Because of $M_{\emptyset}(\alpha_0) = \{m \mid \mathsf{otyp}(m) \le \alpha_0\}$ we obtain by the Boundedness Lemma

$$(\forall n \in \mathsf{field}(\prec))(\exists \alpha_0 < \alpha)[\mathsf{otyp}(n) \le \alpha_0],$$

hence $\mathsf{otyp}(\prec) \le \alpha$.

We prove the Boundedness Lemma by induction on α. Assume

$$\models^{\alpha} \neg Prog(\prec, X), \underline{z}_1 \notin X, \ldots, \underline{z}_n \notin X, \Delta\,. \tag{i}$$

The claim is trivial if $\mathrm{D}(\mathbb{N}) \cap \Delta \neq \emptyset$. If (i) holds by (AxL) then there is a formula $t \in X$ in Δ such that $t^{\mathbb{N}} = z_i$ for some $i \in \{1, \ldots, n\}$. Since $z_i \in M^{\prec}_{\{z_1,\ldots,z_n\}}(\alpha)$ this yields the claim.

If (i) is obtained by the premises

$$\models^{\alpha_i} \neg Prog(\prec, X), \underline{z}_1 \notin X, \ldots, \underline{z}_n \notin X, \Delta_i$$

then we obtain the claim by the induction hypothesis together with the Monotonicity Lemma and the soundness of $\models^{\alpha}$.

The really interesting case is that (i) follows from the premise

$$\models^{\alpha_0} \neg\, Prog(\prec, X), \underline{z} \in \mathsf{field}(\prec) \wedge (\forall y)[\neg y \prec \underline{z} \vee y \in X] \wedge \underline{z} \notin X, \\ \underline{z}_1 \notin X, \ldots, \underline{z}_n \notin X, \Delta \tag{ii}$$

By inversion we get from (ii)

$$\models^{\alpha_0} \neg Prog(\prec, X), \underline{z} \in \mathsf{field}(\prec) \wedge (\forall y)[\neg y \prec \underline{z} \vee y \in X], \underline{z}_1 \notin X, \ldots, \underline{z}_n \notin X, \Delta \tag{iii}$$

and

$$\models^{\alpha_0} \neg Prog(\prec, X), \underline{z} \notin X, \underline{z}_1 \notin X, \ldots, \underline{z}_n \notin X, \Delta\,. \tag{iv}$$

If

$$\mathbb{N} \models \bigvee \Delta[M^{\prec}_{\{z_1,\ldots,z_n\}}(\alpha_0)] \tag{v}$$

then

$$\mathbb{N} \models \bigvee \Delta[M^{\prec}_{\{z_1,\ldots,z_n\}}(\alpha)] \tag{vi}$$

by Monotonicity. If

$$\mathbb{N} \not\models \bigvee \Delta[M^{\prec}_{\{z_1,\ldots,z_n\}}(\alpha_0)] \tag{vii}$$

then the induction hypothesis applied to (iii) gives

$$z \in \mathsf{field}(\prec) \wedge (\forall y \prec z)[y \in M^{\prec}_{\{z_1,\ldots,z_n\}}(\alpha_0)]. \tag{viii}$$

From (iv) we get by the induction hypothesis

$$\mathbb{N} \models \bigvee \Delta[M^{\prec}_{\{z_1,\ldots,z_n,z\}}(\alpha_0)]. \tag{ix}$$

We claim

$$\mathsf{otyp}_{\prec}(z) \leq \overline{\mathrm{en}}^{\prec}_{\{z_1,\ldots,z_n\}}(\alpha_0 + 1) \leq \overline{\mathrm{en}}^{\prec}_{\{z_1,\ldots,z_n\}}(\alpha). \tag{x}$$

To prove (x) recall that $\mathsf{otyp}_{\prec}(z) := \sup\{\mathsf{otyp}_{\prec}(y) + 1 \mid y \prec z\}$ and observe that for $y \prec z$ we obtain $\mathsf{otyp}_{\prec}(y) \leq \overline{\mathrm{en}}^{\prec}_{\{z_1,\ldots,z_n\}}(\alpha_0)$ or $y \in \{z_1, \ldots, z_n\}$ by (viii). Hence

$$\eta := \sup\{\mathsf{otyp}_{\prec}(y) + 1 \mid y \prec z \wedge y \notin \{z_1, \ldots, z_n\}\} \leq \overline{\mathrm{en}}^{\prec}_{\{z_1,\ldots,z_n\}}(\alpha_0 + 1)$$

and for $y \prec z$ such that $y \in \{z_1, \ldots, z_n\}$ we get $\mathsf{otyp}_\prec(y) < \overline{\mathrm{en}}^\prec_{\{z_1,\ldots,z_n\}}(\alpha_0+1)$ since $\mathsf{otyp}_\prec(y)$ is omitted in the enumeration. Because of

$$\overline{\mathrm{en}}^\prec_{(z_1,\ldots,z_n,z)}(\alpha_0) \leq \overline{\mathrm{en}}^\prec_{\{z_1,\ldots,z_n\}}(\alpha_0+1) \leq \overline{\mathrm{en}}^\prec_{\{z_1,\ldots,z_n\}}(\alpha) \tag{xi}$$

we get by (x) and (xi) $M^\prec_{\{z_1,\ldots,z_n,z\}}(\alpha_0) \subseteq M^\prec_{\{z_1,\ldots,z_n\}}(\alpha)$. This finally yields $\mathbf{N} \models \bigvee \Delta[M^\prec_{\{z_1,\ldots,z_n\}}(\alpha)]$ by (ix) and Monotonicity. □

As a consequence of the Boundedness and the ω–Completeness Theorem we obtain

1.3.9. Corollary. *If $\prec$ is an arithmetical definable well-ordering then* $\mathsf{otyp}(\prec) < \omega_1^{\mathrm{CK}}$.

It should be mentioned that Corollary 1.3.9 can – with just a little effort – be extended to Σ^1_1-definable well-orderings $\prec$ and thus comprises the well-known Boundedness Theorem of recursion theory without referring to the Analytical Hierarchy Theorem (cf. Beckmann and Pohlers [1997]).

Conversely to the Boundedness Theorem we obtain $\mathsf{otyp}(\prec)$ also as an upper bound for $\mathrm{tc}(\mathit{TI}(\prec, X))$. We show

$$\overset{5\cdot(\mathsf{otyp}_\prec(n)+1)}{\models\!\!=\!=} \neg \mathit{Prog}(\prec, X), \underline{n} \in X \tag{20}$$

by induction on $\mathsf{otyp}_\prec(n)$ where – for simplicity – we assume that $\prec$ is a primitive recursively definable relation whose field is all of $\mathbf{N}$.

We have

$$\overset{5\cdot(\mathsf{otyp}_\prec(m)+1)}{\models\!\!=\!=} \neg \mathit{Prog}(\prec, X), \neg \underline{m} \prec \underline{n}, \underline{m} \in X \tag{i}$$

either as an instance of (AxM) or by induction hypothesis. Hence

$$\overset{5\cdot\mathsf{otyp}_\prec(n)+3}{\models\!\!=\!=} \neg \mathit{Prog}(\prec, X), (\forall y)[y \prec \underline{n} \rightarrow y \in X] \tag{ii}$$

by two $(\vee)$ and one $(\forall)$ "inference". By (AxL) we have

$$\overset{0}{\models} \neg \mathit{Prog}(\prec, X), \underline{n} \notin X, \underline{n} \in X\,. \tag{iii}$$

By (ii) and (iii) we obtain

$$\overset{5\cdot\mathsf{otyp}_\prec(n)+4}{\models\!\!=\!=} \neg \mathit{Prog}(\prec, X), (\forall y)[y \prec \underline{n} \rightarrow y \in X] \wedge \underline{n} \notin X, \underline{n} \in X\,. \tag{iv}$$

One additional "inference" $(\exists)$ leads to

$$\overset{5\cdot(\mathsf{otyp}_\prec(n)+1)}{\models\!\!=\!=} \neg \mathit{Prog}(\prec, X), \underline{n} \in X\,. \qquad \square$$

From (20) we obtain by a clause $(\forall)$ and two clauses $(\vee)$ the following theorem.

1.3.10. Theorem. *If $\prec$ is a primitive recursive well-founded relation whose order type is a limit ordinal then* $\mathsf{otyp}(\prec) \leq \mathrm{tc}(\mathit{TI}(\prec, X)) \leq \mathsf{otyp}(\prec) + 2$.

Of course we should read Theorem 1.3.10 as $\mathsf{otyp}(\prec) = \mathsf{tc}\big(\mathit{TI}(\prec, X)\big)$, since the "+2" is just due to the syntactical peculiarities in the definition of $\models^{\alpha} \Delta$. But, since all important ordinals will be limits, this is of no importance. We define

$$\mathrm{spec}_{\Pi^1_1}(\mathbf{N}) := \{\mathsf{tc}(F) \mid F \text{ is a } \Pi^1_1\text{-sentence and } \mathbf{N} \models F\}$$

and call $\mathrm{spec}_{\Pi^1_1}(\mathbf{N})$ the Π^1_1-spectrum of $\mathbf{N}$. Due to Theorems 1.3.4 and 1.3.10 we then have

$$\mathrm{spec}_{\Pi^1_1}(\mathbf{N}) = \omega_1^{\mathrm{CK}}. \tag{21}$$

More generally we define the Π^1_1-spectrum of a theory $\mathbf{Ax}$ in the language of second order arithmetic by

$$\mathrm{spec}_{\Pi^1_1}(\mathbf{Ax}) := \{\mathsf{tc}\Big((\forall \vec{X})F(\vec{X})\Big) \mid (\forall \vec{X})F(\vec{X}) \text{ is a } \Pi^1_1\text{-sentence and } \mathbf{Ax} \vdash F(\vec{X})\}.$$

$$\|\mathbf{Ax}\|_{\Pi^1_1} := \sup(\mathrm{spec}_{\Pi^1_1}(\mathbf{Ax})). \tag{22}$$

For a recursively enumerable theory $\mathbf{Ax}$ the set $\{F \mid \mathbf{Ax} \vdash F\}$ is recursively enumerable, too. Since ω_1^{CK} is recursively regular we get by Theorem 1.3.4 and the fact that for true Π^1_1–sentences F an upper bound for $\mathsf{tc}(F)$ can be effectively computed from F via the depth of its search tree

$$\|\mathbf{Ax}\|_{\Pi^1_1} < \omega_1^{\mathrm{CK}}$$

for all recursively enumerable theories $\mathbf{Ax}$. The *proof theoretic ordinal* of a theory $\mathbf{Ax}$ is commonly defined as

$$\|\mathbf{Ax}\| := \sup\{\mathsf{otyp}(\prec) \mid \prec \text{ is primitive recursively definable and } \mathbf{Ax} \vdash \mathit{TI}(\prec, X)\}.$$

From the Boundedness Theorem we get immediately

$$\|\mathbf{Ax}\| \le \|\mathbf{Ax}\|_{\Pi^1_1} < \omega_1^{\mathrm{CK}} \tag{23}$$

for all recursively enumerable theories $\mathbf{Ax}$. By showing that for every ordinal $\alpha < \|\mathbf{Ax}\|_{\Pi^1_1}$ there is a primitive recursive order relation $\prec$ such that $\alpha \le \mathsf{otyp}(\prec)$ and $\mathbf{Ax} \vdash \mathit{TI}(\prec, X)$ we get from Theorem 1.3.10 together with (23)

$$\|\mathbf{Ax}\| = \|\mathbf{Ax}\|_{\Pi^1_1} \tag{24}$$

for all theories which will be analyzed.

There is, however, also a more general argument for (24) which we are going to sketch roughly. Assume that $\mathbf{Ax}$ is a theory comprising $\mathbf{PA}$ and let $(\forall \vec{Y})F(\vec{Y})$ be a Π^1_1-sentence. Denote by $\prec_{S_{F(\vec{Y})}}$ the Kleene–Brouwer ordering in the search tree $S_{F(\vec{Y})}$ for $F(\vec{Y})$ and assume that $\mathbf{Ax} \not\vdash \mathit{TI}(\prec_{S_{F(\vec{Y})}}, X)$. Then there is a model $\mathfrak{M} \models \mathbf{Ax}$ and an assignment $T \subseteq \mathfrak{M}$ for X such that $\mathfrak{M} \not\models \mathit{TI}(\prec_{S_{F(\vec{Y})}}, X)[T]$. Therefore there is an infinite path, say $P \subseteq \mathfrak{M}$, through $S_{F(\vec{Y})}$ which is definable by a first order formula with parameter T. According to the Semantical Main-Lemma we get assignments $\Phi(Y_i) \subseteq \mathfrak{M}$ for all Y_i belonging to $\vec{Y}$ which are definable by first order formulas with parameter T. Since we have induction in $\mathfrak{M}$ for first order formulas we obtain $\mathfrak{M} \not\models F(\vec{Y})[\Phi]$ as in the proof of the Semantical Main Lemma using a local truth predicate. Hence $\mathbf{Ax} \not\vdash F(\vec{Y})$ and we have shown

$$\mathbf{Ax} \vdash F(\vec{Y}) \Rightarrow \mathbf{Ax} \vdash \mathsf{TI}(\prec_{S_{F(\vec{Y})}}, X).$$

Since $\prec_{S_{F(\vec{Y})}}$ is primitive recursively definable and we have $\mathsf{tc}(F(\vec{Y})) \leq \mathsf{otyp}(\prec_{S_{F(\vec{Y})}}) \leq \|\mathbf{Ax}\|$ if $\mathbf{Ax} \vdash F(\vec{Y})$ this implies $\|\mathbf{Ax}\|_{\Pi^1_1} \leq \|\mathbf{Ax}\|$. Summarizing we get

1.3.11. Theorem. *Let* $\mathbf{PA} \subseteq \mathbf{Ax}$ *then* $\|\mathbf{Ax}\| = \|\mathbf{Ax}\|_{\Pi^1_1}$.

The computation of the ordinal $\|\mathbf{Ax}\|$ is commonly called the *ordinal analysis of* $\mathbf{Ax}$. In view of Theorem 1.3.11 we also talk about a Π^1_1–analysis of $\mathbf{Ax}$. To explain the connection between $\|\mathbf{Ax}\|_{\Pi^1_1}$ and $\|\mathbf{Ax}\|_{\Sigma_1^{\omega_1^{CK}}}$ we use again Theorem 1.3.4 which says

$$\mathbf{N} \models (\forall \vec{X})F(\vec{X}) \;\Leftrightarrow\; (\exists \alpha < \omega_1^{CK})[\models^{\alpha} F(\vec{X})]. \tag{25}$$

Assume that we have coded the language $\mathcal{L}^2_{\mathbf{N}}$ within Set Theory (c.f. Barwise [1975]). The fact that "z is an infinitary proof tree for $F(\vec{X})$ of length $\leq \alpha$" can be expressed by a Δ_0 formula, say $G(\alpha, z, \ulcorner F(\vec{X}) \urcorner)$. It is easy to check that for $\alpha \in \mathbf{On}$ and $z \in \mathbf{L}$ such that $\mathbf{L} \models G(\alpha, z, \ulcorner F(\vec{X}) \urcorner)$ we have $z \in \mathbf{L}_{\alpha+n}$ for some $n < \omega$. If we take into account that $F(\vec{X})$ may contain additional number parameters, say $\vec{\underline{n}}$, (25) turns into

$$(\forall \vec{n}) \Big[\mathbf{N} \models (\forall \vec{X})F(\vec{X}, \vec{\underline{n}}) \;\Leftrightarrow\; (\exists \alpha < \omega_1^{CK})(\exists z \in \mathbf{L}_{\omega_1^{CK}})G(\alpha, z, \ulcorner F(\vec{X}, \vec{\underline{n}}) \urcorner)\Big].$$

This is the well known Hyperarithmetical Quantifier Theorem telling that every Π^1_1-formula is equivalent to a Σ_1-formula over $\mathbf{L}_{\omega_1^{CK}}$. If we put

$$|H|_{\Sigma_1} := \min\{\alpha \mid \mathbf{L}_\alpha \models H\} \tag{26}$$

for Σ_1-sentences H we get

$$\mathsf{tc}\Big((\forall \vec{X})F(\vec{X})\Big) \leq |(\exists \xi)(\exists z)G(\xi, z, \ulcorner F(\vec{X}) \urcorner)|_{\Sigma_1} \leq \mathsf{tc}\Big((\forall \vec{X})F(\vec{X})\Big) + n$$

for some $n < \omega$. Defining

$$\|\mathbf{Ax}\|_{\Sigma_1^{\omega_1^{CK}}} := \sup\{|(\exists \xi)(\exists z)G(\xi, z, \ulcorner F(\vec{X}) \urcorner)|_{\Sigma_1} \mid \mathbf{Ax} \vdash F(\vec{X})\}$$

for a $\mathcal{L}^2_{\mathbf{N}}$–theory $\mathbf{Ax}$ we get

$$\|\mathbf{Ax}\|_{\Sigma_1^{\omega_1^{CK}}} = \|\mathbf{Ax}\|_{\Pi^1_1} = \|\mathbf{Ax}\|. \tag{27}$$

It is evident that the ordinal $\|\mathbf{Ax}\|_{\Sigma_1^{\omega_1^{CK}}}$ for $\mathcal{L}^2_{\mathbf{N}}$-theories is the exact counterpart of the ordinal $\|\mathbf{Ax}\|_{\Sigma_1^{\omega_1^{CK}}}$ which we defined in the previous section for theories in the language $\mathcal{L}(\in)$ of Set Theory.

Analogously to Theorem 1.2.5 we get[2]

$$\|\mathbf{Ax} + F\|_{\Pi_1^1} = \|\mathbf{Ax}\|_{\Pi_1^1} \tag{28}$$

for all true Σ_1^1-sentences F with the same proof.

We mentioned already in the beginning of the section that $\mathcal{L}_{\mathbb{N}}^2$ can be regarded as a sublanguage of $\mathcal{L}(\in)$. Call F a Π_1^1-sentence of $\mathcal{L}$ if it is obtained as a translation of a Π_1^1-sentence of $\mathcal{L}_{\mathbb{N}}^2$, i.e., if it has the form $(\forall x)[x \subseteq \omega \to G(x)]$ where $G(x)$ is a Δ_0-formula whose quantifiers are all restricted to ω or to natural numbers. If $\mathbf{Ax}$ is a theory extending $\mathbf{KP}$ then we can use the familiar unsecured sequences argument to show that $\mathbf{Ax}$ proves that for every Π_1^1-formula $F(\vec{x})$ there is a Δ_0-definable order relation $\prec_{F(\vec{x})}$ such that

$$\mathbf{Ax} \vdash (\forall \vec{x})[F(\vec{x}) \leftrightarrow \mathit{Wf}(\prec_{F(\vec{x})})].$$

We will see later (cf. Section 3.3.3) that for a theory $\mathbf{Ax}$ which proves Axiom β – which says that every well-ordering can be order isomorphically mapped onto an ordinal – we have directly

$$\mathbf{Ax} \vdash (\forall \vec{x})[\mathit{Wf}(\prec_{F(\vec{x})}) \leftrightarrow (\exists \xi < \mathbf{L}_{\omega_1^{CK}}) I_F(\xi, \vec{x})]$$

for a Σ_1-formula $I_F(\xi, \vec{x})$. It is easy to check that $\mathsf{otyp}(\prec_{F(\vec{n})}) \approx |(\exists \xi) I_F(\xi, \vec{n})|_{\Sigma_1}$. Thus we have

$$\|\mathbf{Ax}\|_{\Pi_1^1} = \|\mathbf{Ax}\| = \|\mathbf{Ax}\|_{\Sigma_1^{\omega_1^{CK}}} \tag{29}$$

for sufficiently strong theories in the language of Set Theory.

This rough sketch should suffice to explain that the ordinals $\|\mathbf{Ax}\|$ and $\|\mathbf{Ax}\|_{\Sigma_1^{\omega_1^{CK}}}$ carry the same information. So Π_1^1-analysis and ω_1^{CK}–ordinal analysis are the same things. We will see that indeed a good deal of information is contained in $\|\mathbf{Ax}\|$.

1.4. Methods

Before we come to examples of analyzed theories we want to outline the methods used in ordinal analyses. There are two main steps. The first is to compute upper bounds for the ordinals in $\mathrm{spec}_{\Pi_1^1}(\mathbf{Ax})$, the second to show that these bounds are the best possible ones. We explain the general pattern on the example of an $\mathcal{L}_{\mathbb{N}}^2$-theory $\mathbf{Ax}$.

If $\mathbf{Ax} \vdash F$ then there are formulas $A_1, \ldots, A_n \in \mathbf{Ax}$ such that

$$A_1, \ldots, A_n \vdash F \tag{i}$$

in pure predicate logic. By Gentzen's Hauptsatz we may assume that this derivation is cut-free. It will be quite easy to transform (i) into a truth definition

$$\models^{\alpha} \neg A_1, \ldots, \neg A_n, F \tag{ii}$$

[2]Defining $\mathbf{Ax} + (\exists X)H(X) \vdash G \ :\Leftrightarrow\ \mathbf{Ax} \vdash \neg H(X) \vee G$ this makes sense even without a notion of proof for Second Order Logic.

where α will depend mainly on the complexity of the formulas $A_1, \ldots, A_n$ and F. Then we have to compute upper bounds for $\mathsf{tc}(A)$ for all formulas in $\mathbf{Ax}$. This gives

$$\models^{\alpha_i} A_i \tag{iii}$$

if $\mathsf{tc}(A_i) \leq \alpha_i$ for $i \in \{1, \ldots, n\}$. The problem is now to link (iii) and (ii). This will be achieved by extending the truth definition $\models^{\alpha} \Delta$ into a *semi-formal* calculus $\vdash^{\alpha}_{\rho} \Delta$ by adding the cut rule

(cut) $\vdash^{\beta_1}_{\rho} \Delta, F;\ \vdash^{\beta_2}_{\rho} \Delta, \neg F;\ \beta_1, \beta_2 < \beta$ *and* $\mathsf{rk}(F) < \rho \ \Rightarrow\ \vdash^{\beta}_{\rho} \Delta.$

This will of course destroy its meaning as a truth definition. But we will still have

$$\vdash^{\alpha}_{0} F \ \Leftrightarrow\ \models^{\alpha} F. \tag{iv}$$

By (ii) and (iii) we obtain

$$\vdash^{\beta}_{\rho} F \tag{v}$$

where β and ρ are computable from $\alpha_1, \ldots, \alpha_n$ and α and the problem reduces to the elimination of cut in the semi-formal system. Since the semi-formal system is obviously sound we get

$$\vdash^{\beta}_{\rho} F \ \Rightarrow\ \mathbb{N} \models F \ \Rightarrow\ (\exists \delta < \omega_1^{\mathrm{CK}}) \models^{\delta} F \ \Rightarrow\ (\exists \delta < \omega_1^{\mathrm{CK}}) \vdash^{\delta}_{0} F$$

showing that the Cut-Elimination Theorem holds for the semi-formal calculus. But this is of little help since we do not know how to compute δ from β and ρ. In a moment, however, we will see that for predicative theories we can sharpen Cut-Elimination to

$$\vdash^{\beta}_{\rho} \Delta \ \Rightarrow\ \vdash^{\varphi_\rho \beta}_{0} \Delta. \tag{vi}$$

By (v), (vi) and (iv) we get

$$\mathsf{tc}(F) \leq \varphi_\rho \beta \tag{vii}$$

for $\mathbf{Ax} \vdash F$. Since β and ρ only depend on the formulas in $\mathbf{Ax}$ this will give an upper bound, say ε, for $\|\mathbf{Ax}\|_{\Pi^1_1}$. To show that ε is the best possible bound it suffices to prove that for every $\alpha < \varepsilon$ there is a primitive recursive well-ordering $\prec$ such that $\alpha \leq \mathsf{otyp}(\prec)$ and $\mathbf{Ax} \vdash \mathit{TI}(\prec, X)$.

It will become clear from the following text how this concept has to be modified as to serve also for theories in the language $\mathcal{L}(\in)$ of Set Theory and for impredicative theories. But before that we demonstrate some details on the example on Gentzen's result.

2. First order number theory

2.1. Peano arithmetic

The paradigm for ordinal analysis is still Gentzen's result on Peano Arithmetic. So we opt for it as our first and simplest example. However, rather than to analyze

PA itself we will analyze a conservative extension **NT** of **PA** which allows constants for all primitive recursive functions. We start with an introduction of the theory **NT**.

2.1.1. The axiom system NT

The language $\mathcal{L}_{\mathbb{N}}$ is a first order language which contains set parameters denoted by capital Latin letters X, Y, Z, X_1, ... and constants for 0 and all primitive recursive functions and relations. We assume that the symbols for primitive recursive functions are built up from the symbols C_k^n for the constant function, P_k^n for the projection on the n-th component, S for the successor function by a substitution operator *Sub* and the recursion operator *Rec*.

The theory **NT** comprises the following sentences:

The successor axioms

$$(\forall x)[\neg \underline{0} = Sx]$$
$$(\forall x)(\forall y)[S(x) = S(y) \Rightarrow x = y]$$

The defining axioms for function and relation symbols which are the universal closures of the following formulas

$$C_k^n(x_1, \ldots, x_n) = \underline{k}$$
$$P_k^n(x_1, \ldots, x_n) = x_k$$
$$Sub(g, h_1, \ldots, h_m)(x_1, \ldots, x_n) = g(h_1(x_1, \ldots, x_n)) \cdots (h_m(x_1, \ldots, x_n))$$
$$Rec(g, h)(0, x_1, \ldots, x_n) = g(x_1, \ldots, x_n)$$
$$Rec(g, h)(Sy, x_1, \ldots, x_n) = h(y, Rec(g, h)(y, x_1, \ldots, x_n), x_1, \ldots, x_n)$$
$$(x_1, \ldots, x_n) \in R \leftrightarrow \chi_R(x_1, \ldots, x_n) = 0$$

The scheme of Mathematical Induction

$$F(\underline{0}) \wedge (\forall x)[F(x) \to F(S(x))] \to (\forall x)F(x)$$

for all $\mathcal{L}_{\mathbb{N}}$-formulas $F(u)$.

2.1.2. An upper bound for $\mathrm{spec}_{\Pi^1_1}(\mathbf{NT})$

Following the general pattern as sketched in the previous section we have first to deal with the truth complexity of logically valid sentences. Therefore we have to fix a calculus for pure predicate logic and we opt for a cut free Tait calculus, i.e., one sided sequent calculus, which is given by the following rules:

2.1.2.1. Definition.

(AxL) $\models^m \Delta, A, \neg A$ *for any* m, *if* A *is an atomic formula*

($\vee$) *If* $\models^{m_0} \Delta, A_i$ *for some* $i \in \{1, 2\}$, *then* $\models^m \Delta, A_1 \vee A_2$ *for all* $m > m_0$

($\wedge$) *If* $\models^{m_i} \Delta, A_i$ *and* $m_i < m$ *for all* $i \in \{1, 2\}$, *then* $\models^m \Delta, A_1 \wedge A_2$

($\exists$) *If* $\models^{m_0} \Delta, A(t)$, *then* $\models^{m} \Delta, (\exists x)A(x)$ *for all* $m > m_0$

($\forall$) *If* $\models^{m_0} \Delta, A(u)$ *and* u *not free in* $\Delta, (\forall x)A(x)$, *then* $\models^{m} \Delta, (\forall x)A(x)$ *for all* $m > m_0$.

The identity axioms are the following formulas

$$
\begin{array}{l}
(\forall x)[x = x] \\
(\forall x)(\forall y)[x = y \rightarrow y = x] \\
(\forall x)(\forall y)(\forall z)[x = y \land y = z \rightarrow x = z] \\
(\forall \vec{x})(\forall \vec{y})[x_1 = y_1 \land \cdots \land x_n = y_n \;\rightarrow\; f(x_1, \ldots, x_n) = f(y_1, \ldots, y_n)] \\
(\forall \vec{x})(\forall \vec{y})[x_1 = y_1 \land \cdots \land x_n = y_n \;\rightarrow\; (R(x_1, \ldots, x_n) \rightarrow R(y_1, \ldots, y_n))] \\
(\forall x)(\forall y)[x = y \;\rightarrow\; (x \in X \rightarrow y \in X)]
\end{array}
$$

Due to Gentzen's Hauptsatz we have the following theorem:

2.1.2.2. Theorem. *Let* Δ *be a finite set of formulas such that* $\bigvee \Delta$ *is valid in the sense of first order predicate logic. Then there are finitely many identity axioms* $I_1, \ldots, I_n$ *and an* $m < \omega$ *such that* $\models^{m} \neg I_1, \ldots, \neg I_n, \Delta$.

Let $\vec{u}$ be a list containing all number variables which occur free in Δ. An easy induction on m using the fact

$$\models^{\alpha} \Delta\,(s) \;\; \textit{and} \;\; s^{\mathbb{N}} = t^{\mathbb{N}} \Rightarrow \models^{\alpha} \Delta\,(t) \tag{30}$$

shows

$$\vdash^{m} \Delta(\vec{u}) \;\; \Rightarrow \;\; \models^{m} \Delta\,(\underline{\vec{n}}) \tag{31}$$

for every tuple $\underline{\vec{n}}$ of numerals. We have

$$\models^{5} (\forall x)(\forall y)[x = y \;\rightarrow\; (x \in X \rightarrow y \in X)]$$

and all the other identity and defining axioms for primitive recursive functions and relations are true arithmetical sentences. Thus, using also (16), we have

$$\mathsf{tc}(F) < \omega \tag{32}$$

for all mathematical and identity axioms except induction. What really needs checking is the truth complexity of the scheme of Mathematical Induction. Here we need the following lemma.

2.1.2.3. Tautology Lemma. *For every* $\mathcal{L}_{\mathbb{N}}$*-sentence we have* $\models^{2\cdot\mathsf{rk}(F)} \Delta, \neg F, F$.

The proof is immediate by induction on $\mathsf{rk}(F)$.

The truth complexity for all instances of Mathematical Induction follows from the Induction Lemma.

2.1.2.4. Induction Lemma. *For any natural number n and any $\mathcal{L}_{\mathbb{N}}$-sentence $F(\underline{n})$ we have*

$$\models^{2\cdot[\mathsf{rk}(F(\underline{n}))+n]} \neg F(\underline{0}), \neg(\forall x)[F(x) \to F(S(x))], F(\underline{n})\,. \tag{33}$$

The proof by induction on n is very similar to that of (20). For $n = 0$ we get (33) as an instance of the Tautology Lemma. For the induction step we have

$$\models^{2\cdot[\mathsf{rk}(F(\underline{n}))+n]} \neg F(\underline{0}), \neg(\forall x)[F(x) \to F(S(x))], F(\underline{n}) \tag{i}$$

by the induction hypothesis and obtain

$$\models^{2\cdot\mathsf{rk}(F(\underline{n}))} \neg F(\underline{0}), \neg(\forall x)[F(x) \to F(S(x))], \neg F(\underline{S(n)}), F(\underline{S(n)}) \tag{ii}$$

by the Tautology Lemma. From (i) and (ii) we get

$$\models^{2\cdot[\mathsf{rk}(F(\underline{n}))+n]+1} \neg F(\underline{0}), \neg(\forall x)[F(x) \to F(S(x))], F(\underline{n}) \wedge \neg F(\underline{S(n)}), F(\underline{S(n)})\,. \tag{iii}$$

By a clause ($\exists$) we finally obtain

$$\models^{2\cdot[\mathsf{rk}(F(\underline{n}))+n]+2} \neg F(\underline{0}), \neg(\forall x)[F(x) \to F(S(x))], F(\underline{S(n)})\,. \qquad \square$$

By Lemma 2.1.2.4 we have $\mathsf{tc}(G) \leq \omega + 4$ for all instances G of the Mathematical Induction Scheme. Together with (32) we get

$$\mathsf{tc}(F) \leq \omega + 4 \tag{34}$$

for all identity and non-logical axioms of **NT**.

If $\mathbf{NT} \vdash F$ then there are $\mathcal{L}_{\mathbb{N}}$-sentences $\{F_1, \ldots, F_n\}$ and a natural number m such that

$$\models^{m} \neg F_1, \ldots, \neg F_n, F \tag{35}$$

and for all $i \in \{1, \ldots, n\}$ the formula F_i is either an axiom in **NT** or an identity axiom. For every $\mathcal{L}_{\mathbb{N}}$-sentence F such that $\mathbf{NT} \vdash F$ we thus get by (35) and (31) $\mathcal{L}_{\mathbb{N}}$-sentences $F_1, \ldots, F_n$ such that

$$\models^{\omega} \neg F_1, \ldots, \neg F_n, F \tag{36}$$

and by (34)

$$\models^{\omega+4} F_i \tag{37}$$

for all $i \in \{1, \ldots, n\}$.

As sketched in Section 1.4 the problem of linking (37) and (36) will be solved by introducing a semi-formal calculus.

2.1.2.5. Definition. We define $\vdash^{\alpha}_{\rho} \Delta$ for a finite set Δ of $\mathcal{L}_{\mathbb{N}}$-formulas which contain at most free set–variables inductively by the following clauses:

(AxM) *If $\Delta \cap D(\mathbb{N}) \neq \emptyset$ then $\vdash^{\alpha}_{\rho} \Delta$ for all ordinals α and ρ.*

(AxL) *If* $t^{\mathrm{N}} = s^{\mathrm{N}}$ *then* $\vdash^{\alpha}_{\rho} \Delta, s \notin X, t \in X$ *for all ordinals* α *and* ρ.

($\wedge$) *If* $\vdash^{\alpha_i}_{\rho} \Delta, A_i$ *and* $\alpha_i < \alpha$ *for* $i = 1, 2$ *then* $\vdash^{\alpha}_{\rho} \Delta, A_1 \wedge A_2$.

($\vee$) *If* $\vdash^{\alpha_0}_{\rho} \Delta, A_i$ *and* $\alpha_0 < \alpha$ *for some* $i \in \{1, 2\}$ *then* $\vdash^{\alpha}_{\rho} \Delta, A_1 \vee A_2$.

($\forall$) *If* $\vdash^{\alpha_i}_{\rho} \Delta, A(\underline{i})$ *and* $\alpha_i < \alpha$ *for all* $i \in \mathrm{N}$ *then* $\vdash^{\alpha}_{\rho} \Delta, (\forall x)A(x)$.

($\exists$) *If* $\vdash^{\alpha_0}_{\rho} \Delta, A(\underline{i})$ *and* $\alpha_0 < \alpha$ *for some* $i \in \mathrm{N}$ *then* $\vdash^{\alpha}_{\rho} \Delta, (\exists x)A(x)$.

(cut) *If* $\vdash^{\alpha_1}_{\rho} \Delta, F$, $\vdash^{\alpha_2}_{\rho} \Delta, \neg F$, $\alpha_i < \alpha$ *for* $i \in \{1, 2\}$ *and* $\mathsf{rk}(F) < \rho$ *then* $\vdash^{\alpha}_{\rho} \Delta$.

We have to read $\vdash^{\alpha}_{\rho} \Delta$ as:
"There is an infinitary derivation of Δ *of height* $\leq \alpha$ *whose cut formulas are all of ranks* $< \rho$*."*
We call F the *main formula* of a clause in the definition of $\vdash^{\alpha}_{\rho} \Delta, F$ if F is responsible for Δ, F being an axiom in (AxM) and (AxL) or if the logical symbol introduced in the clause belongs to F. Thus (cut) possesses no main formula.

From the definition we get immediately

$$\models^{\alpha} \Delta \;\Leftrightarrow\; \vdash^{\alpha}_{0} \Delta. \tag{38}$$

From (31) and (38) we have

$$\vdash^{m} \Delta(\vec{u}) \;\Rightarrow\; \vdash^{m}_{0} \Delta(\vec{\underline{z}}) \tag{39}$$

for every tuple $\vec{z}$ of numbers. Another immediate property is

$$\vdash^{\alpha}_{\rho} \Delta, \quad \alpha \leq \beta, \quad \rho \leq \sigma \quad \text{and} \quad \Delta \subseteq \Gamma \;\Rightarrow\; \vdash^{\beta}_{\sigma} \Gamma. \tag{40}$$

The calculus $\vdash^{\alpha}_{\rho}$ is obviously sound. By induction on α one proves easily

$$\vdash^{\alpha}_{\rho} F_1, \ldots, F_n \Rightarrow \mathrm{N} \models (F_1 \vee \cdots \vee F_n)[\Phi] \tag{41}$$

for every assignment Φ: set variables $\longrightarrow \mathsf{Pow}(\mathrm{N})$. Soundness of $\vdash^{\alpha}_{\rho}$ and Completeness of $\models^{\alpha}$ together with (38) show

2.1.2.6. Cut Elimination Theorem. *If* $\vdash^{\alpha}_{\rho} \Delta$ *then there is a* $\gamma < \omega_1^{\mathrm{CK}}$ *such that* $\vdash^{\gamma}_{0} \Delta$.

But, as mentioned before, Theorem 2.1.2.6 does not help us in ordinal analysis. The bound for γ is much too large. What we are looking for is a function which computes a value for γ from the data α and ρ. The key here is the Reduction Lemma which tells us how to avoid a cut of rank ρ for the costs of an increasing length of the derivation.

2.1.2.7. Reduction Lemma. *Assume* $\vdash^{\alpha}_{\rho} \Delta, F$ *and* $\vdash^{\beta}_{\rho} \Gamma, \neg F$ *as well as* $\mathsf{rk}(F) = \rho$. *Then* $\vdash^{\alpha \# \beta}_{\rho} \Delta, \Gamma$.

The proof is by induction on $\alpha \# \beta$. Assume first that F is not the main formula of the last clause in the definition of $\frac{\alpha}{\rho} \Delta, F$. Then either $\frac{\alpha}{\rho} \Delta$, and hence also $\frac{\alpha \# \beta}{\rho} \Delta, \Gamma$, holds by (AxM) or (AxL) or we have $\frac{\alpha_i}{\rho} \Delta_i, F$ for $\alpha_i < \alpha$. But then we get by induction hypothesis $\frac{\alpha_i \# \beta}{\rho} \Delta_i, \Gamma$, and obtain $\frac{\alpha \# \beta}{\rho} \Delta, \Gamma$ by the same clause.The case that $\neg F$ is not the main formula in $\frac{\beta}{\rho} \Gamma, \neg F$ is symmetrical.

We may therefore assume that both, F and $\neg F$, are main formulas. Let us first assume that $\rho = 0$, i.e., that F is atomic. Then both $\frac{\alpha}{\rho} \Delta, F$ and $\frac{\beta}{\rho} \Gamma, \neg F$ are axioms whose main formulas are F and $\neg F$, respectively. This excludes axioms according to (AxM) because we cannot have $\{F, \neg F\} \subseteq \mathrm{D}(\mathbb{N})$. But then we have axioms according to (AxL) and we may assume that F is a formula $t \in X$. But then $\neg F \equiv t \notin X$ and we have a formula $s_1 \notin X$ in Δ and a formula $s_2 \in X$ in Γ such that $s_1^{\mathbb{N}} = t^{\mathbb{N}} = s_2^{\mathbb{N}}$. But then $\frac{\alpha \# \beta}{\rho} \Delta, \Gamma$ holds by (AxL).

Now assume $0 < \rho$. We will only treat the more complicated case that F is a formula $(\forall x)A(x)$. The remaining cases are either simpler or symmetrical. Then $\neg F$ is the formula $(\exists x)\neg A(x)$ and we have the premises

$$\frac{\alpha_i}{\rho} \Delta, F, A(\underline{i}) \tag{i}$$

for all $i \in \mathbb{N}$ and

$$\frac{\beta_0}{\rho} \Gamma, \neg F, \neg A(\underline{i}_0). \tag{ii}$$

From (i) and $\frac{\beta}{\rho} \Gamma, F$ as well as from (ii) and $\frac{\alpha}{\rho} \Delta, F$ we obtain by the induction hypothesis

$$\frac{\alpha_{i_0} \# \beta}{\rho} \Delta, \Gamma, A(\underline{i}_0) \tag{iii}$$

and

$$\frac{\alpha \# \beta_0}{\rho} \Delta, \Gamma, \neg A(\underline{i}_0). \tag{iv}$$

Since $\mathsf{rk}(A(\underline{i}_0)) < \mathsf{rk}(F) = \rho$ and $\alpha_{i_0} \# \beta < \alpha \# \beta$ as well as $\alpha \# \beta_0 < \alpha \# \beta$, we obtain $\frac{\alpha \# \beta}{\rho} \Delta, \Gamma$ from (iii) and (iv) by cut. □

As a first consequence of the Reduction Lemma we obtain the Elimination Lemma.

2.1.2.8. Elimination Lemma. *If* $\frac{\alpha}{\rho+1} \Delta$ *then* $\frac{2^\alpha}{\rho} \Delta$.

The proof is by induction on α. If the last inference is not a cut of complexity ρ we obtain the claim immediately from the induction hypothesis and the fact that $\lambda\xi . 2^\xi$ is order preserving. The critical case is a cut $\frac{\alpha_1}{\rho+1} \Delta, F$; $\frac{\alpha_2}{\rho+1} \Delta, \neg F \;\Rightarrow\; \frac{\alpha}{\rho+1} \Delta$ with $\mathsf{rk}(F) = \rho$. By the induction hypothesis and the Reduction Lemma we obtain $\frac{2^{\alpha_1} \# 2^{\alpha_2}}{\rho} \Delta$ and we have $2^{\alpha_1} \# 2^{\alpha_2} \leq 2^{\max\{\alpha_1, \alpha_2\}} \cdot 2 \leq 2^\alpha$. □

The Elimination Lemma provides the first step in the proof of the following more general lemma.

2.1.2.9. Predicative Elimination Lemma. *If* $\vdash^{\alpha}_{\beta+\omega^\rho} \Delta$ *then* $\vdash^{\varphi_\rho\alpha}_{\beta} \Delta$.

The proof is by induction on ρ with side induction on α. For $\rho = 0$ we obtain $\vdash^{2^\alpha}_{\beta} \Delta$ by the first Elimination Lemma which, since $2^\alpha \le \omega^\alpha = \varphi_0\alpha$, entails the claim. Now assume $\rho > 0$. If the last clause was not a cut of rank $\ge \beta$ we obtain the claim from the induction hypotheses and the fact that the functions φ_ρ are order preserving. Therefore assume that the last inference is

$$\vdash^{\alpha_1}_{\beta+\omega^\rho} \Delta, F \quad \vdash^{\alpha_2}_{\beta+\omega^\rho} \Delta, \neg F \;\Rightarrow\; \vdash^{\alpha}_{\beta+\omega^\rho} \Delta$$

such that $\beta \le \mathsf{rk}(F) < \beta+\omega^\rho$. But then there is an ordinal ϕ such that $\mathsf{rk}(F) = \beta+\phi$ which, writing ϕ in Cantor normal form, means $\mathsf{rk}(F) = \beta+\omega^{\sigma_1}+\cdots+\omega^{\sigma_n} < \beta+\omega^\rho$. Hence $\sigma_1 < \rho$ and, putting $\sigma := \sigma_1$, we get $\mathsf{rk}(F) < \beta + \omega^\sigma \cdot (n+1)$. By the side induction hypothesis we have $\vdash^{\varphi_\rho\alpha_1}_{\beta} \Delta, F$ and $\vdash^{\varphi_\rho\alpha_2}_{\beta} \Delta, \neg F$. By a cut it follows $\vdash^{\varphi_\rho\alpha_1+\varphi_\rho\alpha_2}_{\beta+\omega^\sigma\cdot(n+1)} \Delta$. If we define $\varphi^{(0)}_\sigma\alpha := \alpha$ and $\varphi^{(n+1)}_\sigma\alpha := \varphi_\sigma(\varphi^{(n)}_\sigma\alpha)$ then we obtain from $\sigma < \rho$ by $n+1$-fold application of the main induction hypothesis $\vdash^{\varphi^{(n+1)}_\sigma(\varphi_\rho\alpha_1+\varphi_\rho\alpha_2)}_{\beta} \Delta$. Finally we show $\varphi^{(n)}_\sigma(\varphi_\rho\alpha_1 + \varphi_\rho\alpha_2) < \varphi_\rho\alpha$ by induction on n. For $n = 0$ we have $\varphi^{(0)}_\sigma(\varphi_\rho\alpha_1 + \varphi_\rho\alpha_2) = \varphi_\rho\alpha_1 + \varphi_\rho\alpha_2 < \varphi_\rho\alpha$ since $\alpha_i < \alpha$ and $\varphi_\rho\alpha \in Cr(0)$. For the induction step we have $\varphi^{(n+1)}_\sigma(\varphi_\rho\alpha_1 + \varphi_\rho\alpha_2) = \varphi_\sigma(\varphi^{(n)}_\sigma(\varphi_\rho\alpha_1 + \varphi_\rho\alpha_2)) < \varphi_\rho\alpha$ since $\sigma < \rho$ and $\varphi^{(n)}_\sigma(\varphi_\rho\alpha_1+\varphi_\rho\alpha_2) < \varphi_\rho\alpha$ by the induction hypothesis. Hence $\vdash^{\varphi_\rho\alpha}_{\beta} \Delta$. □

By the Predicative Elimination Lemma we obtain the function which computes an upper bound for the height of the cut free derivation.

2.1.2.10. Elimination Theorem. *Let* $\vdash^{\alpha}_{\rho} \Delta$ *such that* $\rho =_{\mathsf{NF}} \omega^{\rho_1} + \cdots + \omega^{\rho_n}$. *Then* $\vdash^{\varphi_{\rho_1}\varphi_{\rho_2}\cdots\varphi_{\rho_n}\alpha}_{0} \Delta$.

Back to the axiom system **NT**. If $\mathbf{NT} \vdash F$ then we obtain by (36), (37), and (38) $\vdash^{\omega+4+n}_{m} F$ for $m := \max\{\mathsf{rk}(F_1), \ldots, \mathsf{rk}(F_n)\} + 1 < \omega$. By the Elimination Theorem (or even m-fold application of the Elimination Lemma) we obtain $\vdash^{exp^m(\omega,\omega+4+n)}_{0} F$. Hence $\mathsf{tc}(F) \le exp^m(\omega, \omega+4+n) < \varepsilon_0$ and we have

2.1.2.11. Theorem. $\mathrm{spec}_{\Pi^1_1}(\mathbf{NT}) \subseteq \varepsilon_0$.

2.1.3. Lower bounds for $\mathrm{spec}_{\Pi^1_1}(\mathbf{NT})$

We want to show that the bound given in Theorem 2.1.2.11 is the best possible one. By Theorem 1.3.10 it suffices to have Theorem 2.1.3.1 below.

2.1.3.1. Theorem. *For every ordinal* $\alpha < \varepsilon_0$ *there is a primitive recursive well-order* $\prec$ *on the natural numbers of order type* α *such that* $\mathbf{NT} \vdash \mathsf{TI}(\prec, X)$.

The first step in proving Theorem 2.1.3.1 is to represent ordinals below ε_0 by primitive recursive well-orders. This is done by an arithmetization. We simultaneously define a set $\mathsf{On} \subseteq \mathbb{N}$ and a relation $a \prec b$ for $a, b \in \mathsf{On}$ together with an evaluation map $|\cdot|: \mathsf{On} \longrightarrow \mathbf{On}$ such that On and $\prec$ become primitive recursive and $a \prec b \Leftrightarrow |a| < |b|$. We put

- $0 \in \mathsf{On}$ *and* $|0| = 0$
- *If* $z_1, \ldots, z_n \subseteq \mathsf{On}$ *and* $z_1 \succeq \cdots \succeq z_n$ *then* $\langle z_1, \ldots, z_n \rangle \in \mathsf{On}$ *and* $|\langle z_1, \ldots, z_n \rangle| = \omega^{|z_1|} + \cdots + \omega^{|z_n|}$

and

- $a \prec b :\Leftrightarrow a \in \mathsf{On} \wedge b \in \mathsf{On} \wedge [(a = 0 \wedge b \neq 0)$
  $\vee (lh(a) < lh(b) \wedge (\forall i < lh(a))((a)_i = (b)_i))$
  $\vee (\exists i) < \min\{lh(a), lh(b)\})(\forall j) < i)((a)_j = (b)_j \wedge (a)_i \prec (b)_i)]$

Observe that On and $\prec$ are defined by simultaneous course of values recursion and thence are primitive recursive. It is also easy to check that $a \prec b \Leftrightarrow |a| < |b|$. The order $\langle \mathsf{On}, \prec \rangle$ is a well-order of order type ε_0. We may therefore represent every ordinal $\alpha < \varepsilon_0$ by an initial segment $\prec_\alpha$ of the well-order $\prec$. Thus we can talk about ordinals $< \varepsilon_0$ in $\mathcal{L}_{\mathbb{N}}$. We will not distinguish between ordinals and their representations in $\mathcal{L}_{\mathbb{N}}$ and regard formulas $(\forall \alpha)[\cdots]$ as abbreviations for $(\forall x)[x \in \mathsf{On} \to \cdots]$ as well as $(\exists \alpha)[\cdots]$ as abbreviation for $(\exists x)[x \in \mathsf{On} \wedge \cdots]$. We also write $\alpha < \beta$ instead of $\alpha \prec \beta$. We introduce the following formulas:

- $\alpha \subseteq X \;:\Leftrightarrow\; (\forall \xi)[\xi < \alpha \to \xi \in X]$
- $\mathit{Prog}(X) \;:\Leftrightarrow\; (\forall \alpha)[\alpha \subseteq X \to \alpha \in X]$
- $\mathit{TI}(\alpha, X) \;:\Leftrightarrow\; \mathit{Prog}(X) \to \alpha \subseteq X$

Our aim is to show $\mathit{TI}(\alpha, X)$ for all $\alpha < \varepsilon_0$. Since $\mathit{TI}(0, X)$ holds trivially and $\varepsilon_0 = \sup\{\mathit{exp}^n(\omega, 0) \mid n \in \omega\}$, we are done as soon as we succeed in proving

$$\mathbf{NT} \vdash \mathit{TI}(\alpha, X) \Rightarrow \mathbf{NT} \vdash \mathit{TI}(\omega^\alpha, X) \tag{42}$$

because $\mathbf{NT} \vdash \mathit{TI}(\alpha, X)$ and $\beta < \alpha$ obviously entails $\mathbf{NT} \vdash \mathit{TI}(\beta, X)$. The first observation is

$$\mathbf{NT} \vdash F(X) \Rightarrow \mathbf{NT} \vdash F(\{x \mid G(x)\}) \tag{43}$$

for all $\mathcal{L}_{\mathbb{N}}$-formulas G. The formula $F(\{x \mid G(x)\})$ is obtained from $F(X)$ by replacing all occurrences of $t \in X$ by $G(t)$ and those of $t \notin X$ by $\neg G(t)$. To prove (43) assume

$$\mathbf{NT} \vdash F(X) \tag{i}$$

and let $\mathfrak{S}$ be an arbitrary $\mathcal{L}_{\mathbb{N}}$-structure and Φ: set–variables $\longrightarrow \mathsf{Pow}(\mathfrak{S})$ an assignment such that

$$\mathfrak{S} \models \mathbf{NT}[\Phi]. \tag{ii}$$

We have to show

$$\mathfrak{S} \models F(\{x \mid G(x)\})[\Phi]. \tag{iii}$$

Define a new assignment

$$\Psi(Y) := \begin{cases} \Phi(Y) & \text{if } Y \neq X \\ \{n \in \mathfrak{S} \mid \mathfrak{S} \models G(x)[n, \Phi]\} & \text{otherwise.} \end{cases}$$

Then

$$\mathfrak{S} \models F(X)[\Psi] \text{ iff } \mathfrak{S} \models F(\{x \mid G(x)\})[\Phi]. \tag{iv}$$

We claim

$$\mathfrak{S} \models \mathbf{NT}[\Psi]. \tag{v}$$

Then (v) together with (i) and (iv) prove (iii). To check (v) we have only to take care of formulas in **NT** which contain the set variable X. This can only happen in instances of the scheme of Mathematical Induction or in identity axioms. Let

$$I(X) \ :\Leftrightarrow \ H(X, \underline{0}) \wedge (\forall x)[H(X,x) \to H(X, S(x))] \to (\forall x)H(X,x)$$

be an instance of Mathematical Induction. We have

$$\mathfrak{S} \models I(X)[\Psi] \text{ iff } \mathfrak{S} \models I(\{x \mid G(x)\})[\Phi]. \tag{vi}$$

The right formula in (vi), however, holds by (ii) since $H(\{x \mid G(x)\}, x)$ is also a formula in **NT**. Instances of identity axioms are treated analogously. □

The above proof shows the importance of formulating Mathematical Induction as a scheme.

Let

$$\mathcal{J}(X) := \{\alpha \mid (\forall \xi)[\xi \subseteq X \to \xi + \omega^\alpha \subseteq X]\}$$

denote the *jump* of X. Then, if we assume

$$\mathbf{NT} \vdash \mathit{Prog}(X) \to \mathit{Prog}(\mathcal{J}(X)), \tag{i}$$

we obtain

$$\mathbf{NT} \vdash \mathit{TI}(\alpha, \mathcal{J}(X)) \to \mathit{TI}(\omega^\alpha, X). \tag{ii}$$

To prove (ii) assume (working informally in **NT**) $\mathit{TI}(\alpha, \mathcal{J}(X))$, i.e.

$$\mathit{Prog}(\mathcal{J}(X)) \to \alpha \subseteq \mathcal{J}(X) \tag{iii}$$

which entails

$$\mathit{Prog}(\mathcal{J}(X)) \to \alpha \in \mathcal{J}(X). \tag{iv}$$

Choosing $\xi = 0$ in the definition of the jump turns (iv) into

$$\mathit{Prog}(\mathcal{J}(X)) \to \omega^\alpha \subseteq X, \tag{v}$$

which, together with (i), gives

$$\mathit{Prog}(X) \to \omega^\alpha \subseteq X, \tag{vi}$$

which is $\mathit{TI}(\omega^\alpha, X)$. Once we have (ii) we also get (42) because $\mathit{TI}(\alpha, X)$ implies $\mathit{TI}(\alpha, \mathcal{J}(X))$ by (43).

It remains to prove (i). Again we work informally in **NT**. Assume

$$\mathit{Prog}(X). \tag{vii}$$

We want to prove $\mathit{Prog}(\mathcal{J}(X))$ i.e. $(\forall\alpha)[\alpha \subseteq \mathcal{J}(X) \to \alpha \in \mathcal{J}(X)]$. Thus, assuming also

$$\alpha \subseteq \mathcal{J}(X), \tag{viii}$$

we have to show $\alpha \in \mathcal{J}(X)$. i.e. $(\forall\xi)[\xi \subseteq X \to \xi + \omega^\alpha \subseteq X]$. That means that we have to prove $\eta \in X$ under the additional hypotheses

$$\xi \subseteq X \tag{ix}$$

and

$$\eta < \xi + \omega^\alpha. \tag{x}$$

If $\eta < \xi$ we obtain $\eta \in X$ by (ix). Let $\xi \leq \eta < \xi + \omega^\alpha$. If $\alpha = 0$ the $\eta = \xi$ and we obtain $\eta \in X$ by (ix) and (vii). If $\alpha > 0$ then there is a $\sigma < \alpha$ and a natural number (i.e. a numeral in **NT**), such that $\xi < \underbrace{\omega^\sigma + \cdots + \omega^\sigma}_{n-\text{fold}} =: \omega^\sigma \cdot n$ (c.f. the proof of the Predicative Elimination Lemma). We show

$$\sigma < \alpha \to \xi + \omega^\sigma \cdot n \subseteq X \tag{xi}$$

by induction on n. For $n = 0$ this is (ix). For $n := m + 1$ we have

$$\xi + \omega^\sigma \cdot m \subseteq X \tag{xii}$$

by the induction hypothesis. From $\sigma < \alpha$ we obtain $\sigma \in \mathcal{J}(X)$ from (viii). This together with (xii) entails $\xi + \omega^\sigma \cdot n = \xi + \omega^\sigma \cdot m + \omega^\sigma \in X$. This finishes the proof of (i), hence also that of (42) which in turn implies Theorem 2.1.3.1. □

Summing up we have shown

2.1.3.2. Theorem. (Ordinal Analysis of **NT**) $\|\mathbf{NT}\|_{\Pi^1_1} = \varepsilon_0$.

As a corollary of Theorem 2.1.3.2 and (24) we obtain

2.1.3.3. Gentzen's Theorem. *The proof theoretic ordinal of the axiom system* **NT** *is* ε_0.

As another consequence of Theorem 2.1.2.11 and Theorem 2.1.3.1 is

2.1.3.4. Theorem. *There is a* Π^1_1*-sentence* $(\forall X)(\forall x)F(X,x)$ *which is true in the standard structure* $\mathbb{N}$ *such that* $\mathbf{NT} \vdash F(X, \underline{n})$ *for all* $n \in \mathbb{N}$ *but* $\mathbf{NT} \not\vdash (\forall x)F(X,x)$.

To prove the theorem choose $F(X,x)\ :\Leftrightarrow\ \mathit{Prog}(X) \to x \in \mathsf{On} \to x \in X.$ □

Theorem 2.1.3.4 is a weakened form of Gödel's Theorem. The general form of Gödel's Theorem says that Theorem 2.1.3.4 holds already for a Π^0_1-sentence $(\forall x)F(x)$.

2.1.4. Computational complexity of Π^0_2-sentences

Heuristics. From the point of view of minimal models, Gentzen's result appears surprisingly complicated. Shouldn't it be trivial that the minimal model for **PA** is obtained at stage ω? Consequently there have been comments stating – faintly joking – that Gentzen tried to secure induction up to ω by transfinite induction up to ε_0. However, (27) tells us that this is not at all the case. Since we have $\mathsf{tc}(F) < \omega$ for all arithmetical sentences we get the least Π^∞_0-model for **PA** at $\mathbf{L}_\omega$ for more or less trivial reasons. But it is not at all trivial where to locate the least model for the $\Sigma^{\omega_1^{CK}}_1$-sentences which correspond to the provable Π^1_1-sentences of **PA** according to the Hyperarithmetical Quantifier Theorem. All we know without proof theory is that it must be at some ordinal $\le \omega_1^{CK}$. Since Gentzen's Theorem tells us $\mathsf{tc}(F) < \varepsilon_0$ for all provable Π^1_1-sentences we know that this will be at $\mathbf{L}_{\varepsilon_0}$.

Nevertheless it is an unsatisfactory situation that the Π^1_1-spectrum tells us nothing about the provable arithmetical sentences of **Ax**. By Gödel's Theorem we know that there are arithmetical (even Π^0_1-) sentences which are not provable from **Ax** while by (14) we have $\mathsf{tc}(A) < \omega$ for all true arithmetical sentences. Thus we are far from having $\mathsf{tc}(F) < \|\mathbf{Ax}\| \ \Rightarrow\ \mathbf{Ax} \vdash F$. After all this work, however, one has the feeling to deserve a stronger result. The aim of this section is to show that we can obtain much more information with just a little more effort. This has been quite recently detected by Weiermann [1996] and it is appealing to include his result here because it is inspired by the collapsing techniques of impredicative proof theory on which we want to put the emphasis of this contribution.

The basic idea is quite simple. Recapitulating the computation of the upper bound ε_0 for $\mathrm{spec}_{\Pi^1_1}(\mathbf{Ax})$ we see that we started with a derivation $\frac{m}{0}\!\!\!\!\vdash \neg F_1, \ldots, \neg F_n, F$ and ended up with an infinitary derivation $\frac{\alpha}{0}\!\!\!\!\vdash F$ for some $\alpha < \varepsilon_0$. Call this derivation π. If we assume that F is a Σ^0_1-formula $(\exists y)A(y)$ then π cannot contain applications of the $(\forall)$-rule. Hence π is finite and we may hope that the depth of π can tell something about the size of a witness for the existential quantifier in $(\exists x)F(x)$. So there are two things to take care of:

1. *Assure that the depth of a derivation majorizes witnesses for existential formulas*

2. *Find a method to extract a bound for the length of the finite derivation of a Σ^0_1-formula from its assigned ordinal.*

The realization of *2.* requires the possibility to *collapse* ordinals bigger than ω into finite ones. This technique, however, is known from impredicative proof theory where we have to collapse (recursively) uncountable ordinals into recursively countable ones. Once we have solved *2.* is easy to realize also *1.* All we have to require is that the witnesses in (∃)-clauses are majorized by the collapse of the derivation length. Later we will see that in more complex systems this becomes an even more natural requirement.

While functions which collapse ordinals of transfinite number classes into lower transfinite number classes occur quite naturally in the development of notations for impredicative ordinals, i.e., ordinals above Γ_0, it had for long been unclear which are the right finitary collapsing functions, i.e., functions which collapse transfinite ordinals into finite ones.

It follows from work of Buchholz and Wainer that finitary collapsing functions must be connected to functions in the sub-recursive hierarchy. There is a result that for a certain choice of an infinitary calculus we have (roughly speaking)

$$\vdash_0^{\alpha} (\exists x)F(\underline{n}, x) \Rightarrow \mathbb{N} \models (\exists x < H_\alpha(n))F(n, x)$$

for quantifier free formulas $F(x, y)$, where H_α is the α-th function in the Hardy hierarchy (c.f. the article by Fairtlough-Wainer in this volume). The infinitary system is tailored in such a way that it gives a nice characterization of the Skolem functions for Π_2^0-sentences which are provable in Peano arithmetic. It turned out, however, that the generalization to stronger, especially to impredicative axiom systems is by far not straightforward.

But there are finitary collapsing functions which work smoothly together with the impredicative collapsing functions. The basic idea in their definition is strikingly simple. As soon as we have term notations for ordinals we may define the *norm* $N(\alpha)$ of an ordinal α as the number of symbols in its term notation. The norm is thus a finite ordinal with the property that for every natural number k the set $\{\alpha \mid N(\alpha) < k\}$ is finite. For any *starting function* $G: \mathbb{N} \to \mathbb{N}$ the function

$$\psi_\omega^G(\alpha) := \sup\{\psi_\omega^G(\beta) + 1 \mid \beta < \alpha \textit{ and } N(\beta) \leq G(N(\alpha))\} \cup \{0\} \qquad (44)$$

is then a finitary collapsing function for ordinals possessing a term notation.

Computational complexity. In order to obtain a more refined complexity for arithmetical sentences we refine the truth definition given in Definition 1.3.1. Here a crucial part is played by the function ψ_ω^G. To specify it we have to choose the starting function G.

There are symbols for all primitive recursive functions and relations ins $\mathcal{L}_{\mathbb{N}}$. The truth of any atomic sentence $R(t_1, \ldots, t_n)$ is thus decidable in just one step which means that the starting point in the measurement of computational complexity must be above the primitive recursive functions. Therefore we introduce a function P – a variant of the Ackermann–Peter function – which majorizes all primitive recursive functions and choose P as the starting function in the definition of the finitary collapsing function $\psi_\omega := \psi_\omega^P$.

2.1.4.1. Definition. We put

$$P_0(x) := 2^x$$
$$P_{n+1}(x) := P_n^{(x+2)}(x)$$

and finally

$$P(x) := P_x(x).$$

Then we have the obvious properties

$$x < P_n(x) \text{ for any } n, \tag{45}$$

$$x < y \Rightarrow P_n(x) < P_n(y) \textit{ for all } n \tag{46}$$

as well as

$$m < n \Rightarrow P_m(x) < P_n(x) \textit{ for all } x. \tag{47}$$

Every primitive recursive function $\boldsymbol{f}$ is eventually majorized by P. To see this we assume that $\boldsymbol{f}$ is represented by the constant f and show by induction on the definition of "*f is a constant for primitive recursive function*" the existence of a number e_f such that

$$\boldsymbol{f}(x_1,\ldots,x_{\#f}) < P_{e_f}(\max\{x_1,\ldots,x_{\#f}\}) \tag{48}$$

holds for all $\#f$-tuples $(x_1,\ldots,x_{\#f})$. This is obvious for $f \in \{C_k^n, P_k^n, S\}$. For $f = Sub(g,h_1,\ldots,h_m)$ we put $e := \max\{e_g, e_{h_1},\ldots,e_{h_m}\}$ and obtain by the induction hypothesis and (45) – (47)

$$\boldsymbol{f}(\vec{x}) < P_e(P_e(\max\{x_1,\ldots,x_n\})) \leq P_{e+1}(\max\{x_1,\ldots,x_n\}).$$

Thus let $e_f := e+1$. If $f = Rec(g,h)$ we put $e := \max\{e_g, e_h\}$ and obtain

$$\boldsymbol{f}(m,x_1,\ldots,x_n) < P_e^{(m+1)}(\max\{x_1,\ldots,x_n,m\}) \leq P_{e+1}(\max\{x_1,\ldots,x_n,m\})$$

by induction on m. Therefore let $e_f := e+1$. □

We define $N(0) := 0$ and for an ordinal $\alpha =_{\mathsf{NF}} \varphi_{\xi_1}\eta_1 + \cdots + \varphi_{\xi_n}\eta_n$ such that $n \geq 1$

$$N(\alpha) := \sum_{i=1}^{n}(N(\xi_i) + N(\eta_i) + 1). \tag{49}$$

This defines a norm for all ordinals $< \Gamma_0$. For $\alpha < \Gamma_0$ the finitary collapsing function is thus defined by

$$\psi_\omega(\alpha) := \max\{\psi_\omega(\beta)+1 \mid \beta < \alpha \textit{ and } N(\beta) \leq P(N(\alpha))\} \cup \{0\}, \tag{50}$$

i.e., $\psi_\omega := \psi_\omega^P$ in the sense of (44). We introduce the relation

$$\alpha \ll_\xi^1 \beta :\Leftrightarrow \alpha < \beta \textit{ and } N(\alpha) \leq P(N(\beta) + N(\xi)).$$

We obtain the transitive closure $\ll_\xi$ of $\ll_\xi^1$ by putting

$$\alpha \ll_\xi^{n+1} \beta :\Leftrightarrow (\exists\eta)[\alpha \ll_\xi^n \eta \ll_\xi^1 \beta]$$

and

$$\alpha \ll_\xi \beta :\Leftrightarrow (\exists n)[\alpha \ll_\xi^n \beta].$$

Instead of $\alpha \ll_0 \beta$ we write shortly $\alpha \ll \beta$ and call α *collapsibly less than* β. This is justified by

$$\alpha \ll \beta \;\Rightarrow\; \psi_\omega(\alpha) < \psi_\omega(\beta). \tag{51}$$

It is perhaps noteworthy that

$$\psi_\omega(\alpha) = \max\{k \mid 0 \ll_0^k \alpha\}.$$

We moreover have $N(\xi) = \xi$ for $\xi < \omega$. Hence

$$\beta < \omega \;\Rightarrow\; (\alpha \ll \beta \;\Leftrightarrow\; \alpha < \beta). \tag{52}$$

Using the collapsibly less relation we define a refinement $\Vdash_\rho^\alpha \Delta$ of $\vdash_\rho^\alpha \Delta$ for ordinals $\alpha, \rho < \Gamma_0$ as follows:

2.1.4.2. Definition.

(AxM) *If* $\Delta \cap D(\mathrm{N}) \neq \emptyset$ *then* $\Vdash_\rho^\alpha \Delta$ *for all ordinals* α *and* ρ.

(AxL) *If* $t^{\mathrm{N}} = s^{\mathrm{N}}$ *then* $\Vdash_\rho^\alpha \Delta, s \notin X, t \in X$ *for all ordinals* α *and* ρ.

($\wedge$) *If* $\Vdash_\rho^{\alpha_i} \Delta, A_i$ *and* $\alpha_i \ll \alpha$ *for* $i = 1,2$ *then* $\Vdash_\rho^\alpha \Delta, A_1 \wedge A_2$.

($\vee$) *If* $\Vdash_\rho^{\alpha_0} \Delta, A_i$ *and* $\alpha_0 \ll \alpha$ *for some* $i \in \{1,2\}$ *then* $\Vdash_\rho^\alpha \Delta, A_1 \vee A_2$.

($\forall$) *If* $\Vdash_\rho^{\alpha_i} \Delta, A(\underline{i})$ *and* $\alpha_i \ll_i \alpha$ *for all* $i \in \mathrm{N}$ *then* $\Vdash_\rho^\alpha \Delta, (\forall x)A(x)$.

($\exists$) *If* $\Vdash_\rho^{\alpha_0} \Delta, A(\underline{i})$ *and* $\alpha_0 \ll \alpha$ *and* $i+1 \ll \alpha$ *for some* $i \in \mathrm{N}$ *then* $\Vdash_\rho^\alpha \Delta, (\exists x)A(x)$.

(cut) *If* $\Vdash_\rho^{\alpha_1} \Delta, F$, $\Vdash_\rho^{\alpha_2} \Delta, \neg F$, $\alpha_i \ll \alpha$ *for* $i \in \{1,2\}$ *and* $\mathsf{rk}(F) < \rho$ *then* $\Vdash_\rho^\alpha \Delta$.

As a word of warning we want to emphasize that the definition of the $\ll$ relation and thus also the definition of the relation $\Vdash_\rho^\alpha \Delta$ depends on the term notation for ordinals (actually only on the norm function $N(\alpha)$) and thus is only defined for ordinals below Γ_0. But it is obvious how to extend $\Vdash_\rho^\alpha \Delta$ as soon as we have term notations for larger ordinals.

From Definition 2.1.4.2 we obtain

$$\Vdash_\rho^\alpha \Delta, \quad \alpha \leqslant \beta, \quad \rho \le \sigma \quad and \quad \Delta \subseteq \Gamma \quad \Rightarrow \quad \Vdash_\sigma^\beta \Gamma \tag{53}$$

immediately by induction on α.

The next observations will help us to a better understanding of the relation $\Vdash_\rho^\alpha \Delta$. First we observe

$$\Vdash_0^\alpha \Delta \;\Rightarrow\; \vdash_0^\alpha \Delta \;\Leftrightarrow\; \models^\alpha \Delta. \tag{54}$$

In contrast to the calculus $\vdash_\rho^\alpha$ the refined calculus has the following collapsing property:

$$\Vdash_{0}^{\alpha} \Delta \text{ and } \Delta \subseteq \Sigma_1^0 \;\Rightarrow\; \Vdash_{0}^{\psi_\omega(\alpha)} \Delta. \tag{55}$$

Observation (55) is immediate by induction on α. The crucial point is that a derivation of a set of Σ_1^0-formulas which is cut free cannot contain applications of an $(\forall)$-rule. Hence all ordinal assignments ξ in this derivation increase in the sense of the collapsibly less relation and may – by (51) and (52) – thus be replaced by $\psi_\omega(\xi)$.

On the other hand we also have

$$\Vdash_{0}^{n} (\exists x)F(x) \text{ for an } \Sigma_1^0\text{-sentence } (\exists x)F(x) \;\Rightarrow\; \mathrm{N} \models (\exists x < n)F(x). \tag{56}$$

The proof of (56) is by induction on n and needs

$$\Vdash_{\rho}^{\alpha} \Delta, F \text{ and } \mathrm{N} \not\models F \text{ for } F \text{ quantifier free} \;\Rightarrow\; \Vdash_{\rho}^{\alpha} \Delta. \tag{57}$$

We prove (57) by induction on α. If F does not belong to the main formula of the last inference in $\Vdash_{\rho}^{\alpha} \Delta, F$ the claim follows immediately from the induction hypothesis. If F is the main formula of the last inference it can neither be an axiom (AxL) (because F is a sentence) nor an axiom according to (AxM) (since $F \notin \mathrm{D(N)}$). Thus F is either a conjunction $F_1 \wedge F_2$ and we have the premises $\Vdash_{\rho}^{\alpha_i} \Delta, F, F_i$ for $i \in \{1,2\}$ and $\mathrm{N} \not\models F_i$ for some $i \in \{1,2\}$ or F is a disjunction $F_1 \vee F_2$ and we have the premise $\Vdash_{\rho}^{\alpha_i} \Delta, F, F_i$ and $\mathrm{N} \not\models F_i$. In both cases we obtain $\Vdash_{\rho}^{\alpha} \Delta$ by two fold application of the induction hypothesis and (53). □

It is a good exercise to try to generalize (57) to sentences F containing quantifiers. What goes wrong if F contains universal quantifiers?

Back to the proof of (56). The only possibility to obtain $\Vdash_{0}^{n} (\exists x)F(x)$ is a clause $(\exists)$ with the premise

$$\Vdash_{0}^{m} (\exists x)F(x), F(\underline{i}) \tag{i}$$

for $m < n$ and $i + 1 < n$. If $\mathrm{N} \models F(\underline{i})$ we are done because of $i < n$, otherwise we obtain $\Vdash_{0}^{m} (\exists x)F(x)$ by (57) and then $\mathrm{N} \models (\exists x < m)F(x)$ by the induction hypothesis. Since $m < n$ this yields the claim. □

By (55) and (56) we see how to obtain upper bounds for the witness of a Σ_1^0-sentence F once we succeed in getting $\Vdash_{0}^{\alpha} F$. To obtain (55) it is important to replace – as far as possible – the natural order relation on the ordinals by the collapsible less relation on the term notations. The crucial condition for (56) is to majorize the witnesses of $(\exists)$-clauses. What still needs explanation is the ordinal assignment to $(\forall)$-clauses. There are only finitely many ordinals $\beta \ll \alpha$. So we cannot require the ordinals to increase in the sense of the $\ll$–relation. We will come back to that point.

In analogy to the definition of the truth complexity for Π_1^1-sentences we define the *computational complexity* of a sentence F by

$$\mathsf{cc}(F) := \min\{\alpha \mid \Vdash_{0}^{\alpha} F\}. \tag{58}$$

By (54) we then have

$$\mathsf{tc}(F) \le \mathsf{cc}(F) \tag{59}$$

for all sentences F. One is tempted to define

$$\operatorname{spec}_{\Sigma_1^0}(\mathbf{Ax}) := \{\operatorname{cc}(F) \mid F \text{ is a } \Sigma_1^0\text{-sentence and } \mathbf{Ax} \vdash F\}$$

in analogy to $\operatorname{spec}_{\Pi_1^1}(\mathbf{Ax})$. However, we easily get $\Vdash_0^{n+2} (\exists y)F(y)$ if $\mathrm{N} \models F(\underline{n})$ which entails $\operatorname{cc}((\exists y)F(y)) = \min\{n+2 \mid \mathrm{N} \models F(\underline{n})\}$ for a Σ_1^0-sentence $(\exists y)F(y)$. Therefore we have $\operatorname{spec}_{\Sigma_1^0}(\mathbf{Ax}) = \omega$ for any axiom system $\mathbf{Ax}$ which at least contains the successor axioms which shows that the Σ_1^0–spectrum $\operatorname{spec}_{\Sigma_1^0}(\mathbf{Ax})$ carries no information about $\mathbf{Ax}$. The analogy to $\operatorname{spec}_{\Pi_1^1}(\mathbf{Ax})$ is apparently the wrong one. But recall that Π_1^1 corresponds to $\Sigma_1^{\omega_1^{CK}}$ on the side of sub–systems of Set Theory and $\Sigma_1^{\omega_1^{CK}}$–models are the same as $\Pi_2^{\omega_1^{CK}}$–models (cf. Lemma 1.2.3). The computational aspect of an axiom system is however better reflected in its $\Pi_2^{\omega_1^{CK}}$–model because it says something about its provably total $\Sigma_1^{\omega_1^{CK}}$–functions. Pulling this down to ω one should rather look at Π_2^0–sentences and try to define something as the Π_2^0–spectrum of an axiom system.

As a first observation in that direction we show

$$\Vdash_\rho^\alpha \Delta, (\forall x)F(x) \;\Rightarrow\; \Vdash_\rho^{\alpha+i+1} \Delta, F(\underline{i}) \tag{60}$$

for all $i \in \mathrm{N}$ by induction on α.

If the main formula of the last inference is different from $(\forall x)F(x)$ then the claim follows directly from the induction hypothesis and the fact

$$\alpha_\xi \ll_j \alpha \;\Rightarrow\; \alpha_\xi + i + 1 \ll_j \alpha + i + 1 \tag{61}$$

for all j and i. The crucial case is that of a clause $(\forall)$ with main formula $(\forall x)F(x)$. There we have the premises

$$\Vdash_\rho^{\alpha_j} \Delta, (\forall x)F(x), F(\underline{j})$$

with $\alpha_j \ll_j \alpha$ for all $j \in \mathrm{N}$. By the induction hypothesis we obtain

$$\Vdash_\rho^{\alpha_j+i+1} \Delta, F(\underline{i}). \tag{i}$$

To get the claim from (i) it suffices to check

$$\alpha_j \ll_i \alpha \;\Rightarrow\; \alpha_j + i + 1 \ll \alpha + i + 1. \tag{62}$$

From $\alpha_j \ll_i^1 \alpha$ we get $\alpha_j + i < \alpha + i$ and $N(\alpha_j) \leq P(N(\alpha) + i)$ which entails also $N(\alpha_j + i + 1) = N(\alpha_j) + i + 1 \leq P(N(\alpha + i + 1))$. Hence $\alpha_j + i + 1 \ll^1 \alpha + i + 1$. Iterating the procedure we get $\alpha_j + i + 1 \ll^k \alpha + i + 1$ for all k which implies (62). □

2.1.4.3. Theorem. *If $\operatorname{cc}((\forall x)(\exists y)F(x,y)) = \alpha$ then, for every $i \in \mathrm{N}$, $\mathrm{N} \models (\exists y < \psi_\omega(\alpha + i + 1))F(\underline{i}, y)$.*

To prove the theorem let $(\forall x)(\exists y)F(x,y)$ be a Π_2^0-sentence and put $\alpha := \operatorname{cc}((\forall x)(\exists y)F(x,y))$. Then $\Vdash_0^\alpha (\forall x)(\exists y)F(x,y)$ and we get by (60)

$\|\frac{\alpha+i+1}{0} (\exists y)F(\underline{i}, y)$ for all $i \in \mathbf{N}$. Hence $\mathbf{N} \models (\exists y < \psi_\omega(\alpha+i+1))F(\underline{i}, y)$ by (55) and (56). □

According to Theorem 2.1.4.3 we define

$$\mathsf{W}_\alpha(i) := \psi_\omega(\alpha+i+1). \tag{63}$$

By Theorem 2.1.4.3 it follows that W_α majorizes a Skolem function of a Π^0_2-sentence F if $\mathrm{cc}(F) \leq \alpha$. If we put

$$|(\forall x)(\exists y)F(x, y)|_{\Pi^0_2} := \min\{\alpha \mid (\forall i)[\mathbf{N} \models (\exists y < \mathsf{W}_\alpha(i))F(\underline{i}, y)]\} \tag{64}$$

for a Π^0_2–sentence $(\forall x)(\exists y)F(x, y)$ then we obtain for Π^0_2–sentences G

$$|G|_{\Pi^0_2} \leq \mathrm{cc}(G). \tag{65}$$

We define the Π^0_2–spectrum of a theory $\mathbf{Ax}$ as

$$\mathrm{spec}_{\Pi^0_2}(\mathbf{Ax}) = \{|F|_{\Pi^0_2} \mid F \textit{ is a } \Pi^0_2\textit{-sentence and } \mathbf{Ax} \vdash F\}$$

provided that $|F|_{\Pi^0_2}$ is defined for all such F as well as

$$\|\mathbf{Ax}\|_{\Pi^0_2} := \sup(\mathrm{spec}_{\Pi^0_2}(\mathbf{Ax}))$$

and obtain

$$\|\mathbf{Ax}\|_{\Pi^0_2} \leq \sup\{\mathrm{cc}(F) \mid \mathbf{Ax} \vdash F\}. \tag{66}$$

We call the computation of $\|\mathbf{Ax}\|_{\Pi^0_2}$ a Π^0_2–analysis of $\mathbf{Ax}$.

The definition of the Π^0_2-spectrum is admittedly less intrinsic than that of the Π^1_1-spectrum. We want, however, give some reasons why we think that Π^0_2-spectra do have an intrinsic meaning. The Π^0_2-spectrum depends on the function ψ^P_ω which in turn depends on the starting function P and on the term notation of the ordinals in the Π^0_2-spectrum. Indeed the Π^0_2-spectrum is rather a set of ordinal terms than a set of ordinals. We have already argued that the dependence on a starting function P has natural reasons. P has to majorize all the functions for which there are function symbols in the language (c.f. also the proof of Lemma 2.1.4.6 below). With another choice of the language, e.g. Peano arithmetic with the only function symbols for addition and multiplication or even the language of Set Theory with no functions symbols, a weaker starting function, e.g. something like $\lambda x.\, 3^x$, will do the same job. Also the hierarchy of functions $\mathsf{W}^P_\alpha := \lambda i.\, \psi^P_\omega(\alpha+i+1)$ does not depend too much on the starting function, i.e., for not too different starting functions P and G we will obtain a comparatively small ordinal α such that $\mathsf{W}^P_\alpha \approx \mathsf{W}^G_\alpha$ which means that W^P_α is elementary in W^G_α and vice versa.

More serious is the dependence on the term notations for the ordinals. However, it is hard to imagine an explicit term notation which could alter the Π^0_2-spectrum. Weiermann has shown that the fast growing subrecursive hierarchies – and the hierarchy W_α is fast growing – are very stable against alterations of the term notations (which is not true for the so called slow growing hierarchies). We believe that at least every term notation usable in ordinal analysis will lead to the same finitary collapsing function ψ_ω and thus to the same Π^0_2-spectrum – although we have to admit to see no way of proving this. The lack of a general and "natural" term notation system for all recursive ordinals (or equivalently the lack of a natural subrecursive hierarchy for all recursive functions), **hinders** us from defining generally

$$\mathrm{spec}_{\Pi^0_2}(\mathbb{N}) := \{|F|_{\Pi^0_2} \mid F \textit{ a } \Pi^0_2\textit{-sentence and } \mathbb{N} \models F\}$$

although we conjecture that any possible definition should lead to $\mathrm{spec}_{\Pi^0_2}(\mathbb{N}) = \mathrm{spec}_{\Pi^1_1}(\mathbb{N}) = \omega_1^{\mathrm{CK}}$ (which is motivated by the fact that we have $\mathrm{spec}_{\Pi^0_2}(\mathbf{Ax}) = \mathrm{spec}_{\Pi^1_1}(\mathbf{Ax})$ for all "regular" axiom systems which are so far analyzed).

Anyway, the Π^0_2-spectrum of an axiom system has pleasant properties. Every Π^0_2-sentence $(\forall x)(\exists y)F(x,y)$ defines a partial recursive function $\boldsymbol{f}_F := \mu y\,.\,F(x,y)$ and we call $\boldsymbol{f}_F$ provably recursive in $\mathbf{Ax}$ iff $\mathbf{Ax} \vdash (\forall x)(\exists y)F(x,y)$, i.e., iff $\mathbf{Ax}$ proves that $\boldsymbol{f}_F$ is total. If $\boldsymbol{f}_F$ is provably recursive in $\mathbf{Ax}$ then there is an $\alpha \in \mathrm{spec}_{\Pi^0_2}(\mathbf{Ax})$ such that $\boldsymbol{f}_F = \mu y < \mathsf{W}_\alpha(x)\,.\,F(x,y)$. Therefore all provably recursive functions of $\mathbf{Ax}$ are primitive recursive (even elementary) in W_α for some $\alpha \in \mathrm{spec}_{\Pi^0_2}(\mathbf{Ax})$. By induction on α we obtain $(\forall x)(\exists y)[\mathsf{W}_\alpha(x) = y]$, i.e., we have

$$\alpha < \|\mathbf{Ax}\| \;\Rightarrow\; \mathbf{Ax} \vdash (\forall x)(\exists y)[\mathsf{W}_\alpha(x) = y] \tag{67}$$

for all axiom systems $\mathbf{Ax}$ which allow the definition of the functions W_α. (For this it certainly suffices that $\mathbf{Ax}$ allows the definition of all primitive recursive functions. In the rest of the paper we tacitly assume that this is true for all axiom systems considered. Weaker systems need more subtle considerations which are outside the scope of this contribution). For axiom systems satisfying this assumption we obtain as a corollary of Theorem 2.1.4.3

2.1.4.4. Lemma. *If* $\mathsf{cc}\,\big((\forall x)(\exists y)F(x,y)\big) < \|\mathbf{Ax}\|$ *then there is a provably recursive function* $\boldsymbol{f}$ *of* $\mathbf{Ax}$ *such that* $\mathbb{N} \models (\forall x)F(x, \boldsymbol{f}(x))$.

Because of $|(\forall x)(\exists y)[\mathsf{W}_\alpha(x) = y]|_{\Pi^0_2} = \alpha + 1$ we obtain from (67) also

$$\alpha < \|\mathbf{Ax}\| \;\Rightarrow\; \alpha < \|\mathbf{Ax}\|_{\Pi^0_2}. \tag{68}$$

Hence

$$\|\mathbf{Ax}\|_{\Pi^1_1} = \|\mathbf{Ax}\| \le \|\mathbf{Ax}\|_{\Pi^0_2}. \tag{69}$$

In general the inequality in (69) is proper. For $\alpha := \|\mathbf{Ax}\|_{\Pi^0_2}$, for instance, we get by (28) the inequalities $\|\mathbf{Ax} + (\forall x)(\exists y)[\mathsf{W}_\alpha(x) = y]\|_{\Pi^1_1} = \|\mathbf{Ax}\|_{\Pi^1_1} = \|\mathbf{Ax}\| \le \alpha < \|\mathbf{Ax} + (\forall x)(\exists y)[\mathsf{W}_\alpha(x) = y]\|_{\Pi^0_2}$. In most cases, however, we obtain – spending a little more care on the ordinal assignment – $\sup\{\mathsf{tc}(F) \mid \mathbf{Ax} \vdash F\} = \sup\{\mathsf{cc}\,(F) \mid \mathbf{Ax} \vdash F\}$. This then entails $\|\mathbf{Ax}\|_{\Pi^1_1} = \sup\{\mathsf{tc}(F) \mid \mathbf{Ax} \vdash F\} = \sup\{\mathsf{cc}\,(F) \mid \mathbf{Ax} \vdash F\}$. Together with (69) we then obtain

$$\|\mathbf{Ax}\| = \|\mathbf{Ax}\|_{\Pi^1_1} \le \|\mathbf{Ax}\|_{\Pi^0_2} \le \sup\{\mathsf{cc}\,(F) \mid \mathbf{Ax} \vdash F\} = \|\mathbf{Ax}\|_{\Pi^1_1}, \tag{70}$$

i.e., $\|\mathbf{Ax}\|_{\Pi^1_1} = \|\mathbf{Ax}\|_{\Pi^0_2}$. We are going to call axioms systems for which we have $\|\mathbf{Ax}\|_{\Pi^1_1} = \|\mathbf{Ax}\|_{\Pi^0_2}$ *regular*. Another consequence of (67) is the following theorem.

2.1.4.5. Theorem. *Let* **Ax** *be a regular axiom system, i.e., let* $\|\mathbf{Ax}\| = \|\mathbf{Ax}\|_{\Pi^0_2}$. *Then the provably recursive functions of* **Ax** *are exactly the functions which are primitive recursive (even elementary) in some* W_α *for* $\alpha < \|\mathbf{Ax}\|$.

Without further hint we just remark that the functions W^P_α are closely connected to the Hardy-functions H_α. A detailed study is in Buchholz, Cichon and Weiermann [1994].

We want to close this section with the remark that there is also a Π^0_1 ordinal for theories, whose intention is to express the order type of the shortest primitive recursive well–ordering which is needed to prove the consistency of the theory within a finitistic framework. Due to certain pathologies (cf. Remark 7.1.9. in Girard [1987] which exposes an example due to Kreisel) the definition of the Π^0_1 ordinal is not completely straightforward. We omit a discussion since we believe that the known concepts are still too far from a final form and need further research.

Computational complexity of NT. As an example we want to compute $\mathrm{spec}_{\Pi^0_2}(\mathbf{NT})$. The first step consists in computing the computational complexities of the axioms of **NT**. We observe that

$$\Vdash^{n}_{0} (\forall x_1)\cdots(\forall x_n)G(x_1,\ldots,x_n) \tag{71}$$

holds for all true sentences $(\forall x_1)\cdots(\forall x_n)G(x_1,\ldots,x_n)$ where $G(u_1,\ldots,u_n)$ is a quantifier free formula.

The proof is simple. For every n-tuple $(z_1,\ldots,z_n)$ we have $G(z_1,\ldots,z_n) \in \mathrm{D}(\mathbf{N})$. Hence $\Vdash^{n}_{0} (\forall x_1)\cdots(\forall x_n)G(x_1,\ldots,x_n)$ by n-fold application of a clause $(\forall)$. □

All mathematical axioms of **NT**, except the induction scheme, are Π-sentences of the form $(\forall x_1)\cdots(\forall x_n)G(x_1,\ldots,x_n)$ with $G(u_1,\ldots,u_n)$ quantifier-free. Thus (71) gives us bounds for the computational complexity of all these axioms. To compute the computational complexity of the scheme of Mathematical Induction we first prove

$$\Vdash^{\omega\cdot\mathsf{rk}(F)}_{0} \Delta, F, \neg F \tag{72}$$

by induction on $\mathsf{rk}(F)$. The proof is essentially that of the Tautology Lemma (Lemma 2.1.2.3). A bit more care is needed for the case that F is a formula $(\forall x)G(x)$. There we have

$$\Vdash^{\omega\cdot\mathsf{rk}(G(\underline{z}))}_{0} \Delta, G(\underline{z}), \neg G(\underline{z}) \tag{i}$$

for all ∈ **N** by the induction hypothesis and obtain

$$\Vdash^{\omega\cdot\mathsf{rk}(G(\underline{z}))+z}_{0} \Delta, G(\underline{z}), (\exists x)\neg G(x) \tag{ii}$$

for every $z \in \mathbf{N}$ by a clause $(\exists)$. But $\omega\cdot\mathsf{rk}(G(\underline{z})) + z \ll_z \omega\cdot(\mathsf{rk}(G(z)) + 1)$ holds for all $z \in \mathbf{N}$ and we obtain

$$\Vdash^{\omega\cdot\mathsf{rk}((\forall x)G(x))}_{0} \Delta, (\forall x)G(x), \neg(\forall x)G(x)$$

by a clause $(\forall)$. □

The computation of the computational complexity of instances of the scheme of Mathematical Induction is obtained as in Lemma 2.1.2.4 with some extra care on the ordinal assignment. We prove

$$\Vdash_0^{\omega\cdot\mathsf{rk}(F(\underline{0}))+2\cdot n} \neg F(\underline{0}), \neg(\forall x)[F(x) \to F(Sx)], F(\underline{n}) \tag{73}$$

by induction on n. For $n = 0$ this is (72). For the induction step we get by (72) and the induction hypothesis by an inference $(\wedge)$

$$\Vdash_0^{\omega\cdot\mathsf{rk}(F(\underline{0}))+2n+1} \neg F(\underline{0}), \neg(\forall x)[F(x) \to F(Sx)], F(\underline{n}) \wedge \neg F(S\underline{n}), F(S\underline{n}). \tag{i}$$

We have $n \ll \omega\cdot\mathsf{rk}(F(\underline{0}))+2\cdot(n+1)$ and $\omega\cdot\mathsf{rk}(F(\underline{0}))+2\cdot n+1 \ll \omega\cdot\mathsf{rk}(F(\underline{0}))+2\cdot(n+1)$ and obtain thus from (i)

$$\Vdash_0^{\omega\cdot\mathsf{rk}(F(\underline{0}))+2(n+1)} \neg F(\underline{0}), \neg(\forall x)[F(x) \to F(Sx)], F(S\underline{n}) \tag{ii}$$

by a clause $(\exists)$. □

Noticing that $\omega\cdot\mathsf{rk}(F(\underline{0}))+2n \ll_n \omega\cdot(\mathsf{rk}(F(\underline{0}))+1)$ holds for all $n \in \mathbf{N}$ we obtain from (73) by a clause $(\forall)$

$$\Vdash_0^{\omega\cdot(\mathsf{rk}(F(\underline{0}))+1)} \neg F(\underline{0}), \neg(\forall x)[F(x) \to F(Sx)], (\forall x)F(x). \tag{74}$$

This gives an estimate for the computational complexities of instances of Mathematical Induction. From (72) we obtain also

$$\Vdash_0^{\omega\cdot\mathsf{rk}(F)} \underline{z}_1 \neq \underline{k}_1, \ldots, \underline{z}_n \neq \underline{k}_n, \neg F(\underline{z}_1, \ldots, \underline{z}_n), F(\underline{k}_1, \ldots, \underline{k}_n) \tag{75}$$

for all n-tuples $(z_1, \ldots, z_n)$ and $(k_1, \ldots, k_n)$ of natural numbers. This eventually gives $\omega\cdot\mathsf{rk}(G)$ as an upper bound for the computational complexities of all instances G of identity axioms. Pulling together (71), (74) and (75) we obtain

$$\Vdash_0^{\omega\cdot(\mathsf{rk}(A)+1)} \Delta, A \tag{76}$$

for every non-logical and identity axiom A of **NT**.

The adaption of (39) needs a bit more care. It will, however, clarify the role of the starting function P in the definition of the collapsing function ψ_ω. We show

2.1.4.6. Lemma. *If* $\vdash^m \Delta(u_1, \ldots, u_n)$ *and all free number variables of* $\Delta(u_1, \ldots, u_n)$ *occur in the list* $u_1, \ldots, u_n$ *then there is a finite ordinal* k *such that* $\Vdash_0^{\omega\cdot k+z_1+\cdots+z_n} \Delta(\underline{z}_1, \ldots, \underline{z}_n)$ *holds for all* n*-tuples* $(z_1, \ldots, z_n)$ *of natural numbers.*

The proof of Lemma 2.1.4.6 is by induction on m. As prerequisite it needs

$$\Vdash_\rho^\alpha \Delta(s) \text{ and } s^{\mathbf{N}} = t^{\mathbf{N}} \Rightarrow \Vdash_\rho^\alpha \Delta(t) \tag{77}$$

which follows straightforwardly by induction on α. The only cases in the proof of Lemma 2.1.4.6 which are not immediate from the induction hypotheses are clauses $(\exists)$ and $(\forall)$. Let us start with the $(\exists)$ case. It is of special interest because it needs the starting function P. We have

$$\vdash_0^{m_0} \Delta_0(\vec{u}), G(\vec{u}, t(\vec{u})) \Rightarrow \vdash_0^{m} \Delta_0(\vec{u}), (\exists x)G(\vec{u}, x) \tag{i}$$

and obtain by the induction hypothesis a finite ordinal k_0 such that

$$\Vdash_0^{\omega \cdot k_0 + \vec{z}} \Delta_0(\underline{\vec{z}}), G(\underline{\vec{z}}, t(\underline{\vec{z}})). \tag{ii}$$

The value $t(\underline{\vec{z}})^{\mathbb{N}}$ is computed primitive recursively from $\vec{z}$. By (48) we thus have a natural number k_t such that $t(\underline{\vec{z}})^{\mathbb{N}} < P_{k_t}(\vec{z})$. Taking $k := \max\{k_0 + 1, k_t\}$ we get

$$\omega \cdot k_0 + \vec{z} \ll \omega \cdot k + \vec{z} \text{ and } t(\underline{\vec{z}})^{\mathbb{N}} \ll \omega \cdot k + \vec{z}. \tag{iii}$$

Hence

$$\Vdash_0^{\omega \cdot k + \vec{z}} \Delta_0(\underline{\vec{z}}), (\exists x)G(\underline{\vec{z}}, x) \tag{iv}$$

from (ii), (77) and (iii) by a clause $(\exists)$.

In the case of an $(\forall)$ inference

$$\vdash_0^{m_0} \Delta_0(\vec{u}), G(\vec{u}, v) \Rightarrow \vdash_0^{m} \Delta_0(\vec{u}), (\forall x)G(\vec{u}, x) \tag{v}$$

we have a finite ordinal k_0 such that

$$\Vdash_0^{\omega \cdot k_0 + \vec{z} + i} \Delta_0(\underline{\vec{z}}), G(\underline{\vec{z}}, \underline{i}) \tag{vi}$$

for all $i \in \mathbb{N}$. Putting $k := k_0 + 1$ we get

$$\omega \cdot k_0 + \vec{z} + i \ll_i \omega \cdot k + \vec{z} \tag{vii}$$

for all $i \in \mathbb{N}$ and obtain

$$\Vdash_0^{\omega \cdot k + \vec{z}} \Delta_0(\underline{\vec{z}}), (\forall x)G(\underline{\vec{z}}, x)$$

from (vi) and (vii) by a clause $(\forall)$. □

Lemma 2.1.4.6 together with (76) yield

2.1.4.7. Lemma. *If* $\mathbf{NT} \vdash F$ *for an* $\mathcal{L}_{\mathbb{N}}$*-sentence* F *then there are natural numbers* m *and* n *such that* $\Vdash_n^{\omega^2 + m} F$.

What is still lacking is a proof that the cut elimination procedure also works for the refined calculus $\Vdash_0^{\alpha} \Delta$. We are going to check this step by step. The first step is checking the Reduction Lemma. We claim

$$\Vdash_\rho^{\alpha} \Delta, F \;, \quad \Vdash_\rho^{\beta} \Gamma, \neg F \text{ and } \mathsf{rk}(F) \leq \rho \Rightarrow \Vdash_\rho^{\alpha \# \beta} \Delta, \Gamma. \tag{78}$$

The proof is by induction on $\alpha \# \beta$ and follows the pattern of the proof of the Reduction Lemma (Lemma 2.1.2.7). There is, however, a subtle point in it which will clarify the ordinal assignment in the case of $(\forall)$-clauses. We treat only this case. All other cases follow easily from the induction hypotheses and

$$\alpha \ll_i \beta \Rightarrow \alpha \# \gamma \ll_i \beta \# \gamma. \tag{79}$$

Assume that F is a formula $(\exists x)G(x)$ which is the main formula of an inference

$$\Vdash_{\rho}^{\alpha_0} \Delta, G(\underline{i}), (\exists x)G(x) \Rightarrow \Vdash_{\rho}^{\alpha} \Delta, (\exists x)G(x). \tag{i}$$

Then $\alpha_0 \ll \alpha$ and $i+1 \ll \alpha$. From (i) and the second hypothesis

$$\Vdash_{\rho}^{\beta} \Gamma, (\forall x)\neg G(x) \tag{ii}$$

we obtain

$$\Vdash_{\rho}^{\alpha_0 \# \beta} \Delta, \Gamma, G(\underline{i}) \tag{iii}$$

by the induction hypothesis. On the other hand we obtain by (60) and (53) from (ii) also

$$\Vdash_{\rho}^{\beta+i+1} \Delta, \Gamma, \neg G(\underline{i}). \tag{iv}$$

We have

$$\alpha_0 \# \beta \ll \alpha \# \beta \tag{v}$$

and obtain from $i+1 \ll \alpha$ also

$$\beta + i + 1 \ll \alpha \# \beta. \tag{vi}$$

Because of (v), (vi) and $\mathsf{rk}(G(\underline{i})) < \mathsf{rk}(F) = \rho$ we get

$$\Vdash_{\rho}^{\alpha \# \beta} \Delta, \Gamma$$

from (iii) and (iv) by cut. □

By (78) we obtain

$$\Vdash_{\rho+1}^{\alpha} \Delta \Rightarrow \Vdash_{\rho}^{2^{\alpha+1}} \Delta \tag{80}$$

by induction on α. The proof is that of the Elimination Lemma (Lemma 2.1.2.8). All we need to adapt the proof are

$$\alpha \ll_i \beta \Rightarrow 2^{\alpha+1} \ll_i 2^{\beta+1} \tag{81}$$

and

$$\alpha_1, \alpha_2 \ll \alpha \Rightarrow 2^{\alpha_1+1} \# 2^{\alpha_2+1} \lll 2^{\alpha+1}. \tag{82}$$

These properties, however, follow easily from the fact that

$$N(2^\alpha) \leq 2^{N(\alpha)} \tag{83}$$

which in turn is nearly immediate since for $\alpha = \omega \cdot \nu + n$ we obtain $N(2^\alpha) = N(\omega^\nu \cdot 2^n) \leq 2^n \cdot (N(\nu)+1) \leq 2^{N(\nu)+n} \leq 2^{N(\alpha)}$. □

One should observe that (81) and (82) force us to use a fixed point free formulation of the Elimination Lemma (i.e., $2^{\alpha+1}$ instead of 2^α). The reason is that in case that $\alpha < 2^\alpha$, $\alpha \ll \beta$ and $\beta = 2^\beta$ we do have $2^\alpha < 2^\beta$ but not necessarily $N(2^\alpha) \leq P(N(2^\beta))$.

Now we have all the data for the computation of the Π_2^0-spectrum of **NT**. From Lemma 2.1.4.7, (80) and the fact that $exp^n(2, \omega^2 + m) < \varepsilon_0$ we obtain

$$\mathbf{NT} \vdash F \Rightarrow \Vdash_0^{\alpha} F \tag{84}$$

for all $\mathcal{L}_{\mathbf{N}}$-sentences F and some ordinal $\alpha < \varepsilon_0$. Together with (66) we have

$$\|\mathbf{NT}\|_{\Pi_2^0} \le \varepsilon_0 = \|\mathbf{NT}\| = \|\mathbf{NT}\|_{\Pi_1^1} \le \|\mathbf{NT}\|_{\Pi_2^0}. \tag{85}$$

Combining (85) and Theorem 2.1.4.5 we obtain

2.1.4.8. Theorem. $\mathrm{spec}_{\Pi_2^0}(\mathbf{NT}) = \varepsilon_0 = \|\mathbf{NT}\|$ *and the provably recursive functions of* $\mathbf{NT}$ *are exactly the functions which are primitive recursive in some function* W_α *for* $\alpha < \varepsilon_0$.

Taking $F(x) \;:\Leftrightarrow\; (\exists y)[(x)_0 \notin \mathbf{On} \vee \mathsf{W}_{(x)_0}((x)_1) = y]$ we get $\mathbf{NT} \vdash F(\underline{n})$ for all $n \in \mathbf{N}$ but $\mathbf{NT} \not\vdash (\forall x)F(x)$ since $\mathsf{cc}\big((\forall x)F(x)\big) = \varepsilon_0$. This sharpens Theorem 2.1.3.4 to

2.1.4.9. Theorem. *There is a true* Π_2^0*-sentence* $(\forall x)F(x)$ *such that* $\mathbf{NT} \vdash F(\underline{n})$ *for all* $n \in \mathbf{N}$ *but* $\mathbf{NT} \not\vdash (\forall x)F(x)$.

2.1.5. Computational complexity of Π_2^0-sentences revisited

Heuristics. We have seen in (65) that $|F|_{\Pi_2^0} \le \mathsf{cc}\,(F)$. Proving also the opposite inequality would be a good argument for the naturalness of the computational complexity. However, if $|(\forall x)(\exists y)F(x,y)|_{\Pi_2^0} = \alpha$ we get easily

$$\Vdash_0^{\mathsf{W}_\alpha(i)} (\exists y)F(\underline{i}, y) \tag{86}$$

for all $i \in \mathbf{N}$ but there is no ordinal β such that $\mathsf{W}_\alpha(i) \ll_i \beta$ for all $i \in \mathbf{N}$. This shows that the $\ll_i$–relation cannot yet be the natural one. There is a possibility to redefine the $\ll_i$–relation by $\alpha \ll_i \beta \;:\Leftrightarrow\; \alpha < \beta \wedge N(\alpha) \le \mathsf{W}^*_\beta(i)$ for slightly modified functions W^*_α (a hierarchy which is close to the so called fast growing iteration hierarchy F_α will suffice). The modification is needed to preserve the collapsing property $\alpha \ll \beta \;\Rightarrow\; \mathsf{W}^*_{\alpha+1}(0) < \mathsf{W}^*_{\beta+1}(0)$. Since we have $\mathsf{W}^*_\alpha(i) \ll_i \alpha$ for all infinite α in the sense of the refined relation $\ll_i$ we get from (86)

$$\Vdash_0^{\alpha} (\forall x)(\exists y)F(x,y). \tag{87}$$

for infinite α. Hence $\omega \le |F|_{\Pi_2^0} \;\Rightarrow\; \mathsf{cc}\,(F) \le |F|_{\Pi_2^0}$. The collapsing technique of the previous sections is based on the idea of *local predicativity* which has been originally developed for the analysis of impredicative axiom systems (cf. Pohlers [1978,1981, 1991] and Buchholz et al. [1981]). It is hardly surprising that the redefinition of $\ll_i$ is exactly the transfer of the definition which we knew before from local predicativity. But there is an improvement of this method by (Buchholz [1992]) which he calls *operator controlled derivations.* Quite recently Weiermann (cf. Blankertz and Weiermann [1996]) adapted this method also to predicative systems. We prefer to include this approach instead of introducing the refined version of the $\ll_i$ relations because it shows very vividly the dynamical aspect of cut–elimination in contrast to the more statical aspect of cut–freeness as e.g. in Beckmann and Pohlers [1997]. The fact that it also needs no alteration of the functions W_α indicates the naturalness of Buchholz' approach. In the following section we introduce the concept of operator controlled derivations. We will demonstrate that the computational content of the cut–elimination procedure is measured by the controlling operator.

Operator controlled derivations. In this section we concentrate on arithmetical sentences and dispense therefore with set variables. We call this language $\mathcal{L}^0_{\mathbb{N}}$ which we assume to be formulated as Tait-language (cf. Section 1.3). There are two types of arithmetical sentences

- *sentences of $\bigwedge$-type, which are true atomic sentences, i.e., sentences in $D(\mathbb{N})$, and sentences of the form $(A \land B)$ or $(\forall x)F(x)$*

and

- *sentences of $\bigvee$-type, which are false atomic sentences and sentences of the form $(A \lor B)$ or $(\exists x)F(x)$.*

For every sentence we associate a characteristic set $\mathcal{C}(F)$ of sentences such that

$$\mathbb{N} \models F \;\Leftrightarrow\; \begin{cases} (\forall G \in \mathcal{C}(F))[\mathbb{N} \models G] & \text{if } F \in \bigwedge\text{-type} \\ (\exists G \in \mathcal{C}(F))[\mathbb{N} \models G] & \text{if } F \in \bigvee\text{-type.} \end{cases} \tag{88}$$

We define

$$\mathcal{C}(F) := \begin{cases} \emptyset & \text{if } F \text{ is atomic} \\ \{A, B\} & \text{if } F \text{ is of the shape } A \circ B \\ \{G(\underline{n}) \mid n \in \omega\} & \text{if } F \text{ is a sentence } (\mathsf{Q}x)G(x), \end{cases}$$

where $\circ \in \{\land, \lor\}$ and $\mathsf{Q} \in \{\forall, \exists\}$.

To each formula $G \in \mathcal{C}(F)$ we associate a finite ordinal $o_F(G)$ which is 0 if F is of the shape $A \circ B$ and n if F has the form $(\mathsf{Q}x)H(x)$ and G is $H(\underline{n})$. In order to obtain a more refined truth complexity for arithmetical sentences we introduce *operator controlled infinitary derivations.*

For a function $\mathsf{F}\colon \mathbb{N} \longrightarrow \mathbb{N}$ and a finite set $M \subseteq \mathbb{N}$ we define the function

$$\mathsf{F}[M](n) := \mathsf{F}(\max(M \cup \{n\})).$$

We may interpret F as an operator with values in the collection of finite sets of ordinals by defining

$$\alpha \in \mathsf{F}(n) \;:\Leftrightarrow\; N(\alpha) < \mathsf{F}(n).$$

Instead of $\alpha \in \mathsf{F}(0)$ we write shortly $\alpha \in \mathsf{F}$. To simplify notations further we write $\mathsf{F}[n]$ for $\mathsf{F}[\{n\}]$ and $G \in \mathsf{F}$ and $\mathsf{F}[G]$ instead of $o_F(G) \in \mathsf{F}$ and $\mathsf{F}[o_F(G)]$, respectively, whenever it is clear from the context to which characteristic set $\mathcal{C}(F)$ the sentence G belongs.

2.1.5.1. Definition. Let $\mathsf{F}\colon \mathbb{N} \longrightarrow \mathbb{N}$ be an increasing function. For a finite set of arithmetical sentences we define the proof relation $\mathsf{F} \mid\!\frac{\alpha}{\rho}\, \Delta$ inductively by the following clauses.

($\bigwedge$) *If* $F \in \Delta \cap \bigwedge$*-type,* $\alpha \in \mathsf{F}$ *and* $\mathsf{F}[G] \mid\!\frac{\alpha_G}{\rho}\, \Delta, G$ *as well* $\alpha_G < \alpha$ *for all* $G \in C(F)$ *then* $\mathsf{F} \mid\!\frac{\alpha}{\rho}\, \Delta$

($\bigvee$) *If* $F \in \Delta \cap \bigvee$*-type,* $\alpha \in \mathsf{F}$, $G \in \mathsf{F}$, $\mathsf{F} \mid\!\frac{\alpha_G}{\rho}\, \Delta, G$ *and* $\alpha_G < \alpha$ *for some* $G \in C(F)$ *then* $\mathsf{F} \mid\!\frac{\alpha}{\rho}\, \Delta$

We call F the *main part* in instances of ($\bigwedge$) and ($\bigvee$).

(cut) *If* $\alpha \in \mathsf{F}$ *and* $\mathsf{F} \mid\!\frac{\alpha_0}{\rho}\, \Delta, A$ *as well as* $\mathsf{F} \mid\!\frac{\alpha_0}{\rho}\, \Delta, \neg A$ *for some* $\alpha_0 < \alpha$ *and some* A *such that* $\mathsf{rk}(A) < \rho$ *then* $\mathsf{F} \mid\!\frac{\alpha}{\rho}\, \Delta$

It is then obvious from (88) and the soundness of the cut rule that this infinitary calculus is sound, i.e., we have

$$\mathsf{F} \mid\!\frac{\alpha}{\rho}\, \Delta \;\Rightarrow\; \mathbb{N} \models \bigvee\{F \mid\ F \in \Delta\}. \tag{89}$$

We say that an operator G extends the operator F, written as $\mathsf{F} \subseteq \mathsf{G}$, if $\mathsf{F}(n) \leq \mathsf{G}(n)$ holds for all n. By a straightforward induction on α we easily obtain

2.1.5.2. Lemma. *If* $\alpha \leq \beta$, $\rho \leq \sigma$, $\mathsf{F} \subseteq \mathsf{G} \ni \beta$, $\Delta \subseteq \Gamma$ *and* $\mathsf{F} \mid\!\frac{\alpha}{\rho}\, \Delta$ *then* $\mathsf{G} \mid\!\frac{\beta}{\sigma}\, \Gamma$.

Another easy observation is the following *Detachment Lemma* whose proof is again a straightforward induction on α.

2.1.5.3. Detachment Lemma. *If* $\mathsf{F} \mid\!\frac{\alpha}{\rho}\, \Delta, A$ *and* $\neg A \in D(\mathbb{N})$ *then* $\mathsf{F} \mid\!\frac{\alpha}{\rho}\, \Delta$.

One purpose of the operator is to control the witnesses of existential sentences. This is manifested in the following *Witnessing Lemma.*

2.1.5.4. Witnessing Lemma. *Let* $(\exists y)F(y)$ *be a* Σ^0_1*-sentence such that* $\mathsf{F} \mid\!\frac{\alpha}{0}\, (\exists y)F(y)$. *Then* $\mathbb{N} \models (\exists y < \mathsf{F}(0))F(y)$.

The proof is by induction on α. The only possible premise is $\mathsf{F} \mid\!\frac{\alpha_0}{0}\, (\exists y)F(y), F(\underline{n})$ with $\alpha_0 < \alpha$ and $n = o_{(\exists y)F(y)}(F(\underline{n})) \in \mathsf{F}$, i.e. $n < \mathsf{F}(0)$. If $F(\underline{n}) \in \mathrm{D}(\mathbb{N})$ we are done. Otherwise we have $\mathsf{F} \mid\!\frac{\alpha_0}{0}\, (\exists y)F(y)$ by the Detachment Lemma and obtain the claim by the induction hypothesis. □

The more important aspect of operator controlled derivations, however, is the control they give on functions as stated in Theorem 2.1.5.6 below. The key property here is the following Inversion Lemma.

2.1.5.5. Inversion Lemma. *Let F be a sentence of $\bigwedge$-type such that $\mathsf{F}\,|\!\frac{\alpha}{\rho}\,\Delta, F$. Then we get $\mathsf{F}[G]\,|\!\frac{\alpha}{\rho}\,\Delta, G$ for all $G \in C(F)$.*

The proof is by induction on α. In case of an inference according to $(\bigvee)$ we get – due to $\mathsf{F} \subseteq \mathsf{F}[G]$ – the claim immediately from the induction hypothesis. In case of an inference $(\bigwedge)$ we have the premise $\mathsf{F}[G]\,|\!\frac{\alpha_G}{\rho}\,\Delta, F, G$ for some $\alpha_G < \alpha$ and get the claim from the induction hypothesis, Lemma 2.1.5.2 and the fact that $\mathsf{F}[G][G] = \mathsf{F}[G]$. □

Combining the Inversion and Witnessing Lemmas we get the following theorem.

2.1.5.6. Theorem. *Let $(\forall x)(\exists y)F(x,y)$ be a Π^0_2-sentence such that $\mathsf{F}\,|\!\frac{\alpha}{0}\,(\forall x)(\exists y)F(x,y)$. Then the associated recursive function $\boldsymbol{f}(x) = \mu y\,.\,F(x,y)$ is majorized by F, i.e., we have $\boldsymbol{f}(n) < \mathsf{F}(n)$ for all $n \in \omega$.*

Proof. Pick $n \in \omega$ and apply the Inversion Lemma to get

$$\mathsf{F}[n]\,|\!\frac{\alpha}{0}\,(\exists y)F(\underline{n}, y). \tag{i}$$

Then use the Witnessing Lemma to see that $\mu y\,.\,F(n,y) < F[n](0) = F(n)$. □

We have to study the behavior of the controlling operators during the cut–elimination process.

2.1.5.7. Reduction Lemma. *Let F be a sentence of $\bigvee$-type and $\rho := \mathsf{rk}(F)$. Let F and G be increasing functions such that $2 \cdot m \le \mathsf{F}(m) \le \mathsf{G}(m)$. If $\mathsf{F}\,|\!\frac{\alpha}{\rho}\,\Gamma, \neg F$ and $\mathsf{G}\,|\!\frac{\beta}{\rho}\,\Delta, F$ then $(\mathsf{F} \circ \mathsf{G})\,|\!\frac{\alpha+\beta}{\rho}\,\Gamma, \Delta$.*

We prove the lemma by induction on β. First we observe that $\alpha \in \mathsf{F}$ and $\beta \in \mathsf{G}$ imply $\alpha + \beta \in \mathsf{F} \circ \mathsf{G}$ since $N(\alpha + \beta) \le N(\alpha) + N(\beta) \le 2 \cdot \mathsf{G}(0) \le \mathsf{F}(\mathsf{G}(0))$. We also have $\mathsf{G} \subseteq \mathsf{F} \circ \mathsf{G}$. If the last inference is a cut or an inference according to $(\bigvee)$, whose main part is different from F, we get the claim immediately by the induction hypothesis and $\mathsf{F} \subseteq (\mathsf{F} \circ \mathsf{G})$. If the last inference is

$$(\bigwedge) \quad \mathsf{G}[H]\,|\!\frac{\beta_H}{\rho}\,\Delta, G, H \text{ for all } H \in C(G) \;\Rightarrow\; \mathsf{G}\,|\!\frac{\beta}{\rho}\,\Delta, G$$

then we get by induction hypothesis

$$(\mathsf{F} \circ \mathsf{G}[H])\,|\!\frac{\alpha+\beta_H}{\rho}\,\Gamma, \Delta, H \tag{i}$$

for all $H \in C(G)$ and obtain

$$(\mathsf{F} \circ \mathsf{G})\,|\!\frac{\alpha+\beta}{\rho}\,\Gamma, \Delta \tag{ii}$$

from (i) by a clause $(\bigwedge)$ since $(\mathsf{F} \circ \mathsf{G}[H]) = (\mathsf{F} \circ \mathsf{G})[H]$.

The interesting case is an inference according to $(\bigvee)$

$$\mathsf{G}\,\vdash^{\beta_0}_{\rho}\,\Gamma, F, G \textit{ for some } G \in \mathcal{C}(F) \;\Rightarrow\; \mathsf{G}\,\vdash^{\beta}_{\rho}\,\Gamma, F \tag{iii}$$

whose main part is F. Then we have $G \in \mathsf{G}$, $\beta_0 < \beta$ and obtain

$$(\mathsf{F} \circ \mathsf{G})\,\vdash^{\alpha+\beta_0}_{\rho}\,\Gamma, \Delta, G \tag{iv}$$

by induction hypothesis. From the hypothesis

$$\mathsf{F}\,\vdash^{\alpha}_{\rho}\,\Delta, \neg F \tag{v}$$

we obtain

$$\mathsf{F}[G]\,\vdash^{\alpha}_{\rho}\,\Delta, \neg G \tag{vi}$$

by the Inversion Lemma (Lemma 2.1.5.5). But $G \in \mathsf{G}$ means $o_F(G) < \mathsf{G}(0)$ which in turn entails $\mathsf{F}[G](n) = \mathsf{F}(\max\{o_F(G), n\}) \leq \mathsf{F}(\max\{\mathsf{G}(0), n\}) \leq \mathsf{F}(\mathsf{G}(n))$, i.e. $\mathsf{F}[G] \subseteq (\mathsf{F} \circ \mathsf{G})$. Hence

$$(\mathsf{F} \circ \mathsf{G})\,\vdash^{\alpha+\beta_0}_{\rho}\,\Gamma, \Delta, \neg G \tag{vii}$$

from (vi) and Lemma 2.1.5.2 and we obtain

$$(\mathsf{F} \circ \mathsf{G})\,\vdash^{\alpha+\beta}_{\rho}\,\Gamma, \Delta \tag{viii}$$

from (iv) and (vii) by cut. □

The Reduction Lemma illustrates that avoiding one cut means composing the controlling operator. If we have a derivation $\mathsf{F}\,\vdash^{\alpha}_{\rho+1}\,\Delta$ and we want to reduce the cut rank by 1 we have to iterate the Reduction Lemma α times. This causes no problems as long as α is finite. For transfinite ordinals α, however, we have to decide what we understand by transfinite iterations of operators. There is a big variety of possibilities, e.g., attaching limit ordinals with fundamental sequences and diagonalizing at limit points. However, we already defined a hierarchy of strictly increasing functions by introducing the functions W_α which in turn are based on the collapsing functions ψ^P_ω. First we observe that $2m \leq \mathsf{W}_\alpha(m)$. We claim that we also have

$$\mathsf{W}_\alpha \circ \mathsf{W}_\beta \subseteq \mathsf{W}_{\alpha \,\#\, \beta+1}. \tag{90}$$

This turns the members of the hierarchy W_α into good candidates for the controlling operators. To prove (90) we first observe

$$n < \omega \;\Rightarrow\; \psi^P_\omega(n) = n. \tag{91}$$

Then we show

$$\psi^P_\omega(\alpha + \psi^P_\omega(\beta)) \leq \psi^P_\omega(\alpha \,\#\, \beta) \tag{92}$$

by induction on β. For $\alpha = 0$ this is (91). So assume $\alpha \neq 0$. For $\beta = 0$ the claim follows trivially from (91). If $\beta \neq 0$ we have $\psi^P_\omega(\beta) = \psi^P_\omega(\gamma) + 1$ for some $\gamma \ll \beta$. Since $\alpha \neq 0$ it follows $\alpha + 1 \,\#\, \gamma \lessdot \alpha \,\#\, \beta$ and we obtain

$$\psi_\omega^P(\alpha+\psi_\omega^P(\beta)) = \psi_\omega^P(\alpha+1+\psi_\omega^P(\gamma))$$
$$\leq \psi_\omega^P(\alpha+1 \mathbin{\#} \gamma) \leq \psi_\omega^P(\alpha \mathbin{\#} \beta).$$

Hence $\mathsf{W}_\alpha(\mathsf{W}_\beta(n)) = \psi_\omega^P(\alpha+\psi_\omega^P(\beta+n+1)+1) \leq \psi_\omega^P(\alpha \mathbin{\#} \beta+n+2) = \mathsf{W}_{\alpha \mathbin{\#} \beta+1}(n)$ and we have (90). □

2.1.5.8. Elimination Lemma. *If* $\mathsf{W}_\beta \mathrel{|}\!\frac{\alpha}{\rho+1} \Delta$ *then* $\mathsf{W}_{\omega^\beta \mathbin{\#} \alpha+1} \mathrel{|}\!\frac{\omega^{\alpha+1}}{\rho} \Delta$.

The proof is by induction on α and the crucial case is that of a cut of rank ρ whose premises are

$$\mathsf{W}_\beta \mathrel{|}\!\frac{\alpha_1}{\rho+1} \Delta, A \;\; and \;\; \mathsf{W}_\beta \mathrel{|}\!\frac{\alpha_2}{\rho+1} \Delta, \neg A$$

for some $\alpha_1, \alpha_2 \in \mathsf{W}_\beta$ such that $\alpha_i < \alpha$ for $i = 1, 2$. From the induction hypothesis we obtain

$$\mathsf{W}_{\omega^\beta \mathbin{\#} \alpha_1+1} \mathrel{|}\!\frac{\omega^{\alpha_1+1}}{\rho} \Delta, A \;\; and \;\; \mathsf{W}_{\omega^\beta \mathbin{\#} \alpha_2+1} \mathrel{|}\!\frac{\omega^{\alpha_2+1}}{\Delta}, \neg A \tag{i}$$

which by the Reduction Lemma and (90) entails

$$\mathsf{W}_{\omega^\beta \mathbin{\#} \alpha_1+1 \mathbin{\#} \omega^\beta \mathbin{\#} \alpha_2+1+1} \mathrel{|}\!\frac{\omega^{\alpha_1+1}+\omega^{\alpha_2+1}}{\rho} \Delta. \tag{ii}$$

Next we show

$$\alpha_1, \alpha_2 < \alpha \wedge \alpha_1, \alpha_2 \in \mathsf{W}_\beta \;\Rightarrow\; \mathsf{W}_{\omega^\beta \mathbin{\#} \alpha_1+1 \mathbin{\#} \omega^\beta \mathbin{\#} \alpha_2+1+1} \subseteq \mathsf{W}_{\omega^\beta \mathbin{\#} \alpha+1}. \tag{93}$$

Since $\alpha_1, \alpha_2 \in \mathsf{W}_\beta$ means $N(\alpha_i) < \mathsf{W}_\beta(0) = \psi_\omega^P(\beta+1)$ we have

$$\omega^{\beta \mathbin{\#} \alpha_1+1} \mathbin{\#} \omega^{\beta \mathbin{\#} \alpha_2+1} \ll \omega^{\beta \mathbin{\#} \alpha} \cdot 2 + \psi_\omega^P(\beta+1) \cdot 2. \tag{iii}$$

This entails by (51) and (92)

$$\begin{aligned} \mathsf{W}_{\omega^\beta \mathbin{\#} \alpha_1+1 \mathbin{\#} \omega^\beta \mathbin{\#} \alpha_2+1+2}(n) &= \psi_\omega^P(\omega^{\beta \mathbin{\#} \alpha_1+1} \mathbin{\#} \omega^{\beta \mathbin{\#} \alpha_2+1}+n+2) \\ &< \psi_\omega^P(\omega^{\beta \mathbin{\#} \alpha} \cdot 2 + \psi_\omega^P(\beta+1) \cdot 2+n+2) \\ &\leq \psi_\omega^P(\omega^{\beta \mathbin{\#} \alpha} \cdot 2 \mathbin{\#} \beta \cdot 2+n+4) \\ &\leq \psi_\omega^P(\omega^{\beta \mathbin{\#} \alpha+1}+n+1) = \mathsf{W}_{\omega^\beta \mathbin{\#} \alpha+1}(n). \end{aligned} \tag{iv}$$

From (ii) and (93) we finally obtain

$$\mathsf{W}_{\omega^\beta \mathbin{\#} \alpha+1} \mathrel{|}\!\frac{\omega^{\alpha+1}}{\rho} \Delta.$$

The remaining cases are immediate from the induction hypothesis and the fact that $\alpha \in \mathsf{W}_\beta$ implies $\omega^{\beta \mathbin{\#} \alpha+1} \in \mathsf{W}_{\omega^\beta \mathbin{\#} \alpha+1}$. □

To reobtain $\mathrm{spec}_{\Pi_2^0}(\mathbf{NT})$ in terms of the indices of functions W_α majorizing the Π_2^0–sentences provable in **NT** by operator controlled derivations, we have to check which operators control the provable sentences of **NT**. We introduce

$$\mathsf{par}(F(z_1,\ldots,z_n)) := \{z_1,\ldots,z_n\} \tag{94}$$

if $F(z_1,\ldots,z_n)$ is a sentence with only the shown number parameters $z_1,\ldots,z_n$. Then we prove

$$\mathsf{W}_{2\cdot\mathsf{rk}(F)}[\mathsf{par}(F)] \,\Big|\!\frac{2\cdot\mathsf{rk}(F)}{0}\; \Delta, F, \neg F \tag{95}$$

by induction on $\mathsf{rk}(F)$. First observe that according to (91) we have $\mathsf{W}_0(n) = n+1$. For symmetry reasons we may assume that F has $\bigwedge$-type. If F is atomic, then $\mathcal{C}(F) = \emptyset$ and we obtain $\mathsf{W}_0[\mathsf{par}(F)] \,|\!\frac{0}{0}\, \Delta, F, \neg F$ by a clause $\bigwedge$. Otherwise it is $F \equiv G_0 \wedge G_1$ or $F \equiv (\forall x)G(x)$. In the second case we get

$$\mathsf{W}_{2\cdot\mathsf{rk}(G(\underline{z}))}[\mathsf{par}(G(\underline{z}))] \,\Big|\!\frac{2\cdot\mathsf{rk}(G(\underline{z}))}{0}\; \Delta, G(\underline{z}), \neg G(\underline{z}) \text{ for all } z \in \omega \tag{i}$$

by induction hypothesis. Since

$$z \in \mathsf{W}_{2\cdot\mathsf{rk}(G(\underline{z}))}[\mathsf{par}(G(\underline{z}))] \subseteq \mathsf{W}_{2\cdot\mathsf{rk}(F)}[\mathsf{par}(F)][z] \ni 2\cdot\mathsf{rk}(G(\underline{z}))+1$$

we obtain from (i)

$$\mathsf{W}_{2\cdot\mathsf{rk}(F)}[\mathsf{par}(F)][z] \,\Big|\!\frac{2\cdot\mathsf{rk}(G(\underline{z}))+1}{0}\; \Delta, G(\underline{z}), \neg F \tag{ii}$$

for all natural numbers z by a clause $\bigvee$ and from (ii) finally

$$\mathsf{W}_{2\cdot\mathsf{rk}(F)}[\mathsf{par}(F)] \,\Big|\!\frac{2\cdot\mathsf{rk}(F)}{0}\; \Delta, F, \neg F$$

by a clause $\bigwedge$. The first case is similar but simpler. □

Since we have $\mathsf{W}_0[z_1,\ldots,z_n] \,|\!\frac{0}{0}\, G(\underline{z}_1,\ldots,\underline{z}_n)$ for every true atomic formula $G(\underline{z}_1,\ldots,\underline{z}_n)$ we obtain

$$\mathsf{W}_n \,|\!\frac{n}{0}\, (\forall x_1)\cdots(\forall x_n)G(x_1,\ldots,x_n) \tag{96}$$

for every mathematical axiom $(\forall x_1)\cdots(\forall x_n)G(x_1,\ldots,x_n)$ by n–fold applications of $(\bigwedge)$.

From the method of (95) we obtain also the equality axioms

$$\mathsf{W}_{2\cdot\mathsf{rk}(F(\vec{x}))+2n+2}[\vec{z}] \,\Big|\!\frac{2\cdot\mathsf{rk}(F(\vec{x}))+2n+2}{0}\; (\forall\vec{x})(\forall\vec{y})[\vec{x}=\vec{y} \;\rightarrow\; F(\vec{x}) \;\rightarrow\; F(\vec{y})] \tag{97}$$

where n is the length of the tuple $\vec{x}$ and $\{\vec{z}\} := \{z_1,\ldots,z_m\} = \mathsf{par}((\forall\vec{x})F(\vec{x}))$.

By the same proof as for the Induction Lemma (Lemma 2.1.2.4) we obtain

$$\mathsf{W}_{2\cdot(\mathsf{rk}(F(0))+n)}[\mathsf{par}(F(0))][n] \,\Big|\!\frac{2\cdot(\mathsf{rk}(F(0))+n)}{0}\; \neg F(\underline{0}), \neg(\forall x)[F(x) \rightarrow F(S(x))], F(\underline{n}). \tag{98}$$

But (98) in turn gives

$$\mathsf{W}_{\omega\cdot\mathsf{rk}(F(0))}[\mathsf{par}(F(0))] \,\Big|\!\frac{\omega+3}{0}\; F(\underline{0}) \wedge (\forall x)[F(x) \rightarrow F(x+1)] \;\rightarrow\; (\forall x)F(x). \tag{99}$$

It remains to adapt Lemma 2.1.4.6, i.e., to show that there is a natural number k such that

$$\Big|\!\frac{m}{}\; \Delta(x_1,\ldots,x_n) \;\Rightarrow\; \mathsf{W}_{\omega\cdot k+m}[z_1,\ldots,z_n] \,\Big|\!\frac{m}{0}\; \Delta(\underline{z}_1,\ldots,\underline{z}_n) \tag{100}$$

for all tuples $z_1,\ldots,z_n$. Since $N(\omega\cdot m) = 2\cdot m$ we obtain $P(m) \ll \omega\cdot m$. Hence

$$P(m) = \psi^P_\omega(P(m)) < \psi^P_\omega(\omega\cdot m) < \mathsf{W}_{\omega\cdot m}(0). \tag{101}$$

We prove (100) by induction on m and, as in the proof of Lemma 2.1.4.6, the only critical case is an inference according to $(\exists)$. We proceed as in that proof and obtain from the induction hypothesis for the premise $\vdash^{m_0} \Delta_0(\vec{u}), G(\vec{u}, t(\vec{u}))$ a natural number k_0 such that

$$\mathsf{W}_{\omega\cdot k_0+m_0}[\vec{z}] \vdash^{m_0}_{0} \Delta(\vec{z}), G(\vec{z}, t(\vec{z})). \tag{i}$$

Because $t(\underline{\vec{z}})^{\mathbf{N}}$ is computed primitive recursively from $\vec{z}$ we find by (48) a natural number $k \geq k_0$ such that $t(\underline{\vec{z}})^{\mathbf{N}} < P(k) < \mathsf{W}_{\omega\cdot k+m}[\vec{z}](0)$. Applying an inference $(\bigvee)$ to (i) we therefore get

$$\mathsf{W}_{\omega\cdot k+m}[\vec{z}] \vdash^{m}_{0} \Delta_0(\vec{z}), (\exists x)G(\vec{z}, x).$$

The other cases follow immediately from the induction hypothesis. □

By (96), (97), (99) and (100) we finally obtain

$$\mathbf{NT} \vdash F \;\Rightarrow\; (\exists k)(\exists m)(\exists r)\left[\mathsf{W}_{\omega\cdot k} \vdash^{m}_{r} F\right] \tag{102}$$

for arithmetical sentences F. This yields a Π^0_2–analysis as follows: If

$$\mathbf{NT} \vdash (\forall x)(\exists y)F(x,y) \tag{i}$$

for a Π^0_2–sentence $(\forall x)(\exists y)F(x,y)$ then

$$\mathsf{W}_{\omega\cdot k} \vdash^{m}_{r} (\forall x)(\exists y)F(x,y) \tag{ii}$$

by (102). Defining $\bar{\omega}_n := (\lambda\xi.\, exp(\omega, \xi+1))^{(n)}$ and $\bar{\bar{\omega}}_n(\xi,\eta)$ recursively by $\bar{\bar{\omega}}_0(\xi,\eta) := \xi$ and $\bar{\bar{\omega}}_{n+1}(\xi,\eta) := exp(\omega, \bar{\bar{\omega}}_n(\xi,\eta) \# \bar{\omega}_n(\eta)+1)$ we get from (ii) by r–fold application of the Elimination Lemma (Lemma 2.1.5.8)

$$\mathsf{W}_{\bar{\bar{\omega}}_r(\omega\cdot k, m)} \vdash^{\bar{\omega}_r(m)}_{0} (\forall x)(\exists y)F(x,y). \tag{iii}$$

Putting $\alpha := \bar{\bar{\omega}}_r(\omega\cdot k, m) < \varepsilon_0$ we obtain from (iii) using the Inversion and Witnessing Lemmas (Lemmas 2.1.5.5 and 2.1.5.3)

$$(\forall x \in \mathbf{N})(\exists y < \mathsf{W}_\alpha(x))F(x,y) \tag{iv}$$

which shows

$$|(\forall x)(\exists y)F(x,y)|_{\Pi^0_2} \leq \alpha.$$

On the other hand, if $\alpha := |(\forall x)(\exists y)F(x,y)|_{\Pi^0_2}$ then we get $\mathsf{W}_\alpha[i] \vdash^{1}_{0} (\exists y)F(\underline{i}, y)$ for all $i \in \mathbf{N}$ and thus also $\mathsf{W}_\alpha \vdash^{2}_{0} (\forall y)(\exists y)F(x,y)$. Hence

$$|F|_{\Pi^0_2} = \min\{\alpha \mid\; (\exists\beta)\left[\mathsf{W}_\alpha \vdash^{\beta}_{0} F\right]\} \tag{103}$$

for Π^0_2–sentences F. □

Equation (103) can be taken as evidence for the naturalness of the concept of operator controlled infinitary derivations.

2.2. Peano arithmetic with additional transfinite induction

2.2.1. The theories $\mathbf{NT}_{\prec\restriction}$ and their Π^0_2-spectra

Let $\prec$ be a primitive recursive well-ordering. For simplicity assume that $\mathsf{field}(\prec) = \mathbb{N}$. By $\mathit{TI}(\prec\restriction\underline{m})$ we denote the scheme

$$\mathit{Prog}(\prec, F) \;\to\; (\forall y \prec \underline{m})F(y)$$

for $\mathcal{L}^2_{\mathbb{N}}$–formulas $F(y)$ expressing transfinite induction along $\prec\restriction\underline{m}$. Then

$$\mathit{TI}(\prec\restriction) := \bigcup_{m \in \mathbb{N}} \mathit{TI}(\prec\restriction\underline{m})$$

denotes the scheme of transfinite induction along all proper initial segments of $\prec$. Let

$$\mathbf{NT}_{\prec\restriction} := \mathbf{NT} + \mathit{TI}(\prec\restriction).$$

The aim of the following section is to compute the Π^0_2-spectra of the theories $\mathbf{NT}_{\prec\restriction}$. We will do this using the technique of Section 2.1.5. The importance of the theories $\mathbf{NT}_{\prec\restriction}$ will become clear in a moment.

To extend the calculus $\Vert\!\frac{\alpha}{\rho}\, \Delta$ of Definition 2.1.4.2 let ε be an ε-number and assume that there is a norm

$$N\colon \varepsilon \longrightarrow \omega$$

satisfying:

(N1) $N(0) = 0$.

(N2) $N(\alpha \# \beta) = N(\alpha) + N(\beta)$.

(N3) $\alpha \neq \omega^\alpha \;\Rightarrow\; N(\omega^\alpha) = N(\alpha) + 1$.

(N4) *For all $n \in \omega$ the set $\{\alpha < \varepsilon \mid N(\alpha) < n\}$ is finite.*

Observe that conditions (N1) – N(4) are always satisfiable as soon as we have term notations for the ordinals below ε.

Using the norm N and the starting function P as defined in Definition 2.1.4.1 we may extend the collapsing function ψ_ω and the collapsibly-less relation $\ll_\xi$ to all ordinals below ε. Therefore Definition 2.1.4.2 extends to the ordinals below ε and we get the same results as in Section 2.1.4. Therefore we need only to know $\mathsf{cc}\big(\mathit{TI}(\prec\restriction)\big)$ in order to compute $\mathrm{spec}_{\Pi^0_2}(\mathbf{NT}_{\prec\restriction})$. To get this, however, we need to know a little bit more about the relation $\prec$.

Call a well-ordering $\prec$ a *good representation* for ε if its order type is ε and there is an order preserving mapping

$$o\colon \mathsf{field}(\prec) \longrightarrow \varepsilon$$

satisfying:

(o1) $(\forall m)[o(m) \in \mathsf{Lim}]$.

(o2) $(\forall m < \omega)[N(o(m)) < P(m) \wedge m \le P(N(o(m)))]$.

Let $\prec$ be a good representation for ε. We want to compute $\mathsf{cc}\big(\mathit{TI}(\prec\restriction, F)\big)$ for an $\mathcal{L}_{\mathbf{N}}$-formula F. Let $k := \mathsf{rk}(F)$ and put

$$\alpha_n := \omega \cdot (k+1) \# o(n).$$

Since $m \prec n$ implies $\alpha_m < \alpha_n \in \mathsf{Lim}$ and $N(\alpha_m + 4) = N(\alpha_m) + 4 = 2 \cdot (k+1) + N(o(m)) + 4 < 2 \cdot (k+1) + P(m) + 4 \le P(2 \cdot (k+1) + m) \le P(N(\alpha_n) + m)$ we get

$$m \prec n \;\Rightarrow\; \alpha_m + 4 \ll^1_m \alpha_n. \tag{104}$$

Because of $\omega \le \alpha_n$ and $N(n+1) = n+1 \le P(o(n)) + 1 \le P(\alpha_n)$ we also have

$$n + 1 \ll_0 \alpha_n. \tag{105}$$

We use (104) and (105) in proving

$$\Vdash^{\alpha_n}_0 \neg \mathit{Prog}(\prec, F), (\forall y \prec \underline{n}) F(y) \tag{106}$$

by $\prec$-induction on n. For any $m \in \mathbf{N}$ we have

$$\Vdash^{\alpha'_m}_0 \neg \mathit{Prog}(\prec, F), (\forall y \prec \underline{m}) F(y), \neg \underline{m} \prec \underline{n} \tag{i}$$

either by (AxM) with $\alpha'_m = 0$ or by induction hypothesis with $\alpha'_m = \alpha_m$ if $m \prec n$. By (72) we also have

$$\Vdash^{\omega \cdot k}_0 \neg F(\underline{m}), F(\underline{m}) \tag{ii}$$

and obtain

$$\Vdash^{\alpha_m + 1}_0 \neg \mathit{Prog}(\prec, F), (\forall y \prec \underline{m}) F(y) \wedge \neg F(\underline{m}), \neg \underline{m} \prec \underline{n}, F(\underline{m}) \tag{iii}$$

from (i) and (ii) by an inference $\wedge$. By (105) we get

$$\Vdash^{\alpha_m + 2}_0 \neg \mathit{Prog}(\prec, F), \neg \underline{m} \prec \underline{n}, F(\underline{m}) \tag{iv}$$

from (iii) by an inference $\exists$ and

$$\Vdash^{\alpha_m + 4}_0 \neg \mathit{Prog}(\prec, F), \neg \underline{m} \prec \underline{n} \vee F(\underline{m}) \tag{v}$$

from (v) by two inferences $(\vee)$. Using (104) we finally get

$$\Vdash^{\alpha_n}_0 \neg \mathit{Prog}(\prec, F), (\forall y \prec \underline{n}) F(y) \tag{vi}$$

from (v) by an inference $\forall$. □

Together with the previous section (107) yields that for every sentence F in the theory $\mathbf{NT}_{\prec\restriction}$ there is an ordinal $\beta < \varepsilon$ such that

$$\Vdash^{\beta}_0 F.$$

Together with Lemma 2.1.4.6 this means

$$\mathbf{NT}_{\prec\restriction} \vdash F \;\Rightarrow\; (\exists\alpha<\varepsilon)(\exists m<\omega)[\,\Vdash^{\alpha}_{m} F]$$

from which we get by cut-elimination and the fact that ε is an ε-number

$$\mathbf{NT}_{\prec\restriction} \vdash F \;\Rightarrow\; (\exists\alpha<\varepsilon)[\,\Vdash^{\alpha}_{0} F]. \tag{107}$$

Hence $\mathrm{spec}_{\Pi^0_2}(\mathbf{NT}_{\prec\restriction}) \subseteq \varepsilon$. On the other hand we have $\varepsilon \subseteq \|\mathbf{NT}_{\prec\restriction}\|$ which by (68) entails

$$\sup(\mathrm{spec}_{\Pi^0_2}(\mathbf{NT}_{\prec\restriction})) = \varepsilon = \|\mathbf{NT}_{\prec\restriction}\| = \sup(\mathrm{spec}_{\Pi^1_1}(\mathbf{NT}_{\prec\restriction})). \tag{108}$$

Applying Theorem 2.1.4.5 we get the following theorem.

2.2.1.1. Theorem. *Let ε be an ε-number and $\prec$ a good representation for ε. Then the provably recursive functions of $\mathbf{NT}_{\prec\restriction}$ are exactly the functions which are elementary in W_α – as defined in (63) – for some $\alpha < \varepsilon$.*

2.2.2. Significance of the theories $\mathbf{NT}_{\prec\restriction}$

An ordinal analysis of a theory **Ax** yields – among others – the ordinal $\|\mathbf{Ax}\| < \omega_1^{\mathrm{CK}}$. We call an ordinal analysis for a theory $\mathbf{Ax} \supseteq \mathbf{NT}$ *profound* if it not only computes $\|\mathbf{Ax}\|$ but also provides a primitive recursive well-ordering, say $\prec$, which is a good representation for $\|\mathbf{Ax}\|$ such that

$$\mathbf{Ax} \vdash F \;\Leftrightarrow\; \mathbf{NT}_{\prec\restriction} \vdash F \tag{109}$$

holds for all arithmetical formulas. If we have a profound ordinal analysis of **Ax** we know by (108) its Π^0_2-spectrum and by Theorem 2.2.1.1 also its provably recursive functions.

All known ordinal analyses are profound. The general reason for that can be roughly sketched. Recall from Section 1.4 the main steps in an ordinal analysis which are:

- *Designing a semi-formal calculus $\vdash^{\alpha}_{\rho} \Delta$ which commonly needs a term notation for ordinals.*
- *Transforming a formal derivation $\mathbf{Ax} \vdash F$ into an infinite semi-formal derivation $\vdash^{\alpha}_{\rho} F$.*
- *Cut elimination for the semi-formal calculus, yielding $\vdash^{\alpha}_{\rho} F \Rightarrow \vdash^{\beta}_{0} F$.*

Arithmetizing the term notation gives in general a primitive recursive well-ordering $\prec$ which is a good representation for $\|\mathbf{Ax}\|$.

Unravelling a formal derivation into an infinite one results in a recursive infinite tree. Therefore we may restrict the semi-formal calculus to recursive proof trees. Then there is a recursive predicate, say $\mathit{Proof}_\infty(x,y,z,u)$, such that $\mathit{Proof}_\infty(\underline{e}, \ulcorner\alpha\urcorner, \ulcorner\rho\urcorner, \ulcorner\Delta\urcorner)$ expresses that

"*e is the code of an infinite recursive tree tagged with ordinal notations (i.e., elements in the field of $\prec$,) and codes for finite formula sets which is locally correct with respect to the axioms and rules of the semi-formal calculus witnessing $\vdash^{\alpha}_{\rho} \Delta$.*"

If we assume $Proof_{\mathbf{Ax}}(\underline{e}, \ulcorner F \urcorner)$ then the embedding procedure yields a recursive function g such that

$$Proof_\infty(g(\underline{e}), \underline{n}, \underline{r}, \ulcorner F \urcorner)$$

where n and r are computable from e. This can be done within **NT**. If we secure that all the manipulations which are done to an infinite proof tree during the cut elimination procedure are locally recursive, we can use the Recursion Lemma to obtain a recursive function, say h, such that

$$Proof_\infty(h(g(\underline{e})), \ulcorner \beta \urcorner, 0, \ulcorner F \urcorner).$$

Besides **NT** the Recursion Lemma needs transfinite induction along $\prec\restriction\ulcorner\beta\urcorner$. Therefore this step can be done within $\mathbf{NT}_{\prec\restriction}$. Using the sub-formula property of cut free infinite derivations we obtain

$$Proof_\infty(\underline{e}, \underline{n}, \underline{0}, \ulcorner F \urcorner) \;\rightarrow\; True_k(\ulcorner F \urcorner)$$

by induction on $\prec\restriction n$ where $True_k$ denotes a partial truth predicate for formulas of complexities $\leq k$. So this can be done in $\mathbf{NT}_{\prec\restriction}$, too. Because of

$$\mathbf{NT} \vdash True_k(\ulcorner F \urcorner) \rightarrow F$$

we get, summing up:

$$\begin{aligned}
\mathbf{Ax} \vdash F \;&\Rightarrow\; \mathbf{NT} \vdash Proof_{\mathbf{Ax}}(\underline{e}, F)\\
&\Rightarrow\; \mathbf{NT} \vdash Proof_\infty(g(\underline{e}), n, r, F)\\
&\Rightarrow\; \mathbf{NT}_{\prec\restriction} \vdash Proof_\infty(h(g(\underline{e})), m, 0, F)\\
&\Rightarrow\; \mathbf{NT}_{\prec\restriction} \vdash True_k(\ulcorner F \urcorner)\\
&\Rightarrow\; \mathbf{NT}_{\prec\restriction} \vdash F
\end{aligned}$$

Since we have $\mathbf{NT} \subseteq \mathbf{Ax}$ and $\mathbf{Ax} \vdash \mathsf{TI}(\prec\restriction)$ we also have the opposite direction

$$\mathbf{NT}_{\prec\restriction} \vdash F \;\Rightarrow\; \mathbf{Ax} \vdash F$$

for arithmetical formulas and the ordinal analysis is profound.

Having a profound ordinal analysis for a theory **Ax**, we can try to sharpen (109) by replacing $\mathbf{NT}_{\prec\restriction}$ by a more constructive system. If we restrict the scheme of Mathematical Induction in **NT** to Σ^0_1-formulas we get the theory Σ^0_1-IND. Primitive Recursive Analysis – as a second order theory – is a conservative extension of Σ^0_1-IND.

Let $\prec$ be a primitive recursively definable order relation. For $m \in \mathsf{field}(\prec)$ we have the scheme

$$(\mathsf{PRWO}(\prec\restriction m)) \quad (\forall \vec{x})[(\forall y)(f(\vec{x}, y) \prec m) \rightarrow (\exists y)(\neg f(\vec{x}, \mathsf{S}y) \prec f(\vec{x}, y))]$$

where f varies over constants for primitive recursive functions. Thus $\mathsf{PRWO}(\prec\restriction m)$ expresses that there are no infinite primitive recursive $\prec$-descendent sequences in $\prec\restriction m$. We put

$$\mathsf{PRWO}(\prec\restriction) := \bigcup_{m \in \mathsf{field}(\prec)} \mathsf{PRWO}(\prec\restriction m)$$

and want to replace $\mathbf{NT}_{\prec\restriction}$ by Σ^0_1-IND + PRWO($\prec\restriction$) in (109). This does not work for arbitrary arithmetical formulas but only for Π^0_2-sentences. The proof needs Mints' continuous cut elimination theorem. Let $Proof^{pr}_\infty(e, \underline{m}, \underline{r}, \ulcorner\Delta\urcorner)$ express that *"e is the index of a primitive recursive tree tagged with members of* field($\prec$) *(serving as ordinal notations), numbers (for the cut - rank) and finite formula sets, which is locally correct with respect to the axioms and rules of the semi-formal system (augmented by a replication rule whose premise and conclusion are identical) such that its bottom node is tagged with m (coding the height of the tree), r (coding its cut rank) and Δ".* By Mints' continuous cut elimination there exists a primitive recursive function, say H, such that

$$\Sigma^0_1\text{-IND} \vdash Proof^{pr}_\infty(e, \underline{m}, \underline{r}, \ulcorner\Delta\urcorner) \;\to\; Proof^{pr}_\infty(H(e), \underline{k}, \underline{0}, \ulcorner\Delta\urcorner)$$

where k is computable from $H(e)$ in such a way that, provided that $\prec$ is a well-ordering, $\prec\restriction k$ has order type $exp^r(2, \mathsf{otyp}_\prec(m))$. Giving a sketch of the proof of Mints' theorem would lead us far outside the scope of this contribution. But we want to give a kind of flow chart how to use it in sharpening (109). Thus assume that we have a profound ordinal analysis of $\mathbf{Ax}$ and let $\prec$ be a good representation for $\|\mathbf{Ax}\|$. Let $(\forall x)(\exists y)F(x, y)$ be a Π^0_2-sentence such that

$$\mathbf{Ax} \vdash (\forall x)(\exists y)F(x, y). \tag{i}$$

By (109) we thus obtain

$$\mathbf{NT}_{\prec\restriction} \vdash (\forall x)(\exists y)F(x, y). \tag{ii}$$

By Theorem 2.1.2.2 and (39) there are formulas $F_1, \ldots, F_l$ which either belong to $\mathbf{NT}_{\prec\restriction}$ or are identity axioms such that

$$\vdash^{m}_{0} \neg F_1, \ldots, \neg F_l, (\exists y)F(\underline{k}, y) \tag{iii}$$

for every number k. Looking more carefully at the embedding procedure we observe that the resulting infinitary proof tree is primitive recursive and that an index for that tree can be computed from the formal proof. Since the provably recursive functions of Σ^0_1-IND are exactly the primitive recursive ones, this embedding procedure can be formalized within Σ^0_1-IND. In the next step we observe that all axioms in $\mathbf{NT}_{\prec\restriction}$ and all identity axioms have primitive recursive proof trees in the semi-formal calculus. The only case in which this is not completely obvious is an instance of $\mathit{Prog}(\prec, G) \to (\forall y \prec \underline{n})G(y)$. The proof of (106) shows how to construct the tree. Instead of using induction on $\prec$ — Σ^0_1-IND + PRWO($\prec\restriction$) does not know that $\prec$ is well-ordered — we start with the bottom node and enumerate all possible premises. This gives

$$\frac{\cdots, \neg \mathit{Prog}(\prec, G), \neg \underline{m} \prec \underline{n} \vee G(\underline{m}), \cdots}{\neg \mathit{Prog}(\prec, G), (\forall y \prec \underline{n})G(y)}$$

Above any of these nodes we decide primitive recursively whether $\neg m \prec n$. If this is true then we add $\neg \mathit{Prog}(\prec, G), \neg \underline{m} \prec \underline{n}$ as top node. Otherwise we construct

$$\dfrac{\dfrac{\dfrac{\neg \mathit{Prog}(\prec, G), (\forall y \prec \underline{m})G(y) \qquad\qquad \neg G(\underline{m}), G(\underline{m})}{\neg \mathit{Prog}(\prec, G), (\forall y \prec \underline{m})G(y) \wedge \neg G(\underline{m}), G(m)}}{\neg \mathit{Prog}(\prec, G), G(\underline{m})}}{\neg \mathit{Prog}(\prec, G), \neg \underline{m} \prec \underline{n} \vee G(m)}$$

and repeat the procedure above $\neg \mathit{Prog}(\prec, G), (\forall y \prec \underline{m})G(y)$.

Summing up we obtain a primitive recursive function h such that

$$\Sigma^0_1\text{-IND} \vdash (\forall x)[\mathit{Proof}_{\mathbf{Ax}}(e, \ulcorner(\exists y)F(\dot{x}, y)\urcorner) \;\rightarrow\; \mathit{Proof}^{pr}_{\infty}(h(e), \underline{m}, \underline{r}, \ulcorner(\exists y)F(\dot{x}, y)\urcorner)]. \quad \text{(iv)}$$

Together with Mints' Theorem this yields

$$\mathbf{Ax} \vdash (\forall x)(\exists y)F(x, y) \;\Rightarrow\; \Sigma^0_1\text{-IND} \vdash (\forall x)[\mathit{Proof}^{pr}_{\infty}(H(h(e)), \underline{m}, \underline{0}, \ulcorner(\exists y)F(\dot{x}, y)\urcorner)] \quad \text{(v)}$$

But $(\exists y)F(x, y)$ is a Σ^0_1-formula. A cut free infinitary proof of $(\exists y)F(x, y)$ cannot contain instances of a $\forall$-rule and is thus finite. Every path in the proof tree is primitive recursive and we may therefore use $\mathsf{PRWO}(\prec\restriction)$ to deduce

$$\Sigma^0_1\text{-IND} + \mathsf{PRWO}(\prec\restriction) \vdash (\forall x)\,\mathit{True}_1(\ulcorner(\exists y)F(\dot{x}, y)\urcorner) \quad \text{(vi)}$$

from (v) which in turn entails

$$\Sigma^0_1\text{-IND} + \mathsf{PRWO}(\prec\restriction) \vdash (\forall x)(\exists y)F(x, y). \quad \text{(vii)}$$

By (i) and (vii) we have

$$\mathbf{Ax} \vdash (\forall x)(\exists y)F(x, y) \;\Leftrightarrow\; \Sigma^0_1\text{-IND} + \mathsf{PRWO}(\prec\restriction) \vdash (\forall x)(\exists y)F(x, y) \quad (110)$$

for Π^0_2-sentences $(\forall x)(\exists y)F(x, y)$ since the opposite implication holds obviously. □

Call a function $\boldsymbol{f}$ $\prec$-descendent recursive if it is represented by a function term which is built up from C^n_m, P^n_m and S by *Sub*, *Rec* and the search operator $\mu_\prec$ which is defined by

$$(\mu_\prec f)(\vec{x}) := \min\{y \mid \neg f(\vec{x}, Sy) \prec f(\vec{x}, y)\}.$$

It is not very difficult to show that the provably recursive functions of Σ^0_1-IND + $\mathsf{PRWO}(\prec\restriction)$ are exactly the $\prec$-descendent recursive functions (cf. e.g.,Pohlers [1992] for a proof). Together with Weiermann's result this shows that for a good representation $\prec$ for $\|\mathbf{Ax}\|$ a function $\boldsymbol{f}$ is $\prec$-descendent iff it is primitive recursive in W_α for some $\alpha < \|\mathbf{Ax}\|$. A result which can also be proved directly, even under weaker conditions on $\prec$ (cf. Buchholz, Cichon and Weiermann [1994]). A comprehensive study on $\prec$-descendence and proof theory can be found in Friedman and Sheard [1995]. A completely worked out proof of Mints' theorem is in Buchholz [1991].

3. Impredicative systems

The aim of this chapter is to give upper bounds for the proof-theoretical ordinals of some impredicative axiom systems of Number Theory and Set Theory. We will restrict ourselves to Π^1_1 analyses which are already sufficiently complicated. Moreover, we will also not demonstrate the latest state of the art but restrict ourselves to three axiom systems for Set Theory, $\mathbf{KP}\boldsymbol{\omega}$, axiomatizing an admissible universe, **KPl**, axiomatizing a union of admissible universes and **KPi** axiomatizing an admissible union of admissible universes, and the corresponding axiom systems for Number Theory. Today we know also how to analyse axiom–systems for Mahlo–universes, Π_n–reflection and – though I have not yet seen the proofs – even for Σ_1–separation. T. Arai has announced the analysis of even stronger systems. He uses, however, a different technique which is based on G. Takeuti's methods.

3.1. Some remarks on predicativity and impredicativity

The focus of this contribution is on the ordinal analysis of impredicative systems. In order to distinguish impredicative theories from predicative ones we need a short discussion on predicativity and impredicativity. Limitations of space force us to be rather sketchy.

There are two ordinals which characterize a transitive $\mathcal{L}(\in)$–structure $\mathfrak{M}$. The ordinal $o(\mathfrak{M}) := \min\{\alpha \in \mathbf{On} |\ \alpha \notin \mathfrak{M}\}$ and the least ordinal which cannot be pinned down in $\mathfrak{M}$ (cf. Barwise [1975,III.7 and VII.3]). We need not to repeat the definition of "Pinning down ordinals" because we are going to refine it in the following way. Assume that the language $\mathcal{L}_{\infty,\omega}$ is coded as sets as e.g. in Barwise [1975]. Introduce the notion of an infinitary proof within a semi-formal system for Set Theory as sketched in Section 1.4. Denote by $T \vdash F$ that T is an infinitary proof tree for the formula F. We say that a countable ordinal α is provably pinned down in a transitive $\mathcal{L}(\in)$–structure $\mathfrak{M}$ if there is a well–ordering $\prec$ on ω of order type α in $\mathfrak{M}$, a (possibly infinitary) formula (cf. Pohlers [1989,§19] for examples of such formulas) $\mathit{Found}(\prec)$ in $\mathfrak{M}$ which expresses the well–foundedness of $\prec$ and an infinitary proof T in $\mathfrak{M}$ such that $T \vdash \mathit{Found}(\prec)$. Define $h(\mathfrak{M}) := \min\{\alpha \in \mathbf{On} |\ \alpha$ cannot be provably pinned down by $\mathfrak{M}\}$. Of course we always have $o(\mathfrak{M}) \leq h(\mathfrak{M})$. Now let $\mathfrak{M}$ be the initial part $\mathbf{L}_\alpha$ of the constructible hierarchy. Then $o(\mathbf{L}_\alpha) = \alpha$ and we put $h(\alpha) := h(\mathbf{L}_\alpha)$. Then $\alpha \leq h(\alpha)$. We call an ordinal α *autonomously inaccessible* if $\alpha = h(\alpha)$. For an autonomously accessible ordinal we have $\alpha < h(\alpha)$ which means that α can be provably pinned down by $\mathbf{L}_\alpha$. Then we have a formula $\mathit{Found}(\alpha) \in \mathbf{L}_\alpha$, expressing the well-foundedness of an well–ordering of order type α and a proof $T \in \mathbf{L}_\alpha$ such that $T \vdash \mathit{Found}(\alpha)$. If we denote again by $T \overset{\beta}{\underset{\rho}{\vdash}} F$ that β is an upper bound for the height of T and the complexity of all formulas occurring in T and ρ a strict upper bound for the cut formulas occurring in T then there are ordinals β and ρ less than α such that $T \overset{\beta}{\underset{\rho}{\vdash}} \mathit{Found}(\alpha)$. If we anticipate that we can construct $\mathbf{L}_\alpha$ whenever we have the ordinal α we can interpret autonomously accessible ordinals as ordinals which can be secured by smaller ordinals (cf. Schlüter [1990] for a fully worked out version of these ideas). The notion of autonomously accessible and inaccessible ordinals is due to Feferman (cf. Feferman [1964]).

The Elimination Lemma (Lemma 2.1.2.8) and the Predicative Elimination Lemma (Lemma 2.1.2.9) as well as the Boundedness Theorem (Theorem 1.3.6) carry over. So we get

$$T \overset{\beta}{\underset{n}{\vdash}} \mathit{Found}(\prec) \;\Rightarrow\; \mathsf{otyp}(\prec) \leq exp^n(2, \beta) \tag{111}$$

for $n < \omega$ and

$$T \overset{\beta}{\underset{\rho}{\vdash}} \mathit{Found}(\prec) \;\Rightarrow\; \mathsf{otyp}(\prec) \leq \varphi_\rho \beta. \tag{112}$$

It follows from (111) that ω and from (112) that all strongly critical ordinals are autonomously inaccessible. This has first been observed by Feferman [1964] and independently by Schütte [1965a] who both could also show that these are the only autonomously inaccessible ordinals (cf. Schütte [1965b]). A proof of this fact which is in the spirit of the above sketch can be found in Pohlers [1989].

In some sense the notion of autonomous accessibility captures the idea of predicativity. First we see that without accepting the ordinal ω we stay within the hereditarily finite world. Once we have accepted ω as a set we can look for the ordinals α which are provably pinned down in $\mathbf{L}_{\omega+1}$. Then we construct $\mathbf{L}_\alpha$, look for ordinals provably pinned down in $\mathbf{L}_\alpha$ and so on. This process will stop at the first strongly critical ordinal, i.e., at Γ_0. On the other hand $\mathbf{L}_{\Gamma_0}$ is also exhausted by this procedure. In that sense Γ_0 is known to bound predicativity. We stick to that notation in a very technical manner and call theories whose Π^1_1–ordinals are below Γ_0 *predicative* without further reflection whether there are also possibly stronger principles which can be predicatively justified.

But we will see in the following section that there is a completely novel feature in the ordinal analysis of impredicative (i.e., non predicative) systems, collapsing. The simplest theory which needs a collapsing argument in its Π^1_1–analysis is the theory of non-iterated inductive definitions which is introduced in the next section. Its ordinal is $\psi_{\omega_1}(\varepsilon_{\omega_1+1})$, an ordinal which already has been described by H. Bachmann. There are, however, theories whose Π^1_1–ordinals are between Γ_0 and $\psi_{\omega_1}(\varepsilon_{\omega_1+1})$, e.g., the theory ATR introduced by Friedman which axiomatizes autonomous transfinite recursion (which is the axiom (Aut-Π^0_1) introduced on page 276 together with the full scheme of Mathematical Induction). Its ordinal is Γ_{ε_0}. Most recently many theories between Γ_0 and $\psi_{\omega_1}(\varepsilon_{\omega_1+1})$ have been analyzed. G. Jäger calls these theories *meta-predicative.*

A good summary on predicative theories can be found in the booklet Jäger [1986]. A sample of papers treating meta-predicative theories is Jäger et al. [n.d.], Jäger and Strahm [n.d.], Jäger [1980], Palmgren [n.d.], Strahm [n.d.] and Kahle [1997]. This list has been communicated to me by T. Strahm.

3.2. Axiom systems for number theory

In the present section we will introduce some impredicative axiom systems for Number Theory. We will not give an ordinal analysis for these systems directly – which would be possible in all demonstrated cases – but show that all these systems can be embedded into axioms systems for Set Theory. The ordinal analysis for the number-theoretic systems will then be obtained via an ordinal analysis of the set-theoretic systems. We start with the most simple example of an impredicative axiom system.

3.2.1. The theory $\mathbf{ID}_1$

By a monotone inductive definition on natural numbers we usually understand a monotone operator

$$\Gamma\colon \mathsf{Pow}(\mathbb{N}) \longrightarrow \mathsf{Pow}(\mathbb{N}),$$

i.e., an operator for which we have

$$\boldsymbol{S} \subseteq \boldsymbol{T} \;\Rightarrow\; \Gamma(\boldsymbol{S}) \subseteq \Gamma(\boldsymbol{T}).$$

A set $\boldsymbol{S} \subseteq \mathbb{N}$ is called Γ*-closed* iff $\Gamma(\boldsymbol{S}) \subseteq \boldsymbol{S}$. We obtain the least fixed-point $\mathbf{I}_\Gamma$ of Γ – often called *the fixed-point* of Γ – as the intersection of all Γ-closed subsets of $\mathbb{N}$, i.e., $\mathbf{I}_\Gamma = \bigcap\{\boldsymbol{S} \mid \Gamma(\boldsymbol{S}) \subseteq \boldsymbol{S}\}$. A set $\boldsymbol{P} \subseteq \mathbb{N}^n$ is *inductively definable* iff it is primitive recursive in the fixed point of some inductive definition. An operator is *arithmetically definable* iff there is an $\mathcal{L}_\mathsf{N}$ formula $A(X, x)$ such that $\Gamma(\boldsymbol{S}) = \{x \in \mathbb{N} \mid \mathbb{N} \models A(\boldsymbol{S}, x)\}$. If $A(X, x)$ is an X–positive formula, i.e., if its translation into the Tait–language contains no occurrences of $t \notin X$, then $A(X, x)$ defines a monotonic operator, i.e., an inductive definition.

We do not want to go into the theory of inductively defined sets (cf. Moschovakis [1974] and Barwise [1975] for a profound study). All we want to say here is that the fixed-point $\mathbf{I}_\Gamma$ of an monotone inductive definition comes in stages $\mathbf{I}_\Gamma^\xi$ which are defined by $\mathbf{I}_\Gamma^\xi := \Gamma(\mathbf{I}_\Gamma^{<\xi})$ where $\mathbf{I}_\Gamma^{<\xi} := \bigcup_{\zeta<\xi} \mathbf{I}_\Gamma^\zeta$. By cardinality reasons there is a countable ordinal σ such that $\mathbf{I}_\Gamma^\sigma = \mathbf{I}_\Gamma^{<\sigma}$. One defines

$$|\Gamma| := \min\{\sigma \mid \mathbf{I}_\Gamma^\sigma = \mathbf{I}_\Gamma^{<\sigma}\}$$

and calls $|\Gamma|$ the *closure ordinal* of Γ. Thus $\mathbf{I}_\Gamma = \mathbf{I}_\Gamma^{|\Gamma|}$. For every element $\boldsymbol{s} \in \mathbf{I}_\Gamma$ we may introduce its *inductive norm*

$$|\boldsymbol{s}|_\Gamma := \min\{\xi \mid \boldsymbol{s} \in \mathbf{I}_\Gamma^\xi\}.$$

We then obtain $|\Gamma| = \sup\{|\boldsymbol{s}|_\Gamma + 1 \mid \boldsymbol{s} \in \mathbf{I}_\Gamma\}$. For arithmetically definable operators Γ we have $|\Gamma| \le \omega_1^{\mathsf{CK}}$.

The theory $\mathbf{ID}_1$ axiomatizes the existence of least fixed-points for positively definable arithmetical inductive definitions. Recall the language $\mathcal{L}_\mathsf{N}$ of Number Theory. To obtain the language $\mathcal{L}_{\mathsf{ID}_1}$ we add a set constant I_A for every X–positive formula $A(X, x)$ in the language $\mathcal{L}_\mathsf{N}$ which contains only the shown free variables. We extend the scheme of Mathematical Induction to all $\mathcal{L}_{\mathsf{ID}_1}$ formulas and augment the axioms of $\mathbf{NT}$ by the schemes

$(\mathsf{ID}_1)^1$ $(\forall x)[A(\mathsf{I}_A, x) \;\rightarrow\; x \in \mathsf{I}_A]$

and

$(\mathsf{ID}_1)^2$ $(\forall x)[A(B,x) \to B(x)] \;\to\; (\forall x)[x \in \mathsf{I}_A \to B(x)]$

where $B(x)$ is an arbitrary $\mathcal{L}_{\mathsf{ID}_1}$ formula. While scheme $(\mathsf{ID}_1)^1$ expresses that I_A is $\Gamma_{A(X,x)}$–closed, scheme $(\mathsf{ID}_1)^2$ expresses that it is the least $\Gamma_{A(X,x)}$–closed set. The standard semantics for $\mathcal{L}_{\mathsf{ID}_1}$ is obtained interpreting I_A by $\mathbf{I}_A$.

Instead of giving a direct ordinal analysis for $\mathbf{ID_1}$ – which is possible e.g. cf. Pohlers [1989] – we will show that it can be easily embedded into axiom systems for Set Theory.

3.2.2. Iterated inductive definitions

The expressive power of first order logic with free set parameters is of course not exhausted by the axiom system $\mathbf{ID_1}$. As soon as we have fixed-points of inductive definitions we may use them in the definition of new operators. We are then leaving the realm of 'elementary inductive definitions' on the structure $\mathbf{N}$ in the sense of Moschovakis [1974]. To formalize the iteration let $\prec$ be a well-ordering of order type ν and associate a binary predicate constant J_A to every X-positive $\mathcal{L}_{\mathbf{N}}$ formula $A(X,Y,x,y)$ which contains at most the shown free variables. We are going to write $a \in \mathsf{J}_A^b$ instead of $\langle a,b\rangle \in \mathsf{J}_A$; by $a \in \mathsf{J}_A^{\prec b}$ we abbreviate the formula $(\exists y \prec b)[a \in \mathsf{J}_A^y]$. The language $\mathcal{L}_{\mathsf{ID}_\nu}$ is obtained by augmenting the language $\mathcal{L}_{\mathbf{N}}$ by a constant for $\prec$ and all constants J_A. Denote by $LO(\prec)$ the formula which is saying that $\prec$ is a linear ordering. We obtain the axiom system $\mathbf{ID}_\nu$ by taking all the axioms of $\mathbf{NT}$ and adding

(TI_ν) $\quad LO(\prec) \wedge TI(\prec, F)$

$(\mathsf{ID}_\nu)^1$ $\quad (\forall y \in \mathsf{field}(\prec))[(\forall x)[A(\mathsf{J}_A^y, \mathsf{J}_A^{\prec y}, x, y) \;\to\; x \in \mathsf{J}_A^y]]$

and

$(\mathsf{ID}_\nu)^2$ $\quad (\forall y \in \mathsf{field}(\prec))[(\forall x)[A(B, \mathsf{J}_A^{\prec y}, x, y) \to B(x)] \;\to\; (\forall x)[x \in \mathsf{J}_A^y \to B(x)]]$

where F and $B(x)$ are arbitrary $\mathcal{L}_{\mathsf{ID}_\nu}$–formulas. Since one inductive definition corresponds to one hyperjump the axiom system $\mathbf{ID}_\nu$ may also be interpreted as the system for ν-fold iterated hyperjumps. The constructive number classes $\mathcal{O}_\mu$ for $\mu \le \nu$ can be defined in $\mathbf{ID}_\nu$ and their basic properties can also be proved there. We define

$$\mathbf{ID}_{<\nu} := \bigcup_{\xi<\nu} \mathbf{ID}_\xi$$

where every $\xi < \nu$ is represented by a proper initial segment of $\prec$. One may also combine the axioms $\mathbf{ID_1}$ and the systems $\mathbf{ID}_\nu$ to obtain the system $\mathbf{ID}_{\prec}$. axiomatizing the iteration of inductive definitions along the accessible part of a linear order $\prec$. This is done by taking some arithmetically definable order relation $\prec$ and choosing the X–positive formula

$$A_\prec(X,x) \;:\Leftrightarrow\; (\forall y)[y \prec x \to y \in X].$$

Its fixed-point $\mathsf{I}_{A_\prec}$ is called the *accessible part* of $\prec$ and usually denoted by $\mathsf{Acc}_\prec$. The axioms of $\mathbf{ID}_{\prec^*}$ are

$(\mathsf{Acc})^1$ $(\forall x)[A_\prec(\mathsf{Acc}_\prec, x) \;\rightarrow\; x \in \mathsf{Acc}_\prec]$

(corresponding to $(\mathsf{ID}_1)^1$)

$(\mathsf{Acc})^2$ $(\forall x)[A_\prec(B, x) \rightarrow B(x)] \;\rightarrow\; (\forall x)[x \in \mathsf{Acc}_\prec \rightarrow B(x)]$

(corresponding to $(\mathsf{ID}_1)^2$).

$(\mathsf{ID}_{\prec^*})^1$ $(\forall y \in \mathsf{Acc}_\prec)[(\forall x)[A(\mathsf{J}_A^y, \mathsf{J}_A^{\prec y}, x, y) \;\rightarrow\; x \in \mathsf{J}_A^y]]$

modifying $(\mathsf{ID}_\nu)^1$ and

$(\mathsf{ID}_{\prec^*})^2$ $(\forall y \in \mathsf{Acc}_\prec)[(\forall x)[A(B, \mathsf{J}_A^{\prec y}, x, y) \rightarrow B(x)] \;\rightarrow\; (\forall x)[x \in \mathsf{Acc}_\prec \rightarrow B(x)]]$

modifying $(\mathsf{ID}_\nu)^2$. All these systems of iterated inductive definitions have been introduced by Feferman.

There are even stronger iterations of inductive definitions which, however, can be more elegantly formulated within the framework of Second Order Number Theory.

3.2.3. Iterated inductive definitions in second order

In full Second Order Logic we do not have a recursively enumerable notion of provability. Therefore we have to fix a calculus and regard the proof strength of an axiom system relatively to that calculus. This means that we rather use two sorted first order logic than full second order. To fix a calculus we assume that we have a second order Tait language as introduced in Section 1.3. We extend the calculus of Definition 2.1.2.1 by the following second order rules:

$(\exists^2)$ *If* $\vdash^{m_0} \Delta, A(X)$, *then* $\vdash^{m} \Delta, (\exists Y)A(Y)$ *for all* $m > m_0$

$(\forall^2)$ *If* $\vdash^{m_0} \Delta, A(X)$ *and* X *not free in any of the formulas in* $\Delta, (\forall Y)A(Y)$, *then* $\vdash^{m} \Delta, (\forall Y)A(Y)$ *for all* $m > m_0$.

We say that a formula F is provable from an axiom system $\mathbf{Ax}$ iff there are finitely many instances of identity axioms $G_1, \ldots, G_m$ and finitely many sentences $\{A_1, \ldots, A_n\} \subseteq \mathbf{Ax}$ such that $\vdash^{m} \neg G_1, \ldots, \neg G_m, \neg A_1, \ldots, \neg A_n, F$ for some m.

The strongest axiom system for Number Theory is $\mathbf{NT_2}$ which comprises all the axioms of $\mathbf{NT}$ together with the axiom schemes

(CA) $(\exists X)[(\forall x)(x \in X \leftrightarrow F(x))]$

of *comprehension* and

(AC) $(\forall x)(\exists X)F(x, X) \rightarrow (\exists Y)(\forall x)F(x, Y^x)$

of *choice*. We put $y \in Y^x \;:\Leftrightarrow\; \langle y, x\rangle \in Y$ and assume tacitly that $F(x)$ and $F(x, Y)$ must not contain the variable X.

If the formulas in the schemes (CA) or (AC) are restricted to a complexity class $\mathcal{F}$ we talk about ($\mathcal{F}$-CA) and ($\mathcal{F}$-AC), respectively. By **($\mathcal{F}$-CA)** we denote the axiom systems which comprises all the axioms of **NT** extended to the second order language together with the scheme ($\mathcal{F}$-CA). Analogously we denote by **($\mathcal{F}$-AC)** the axioms of **NT** together with ($\mathcal{F}$-AC).

We will also regard axiom systems which are closed under rules. A rule has the form

(R) $\quad F_1, \ldots, F_n \;\; \vdash F$

and we say that a theory **Ax** is *closed under the rule* (R) if $\mathbf{Ax} \vdash F_i$ for $i = 1, \ldots, n$ implies $F \in \mathbf{Ax}$. For a given rule (R) we define

(R) $\quad$ *The least* $\mathcal{L}^2_{\mathbb{N}}$*–theory which comprises* $\mathbf{NT} + (\Pi^1_0\text{-CA})$ *and is closed under the rule* (R).

Observe that the theory $(\mathbf{R})$ is the union of the theories $(\mathbf{R}_n)$ where $(\mathbf{R}_0) = \mathbf{NT} + (\Pi^1_0\text{-CA})$ and $(\mathbf{R}_{n+1})$ is obtained by closing $(\mathbf{R}_n)$ under all applications of the rule (R).

In $\mathbf{NT_2}$ we may replace the scheme of Mathematical Induction by a single axiom

$$(\forall X)[\underline{0} \in X \wedge (\forall y)(y \in X \to y+1 \in X) \;\; \to \;\; (\forall x)(x \in X)].$$

This is no longer true if we regard axiom systems with restricted comprehension scheme. Therefore we introduce also the axiom systems $(\boldsymbol{\mathcal{F}}\text{- }\mathbf{CA})_0$ and $(\boldsymbol{\mathcal{F}}\text{- }\mathbf{AC})_0$ in which the scheme of Mathematical Induction is replaced by the single axiom.

The formula

$$\mathit{Wf}(\prec) \;\; :\Leftrightarrow \;\; (\forall X)[\mathit{Prog}(\prec, X) \to (\forall x)(x \in X)]$$

expresses the well–foundedness of the relation $\prec$. We have

$$\mathit{Wf}(\prec) \;\; \Leftrightarrow \;\; (\forall X)[(\exists x)(x \in X) \to (\exists x \in X)(\forall y \prec x)(y \notin X)]. \tag{113}$$

In $\mathbf{NT_2}$ the sentence $\mathit{Wf}(\prec)$ entails the scheme $\mathit{TI}(\prec, F)$ for arbitrary $\mathcal{L}^2_{\mathbb{N}}$ formulas $F(x)$. This, too, is not longer true for restricted comprehension. Therefore we introduce the scheme

(BI) $\quad \mathit{Wf}(\prec) \to \mathit{TI}(\prec, F)$

of *Bar Induction* for definable relations $\prec$ and arbitrary $\mathcal{L}^2_{\mathbb{N}}$-formulas $F(x)$. A relation $\prec$ is definable iff there is an $\mathcal{L}^2_{\mathbb{N}}$-formula $G(x,y)$ such that $x \prec y \;\leftrightarrow\; G(x,y)$. The formula $G(x,y)$ may contain additional parameters. We will sometimes emphasize this by writing

$$x \prec_{G,\vec{x},\vec{X}} y \;\; :\Leftrightarrow \;\; G(x,y,\vec{x},\vec{X}).$$

Roughly speaking Bar Induction says that a relation which is well–founded with respect to sets is also well–founded for classes. If the defining formulas for the relation $\prec$ in the scheme (BI) are restricted to the complexity class $\mathcal{F}$ then we talk about ($\mathcal{F}$-BI). If also the complexity of the allowed classes is restricted to another complexity class $\mathcal{F}_2$ we notate that as $(\mathcal{F}\text{-BI})\restriction\mathcal{F}_2$.

If X is a set parameter we may define the binary relation

$$x \prec_X y \ :\Leftrightarrow \ \langle x, y\rangle \in X.$$

Sometimes Bar Induction is formulated as the single axiom

(Bi) $\quad (\forall X)[\mathit{Wf}(\prec_X) \to \mathit{TI}(\prec_X, F)]$

which in the presence of (Π^1_0-CA) has the strength of (Π^1_0-BI). Weaker than (BI) is

(BR) $\quad \mathit{Wf}(\prec) \ \vdash \ \mathit{TI}(\prec, F)$

which is known as *Bar Rule.* M. Rathjen has shown in Rathjen [1991] that the Bar Rule is of the same strength as parameter free Bar Induction, i.e., the axiom of Bar Induction in which the defining formula for the relation $\prec$ must not contain parameters (not even individual parameters).

As a basis for nearly all our second order axiom systems we will use **(Π^1_0-CA)**, i.e., the scheme for arithmetical comprehension. This system is also known as **(ACA)**. We use both notions interchangeable. Observe that **(ACA)** proves that a relation $\prec$ is well-founded iff it does not contain an infinite $\prec$–descending sequence, i.e.,

$$\mathit{Wf}(\prec) \ \Leftrightarrow \ \neg[(\exists X)[(\exists x)(x \in X) \land \mathit{LO}(\prec \restriction X) \land (\forall x \in X)(\exists y \in X)(y \prec x)]]. \quad (114)$$

The axiom system **(ACA)** proves also the equivalence of the schemes (Bi), (Π^1_0-BI) and the following quantifier scheme (cf. Feferman [1970]).

(QS) $\quad (\forall X)A(X) \to A(F)$ *for arithmetical* $A(X)$ *and arbitrary* $F(x)$.

Because of this equivalence it has become common to call (QS) also (Π^1_0-BI).

The iteration of inductive definitions is quite elegantly expressed in a second order language. For an X–positive formula $A(X, Y, x, y)$ we introduce the abbreviation

$$\mathit{Cl}_A(X, Y, y) \ :\Leftrightarrow \ (\forall x)[A(X, Y, x, y) \to x \in X]$$

and define

$$\begin{aligned} \mathit{IT}_A(\prec, X) \ :\Leftrightarrow \ & (\forall y \in \mathsf{field}(\prec))[\mathit{Cl}_A(X^y, X^{\prec y}, y) \\ & \land (\forall Y)(\mathit{Cl}_A(Y, X^{\prec y}, y) \ \to \ (\forall x)(x \in X^y \to x \in Y))]. \end{aligned}$$

Then $\mathit{IT}_A(\prec, X)$ says 'X is an hierarchy of fixed-points for $A(X, Y, x, y)$ iterated along $\prec$'.

3.2.3.1. Definition. Let $\prec$ be a primitive recursive well-ordering of order type ν. For an X–positive arithmetical formula $A(X, Y, x, y)$ we introduce the schemes

(IT_ν) $\quad (\exists X)\mathit{IT}_A(\prec, X)$

and

(B-IT_ν) $\quad (\forall Y)(\forall y \in \mathsf{field}(\prec))[\mathit{IT}_A(\prec, Y) \land \mathit{Cl}_A(B, Y^{\prec y}, y) \ \to \ (\forall x)(x \in Y^y \to B(x))]$

where $B(x)$ is an arbitrary $\mathcal{L}^2_{\mathsf{N}}$–formula. We define the theories

$$\mathbf{ID}^2_\nu := \mathbf{(ACA)} + \mathit{WO}(\prec) + (\mathsf{IT}_\nu)$$

and

$$\mathbf{BID}^2_\nu := (\mathbf{ACA}) + \mathit{WO}(\prec) + (\mathsf{IT}_\nu) + (\mathsf{B\text{-}IT}_\nu)$$

where $\mathit{WO}(\prec)$ stands for $\mathit{LO}(\prec) \wedge \mathit{Wf}(\prec)$ and A varies over all X–positive formulas $A(X,Y,x,y)$ which contain at most the shown variables free. Denote by

$$(\mathsf{ITA}) \qquad (\forall x)[\mathit{WO}(\prec_{F,x}) \to (\exists X)\mathit{IT}_A(\prec_{F,x}, X)]$$

the scheme in which $F(u,v,x)$ and $A(X,Y,x,y)$ are supposed to vary over arithmetical formulas without further parameters. We define

$$\mathbf{ID}^{2*} := (\mathbf{ACA}) + (\mathsf{ITA}).$$

Finally put

$$(\mathsf{B\text{-}ITA}) \quad (\forall x)(\forall y)(\forall X)[\mathit{WO}(\prec_{F,x}) \wedge \mathit{IT}_A(\prec_{F,x}, X) \wedge \mathit{Cl}_A(B, X^{\prec_{F,x}y}, y) \\ \to \; (\forall x)(x \in X^y \to B(x))]$$

where $F(u,v,x)$ and $A(X,Y,x,y,)$ are as above and $B(x)$ is an arbitrary $\mathcal{L}^2_{\mathrm{N}}$–formula. The schemes $(\mathsf{B\text{-}IT}_\nu)$ and $(\mathsf{B\text{-}ITA})$ have the flavor of Bar Induction; therefore the B in their identifiers. Let

$$\mathbf{BID}^{2*} := (\mathbf{ACA}) + (\mathsf{ITA}) + (\mathsf{B\text{-}ITA}).$$

One should observe that $\mathbf{BID}^{2*}$ allows only iterations along arithmetically definable well-orderings, i.e., along well-orderings of length $< \omega_1^{\mathrm{CK}}$. So it may appear weaker than $\mathbf{ID}_{\prec^*}$ which allows iterations along accessible parts of arithmetically definable orderings, i.e., along well-orderings of length $\leq \omega_1^{\mathrm{CK}}$. However, we will sketch in a moment that $\mathbf{BID}^{2*}$ comprises $\mathbf{ID}_{\prec^*}$.

If we drop the restriction to arithmetically definable well–orderings we obtain the schemes

$$(\mathsf{IT}) \qquad (\forall X)[\mathit{WO}(\prec_X) \to (\exists Y)\mathit{IT}_A(\prec_X, Y)]$$

and

$$(\mathsf{B\text{-}IT}) \quad (\forall y)(\forall Y)[\mathit{WO}(\prec_F) \wedge \mathit{IT}_A(\prec_F, Y) \wedge \mathit{Cl}_A(B, Y^{\prec_F y}, y) \\ \to \; (\forall x)(x \in Y^y \to B(x))],$$

respectively, where $F(u,v)$, and $B(x)$ may now be arbitrary $\mathcal{L}^2_{\mathrm{N}}$–formulas and $G(X,Y,x,y)$ is an X-positive arithmetical formula. All formulas may contain additional parameters. We put

$$\mathbf{Aut\text{-}ID} := (\mathbf{ACA}) + (\mathsf{IT})$$

and

$$(\mathbf{Aut\text{-}BID}) := (\mathbf{ACA}) + (\mathsf{IT}) + (\mathsf{B\text{-}IT}).$$

One easily shows

$$(\mathbf{\Pi^1_0\text{-}CA}) \vdash \mathit{WO}(\prec) \wedge \mathit{IT}_A(\prec, Y) \wedge \mathit{IT}_A(\prec, Z) \;\to\; (\forall x \in \mathsf{field}(\prec))(Y^x = Z^x). \quad (115)$$

There is an canonical embedding $F \mapsto F^*$ from the language of $\mathbf{ID}_\nu$ into the language of $\mathbf{ID}^2_\nu$. We replace every occurrence of $s \in \mathsf{J}^t_A$ by $(\exists X)[\mathit{IT}_A(\prec, X) \wedge s \in X^t]$. The theory $\mathbf{ID}^2_\nu$ shows that there is a set W_A such that $IT_A(\prec, W_A)$ and by (115) we obtain that W^t_A is uniquely defined for $t \prec \nu$. Hence $(s \in \mathsf{J}^t_A)^*$ iff $s \in W^t$ for all $t \in \mathsf{field}(\prec)$ which entails $\mathsf{J}^{t\,*}_A = W^t$ and $\mathsf{J}^{\prec t*}_A = W^{\prec t}$. Since we have $Cl_A(W^t_A, W^{\prec t}_A, t)$ we get $(\mathsf{ID}_\nu)^{1\,*}$ and from $\mathsf{B\text{-}IT}_\nu$ also $(\mathsf{ID}_\nu)^2$. So we have shown

$$\mathbf{ID}_\nu \subseteq \mathbf{BID}^2_\nu. \tag{116}$$

An embedding $F \mapsto F^*$ of the language of $\mathsf{ID}_{\prec^*}$ into the second order language is obtained similarly. Define

$$IT^y_A(\prec, X) \;:\Leftrightarrow\; (\forall x \preceq y)[Cl_A(X^x, X^{\prec x}, x) \wedge (\forall Y)(Cl_A(Y, X^{\prec x}, x) \;\rightarrow\; (\forall z)(z \in X^x \rightarrow z \in Y))]$$

and let $X \subseteq Y$ stand for $(\forall x)[x \in X \rightarrow x \in Y]$. Define

$$\mathsf{Acc}(\prec, X) \;:\Leftrightarrow\; Prog(\prec, X) \wedge (\forall Z)[Prog(\prec, Z) \rightarrow X \subseteq Z]$$

and replace all occurrences of $s \in \mathsf{Acc}_\prec$ by $(\exists X)[\mathsf{Acc}(\prec, X) \wedge s \in X]$ and finally replace all occurrences of $s \in \mathsf{J}^t_A$ by $(\exists X)[IT^t_A(\prec, X) \wedge s \in X^t]$. We indicate that the translations $(\mathsf{Acc})^{1*}$, $(\mathsf{Acc})^{2*}$, $(\mathsf{ID}_{\prec^*})^{1*}$ and $(\mathsf{ID}_{\prec^*})^{2*}$ are all provable in $\mathbf{BID}^{2*}$. We argue informally in $\mathbf{BID}^{2*}$. Let $x \prec^s y \;:\Leftrightarrow\; x \prec y \preceq s$. Put $A_0(X, Y, x, y) \;:\Leftrightarrow\; (\forall z)[z \prec x \rightarrow z \in X]$. Then there is a set T such that $IT_{A_0}(\prec^0, T)$ if 0 denotes the least element in $\prec$. We then have $Prog(\prec, T^0)$ and define $S := T^0$. Moreover we have $Prog(\prec, X) \rightarrow S \subseteq X$ and thus $\mathsf{Acc}(\prec, S)$. So $\mathsf{Acc}^*_\prec = S$ and we get $(\mathsf{Acc})^{1*}$ by $Prog(\prec, S)$ and $(\mathsf{Acc})^{2*}$ from (B-ITA). To prove $(\mathsf{ID}_{\prec^*})^{1*}$ we assume $s \in \mathsf{Acc}^*_\prec$, i.e., $s \in S$. Then we obtain $Wf(\prec^s)$ because otherwise according to (114) there would be a non empty set $V \subseteq \mathsf{field}(\prec^s)$ containing s and an infinite $\prec$-descending sequence. Since $\prec$ is a linear order this entails that if all $\prec$-predecessors of an element x do not belong to V then x cannot belong to V. That means $Prog(\prec, \neg V)$ where $\neg V$ denotes the complement of V. But then $S \subseteq \neg V$ which contradicts $s \in V$. Now choose any X-positive formula $A(X, Y, x, y)$. By ITA there exists a set W such that $IT_A(\prec^s, W)$. By the uniqueness property (115) we obtain $(\forall y)[y \preceq s \rightarrow (\forall x)(x \in \mathsf{J}^y_A)^* \leftrightarrow x \in W^y]$. Hence $(\mathsf{J}^s_A)^* = W^s$ and $(\mathsf{J}^{\prec s}_A)^* = W^{\prec s}$ and we get $(\mathsf{ID}_{\prec^*})^{1*}$ from $IT_A(\prec\restriction s, W)$ and $(\mathsf{ID}_{\prec^*})^{2*}$ from B-ITA.

So we have

$$\mathbf{ID}_{\prec^*} \subseteq \mathbf{BID}^{2*}. \tag{117}$$

As remarked before the schemes $(\mathsf{B\text{-}IT}_\nu)$, (B-ITA) and (B-IT) have the flavor of Bar Induction. We are going to substantiate this remark. First we show

$$\mathbf{Ax} \in \{\mathbf{BID}^2_\nu, \mathbf{BID}^{2*}, \mathbf{Aut\text{-}BID}\} \;\Rightarrow\; \mathbf{Ax} \vdash \mathsf{Bi}. \tag{118}$$

Assume $Wf(\prec_X)$ and $Prog(\prec_X, B)$ for some formula $B(x)$. We have to show $(\forall x)B(x)$. For $A(Z, Y, x, y) \;:\Leftrightarrow\; (\forall z \prec_X x)(z \in Z)$ we obtain by IT_ν, ITA or IT, respectively, a set T such that $IT_A(\prec^0_X, T)$. Since $\prec^0_X$ is trivially well-founded we get $Prog(\prec_X, B) \rightarrow (\forall x)B(x)$ instantiating y by 0, $\prec_X$ by $\prec^0_X$ and Y by T in $\mathsf{B\text{-}IT}_\nu$, B-ITA or B-IT, respectively. Hence $(\forall x)B(x)$. □

But the opposite is also true. We prove

$$\begin{aligned} &\mathbf{ID}^2_\nu + (\Pi^1_0\text{-}\mathsf{BI}) \vdash (\mathsf{B\text{-}IT}_\nu) \\ &\mathbf{ID}^{2*} + (\Pi^1_0\text{-}\mathsf{BI}) \vdash (\mathsf{B\text{-}ITA}) \qquad (119)\\ &\mathbf{Aut\text{-}ID} + (\Pi^1_0\text{-}\mathsf{BI}) \vdash (\mathsf{B\text{-}IT}). \end{aligned}$$

Assume $\mathit{Wf}(\prec)$, $\mathit{IT}_A(\prec, X)$ and $\mathit{Cl}_A(B, X^{\prec y}, y)$. We have to show $(\forall x \in X^y)B(x)$. For the arithmetical formula $C(Y) \;:\Leftrightarrow\; \mathit{Cl}_A(Y, X^{\prec y}, y) \to (\forall x \in X^y)(x \in Y)$ we get from IT_ν, ITA or IT, respectively, $(\forall Y)C(Y)$. Using (QS) which is equivalent to $(\Pi^1_0\text{-}\mathsf{BI})$ we obtain $C(B)$. Hence $(\forall x \in X^y)B(x)$. □

So we have the following theorem

3.2.3.2. Theorem. $\mathbf{BID}^2_\nu = \mathbf{ID}^2_\nu + (\mathsf{Bi})$, $\mathbf{BID}^{2*} = \mathbf{ID}^{2*} + (\mathsf{Bi})$ *and* $\mathbf{Aut\text{-}BID} = \mathbf{Aut\text{-}ID} + (\mathsf{Bi})$, *where* $=$ *means that both theories prove the same theorems.*

3.2.4. Π^1_1-comprehension and beyond

We mentioned already that the fixed-point of an monotonic operator Γ can be obtained as the intersection of all Γ-closed sets. For an operator which is arithmetically defined by a formula $A(X, x)$ this means that we have

$$\begin{aligned} \mathbf{I}_A &= \{x \mid (\forall X)[(\forall y)(A(X, y) \to y \in X) \;\to\; x \in X]\} \qquad (120)\\ &= \{x \mid (\forall X)[\mathit{Cl}_A(X) \to x \in X]\} \end{aligned}$$

Vice versa, every Π^1_1-set can be shown to be an inductive set, i.e., a set which is primitive recursive in some fixed-point. So $(\Pi^1_1\text{-CA})$ and inductive definitions are canonically connected. To iterate $(\Pi^1_1\text{-CA})$ we introduce the following notations

3.2.4.1. Definition. Let $H(X, x, y)$ be an $\mathcal{L}^2_{\mathrm{N}}$-formula which contains only the shown parameters. We define

$$J_H(\prec, X) \;:\Leftrightarrow\; (\forall x)(\forall y)[x \in X^y \leftrightarrow H(X^{\prec y}, x, y)]$$

and call X the *jump hierarchy* based on $H(X, x, y)$ along $\prec$.

For a primitive recursive well-ordering of order type ν we introduce the scheme

$(\Pi^1_1\text{-CA}_\nu)$ $\;(\exists Z)J_H(\prec, Z)$ *for* $H(X, x, y) \in \Pi^1_1$

of ν–fold iterated Π^1_1-comprehension. We sometimes express this sloppily by $(\exists Z)J_H(\nu, Z)$ if we do not want to emphasize the order–relation but its order type. Transfinitely iterated Π^1_1-comprehensions are axiomatized by

$(\mathsf{Aut}\text{-}\Pi^1_1)$ $\;(\forall X)[\mathit{Wf}(\prec_X) \;\to\; (\exists Z)J_H(\prec_X, Z)]$ *for* $H(X, x, y) \in \Pi^1_1$.

For a primitive recursive ordering $\prec$ of order type ν we define

$$(\mathbf{\Pi^1_1\text{-}CA_\nu}) := (\mathbf{ACA}) + \mathit{WO}(\prec) + (\Pi^1_1\text{-CA}_\nu),$$

$$(\mathbf{\Pi^1_1\text{-}CA_{<\nu}}) := \bigcup_{\xi<\nu} (\mathbf{\Pi^1_1\text{-}CA_\xi})$$

and

$$(\mathbf{Aut\text{-}\Pi^1_1}) := (\mathbf{ACA}) + (\mathsf{Aut\text{-}\Pi^1_1}).$$

Notice that $(\Pi^1_1\text{-CA})$ and $(\Pi^1_1\text{-CA}_\nu)$ have different meanings. Due to the possible presence of set parameters in Π^1_1-comprehension formulas we have

$$(\mathbf{\Pi^1_1\text{-}CA}) = (\mathbf{\Pi^1_1\text{-}CA}_{<\omega})$$

though notationally this looks strange. Recall that $(\mathbf{\Pi^1_1\text{-}\ CA})_0$ means $(\Pi^1_1\text{-CA}) + (\mathbf{ACA})_0$, i.e., $(\Pi^1_1\text{-CA})$ together with the axiom of Mathematical Induction. The theories $(\mathbf{\Pi^1_1\text{-}CA}_\nu)_0$ are defined analogously.

We will see that the theories $(\mathbf{\Pi^1_1\text{-}CA}_\nu)$ and $\mathbf{ID}^2_\nu$ are equivalent. In a first step we show

$$(\mathbf{ID}^2_\nu)_0 \subseteq (\mathbf{\Pi^1_1\text{-}CA}_\nu)_0 \quad \textit{and} \quad (\mathbf{Aut\text{-}ID})_0 \subseteq (\mathbf{Aut\text{-}\Pi^1_1})_0. \tag{121}$$

Let $\prec$ be a well-ordering of order type ν or assume $\mathit{Wf}(\prec)$. Let $A(X, Y, x, y)$ be an X–positive arithmetical formula and put

$$H(X, x, y) \;:\Leftrightarrow\; (\forall Z)[\mathit{Cl}_A(Z, X, y) \to x \in Z]. \tag{i}$$

Then $H(X, x, y) \in \Pi^1_1$ and by $(\Pi^1_1\text{-CA})_\nu$ or $(\mathsf{Aut\text{-}\Pi^1_1})$ there is a set S such that $J_H(\prec, S)$. Hence

$$S^y = \{x \mid H(S^{\prec y}, x, y)\} = \{x \mid (\forall X)[\mathit{Cl}_A(X, S^{\prec y}, y) \to x \in X]\} \tag{ii}$$

which in turn implies

$$(\forall X)[\mathit{Cl}_A(X, S^{\prec y}, y) \to S^y \subseteq X]. \tag{iii}$$

From (iii) we obtain

$$\begin{aligned} A(S^y, S^{\prec y}, x, y) \to \mathit{Cl}_A(X, S^{\prec y}, y) &\to S^y \subseteq X \\ &\to A(X, S^{\prec y}, x, y) \\ &\to x \in X \end{aligned} \tag{iv}$$

by the monotonicity of the X-positive formula $A(X, Y, x, y)$. But (iv) means

$$A(S^y, S^{\prec y}, x, y) \to x \in S^y. \tag{v}$$

Pulling (iii) and (v) together we obtain $\mathit{IT}_A(\prec, S)$. □

To prove also the other inclusion in (121) we use the fact that already $(\mathbf{ACA})_0$ proves that every Π^1_1-formula $H(Y, y, \vec{z})$ is equivalent to the well–foundedness of its associated tree of unsecured sequences, i.e., that there is an arithmetical formula $T_H(Y, x, y, \vec{z})$ such that

$$(\mathbf{ACA})_0 \vdash H(Y, y, \vec{z}) \;\leftrightarrow\; \{x \mid T_H(Y, x, y, \vec{z})\} \;\; \textit{is a well–founded tree.}$$

Defining $A(X, Y, x, \vec{z}) \;:\Leftrightarrow\; (\forall y)[T_H(Y, (x)_0 \frown \langle y \rangle, (x)_1, \vec{z}) \to \langle (x)_0 \frown \langle y \rangle, (x)_1 \rangle \in X]$ we have an X-positive arithmetical formula such that

$$H(Y,y,\vec{z}) \leftrightarrow \{x \mid T_H(Y,x,y,\vec{z})\} \text{ is well-founded} \\ \leftrightarrow \langle\langle\rangle, y\rangle \in \mathsf{I}_{A(Y,\vec{z})} \\ \leftrightarrow (\forall Z)[Cl_A(Z,Y,\vec{z}) \to \langle\langle\rangle, y\rangle \in Z]. \tag{122}$$

The last equivalence is provable in $(\mathbf{ACA})_0$.

To obtain therefrom the opposite inclusion in (121) let $H(Y,x,y)$ be a Π^1_1-formula and $A(X,Y,x,y)$ the arithmetical formula such that according to (122)

$$H(Y,x,y) \quad \leftrightarrow \quad (\forall Z)[Cl_A(Z,Y,y) \to \langle\langle\rangle, x\rangle \in Z]. \tag{i}$$

Define $A'(X,Y,x,y) \;:\Leftrightarrow\; A(X,\{z \mid \langle\langle\rangle, z\rangle \in Y\},x,y)$ and let $\prec$ either by a primitive recursive well-ordering of order type ν or assume $WO(\prec)$. Then either $\mathbf{ID}^2_\nu$ or **Aut-ID** prove the existence of a set S such that $IT_{A'}(\prec, S)$, i.e.,

$$Cl_A(S^y, \{z \mid \langle\langle\rangle, z\rangle \in S^{\prec y}\}, y) \tag{ii}$$

and

$$(\forall Z)[Cl_A(Z, \{z \mid \langle\langle\rangle, z\rangle \in S^{\prec y}\}, y) \to S^y \subseteq Z]. \tag{iii}$$

Hence

$$\langle\langle\rangle, x\rangle \in S^y \to (\forall Z)[Cl_A(Z, \{z \mid \langle\langle\rangle, z\rangle \in S^{\prec y}\}, y) \to \langle\langle\rangle, x\rangle \in Z] \\ \to H(\{z \mid \langle\langle\rangle, z\rangle \in S^{\prec y}\}, x, y) \tag{iv}$$

and

$$H(\{z \mid \langle\langle\rangle, z\rangle \in S^{\prec y}\}, x, y) \to (\forall Z)[Cl_A(Z, \{z \mid \langle\langle\rangle, z\rangle \in S^{\prec y}\}, y) \to \langle\langle\rangle, x\rangle \in Z] \\ \to \langle\langle\rangle, x\rangle \in S^y. \tag{v}$$

So we have

$$\langle\langle\rangle, x\rangle \in S^y \quad \leftrightarrow \quad H(\{x \mid \langle\langle\rangle, x\rangle \in S^{\prec y}\}, x, y) \tag{vi}$$

and putting $T \;:=\; \{\langle x,y\rangle \mid \langle\langle\rangle, x\rangle \in S^y\}$ we obtain $J_H(\prec, T)$. Therefore $(\exists Z)J_H(\prec, Z)$ is a theorem of $\mathbf{ID}^2_\nu$ or **Aut-ID**, respectively. □

So we have together with (121) and Theorem 3.2.3.2 the following theorem.

3.2.4.2. Theorem. $(\mathbf{\Pi^1_1\text{-}CA}_\nu)_0 = (\mathbf{ID}^2_\nu)_0$, $(\mathbf{Aut\text{-}\Pi^1_1})_0 = (\mathbf{Aut\text{-}ID})_0$, $(\mathbf{\Pi^1_1\text{-}CA}_\nu) = \mathbf{ID}^2_\nu$, $(\mathbf{\Pi^1_1\text{-}CA}_\nu) + (\mathsf{Bi}) = \mathbf{ID}^2_\nu + (\mathsf{Bi}) = \mathbf{BID}^2_\nu$, $\mathbf{Aut\text{-}ID} = (\mathbf{Aut\text{-}\Pi^1_1})$ *and* $\mathbf{Aut\text{-}BID} = \mathbf{Aut\text{-}ID} + (\mathsf{Bi}) = (\mathbf{Aut\text{-}\Pi^1_1}) + (\mathsf{Bi})$.

Regarding (116) we obtain the following chain

$$\mathbf{ID}_\nu \subseteq \mathbf{BID}^2_\nu = (\mathbf{\Pi^1_1\text{-}CA}_\nu) + (\Pi^1_0\text{-}\mathsf{BI}). \tag{123}$$

Feferman [1970] has shown that for $\nu = \omega^\rho$ with $\rho > 0$ the theory $\mathbf{ID}_\nu$ proves the existence of an ω-model for $(\mathbf{\Pi^1_1\text{-}CA}_{<\nu})$ $+(\Pi^1_0\text{-}\mathsf{BI})$. This shows that $(\mathbf{\Pi^1_1\text{-}CA}_{<\nu})$ $+(\Pi^1_0\text{-}\mathsf{BI})$ is proof theoretical reducible to $\mathbf{ID}_{<\nu}$ for $\nu = \omega^\rho$, $\rho \in \mathsf{Lim}$ in the sense that $\mathbf{ID}_{<\nu}$ proves the existence of an ω-model for every finite subtheory of $(\mathbf{\Pi^1_1\text{-}CA}_{<\nu})$ $+(\Pi^1_0\text{-}\mathsf{BI})$. On the other hand the theory $(\mathbf{\Pi^1_1\text{-}CA}_{<\nu})$ proves the existence of an

ω–model for all $\mathbf{ID}_\xi$, $\xi < \nu$. The scheme (Π^1_0-BI) is not needed here since the translation of every $\mathcal{L}_{\mathsf{ID}_\xi}$–formula F is arithmetical in some X^y with $J_H(\prec, X)$. The translation of $\mathsf{B\text{-}IT}_\nu$ is thus obtained by (Π^1_0-CA). Writing $\leq$ for proof theoretical reducibility we get for $\nu = \omega^\rho$, $\rho \in \mathsf{Lim}$ the following chain.

$$\mathbf{ID}_{<\nu} \subseteq \mathbf{BID}^2_{<\nu} = (\mathbf{\Pi^1_1\text{-}CA}_{<\nu}) + (\Pi^1_0\text{-BI}) \leq \mathbf{ID}_{<\nu} \leq (\mathbf{\Pi^1_1\text{-}CA}_{<\nu}) \leq \mathbf{ID}_{<\nu}. \quad (124)$$

This shows that all these theories are proof theoretically equivalent.

To close the section we mention the results of H. Friedman [1970] who showed

$$(\mathbf{\Delta^1_2\text{-}CA}) \equiv (\mathbf{\Pi^1_1\text{-}CA}_{<\varepsilon_0}) \equiv (\mathbf{\Sigma^1_2\text{-}AC}) \quad (125)$$

where

(Δ^1_2-CA) $\quad (\forall x)[A(x) \leftrightarrow B(x)] \;\rightarrow\; (\exists X)(\forall x)[x \in X \leftrightarrow A(x)]$

for $A(x) \in \Pi^1_2$ and $B(x) \in \Sigma^1_2$ is the scheme of Δ^1_2–comprehension.

A simpler argument than that in H. Friedman's results is given in Feferman [1970] to characterize the Δ^1_2–comprehension rule. To define the rule let the class of 'essentially' Π^1_1 formulas be the smallest class of formulas which contains the arithmetical formulas and is closed under the positive boolean operations $\wedge$ and $\vee$, first order quantification and second order $\forall$–quantification. Dually the class of essentially Σ^1_1–formulas is the class of formulas whose negation is logically equivalent to an essentially Π^1_1-formula. Analogously we obtain the class of essentially Π^1_2–formulas when we start with the essentially Σ^1_1–formulas instead of arithmetical formulas and the class of essentially Σ^1_2–formulas as its dual class. The Δ^1_2 comprehension rule is defined as follows.

(Δ^1_2–CR) $\quad (\forall x)[A(x) \leftrightarrow B(x)] \;\vdash (\exists X)(\forall x)[x \in X \leftrightarrow A(x)] \in \mathbf{Ax}$ *for* $A(x)$ *essentially* Π^1_2 *and* $B(x)$ *essentially* Σ^1_2.

Feferman shows

$$(\mathbf{\Delta^1_2\text{-}CR}) \equiv (\mathbf{\Pi^1_1\text{-}CA}_{<\omega^\omega}). \quad (126)$$

3.3. Axiom systems for set theory

3.3.1. The axiom system KPω

We introduced the axiom system $\mathbf{KP}\omega$ already in Section 1.2. For ordinal analysis it will be more convenient to restate the axioms of $\mathbf{KP}\omega$ in a more parsimonious way. We keep (Ext) and modify the pairing axiom (Pair) to

(Pair') $\quad (\forall x)(\forall y)(\exists z)[x \in z \wedge y \in z]$.

It is obvious that (Pair) follows from (Pair') by Δ_0–separation. In a similar way we modify the axiom of union to

(Union') $\quad (\forall u)(\exists w)(\forall y \in u)(\forall z \in y)[z \in w]$

an axiom which requires only the existence of a superset of the union. Again it is clear that we obtain (Union) form (Union') by Δ_0–separation. Similarly we modify the axiom of infinity to

(Inf') $(\exists u)[u \neq \emptyset \wedge (\forall x \in u)(\exists v \in u)(x \in v)]$.

For convenience we introduce an axiom system **BST** for Basic Set Theory. It comprises the axioms (Ext), (Pair'), (Union'), (Δ_0–Separation), and the scheme (FOUND) of foundation. Adding (Inf') to **BST** we obtain the system **BSTω**. All systems we will regard here are based on **BST**. If **Ax** is such a system we denote by $\mathbf{Ax}^r$ the system which is obtained by restricting the foundations scheme (FOUND) to Δ_0-formulas. The system W–**Ax** is the intermediate system between $\mathbf{Ax}^r$ and **Ax** in which we have (Δ_0-FOUND) but allow full Mathematical Induction, i.e., the scheme $F(0) \wedge (\forall x \in \omega)(F(x) \rightarrow F(x+1)) \rightarrow (\forall x \in \omega)F(x)$ for arbitrary formulas F.

Adding the scheme (Δ_0–Collection) to the axioms in **BST** we obtain the axiom system **KP**, adding it to **BSTω** the system **KPω** These systems are thoroughly studied in Barwise [1975]. We list some of the most important properties of the system **KPω** without giving proofs. All proofs can be found in Barwise [1975]. Many of these properties are already provable from axioms which are weaker than $\mathbf{KP}\omega^r$. However, we do not have enough space to go into more details here.

We use class–terms of the form $\{x|\ A(x)\}$ freely though they are not regarded as terms of the language. The formula $z \in \{x|\ A(x)\}$ is an 'abbreviation' for $A(z)$.

3.3.1.1. Σ-Persistency Lemma. *Let F be a Σ-formula. Then $\mathbf{KP}^r$ proves $F^a \wedge a \subseteq b \rightarrow F^b$ and $F^a \rightarrow F$.*

3.3.1.2. Σ-Reflection Theorem. *For every Σ-formula F we have $\mathbf{KP}^r \vdash F \rightarrow (\exists a)F^a$.*

As a consequence of Σ–Reflection we obtain that in $\mathbf{KP}^r$ every Σ formula is provably equivalent to a Σ_1 formula.

3.3.1.3. Σ-Collection Theorem. *For every Σ-formula $F(x,y)$ we have*

$$\mathbf{KP}^r \vdash (\forall x \in a)(\exists y)F(x,y) \rightarrow (\exists z)(\forall x \in a)(\exists y \in z)F(x,y).$$

3.3.1.4. Σ-Replacement Theorem. *For every Σ-formula $A(\vec{x},y)$ we have*

$$\mathbf{KP}^r \vdash (\forall \vec{x} \in u)(\exists! y)A(\vec{x},y) \rightarrow (\exists f)[\mathit{Fun}(f) \wedge \mathit{dom}(f) = u \wedge (\forall \vec{x} \in u)A(\vec{x}, f(\vec{x}))].$$

3.3.1.5. Δ-Separation Theorem. *Let $A(x)$ be a Π and $B(x)$ be a Σ formula. Then*

$$\mathbf{KP}^r \vdash (\forall a)[(\forall x \in a)[A(x) \leftrightarrow B(x)] \rightarrow (\exists z)(\forall x \in a)[x \in z \leftrightarrow x \in a \wedge B(x)]].$$

There are many basic relations which are Δ_0–definable (cf. Barwise [1975] for details). The fact that α is an ordinal, e.g., can be expressed by $\mathit{Tran}(\alpha) \wedge (\forall x \in \alpha)\, \mathit{Tran}(x)$.

Similarly we can express by $\mathbf{On}(\alpha) \wedge (\exists x \in \alpha)[x \in \alpha] \wedge (\forall x \in \alpha)(\exists y \in \alpha)[x \in y]$ that α is a limit ordinal. The usual basic notations as Rel(r) (r is a relation), Fun(f) (f is a function) are Δ_0 definable. See Barwise [1975,pp.14-29] for a more complete list.

If $F(x_1, \ldots, x_n)$ is a Δ–formula of $\mathbf{KP}\boldsymbol{\omega}$ then we may introduce a new relation symbol R together with its defining axiom

$$(\forall x_1) \cdots (\forall x_n)[\mathsf{R}(x_1, \ldots, x_n) \leftrightarrow F(x_1, \ldots, x_n)].$$

Adding defined Δ relation–symbols to the language $\mathcal{L}(\in)$ will not alter the class of Σ– and Δ–formulas. If $F(x_1, \ldots, x_n, y)$ is a Σ–formula such that

$$\mathbf{KP} \vdash (\forall x_1) \cdots (\forall x_n)(\exists! y) F(x_1, \ldots, x_n, y)$$

then we may add an n–ary function symbol F to the language of **KP** together with its defining axiom

$$(\forall x_1) \cdots (\forall x_n)(\forall y)[\mathsf{F}(x_1, \ldots, x_y) = y \leftrightarrow F(x_1, \ldots, x_n, y)].$$

Extensions by definitions of Σ function–symbols will also not alter the class of Δ– and Σ–formulas of $\mathbf{KP}\boldsymbol{\omega}$. Details about 'Adding Defined Symbols to **KP**' can be found in Chapter I 5 of Barwise [1975].

One of the most important theorems of **KP** is the following Σ–Recursion Theorem.

3.3.1.6. Σ-Recursion Theorem. *Let* G *by an $n+2$–ary Σ function–symbol of* **KP**. *Then there is an $n+1$–ary Σ–function symbol* F *of* **KP** *such that*

$$\mathbf{KP} \vdash \mathsf{F}(\vec{a}, \alpha) = \mathsf{G}(\vec{a}, \alpha, \bigcup_{\xi<\alpha} \mathsf{F}(\vec{a}, \xi)).$$

The above Σ–Recursion Theorem is a special case of the more general Σ–Recursion Theorem as stated in Barwise [1975].

Because of the axiom of infinity we obtain $\mathbf{KP}\boldsymbol{\omega} \vdash (\exists \alpha)\mathsf{Lim}(\alpha)$ and thus a Δ_0–definition of ω as a point by $\mathsf{Lim}(\omega) \wedge (\forall x \in \omega)[\neg \mathsf{Lim}(x)]$.

Stronger than Σ–reflection is the Π_2–reflection scheme

(Π_2-Ref) $\quad F \rightarrow (\exists a)[a \neq \emptyset \wedge F^a]$ *for Π_2 formulas F.*

Observe that any model of $\mathbf{KP}\boldsymbol{\omega}$ in the constructible hierarchy already satisfies Π_2–reflection. To see that let $F \equiv (\forall y)(\exists y)A(x, y)$ be a Π_2–sentence and assume $\mathbf{L}_\alpha \models \mathbf{KP}\boldsymbol{\omega}$. For $a \in \mathbf{L}_\alpha$ there is a least $\beta_0 < \alpha$ such that $a \in \mathbf{L}_{\beta_0}$ and we define β_{n+1} to be the least ordinal such that $\mathbf{L}_\alpha \models (\forall x \in \mathbf{L}_{\beta_n})(\exists y \in \mathbf{L}_{\beta_{n+1}})A(x, y)$. The ordinal β_{n+1} exists by Σ–Reflection. This defines a sequence $\langle \beta_n; n \in \omega \rangle$ which is Σ–definable in $\mathbf{L}_\alpha$. Hence $\beta := \sup \{\beta_n | \ n \in \omega\} < \alpha$ and we have $\mathbf{L}_\alpha \models (\forall x \in \mathbf{L}_\beta)(\exists y \in \mathbf{L}_\beta)A(x, y)$.

Due to the Σ–Recursion Theorem we can prove the existence of the stages of inductively defined sets in $\mathbf{KP}\boldsymbol{\omega}$. Let S be an additional unary predicate symbol. We write $a \in S$ instead of $S(a)$. We obtain the stages of an inductive definition as stated in the following theorem.

3.3.1.7. Theorem. *Let $B(\vec{x}, y, S)$ be a Δ-formula of* **KPω**. *Then there is a Σ function–symbol I_B such that*

$$\mathbf{KP}\omega \vdash \mathsf{I}_B(\alpha, \vec{x}) = \{y \in x_1 |\ B(\vec{x}, y, \{z \in x_1 |\ (\exists \xi \in \alpha)[z \in \mathsf{I}_B(\xi, \vec{x})]\})\}.$$

Putting $\mathsf{G}(\vec{x}, y, S) := \{y \in x_1 |\ B(\vec{x}, y, S)\}$ we observe that G possesses a Σ definition and get the theorem immediately from the Σ–Recursion Theorem. □

A Δ–formula $B(\vec{x}, y, S)$ and a tuple $\vec{b}$ of sets induce an operator

$$\Gamma_{B,\vec{b}}\colon \mathsf{Pow}(b_1) \longrightarrow \mathsf{Pow}(b_1)$$
$$\Gamma_{B,\vec{b}}(S) := \{y \in b_1 |\ B(\vec{b}, y, S)\}$$

which depends on the parameter list $\vec{b}$. Again we say that S *occurs positively* in a formula $F(S)$ if the formula corresponding to $F(S)$ in the Tait–language (cf. Section 1.3) does not have occurrences of the form $t \notin S$. We sometimes denote this by $F(S^+)$. For S–positive formulas $B(\vec{x}, y, S)$ the associated operator $\Gamma_{B,\vec{b}}$ is monotonic, i.e., we have $S \subseteq T \Rightarrow \Gamma_{B,\vec{b}}(S) \subseteq \Gamma_{B,\vec{b}}(T)$. We say that a set is *closed* under an operator Γ if we have $\Gamma(S) \subseteq S$ for all S. For an monotonic operator $\Gamma\colon \mathsf{Pow}(b) \longrightarrow \mathsf{Pow}(b)$ we obtain its least fixed–point as the intersection of all Γ–closed subsets of b.

The fact that a class $T := \{x \in b_1 |\ A(x)\} \in \mathsf{Pow}(b_1)$ is $\Gamma_{B,\vec{b}}$ -closed can be expressed by the formula

$$Cl_B(\vec{b}, T) :\equiv (\forall y \in b_1)[B(\vec{b}, y, T) \to y \in T]. \tag{127}$$

Defining $\mathsf{I}_B(\vec{b}) := \{x \in b_1 |\ (\exists \xi)[x \in \mathsf{I}_B(\xi, \vec{b})]\}$ we obtain the following theorem.

3.3.1.8. Theorem. *Let $B(\vec{x}, y, S)$ by an S-positive Δ formula of* **KPω**. *Then*

$$\mathbf{KP}\omega \vdash Cl_B(\vec{b}, \mathsf{I}_B(\vec{b}))$$

and

$$\mathbf{KP}\omega \vdash Cl_B(\vec{b}, A) \ \to\ (\forall \xi)(\forall x)[x \in \mathsf{I}_B(\xi, \vec{b}) \to A(x)].$$

It follows from Theorem 3.3.1.8 that $\mathsf{I}_B(\vec{b})$ is the least fixed–point of the operator $\Gamma_{B,\vec{b}}$.

To prove the theorem pick a tuple $\vec{b}$, $c \in b_1$ and assume $B(\vec{b}, c, \mathsf{I}_B(\vec{b}))$. Since $B(\vec{x}, y, S)$ is S–positive the formula $B(\vec{b}, c, \mathsf{I}_B(\vec{b}))$ is still a Σ–formula and by Σ–Reflection we obtain a set d such that $B(\vec{b}, c, \{x \in b_1 |\ (\exists \xi \in d)(x \in \mathsf{I}_B(\xi, \vec{b}))\})$. Now define $\beta := \bigcup \{\xi \in d |\ \xi \in \mathbf{On}\}$ and $\alpha := \beta \cup \{\beta\}$. Then α is a set by Δ_0–Separation, Union and Pair such that $\{\xi \in d |\ \xi \in \mathbf{On}\} \subseteq \alpha$. By Σ–Persistency it follows $B(\vec{b}, c, \{x \in b_1 |\ (\exists \xi \in \alpha)(x \in \mathsf{I}_B(\xi, \vec{b}))\})$ and by Theorem 3.3.1.7 we obtain $c \in \mathsf{I}_B(\alpha, \vec{b})$, i.e., $c \in \mathsf{I}_B(\vec{b})$. This proves the first part of the theorem.

For the second part we show

$$Cl_B(\vec{b}, S) \ \Rightarrow\ \mathsf{I}_B(\xi, \vec{b}) \subseteq S \tag{i}$$

by induction on ξ. Assuming the hypothesis $Cl_B(\vec{b}, S)$ we obtain by induction hypothesis

$$\bigcup_{\zeta<\xi} \mathsf{I}_B(\zeta, \vec{b}) \subseteq S. \tag{ii}$$

The monotonicity of the induced operator therefore implies

$$(\forall x \in b_1)\Big[B(\vec{b}, x, \bigcup_{\zeta<\xi} \mathsf{I}_B(\zeta, \vec{b})) \;\Rightarrow\; B(\vec{b}, x, S)\Big]. \tag{iii}$$

By Theorem 3.3.1.7 and the hypothesis $Cl_B(\vec{b}, S)$ we get (i) from (iii). □

It follows from Barwise [1975] that every primitive recursive function has a Δ_1–definition in **KPω**. Therefore we may add function symbols for all primitive recursive functions to the language of **KPω**. So we may regard the second order language of $\mathbf{NT_2}$ as sublanguage of **KPω**. We mentioned that already in Section 1.3. Interpreting the language $\mathcal{L}^2_{\mathbb{N}}$ as a sublanguage of $\mathcal{L}(\in)$ augmented by Σ function symbols (call this language $\mathcal{L}(\in, \ldots)$) turns first order formulas with set parameters into Δ_0 formulas (which in turn are Δ for the theory **KPω**). It is obvious that the scheme of Mathematical Induction can be easily derived from the Foundation Scheme. All defining axioms for primitive recursive functions are provable in **KPω**. By Theorem 3.3.1.8 we may also interpret the additional constants I_B. So we get the following theorem.

3.3.1.9. Theorem. *The theory* $\mathbf{ID_1}$ *viewed as a theory in the language* $\mathcal{L}(\in, \ldots)$ *is a subtheory of* **KPω**.

3.3.2. The theory KPl

We are now introducing the theory **KPl** which axiomatizes a set universe which is the union of admissible universes. Therefore we augment the language of Set Theory by an additional constant Ad whose intended interpretation is that of an admissibility predicate. The defining axioms for Ad are

(Ad1) $(\forall u)[\mathsf{Ad}(u) \to \omega \in u \wedge \mathit{Tran}(u)]$

(Ad2) $(\forall x)(\forall y)[\mathsf{Ad}(x) \wedge \mathsf{Ad}(y) \to x \in y \vee x = y \vee y \in x]$

(Ad3) $(\forall x)[\mathsf{Ad}(x) \to (\text{Pair'})^x \wedge (\text{Union'})^x \wedge (\Delta_0\text{-Separation})^x \wedge (\Delta_0\text{-Collection})^x]$

3.3.2.1. Definition. The theory **KPl** is the system **BSTω** together with the axioms (Ad1) - (Ad3) and the axiom

(Lim) $(\forall x)(\exists u)[\mathsf{Ad}(u) \wedge x \in u]$.

Let $\lambda\xi.\,\Omega_\xi$ enumerate the class $\overline{\mathbf{Reg}} \cup \{0\}$ of admissible ordinals (augmented by 0) and their limits. Then the smallest constructible model of **KPl** is $\mathbf{L}_{\Omega_\omega}$, i.e., $\|\mathbf{KPl}\|_\infty = \Omega_\omega$. Since we have $\mathbf{KPl} \vdash \mathsf{Ad}(u) \to F^u$ for every sentence $F \in \mathbf{KP}\omega$ we obtain

3.3.2.2. Lemma. $\mathbf{KP}\omega^{\mathbf{r}} \vdash F \Rightarrow \mathbf{KPl}^{\mathbf{r}} \vdash \mathsf{Ad}(u) \to F^u$. *Also,* $\mathbf{KP}\omega \vdash F \Rightarrow \mathbf{KPl} \vdash \mathsf{Ad}(u) \to F^u$.

As in $\mathbf{KP}\omega$ the most important theorem in **KPl** will be the Σ–recursion theorem. But we do not have full Δ_0–collection in in **KPl**. Therefore we will obtain Σ–recursion only in a relativized version. Let $\mathbb{F}$ be a collection of new function symbols. Denote by $\mathbf{KPl}(\mathbb{F})$ the theory **KPl** formulated in the language $\mathcal{L}(\in, \mathbb{F})$ together with defining axioms for the function symbols in $\mathbb{F}$. Let $A(\vec{x}, y)$ be a Σ–formula such that $\mathbf{KPl}(\mathbb{F}) \vdash (\forall \vec{x})(\exists! y) A(\vec{x}, y)$ and $\mathbf{KPl}(\mathbb{F}) \vdash (\forall u)[\mathsf{Ad}(u) \to (\forall \vec{x} \in u)(\exists y \in u) A(\vec{x}, y)]$. Then we introduce a new function symbol G and its defining axiom

$$(D_G) \qquad (\forall \vec{x})(\forall y)[G(\vec{x}) = y \leftrightarrow A(\vec{x}, y)]$$

and call G a *relative Σ–function symbol* of $\mathbf{KPl}(\mathbb{F})$. Let $\mathbf{KPl}(\mathbb{G})$ be the $\mathrm{L}(\in, \mathbb{F} \cup \{G\})$-theory $\mathbf{KPl}(\mathbb{F}) + D_G$. By the common techniques we obtain that $\mathbf{KPl}(\mathbb{G})$ is an extension by definitions of $\mathbf{KPl}(\mathbb{F})$ in the following strong sense.

3.3.2.3. Lemma. *For every formula $F(\vec{x})$ in the language $\mathcal{L}(\in, \mathbb{G})$ there is a formula $F_0(\vec{x})$ in the language $\mathcal{L}(\in, \mathbb{F})$ such that*

$$\mathbf{KPl}(\mathbb{G}) \vdash (\forall \vec{x})[F(\vec{x}) \leftrightarrow F_0(\vec{x})].$$

If $F(\vec{x})$ is a Σ-formula of $\mathcal{L}(\in, \mathbb{G})$ then $F_0(\vec{x})$ is a Σ-formula of $\mathcal{L}(\in, \mathbb{F})$ such that

$$\mathbf{KPl}(\mathbb{F}) \vdash (\forall u)[\mathsf{Ad}(u) \to (\forall \vec{x} \in u)(F_0(\vec{x}) \leftrightarrow F_0(\vec{x})^u)].$$

If $F(\vec{x})$ is a Δ_0-formula of $\mathcal{L}(\in, \mathbb{G})$ then there is moreover also a Π-formula $F_1(\vec{x})$ such that

$$\mathbf{KPl}(\mathbb{G}) \vdash (\forall \vec{x})[F(\vec{x}) \leftrightarrow F_1(x)]$$
$$\mathbf{KPl}(\mathbb{F}) \vdash (\forall u)[\mathsf{Ad}(u) \to (\forall \vec{x} \in u)(F_0(\vec{x}) \leftrightarrow F_0(\vec{x})^u \leftrightarrow F_1(\vec{x})^u)].$$

Iterating Lemma 3.3.2.3 we see that we can identify the theory **KPl** with its closure under extensions by definitions of relative Σ–function symbols. The Σ–Recursion Theorem can now be modified in the following way.

3.3.2.4. Relativized Σ–Recursion Theorem. *Let G be an $n+2$–ary relative Σ–function symbol of $\mathbf{KPl}^{\mathbf{r}}$. Then there exists an $n+1$–ary relative Σ–function symbol F such that*

$$\mathbf{KPl}^{\mathbf{r}} \vdash \mathsf{F}(\vec{x}, \alpha) = \mathsf{G}(\vec{x}, \alpha, \bigcup_{\xi < \alpha} \mathsf{F}(\vec{x}, \xi))$$

To prove the theorem we follow Barwise [1975] as far as possible. Let $C(\vec{x}, \alpha, v, f)$ be the Δ_0–formula $(\alpha \notin \mathbf{On} \wedge f = \emptyset \wedge v = \emptyset) \vee (\alpha \in \mathbf{On} \wedge \mathit{Fun}(f) \wedge \mathit{dom}(f) = \alpha \wedge (\forall \xi \in \alpha)[f(\xi) = \mathsf{G}(\vec{x}, \xi, \bigcup_{\zeta < \xi} f(\zeta)) \wedge v = \mathsf{G}(\vec{x}, \alpha, \bigcup_{\zeta \in \alpha} f(\zeta))])$. We then show

$$\mathbf{KPl}^{\mathbf{r}} \vdash (\forall \vec{x})(\forall \alpha)(\exists! z)(\exists f) C(\vec{x}, \alpha, z, f) \qquad \text{(i)}$$

and

$$\mathbf{KPl^r} \vdash (\forall u)[\mathsf{Ad}(u) \to (\forall \vec{x} \in u)(\forall \alpha \in u)(\exists z \in u)(\exists f \in u)C(\vec{x}, \alpha, z, f)] \quad \text{(ii)}$$

and introduce a new relative Σ–function symbol F and its defining axiom

$$(\forall \vec{x})(\forall \alpha)(\forall z)[\mathsf{F}(\vec{x}, \alpha) = z \leftrightarrow (\exists f)C(\vec{x}, \alpha, z, f)]. \quad \text{(iii)}$$

To prove (i) and (ii) it suffices to show that $\mathbf{KPl^r}$ proves

$$C(\vec{x}, \alpha, z, f) \wedge C(\vec{x}, \alpha, z', f') \to z = z' \wedge f = f' \quad \text{(iv)}$$

and

$$\mathsf{Ad}(u) \wedge \vec{x} \in u \wedge \alpha \in u \to (\exists z \in u)(\exists f \in u)C(\vec{x}, \alpha, z, f). \quad \text{(v)}$$

We prove (iv) and (v) by induction on α. Since $C(\vec{x}, \alpha, z, f)$ is Δ_0 we can formalize this in $\mathbf{KPl^r}$ where we have (Δ_0-FOUND). The proof of (iv) is exactly the same as in Barwise [1975]. For the proof of (v) we also follow Barwise [1975] but use the additional observation that $\mathsf{G}(\vec{x}, \alpha, f) \in u$ whenever u is admissible and $\vec{x}, \alpha, f \in u$. We do not have full Σ–replacement in $\mathbf{KPl^r}$ but we may use its relativized version according to Theorem 3.3.1.4 and Lemma 3.3.2.2 whenever it is used in Barwise [1975]. Once we have the relative function symbol F we may proceed literally as in Barwise [1975]. □

There is a relativized version of Theorem 3.3.1.7. Let $B(\vec{x}, v, S)$ be a Δ_0-formula. Then we obtain a relative Σ–function symbol G such that $G(\vec{x}, v, s) = \{y \in x_1 \mid B(\vec{x}, y, s)\}$ and we apply the relativized Σ–recursion Theorem (Theorem 3.3.2.4) to obtain a relative Σ–function symbol I_B such that

$$\mathsf{I}_B(\alpha, \vec{x}) = \{y \in x_1 \mid B(\vec{x}, y, \{z \in x_1 \mid (\exists \xi \in \alpha)(z \in \mathsf{I}_B(\xi, \vec{x}))\})\}$$

is provable in $\mathbf{KPl^r}$. Defining $\mathsf{I}_B^u(\vec{x}) := \bigcup_{\xi \in u} \mathsf{I}_B(\xi, \vec{x})$ we obtain the following analogue of Theorem 3.3.1.8.

3.3.2.5. Theorem. (Inductive Definitions in $\mathbf{KPl}$ and $\mathbf{KPl^r}$) *Let $B(\vec{x}, y, S)$ be an S–positive Δ_0–formula. Then there is a relative Σ–function symbol I_B such that*

$$\mathsf{Ad}(u) \wedge \vec{x} \in u \to \mathit{Cl}_B(\vec{x}, \mathsf{I}_B^u(\vec{x}))$$

and

$$\mathsf{Ad}(u) \wedge \vec{x} \in u \wedge \mathit{Cl}_B(\vec{x}, A) \to \mathsf{I}_B^u(\vec{x}) \subseteq \{x \in u \mid A(x)\}$$

are provable in $\mathbf{KPl}$ *for arbitrary formulas $A(x)$ and in* $\mathbf{KPl^r}$ *for Δ_0–formulas $A(x)$.*

The proof is essentially that of Theorem 3.3.1.8. We fix an admissible u such that $\vec{x} \in u$ and repeat the proof substituting Σ–reflection by its relativized version. If we only consider Δ_0–formulas $A(x)$ then only (Δ_0-FOUND) is needed. Otherwise we need the full strength of $\mathbf{KPl}$. □

Observe that Theorem 3.3.2.5 does not immediately follow from Theorem 3.3.1.8 by Lemma 3.3.2.2 since the definitions of the function symbols I_B differ slightly.

3.3.2.6. Corollary. *Let $B(\vec{x}, y, S)$ be an S-positive Δ_0-formula. Then* $\mathbf{KPl^r}$ *proves that for every admissible set $\boldsymbol{a}$ containing the parameters $\vec{b}$ and every admissible set $\boldsymbol{u}$ containing $\boldsymbol{a}$ the class $\mathsf{I}_B^a(\vec{b})$ is a set in $\boldsymbol{u}$ which is the least fixed point of the monotone operator $\Gamma_{B,\vec{b}}$ induced by $B(\vec{b}, y, S)$.*

Proof. The function symbol I_B is a relative Σ–function symbol. and we have $\mathsf{I}_B^a(\vec{b}) \subseteq \boldsymbol{a} \in \boldsymbol{c}$. So by Lemma 3.3.2.3 and Δ_0–Separation relativized to $\boldsymbol{u}$ we obtain $\mathsf{I}_B^a(\vec{b}) \in \boldsymbol{u}$. That $\mathsf{I}_B^a(\vec{b})$ is the least fixed–point of $\Gamma_{B,\vec{b}}$ follows from Theorem 3.3.2.5. □

3.3.3. The quantifier theorem and axiom β

The most important tool for embedding subsystems of $\mathbf{NT_2}$ into subsystems of Set Theory is the Quantifier Theorem which we are going to present in this section. It is based on a theorem which is commonly known as Spector-Gandy Theorem.

First we fix the following notations:

$$\mathit{Prog}^a(\prec, T) \;:\Leftrightarrow\; (\forall x \in a)[(\forall y \in a)(y \prec x \to y \in T) \to x \in T]$$

where T may be a set or a class–term,

$$\mathit{TI}^a(\prec, T) \;:\Leftrightarrow\; \mathit{Prog}^a(\prec, T) \to (\forall x \in a)(x \in T),$$

$$\mathit{Wf}^a(\prec) \;:\Leftrightarrow\; (\forall x)\, \mathit{TI}^a(\prec, x),$$

$$\mathit{WO}^a(\prec) \;:\Leftrightarrow\; a = \mathsf{field}(\prec) \wedge \mathit{LO}(\prec) \wedge \mathit{Wf}^a(\prec)$$

where $\mathit{LO}(\prec)$ says that $\prec$ is a linear ordering. For a Δ_0-formula $A(\vec{x}, x, y)$ we define a relation $x \prec_{\vec{x}} y \;:\Leftrightarrow\; A(\vec{x}, x, y)$ and call it a Δ_0-relation.

3.3.3.1. Lemma. *Let $\prec_{\vec{x}}$ be a Δ_0-relation and define $A(b, \vec{x}, x, S)$ as $(\forall y \in b)[y \prec_{\vec{x}} x \to y \in S]$. Then*

$$\mathbf{KPl^r} \vdash \mathsf{Ad}(a) \wedge b, \vec{x} \in a \;\to\; (\mathit{Wf}^b(\prec_{\vec{x}}) \leftrightarrow (\forall y \in b)(\exists \xi \in a)(y \in \mathsf{I}_A^a(\xi, b, \vec{x}))).$$

To sketch the proof we work informally in $\mathbf{KPl^r}$. We have $\mathit{Prog}^b(\prec_{\vec{x}}, x) \leftrightarrow \mathit{Cl}_A(b, \vec{x}, x)$. Let $\boldsymbol{a}$ be an admissible set containing b and all members of $\vec{x}$ as elements. By Corollary 3.3.2.6 we get $d := \{y \in \boldsymbol{a} \mid (\exists \xi \in \boldsymbol{a})(y \in \mathsf{I}_A^a(\xi, b, \vec{x}))\}$ as a set provable in $\mathbf{KPl^r}$. We have to show

$$\mathit{Wf}^b(\prec_{\vec{x}}) \leftrightarrow b \subseteq d. \tag{i}$$

For the direction from left to right observe that $\mathit{Wf}^b(\prec_{\vec{x}})$ implies $\mathit{Prog}^b(\prec_{\vec{x}}, d) \to b \subseteq d$. But by Theorem 3.3.2.5 we have $\mathit{Cl}_A(b, \vec{x}, d)$ which is equivalent to $\mathit{Prog}^b(\prec_{\vec{x}}, d)$. For the opposite direction we use the second part of Theorem 3.3.2.5 to get $\mathit{Cl}_A(b, \vec{x}, x) \to d \subseteq x$ for any x. Together with $b \subseteq d$ this implies $\mathit{Prog}^b(\prec_{\vec{x}}, x) \to b \subseteq x$ which is $\mathit{Wf}^b(\prec_{\vec{x}})$. □

The following notion is motivated by the Π^1_1-completeness of well-foundedness.

3.3.3.2. Definition. Let **Ax** be a theory in the language of Set Theory. We say that a formula $A(\vec{x})$ is $\Pi^1(\mathbf{Ax})$ iff there is a Δ_0–relation $\prec_{\vec{x}} \subseteq \omega \times \omega$ such that

$$\mathbf{Ax} \vdash (\forall \vec{x})[A(\vec{x}) \leftrightarrow \mathsf{Wf}(\prec_{\vec{x}})].$$

Observe that the folklore fact that every Π^1_1–formula is equivalent to the well–foundedness of its associated tree of unsecured sequences can be proved in $\mathbf{BST^r}$. Therefore every Π^1_1–formula is a $\Pi^1(\mathbf{BST^r})$–formula such that in the defining formula for its corresponding Δ_0–relation all quantifier are restricted to ω.

3.3.3.3. Theorem. *For every $\Pi^1(\mathbf{KPl^r})$-formula $A(\vec{x})$ there is a Σ-formula $C(\vec{x})$ such that $\mathbf{KPl^r} \vdash \mathsf{Ad}(a) \wedge \vec{x} \in a \;\rightarrow\; (A(\vec{x}) \leftrightarrow C(\vec{x})^a)$.*

We indicate the proof. Let $\prec_{\vec{x}} \subseteq \omega \times \omega$ the Δ_0–relation such that $A(\vec{x}) \leftrightarrow \mathsf{Wf}(\prec_{\vec{x}})$. Choose an admissible $\boldsymbol{a}$ such that $\vec{x}, \omega \in \boldsymbol{a}$ and define $B(\vec{x}, x, S) \;:\Leftrightarrow\; (\forall y \in \omega)[y \prec_{\vec{x}} x \;\rightarrow\; y \in S]$. Then $A(\vec{x}) \leftrightarrow \mathsf{Wf}^{\omega}(\prec_{\vec{x}}) \leftrightarrow (\forall y \in \omega)(\exists \xi \in \boldsymbol{a})(y \in \mathsf{I}_B(\xi, \vec{x}))$ by Lemma 3.3.3.1. Since I_B is a relative Σ–function symbol, $\mathsf{Ad}(\boldsymbol{a})$ and $\vec{x} \in \boldsymbol{a}$ we get $A(\vec{x}) \leftrightarrow C(\vec{x})^{\boldsymbol{a}}$ for $C(\vec{x}) \;:\Leftrightarrow\; (\forall y \in \omega)(\exists \xi)(y \in \mathsf{I}_B(\xi, \vec{x}))$. □

3.3.3.4. Quantifier Theorem. *For every Π^1_1-formula $A(\vec{X}, \vec{x})$ in the language $\mathcal{L}^2_{\mathrm{N}}$ there is a Σ-formula $C(\vec{X}, \vec{x})$ so that $\mathsf{Ad}(a) \rightarrow (\forall \vec{X} \in a)(\forall \vec{x} \in \omega)[A(\vec{X}, \vec{x}) \leftrightarrow C(\vec{X}, \vec{x})^a]$ is provable in $\mathbf{KPl^r}$. For every Σ^1_2 formula $A(\vec{X}, \vec{x})$ there is a Σ-formula $C(\vec{X}, \vec{x})$ such that $A(\vec{X}, \vec{x}) \leftrightarrow C(\vec{X}, \vec{x})$ is provable in $\mathbf{KPl^r}$. Dually for every Π^1_2-formula $A(\vec{X}, \vec{x})$ there is a Π-formula $C(\vec{X}, \vec{x})$ such that $A(\vec{X}, \vec{x}) \leftrightarrow C(\vec{X}, \vec{x})$ is provable in $\mathbf{KPl^r}$.*

Recall that we regard $\mathcal{L}^2_{\mathrm{N}}$ as a sublanguage of $\mathcal{L}(\in, \ldots)$ by restricting all first order quantifiers to ω and all second order quantifiers to subsets of ω. As already remarked every Π^1_1–formula is a $\Pi^1(\mathbf{BST^r})$ and hence also a $\Pi^1(\mathbf{KPl^r})$–formula. Assume $\mathsf{Ad}(\boldsymbol{a})$ and $\vec{X} \in \boldsymbol{a}$. By Theorem 3.3.3.3 we obtain a Σ-formula $C(\vec{X}, x)$ such that $A(\vec{X}, \vec{x}) \leftrightarrow C(\vec{X}, \vec{x})^{\boldsymbol{a}}$. For the second claim assume $A(\vec{X}, \vec{x}) \;\Leftrightarrow\; (\exists Y)[Y \subseteq \omega \wedge B(\vec{X}, Y, \vec{x})]$ for a Π^1_1–formula $B(\vec{X}, Y, \vec{x})$. By the first claim there is a Σ–formula $C'(\vec{X}, Y, \vec{x})$ such that for $\mathsf{Ad}(\boldsymbol{a})$ and $\vec{X}, Y \in \boldsymbol{a}$ we get $B(\vec{X}, Y, \vec{x}) \leftrightarrow C'(\vec{X}, Y, \vec{x})^{\boldsymbol{a}}$ provable in $\mathbf{KPl^r}$. By axiom (Lim) we always find such an $\boldsymbol{a}$ and thus obtain the claim by defining $C(\vec{X}, \vec{x}) \;:\Leftrightarrow\; (\exists Z)[\mathsf{Ad}(Z) \wedge \vec{X} \in Z \wedge (\exists Y \in Z)(Y \subseteq \omega \wedge C'(\vec{X}, Y, \vec{x})^Z]$. The last claim follows from the second by taking negations. □

As a corollary of Theorem 3.3.3.4 we get Π^1_1–comprehension in $\mathbf{KPl^r}$.

3.3.3.5. Π^1_1–Comprehension Theorem. *Let $H(\vec{X}, \vec{x}, x)$ be a Π^1_1-formula. Then $\mathbf{KPl^r}$ proves $(\forall \vec{X})(\forall \vec{x} \in \omega)(\exists y)[y = \{z \in \omega \mid H(\vec{X}, \vec{x}, z)\}]$.*

Proof. By Theorem 3.3.3.4 there is a Σ–formula $C(\vec{X}, \vec{x}, x)$ such that $\mathbf{KPl^r} \vdash \mathsf{Ad}(a) \wedge \vec{X} \in a \wedge \vec{x}, x \in \omega \rightarrow (H(\vec{X}, \vec{x}, x) \leftrightarrow C(\vec{X}, \vec{x}, x)^a)$. Using axiom

(Lim) we find such an admissible $\boldsymbol{a}$ and another admissible $\boldsymbol{u}$ such that $\boldsymbol{a} \in \boldsymbol{u}$. Now we apply Δ_0–Separation relativized to $\boldsymbol{u}$ to obtain $d := \{x \in \omega \mid C(\vec{X}, \vec{x}, x)^{\boldsymbol{a}}\}$ as a set. Obviously d is a witness for y in the claim. □

The Quantifier Theorem is a good example for reducing the complexity of 'analytical' formulas by translating them into the language of Set Theory. (The reason for this reduction is the presence of the axiom (Found) in Set Theory). Another example is Axiom β which turns the Π_1 notion of well–foundedness into a Δ–notion. We will show that Axiom β is provable in $\mathbf{KPl^r}$. This needs some notations. Define

$$\mathsf{Found}(a, r) \;:\Leftrightarrow\; (\forall x)[x \subseteq a \wedge x \neq \emptyset \rightarrow (\exists z \in x)(\forall y \in x)((y, z) \notin r)] \quad (128)$$

expressing that r is a well–founded relation on the set a. To increase readability we use the infix notation for relations, i.e., $x\ r\ y$ instead of $(x, y) \in r$. Recall that $\mathsf{Found}(a, r)$ entails $(\forall x \in a)(x \not r\, x)$.

$$(Ax\beta) \quad (\forall x)(\forall r)[\mathsf{Found}(x, r) \;\rightarrow\; (\exists f)(\exists \alpha)(\mathit{Fun}(f) \wedge \mathit{dom}(f) = x \wedge \mathit{rng}(f) = \alpha \wedge (\forall u \in x)(\forall v \in x)(u\ r\ v \rightarrow f(u) < f(v)))].$$

3.3.3.6. Theorem. *Axiom β is a theorem of* $\mathbf{KPl^r}$.

Proof. We start with the obvious observation

$$\mathbf{KPl^r} \vdash \mathsf{Found}(b, r) \rightarrow \mathsf{Wf}^b(r). \quad \text{(i)}$$

Then we obtain from Lemma 3.3.3.1

$$\mathbf{KPl^r} \vdash \mathsf{Ad}(a) \wedge b, r \in a \rightarrow [\mathsf{Found}(b, r) \leftrightarrow (\forall y \in b)(\exists \xi \in a)(y \in \mathsf{I}^a_B(\xi, b, r))] \quad \text{(ii)}$$

for $B(x, S) \;:\Leftrightarrow\; (\forall y)[y\ r\ x \rightarrow y \in S]$. Let b and r be given. By (Lim) we choose an admissible $\boldsymbol{a}$ such that $b, r \in \boldsymbol{a}$. For $x \in b$ there is by (ii) a $\xi \in \boldsymbol{a}$ such that $x \in \mathsf{I}_B(\xi, b, r)$ and we define $f(x) := \min\{\xi \in a \mid x \in \mathsf{I}^a_B(\xi, b, r)\}$. This defines a function f with $\mathit{dom}(f) = b$. Defining $\alpha := \sup\{f(x) + 1 \mid x \in b\}$ we see that f is an $(r, \in)$–homomorphisms from b onto α. □

Observe that the function f whose existence is required by $(Ax\beta)$ is uniquely determined and has for transitive well–founded r the property that $f(x) = \{f(y) \mid y\ r\ x\}$. We often denote this function by otyp_r and define $\mathsf{otyp}(r) := \mathit{rng}(\mathsf{otyp}_r)$. As a corollary of the proof of Theorem 3.3.3.6 we obtain

$$\mathbf{KPl^r} \vdash \mathsf{Ad}(a) \wedge \mathsf{Found}(b, r) \wedge b, r \in a \rightarrow \mathsf{otyp}_r \in a. \quad (129)$$

As soon as we have $(Ax\beta)$ we see that well–foundedness for sets is in $\mathbf{KPl}$ extensible to well-foundedness for classes.

3.3.3.7. Theorem. *Let $\prec_{\vec{x}}$ be a Δ_0-relation. Then* $\mathbf{KPl} \vdash \mathsf{Wf}^a(\prec_{\vec{x}}) \rightarrow \mathsf{TI}^a(\prec_{\vec{x}}, T)$ *for any class–term T. For Δ_0-classes T this is already provable in* $\mathbf{KPl^r}$.

Proof. Assume $\mathsf{Wf}^a(\prec_{\vec{x}})$. We have to show that from the hypothesis $\mathsf{Prog}^a(\prec_{\vec{x}}, T)$ we obtain $a \subseteq T$. Define $r := \{(x, y) \mid x \in a \wedge y \in a \wedge x \prec_{\vec{x}} y\}$. Then r is a set by Δ_0–separation and we obtain $\mathit{Found}(a,r)$ from $\mathit{Wf}^a(\prec_{\vec{x}})$. By $(Ax\beta)$ there is an ordinal α and an order preserving mapping $f\colon a \overset{\text{onto}}{\longrightarrow} \alpha$. Put $B(\xi) \;:\Leftrightarrow\; \xi < \alpha \to (\forall y \in a)[f(y) = \xi \to y \in T]$. Since we have the scheme (FOUND) we have induction on ordinals, i.e., especially

$$(\forall\xi)[(\forall\zeta < \xi)B(\zeta) \to B(\xi)] \;\;\to\;\; (\forall\xi)B(\xi). \tag{i}$$

But from $(\forall\xi)B(\xi)$ we immediately obtain $(\forall y \in a)[y \in T]$, i.e., $a \subseteq T$. So assume $(\forall\zeta < \xi)B(\zeta)$ for $\xi < \alpha$. We have to show $B(\xi)$. Let $x \in a$ such that $\xi = f(x)$. For all $y \in a$ such that $y\ r\ x$ we get $f(y) < \xi$ and thus $B(f(y))$. Hence $(\forall y \in a)(y\ r\ x \to y \in T)$ which by $\mathit{Prog}^a(\prec_{\vec{x}}, T)$ implies $x \in T$. Hence $B(\xi)$ and we are done. If T is a Δ_0–class then (Δ_0–FOUND) suffices to have (i) which allows to formalize the proof in **KPl^r**. □

3.3.3.8. Corollary. *The schemes* (Π^1_0-BI) *as well as* (Bi) *are theorems of* **KPl**.

Proof. Let $\prec_{\vec{X},\vec{x}}$ be arithmetically definable in $\mathcal{L}^2_{\mathbb{N}}$. Then its defining formula is Δ_0 in the sense of $\mathcal{L}(\in,\ldots)$. The formula $\mathit{Wf}(\prec_{\vec{X},\vec{x}})$ as $\mathcal{L}(\in,\ldots)$–formula becomes $\mathit{Wf}^\omega(\prec_{\vec{X},\vec{x}})$. Hence $\mathbf{KPl} \vdash \mathit{Wf}^\omega(\prec_{\vec{X},\vec{x}}) \to \mathit{TI}^\omega(\prec_{\vec{X},\vec{x}}, T)$ for every class term T by Theorem 3.3.3.7 which implies (Π^0_1-BI). As a special case we obtain the provability of $\mathit{Wf}^\omega(\prec_X) \to \mathit{TI}^\omega(\prec_X, T)$ which entails (Bi). □

Summarizing the work of this section we have the following results.

3.3.3.9. Theorem. *The theory* $(\mathbf{\Pi^1_1}\text{-}\mathbf{CA})_0$ *is a subtheory of* **KPl^r**. *The theory* $(\mathbf{\Pi^1_1}\text{-}\mathbf{CA})$ *is a subtheory of W-*__KPl__ *and* $(\mathbf{\Pi^1_1}\text{-}\mathbf{CA})$ + (Bi) *a subtheory of* **KPl**.

3.3.4. The theories KPi and KPβ

Much stronger than the theory **KPl** is the theory **KPi** which we are going to study in this section.

3.3.4.1. Definition. The theory **KPi** is the union of the axioms in **KP**$\boldsymbol{\omega}$ and **KPl**.

The axioms in **KPi** describe a universe which is admissible and simultaneously the union of admissible universes. So we obtain $\|\mathbf{KPi}\|_\infty = I$ where I denotes the first recursively inaccessible ordinal, i.e., the least ordinal which is admissible and the limit of admissible ordinals. Most of the properties of the theory **KPi** follow from the previous sections.

3.3.4.2. Theorem. (Δ^1_2-comprehension in **KPi^r**) *Suppose that* $A(\vec{X},\vec{x},y)$ *is a* Π^1_2- *and* $B(\vec{X},\vec{x},y)$ *a* Σ^1_2*–formula. Then* $(\forall y \in \omega)(A(\vec{X},\vec{x},y) \leftrightarrow B(\vec{X},\vec{x},y)) \to (\exists z)(z = \{y \in \omega \mid A(\vec{X},\vec{x},y)\})$ *is provable in* **KPi^r**.

Proof. From the hypothesis $(\forall y \in \omega)(A(\vec{X}, \vec{x}, y) \leftrightarrow B(\vec{X}, \vec{x}, y))$ and the Quantifier Theorem 3.3.3 is follows that $A(\vec{X}, \vec{x}, y)$ is a Δ–formula of $\mathbf{KPi^r}$. Therefore $z := \{y \in \omega \mid A(\vec{X}, \vec{x}, y)\}$ is a set by Δ–comprehension. □

As an immediate consequence of Theorem 3.3.4.2 we obtain the following theorem.

3.3.4.3. Theorem. *The theory* $(\mathbf{\Delta^1_2}\text{-}\mathbf{CA})_0$ *is a subtheory of* $\mathbf{KPi^r}$, *the theory* $(\mathbf{\Delta^1_2}\text{-}\mathbf{CA})$ *is a subtheory of W-*$\mathbf{KPi}$ *and the theories* $(\mathbf{\Delta^1_2}\text{-}\mathbf{CA})+(\mathsf{Bi})$ *and* $(\mathbf{\Delta^1_2}\text{-}\mathbf{CA})+(\mathsf{Bi})$ *are subtheories of* $\mathbf{KPi}$.

We have seen in the previous section that $\mathbf{KPl}$ proves $(Ax\beta)$. We will now show that $\mathbf{KP\omega}$ cannot prove $(Ax\beta)$. It will moreover become clear that augmenting $\mathbf{KP\omega}$ by $(Ax\beta)$ will give a theory of the strength of $\mathbf{KPi}$.

3.3.4.4. Definition. Let $\mathbf{KP\beta}$ be the theory $\mathbf{KP\omega} + (Ax\beta)$.

Since we have already shown that $\mathbf{KPl^r} \vdash (Ax\beta)$ we get immediately

3.3.4.5. Theorem. *The theory* $\mathbf{KP\beta^r}$ *is a subtheory of* $\mathbf{KPi^r}$. *The theory W-*$\mathbf{KP\beta}$ *is a subtheory of W-*$\mathbf{KPi}$ *and* $\mathbf{KP\beta}$ *is a subtheory of* $\mathbf{KPi}$.

We will show that conversely $\mathbf{KP\beta}$ is of the same proof theoretical strength as $\mathbf{KPi}$. That, however, does not mean that both theories coincide. Though $\|\mathbf{KP\beta}\|_\infty = I = \|\mathbf{KPi}\|_\infty$ there are ordinals α such that $\mathbf{L}_\alpha \models \mathbf{KP\beta}$ but $\mathbf{L}_\alpha \not\models \mathbf{KPi}$ (e.g., $\alpha = \aleph_1^{\mathbf{L}+}$ cf. Platek [1966] 5.11). We will first check that $\mathbf{KP\beta}$ allows the embedding of the same $\mathcal{L}^2_{\mathbb{N}}$–theories as $\mathbf{KPi}$. Therefore we need an equivalent for the Quantifier Theorems of $\mathbf{KPl}$.

3.3.4.6. Lemma. *For every* $\Pi^1(\mathbf{KP\beta^r})$*-formula* $A(\vec{x})$ *there is a* Σ*-formula* $B(\vec{x})$ *such that* $\mathbf{KP\beta^r} \vdash (\forall \vec{x})(A(\vec{x}) \leftrightarrow B(\vec{x}))$. *For every* Σ^1_2*-formula* $A(\vec{X}, \vec{x})$ *there is a* Σ*-formula* $B(\vec{X}, \vec{x})$ *and dually for every* Π^1_2*-formula, a* Π*-formula such that* $\mathbf{KP\beta^r} \vdash (\forall \vec{X})(\forall \vec{x} \in \omega)[\vec{X} \subseteq \omega \to (A(\vec{X}, \vec{x}) \leftrightarrow B(\vec{X}, \vec{x}))]$.

Proof. Let $\prec_{\vec{X},\vec{x}} \subseteq \omega \times \omega$ be a Δ_0–relation such that $\mathbf{KP\beta^r} \vdash \mathit{Wf}(\prec_{\vec{X},\vec{x}}) \leftrightarrow A(\vec{X}, \vec{x})$. Then $r := \{(x, y) \mid x \in \omega \wedge y \in \omega \wedge x \prec_{\vec{X},\vec{x}} y\}$ is a set and we have $\mathit{Wf}(\prec_{\vec{X},\vec{x}}) \leftrightarrow \mathit{Found}(\omega, r)$. By $(Ax\beta)$ we obtain

$$\mathit{Found}(\omega, r) \leftrightarrow (\exists f)(\exists \alpha)[f\colon \mathsf{field}(r) \xrightarrow{\text{onto}} \alpha \text{ order preserving}].$$

The right hand side, however, is Σ_1.

For the second claim assume $A(\vec{X}, \vec{x}) \Leftrightarrow (\exists Y)[Y \subseteq \omega \wedge A_0(\vec{X}, Y, \vec{x})]$ where $A_0(\vec{X}, Y, \vec{x})$ is Π^1_1. The 'unsecured sequences' argument is of course formalizable in $\mathbf{KP\beta^r}$ (it needs only arithmetical comprehension which is covered by Δ_0–separation) and we therefore get $A_0(\vec{X}, Y, \vec{x})$ as a $\Pi^1(\mathbf{KP\beta^r})$–formula. By the first claim we therefore have a Σ–formula, say $C_0(\vec{X}, Y, \vec{x})$, which is in $\mathbf{KP\beta^r}$ equivalent to $A_0(\vec{X}, Y, \vec{x})$. Defining $C(\vec{X}, \vec{x}) :\Leftrightarrow (\exists Y)[Y \subseteq \omega \wedge C_0(\vec{X}, Y, \vec{x})]$ we get a Σ–formula which is in $\mathbf{KP}\boldsymbol{\beta}^{\mathbf{r}}$ equivalent to $A(\vec{X},\vec{x})$. The dual claim follows by taking negations. □

Now having also a Quantifier Theorem in $\mathbf{KP}\boldsymbol{\beta}^{\mathbf{r}}$ we get with the same proof the equivalent of Theorem 3.3.4.2.

3.3.4.7. Theorem. (Δ^1_2–Comprehension in $\mathbf{KP}\boldsymbol{\beta}^{\mathbf{r}}$) *Let $A(\vec{X},\vec{x},y)$ be a Π^1_2- and $B(\vec{X},\vec{x},y)$ a Σ^1_2-formula. Then*

$$(\forall y\in\omega)(A(\vec{X},\vec{x},y)\leftrightarrow B(\vec{X},\vec{x},y)) \;\rightarrow\; (\exists z)(z=\{y\in\omega \mid A(\vec{X},\vec{x},y)\})$$

is provable in $\mathbf{KP}\boldsymbol{\beta}^{\mathbf{r}}$.

As in Theorem 3.3.3.7 we get from $(Ax\beta)$ the well–foundedness for classes for every Δ_0–relation which is well–founded for sets. I.e., we have with the same proof

3.3.4.8. Theorem. *Let $\prec_{\vec{x}}$ be a Δ_0-relation. Then $\mathbf{KP}\boldsymbol{\beta} \vdash \mathit{Wf}^a(\prec_{\vec{x}}) \rightarrow \mathit{TI}(\prec,T)$ for any class–term T. For Δ_0-classes T this is already provable in* $\mathbf{KP}\boldsymbol{\beta}^{\mathbf{r}}$.

Summing up we obtain

3.3.4.9. Theorem. *The theory $(\boldsymbol{\Delta}^1_2\text{-}\mathbf{CA})_0$ is a subtheory of $\mathbf{KP}\boldsymbol{\beta}^{\mathbf{r}}$, the theory $(\boldsymbol{\Delta}^1_2\text{-}\mathbf{CA})$ is a subtheory of W-$\mathbf{KP}\boldsymbol{\beta}$ and the theories $(\boldsymbol{\Delta}^1_2\text{-}\mathbf{CA})+(\mathsf{Bi})$ and $(\boldsymbol{\Delta}^1_2\text{-}\mathbf{CA})+(\mathsf{Bi})$ are subtheories of* $\mathbf{KP}\boldsymbol{\beta}$.

It follows from Theorem 3.3.4.9 that for every ordinal α for which we have $\mathbf{L}_\alpha \models \mathbf{KP}\boldsymbol{\beta}$ we also have $\mathbf{L}_\alpha \cap \mathsf{Pow}(\omega) \models (\boldsymbol{\Delta}^1_2\text{-}\mathbf{CA})+(\mathsf{Bi})$. But again the opposite claim is not true. S. Simpson in a private communication to G. Jäger constructed the following counterexample. Let α be the least admissible ordinal such that $\mathbf{L}_\alpha \models (\exists\xi)[\xi \text{ is uncountable}]$. Then $\alpha = (\aleph_1^{\mathbf{L}_\alpha})^+$. While $\mathbf{L}_\alpha \models \mathbf{NT_2}$ and therefore also $\mathbf{L}_\alpha \models (\boldsymbol{\Delta}^1_2\text{-}\mathbf{CA})+(\mathsf{Bi})$ we have $\mathbf{L}_\alpha \not\models (Ax\beta)$.

3.3.5. Theories of iterated admissibility

There is a tremendous gap between **KPl** –even $\mathbf{KPi}^{\mathbf{r}}$ – and **KPi**. Within this gap there is a whole zoo of theories. M. Rathjen in his thesis [1988] studied these theories exhaustively. Unfortunately his thesis never appeared in English. But there is not enough space to introduce all these theories. As examples, however, we will introduce some theories for iterated admissibility, i.e., theories which axiomatize universes containing a certain number of admissibles. As in Barwise [1975] we denote by τ_ξ the ξth admissible, i.e., $\lambda\xi\,.\,\tau_\xi$ enumerates the class of admissible ordinals including ω. So we have $\tau_0=\omega$, $\tau_1=\omega_1^{\mathrm{CK}}=\Omega_1$, $\tau_\omega=\Omega_{\omega+1}$ etc.

3.3.5.1. Definition. Let

$$\begin{aligned}\mathit{ItAd}(\alpha,f) \;:\Leftrightarrow\;& \mathit{Fun}(f)\wedge\alpha\in\mathbf{On}\wedge \mathit{dom}(f)=\alpha\wedge(\forall\xi<\alpha)[\mathsf{Ad}(f(\xi))\\ &\wedge(\forall\zeta<\xi)(f(\zeta)\in f(\xi))\wedge(\forall x\in f(\xi))(\mathsf{Ad}(x)\rightarrow(\exists\zeta<\xi)(x=f(\zeta)))]\end{aligned}$$

express that there are at least α many admissibles. We define

$$\mathbf{KPl}_\nu := \mathbf{KPl} + (\forall \xi < \nu)(\exists f)\mathit{ItAd}(\xi, f).$$

If $\prec$ is a well-ordering we put

$$\mathbf{KPl}_\prec := \mathbf{KPl} + \mathit{WO}(\prec) + (\forall \xi \in \mathsf{otyp}(\prec))(\exists f)\mathit{ItAd}(\xi, f).$$

The theory

$$\mathbf{Aut\text{–}KPl} := \mathbf{KPl} + (\forall \alpha)(\exists f)\mathit{ItAd}(\alpha, f)$$

axiomatizes a universe which contains as many admissibles as ordinals. Observe that we can define $x = \mathbf{L}_{\Omega_1} \;:\Leftrightarrow\; \mathsf{Ad}(x) \wedge (\forall y \in x)\neg\mathsf{Ad}(y)$. The formula $x \in \mathbf{L}_{\Omega_1}$ is a Δ–formula since $x \in \mathbf{L}_{\Omega_1} \;\Leftrightarrow\; (\forall y)[\mathsf{Ad}(y) \rightarrow x \in y] \;\Leftrightarrow\; (\exists y)[y = \mathbf{L}_{\Omega_1} \wedge x \in y]$. We define the theory

$$\mathbf{KPl}^* := \mathbf{KPl} + (\forall \xi \in \mathbf{L}_{\Omega_1})(\exists f)\mathit{ItAd}(\xi, f).$$

It follows by (Δ_0-FOUND) that the enumerating function of the admissibles is uniquely determined, i.e., we have

$$\mathbf{KPl}^r \vdash \mathit{ItAd}(\alpha, f) \wedge \mathit{ItAd}(\alpha, g) \rightarrow f = g. \tag{130}$$

As a consequence of (130) we obtain $\mathbf{KPl}^* \vdash (\forall \alpha \in \mathbf{L}_{\Omega_1})(\exists! f)\mathit{ItAd}(\alpha, f)$. The formula

$$\begin{aligned} \mathit{RItAd}(\alpha, f, a) \;:\Leftrightarrow\; & \alpha \in \mathbf{On} \wedge \mathsf{Ad}(a) \wedge \mathit{Fun}(f) \wedge \mathit{dom}(f) = \alpha \wedge f(0) = a \wedge \\ & (\forall \xi < \alpha)[\mathsf{Ad}(f(\xi)) \wedge (\forall \zeta < \xi)(f(\zeta) \in f(\xi)) \\ & \wedge (\forall x \in f(\xi))(\mathsf{Ad}(x) \wedge a \in x \rightarrow (\exists \zeta < \xi)(x = f(\zeta))))] \end{aligned}$$

expresses α–fold iteration of admissibility relative to a start–admissible a. It is easy to show that

$$\mathbf{Aut\text{-}KPl}^r \vdash \mathsf{Ad}(a) \rightarrow (\forall \alpha)(\exists! f)\mathit{RItAd}(\alpha, f, a). \tag{131}$$

The uniqueness follows again by (Δ_0-FOUND). To prove the existence of the function f we choose by axiom (Lim) two admissible sets $\boldsymbol{b}$ and $\boldsymbol{u}$ such that $\alpha, a \in \boldsymbol{b} \in \boldsymbol{u}$. By Δ_0–separation relativized to $\boldsymbol{u}$ we obtain $\beta := \boldsymbol{b} \cap \mathbf{On} \in \boldsymbol{u}$. Then $\beta \notin \boldsymbol{b}$ and thus $\beta \notin a$. There is a function g such that $\mathit{ItAd}(\beta + \beta, g)$. By ($\Delta_0$-FOUND) we obtain $(\forall \xi < \beta + \beta)(\xi \in g(\xi + 1))$. Because of $\beta \in g(\beta + 1)$, $\beta \notin a$ and axiom (Ad2) we get $a \in g(\beta + 1)$. Therefore there is a $\rho \leq \beta$ such that $a = g(\rho)$. We may now define $f(\xi) := g(\rho + \xi)$ for $\xi < \alpha < \beta$ and easily check $\mathit{RItAd}(\alpha, f, a)$.

We will now show that the theories for iterated inductive definitions can be embedded into the theories of iterated admissibility. In the following lemma we use the same notational conventions as in Section 3.2.3. So $x \in X^y$ stands for $(x, y) \in X$ and for a relation r we write $x \in X^{ry}$ for $(\exists z)(z \; r \; y \wedge x \in X^z)$. The capital letters serve only to improve readability and to emphasize the close connection to Section 3.2.3. Their meaning is that of ordinary variables for sets.

3.3.5.2. Lemma. *Let $A(S,T,x,y,\vec{z})$ be an S–positive Δ_0-formula, $r \subseteq \omega \times \omega$ and define $Cl^{\omega}_{A}(S,T,y,\vec{z}) \;:\Leftrightarrow\; (\forall x \in \omega)[A(S,T,x,y,\vec{z}) \;\rightarrow\; x \in S]$ and*

$$IT^{\omega}_{A}(r,X,\vec{z}) \;:\Leftrightarrow\; r \subseteq \omega \times \omega \wedge (\forall y \in \omega) Cl^{\omega}_{A}(X^{y}, X^{ry}, y, \vec{z}) \\ \wedge (\forall Y)(Y \subseteq \omega \rightarrow Cl^{\omega}_{A}(Y, X^{ry}, y, \vec{z}) \rightarrow X^{y} \subseteq Y).$$

Then we obtain

$$\mathbf{KPl}_{\prec} \vdash (\exists X) IT^{\omega}_{A}(\prec, X, \vec{z}) \tag{132}$$

$$\mathbf{KPl}^{*} \vdash WO^{\omega}(r) \wedge r, \vec{z} \in \mathbf{L}_{\Omega_1} \rightarrow (\exists X) IT^{\omega}_{A}(r, X, \vec{z}) \tag{133}$$

and

$$\mathbf{Aut\text{–}KPl} \vdash WO^{\omega}(r) \rightarrow (\exists X) IT^{\omega}_{A}(r, X, \vec{z}). \tag{134}$$

The proof of all three claims is essentially the same. We sketch the most complicated case of (134). Assume $WO^{\omega}(r)$ and chose by (Lim) an admissible $\boldsymbol{a}$ such that $r, \vec{z} \in \boldsymbol{a}$. Then $\mathsf{field}(r) \cap \omega \in \boldsymbol{a}$ and we get by (129) $\mathsf{otyp}_r \in \boldsymbol{a}$. Hence also $\alpha := \mathsf{otyp}(r) \in \boldsymbol{a}$. By (131) there is function f such that $RltAd(\alpha, f, \boldsymbol{a})$. By (Lim) we obtain an admissible set $\boldsymbol{u}$ such that $\boldsymbol{a}, f, \alpha \in \boldsymbol{u}$. For $\xi < \alpha$ let $\boldsymbol{u}_{\xi} := f(\xi)$, so $\boldsymbol{u}_0 = \boldsymbol{a}$. By Σ–Recursion relativized to $\boldsymbol{u}$ we obtain functions g and h_y for $y \in \omega$ satisfying

$$h_y(\eta) = \begin{cases} \{x \in \omega \mid A(\bigcup\{h_y(\zeta) \mid \zeta < \eta\}, \emptyset, x, y, \vec{z})\} \text{ and } dom(h_y) = \mathbf{On} \cap \boldsymbol{a} \\ \hfill \text{if } y \notin \mathsf{field}(r) \\ \{x \in \omega \mid A(\bigcup\{h_y(\zeta) \mid \zeta < \eta\}, \bigcup\{g(\zeta) \mid \zeta < \mathsf{otyp}_r(y)\}, x, y, \vec{z})\} \\ \quad \text{and } dom(h_y) = \mathbf{On} \cap \boldsymbol{u}_{\mathsf{otyp}_r(y)} \text{ and } g(\mathsf{otyp}_r(y)) = \bigcup rng(h_y) \\ \hfill \text{if } y \in \mathsf{field}(r). \end{cases} \tag{i}$$

Put

$$S := \{(x,y) \mid y \in \mathsf{field}(r) \wedge x \in g(\mathsf{otyp}_r(y))\} \\ \cup \{(x,y) \mid y \in \omega \wedge y \notin \mathsf{field}(r) \wedge x \in \bigcup rng(h_y)\}.$$

By construction we have $S, h_y, g, \{(h_y, y) \mid y \in \omega\} \in \boldsymbol{u}$. We show

$$Cl^{\omega}_{A}(S^{y}, S^{ry}, y, \vec{z}) \tag{ii}$$

and

$$(\forall X)[X \subseteq \omega \rightarrow Cl^{\omega}_{A}(X, S^{ry}, y, \vec{z}) \rightarrow S^{y} \subseteq X]. \tag{iii}$$

First assume $y \notin \mathsf{field}(r)$. Then $S^{ry} = \emptyset$. The function h_y is Σ–definable in the admissible set $\boldsymbol{a}$. We have $S^{y} = \{x \mid (\exists \zeta \in \boldsymbol{a})(x \in h_y(\zeta))\}$. Thus S^{y} corresponds to the class $\mathsf{I}_A(\vec{z})$ of Section 3.3.1 and we show (ii) and (ii) as in the proof of Theorem 3.3.1.8.

Now assume $y \in \mathsf{field}(r)$. Let $\xi := \mathsf{otyp}_r(y)$. We show

$$g \restriction \xi \in \boldsymbol{u}_{\xi} \tag{iv}$$

by induction on ξ. From the induction hypothesis we obtain $g\restriction\zeta \in \boldsymbol{u}_\zeta \in \boldsymbol{u}_\xi$ for all $\zeta < \xi$. The function $\xi > \zeta \mapsto g\restriction\zeta$ is thus Σ-definable in $\boldsymbol{u}_\xi$ and because of $\xi \in \boldsymbol{a} \in \boldsymbol{u}_\xi$ we get $\{(\zeta, g\restriction\zeta) \mid \zeta < \xi\} \in \boldsymbol{u}_\xi$ by Σ–replacement relativized to $\boldsymbol{u}_\xi$. This proves (iv) in case that $\xi \in \mathsf{Lim}$. If $\xi = \zeta + 1$ then $\zeta = \mathsf{otyp}_r(z)$ for some $z \in \mathsf{field}(r)$ and $dom(h_z) = \boldsymbol{u}_\zeta$ as well as $g\restriction\zeta$ are elements of $\boldsymbol{u}_\xi$. By Σ–recursion relativized to $\boldsymbol{u}_\xi$ we obtain $h_z \in \boldsymbol{u}_\xi$. Therefore $g\restriction\xi = g\restriction\zeta \cup \{(\zeta, \bigcup rng(h_z))\} \in \boldsymbol{u}_\xi$. This proves (iv). By (iv) and the definition of S we obtain $S^{ry} \in \boldsymbol{u}_\xi$ and $S^y = \bigcup rng(h_y)$. For $\eta \in \boldsymbol{u}_\xi$ we have $h_y(\eta) = \{x \in \omega \mid A(\bigcup_{\zeta<\eta} h_y(\zeta), S^{ry}, x, y, \vec{z})\}$ which shows that h_y is definable in $\boldsymbol{u}_\xi$ by Σ–Recursion. Therefore the formula $A(S^y, S^{ry}, x, y, \vec{z})$ is still Σ and we obtain $(\exists \eta \in \boldsymbol{u}_\xi) A(\bigcup_{\zeta<\eta} h_y(\zeta), S^{ry}, x, y, \vec{z})$ by Σ-reflection relativized to $\boldsymbol{u}_\xi$. Hence $x \in S^y$. This proves (ii). To prove also (iii) we show

$$Cl^\omega_A(X, S^{ry}, y, \vec{z}) \to h_y(\eta) \subseteq X \tag{v}$$

by induction on $\eta \in \boldsymbol{u}_\xi$. From the induction hypothesis we have $\bigcup_{\zeta<\eta} h_y(\zeta) \subseteq X$ which implies by X–positivity $A(\bigcup_{\zeta<\eta} h_y(\zeta), S^{ry}, x, y, \vec{z}) \to A(X, S^{ry}, x, y, \vec{z})$. Together with the hypothesis $Cl^A_\omega(X, S^{ry}, y, \vec{z})$ this yields $h_y(\eta) \subseteq X$. □

The following theorem is a consequence of Lemma 3.3.5.2.

3.3.5.3. Theorem.

(i) $(\mathbf{\Pi^1_1\text{-}CA_\nu})_0 = (\mathbf{ID}^2_\nu)_0 \subseteq \mathbf{KPl}^r_\nu$

(ii) $(\mathbf{\Pi^1_1\text{-}CA_\nu}) = \mathbf{ID}^2_\nu \subseteq \text{W-}\mathbf{KPl}_\nu$

(iii) $\mathbf{ID}_\nu \subseteq (\mathbf{\Pi^1_1\text{-}CA_\nu}) + (\mathsf{Bi}) = \mathbf{BID}^2_\nu \subseteq \mathbf{KPl}_\nu$

(iv) $(\mathbf{Aut\text{-}\Pi^1_1})_0 = (\mathbf{Aut\text{-}ID})_0 \subseteq \mathbf{Aut\text{-}KPl^r}$

(v) $(\mathbf{Aut\text{-}\Pi^1_1}) = \mathbf{Aut\text{-}ID} \subseteq \text{W-}\mathbf{Aut\text{–}KPl}$

(vi) $(\mathbf{Aut\text{-}\Pi^1_1}) + (\mathsf{Bi}) = \mathbf{Aut\text{-}BID} \subseteq \mathbf{Aut\text{–}KPl}$

(vii) $\mathbf{ID}_{\prec^*} \subseteq \mathbf{BID}^* \subseteq \mathbf{KPl}^*$

Proof. Claims (i), (ii), (iv), (v) and (vii) follow immediately from Lemma 3.3.5.2 and Theorem 3.2.4.2. Claims (iii) and (vi) follow from Lemma 3.3.5.2, Theorems 3.2.4.2 and 3.2.3.2 and Corollary 3.3.3.8. □

The opposite inclusions in claims (iv), (v) and (vi) are also true. In (i), (ii) and (iii) this can of course only be true for limit ordinals ν. And this is the case.

3.4. Ordinal analysis for set-theoretic axioms systems

3.4.1. Ramified set theory

We introduce the language $\mathcal{L}_{\mathrm{RS}}$ and of *Ramified Set Theory*. The basic symbols are $\in$, $\notin$, Ad, $\neg\mathsf{Ad}$ and a constant L_α for every ordinal α.

3.4.1.1. Definition. (Inductive definition of the set terms of $\mathcal{L}_{\mathrm{RS}}$) Every constant L_α is an *atomic set term* of stage α. If $a_1, \ldots, a_n$ are set terms of stages $< \alpha$ and $F(x, x_1, \ldots, x_n)$ is an $\mathcal{L}$ formula without further free variables then

$$\{x \in \mathsf{L}_\alpha | \, F(x, a_1, \ldots, a_n)^{\mathsf{L}_\alpha}\}$$

is a *composed set term* of stage α. We denote the stage of a set term s by $\mathsf{stg}\,(s)$.

There are only sentences in $\mathcal{L}_{\mathrm{RS}}$. An $\mathcal{L}_{\mathrm{RS}}$–sentence is obtained from an $\mathcal{L}(\in, \mathsf{Ad})$–formula (in 'Tait'-style) by replacing all free variables by $\mathcal{L}_{\mathrm{RS}}$–terms and restricting all unbounded quantifiers to $\mathcal{L}_{\mathrm{RS}}$–terms. To have a uniform notation we refer to $\mathcal{L}_{\mathrm{RS}}$–terms and –sentences as $\mathcal{L}_{\mathrm{RS}}$*–expressions.* Let

$$\mathcal{T}_\alpha := \{t | \; \mathsf{stg}(t) < \alpha\}. \tag{135}$$

We do not count the equality symbol among the basic symbols of $\mathcal{L}_{\mathrm{RS}}$ but define

$$s = t \;\; :\Leftrightarrow \;\; (\forall x \in s)[x \in t] \wedge (\forall x \in t)[s \in s].$$

We transfer the Levy hierarchy to the language $\mathcal{L}_{\mathrm{RS}}$.

3.4.1.2. Definition. Let $\mathcal{F}$ be a complexity class in the Levy hierarchy. We call an $\mathcal{L}_{\mathrm{RS}}$–sentence F an $\mathcal{F}^\alpha$–sentence if there is a $\mathcal{F}$–formula $G(\vec{x})$ in the language $\mathcal{L}$ which has only the shown free variables and a tuple $\vec{a}$ of $\mathcal{L}_{\mathrm{RS}}$–terms of stages less than α such that $F \equiv G(\vec{a})^{\mathsf{L}_\alpha}$.

If F^{L_κ} is a $\mathcal{F}^\kappa$–sentence, we denote by F^z the sentence which is obtained by replacing all quantifier restrictions $\in \mathsf{L}_\kappa$ by $\in z$.

The standard interpretation of $\mathcal{L}_{\mathrm{RS}}$ is given by

$$\mathsf{L}_\alpha{}^{\mathbf{L}} = \mathbf{L}_\alpha$$
$$\{x \in \mathsf{L}_\alpha | \, F(x, a_1, \ldots, a_n)^{\mathsf{L}_\alpha}\}^{\mathbf{L}} = \{x \in \mathbf{L}_\alpha | \; \mathbf{L}_\alpha \models F(x, a_1^{\mathbf{L}}, \ldots, a_n^{\mathbf{L}})\}.$$

It is obvious that for every set term s of stage α we have $s^{\mathbf{L}} \in \mathbf{L}_{\alpha+1}$. We have

$$\mathbf{L} \models s \in \mathsf{L}_\alpha \;\; \Leftrightarrow \;\; \mathbf{L} \models s = t \;\; \text{for some set term } t \text{ with } \mathsf{stg}\,(t) < \alpha, \tag{136}$$

$$\mathbf{L} \models \mathsf{Ad}(t) \;\; \Leftrightarrow \;\; \mathbf{L} \models t = \mathsf{L}_\kappa \;\; \text{for some } \kappa \in \mathbf{Reg} \text{ such that } \kappa \le \mathsf{stg}(t) \tag{137}$$

and

$$\mathbf{L} \models s \in \{x \in \mathsf{L}_\alpha | \, F(x)\} \;\; \Leftrightarrow \;\; \mathbf{L} \models t = s \wedge F(t) \tag{138}$$

for some set term t with $\mathsf{stg}(t) < \alpha$. This motivates the following definition.

3.4.1.3. Definition. We say that $\mathcal{L}_{\mathrm{RS}}$-sentences of the shape $s \in r$, $\mathsf{Ad}(t)$, $A \vee B$ and $(\exists x \in r)G(x)$ have $\bigvee$-type. Dually sentences of the shape $s \notin t$, $\neg\mathsf{Ad}(t)$, $A \wedge B$ and $(\forall x \in r)G(x)$ are said to have $\bigwedge$-type.

We put

$$\mathcal{C}(s \in r) := \begin{cases} \{t = s \mid \mathsf{stg}(t) < \alpha\} & \text{if } r = \mathbf{L}_\alpha \\ \{t = s \wedge F(t) \mid \mathsf{stg}(t) < \alpha\} & \text{if } r = \{x \in \mathsf{L}_\alpha \mid F(x)\}, \end{cases}$$
$$\mathcal{C}(\mathsf{Ad}(t)) := \{\mathsf{L}_\kappa = t \mid \kappa \in \mathbf{Reg} \wedge \kappa \leq \mathsf{stg}(t)\},$$
$$\mathcal{C}(A \vee B) := \{A, B\}$$

and

$$\mathcal{C}((\exists x \in r)G(x)) := \begin{cases} \{G(t) \mid \mathsf{stg}(t) < \alpha\} & \text{if } r = \mathbf{L}_\alpha \\ \{F(t) \wedge G(t) \mid \mathsf{stg}(t) < \alpha\} & \text{if } r = \{x \in \mathsf{L}_\alpha \mid F(x)\}. \end{cases}$$

This defines the *characteristic sub-sentences-set* $\mathcal{C}(F)$ for all sentences F of $\bigvee$-type. Dually we define

$$\mathcal{C}(F) := \{\neg G \mid G \in \mathcal{C}(\neg F)\}$$

for sentences F of $\bigwedge$-type.

If F is not a conjunction or disjunction then every $G \in \mathcal{C}(F)$ is of the form $H(t)$ for some characteristic set term t. We define $t_F(G) := t$ and $o_F(G) := \mathsf{stg}(t)$. For $F = A_0 \circ A_1$, $\circ \in \{\vee, \wedge\}$ and $G \in \mathcal{C}(F)$ we put $t_F(G) := \mathsf{L}_i$ and $o_F(G) := i$ if $G = A_i$ for $i \in \{0, 1\}$.

The following lemma is an immediate consequence of Definition 3.4.1.3 and its preceding remarks.

3.4.1.4. Lemma. *For every sentence F of $\bigvee$-type we have*

$$\mathbf{L} \models F \;\Leftrightarrow\; \mathbf{L} \models \bigvee_{G \in \mathcal{C}(F)} G$$

and dually

$$\mathbf{L} \models F \;\Leftrightarrow\; \mathbf{L} \models \bigwedge_{G \in \mathcal{C}(F)} G$$

for sentences of $\bigwedge$-type.

Lemma 3.4.1.4 is the basis for the following infinitary calculus.

3.4.1.5. Definition. We define the relation $\models^{\alpha} \Delta$ for finite sets of $\mathcal{L}_{\mathrm{RS}}$-sentences Δ inductively by the following two clauses:

($\bigvee$) *If F is of $\bigvee$-type and $\models^{\alpha_0} \Delta, G$ as well as $\alpha_0 < \alpha$ and $o_F(G) < \alpha$ hold for some $G \in \mathcal{C}(F)$ then $\models^{\alpha} \Delta, F$*

and

($\bigwedge$) *If F is of $\bigwedge$-type and $\models^{\alpha_G} \Delta, G$ as well as $\alpha_G < \alpha$ hold for all $G \in \mathcal{C}(F)$ then $\models^{\alpha} \Delta, F$.*

From Definition 3.4.1.5 we get for $\mathcal{L}_{\mathrm{RS}}$-sentences F

$$\mathbf{L} \models F \;\Leftrightarrow\; (\exists\alpha) \models^{\alpha} F \tag{139}$$

and for a Σ_1-sentence F and an ordinal γ we have

$$\models^{\alpha} F^{\mathsf{L}_\gamma} \;\Rightarrow\; \mathbf{L}_\alpha \models F. \tag{140}$$

If we put

$$\mathsf{tc}(F) := \begin{cases} \min\{\alpha \mid\; \models^{\alpha} F\} & \text{if } \mathbf{L} \models F \\ \infty & \text{otherwise} \end{cases} \tag{141}$$

for $\mathcal{L}_{\mathrm{RS}}$–sentences F we obtain from (140)

$$|F|_{\Sigma_1} \leq \mathsf{tc}(F^{\mathsf{L}_\gamma}) \tag{142}$$

for all Σ_1–sentences F and all ordinals γ.

We are going to define a *rank* function for $\mathcal{L}_{\mathrm{RS}}$–expressions in such a way that all sentences in the characteristic sub-sentences set of a sentence F get lower rank.

3.4.1.6. Definition. For an $\mathcal{L}_{\mathrm{RS}}$–expression E we define $\mathsf{rk}(E)$ by the following clauses:

$$\mathsf{rk}(\mathsf{L}_\alpha) := \omega \cdot \alpha$$

$$\mathsf{rk}(\{x \in \mathsf{L}_\alpha \mid F(x)\}) := \max\{\mathsf{rk}(\mathsf{L}_\alpha) + 1, \mathsf{rk}(F(\mathsf{L}_0)) + 2\}$$

$$\mathsf{rk}(\mathsf{Ad}(t)) := \mathsf{rk}(\neg\mathsf{Ad}(t)) := \mathsf{rk}(t) + 5$$

$$\mathsf{rk}(s \in t) := \mathsf{rk}(s \notin t) := \max\{\mathsf{rk}(s) + 6, \mathsf{rk}(t) + 1\}$$

$$\mathsf{rk}(A \vee B) := \mathsf{rk}(A \wedge B) := \max\{\mathsf{rk}(A), \mathsf{rk}(B)\} + 1$$

$$\mathsf{rk}((\exists x \in s)F(x)) := \mathsf{rk}((\forall x \in s)F(x)) := \max\{\mathsf{rk}(s), \mathsf{rk}(F(\mathsf{L}_0)) + 2\}$$

The crucial property of the rank function is stated in the following theorem.

3.4.1.7. Theorem. *For* $G \in \mathcal{C}(F)$ *we have* $\mathsf{rk}(G) < \mathsf{rk}(F)$.

The proof of the theorem is a bit lengthy and we will not give all details. Let a, b and c be $\mathcal{L}_{\mathrm{RS}}$–expressions. First we show

$$\mathsf{stg}(a) < \mathsf{stg}(b(\mathsf{L}_0)) \;\Rightarrow\; \mathsf{rk}(b(a)) < \mathsf{rk}(b(\mathsf{L}_0)) \tag{i}$$

by an easy induction on $\mathsf{rk}(b(\mathsf{L}_0))$. The next step is to show

$$\mathsf{stg}(c) < \alpha \;\Rightarrow\; \mathsf{rk}(b(c)) < \max\{\omega \cdot \alpha, \mathsf{rk}(b(\mathsf{L}_0)) + 1\} \tag{ii}$$

by induction on $\mathsf{rk}(b(\mathsf{L}_0))$. From (ii) we get easily

$$\mathsf{stg}(c) < \alpha \;\Rightarrow\; \mathsf{rk}(F(c)) + 1 < \mathsf{rk}(s \in \{x \in \mathsf{L}_\alpha \mid F(x)\}). \tag{iii}$$

Now we compute

$$\mathsf{rk}(a = b) = \max\{\mathsf{rk}(a), \mathsf{rk}(b)\} + 4. \tag{iv}$$

Finally we show

$$G \in \mathcal{C}(F) \;\Rightarrow\; \mathsf{rk}(G) < \mathsf{rk}(F) \tag{v}$$

by distinguishing cases on the shape of F. We only consider sentences of $\bigvee$-type. The case of $\bigwedge$-type is dual.

If $F \equiv \mathsf{Ad}(t)$ then $G \equiv (t = \mathsf{L}_\kappa)$ for some $\kappa \in \mathbf{Reg} \cap \mathsf{stg}(t) + 1$. Hence $\mathsf{rk}(G) = \max\{\mathsf{rk}(t), \kappa\} + 4 < \mathsf{rk}(t) + 5 = \mathsf{rk}(F)$.

If $F \equiv (s \in \mathsf{L}_\alpha)$ then $G \equiv (t = s)$ for some t such that $\mathsf{stg}(t) < \alpha$. But then $\mathsf{rk}(G) = \max\{\mathsf{rk}(t), \mathsf{rk}(s)\} + 4 < \max\{\omega \cdot \alpha + 1, \mathsf{rk}(s) + 6\} = \mathsf{rk}(F)$.

If $F \equiv (s \in \{x \in \mathsf{L}_\alpha | H(x)\})$ then $G \equiv (s = t \wedge H(t))$ for some t such that $\mathsf{stg}(t) < \alpha$. Hence $\mathsf{rk}(G) = \max\{\mathsf{rk}(s = t), \mathsf{rk}(H(t))\} + 1 = \max\{\mathsf{rk}(s) + 5, \mathsf{rk}(t) + 5, \mathsf{rk}(H(t)) + 1\}$. But $\mathsf{rk}(s) + 5 < \mathsf{rk}(s) + 6 \leq \mathsf{rk}(F)$, $\mathsf{rk}(t) + 5 < \omega \cdot \alpha \leq \mathsf{rk}(F)$ and by (iii) also $\mathsf{rk}(H(t)) + 1 < \mathsf{rk}(F)$.

The claim is obvious for $F \equiv (A \vee B)$.

If $F \equiv (\exists x \in \mathsf{L}_\alpha)H(x)$ then $G \equiv H(t)$ for some t with $\mathsf{stg}(t) < \alpha$. Then by (ii) $\mathsf{rk}(G) < \max\{\omega \cdot \alpha, \mathsf{rk}(H(\mathsf{L}_0)) + 1\} \leq \max\{\omega \cdot \alpha, \mathsf{rk}(H(\mathsf{L}_0)) + 2\} = \mathsf{rk}(F)$.

If $F \equiv (\exists x \in \{y \in \mathsf{L}_\alpha | H(y)\})K(x)$ then $G \equiv H(t) \wedge K(t)$ for some t such that $\mathsf{stg}(t) < \alpha$. Then by (ii) $\mathsf{rk}(G) = \max\{\mathsf{rk}(H(t)), \mathsf{rk}(K(t))\} + 1 < \max\{\omega \cdot \alpha, \mathsf{rk}(H(\mathsf{L}_0)) + 2, \mathsf{rk}(K(\mathsf{L}_0)) + 2\} = \max\{\mathsf{rk}(\{x \in \mathsf{L}_\alpha | H(x)\}), \mathsf{rk}(K(\mathsf{L}_0)) + 2\} = \mathsf{rk}(F)$. □

Because of $\mathsf{rk}(G) < \mathsf{rk}(F)$ for all $G \in \mathcal{C}(F)$ we get by induction on $\mathsf{rk}(F)$

$$\mathbf{L} \models F \;\Rightarrow\; \overset{\mathsf{rk}(F)}{\models\!\!=} F. \tag{143}$$

Hence

$$\mathsf{tc}(F) \leq \mathsf{rk}(F) \textit{ for all } \mathcal{L}_{\mathrm{RS}}\textit{-sentences } F. \tag{144}$$

3.4.2. A semi-formal calculus for ramified set theory

It follows by (139) that the rules

(cut) $\models^{\alpha} \Delta, A$ *and* $\models^{\beta} \Delta, \neg A \;\Rightarrow\; \models^{\delta} \Delta$

and

(Ref) $\models^{\alpha} \Delta, F^{\mathsf{L}_\kappa} \;\Rightarrow\; \models^{\delta} \Delta, (\exists z \in \mathsf{L}_\kappa)[z \neq \emptyset \wedge F^z]$ *for* $\kappa \in \mathbf{Reg}$, F^{L_κ} *a* Π_2^κ*-sentence*

are admissible for some ordinal δ. However, we do not yet know how to compute δ from α or α and β respectively. Therefore we design a semi-formal system having these rules as basic inferences.

3.4.2.1. Definition. Let Δ be a finite set of $\mathcal{L}_{\mathrm{RS}}$–sentences and α and ρ ordinals. We define the relation $\vdash^{\alpha}_{\rho} \Delta$ by the following clauses:

($\bigwedge$) *If* $F \in \Delta \cap \bigwedge$*-type,* $\vdash_{\rho}^{\alpha_G} \Delta, G$ *and* $\alpha_G < \alpha$ *for all* $G \in \mathcal{C}(F)$ *then* $\vdash_{\rho}^{\alpha} \Delta$.

($\bigvee$) *If* $F \in \Delta \cap \bigvee$*-type,* $\vdash_{\rho}^{\alpha_G} \Delta, G$ *and* $o_F(G), \alpha_G < \alpha$ *for some* $G \in \mathcal{C}(F)$ *then* $\vdash_{\rho}^{\alpha} \Delta$.

(Ref_κ) *If* $\kappa \in \mathbf{Reg}$, $F^{\mathsf{L}_\kappa} \in \Pi_2^\kappa$, $(\exists z \in \mathsf{L}_\kappa)[z \neq \emptyset \wedge F^z] \in \Delta$, $\vdash_{\rho}^{\alpha_0} \Delta, F^{\mathsf{L}_\kappa}$ *and* $\kappa, \alpha_0 + 1 < \alpha$ *then* $\vdash_{\rho}^{\alpha} \Delta$.

We call F the *main part* in instances of ($\bigwedge$) and ($\bigvee$) and $(\exists z \in \mathsf{L}_\kappa)[z \neq \emptyset \wedge F^z]$ the main part in an instance of (Ref_κ).

(cut) *If* $\vdash_{\rho}^{\alpha_0} \Delta, A$, $\vdash_{\rho}^{\alpha_0} \Delta, \neg A$ *for some* $\alpha_0 < \alpha$ *and some* A *such that* $\mathsf{rk}(A) < \rho$ *then* $\vdash_{\rho}^{\alpha} \Delta$.

We say that Δ is semi-formal derivable in $\mathcal{L}_{\text{RS}}$ if there are ordinals α and ρ such that $\vdash_{\rho}^{\alpha} \Delta$.

As an immediate consequence of the soundness of all rules we get the soundness of the semi-formal calculus, i.e.,

$$\vdash_{\rho}^{\alpha} \Delta \;\Rightarrow\; \mathbf{L} \models \bigvee \Delta. \tag{145}$$

Since $\vdash_{\rho}^{\alpha} \Delta$ is a correct calculus which derives sentences we get cut elimination nearly for free. We prove:

$$\textit{If } \vdash_{\rho}^{\alpha} \Delta, \Gamma \textit{ and } \mathbf{L} \not\models F \textit{ for all } F \in \Gamma \textit{ then } \models^{\alpha} \Delta \tag{146}$$

by induction on α. The claim is immediate from the induction hypothesis if the last inference is according to ($\bigwedge$) or ($\bigvee$) and its main part does not belong to Γ. If the main part is in Γ we have in the case of an inference according to ($\bigvee$) an $F \in \Gamma \cap \bigvee$-type and the premise $\vdash_{\rho}^{\alpha_0} \Delta, \Gamma, G$ for some $G \in \mathcal{C}(F)$. Then $\mathbf{L} \not\models G$ and $\models^{\alpha} \Delta$ follows by induction hypothesis. In the case of an inference according to ($\bigwedge$) there is an $F \in \Gamma \cap \bigwedge$-type and therefore a $G \in \mathcal{C}(F)$ such that $\mathbf{L} \not\models G$. But then there is a premise $\vdash_{\rho}^{\alpha_G} \Delta, \Gamma, G$ with $\alpha_G < \alpha$ and we obtain $\models^{\alpha} \Delta$ by the induction hypothesis. If the last inference is a cut with cut–sentence F then either $\mathbf{L} \not\models F$ or $\mathbf{L} \not\models \neg F$. We pick the corresponding premise and obtain the claim by induction hypothesis. If the last inference is according to (Ref_κ) we distinguish the cases that for its main part we have $\mathbf{L} \models (\exists z \in \mathsf{L}_\kappa) F^z$ or not. In the second case we also have $\mathbf{L} \not\models F^{\mathsf{L}_\kappa}$ and obtain the claim by the induction hypothesis. In the first case we have $(\exists z \in \mathsf{L}_\kappa)[z \neq \emptyset \wedge F^z] \in \Delta$ and get $\models^{\alpha} \Delta$ from $\mathsf{rk}((\exists z \in \mathsf{L}_\kappa) F^z) = \kappa < \alpha$ and (143). □

Summing up and taking Γ as the empty set in (146) we get the following theorem.

3.4.2.2. Theorem. *The semi-formal calculus is sound and allows cut elimination. Especially we get* $\models^{\alpha} \Delta$ *from* $\vdash_{\rho}^{\alpha} \Delta$.

3.4.3. Operator–controlled derivations

It follows from Theorem 3.4.2.2 that cut–elimination alone cannot be crucial for the ordinal analysis of theories **Ax** for which we have

$$\mathbf{Ax} \vdash \mathit{Wf}(\prec) \;\Leftrightarrow\; \mathbf{Ax} \vdash (\exists \xi \in \mathsf{L}_{\omega_1^{CK}}) H_{\prec}(\xi),$$

i.e. for theories in which the well-foundedness of a Δ_0–definable ordering can be expressed by an $\Sigma_1^{\omega_1^{CK}}$–sentence. This is true for all theories comprising $\mathbf{KP}\boldsymbol{\omega}$. The main problem there is to collapse the ordinals which arise canonically in the embedding procedure of $\Sigma_1^{\omega_1^{CK}}$–sentences into ordinals below ω_1^{CK}. Collapsing is therefore the Leitmotiv of Impredicative Proof–Theory (but we will see that cut–elimination will be needed for collapsing). We will use the technique of *local predicativity*, first introduced in Pohlers [1982a,1982b], Buchholz et al. [1981], but we are going to use an essential simplification of the original technique which has been introduced by Buchholz [1992]. We already used collapsing techniques in the Π_2^0–analyses of Sections 2.1.4 and 2.1.5. We will, however, not give Π_2^0–analyses for impredicative theories. Already for Π_1^1–analyses the matter is sufficiently complicated. We just beg the reader to believe that the refinement to Π_2^0–analyses can be done by modifying the techniques of Section 2.1.5 (cf. Blankertz and Weiermann [1996], Blankertz [1997] for more details). Another way to extend the following analyses to Π_2^0–analyses is to apply Section 2.2.2.

Our presentation will follow quite closely that of Buchholz in Buchholz [1992]. (Those who have tried know that it is hardly possible to improve Buchholz's presentations.)

3.4.3.1. Definition. An *(ordinal-)operator* is a map

$$\mathcal{H}\colon \mathsf{Pow}(\mathbf{On}) \longrightarrow \mathsf{Pow}(\mathbf{On}).$$

We introduce the abbreviations

$$\begin{aligned} \alpha \in \mathcal{H} \;&:\Leftrightarrow\; \alpha \in \mathcal{H}(\emptyset) \\ M \subseteq \mathcal{H} \;&:\Leftrightarrow\; M \subseteq \mathcal{H}(\emptyset) \\ \mathcal{H} \subseteq M \;&:\Leftrightarrow\; \mathcal{H}(\emptyset) \subseteq M \\ \mathcal{H} \subseteq \mathcal{H}' \;&:\Leftrightarrow\; (\forall X)[\mathcal{H}(X) \subseteq \mathcal{H}'(X)] \end{aligned} \tag{147}$$

and call an operator $\mathcal{H}$ closed under a function $f\colon \mathbf{On}^n \longrightarrow \mathbf{On}$ if

$$(\forall X \in \mathsf{Pow}(\mathbf{On}))(\forall \xi_1)\cdots(\forall \xi_n)[f(\xi_1,\ldots,\xi_n) \in \mathcal{H}(X) \;\Leftrightarrow\; \{\xi_1,\ldots,\xi_n\} \subseteq \mathcal{H}(X)].$$

In case that $\mathcal{H}$ is closed under $f(\alpha_1,\ldots,\alpha_n) := \omega^{\alpha_1} \# \cdots \# \omega^{\alpha_n}$ we call it *Cantorian–closed.*

A set $M \subseteq \mathbf{On}$ and an operator $\mathcal{H}$ induce a new operator $\mathcal{H}[M]$ by

$$\mathcal{H}[M](X) := \mathcal{H}(M \cup X). \tag{148}$$

For an $\mathcal{L}_{\mathrm{RS}}$–expression E let

$$\mathsf{par}(E) := \{\alpha \mid \mathsf{L}_\alpha \text{ occurs in } E\}. \tag{149}$$

If Θ is a set of $\mathcal{L}_{RS}$–expressions and $\mathcal{H}$ an operator we define $\mathcal{H}[\Theta] := \mathcal{H}[\mathsf{par}(\Theta)]$.

3.4.3.2. Definition. An operator $\mathcal{H}$ is *acceptable* if it satisfies the following conditions:

$$\begin{aligned} & 0 \in \mathcal{H}(\emptyset) \\ & \mathcal{H} \text{ is Cantorian–closed} \\ & (\forall X \in \mathsf{Pow}(\mathbf{On}))[X \subseteq \mathcal{H}(X)] \\ & (\forall X \in \mathsf{Pow}(\mathbf{On}))(\forall Y \in \mathsf{Pow}(\mathbf{On}))[X \subseteq \mathcal{H}(Y) \Rightarrow \mathcal{H}(X) \subseteq \mathcal{H}(Y)]. \end{aligned} \tag{150}$$

3.4.3.3. Definition. Let $\mathcal{H}\colon \mathsf{Pow}(\mathbf{On}) \longrightarrow \mathsf{Pow}(\mathbf{On})$ be an operator. For a finite set Δ of $\mathcal{L}_{RS}$–sentences we define $\mathcal{H} \mid\!\frac{\alpha}{\rho}\, \Delta$ iff $\mathsf{par}(\Delta) \cup \{\alpha\} \subseteq \mathcal{H}$ and one of the following conditions is satisfied:

($\bigwedge$) *There is a sentence $F \in \Delta \cap \bigwedge$-type such that $\mathcal{H}[t_F(G)] \mid\!\frac{\alpha_G}{\rho}\, \Delta, G$ and $\alpha_G < \alpha$ for all $G \in \mathcal{C}(F)$.*

($\bigvee$) *There is a sentence $F \in \Delta \cap \bigvee$-type such that $\mathcal{H} \mid\!\frac{\alpha_G}{\rho}\, \Delta, G$ and $o_F(G), \alpha_G < \alpha$ for some $G \in \mathcal{C}(F)$.*

(Ref$_\kappa$) *There is a Π_2^κ–sentence F^{L_κ} such that $(\exists z \in \mathsf{L}_\kappa)[z \neq \emptyset \wedge F^z] \in \Delta$, $\mathcal{H} \mid\!\frac{\alpha_0}{\rho}\, \Delta, F^{\mathsf{L}_\kappa}$ and $\kappa, \alpha_0 + 1 < \alpha$.*

(cut) *There is a sentence A such that $\mathsf{rk}(A) < \rho$, $\mathcal{H} \mid\!\frac{\alpha_0}{\rho}\, \Delta, A$ and $\mathcal{H} \mid\!\frac{\alpha_0}{\rho}\, \Delta, \neg A$ for some $\alpha_0 < \alpha$.*

We say that Δ is *operator controlled derivable* if there are an acceptable operator $\mathcal{H}$ and ordinals α and ρ such that $\mathcal{H} \mid\!\frac{\alpha}{\rho}\, \Delta$.

From now on we will only regard acceptable operators without mentioning it explicitly.

The following properties of operator–controlled derivations are immediate consequences of Definition 3.4.3.3.

$$\text{If } \mathcal{H} \mid\!\tfrac{\alpha}{\rho}\, \Delta,\ \mathcal{H} \subseteq \mathcal{H}',\ \Delta \subseteq \Gamma,\ \alpha \le \beta \in \mathcal{H}',\ \rho \le \sigma \text{ and } \mathsf{par}(\Gamma) \subseteq \mathcal{H}' \text{ then } \mathcal{H}' \mid\!\tfrac{\beta}{\sigma}\, \Gamma. \tag{151}$$

$$\text{If } \mathcal{H} \mid\!\tfrac{\alpha}{\rho}\, \Delta, (\exists x \in \mathsf{L}_\delta)F(x) \text{ and } \delta \le \beta \in \mathcal{H} \text{ then } \mathcal{H} \mid\!\tfrac{\alpha}{\rho}\, \Delta, (\exists x \in \mathsf{L}_\beta)F(x). \tag{152}$$

$$\text{If } \mathcal{H} \mid\!\tfrac{\alpha}{\rho}\, \Delta, (\forall x \in \mathsf{L}_\kappa)F(x) \text{ and } \beta \in \kappa \cap \mathcal{H} \text{ then } \mathcal{H} \mid\!\tfrac{\alpha}{\rho}\, \Delta, (\forall x \in \mathsf{L}_\beta)F(x). \tag{153}$$

$$\text{If } \mathcal{H} \mid\!\tfrac{\alpha}{\rho}\, \Delta, F \vee G \text{ then } \mathcal{H} \mid\!\tfrac{\alpha}{\rho}\, \Delta, F, G. \tag{154}$$

We refer to (151) as *Structural Rule*, to (152) as *Upward Persistency*, to (153) as *Downward Persistency* and to (154) as $\vee$*–exportation*. All these properties are easily proved by induction on α.

The predicative cut–elimination procedure works also for operator controlled derivations. First we prove

3.4.3.4. Inversion Lemma. *If* $\mathcal{H} \mid\!\frac{\alpha}{\rho}\, \Delta, F$ *and* $F \in \bigwedge$*-type then* $\mathcal{H}[t_F(G)] \mid\!\frac{\alpha}{\rho}\, \Delta, G$ *for all* $G \in \mathcal{C}(F)$.

The proof is a straightforward induction on α.

3.4.3.5. Reduction Lemma. *If* $F \in \bigvee$*-type,* $\mathsf{rk}(F) = \rho \notin \mathbf{Reg}$ *and* $\mathcal{H} \mid\!\frac{\alpha}{\rho}\, \Delta, \neg F$ *as well as* $\mathcal{H} \mid\!\frac{\beta}{\rho}\, \Gamma, F$ *then* $\mathcal{H} \mid\!\frac{\alpha+\beta}{\rho}\, \Delta, \Gamma$.

The proof is that of the Reduction Lemma (Lemma 2.1.5.7) of Section 2.1.5. Since we restrict ourselves to Π^1_1–ordinal analysis we consider finite ordinals as trivial. Therefore we don't have to compose the controlling operators. This makes the proof simpler. The hypothesis $\rho \notin \mathbf{Reg}$ is needed to exclude the case that F is the main part of an inference according to (Ref).

As in Section 2.1.2 we obtain the Predicative Elimination Lemma as a straightforward consequence of the Reduction Lemma.

3.4.3.6. Predicative Elimination Lemma. *Let* $\mathcal{H}$ *be an operator which is closed under the function* $\alpha, \beta \mapsto \varphi_\alpha \beta$, $\mathcal{H} \mid\!\frac{\alpha}{\beta+\omega^\rho}\, \Delta$, $[\beta, \beta+\omega^\rho) \cap \mathbf{Reg} = \emptyset$ *and* $\rho \in \mathcal{H}$. *Then* $\mathcal{H} \mid\!\frac{\varphi_\rho \alpha}{\beta}\, \Delta$.

The proof is that of the Predicative Elimination Lemma (Lemma 2.1.2.9) with some extra care on the controlling operator. As an example we treat the case of a cut. There we have the premises

$$\mathcal{H} \mid\!\frac{\alpha_0}{\beta+\omega^\rho}\, \Delta, F \text{ and } \mathcal{H} \mid\!\frac{\alpha_0}{\beta+\omega^\rho}\, \Delta, \neg F. \tag{i}$$

Using the induction hypothesis we obtain

$$\mathcal{H} \mid\!\frac{\varphi_\rho \alpha_0}{\beta}\, \Delta, F \quad \text{and} \quad \mathcal{H} \mid\!\frac{\varphi_\rho \alpha_0}{\beta}\, \Delta, \neg F. \tag{ii}$$

If $\mathsf{rk}(F) < \beta$ we obtain

$$\mathcal{H} \mid\!\frac{\varphi_\rho \alpha}{\beta}\, \Delta$$

by a cut since $\varphi_\rho \alpha_0 < \varphi_\rho \alpha \in \mathcal{H}$. If $\beta \le \mathsf{rk}(F) =_{\mathsf{NF}} \beta + \omega^{\rho_1} + \cdots + \omega^{\rho_n} < \beta + \omega^\rho$ we first obtain

$$\mathcal{H} \mid\!\frac{\varphi_\rho \alpha_0 \cdot 2}{\mathsf{rk}(F)}\, \Delta \tag{iii}$$

by the Reduction Lemma (Lemma 3.4.3.5). Since $\mathsf{par}(F) \subseteq \mathcal{H}$ we also get $\mathsf{rk}(F) \in \mathcal{H}$ and therefore $\{\rho_1, \ldots, \rho_n\} \subseteq \mathcal{H}$. From (iii) we first get

$$\mathcal{H} \mid\!\frac{\varphi_\rho \alpha}{\mathsf{rk}(F)}\, \Delta \tag{iv}$$

and finally

$$\mathcal{H} \mid\!\frac{\varphi_\rho \alpha}{\beta}\, \Delta$$

by n-fold application of the main induction hypothesis since $\varphi_{\rho_i}\varphi_\rho\alpha = \varphi_\rho\alpha$ for $i = 1, \ldots n$. □

3.4.3.7. Boundedness Theorem. *Let* $\mathcal{H} \mathrel{|\!\frac{\alpha}{\rho}} \Delta, F^{\mathsf{L}_\kappa}$ *for a* Σ^κ*–sentence* F^{L_κ}. *Then* $\mathcal{H} \mathrel{|\!\frac{\alpha}{\rho}} \Delta, F^{\mathsf{L}_\beta}$ *for all* $\beta \in [\alpha, \kappa) \cap \mathcal{H}$.

The proof is by induction on α. The claim follows immediately from the induction hypothesis if the main part of the last inference is different from F^{L_κ}. So assume that F^{L_κ} is the main part of the last inference. If $F^{\mathsf{L}_\kappa} \in \bigwedge$-type then every member of $\mathcal{C}(F^{\mathsf{L}_\kappa})$ is a Σ^κ–sentence of the form G^{L_κ} and we have the premise

$$\mathcal{H}[t_{F^{\mathsf{L}_\kappa}}(G^{\mathsf{L}_\kappa})] \mathrel{|\!\frac{\alpha_G}{\rho}} \Delta, F^{\mathsf{L}_\kappa}, G^{\mathsf{L}_\kappa}. \tag{i}$$

Hence

$$\mathcal{H}[t_{F^{\mathsf{L}_\kappa}}(G^{\mathsf{L}_\kappa})] \mathrel{|\!\frac{\alpha_G}{\rho}} \Delta, F^{\mathsf{L}_\beta}, G^{\mathsf{L}_\beta} \text{ for all } G^{\mathsf{L}_\kappa} \in \mathcal{C}(F^{\mathsf{L}_\kappa}) \tag{ii}$$

by induction hypothesis. But since F^{L_κ} is a Σ^κ–sentence we have $t_{F^{\mathsf{L}_\kappa}}(G^{\mathsf{L}_\kappa}) = t_{F^{\mathsf{L}_\beta}}(G^{\mathsf{L}_\beta})$ and get the claim from (ii) by an inference according to $(\bigwedge)$.

If $F^{\mathsf{L}_\kappa} \in \bigvee$-type and $F^{\mathsf{L}_\kappa} \not\equiv (\exists x \in \mathsf{L}_\kappa)G(x)$ then every member of $\mathcal{C}(F^{\mathsf{L}_\kappa})$ is again a Σ^κ–sentence of the form G^{L_κ} and we have the premise

$$\mathcal{H} \mathrel{|\!\frac{\alpha_0}{\rho}} \Delta, F^{\mathsf{L}_\kappa}, G^{\mathsf{L}_\kappa} \tag{iii}$$

for some $\alpha_0 < \alpha$. Applying the induction hypothesis to (iii) we get the claim by an inference $(\bigvee)$.

If $F^{\mathsf{L}_\kappa} \equiv (\exists x \in \mathsf{L}_\kappa)G(x)^{\mathsf{L}_\kappa}$ then we are either in the case of an inference $(\bigvee)$ whose premise is

$$\mathcal{H} \mathrel{|\!\frac{\alpha_0}{\rho}} \Delta, F^{\mathsf{L}_\kappa}, G(t)^{\mathsf{L}_\kappa} \tag{iv}$$

with $\alpha_0 < \alpha$ and $\mathbf{stg}(t) < \alpha$ or in the case of an inference (Ref_κ)

$$\mathcal{H} \mathrel{|\!\frac{\alpha_0}{\rho}} \Delta, F^{\mathsf{L}_\kappa}, (\forall x \in \mathsf{L}_\kappa)(\exists y \in \mathsf{L}_\kappa)G(x,y) \quad \Rightarrow \quad \mathcal{H} \mathrel{|\!\frac{\alpha}{\rho}} \Delta, (\exists z \in \mathsf{L}_\kappa)(\forall x \in z)(\exists y \in x)G(x,y) \tag{v}$$

for $\alpha_0 + 1 < \alpha$. In the first case we get

$$\mathcal{H} \mathrel{|\!\frac{\alpha_0}{\rho}} \Delta, F^{\mathsf{L}_\beta}, G(t)^{\mathsf{L}_\beta} \tag{vi}$$

by induction hypothesis. We have $G(t)^{\mathsf{L}_\beta} \in \mathcal{C}((\exists x \in \mathsf{L}_\beta)G(x)^{\mathsf{L}_\beta})$ because of $\mathbf{stg}(t) < \alpha \le \beta$ and we get the claim from (iv) by an inference $(\bigvee)$. In the second case we get by the Inversion Lemma and the induction hypothesis

$$\mathcal{H}[t] \mathrel{|\!\frac{\alpha_0}{\rho}} \Delta, F^{\mathsf{L}_\beta}, (\exists y \in \mathsf{L}_{\alpha_0})G(t,y) \tag{vii}$$

for all $\mathcal{L}_{\mathrm{RS}}$–terms t such that $\mathbf{stg}(t) < \kappa$. Since $\alpha_0 < \alpha \le \beta < \kappa$ we obtain

$$\mathcal{H} \mathrel{\Big|\!\frac{\alpha_0+1}{\rho}} \Delta, F^{\mathsf{L}_\beta}, (\forall x \in \mathsf{L}_{\alpha_0})(\exists y \in \mathsf{L}_{\alpha_0})G(x,y) \tag{viii}$$

from (vii) by an inference $(\bigwedge)$. But

$$(\forall x \in \mathsf{L}_{\alpha_0})(\exists y \in \mathsf{L}_{\alpha_0})G(x,y) \in \mathcal{C}\big((\exists z \in \mathsf{L}_\beta)(\forall x \in z)(\exists y \in z)G(x,y)\big)$$

and we obtain the claim from (viii) by an inference $(\bigvee)$. □

3.4.4. Collapsing functions

In order to define operators which allow the collapsing of derivations we need to know more about ordinals. This theory turns out to be very complicated. To simplify things we are going to use an abstraction. Instead of using admissible, i.e., recursively regular ordinals, we will develop the theory on the basis of just regular ordinals. This has the advantage not to have to bother about the complexity of the ordinal functions which we are going to define. Complexity arguments will be replaced by simple cardinality arguments. The disadvantage, however, is that the replacement of regular ordinals by recursively regular ones is not at all easy. See Rathjen [1995] and Schlüter [1993,n.d.] for details. It is outside the scope of this contribution even to give a hint how this can be performed. All we can indicate is that the segment below ω_1, the first regular ordinal, is recursive and therefore already a segment below ω_1^{CK}. This yields at least a correct computation of the $\Sigma_1^{\omega_1^{\mathrm{CK}}}$–ordinals of the analyzed axiom systems.

In this section we denote by **Reg** the class of regular ordinals above ω. By $\overline{\mathbf{Reg}}$ we denote its topological closure, i.e., the class of uncountable cardinals. Let $\lambda\xi\,.\,\Omega_\xi$ denote the enumerating function of $\overline{\mathbf{Reg}} \cup \{0\}$. Then $\Omega_0 = 0$, $\Omega_1 = \aleph_1 = \omega_1$, $\Omega_\omega = \aleph_\omega$ etc. We reserve the letters $\kappa, \pi, \kappa_1, \ldots, \pi_1, \ldots$ to denote members of **Reg** exclusively. We put

$$\alpha^+ := \min\{\kappa \in \mathbf{Reg} \mid \alpha < \kappa\}.$$

For the following we assume that there exist a weakly inaccessible ordinal. Let I be the least such ordinal, i.e. I is a regular ordinal for which we have $\Omega_I = I$ and $\sigma < \Omega_\sigma$ for all $\sigma \in \mathbf{Reg} \cap I$. Then we have $\kappa = \Omega_{\sigma+1}$ for all $\kappa \in \mathbf{Reg} \cap I$.

3.4.4.1. Definition. We define sets $Cl(\alpha, \beta)$ and functions

$$\psi_\kappa \alpha := \min\{\beta \mid \kappa \in Cl(\alpha,\beta) \wedge Cl(\alpha,\beta) \cap \kappa \subseteq \beta\}$$

by recursion on α.

The set $Cl(\alpha,\beta)$ is the least set which contains $\beta \cup \{0, I\}$ and is closed under ordinal addition $+$, the Veblen–function $\lambda\xi\,.\,\lambda\eta\,.\,\varphi_\xi\eta$, the enumerating function $\lambda\xi\,.\,\Omega_\xi$ of the class $\overline{\mathbf{Reg}}$ and the function $\lambda\xi < \alpha\,.\,\lambda\pi\,.\,\psi_\pi\xi$.

We call $Cl(\alpha,\beta)$ the α*–iterated closure* of β. There are some immediate consequences of Definition 3.4.4.1.

$$\alpha_0 \le \alpha \text{ and } \beta_0 \le \beta \;\Rightarrow\; Cl(\alpha_0, \beta_0) \subseteq Cl(\alpha, \beta) \tag{155}$$

$$\beta \in \mathsf{Lim} \;\Rightarrow\; Cl(\alpha, \beta) = \bigcup_{\eta \in \beta} Cl(\alpha, \eta) \tag{156}$$

$$|Cl(\alpha, \beta)| < \beta^+ \tag{157}$$

$$Cl(\alpha, \psi_\kappa \alpha) \cap \kappa = \psi_\kappa \alpha. \tag{158}$$

Since $\sigma < \Omega_{\sigma+1} = \kappa \subseteq Cl(\alpha, \kappa)$ for $\kappa < I$ we get

$$\kappa \in Cl(\alpha, \kappa) \text{ for all } \alpha. \tag{159}$$

A little more effort is needed to show

$$\psi_\kappa \alpha < \kappa \text{ and } \psi_\kappa \alpha \notin Cl(\alpha, \psi_\kappa \alpha). \tag{160}$$

By (156) and (159) we have $\eta_0 := \min\{\xi \mid \kappa \in Cl(\alpha, \xi)\} < \kappa$. Putting

$$\eta_{n+1} := \min\{\xi \mid Cl(\alpha, \eta_n) \cap \kappa \subseteq \xi\}$$

we obtain $\eta_n < \kappa$ from (157) by induction on n. Hence $\eta := \sup_{n \in \omega} \eta_n < \kappa$ and $Cl(\alpha, \eta) \cap \kappa = \bigcup_{n \in \omega} Cl(\alpha, \eta_n) \cap \kappa \subseteq \sup_{n \in \omega} \eta_{n+1} = \eta$. This shows $\psi_\kappa \alpha \le \eta < \kappa$ and, since $Cl(\alpha, \psi_\kappa \alpha) \cap \kappa = \psi_\kappa \alpha$, also $\psi_\kappa \alpha \notin Cl(\alpha, \psi_\kappa \alpha)$.

Putting $I^\Gamma := \min\{\gamma \in \mathsf{SC} \mid I < \gamma\}$ it follows from (160) that $Cl(\alpha, \beta) \subseteq I^\Gamma$ for $\beta \subseteq I^\Gamma$.

$$\alpha_0 < \alpha \text{ and } \alpha_0 \in Cl(\alpha, \psi_\kappa \alpha) \;\Rightarrow\; \psi_\kappa \alpha_0 < \psi_\kappa \alpha \tag{161}$$

follows since $\psi_\kappa \alpha_0 \in Cl(\alpha, \psi_\kappa \alpha) \cap \kappa$.

We obtain

$$\psi_\kappa \alpha \notin \{\Omega_\sigma \mid \sigma < \Omega_\sigma\} \tag{162}$$

since the assumption $\psi_\kappa \alpha = \Omega_\sigma > \sigma \in Cl(\alpha, \psi_\kappa \alpha)$ leads to $\psi_\kappa \alpha \in Cl(\alpha, \psi_\kappa \alpha)$ which contradicts (160).

We have

$$\psi_\kappa \alpha \in \mathsf{SC} \tag{163}$$

since $\xi, \eta < \psi_\kappa \alpha \subseteq Cl(\alpha, \psi_\kappa \alpha)$ entails $\varphi_\xi \eta \in Cl(\alpha, \psi_\kappa \alpha) \cap \kappa = \psi_\kappa \alpha$.

By definition we have $\Omega_\sigma \in Cl(\alpha, \beta)$ for $\sigma \in Cl(\alpha, \beta)$. This extends to

$$\Omega_\sigma \in Cl(\alpha, \beta) \;\Leftrightarrow\; \sigma \in Cl(\alpha, \beta) \tag{164}$$

which is obvious for $\sigma = \Omega_\sigma$. If we assume $\Omega_\sigma > \sigma \notin Cl(\alpha, \beta)$ then we get $\Omega_\sigma \notin \beta \cup \{0, I\}$ as well as $\Omega_\sigma \neq \psi_\kappa \eta$ for all κ and η. If $\Omega_\sigma = \xi + \eta$ or $\Omega_\sigma = \varphi_\xi \eta$ for some ξ and η then $\Omega_\sigma \in \{\xi, \eta\}$. This shows that the set $Cl(\alpha, \beta) \setminus \{\Omega_\sigma\}$ contains $\beta \cup \{0, I\}$ and has the same closure properties as $Cl(\alpha, \beta)$. Hence $Cl(\alpha, \beta) = Cl(\alpha, \beta) \setminus \{\Omega_\sigma\}$, i.e., $\Omega_\sigma \notin Cl(\alpha, \beta)$.

We define the set of *strongly critical components* $\mathsf{SC}(\alpha)$ of an ordinal α by

$$\mathsf{SC}(\alpha) := \begin{cases} \emptyset & \text{if } \alpha = 0 \\ \{\alpha\} & \text{if } \alpha \in \mathsf{SC} \\ \mathsf{SC}(\xi) \cup \mathsf{SC}(\eta) & \text{if } \alpha =_{\mathsf{NF}} \varphi_\xi \eta \\ \mathsf{SC}(\alpha_1) \cup \cdots \mathsf{SC}(\alpha_n) & \text{if } \alpha =_{\mathsf{NF}} \alpha_1 + \cdots + \alpha_n. \end{cases} \tag{165}$$

Then obviously $\mathsf{SC}(\alpha) \subseteq \mathsf{SC}$ and $\mathsf{SC}(\alpha) \subseteq \alpha + 1$. Now we get

$$\eta \in Cl(\alpha, \beta) \;\Leftrightarrow\; \mathsf{SC}(\eta) \subseteq Cl(\alpha, \beta). \tag{166}$$

The claim is trivial for $\eta \in \mathsf{SC}$. So assume $\eta \notin \mathsf{SC}$. The direction "$\Leftarrow$" is clear by definition. We show the opposite direction by induction on η. Assume $\mathsf{SC}(\eta) \nsubseteq Cl(\alpha, \beta)$. Then $\eta \notin \beta \cup \{0, I\}$. If $\eta =_{\mathsf{NF}} \varphi_{\xi_1}\xi_2$ or $\eta = \xi_1 + \xi_2$ for $\xi_i < \eta$ we get $\mathsf{SC}(\eta) \subseteq \mathsf{SC}(\xi_1) \cup \mathsf{SC}(\xi_2)$ and $\mathsf{SC}(\xi_i) \nsubseteq Cl(\alpha, \beta) \Rightarrow \xi_i \notin Cl(\alpha, \beta)$ by induction hypothesis. Therefore we can't have $\eta = \varphi_{\xi_1}\xi_2$ or $\eta = \xi_1 + \xi_2$ such that $\xi_i \in Cl(\alpha, \beta)$ for $i = 1, 2$. Thus $Cl(\alpha, \beta) \setminus \{\eta\}$ contains $\beta \cup \{0, I\}$ and satisfies the same closure conditions as $Cl(\alpha, \beta)$, i.e., $\eta \notin Cl(\alpha, \beta)$.

If $\kappa = \Omega_{\sigma+1}$ then $\sigma \in Cl(\alpha, \psi_\kappa\alpha)$ by the definition of $\psi_\kappa\alpha$ and (164). Hence $\Omega_\sigma \in Cl(\alpha, \psi_\kappa\alpha) \cap \kappa = \psi_\kappa\alpha < \Omega_{\sigma+1}$ and we have

$$\kappa = \Omega_{\sigma+1} \;\Rightarrow\; \Omega_\sigma < \psi_\kappa\alpha < \Omega_{\sigma+1}. \tag{167}$$

Let $\xi < \alpha$. By (167) we obtain $\kappa \in Cl(\xi, \psi_\kappa\alpha)$. Since $Cl(\xi, \psi_\kappa\alpha) \cap \kappa \subseteq Cl(\alpha, \psi_\kappa\alpha) \cap \kappa = \psi_\kappa\alpha$ we get $\psi_\kappa\xi \le \psi_\kappa\alpha$ and therefore

$$\xi \le \alpha \;\Rightarrow\; \psi_\kappa\xi \le \psi_\kappa\alpha \wedge Cl(\xi, \psi_\kappa\xi) \subseteq Cl(\alpha, \psi_\kappa\alpha). \tag{168}$$

Together with (161) this gives

$$\alpha < \beta \wedge \alpha \in Cl(\gamma, \psi_\kappa\gamma) \textit{ for some } \gamma \le \beta \;\Rightarrow\; \psi_\kappa\alpha < \psi_\kappa\beta. \tag{169}$$

It follows from (167) that for $\kappa < I$ the ordinal $\psi_\kappa\alpha$ is not in $\overline{\mathbf{Reg}}$. For $\kappa = I$ the situation is different. There we have

$$\Omega_{\psi_I\alpha} = \psi_I\alpha \tag{170}$$

showing that $\psi_I\alpha$ is a limit of regular ordinals. To prove (170) choose σ such that $\Omega_\sigma \le \psi_I\alpha < \Omega_{\sigma+1}$. Then $\Omega_{\sigma+1} < I$ which entails $\Omega_{\sigma+1} \notin Cl(\alpha, \psi_I\alpha)$. Hence $\sigma \notin Cl(\alpha, \psi_I\alpha)$. But then $\psi_I\alpha \le \sigma \le \Omega_\sigma \le \psi_I\alpha$.

Since $\Omega_\sigma \in \mathsf{SC}$ and $\Omega_\sigma \ne \psi_\pi\eta$ for $\pi \ne I$ by (167) we also get

$$\Omega_\sigma = \sigma \in Cl(\alpha, \beta) \;\Rightarrow\; \sigma = I \textit{ or } \sigma = \psi_I\eta \textit{ for some } \eta. \tag{171}$$

So I and the ordinals of the form $\psi_I\xi$ are the only ordinals in $Cl(\alpha, \beta)$ which are fixed points of $\lambda\xi\,.\,\Omega_\xi$.

$$[\Omega_\sigma, \Omega_{\sigma+1}) \cap Cl(\alpha, \beta) \ne \emptyset \;\Rightarrow\; \Omega_\sigma \in Cl(\alpha, \beta). \tag{172}$$

We prove (172) by contraposition. Assume that $\Omega_\sigma \notin Cl(\alpha, \beta)$. We are going to show that then $M := Cl(\alpha, \beta) \setminus [\Omega_\sigma, \Omega_{\sigma+1})$ satisfies the same closure condition as $Cl(\alpha, \beta)$. Then $M = Cl(\alpha, \beta)$ which shows $Cl(\alpha, \beta) \cap [\Omega_\sigma, \Omega_{\sigma+1}) = \emptyset$. First we have $\beta \cup \{0, I\} \subseteq M$. If $\xi, \eta \notin [\Omega_\sigma, \Omega_{\sigma+1})$ we get $\xi + \eta \notin [\Omega_\sigma, \Omega_{\sigma+1})$ as well as

$\varphi_\xi\eta \notin [\Omega_\sigma, \Omega_{\sigma+1})$. If $\psi_\kappa\xi \in [\Omega_\sigma, \Omega_{\sigma+1})$ for $\{\xi, \kappa\} \subseteq Cl(\alpha,\beta)$ but $\kappa, \xi \notin [\Omega_\sigma, \Omega_{\sigma+1})$ then we have $\kappa = \Omega_{\sigma+1}$ by (167). Hence $\Omega_\sigma \in Cl(\alpha,\beta)$ by (164). A contradiction. That M is closed under $\lambda\xi\,.\,\Omega_\xi$ is obvious.

Next we observe

$$Cl(\alpha,\Omega_\sigma) = Cl(\alpha, \psi_{\Omega_{\sigma+1}}\alpha). \tag{173}$$

To prove (173) put $\xi := \min\{\eta|\ \eta \notin Cl(\alpha,\Omega_\sigma)\}$. Then $\xi \subseteq Cl(\alpha,\Omega_\sigma) \cap \Omega_{\sigma+1}$ which entails $Cl(\alpha,\Omega_\sigma) = Cl(\alpha,\xi)$. Therefore we have $\xi \notin Cl(\alpha,\xi)$ and $\xi < \Omega_{\sigma+1}$ which implies $\psi_{\Omega_{\sigma+1}}\alpha \leq \xi$ because $\Omega_{\sigma+1} \in Cl(\alpha,\Omega_\sigma) \subseteq Cl(\alpha,\xi)$. Hence $Cl(\alpha,\Omega_\sigma) \subseteq Cl(\alpha,\psi_{\Omega_{\sigma+1}}\alpha) \subseteq Cl(\alpha,\xi) = Cl(\alpha,\Omega_\sigma)$ by(167).

From (173) we obtain especially

$$Cl(I^\Gamma, 0) \cap \Omega_1 = \psi_{\omega_1}(I^\Gamma). \tag{174}$$

It follows from (174) that all ordinals below $\psi_{\omega_1}I^\Gamma$ can be represented by terms which are built up from 0, I, by the unary function $\lambda\xi\,.\,\Omega_\xi$ and the binary functions $\lambda\xi\,.\,\lambda\eta\,.\,\xi+\eta$, $\lambda\xi\,.\,\lambda\eta\,.\,\varphi_\xi\eta$ and $\lambda\pi\,.\,\lambda\eta\,.\,\psi_\pi\eta$ solely. We already defined the notions $\alpha =_{\mathsf{NF}} \alpha_1 + \cdots + \alpha_n$ and $\alpha =_{\mathsf{NF}} \varphi_\xi\eta$. This gives unique term notations for terms not in SC. If we define

$$\alpha =_{\mathsf{NF}} \Omega_\sigma \quad :\Leftrightarrow \quad \alpha = \Omega_\sigma \wedge \sigma < \alpha$$

and

$$\alpha =_{\mathsf{NF}} \psi_\kappa\xi \quad :\Leftrightarrow \quad \alpha = \psi_\kappa\xi \wedge \xi \in Cl(\xi, \psi_\kappa\xi)$$

we get

$$\alpha =_{\mathsf{NF}} \Omega_\xi \wedge \alpha =_{\mathsf{NF}} \Omega_\eta \;\Rightarrow\; \xi = \eta \tag{175}$$

as well as

$$\alpha =_{\mathsf{NF}} \psi_\kappa\xi \wedge \beta =_{\mathsf{NF}} \psi_\kappa\eta \;\Rightarrow\; (\alpha = \beta \;\Leftrightarrow\; \xi = \eta). \tag{176}$$

While (175) holds obviously we get (176) because $\xi < \eta$ and $\xi \in Cl(\xi, \psi_\kappa\xi)$ implies $\psi_\kappa\xi \leq \psi_\kappa\eta$ by (168) and therefore $\xi \in Cl(\eta, \psi_\kappa\eta) \cap \eta$. Hence $\alpha < \beta$ by (161). This entails also the opposite direction.

As a consequence of (176), (167) and (170) we obtain that $\alpha =_{\mathsf{NF}} \psi_\pi\eta$ determines π and η uniquely. In order to decide whether $\psi_\kappa\alpha$ is in normal form, we have to decide $\alpha \in Cl(\alpha, \psi_\kappa\alpha)$. By (173), however, it suffices to decide $\alpha \in Cl(\alpha, |\psi_\kappa\alpha|)$ where $|\psi_\kappa\alpha|$ denotes the cardinality of $\psi_\kappa\alpha$. For $\mu \in \overline{\mathbf{Reg}}$ we get more generally $\beta \in Cl(\alpha,\mu)$ if and only if one of the following conditions is satisfied:

$$\begin{array}{l} \beta \in \mu \cup \{0, I\} \\ \beta \notin \mathsf{SC} \textit{ and } \mathsf{SC}(\beta) \subseteq Cl(\alpha,\mu) \\ \mu \leq \beta = \psi_\pi\eta,\ \eta < \alpha \textit{ and } \{\pi,\eta\} \subseteq Cl(\alpha,\mu) \\ \mu \leq \beta = \Omega_\sigma \textit{ and } \sigma \in Cl(\alpha,\mu). \end{array} \tag{177}$$

This gives raise for the following definition.

3.4.4.2. Definition. For ordinals $\alpha, \mu \in Cl(I^\Gamma, 0)$ we define finite sets $K_\mu(\alpha)$ by

$$K_\mu(\alpha) := \begin{cases} \bigcup\{K_\mu(\beta) \mid \beta \in \mathsf{SC}(\alpha)\} & \text{if } \alpha \notin \mathsf{SC} \\ \emptyset & \text{if } \alpha \in \mu \cup \{I, 0\} \\ \{\xi\} \cup K_\mu(\xi) \cup K_\mu(\pi) & \text{if } \mu \leq \alpha = \psi_\pi \xi \\ K_\mu(\sigma) & \text{if } \mu \leq \alpha = \Omega_\sigma. \end{cases}$$

It does not matter that the above definition is not deterministic in the case that α is an ordinal below μ which is not strongly critical. We get $K_\mu(\alpha) = \emptyset$ regardless of the clause we apply.

From (177) and Definition 3.4.4.2 we get $\beta \in Cl(\alpha, \mu) \;\Leftrightarrow\; (\forall \xi \in K_\mu(\beta))[\xi < \alpha]$. Putting $K_\mu(\alpha) < \beta \;:\Leftrightarrow\; (\forall \xi \in K_\mu(\alpha))[\xi < \beta]$ we get

$$\alpha =_{\mathsf{NF}} \psi_\pi \eta \;\Leftrightarrow\; \alpha = \psi_\pi \eta \textit{ and } K_{|\psi_\pi \eta|}(\eta) < \eta. \tag{178}$$

Observe that for ordinals $\alpha \in Cl(I^\Gamma, 0)$ the cardinality is determined by

$$|\alpha| = \begin{cases} \alpha & \text{if } \alpha = \Omega_\sigma \text{ or } \alpha = \psi_I \eta \\ \max\{|\alpha_1|, \ldots, |\alpha_n|\} & \text{if } \alpha = \alpha_1 + \cdots + \alpha_n \\ |\eta| & \text{if } \alpha = \varphi_\xi \eta \\ \Omega_\sigma & \text{if } \alpha = \psi_\pi \eta \text{ and } I > \pi = \Omega_{\sigma+1}. \end{cases} \tag{179}$$

All that opens the possibility to define simultaneously a term-system T together with an evaluation function $|\;|_O \colon T \longrightarrow \mathbf{On}$ and a "less than" relation $<$ on the ordinal–terms such that $a < b \;\Leftrightarrow\; |a|_O < |b|_O$ and the "less–than" relation on the ordinal terms becomes primitive recursive. We will not do this in all details but only indicate the essential steps. There are the following sets of ordinal–terms

- *the set T comprising all ordinal terms*
- *the set P of principal terms denoting additively indecomposable ordinals*
- *the set SC denoting strongly critical ordinals in* SC
- *the set K of cardinal terms denoting ordinals in* $\overline{\mathbf{Reg}}$
- *the set F of fixed–point terms denoting ordinals which are fixed–points of the enumerating function of* $\overline{\mathbf{Reg}}$
- *the set R of regular–terms denoting ordinals in* $\mathbf{Reg}$

which are defined by:

- $R \subseteq K \subseteq SC \subseteq P \subseteq T$
- $F \subseteq K$
- $\underline{0} \in T$, $|\underline{0}|_O := 0$
- $\underline{I} \in R \cap F$, $|\underline{I}|_O := I$
- $a_1, \ldots, a_n \in P$ *and* $a_1 \geq \cdots \geq a_n \;\Rightarrow\; a_1 + \cdots + a_n \in T$, $|a_1 + \cdots + a_n|_O := |a_1|_O + \cdots + |a_n|_O$

- $a, b \in T \Rightarrow \overline{\varphi}_a b \in P$, $|\overline{\varphi}_a b|_{\mathcal{O}} := \overline{\varphi}_{|a|_{\mathcal{O}}} |b|_{\mathcal{O}}$

where $\overline{\varphi}$ is the fixed–point free version of the function φ, i.e.

$$\overline{\varphi}_\alpha \beta := \begin{cases} \varphi_\alpha(\beta+1) & \text{if } \beta = \gamma + n \text{ for some } n < \omega \text{ and } \gamma \text{ such that } \varphi_\alpha \gamma = \gamma \\ \varphi_\alpha \beta & \text{otherwise,} \end{cases}$$

- *If* $p \in R$, $a \neq \underline{I}$, $a \in T$ *and* $K_{\|\psi_p a\|}(a) < a$ *then* $\psi_p a \in SC$ *and* $|\psi_p a|_{\mathcal{O}} := \psi_{|p|_{\mathcal{O}}} |a|_{\mathcal{O}}$
- *If* $a \in T$ *and* $K_{\|\psi_{\underline{I}} a\|}(a) < a$ *then* $\psi_{\underline{I}} a \in F$ *and* $|\psi_{\underline{I}} a|_{\mathcal{O}} := \psi_I |a|_{\mathcal{O}}$
- *If* $a \in T \setminus F$ *then* $\Omega_a \in K$ *and* $\Omega_{a+\underline{1}} \in R$ *and* $|\Omega_a|_{\mathcal{O}} := \Omega_{|a|_{\mathcal{O}}}$

where $\underline{1} := \varphi_{\underline{0}} \underline{0}$.

The definition of the sets $K_p(a)$ for $p \in K$ and $a \in T$ should be obvious from Definition 3.4.4.2. Similarly obvious is the definition of the "cardinality" $\|a\| = ||a|_{\mathcal{O}}|$ of an ordinal term a from (179).

Finally we have (omitting some obvious cases)

$$a < b \Leftrightarrow \begin{cases} a = 0 \wedge b \neq 0, \text{ or } \|a\| < \|b\|, \text{ or } \|a\| = \|b\| \text{ and one of the following conditions is satisfied} \\ a = \psi_p c \wedge b = \psi_p d \wedge c < d \\ a = \overline{\varphi}_c d \wedge b \in SC \wedge c, d < b \\ a \in SC \wedge b = \overline{\varphi}_c d \wedge (a \leq c \vee a \leq d) \\ a = \overline{\varphi}_c d \wedge b = \overline{\varphi}_e f \wedge \begin{cases} c < e \wedge d < b \\ c = e \wedge d < f \\ e < c \wedge a < f \end{cases} \\ a = a_1 + \cdots + a_m \wedge b = b_1 + \cdots + b_n \\ \quad \wedge (\exists k \leq n)[(\forall i < k)[a_i = b_i] \wedge [(k \leq m \wedge a_k < b_k) \vee m < k]] \\ \quad \text{for } m, n \geq 1 \end{cases}$$

This shows that the set T together with the sets $K_p(a)$, the cardinality $\|a\|$ and the relation $a < b$ can be simultaneously defined by course–of–values recursion. It follows that T and the $<$–relation on T are primitive recursive and it is pretty easy to realize that

$$\{|a|_{\mathcal{O}} \mid a \in T\} = Cl(I^\Gamma, 0).$$

Since $Cl(I^\Gamma, 0) \cap \Omega_1$ is a segment of the ordinals we obtain $Cl(I^\Gamma, 0) \cap \Omega_1 \subseteq \omega_1^{CK}$ which entails $\psi_{\omega_1} I^\Gamma < \omega_1^{CK}$ by (174). This shows that we at least may replace Ω_1 by ω_1^{CK}. It needs, however, considerably more effort to show that we can replace all regular ordinals by recursively regular ones without changing the segment $Cl(I^\Gamma, 0) \cap \omega_1^{CK}$. Nevertheless, we will pretend that we have done that and interpret the ordinals in **Reg** as recursively regular, i.e. admissible ordinals.

3.4.5. The Collapsing Theorem

We are now prepared to define the controlling operator which allows to collapse the controlled derivation.

3.4.5.1. Definition. Put

$$\mathcal{H}_\gamma(X) := \bigcap\{Cl(\alpha,\beta) \mid X \subseteq Cl(\alpha,\beta) \wedge \gamma < \alpha\}.$$

We certainly have $0 \in \mathcal{H}_\gamma$ for all γ and obtain from (166) that all $\mathcal{H}_\gamma$ are Cantorian closed and closed under φ. By definition we get $X \subseteq \mathcal{H}_\gamma(X)$ for all $X \subseteq \mathbf{On}$. If we assume $X \subseteq \mathcal{H}_\gamma(Y)$, $\xi \in \mathcal{H}_\gamma(X)$ and $Y \subseteq Cl(\alpha,\beta)$ for some $\gamma < \alpha$ then we obtain also $X \subseteq Cl(\alpha,\beta)$ and therefore also $\xi \in Cl(\alpha,\beta)$. Hence $\xi \in \mathcal{H}_\gamma(Y)$ and we have $\mathcal{H}_\gamma(X) \subseteq \mathcal{H}_\gamma(Y)$. Pulling this together we have the following lemma.

3.4.5.2. Lemma. *The operators $\mathcal{H}_\gamma$ are all acceptable and closed under the Veblen–function φ and the function $\lambda\xi\,.\,\Omega_\xi$.*

But we also get the closure of the operators $\mathcal{H}_\gamma$ under the functions ψ_ξ in the following sense.

$$\xi \le \gamma \wedge \xi, \kappa \in \mathcal{H}_\gamma(X) \;\Rightarrow\; \psi_\kappa\xi \in \mathcal{H}_\gamma(X). \tag{180}$$

From (172) and (164) we get

$$[\Omega_\sigma, \Omega_{\sigma+1}] \cap \mathcal{H}_\gamma(X) \neq \emptyset \;\Rightarrow\; \{\Omega_\sigma, \Omega_{\sigma+1}\} \subseteq \mathcal{H}_\gamma(X). \tag{181}$$

The operators are also cumulative. We have

$$\gamma \le \delta \;\Rightarrow\; \mathcal{H}_\gamma \subseteq \mathcal{H}_\delta. \tag{182}$$

The aim of this section is to prove the Collapsing Theorem, i.e., to show that every derivation of a set of Σ^κ–sentences can be collapsed into a derivation whose derivation length is less than κ. Obviously such an derivation must not contain cuts whose cut–sentence has a complexity above κ. Therefore collapsing below κ has to come together with the elimination of all cuts of complexities above κ. Cut elimination as stated in Theorem 3.4.2.2 is of no help since it doesn't say anything about the controlling operators. However, we know already from the Predicative Elimination Lemma 3.4.3.6 that we can eliminate cuts whose cut-ranks are in $(\Omega_\lambda, \Omega_{\lambda+1}]$ for $\lambda \in \mathsf{Lim}$ or in $(\Omega_\sigma + 1, \Omega_{\sigma+1}]$ without losing information about the controlling operator. Therefore we introduce the class of left initial points of these intervals, i.e., we put

$$\widehat{\mathbf{Reg}} := \{I+1\} \cup \{\Omega_\sigma \mid \sigma \in I \cap \mathsf{Lim}\} \cup \{\Omega_\sigma + 1 \mid \sigma \in I \setminus \mathsf{Lim}\}$$

The crucial situation which is not yet covered by the Predicative Elimination Lemma are derivations of the form

$$\mathcal{H} \,\Big|\!\frac{\alpha}{\mu}\, \Delta \;\; \textit{for } \mu \in \widehat{\mathbf{Reg}}.$$

In the case that Δ is a set of Σ^κ–sentences we want to collapse this derivation below κ. We are going to show the following theorem.

3.4.5.3. Collapsing Theorem. *Let Δ be a set of Σ^κ–sentences, $\mu \in \widehat{\mathbf{Reg}}$, $\{\kappa, \gamma, \mu\} \subseteq \mathcal{H}_\gamma$ and assume $\mathcal{H}_\gamma \mathrel{|\!\frac{\alpha}{\mu}} \Delta$. Then this derivation is collapsed to $\mathcal{H}_{\gamma+\omega^{\mu+\alpha}} \mathrel{|\!\frac{\psi_\kappa(\gamma+\omega^{\mu+\alpha})}{\psi_\kappa(\gamma+\omega^{\mu+\alpha})}} \Delta$.*

To make the induction work we assume that Θ is a set of $\mathcal{L}_{\mathrm{RS}}$–terms and prove the more general claim

$$\left.\begin{array}{l} \Delta \subseteq \Sigma^\kappa,\ \mu \in \widehat{\mathbf{Reg}} \\ \mathsf{par}(\Theta) \subseteq \bigcap\{Cl(\gamma+1, \psi_\tau(\gamma+1))\mid \tau \geq \kappa\} \\ \kappa, \gamma, \mu \in \mathcal{H}_\gamma[\Theta] \\ \mathcal{H}_\gamma[\Theta] \mathrel{|\!\frac{\alpha}{\mu}} \Delta \end{array}\right\} \Rightarrow \mathcal{H}_{\gamma+\omega^{\mu+\alpha}}[\Theta] \mathrel{|\!\frac{\psi_\kappa(\gamma+\omega^{\mu+\alpha})}{\psi_\kappa(\gamma+\omega^{\mu+\alpha})}} \Delta \quad \text{(i)}$$

by main induction on μ with side induction on α. To simplify notations we abbreviate the first three lines in the assumptions of claim (i) by $\mathsf{Asmp}(\Delta; \Theta; \mu; \kappa; \gamma)$. So (i) becomes

$$\mathsf{Asmp}(\Delta; \Theta; \mu; \kappa; \gamma) \wedge \mathcal{H}_\gamma[\Theta] \mathrel{|\!\frac{\alpha}{\mu}} \Delta \Rightarrow \mathcal{H}_{\gamma+\omega^{\mu+\alpha}}[\Theta] \mathrel{|\!\frac{\psi_\kappa(\gamma+\omega^{\mu+\alpha})}{\psi_\kappa(\gamma+\omega^{\mu+\alpha})}} \Delta. \quad \text{(ii)}$$

To prepare the induction we first observe that by Lemma 3.4.5.2 we have

$$\mathsf{Asmp}(\Delta; \Theta; \mu; \kappa; \gamma) \wedge \alpha \in \mathcal{H}_\gamma[\Theta] \Rightarrow \gamma + \omega^{\mu+\alpha} \in \mathcal{H}_\gamma[\Theta]. \quad \text{(iii)}$$

From (iii), (182), $\gamma \leq \gamma + \omega^{\mu+\alpha}$ and (180)we then obtain

$$\mathsf{Asmp}(\Delta; \Theta; \mu; \kappa; \gamma) \wedge \alpha \in \mathcal{H}_\gamma[\Theta] \Rightarrow \psi_\kappa(\gamma + \omega^{\mu+\alpha}) \in \mathcal{H}_{\gamma+\omega^{\mu+\alpha}}[\Theta]. \quad \text{(iv)}$$

Most important is the following collapsing property of the function $\psi_\kappa(\gamma + \omega^{\mu+\alpha})$.

$$\mathsf{Asmp}(\Delta; \Theta; \mu; \kappa; \gamma) \wedge \alpha \in \mathcal{H}_\gamma[\Theta] \wedge \gamma+\omega^{\mu+\alpha} < \beta \Rightarrow \psi_\kappa(\gamma + \omega^{\mu+\alpha}) < \psi_\kappa\beta. \quad \text{(v)}$$

To obtain (v) it suffices by (169) to find some $\delta \leq \beta$ such that $\gamma+\omega^{\mu+\alpha} \in Cl(\delta, \psi_\kappa\delta)$. But we have $\gamma+1 \leq \gamma+\omega^{\mu+\alpha}$ and $\gamma+\omega^{\mu+\alpha} \in \mathcal{H}_\gamma[\Theta] \subseteq Cl(\gamma+1, \psi_\kappa(\gamma+1))$ by (iii) and the assumption $\mathsf{par}(\Theta) \subseteq Cl(\gamma+1, \psi_\kappa(\gamma+1))$. So we may choose $\delta := \gamma+1$.

To prove claim (i) we run through the cases. If the last inference was by $(\bigwedge)$ then there is a sentence $F \in \Delta \cap \bigwedge$-type and we have the premises

$$\mathcal{H}_\gamma[\Theta \cup t_F(G)] \mathrel{|\!\frac{\alpha_G}{\mu}} \Delta, G \ \textit{and}\ \alpha_G < \alpha \ \textit{for all}\ G \in C(F). \quad \text{(vi)}$$

Since $F \in \Sigma^\kappa \cap \bigwedge$-type there is a $\delta \in \mathsf{par}(F) \cap \kappa \subseteq \mathcal{H}_\gamma[\Theta] \cap \kappa$ such that $\mathsf{par}(t_F(G)) \subseteq \delta$ for all $G \in C(F)$. For $\tau \geq \kappa$ we get $\mathcal{H}_\gamma[\Theta] \cap \tau \subseteq Cl(\gamma+1, \psi_\tau(\gamma+1)) \cap \tau = \psi_\tau(\gamma+1)$. Hence $\mathsf{par}(t_F(G)) \subseteq \delta < \psi_\tau(\gamma+1) \subseteq Cl(\gamma+1, \psi_\tau(\gamma+1))$ which shows $\mathsf{par}(\Theta \cup t_F(G)) \subseteq \bigcap_{\tau \geq \kappa} Cl(\gamma+1, \psi_\tau(\gamma+1))$ for all $G \in C(F)$. So we have $\mathsf{Asmp}(\Delta, G; \Theta \cup t_F(G); \mu; \kappa; \gamma)$ for all $G \in C(F)$ and get

$$\mathcal{H}_{\gamma+\omega^{\mu+\alpha_G}}[\Theta \cup t_F(G)] \mathrel{|\!\frac{\psi_\kappa(\gamma+\omega^{\mu+\alpha_G})}{\psi_\kappa(\gamma+\omega^{\mu+\alpha_G})}} \Delta, G \quad \text{(vii)}$$

for all $G \in C(F)$ by the side induction hypothesis. By (v) we have $\psi_\kappa(\gamma + \omega^{\mu+\alpha_G}) < \psi_\kappa(\gamma + \omega^{\mu+\alpha})$ and $\psi_\kappa(\gamma + \omega^{\mu+\alpha}) \in \mathcal{H}_{\gamma+\omega^{\mu+\alpha}}[\Theta]$ by (iv). Therefore we obtain

$$\mathcal{H}_{\gamma+\omega^{\mu+\alpha}}[\Theta] \,\Big|\frac{\psi_\kappa(\gamma+\omega^{\mu+\alpha})}{\psi_\kappa(\gamma+\omega^{\mu+\alpha})}\, \Delta$$

from (vii) by an inference $(\bigwedge)$.

In the case of an inference $(\bigvee)$ there is a sentence $F \in \Delta \cap \bigvee$-type, a sentence $G \in \mathcal{C}(F)$ and some $\alpha_0 < \alpha$ such that

$$\mathcal{H}_\gamma[\Theta] \,\Big|\frac{\alpha_0}{\mu}\, \Delta, G \ \textit{and}\ o_F(G) < \alpha. \tag{viii}$$

By side induction hypothesis we obtain

$$\mathcal{H}_{\gamma+\omega^{\mu+\alpha_0}}[\Theta] \,\Big|\frac{\psi_\kappa(\gamma+\omega^{\mu+\alpha_0})}{\psi_\kappa(\gamma+\omega^{\mu+\alpha_0})}\, \Delta, G. \tag{ix}$$

From $\gamma + \omega^{\mu+\alpha_0} < \gamma + \omega^{\mu+\alpha}$ and $\alpha_0 \in \mathcal{H}_\gamma[\Theta]$ we obtain $\psi_\kappa(\gamma + \omega^{\mu+\alpha_0}) < \psi_\kappa(\gamma + \omega^{\mu+\alpha})$ by (v). Since $o_F(G) \in \mathsf{par}(\Delta, G) \cap \kappa \subseteq \mathcal{H}_\gamma[\Theta] \cap \kappa \subseteq \psi_\kappa(\gamma + 1) \leq \psi_\kappa(\gamma + \omega^{\mu+\alpha})$ we obtain

$$\mathcal{H}_{\gamma+\omega^{\mu+\alpha}}[\Theta] \,\Big|\frac{\psi_\kappa(\gamma+\omega^{\mu+\alpha})}{\psi_\kappa(\gamma+\omega^{\mu+\alpha})}\, \Delta$$

from (ix) by an inference $(\bigvee)$.

In the case of an inference (Ref_κ) there is a Π_2^κ sentence $(\forall x \in \mathsf{L}_\kappa)(\exists y \in \mathsf{L}_\kappa)F(x, y)$ such that $(\exists z \in \mathsf{L}_\kappa)[z \neq \emptyset \wedge (\forall x \in z)(\exists y \in z)F(x, y)] \in \Delta$, an ordinal α_0 such that $\kappa, \alpha_0 + 1 < \alpha$ and

$$\mathcal{H}_\gamma[\Theta] \,\Big|\frac{\alpha_0}{\mu}\, \Delta, (\forall x \in \mathsf{L}_\kappa)(\exists y \in \mathsf{L}_\kappa)F(x, y). \tag{x}$$

By inversion we obtain from (x)

$$\mathcal{H}_\gamma[\Theta, t] \,\Big|\frac{\alpha_0}{\mu}\, \Delta, (\exists y \in \mathsf{L}_\kappa)F(t, y) \tag{xi}$$

for all $t \in \mathcal{T}_\kappa$. Let $\eta := \psi_\kappa(\gamma + \omega^{\mu+\alpha_0+1})$ and $\gamma_n := \gamma + \omega^{\mu+\alpha_0} \cdot n$. By $\mu, \gamma, \alpha_0 \in \mathcal{H}_\gamma[\Theta]$ we get $\gamma_n \in \mathcal{H}_\gamma[\Theta] \subseteq \mathcal{H}_{\gamma_n}[\Theta, t]$ for all $n \in \omega$. For $t \in \mathcal{T}_\eta$ there is an $n \in \omega$ such that $\mathsf{stg}(t) < \psi_\kappa\gamma_n$. We have $\mathsf{par}(\Theta) \subseteq Cl(\gamma + 1, \psi_\tau(\gamma + 1)) \subseteq Cl(\gamma_n + 1, \psi_\tau(\gamma_n + 1))$ and obtain $\mathsf{par}(t) \subseteq \mathsf{stg}(t) < \psi_\kappa\gamma_n \in \mathcal{H}_{\gamma_n}[\Theta] \cap \tau \subseteq Cl(\gamma_n + 1, \psi_\tau(\gamma_n + 1)) \cap \tau = \psi_\tau(\gamma_n + 1)$ for all $\tau \geq \kappa$. Hence $\mathsf{par}(t) \subseteq Cl(\gamma_n + 1, \psi_\tau(\gamma_n + 1))$ for all $\tau \geq \kappa$. So we have

$$\mathsf{Asmp}(\Delta, (\exists y \in \mathsf{L}_\kappa)F(t, y); \Theta, t; \mu; \kappa; \gamma_n) \tag{xii}$$

for all $t \in \mathcal{T}_\eta$. Observing that $\gamma_n + \omega^{\mu+\alpha_0} = \gamma_{n+1}$ the induction hypothesis applied to (xi) yields

$$\mathcal{H}_{\gamma_{n+1}}[\Theta, t] \,\Big|\frac{\psi_\kappa(\gamma_{n+1})}{\psi_\kappa(\gamma_{n+1})}\, \Delta, (\exists y \in \mathsf{L}_\kappa)F(t, y) \tag{xiii}$$

for all $t \in \mathcal{T}_\eta$. Using the Boundedness Theorem 3.4.3.7 we obtain

$$\mathcal{H}_{\gamma_{n+1}}[\Theta, t] \,\Big|\frac{\psi_\kappa(\gamma_{n+1})}{\psi_\kappa(\gamma_{n+1})}\, \Delta, (\exists y \in \mathsf{L}_\eta)F(t, y) \tag{xiv}$$

for all $t \in \mathcal{T}_\eta$. Since $\gamma_n < \gamma + \omega^{\mu+\alpha_0+1} < \gamma + \omega^{\mu+\alpha}$ we get

$$\mathcal{H}_{\gamma+\omega^{\mu+\alpha_0+1}}[\Theta] \,\Big|\frac{\psi_\kappa(\omega^{\mu+\alpha_0+1})}{\psi_\kappa(\omega^{\mu+\alpha_0+1})}\, \Delta, (\forall x \in \mathsf{L}_\eta)(\exists y \in \mathsf{L}_\eta)F(x, y) \tag{xv}$$

by an inference $(\bigwedge)$. Since $\mathcal{H}_0 \overset{\delta}{\underset{0}{\vdash}} (\exists x \in \mathsf{L}_\eta)(x \in \mathsf{L}_\eta)$ for some $\delta < \psi_\kappa(\gamma + \omega^{\mu+\alpha})$ (cf. (191) below) we obtain

$$\mathcal{H}_{\gamma+\omega^{\mu+\alpha}}[\Theta] \overset{\psi_\kappa(\gamma+\omega^{\mu+\alpha})}{\underset{\psi_\kappa(\gamma+\omega^{\mu+\alpha})}{\vdash}} \Delta, (\exists z \in \mathsf{L}_\kappa)[z \neq \emptyset \wedge (\forall x \in z)(\exists y \in z)F(x,y)]$$

from (xv) by inferences $(\bigwedge)$ and $(\bigvee)$.

In the case of an inference (Ref_π) with $\pi < \kappa$ we obtain the claim directly from the side induction hypotheses.

The real crucial case is a cut. There we have a sentence A with $\mathsf{rk}(A) \leq \mu$ and an ordinal α_0 such that

$$\mathcal{H}_\gamma[\Theta] \overset{\alpha_0}{\underset{\mu}{\vdash}} \Delta, A \ \textit{and}\ \mathcal{H}_\gamma[\Theta] \overset{\alpha_0}{\underset{\mu}{\vdash}} \Delta, \neg A. \tag{xvi}$$

The simple case is $\mathsf{rk}(A) < \kappa$. Since $\mathcal{H}_\gamma[\Theta]$ is Cantorian closed and $\mathsf{par}(A) \subseteq \mathcal{H}_\gamma[\Theta]$ we get $\mathsf{rk}(A) \in \mathcal{H}_\gamma[\Theta] \cap \kappa \subseteq \psi_\kappa(\gamma+1) \leq \psi_\kappa(\gamma + \omega^{\mu+\alpha})$. This together with the side induction hypotheses applied to (xvi) yields the claim by a cut.

Now assume $\kappa \leq \mathsf{rk}(A) \leq \mu$. First we consider the sub-case that $\mathsf{rk}(A) \notin$ **Reg**. Then $\kappa \leq \mathsf{rk}(A) < \mathsf{rk}(A)^+ =: \pi \leq \mu$. As before we have $\mathsf{rk}(A) \in \mathcal{H}_\gamma[\Theta]$ and thus also $\pi \in \mathcal{H}_\gamma[\Theta]$ by (199) and trivially $\Delta \cup \{A\} \cup \{\neg A\} \subseteq \Sigma^\pi$. Hence $\mathsf{Asmp}(\Delta, (\neg)A; \Theta; \mu; \pi; \gamma)$ and the induction hypothesis applied to (xvi) yields

$$\mathcal{H}_{\gamma+\omega^{\mu+\alpha_0}}[\Theta] \overset{\psi_\pi(\gamma+\omega^{\mu+\alpha_0})}{\underset{\psi_\pi(\gamma+\omega^{\mu+\alpha_0})}{\vdash}} \Delta, A \ \textit{and}\ \mathcal{H}_{\gamma+\omega^{\mu+\alpha_0}}[\Theta] \overset{\psi_\pi(\gamma+\omega^{\mu+\alpha_0})}{\underset{\psi_\pi(\gamma+\omega^{\mu+\alpha_0})}{\vdash}} \Delta, \neg A. \tag{xvii}$$

In this situation we want to make a cut, apply the Predicative Elimination Lemma (Lemma 3.4.3.6) and then use the main induction hypothesis. Since the same situation will return we are going to state this in a more general form.

$$\begin{array}{l} \textit{Let}\ \gamma \leq \eta < \gamma + \omega^{\mu+\alpha},\ \eta \in \mathcal{H}_\eta[\Theta],\ \mathsf{rk}(A) < \pi \leq \mu,\ \mathcal{H}_\eta[\Theta] \overset{\beta}{\underset{\beta}{\vdash}} \Delta, A\ \textit{and} \\ \mathcal{H}_\eta[\Theta] \overset{\beta}{\underset{\beta}{\vdash}} \Delta, \neg A\ \textit{for some}\ \beta < \pi.\ \textit{Then}\ \mathcal{H}_{\gamma+\omega^{\mu+\alpha}}[\Theta] \overset{\psi_\kappa(\gamma+\omega^{\mu+\alpha})}{\underset{\psi_\kappa(\gamma+\omega^{\mu+\alpha})}{\vdash}} \Delta. \end{array} \tag{xviii}$$

Applying (xviii) with $\eta = \gamma + \omega^{\mu+\alpha_0}$ and $\beta = \psi_\pi(\gamma + \omega^{\mu+\alpha_0})$ to (xvii) will then finish the case.

To prove (xviii) put $\delta := \max\{\mathsf{rk}(A), \beta\} + 1 < \pi$ and choose $\rho \in \overline{\mathbf{Reg}}$ such that $\rho < \delta < \rho^+$. Defining $\hat{\rho} := \rho$ if $\rho \notin \mathbf{Reg}$ and $\hat{\rho} := \rho + 1$ otherwise we get $[\hat{\rho}, \hat{\rho} + \omega^\delta) \cap \mathbf{Reg} = \emptyset$. From the hypotheses we obtain

$$\mathcal{H}_\eta[\Theta] \overset{\beta+1}{\underset{\hat{\rho}+\omega^\delta}{\vdash}} \Delta \tag{xix}$$

by a cut. Since $\mathsf{par}(A) \subseteq \mathcal{H}_\eta[\Theta]$ and $\beta \in \mathcal{H}_\eta[\Theta]$ we have $\delta \in \mathcal{H}_\eta[\Theta]$. Using the Predicative Elimination Lemma (Lemma 3.4.3.6) we therefore obtain

$$\mathcal{H}_\eta[\Theta] \overset{\varphi_\delta(\beta+1)}{\underset{\hat{\rho}}{\vdash}} \Delta. \tag{xx}$$

By (199) we get $\hat{\rho} \in \mathcal{H}_\eta[\Theta]$. From $\gamma \leq \eta$ we also obtain $\mathsf{par}(\Theta) \subseteq \bigcap_{\tau \geq \kappa} Cl(\gamma+1, \psi_\tau(\gamma+1)) \subseteq \bigcap_{\tau \geq \kappa} Cl(\eta+1, \psi_\tau(\eta+1))$. So we have $\mathsf{Asmp}(\Delta; \Theta; \hat{\rho}; \kappa; \eta)$ and $\hat{\rho} < \mu$. The main hypothesis applied to (xx) thus yields

$$\mathcal{H}_{\eta+\omega^{\hat\rho+\varphi_\delta(\beta+1)}}[\Theta] \;\Big|\frac{\psi_\kappa(\eta+\omega^{\hat\rho+\varphi_\delta(\beta+1)})}{\psi_\kappa(\eta+\omega^{\hat\rho+\varphi_\delta(\beta+1)})}\, \Delta. \tag{xxi}$$

Since $\hat\rho + \varphi_\delta(\beta+1) < \pi \le \mu$ we have $\omega^{\hat\rho+\varphi_\delta(\beta+1)} < \omega^\mu$. From $\eta < \gamma + \omega^{\mu+\alpha}$ we either get $\eta \le \gamma$ and thus also $\eta + \omega^{\hat\rho+\varphi_\delta(\beta+1)} < \gamma + \omega^\mu \le \gamma + \omega^{\mu+\alpha}$ or $\eta = \gamma + \zeta$ with $\zeta < \omega^{\mu+\alpha}$ which entails $\eta + \omega^{\hat\rho+\varphi_\delta(\beta+1)} < \gamma + \zeta + \omega^\mu < \gamma + \omega^{\mu+\alpha}$. Hence $\psi_\kappa(\eta + \omega^{\hat\rho+\varphi_\delta(\beta+1)}) \le \psi_\kappa(\gamma + \omega^{\mu+\alpha})$ and we obtain the claim from (xxi) by a structural rule.

Now we consider the sub-case that $\kappa \le \mathsf{rk}(A) =: \pi \in \mathbf{Reg}$. Then either A or $\neg A$ has the shape $(\exists x \in \mathsf{L}_\pi)G(x)$. Since $\kappa \le \pi$ we easily check $\mathsf{Asmp}(\Delta, (\exists x \in \mathsf{L}_\pi)G(x); \Theta; \mu; \pi; \gamma)$ and obtain thus by the induction hypothesis

$$\mathcal{H}_{\gamma+\omega^{\mu+\alpha_0}}[\Theta] \;\Big|\frac{\psi_\pi(\gamma+\omega^{\mu+\alpha_0})}{\psi_\pi(\gamma+\omega^{\mu+\alpha_0})}\, \Delta, (\exists x \in \mathsf{L}_\pi)G(x). \tag{xxii}$$

Defining $\zeta := \psi_\pi(\mu + \omega^{\gamma+\alpha_0})$ we have

$$\zeta \in \mathcal{H}_{\gamma+\omega^{\mu+\alpha_0}}[\Theta] \cap \pi \tag{xxiii}$$

by (iv) and get from (xxii) by the Boundedness Theorem (Theorem 3.4.3.7)

$$\mathcal{H}_{\gamma+\omega^{\mu+\alpha_0}}[\Theta] \;\Big|\frac{\psi_\pi(\gamma+\omega^{\mu+\alpha_0})}{\psi_\pi(\gamma+\omega^{\mu+\alpha_0})}\, \Delta, (\exists x \in \mathsf{L}_\zeta)G(x). \tag{xxiv}$$

Respecting (xxiii) we may apply Downward Persistency (153) to the second premise in (xvi) and obtain

$$\mathcal{H}_{\gamma+\omega^{\mu+\alpha_0}}[\Theta] \;\Big|\frac{\alpha_0}{\mu}\, \Delta, (\forall x \in \mathsf{L}_\zeta)\neg G(x). \tag{xxv}$$

Since $\gamma < \gamma + \omega^{\mu+\alpha_0} \in \mathcal{H}_\gamma[\Theta]$ we obtain $\mathsf{Asmp}(\Delta, (\forall x \in \mathsf{L}_\zeta)\neg G(x); \Theta; \mu; \pi; \gamma + \omega^{\mu+\alpha_0})$ and may therefore apply the side induction hypothesis to (xxv). This yields

$$\mathcal{H}_{\gamma+\omega^{\mu+\alpha_0}+\omega^{\mu+\alpha_0}}[\Theta] \;\Big|\frac{\psi_\pi(\gamma+\omega^{\mu+\alpha_0}+\omega^{\mu+\alpha_0})}{\psi_\pi(\gamma+\omega^{\mu+\alpha_0}+\omega^{\mu+\alpha_0})}\, \Delta, (\forall x \in \mathsf{L}_\zeta)\neg G(x). \tag{xxvi}$$

Putting $\eta := \gamma + \omega^{\mu+\alpha_0} + \omega^{\mu+\alpha_0}$ and $\beta := \psi_\pi(\gamma + \omega^{\mu+\alpha_0} + \omega^{\mu+\alpha_0})$ we obtain from (xxiv) and (xxvi)

$$\mathcal{H}_\eta[\Theta] \;\Big|\frac{\beta}{\beta}\, \Delta, (\exists x \in \mathsf{L}_\zeta)G(x) \;\; \textit{and} \;\; \mathcal{H}_\eta[\Theta] \;\Big|\frac{\beta}{\beta}\, \Delta, (\forall x \in \mathsf{L}_\zeta)\neg G(x). \tag{xxvii}$$

We realize that $\mathsf{rk}((\exists x \in \mathsf{L}_\zeta)G(x)) < \pi$, $\beta < \pi$, $\gamma \le \eta < \gamma + \omega^{\mu+\alpha}$ and obtain by (xviii) the claim. □

One should notice that the collapsing procedure not only collapses the derivations but also removes applications of the rules (Ref_κ).

3.4.6. Controlling operators for axiom systems of set theory

The aim of the following section is to determine controlling operators for different axiom systems for Set Theory. We will see that this is a fairly straightforward procedure which parallels the last part of Section 2.1.5. Due to extensionality of sets, however, it will turn out to be more painstaking. We start with pure logic.

Controlling operators for pure logic. First we show an analogue of (95)

3.4.6.1. Lemma. *Let Δ be a finite set of $\mathcal{L}_{\mathrm{RS}}$–sentences and F be a sentence such that $\{F, \neg F\} \subseteq \Delta$. Then $\mathcal{H}[\mathsf{par}(\Delta)] \mathrel{|\!\frac{2\cdot\mathsf{rk}(F)}{0}} \Delta$ holds for every acceptable operator $\mathcal{H}$.*

The proof is by induction on $\mathsf{rk}(F)$. Without loss of generality we assume $F \in \bigvee$-type. Then we have

$$\mathcal{H}[\mathsf{par}(\Delta, G)] \mathrel{|\!\frac{2\cdot\mathsf{rk}(G)}{0}} \Delta, G, \neg G \tag{i}$$

for all $G \in \mathcal{C}(F)$ by the induction hypothesis and obtain first

$$\mathcal{H}[\mathsf{par}(\Delta, G)] \mathrel{|\!\frac{2\cdot\mathsf{rk}(G)+1}{0}} \Delta, F, \neg G \tag{ii}$$

for all $G \in \mathcal{C}(F)$ by an inference $(\bigvee)$ and, since $2 \cdot \mathsf{rk}(G) + 1 < 2 \cdot \mathsf{rk}(F)$ and $\mathsf{par}(\Delta, G) \subseteq \mathsf{par}(\Delta \cup t_F(G))$, from (ii) finally

$$\mathcal{H}[\mathsf{par}(\Delta)] \mathrel{|\!\frac{2\cdot\mathsf{rk}(F)}{0}} \Delta, F, \neg F$$

by an inference $(\bigwedge)$. □

For a finite set Δ of $\mathcal{L}_{\mathrm{RS}}$–sentences we define

$$\mathsf{rk}(\Delta) := \max\{\mathsf{rk}(F) \mid F \in \Delta\}. \tag{183}$$

Now recall the cut–free Tait–calculus introduced in Section 2.1.2. We obtain the following lemma.

3.4.6.2. Lemma. *Let $\Delta(x_1, \ldots, x_n)$ be a finite set of $\mathcal{L}(\in, \mathsf{Ad})$–formulas whose free variables occur all in the list $\{x_1, \ldots, x_n\}$ such that $\mathrel{|\!\frac{m}{}} \Delta(x_1, \ldots, x_n)$. For any ordinal λ, any n–tuple $(a_1, \ldots, a_n)$ of $\mathcal{L}_{\mathrm{RS}}$–terms of stages less than λ and any acceptable operator $\mathcal{H}$ we obtain*

$$\mathcal{H}[\mathsf{par}(\Delta(a_1, \ldots, a_n)^{\mathsf{L}_\lambda})] \mathrel{|\!\frac{2\cdot(\alpha+m)}{0}} \Delta(a_1, \ldots, a_n)^{\mathsf{L}_\lambda}$$

for $\alpha := \mathsf{rk}(\Delta(a_1, \ldots, a_n)^{\mathsf{L}_\lambda})$.

The proof is by induction on m. We abbreviate $\Delta(a_1, \ldots, a_n)^{\mathsf{L}_\lambda}$ by Δ'. In the case of an axiom (AxL) we obtain the claim from Lemma 3.4.6.1. In the case of an inference $(\vee)$ there is a sentence $A_0(a_1, \ldots, a_n)^{\mathsf{L}_\lambda} \vee A_1(a_1, \ldots, a_n)^{\mathsf{L}_\lambda} \in \Delta'$. Then $\mathsf{par}(A_i(a_1, \ldots, a_n)^{\mathsf{L}_\lambda}) \subseteq \mathsf{par}(\Delta')$ and $\mathsf{rk}(A_i(a_1, \ldots, a_n)^{\mathsf{L}_\lambda}) < \mathsf{rk}(\Delta')$ and we get from the induction hypothesis $\mathcal{H}[\mathsf{par}(\Delta')] \mathrel{|\!\frac{2\cdot(\alpha+m_0)}{0}} \Delta', A_i(a_1, \ldots, a_n)^{\mathsf{L}_\lambda}$ for some $i \in \{0, 1\}$. By an inference $(\bigvee)$ we finally obtain $\mathcal{H}[\mathsf{par}(\Delta')] \mathrel{|\!\frac{2\cdot(\alpha+m)}{0}} \Delta'$.

The case of an inference $(\wedge)$ is treated analogously.

In the case of an inference $(\exists)$ there are two subcases. First assume that we are in the situation of an unrestricted quantifier. Then there is a sentence $(\exists x \in \mathsf{L}_\lambda)F(x, \vec{a})^{\mathsf{L}_\lambda} \in \Delta'$. For $a \in \mathcal{T}_\lambda$ we have $F(a, \vec{a})^{\mathsf{L}_\lambda} \in \mathcal{C}((\exists x \in \mathsf{L}_\lambda)F(x, \vec{a})^{\mathsf{L}_\lambda})$, $o_{(\exists x \in \mathsf{L}_\lambda)F(x,\vec{a})^{\mathsf{L}_\lambda}}(F(a, \vec{a})^{\mathsf{L}_\lambda}) = \mathsf{stg}(a) < \lambda \le \alpha$ and $\mathsf{rk}(F(a, \vec{a})) < \alpha$. Moreover, we have the premise $\models^{m_0} \Delta(\vec{x}), F(y,\vec{x})$. If $y \in \{x_1,\dots,x_n\}$ then $\mathsf{par}(F(a_i,\vec{a})^{\mathsf{L}_\lambda}) \subseteq \mathsf{par}(\Delta')$. Otherwise we replace y by L_0. In both cases we get $\mathcal{H}[\mathsf{par}(\Delta')] \mathrel{|}\!\frac{2\cdot(\alpha+m_0)}{0}\ \Delta', F(a,\vec{a})^{\mathsf{L}_\lambda}$ by the induction hypothesis and obtain the claim by an inference $(\bigvee)$.

In the case of an restricted quantifier there is a sentence $(\exists x \in a_i)F(x,\vec{a})^{\mathsf{L}_\lambda}$ in Δ'. Assume that $a_i = \{x \in \mathsf{L}_\delta \mid G(x,\vec{a})\}$. From the premise $\models^{m_0} \Delta, y \in \vec{x}_i \wedge F(y,\vec{x})$ we obtain by the induction hypothesis $\mathcal{H}[\mathsf{par}(\Delta')] \mathrel{|}\!\frac{2\cdot(\alpha+m_0)}{0}\ \Delta', a \in a_i \wedge F(a,\vec{a})$ for $a \in \mathcal{T}_\lambda$ as in the previous subcase. By $\bigwedge$-inversion this implies $\mathcal{H}[\mathsf{par}(\Delta')] \mathrel{|}\!\frac{2\cdot(\alpha+m_0)}{0}\ \Delta', a \in a_i$ and $\mathcal{H}[\mathsf{par}(\Delta')] \mathrel{|}\!\frac{2\cdot(\alpha+m_0)}{0}\ \Delta', F(a,\vec{a})$. We easily prove

$$\mathcal{H} \mathrel{|}\!\frac{\alpha}{\rho}\ \Delta, a \in \{x \in \mathsf{L}_\delta \mid G(x)\} \ \Rightarrow\ \mathcal{H} \mathrel{|}\!\frac{\alpha}{\rho}\ \Delta, G(a) \tag{184}$$

by induction on α. Using (184) we obtain $\mathcal{H}[\mathsf{par}(\Delta')] \mathrel{|}\!\frac{2\cdot(\alpha+m_0)+1}{0}\ \Delta', G(a,\vec{a}) \wedge F(a,\vec{a})$ and obtain the claim by an inference $(\bigvee)$.

In the case of an inference $(\forall)$ we have the premise $\models^{m_0} \Delta(\vec{x}), F(y,\vec{x})$ with $y \notin \{x_1,\dots,x_n\}$. The induction hypothesis implies $\mathcal{H}[\mathsf{par}(\Delta'), b] \mathrel{|}\!\frac{2\cdot\alpha+m_0}{0}\ \Delta', F(b,\vec{a})^{\mathsf{L}_\lambda}$ for all $b \in \mathcal{T}_\lambda$. Using an inference $(\bigwedge)$ we obtain $\mathcal{H}[\mathsf{par}(\Delta')] \mathrel{|}\!\frac{2\cdot\alpha+m}{0}\ \Delta'$. □

Regarding identity axioms as part of Pure Logic the next step is to deal with these axioms. We already mentioned that, because of the extensionality of sets, this is by far not simple. The tedious point is the bookkeeping of derivation lengths. To obtain precise bounds we are forced to derive all axioms step by step which is a bore. However, we do not need absolutely exact bounds. The collapsing procedure will equalize too precise bounds anyway. We already observed that the rank of a sentence is always an upper bound for its truth complexity. So we will introduce a more liberal derivation calculus $\vdash \Delta$ and show afterwards that $\vdash \Delta$ entails $\mathcal{H}[\mathsf{par}(\Delta)] \mathrel{|}\!\frac{\alpha}{0}\ \Delta$ where α is computable from $\mathsf{rk}(\Delta)$.

For a set $\Theta \subseteq \mathbf{On}$ we define $\overline{\Theta}$ to be the closure of $\Theta \cup \{\omega\}$ under successor, i.e., $\xi \mapsto \xi+1$ and regular successor, i.e., $\xi \mapsto \xi^+$. We define the relation $\vdash \Delta$ by the rules

$(\bigwedge')$ $\vdash \Delta, G$ *for all* $G \in \mathcal{C}(F) \ \Rightarrow\ \vdash \Delta, F$

and

$(\bigvee')$ $\vdash \Delta, \Gamma,\ \Gamma \subseteq \mathcal{C}(F)$ *and* $\mathsf{par}(\Delta,\Gamma) \subseteq \overline{\mathsf{par}(\Delta,F)} \ \Rightarrow\ \vdash \Delta, F.$

Here we want Δ, ..., to denote multi-sets, i.e., sequences which are independent from the order of their elements but count their multiplicity.

To avoid distinctions by cases we introduce for $a \in \mathcal{T}_{\mathsf{stg}(b)}$ the relation

$$a \,\epsilon\, b \wedge G(s) \ :\Leftrightarrow\ \begin{cases} F(a) \wedge G(s) & \text{if } b = \{x \in \mathsf{L}_\alpha \mid F(x)\} \\ G(s) & \text{if } b = \mathsf{L}_\alpha. \end{cases} \tag{185}$$

Dually we put

$$a \not\epsilon b \vee G(s) \;:\Leftrightarrow\; \begin{cases} \neg F(a) \vee G(s) & \text{if } b = \{x \in \mathsf{L}_\alpha \mid F(x)\} \\ G(s) & \text{if } b = \mathsf{L}_\alpha. \end{cases} \tag{186}$$

For multi-sets we define analogously

$$\{\ldots, a \not\epsilon b, G(s), \ldots\} := \begin{cases} \{\ldots, \neg F(a), G(s), \ldots\} & \text{if } b = \{x \in \mathsf{L}_\alpha \mid F(x)\} \\ \{\ldots, G(s), \ldots\} & \text{if } b = \mathsf{L}_\alpha. \end{cases}$$

This has the notational advantage that $\mathcal{C}(a \in b) = \{t \,\epsilon\, b \wedge t = a \mid \mathsf{stg}(t) < \mathsf{stg}(b)\}$ and $\mathcal{C}((\exists x \in b)F(x)) = \{t \,\epsilon\, b \wedge F(t) \mid \mathsf{stg}(t) < \mathsf{stg}(b)\}$ independent of the shape of b.

There is a number of inference rules which are derivable or admissible within the calculus $\vdash$. We list the most important ones

(Str) $\vdash \Delta$ *and* $\Delta \subseteq \Gamma \;\Rightarrow\; \vdash \Gamma$

(Taut) $\vdash A, \neg A$

(Sent) $\vdash \Delta, A \;\Rightarrow\; \vdash \Delta, \neg B, A \wedge B$

($\in$) $\vdash \Delta, t \,\epsilon\, b \wedge t = a$ *for some* $t \in \mathcal{T}_{\mathsf{stg}(b)} \;\Rightarrow\; \vdash \Delta, a \in b$

($\notin$) $\vdash \Delta, t \not\epsilon b, t \neq a$ *for all* $t \in \mathcal{T}_{\mathsf{stg}(b)} \;\Rightarrow\; \vdash \Delta, a \notin b$

($\forall^\alpha$) $\vdash \Delta, F(t)$ *for all* $t \in \mathcal{T}_\alpha \;\Rightarrow\; \vdash \Delta, (\forall x \in \mathsf{L}_\alpha)F(x)$

($\exists^\alpha$) $t \in \mathcal{T}_\alpha$, $\mathsf{par}(t) \subseteq \overline{\mathsf{par}(\Delta, F(\mathsf{L}_0))}$ *and* $\vdash \Delta, F(t) \;\Rightarrow\; \vdash \Delta, (\exists x \in \mathsf{L}_\alpha)F(x)$

($\forall^b$) $\vdash \Delta, t \not\epsilon b, F(t)$ *for all* $t \in \mathcal{T}_{\mathsf{stg}(b)} \;\Rightarrow\; \vdash \Delta, (\forall x \in b)F(x)$

($\exists^b$) $t \in \mathcal{T}_{\mathsf{stg}(b)}$, $\mathsf{par}(t) \subseteq \overline{\mathsf{par}(\Delta, F(\mathsf{L}_0))}$ *and* $\vdash \Delta, t \,\epsilon\, b \wedge F(t)$
$\Rightarrow\; \vdash \Delta, (\exists x \in b)F(x)$

We refer to (Str) as *Structural Rule*, to (Taut) as *Tautology Rule*, to (Sent) as *Sentential Rule* , to ($\in$) , ($\notin$) as $\in$*–rule* or $\notin$*–rule*, etc. The proofs are all obvious.

For a multi-set Δ of $\mathcal{L}_{\mathrm{RS}}$–sentences we define

$$\#\Delta := \sum_{F \in \Delta}^{\#} \omega^{\mathsf{rk}(F)}$$

and observe that in rules according to $(\bigwedge')$ and $(\bigvee')$ we always have $\#\Delta_p < \#\Delta_c$ if Δ_p denotes a premise and Δ_c the conclusion of the rule. This will be the main argument in showing

3.4.6.3. Lemma. *Let $\mathcal{H}$ be an acceptable operator which is closed under $\xi \mapsto \xi^+$. Then $\vdash \Delta$ implies $\mathcal{H}[\mathsf{par}(\Delta)] \,\big|\!\frac{\#\Delta}{0}\, \Delta$.*

The proof is by induction on the definition of $\vdash \Delta$. In the case of an inference $(\bigwedge')$ we get the claim immediately from the induction hypothesis, the previous remark and the fact that for $G \in \mathcal{C}(F)$ we have $\mathsf{par}(G) \subseteq \mathsf{par}(F \cup t_F(G))$. In the case of an inference $(\bigvee')$ we have $\mathcal{H}[\mathsf{par}(\Delta, \Gamma)] \,\big|\!\frac{\#(\Delta,\Gamma)}{0}\, \Delta, \Gamma$, $\Gamma \subseteq \mathcal{C}(F)$ and

$\mathsf{par}(\Delta,\Gamma) \subseteq \overline{\mathsf{par}(\Delta,F)}$. Since $\mathcal{H}$ is Cantorian closed and closed under $\xi \mapsto \xi^+$ the latter implies $\mathcal{H}[\mathsf{par}(\Delta,\Gamma)] \subseteq \mathcal{H}[\mathsf{par}(\Delta,F)]$. So we get $\mathcal{H}[\mathsf{par}(\Delta,F)] \mathrel{|}\!\frac{\#(\Delta,\Gamma)}{0} \Delta,\Gamma$ which entails $\mathcal{H}[\mathsf{par}(\Delta,F)] \mathrel{|}\!\frac{\#(\Delta,\Gamma)+|\Gamma|}{0} \Delta,F$ where $|\Gamma|$ denotes the cardinality of the finite set Γ. Since $\mathsf{rk}(A) < \mathsf{rk}(F)$ for all $A \in \mathcal{C}(F)$ we get $\sum\limits_{A\in\Gamma}^{\#} \omega^{\mathsf{rk}(A)} + |\Gamma| < \omega^{\mathsf{rk}(F)}$, hence $\#(\Delta,\Gamma) + |\Gamma| < \#(\Delta,F)$ and this yields the claim. □

We are now going to derive a series of sentences which will be needed in the computation of the truth complexities of identity and non–logical axioms of Set Theory.

$$\vdash a \notin a \textit{ for all } \mathcal{L}_{\mathrm{RS}}\textit{-terms } a. \tag{187}$$

The proof is by induction on $\mathsf{rk}(a)$. We obtain $\vdash b \notin b$ for all $b \in \mathcal{T}_{\mathsf{stg}(a)}$ by the induction hypothesis. Hence $\vdash b \not\varepsilon a, b \varepsilon a \wedge b \notin b$ by (Sent) which in turn gives $\vdash b \not\varepsilon a, (\exists x \in a)[x \notin b]$ by $(\exists^b)$. This implies $\vdash b \not\varepsilon a, b \neq a$ for all $b \in \mathcal{T}_{\mathsf{stg}(a)}$. Hence $\vdash a \notin a$ by $(\notin)$. □

$$\vdash a \subseteq a \textit{ hence also } \vdash a = a \textit{ for all } \mathcal{L}_{\mathrm{RS}}\textit{-terms } a \tag{188}$$

The proof is by in induction on $\mathsf{rk}(a)$. We get $\vdash b \subseteq b$ for all $a \in \mathcal{T}_{\mathsf{stg}(a)}$ by the induction hypothesis. For symmetry reasons this implies $\vdash b = b$. So $\vdash b \not\varepsilon a, b \varepsilon a \wedge b = b$ by (Sent) and $\vdash b \not\varepsilon a, b \in a$ for all $b \in \mathcal{T}_{\mathsf{stg}(a)}$ by $(\in)$ which implies $\vdash (\forall x \in a)[x \in a]$ by $(\forall^a)$. □

As a corollary of the proof of (188) we obtain

$$\vdash b \not\varepsilon a, b \in a \textit{ for all } b \in \mathcal{T}_{\mathsf{stg}(a)}. \tag{189}$$

By (Sent) we have $\vdash (\exists x \in a)[x \notin b], (\exists x \in b)[x \notin a], (\forall x \in b)[x \in a] \wedge (\forall x \in a)[x \in b]$ which entails

$$\vdash a \neq b, b = a. \tag{190}$$

Since we have $\vdash b = b$ for all $b \in \mathcal{T}_\alpha$ we get

$$\vdash b \in \mathsf{L}_\alpha \textit{ for all } b \in \mathcal{T}_\alpha. \tag{191}$$

Now we show

$$\vdash \mathit{Tran}(\mathsf{L}_\alpha). \tag{192}$$

For $a \in \mathcal{T}_\alpha$ and $b \in \mathcal{T}_{\mathsf{stg}(a)}$ we get $\mathsf{stg}(b) < \alpha$. Hence by (191) $\vdash b \not\varepsilon a, b \in \mathsf{L}_\alpha$ which in turn gives $\vdash (\forall y \in a)[y \in \mathsf{L}_\alpha]$ for all $a \in \mathcal{T}_\alpha$. Thus $\vdash (\forall x \in \mathsf{L}_\alpha)(\forall y \in x)[y \in \mathsf{L}_\alpha]$. □

We prove the next item

$$\mathsf{stg}(b) < \delta \;\Rightarrow\; \vdash a \neq b, \mathsf{L}_\delta \neq a \tag{193}$$

by induction on δ. Put $\beta := \mathsf{stg}(b)$. For $t \in \mathcal{T}_\beta$ and $s \in \mathcal{T}_{\mathsf{stg}(a)}$ we obtain:

$$\vdash s \neq t, \mathsf{L}_\beta \neq s \quad \text{by induction hypothesis} \tag{i}$$

$$\vdash t \notin b, t \neq s, \mathsf{L}_\beta \neq s \quad \textit{for all } t \in \mathcal{T}_\beta \quad \text{from (i) by (Str)} \tag{ii}$$

$$\vdash s \notin b, \mathsf{L}_\beta \neq s \quad \text{from (ii) by } (\notin) \tag{iii}$$

$$\vdash s \,\epsilon\, a \wedge s \notin b, s \not\epsilon\, a, \mathsf{L}_\beta \neq s \quad \text{from (iii) by (Sent)} \tag{iv}$$

$$\vdash (\exists x \in a)[x \notin b], s \not\epsilon\, a, \mathsf{L}_\beta \neq s \quad \text{from (iv) by } (\exists^a) \tag{v}$$

Since $a = b \equiv a \subseteq b \wedge b \subseteq a \equiv (\forall x \in a)[x \in b] \wedge (\forall x \in b)[x \in a]$ we may continue

$$\vdash a \neq b, s \not\epsilon\, a, \mathsf{L}_\beta \neq s \quad \textit{for all } s \in \mathcal{T}_{\mathsf{stg}(a)} \quad \text{from (v) by } (\textstyle\bigvee') \tag{vi}$$

$$\vdash a \neq b, \mathsf{L}_\beta \notin a \quad \text{from (vi) by } (\notin) \tag{vii}$$

$$\vdash a \neq b, (\exists x \in \mathsf{L}_\delta)[x \notin a] \quad \text{from (vii) by } (\exists^\delta) \tag{viii}$$

$$\vdash a \neq b, \mathsf{L}_\delta \neq a \quad \text{from (vii) by } (\textstyle\bigvee'). \qquad \square$$

The most painstaking part is the proof of the following theorem

3.4.6.4. Equality Theorem. *For every $\mathcal{L}_{\mathrm{RS}}$-sentence $F(\mathsf{L}_0)$ and all $\mathcal{L}_{\mathrm{RS}}$-terms a and b we have*

$$\vdash a \neq b, \neg F(a), F(b).$$

To prepare the proof we show a more general lemma which entails the Equality Theorem. Let $a \neq' b$ denote the multi-set $\neg a \subseteq b, \neg b \subseteq a$.

3.4.6.5. Lemma. *Assume that $A(x_1, \ldots, x_n)$ is a Δ_0-formula in $\mathcal{L}(\in, \mathsf{Ad})$ in which every variable $x_1, \ldots, x_n$ occurs at most once and let $a_1, \ldots, a_n$ and $b_1, \ldots, b_n$ be $\mathcal{L}_{\mathrm{RS}}$-terms. Then*

$$\vdash a_1 \neq' b_1, \ldots, a_n \neq' b_n, \neg A(a_1, \ldots, a_n), A(b_1, \ldots, b_n).$$

We prove the lemma by induction on $\mathsf{rk}(A(a_1, \ldots, a_n)) \# \mathsf{rk}(A(b_1, \ldots, b_n))$ and distinguish the following cases:

$A(x_1, \ldots, x_n) \equiv x_1 \in x_2$. For $\mathsf{stg}(a) < \mathsf{stg}(a_2)$ and $\mathsf{stg}(b) < \mathsf{stg}(b_2)$ we have $\mathsf{rk}(a = a_1) < \mathsf{rk}(a_1 \in a_2)$ and $\mathsf{rk}(b = b_1) < \mathsf{rk}(b_1 \in b_2)$. So we get for $a \in \mathcal{T}_{\mathsf{stg}(a_2)}$ and $b \in \mathcal{T}_{\mathsf{stg}(b_2)}$

$$\vdash a_1 \neq' b_1, a \neq a_1, a \neq' b, b = b_1 \quad \text{by induction hypothesis} \tag{i}$$

$$\vdash a_1 \neq' b_1, a \neq a_1, b \not\epsilon\, b_2, a \neq' b, b \,\epsilon\, b_2 \wedge b = b_1 \quad \text{from (i) by (Sent)} \tag{ii}$$

$$\vdash a_1 \neq' b_1, a \neq a_1, b \not\epsilon\, b_2, a \neq b, b_1 \in b_2 \quad \text{from (ii) by } (\in) \text{ and } (\textstyle\bigvee') \tag{iii}$$

$$\vdash a_1 \neq' b_1, a \neq a_1, a \notin b_2, b_1 \in b_2 \quad \text{from (iii) by } (\notin) \tag{iv}$$

$$\vdash a_1 \neq' b_1, a \not\epsilon\, a_2, a \neq a_1, a \,\epsilon\, a_2 \wedge a \notin b_2, b_1 \in b_2 \quad \text{from (iv) by (Sent)} \tag{v}$$

$$\vdash a_1 \neq' b_1, a \not\epsilon\, a_2, a \neq a_1, \neg(a_2 \subseteq b_2), b_1 \in b_2 \quad \text{from (v) by } (\exists^{a_2}) \tag{vi}$$

$$\vdash a_1 \neq' b_1, a_2 \neq' b_2, a_1 \notin a_2, b_1 \in b_2 \quad \text{from (vi) by } (\notin).$$

$A(x_1, \ldots, x_n) \equiv \mathsf{Ad}(x_1)$. For $\kappa \leq \min\{\mathsf{stg}(a_1), \mathsf{stg}(b_1)\} =: \beta$ we have $\mathsf{rk}(\mathsf{L}_\kappa = a_1) < \mathsf{rk}(\mathsf{Ad}(a_1))$ and $\mathsf{rk}(\mathsf{L}_\kappa = b_1) < \mathsf{rk}(\mathsf{Ad}(b_1))$. Therefore we obtain:

$$\vdash a_1 \neq' b_1, \mathsf{L}_\kappa \neq a_1, \mathsf{L}_\kappa = b_1 \textit{ for all } \kappa \leq \beta \quad \text{by induction hypothesis} \tag{vii}$$

$$\vdash a_1 \neq' b_1, \mathsf{L}_\kappa \neq a_1, \mathsf{Ad}(b_1) \quad \text{from (vii) by } (\bigvee{}') \tag{viii}$$

$$\vdash a_1 \neq' b_1, \mathsf{L}_\kappa \neq a_1, \mathsf{Ad}(b_1) \textit{ for all } \kappa \leq \mathsf{stg}(a_1) \quad \text{from (viii) and (193)} \tag{ix}$$

$$\vdash a_1 \neq' b_1, \neg\mathsf{Ad}(a_1), \mathsf{Ad}(b_1) \quad \text{from (ix) by } (\bigwedge{}').$$

For $A(x_1, \ldots, x_n) \equiv A_0(x_1, \ldots, x_n) \wedge A_1(x_1, \ldots, x_n)$ we obtain the claim easily from the induction hypothesis.

$A \equiv (\exists y \in x_1) B(x_2, \ldots, x_n, y)$. Putting $\vec{a_2} := a_2, \ldots, a_n$ and defining $\vec{b_2}$ analogously we obtain for $a \in \mathcal{T}_{\mathsf{stg}(a_1)}$ and $b \in \mathcal{T}_{\mathsf{stg}(b_1)}$:

$$\vdash \vec{a_2} \neq' \vec{b_2}, a \neq' b, \neg B(\vec{a_2}, a), B(\vec{b_2}, b) \quad \text{by induction hypothesis} \tag{x}$$

$$\vdash \vec{a_2} \neq' \vec{b_2}, b \not\epsilon\, b_1, a \neq' b, \neg B(\vec{a_2}, a), b \,\epsilon\, b_1 \wedge B(\vec{b_2}, b) \quad \text{from (x) by (Sent)} \tag{xi}$$

$$\vdash \vec{a_2} \neq' \vec{b_2}, b \not\epsilon\, b_1, a \neq b, \neg B(\vec{a_2}, a), (\exists y \in b_1) B(\vec{b_2}, y) \quad \text{from (xi) by } (\exists^{b_1}) \text{ and } (\bigvee{}') \tag{xii}$$

$$\vdash \vec{a_2} \neq' \vec{b_2}, a \notin b_1, \neg B(\vec{a_2}, a), (\exists y \in b_1) B(\vec{b_2}, y) \quad \text{from (xii) by } (\notin) \tag{xiii}$$

$$\vdash \vec{a_2} \neq' \vec{b_2}, a \,\epsilon\, a_1 \wedge a \notin b_1, a \not\epsilon\, a_1, \neg B(\vec{a_2}, a), (\exists y \in b_1) B(\vec{b_2}, y) \quad \text{from (xiii) by (Sent)} \tag{xiv}$$

$$\vdash \vec{a_2} \neq' \vec{b_2}, \neg(a_1 \subseteq b_1), a \not\epsilon\, a_1, \neg B(\vec{a_2}, a), (\exists y \in b_1) B(\vec{b_2}, y) \quad \text{from (xiv) by } (\exists^{a_1}) \tag{xv}$$

$$\vdash \vec{a_2} \neq' \vec{b_2}, a_1 \neq' b_1, (\forall y \in a_1) \neg B(\vec{a_2}, y), (\exists y \in b_1) B(\vec{b_2}, y) \quad \text{from (xv) by } (\forall^{a_1}) \text{ and (Str).}$$

Since the claim is symmetrical in $\neg A(a_1, \ldots, a_n)$ and $A(b_1, \ldots, b_n)$ the distinction by cases is complete. □

To infer the Equality Theorem from Lemma 3.4.6.5 let $G(x_1, \ldots, x_n)$ be an Δ_0-formula in which every variable x_i occurs at most once such that $F(a) \equiv G(a, \ldots, a)$. Then apply the lemma to $G(x_1, \ldots, x_n)$ and infer the theorem by an inference $(\bigvee{}')$.

Observe that by (188), (190) and the Equality Theorem we have

$$\vdash A^{\mathsf{L}_\lambda} \textit{ for all identity axioms } A. \tag{194}$$

If $\mathsf{F}(x_1, \ldots, x_n)$ is an $\mathcal{L}(\in, \mathsf{Ad})$ formula and $a_1, \ldots, a_n$ are all $\mathcal{L}_{\mathrm{RS}}$-terms of stages $< \lambda$ we obtain that $\mathsf{rk}(F(a_1, \ldots, a_n)) < \omega \cdot \lambda + n$ for some $n < \omega$. So putting together Lemma 3.4.6.2 and (194) we obtain the following theorem.

3.4.6.6. Theorem. *Let $F(x_1,\ldots,x_n)$ be an $\mathcal{L}(\in,\mathsf{Ad})$-formula which is valid in Pure Logic with identity and $\mathcal{H}$ an acceptable operator. Then there is an $m<\omega$ such that*

$$\mathcal{H}\left[\bigcup_{i=1}^{n}\mathsf{par}(a_i)\cup\{\lambda\}\right] \,\Big|\!\frac{\omega^{\omega\cdot\lambda+m}}{\omega\cdot\lambda+m}\, F(a_1,\ldots,a_n)^{\mathsf{L}_\lambda}.$$

Controlling operators for set-theoretic axioms. Having established controlling operators and derivation lengths for logically valid sentences we will now determine controlling operators and derivation length for axioms of Set Theory. This is fairly easy for the axiom of extensionality. Since we defined $a=b$ to stand for $a\subseteq b\wedge b\subseteq a$ we obtain

$$(\forall x\in\mathsf{L}_\lambda)(\forall y\in\mathsf{L}_\lambda)[x=y\leftrightarrow(\forall z\in x)(z\in y)\wedge(\forall z\in y)(z\in x)]. \tag{195}$$

We derive the remaining axioms in their modified forms as introduced in Section 3.3. We prove

$$\vdash(\text{Pair'})^{\mathsf{L}_\lambda}\quad \textit{for}\quad \lambda\in\mathsf{Lim}. \tag{196}$$

Let $a,b\in\mathcal{T}_\lambda$ and $\beta:=\max\{\mathsf{stg}(a),\mathsf{stg}(b)\}+1$. Then

$$\vdash a\in\mathsf{L}_\beta\wedge b\in\mathsf{L}_\beta\quad\text{from (191).} \tag{i}$$

Since $\beta\in\overline{\mathsf{par}(a)\cup\mathsf{par}(b)}$ and $\beta<\lambda$ we get from (i) by $(\bigvee')$

$$\vdash(\exists z\in\mathsf{L}_\lambda)[a\in z\wedge b\in z]. \tag{ii}$$

Hence by twofold $(\forall^\lambda)$

$$\vdash(\forall x\in\mathsf{L}_\lambda)(\forall y\in\mathsf{L}_\lambda)(\exists z\in\mathsf{L}_\lambda)[x\in z\wedge y\in z]. \qquad\square$$

Next we show

$$\vdash(\text{Union'})^{\mathsf{L}_\lambda}\quad \textit{for}\quad \lambda\in\mathsf{Lim}. \tag{197}$$

Let $a\in\mathcal{T}_\lambda$ and $\alpha:=\mathsf{stg}(a)$. For $t\in\mathcal{T}_\alpha$ and $s\in\mathcal{T}_{\mathsf{stg}(t)}$ we obtain:

$$\vdash s\notin t, s\in\mathsf{L}_\alpha\quad\text{from (190) by (Str)} \tag{i}$$

$$\vdash(\forall x\in t)[x\in\mathsf{L}_\alpha]\quad\text{from (i) by }(\forall^t) \tag{ii}$$

$$\vdash t\notin a,(\forall x\in t)[x\in\mathsf{L}_\alpha]\quad\text{from (ii) by (Str)} \tag{iii}$$

$$\vdash(\forall y\in a)(\forall x\in y)[x\in\mathsf{L}_\alpha]\quad\text{from (iii) by }(\forall^a) \tag{iv}$$

$$\vdash(\exists w\in\mathsf{L}_\lambda)(\forall y\in a)(\forall x\in y)[x\in w]\quad\text{from (iv) by }(\exists^{\mathsf{L}_\lambda}) \tag{v}$$

$$\vdash(\forall u\in\mathsf{L}_\lambda)(\exists w\in\mathsf{L}_\lambda)(\forall y\in u)(\forall x\in y)[x\in w]\quad\text{from (v) by }(\forall^{\mathsf{L}_\lambda}). \qquad\square$$

We prove the *set existence axiom-schemes* of Separation and Collection in the form

$(\Delta_0$-Sep$)$ $\quad(\forall\vec{v})(\forall a)(\exists z)[(\forall x\in z)(x\in a\wedge F(x,\vec{v}))\wedge(\forall x\in a)(F(x,\vec{v})\ \rightarrow\ x\in z)]$

(Δ_0-Col) $(\forall\vec{v})(\forall u)[(\forall x \in u)(\exists y)F(x,y,\vec{v}) \to (\exists z)(\forall x \in u)(\exists y \in z)F(x,y,\vec{v})]$

for Δ_0–formulas $F(x,\vec{v})$ and $F(x,y,\vec{v})$, respectively. We first prove

$$\vdash (\Delta_0\text{-Sep})^{\mathsf{L}_\lambda} \quad for \quad \lambda \in \mathsf{Lim}. \tag{198}$$

Let $\{a, a_1, \ldots, a_n\} \subseteq \mathsf{L}_\lambda$ and $\alpha := \max\{\mathsf{stg}(a), \mathsf{stg}(a_1), \ldots, \mathsf{stg}(a_n)\} + 1$. Define

$$b := \{x \in \mathsf{L}_\alpha | \, x \in a \wedge F(x, a_1, \ldots, a_n)\}. \tag{i}$$

Then we obtain for $t \in \mathcal{T}_\alpha$:

$$\vdash t \notin b, t \in a \wedge F(t, a_1, \ldots, a_n) \quad \text{by (Taut) and (Sent)} \tag{ii}$$

$$\vdash (\forall x \in b)[x \in a \wedge F(x, a_1, \ldots, a_n)] \quad \text{from (ii) by } (\forall^b) \tag{iii}$$

$$\vdash t \notin a, \neg F(t,\vec{a}), t \in a \wedge F(t,\vec{a}) \wedge t = t \quad for \quad t \in \mathcal{T}_{\mathsf{stg}(a)} \quad \text{from (188) and (189) by (Sent)} \tag{iv}$$

$$\vdash t \notin a, \neg F(t,\vec{a}), t \,\epsilon\, b \wedge t = t \quad \text{reformulation of (iv)} \tag{v}$$

$$\vdash t \notin a, \neg F(t,\vec{a}) \vee t \in b \quad \text{from (v) by } (\in) \text{ and } (\bigvee{}') \tag{vi}$$

$$\vdash (\forall x \in a)[F(x,\vec{a}) \to x \in b] \quad \text{from (vi) by } (\forall^a) \tag{vii}$$

$$\vdash (\exists z \in \mathsf{L}_\lambda)[(\forall x \in z)(x \in a \wedge F(x,\vec{a})) \wedge (\forall x \in a)(F(x,\vec{a}) \to x \in z)] \quad \text{from (iii) and (vii) by } (\bigwedge{}') \text{ and } (\exists^\lambda). \tag{viii}$$

From (viii) we finally obtain $(\Delta_0\text{-Sep})^{\mathsf{L}_\lambda}$ by inferences $(\forall^\lambda)$. □

We show

$$\vdash (\text{Inf'})^{\mathsf{L}_\lambda} \quad for \quad \omega < \lambda \in \mathsf{Lim}. \tag{199}$$

Let $a \in \mathcal{T}_\omega$ and $\alpha := \mathsf{stg}(a) + 1$. Then $\vdash a \in \mathsf{L}_\alpha$ and $\vdash a \in \mathsf{L}_\omega$ by (191). This entails $\vdash (\exists y \in \mathsf{L}_\omega)[y \in \mathsf{L}_\omega]$ and $\vdash (\exists z \in \mathsf{L}_\omega)[a \in z]$ for all $t \in \mathcal{T}_\omega$. Hence $\vdash \mathsf{L}_\omega \neq \emptyset \wedge (\forall x \in \mathsf{L}_\omega)(\exists z \in \mathsf{L}_\omega)[x \in z]$. Since $\omega < \lambda$ we get the claim by an inference $(\bigvee{}')$.

We summarize Lemma 3.4.6.3, (194), (196), (197), (198) and (199).

3.4.6.7. Lemma. *Let λ be a limit ordinal and A be one of the axioms* (Ext), (Pair'), (Union'), (Δ_0–Sep) *or* (Inf') *and $\mathcal{H}$ an acceptable operator. Then there is an $n < \omega$ such that $\mathcal{H}[\{\lambda\}] \,\big|\!\frac{\omega(\lambda+n)}{0}\, A^{\mathsf{L}_\lambda}$*

Next we prove that for acceptable $\mathcal{H}$ and $\kappa \in \mathbf{Reg}$ there is an $n < \omega$ such that

$$\mathcal{H}[\{\kappa\}] \,\big|\!\frac{\kappa+n}{0}\, (\Delta_0\text{-Col})^\kappa. \tag{200}$$

Let $\kappa \in \mathbf{Reg}$ and $\{a, a_1, \ldots, a_n\} \subseteq \mathcal{T}_\kappa$. By Lemma 3.4.6.1 we obtain

$$\mathcal{H}[\mathsf{par}(\vec{a}, a) \cup \{\kappa\}] \,\big|\!\frac{\kappa+4}{0}\, \neg(\forall x \in a)(\exists y \in \mathsf{L}_\kappa)F(x,y,\vec{a}), (\forall x \in a)(\exists y \in \mathsf{L}_\kappa)F(x,y,\vec{a}). \tag{i}$$

By $(\mathrm{Ref})_\kappa$ this implies

$$\mathcal{H}[\mathsf{par}(\vec{a},a)\cup\{\kappa\}] \,\Big|\!\frac{\kappa+6}{0}\; \neg(\forall x\in a)(\exists y\in \mathsf{L}_\kappa)F(x,y,\vec{a}), \quad (\exists z\in \mathsf{L}_\kappa)(\forall x\in a)(\exists y\in z)F(x,y,\vec{a}). \tag{ii}$$

By two clauses $(\bigvee)$ and some clauses $(\bigwedge)$ this entails

$$\mathcal{H}[\{\kappa\}] \,\Big|\!\frac{\kappa+n}{0}\; (\forall\vec{v}\in \mathsf{L}_\kappa)(\forall u\in \mathsf{L}_\kappa)(\forall x\in u)(\exists y\in \mathsf{L}_\kappa)F(x,y,\vec{v}) \rightarrow (\exists z\in \mathsf{L}_\kappa)(\forall x\in u)(\exists y\in z)F(x,y,\vec{v}).$$

□

To deal with the foundation scheme (and axiom) we prove a lemma.

3.4.6.8. Foundation Lemma. *Let* $\{\lambda\}\cup\mathsf{par}(F(\mathsf{L}_0)^{\mathsf{L}_\lambda})\subseteq\mathcal{H}$ *for an acceptable operator* $\mathcal{H}$. *Then*

$$\mathcal{H}[\mathsf{par}(a)] \,\Big|\!\frac{2\cdot\mathsf{rk}(F(a)^{\mathsf{L}_\lambda})+3\cdot(\mathsf{stg}(a)+1)}{0}\; \neg F(a)^{\mathsf{L}_\lambda}, (\exists x\in \mathsf{L}_\lambda)[F(x)^{\mathsf{L}_\lambda}\wedge(\forall y\in x)\neg F(y)^{\mathsf{L}_\lambda}]$$

for all $a\in\mathcal{T}_\lambda$.

We prove the lemma by induction on $\mathsf{stg}(a)$ and get

$$\mathcal{H}[\mathsf{par}(b)] \,\Big|\!\frac{2\cdot\mathsf{rk}(F(b)^{\mathsf{L}_\lambda})+3\cdot(\mathsf{stg}(b)+1)}{0}\; \neg F(b)^{\mathsf{L}_\lambda}, (\exists x\in \mathsf{L}_\lambda)[F(x)^{\mathsf{L}_\lambda}\wedge(\forall y\in x)\neg F(y)^{\mathsf{L}_\lambda}] \tag{i}$$

for all $b\in\mathcal{T}_{\mathsf{stg}(a)}$. By the structural rule (151) this entails

$$\mathcal{H}[\mathsf{par}(a,b)] \,\Big|\!\frac{2\cdot\mathsf{rk}(F(b)^{\mathsf{L}_\lambda})+3\cdot(\mathsf{stg}(b)+1)}{0}\; b\notin a, \neg F(b)^{\mathsf{L}_\lambda}, \quad (\exists x\in \mathsf{L}_\lambda)[F(x)^{\mathsf{L}_\lambda}\wedge(\forall y\in x)\neg F(y)^{\mathsf{L}_\lambda}]. \tag{ii}$$

Using $(\bigwedge)$ this implies

$$\mathcal{H}[\mathsf{par}(a)] \,\Big|\!\frac{2\cdot\mathsf{rk}(F(b)^{\mathsf{L}_\lambda})+3\cdot(\mathsf{stg}(a))+1}{0}\; (\forall z\in a)\neg F(z)^{\mathsf{L}_\lambda}, \quad (\exists x\in \mathsf{L}_\lambda)[F(x)^{\mathsf{L}_\lambda}\wedge(\forall y\in x)\neg F(y)^{\mathsf{L}_\lambda}]. \tag{iii}$$

By Lemma 3.4.6.1 we have

$$\mathcal{H}[\mathsf{par}(a)] \,\Big|\!\frac{2\cdot\mathsf{rk}(F(a)^{\mathsf{L}_\lambda})}{0}\; F(a)^{\mathsf{L}_\lambda}, \neg F(a)^{\mathsf{L}_\lambda}, (\exists x\in \mathsf{L}_\lambda)[F(x)^{\mathsf{L}_\lambda}\wedge(\forall y\in x)\neg F(y)^{\mathsf{L}_\lambda}]. \tag{iv}$$

Putting $\alpha := 2\cdot\mathsf{rk}(F(a)^{\mathsf{L}_\lambda})$ we obtain

$$\mathcal{H}[\mathsf{par}(a)] \,\Big|\!\frac{\alpha+3\cdot(\mathsf{stg}(a))+2}{0}\; F(a)^{\mathsf{L}_\lambda}\wedge(\forall z\in a)\neg F(z)^{\mathsf{L}_\lambda}, \neg F(a), \quad (\exists x\in \mathsf{L}_\lambda)[F(x)^{\mathsf{L}_\lambda}\wedge(\forall y\in x)\neg F(y)^{\mathsf{L}_\lambda}] \tag{v}$$

from (iii) and (iv). Hence

$$\mathcal{H}[\mathsf{par}(a)] \,\Big|\!\frac{\alpha+3\cdot(\mathsf{stg}(a))+3}{0}\; \neg F(a)^{\mathsf{L}_\lambda}, (\exists x\in \mathsf{L}_\lambda)[F(x)^{\mathsf{L}_\lambda}\wedge(\forall y\in x)\neg F(y)^{\mathsf{L}_\lambda}] \tag{vi}$$

by $(\bigvee)$. □

We get the following theorem as a corollary of the Foundation Lemma.

3.4.6.9. Foundation Theorem. *Let $F(x, \vec{x})$ be an $\mathcal{L}(\in, \mathsf{Ad})$ formula without further free variables, $\lambda \in \mathsf{Lim}$ and $\mathcal{H}$ an acceptable operator with $\lambda \in \mathcal{H}$. Then there is an $n < \omega$ such that*

$$\mathcal{H} \overset{\omega\cdot\lambda+\lambda+n}{\underset{0}{\vdash}} (\forall \vec{x} \in \mathsf{L}_\lambda)[(\exists x \in \mathsf{L}_\lambda)F(x,\vec{x})^{\mathsf{L}_\lambda} \to (\exists x \in \mathsf{L}_\lambda)[F(x,\vec{x})^{\mathsf{L}_\lambda} \wedge (\forall y \in x)\neg F(y,\vec{x})^{\mathsf{L}_\lambda}]].$$

To prove the theorem we introduce the abbreviation $\mathrm{Found}(\{x \in \mathsf{L}_\lambda | F(x,\vec{x})^{\mathsf{L}_\lambda}\}) :\Leftrightarrow$

$$(\exists x \in \mathsf{L}_\lambda)F(x,\vec{x})^{\mathsf{L}_\lambda} \to (\exists x \in \mathsf{L}_\lambda)[F(x,\vec{x})^{\mathsf{L}_\lambda} \wedge (\forall y \in x)\neg F(y,\vec{x})^{\mathsf{L}_\lambda}].$$

Observe that for $\{a, a_1, \ldots, a_n\} \subseteq \mathcal{T}_\lambda$ we have $\mathsf{rk}(F(a, a_1, \ldots, a_n)^{\mathsf{L}_\lambda}) \leq \omega \cdot \lambda + k$ for some $k < \omega$. From the Foundation Lemma we obtain

$$\mathcal{H}[\mathsf{par}(a_1, \ldots, a_n)] \overset{\omega\cdot\lambda+\lambda}{\underset{0}{\vdash}} \mathrm{Found}(\{x \in \mathsf{L}_\lambda | F(x,\vec{a})^{\mathsf{L}_\lambda}\})$$

by an inference $(\bigwedge)$. Applying n inferences $(\bigwedge)$ we finally obtain the claim. □

So far we are already prepared to compute an upper bound for $\|\mathbf{KP}\boldsymbol{\omega}\|_{\omega_1^{\mathrm{CK}}}$. If $\mathbf{KP}\boldsymbol{\omega} \vdash F$ then we have finitely many axioms $A_1, \ldots, A_n$ of $\mathbf{KP}\boldsymbol{\omega}$, such that $A_1 \to \cdots \to A_n \to F$ is valid in first order logic with identity. Hence $\mathcal{H}_0 \overset{\Omega\cdot\omega^m}{\underset{\Omega+n_0}{\vdash}} A_1^{\mathsf{L}_\Omega} \to \cdots \to A_n^{\mathsf{L}_\Omega} \to F^{\mathsf{L}_\Omega}$ for $\Omega := \omega_1^{\mathrm{CK}}$ by Theorem 3.4.6.6. By Lemma 3.4.6.7, (200) and the Foundation Theorem (Theorem 3.4.6.9) we therefore obtain $\mathcal{H}_0 \overset{\Omega\cdot 2+\omega}{\underset{\Omega+n}{\vdash}} F^{\mathsf{L}_\Omega}$ by cuts. Applying the Predicative Elimination Lemma (Lemma 3.4.3.6) we then obtain $\mathcal{H}_0 \overset{\varphi_0^{(n-1)}(\Omega\cdot 2+\omega)}{\underset{\Omega+1}{\vdash}} F^{\mathsf{L}_\Omega}$. Assuming that F is a Σ_1–formula and putting $\alpha := \varphi_0^{(n-1)}(\Omega \cdot 2 + \omega) < \varepsilon_{\Omega+1}$ we get by the Collapsing Theorem (Theorem 3.4.5.3) $\mathcal{H}_{\omega^{\Omega+\alpha}} \overset{\psi_\Omega(\omega^{\Omega+\alpha})}{\underset{\psi_\Omega(\omega^{\Omega+\alpha})}{\vdash}} F^{\mathsf{L}_\Omega}$. Finally applying Theorem 3.4.2.2 we obtain $\overset{\psi_\Omega(\omega^{\Omega+\alpha})}{\models} F^{\mathsf{L}_\Omega}$. Therefore we have shown

3.4.6.10. Theorem. *For $\Omega := \omega_1^{\mathrm{CK}}$ we have $\|\mathbf{KP}\boldsymbol{\omega}\|_\Omega \leq \psi_\Omega(\varepsilon_{\Omega+1})$.*

To obtain also an ordinal analysis for the theories **KPl** and **KPi** we need controlling operators for axioms (Ad1), (Ad2), (Ad3) and (Lim). Let $\mathcal{H}$ be an acceptable operator which is closed under regular successors and λ an ordinal satisfying $(\forall \xi < \lambda)(\exists \kappa \in \lambda)[\kappa \in \mathbf{Reg} \wedge \xi < \kappa]$ and $\kappa \in \mathbf{Reg} \cap \lambda$. Then $\omega = \{x \in \mathsf{L}_\omega | x \in \mathbf{On}\} \in \mathcal{T}_\kappa$ and by (191) and (192) we obtain $\vdash \omega \in \mathsf{L}_\kappa \wedge \mathit{Tran}(\mathsf{L}_\kappa)$. By the Equality Theorem 3.4.6.4 we have $\vdash a \neq \mathsf{L}_\kappa, \neg(\omega \in \mathsf{L}_\kappa \wedge \mathit{Tran}(\mathsf{L}_\kappa)), \omega \in a \wedge \mathit{Tran}(a)$ for every term a. By Lemma 3.4.6.3, a cut, an inference $(\bigwedge)$ and two inferences $(\bigvee)$ we therefore obtain

$$\mathcal{H}[\mathsf{par}(a) \cup \{\lambda\}] \overset{\alpha}{\underset{\lambda}{\vdash}} \mathsf{Ad}(a) \to \omega \in a \wedge \mathit{Tran}(a) \tag{i}$$

for some $\alpha < \lambda$ depending on a. By an inference $(\bigwedge)$ we obtain from (i)

$$\mathcal{H}[\{\lambda\}] \overset{\lambda}{\underset{\lambda}{\vdash}} (\forall x \in \mathsf{L}_\lambda)[\mathsf{Ad}(x) \to \omega \in x \wedge \mathit{Tran}(x)], \text{ i.e., } \mathcal{H}[\{\lambda\}] \overset{\lambda}{\underset{\lambda}{\vdash}} (\mathrm{Ad1})^{\mathsf{L}_\lambda}. \tag{201}$$

Similarly we obtain $\vdash \mathsf{L}_\kappa \in \mathsf{L}_\mu \vee \mathsf{L}_\kappa = \mathsf{L}_\mu \vee \mathsf{L}_\mu \in \mathsf{L}_\kappa$ by (191) and (188) and $\vdash a \neq \mathsf{L}_\mu, b \neq \mathsf{L}_\kappa, \neg(\mathsf{L}_\kappa \in \mathsf{L}_\mu \vee \mathsf{L}_\kappa = \mathsf{L}_\mu \vee \mathsf{L}_\mu \in \mathsf{L}_\kappa), a \in b \vee a = b \vee b \in a$ for all $\mu, \kappa \in \lambda \cap \mathbf{Reg}$. As in the previous case this implies

$$\mathcal{H}[\{\lambda\}] \,\vdash^{\lambda}_{\lambda}\, (\mathrm{Ad2})^{\mathsf{L}_\lambda}. \tag{202}$$

Now let A be one of the axioms (Pair'), (Union'), (Δ_0–Separation) or (Δ_0–Collection). For any $\kappa \in \lambda \cap \mathbf{Reg}$ we get by Lemma 3.4.6.7 and (200) $\mathcal{H}[\{\kappa\}] \,\vdash^{\alpha}_{0}\, A^{\mathsf{L}_\kappa}$ for some $\alpha < \kappa^+$. As in the previous cases this implies

$$\mathcal{H}[\{\lambda\}] \,\vdash^{\lambda}_{\lambda}\, (\mathrm{Ad3})^{\mathsf{L}_\lambda}. \tag{203}$$

For $b \in \mathcal{T}_\lambda$ we have $\kappa := \mathsf{stg}(b)^+ < \lambda$. By (188) and (191) it follows $\vdash \mathsf{Ad}(\mathsf{L}_\kappa) \wedge b \in \mathsf{L}_\kappa$. By $(\bigvee')$ it follows $\vdash (\exists y \in \mathsf{L}_\lambda)[\mathsf{Ad}(y) \wedge b \in y]$ and by $(\bigwedge')$ we finally obtain $\vdash (\forall x \in \mathsf{L}_\lambda)(\exists y \in \mathsf{L}_\lambda)[\mathsf{Ad}(y) \wedge x \in y]$. Hence

$$\mathcal{H}[\{\lambda\}] \,\vdash^{\omega^{\lambda+n}}_{0}\, (\mathrm{Lim})^{\mathsf{L}_\lambda}. \tag{204}$$

Since Ω_ω satisfies $(\forall x \in \Omega_\omega)(\exists \kappa \in \Omega_\omega)[\kappa \in \mathbf{Reg} \wedge x \in \kappa]$ we obtain from Lemma 3.4.6.2, Lemma 3.4.6.7, Lemma 3.4.6.3, the Foundation Theorem 3.4.6.9 and (201) through (204) that for every sentence F there is an $m < \omega$ such that

$$\mathbf{KPl} \vdash F \;\Rightarrow\; \mathcal{H}_0 \,\vdash^{\Omega_\omega \cdot 2+\omega}_{\Omega_\omega + m}\, F^{\mathsf{L}_{\Omega_\omega}}. \tag{205}$$

If we assume that F in (205) is a $\Sigma_1^{\omega_1^{CK}}$–sentence we can apply the Predicative Elimination Lemma 3.4.3.6, the Collapsing Theorem 3.4.5.3 and Theorem 3.4.2.2 to obtain

$$\mathbf{KPl} \vdash F \text{ and } F \in \Sigma_1^{\omega_1^{CK}} \;\Rightarrow\; \models^{\alpha} F \text{ for some } \alpha < \psi_\Omega(\varepsilon_{\Omega_\omega+1}).$$

So we have shown the following theorem

3.4.6.11. Theorem. $\|\mathbf{KPl}\|_\Omega \leq \psi_\Omega(\varepsilon_{\Omega_\omega+1})$.

The ordinal I not only satisfies $(\forall x \in \Omega_\omega)(\exists \kappa \in \Omega_\omega)[\kappa \in \mathbf{Reg} \wedge x \in \kappa]$ but also $I \in \mathbf{Reg}$. So we obtain by (200), Lemma 3.4.6.2, Lemma 3.4.6.7, Lemma 3.4.6.3, the Foundation Theorem (Theorem 3.4.6.9) and (201) through (204)

$$\mathbf{KPi} \vdash F \;\Rightarrow\; \mathcal{H}_0 \,\vdash^{I\cdot 2+\omega}_{I+n}\, F^{\mathsf{L}_I}$$

for some $n < \omega$. As before this implies that there is an $\alpha < \psi_\Omega(\varepsilon_{I+1})$ such that $\models^{\alpha} F$ for Σ_1^Ω-sentences which are provable in **KPi**. So we have

3.4.6.12. Theorem. $\|\mathbf{KPi}\|_\Omega \leq \psi_\Omega(\varepsilon_{I+1})$.

To obtain also an analysis of the theory **KPl^r** we have to do a bit more. First we show

$$\vdash^{k} \neg\mathrm{Lim}, \Delta(\vec{x}) \;\Rightarrow\; (\exists n < \omega)(\forall \mu < \lambda)(\forall \vec{a} \in \mathcal{T}_{\Omega_\mu})\mathcal{H}[\vec{a}] \,\vdash^{\Omega_{\mu+n}+2k}_{\Omega_{\mu+n}+1}\, \Delta(\vec{a})^{\mathsf{L}_{\Omega_{\mu+n}}} \tag{206}$$

for a limit ordinal λ and a finite set $\Delta(\vec{x})$ of Σ–formulas which contain only the shown free variables and an acceptable operator $\mathcal{H}$ which is closed under $\xi \mapsto \xi^+$.

The proof is by induction on k and runs essentially as that of Lemma 3.4.6.2. But there is the additional (and critical) case that the main formula of the last inference is $\neg$Lim. Then we have the premise $\mid\!\frac{k_0}{}\ \neg\mathrm{Lim}, (\forall y)[\neg\mathsf{Ad}(y) \vee x_i \notin y], \Delta(\vec{x})$ which by inversion yields $\mid\!\frac{k_0}{}\ \neg\mathrm{Lim}, \neg\mathsf{Ad}(y) \vee x_i \notin y, \Delta(\vec{x})$ for a free variable y not occurring in $\Delta(\vec{x})$. By induction hypothesis there is an $n_0 < \omega$ such that

$$\mathcal{H}[\vec{a}, b]\ \Big|\frac{\Omega_{\mu+n_0+1}+2k_0}{\Omega_{\mu+n_0+1}+1}\ \neg\mathsf{Ad}(b) \vee a_i \notin b, \Delta(\vec{a})^{\mathsf{L}_{\Omega_{n_0+m+1}}}$$

for all $\vec{a} \in \mathcal{T}_{\Omega_\mu}$ and all $b \in \mathcal{T}_{\Omega_{\mu+1}}$. So we obtain

$$\mathcal{H}[\vec{a}]\ \Big|\frac{\Omega_{\mu+n_0+1}+2k_0+1}{\Omega_{\mu+n_0+1}+1}\ (\forall z \in \mathsf{L}_{\Omega_{\mu+1}})[\neg\mathsf{Ad}(z) \vee a_i \notin z], \Delta(\vec{a})^{\mathsf{L}_{\Omega_{\mu+n_0+1}}}.$$

By (188) and (191) we have $\mathcal{H}[\{a_i\}]\ \Big|\frac{\Omega_{\mu+1}}{0}\ (\exists z \in \mathsf{L}_{\Omega_{\mu+1}})[\mathsf{Ad}(z) \wedge a_i \in z]$ and obtain

$$\mathcal{H}[\vec{a}]\ \Big|\frac{\Omega_{\mu+n_0+1}+2k}{\Omega_{\mu+n_0+1}+1}\ \Delta(\vec{a})^{\mathsf{L}_{\Omega_{\mu+n_0+1}}}$$

by cut.

Next we observe that the axioms (Pair'), (Union') and (Δ_0–Separation) are dispensable. They can be derived from (Ad3) together with (Lim). The same is true for (Inf'). The scheme of Δ_0-Foundation can be formulated as

$$(\Delta_0\text{-Found})\quad (\forall u)[\mathit{Tran}(u) \wedge (\forall x \in u)[(\forall y \in x)F(y) \to F(x)] \to (\forall x \in u)F(x)]$$

where $F(x)$ is a Δ_0–formula. So if $\mathbf{KPl^r} \vdash \Delta(\vec{x})$ for a finite set $\Delta(\vec{x})$ of Σ–formulas then there are finitely many instances of axioms $A_1, \ldots, A_k$ such that

$$\mid\!\frac{m}{}\ \neg\mathrm{Lim}, \neg A_1, \ldots, \neg A_k, \Delta(\vec{x}).$$

With the exception of axiom (Lim) every of these formulas $\neg A_i$ is a Σ_1–formula. For a limit ordinal λ and a tuple $\vec{a} \in \mathbf{L}_{\Omega_\lambda}$ we get by (206) $\mu < \lambda$ and $k < \omega$ such that

$$\mathcal{H}[\vec{a}]\ \Big|\frac{\Omega_\mu+k}{\Omega_\mu+1}\ \neg A_1^{\mathsf{L}_{\Omega_\mu}}, \ldots, \neg A_n^{\mathsf{L}_{\Omega_\mu}}, \Delta(\vec{a})^{\mathsf{L}_{\Omega_\mu}}.$$

Applying some cuts this implies $\mathcal{H}[\vec{a}]\ \Big|\frac{\Omega_\mu+k}{\Omega_\mu+1}\ \Delta(\vec{a})^{\mathsf{L}_{\Omega_\mu}}$. So we have shown

$$\mathbf{KPl^r} \vdash \Delta(\vec{u}) \Rightarrow (\forall \lambda \in \mathsf{Lim})(\forall \vec{a} \in \Omega_\lambda)(\exists \mu < \lambda)(\exists k < \omega)[\mathcal{H}_0[\vec{a}]\ \Big|\frac{\Omega_\mu+k}{\Omega_\mu+1}\ \Delta(\vec{a})^{\mathsf{L}_{\Omega_\mu}}] \quad (207)$$

For a Σ-sentence F this entails

$$\mathbf{KPl^r} \vdash F \ \Rightarrow\ (\exists m \in \omega)\left[\mathcal{H}_0\ \Big|\frac{\Omega_m+\omega}{\Omega_m+1}\ F^{\mathsf{L}_{\Omega_\omega}}\right] \quad (208)$$

which for a Σ_1^Ω-sentence F by the Collapsing Theorem and Theorem 3.4.2.2 implies $\Big|\!\!\frac{\psi_\Omega(\Omega_m^2 \cdot \omega^\omega)}{}\ F$. Since $\lim_{m \in \omega} \psi_\Omega(\Omega_m^2 \cdot \omega^\omega) = \psi_\Omega(\Omega_\omega)$ we have

3.4.6.13. Theorem. $\|\mathbf{KPl^r}\|_\Omega \le \psi_\Omega(\Omega_\omega)$.

Though there is a tremendous gap between **KPl** and **KPi** there is no difference between $\mathbf{KPl^r}$ and $\mathbf{KPi^r}$. If we try the same analysis as we did for $\mathbf{KPl^r}$ the axiom of Δ_0–Collection will spoil the argument. This, however, can be remedied by augmenting the logical calculus by the following non-logical rule.

$$(\Delta_0\text{-Collection Rule}) \qquad \vdash^{m} \Delta, (\forall x \in a)(\exists y)F(x,y)$$
$$\Rightarrow \quad \vdash^{m+1} \Delta, (\exists z)(\forall x \in a)(\exists y \in z)F(x,y).$$

By $[\Delta_0] \vdash^{m} \Delta$ we denote that Δ is provable in the extended calculus. We obtain for a finite set $\Delta(\vec{u})$ of Σ– formulas and an ordinal λ

$$[\Delta_0] \vdash^{m} \neg\mathrm{Lim}, \Delta(\vec{u}) \;\Rightarrow\; (\forall \mu \in \lambda)(\forall \vec{a} \in \mathcal{T}_{\Omega_\mu})[\mathcal{H}[\vec{a}] \,\vdash^{\Omega_{\mu+2m}}_{\Omega_{\mu+2m+1}} \Delta(\vec{a})^{\mathsf{L}_{\Omega_{\mu+2m}}}]. \quad (209)$$

The proof parallels that of (206). The additional case is an application of the Δ_0–Collection Rule. By the induction hypothesis we then have

$$\mathcal{H}[\vec{a}] \,\vdash^{\Omega_{\mu+2m_0}}_{\Omega_{\mu+2m_0}+1} \Delta(\vec{a})^{\Omega_{\mu+2m_0}}, (\forall x \in a_i)(\exists y \in \mathsf{L}_{\Omega_{\mu+2m_0}})F(x,y).$$

By an inference $(\mathrm{Ref}_{\Omega_{\mu+2m_0+1}})$ and Upward Persistency this implies

$$\mathcal{H}[\vec{a}] \,\vdash^{\Omega_{\mu+2m_0+2}}_{\Omega_{\mu+2m_0+2}+1} \Delta(\vec{a})^{\Omega_{\lambda+2m_0+2}}, (\exists z \in \mathsf{L}_{\Omega_{\mu+2m_0+2}})(\forall x \in a_i)(\exists y \in z)F(x,y).$$

Now replacing (206) by (209) we obtain with the same strategy as in the case of $\mathbf{KPl^r}$

$$\mathbf{KPi^r} \vdash F \;\Rightarrow\; (\exists n \in \omega)\left[\mathcal{H}_0 \,\vdash^{\Omega_n}_{\Omega_n+1} F\right]$$

for Σ_1^Ω sentences F. So we have

3.4.6.14. Theorem. $\|\mathbf{KPi^r}\|_\Omega \le \psi_\Omega(\Omega_\omega)$.

We will roughly sketch how an analysis for W–**KPl** and W–**KPi** can be obtained without going too much into details. The stumbling block here is the scheme of Mathematical Induction. The remedy is to augment the calculi $\vdash^{m}$ and $[\Delta_0] \vdash^{m}$, respectively, by an ω– and a cut–rule. Call these extended calculi $[\omega] \vdash^{m}_{r}$ and $[\Delta_0, \omega] \vdash^{m}_{r}$, respectively. Then if W–**KPl** $\vdash F$ or W–**KPi** $\vdash F$ we obtain $[(\Delta_0)] \vdash^{m} \neg A_1, \ldots, \neg A_k, \neg MI_1, \ldots, \neg MI_l, F$ for finitely many instances MI_i of Mathematical Induction. Since $[(\Delta_0), \omega] \vdash^{\omega+n}_{0} MI_i$ we obtain by cut $[(\Delta_0), \omega] \vdash^{\omega+\omega}_{r} \neg A_1, \ldots, \neg A_k, F$ which by the usual cut-elimination for ω–logic (cf. Lemma 2.1.2.9) entails $[(\Delta_0), \omega] \vdash^{\alpha}_{0} \neg A_1, \ldots, \neg A_k, F$ for some $\alpha < \varepsilon_0$. Now we are in a situation similar to that in the analyses of $\mathbf{KPl^r}$ and $\mathbf{KPi^r}$. While (209) modifies directly to

$$[\Delta_0, \omega] \vdash^{\alpha}_{0} \neg\mathrm{Lim}, \Delta(\vec{u}) \;\Rightarrow\; (\forall \vec{a} \in \mathcal{T}_{\Omega_\lambda})[\mathcal{H}[\vec{a}] \,\vdash^{\Omega_{\lambda+2\alpha}}_{\Omega_{\lambda+2\alpha}+1} \Delta(\vec{a})^{\mathsf{L}_{\Omega_{\lambda+2\alpha}}}], \quad (210)$$

we cannot directly adapt (206) to the calculus $[\omega] \overset{\alpha}{\underset{0}{\vdash}}$ because we have infinite derivations. We have to refine the argument and prove

$$[\omega] \overset{\alpha}{\underset{0}{\vdash}} \neg\mathrm{Lim}, \Delta(\vec{u}) \;\Rightarrow\; (\forall\lambda \in \mathsf{Lim})(\forall \vec{a}, b \in \mathcal{T}_{\Omega_\lambda})[\Delta(\vec{a})^b \subseteq \Sigma^\kappa \Rightarrow \mathcal{H}_{\omega^{\Omega_\lambda+3\alpha}}[\vec{a}, b] \overset{\Omega_\lambda+3\alpha}{\underset{\kappa+1}{\vdash}} \Delta(\vec{a})^b] \tag{211}$$

for a finite set of Σ–formulas $\Delta(\vec{u})$ by induction on α. The crucial case is again that the main formula of the last inference is $\neg$Lim. Let $\vec{a}, b$ such that $\Delta(\vec{a})^b \subseteq \Sigma^\kappa$. From the induction hypothesis we obtain $\mathcal{H}_{\omega^{\Omega_\lambda+3\alpha_0}}[\vec{a}, b, c] \overset{\Omega_\lambda+3\alpha_0}{\underset{\kappa^++1}{\vdash}} \neg\mathsf{Ad}(c) \vee a_i \notin c, \Delta(\vec{a})^b$ for all $c \in \mathcal{T}_{\kappa^+}$ which by an inference $(\bigwedge)$ implies $\mathcal{H}_{\omega^{\Omega_\lambda+3\alpha_0}}[\vec{a}, b] \overset{\Omega_\lambda+3\alpha_0+1}{\underset{\kappa^++1}{\vdash}} \neg(\exists z \in \mathsf{L}_{\kappa^+})[\mathsf{Ad}(z) \wedge a_i \in z], \Delta(\vec{a})^b$. By cut we obtain $\mathcal{H}_{\omega^{\Omega_\lambda+3\alpha_0}}[\vec{a}, b] \overset{\Omega_\lambda+3\alpha_0+2}{\underset{\kappa^++1}{\vdash}} \Delta(\vec{a})^b$. Since $\omega^{\Omega_\lambda+3\alpha_0} + \omega^{\kappa^++\Omega_\lambda+3\alpha_0+2} < \omega^{\Omega_\lambda+3\alpha}$ we obtain $\mathcal{H}_{\omega^{\Omega_\lambda+3\alpha}}[\vec{a}, b] \overset{\Omega_\lambda+3\alpha}{\underset{\kappa+1}{\vdash}} \Delta(\vec{a})^b$ by the Collapsing Theorem.

By the now familiar technique we obtain from (210)

$$\text{W–}\mathbf{KPi} \vdash F \;\Rightarrow\; (\exists\alpha \in \varepsilon_0)\left[\mathcal{H}_0 \overset{\Omega_\alpha}{\underset{\Omega_\alpha+1}{\vdash}} F^{\mathsf{L}_{\Omega_\alpha}}\right] \tag{212}$$

for Σ_1–sentences F. By the Collapsing Theorem and Theorem 3.4.2.2 this implies

3.4.6.15. Theorem. $\|\text{W–}\mathbf{KPi}\|_\Omega \leq \psi_\Omega(\Omega_{\varepsilon_0})$.

From (211) we obtain as in the proof of (208)

$$\text{W–}\mathbf{KPl} \vdash F \;\Rightarrow\; (\exists\alpha \in \varepsilon_0)\left[\mathcal{H}_{\omega^{\Omega_\omega+\alpha}} \overset{\Omega_\omega+\alpha}{\underset{\Omega+1}{\vdash}} F^{\mathsf{L}_\Omega}\right] \tag{213}$$

for Σ_1–sentences F. Since $\psi_\Omega(\omega^{\Omega+\Omega_\omega+\alpha}) = \psi_\Omega(\Omega_\omega \cdot \omega^\alpha) < \psi_\Omega(\Omega_\omega \cdot \varepsilon_0)$ for $\alpha < \varepsilon_0$ we get from (213) by the Collapsing Theorem and Theorem 3.4.2.2

3.4.6.16. Theorem. $\|\text{W–}\mathbf{KPl}\|_\Omega \leq \psi_\Omega(\Omega_\omega \cdot \varepsilon_0)$.

The analysis of theories for iterated admissibles needs serious extra work which has first been done by M. Rathjen. We have to show that there are operator controlled derivations for the axioms $\mathit{ItAd}(\alpha, f)$. This is prepared by a Lemma which is provable in $\mathbf{KPl^r}$.

3.4.6.17. Lemma. ($\mathbf{KPl^r}$) *Let $\boldsymbol{a}$ and $\boldsymbol{u}$ be admissible sets such that $\boldsymbol{a} \in \boldsymbol{u}$ and let $\alpha \in \boldsymbol{a}$. Then $(\forall\xi < \alpha)(\exists f \in \boldsymbol{a})\mathit{ItAd}(\xi, f)$ implies $(\exists g \in \boldsymbol{u})\mathit{ItAd}(\alpha, g)$.*

Proof. Since $\mathit{ItAd}(\xi, f)$ is a Δ_0–formula we get from the hypothesis by Δ_0–Collection relative to $\boldsymbol{a}$ a set $b \in \boldsymbol{a}$ such that $(\forall\xi < \alpha)(\exists f \in b)\mathit{ItAd}(\xi, f)$. By (130) for every $\xi < \alpha$ there is exactly one such $f_\xi \in b$. By Δ_0–Separation relative to $\boldsymbol{a}$ we get $h := \bigcup_{\xi<\alpha} f_\xi \in \boldsymbol{a}$. Clearly h is a function. Let c be an $\in$–least element of $\{x \in \boldsymbol{u} \mid \mathsf{Ad}(x) \wedge x \notin rng(h)\}$. Since $\mathsf{Ad}(\boldsymbol{a}) \wedge \boldsymbol{a} \in \boldsymbol{u}$ the set c exists by Δ_0-Foundation and we define

$$g := \begin{cases} h & \text{if } \alpha \in \mathsf{Lim} \cup \{0\} \\ h \cup \{(\gamma, c)\} & \text{if } \alpha = \gamma + 1. \end{cases}$$

Then g is a function with domain α and we easily check $\mathit{ItAd}(\alpha, g)$. □

Recall the function $\lambda\xi\,.\,\tau_\xi$ which enumerates the admissible ordinals. We have

$$\tau_\alpha = \begin{cases} \omega & \text{if } \alpha = 0 \\ \Omega_\alpha & \text{for } 0 < \alpha < \omega \\ \Omega_{\alpha+1} & \text{for } \omega < \alpha < I \\ I & \text{for } \alpha = I \end{cases}$$

From Lemma 3.4.6.17 and (207) we get for a limit ordinal ν and $\alpha < \nu$ a $k < \omega$ such that

$$\mathcal{H}_0[\{\alpha\}] \mathrel{|}\!\frac{\tau_{\alpha+k}+m}{\tau_{\alpha+k}+1} \neg\mathsf{Ad}(\mathsf{L}_{\tau_\alpha}), \neg\mathsf{Ad}(\mathsf{L}_{\tau_{\alpha+1}}), \alpha \notin \mathsf{L}_{\tau_\alpha}, \neg(\forall\xi \in \alpha)(\exists f \in \mathsf{L}_{\tau_\alpha})[\mathit{ItAd}(\xi, f)],$$
$$\mathsf{L}_{\tau_\alpha} \notin \mathsf{L}_{\tau_{\alpha+1}}, (\exists g \in \mathsf{L}_{\tau_{\alpha+1}})[\mathit{ItAd}(\alpha, g)].$$

For $\alpha < I$ we have $\mathcal{H}_0[\{\alpha\}] \mathrel{|}\!\frac{\tau_{\alpha+1}}{\tau_{\alpha+1}} \mathsf{Ad}(\mathsf{L}_{\tau_\alpha})$, $\mathcal{H}_0[\{\alpha\}] \mathrel{|}\!\frac{\tau_{\alpha+2}}{\tau_{\alpha+2}} \mathsf{Ad}(\mathsf{L}_{\tau_{\alpha+1}})$, $\vdash \alpha \in \mathsf{L}_{\tau_\alpha}$ and $\vdash \mathsf{L}_{\tau_\alpha} \in \mathsf{L}_{\tau_{\alpha+1}}$. So we get by some cuts

$$\mathcal{H}_0[\{\alpha\}] \mathrel{|}\!\frac{\tau_{\alpha+k}+m'}{\tau_{\alpha+k}+n'} \neg(\forall\xi \in \alpha)(\exists f \in \mathsf{L}_{\tau_\alpha})[\mathit{ItAd}(\xi, f)], (\exists g \in \mathsf{L}_{\tau_{\alpha+1}})[\mathit{ItAd}(\alpha, g)]. \quad (214)$$

Now we show by induction on $\alpha < I$

$$\mathcal{H}_0[\{\alpha\}] \mathrel{|}\!\frac{\Omega_{\alpha+\omega}+4(\alpha+1)}{\Omega_{\alpha+\omega}} (\exists f \in \mathsf{L}_{\tau_{\alpha+1}})\mathit{ItAd}(\alpha, f). \quad (215)$$

By the induction hypothesis, $\mathcal{H}_0[\{\alpha, \xi\}] \mathrel{|}\!\frac{\Omega_{\xi+\omega}+4\cdot(\xi+1)}{\Omega_{\xi+\omega}} \alpha \notin \mathbf{On}, (\exists f \in \mathsf{L}_{\tau_{\xi+1}})\mathit{ItAd}(\xi, f)$ for all $\xi < \alpha$. This implies $\mathcal{H}_0[\{\alpha\}] \mathrel{|}\!\frac{\Omega_{\alpha+\omega}+4\alpha+3}{\Omega_{\alpha+\omega}} (\forall\xi < \alpha)(\exists f \in \mathsf{L}_{\tau_\alpha})\mathit{ItAd}(\xi, f)$ by two inferences $(\bigvee)$, an inference $(\bigwedge)$ and Upward Persistency. From this and (214) we obtain the claim by a cut. □

Since $\xi < \nu$ implies $\xi + \omega \le \nu$ for $\nu \in \mathsf{Lim}$ we obtain from (215)

$$\mathcal{H}_0[\{\nu\}] \mathrel{|}\!\frac{\Omega_\nu+\nu}{\Omega_\nu} (\forall\xi < \nu)(\exists f \in \mathsf{L}_{\Omega_\nu})\mathit{ItAd}(\xi, f) \quad (216)$$

for $\nu \in \mathsf{Lim}$. The strategy we applied for the ordinal analysis of **KPl** together with (216) now yields

$$\mathbf{KPl}_\nu \vdash F(\vec{u}) \;\Rightarrow\; \mathcal{H}_0[\vec{a}] \mathrel{|}\!\frac{\Omega_\nu\cdot 2+\omega}{\Omega_\nu+m} F(\vec{a})^{\mathsf{L}_{\Omega_\nu}} \quad (217)$$

for a limit ordinal ν and all $\vec{a} \in \mathcal{T}_{\Omega_\nu}$. By the meanwhile familiar argument this entails

3.4.6.18. Theorem. $\|\mathbf{KPl}_\nu\|_\Omega \le \psi_\Omega(\varepsilon_{\Omega_\nu+1})$

An ordinal analysis of the theory $\mathbf{KPl}^{\mathbf{r}}_\nu$ for limit ordinals ν is obtained in in a similar way as that of $\mathbf{KPl}^{\mathbf{r}}$. We modify (206) to

$$\mathrel{|}\!\frac{k}{}\; \neg\text{Lim}, \neg(\forall\xi < \nu)(\exists f)\mathit{ItAd}(\xi, f), \Delta(\vec{u}) \;\Rightarrow \quad (218)$$
$$(\forall\alpha < \nu)(\forall\vec{a} \in \mathcal{T}_{\tau_\alpha}) \left[\mathcal{H}[\vec{a}, \{\alpha\}] \mathrel{|}\!\frac{\Omega_{\alpha+\omega}+4(\alpha+k)}{\Omega_{\alpha+\omega}} \Delta(\vec{a})^{\mathsf{L}_{\Omega_{\alpha+\omega}}}\right]$$

for a finite set $\Delta(\vec{u})$ of Σ–formulas and an acceptable operator which is closed under $\lambda\xi.\Omega_\xi$. The proof is analogous to that of (206) with the additional case that the main formula of the last inference is $\neg(\forall\xi<\nu)(\exists f)\mathit{ItAd}(\xi,f)$. Then we have the premise

$$\vdash^{k_0} \neg\mathrm{Lim}, \neg(\forall\xi<\nu)(\exists f)\mathit{ItAd}(\xi,f), u_i\in\mathbf{On}\wedge u_i<\nu\wedge(\forall f)\neg\mathit{ItAd}(u_i,f).$$

Applying inversion we obtain for $\alpha, a_i<\nu$, $\vec{a}\in\mathcal{T}_{T_\alpha}$ and $f\in\mathcal{T}_{T_{\alpha+1}}$ by induction hypothesis $\mathcal{H}[\vec{a},\{\alpha\},f] \mid\!\frac{\Omega_{\alpha+\omega}+4(\alpha+1+k_0)}{\Omega_{\alpha+\omega}} \neg\mathit{ItAd}(a_i,f),\Delta(\vec{a})^{\mathsf{L}_{\Omega_{\alpha+\omega}}}$. By an inference $(\bigwedge)$ this yields $\mathcal{H}[\vec{a},\{\alpha\}] \mid\!\frac{\Omega_{\alpha+\omega}+4(\alpha+k_0+1)+1}{\Omega_{\alpha+\omega}} \neg(\exists f\in\mathsf{L}_{T_{\alpha+1}})\mathit{ItAd}(\alpha,f),\Delta(\vec{a})^{\mathsf{L}_{\Omega_{\alpha+\omega}}}$ which together with (215) entails the claim by a cut.

If $\mathbf{KPl}^r_\nu \vdash F$ for a Σ–sentence F then there are Π_1 sentences $A_1,\dots,A_n$ – the axioms different from Lim and the iteration axiom – such that by (218) we obtain

$$\mathcal{H}_0[\{\alpha\}] \mid\!\frac{\Omega_{\alpha+\omega}+\Omega_\alpha}{\Omega_{\alpha+\omega}} \neg A_1^{\mathsf{L}_{\Omega_{\alpha+\omega}}},\dots,A_n^{\mathsf{L}_{\Omega_{\alpha+\omega}}},F^{\mathsf{L}_{\Omega_{\alpha+\omega}}}$$

for some $\alpha<\nu$. Applying some cuts we get

$$\mathbf{KPl}^r_\nu \vdash F \;\Rightarrow\; (\exists\alpha<\nu)\left[\mathcal{H}_0[\nu] \mid\!\frac{\Omega_{\alpha+\omega}\cdot 2}{\Omega_{\alpha+\omega}} F^{\mathsf{L}_{\Omega_{\alpha+\omega}}}\right] \tag{219}$$

for Σ–sentences F. If we assume that $\omega<\nu$ and ν is additively indecomposable then $\alpha<\nu$ implies $\alpha+\omega<\nu$. So we obtain from (219)

3.4.6.19. Theorem. *$\|\mathbf{KPl}^r_\nu\|_\Omega\le\psi_\Omega(\Omega_\nu)$ for additively indecomposable ordinals ν.*

We can also say something for limit ordinals which are additively decomposable. We generalize (211) to

$$\begin{aligned}[\omega] \mid\!\frac{\alpha}{0} \neg\mathrm{Lim}, \neg(\forall\xi<\nu)(\exists f)\mathit{ItAd}(\xi,f)\Delta(\vec{u}) \;\Rightarrow\; \\ (\forall\vec{a},b\in\mathcal{T}_{\Omega_\nu})[\Delta(\vec{a})^b\subseteq\Sigma^\kappa \Rightarrow \mathcal{H}_{\omega^{\Omega_\nu+4\alpha}}[\vec{a},b] \mid\!\frac{\Omega_\nu+4\alpha}{\kappa+1} \Delta(\vec{a})^b].\end{aligned} \tag{220}$$

Therefore we obtain from $\mathbf{KPl}^r_\nu \vdash F$ for a Σ–sentence F a $k<\omega$ such that $\mathcal{H}_{\omega^{\Omega_\nu+k}} \mid\!\frac{\Omega_\nu+k}{\Omega+1} \mathsf{F}^{\mathsf{L}_\Omega}$. This shows

$$\|\mathbf{KPl}^r_\nu\|_\Omega\le\psi_\Omega(\Omega_\nu\cdot\omega^\omega)$$

for limit ordinals which are additively decomposable. I have, however, never checked whether this ordinal is the precise one or in other way meaningful. In all relevant applications the ordinal ν is additively indecomposable.

We also obtain an analysis of W–$\mathbf{KPl}_\nu$. If W–$\mathbf{KPl}_\nu \vdash F$ for a Σ–sentence F we obtain similarly as in the case of W–$\mathbf{KPl}$ an $\alpha<\varepsilon_0$ such that by (220) $\mathcal{H}_{\omega^{\Omega_\nu+\alpha}}[\{\alpha\}] \mid\!\frac{\Omega_\nu+\alpha}{\Omega+1} F^{\mathsf{L}_\Omega}$. Applying Collapsing and Theorem 3.4.2.2 we obtain

3.4.6.20. Theorem. *For limit ordinals ν we have $\|$W–$\mathbf{KPl}_\nu\|_\Omega\le\psi_\Omega(\Omega_\nu\cdot\varepsilon_0)$.*

Recall from (169) and (170) that $\psi_I 0$ is the least fixed point of the function $\lambda\xi.\Omega_\xi$. By (215) we get therefore

$$\mathcal{H}_0 \left|\frac{\psi_I 0}{\psi_I 0}\right. (\forall\alpha \in \mathsf{L}_{\psi_I 0})(\exists f \in \mathsf{L}_{\psi_I 0})\, ItAd(\alpha, f). \tag{221}$$

By the now familiar procedures we obtain from (221)

$$\textbf{Aut–KPl} \vdash F \;\Rightarrow\; \mathcal{H}_0 \left|\frac{\psi_I 0\cdot 2+\omega}{\psi_I 0+n}\right. F^{\mathsf{L}_{\psi_I 0}} \tag{222}$$

$$\textbf{Aut-KPl}^{\mathbf{r}} \vdash F \;\Rightarrow\; (\exists\xi < \psi_I 0)\left[\mathcal{H}_0[\{\xi\}] \left|\frac{\Omega_\xi}{\Omega_\xi+1}\right. F^{\mathsf{L}_{\psi_I 0}}\right] \tag{223}$$

$$\text{W–}\textbf{Aut–KPl} \vdash F \;\Rightarrow\; (\exists\alpha < \varepsilon_0)(\forall b \in \mathsf{L}_{\psi_I 0})\left[\mathcal{H}_{\psi_\Omega I\cdot\alpha}[\{\alpha\}] \left|\frac{\psi_I 0+\alpha}{\Omega+1}\right. F^{\mathsf{L}_\Omega}\right] \tag{224}$$

for Σ–sentences F. Therefore we obtain

3.4.6.21. Theorem.

(i) $\|\textbf{Aut-KPl}^{\mathbf{r}}\|_\Omega \leq \psi_\Omega(\psi_I 0)$

(ii) $\|\text{W–}\textbf{Aut-KPl}^{\mathbf{r}}\|_\Omega \leq \psi_\Omega(\psi_I 0 \cdot \varepsilon_0)$

(i) $\|\textbf{Aut–KPl}\|_\Omega \leq \psi_\Omega(\varepsilon_{\psi_I 0+1})$

From (215) we get also $\mathcal{H}_0 \left|\frac{\Omega_\Omega}{\Omega_\Omega}\right. (\forall\xi \in \mathsf{L}_{\Omega_\Omega})(\exists f \in \mathsf{L}_{\Omega_\Omega})\, ItAd(\xi, f)$ which together with our previous work shows

3.4.6.22. Theorem. $\|\textbf{KPl}^{*}\|_\Omega \leq \psi_\Omega(\varepsilon_{\Omega_\Omega+1})$.

These examples should suffice to demonstrate the technique of ordinal analysis. We already mentioned that it is not the most recent state of the art. On the one hand the analyses of much stronger axiom systems such as Π_3 reflection and even Σ_1–Separation – which corresponds to Π^1_2–comprehension – are known today. The collapsing functions needed there are based on a Π^0_3–reflecting ordinal which – when we stick to the simplification of using regular cardinals instead of recursively regular ones – corresponds to the first weakly compact cardinal. The cardinals which are in this way connected to collapsing functions for Σ_1–separation are considerably larger. On the other hand we also omitted a whole zoo of theories which are between $(\mathbf{\Delta^1_2}$**-CA**$)$ and $(\mathbf{\Delta^1_2}$**-CA**$)$ + (Bi), i.e., between $\mathbf{KPl}_{\varepsilon_0}$ and **KPi**. Examples for such omitted systems are W–**KPi** + (Σ–FOUND) which on the side of subsystems of Number Theory corresponds to autonomously iterated Δ^1_2 comprehension (**Aut–**$\mathbf{\Delta^1_2}$) (and whose ordinal is $\psi_\Omega(\psi_I(\varepsilon_{I+1}))$).

We arranged the theories in Table 1 in such a way that in every row the theories are embeddable from left to right. We have shown that the ordinal in the column $\|\mathbf{Ax}\|_{\omega_1^{CK}}$ is an upper bound for the "strongest" theory, i.e., the Set Theoretic theory. To show that the bounds are precise it suffices to arithmetize the notation system (whose construction is indicated on page 308) and to give a well-ordering proof for the segment below $\|\mathbf{Ax}\|_{\omega_1^{CK}}$ within the theory in the leftmost column. All the shown

Theories for Inductive Definitions		Restricted Comprehension and Choice in $\mathbf{NT_2}$	Set Theoretic Theories	$\|\mathbf{Ax}\|_{\omega_1^{CK}}$
1st–order	2nd–order			
$\mathbf{ID}_1$	$\mathbf{ID}_1^2$	$(\mathbf{\Pi}_1^1\text{-}\mathbf{CA})^-$	$\mathbf{KP}\omega$	$\psi_\Omega(\varepsilon_{\Omega+1})$
$\mathbf{ID}_{<\omega}$	$(\mathbf{ID}^2_{<\omega})_0$	$(\mathbf{\Pi}_1^1\text{- }\mathbf{CA})_0$ $(\mathbf{\Delta}_2^1\text{- }\mathbf{CA})_0$	$\mathbf{KPl}^r$ $\mathbf{KPi}^r$ $\mathbf{KP}\beta^r$	$\psi_\Omega(\Omega_\omega)$
W–$\mathbf{ID}_\omega$†	$\mathbf{ID}^2_{<\omega}$	$(\mathbf{\Pi}_1^1\text{-}\mathbf{CA})$	W–$\mathbf{KPl}$	$\psi_\Omega(\Omega_\omega \cdot \varepsilon_0)$
$\mathbf{ID}_\omega$	$\mathbf{BID}^2_\omega$	$(\mathbf{\Pi}_1^1\text{-}\mathbf{CA}) + (\mathsf{Bi})$	$\mathbf{KPl}$	$\psi_\Omega(\varepsilon_{\Omega_\omega+1})$
$\mathbf{ID}_{<\omega^\omega}$	$\mathbf{BID}^2_{<\omega^\omega}$ $\mathbf{ID}^2_{<\omega^\omega}$	$(\mathbf{\Pi}_1^1\text{-}\mathbf{CA}_{<\omega^\omega})$ $(\mathbf{\Delta}_2^1\text{-}\mathbf{CR})$	$\mathbf{KPl}^r_{\omega^\omega}$	$\psi_\Omega(\Omega_{\omega^\omega})$
$\mathbf{ID}_{<\varepsilon_0}$	$\mathbf{BID}^2_{<\varepsilon_0}$ $\mathbf{ID}^2_{<\varepsilon_0}$	$(\mathbf{\Pi}_1^1\text{-}\mathbf{CA}_{<\varepsilon_0})$ $(\mathbf{\Delta}_2^1\text{-}\mathbf{CA})$, $(\mathbf{\Sigma}_2^1\text{-}\mathbf{AC})$	$\mathbf{KPl}^r_{\varepsilon_0}$ W–$\mathbf{KPi}$ W–$\mathbf{KP}\beta$	$\psi_\Omega(\Omega_{\varepsilon_0})$
$\mathbf{ID}_{<\nu}$	$\mathbf{BID}^2_{<\nu}$ $\mathbf{ID}^2_{<\nu} + (\mathsf{Bi})$	$(\mathbf{\Pi}_1^1\text{-}\mathbf{CA}_{<\nu})$ $(\mathbf{\Pi}_1^1\text{-}\mathbf{CA}_{<\nu}) + (\mathsf{Bi})$	$\mathbf{KPl}^r_\nu$	$\psi_\Omega(\Omega_\nu)$‡
	$(\mathbf{ID}^2_\nu)_0$	$(\mathbf{\Pi}_1^1\text{-}\mathbf{CA}_\nu)_0$	$\mathbf{KPl}^r_\nu$	$\psi_\Omega(\Omega_\nu)$‡‡
W–$\mathbf{ID}_\nu$†	$\mathbf{ID}^2_\nu$	$(\mathbf{\Pi}_1^1\text{-}\mathbf{CA}_\nu)$	W–$\mathbf{KPl}_\nu$	$\psi_\Omega(\Omega_\nu \cdot \varepsilon_0)$††
$\mathbf{ID}_\nu$	$\mathbf{BID}^2_\nu$ $\mathbf{ID}^2_\nu + (\mathsf{Bi})$	$(\mathbf{\Pi}_1^1\text{-}\mathbf{CA}_\nu) + (\mathsf{Bi})$	$\mathbf{KPl}_\nu$	$\psi_\Omega(\varepsilon_{\Omega_\nu+1})$††
$\mathbf{ID}_{\prec^*}$	$\mathbf{BID}^{2*}$ $\mathbf{ID}^{2*} + (\mathsf{Bi})$		$\mathbf{KPl}^* = \mathbf{KPl}^r_\Omega$	$\psi_\Omega(\varepsilon_{\Omega_\Omega+1})$
	$(\mathbf{Aut\text{-}ID})_0$	$(\mathbf{Aut\text{-}\Pi}_1^1)_0$	$\mathbf{Aut\text{-}KPl}^r$	$\psi_\Omega(\psi_I 0)$
	$\mathbf{Aut\text{-}ID}$	$(\mathbf{Aut\text{-}\Pi}_1^1)$	W–$\mathbf{Aut}$–$\mathbf{KPl}$	$\psi_\Omega(\psi_I 0 \cdot \varepsilon_0)$
	$\mathbf{Aut\text{-}BID}$	$(\mathbf{Aut\text{-}\Pi}_1^1)$+$(\mathsf{Bi})$	$\mathbf{Aut}$–$\mathbf{KPl}$	$\psi_\Omega(\varepsilon_{\psi_I 0+1})$
		$(\mathbf{\Delta}_2^1\text{-}\mathbf{CA}) + (\mathsf{Bi})$ $(\mathbf{\Sigma}_2^1\text{-}\mathbf{AC}) + (\mathsf{Bi})$	$\mathbf{KPi}$ $\mathbf{KP}\beta$	$\psi_\Omega(\varepsilon_{I+1})$

Fig. 1: Π_1^1–ordinals of some impredicative theories. Notice that $\Omega := \omega_1^{CK}$

†We did not define this theory. For details cf. Buchholz et al. [1981]
‡For $\nu = \omega^\rho$, $\rho \in \mathsf{Lim}$ ††For $\nu \in \mathsf{Lim}$ ‡‡For additively indecomposable ν

upper bounds are precise ones. Unfortunately there is not enough space to indicate the well–ordering proofs. Details for the well-ordering proof for ID_ν are in Buchholz and Pohlers [1978] and Buchholz et al. [1981]. More well-orderings proof are carried through in Rathjen [1988].

Notice that the remark in the beginning of Section 2.2.2 applies to all the ordinal analyses given in this article. So these analyses are all profound and can therefore be extended to Π^0_2–analyses.

References

P. H. G. Aczel, H. Simmons, and S. S. Wainer

[1992] eds., *Proof Theory*, Cambridge, July, Cambridge University Press.

J. Barwise

[1975] *Admissible Sets and Structures*, Perspectives in Mathematical Logic, Springer-Verlag, Berlin.

A. Beckmann and W. Pohlers

[1997] *Application of cut-free infinitary derivations to generalized recursion theory.* To appear in Annals of Pure and Applied Logic.

B. Blankertz

[1997] *Beweistheoretische Techniken zur Bestimmung von Π^0_2-Skolem Funktionen*, Dissertation, Westfälische Wilhelms-Universität, Münster.

B. Blankertz and A. Weiermann

[1996] How to characterize provably total functions by the Buchholz' operator method, in: *Gödel '96: Logical Foundations of Mathematics, Computer Science and Physics – Kurt Gödel's Legacy*, P. Hájek, ed., Lecture Notes in Logic #6, Springer-Verlag, Berlin, pp. 205–213.

W. Buchholz

[1991] Notation systems for infinitary derivations, *Archive for Mathematical Logic*, 30, pp. 277–296.

[1992] A simplified version of local predicativity, in: Aczel, Simmons and Wainer [1992], pp. 115–147.

W. Buchholz, E. A. Cichon, and A. Weiermann

[1994] A uniform approach to fundamental sequences and hierarchies, *Mathematical Logic Quarterly*, 40, pp. 273–286.

W. Buchholz, S. Feferman, W. Pohlers, and W. Sieg

[1981] eds., *Iterated Inductive Definitions and Subsystems of Analysis: Recent Proof-Theoretical Studies*, Lecture Notes in Mathematics #897, Springer-Verlag, Berlin.

W. Buchholz and W. Pohlers

[1978] Provable well orderings of formal theories for transfinitely iterated inductive definitions, *Journal of Symbolic Logic*, 43, pp. 118–125.

S. Feferman

[1964] Systems of predicative analysis, *Journal of Symbolic Logic*, 29, pp. 1–30.

[1970] Formal theories for transfinite iteration of generalized inductive definitions and some subsystems of analysis, in: Kino, Myhill and Vesley [1970], pp. 303–326.

H. M. Friedman

[1970] Iterated inductive definitions and Σ^1_2-AC, in: Kino, Myhill and Vesley [1970], pp. 435–442.

H. M. Friedman and M. Sheard

[1995] Elementary descent recursion and proof theory, *Annals of Pure and Applied Logic*, 71, pp. 1–45.

J.-Y. Girard

[1987] *Proof Theory and Logical Complexity*, vol. 1, Bibliopolis, Naples.

G. Jäger

[1980] *Theories for iterated jumps.* Handwritten notes.

[1986] *Theories for Admissible Sets. A Unifying Approach to Proof Theory*, Studies in Proof Theory, Lecture Notes, #2, Bibliopolis, Naples.

G. Jäger, R. Kahle, A. Setzer, and T. Strahm

[n.d.] *The proof-theoretic analysis of transfinitely iterated fixed point theories.* To appear in Journal of Symbolic Logic.

G. Jäger and W. Pohlers

[1983] Eine beweistheoretische Untersuchung von (Δ^1_2-CA) + (BI) und verwandter Systeme, *Bayerische Akademie der Wissenschaften, Sitzungsberichte 1982*, pp. 1–28.

G. Jäger and T. Strahm

[n.d.] *Fixed point theories and dependent choice.* Submitted for publication.

R. Kahle

[1997] *Applicative Theories and Frege Structures*, Dissertation, Institut für Informatik und angewandte Mathematik, Universität Bern, Bern.

A. Kino, J. Myhill, and R. E. Vesley

[1970] eds., *Intuitionism and Proof Theory*, Studies in Logic and the Foundations of Mathematics, Amsterdam, North-Holland.

Y. N. Moschovakis

[1974] *Elementary Induction on Abstract Structures*, Studies in Logic and the Foundations of Mathematics #77, North-Holland, Amsterdam.

E. Palmgren

[n.d.] *On universes in type theory.* To appear.

R. Platek

[1966] *Foundations of Recursion Theory II*, dissertation, Stanford University.

W. Pohlers

[1978] Ordinals connected with formal theories for transfinitely iterated inductive definitions, *Journal of Symbolic Logic*, 43, pp. 161–182.

[1981] Proof-theoretical analysis of ID_ν by the method of local predicativity, in: Buchholz et al. [1981], pp. 261–357.

[1982a] Admissibility in proof theory; a survey, in: *Logic, Methodology and Philosophy of Science VI*, L. J. Cohen, J. Łos, H. Pfeiffer, and K.-P. Podewski, eds., Studies in Logic and the Foundations of Mathematics #104, North-Holland, Amsterdam, Aug., pp. 123–139.

[1982b] Cut elimination for impredicative infinitary systems II. Ordinal analysis for iterated inductive definitions, *Archiv für Mathematische Logik und Grundlagenforschung*, 22, pp. 69–87.

[1989] *Proof Theory. An Introduction*, Lecture Notes in Mathematics #1407, Springer-Verlag, Berlin.

[1991] Proof theory and ordinal analysis, *Archive for Mathematical Logic*, 30, pp. 311–376.

[1992] A short course in ordinal analysis, in: Aczel, Simmons and Wainer [1992], pp. 27–78.

M. RATHJEN

[1988] *Untersuchungen zu Teilsystemen der Zahlentheorie zweiter Stufe und der Mengenlehre mit einer zwischen Δ^1_2-CA und Δ^1_2-CA+BI liegenden Beweisstärke*, Dissertation, Westfälische Wilhelms-Universität, Münster.

[1991] The role of parameters in bar rule and bar induction, *Journal of Symbolic Logic*, 56, pp. 715–730.

[1995] Recent advances in ordinal analysis: Π^1_2–**CA** and related systems, *Bulletin of Symbolic Logic*, 1, pp. 468–485.

A. SCHLÜTER

[n.d.] *What is provable using first order reflection.* To appear in Annal of Pure and Applied Logic.

A. SCHLÜTER

[1990] *Autonom erreichbare Mengen*, Diplomarbeit, Westfälische Wilhelms-Universität, Münster.

A. SCHLÜTER

[1993] *Zur Mengenexistenz in formalen Theorien der Mengenlehre*, Dissertation, Westfälische Wilhelms-Universität, Münster.

K. SCHÜTTE

[1960] *Beweistheorie*, Springer-Verlag, Berlin. Translated into English as Schütte [1977].

[1965a] Eine Grenze für die Beweisbarkeit der transfiniten Induktion in der verzweigten Typenlogik, *Archiv für Mathematische Logik und Grundlagenforschung*, 7, pp. 45–60.

[1965b] Predicative well-orderings, in: *Formal Systems and Recursive Functions*, J. N. Crossley and M. A. E. Dummett, eds., Studies in Logic and the Foundations of Mathematics, North-Holland, Amsterdam, July, pp. 280–303.

[1977] *Proof Theory*, Springer-Verlag, Berlin.

T. STRAHM

[n.d.] *First steps into metapredicativity in explicit mathematics.* In preparation.

G. TAKEUTI

[1975] *Proof Theory*, North-Holland, Amsterdam.

A. WEIERMANN

[1996] How to characterize provably total functions by local predicativity, *Journal of Symbolic Logic*, 61, pp. 52–69.

CHAPTER V

Gödel's Functional ("Dialectica") Interpretation

Jeremy Avigad
Department of Philosophy, Carnegie Mellon University
Pittsburgh, PA 15213

Solomon Feferman
Departments of Mathematics and Philosophy, Stanford University
Stanford, CA 94305

Contents

HANDBOOK OF PROOF THEORY
Edited by S. R. Buss

1. Introduction

1.1. Functional interpretations

In 1958, Kurt Gödel published in the journal *Dialectica* an interpretation of intuitionistic arithmetic in a quantifier-free theory of functionals of finite type, an interpretation which has since come to be known as Gödel's functional or *Dialectica* interpretation. When combined with Gödel's *double-negation* interpretation, which reduces classical arithmetic to intuitionistic arithmetic, the Dialectica interpretation (also referred to below as the D-interpretation) yields a reduction of the classical theory as well. This approach has since been extended and adapted to other theories, but the pattern usually follows Gödel's original example:

- first, one reduces a classical theory C to a variant I based on intuitionistic logic;
- then one reduces the theory I to a quantifier-free functional theory F.

Functional interpretations of this form can be interesting for a number of reasons. To begin with, the work can be seen as a contribution to a modified form of Hilbert's program, since, from a foundational point of view, the consistency of C is thereby reduced to the consistency of F. Subsequent analyses of F often lead to further gains, yielding, for example, reductions of *(prima facie)*

- infinitary systems to finitary ones,
- non-constructive systems to constructive ones, and
- impredicative systems to predicative ones.

Secondly, functional interpretation provides a way of extracting (or "unwinding") constructive information from proofs in I or C. For example, as a direct consequence of the interpretation one usually obtains the result that any recursive function whose totality can be proven either in I or in C is represented by a term of F. Via an additional interpretation of F in I, this characterization is in fact usually shown to be exact. It often turns out that the terms of F represent a natural class of functions, such as the primitive recursive or polynomial-time computable functions. In other cases, the theory F embodies independently interesting computational constructs, such as bar-recursion or polymorphism, which are discussed in this chapter.

Finally, functional interpretations often provide a useful stepping-stone to other goals. For example, the analyses of Gödel's functional calculus T due to Tait and Howard provide an alternative means to the ordinal analysis of classical arithmetic, and non-constructive interpretations due to the second author yield consequences for various subsystems of second-order arithmetic. Many of these results can also be obtained using Herbrand-Gentzen methods of syntactic transformation, and in certain domains (for example, in the ordinal analysis of strong subsystems of analysis and set-theory) these latter methods are the only ones currently known.[1] Nonetheless,

[1] Whether there is a fundamental obstacle against the use of D-interpretations for such purposes is a matter of methodological interest.

functional interpretations have proven to be a relatively powerful and versatile tool, with distinct advantages that will be illustrated below.

1.2. Historical background

Although the Dialectica interpretation was not published until 1958, Gödel began to develop these ideas in the latter part of the 1930's as a possible modification of Hilbert's program (cf. Sieg and Parsons [1994]), and the D-interpretation itself was arrived at by 1941 and presented in a lecture to the Mathematics and Philosophy Clubs at Yale University. Full notes for that lecture are found in Gödel's *Nachlass* and have been made available in Volume III of his *Collected Works* as Gödel [1941].

The interpretation was first brought to the attention of the logic community at large in a lecture by Georg Kreisel at the Summer Institute in Symbolic Logic held at Cornell University in 1957 (cf. the notes Kreisel [1957]). The publication Gödel [1958], in German, was for an issue of *Dialectica* in honor of Paul Bernays' 70th birthday. Gödel worked on a translation and expansion of that article for another issue in honor of Bernays a decade later, but though it reached the stage of proof sheets it never appeared in print until it was retrieved from the *Nachlass* for publication in Volume II of his *Collected Works* as Gödel [1972].

Subsequent work on functional interpretations was carried out in the 1960s through the early 1980s, following the initial developments by Kreisel in the late 1950s. After a lapse in the latter part of the 1980s, there has been a resurgence of interest in these methods in the 1990s, yielding a number of new applications. Some of the different general directions of work may be indicated roughly as follows (more or less along the lines of Troelstra [1990,pp. 236–239], which should be consulted for publication information concerning items not found in the references to this chapter).

1. The functionals in Gödel's interpretation are defined by schemata for explicit definition and a natural extension of primitive recursion to finite types, and are therefore called *primitive recursive functionals of finite type.*[2] There have been a number of investigations of this class of functionals, which have set-theoretic, recursion-theoretic and term models. Prominent in the study of the latter are various methods of normalization and related assignments of ordinals. Among the contributors here that one should mention are S. Hinata, J. Diller, W. Tait and W. Howard.

2. Next, Gödel's interpretation has been adapted and extended both to stronger and weaker theories. Here, briefly, in semi-chronological order, are some of the kinds of systems to which the D-interpretation has been extended, either directly in the case of intuitionistic systems, or indirectly by combination with the negative translation in the case of classical systems. (We also indicate some of the main contributors to each):

[2]This will be distinguished below from the class of functionals introduced by Kleene [1959b] using a weaker predicative extension of primitive recursion to finite types.

   (a) Intuitionistic arithmetic with principles of transfinite induction (G. Kreisel).
   (b) Impredicative ("full") classical analysis formulated with function variables (G. Kreisel, C. Spector, W. Howard, H. Luckhardt).
   (c) Subsystems of classical arithmetic (C. Parsons).
   (d) Impredicative systems of classical analysis formulated with set variables (J.-Y. Girard).
   (e) Intuitionistic and classical theories of ordinals (W. Howard, S. Feferman).
   (f) Predicative systems of classical analysis (W. Maass, S. Feferman).
   (g) Classical analysis with a game quantifier (W. Friedrich).
   (h) Systems of feasible arithmetic (S.A. Cook and A. Urquhart).
   (i) Iterated arithmetical fixed point theories (J. Avigad).
3. Furthermore, the interpretations have been applied towards a number of interesting proof theoretic ends. These applications include:
   (a) The no-counterexample interpretation for Peano Arithmetic (G. Kreisel).
   (b) Closure and conservation results for intuitionistic systems (A.S. Troelstra).
   (c) Conservation results for classical systems and characterization of the provably recursive functionals (C. Parsons, S. Feferman, U. Kohlenbach).
4. Finally, useful variants of Gödel's original interpretation have also been developed, among them those due to J. Shoenfield, J. Diller and W. Nahm, M. Beeson, U. Kohlenbach.

These lists, as well as the treatment below, are not comprehensive. For more information we refer the reader to the surveys Troelstra [1990] and Feferman [1993], the encyclopedic treatment of Troelstra [1973], and the related articles Feferman [1977] and Troelstra [1977].[3]

1.3. An overview of this chapter

In this chapter we try to give a broad and self-contained survey of the D-interpretation and its applications. First we provide the details of the interpretation of arithmetic, explicitly presenting the relevant axioms and the functional theory T. In section 3 we present some of the useful information that can be gleaned from the interpretation, and take a broader look at the general form of the interpretation in order to understand better how it might be adapted to other contexts.

The presentation of a functional theory raises the issue of what its models look like. In the case of Gödel's T, that issue is addressed in section 4. In section 5 we

[3]Unfortunately, notation in literature is not uniform, and here we have struck some compromises. For example, though the use of PR^{ω} to denote Gödel's theory of the primitive recursive functionals of finite type in Feferman [1977,1990] is more descriptive, here we follow Gödel's original use of the name T.

show how to weaken T in order to obtain useful characterizations of the provably total recursive functions of certain fragments of arithmetic, namely $I\Sigma_1$ and S^1_2.

In section 6 we go in the opposite direction and consider a strengthening of T that suffices to interpret full second-order arithmetic ("analysis"). In the following three sections we then consider ways in which the D-interpretation can also be used to obtain information regarding a number of interesting subsystems of analysis.

Finally, in the last section we show how functional theories based on polymorphic types arise in a natural way from a functional interpretation of full analysis, formulated using predicate variables instead of function variables.

The authors are very much indebted to Ulrich Kohlenbach for numerous comments and suggestions, as well as corrections to a draft of this chapter.

2. The Dialectica interpretation of arithmetic

2.1. Theories of arithmetic and the double-negation interpretation

The first-order theory Peano arithmetic, or *PA*, has already been discussed in Chapter II. Peano arithmetic has its intuitionistic analogue in a theory known as Heyting arithmetic, or *HA*, which differs from the former only in that it uses intuitionistic axioms and rules as the underlying predicate logic. For concreteness, we take the following list of axioms and rules, which is that used by Gödel [1958] (cf. also Troelstra [1973,1977]):

1. From φ, $\varphi \to \psi$ conclude ψ
2. From $\varphi \to \psi$, $\psi \to \theta$ conclude $\varphi \to \theta$
3. $\varphi \vee \varphi \to \varphi$, $\varphi \to \varphi \wedge \varphi$
4. $\varphi \to \varphi \vee \psi$, $\varphi \wedge \psi \to \varphi$
5. $\varphi \vee \psi \to \psi \vee \varphi$, $\varphi \wedge \psi \to \psi \wedge \varphi$
6. From $\varphi \to \psi$ conclude $\theta \vee \varphi \to \theta \vee \psi$
7. From $\varphi \to (\psi \to \theta)$ conclude $\varphi \wedge \psi \to \theta$, and conversely
8. $\bot \to \theta$
9. From $\varphi \to \psi$ conclude $\varphi \to \forall x\, \psi$, assuming x is not free in φ
10. $\forall x\, \varphi \to \varphi[t/x]$, assuming t is free for x in φ
11. $\varphi[t/x] \to \exists x\, \varphi$, assuming t is free for x in φ
12. From $\varphi \to \psi$ conclude $\exists x\, \varphi \to \psi$, assuming x is not free in ψ

Here $\varphi[t/x]$ denotes the result of replacing all free occurrences of the variable x by t in the formula φ. It is common in intuitionistic systems to define negation by

$$\neg A = A \to \bot$$

where $\bot$ is an identically false statement, or "contradiction"; $\bot$ may be taken to be a closed atomic formula, or identified with $0 = 1$. We take the equality axioms to be given by

1. $x = x$
2. $x = y \to (\varphi[x/z] \to \varphi[y/z])$, where φ is atomic.

Finally, classical logic is obtained by adding to this list *tertium non datur*, the law of excluded middle:

$$\varphi \vee \neg\varphi.$$

Classical predicate logic can be reduced in a simple way to intuitionistic predicate logic via the so-called *double-negation* (or *negative*) translation due (independently) to Gödel and Gentzen. This is defined as follows:

1. $\varphi^N = \neg\neg\varphi$, for φ atomic
2. $(\varphi \wedge \psi)^N = \varphi^N \wedge \psi^N$
3. $(\varphi \vee \psi)^N = \neg(\neg\varphi^N \wedge \neg\psi^N)$
4. $(\varphi \to \psi)^N = \varphi^N \to \psi^N$
5. $(\forall x\, \varphi(x))^N = \forall x\, \varphi(x)^N$
6. $(\exists x\, \varphi(x))^N = \neg\forall x\, \neg\varphi(x)^N$

The "double negation" appellation is due not only to clause 1, but also the fact that $(\varphi \vee \psi)^N \leftrightarrow \neg\neg(\varphi^N \vee \psi^N)$ and $(\exists x\, \varphi)^N \leftrightarrow \neg\neg\exists x\, \varphi^N$ are provable intuitionistically.

Clearly, from a classical point of view every formula is equivalent to its N-interpretation. Moreover, one has the following

2.1.1. Theorem. *Suppose a set of axioms S proves a formula φ using classical logic. Then S^N proves φ^N using intuitionistic logic.*

For a proof of this and more general results, see Troelstra [1973] or Troelstra [1977, section 3.8]. For the case at hand, the preceding theorem provides the following useful

2.1.2. Corollary. *Suppose PA proves a formula φ. Then HA proves φ^N.*

2.1.3. Proof. We only need to verify that *HA* proves the N-interpretation of each axiom and rule of *PA*. Since *HA* proves $x = y \vee \neg(x = y)$ (using a double induction on x and y), the N-interpretations of the quantifier-free axioms of *PA* follow from their *HA* counterparts. Finally, the N-interpretation of an instance of the induction scheme is again an instance of the induction scheme. □

2.2. The primitive recursive functionals of finite type

The Dialectica interpretation reduces *HA* to a theory T which axiomatizes a class of functionals that Gödel called the "primitive recursive functionals of finite type."[4] While T is quantifier-free, its language is many-sorted, in that each term is assigned a *type symbol*, or *type* for short. The set of types is generated inductively by the following rules:

[4]For a variant of the D-interpretation that applies directly to *PA*, see Shoenfield [1967].

1. 0 is a type.
2. If σ and τ are types then so is $\sigma \to \tau$[5]

The intended use is that objects of type 0 are considered to be natural numbers and objects of type $\sigma \to \tau$ are considered to be *functions* from objects of type σ to objects of type τ. The latter may be interpreted as constructive functions in some sense or other, or set-theoretically as all functions of the specified type. By convention we interpret

$$\tau_1 \to \tau_2 \to \ldots \to \tau_n$$

by associating parentheses to the right, i.e. as

$$\tau_1 \to (\tau_2 \to (\ldots \to \tau_n) \ldots).$$

Objects of a type $(\sigma \to \tau) \to \rho$ have function arguments and are usually called *functionals.* An interpretation $\langle M_\sigma : \sigma \text{ a type}\rangle$ of such a typed language is called a *functional type structure* or simply a *type structure.*

One might also wish to include on the preceding list the following additional closure condition:

3. If σ and τ are types then so is $\sigma \times \tau$.

Here $\sigma \times \tau$ denotes the set of ordered pairs of objects $\langle s, t\rangle$ with s an object of type σ and t an object of type τ. This closure condition is eliminable in favor of 2 by "currying," that is, interpreting the type $(\rho \times \sigma) \to \tau$ as the type $\rho \to (\sigma \to \tau)$ and adopting the term-reading conventions described below. (Always in such choices there are trade-offs: fewer closure conditions on type symbols simplifies description of models, normalization of terms, etc., but more closure conditions provide added flexibility and naturalness of formulation.)

To each type σ we can assign a natural number $\text{lev}(\sigma)$ as its *type level*, by:

1. $\text{lev}(0) = 0$
2. $\text{lev}(\sigma \to \tau) = \max(\text{lev}(\sigma) + 1, \text{lev}(\tau))$.

The language of T is said to be of *finite type* since every type is assigned a finite level by this convention. The *pure types* (n) are defined by

1. $(0) = 0$
2. $(n + 1) = (n) \to 0$.

Where the context determines that we are dealing with type symbols, we drop the parentheses around symbols for pure types. Then $\text{lev}(n) = n$ for each $n < \omega$.

We now define the set of terms of T, as well as the relation $t : \tau$ (read "term t has type τ"), inductively via the following rules:

1. There are infinitely many variables $x^\tau, y^\tau, z^\tau, \ldots$ of each type τ.
2. If s is a term of type σ and t a term is of type $\sigma \to \tau$ then $t(s)$ is a term of type τ.
3. 0 is a constant of type 0.

[5]Gödel [1958] used (τ, σ) for our $\sigma \to \tau$. There are many alternative notations in use in the literature such as $(\sigma)\tau$ or $\tau(\sigma)$ or τ^σ, etc.; *caveat emptor.*

4. Sc is a constant of type $0 \to 0$.
5. For each pair of types σ, τ, $\mathrm{K}_{\sigma,\tau}$ is a constant of type $\sigma \to \tau \to \sigma$.
6. For each triple of types ρ, σ, τ, $\mathrm{S}_{\rho,\sigma,\tau}$ is a constant of type $(\rho \to \sigma \to \tau) \to (\rho \to \sigma) \to (\rho \to \tau)$.
7. For each type σ, R_σ is a constant of type $\sigma \to (0 \to \sigma \to \sigma) \to 0 \to \sigma$.

The intended interpretation is that 0 denotes the constant zero, $\mathrm{Sc}(t)$ denotes the successor of t (which we will also write t'), and $t(s)$ denotes the result of applying the function(al) t to the argument s. The intuitive meanings of $\mathrm{K}_{\sigma,\tau}$, $\mathrm{S}_{\rho,\sigma,\tau}$, and R_σ will become clear when we present their defining equations below.

Where possible without ambiguity, we will suppress the type superscripts on the constants K, S, R and on variables, and write, for example, $x, y, z, \ldots$. We will sometimes use capital letters $X, Y, Z \ldots$ to denote variables of functional type. To improve readability, if t is a term of type $\rho \to (\sigma \to \tau)$, r is a term of type ρ, and s is a term of type σ, we will write $t(r, s)$ instead of $t(r)(s)$ (which we in turn interpret by associating to the left, yielding $(t(r))(s)$). Similarly $t(r_1, r_2, \ldots, r_n)$ denotes $t(r_1)(r_2)\ldots(r_n)$.[6]

At the risk of confusion, *sequences* of variables (possibly empty) will be indicated by the same font, e.g. $x = (x_1, \ldots, x_n)$ or $X = (X_1, \ldots, X_n)$. If x is such a sequence then the term $t(x)$ should be interpreted as $t(x_1, \ldots, x_n)$ and the prefix $\exists x$ should be interpreted as the quantifier string $\exists x_1 \, \exists x_2 \, \ldots \exists x_n$. In general, we will rely on context to determine whether we are dealing with a single variable or sequence of such. If x and y denote sequences of variables, then x, y denotes their concatenation, as in $t(x, y)$ or $\exists x, y$.

In the intended interpretation, of course, 0 and Sc satisfy

$$x' \neq 0$$

and

$$x' = y' \to x = y.$$

The constants K and S in (5) and (6) above are the usual *typed combinators* whose interpretation is given by

$$\mathrm{K}(s, t) = s \quad \text{for } s : \sigma \text{ and } t : \tau,$$

and

$$\mathrm{S}(r, s, t) = r(t)(s(t)) \quad \text{for } r : (\rho \to \sigma \to \tau),\ s : \rho \to \sigma, \text{ and } t : \rho.$$

The meaning of equality at higher types is a delicate matter which will be taken up in section 2.5 below. For the time being we read these equations naively.

If terms are generated from the variables and constants solely by the operation of application then one obtains *combinatory completeness* as usual, i.e. we can associate

[6]In the literature on functional interpretations, parentheses are often dropped altogether for the sake of brevity, so that one may encounter e.g. $Xxyz$ instead of $X(x, y, z)$. When this is the case, convention and the types of the associated terms and variables dictate the appropriate reading.

with each term t and variable x another term $\lambda x.t$ whose free variables are all those of t other than x, and which satisfies the equation

$$(\lambda x.t)(s) = t[s/x].$$

It follows that if t has free variables among $x_1, \ldots, x_n$, then $(\lambda x_1 \ldots \lambda x_n.t)$ is a closed term with

$$(\lambda x_1 \ldots \lambda x_n.t)(s_1, \ldots, s_n) = t[s_1/x_1, \ldots, s_n/x_n]$$

for all s_i of the same type as x_i for $i = 1, \ldots, n$. Alternatively, we could have taken the λ operation as a basic term-forming operation, where the terms thereby formed have the defining equation above.

Finally, we have the equations for the recursors R, which define a simple form of primitive recursion:

$$\begin{aligned} &\mathrm{R}(f, g, 0) = f \\ &\mathrm{R}(f, g, n') = g(n, \mathrm{R}(f, g, n)). \end{aligned}$$

That is, $R(f, g)$ is a function h of type $0 \to \sigma$, with defining equations

$$\begin{aligned} &h(0) = f \\ &h(n') = g(n, h(n)). \end{aligned}$$

There is no need to mention parameters to h explicitly, since these can be absorbed in the types of f and g. We note that this kind of higher-type iteration is clearly anticipated in Hilbert [1926], and even, to some extent, in Weyl [1918].

The atomic formulas of T consist of assertions of equality between terms of the same type,[7] and more complex formulas are obtained by combining these with the usual propositional connectives. The axioms of T consist of the defining equations of 0, Sc, K, S, and R described above, a rule allowing for the substitution of arbitrary terms for variables of the same type, equality axioms, the axioms of *classical* propositional logic, and the scheme of induction

$$\text{from } \varphi(0) \text{ and } \varphi(x) \to \varphi(x') \text{ conclude } \varphi(t)$$

for arbitrary formulas φ and terms t in the language. Note that if one has included products $\sigma \times \tau$ among the finite types, one also needs to add basic terms denoting pairing and projection operations of the appropriate types, together with their defining equations as well.

In later sections we will consider variants of T which allow for more elaborate types (e.g. "transfinite types") and functionals. For the time being, T is strong enough to interpret HA, as we now show.

[7]In one treatment of equality in T, these are only for terms of type 0; cf. section 2.5 below.

2.3. The Dialectica interpretation

To each formula φ in the language of arithmetic we associate its *Dialectica* (or D-) *interpretation* φ^D, which is a formula of the form

$$\varphi^D = \exists x\, \forall y\, \varphi_D$$

where φ_D is a (quantifier-free) formula in the language of T. Here the free variables of φ_D consist of those free in φ, together with the sequences of variables (possibly empty) x and y. However, the free variables of φ are generally suppressed and we also write $\varphi_D(x,y)$ for φ_D. If one or more free variables z of φ are exhibited, as $\varphi(z)$, then we write $\varphi_D(x,y,z)$ for φ_D.

The associations $(\ )^D$ and $(\ )_D$ are defined inductively as follows, where

$$\varphi^D = \exists x\, \forall y\, \varphi_D \quad \text{and} \quad \psi^D = \exists u\, \forall v\, \psi_D.$$

1. For φ an atomic formula, x and y are both empty and $\varphi^D = \varphi_D = \varphi$.
2. $(\varphi \wedge \psi)^D = \exists x, u\, \forall y, v\, (\varphi_D \wedge \psi_D)$.
3. $(\varphi \vee \psi)^D = \exists z, x, u\, \forall y, v\, ((z = 0 \wedge \varphi_D) \vee (z = 1 \wedge \psi_D))$.
4. $(\forall z\, \varphi(z))^D = \exists X\, \forall z, y\, \varphi_D(X(z), y, z)$.
5. $(\exists z\, \varphi(z))^D = \exists z, x\, \forall y\, \varphi_D(x, y, z)$.
6. $(\varphi \to \psi)^D = \exists U, Y\, \forall x, v\, (\varphi_D(x, Y(x,v)) \to \psi_D(U(x), v))$.

The case of $\to$ has been put last here because this requires special explanation below. Since we have defined $\neg\varphi$ to be $\varphi \to \bot$, from 6 we obtain

7. $(\neg\varphi)^D = \exists Y\, \forall x\, \neg\varphi_D(x, Y(x))$.

In clause 1 we assume the obvious identification of the symbols $+$ and $\times$ of *HA* with the terms that represent addition and multiplication in T. The definition of φ^D for atomic φ, of $(\varphi \wedge \psi)^D$ and of $(\exists z\, \varphi(z))^D$ needs no comment. The definition of $(\varphi \vee \psi)^D$ is also clear on a constructive reading: the new parameter z tells which disjunct is being established, according as to whether $z = 0$ or $z = 1$. The definition of $(\forall z\, \varphi(z))^D$ is obtained by prefixing a universal quantifier to φ^D to obtain

$$\forall z\, \exists x\, \forall y\, \varphi_D(x, y, z)$$

and then "skolemizing" the existentially quantified variable.

The motivation behind the definition of $(\varphi \to \psi)^D$ given by Gödel is as follows: from a witness x to the hypothesis φ^D one should be able to obtain a witness u to the conclusion ψ^D, such that from a counterexample v to the conclusion one should be able to find a counterexample y to the hypothesis. In short, one uses equivalences (i–iv) below to bring quantifiers to the front, and then skolemizes the existential variables:

$$\begin{array}{lll}
(\exists x\, \forall y\, \varphi_D(x,y) \to \exists u\, \forall v\, \psi_D(u,v)) & \leftrightarrow & \text{(i)} \\
\forall x\, (\forall y\, \varphi_D(x,y) \to \exists u\, \forall v\, \psi_D(u,v)) & \leftrightarrow & \text{(ii)} \\
\forall x\, \exists u\, (\forall y\, \varphi_D(x,y) \to \forall v\, \psi_D(u,v)) & \leftrightarrow & \text{(iii)} \\
\forall x\, \exists u\, \forall v\, (\forall y\, \varphi_D(x,y) \to \psi_D(u,v)) & \leftrightarrow & \text{(iv)} \\
\forall x\, \exists u\, \forall v\, \exists y\, (\varphi_D(x,y) \to \psi_D(u,v)) & \leftrightarrow & \text{(v)} \\
\forall x\, \exists u, Y_1\, \forall v\, (\varphi_D(x, Y_1(v)) \to \psi_D(u,v)) & \leftrightarrow & \text{(vi)} \\
\exists U, Y\, \forall x, v\, (\varphi_D(x, Y(x,v)) \to \psi_D(U(x), v)). & &
\end{array}$$

The definitions of $(\varphi \wedge \psi)^D$, $(\varphi \vee \psi)^D$, and $(\exists z\ \varphi(z))^D$ are all justified from a constructive as well as classical point of view. The definition of $(\forall z\ \varphi(z))^D$ is justified by an application of the axiom of choice,

$$(AC) \qquad \forall x\ \exists y\ \varphi(x,y) \to \exists Y\ \forall x\ \varphi(x, Y(x))$$

for arbitrary formulas φ. (AC) is usually accepted classically in set theory; it is also accepted by many constructivists, since the reading of the hypothesis is that one has a constructive proof of $\forall x\ \exists y\ \varphi(x,y)$, and such a proof must provide a means Y of constructively associating with each x a solution $y = Y(x)$ of $\varphi(x,y)$.

The analysis of $(\varphi \to \psi)^D$ is more delicate. While equivalences (i–vi) above are all classically justified, only (i), (iii), (v) and (vi) are acceptable from a constructive point of view, the first two by logic and the latter two by (AC). Equivalence (ii), namely,

$$(IP') \qquad (\forall y\ \varphi \to \exists u\ \forall v\ \psi) \to \exists u\ (\forall y\ \varphi \to \forall v\ \psi)$$

is a special case of a principle called *independence of premise.* Though this is valid in classical logic, it is *not* generally accepted constructively, since the constructive reading of the hypothesis $(\theta \to \exists u\ \eta)$ is that we have a constructive means of turning any proof of the premise θ into a proof of η with a witness for the existential quantifier applied to η. In general, the choice of such a u will then depend on the proof of θ, while (IP') tells us that u can be chosen independently of any proof of that premise.

Equivalence (iv) can be justified by a generalization of Markov's principle, namely

$$(MP') \qquad \neg\forall y\ \theta \to \exists y\ \neg\theta$$

in which θ is assumed to be quantifier-free. (Assuming that the law of excluded middle holds for ψ_D, argue thus: if ψ_D is true then (iv) is justified, and if ψ_D is false apply (MP').) The problem with (MP') is that there is no evident way to choose constructively a witness y to $\neg\theta$ from a proof that $\forall y\ \theta$ leads to a contradiction. However if y ranges over the natural numbers one can search for such a y given that one accepts its existence. In this case (MP') boils down to the usual form of Markov's principle

$$(MP) \qquad \forall x\ (\neg\neg\exists y\ \varphi(x,y) \to \exists y\ \varphi(x,y))$$

which is accepted in the Russian school of constructivity for φ quantifier-free.

While the reasoning leading to the form of the D-interpretation is not fully constructive it can still be used as a tool in constructive metamathematics and to derive constructive information. In section 3.1 we'll see that the D-interpretation verifies the three principles (AC), (IP'), and (MP') just discussed, and hence allows one to use the D-interpretation to extract constructive information from non-constructive proofs.

2.4. Verifying the axioms of arithmetic

Gödel's main result is as follows.

2.4.1. Theorem. *Suppose φ is a formula in the language of arithmetic, and HA proves φ. Then there is a sequence of terms t such that T proves $\varphi_D(t, y)$.*

We express this by saying that *HA* is *D-interpreted* in T. Combining Theorem 2.4.1 with Corollary 2.1.2 we obtain

2.4.2. Corollary. *Suppose φ is a formula in the language of arithmetic, such that PA proves φ. Then there is a sequence of terms t such that T proves $(\varphi^N)_D(t, y)$.*

In short, *PA* is *ND-interpreted* in T.

One proves Theorem 2.4.1 by induction on the length of the proof in *HA*. One only has to verify that the claim holds true when φ is an axiom of *HA*, and that it is maintained under rules of inference.

We begin by considering the axioms of rules of intuitionistic logic, listed in section 2.1. For most of these the verification is routine, and we only address a few key examples. For example, consider the rule "from $\varphi \to \psi$ and φ conclude ψ." Given a term a such that T proves $\varphi_D(a, y)$ and terms b and c such that T proves $\varphi_D(x, b(x, v)) \to \psi_D(c(x), v)$, we want a term d such that T proves $\psi_D(d, v)$. By substituting $b(a, v)$ for y in the first hypothesis and a for x in the second, we see that taking $d = c(a)$ works; so, in a sense, modus ponens corresponds to functional application. The reader can verify that, similarly, the axiom "from $\varphi \to \psi$ and $\psi \to \theta$ conclude $\varphi \to \theta$" corresponds to the composition of functions.

Handling the axiom $\varphi \to \varphi \wedge \varphi$ requires a bit more work. If the interpretation of the hypothesis is given by $\exists x \, \forall y \, \varphi_D(x, y)$, the interpretation of the conclusion is

$$\exists x_1, x_2 \, \forall y_1, y_2 \, (\varphi_D(x_1, y_1) \wedge \varphi_D(x_2, y_2)).$$

According to the clause for implication, we need to provide terms $c_1(x)$ and $c_2(x)$ taking a witness x for the hypothesis to witnesses x_1 and x_2 for the conclusion; for this purpose, we can simply take $c_1(x)$ and $c_2(x)$ to be x. But we also need a functional $d(x, y_1, y_2)$ that will take y_1 and y_2 witnessing the failure of $\varphi_D(x, y_1) \wedge \varphi_D(x, y_2)$ to a value d representing the failure of $\varphi_D(x, d)$. This functional must effectively determine which of $\varphi_D(x, y_1)$ and $\varphi_D(x, y_2)$ is false. We need the following

2.4.3. Lemma. *If φ is a formula of arithmetic, there is a term t_φ such that T proves*

$$t_\varphi(x, y) = 0 \leftrightarrow \varphi_D(x, y).$$

The lemma is proved by induction on the size of φ_D; the key instance occurs when φ_D is simply an atomic formula $s_1 = s_2$, for which case the required t can be obtained from an application of primitive recursion.

Another instance of primitive recursion yields a functional Cond such that T proves

$$\mathrm{Cond}(w, u, v) = \begin{cases} u & \text{if } w = 0 \\ v & \text{otherwise.} \end{cases}$$

Taking $d = \text{Cond}(t_\varphi(x, y_1), y_1, y_2)$ above then suffices to complete the proof for $\varphi \to \varphi \wedge \varphi$.

Aside from the two instances just described, the recursors R are not otherwise used to verify the logical axioms. Their primary purpose is to interpret the induction rule, "from $\varphi(0)$ and $\varphi(u) \to \varphi(u')$ conclude $\varphi(u)$." Inductively we are given terms a, b, and c so that T proves $\varphi_D(0, a, y)$ and

$$\varphi_D(u, x, b(u, x, y_1)) \to \varphi_D(u', c(u, x), y_1).$$

We want a term d such that $\varphi_D(u, d(u), y)$. Using the recursors we can define d using primitive recursion, so that

$$\begin{aligned} d(0) &= a \\ d(u') &= c(u, d(u)). \end{aligned}$$

This yields

$$\varphi_D(0, d(0), y)$$

and

$$\varphi_D(u, d(u), b(u, d(u), y)) \to \varphi_D(u', d(u'), y).$$

The following lemma will then allow us to conclude $\varphi_D(u, d(u), y)$, as desired.

2.4.4. Lemma. *From $\psi(0, y)$ and $\psi(u, b(u, y)) \to \psi(u', y)$ one can prove, in T, $\psi(u, y)$.*

The idea is to work backwards: if "$\dot{-}$" denotes truncated (or "cut off") subtraction, note that $\psi(u, y)$ follows from

$$\psi(u \dot{-} 1, b(u \dot{-} 1, y)),$$

which in turn follows from

$$\psi(u \dot{-} 2, b(u \dot{-} 2, b(u \dot{-} 1, y))),$$

and so on, until the first argument is equal to 0. More formally, one defines in T a function $e(y, z)$ by $e(y, 0) = y$ and $e(y, z') = b(u \dot{-} z', e(y, z))$, and then uses induction to prove that $\psi(w, e(y, u \dot{-} w))$ holds for every w less than or equal to u. For details, see Spector [1962] or Troelstra [1973].

This leaves only the quantifier-free axioms regarding 0, Sc, +, and $\times$ in *HA*, which follow immediately from their counterparts in T.

Once more we emphasize that the recursors of T are only essentially needed for the nonlogical axioms. Roughly speaking, it is the combinatory completeness of T that allows us to verify the axioms of intuitionistic predicate logic, whereas the recursors are the functional analogue of induction. This observation allows one to generalize the D-interpretation to other theories, as discussed in section 3.3 below.

2.5. Equality at higher types

We return to the question as to how equality at higher types is to be treated in T. There are two basic choices, the *intensional* formulation taken in Gödel [1958] and the *(weakly) extensional* formulation of Spector [1962]. In Gödel's formulation, we have a decidable equality relation $=_\sigma$ at each type, i.e. all formulas $s =_\sigma t$ with s, t of type σ are taken to be atomic, and the law of excluded middle is accepted for these. Suppressing type subscripts where there is no ambiguity, the equality axioms for T in this version are, as usual,

1. $s = s$
2. $s = t \wedge \varphi[s/x] \rightarrow \varphi[t/x]$.

The axioms for K, S and R at each type may be read as they stand.

In Spector's formulation, the atomic formulas are equations between terms of type 0 only, and the law of excluded middle is accepted only for these. For $\sigma = (\sigma_1 \rightarrow \ldots \rightarrow \sigma_n \rightarrow 0)$, an equation $s =_\sigma t$ between terms of type σ is regarded as an abbreviation for

$$s(x_1, \ldots, x_n) = t(x_1, \ldots, x_n)$$

where the x_i are fresh variables of type σ_i for $i = 1, \ldots, n$. In particular, the axioms for K, S and R are to be read as such abbreviations. Now in this version, the axioms 1, 2 are taken only for terms s, t of type 0; denote these by 1_0 and 2_0 respectively. For s, t of type $\sigma \neq 0$ we must further adjoin a rule

2′. From $s(x_1, \ldots, x_n) = t(x_1, \ldots, x_n)$ and $\varphi[s/x]$ infer $\varphi[t/x]$.

An alternative suggested by Gödel in a footnote to his revision [1972] of [1958] and made explicit by Troelstra [1990] is to follow Spector in taking only equations at type 0 as basic but to assume as axioms, besides 1_0 and 2_0, just the following consequences of 2′ for K, S, and R:

$$s[\mathrm{K}(u, v)/x] = s[u/x], \quad s[\mathrm{S}(u, v, w)/x] = s[u(v, u(w))/x],$$

$$s[\mathrm{R}(u, v, 0)/x] = s[u/x], \quad s[\mathrm{R}(u, v, w')/x] = s[v(w, \mathrm{R}(u, v, w))/x],$$

where $s[x]$ is a term of type 0 and u, v, w are terms of appropriate type. Note that the resulting theory is contained in both Gödel's and Spector's versions.

Most of the results for the functional interpretation are insensitive to which of these three formulations of T (and of other systems like T to be considered below) is taken. When it is necessary to make a distinction, as for example in the next section when considering extensions of T by adjunction of quantifiers, and in section 4.1 when considering models of T, we shall use *WE-T* to denote Spector's version and T_0 to denote the version, just described, elicited by Troelstra, while reserving T itself for Gödel's version. However, one further system must be considered along with the latter. It is implicit in the idea of intensional equality at higher types that we have an effective procedure to decide whether any two objects of the same type are identical. This is made explicit by adjoining a constant E_σ for the characteristic function of $=_\sigma$ at each type σ, with axiom:

$$\mathrm{E}_\sigma(x, y) = 0 \leftrightarrow x =_\sigma y.$$

The system T with these additional constants and axioms is then denoted $I\text{-}T$.

3. Consequences and benefits of the interpretation

3.1. Higher type arithmetic

Before turning to some of the immediate consequences of the Dialectica interpretation, we pause to consider some easy generalizations. First of all, note that the D-translation applies equally well if the source language is also typed. In other words, if we define HA^ω to be a version of T with quantifiers ranging over each finite type, with the axioms and rules of intuitionistic predicate logic and induction for all formulas in the new language, then we can apply the D-translation to this theory as well.

A problem, however, arises in verifying the interpretation of the axiom $\varphi \to \varphi \wedge \varphi$, and this depends on how equality at higher types is treated. For if φ is an equation $s =_\sigma t$, in order to extend the argument for Lemma 2.4.3 we shall need to have a characteristic function E_σ for $=_\sigma$. Otherwise we shall have to limit ourselves to a formulation of HA^ω using only equations of type 0. Thus we are led to consider three versions of HA^ω corresponding to the three versions $I\text{-}T$, $WE\text{-}T$ and T_0 in section 2.5, denoted respectively $I\text{-}HA^\omega$, $WE\text{-}HA^\omega$, and HA^ω_0. To be more precise, in $WE\text{-}HA^\omega$, given the availability of quantification at higher types, we may take equality there to be introduced *by definition* in terms of equality at lower types so as to satisfy the equivalences

$$x =_{\sigma\to\tau} y \leftrightarrow \forall z^\sigma\ (x(z) =_\tau y(z))$$

for each σ, τ. What is a new option now is that a *fully extensional* version $E\text{-}HA^\omega$ can be formulated as follows. In this theory one takes the symbol $=_\sigma$ to be basic at each type, and adopts the full axioms 1, 2 of section 2.5 as in the intensional version. What makes the difference is that the preceding equivalences are now taken as axioms in the fully extensional theory.

Now, with little or no modification, the proof sketched in section 2.4 carries over to show that for H any one of the three systems $I\text{-}HA^\omega$, $WE\text{-}HA^\omega$, or HA^ω_0, the system H is D-interpreted in the corresponding quantifier-free subsystem, i.e. $I\text{-}T$, $WE\text{-}T$, or T_0, resp. This does *not* hold for the system $E\text{-}HA^\omega$, as shown by Howard [1973]. In particular, no functional of that system satisfies the D-interpretation of $\forall u, x, y\ (x =_1 y \to u(x) =_0 u(y))$ where x, y are of type 1 (i.e. $0 \to 0$) and u is of type 2 (i.e. $1 \to 0$). However, $E\text{-}HA^\omega$ may be formally interpreted in HA^ω_0 preserving all formulas all of whose variables are of type 0 or 1, by relativizing the quantifiers to the hereditarily extensional objects in each type; we may then apply the D-interpretation to HA^ω_0 as above. This is the route taken by Luckhardt [1973]; cf. also Feferman [1977,section 4.4.2] for a brief outline of the details involved. Clearly $HA^\omega_0 \subset I\text{-}HA^\omega$ and $HA^\omega_0 \subset WE\text{-}HA^\omega \subset E\text{-}HA^\omega$.

In the following we shall take it that HA^ω is any one of the three systems $I\text{-}HA^\omega$, $WE\text{-}HA^\omega$, and HA^ω_0 for which the D-interpretation works directly as described above, and T is now taken for the corresponding quantifier-free subsystem; however, the reader is free to settle on any one of these three pairs as the preferred one according to taste.[8]

Let us return to the principles (AC), (IP'), and (MP') which figured in the motivation for the definition of $(\varphi \to \psi)^D$ in section 2.3. These make use of the quantified variables of arbitrary type and are thus formulated in the language of HA^ω. Given the discussion in section 2.3 it shouldn't be surprising that, in fact, we have

3.1.1. Theorem. *Over intuitionistic logic, the schemata* $(AC) + (IP') + (MP')$ *and* $\varphi \leftrightarrow \varphi^D$ *are equivalent.*

Henceforth we'll use $HA^{\#}$ to denote the theory HA^ω together with either of these two schemata. The previous discussion tells us that

3.1.2. Theorem. $HA^{\#}$ *is D-interpreted in* T.

What, now, can be said about classical theories of higher types? Take PA^ω to be the classical extension of HA^ω, obtained simply by adjoining the law of excluded middle for all formulas. It is N-interpreted in HA^ω just as PA is N-interpreted in HA. By the preceding theorem, it is natural to consider the system $PA^{\#}$ which is defined to consist of PA^ω together with all instances of $\varphi \leftrightarrow \varphi^{ND}$ for φ in the language of PA^ω. It then follows immediately that

3.1.3. Theorem. $PA^{\#}$ *is ND-interpreted in* T.

Consider now the principles (AC), (IP'), and (MP') on the classical side. Of these, (IP') and (MP') follow from classical logic, but (AC) is not derivable from $PA^{\#}$. Moreover, (AC) is not in general preserved under the ND-interpretation. For, the N-interpretation of $\forall x\, \exists y\, \varphi(x, y) \to \exists Y\, \forall x\, \varphi(x, Y(x))$ is equivalent to

$$\forall x\, \neg\forall y\, \neg\varphi^N(x, y) \to \neg\forall Y\, \neg\forall x\, \varphi^N(x, Y(x))$$

or, equivalently,

$$\forall x\, \neg\neg\exists y\, \varphi^N(x, y) \to \neg\neg\exists Y\, \forall x\, \varphi^N(x, Y(x)),$$

which cannot in general be proved in $HA^{\#}$. It *is*, however, provable for the special case $(QF\text{-}AC)$ of the axiom of choice with quantifier-free matrix φ. For, in that case,

$$\forall x\, \neg\neg\exists y\, \varphi^N(x, y) \leftrightarrow \forall x\, \exists y\, \varphi^N(x, y)$$

[8]But note that Troelstra [1990,p. 351] says that "*WE-HA*$^\omega$ as an intermediate possibility [between *I-HA*$^\omega$ and HA_0^ω] is not very attractive: the deduction theorem does not hold for this theory." Yet another alternative for dealing with the problem of verifying the axioms $\varphi \to \varphi \wedge \varphi$ without restriction on atomic formulas, but without additional E_σ functionals, makes use of a variant of the D-interpretation due to Diller and Nahm [1974], described in Troelstra [1973,pp. 243–245]. We shall not go into that variant in this chapter.

is provable in $HA^{\#}$ using (MP'). Then from (AC) in that system we infer $\exists Y\ \forall x\ \varphi^N(x, Y(x))$, which implies intuitionistically its double negation. The conclusion is that $(QF\text{-}AC)$ is ND-interpreted in T. The following observation by Kreisel [1959,p. 120] then gives the proper analogue to Theorem 3.1.1:

3.1.4. Theorem. *Over classical logic, the schemata* $(QF\text{-}AC)$ *and* $\varphi \leftrightarrow \varphi^{ND}$ *are equivalent.*

3.1.5. Proof. In the forward direction, one need only consider negative formulas φ, i.e. those which do not contain disjunction or existence symbols. For such formulas it is easily proved by induction on φ that φ^{ND} has the form $\exists X\ \forall y\ \psi(X(y), y)$ where ψ is quantifier-free and where $\varphi \leftrightarrow \exists X\ \forall y\ \psi(X(y), y) \leftrightarrow \forall y\ \exists x\ \psi(x, y)$. The reverse direction is immediate. □

Thus $PA^{\#}$ can equally well be thought of as $PA^{\omega} + (QF\text{-}AC)$.

When analyzing classical or intuitionistic theories of first- or second-order arithmetic, it often turns to be useful to embed them in fragments or extensions of $PA^{\#}$ and $HA^{\#}$ respectively; cf. the discussion in section 3.3 below.

3.2. Some consequences of the D-interpretation

Since the D-interpretations of HA and $HA^{\#}$, as well as the ND-interpretations of PA and $PA^{\#}$, are purely syntactic, they can be formalized in a weak theory of arithmetic. This yields the following theorem, which is of foundational importance.

3.2.1. Theorem. *Let* S *be any of the theories* HA, $HA^{\#}$, PA, *or* $PA^{\#}$. *Then a weak base theory proves*

$$\mathrm{Con}(T) \to \mathrm{Con}(S).$$

Of course, the interpretations yield far more information than just the relative consistency of the theories involved. Recall that a formula $\theta(x_1, x_2, \ldots, x_n)$ in the language of arithmetic is Δ^0_0 if all its quantifiers are bounded, in which case it defines a primitive recursive relation on the natural numbers. The characteristic function of this relation can be represented by a term t in the language of T, such that $PA^{\#}$ or $HA^{\#}$ proves $\theta(\vec{x}) \leftrightarrow t(\vec{x}) = 0$. Though it is somewhat an abuse of notation, when we say below that "T proves $\theta(\vec{x})$" for such a θ, we mean that it proves $t(\vec{x}) = 0$.

3.2.2. Theorem. *Let* S *be any of the theories above, and suppose* S *proves the* Π^0_2 *formula*

$$\forall x\ \exists y\ \theta(x, y),$$

where θ *is* Δ^0_0. *Then there is a term* f *such that* T *proves*

$$\theta(x, f(x)).$$

3.2.3. Proof. By embedding *HA* and *PA* in $HA^{\#}$ and $PA^{\#}$ respectively and proving the equivalence just discussed, we can assume that θ is quantifier-free. In that case the D-interpretation of $\forall x\, \exists y\, \theta(x, y)$ is $\exists Y\, \forall x\, \theta(x, Y(x))$, and its ND-interpretation is $\exists Y\, \forall x\, \neg\neg\theta(x, Y(x))$. The conclusion then follows from Theorem 3.1.2 in the case of $HA^{\#}$, and Theorem 3.1.3 in the case of $PA^{\#}$. □

Now suppose h is a recursive function whose graph is defined by a Σ^0_1 formula $\varphi(x, y)$ in the standard model. We say that the theory S proves h to be total if it proves $\forall x\, \exists! y\, \varphi(x, y)$. In section 2.2 we axiomatized the primitive recursive functionals of finite type, without addressing the issue of what exactly the terms denote. Deferring this discussion to section 4 below, for now let us assume that at least the closed type 1 terms denote functions. As a corollary to Theorem 3.2.2 we have

3.2.4. Corollary. *Every provably total recursive function of HA, $HA^{\#}$, PA, or $PA^{\#}$ is denoted by a term of T.*

In fact, in section 4.1 we will see that there are models of T that can be formalized in the language of arithmetic, yielding an interpretation of T in *HA*. This yields the following result, which is interesting in that it makes no mention of T at all:

3.2.5. Corollary. *PA, and hence HA + (MP), is conservative over HA for Π^0_2 sentences.*

When it comes to *PA*, Theorem 3.2.2 is sharp in the following sense. Consider the Π^0_3 sentence "for every x there exists a y, such that either y is a halting computation for the Turing machine with index x, or Turing machine x doesn't halt." Though this statement is provable in *PA*, any function returning such a y for every x cannot be recursive since it solves the halting problem. Later we'll see that, on the other hand, the functions represented by terms of T *are* recursive, so that the analogue of Theorem 3.2.2 does not hold for Π^0_3 formulas.

Nonetheless, one *can* extract a different kind of constructive information from *PA*-proofs of complex formulas. Suppose *PA* (or $PA^{\#}$) proves a formula φ given by

$$\forall x_1\, \exists y_1\, \forall x_2\, \exists y_2\, \ldots \forall x_n\, \exists y_n\, \theta(x_1, x_2, \ldots, x_n, y_1, y_2, \ldots, y_n)$$

where θ is quantifier-free. The N-interpretation of this formula intuitionistically implies

$$\neg\exists x_1\, \forall y_1\, \exists x_2\, \forall y_2\, \ldots \exists x_n\, \forall y_n\, \neg\theta(x_1, x_2, \ldots, x_n, y_1, y_2, \ldots, y_n),$$

whose D-interpretation is the same as that of

$$\forall X_1, X_2, \ldots, X_n\, \exists y_1, y_2, \ldots, y_n$$
$$\neg\neg\theta(X_1, X_2(y_1), \ldots, X_n(y_1, y_2, \ldots, y_{n-1}), y_1, y_2, \ldots, y_n).$$

As a result we have

3.2.6. Corollary. *Suppose PA proves φ as above. Then there are terms*

$$t_1 = F_1(X_1, X_2, \ldots, X_n)$$
$$\vdots$$
$$t_n = F_n(X_1, X_2, \ldots, X_n)$$

such that T proves

$$\theta(X_1, X_2(t_1), \ldots, X_n(t_1, t_2, \ldots, t_{n-1}), t_1, t_2, \ldots, t_n).$$

The observation that the functionals $F_1, F_2, \ldots, F_n$ can be taken to be recursive in their arguments is known as Kreisel's "no-counterexample interpretation" (cf. Kreisel [1951,1959]). To make sense of the name, think of the functions $X_1, X_2, \ldots, X_n$ as trying to provide counterexamples to the truth of φ, by making $\theta(X_1, X_2(y_1), \ldots, X_n(y_1, \ldots, y_{n-1}), y_1, \ldots, y_n)$ false for any given values of $y_1, y_2, \ldots, y_n$; in which case $F_1, F_2, \ldots, F_n$ effectively provide witnesses that foil the purported counterexample.

3.3. Benefits of the D-interpretation

In general, the D-interpretation is a powerful tool when applied to the reduction of an intuitionistic theory I to a functional theory F. As we have seen, the very form of the D-interpretation automatically brings a number of benefits. Since little more than the combinatorial completeness of F is necessary to interpret the logical axioms of I, one only has to worry about interpreting the non-logical axioms of I and tailor the functionals of F accordingly. As an added bonus, I can often be embedded in a higher-type analogue I^ω to which we can add the schemata (AC), (IP'), (MP') at no extra cost. Taken together (AC), (IP'), and (MP') prove the scheme $\varphi \leftrightarrow \varphi^D$, a fact that is often useful in practice since it allows one to pull facts about the D-interpretation back to the theory being interpreted. This observation will be employed in sections 6 and 7 below.

If one is trying to analyze a classical theory C by reducing it to I via a double-negation interpretation, one has, of course, to ensure that I is strong enough to prove the doubly-negated axioms of C. Once again, though, the choice of a D-interpretation for I has some advantages: C can often also be embedded in a higher-type analogue C^ω in a natural way, and the fact that Markov's principle is verified in the interpretation guarantees that one ultimately obtains Skolem terms for provable Π^0_2 sentences. This in turn yields a characterization of C's provably total recursive functions.

These advantages of the D-interpretation are summed up in Feferman [1993]:

> Applied to intuitionistic systems it takes care of the underlying logic once and for all, verifies the Axiom of Choice AC in all types, and interprets various forms of induction by suitably related forms of recursion. This

> then leads for such systems to a perspicuous mathematical characterization of the provably recursive functions and functionals. For application to classical systems, one must first apply the negative translation (again taken care of once and for all). Since the D-interpretation verifies Markov's principle even at higher types ... at least the provably recursive functions and functionals are preserved, as well as QF-AC in all types and induction schemata. The main disadvantage, though, comes with the analysis of other statements whose negative translation may lead to a complicated D-interpretation; special tricks may have to be employed to handle these.

A further distinguishing feature of the D-interpretation is its nice behavior with respect to modus ponens. In contrast to cut-elimination, which entails a global (and computationally infeasible) transformation of proofs, the D-interpretation extracts constructive information through a purely local procedure: when proofs of φ and $\varphi \to \psi$ are combined to yield a proof of ψ, witnessing terms for the antecedents of this last inference are combined to yield a witnessing term for the conclusion. As a result of this modularity, the interpretation of a theorem can be readily obtained from the interpretations of the lemmata used in its proof.

The process of applying the D-interpretation to specific classical theorems can sometimes be used to obtain appropriate constructivizations thereof, or to uncover additional numerical information that is implicit in their classical proofs; cf., for example, Bishop [1970] or Kohlenbach [1993].

4. Models of T, type structures, and normalizability

In section 2.2 we presented the set of terms of the theory T without a discussion of what these terms denote. In this section we exhibit several kinds of functional and term models.

4.1. Functional models

The most obvious model of T considered in either the intensional or extensional sense is the *full (set-theoretic) hierarchy of functionals of finite type*, in which the objects of type 0 are the natural numbers, and each type $\sigma \to \tau$ represents the set of *all* functions from the objects of type σ to those of type τ. The denotations of 0, Sc, K, S, and R are then apparent, as well as the denotation of terms built up through the application of these constants. The equality relation in T is taken to denote true (extensional) equality in this model. By the primitive recursive functionals in the set-theoretic sense we mean those denoted by closed terms of T.

One can obtain a "smaller" type structure in which the elements of each type are indices for recursive functions, as follows. Let φ_e denote a standard enumeration of the recursive functions, say, using Kleene's universal predicate. Define $M_0 = \mathrm{N}$, and

$$M_{\sigma\to\tau} = \{e \mid \forall x \in M_\sigma \ \exists y \in M_\tau \ (\varphi_e(x) \downarrow = y)\}.$$

One can associate recursive indices to constants of T in a natural way, and interpret equality as equality of indices. The resulting model $\mathcal{M}$ of $I\text{-}T$ (and of $I\text{-}HA^\omega$) is known as the *hereditarily recursive operations* (though "hereditarily total recursive operations" might be more accurate) and is usually denoted by *HRO*. Equality is not extensional in this model, since many different indices can represent the same functional. If instead one wants a model of $WE\text{-}T$ (and of $WE\text{-}HA^\omega$ or even $E\text{-}HA^\omega$), one can consider the *hereditarily effective operations* $\mathcal{N}$, usually denoted by *HEO*. For this model, sets N_σ and the equality relation $=_\sigma$ are defined inductively as follows: set $N_0 = \mathbf{N}$ and $=_0$ the usual equality relation for natural numbers,

$$N_{\sigma\to\tau} = \{e \mid \forall x \in N_\sigma\ \exists y \in N_\tau\ (\varphi_e(x)\downarrow= y) \wedge \\ \forall x \in N_\sigma, y \in N_\sigma\ (x =_\sigma y \to \varphi_e(x) =_\tau \varphi_e(y))\},$$

and

$$e =_{\sigma\to\tau} f \equiv \forall z \in N_\sigma\ (\varphi_e(z) =_\tau \varphi_f(z)).$$

Both *HEO* and *HRO* can be formalized in *HA* in the sense that the sets M_σ (resp. N_σ) and the equality relations $=_\sigma$ are defined by formulas in the language of arithmetic, *HA* proves each axiom of T true in the interpretation, and the natural numbers in the model correspond to the natural numbers of *HA*. Notice, however, that the complexity of the formulas defining M_σ and N_σ grow with the level of σ, so that *HA* (or *PA*) cannot prove the consistency of T outright.

By generalizing the notion of a continuous function to higher types, Kleene and Kreisel independently obtained further models of T (cf. also Troelstra [1973]); these will be of significance in section 6.

All of the models described in this subsection contain more than just the primitive recursive functionals (or the indices for such), and hence model extensions of T as well.[9] In contrast, a "minimal" model of T, which only contains objects denoted by terms, is provided by the *term model*, which we now describe.

4.2. Normalization and the term model

The defining equations for the typed combinators K, S, and R in section 2.2 describe a symmetric equality relation. From a computational point of view, it is often more useful to think of these defining equations as describing a *directed* relation, in which terms on the left-hand sides of the equations are "reduced" to more basic ones on the right. For example, the defining equations of the theory T yield the following reduction rules:

1. $\mathrm{K}(s,t) \triangleright s$
2. $\mathrm{S}(r,s,t) \triangleright r(t)(s(t))$
3. $\mathrm{R}(s,t,0) \triangleright s$

[9]Some, like the recursion theoretic models, have generalizations to transfinite types; see, e.g. Beeson [1982]. Category-theoretic methods have also been used to construct models of functional theories, though we will not discuss such models here.

4. $R(s,t,x') \triangleright t(x, R(s,t,x))$.

If one uses lambda terms instead of combinators, the first two clauses can be replaced by the reduction rule $(\lambda x.s)(t) \triangleright t[s/x]$. If one wants to include product types, corresponding reduction rules for the pairing and projection operations can be defined as well.

If s and t are terms, we say that s *reduces to* t *in one step*, written $s \to t$, if t can be obtained by replacing some subterm u of s by a v such that $u \triangleright v$. We say that s *reduces to* t if t can be obtained from s by a finite sequence of one-step reductions. In other words, the reducibility relation $\to^*$ is the reflexive-transitive closure of $\to$.

Such a reducibility relation is an example of a *rewrite system* (cf. Dershowitz and Jouannaud [1990]). The following terminology is standard.

4.2.1. Definition.

1. If s is a term and u is a subterm of s, then u is a *redex* of s if a reduction rule can be applied to u; i.e. there is some v such that $u \triangleright v$.
2. If s has a redex, then s is *reducible*. Otherwise, s is *irreducible*, or in *normal form*.
3. A term is *normalizable* if it can be reduced to one in normal form. A system of reduction rules is *normalizing* if every term is normalizible.
4. A term s is *strongly normalizable* if there are no infinite (one-step) reduction sequences beginning with s; that is, *every* such sequence eventually leads to a term in normal form. A system of reduction rules is *strongly normalizing* if every term is strongly normalizable.
5. A system of reduction rules is *confluent*, or has the *Church-Rosser property*, if whenever $s \to^* u$ and $s \to^* v$ then there is a term t such that $u \to^* t$ and $v \to^* t$.

If a system of rules is confluent and s is any term, then s has at most one normal form. Furthermore, if the system is also normalizing, then s has *exactly* one normal form. Identifying closed terms with their normal forms then provides a *term model* for the defining equations corresponding to the reduction rules. Under this very simple semantics, one can think of each closed term representing nothing more than the "program" it computes, when applied to other closed terms in normal form. Strong normalizability implies that the "programming language" is insensitive to its implementation, in the sense that every reduction sequence terminates regardless of the order in which the reductions are performed.

In all the functional theories we consider in this chapter (together with the associated reduction relations), the only closed irreducible terms of type 0 are in fact numerals. In that case, we can identify type 0 objects of the corresponding term model with natural numbers. Showing that the reduction relation associated with a functional theory is confluent and normalizing then has two benefits:

- it implies that the functional theory is consistent, that is, it cannot prove $0 = 1$; and

- it justifies the intuition that the functional theory describes "computable" entities.

4.3. Strong normalization for T

Given the reduction rules corresponding to the theory T described above, one can apply brute but judicious combinatorial force to verify the following

4.3.1. Lemma. *The reduction relation $\rightarrow^*$ associated with T is confluent. Furthermore, any closed irreducible term of type 0 is a numeral.*

The "shortest known proof" of this, due to Tait and Martin-Löf, can be found in Hindley and Seldin [1986,appendix 1].

Assuming we can prove that the relation $\rightarrow^*$ is also normalizing, the resulting term model will satisfy the axioms of T: the uniqueness of normal forms implies that terms on either side of an equality axiom have the same interpretation under any instantiation of the variables; and the fact that the objects of type 0 in the term model are numerals reduces induction in T to induction in the metatheory.

Proving that $\rightarrow^*$ is normalizing, however, is bound to be tricky. Because it implies the consistency of Peano arithmetic, the proof must somehow go beyond the capabilities of that theory. W. Tait developed an elegant and flexible technique for proving normalization, using appropriate "convertibility" predicates (cf. Tait [1967] and Tait [1971]).

4.3.2. Definition. For each type σ, we define the set of *reducible terms of type σ*, denoted by Red_σ:

1. If t is a term of type 0, then t is in Red_0 if and only if t is normalizing.
2. If t is a term of type $\sigma \rightarrow \tau$, then t is in $\mathrm{Red}_{\sigma\rightarrow\tau}$ if and only if whenever s in Red_σ, $t(s)$ is in Red_τ.

Writing the second clause symbolically, we have that t is in $\mathrm{Red}_{\sigma\rightarrow\tau}$ if and only if

$$\forall s\ (s \in \mathrm{Red}_\sigma \rightarrow t(s) \in \mathrm{Red}_\tau).$$

Notice that the quantifier complexity of the first-order formula expressing "$t \in \mathrm{Red}_\rho$" grows with the complexity of ρ.

The normalization proof proceeds in two steps:

1. One shows, by induction on terms, that every t of type σ is in Red_σ.
2. One shows, by induction on the type σ, that every t in Red_σ is normalizing.

The only aspect of the proof that cannot be carried out in a weak base theory is the verification of clause 2, when t is the recursor R_σ: at this point the argument requires induction on a formula involving the predicate Red_σ. As a result we have

4.3.3. Theorem. *T is normalizing. Moreover, for each term t of T, PA proves that t is normalizable.*

This does not mean that Peano arithmetic proves that "for every term t, t is normalizable," since for various t the corresponding PA-proof can grow increasingly complex. The latter result, however, follows from the soundness of PA.

Another approach to proving normalization involves assigning ordinals (or notations representing ordinals) to terms in such a way that each one-step reduction leads to a decrease in the associated ordinal. For terms of T, this task is carried out by Howard [1970] using notations below the ordinal ε_0. Via a formal treatment of the term model, this yields, as a by-product, an ordinal analysis: over a weak base theory the assertion that "there are no infinite descending sequences of ordinal notations beneath ε_0" implies the consistency of PA.

4.4. Infinitely long terms

Another term model for T which is of special interest was provided by Tait [1965]. This uses infinitely long terms to replace the recursors, and thus produces a system of terms which is closer in character to the ordinary typed λ-calculus. The closure conditions on terms are as follows:

1. There are infinitely many variables $x^\tau, y^\tau, z^\tau, \ldots$ of each type τ.
2. 0 is a constant of type 0.
3. Sc is a constant of type $(0 \to 0)$.
4. If s is a term of type σ and t is a term of type $(\sigma \to \tau)$ then $t(s)$ is a term of type τ.
5. If t is a term of type τ then $\lambda x^\sigma.t$ is a term of type $(\sigma \to \tau)$.
6. If t_n $(n = 0, 1, 2, \ldots)$ is a sequence of terms of type τ then $\langle t_n \rangle$ is a term of type $(0 \to \tau)$.

Write $\mathbf{n}$ for $Sc^n(0)$. Then we translate each term t of T into a term t^+ of this system of infinite terms by taking $t^+ = t$ for t a variable, 0 or Sc,

$$\mathrm{K}^+ = \lambda x \lambda y.x, \quad \mathrm{S}^+ = \lambda x \lambda y \lambda z.x(z, y(z))$$

and

$$\mathrm{R}^+ = \lambda f \lambda g \lambda x.\langle t_n \rangle \text{ where } t_0 = f \text{ and } t_{n+1} = g(\mathbf{n}, t_n),$$

and by requiring $(\cdot)^+$ to preserve application.

Each term t of the infinite system is assigned an ordinal $|t|$ as length in a natural way, with $|t| = 1$ for t a variable or constant, $|\lambda x.t| = |t| + 1$, $|t(s)| = |s| + |t|$ and, finally, $|\langle t_n \rangle| = \sup_{n<\omega}(|t_n| + 1)$. Note that for each of the constants C of T, $|C^+| \leq \omega$, so for each term t of T, $|t^+| < \omega \cdot 2$.

We have three immediate reduction rules for this system of infinite terms:

1. $(\lambda x.t[x])(s) \rhd t[s/x]$
2. $\langle t_n \rangle(\mathbf{m}) \rhd t_m$

3. $(\langle t_n \rangle(r))(s) \triangleright \langle t_n(s) \rangle(r)$, when r is not a numeral and $\langle t_n \rangle(r)$ is not of type 0.
The relation $\rightarrow^*$ is then the least reflexive and transitive relation which extends the $\triangleright$ relation and preserves application. As before, a term t is said to be in normal form if whenever $t \rightarrow^* u$ then t is identical with u.

4.4.1. Theorem. *For each term t of T we can find a term t° in normal form such that $t^+ \rightarrow^* t^\circ$ and $|t^\circ| < \varepsilon_0$.*

The idea of Tait's proof of Theorem 4.4.1 is very much the same as that for the cut-elimination theorem for the extension of Gentzen's classical propositional sequent calculus to that for logic with countably long conjunctions Π and disjunctions Σ. Derivations in *PA* are translated into derivations in this calculus, by first translating formulas φ into propositional formulas φ^+, using $(\forall x\ \varphi[x])^+ = \Pi_{n<\omega}\varphi^+[\mathbf{n}/x]$. Then each derivation d from PA is translated into an infinite propositional derivation d^+ with finite cut-rank and $|d^+| < \omega \cdot 2$. Each derivation whose cut-rank is $\leq m+1$ and ordinal length is $\leq \alpha$ is effectively reduced to a derivation of the same end-formula whose cut-rank is $\leq m$ and ordinal length is $\leq \omega^\alpha$. So for each derivation d of T we eventually reduce d^+ to a derivation d° of length $< \varepsilon_0$. Similarly, we can assign a "cut-rank" to reducible infinite terms, and lower cut-complexity at the same exponential cost of increasing ordinal bounds. See Tait [1968], Schwichtenberg [1977], or Chapters III IV in this volume for more details concerning cut-elimination for sequent calculi for infinitary languages, and Tait [1965] or Feferman [1977] for details concerning normalization for infinitary term calculi.

More information can be extracted from these procedures as follows. Schwichtenberg [1977,section 4.2.2] shows in detail how the infinitary derivations generated from those in PA, as described above, may be coded by indices for primitive recursive functions. For each derivation d' in the sequence of reductions from d^+ to d°, the code of d' both determines the structure of d' as a tree and contains a bound on its cut-rank ($< \omega$) and on its length ($< \varepsilon_0$, in a primitive recursive notation system for ε_0). Exactly the same kind of thing can be done for each infinite term t' in the reduction sequence from t^+ to t°. This allows one to define an effective valuation function on t° when the initial t is a closed term of type 1 (i.e. $0 \rightarrow 0$), and that leads one to

4.4.2. Theorem. *The functions of type 1 generated by the primitive recursive functionals may be defined by schemata of effective transfinite recursion on ordinals $< \varepsilon_0$.*

The schemata referred to define, for any given ordinal $\alpha < \varepsilon_0$, a function $F(x, y)$ of numbers x, y by $F(0, y) = G(y)$ and for $x \neq 0$, $F(x, y)$ in terms of $F(H_i(x, y), y)$ where the H_i are one or more functions with $H_i(x, y) <_\alpha x$. The collection of F definable using such schemata coupled with the usual schemata for primitive recursive definition is denoted REC($<\varepsilon_0$). In this way one obtains the following result due to Kreisel [1951,1952] from Corollary 3.2.4 and Theorem 4.4.2:

4.4.3. Theorem. *The provably recursive functions of PA (and hence also of HA) are exactly those in* REC($<\varepsilon_0$).

A similar characterization can be obtained of the functionals which are asserted to exist in Kriesel's no-counterexample interpretation of *PA*, here via the ND-interpretation of *PA* in *T* (Corollary 3.2.6).

Returning to Theorem 4.4.2, for its proof one makes use of the following simple description of the normal terms $t[x_1, \ldots, x_n]$ of type 0 in which all the free variables x_i are of type 0: either

1. $t = 0$ or
2. $t = \mathrm{Sc}(s)$ where s is normal or
3. t is a variable of type 0 or
4. $t = \langle t_n \rangle (s)$ where each t_n and s is normal of type 0, but s is not a numeral.

It follows that the only closed terms of type 0 are numerals. The closed normal terms of type 1 are just those of the form $\lambda x^0.t[x]$ where t is normal of type 0.

5. The interpretation of fragments of arithmetic

5.1. $I\Sigma_1$ and the primitive recursive functions

In the interpretation of *PA* and *HA* the recursors R_σ were used to interpret the induction axioms, and it should not be surprising that weaker forms of recursion can be used to interpret weaker forms of induction. Let $I\Sigma_1$ be the fragment of *PA* in which induction is restricted to Σ_1^0 formulas. A nice application of the D-interpretation due to Parsons (cf. [1970,1972]) shows that any provably total recursive function of $I\Sigma_1$ is in fact primitive recursive.

The definition of the recursors R_σ in section 2.2 for $\sigma \neq 0$ is "impredicative," in that the evaluation of $\mathrm{R}_\sigma(f, g, n')$ at a given argument x presumes that $\mathrm{R}_\sigma(f, g, n)$ has already been defined for arbitrary arguments z. A "predicative" restriction of this scheme is given by the recursors $\hat{\mathrm{R}}_\sigma$ due to Kleene, which have the defining schemata

$$\hat{\mathrm{R}}_\sigma(f, g, 0, b) = f(b)$$
$$\hat{\mathrm{R}}_\sigma(f, g, n', b) = g(n, \hat{\mathrm{R}}_\sigma(f, g, n, b), b).$$

Note that each type σ is uniquely of the form $(\sigma_1, \ldots, \sigma_k) \to 0$ for some sequence $(\sigma_1, \ldots, \sigma_k)$. In the equations above, then, f is of type σ, g is of type $0 \to 0 \to \sigma$, and $b = (b_1^{\sigma_1}, \ldots, b_k^{\sigma_k})$ is a sequence of variables chosen so that $\hat{\mathrm{R}}_\sigma(f, g, n, b)$ is of type 0.[10] We let $\widehat{T}$ denote the restriction of T which only allows this type of recursion.[11]

---

[10] In contrast to Gödel's recursors, each $\hat{\mathrm{R}}_\sigma$ can in fact be defined from $\hat{\mathrm{R}}_0$ by the equation $\hat{\mathrm{R}}_\sigma = \lambda f, g, n, b\ \hat{\mathrm{R}}_0(f(b), \lambda k, l\ g(k, l, b), n)$.

[11] In the versions $\widehat{T_0}$ and $I\text{-}\widehat{T}$ we need to add, as in Parsons [1972], conditional functionals $\mathrm{Cond}_\rho : (0, \rho, \rho) \to 0$. These functionals have defining equations $\mathrm{Cond}_\rho(w, u, v) = u$ if $w = 0$ and $\mathrm{Cond}_\rho(w, u, v) = v$ otherwise; in these theories they are no longer definable using the $\hat{\mathrm{R}}$ recursors.

5.1.1. Theorem. *The closed level 1 terms of $\widehat{T}$ denote primitive recursive functions. Moreover, there is a natural translation of type 0 terms t of $\widehat{T}$ whose only free variables are of type 0 to terms t^{PRA} of PRA, such that if $\widehat{T}$ proves $t = s$, then PRA proves $t^{PRA} = s^{PRA}$.*

5.1.2. Proof. The proof of the first assertion is by adaptation of Kleene [1959b, sections 1.5–1.7] to the present framework.[12] The idea is to define a class KL of functionals $F(b)$ of lists of arguments b of arbitrary finite type but with values of type 0 only, such that each F in KL is defined by a term in $\widehat{T}$, while each term t in $\widehat{T}$ is represented by an F in KL either directly (if it is of type 0) or by abstraction on some of the variables of F (otherwise).

In defining KL and in the proof of these facts we modify the conventions of section 2.2 as follows. We will use $\underline{\sigma}$ to denote (possibly empty) sequences of types $(\sigma_1, \ldots, \sigma_k)$, and by $(\underline{\sigma} \to 0)$ we mean 0 if $k = 0$ and $(\sigma_1 \to \ldots \to \sigma_k \to 0)$ otherwise. We will use $u^{\underline{\sigma}}$ to denote a sequence of terms $(u_1^{\sigma_1}, \ldots, u_k^{\sigma_k})$, and $\lambda u^{\underline{\sigma}}.F(u, \ldots)$ means $F(\ldots)$ when $n = 0$ and $\lambda u_1^{\sigma_1} \ldots \lambda u_k^{\sigma_k}.F(u_1, \ldots, u_k, \ldots)$ otherwise. So each type τ is uniquely of the form $(\underline{\sigma} \to 0)$ for some $\underline{\sigma}$, and $lev(\tau) = \max_{1 \le i \le k}(lev(\sigma_i) + 1)$.

The functionals F in KL are generated by the following schemata, in which n is a variable of type 0 and $b = (b_1^{\tau_1}, \ldots, b_\ell^{\tau_\ell})$:

1. $F(b) = 0$
2. $F(n, b) = n$
3. $F(n, b) = n'$
4. $F(a^\tau, b) = a(t_1, \ldots, t_k)$ where $\tau = (\underline{\sigma} \to 0)$, $\tau \neq 0$, $\sigma_i = (\underline{\rho}_i \to 0)$ (possibly 0) and $t_i = \lambda w_i^{\underline{\rho}_i} G_i(w_i, a, b)$ for $i = 1, \ldots, k$
5. $F(b) = G(H(b), b)$ where the first argument of G is of type 0
6. $F(b) = G(b_\pi)$ where b_π is a permutation of b by π
7. $F(0, b) = G(b)$, $F(k', b) = H(k, b, F(k, b))$

Then the following facts are established:

1. The functionals F of KL are closed under substitution of one or more terms t of type τ for variables a^τ of F where $\tau = (\underline{\sigma} \to 0)$, $\tau \neq 0$ and $t = \lambda u^{\underline{\sigma}}.H(u, \ldots)$.
2. For each $F(b)$ in KL we can find a term t of $\widehat{T}$ with free variables b, which defines F.
3. If t is a term of $\widehat{T}$ with free variables b, and t is of type $(\underline{\sigma} \to 0)$ (possibly 0), then we can find a functional $F(u, b)$ of KL such that $t(u) = F(u, b)$ for all u, b.
4. If $F(b)$ has all its variables $b_1, \ldots, b_\ell$ of type 0, then F is primitive recursive.

Fact 1 corresponds to the Full Substitution Theorem of Kleene [1959b,section 1.6]. It is proved by induction on the maximum m of the levels τ of the variables a^τ being substituted for and, for given m, by induction on the generation of F in KL. Facts 2 and 3 are straightforward, using 1 for the application step in 3. Finally, fact 4

[12]Here we do not consider the version I-$\widehat{T}$, with its additional functionals E_ρ.

follows by observation that scheme 4 of the definition of KL can never be applied in the generation of functionals all of whose arguments are of type 0.

This shows that the type 1 terms of $\widehat{T}$ denote primitive recursive functions. The conservation result of Theorem 5.1.1 is obtained by using the argument above to transform proofs in $\widehat{T}$ to proofs from schemata 1–7, and then using fact 4 to transform this to a proof in PRA.

(Remark. One can think of the proof of fact 1 as a normalization argument for a system of terms generated from schemata 1–7 together with a scheme of full substitution in place of scheme 4. Another route to the proof of Theorem 5.1.1 should be possible by normalization of the terms of $\widehat{T}$, but that would apparently require more work. For a proof of an analogous result for a system of feasible functionals of finite type, see the reference after Theorem 5.2.1 below.) □

Let $\widehat{HA}^{\#}$ and $\widehat{PA}^{\#}$ denote variants of $HA^{\#}$ and $PA^{\#}$ respectively, in which induction is restricted to existential formulas. The reader can verify that the recursors $\hat{\mathrm{R}}$ are sufficient to interpret induction for these formulas, yielding

5.1.3. Theorem. *$\widehat{HA}^{\#}$ is D-interpreted in $\widehat{T}$, and $\widehat{PA}^{\#}$ is ND-interpreted in $\widehat{T}$.*

Since $\widehat{PA}^{\#}$ proves that any Σ^0_1 formula is equivalent to an existential one, $I\Sigma_1$ can be embedded in this theory, and Theorems 5.1.1 and 5.1.3 yield

5.1.4. Theorem. *$I\Sigma_1$ is conservative over PRA for Π^0_2 formulas, in the sense that if $I\Sigma_1$ proves $\forall x\, \exists y\, \varphi(x,y)$ for a Δ^0_0 formula φ, then there is a term f such that PRA proves $\varphi(x, f(x))$. Every provably total recursive function of $I\Sigma_1$ is primitive recursive.*

Since $I\Sigma_1$ can prove each primitive recursive function to be total, this last characterization is exact.

5.2. S^1_2 and the polynomial-time computable functions

When it comes to bounded arithmetic (cf. Chapter II), $I\Sigma_1$ is analogous to Buss' theory S^1_2, and PRA is analogous to Cook's theory PV, which axiomatizes the polynomial-time computable functions as characterized in section 1.3.7 of Chapter II. In Buss [1986] it is shown that S^1_2 is conservative over PV in the sense of Theorem 5.1.4. This result has been reobtained using a D-interpretation in Cook and Urquhart [1993], and it is this presentation that we now sketch (cf. also Feferman [1990]).

Cook and Urquhart start by defining a higher-type version PV^ω of the theory PV. Aside from a careful choice of initial functions, which hinge on the fact that one is supposed to think of the natural numbers in terms of their binary representations, the computational strength of PV^ω comes from recursors $\tilde{\mathrm{R}}$, which allow for "higher-type limited recursion on notation." These are described by the equations

$$\tilde{\mathrm{R}}(f,g,h,0,b) = f(b)$$

$$\tilde{\mathrm{R}}(f,g,h,n',b) = \begin{cases} g(n,\tilde{\mathrm{R}}(f,g,h,\lfloor n/2 \rfloor,b),b) & \text{if this has length less than} \\ & \text{that of } h(n,b) \\ h(n,b) & \text{otherwise.} \end{cases}$$

Here b once again denotes a sequence of variables chosen so that $\tilde{R}(f,g,h,0,b)$ is of type 0, h is a type 1 function which provides a bound on the growth of the function being defined, $\lfloor \cdot/2 \rfloor$ is among the initial functions of PV^ω, and additional initial functions representing "length" subtraction and a conditional are used to express the second equation.[13] The following theorem is analogous to Theorem 5.1.1.

5.2.1. Theorem. *The level 1 terms of PV^ω denote polynomial-time computable functions. Moreover, there is a natural translation of type 0 terms t of PV^ω whose only free variables are of type 0 to terms t^{PV} of PV, so that if PV^ω proves $t = s$, then PV proves $t^{PV} = s^{PV}$.*

Full details are provided in Cook and Urquhart [1993,pp. 140–146].

Finally, IPV^ω and CPV^ω are defined to be quantifier versions of PV^ω based on intuitionistic and classical logic respectively, where only type 0 equality is allowed. In these theories, induction is allowed for "NP-predicates," that is, formulas of the form $\exists y \leq t\ (r = s)$ where all the free variables of t have type 0. Since the recursors $\tilde{\mathrm{R}}$ are sufficient to interpret this form of induction, we have, in analogy to Theorem 5.1.3,

5.2.2. Theorem. *$IPV^\omega + (MP)$ is D-interpreted in PV^ω, and CPV^ω is ND-interpreted in PV^ω.*

S^1_2 can be embedded in CPV^ω, since the latter theory can prove any Σ^b_1 formula equivalent to an NP-predicate as described above. Hence we have the following

5.2.3. Theorem. *S^1_2 is conservative over PV for $\forall\Sigma^b_1$ sentences, in the sense that if S^1_2 proves $\forall x\, \exists y\, \varphi(x,y)$ for φ a Σ^b_1 formula, then there is a term f such that PV proves $\varphi(x, f(x))$. Every Σ^b_1-definable function of S^1_2 is polynomial-time computable.*

By Theorem 1.3.4.1 of Chapter II this last characterization is exact.

6. The interpretation of analysis

6.1. Towards the interpretation of stronger theories

At the end of his 1958 paper, Gödel made the following suggestions regarding the functional interpretation of stronger theories.

[13]In Cook and Urquhart [1993,section 6] only the type 0 recursors are taken to be basic, with the others defined as in footnote 10.

> It is clear that, starting from the same basic idea, one can also construct systems that are much stronger than T, for example by admitting transfinite types or the sort of inference that Brouwer used in proving the "fan theorem."

In section 10 we will address the first proposal, expanding the notion of "type" to the transfinite. In this section we explore the second proposal. The fan theorem that Gödel refers to is, from a classical point of view, equivalent to weak König's lemma, to be dealt with in section 7. Brouwer was able to prove the fan theorem using a principle of "bar induction" which he felt was justified by a constructive interpretation of the terms involved (cf. Beeson [1985], Troelstra and van Dalen [1988]). In a posthumous paper, C. Spector [1962] used a generalization of this principle to justify a computational scheme which he dubbed "bar recursion." With this scheme Spector was able to provide a functional interpretation of full second-order arithmetic, that is, the theory $PA^2 + (CA)$ defined in section 6.2 below.[14]

While Spector used bar induction to justify bar recursion, he apparently intended to show that, conversely, bar recursion could be used to obtain a functional interpretation of bar induction. This task was in fact carried out by Howard [1968], and is the approach we outline here.

6.2. Analysis and higher type extensions

The language of second-order arithmetic extends the language of Peano arithmetic with variables that range over sets of numbers, and a binary membership relation $x \in Y$. In this language we only allow first-order equality, defining the assertion $Y = Z$ to mean

$$\forall x\, (x \in Y \leftrightarrow x \in Z).$$

The theory of second-order arithmetic, $PA^2 + (CA)$, extends Peano arithmetic with induction for formulas in the expanded language, and the comprehension scheme

$$(CA) \qquad \exists Y\, \forall x\, (x \in Y \leftrightarrow \varphi(x))$$

for arbitrary formulas φ. (As usual we denote the scheme in which φ is restricted to formulas in Γ by $(\Gamma\text{-}CA)$.) The theory $PA^2 + (CA)$ is often called "analysis" due to the observation that, via the coding of real numbers and continuous functions as sets of natural numbers, one can develop a good theory of the continuum from these axioms. (In fact, by the work of Weyl [1918,1994], first-order arithmetic comprehension with parameters suffices for most practical purposes.)

In order to embed $PA^2 + (CA)$ in an extension of $PA^{\#}$, we identify sets Y with their characteristic functions χ_Y, and read $t \in Y$ as $\chi_Y(t) \neq 0$. This provides a

[14] We should mention in this connection Kreisel [1959], in which it is shown that if second-order arithmetic proves a formula φ, the witnessing functions can be taken to be "recursively continuous," in a sense defined, independently, by Kreisel [op. cit.] and Kleene [1959a] (cf. also Feferman [1993]). For an interesting extension of bar recursion to transfinite types, see Friedrich [1985].

natural way of interpreting the set variables of $PA^2 + (CA)$ with function variables of type 1.

Recall the axiom of choice

$$(AC) \qquad \forall x \, \exists y \, \varphi(x,y) \rightarrow \exists f \, \forall x \, \varphi(x, f(x)).$$

We will denote the scheme in which the variables x and y are restricted to be of type σ and τ respectively by $(AC_{\sigma\tau})$. We have seen that adding the axiom of choice (AC) to HA^ω results in a theory that is no stronger than Heyting arithmetic. In contrast, the addition of even (AC_{00}) to PA^ω results in a theory that includes (CA), since one can apply this choice principle to the classically valid formula

$$\forall x \, \exists y \, ((y = 0 \wedge \neg\varphi(x)) \vee (y = 1 \wedge \varphi(x))).$$

More generally, assuming that Γ satisfies some basic closure conditions (including closure under negations), $(\Gamma\text{-}CA)$ follows classically from $(\Gamma\text{-}AC_{00})$. So, to interpret analysis, it clearly suffices to interpret the theory $PA^\omega + (AC_{00})$.

6.3. The principle of bar induction

In the following statement of the principle of bar induction used in the Spector-Howard interpretation, the variable c represents a finite sequence $\langle c_0, c_1, \ldots, c_{k-1}\rangle$ of objects of type σ (suitably coded), while the variable f ranges over infinite sequences of objects of type σ, which is to say that f is a functional of type $0 \rightarrow \sigma$. Given such an f, let $\overline{f}(k)$ to denote the initial segment of f of length k, i.e. the finite sequence

$$\langle f(0), f(1), \ldots, f(k-1)\rangle.$$

Finally, if u is an element of type σ, $c\hat{\ }u$ denotes the sequence obtained by appending u to c.

6.3.1. Principle of bar induction at type σ. *Suppose φ and ψ are predicates of finite sequences of objects of type σ, with the following four properties:*

1. $\forall f \, \exists k \, \psi(\overline{f}(k))$
2. $\forall c \, (\psi(c) \vee \neg\psi(c))$
3. $\forall c \, (\psi(c) \rightarrow \varphi(c))$
4. $\forall c \, (\forall u \, \varphi(c\hat{\ }u) \rightarrow \varphi(c))$

Then φ holds at the empty sequence, $\langle\rangle$.

To make sense of this principle, imagine the tree of all finite sequences of objects of type σ. Clauses (1–3) imply that every path through the tree passes through a node c where ψ holds, and hence φ holds as well. The first nodes c where $\psi(c)$ hold form, in Brouwer's terminology, a "bar." Clause (4) asserts that if φ holds at every child of a node, then it holds at the node as well, allowing the property φ to "percolate" towards the root.

The principle of bar induction may be read as a statement of induction over a well-founded relation. Classically, it follows from an appropriate axiom of dependent choices. Assuming Clause (4), the failure of φ to hold at the empty sequence would allow one to construct an infinite sequence

$$c_0, c_1, c_2, \dots$$

with the property that φ does not hold at any initial segment $\langle c_0, \dots c_{k-1}\rangle$. Defining the function f by

$$f(k) =_{\text{def}} c_k$$

would then provide a counterexample to (1) and (3).

In the special case in which σ is the type of natural numbers, Principle 6.3.1 is Kleene's exposition of Brouwer's "bar theorem" (cf. Kleene and Vesley [1965]). The generalization to higher types is due to Spector.

To express this principle in the language of $HA^{\#}$, note that we may identify each sequence $\langle c_0, c_1, \dots, c_{k-1}\rangle$ with the pair C, k, where C is of type $0 \to \sigma$ and

$$C(x) = \begin{cases} c_x & \text{if } x < k \\ 0^\sigma & \text{otherwise.} \end{cases}$$

(Inductively one can define for each σ a constant "zero" functional denoted by 0^σ.) References to such c can then be taken as shorthand for references to such pairs C, k. For this representation we will use $\hat{c}$ and length(c) to denote C and k respectively.

Let (BI_σ) denote the principle of bar induction for sequences of objects of type σ, and let (BI) denote the same principle for arbitrary σ. We will see in section 6.5 that the theory $HA^{\#} + (BI)$ is in fact strong enough to interpret $PA^\omega + (AC_{00})$, and hence analysis. First, however, we show that bar induction has a computational analogue in a form of recursion, much in the way that arithmetic induction has its computational analogue in the recursors R of section 2.2.

6.4. The interpretation of bar induction using bar recursion

For each σ and τ, the principle of bar recursion uniformly associates with given functionals G, H, and Y a new functional F which maps finite sequences c of objects of type σ to objects of type τ, according to the defining equation

$$F(c) = \begin{cases} G(c) & \text{if } Y(\hat{c}) < \text{length}(c) \\ H(\lambda u.F(c\hat{\ }u), c) & \text{if } Y(\hat{c}) \geq \text{length}(c). \end{cases}$$

To make the uniformity clear, and in analogy to the recursions in section 2.2, one introduces a single functional $\mathrm{B}_{\sigma\tau}$ for each σ and τ and replaces F above by $\mathrm{B}_{\sigma\tau}(G, H, Y)$.

From the definition of F we see that if $Y(\hat{c}) <$ length(c) then $F(c)$ is defined outright, and otherwise $F(c)$ depends on the values $F(c\hat{\ }u)$, for arbitrary u. One should think of the values of F as being computed by recursion along a well-founded tree of sequences determined by Y, in a way that will be made more explicit below. Spector showed that the principle of bar induction can be used to justify bar recursion, when the argument Y is taken to be a continuous functional (in the Kleene-Kreisel sense referred to in section 4.1 and footnote 14).

The following informal argument gives a hint as to how bar induction comes into play. Given the functional F described above, we would like to see that the value $F(\langle\rangle)$ is "defined." Let $\psi(c)$ denote the property $Y(\hat{c}) < \text{length}(c)$, and let $\varphi(c)$ denote the property that $F(c)$ is defined. Clauses (3–4) of Principle 6.3.1 are clearly seen to hold, and in an appropriate formalization clause (2) can be justified intuitionistically as well.

Seeing that the well-foundedness condition (1) holds requires verifying that for every function f, there is a finite initial segment $c = \overline{f}(k)$ so that $Y(\hat{c}) < k$. Suppose $Y(f) = n$. The computation of Y on f can only depend on finitely many values of f, contained among the list $f(0), f(1), \ldots, f(m)$ for some m. In particular, if $k = \max(m, n+1)$ and $c = \overline{f}(k)$, then

$$Y(\hat{c}) = Y(f) = n < k = \text{length}(c),$$

as desired.

The following lemma shows that conversely, bar recursion can be used to justify bar induction. We use $HA^{\#} + (BR)$ to denote $HA^{\#}$ augmented by the principle of bar recursion for arbitrary types.

6.4.1. Lemma. *$HA^{\#} + (BR)$ proves the principle of bar induction, (BI).*

While the proof requires some effort, the underlying idea is straightforward. By Theorem 3.1.1, $HA^{\#}$ proves (BI) equivalent to its D-interpretation, which asserts the existence of various functionals. One shows how to define these functionals explicitly, using bar recursion in a key instance, and employing a trick due to Kreisel to verify that these functionals have the desired properties. We refer the reader to Howard [1968] for details.

On the other hand, by Theorem 3.1.2, $HA^{\#} + (BR)$ is clearly D-interpreted in $T + (BR)$. Combining this observation with the previous lemma yields

6.4.2. Theorem. *$HA^{\#} + (BI)$ is D-interpreted in $T + (BR)$.*

6.5. Interpreting $PA^{\omega} + (AC_{00})$

We have seen that full second-order arithmetic can be embedded in the theory $PA^{\omega} + (AC_{00})$. Spector's interpretation applies not only to this but more generally to $(AC_{0\sigma})$ and an even stronger axiom scheme, (DC_{σ}), which asserts the existence of sequences formed by making dependent choices:

$$\forall x, a\, \exists b\, \varphi(x, a, b) \rightarrow \exists f\, \forall x\, \varphi(x, f(x), f(x+1)),$$

where x is of type 0 and a and b are of type σ. Let (DC) denote the union of the (DC_σ). In section 6.3 we pointed out that, classically, (DC) can be used to justify bar induction. The following theorem represents a kind of converse, and shows the full strength of Spector's interpretation.

6.5.1. Theorem. *$PA^\omega + (DC)$ is N-interpreted in $HA^\# + (BI)$.*

By Theorem 3.1.1 this is tantamount to the assertion that the double-negation interpretation of (DC) is provable in the latter theory. Howard was able to obtain this result by first reducing (DC) to a special case of bar induction in an appropriate extension of PA^ω (which is plausible when one considers the contrapositive of (DC)), and then deriving the double negation of this special case in $HA^\# + (BI)$. The entire proof would take us too far afield, and so once again we refer the reader to Howard [1968] for details.

Together with Theorem 6.4.2 this yields

6.5.2. Corollary. *$PA^\omega + (DC)$ is ND-interpreted in $T + (BR)$.*

Since $PA^\omega + (AC_{00})$ is contained in $PA^\omega + (DC)$, we have

6.5.3. Corollary. *The consistency of $T + (BR)$ implies the consistency of $PA^2 + (CA)$. Moreover, every provably total recursive function of $PA^2 + (CA)$ is represented by a type 1 term of $T + (BR)$.*

We have both a functional and a term model for $T + (BR)$. The former is given by the continuous functionals of Kleene and Kreisel indicated in section 4.1 and footnote 14, with the constants interpreted by recursively continuous functionals. The latter is established by work of Tait [1971] and independently Luckhardt [1973], which shows that $T + (BR)$ is normalizing and confluent. Both models can be formalized in $PA^2 + (CA)$, thus proving

6.5.4. Theorem. *The provably total recursive functions of $PA^2 + (CA)$ are exactly the ones represented by bar-recursive terms.*

6.6. Evaluation of Spector's interpretation

Spector was careful not to claim that the generalization of bar induction to higher types, which he used to justify bar recursion for continuous functionals, should be accepted on intuitionistic grounds. In fact, he offers the following caveat:

> The author believes that the bar theorem is itself questionable, and that until the bar theorem can be given a suitable foundation, the question of whether bar induction is intuitionistic is premature.

The question of whether bar recursion can be justified on constructive grounds was taken up in a seminar on the foundations of analysis led by G. Kreisel at Stanford in the summer of 1963. The seminar's conclusion, summarized by Kreisel in an ensuing report, was that

> ... the answer is negative by a wide margin, since not even bar recursion of type 2 can be proved consistent [by constructively accepted principles].

Despite this disappointing assessment, we feel that Spector's reduction of all of classical analysis to the *prima facie* simple computational construct of bar recursion is, in the end, rather remarkable.

7. Conservation results for weak König's lemma

7.1. The theory WKL_0

In this section we study a subsystem of second-order arithmetic known as WKL_0. WKL_0 is an interesting theory because it is just strong enough to prove, among other things, the Heine-Borel theorem, and the completeness and compactness of first-order logic (cf. Simpson [1987]). These facts make it all the more surprising that from a proof-theoretic standpoint the theory is fairly weak.

Formally WKL_0 is the fragment of $PA^2 + (CA)$ in which induction is restricted to Σ_1^0 formulas (set parameters are allowed), and instead of full comprehension the only set existence principles are given by a recursive comprehension scheme, (RCA), which asserts the existence of Δ_1^0-definable sets, and a weak version of König's lemma. The latter, denoted (WKL), asserts that every infinite binary tree has a path.

In order to express (WKL), let $\{0,1\}^k$ (resp. $\{0,1\}^{<\omega}$, $\{0,1\}^\omega$) denote the set of length-k (resp. finite, infinite) binary sequences, and fix a reasonable encoding of finite binary sequences as natural numbers. If s and t are sequences so coded, let $\text{length}(s)$ denote the length of s and let the primitive recursive predicate $t \subseteq s$ assert that t is an initial segment of s. If b is an element of $\{0,1\}^\omega$, let $\bar{b}$ denote the initial segment function

$$\bar{b}(x) =_{\text{def}} \langle b(0), b(1), \ldots, b(x-1) \rangle.$$

Finally, define the predicates

$$\text{BinFunc}(b) \equiv_{\text{def}} \forall x\ (b(x) = 0 \vee b(x) = 1)$$

which asserts that b is an element of $\{0,1\}^\omega$,

$$\text{BinTree}(f) \equiv_{\text{def}} \forall s \in f\ (s \in \{0,1\}^{<\omega} \wedge \forall t \subseteq s\ (t \in f))$$

which asserts that f is a binary tree, that is, a set of binary sequences closed under initial segments, and

$$\text{Bounded}(f, k) \equiv_{\text{def}} \forall s \in \{0,1\}^k\ (s \notin f)$$

which asserts that the height of the binary tree f is less than or equal to k. Weak König's lemma can be now be written

$$\forall f\ (\text{BinTree}(f) \wedge \forall k\ \neg \text{Bounded}(f, k) \rightarrow \exists b\ (\text{BinFunc}(b) \wedge \forall k\ \bar{b}(k) \in f)).$$

In words, if f is a binary tree with branches at every level, then there is a path b through f.

Although the predicate Bounded is primitive recursive (since there are only finitely many binary sequences of length k), both BinFunc and BinTree require a universal quantifier. To avoid these, it turns out to be useful to employ the following trick. Given a function b, define b^{bin} by

$$b^{\text{bin}} =_{\text{def}} \begin{cases} 0 & \text{if } b(x) = 0 \\ 1 & \text{otherwise,} \end{cases}$$

denoting the casting of b to a binary function. (Formally, bin is a functional of type $(0 \to 0) \to (0 \to 0)$.) Similarly, define f^{tree} so that for any s

$$s \in f^{\text{tree}} \leftrightarrow (s \in \{0,1\}^{<\omega} \wedge \forall t \subseteq s \ (t \in f)),$$

so that f^{tree} is a binary tree obtained from f by pruning away extraneous sequences. The theory $\widehat{HA}^{\omega}$ can then prove

$$\text{BinFunc}(b^{\text{bin}}) \wedge (\text{BinFunc}(b) \to b = b^{\text{bin}})$$

and, similarly,

$$\text{BinTree}(f^{\text{tree}}) \wedge (\text{BinTree}(f) \to f = f^{\text{tree}}).$$

As a result, when we are working in the language of $\widehat{HA}^{\omega}$, we can just as well take (*WKL*) to be given by

$$\forall f \ (\forall k \ \neg\text{Bounded}(f^{\text{tree}}, k) \to \exists b \ \forall k \ (\overline{b^{\text{bin}}}(k) \in f^{\text{tree}})).$$

Since Δ^0_1 comprehension follows from (*QF-AC*), we can take WKL_0 to be contained in

$$\widehat{PA}^{\#} + (WKL).$$

The fact that WKL_0 is proof-theoretically weak is shown by the following celebrated theorem of H. Friedman.

7.1.1. Theorem. *WKL_0 is conservative over PRA for Π^0_2 formulas.*

While Friedman's original proof of Theorem 7.1.1 was model-theoretic, Kohlenbach [1992] showed how to use a D-interpretation to obtain the same result. In fact, Kohlenbach's work was dedicated to somewhat more general results; the approach we present here is a simplification of his methods applied to the specific case at hand. The details are manageable enough so that we can present them here in full. (The realization that hereditary majorizability, defined in the next section, can be used to obtain Theorem 7.1.1 is due to Sieg [1985], who used it with Herbrand methods. For other proof-theoretic approaches, as well as Harrington's strengthening of this theorem, see Hájek [1993], Avigad [1996a].)[15]

[15] A number of other interesting applications of the technique of majorization in combination with functional interpretation have been made by Kohlenbach in a series of papers; cf. Kohlenbach [1996a,1996b] among others.

7.2. Hereditary majorizability

If one puts the discrete topology on the set $\{0,1\}$, weak König's lemma expresses the compactness of $\{0,1\}^\omega$ under the associated product topology. Recall that $\widehat{T}$ has models in which functionals F of type $(0 \to 0) \to 0$ are computable, and hence continuous, since, for any g, the computation of $F(g)$ depends on only finitely many values of g. As a result, the compactness of $\{0,1\}^\omega$ implies that any such F is necessarily bounded when restricted to functions in this space.

The notion of hereditary majorizability, due to Howard [1973], is an effective generalization of this observation.

7.2.1. Definition. For each type σ, we define a relation $a \leq_* b$ for terms a and b of type σ, as follows.

1. If $\sigma = 0$, $a \leq_* b$ is just $a \leq b$.
2. If $\sigma = (\tau \to \rho)$, then $a \leq_* b$ if and only if

$$\forall x, y\ (x \leq_* y \to a(x) \leq_* b(y))$$

where the variables x and y are of type τ.

The relation $a \leq_* b$ is read "a is hereditarily majorized by b."

Notice that

$$\forall f\ (\text{BinFunc}(f) \leftrightarrow f \leq_* \lambda x.1).$$

It is not difficult to majorize type 1 functions:

7.2.2. Proposition. *Given a type 1 term f, define*

$$f^*(x) =_{\text{def}} \max_{y \leq x} f(y).$$

Then $f \leq_* f^*$.

Though we cannot prove in $\widehat{T}$ that for every functional F there is another functional G that majorizes it, we *can* majorize *closed* terms.

7.2.3. Proposition. *For every closed term F in the language of $\widehat{T}$ there is another closed term G such that $\widehat{HA}^\omega$ proves $F \leq_* G$.*

7.2.4. Proof. Inductively, for each term $F[x_1, \ldots, x_n]$ with free variables shown, one constructs a term $G[y_1, \ldots, y_n]$ such that $\widehat{T}$ proves

$$x_1 \leq_* y_1 \wedge \ldots \wedge x_n \leq_* y_n \to F \leq_* G.$$

The main case is for $F = \hat{\text{R}}$. Here we can take G defined by

$$\begin{aligned} G(f, g, 0, b) &= f(b) \\ G(f, g, n', b) &= \max(G(f, g, n, b), g(n, G(f, g, n, b), b)). \end{aligned}$$

See Howard [1973] for further details. □

Notice that if F is a closed term of type $(0 \to 0) \to 0$, $F \leq_* G$, and $k = G(\lambda x.1)$, then Definition 7.2.1 and the remark immediately following imply that for every b in $\{0,1\}^\omega$, we have that $F(b) < k$.

The following technical lemma will be needed in section 7.4.

7.2.5. Lemma. *If the term B is of type $(0 \to 0) \to (0 \to 0)$, then $\widehat{T}$ proves*

$$\forall f \; \mathrm{BinFunc}(B(f)) \leftrightarrow B \leq_* \lambda f \lambda x.1.$$

7.2.6. Proof. $B \leq_* \lambda f \lambda x.1$ is equivalent to

$$\forall f, g \; (f \leq_* g \to B(f) \leq_* \lambda x.1),$$

that is,

$$\forall f, g \; (f \leq_* g \to \mathrm{BinFunc}(B(f))). \tag{1}$$

Taking $g = f^*$ from Proposition 7.2.2 we see that (1) implies $\forall f \; \mathrm{BinFunc}(B(f))$, and the converse is trivial. □

7.3. Reducing WKL_0 to $\widehat{HA}^{\#} + (WKL')$

Since WKL_0 is contained in $\widehat{PA}^{\#} + (WKL)$, to obtain Theorem 7.1.1 it is sufficient, by Theorems 5.1.3 and 5.1.1, to show that the latter theory is conservative over $\widehat{HA}^{\#}$ for Π^0_2 sentences. Our first step is to reduce it to an intuitionistic variant.

7.3.1. Lemma. *The theory $\widehat{PA}^{\#} + (WKL)$ is N-interpreted in $\widehat{HA}^{\#} + (WKL)$.*

7.3.2. Proof. $\widehat{PA}^{\#}$ is N-interpreted in $\widehat{HA}^{\#}$, and the double-negation interpretation of (WKL), given by

$$\forall f \; (\forall k \; \neg \mathrm{Bounded}(f^{\mathrm{tree}}, k) \to \neg\neg \exists b \; \forall k \; (\overline{b^{\mathrm{bin}}}(k) \in f^{\mathrm{tree}})),$$

follows from (WKL) since the double-negation only weakens the conclusion. □

We define a convenient variant of (WKL) as follows:

(WKL') $\qquad \forall f \; \exists b \; \forall k \; (\neg \mathrm{Bounded}(f^{\mathrm{tree}}, k) \to \overline{b^{\mathrm{bin}}}(k) \in f^{\mathrm{tree}}).$

Proving the following lemma is a simple exercise in intuitionistic logic.

7.3.3. Lemma. *Over intuitionistic logic, (WKL) is implied by (WKL').*

In fact, the two principles are equivalent over $\widehat{HA}^{\#}$, but the converse direction requires more work and is not needed below.

7.4. Reducing $\widehat{HA}^{\#} + (WKL')$ to $\widehat{HA}^{\#}$

Though $\widehat{HA}^{\#}$ doesn't prove (WKL), we can show that adding (WKL) doesn't allow $\widehat{HA}^{\#}$ to prove any new Π^0_2 sentences. The main avenue to this result is abstracted in the following

7.4.1. Lemma. *Suppose $\widehat{HA}^{\#}$ proves*

$$\exists a\, \forall b, c\; S(a,b,c) \to \forall x\, \exists y\; R(x,y). \tag{2}$$

Then there is a specific term $\tilde{c}(a,x)$ (whose other free variables are among those of R and S) such that $\widehat{HA}^{\#}$ proves

$$\forall x\, \exists a\, \forall b\; S(a,b,\tilde{c}(a,x)) \to \forall x\, \exists y\; R(x,y).$$

The proof is straightforward, using the scheme $\varphi \leftrightarrow \varphi^D$ of $\widehat{HA}^{\#}$ to convert (2) with "$\forall x$" deleted to its D-translation, applying Theorem 5.1.3 to extract a term $\tilde{c}$ of $\widehat{T}$, and then manipulating quantifiers. The upshot is that to eliminate the assumption $\exists a\, \forall b, c\; S(a,b,c)$ from a proof of $\forall x\, \exists y\; R(x,y)$ in $\widehat{HA}^{\#}$, it suffices to show that one can prove

$$\exists a\, \forall b\; S(a,b,\tilde{c}(a,x))$$

for any specific term $\tilde{c}$.

We now apply this to the situation at hand.

7.4.2. Lemma. *$\widehat{HA}^{\#} + (WKL)$ is conservative over $\widehat{HA}^{\#}$ for Π^0_2 sentences.*

7.4.3. Proof. By Lemma 7.3.3 and the deduction theorem, if $\widehat{HA}^{\#} + (WKL)$ proves

$$\forall x\, \exists y\; R(x,y),$$

then $\widehat{HA}^{\#}$ proves

$$\forall f\, \exists b\, \forall k\, (\neg \mathrm{Bounded}(f^{\mathrm{tree}}, k) \to \overline{b^{\mathrm{bin}}}(k) \in f^{\mathrm{tree}}) \to \forall x\, \exists y\; R(x,y).^{16}$$

Applying (AC) to the hypothesis, we obtain

$$\exists B\, \forall f, k\, (\neg \mathrm{Bounded}(f^{\mathrm{tree}}, k) \to \overline{B(f)^{\mathrm{bin}}}(k) \in f^{\mathrm{tree}}) \to \forall x\, \exists y\; R(x,y).$$

Applying Lemma 7.4.1 with k in place of c, we are reduced to showing that $HA^{\#}$ proves

$$\exists B\, \forall f\, (\neg \mathrm{Bounded}(f^{\mathrm{tree}}, \tilde{k}(B,x)) \to \overline{B(f)^{\mathrm{bin}}}(\tilde{k}(B,x)) \in f^{\mathrm{tree}}) \tag{3}$$

[16] Here we need to use $\widehat{HA}^{\#}_0$ or $I\text{-}\widehat{HA}^{\#}$ instead of $WE\text{-}\widehat{HA}^{\#}$, since the deduction theorem fails for the latter theory; cf. 3.1. However, for a way around this, see Kohlenbach [1992], p. 1246.

for any closed term $\tilde{k}$.

We now bring in the notion of hereditary majorizability to bound the value of $\tilde{k}(B,x)$. By Proposition 7.2.3, we can find a term k^* that hereditarily majorizes $\tilde{k}$. Define

$$k'(x) =_{\text{def}} k^*(\lambda x \lambda f.1, x).$$

Since $x \leq_* x$, from Lemma 7.2.5 we see that $\widehat{HA}^{\#}$ proves

$$\forall f \ \text{BinFunc}(B(f)) \to \tilde{k}(B,x) \leq k'(x). \tag{4}$$

To verify (3), we only need construct an appropriate B. Define $B'(x,f)$ as follows. Let l be the maximum value less than or equal to $k'(x)$ such that f^{tree} has an element of length l, and let s be any (e.g. the leftmost) such element. Let

$$B'(x,f)(y) =_{\text{def}} \begin{cases} s_y & \text{if } y < l \\ 0 & \text{otherwise,} \end{cases}$$

so that $B'(x,f)$, as an element of $\{0,1\}^\omega$, consists of s followed by a string of zeros. In particular, $\widehat{HA}^{\#}$ proves

$$l \leq k'(x) \wedge \neg\text{Bounded}(f^{\text{tree}}, l) \to \overline{B'(x,f)}(l) \in f^{\text{tree}}. \tag{5}$$

Finally, let $B(x) =_{\text{def}} \lambda f.B'(x,f)$. Then $\widehat{HA}^{\#}$ proves

$$\forall f \ \text{BinFunc}(B(x,f))$$

and so, using (4),

$$\forall x \ (\tilde{k}(B(x),x) \leq k'(x)).$$

By (5) we then have

$$\forall f \ (\neg\text{Bounded}(f^{\text{tree}}, \tilde{k}(B(x),x)) \to \overline{B(x,f)^{\text{bin}}}(\tilde{k}(B(x),x)) \in f^{\text{tree}}).$$

So $B(x)$ witnesses the existential quantifier in (3). □

Friedman's Theorem 7.1.1 now follows from Lemmas 7.3.1 and 7.4.2, together with Theorems 5.1.3 and 5.1.1.

Harrington's theorem states that *WKL₀* is conservative over its subtheory *RCA₀* (which omits (*WKL*)) for Π^1_1 sentences. Since *RCA₀* is easily interpretable in $I\Sigma_1$, this, together with Theorem 5.1.4, yields another proof of Friedman's result. However, the question as to whether a functional interpretation can be used to obtain Harrington's strengthening is still open.

The proof above can be adapted to yield similar conservation results for weaker fragments of second-order arithmetic, such as "elementary analysis," whose provably total recursive functions are all elementary recursive (cf. also Kohlenbach [1996b]). It is reasonable to conjecture that functional interpretations can be used to obtain similar results for systems of polynomial-time-computable arithmetic, especially in view of the conservation result of Ferreira [1988,1994], which showed that a suitable version of (WKL) is conservative over S^1_2 for Π^0_2 formulas (cf. also Cantini [1996]). However, the functional B′ and the relation Bounded(f, k) from the proof above cannot be defined in, say, PV^ω, so new considerations seem to be necessary. (If the reader is concerned with polynomial *bounds* rather than a polynomial-time-computable Skolem function, he or she should consult Corollary 4.28 of Kohlenbach [1996b].)

8. Non-constructive interpretations and applications

8.1. Overview of the section; general pattern.

This section shows how the D-interpretation can be extended by inclusion of certain non-constructive functional operators in order to obtain conservation results of various finite type systems contained in $PA^\omega + (AC)$ over related second-order systems. As we have noted in section 3.1, (QF-AC) is automatically preserved under the ND-interpretation. This can be extended to (Γ-AC) for some larger classes of formulas Γ by adjunction of suitable (necessarily non-constructive) Skolem functionals with associated axioms which imply that each formula in Γ is equivalent to a QF- (i.e. quantifier-free) formula.

The paradigm case is given by adjunction of the non-constructive minimum operator μ, which allows us to reduce every arithmetical formula to QF form. Then axioms (μ) for this with $PA^\omega + (QF\text{-}AC)$ imply the second-order system Σ^1_1-DC, and it is shown that the system $PA^\omega + (\mu) + (QF\text{-}AC)$ is ND-interpreted in $T + (\mu)$. Normalization of a system of infinite terms for the latter, just as described in section 4.4 above for T by the methods of Tait [1965], shows that the type 1 functions thus generated lie in the hierarchy of hyperarithmetic functions H_α for $\alpha < \varepsilon_0$. Then formalization of this model can be carried out in a theory $(\Pi^0_1\text{-}CA)_{<\varepsilon_0}$ of relative arithmetical comprehension iterated up to ε_0, yielding a number of conservation results, including a well-known one due to Friedman [1970]. The arguments for this are detailed in sections 8.2 and 8.3. It is also shown that when the Bar Rule is adjoined to the above finite type system, we obtain conservation over the full predicative system $(\Pi^0_1\text{-}CA)_{<\Gamma_0}$.

Analogous results to those in sections 8.2–8.3 are sketched in section 8.4 for adjunction of the Kleene basis operator which transforms every Σ^1_1 formula into an equivalent Π^0_1 formula and thence a QF-formula using μ. This allows us to lift the preceding results throughout one step up in the analytical hierarchy. The results of sections 8.2–8.4 are due to Feferman [1971,1977,1979].

8.2. Finite type systems with non-constructive μ-operator, and related second-order systems

8.2.1. Subsystems of PA^ω

Besides the classical finite type theory PA^ω which includes the constants K, S and R in all types and full induction for all formulas, we shall consider also its subsystems $\widehat{PA}^\omega$ in which R is replaced by the predicative recursion operator $\widehat{\mathrm{R}}$ introduced in section 5.1, and subsystems of both, obtained by restricting induction as follows:

$$(\mathit{Res\text{-}Ind}) \qquad 0 \in X \wedge \forall x\,(x \in X \to x' \in X) \to \forall x\,(x \in X).$$

Here X ranges over sets of type 1 (identified with characteristic functions as in section 6.3). When we add principles which imply (Γ-CA) for various classes of formulas Γ, we can infer from ($\mathit{Res\text{-}Ind}$) all substitution instances for X by formulas φ in Γ, but not full induction in general. If S is any system of second or higher order which includes the scheme of full induction, we shall denote by $\mathit{Res\text{-}S}$ the system obtained by replacing that scheme by the axiom ($\mathit{Res\text{-}Ind}$) above. (In the literature this is also denoted $S\!\restriction$ or S_0.) The higher type systems which will figure prominently in the following are PA^ω, $\mathit{Res\text{-}PA}^\omega$, $\widehat{PA}^\omega$ and $\mathit{Res\text{-}}\widehat{PA}^\omega$, together with ($QF$-$AC$) and some special axioms for non-constructive functionals.

8.2.2. Functionals for numerical quantification and choice

The functional E_0 of type 2 with values 0 and 1 only is defined by

$$(\mathrm{E}_0) \qquad \mathrm{E}_0(f) = 0 \leftrightarrow \exists x\,(f(x) = 0).$$

(This functional is also denoted ${}^2E$ in the recursion-theoretic literature.) Using this axiom, every arithmetical formula is equivalent to a QF-formula. If we are to use it with an ND-interpretation, we should consider the two implications:

$$f(x) = 0 \to \mathrm{E}_0(f) = 0 \quad \text{and} \quad \mathrm{E}_0(f) = 0 \to \exists x\,(f(x) = 0).$$

The ND-interpretation preserves the first of these but requires for the second a functional X such that $\mathrm{E}_0(f) = 0 \to f(X(f)) = 0$. Combining this with the first implication gives $f(x) = 0 \to f(X(f)) = 0$. Such an X is then a choice or Skolem functional for type 0 quantification. We shall use the new symbol μ for such a functional and take as its axiom

$$(\mu) \qquad f(x) = 0 \to f(\mu(f)) = 0.$$

Then E_0 may be defined simply in terms of μ by $\mathrm{E}_0(f) = 0$ if $f(\mu(f)) = 0$, otherwise $\mathrm{E}_0(f) = 1$. The advantage of using μ in place of E_0 is that its axiom is preserved directly under the ND-interpretation without requiring any supplementary functionals. Note that the non-constructive minimum operator in its usual sense, defined as the least x such that $f(x) = 0$ if $\exists x\,(f(x) = 0)$ and 0 otherwise, is also definable in terms of μ and the primitive recursive bounded minimum operator.

8.2.3. Second-order forms of (CA), (AC) and (DC)

We shall consider various restricted forms of these principles for the usual arithmetical (Π^0_n and Σ^0_n) and analytical (Π^1_n and Σ^1_n) hierarchies. That is, for Γ one of these classes of formulas, we shall consider the schemata

$$(\Gamma\text{-}CA) \qquad \exists Y\, \forall x\, (x \in Y \leftrightarrow \varphi(x))$$

$$(\Gamma\text{-}AC_{01}) \qquad \forall x\, \exists f\, \varphi(x, f) \rightarrow \exists g\, \forall x\, \varphi(x, g_x)$$

$$(\Gamma\text{-}DC_{1}) \qquad \forall x, f\, \exists g\, \varphi(x, f, g) \rightarrow \forall f\, \exists g\, (g_0 = f \wedge \forall x\, \varphi(x, g_x, g_{x+1}))$$

where in each case φ ranges over formulas in Γ, and in the two choice principles g_x is written for $\lambda y.g(x,y)$. In addition, we shall consider the scheme

$$(\Delta_\Gamma\text{-}CA) \qquad \forall x\, (\varphi(x) \leftrightarrow \neg\psi(x)) \rightarrow \exists Y\, \forall x\, (x \in Y \leftrightarrow \varphi(x))$$

where φ and ψ again range over formulas in Γ. In the context of second-order systems, we omit the subscripts from the (AC) and (DC) principles. Also we use these principles to name the second-order system S obtained by adjoining them to PA together with full induction for second-order formulas. Then, in accordance with 8.2.1, we use $Res\text{-}S$ to name the same system with restricted induction. The systems of particular concern to us in this section and the next are: $\Pi^0_1\text{-}CA$, $\Delta^1_1\text{-}CA$, $\Sigma^1_1\text{-}AC$, $\Sigma^1_1\text{-}DC$, and their restricted versions, while in section 8.4 we shall take up the corresponding systems one level up in the analytic hierarchy. Note that the system $\Pi^0_1\text{-}CA$ proves the same theorems as the system ACA based on the arithmetical comprehension axiom, i.e. $(\Gamma\text{-}CA)$ where Γ is the class of arithmetical formulas. For, we can derive closure under complement from $(\Pi^0_1\text{-}CA)$ and hence the principle $(\Sigma^0_1\text{-}CA)$, and then by iterating these we obtain (ACA) in general. We note this for the record in the following; the further relationships below are established just as easily.

8.2.4. Theorem.

1. $\Pi^0_1\text{-}CA = ACA$
2. $\Pi^0_1\text{-}AC = \Sigma^1_1\text{-}AC$
3. $\Pi^0_1\text{-}DC = \Sigma^1_1\text{-}DC$
4. $\Pi^0_1\text{-}CA \subseteq \Delta^1_1\text{-}CA \subseteq \Sigma^1_1\text{-}AC \subseteq \Sigma^1_1\text{-}DC$
5. *The same results hold for the corresponding theories with* $(Res\text{-}Ind)$.

Apropos of this last, it is a familiar result that $Res\text{-}\Pi^0_1\text{-}CA$ or ACA_0 is a conservative extension of PA; a model-theoretic argument gives the quickest proof of that.

8.2.5. Transfinite induction below ε_0.

In the following we shall use variables $\alpha, \beta, \gamma, \ldots$ to range both over ordinals and over terms in a natural recursive ordinal representation system for ordinals up to ε_0. We write $<_\alpha$ for the initial segment determined in this ordering by α. The scheme of transfinite induction up to α for a formula $\varphi(x)$ is given by

$$(TI_\alpha(\varphi)) \qquad \forall x\, (\forall y\, (y <_\alpha x \to \varphi(y)) \to \varphi(x)) \to \forall x\, \varphi(x).$$

It was proved by Gentzen that this scheme is derivable in *PA* for each formula φ of arithmetic and each $\alpha < \varepsilon_0$. Analysis of his argument actually shows more:

8.2.6. Theorem. *Suppose S is any system whose language contains that of PA and which contains full induction. Then for each formula φ of S and each $\alpha < \varepsilon_0$ we can prove $(TI_\alpha)(\varphi)$ in S.*

The argument for this proceeds by showing that transfinite induction up to ω^α with respect to φ can be reduced to transfinite induction up to α with respect to a formula φ^* which is built by propositional operations and numerical (first-order) quantification from φ. (For a simple choice of φ^* due to Schütte, cf. Feferman [1977,p. 946].)

8.2.7. Transfinitely iterated arithmetical comprehension

The hyperarithmetic hierarchy up to any α and relative to any initial set X is the sequence of sets $\langle H^X_\beta \rangle_{\beta<\alpha}$ for which $H^X_0 = X$, $(H^X_{\beta+1}) = (H^X_\beta)'$, and for limit β, H^X_β is the join of the sequence H^X_γ for $\gamma < \beta$. (Here Y' represents the set obtained by applying the Turing jump operator to Y.) Such a sequence may be represented by its join, which is a single set $H^X(\alpha)$ consisting of the pairs $\langle \beta, x \rangle$ such that $\beta < \alpha$ and $x \in H^X_\beta$. The assertion that $H^X(\alpha)$ exists for each X is denoted $(\Pi^0_1\text{-}CA)_\alpha$, since the successor step from β to $\beta + 1$ may be considered as being obtained by one application of $(\Pi^0_1\text{-}CA)$. We use $(\Pi^0_1\text{-}CA)_{<\varepsilon_0}$ to denote the collection of all $(\Pi^0_1\text{-}CA)_\alpha$ for $\alpha < \varepsilon_0$, as well as the system based on this collection. In these terms we can strengthen the first inclusion in part 4 of the Theorem 8.2.4 to the following

8.2.8. Theorem. $(\Pi^0_1\text{-}CA)_{<\varepsilon_0} \subseteq \Delta^1_1\text{-}CA$.

The proof consists in showing, for each $\alpha < \varepsilon_0$, the existence of $H^X(\beta)$ for $\beta < \alpha$ by the scheme of transfinite induction up to α. When passing to the limit β one makes use of the fact that if $H^X(\gamma)$ exists then it is uniquely determined by its recursive defining conditions, so the set $H^X(\beta)$ can be defined in Δ^1_1 form (cf. Feferman [1977,pp. 944–947] for further details).

Note that what is essential for this proof is availability of transfinite induction for Σ^1_1 formulas in the system $\Delta^1_1\text{-}CA$, which we do not have in its restricted version. Our next step is to put these second-order systems together with the finite-type systems of 8.2.1. This leads to two chains of interest:

8.2.9. Theorem.

1. $(\Pi^0_1\text{-}CA)_{<\varepsilon_0} \subseteq \Delta^1_1\text{-}CA \subseteq \Sigma^1_1\text{-}AC \subseteq \Sigma^1_1\text{-}DC \subseteq PA^\omega + (\mu) + (QF\text{-}AC)$
2. $PA \subseteq Res\text{-}\Pi^0_1\text{-}CA \subseteq Res\text{-}\Delta^1_1\text{-}CA \subseteq Res\text{-}\Sigma^1_1\text{-}AC$
$\subseteq Res\text{-}\widehat{PA}^\omega + (\mu) + (QF\text{-}AC)$.

What's added here to the information from above are the final inclusions. In 1, from $\forall x, f\ \exists g\ \varphi(x, f, g)$ with φ arithmetical, one uses (μ) and $(QF\text{-}AC)$ to infer

$$\exists G\ \forall x, f\ \varphi(x, f, G(f)).$$

Then, given any initial f, a sequence $\langle g_x \rangle$ with $g_0 = f \wedge \forall x\ \varphi(x, g_x, g_{x+1})$ is defined by the recursion $g_{x+1} = G(g_x)$ using the recursor R with values at type 1. The reason the same sort of argument cannot be carried out in 2 is that there we only have availability of the recursor $\widehat{\mathrm{R}}$ whose values are confined to type 0. Recursion is not needed though to infer the principle $(\Sigma^1_1\text{-}AC)$ from (μ) and $(QF\text{-}AC)$.

The next section is devoted to showing how these chains of inclusions can be closed proof-theoretically.

8.3. Functional interpretations using the (μ) operator

As noted above, the (μ) axiom is preserved under the ND-interpretation; in combination with the Theorems 3.1.3 and 3.1.4, this immediately yields:

8.3.1. Theorem. *$PA^\omega + (\mu) + (QF\text{-}AC)$ is ND-interpreted in $T + (\mu)$.*

The next step is to show how $T + (\mu)$ can be modeled in $(\Pi^0_1\text{-}CA)_{<\varepsilon_0}$. This is by an extension to μ, as a type 2 constant symbol, of Tait's infinite term model of T from section 4.4. No new arguments are needed for normalization and so, just as before, every term of $T + (\mu)$ translates into a (possibly) infinitely long term which, in turn, reduces effectively to a normal term t with $|t| < \varepsilon_0$. Only the description of normal infinitely long terms t at type 0 needs to be modified. For this purpose, one examines such t which may contain free variables x of type 0 and/or f of type 1, and arrives at the following possibilities (cf. the end of 4.4): either

1. $t = 0$ or
2. $t = Sc(s)$ where s is normal or
3. t is a variable of type 0 or
4. $t = \langle t_n \rangle(s)$ where each t_n is normal and s is normal or
5. $t = f(s)$ where f is a type 1 variable and s is normal or
6. $t = \mu(\lambda x.s)$ where s is normal or
7. $t = \mu(\langle t_n \rangle)$ where each t_n is normal.

Every closed term of type 1 in $T + (\mu)$ then reduces effectively to a normal term of the form $\lambda x.t$ where t is normal of type 0 or of the form $\langle t_n \rangle$ where each t_n is normal. What these represent as functions is the effective transfinite iteration of the μ operator, or equivalently the jump (E_0) operator, up to ordinals less than ε_0. Thus each closed term t of type 1 in $T + (\mu)$ represents a function which is recursive in H_α for some $\alpha < \varepsilon_0$. More generally, if t is a term with a free variable f of type 1, then t represents a function recursive uniformly in H_α^f for some $\alpha < \varepsilon_0$. Formalization of this argument leads finally to the following:

8.3.2. Theorem. *$PA^\omega + (\mu) + (QF\text{-}AC)$ is a conservative extension of $(\Pi_1^0\text{-}CA)_{<\varepsilon_0}$ for Π_2^1 sentences.*

For, given a provable Π_2^1 sentence $\forall f\ \exists g\ \varphi(f,g)$ with φ arithmetical, we use 8.3.1 to obtain a term $t[f]$ of type 1 such that $T + (\mu)$ proves $\varphi'(f, t[f])$, where φ' is quantifier- free and equivalent to φ under (μ), and then apply the preceding modeling argument.

8.3.3. Corollary. *$\Sigma_1^1\text{-}DC$ is a conservative extension of $(\Pi_1^0\text{-}CA)_{<\varepsilon_0}$ for Π_2^1 sentences.*

This result is due originally to Friedman [1970], who established it by model-theoretic techniques. Subsequently to the appearance of the above approach in the publications referred to in section 8.1, Feferman and Sieg [1981] used Herbrand-Gentzen methods with the μ operator to the same end.

Now turning to the restricted systems, the main result is

8.3.4. Theorem. *$Res\text{-}\widehat{PA}^\omega + (\mu) + (QF\text{-}AC)$ is a conservative extension of $Res\text{-}\Pi_1^0\text{-}CA$ for Π_2^1 sentences, and hence of PA for arithmetical sentences.*

Here, for the proof, one makes use of the ND-interpretation of our higher type system in $\widehat{T} + (\mu)$. The recursor $\widehat{\mathrm{R}}$ can then be eliminated from the picture, since it can be defined arithmetically, hence by means simply of $+$, $\cdot$, and μ. Terms of the resulting system then normalize without passing to infinite terms, and so the functions represented by closed terms of type 1 are also arithmetical. The same holds for terms $t[f]$ of type 1 uniformly in a type 1 variable f.

8.3.5. Corollary. *$Res\text{-}\Sigma_1^1\text{-}AC$ is a conservative extension of $Res\text{-}\Pi_1^0\text{-}CA$ for Π_2^1 sentences, and hence of PA for arithmetical sentences.*

This was first established by model-theoretic methods by Barwise and Schlipf [1975] and independently Friedman [1976]. Again, Feferman and Sieg [1981] used Herbrand-Gentzen methods with μ to obtain the same result. Note that $Res\text{-}\Sigma_1^1\text{-}DC$ is stronger than $Res\text{-}\Pi_1^0\text{-}CA$ because it proves the existence of H_ω.

8.3.6. Predicatively provable ordinals

By formalizing work of Kleene, the systems $(\Pi^0_1\text{-}CA)_{<\alpha}$ are another form of ramified analysis RA_α when $\alpha = \omega \cdot \alpha$. The proof-theoretic ordinals[17] of these were established by Schütte [1960] (cf. also Schütte [1977]) in terms of the Veblen hierarchy of normal functions φ_α of ordinals, defined by:

1. $\varphi_0(\xi) = \omega^\xi$
2. For $\alpha \neq 0$, φ_α enumerates $\{\xi : \varphi_\beta(\xi) = \xi \text{ for all } \beta < \alpha\}$.

Each $\varphi_{\alpha+1}$ is then the critical function of φ_α, i.e. it enumerates the set of fixed points $\{\xi : \varphi_\alpha(\xi) = \xi\}$. In particular, $\varphi_1(\xi) = \varepsilon_\xi$, etc. The diagonal function $\lambda\xi.\varphi_\xi(0)$ is also normal and its least fixed point is denoted Γ_0. There is a natural notation system for ordinals up to Γ_0 (Feferman [1968b]), and we shall now use $\alpha, \beta, \gamma, \ldots$ to range over that system as well as over ordinals in the usual sense. It follows from Schütte's work referenced above that

8.3.7. Theorem. *For $\alpha = \omega \cdot \alpha$ (α not 0), the proof-theoretic ordinal of RA_α or, equivalently, of $(\Pi^0_1\text{-}CA)_{<\alpha}$, is $\varphi_\alpha(0)$.*

Now, the so-called *autonomous ordinals* of ramified analysis were defined to be those generated by the following "boot-strap" procedure:

1. 0 is an autonomous ordinal.
2. If β is an autonomous ordinal and α is a provably recursive ordinal of RA_β then α is an autonomous ordinal.

The systems RA_α for α autonomous were proposed by Kreisel as a characterization of predicative analysis, and so the autonomous ordinals are identified with the predicatively provable ordinals. Using 8.3.7, it was established independently by Schütte and Feferman in the mid 1960s that Γ_0 is the least non-autonomous ordinal, hence the least impredicative ordinal. In order to investigate the extent of predicativity in ordinary, unramified, analysis, Feferman [1964] described an unramified second-order system with finitely many schemata whose proof-theoretic ordinal is Γ_0. This was strengthened later in Feferman [1979] to the following:

8.3.8. Theorem. *The system $\Sigma^1_1\text{-}DC$ + (Bar-Rule) has proof-theoretic ordinal Γ_0; it is conservative over $(\Pi^0_1\text{-}CA)_{<\Gamma_0}$ for Π^1_2 sentences.*

The Bar-Rule, which is related to the principle of Bar Induction described in section 6.3 above, allows one to infer the scheme of transfinite induction on a recursive ordering when its well-foundedness has been established. The idea of the proof of Theorem 8.3.8 sketched op. cit. is to apply a functional interpretation to the still larger system $PA^\omega + (\mu) + (QF\text{-}AC) + (Bar\text{-}Rule)$, carrying it into $T + (\mu) + (Bar\text{-}Rule)$,

[17] The proof-theoretic ordinal of a system S is defined in several ways, e.g. as the least ordinal α which is not the order-type of a provably recursive well-ordering of the system, or as the least α (in a standard notation system) for which the consistency of S can be proved by transfinite induction up to α over a weak base system. These are equivalent in practice.

and then to model the latter in a boot-strap fashion in a system of infinitely long terms of length less than Γ_0. The final result is then obtained by formalization of the type 1 functions of that model as for Theorem 8.3.2 above. The details of this are rather complicated, and in its place a more digestible proof of Theorem 8.3.8 was subsequently obtained by Feferman and Jäger [1983] by Herbrand-Gentzen methods with the μ operator. But the functional interpretation served an important intermediate, partly heuristic role in arriving at this result.

8.4. Functional interpretations using the Kleene basis operator

8.4.1. Testing for (non-)well-foundedness as a functional

As in sections 6 and 7, we write $\overline{g}(x)$ for the sequence number coding the sequence $\langle g(0), \ldots, g(x-1)\rangle$ when g is of type 1. Then a type 1 function f represents a well-founded tree if $\forall g\ \exists x\ f(\overline{g}(x)) \neq 0$, where the tree consists of all sequence numbers s with $f(s) = 0$. Kleene's basis theorem for Σ^1_1 predicates provides a functional μ_1 which chooses a descending branch in this tree if it is not well-founded, recursive in the Suslin quantifier E_1 with values 0 and 1, given by

$$(\mathrm{E}_1) \qquad E_1(f) = 0 \leftrightarrow \exists g\ \forall x\ f(\overline{g}(x)) = 0.$$

Specifically, when $\mathrm{E}_1(f) = 0$, μ_1 is the left-most descending branch in this tree, with the successive values of $\mu_1(f) = g$ defined recursively by taking $g(x)$ to be the least y such that there is an infinite descending branch in the f-tree extending $\overline{g}(x)\hat{\ }\langle y\rangle$. Thus μ_1 satisfies the axiom

$$(\mu_1) \qquad \forall x\ f(\overline{g}(x)) = 0 \rightarrow \forall x\ f(\overline{\mu_1(f)}(x)) = 0.$$

Of course then E_1 is definable in terms of μ_1. This axiom is preserved under the N-interpretation, and in the presence of the axiom (μ) is equivalent to a quantifier-free statement,which is further preserved under the D-interpretation. So this leads us to consider the systems of 8.2.1 augmented by the axioms for both μ and μ_1. Then we obtain conservation results for these over second-order systems involving Π^1_1-CA, Σ^1_2-AC and Σ^1_2-DC. In addition, we have to consider the following.

8.4.2. Transfinitely iterated (Π^1_1-CA)

The explanation of the systems $(\Pi^1_1\text{-}CA)_\alpha$ and $(\Pi^1_1\text{-}CA)_{<\alpha}$ is analogous to that for the systems of iterated arithmetical comprehension. These formalize the existence of relative hierarchies based on iteration of the hyperjump operator, definable by E_1. Then in analogy to Theorem 8.2.9, we easily obtain the following chains of inclusions:

8.4.3. Theorem.

1. $(\Pi^1_1\text{-}CA)_{<\varepsilon_0} \subseteq \Delta^1_2\text{-}CA \subseteq \Sigma^1_2\text{-}AC \subseteq \Sigma^1_2\text{-}DC \subseteq PA^\omega + (\mu) + (\mu_1) + (QF\text{-}AC)$

2. $Res\text{-}\Pi^1_1\text{-}CA \subseteq Res\text{-}\Delta^1_2\text{-}CA \subseteq Res\text{-}\Sigma^1_2\text{-}AC$
$$\subseteq Res\text{-}\widehat{PA}^{\omega} + (\mu) + (\mu_1) + (QF\text{-}AC)^{18}$$

In analogy to 8.3.2, using the ND-interpretation of $PA^{\omega} + (\mu) + (\mu_1) + (QF\text{-}AC)$ in $T + (\mu) + (\mu_1)$ we can also obtain:

8.4.4. Theorem. *$PA^{\omega} + (\mu) + (\mu_1) + (QF\text{-}AC)$ is a conservative extension of $(\Pi^1_1\text{-}CA)_{<\varepsilon_0}$ for Π^1_3 statements.*

8.4.5. Corollary. *$\Sigma^1_2\text{-}DC$ is a conservative extension of $(\Pi^1_1\text{-}CA)_{<\varepsilon_0}$ for Π^1_3 statements.*

Similarly we obtain a conservation result for $Res\text{-}\Pi^1_2\text{-}AC$ over $Res\text{-}\Pi^1_1\text{-}CA$. Corollary 8.4.5 is again a result first obtained by Friedman [1970]. For more details of the arguments using the approach sketched here through functional interpretations, see Feferman [1977,section 8].

8.4.6. Autonomously iterated Π^1_1-CA

Just as for the case of autonomously iterated ramified analysis or $\Pi^0_1\text{-}CA$, we may explain what are the autonomous ordinals for iteration of $(\Pi^1_1\text{-}CA)$. A much larger recursive ordinal notation system is needed to determine these, using the so-called Bachmann hierarchies of ordinal functions (subsequently simplified through work of Feferman, Aczel and Buchholz — cf. Schütte [1977,Chapter IX]). We shall not try to describe these here, except to take up the first ordinal of proof-theoretical interest from that hierarchy, the so-called Howard ordinal, in section 9.7 below. We content ourselves instead with the statement of an analogue of Theorem 8.3.8 in the following form.

8.4.7. Theorem. *$\Sigma^1_2\text{-}DC$ + (Bar-Rule) is a conservative extension of autonomously iterated $\Pi^1_1\text{-}CA$ for Π^1_3 statements.*

This can be proved by an extension of the functional interpretation methods indicated above, thus also giving conservation for the finite type system $PA^{\omega} + (\mu) + (\mu_1) + (QF\text{-}AC) + (Bar\text{-}Rule)$ over autonomously iterated $\Pi^1_1\text{-}CA$. Theorem 8.4.7 has also been proved by Herbrand-Gentzen methods with the functionals μ and μ_1 in Feferman and Jäger [1983].

8.4.8. Further results and methodological discussion

Friedman [1970] obtained further analogues to Corollaries 8.3.5 and 8.4.5, of the following form:

[18] Actually, it is known from the Kondô-Addison theorem that we can strengthen two of the inclusions in 1 by $\Delta^1_2\text{-}CA = \Sigma^1_2\text{-}AC = \Sigma^1_2\text{-}DC$.

8.4.9. Theorem. *For $n \geq 2$, Σ^1_{n+1}-DC is a conservative extension of $(\Pi^1_n\text{-}CA)_{<\varepsilon_0}$ for Π^1_4 statements.*

This can be improved to conservation results for finite type extensions of Σ^1_{n+1}-DC by means of additional axioms for suitable Skolem functionals. However, those functionals do not occur naturally in practice, unlike the non-constructive minimum operator and the Kleene basis operator, so such results would be somewhat artificial to state. Friedman's own proof of this theorem was by model-theoretic methods, and later Feferman and Sieg [1981] gave a proof by Herbrand-Gentzen methods. Subsequently Feferman and Jäger [1983], using the same methods, obtained the related analogues of Theorem 8.4.7:

8.4.10. Theorem. *For $n \geq 2$, Σ^1_{n+1}-DC + (Bar-Rule) is a conservative extension of autonomously iterated Π^1_n-CA for Π^1_4 statements.*

9. The interpretation of theories of ordinals

9.1. A classical theory of countable tree ordinals

We now extend the methods of the previous section to a finite type theory OR^ω_1 of countable (tree) ordinals including the μ operator, in order to obtain a conservation result over a classical theory ID_1 of one arithmetical inductive definition.[19] This work is compared with that of Howard [1972] on analogous constructive theories; it is an interesting open question how these results may be combined. The work described here is based on an unpublished paper, Feferman [1968a].

The system OR^ω_1, to which the ND-interpretation is to be applied, is a variant of that used in Feferman [1968a]. It extends $PA^\omega + (\mu)$ as follows. We expand the type structure by an additional ground type, denoted Ω. Then types $\sigma, \tau, \ldots$ are generated from the ground types 0 and Ω by closing under $(\sigma \to \tau)$ as before. We have infinitely many variables of each type; we use the letters $\alpha, \beta, \gamma, \ldots, \xi, \eta, \zeta$ for variables of the new type Ω. These are informally understood to range over countable tree ordinals, which are closed under formation of suprema in the sense that if f is of type $0 \to \Omega$ then $\mathrm{Sup}(f) \in \Omega$ and $\mathrm{Sup}(f)$ represents the tree whose immediate subtrees are given by the sequence of values $f(x)$ for x a natural number. Formally, the constants of OR^ω_1 besides those of $PA^\omega + (\mu)$ are as follows:

1. 0_Ω is a constant of type Ω
2. Sup is a constant of type $(0 \to \Omega) \to \Omega$
3. Sup^{-1} is a constant of type $\Omega \to (0 \to \Omega)$
4. For each σ, $\mathrm{R}_{\Omega,\sigma}$ is a constant of type $(\Omega, (0 \to \sigma) \to \sigma), \sigma, \Omega \to \sigma$.

[19] Recent work of Burr and Hartung [n.d.] and Burr [1997] extends these results with an interpretation of $KP\omega$ (essentially a set-theoretic analogue of ID_1) in a theory of primitive recursive *set* functionals of finite type, instead of the ordinal functionals of OR^ω.

We shall omit the subscript 'σ' from the ordinal recursor whenever there is no ambiguity. In the following we shall write α_x for $(\mathrm{Sup}^{-1}(\alpha))(x)$ for x of type 0; for α non-zero, this represents the immediate subtree at position x. Just as we do not assume extensionality for functions, we do not assume extensionality for ordinals, i.e. it does *not* follow from $\forall x\ (\alpha_x = \beta_x)$ that $\alpha = \beta$. Nevertheless we shall assume there is a one-one correspondence between non-zero ordinals and functions of which they are the suprema, so that Sup^{-1} will indeed be the inverse of the Sup operation. This is not necessary, but simplifies some points.

The logic taken for OR_1^ω is full classical quantificational logic in its finite type language. The axioms of OR_1^ω consist of:

1. The axioms of $PA^\omega + (\mu)$ with the induction scheme extended to all formulas of the language
2. $\mathrm{Sup}(f) \neq 0_\Omega \wedge \mathrm{Sup}^{-1}(\mathrm{Sup}(f)) = f$, for f of type $(0 \to \Omega)$
3. $\alpha \neq 0_\Omega \to \mathrm{Sup}(\mathrm{Sup}^{-1}(\alpha)) = \alpha$
4. $(0_\Omega)_x = 0_\Omega$
5. (a) $\mathrm{R}_\Omega(f, a, 0_\Omega) = a$
   (b) $\alpha \neq 0_\Omega \to \mathrm{R}_\Omega(f, a, \alpha) = f(\alpha, \lambda x.\mathrm{R}_\Omega(f, a, \alpha_x))$ for each type σ, where the variable a is of type σ, and the variable f is of type $\Omega, (0 \to \sigma) \to \sigma$
6. $\varphi(0_\Omega) \wedge \forall\alpha\ (\alpha \neq 0_\Omega \wedge \forall x\ \varphi(\alpha_x) \to \varphi(\alpha)) \to \forall\alpha\ \varphi(\alpha)$
7. $(QF\text{-}AC)$

The quantifier-free subsystem of OR_1^ω may be axiomatized as follows.

1′. $T + (\mu)$, with induction rule extended to all quantifier- free formulas of OR_1^ω

2′.-5′. The same as 2-5 in OR_1^ω

6′. The rule of induction on ordinals, i.e. from $\varphi(0_\Omega)$ and $(\alpha \neq 0_\Omega \wedge \forall x\ \varphi(\alpha_x) \to \varphi(\alpha))$ infer $\varphi(\alpha)$ for all quantifier-free φ.

(This last is to be expressed in quantifier-free form using the μ operator.) We denote this system by $T_\Omega + (\mu)$. Just as for Theorem 8.3.1 we readily obtain:

9.1.1. Theorem. *OR_1^ω is ND-interpreted in $T_\Omega + (\mu)$.*

9.2. The classical systems ID_1 of one arithmetical inductive definition

We remind the reader briefly of these relatively familiar systems. Here $\theta(P^+, x)$ denotes a formula in the language of arithmetic with one additional predicate or set symbol P that has only positive occurrences in θ. This determines a monotonic operator from sets P to sets $\{x : \theta(P, x)\}$, which thus has a least fixed point. The theory $ID_1(\theta)$ associated with θ is given by the following axioms expressing that P is such a least fixed point, to the extent possible within the given first-order language:

1. The axioms of *PA* with induction expanded to include all formulas containing the symbol P.
2. $\theta(P, x) \to P(x)$
3. $\forall x\ (\theta(\psi/P, x) \to \psi(x)) \to \forall x\ (P(x) \to \psi(x))$, for each formula ψ.

Here $\theta(\psi/P, x)$ denotes the result of substituting any occurrence of the form $P(t)$ in θ by $\psi(t/x)$. The logic of $ID_1(\theta)$ is, of course, classical. When referring to any system described in this way, we simply write ID_1.

We will use ID_1^i to denote the corresponding theories based on intuitionistic logic. Here, however, we need to specify how the positivity requirement on θ is to be interpreted. We will say that θ is *weakly positive* if it is positive in the classical sense, and *strongly* (or *strictly*) *positive* if there are no occurrences of P in the antecedent of an implication, where the basic logical connectives are taken to be those of section 2.1. An even more restrictive requirement on θ is that it be an *accessibility* inductive definition, as described in section 9.6; cf. section 9.8 for a discussion.

9.3. Translation of ID_1 into OR_1^ω

The obvious way to carry out the translation of $ID_1(\theta)$ into OR_1^ω is to define the approximations to the least fixed point from below. For this purpose, we need a form of ordinal recursion which defines a function at an ordinal α in terms of its values at all smaller ordinals β. The appropriate less-than relation for tree ordinals is introduced as follows. We use the letter 's' to range over numbers of finite sequences of natural numbers. The number 0 is chosen to code the empty sequence, and the extension of a sequence s by one new term x is denoted $s\hat{}\langle x\rangle$.
For an ordinal α, the predecessor α_s of α "down along s" is defined recursively by:

1. $\alpha_0 = \alpha$
2. $\alpha_{s\hat{}\langle x\rangle} = (\alpha_s)_x$.

Then we define

3. $\beta < \alpha \leftrightarrow \alpha \neq 0_\Omega \wedge \exists s\, (s \neq 0 \wedge \beta = \alpha_s)$.

9.3.1. Lemma.

1. *The $<$ relation between ordinals is transitive.*
2. $\forall \alpha, \beta\, \exists \gamma\, (\alpha < \gamma \wedge \beta < \gamma)$.
3. *("Ω-upper bounds", or Quantifier-free collection) For every quantifier-free formula φ, we have $\forall x\, \exists \beta\, \varphi(x, \beta) \rightarrow \exists \alpha\, \forall x\, \exists \beta < \alpha\, \varphi(x, \beta)$.*

9.3.2. Proof. Claim 1 is immediate by definition. For claim 2, define f with $f(0) = \alpha$, $f(x') = \beta$, and take $\gamma = \mathrm{Sup}(f)$. Finally, under the hypothesis of claim 3, we have by $(QF\text{-}AC)$, existence of an f such that $\forall x\ \varphi(x, f(x))$; then take $\alpha = \mathrm{Sup}(f)$. □

Given a function f of type $\Omega \rightarrow \sigma$, the restriction of f to α for $\alpha \neq 0_\Omega$ is definable as $\lambda s, x.\, f(\alpha_{s\hat{}\langle x\rangle})$, which we also write both as $\lambda s \neq 0.\, f(\alpha_s)$ and as $\lambda\beta < \alpha.\, f(\beta)$. Then one can derive by use of our recursor R_Ω the following more general form of recursion with values in any type σ, given any a of type σ and G of type $\Omega, (0 \rightarrow \sigma) \rightarrow \sigma$:

$$(<\text{-}Rec_\Omega) \qquad F(0_\Omega) = a, \text{ and for } \alpha \neq 0_\Omega,\ F(\alpha) = G(\alpha, \lambda\beta < \alpha.\, F(\beta)).$$

Correspondingly we can derive the following more general form of induction on Ω:

$(<\text{-}Ind_\Omega)$ $\qquad \varphi(0_\Omega) \wedge \forall\alpha\,(\alpha \neq 0_\Omega \wedge \forall\beta < \alpha\,\varphi(\beta) \rightarrow \varphi(\alpha)) \rightarrow \forall\alpha\,\varphi(\alpha)$

9.3.3. Lemma. *Given arithmetical $\theta(P, x)$, we can define a function F of type $\Omega, 0 \rightarrow 0$ in OR_1^ω satisfying: $F(\alpha, x) = 0 \leftrightarrow \theta(\{y \mid \exists\beta < \alpha\, F(\beta, y) = 0\}, x)$ for all α.*

The proof of this depends essentially on the use of μ to eliminate all the quantifiers in θ and to eliminate the quantifier '$\exists\beta < \alpha$', which is just a quantifier over non-zero sequence numbers, so that we can then apply the principle $(<\text{-}Rec_\Omega)$ above.

Now suppose that $\theta(P, x)$ has just positive occurrences of P. Using the function F from the preceding lemma, define

$$P(x) \leftrightarrow \exists\alpha\, P(\alpha, x), \text{ where } P(\alpha, x) \leftrightarrow F(\alpha, x) = 0, \tag{6}$$

so that

$$P(\alpha, x) \leftrightarrow \theta(\{y \mid \exists\beta < \alpha\, P(\beta, y)\}, x), \text{ for all } \alpha. \tag{7}$$

9.3.4. Theorem. *The predicate P thus defined provably satisfies the axioms of $ID_1(\theta)$ in OR_1^ω.*

9.3.5. Proof. First, to show $\theta(P, x) \rightarrow P(x)$, we must show

$$\theta(\{y \mid \exists\beta\, P(\beta, y)\}, x) \rightarrow \exists\alpha\, P(\alpha, x).$$

Using the positivity of P in θ, we may bring the hypothesis of this implication to prenex normal form

$$Q_1 z_1 \ldots Q_n z_n\, \exists\beta_1 \ldots \exists\beta_m\, \theta_0(x, z_1, \ldots, z_n, \beta_1, \ldots, \beta_m),$$

where Q_i is $\forall$ or $\exists$ and θ_0 is quantifier-free. Then by successive application of parts 2 and 3 of Lemma 9.3.1 we obtain

$$\exists\alpha\, Q_1 z_1 \ldots Q_n z_n\, \exists\beta_1 < \alpha \ldots \exists\beta_m < \alpha\, \theta_0(x, z_1, \ldots, z_n, \beta_1, \ldots, \beta_m),$$

which is equivalent back to

$$\exists\alpha\, \theta(\{y \mid \exists\beta < \alpha\, P(\beta, y)\}, x).$$

Hence by (6) and (7) we may conclude $\exists\alpha\, P(\alpha, x)$, i.e. $P(x)$.

Next, to show

$$\forall x\,(\theta(\psi/P, x) \rightarrow \psi(x)) \rightarrow \forall x\,(P(x) \rightarrow \psi(x)),$$

we simply prove by use of the principle $(<\text{-}Ind_\Omega)$ that, under the hypothesis, we have $\forall\alpha\,\forall x\,(P(\alpha, x) \rightarrow \psi(x))$. This proceeds as usual. □

9.4. Models of $T_\Omega + (\mu)$

We can build models of this system which parallel those of T in section 4.1. First of all, a full, extensional, set-theoretical model $\langle M_\sigma \rangle$ is definable as follows:

1. $M_0 = \mathbb{N}$
2. M_Ω is the smallest set X which contains 0 and which is such that whenever $f : N \to X$, then $\langle 1, f \rangle \in X$.
3. $M_{\sigma \to \tau} = \{ f \mid f \text{ maps } M_\sigma \text{ into } M_\tau \}$.

Then we define

4. $0_\Omega = 0$, $\mathrm{Sup}(f) = \langle 1, f \rangle$ for $f : N \to M_\Omega$, and $\mathrm{Sup}^{-1}(\langle 1, f \rangle) = f$.

So for $\alpha = \mathrm{Sup}(f)$ we have $\alpha_x = f(x)$. With each $\alpha \in M_\Omega$ is associated an ordinal $|\alpha|$, in the usual set-theoretical sense, by

5. $|0| = 0$ and $|\mathrm{Sup}(f)| = \sup \{ |f(x)| + 1 \mid x \in N \}$.

Then $|\alpha_x| < |\alpha|$ whenever $\alpha \neq 0$, and thus R_Ω is definable by classical ordinal recursion so as to satisfy axiom 5 of $T_\Omega + (\mu)$.

For an intensional recursion-theoretic model analogous to *HRO* we make use of recursion in the μ operator as a type 2 functional, which is a special case of the Kleene [1959b] development of recursion in finite type objects. In the following we shall use $f, g, h, \ldots$ to range over $\mathbb{N}$ and write $f(x_1, \ldots, x_n)$ for Kleene's $\{f\}(\mu, x_1, \ldots, x_n)$ whenever it is defined. Now the model $\langle N_\sigma \rangle$ is defined by

1. $N_0 = \mathbb{N}$
2. N_Ω = the smallest set $X \subseteq \mathbb{N}$ such that $0 \in X$ and whenever $f : N \to X$ then $\langle 1, f \rangle \in X$ (where the pairing function $\langle x, y \rangle$ is assumed to be from $\mathbb{N}^2$ to $\mathbb{N} - \{0\}$).
3. $N_{\sigma \to \tau} = \{ f \mid \forall x \, (x \in N_\sigma \to f(x) \in N_\tau) \}$.

Then, just as above, we take

4. $0_\Omega = 0$, $\mathrm{Sup}(f) = \langle 1, f \rangle$ for $f \in N_{0 \to \Omega}$, and $\mathrm{Sup}^{-1}(\langle 1, f \rangle) = f$.

Thus, again, $\alpha_x = f(x)$ for $\alpha = \mathrm{Sup}(f)$, and we may assign set-theoretical ordinals $|\alpha|$ to members α of N_Ω by the same definition as in 5 above. Finally, we may define the interpretations of the ordinal recursors $\mathrm{R}_{\Omega,\sigma}$ as functionals partial recursive in μ for each type σ, by means of the recursion theorem, which is used to produce numbers r satisfying

$$r(f, a, 0) = a, \quad \text{and} \quad r(f, a, \langle 1, g \rangle) \simeq f(g, \lambda x.r(f, a, g(x))).$$

Then induction on $|\alpha|$ shows that $r(f, a, \alpha)$ is defined on the objects f, a of appropriate type, for all $\alpha \in N_\Omega$.

Finally, one can define an analogue of the hereditarily extensional recursion-theoretic *HRE* model by first defining the notion $\alpha =_\Omega \beta$ inductively, and then defining $=_{\sigma \to \tau}$ in terms of $=_\sigma$ and $=_\tau$ as for *HRE*, with the objects at each type being those which preserve the defined equality relations at each type.

9.5. Interpreting OR_1^ω in ID_1

We can interpret OR_1^ω in $ID_1(O)$, where O is the set of Church-Kleene constructive ordinal notations, by first applying the ND-interpretation of OR_1^ω in its quantifier-free subtheory $T_\Omega + (\mu)$ and then formalizing the HRO model of the latter in ID_1. For this, though, instead of making use of Kleene's definition of recursion in μ as a special case of recursion in finite type objects, one takes a more concrete version which is possible from the fact that the partial-recursive-in-μ functions are exactly the Π_1^1 partial functions; those can be enumerated by uniformizing the Π_1^1 relations in a standard enumeration. Since O is a complete Π_1^1 predicate, this generalized recursion theory can be formalized in $ID_1(O)$ and, further, the definitions of the N_σ can be given for each σ in that theory. The resulting interpretation preserves arithmetical statements, and is such that with each closed term t of type Ω is associated a number n_t for which $ID_1(O)$ proves $n_t \in O$, and such that $|t| \leq |n_t|$. It follows that the provable ordinals of OR_1^ω are the same as those of $ID_1(O)$, and that is the same as its proof-theoretic ordinal. The latter will be described in section 9.8.

9.6. Functional interpretation of a constructive theory of countable tree ordinals

Howard [1972] introduced a first-order theory U that he called a system of abstract constructive ordinals, with just two sorts of variables, numbers and ordinals. In place of $(QF\text{-}AC)$ this made use of the principle of ω-upper bounds (Lemma 9.3.1 above, clause 3). Howard's system can be translated directly into a finite type theory U^ω which is the same as OR_1^ω, except that we omit the μ operator, and base it on intuitionistic logic in place of classical logic. Then U^ω has a D-interpretation in the system T_Ω of section 9.1, again without the μ operator; that system is just another version of Howard's quantifier-free finite-type theory V of abstract constructive (tree) ordinals op. cit. Howard gave a term model of V which is an extension of Tait's term model of T using infinitely long terms (cf. section 4.4), and used that to obtain an upper bound to the proof-theoretic ordinal of U as the least ordinal greater than $|t|$ for t a normal term of type Ω. The so-called *(Bachmann-)Howard ordinal* thus obtained will be described in the next section. Howard also showed how to translate intuitionistic $ID_1^i(\theta)$ for θ of a certain special form into his system U. This special form includes "accessibility" inductive definitions, where $\theta(P, x)$ is of the form $\forall y\ (y <_t x \to P(y))$ and t is a closed term which determines $y <_t x$ by definition as $t(x, y) = 0$. In the final part of Howard [1972] it was shown how to use work of Gerber [1970] to establish transfinite induction up to each ordinal smaller than the Howard ordinal in such a theory of an accessibility inductive definition. Thus the Howard ordinal is the proof-theoretic ordinal of U and is a least upper bound for the ordinals of all systems $ID_1^i(\theta)$ of the special form considered by Howard.

9.7. The Howard ordinal

The original description of this ordinal made use of an extension of the Veblen hierarchy of ordinal functions in a form due to Bachmann [1950], which gives sense to φ_α for suitable uncountable α. For the specific purposes of 9.6, it is sufficient to tell how this is to be done for α up to the first epsilon number greater than the least uncountable ordinal ω_1, namely ε_{ω_1+1}.[20] Roughly speaking one first assigns to each term α for a limit ordinal in a notation system for ordinals up to ε_{ω_1+1} a fundamental sequence of order type $\leq \omega_1$ in a reasonably canonical way. If the cofinality type of this sequence is countable, one proceeds to define φ_α in terms of the simultaneous fixed points of the φ_β for the terms $\beta = \alpha_\nu$ in that fundamental sequence as with the Veblen hierarchy; if its fundamental sequence is of length ω_1 then one diagonalizes, i.e. takes $\varphi_\alpha(\nu)$ to be $\varphi_{\alpha_\nu}(0)$. The Howard ordinal is then defined to be $\varphi_\alpha(0)$ for $\alpha = \varepsilon_{\omega_1+1}$.[21]

9.8. Discussion

From a foundational standpoint, it is desirable to have a reduction of classical ID_1 to a constructive theory. Although a slight variant of the double-negation interpretation serves to reduce ID_1 to its formally intuitionistic counterpart ID_1^i (cf. Buchholz et al. [1981,p. 56]), the latter theory is not evidently constructive, in the sense that there is no direct constructive justification of axioms 2 and 3 of section 9.2 for θ in which the predicate symbol P occurs only in a weakly positive way. (An indirect justification is provided by the intuitionistic theory of species — i.e. the formal counterpart of second-order analysis — whose constructivity is, however, a matter of dispute; cf. the discussion in Feferman [1982b,pp. 77–78].) What we really desire is a reduction of ID_1 to an intuitionistic theory of accessibility inductive definitions, whose very form provides a clear picture of how the corresponding sets are generated from the "bottom up"; or, alternatively, a reduction of ID_1 to the constructive theory U^ω without the μ operator.

In fact, the first type of reduction has been given by work of Pohlers and Buchholz in a series of steps beginning in the mid-1970s using interesting (prima-facie) uncountably infinitary extensions of Gentzen-Schütte style proof theory; cf. the reports in Buchholz et al. [1981]. Indeed, they have succeeded in determining the proof-theoretic ordinals of classical theories of iterated inductive definitions ID_α and in reducing them to corresponding intuitionistic theories of iterated accessibility inductive definitions, thus showing the proof-theoretic ordinals to be the same. In particular, in the special case of ID_1, the result of their work yields the Howard ordinal as the proof-theoretic ordinal of the system whether taken in its classical or restricted intuitionistic form.

---

[20] Here, ordinals are treated set-theoretically.

[21] The Bachmann approach has since been superseded by the Feferman-Aczel-Buchholz approach described in Schütte [1977,Chapter IX]: the latter is simpler in not requiring a prior assignment of fundamental sequences.

We do not, however, know how to achieve this same result via the method of functional interpretation, nor do we have a direct reduction of ID_1 to U^ω. In addition, no one has yet extended the method of functional interpretation to iterated ID_α, either classical or intuitionistic, in an informative way specific to those systems.[22] It would be of interest to know whether there is some fundamental methodological obstacle for doing so, or if it is simply for lack of a new idea — or simply lack of trying hard enough.

10. Interpretations based on polymorphism

10.1. Transfinite types and polymorphism

In this section we address strengthenings of T that provide mechanisms for defining "transfinite" types. For example, recall the types (n) from section 2.2, defined by $(0) = 0$ and $(n+1) = (n) \to 0$. One might want to define a function f that, for each natural number n, returns an object of type (n). Such a function f is an element of the product type

$$\Pi_{n \in 0}\,(n),$$

and clearly goes beyond the finite-type capabilities of T. The function f is also "polymorphic" in the sense that, for each n, the type of $f(n)$ depends on its argument.

A down-to-earth example of polymorphism arises in the context of writing a sorting algorithm. Instead of writing separate routines that sort lists of integers, real numbers, strings, and so on, one would prefer to write a general routine that, given a type X and a comparison function in $X \times X \to 0$, sorts lists of objects of type X. Assuming such lists are represented by the type List(X), for each type X we want Sort(X) to have the type

$$(X \times X \to 0) \times \text{List}(X) \to \text{List}(X).$$

The type of the function Sort itself can then be written

$$\Pi_X\,((X \times X \to 0) \times \text{List}(X) \to \text{List}(X))$$

where the product Π_X now ranges over (some collection of) types.

In this section we will consider two different kinds of polymorphism. In the first, the variable X in the preceding example is allowed to range over *all* types, include the type of Sort itself. This scheme is known as *impredicative polymorphism* and was discovered independently by J.-Y. Girard, who was looking for a D-interpretation for second-order arithmetic (cf. Girard [1971]), and J. Reynolds, who was exploring

[22]For α a provably recursive ordinal of analysis we can, in principle, treat ID_α as a subsystem of analysis and then apply Spector's interpretation (section 6) or Girard's interpretation (section 10). But it would appear very difficult to extract meaningful descriptions of the associated proof-theoretic ordinals and reductions to the corresponding intuitionistic systems via those interpretations. At any rate this does not look at all like a practical possibility.

type-theoretic constructs from a computational point of view (cf. Reynolds [1974]). Given these independent motivations, Girard's main theorem is quite satisfying: the provably total recursive functions of second-order arithmetic are exactly the ones computable in the Girard-Reynolds framework.

The circularity of allowing the variable X in a type $\Pi_X\ T(X)$ to range over all types, including $\Pi_X\ T(X)$ itself, might seem disconcerting. *Predicative polymorphism* is more benign in that regard, since in this framework the variable X is restricted to range over a pre-set universe of "smaller" types. For example, in the type

$$\Pi_{X \in U_0}\ T(X)$$

the variable X takes values in the fixed universe U_0. This kind of polymorphism was developed by Martin-Löf as a framework for constructive mathematics (cf. Martin-Löf [1973]), and has been implemented in the underlying language of the Nuprl proof-development system (cf. Chapter X or Constable et al. [1986]). Once again, there is a result that nicely characterizes the axiomatic strength of this kind of polymorphism: the provably total recursive functions of predicative analysis (cf. section 8.3.6) are exactly the ones that can be defined using nested universes of types.

Here we will focus on the interpretative strength of polymorphism. For a detailed discussion of the various computational aspects of the subject, we refer the reader to Mitchell [1990] and Gallier [1990].

10.2. The second-order polymorphic lambda calculus, F

We now define Girard's theory F, a polymorphic extension of T strong enough to interpret second-order arithmetic. This allows for terms which can be applied to (terms representing) types, and abstraction across types. It is simpler here to take abstraction as basic operators rather than defined in terms of combinators. The types of F are defined inductively, as follows:

1. There are infinitely many type variables $X, Y, Z, \ldots$
2. 0 is a type.
3. If σ and τ are types, so are $\sigma \rightarrow \tau$ and $\sigma \times \tau$.
4. If σ is a type and X is a type variable, then $\Pi_X\ \sigma$ and $\Sigma_X\ \sigma$ are also types.

A term of type $\Pi_X\ \sigma$ denotes a polymorphic function that, for each type τ, returns an object of type $\sigma[\tau/X]$. A term of type $\Sigma_X\ \sigma$ denotes a pair $\langle \tau, t\rangle$, where τ is a type and t is an object of type $\sigma[\tau/X]$.

The terms of F are also defined inductively, as follows:

1. For every type σ there are infinitely many variables, x^σ, y^σ, z^σ, ...
2. 0, Sc, the recursors R_σ, pairing operators $\langle \cdot, \cdot \rangle$, and projection operators π_0, π_1 are terms of appropriate type.
3. If t is a term of type τ and x is a variable of type σ, then $\lambda x.t$ is a term of type $\sigma \rightarrow \tau$.
4. If t is a term of type $\sigma \rightarrow \tau$ and s is a term of type σ, then $t(s)$ is of type τ.

5. If t is a term of type τ, and the type variable X is not free in the type of a free variable of t, then $\Lambda X.t$ is a term of type $\Pi_X \tau$.
6. If t is a term of type $\Pi_X \tau$ and σ is a type, then $t(\sigma)$ is a term of type $\tau[\sigma/X]$.
7. If σ is a type and t is a term of type $\tau[\sigma/X]$, then $\langle\sigma, t\rangle$ is a term of type $\Sigma_X \tau$.
8. If $t[x^\tau]$ is a term of type σ and the type variable X is only free in the type of x^τ, then
$$\nabla\langle X, x^\tau\rangle.t$$
is a term of type $\Sigma_X \tau \to \sigma$.

While clauses (1–4) are essentially imported from T, clauses (5–8) provide the new polymorphic capabilities. The choice of lambda terms instead of combinators in clause (3) conforms with the majority of the literature on this subject.

The term $\Lambda X.t$ in clause (6) denotes a function that associates to every type σ an object of type $\tau[\sigma/X]$, with defining equation

$$(\Lambda X.t)(\sigma) = t[\sigma/X].$$

The requirement that X is not free in the type of any free variable of t precludes terms like

$$\Lambda X.x^X,$$

in which the free variable x cannot be assigned any reasonable type. On the other hand, it does allow constructions such as the polymorphic identity function

$$\Lambda X.\lambda x^X.x^X$$

and the Sort function defined in section 10.1. Strictly speaking, the application operation $(\Lambda X.t)(\sigma)$ in clause (6) is different from the application operation $(\lambda x.t)(s)$ in clause (4), though we will use the same notation.

The function $\nabla\langle X, x^\tau\rangle.t$ in clause (8) is a bit trickier: it takes a pair $\langle\sigma, s\rangle$, for which s is of type $\tau[\sigma/X]$, and returns the value of t with X replaced by σ and $x^{\tau[\sigma/X]}$ replaced by s. In other words, this term has the defining equation

$$(\nabla\langle X, x\rangle.t)\langle\sigma, s\rangle = t[\sigma/X][s/x^{\tau[\sigma/X]}].$$

As usual, the variable X becomes bound in clause (6), and both variables X and x become bound in clause (8).

For technical reasons the interpretation in the next section also requires a constant zero functional 0_σ of each type. We omit the rules defining its behavior, though they can be found in Girard [1971].

10.3. The interpretation of analysis

Whereas in Spector's interpretation of analysis the second-order objects of $PA^2 + (CA)$ become identified with function variables, in Girard's interpretation it is more natural to treat them as predicates. (Moreover, the interpretation extends to a higher-type version of this system, with predicates instead of functions at each level.) As in the interpretation of *PA*, we first apply the double-negation interpretation to reduce $PA^2 + (CA)$ to its intuitionistic variant, $HA^2 + (CA)$. Unlike the case of (AC) in Spector's interpretation, $(CA)^N$ follows directly from (CA) in HA^2.

The D-interpretation we are about to define translates formulas φ in the language of second-order arithmetic to formulas φ^D of the form

$$\exists x^\sigma \, \forall y^\tau \, F_\varphi(x,y) = 0 \tag{8}$$

where φ_D is quantifier-free with only type 0 equality. Adding dummy quantifiers as necessary and using the pairing and projection operations to combine quantified variables, we can assume that there is always exactly one variable after each quantifier.

The first new step needed in defining φ^D beyond first-order formulas is to find a suitable translation for formulas of the form $t \in Z$. Since we are aiming to interpret the comprehension axioms, we want the variable Z to "range" over arbitrary formulas φ. We define $(t \in Z)^D$ to be

$$\exists x^X \, \forall y^Y \, f(x,y,t) = 0 \tag{9}$$

where now X and Y are type variables and f is a function variable of type

$$X \times Y \times 0 \to 0.$$

Intuitively, $(t \in Z)^D$ represents an "arbitrary" formula of the form (8).

The translation is extended to the logical connectives and first-order quantifiers as before. To define $(\exists Z \, \varphi(Z))^D$, suppose $\varphi(Z)^D$ is given by

$$\exists x^{\sigma[X,Y]} \, \forall y^{\tau[X,Y]} \, F(x,y,f) = 0$$

where X, Y, and f are the variables we have associated with Z in the preceding paragraph. The formula $\exists Z \, \varphi(Z)$ should then translate to

$$\exists X, Y, f, x^{\sigma[X,Y]} \, \forall y^{\tau[X,Y]} \, F(x,y,f) = 0.$$

But now the type of y depends on the existentially quantified types X and Y, which is problematic. We remedy this by replacing y with an element of the appropriate product type to obtain

$$\exists X, Y, f, x^{\sigma[X,Y]} \, \forall y^{\Pi_{X,Y} \, \tau[X,Y]} \, F(x, y(X,Y), f) = 0.$$

Finally, we replace the existentially quantified variables X, Y, and f by a single variable u of type $\Sigma_X \, \Sigma_Y \, (X \times Y \times 0 \to 0)$. Then u may be paired with x to put $(\exists Z \, \varphi(Z))^D$ again in the form (8).

Similar manipulations are used to define the translation of the formula $\forall Z \, \varphi(Z)$. When this is done, the interpretation of HA^2 is verified just as in section 2.4. The comprehension axioms (CA) are also easily dealt with, as a consequence of the translation we have chosen for $x \in Z$. Details can be found in Girard [1971].[23]

This yields

[23]There, and in the literature, F is often used more specifically to denote the set of *reduction*

10.3.1. Theorem. *$PA^2 + (CA)$ is ND-interpreted in F.*

10.4. Strong normalization for F

In his doctoral dissertation Girard [1972] presented a powerful generalization of Tait's convertibility methods (cf. section 4.3), and used it to show that F is strongly normalizing.

Since the normalization of F implies the consistency of second-order arithmetic, the full argument cannot be formalized in that theory. To appreciate the difficulties that arise in trying to adapt the argument of section 4.3, consider the problem of extending the reducibility predicate defined there to a term t of type $\Pi_X\ \tau$. We would like to say that t is reducible if for every type σ the term $t(\sigma)$ is reducible. But σ is arbitrary, and could very well be the type $\Pi_X\ \tau$ itself. In that case, determining the reducibility of $t(\sigma)$ will require some knowledge of what it means for terms of type $\Pi_X\ \tau$ to be reducible— which is exactly the notion we are trying to define.

Girard's clever dodge is to define the notion of a "reducibility candidate," which is a predicate of terms which satisfies certain closure conditions. The reducibility predicate of section 4.3 is an example of a reducibility candidate, and, in fact, it is just these closure conditions that allow one to carry out the necessary induction on terms. Girard then declared $t \in \Pi_X\ \tau$ to be reducible if for *every* reducibility candidate C and every type σ, $t(\sigma)$ is reducible of type $\tau(\sigma)$, where now reducibility for $t(\sigma)$ is defined in terms of C.

Since the definition of reducibility for terms of type $\Pi_X\ \tau$ involves prefixing a second-order quantifier to a formula involving the definition of reducibility for terms of type τ, for arbitrary polymorphic types the definition requires second-order formulas of arbitrary complexity. The net result is the following

10.4.1. Theorem. *For each term t of F, $PA^2 + (CA)$ proves that t is strongly normalizing. In addition, $PA^2 + (CA)$ proves the confluence of F.*

Thus F can be interpreted in $PA^2 + (CA)$ via its model in the normal terms. This yields the following

10.4.2. Theorem. *The provably total recursive functions of $PA^2 + (CA)$ are exactly the ones that are represented by terms of F.*

The method of reducibility candidates extends to stronger functional theories, including a typed extension of F also introduced in Girard's dissertation, and Co-

---

rules corresponding to the defining equations of the theory described above; we have chosen to blur this distinction. A more minimal version of F which omits the base type 0 and the sum type — essentially the system $\lambda^{\rightarrow,\forall}$ of Mitchell [1990], Gallier [1990] — is discussed in Girard, Lafont and Taylor [1989].

The theory Y of Girard [1971] is essentially a logic-free version of our theory F. More precisely, if F proves a quantifier-free formula φ involving only type 0 equality, then Y proves the logic-free translation of φ when free variables are instantiated by any closed terms.

quand's "Calculus of Constructions." In fact, by identifying natural deductions with terms of an appropriate functional calculus, Girard [1971] was also able to use these methods to give a new proof of "Takeuti's conjecture," an extension of Gentzen's Hauptsatz to higher-order deductive sytems; this was realized independently by Martin-Löf [1971] and Prawitz [1971]. See Gallier [1990], Coquand [1990], Girard, Lafont and Taylor [1989] for more details.

10.5. Theories based on predicative polymorphism

A more restrictive, "predicative," method of extending the theory T polymorphically is represented by a sequence of theories P_n, based on Martin-Löf's theories ML_n. The theory P_0 is equivalent to T, while the theory P_{n+1} adds $n+1$ "universes" of types, U_0 through U_n, each of which will itself be a type.

Like the theory F, the theories P_n have product and sum types, but rather than taking products over *types*, one takes products over *terms* of a given type. That is, one has the following type formation rules:

1. 0 is a type.
2. If σ and $\tau[x^\sigma]$ are types, then so are $\Pi_{x\in\sigma}\ \tau$ and $\Sigma_{x\in\sigma}\ \tau$.

Terms of type $\Pi_{x\in\sigma}\ \tau$ denote functions f which take elements a of type σ to elements $f(a)$ of type $\tau[a/x]$, and terms of type $\Sigma_{x\in\sigma}\ \tau$ denote pairs $\langle a, b\rangle$, where a is an object of type σ and b is an object of type $\tau[a/x]$. The $\rightarrow$ and $\times$ constructions can be seen as special cases of the Π and Σ constructions respectively, in which the type τ doesn't depend on the variable x^σ.

One has to modify the usual formation rules for terms to accommodate these new dependent types. For example, the new rule of explicit definition takes the form

$$\text{If } t[x^\sigma] \text{ is a term of type } \tau[x^\sigma], \text{ then } \lambda x.t \text{ is a term of type } \Pi_{x\in\sigma}\tau.$$

Similar rules take the place of those of T regarding application, pairing, projection, and the recursors.

We haven't yet provided a mechanism for defining types that depend on variables. What makes the systems P_n polymorphic is that *types can be represented by terms*, since elements of type U_i are themselves taken to denote types. For example, P_1 has the following rules:

1. U_0 is a type
2. 0 (the type of the natural numbers) is an element of U_0
3. If σ and $\tau[x^\sigma]$ are elements of U_0, then so are $\Pi_{x\in\sigma}\tau$ and $\Sigma_{x\in\sigma}\tau$.
4. If t is a term of type U_0, then t is a type.

The last clause places the theory P_1 in sharp distinction to T, where there is no interplay between terms and types. In P_1, one can define a term t of type U_0, conclude that t is a type, and then define another term s of type t. For example, the types (n) of section 10.1 can be defined by a simple instance of primitive recursion with range U_0, whereupon the term $\Pi_{n\in 0}(n)$ is an element of U_0 and hence a type. In general, each theory P_n has n universes $U_0, \ldots, U_{n-1}$, each of which contains terms denoting all "smaller" universes.

10.6. The interpretation of predicative theories of analysis

The theory $\widehat{ID}_1$ is a weakening of the theory ID_1 defined in section 9.2. Here, rather than assert that P is a *least* fixed point of the positive arithmetic operator given by θ, one simply asserts that P is *some* fixed point; that is, one replaces 2 and 3 from section 9.2 by the single axiom

$$\forall x\ (\theta(P, x) \leftrightarrow P(x)).$$

In general, each theory $\widehat{ID}_n$ allows n nested inductive definitions, by allowing one to use any predicate of $\widehat{ID}_i$ to define a fixed-point in $\widehat{ID}_{i+1}$. Also we set

$$\widehat{ID}_{<\omega} = \bigcup_{n<\omega} \widehat{ID}_n.$$

It is known (cf. Feferman [1982a], Friedman, McAloon and Simpson [1982], Avigad [1996b]) that $\widehat{ID}_{<\omega}$ has strength Γ_0, and hence proves the same arithmetic formulas as predicative analysis (cf. the discussion preceding Theorem 8.3.8) as well as an important second-order theory known as ATR_0.[24] The following theorem is due to Avigad [n.d.].

10.6.1. Theorem. *Each theory $\widehat{ID}_n$ is ND-interpreted in P_n.*[25]

The interpretation is somewhat tortuous, so we only provide a rough outline here. The first reduction relies on the following lemma, due to P. Aczel (cf. Feferman [1982a]).

10.6.2. Lemma. *$\widehat{ID}_1$ is interpretable in Σ^1_1-AC. More generally, each theory $\widehat{ID}_{n+1}$ is interpretable in Σ^1_1-$AC(\widehat{ID}_n)$.*

Here the theory Σ^1_1-$AC(\widehat{ID}_n)$ is a second-order version of $\widehat{ID}_n$ together with a choice scheme for Σ^1_1 formulas in the expanded language. Given any arithmetic (or even Σ^1_1) formula $\varphi(x, Y)$ in which the predicate Y occurs positively, one can obtain a Σ^1_1 formula ψ defining a fixed-point of the corresponding set operation. (The proof is similar to that of Gödel's fixed-point lemma, while the scheme (Σ^1_1-AC) is needed to show that that formula ψ works as advertised.) One then uses these formulas ψ to interpret the fixed-point constants of $\widehat{ID}_{n+1}$.

By Theorem 8.3.1, one can interpret Σ^1_1-AC in the theory $T + (\mu)$, in which formulas in the language of Peano arithmetic translate to formulas that are quantifier-free. We can instead interpret Σ^1_1-AC in P_1 if we interpret formulas $t \in Z$ by equation (9) of section 10.3, where now the type variables range over the universe U_0 instead of all types. Iterating this idea leads to Theorem 10.6.1.

---

[24]The first functional interpretation of predicative ramified analysis was given by Maass [1976] via a specialization of Girard's interpretation described in section 10.3 above.

[25]The theories P_n of Avigad [n.d.] are logic-free subsystems of the theories P_n defined here.

One can show (cf. Coquand [n.d.], Martin-Löf [1973]) that terms of $P_{<\omega}$ ($= \bigcup P_n$) are normalizing. Since each $\widehat{ID}_\omega$ can define a recursion-theoretic model for P_n, we obtain

10.6.3. Corollary. *The provably total recursive functions of each $\widehat{ID}_n$ are exactly the ones represented by terms of P_n. The provably total recursive functions of $\widehat{ID}_{<\omega}$, and hence predicative analysis and ATR_0, are exactly the ones represented by terms of $P_{<\omega}$.*

10.7. Final comments and questions

The functional interpretation of a classical theory provides a nice interplay between logic and theoretical computer science. Given a set of classical axioms, the corresponding computational schemata provide a constructive understanding of their strength, and normalization proofs provide evidence that these abstract axioms have interesting computational consequences. Conversely, given some computational schemata, the calibration of their classical axiomatic strength helps round out our understanding of their capabilities.

Martin-Löf has described extensions of the theories ML_n with "elimination rules" for the universes, yielding the same proof-theoretic strength as the impredicative theories ID_n (cf. Griffor and Rathjen [1994,Theorem 4.14]). Can such elimination rules be used to give the theories ID_n a direct functional interpretation?

The interpretations of Spector and Girard show that bar-recursion and impredicative polymorphism each exactly capture the provably total recursive functions of second-order arithmetic. Are there natural characterizations for interesting fragments and extensions of the latter theory, other than the ones we have already described in this chapter?

References

J. Avigad

[n.d.] *Predicative functionals and an interpretation of $\widehat{ID}_{<\omega}$.* To appear in the *Annals of Pure and Applied Logic.*

[1996a] Formalizing forcing arguments in subsystems of second-order arithmetic, *Annals of Pure and Applied Logic*, 82, pp. 165–191.

[1996b] On the relationship between ATR_0 and $\widehat{ID}_{<\omega}$, *Journal of Symbolic Logic*, 61, pp. 768–779.

H. Bachmann

[1950] Die Normalfunktionen und das Problem der ausgezeichneten Folgen von Ordnungzahlen, *Vierteljahresschr. Nat. Ges., Zürich*, 95, pp. 5–37.

J. Barwise

[1977] ed., *The Handbook of Mathematical Logic*, North-Holland, Amsterdam.

J. Barwise and J. S. Schlipf

[1975] On recursively saturated models of arithmetic, in: *Model theory and algebra: a memorial tribute to Abraham Robinson*, D. Saracino and V. B. Weispfennig, eds., Lecture Notes in Mathematics #498, Springer-Verlag, Berlin, pp. 42–55.

M. J. Beeson

[1978] A type-free Gödel interpretation, *Journal of Symbolic Logic*, 43, pp. 213–227.

[1982] Recursive models for constructive set theories, *Annals of Mathematical Logic*, 23, pp. 127–178.

[1985] *Foundations of Constructive Mathematics*, Springer-Verlag, Berlin.

E. Bishop

[1970] Mathematics as a numerical language, in: Kino, Myhill and Vesley [1970], pp. 53–71.

W. Buchholz, S. Feferman, W. Pohlers, and W. Sieg

[1981] *Iterated Inductive Definitions and Subsystems of Analysis: Recent Proof-Theoretical Studies*, Lecture Notes in Mathematics #897, Springer-Verlag, Berlin.

W. Burr

[1997] *A Diller-Nahm-style functional interpretation of* $KP\omega$. Preliminary draft.

W. Burr and V. Hartung

[n.d.] *A characterization of the* Σ_1 *definable functions of* $KP\omega + (uniform\ AC)$. To appear in the *Archive for Mathematical Logic.*

S. R. Buss

[1986] *Bounded Arithmetic*, Bibliopolis, Naples. A reprinting of the author's 1985 Princeton University dissertation.

A. Cantini

[1996] Asymmetric interpretation for bounded theories, *Mathematical Logic Quarterly*, 42, pp. 270–288.

R. L. Constable et al.

[1986] *Implementing Mathematics with the Nuprl Proof Development System*, vol. 37 of Graduate Texts in Mathematics, Prentice-Hall, Englewood Cliffs, NJ.

S. A. Cook and A. Urquhart

[1993] Functional interpretations of feasibly constructive arithmetic, *Annals of Pure and Applied Logic*, 63, pp. 103–200.

C. Coquand

[n.d.] A normalization proof for Martin-Löf's type theory, in: *25 Years of Constructive Type Theory*, G. Sambin and J. M. Smith, eds., Oxford University Press. To appear.

T. Coquand

[1990] Metamathematical investigations of a calculus of constructions, in: Odifreddi [1990], pp. 91–122.

N. Dershowitz and J.-P. Jouannaud

[1990] Rewrite systems, in: van Leeuwen [1990], pp. 243–320.

J. Diller and W. Nahm

[1974] Eine Variante zur Dialectica Interpretation der Heyting Arithmetik endlicher Typen, *Archive für Mathematische Logik und Grundlagenforschung*, 16, pp. 49–66.

S. Feferman

[1964] Systems of predicative analysis, *Journal of Symbolic Logic*, 29, pp. 1–30.

[1968a] *Ordinals associated with theories for one inductively defined set.* Unpublished paper.

[1968b] Systems of predicative analysis II: representations of ordinals, *Journal of Symbolic Logic*, 33, pp. 193–220.

[1971] Ordinals and functionals in proof theory, in: *Proceedings of the International Congress of Mathematicians, Nice*, vol. 1, Gauthier-Villars, Paris, pp. 229–233.

[1977] Theories of finite type related to mathematical practice, in: Barwise [1977], pp. 913–971.

[1979] A more perspicuous formal system for predicativity, in: *Konstruktionen versus Positionen I*, K. Lorenz, ed., de Gruyter, Berlin, pp. 87–139.

[1982a] Iterated inductive fixed-point theories: application to Hancock's conjecture, in: Metakides [1982].

[1982b] Monotone inductive definitions, in: *The L. E. J. Brouwer Centenary Symposium*, A. S. Troelstra and D. van Dalen, eds., North-Holland, Amsterdam, pp. 77–89.

[1990] *Milking the Dialectica interpretation.* Unpublished notes.

[1993] Gödel's Dialectica interpretation and its two-way stretch, in: *Computational Logic and Proof Theory*, G. Gottlob, A. Leitsch, and D. Mundici, eds., Lecture Notes in Computer Science #713, Springer-Verlag, Berlin, pp. 23–40.

S. Feferman and G. Jäger

[1983] Choice principles, the bar rule, and autonomously iterated comprehension schemes in analysis, *Journal of Symbolic Logic*, 48, pp. 63–70.

S. Feferman and W. Sieg

[1981] Proof-theoretic equivalences between classical and constructive theories for analysis, in: Buchholz et al. [1981], pp. 78–142.

J. E. Fenstad

[1971] ed., *Proceedings of the Second Scandinavian Logic Symposium*, North-Holland, Amsterdam.

F. Ferreira

[1988] *Polynomial time computable arithmetic and conservative extensions*, PhD thesis, Pennsylvania State University.

[1994] A feasible theory for analysis, *Journal of Symbolic Logic*, 59, pp. 1001–1011.

H. M. Friedman

[1970] Iterated inductive definitions and $\Sigma^1_2{-}AC$, in: Kino, Myhill and Vesley [1970], pp. 435–442.

[1976] Systems of second-order arithmetic with restricted induction I, II (abstract), *Journal of Symbolic Logic*, 41, pp. 557–559.

H. M. Friedman, K. McAloon, and S. G. Simpson

[1982] A finite combinatorial principle which is equivalent to the 1-consistency of predicative analysis, in: Metakides [1982], pp. 197–230.

W. Friedrich

[1985] Gödelsche Funktionalinterpretation für eine Erweiterung der Klassischen Analysis, *Zeitschrift für mathematische Logik und Grundlagen der Mathematik*, 31, pp. 3–29.

J. H. Gallier

[1990] On Girard's 'Candidats de Reductibilité', in: Odifreddi [1990], pp. 123–203.

H. Gerber

[1970] Brouwer's bar theorem and a system of ordinal notations, in: Kino, Myhill and Vesley [1970], pp. 327–338.

J.-Y. Girard

[1971] Une extension de l'interpretation de Gödel à l'analyse, et son application à l'élimination des coupures dans l'analyse et dans la théorie des types, in: Fenstad [1971], pp. 63–92.

[1972] *Interprétation fonctionelle et élimination de coupures de l'arithmétique d'ordre supérieur*, PhD thesis, Université de Paris VII.

J.-Y. Girard, Y. Lafont, and P. Taylor

[1989] *Proofs and Types*, Cambridge University Press.

K. Gödel

[1941] In what sense is intuitionistic logic constructive?, in: Gödel [1994], pp. 189–200.

[1958] Über eine bisher noch nicht benüzte Erweiterung des finiten Standpunktes, *Dialectica*, 12, pp. 280–287. Reproduced with English translation in Gödel [1990], pp. 241–251.

[1972] On an extension of finitary methods which has not yet been used, in: Gödel [1994], pp. 271–280.

[1990] *Collected Works*, vol. II, Oxford University Press, New York. Solomon Feferman et al. eds.

[1994] *Collected Works*, vol. III, Oxford University Press, New York. Solomon Feferman et al. eds.

E. Griffor and M. Rathjen

[1994] The strength of some Martin-Löf type theories, *Archive for Mathematical Logic*, 33, pp. 347–385.

P. Hájek

[1993] Interpretability and fragments of arithmetic, in: *Arithmetic, Proof Theory, and Computational Complexity*, P. Clote and J. Krajicek, eds., Oxford University Press, Oxford.

J. van Heijenoort

[1967] *From Frege to Gödel: A sourcebook in mathematical logic, 1879-1931*, Harvard University Press.

A. Heyting

[1959] ed., *Constructivity in Mathematics*, North-Holland, Amsterdam.

D. Hilbert

[1926] Über das Unendliche, *Mathematische Annalen*, 95, pp. 161–190. English translation in van Heijenoort [1967], pp.367–392.

J. R. Hindley and J. P. Seldin

[1986] *Introduction to Combinators and λ-calculus*, Cambridge University Press, Cambridge.

W. A. Howard

[1968] Functional interpretation of bar induction by bar recursion, *Compositio Mathematica*, 20, pp. 107–124.

[1970] Assignment of ordinals to terms for primitive recursive functionals of finite type, in: Kino, Myhill and Vesley [1970], pp. 453–468.

[1972] A system of abstract constructive ordinals, *Journal of Symbolic Logic*, 37, pp. 355–374.

[1973] Hereditarily majorizable functionals of finite type, in: Troelstra [1973], pp. 454–461.

A. Kino, J. Myhill, and R. E. Vesley

[1970] eds., *Intuitionism and Proof Theory*, North-Holland, Amsterdam.

S. C. Kleene

[1959a] Countable functionals, in: Heyting [1959], pp. 81–100.

[1959b] Recursive functionals and quantifiers of finite types, I, *Transactions of the American Mathematics Society*, 91, pp. 1–52.

S. C. Kleene and R. E. Vesley

[1965] *Foundations of Intuitionistic Mathematics*, North-Holland, Amsterdam.

U. W. Kohlenbach

[1992] Effective bounds from ineffective proofs in analysis: an application of functional interpretation and majorization, *Journal of Symbolic Logic*, 57, pp. 1239–1273.

[1993] Effective moduli from ineffective uniqueness proofs: an unwinding of de La Vallée Poussin's proof for Chebycheff approximation, *Annals of Pure and Applied Logic*, 64, pp. 27–94.

[1996a] Analyzing proofs in analysis, in: *Logic: From Foundations to Applications: European Logic Colloquium '93*, W. Hodges et al., eds., Clarendon Press, Oxford, pp. 225–260.

[1996b] Mathematically strong subsystems of analysis with low rate of growth of provably recursive functionals, *Archive for Mathematical Logic*, 36, pp. 31–71.

G. KREISEL

[1951] On the interpretation of non-finitist proofs, part I, *Journal of Symbolic Logic*, 16, pp. 241–267.

[1952] On the interpretation of non-finitist proofs, part II: Interpretation of number theory, applications, *Journal of Symbolic Logic*, 17, pp. 43–58.

[1957] Gödel's interpretation of Heyting's arithmetic, in: *Summaries of talks, Summer Institute for Symbolic Logic, Cornell University*, Institute for Defense Analyses.

[1959] Interpretation of analysis by means of constructive functionals of finite type, in: Heyting [1959], pp. 101–128.

J. VAN LEEUWEN

[1990] ed., *The Handbook of Theoretical Computer Science*, vol. B, MIT Press (Cambridge, MA) / Elsevier (Amsterdam).

H. LUCKHARDT

[1973] *Extensional Gödel Functional Interpretation*, Lecture Notes in Mathematics #306, Springer-Verlag, Berlin.

W. MAASS

[1976] Eine Funktionalinterpretation der predikativen Analysis, *Archiv für Mathematische Logik und Grundlagenforschung*, 18, pp. 27–46.

P. MARTIN-LÖF

[1971] Hauptsatz for the theory of species, in: Fenstad [1971], pp. 217–233.

[1973] An intuitionistic theory of types: predicative part, in: *Logic Colloquium '73*, H. E. Rose and J. C. Shepherdson, eds., North-Holland, Amsterdam.

G. METAKIDES

[1982] ed., *Patras Logic Symposion*, North-Holland, Amsterdam.

J. C. MITCHELL

[1990] Type systems for programming languages, in: van Leeuwen [1990], pp. 365–458.

P. ODIFREDDI

[1990] ed., *Logic and Computer Science*, Academic Press, London.

C. PARSONS

[1970] On a number-theoretic choice scheme and its relation to induction, in: Kino, Myhill and Vesley [1970], pp. 459–473.

[1972] On n-quantifier induction, *Journal of Symbolic Logic*, 37, pp. 466–482.

D. PRAWITZ

[1971] Ideas and results in proof theory, in: Fenstad [1971], pp. 235–307.

J. C. REYNOLDS

[1974] Towards a theory of type structure, in: *Programming Symposium*, B. Robinet, ed., Lecture Notes in Computer Science #19, Springer-Verlag, Berlin, pp. 408–425.

K. SCHÜTTE

[1960] *Beweistheorie*, Springer-Verlag, Berlin.

[1977] *Proof Theory*, Springer-Verlag, Berlin.

H. SCHWICHTENBERG

[1977] Proof theory: Some aspects of cut-elimination, in: Barwise [1977], pp. 867–895.

J. R. SHOENFIELD

[1967] *Mathematical Logic*, Addison Wesley, Reading, MA.

W. SIEG

[1985] Fragments of arithmetic, *Annals of Pure and Applied Logic*, 28, pp. 33–72.

W. SIEG AND C. PARSONS

[1994] Introductory note to 1938a, in: Gödel [1994], pp. 62–85.

S. G. SIMPSON

[1987] Subsystems of Z_2 and reverse mathematics, in: appendix to Takeuti [1987], pp. 432–446.

C. SPECTOR

[1962] Provably recursive functionals of analysis: a consistency proof of analysis by an extension of principles formulated in current intuitionistic mathematics, in: *Recursive Function Theory: Proc. Symposia in Pure Mathematics*, J. C. E. Dekker, ed., vol. 5, American Mathematical Society, Providence, Rhode Island, pp. 1–27.

W. W. TAIT

[1965] Infinitely long terms of transfinite type, in: *Formal Systems and Recursive Functions*, J. N. Crossley and M. A. E. Dummett, eds., North-Holland, Amsterdam, pp. 176–185.

[1967] Intensional interpretations of functionals of finite type, I, *Journal of Symbolic Logic*, 32, pp. 198–212.

[1968] Normal derivability in classical logic, in: *The Syntax and Semantics of Infinitary Logic*, J. Barwise, ed., Lecture Notes in Mathematics #72, Springer-Verlag, Berlin, pp. 204–236.

[1971] Normal form theorem for bar recursive functions of finite type, in: Fenstad [1971], pp. 353–367.

G. TAKEUTI

[1987] *Proof Theory*, North-Holland, Amsterdam, second ed.

A. S. TROELSTRA

[1973] *Metamathematical Investigation of Intuitionistic Arithmetic and Analysis*, Lecture Notes in Mathematics #344, Springer-Verlag, Berlin.

[1977] Aspects of constructive mathematics, in: Barwise [1977], pp. 973–1052.

[1990] Introductory note to 1958 and 1972, in: Gödel [1990], pp. 217–241.

A. S. TROELSTRA AND D. VAN DALEN

[1988] *Constructivism in Mathematics: An Introduction*, vol. 1, North-Holland, Amsterdam.

H. WEYL

[1918] *Das Kontinuum. Kritische Untersuchungen über die Grundlagen der Analysis*, Veit, Leipzig. Second edition (1932).

[1994] *The Continuum. A Critical Examination of the Foundation of Analysis*, Dover, New York. English translation of Weyl [1918].

CHAPTER VI

Realizability

A. S. Troelstra[1]

Institute for Language, Logic and Computer Science, University of Amsterdam
NL-1018 TV Amsterdam, The Netherlands

Contents

HANDBOOK OF PROOF THEORY
Edited by S. R. Buss

[1]S. Buss, U. Kohlenbach, H. Luckhardt, J.R. Moschovakis and J. van Oosten have commented on earlier drafts of this paper. Van Oosten also provided a sketch for section 8 which has been used in composing the final version.

1. Numerical realizability

1.1. Introduction

There is not just one single notion of realizability, but a whole family of notions, which of course resemble each other in certain respects. This section is devoted to a fairly detailed discussion of the earliest and most basic notion of realizability, S.C. Kleene's realizability by numbers. In later sections we discuss more briefly variations of the basic notion. We do not aim at an exhaustive description of all possible proof-theoretic applications of realizability, but rather aim at presenting illustrative examples. Most of the sections are followed by "Notes", containing suggestions for further reading, some historical comments, etc. The historical comments concern mainly the period *after* 1972, since the history up till 1972 is fairly completely documented in Troelstra [1973a].

Realizability by numbers was introduced by Kleene [1945] as a semantics for intuitionistic arithmetic, by defining for arithmetical sentences A a notion "the number $\mathbf{n}$ realizes A", intended to capture some essential aspects of the intuitionistic meaning of A. Here $\mathbf{n}$ is not a term of the arithmetical formalism, but an element of the natural numbers $\mathbb{N}$. The definition is by induction on the complexity of A:

- $\mathbf{n}$ realizes $t = s$ iff $t = s$ holds;
- $\mathbf{n}$ realizes $A \wedge B$ iff $\mathbf{p}_0\mathbf{n}$ realizes A and $\mathbf{p}_1\mathbf{n}$ realizes B;
- $\mathbf{n}$ realizes $A \vee B$ iff $\mathbf{p}_0\mathbf{n} = 0$ and $\mathbf{p}_1\mathbf{n}$ realizes A or $\mathbf{p}_0\mathbf{n} = 1$ and $\mathbf{p}_1\mathbf{n}$ realizes B;
- $\mathbf{n}$ realizes $A \to B$ iff for all $\mathbf{m}$ realizing A, $\mathbf{n}\bullet\mathbf{m}$ is defined and realizes B;
- $\mathbf{n}$ realizes $\neg A$ if for no $\mathbf{m}$, $\mathbf{m}$ realizes A;
- $\mathbf{n}$ realizes $\exists y\, A$ iff $\mathbf{p}_1\mathbf{n}$ realizes $A[y/\overline{\mathbf{p}_0\mathbf{n}}]$.
- $\mathbf{n}$ realizes $\forall y\, A$ iff $\mathbf{n}\bullet\mathbf{m}$ is defined and realizes $A[y/\overline{\mathbf{m}}]$, for all $\mathbf{m}$.

Here $\mathbf{p}_1$ and $\mathbf{p}_0$ are the inverses of some standard primitive recursive pairing function $\mathbf{p}$ coding $\mathbb{N}^2$ onto $\mathbb{N}$, and $\overline{\mathbf{m}}$ is the standard term $S^{\mathbf{m}}0$ (numeral) in the language of intuitionistic arithmetic corresponding to $\mathbf{m}$; $\bullet$ is partial recursive function application, i.e. $\mathbf{n}\bullet\mathbf{m}$ is the result of applying the function with code $\mathbf{n}$ to $\mathbf{m}$. (Later on we also use $\bar{m}, \bar{n}, \ldots$ for numerals.)

The definition may be extended to formulas with free variables by stipulating that $\mathbf{n}$ realizes A if $\mathbf{n}$ realizes the universal closure of A.

Reading "there is a number realizing A" as "A is constructively true", we see that a realizing number provides witnesses for the constructive truth of existential quantifiers and disjunctions, and in implications carries this type of information from premise to conclusion by means of partial recursive operators. In short, realizing numbers "hereditarily" encode information about the realization of existential quantifiers and disjunctions.

Realizability, as an interpretation of "constructively true" is reminiscent of the well-known Brouwer-Heyting-Kolmogorov explanation (BHK for short) of the intuitionistic meaning of the logical connectives. BHK explains "p proves A" for compound A in terms of the provability of the components of A. For prime formulas the notion of proof is supposed to be given. Examples of the clauses of BHK are:

- p proves $A \to B$ iff p is a construction transforming any proof c of A into a proof $p(c)$ of B;
- p proves $A \wedge B$ iff $p = (p_0, p_1)$ and p_0 proves A, p_1 proves B;
- p proves $A \vee B$ iff $p = (p_0, p_1)$ with $p_0 \in \{0, 1\}$, and p_1 proves A if $p_0 = 0$, p_1 proves B if $p_0 \neq 0$.

Realizability corresponds to BHK if (a) we concentrate on (numerical) information concerning the realizations of existential quantifiers and the choices for disjunctions, and (b) the constructions considered for $\forall, \to$ are encoded by (partial) recursive operations.

Realizability gives a classically meaningful definition of intuitionistic truth; the set of realizable statements is closed under deduction and must be consistent, since 1=0 cannot be realizable. It is to be noted that decidedly non-classical principles are realizable, for example

$$\neg\forall x[\exists y Txxy \vee \forall y \neg Txxy]$$

is easily seen to be realizable. (T is Kleene's T-predicate, which is assumed to be available in our language; $Txyz$ is primitive recursive in x, y, z and expresses that the algorithm with code x applied to argument y yields a computation with code z; U is a primitive recursive function extracting from a computation code z the result Uz.) For $\neg A$ is realizable iff no number realizes A, and realizability of $\forall x[\exists y Txxy \vee \forall y \neg Txxy]$ requires a total recursive function deciding $\exists y Txxy$, which does not exist (more about this below). In this way realizability shows how in constructive mathematics principles may be incorporated which cause it to diverge from the corresponding classical theory, instead of just being included in the classical theory.

Some notational habits adopted in this paper are: dropping of distinguishing sub- and superscripts where the context permits; saving on parentheses, e.g. for a binary predicate R applied to x, y we often write Rxy instead of $R(x, y)$ (this habit has just been demonstrated above). The symbol $\equiv$ is used for literal identity of expressions modulo renaming of bound variables. $\Rightarrow$ is used as metamathematical consequence relation, and in particular $\mathcal{A}, \mathcal{B} \Rightarrow \mathcal{C}$ expresses a rule which derives $\mathcal{C}$ from premises $\mathcal{A}, \mathcal{B}$. FV($\mathcal{A}$) is the set of free variables of expression $\mathcal{A}$.

1.2. Formalizing realizability in HA

In order to exploit realizability proof-theoretically, we have to formalize it. Let us first discuss its formalization in ordinary intuitionistic first-order arithmetic **HA** ("Heyting's Arithmetic"), based on intuitionistic predicate logic with equality, and containing symbols for all primitive recursive functions, with their recursion equations as axioms.

$x, y, z, \ldots$ are numerical variables, S is successor. We use the notation $\bar{n}$ for the term $S^n 0$; such terms are called *numerals*. $\mathbf{p}_0, \mathbf{p}_1$ bind stronger than infix binary operations, i.e. $\mathbf{p}_0 t + s$ is $(\mathbf{p}_0 t) + s$. For primitive recursive predicates R, $Rt_1 \ldots t_n$

may be treated as a prime formula since the formalism contains a symbol for the characteristic function χ_R.

Now we are ready for a formalized definition of "x realizes A" in **HA**.

Definition. By recursion on the complexity of A we define $x\,\underline{\mathrm{rn}}\,A$, $x \notin \mathrm{FV}(A)$, "x numerically realizes A" :

$$\begin{array}{ll} x\,\underline{\mathrm{rn}}\,(t=s) & := (t=s) \\ x\,\underline{\mathrm{rn}}\,(A \wedge B) & := (\mathbf{p}_0 x\,\underline{\mathrm{rn}}\,A) \wedge (\mathbf{p}_1 x\,\underline{\mathrm{rn}}\,B), \\ x\,\underline{\mathrm{rn}}\,(A \to B) & := \forall y(y\,\underline{\mathrm{rn}}\,A \to \exists z(Txyz \wedge Uz\,\underline{\mathrm{rn}}\,B)), \\ x\,\underline{\mathrm{rn}}\,\forall y\,A & := \forall y \exists z(Txyz \wedge Uz\,\underline{\mathrm{rn}}\,A), \\ x\,\underline{\mathrm{rn}}\,\exists y\,A & := \mathbf{p}_1 x\,\underline{\mathrm{rn}}\,A[y/\mathbf{p}_0 x]. \end{array}$$

Note that $\mathrm{FV}(x\,\underline{\mathrm{rn}}\,A) \subset \{x\} \cup \mathrm{FV}(A)$. □

Remarks. (i) We have omitted clauses for negation and disjunction, since in arithmetic we can take $\neg A := A \to 1 = 0$, $A \vee B := \exists x((x = 0 \to A) \wedge (x \neq 0 \to B))$. If we spell out $x\,\underline{\mathrm{rn}}\,(A \vee B)$ on the basis of this definition, we find

$$x\,\underline{\mathrm{rn}}\,(A \vee B) := (\mathbf{p}_0 x = 0 \to (\mathbf{p}_0\mathbf{p}_1 x)0\,\underline{\mathrm{rn}}\,A) \wedge (\mathbf{p}_0 x \neq 0 \to (\mathbf{p}_1\mathbf{p}_1 x)0\,\underline{\mathrm{rn}}\,B),$$

(ii) The definition of realizability permits slight variations, e.g. for the first clause we might have taken

$$x\,\underline{\mathrm{rn}}'(t=s) := (x = t \wedge t = s).$$

However, it is routine to see that this variant $\underline{\mathrm{rn}}'$-realizability is *equivalent* to $\underline{\mathrm{rn}}$-realizability in the following sense: for each formula A there are two partial recursive functions ϕ_A and ψ_A such that

$$\vdash x\,\underline{\mathrm{rn}}\,A \to \phi_A(x)\,\underline{\mathrm{rn}}'A$$
$$\vdash x\,\underline{\mathrm{rn}}'A \to \psi_A(x)\,\underline{\mathrm{rn}}\,A.$$

(If in the future we shall call two versions of a realizability notion equivalent, it will always be in this or a similar sense.) Similarly, if we treat $\vee$ as a primitive, the clause for $x\,\underline{\mathrm{rn}}\,(A \vee B)$ given above may be simplified to

$$x\,\underline{\mathrm{rn}}\,(A \vee B) := (\mathbf{p}_0 x = 0 \wedge \mathbf{p}_1 x\,\underline{\mathrm{rn}}\,A) \vee (\mathbf{p}_0 x \neq 0 \wedge \mathbf{p}_1 x\,\underline{\mathrm{rn}}\,B),$$

which yields an equivalent notion of realizability.

(iii) In terms of partial recursive function application $\bullet$ and the definedness predicate $\downarrow$ ($t\downarrow$ means "t is defined"), we can write more succinctly:

$$\begin{array}{ll} x\,\underline{\mathrm{rn}}\,(A \to B) & := \forall y(y\,\underline{\mathrm{rn}}\,A \to x{\bullet}y\downarrow \wedge\, x{\bullet}y\,\underline{\mathrm{rn}}\,B), \\ x\,\underline{\mathrm{rn}}\,\forall y\,A & := \forall y(x{\bullet}y\downarrow \wedge\, x{\bullet}y\,\underline{\mathrm{rn}}\,B). \end{array}$$

where $t\downarrow$ expresses that t is defined (cf. next subsection). Of course, the partial operation $\bullet$ and the definedness predicate $\downarrow$ are not part of the language, but expressions containing them may be treated as abbreviations, using the following equivalences:

$$\begin{array}{l} t_1 = t_2 \leftrightarrow \exists x(t_1 = x \wedge t_2 = x), \\ t_1{\bullet}t_2 = x \leftrightarrow \exists yzu(t_1 = y \wedge t_2 = z \wedge Tyzu \wedge Uu = x), \\ t\downarrow\, \leftrightarrow \exists z(t = z). \end{array}$$

(t_1, t_2 terms containing $\bullet$, x, y, z, u not free in t_1, t_2). However, note that the logical complexity of $A(t)$, where t is an expression containing $\bullet$, depends on the complexity of t! (On the other hand, $t\downarrow$ is always expressible in Σ^0_1-form.) For metamathematical investigations it is therefore more convenient to formalize realizability in a conservative extension $\mathbf{HA}^*$ of $\mathbf{HA}$ in which we can treat "$\bullet$" as a primitive. Treating $t_1 = t_2$ for partially defined t_1, t_2 as an abbreviation in a rigorous way is possible, but involves a good deal of lengthy inductions, as demonstrated in Kleene [1969]. Since ordinary logic deals with total functions only, we first need to extend our logic to the (intuitionistic) logic of partial terms LPT, or intuitionistic E^+-logic, in the terminology of Troelstra and van Dalen[1988,2.2.3]. LPT first appeared in Beeson [1981].

1.3. Intuitionistic predicate logic with partial terms LPT

Variables are supposed to range over the objects of the domain considered, so always denote; arbitrary terms need not denote, so we need a predicate $\mathbf{E}$, expressing definedness; $\mathbf{E}t$ reads "t denotes" or "t is defined". Instead of $\mathbf{E}t$ we shall write $t\downarrow$, in the notation commonly used in recursion theory.

If we also have equality in our logic, and read $t = s$ as "t and s are both defined and equal", we can express $t\downarrow$ as $t = t$.

The following axiomatization is a convenient (but not canonical) choice for arguments proceeding by induction on the length of formal deductions:

L1	$A \to A$,
L2	$A,\ A \to B \Rightarrow B$,
L3	$A \to B,\ B \to C \Rightarrow A \to C$,
L4	$A \wedge B \to A,\ A \wedge B \to B$,
L5	$A \to B,\ A \to C \Rightarrow A \to B \wedge C$,
L6	$A \to A \vee B,\ B \to A \vee B$,
L7	$A \to C,\ B \to C \Rightarrow A \vee B \to C$,
L8	$A \wedge B \to C \Rightarrow A \to (B \to C)$,
L9	$A \to (B \to C) \Rightarrow A \wedge B \to C$,
L10	$\bot \to A$,
L11	$B \to A \Rightarrow B \to \forall x\, A \ \ (x \notin \mathrm{FV}(B))$,
L12	$\forall x\, A \wedge t\downarrow \to A[x/t]$,
L13	$A[x/t] \wedge t\downarrow \to \exists x\, A$,
L14	$A \to B \Rightarrow \exists x\, A \to B \ \ (x \notin \mathrm{FV}(B))$

where $t\downarrow := t = t$. For equality we have (F function symbol, R relation symbol of the language):

EQ
$$\begin{cases} \forall xy(x = y \to y = x), & \forall xyz(x = y \wedge y = z \to x = z), \\ \forall \vec{x}\vec{y}(\vec{x} = \vec{y} \wedge F\vec{x}\downarrow \to F\vec{x} = F\vec{y}), & \forall \vec{x}\vec{y}(R\vec{x} \wedge \vec{x} = \vec{y} \to R\vec{y}) \end{cases}$$

Basic predicates and functions of the language are assumed to be strict:

STR $\quad F(t_1, \ldots, t_n)\downarrow \to t_i\downarrow, \quad R(t_1, \ldots, t_n) \to t_i\downarrow$

Note that this logic reduces to ordinary first-order intuitionistic logic if all functions are total, i.e. $\forall \vec{x}(f\vec{x}\downarrow)$, since then $t\downarrow$ for all terms t.

For the notion *"equally defined and equal if defined"* introduced by

$$t \simeq s := (t{\downarrow} \vee s{\downarrow}) \to t = s,$$

we can prove the replacement schema for arbitrary formulas A

$$t \simeq s \wedge A[x/t] \to A[x/s].$$

1.4. Conservativeness of defined functions

Relative to the logic of partial terms, the following conservative extension result is easily proved. Let Γ be a theory based on LPT, such that

$$\Gamma \vdash A(\vec{x}, y) \wedge A(\vec{x}, z) \to y = z.$$

Then we may introduce a symbol ϕ_A for a partial function with axiom

$$\mathrm{Ax}(\phi_A) \qquad A(\vec{x}, y) \leftrightarrow y = \phi_A(\vec{x}).$$

The conservativeness of this addition can be proved in a straightforward syntactic way; the easiest method, however, uses completeness for Kripke models, see Troelstra and van Dalen [1988,2.7].

Let Γ^* consist of Γ and all substitution instances of the axiom schemata w.r.t. the extended language, and let $\phi(\Gamma^*)$ be the result of systematically eliminating the function symbol ϕ_A from the elements of Γ, and assume $\phi(\Gamma^*)$ to be provable from Γ, then the conservative extension result still holds in the form: "$\Gamma^* + \mathrm{Ax}(\phi_A)$ is conservative over Γ".

This extended result applies to $\mathbf{HA}^*$ defined below, since eliminating the symbol for partial recursive function application from instances of induction yields instances of induction in the language of **HA**.

1.5. Formalizing elementary recursion theory in $\mathbf{HA}^*$

$\mathbf{HA}^*$ is the conservative extension of **HA**, formulated in the intuitionistic logic of partial terms, with a primitive binary partial operation $\bullet$ of partial recursive function application. $t_1 \bullet t_2 \bullet t_3 \ldots$ abbreviates $(\ldots((t_1 \bullet t_2) \bullet t_3)\ldots)$ (association to the left).

Note that strictness entails in particular $t \bullet t'{\downarrow} \to t{\downarrow} \wedge t'{\downarrow}$ for the application operation. Of course we have to require totality for the primitive recursive functions; it suffices to demand $0{\downarrow}$, $Sx{\downarrow}$. In all other cases the primitive recursive functions satisfy equations with $=$, characterizing them inductively in terms of functions introduced before (e.g. $x + 0 = x$, $x + Sy = S(x + y)$). By induction one can then prove $Fx_1 \ldots x_n{\downarrow}$ for each primitive recursive function symbol F.

A smooth formalization of elementary recursion theory in $\mathbf{HA}^*$ can be given by using Kleene's index method in combination with the theory of elementary inductive definitions in arithmetic Troelstra and van Dalen [1988,3.6, 3.7]. In particular we obtain the smn-theorem, the recursion theorem (Kleene's fixed-point theorem), the Kleene normal form theorem, etc. Moreover, by the normal form theorem, every partial recursive function is definable by a term of the language of $\mathbf{HA}^*$.

Notation. If t is a term in the language of $\mathbf{HA}^*$, then $\Lambda x.t$ is a canonically chosen code number for t as a partial recursive function of x, uniformly in the other free variables; by the smn-theorem we may therefore assume $\Lambda x.t$ to be primitive recursive in $\mathrm{FV}(t) \setminus \{x\}$. $\Lambda x_1 \dots x_n.t$ abbreviates $\Lambda x_1(\Lambda x_2 \dots (\Lambda x_n.t) \dots)$. □

We note the following

Lemma. *In* $\mathbf{HA}^*$ *the* Σ^0_1*-formulas of* $\mathbf{HA}$ *are equivalent to prime formulas of the form* $t = t$ *for suitable* t, *and each formula* $t = s$ *is equivalent to a* Σ^0_1*-formula of* $\mathbf{HA}$.

Proof. Systematically using the equivalences mentioned above transforms any formula $t = s$ of $\mathbf{HA}^*$ into a Σ^0_1-formula of $\mathbf{HA}$. Conversely, let a Σ^0_1-formula be given; by the normal form results of recursion theory, we can write this in the form $\exists z T(\bar{n}, \langle \vec{x} \rangle, z)$ for a numeral $\bar{n}$; this is equivalent to $\bar{n}\bullet\langle \vec{x} \rangle = \bar{n}\bullet\langle \vec{x} \rangle$. □

We are now ready to formalize $x\,\underline{\mathrm{rn}}\,A$ directly in $\mathbf{HA}^*$.

1.6. Formalizing $\underline{\mathrm{rn}}$-realizability in HA*

Definition. $x\,\underline{\mathrm{rn}}\,A$ is defined by induction on the complexity of A, $x \notin \mathrm{FV}(A)$.

$$
\begin{aligned}
x\,\underline{\mathrm{rn}}\,P &:= P \wedge x\downarrow \text{ for } P \text{ prime},\\
x\,\underline{\mathrm{rn}}\,(A \wedge B) &:= \mathbf{p}_0 x\,\underline{\mathrm{rn}}\,A \wedge \mathbf{p}_1 x\,\underline{\mathrm{rn}}\,B,\\
x\,\underline{\mathrm{rn}}\,(A \to B) &:= \forall y(y\,\underline{\mathrm{rn}}\,A \to x\bullet y\,\underline{\mathrm{rn}}\,B) \wedge x\downarrow,\\
x\,\underline{\mathrm{rn}}\,\forall y\,A &:= \forall y(x\bullet y\,\underline{\mathrm{rn}}\,A),\\
x\,\underline{\mathrm{rn}}\,\exists y\,A &:= \mathbf{p}_1 x\,\underline{\mathrm{rn}}\,A[y/\mathbf{p}_0 x].
\end{aligned}
$$

We also define a combination of realizability with truth, $x\,\underline{\mathrm{rnt}}\,A$; the clauses are the same as for $\underline{\mathrm{rn}}$, the clause for implication excepted, which now reads:

$$x\,\underline{\mathrm{rnt}}\,(A \to B) := \forall y(y\,\underline{\mathrm{rnt}}\,A \to x\bullet y\,\underline{\mathrm{rnt}}\,B) \wedge x\downarrow \wedge (A \to B). \quad \square$$

Remarks. (i) $t\,\underline{\mathrm{rn}}\,A$ is $\exists$-free (i.e. does not contain $\exists$) for all A. Note that, by our definition of $\vee$ in terms of the other operators, $\exists$-free implies $\vee$-free.

(ii) The clauses "$\wedge\, x\downarrow$" have been added for the cases of prime formulas and implications, in order to guarantee the truth of part (i) of the following lemma.

(iii) For negations we have $x\,\underline{\mathrm{rn}}\,\neg A \leftrightarrow \forall y(\neg y\,\underline{\mathrm{rn}}\,A) \wedge x\downarrow$, and $x\,\underline{\mathrm{rn}}\,\neg\neg A \leftrightarrow \forall y(\neg y\,\underline{\mathrm{rn}}\,\neg A) \wedge x\downarrow \leftrightarrow \forall y \neg\forall z \neg(z\,\underline{\mathrm{rn}}\,A) \wedge x\downarrow \leftrightarrow \neg\neg\exists z(z\,\underline{\mathrm{rn}}\,A) \wedge x\downarrow$.

The following lemmas are easily proved by induction on A.

Lemma. *(Definedness of realizing terms; Substitution Property) For* $\mathrm{R} \in \{\underline{\mathrm{rn}}, \underline{\mathrm{rnt}}\}$

(i) $\vdash t\,\mathrm{R}\,A \to t\downarrow$,

(ii) $(x\,\mathrm{R}\,A)[y/t] \equiv x\,\mathrm{R}\,(A[y/t])$ $(x \notin \mathrm{FV}(A) \cup \mathrm{FV}(t), y \not\equiv x)$.

Proof. By induction on the complexity of A. Let e.g. $t\,\underline{\mathrm{rn}}\,\exists y A$, then $\mathbf{p}_1 t\,\underline{\mathrm{rn}}\,A[y/\mathbf{p}_0 t]$, hence by induction hypothesis $\mathbf{p}_1 t\downarrow$, and so by strictness $t\downarrow$. □

Lemma. $\mathbf{HA}^* \vdash t\,\underline{\mathrm{rnt}}\,A \to A$.
A similar lemma holds for all combinations of realizability with truth (i.e. realizabilities with $\underline{\mathrm{t}}$ in their mnemonic code) we shall encounter in the sequel; we shall not bother to state it explicitly in the future. We can readily prove that realizability is sound for $\mathbf{HA}^*$:

1.7. Theorem. *(Soundness theorem)*

$$\mathbf{HA}^* \vdash A \Rightarrow \mathbf{HA}^* \vdash t\,\underline{\mathrm{rn}}\,A \wedge t\,\underline{\mathrm{rnt}}\,A$$

for a suitable term t with $\mathrm{FV}(t) \subset \mathrm{FV}(A)$.

Proof. The proof proceeds by induction on the length of derivations; that is to say, we have to find realizing terms for the axioms, and for the rules we must show how to find a realizing term for the conclusion from realizing terms for the premises. We check some cases.

L5. Assume $t\,\underline{\mathrm{rn}}\,(A \to B)$, $t'\,\underline{\mathrm{rn}}\,(A \to C)$, and $x\,\underline{\mathrm{rn}}\,A$; then $\mathbf{p}(t{\bullet}x, t'{\bullet}x)\,\underline{\mathrm{rn}}\,(B \wedge C)$, so $\Lambda x.\mathbf{p}(t{\bullet}x, t'{\bullet}x)\,\underline{\mathrm{rn}}\,(A \to B \wedge C)$.

L14. Assume $t\,\underline{\mathrm{rn}}\,(A \to B)$, $x \notin \mathrm{FV}(B)$, and let $y\,\underline{\mathrm{rn}}\,\exists x\, A$, then $\mathbf{p}_1 y\,\underline{\mathrm{rn}}\,A[x/\mathbf{p}_0 y]$, hence $t[x/\mathbf{p}_0 y]{\bullet}(\mathbf{p}_1 y)\,\underline{\mathrm{rn}}\,B$, so $\Lambda y.t[x/\mathbf{p}_0 y]{\bullet}(\mathbf{p}_1 y)\,\underline{\mathrm{rn}}\,(\exists x\, A \to B)$.

Of the non-logical axioms, only induction requires attention. Suppose

$$x\,\underline{\mathrm{rn}}\,(A[y/0] \wedge \forall y(A \to A[y/Sy])).$$

Then

$$\mathbf{p}_0 x\,\underline{\mathrm{rn}}\,A[y/0], \quad z\,\underline{\mathrm{rn}}\,A \to (\mathbf{p}_1 x){\bullet}y{\bullet}z\,\underline{\mathrm{rn}}\,A[y/Sy].$$

So let t be such that

$$t{\bullet}0 \simeq \mathbf{p}_0 x, \quad t{\bullet}(Sy) \simeq (\mathbf{p}_1 x){\bullet}y{\bullet}(t{\bullet}y).$$

The existence of t follows either by an application of the recursion theorem, or is immediate if closure under recursion has been built directly into the definition of recursive function. It is now easy to prove by induction that t realizes induction for A. □

A statement weaker than soundness is $\vdash A \Rightarrow \vdash \exists x(x\,\underline{\mathrm{rn}}\,A)$; we might call this *weak soundness.* We can also prove a stronger version of soundness:

1.8. Theorem. *(Strong Soundness Theorem) For closed A*

$$\mathbf{HA}^* \vdash A \Rightarrow \mathbf{HA}^* \vdash \bar{n}\,\underline{\mathrm{rn}}\,A \wedge \bar{n}\,\underline{\mathrm{rnt}}\,A \quad \textit{for some numeral } \bar{n}.$$

Proof. Let $\mathbf{HA}^* \vdash A$; from the soundness theorem we find a term t such that

$$t\,\underline{\mathrm{rn}}\,A, \text{ hence } t\!\downarrow.$$

$t\!\downarrow$, i.e. $t = t$ is equivalent to a Σ^0_1-formula of $\mathbf{HA}$, say $\exists x(s = 0)$, and $\mathbf{HA}$ proves only true Σ^0_1-formulas, from which we see that $t = \bar{n}$ must be provable in $\mathbf{HA}^*$ for some numeral $\bar{n}$. Similarly for $\underline{\mathrm{rnt}}$. □

1.9. Remark. If one formalizes the proof of the soundness theorem, it is easy to see that there are primitive recursive functions ψ, ϕ such that

$$\mathbf{HA} \vdash \mathrm{Prf}(x, \ulcorner A \urcorner) \to \mathrm{Prf}(\phi(x), \mathrm{Sub}(\ulcorner y\, \underline{\mathrm{rn}}\, A \urcorner, y, \psi(x)))$$

where "Prf" is the formalized proof-predicate of $\mathbf{HA}^*$, $\ulcorner \xi \urcorner$ is the gödelnumber of expression ξ, and $\mathrm{Sub}(\ulcorner B \urcorner, x, \ulcorner s \urcorner)$ is the gödelnumber of $B[x/s]$.

In fact, the whole implication is provable even in primitive recursive arithmetic. But the statement expressing a formalized version of the *strong* completeness theorem:

$$\mathrm{Prf}(x, \ulcorner A \urcorner) \to \mathrm{Prf}(\phi(x), \ulcorner \overline{\psi(x)}\, \underline{\mathrm{rn}}\, A \urcorner)$$

(A closed, for suitable provably recursive ϕ, ψ) is not provable in $\mathbf{HA}$ (see section 1.16).

1.10. Lemma. *(Self-realizing formulas) For $\exists$-free formulas, canonical realizers exist, that is to say for each $\exists$-free A we have in* $\mathbf{HA}^*$

(i) $\vdash \exists x(x\, \underline{\mathrm{rn}}\, A) \to A$,

(ii) $\vdash A \to t_A\, \underline{\mathrm{rn}}\, A$ *for some term* t_A *with* $\mathrm{FV}(t_A) \subset \mathrm{FV}(A)$.

(iii) A formula A is provably equivalent to its own realizability, i.e. $A \leftrightarrow \exists x(x\, \underline{\mathrm{rn}}\, A)$*), iff A is provably equivalent to an existentially quantified $\exists$-free formula.*

(iv) Realizability is idempotent, i.e. $\exists x(x\, \underline{\mathrm{rn}}\, \exists y(y\, \underline{\mathrm{rn}}\, A)) \leftrightarrow \exists x(x\, \underline{\mathrm{rn}}\, A)$*; in fact, even* $\exists x(x\, \underline{\mathrm{rn}}\, (A \leftrightarrow \exists y(y\, \underline{\mathrm{rn}}\, A)))$ *holds.*

Proof. Take $t_{s=s'} := 0$, $t_{A \wedge B} := \mathbf{p}(t_A, t_B)$, $t_{\forall x A} := \Lambda x.t_A$, $t_{A \to B} := \Lambda x.t_B$ ($x \notin \mathrm{FV}(t_B)$), and prove (i) and (ii) by simultaneous induction on A. (iii) and (iv) are immediate corollaries. □

Remark. An observation of practical usefulness is the following. For any definable predicate with canonical realizers (i.e. a predicate A definable by an $\exists$-free formula) we obtain an equivalent realizability if we read restricted quantifiers $\forall x(A(x) \to \ldots)$ and $\exists x(A(x) \wedge \ldots)$ as quantifiers $\forall x{\in}A$, $\exists x{\in}A$ over a new domain with realizability clauses copied from numerical quantification, i.e.

$$x\, \underline{\mathrm{rn}}\, \forall y{\in}A.B := \forall y{\in}A(x{\bullet}y\, \underline{\mathrm{rn}}\, B) \wedge x{\downarrow},$$
$$x\, \underline{\mathrm{rn}}\, \exists y{\in}A.B := \mathbf{p}_1 x\, \underline{\mathrm{rn}}\, B[x/\mathbf{p}_0 x] \wedge A(\mathbf{p}_0 x).$$

In short, we may simply forget about the canonical realizers.

1.11. Axiomatizing provable realizability

As we have seen already in the introduction, realizability validates more than what is provable in $\mathbf{HA}$; in fact, we can formally prove realizability of in $\mathbf{HA}^*$ an intuitionistic version of Church's thesis:

CT_0 $\quad \forall x \exists y\, A(x,y) \to \exists z \forall x(A(x, z{\bullet}x) \wedge z{\bullet}x{\downarrow})$.

CT_0 is certainly not *provable in* $\mathbf{HA}$, since it is in fact refutable in classical arithmetic. This version of Church's thesis is in fact a combination of the well-known version which states "Each humanly computable function is recursive" and the intuitionistic reading of $\forall x \exists y A(x,y)$ which states that there is a method for constructing, for each given x, a y such that $A(x,y)$. Such a method describes a humanly computable function.

We now ask ourselves: is there a reasonably simple axiomatization (by a few axiom schemata say) of the formulas provably realizable in **HA**? The answer is yes, the provably realizable formulas can be axiomatized by a generalization of CT_0, namely *"Extended Church's Thesis"*:

ECT_0 $\quad \forall x(Ax \to \exists y\, Bxy) \to \exists z \forall x(Ax \to z{\bullet}x{\downarrow} \wedge B(x, z{\bullet}x))$ (A $\exists$-free).

Lemma. *Each instance of* ECT_0 *is* $\mathbf{HA}^*$*-realizable.*

Proof. Suppose

$$u \,\underline{\mathrm{rn}}\, \forall x(Ax \to \exists y Bxy)$$

Then $\forall xv(v \,\underline{\mathrm{rn}}\, Ax \to u{\bullet}x{\bullet}v \,\underline{\mathrm{rn}}\, \exists y Bxy)$, and since A is $\exists$-free, in particular $\forall x(Ax \to u{\bullet}x{\bullet}t_A \,\underline{\mathrm{rn}}\, \exists y Bxy)$, so $\forall x(Ax \to \mathbf{p}_1(u{\bullet}x{\bullet}t_A) \,\underline{\mathrm{rn}}\, B(x, \mathbf{p}_0(u{\bullet}x{\bullet}t_A)))$. Then it is straightforward to see that

$$\mathbf{p}(\Lambda x.\mathbf{p}_0(u{\bullet}x{\bullet}t_A), \Lambda xv.\mathbf{p}(0, \mathbf{p}_1(u{\bullet}x{\bullet}t_A)))$$

realizes the conclusion. □

Remark. The condition "A is $\exists$-free" in ECT_0 cannot be dropped: applying unrestricted ECT_0 to $Ax := \exists z Txxz \vee \neg\exists z Txxz$, $Bxy := (y = 0 \wedge \exists z Txxz) \vee (y = 1 \wedge \neg\exists z Txxz)$ yields a contradiction. In fact, this example can be used to show that even unrestricted $\mathrm{ECT}_0!$ fails ($\mathrm{ECT}_0!$ is like ECT_0 except that $\exists y$ in the premise is replaced by $\exists! y$; $\exists! y$ means "there is a unique y such that").

Theorem. *(Characterization Theorem for* $\underline{\mathrm{rn}}$*-realizability)*

(i) $\mathbf{HA}^* + \mathrm{ECT}_0 \vdash A \leftrightarrow \exists x(x \,\mathrm{R}\, A)$ *for* $\mathrm{R} \in \{\underline{\mathrm{rn}}, \underline{\mathrm{rnt}}\}$,

(ii) For closed A, $\mathbf{HA}^* + \mathrm{ECT}_0 \vdash A \Leftrightarrow \mathbf{HA}^* \vdash \bar{n} \,\underline{\mathrm{rn}}\, A$ *for some numeral* $\bar{n}$.

Proof. (i) is proved by a straightforward induction on A. The crucial case is $A \equiv B \to C$; then $B \to C \leftrightarrow (\exists x(x \,\underline{\mathrm{rn}}\, B) \to \exists y(y \,\underline{\mathrm{rn}}\, C))$ (by the induction hypothesis) $\leftrightarrow \forall x(x \,\underline{\mathrm{rn}}\, B \to \exists y(y \,\underline{\mathrm{rn}}\, C))$ (by pure logic) $\leftrightarrow \exists z \forall x(x \,\underline{\mathrm{rn}}\, B \to z{\bullet}x \,\underline{\mathrm{rn}}\, C)$ (by ECT_0, since $x \,\underline{\mathrm{rn}}\, B$ is $\exists$-free) $\equiv \exists z(z \,\underline{\mathrm{rn}}\, (B \to C))$.

(ii). The direction $\Rightarrow$ follows from the strong soundness theorem plus the lemma; $\Leftarrow$ is an immediate consequence of (i). □

Curiosity prompts us to ask which formulas are classically provably realizable, i.e. provably realizable in first-order Peano Arithmetic **PA**, which is just **HA** with classical logic. The answer is contained in the following

Proposition. $\mathbf{PA} \vdash \exists x(x \,\underline{\mathrm{rn}}\, A) \Leftrightarrow \mathbf{HA} + \mathrm{M} + \mathrm{ECT}_0 \vdash \neg\neg A$,
where M *is* Markov's principle*:*

M $\quad \forall x(A \vee \neg A) \wedge \neg\neg\exists x\, A \to \exists x\, A.$

Proof. Let $\mathbf{PA} \vdash \exists x(x \,\underline{\mathrm{rn}}\, A)$, and let B be a negative formula (i.e. a formula in the $\wedge, \forall, \rightarrow$-fragment) such that $\mathbf{HA}+\mathrm{M} \vdash x \,\underline{\mathrm{rn}}\, A \leftrightarrow B(x)$. Then $\mathbf{PA} \vdash \neg\forall x \neg(x \,\underline{\mathrm{rn}}\, A)$, and since **PA** is conservative over **HA** for negative formulas (in consequence of Gödel's negative translation), also $\mathbf{HA} \vdash \neg\forall x\neg B$, i.e. $\mathbf{HA}+\mathrm{M} \vdash \neg\neg\exists x(x \,\underline{\mathrm{rn}}\, A)$, and thus it follows that $\mathbf{HA}+\mathrm{M}+\mathrm{ECT}_0 \vdash \neg\neg A$. The converse is simpler. □

1.12. Extensions of HA*

For suitable sets Γ of extra axioms, we may replace $\mathbf{HA}^*$ in the soundness and characterization theorem by $\mathbf{HA}^* + \Gamma$. Weak soundness and the characterization theorem require for all $A \in \Gamma$

(1) $\quad \mathbf{HA}^* + \Gamma \vdash \exists x(x \,\underline{\mathrm{rn}}\, A)$.

Soundness requires for all $A \in \Gamma$

(2) $\quad \mathbf{HA}^* + \Gamma \vdash t \,\underline{\mathrm{rn}}\, A$ for some term t,

and strong soundness requires (2) and in addition: $\mathbf{HA}^* + \Gamma$ proves only true Σ^0_1-formulas.

Examples

(a) For Γ any set of $\exists$-free formulas soundness and the characterization theorem extend. If $\mathbf{HA}^* + \Gamma$ proves only true Σ^0_1-formulas, strong soundness holds. The next two examples permit characterization and strong soundness.

(b) Let $\prec$ be a primitive recursive well-ordering of $\mathbb{N}$, provably total and linear in $\mathbf{HA}^*$; for Γ we take all instances of *transfinite induction over* $\prec$:

$\mathrm{TI}(\prec) \qquad \forall y(\forall x{\prec}y\, A \rightarrow A[x/y]) \rightarrow \forall x A.$

(c) Γ is the set of instances of Markov's principle (cf. the last proposition in 1.11). In fact, in the presence of CT_0, which is valid under realizability, Γ may be replaced by a single axiom:

$$\forall xy(\neg\neg\exists z Txyz \rightarrow \exists z Txyz).$$

It is also worth noting that in the presence of M, we can use the following variant of ECT_0 which is equivalent to ECT_0:

$\mathrm{ECT}_0' \qquad \forall x(\neg A \rightarrow \exists y Bxy) \rightarrow \exists z \forall x(\neg A \rightarrow z{\bullet}x{\downarrow} \wedge B(x, z{\bullet}y)).$

(d) An extension of another kind is obtained if we enrich the language with constants for inductively defined predicates, e.g. the tree predicate Tr. Intuitively, Tr is the least set containing the (code of the) single-node tree (i.e. $\langle\ \rangle \in \mathrm{Tr}$), and with every recursive sequence of tree codes $n{\bullet}0, n{\bullet}1, \ldots, n{\bullet}m, \ldots$ in Tr, Tr also contains a code for the infinite tree having the trees with codes $n{\bullet}m$ as immediate subtrees, namely $\mathbf{p}(1,n)$. Thus if

$$A(X,x) := (x = 0) \vee (\mathbf{p}_0 x = 1 \wedge \forall m(\mathbf{p}_1 x{\bullet}m \in X))$$

we have

$$A(\mathrm{Tr}, x) \rightarrow x \in \mathrm{Tr},$$
$$\forall x(A(\lambda y.B, x) \rightarrow B[y/x]) \rightarrow \forall x \in \mathrm{Tr}.B[y/x]$$

for all B in the language extended with the new primitive predicate Tr. Then we can extend $\underline{\mathrm{rn}}$-realizability simply by putting

$$x\,\underline{\mathrm{rn}}\,(t \in \mathrm{Tr}) := t \in \mathrm{Tr}.$$

Let us check that the soundness theorem extends. $A(\mathrm{Tr}, x)$ is equivalent to an $\exists$-free formula, so its realizability implies its truth, and $x \in \mathrm{Tr}$ follows. As to the schema, assume

$$u\,\underline{\mathrm{rn}}\,\forall x(A(\lambda y.B, x) \to B[y/x]),\ \text{or}$$
$$u\,\underline{\mathrm{rn}}\,\forall x[(x = 0 \to B(0)) \wedge (\mathbf{p}_0 x = 1 \wedge \forall y B(\mathbf{p}_1 x\bullet y) \to Bx)].$$

So

$$\mathbf{p}_0(u\bullet 0)\bullet(0,0)\,\underline{\mathrm{rn}}\,B(0),$$
$$\mathbf{p}_1(u\bullet x)\bullet v\,\underline{\mathrm{rn}}\,B(x) \text{ if } \mathbf{p}_0 x = 1 \ \text{ and } \ v\,\underline{\mathrm{rn}}\,(\mathbf{p}_0 x = 1 \wedge \forall y B(\mathbf{p}_1 x\bullet y).$$

Assume $\forall y(e\bullet(\mathbf{p}_1 x\bullet y)\,\underline{\mathrm{rn}}\,B(\mathbf{p}_1 x\bullet y))$, $\mathbf{p}_0 x = 1$. Then

$$v = \mathbf{p}(0, \Lambda y.e\bullet(\mathbf{p}_1 x\bullet y))\,\underline{\mathrm{rn}}\,(\mathbf{p}_0 x = 1 \wedge \forall y B(\mathbf{p}_1 x\bullet y)).$$

Therefore

$$\text{if } \mathbf{p}_0 x = 1 \ \text{ and } \ \forall y(e\bullet(\mathbf{p}_1 x\bullet y)\,\underline{\mathrm{rn}}\,B(\mathbf{p}_1 x\bullet y))$$
$$\text{then } \mathbf{p}_1(u\bullet x)\bullet(0, \Lambda y.e\bullet(\mathbf{p}_1 x\bullet y))\,\underline{\mathrm{rn}}\,B(x).$$

Now we construct by the recursion theorem an e such that

$$e\bullet x \simeq \begin{cases} \mathbf{p}_0(u\bullet 0)\bullet 0 \text{ if } x = 0, \\ \mathbf{p}_1(u\bullet x)\bullet\mathbf{p}(0, \Lambda y.e\bullet(\mathbf{p}_1 x\bullet y)) \text{ if } \mathbf{p}_0 x = 1, \\ \text{undefined otherwise.} \end{cases}$$

We then prove by induction on Tr that $\forall x \in \mathrm{Tr}(e\bullet x\,\underline{\mathrm{rn}}\,B(x))$. This is straightforward. This example is capable of considerable generalization, namely to arithmetic enriched with constants for predicates introduced by iterated inductive definitions of higher level; see e.g. Buchholz et al. [1981,IV, section 6].

The examples just mentioned also permit extension of $\underline{\mathrm{rnt}}$-realizability.

We end the section with some applications of $\underline{\mathrm{rn}}$- and $\underline{\mathrm{rnt}}$-realizability.

1.13. Proposition. *(Consistency and inconsistency results)*

(i) $\mathbf{HA}^* + \mathrm{ECT}_0$ *is consistent relative to* $\mathbf{HA}^*$ *(and hence also relative to* $\mathbf{PA}$*).*

(ii) $\neg\forall x(A \vee \neg A)$, $\neg(\forall x\neg\neg B \to \neg\neg\forall x B)$ *are consistent with* $\mathbf{HA}^*$ *for certain arithmetical* A, B.

(iii) The schema "Independence of Premise"

IP $\qquad (\neg A \to \exists z B) \to \exists z(\neg A \to B)$

is not derivable in $\mathbf{HA}^* + \mathrm{CT}_0 + \mathrm{M}$*; in fact,* $\mathbf{HA}^* + \mathrm{IP} + \mathrm{CT}_0 + \mathrm{M} \vdash 1 = 0$.

Proof. (i) Immediate from the characterization theorem.

(ii) is a corollary of the realizability of CT_0: take $A \equiv \exists y Txxy$, $B \equiv \exists y Txxy \vee \neg\exists y Txxy$.

(iii) By M, $\neg\neg\exists y Txxy \to \exists z Txxz$; apply IP to obtain $\forall x \exists z(\neg\neg\exists y Txxy \to Txxz)$, then by CT_0 there is a total recursive F such that $\neg\neg\exists y Txxy \to T(x, x, Fx)$, and this would make $\exists y Txxy$ recursive in x. □

We next give an example of a conservative extension result.

1.14. Definition. CC($\underline{\text{rn}}$) (the $\underline{\text{rn}}$-Conservative Class) is the class of formulas A such that whenever $B \to C$ is a subformula of A, then B is $\exists$-free. □

Lemma. *For* $A \in \mathrm{CC}(\underline{\text{rn}})$ *we have* $\vdash \exists x(x\,\underline{\text{rn}}\,A) \to A$.

Proof. By induction on the structure of A. Consider the case $A \equiv B \to C$; then B is $\exists$-free, so there is a t_B such that $\vdash B \to t_B\,\underline{\text{rn}}\,B$. Assume B and $x\,\underline{\text{rn}}\,(B \to C)$, then $x{\bullet}t_B{\downarrow} \;\wedge\; x{\bullet}t_B\,\underline{\text{rn}}\,C$, hence by the induction hypothesis C; therefore $(x\,\underline{\text{rn}}\,(B \to C)) \to (B \to C)$. □
The lemma in combination with the characterization theorem yields

Proposition. $\mathbf{HA}^* + \mathrm{ECT}_0$ *is conservative over* $\mathbf{HA}^*$ *w.r.t. formulas in* CC($\underline{\text{rn}}$)*:*

$$(\mathbf{HA}^* + \mathrm{ECT}_0) \cap \mathrm{CC}(\underline{\text{rn}}) = \mathbf{HA}^* \cap \mathrm{CC}(\underline{\text{rn}}).$$

The following proposition follows from $\underline{\text{rnt}}$-realizability.

1.15. Proposition. *(Derived rules) In* $\mathbf{HA}^*$

(i) For sentences $\vdash A \vee B \Rightarrow \vdash A$ *or* $\vdash B$ *(Disjunction property* DP*),*

(ii) For sentences $\vdash \exists x A \Rightarrow \vdash A[x/\bar{n}]$ *for some numeral* $\bar{n}$ *(Explicit Definability for Numbers* EDN*),*

(iii) Extended Church's Rule: for $\exists$*-free* A

ECR $\quad \vdash \forall x(A \to \exists y Bxy) \Rightarrow \vdash \exists z \forall x(A \to z{\bullet}x{\downarrow} \wedge B(x, z{\bullet}x))$.

Proof. (i) follows from (ii) (actually, (i) and (ii) are equivalent for systems containing a minimum of arithmetic, see Friedman [1975]). As to (ii), let $\vdash \exists x A$, then by the strong soundness for $\underline{\text{rnt}}$-realizability $\vdash \bar{m}\,\underline{\text{rnt}}\,\exists x A$ for some numeral $\bar{m}$, so $\vdash \mathbf{p}_1\bar{m}\,\underline{\text{rnt}}\,A[x/\mathbf{p}_0\bar{m}]$, and hence $\vdash A[x/\mathbf{p}_0\bar{m}]$.

(iii) Assume $\vdash \forall x(A \to \exists y Bxy)$, then for a suitable t $\vdash t\,\underline{\text{rnt}}\,\forall x(A \to \exists y Bxy)$, i.e.

$$\vdash \forall x \forall z(z\,\underline{\text{rnt}}\,A \to \mathbf{p}_1(t{\bullet}x{\bullet}z)\,\underline{\text{rnt}}\,B(x, \mathbf{p}_0(t{\bullet}x{\bullet}z))).$$

Since $t_A\,\underline{\text{rnt}}\,A$,

$$\vdash \forall x(A \to \mathbf{p}_1(t{\bullet}x{\bullet}t_A)\,\underline{\text{rnt}}\,B(x, \mathbf{p}_0(t{\bullet}x{\bullet}t_A))),$$

and therefore $\vdash \forall x(A \to B(x, \mathbf{p}_0(t{\bullet}x{\bullet}t_A)))$. So we can take $z = \Lambda x.\mathbf{p}_0(t{\bullet}x{\bullet}t_A)$. □

1.16. Remark. The DP cannot be formalized in any consistent extension of **HA** itself (Myhill [1973a], Friedman [1977a]). We sketch Myhill's argument (the result of Friedman is even stronger). Assume that there is a provably recursive function f satisfying

$$\vdash \mathrm{Prf}(x, \ulcorner A \vee B \urcorner) \to ((fx = 0 \wedge \mathrm{Pr}(\ulcorner A \urcorner)) \vee ((fx = 1 \wedge \mathrm{Pr}(\ulcorner B \urcorner))).$$

So $f = \{\bar{p}\}$, and $\vdash \forall x \exists y T\bar{n}xy$. Let F enumerate all primitive recursive functions, i.e. $\lambda n.F(i, n)$ is the i-th primitive recursive function. Put

$$D(n) := \bar{p}{\bullet}F(n, n) \neq 0,$$

then $\vdash \forall n(Dn \vee \neg Dn)$ (i.e. $\mathrm{Prf}(\bar{k}, \ulcorner\forall n(Dn \vee \neg Dn)\urcorner)$ for a specific $\bar{k}$), from which we can find a particular primitive recursive $\lambda n.F(\bar{m}, n)$ such that $\vdash \mathrm{Prf}(F(\bar{m}, \bar{n}), \ulcorner D\bar{n} \vee \neg D\bar{n}\urcorner)$. Then $D\bar{m} \to \bar{p}{\bullet}F(\bar{m}, \bar{m}) \neq 0 \to \mathrm{Prf}(F(\bar{m}, \bar{m}), \ulcorner D\bar{m} \vee \neg D\bar{m}\urcorner) \wedge \mathrm{Pr}(\ulcorner\neg D\bar{m}\urcorner)$, hence $\neg D\bar{m}$ follows, since $\mathbf{HA}^*$ is consistent. If we start assuming $\neg D\bar{m}$, we similarly obtain a contradiction.

From this we see that DP cannot be proved in $\mathbf{HA}^*$ itself; for if DP were provable in $\mathbf{HA}^*$, then a function f as above would be given by

$f(x) := \mathbf{p}_0$(the least y s.t.(x does not prove a closed disjunction and $y = 0$)
or (for some closed $\ulcorner A \vee B\urcorner$, $\mathrm{Prf}(x, \ulcorner A \vee B\urcorner) \wedge \mathbf{p}_0 y = 0 \wedge \mathrm{Prf}(\mathbf{p}_1 y, \ulcorner A\urcorner)$)
or (for some closed $\ulcorner A \vee B\urcorner$, $\mathrm{Prf}(x, \ulcorner A \vee B\urcorner) \wedge \mathbf{p}_1 y = 1 \wedge \mathrm{Prf}(\mathbf{p}_1 y, \ulcorner B\urcorner)$))).

This in turn implies that the strong soundness theorem is not formalizable in $\mathbf{HA}^*$, since strong soundness for <u>rn</u>-realizability immediately implies EDN for $\mathbf{HA}^* + \mathrm{ECT}_0$.

1.17. Notes

Slash relations. Already in Kleene [1945], a modification of numerical realizability was considered, namely $\Gamma \vdash$-realizability; let us use "R" as a short designation for this kind of realizability. The clauses for $\forall, \wedge$ and for prime formulas are as for ordinary realizability; the clauses for $\vee, \exists, \to$ become:

- $\mathbf{n}\,\mathrm{R}\,(A \vee B)$ iff $\mathbf{p}_0\mathbf{n} = 0$, $\mathbf{p}_1\mathbf{n}\,\mathrm{R}\,A$ and $\Gamma \vdash A$, or $\mathbf{p}_0\mathbf{n} \neq 0$, $\mathbf{p}_1\mathbf{n}\,\mathrm{R}\,B$ and $\Gamma \vdash B$;
- $\mathbf{n}\,\mathrm{R}\,\exists x A$ iff $\mathbf{p}_1\mathbf{n}\,\mathrm{R}\,A[x/\overline{\mathbf{p}_0\mathbf{n}}]$ and $\Gamma \vdash A[x/\overline{\mathbf{p}_0\mathbf{n}}]$;
- $\mathbf{n}\,\mathrm{R}\,(A \to B)$ iff for all $\mathbf{m}$, if $\mathbf{m}\,\mathrm{R}\,A$ and $\Gamma \vdash A$, then $\mathbf{n}{\bullet}\mathbf{m}$ is defined and $\mathbf{n}{\bullet}\mathbf{m}\,\mathrm{R}\,B$.

Kleene [1952,Example 2 on page 510] used this notion to obtain a version of Church's thesis. Later Kleene [1962] observed that by dropping the realizability part and retaining only the provability part, one obtained an inductively defined property of formulas which could be used to obtain quite simple proofs of (generalizations of) the disjunction- and existence properties for logic and arithmetic. For easy reference, let us define $\Gamma|A$ ("Γ slashes A") for arithmetic, treating $\vee, \neg$ as defined, and putting $\Gamma|\vdash A$ as short for "$\Gamma|A$ and $\Gamma \vdash A$":

$\Gamma\|P$	iff $\Gamma \vdash P$ for prime sentences P,
$\Gamma\|(A \wedge B)$	iff $\Gamma\|A$ and $\Gamma\|B$,
$\Gamma\|(A \to B)$	iff $\Gamma\|\vdash A \Rightarrow \Gamma\|B$,
$\Gamma\|\exists x A$	iff $\Gamma\|\vdash A[x/\bar{n}]$ for some numeral $\bar{n}$,
$\Gamma\|\forall x A$	iff $\Gamma\|A[x/\bar{n}]$ for all numerals $\bar{n}$.

$\Gamma|A$ for a formula A is defined as $\Gamma|B$ for some universal closure B of A. (For predicate logic, clauses for $\vee, \bot$ have to be added.)

"|" is sometimes called a realizability, but we think it better to reserve the term realizability for notions where realizing objects appear explicitly. Since the "|" in "$\Gamma|A$" has nothing to do with division, the term "divides" for "|" is also not advisable. Therefore we call the notions derived from, or similar to Kleene's $\Gamma|A$ simply *slash relations* or *slashes*.

In one respect $\Gamma|A$ is not well behaved; it is not closed under deduction, since it may happen that $\Gamma|A$, but not $\Gamma|(A \vee A)$. Aczel [1968] gave a simple modification which overcomes this defect: the deducibility requirements in the clauses for $\vee, \exists$ are dropped, and for implication and universal quantification we require instead

$$\Gamma|(A \to B) \text{ iff } (\Gamma|A \Rightarrow \Gamma|B) \text{ and } \Gamma \vdash A \to B,$$
$$\Gamma|\forall x Ax \text{ iff } \Gamma \vdash \forall x Ax \text{ and } \Gamma|A[x/\bar{n}] \text{ for all } \bar{n}.$$

Now $\Gamma|A \Rightarrow \Gamma \vdash A$ holds for all A, and the modified slash yields the same applications as the original one. In fact, one easily proves by formula induction that $\Gamma|\vdash A$ in the sense of Kleene iff $\Gamma|A$ in the sense of Aczel. The Aczel slash also has an appealing model-theoretic interpretation; see e.g. Troelstra and van Dalen [1988,13.7].

It is also worth noting that $C|C$ is both *necessary* and *sufficient* for the validity of the rule "For all A, $\vdash C \to \exists x A \Rightarrow \vdash C \to A[x/\bar{n}]$ for some $\bar{n}$" (Kleene [1962], Troelstra [1973a,3.1.8]).

Slash operators in many variants have been widely used for obtaining metamathematical results for formalisms based on intuitionistic logic.

The slash as defined above applies to *sentences* only, but the use of partial reflection principles in combination with formalized versions of the slash relation, restricted to formulas of bounded complexity, may be used to deal with free numerical variables, see Troelstra [1973a,3.1.16].

Suitable slash relations for systems beyond arithmetic may be defined by considering conservative extensions with extra "witnessing constants" for existential statements. The explicit definability property for numbers EDN can then be proved by proving soundness of slash for the extended system (a typical example is Moschovakis [1981]).

Friedman [1973] describes the extension of the Kleene slash to higher-order logic. In Scedrov and Scott [1982] it is shown that this extension of the slash is in fact equivalent to a categorical construction on the free topos due to P. Freyd (see Lambek and Scott [1986]).

Friedman and Scedrov [1983] use slash relations *and* numerical realizability combined with truth ($\underline{\mathrm{q}}$-realizability, see below) to obtain the explicit set-existence property (explicit definability property for sets) for intuitionistic second-order arithmetic **HAS** (cf. 7.1) and intuitionistic set theory plus countable choice or relativized dependent choice.

Friedman and Scedrov [1986] use a slash relation to establish a very interesting result: there is a particular number-theoretic property $A(n)$ such that if **HA** proves transfinite induction for a primitive recursive binary relation $\prec$ w.r.t. A, then $\prec$ is well-founded with ordinal less than ϵ_0. (The corresponding result is false for **PA**, cf. Kreisel [1953].) If transfinite induction is proved for $\prec$ w.r.t. A for the theory $\mathbf{HA}^+$ obtained by adding transfinite induction for all recursive-wellorderings, then $\prec$ is well-founded.

Of the many papers discussing or making use of slash relations we further mention: Beeson[1975,1976b,1977a], Beeson and Scedrov [1984], Dragalin[1980,1988], Friedman [1977a], Krol' [1977], Moschovakis [1967], Myhill[1973b,1975], Robinson[1965].

$\underline{q}$-realizability. Since in soundness theorems for formalized realizability we prove deducibility instead of just truth, one can replace deducibility in the definition of $\Gamma\vdash$-realizability by truth; let us use "$\underline{q}$" for this realizability. The clauses for $\exists, \to$ then become:

$$\begin{aligned} x\,\underline{q}\,(A \to B) &:= \forall y(y\,\underline{q}\,A \wedge A \to x{\bullet}y\,\underline{q}\,B) \wedge x{\downarrow}, \\ x\,\underline{q}\,\exists y A &:= \mathbf{p}_1 x\,\underline{q}\,A[y/\mathbf{p}_0 x] \wedge A[y/\mathbf{p}_0 x]. \end{aligned}$$

Such a $\underline{q}$-variant was used in Kleene [1969] to obtain derived rules for intuitionistic analysis with function variables. $\underline{q}$-realizability is also not closed under deducibility (think of an instance A of CT_0 unprovable in **HA**; then A is $\underline{q}$-realizable, but $A \vee A$ is not). Grayson [1981a] observed that an Aczel-style modification could be used instead of $\underline{q}$-realizability; this corresponds to our $\underline{rnt}$-realizability. Friedman and Scedrov [1984] use $\underline{rn}$-realizability and $\underline{q}$-realizability to obtain consistency with Church's thesis, the disjunction property and the numerical existence property for set theories based on intuitionistic logic, with axioms asserting the existence of very large cardinals, thereby demonstrating that the metamathematical properties just mentioned, often regarded as a test for the constructive character of a system, are not affected by assumptions concerning large cardinals.

Shanin's algorithm. In a number of papers Shanin presented a systematic way of making the constructive meaning of arithmetical formulas explicit. His method is logically equivalent to $\underline{rn}$-realizability, as shown by Kleene [1960]. On the one hand Shanin's algorithm is more complicated than realizability, on the other hand it has the advantage of being the identity on $\exists$-free formulas.

2. Abstract realizability and function realizability

2.1. After the leisurely introduction to numerical realizability in the preceding section, we now turn to variations and generalizations. In order to distinguish easily the various concepts of realizability, we shall use a certain mnemonic code:

$\underline{r}$ signifies "realizability",
$\underline{n}$ signifies "numerical" or "by numbers",
$\underline{f}$ signifies "by functions",
$\underline{m}$ signifies "modified",
$\underline{t}$ signifies "combined with truth",
$\underline{l}$ signifies "Lifschitz variant of",
$\underline{e}$ signifies "extensional".

Thus "$\underline{rft}$" refers to "realizability by functions combined with truth" etc. Strictly speaking, the $\underline{r}$ is redundant in many of these mnemonic codes.

A simple generalization of numerical realizability is realizability with a different set of realizing objects and/or different application operator; abstractly, the realizing objects with application have to form a combinatory algebra. We shall first sketch an abstract version of numerical realizability, namely realizability in a combinatory algebra with induction, then consider the interesting special case of function realizability.

2.2. Definition. (*The theory* **APP**) The language is single-sorted, based on LPT. The only non-logical predicate is N (natural numbers). There is an application operation $\bullet$ and constants

0 (zero), S (successor), P (predecessor),
$\mathbf{p}$, $\mathbf{p}_0$, $\mathbf{p}_1$ (pairing with inverses),
$\mathbf{k}$, $\mathbf{s}$ (combinators), $\mathbf{d}$ (numerical definition by cases).

(We have used the same symbols for pairing and inverses as in the case of $\mathbf{HA}^*$, even if there is a slight difference in syntax: $\mathbf{p}(t, t')$ in $\mathbf{HA}^*$ corresponds to $(\mathbf{p}t)t'$ in **APP**.) For $t_1 \bullet t_2$ we simply write $(t_1 t_2)$, and we use association to the left, i.e. $t_1 t_2 \ldots t_n$ is short for $(\ldots((t_1 t_2)t_3)\ldots t_n)$.

Axioms for the constants:

$$N0,\ Nx \to N(Sx),\ Nx \to N(Px),$$
$$P(St) \simeq t,\ P0 = 0,\ 0 \neq St,$$
$$\mathbf{k}x\downarrow,\ \mathbf{k}ty \simeq t,\ \mathbf{s}xy\downarrow,\ \mathbf{s}tt't'' \simeq tt''(t't''),$$
$$\mathbf{p}xy\downarrow,\ \mathbf{p}_0x\downarrow,\ \mathbf{p}_1x\downarrow,\ \mathbf{p}_0(\mathbf{p}tx) = t,\ \mathbf{p}_1(\mathbf{p}xt) = t,$$
$$Nu \wedge Nv \to (u \neq v \to \mathbf{d}xyuv = x) \wedge \mathbf{d}xyuu = y.$$

Observe that by the general LPT-axioms we have $tt'\downarrow \to t\downarrow \wedge t'\downarrow$. Finally we have induction:

$$A[x/0] \wedge \forall x \in \mathrm{N}(A \to A[x/Sx]) \to \forall x \in \mathrm{N}.\, A \quad \Box$$

The combinators $\mathbf{k}, \mathbf{s}$ permit us to have λ-abstraction defined by induction on the construction of terms:

$$\lambda x.t := \mathbf{k}t \text{ for } t \text{ a constant or variable} \not\equiv x,$$
$$\lambda x.x := \mathbf{skk},$$
$$\lambda x.tt' := \mathbf{s}(\lambda x.t)(\lambda x.t').$$

For this definition

$$\mathrm{FV}(\lambda x.t) \equiv \mathrm{FV}(t) \setminus \{x\},$$
$$t'\downarrow \to (\lambda x.t)t' \simeq t[x/t'] \text{ if } t' \text{ is free for } x \text{ in } t,$$
$$\lambda x.t\downarrow \text{ for all } t.$$

It is not generally true that[2]

(1) if $x \notin \mathrm{FV}(t'), y \not\equiv x$ then $\lambda x.(t[y/t']) \simeq (\lambda x.t)[y/t']$,

(consider e.g. $t \equiv y, t' \equiv \mathbf{kk}$) but we do have, for $x \notin \mathrm{FV}(t')$, $y \notin \mathrm{FV}(t'')$, $y \not\equiv x$

(2) $t''\downarrow \to ((\lambda x.t)[y/t'])t'' \simeq t[x/t''][y/t'] \equiv t[y/t'][x/t'']$.

Property (1) can be guaranteed by an alternative definition of abstraction:

$$\lambda' x.x := \mathbf{skk},$$
$$\lambda' x.t := \mathbf{k}t \text{ if } x \notin \mathrm{FV}(t),$$
$$\lambda' x.tt' := \mathbf{s}(\lambda' x.t)(\lambda' x.t') \text{ if } x \in \mathrm{FV}(tt'),$$

[2]This was overlooked in the proofs in Troelstra and van Dalen [1988,section 9.3], but is easily remedied by the use of (2). Strahm [1993] recently showed that this problem might also be overcome by considering instead of **APP** based on combinators, a theory based on lambda-abstraction as a primitive (without a ξ-rule of the form $t = s \Rightarrow \lambda x.t = \lambda x.s$!) and an explicit substitution operator *as part of* the language.

but then we lose the property that $\lambda x.t\downarrow$ for all t. A recursor and a minimum operator may be defined with help of a fixed point operator (see e.g. Troelstra and van Dalen [1988,9.3]) which permits us to define in **APP** all partial recursive functions. It follows that **HA** can be embedded into **APP** in a natural and straightforward way.

Remark. Partial combinatory algebras are structures $(X, \bullet, \mathbf{k}, \mathbf{s})$, $\mathbf{k} \neq \mathbf{s}$, satisfying the relevant axioms above; in such structures we can always define terms forming a copy of $\mathbb{N}$, and appropriate $S, P, \mathbf{p}, \mathbf{p}_0, \mathbf{p}_1$, and we might simply have postulated induction for this particular copy of $\mathbb{N}$. However, in describing models it is more convenient not to be tied to a specific representation of $\mathbb{N}$ relative to the combinators. Also, we want to leave open the possibility that the interpretation of N is non-standard.

2.3. The model of the partial recursive operations PRO

The basic combinatory algebra is

$$(\mathbb{N}, \bullet, \Lambda xy.x, \Lambda xyz.x\bullet z\bullet(y\bullet z));$$

where $\bullet$ is partial recursive function application for $\mathbb{N}$. 0, S, P get their usual interpretation (more precisely, we choose codes $\Lambda x.Sx, \Lambda x.Px$ etc.; for $\mathbf{d}$ take $\Lambda uvxy[u \cdot \mathrm{sg}|x-y| + v(1 \dot{-} |x-y|)]$.

In $\mathbf{HA}^*$ we can prove PRO to be a model of **APP**, in the sense that $\mathbf{APP} \vdash A \Rightarrow \mathbf{HA}^* \vdash [\![A]\!]_{\mathrm{PRO}}$. Here and in the sequel we use "interpretation brackets": given some model $\mathcal{M}$, we use $[\![t]\!]_{\mathcal{M}}$, $[\![A]\!]_{\mathcal{M}}$ to indicate the interpretation of term t, formula A in the model $\mathcal{M}$. Thus "$[\![A]\!]_{\mathcal{M}}$ holds" means the same as $\mathcal{M} \models A$.

2.4. Definition. (*Abstract realizability*) $x\,\underline{\mathrm{r}}\,A$ in **APP** is defined by

$$\begin{array}{ll} x\,\underline{\mathrm{r}}\,P & := P \wedge x\downarrow \text{ for } P \text{ prime}, \\ x\,\underline{\mathrm{r}}\,(A \wedge B) & := (\mathbf{p}_0 x\,\underline{\mathrm{r}}\,A) \wedge (\mathbf{p}_1 x\,\underline{\mathrm{r}}\,B), \\ x\,\underline{\mathrm{r}}\,(A \to B) & := \forall y(y\,\underline{\mathrm{r}}\,A \to x\bullet y\,\underline{\mathrm{r}}\,B) \wedge x\downarrow, \\ x\,\underline{\mathrm{r}}\,\forall y\,A & := \forall y(x\bullet y\,\underline{\mathrm{r}}\,A), \\ x\,\underline{\mathrm{r}}\,\exists y\,A & := \mathbf{p}_1 x\,\underline{\mathrm{r}}\,A[y/\mathbf{p}_0 x]. \quad \Box \end{array}$$

Remark. $x\,\underline{\mathrm{r}}\,\forall y{\in}N.A$ becomes literally $\forall yz(z\,\underline{\mathrm{r}}\,(y \in N) \to (x\bullet y\bullet z)\,\underline{\mathrm{r}}\,A))$. It is easy to see that realizability with a special clause for the relativized quantifier

$$x\,\underline{\mathrm{r}}'\,\forall y{\in}N.A \;:=\; \forall y{\in}N(x\bullet y\,\underline{\mathrm{r}}'\,A)$$

is in fact equivalent.

The $\exists$-free formulas play the same role in **APP** as they do in $\mathbf{HA}^*$, i.e. $\exists$-free formulas have canonical realizing terms, their realizability coincides with their truth, and equivalence of realizability with truth for a formula A means that A is equivalent to a formula $\exists x B$, B $\exists$-free; the schema characterizing $\underline{\mathrm{r}}$-realizability is an *Extended Axiom of Choice*

EAC $\quad \forall x(Ax \to \exists y Bxy) \to \exists z \forall x(Ax \to z\bullet x\downarrow \wedge B(x, z\bullet x))$ (A is $\exists$-free) etc.

We may specialize $\underline{\mathbf{r}}$-realizability to PRO-$\underline{\mathbf{r}}$-realizability by interpreting **APP** in PRO. It is then not difficult to show that the resulting realizability of **HA** (as embedded in the obvious way into **APP**) becomes equivalent to $\underline{\mathbf{rn}}$-realizability; cf. Renardel de Lavalette [1984].

2.5. Partial continuous function application

Another important model of **APP** is the model PCO of functions with partial continuous application. Before we can discuss this, we need some preliminaries.

Notations. From primitive recursive $\mathbf{p}, \mathbf{p}_0, \mathbf{p}_1$ we can construct primitive recursive encodings $\mathbf{p}^n$ of n-tuples of natural numbers, with primitive recursive inverses $\mathbf{p}_i^n$ $(0 \leq i < n)$. We may also assume finite sequences of natural numbers to be coded onto $\mathbb{N}$; we write $\langle n_0, \ldots, n_{x-1} \rangle$ for (the code of) the finite sequence $n_0, \ldots, n_{x-1}$; $\langle\ \rangle$ is the (code of the) empty sequence (which may be assumed to be equal to 0; see below).

If m is (a code of) a sequence, $\mathrm{lth}(m)$ is its length; $*$ is a primitive recursive concatenation function for codes of sequences. We abbreviate

$$\begin{aligned} n \preceq m &:= \exists n'(n * n' = m), \\ n \prec m &:= (n \preceq m \wedge n \neq m), \\ \hat{x} &:= \langle x \rangle. \end{aligned}$$

The primitive recursive inverse function $\lambda xy.(x)_y$ of sequence encoding satisfies

$$m = \langle n_0, \ldots, n_{x-1} \rangle \Rightarrow (m)_y = n_y \text{ for } y < x,\ (m)_y = 0 \text{ for } y \geq x.$$

For reasons of technical convenience we assume monotonicity in the arguments for encodings of pairs, n-tuples and finite sequences:

$$n < n' \rightarrow \mathbf{p}(n,m) < \mathbf{p}(n',m), \quad m < m' \rightarrow \mathbf{p}(n,m) < \mathbf{p}(n,m'),$$

and similarly for p-tuples;

$$n \leq n * m; \quad \mathrm{lth}(n) = \mathrm{lth}(m) \wedge \forall x((n)_x \leq (m)_x) \rightarrow n \leq m.$$

For example, for $\mathbf{p}$ we may take $\mathbf{p}(n,m) = \frac{1}{2}(n+m)(n+m+1)+m$. These monotonicity conditions in fact enforce $\langle\ \rangle = 0$. Encoding of n-tuples $(\alpha_1, \ldots, \alpha_n)$ of sequences is obtained by

$$(\alpha_0, \ldots, \alpha_n) := \lambda x.\mathbf{p}^n(\alpha_1 x, \ldots, \alpha_n x).$$

For initial segments of functions we use

$$\bar{\alpha}0 := \langle\ \rangle, \quad \bar{\alpha}(x+1) := \langle \alpha 0, \ldots, \alpha x \rangle. \quad \square$$

Definition. *Elementary Analysis* **EL** is a conservative extension of **HA** obtained by adding to **HA** variables $(\alpha, \beta, \gamma, \delta, \epsilon)$ and quantifiers for (total) functions from $\mathbb{N}$ to $\mathbb{N}$ (i.e. infinite sequences of natural numbers). There is λ-abstraction for explicit definition of functions, and a recursion-operator Rec such that (t a numerical term, ϕ a function term; $\phi(t,t') := \phi\mathbf{p}(t,t')$)

$$\mathrm{Rec}(t,\phi)(0) = t, \quad \mathrm{Rec}(t,\phi)(Sx) = \phi(x, \mathrm{Rec}(t,\phi)(x)).$$

Induction is extended to all formulas in the new language.

The functions of **EL** are assumed to be closed under "recursive in", which is expressed by including a weak choice axiom for quantifier-free A:

QF-AC $\quad \forall n \exists m A(n,m) \to \exists \alpha \forall n A(n, \alpha n) \quad \Box$

Definition. In **EL** we introduce abbreviations for *partial continuous application*

$$\begin{aligned} \alpha(\beta) = x \quad &:= \quad \exists y(\alpha(\bar{\beta}y) = x+1 \wedge \forall y'{<}y(\alpha(\bar{\beta}y') = 0), \\ \alpha|\beta = \gamma \quad &:= \quad \forall x(\lambda n.\alpha(\hat{x} * n)(\beta) = \gamma x) \wedge \alpha 0 = 0, \text{ or equivalently} \\ &\quad\ \forall x \exists y(\alpha(\hat{x} * \bar{\beta}y) = \gamma x + 1 \wedge \forall y'{<}y(\alpha(\hat{x} * \bar{\beta}y') = 0)) \wedge \alpha 0 = 0. \end{aligned}$$

We may introduce $|$, $\cdot(\cdot)$ as primitive operators in a conservative extension $\mathbf{EL}^*$ based on the logic of partial terms (1.3). $\Box$

Definition. $\mathbf{EL}^*$ is a conservative extension of **EL** based on the logic of partial terms, to which $\lambda\alpha\beta.\alpha|\beta$ and $\lambda\alpha\beta.\alpha(\beta)$ have been added as primitive operations. Numerical lambda-abstraction satisfies:

$$s{\downarrow} \wedge (\lambda x.t){\downarrow} \to (\lambda x.t)s = t[x/s], \quad (\lambda x.t){\downarrow} \leftrightarrow \forall x(t{\downarrow}).$$

For function application we require

$$\phi t{\downarrow} \leftrightarrow \phi{\downarrow} \wedge \phi{\downarrow}.$$

(The implication from left to right must hold since $\phi{\downarrow}$ is supposed to imply totality of the function denoted by ϕ.) For Rec we have

$$\mathrm{Rec}(t,\phi){\downarrow} \leftrightarrow t{\downarrow} \wedge \phi{\downarrow}.$$

$\Box$

2.6. The model of the partial continuous operations PCO

This model has as domain all the total functions from $\mathbb{N}$ to $\mathbb{N}$; the application is $|$ defined above. The elementary theory of the model can be formalized in $\mathbf{EL}^*$. Some work is needed to show that PCO is actually a model of **APP**.

Definition. (*The class of neighbourhood functions*)

$$\alpha \in K^* \quad := \quad \alpha 0 = 0 \wedge \forall nm(\alpha n > 0 \to \alpha n = \alpha(n * m)) \wedge \forall \beta \exists x(\alpha(\bar{\beta}x) > 0).$$

$\Box$

Crucial is the following

Lemma. *To each function term ϕ, and each numerical term t of $\mathbf{EL}^*$, we can construct function terms $\Phi_\phi \in K^*$, $\Phi_t \in K^*$ respectively, such that*

(i) $\Phi_\phi|\alpha \simeq \phi$;

(ii) $(\Phi_t|\alpha){\downarrow}$ *iff* $t{\downarrow}$;

(iii) $t{\downarrow} \to (\Phi_t|\alpha)0 = t$;

(iv) $\mathrm{FV}(\Phi_t) \subset \mathrm{FV}(t) \setminus \{\alpha\}$, $\mathrm{FV}(\Phi_\phi) \subset \mathrm{FV}(\phi) \setminus \{\alpha\}$, Φ_t, Φ_ϕ *primitive recursive in their free variables.*

Proof. (i)–(iv) are proved by simultaneous induction on the construction of numerical and function terms. The reason that we need a function term with partial continuous application $|$ to represent a numerical term (instead of application $\cdot(\cdot)$) is that a numerical term t may contain function-terms as subterms, which all have to be defined by the strictness condition of the logic of partial terms; this is a Π^0_2-condition and cannot be expressed by definedness of a numerical term.

We consider a few typical cases. In all cases we put $\Phi_\phi 0 = 0$, $\Phi_t 0 = 0$.

Case 1. $t \equiv x$. Take $\Phi_t(\hat{z} * \bar{\alpha}n) = x + 1$. Similarly for $t \equiv 0$.

Case 2. $\phi \equiv \alpha$. Take

$$\Phi_\alpha(\hat{z} * \bar{\alpha}n) = \begin{cases} \alpha z + 1 \text{ if } z < n, \\ 0 \text{ otherwise.} \end{cases}$$

Case 3. $t \equiv \phi(\psi)$. Take

$$\Phi_{\phi(\psi)}(\hat{x} * \bar{\alpha}n) = \begin{cases} z + 1 \text{ if } \exists u{<}n \forall y{<}\mathrm{lth}(u)(\Phi_\psi(\hat{y} * \bar{\alpha}n) = (u)_y + 1 \wedge \\ \qquad \Phi_\phi(u) = z + 1) \wedge \Phi_\phi(\hat{x} * \bar{\alpha}n) > 0 \wedge \Phi_\psi(\hat{x} * \bar{\alpha}n) > 0, \\ 0 \text{ otherwise.} \end{cases}$$

Case 4. $\phi \equiv \psi|\xi$. Take

$$\Phi_{\psi|\xi}(\hat{x} * \bar{\alpha}n) = \begin{cases} z + 1 \text{ if } \exists u{<}n \forall y{<}\mathrm{lth}(u)(\Phi_\xi(\hat{y} * \bar{\alpha}n) = (u)_y + 1 \wedge \\ \qquad \Phi_\psi(\hat{x} * u) = z + 1) \wedge \Phi_\psi(\hat{x} * \bar{\alpha}n) > 0 \wedge \Phi_\xi(\hat{x} * \bar{\alpha}n) > 0, \\ 0 \text{ otherwise .} \end{cases}$$

The other cases are left to the reader. □

It is now easy to prove in $\mathbf{EL}^*$ that PCO is a model of **APP** (do not confuse the λ-abstraction in $\mathbf{EL}^*$ with the defined λ-operator in **APP**). For example, an interpretation of $[\![\mathbf{s}]\!]$ is found as follows. If ϕ is a function term of $\mathbf{EL}^*$, let us write $\Lambda\alpha.\phi$ for the Φ_ϕ given by the lemma. Then we put

$$[\![\mathbf{s}]\!]_{\mathrm{PCO}} := \Lambda\alpha\Lambda\beta\Lambda\gamma.(\alpha|\gamma)|(\beta|\gamma).$$

Clearly $([\![\mathbf{s}]\!]|\alpha)|\beta = \Lambda\gamma(\alpha|\gamma)|(\beta|\gamma)$, since all terms Φ_ϕ are total, hence defined. Moreover $(([\![\mathbf{s}]\!]|\alpha)|\beta)|\gamma \simeq (\alpha|\gamma)|(\beta|\gamma)$.

We spell out a definition of realizability for this application which is not literally what one obtains by interpreting $\underline{\mathrm{r}}$-realizability in PCO, but equivalent to it:

2.7. Definition. *(Realizability by functions)* With each formula A of $\mathbf{EL}^*$ we associate $\alpha \,\underline{\mathtt{rf}}\, A$ ($\alpha \notin \mathrm{FV}(A)$) as follows:

$$\begin{array}{ll} \alpha \,\underline{\mathtt{rf}}\, P & := P \wedge \alpha{\downarrow} \;\; (P \text{ prime}), \\ \alpha \,\underline{\mathtt{rf}}\, (A \wedge B) & := (\mathbf{p}_0\alpha \,\underline{\mathtt{rf}}\, A) \wedge (\mathbf{p}_1\alpha \,\underline{\mathtt{rf}}\, B), \\ \alpha \,\underline{\mathtt{rf}}\, (A \to B) & := \forall\beta(\beta \,\underline{\mathtt{rf}}\, A \to \alpha|\beta \,\underline{\mathtt{rf}}\, B) \wedge \alpha{\downarrow}, \\ \alpha \,\underline{\mathtt{rf}}\, \forall x\, A & := \forall x(\alpha|\lambda n.x \,\underline{\mathtt{rf}}\, A), \\ \alpha \,\underline{\mathtt{rf}}\, \forall\beta\, A & := \forall\beta(\alpha|\beta \,\underline{\mathtt{rf}}\, A), \\ \alpha \,\underline{\mathtt{rf}}\, \exists x\, A & := \mathbf{p}_1\alpha \,\underline{\mathtt{rf}}\, A[x/(\mathbf{p}_0\alpha)0], \\ \alpha \,\underline{\mathtt{rf}}\, \exists\beta\, A & := \mathbf{p}_1\alpha \,\underline{\mathtt{rf}}\, A[\beta/\mathbf{p}_0\alpha]. \end{array}$$

$\underline{\mathtt{rft}}$-realizability is defined by modifying $\underline{\mathtt{rf}}$-realizability as before. □

Now the theory runs to a large extent parallel to numerical realizability. The role of ECT_0 is taken over by the following schema of *Generalized Continuity*:

GC $\quad \forall\alpha(A \to \exists\beta B(\alpha,\beta)) \to \exists\gamma\forall\alpha(A \to \gamma|\alpha{\downarrow} \wedge B(\alpha, \gamma|\alpha))$ (A $\exists$-free) where $\exists$-*free* in $\mathbf{EL}^*$ is defined as before; in $\mathbf{EL}$, $\exists$-free formulas correspond to the class of formulas constructed from $t = s$, $\exists x(t = s)$, $\exists \alpha(t = s)$ by means of $\rightarrow, \wedge, \forall$.

2.8. Proposition. *(Examples of applications) For* $\mathbf{EL}^*$ *we have*

(i) $\vdash \forall\alpha(A \rightarrow \exists\beta B(\alpha,\beta)) \Rightarrow\; \vdash \exists\gamma\forall\alpha(A \rightarrow \gamma|\alpha\downarrow \wedge B(\alpha,\gamma|\alpha))$ *for almost negative* A *(Generalized Continuity Rule* GCR*).*

(ii) For $\exists\alpha A\alpha$ *closed,* $\vdash \exists\alpha\, A\alpha \;\Rightarrow$ *there exists some* $\bar{n}$ *such that* $\vdash A(\{\bar{n}\}) \wedge \forall m(\bar{n}\bullet m\downarrow)$, *i.e. if* $\vdash \exists\alpha\, A(\alpha)$, *there is a total recursive function* f *such that* $\vdash A(f)$.

(iii) $\mathrm{CC}(\underline{\mathtt{rf}}) \cap \mathbf{EL}^* = \mathrm{CC}(\underline{\mathtt{rf}}) \cap (\mathbf{EL}^* + \mathrm{GC})$, *where the conservative class* $\mathrm{CC}(\underline{\mathtt{rf}})$ *for* $\underline{\mathtt{rf}}$*-realizability is defined in complete analogy to* $\mathrm{CC}(\underline{\mathtt{rn}})$.

Proof. (Of (ii).) The strong soundness theorem yields in this case a particular function term ϕ such that $\vdash \phi\,\underline{\mathtt{rft}}\,\exists\alpha\, A\alpha$, hence $\vdash \mathbf{p}_1\phi\,\underline{\mathtt{rft}}\,A[\alpha/\mathbf{p}_0\phi]$, and thus $\vdash A[\alpha/\mathbf{p}_0\phi] \wedge \mathbf{p}_0\phi\downarrow$; $\mathbf{p}_0(\phi)$ is a closed function term in the language of $\mathbf{EL}$ which may be written as $\{\bar{n}\}$. ($\{\bar{n}\}$ is short for $\lambda x.(\bar{n}\bullet x)$.) □

2.9. Examples of extensions

Bar Induction for Decidable predicates is an induction principle

$$\mathrm{BI_D} \qquad \forall\alpha\exists x\, P(\bar{\alpha}x) \wedge \forall n(Pn \vee \neg Pn) \wedge \forall n(Pn \rightarrow Qn) \wedge \forall n(\forall m(Q(n * \langle m\rangle) \rightarrow Qn) \rightarrow Q\langle\,\rangle$$

An equivalent principle is BI! with $\forall\alpha\exists x\, P(\bar{\alpha}x) \wedge \forall n(Pn \vee \neg Pn)$ replaced by $\forall\alpha\exists! x\, P(\bar{\alpha}x)$. $\mathrm{BI_D}$ implies the *Fan theorem for Decidable predicates*

$$\mathrm{FAN_D} \qquad \forall\alpha{\le}\beta\exists x\, A(\bar{\alpha}x) \wedge \forall n(An \vee \neg An) \rightarrow \exists z\forall\alpha{\le}\beta\exists x{\le}z\, A(\bar{\alpha}x)$$

where $\alpha \le \beta := \forall x(\alpha x \le \beta x)$. Since $\underline{\mathtt{rf}}$-realizability validates continuity principles, in fact the stronger

$$\mathrm{FAN} \qquad \forall\alpha\exists x\, A(\alpha,x) \rightarrow \exists z\forall\alpha{\le}\beta\exists x{\le}z\, A(\alpha,x)$$

holds.

In the soundness and characterization theorems $\mathbf{EL}^*$ may be replaced by $\mathbf{EL}^* + \Gamma$, where for example Γ can be the set of all instances of one or more of the following schemata: M, $\mathrm{TI}(\prec)$, $\mathrm{FAN_D}$, $\mathrm{BI_D}$.

2.10. Notes

Kleene [1965b] contains the first formalization of $\underline{\mathtt{rf}}$-realizability. In this paper Kleene shows a.o. that formulas such that all their subformulas in the scope of an universal function quantifier are $\exists$-free are true iff $\underline{\mathtt{rf}}$-realizable (provable classically). In Kleene [1969] a thorough formalized treatment of function realizability is given, also of a (version of) $\underline{\mathtt{rft}}$-realizability. Another notion of realizability by functions is found in Moschovakis [1993,1994].

Vesley [1996] considers a notion of function-realizability which combines the idea of $\underline{\mathtt{rf}}$-realizability with the topological model of elementary intuitionistic analysis in Moschovakis [1973] (itself an adaptation of a model due to Scott [1968]).

Barendregt [1973] used an abstract version of realizability to show consistency of an axiom of choice with combinatory logic. Staples[1973,1974] used realizability with combinators for higher-order logic and set theory. Abstract realizability for theories including **APP** was introduced by Feferman[1975,1979].

Of the researches using abstract versions of realizability we further mention Beeson[1977b,1979b,1980,1985], Renardel de Lavalette[1984,1990].

3. Modified realizability

In the case of numerical and function realizability, we started with the concrete and ended with the abstract version.

For modified realizability on the other hand, it is advantageous to start with the abstract setting, and afterwards to specialize to more concrete versions. The abstract setting of modified realizability is not a type-free theory such as **APP**, sketched above, but a system $\mathbf{HA}^\omega$ of intuitionistic finite-type arithmetic.

3.1. Description of intuitionistic finite-type arithmetic $\mathbf{HA}^\omega$

The *set of finite type symbols* $\mathcal{T}$ is generated by the clauses $0 \in \mathcal{T}$ (type of the natural numbers); if $\sigma, \tau \in \mathcal{T}$ then $(\sigma \times \tau) \in \mathcal{T}$ (formation of product types) and $(\sigma \to \tau) \in \mathcal{T}$ (formation of *function types*). We use $\sigma, \sigma', \ldots, \tau, \tau', \ldots, \rho, \rho', \ldots$ for arbitrary type symbols.

As an alternative for $(\sigma \to \tau)$ we write $(\sigma\tau)$; 1 is short for (00), $n+1$ for $(n0)$. Outer parentheses in type symbols are usually omitted. Further saving on parentheses is obtained by the convention of association to the *right*, i.e. $\sigma_0\sigma_1\sigma_2\sigma_3$ abbreviates $(\sigma_0(\sigma_1(\sigma_2\sigma_3)))$; $\sigma_1 \times \sigma_2 \times \cdots \times \sigma_n$ abbreviates $(\cdots((\sigma_1 \times \sigma_2) \times \sigma_3) \cdots \times \sigma_n)$.

The language of of *intuitionistic finite-type arithmetic* $\mathbf{HA}^\omega$ is a many-sorted language with variables $(x^\sigma, y^\sigma, z^\sigma, \ldots)$ of all types; for each $\sigma \in \mathcal{T}$ there is a primitive equality $=_\sigma$, and there are some constants listed below, and an application operation $\mathrm{App}_{\sigma,\tau}$ from $\sigma \to \tau$ and σ to τ. For arbitrary terms we use $t, t', t'', \ldots, s, s', s'', \ldots$. In order to indicate that t is a term of type σ we write $t \in \sigma$ or t^σ. If $t \in \sigma \to \tau$, $t' \in \sigma$, then $\mathrm{App}_{\sigma,\tau}(t, t') \in \tau$. For $\mathrm{App}_{\sigma,\tau}(t, t')$ we simply write (tt') or even tt'; we save on parentheses by association to the *left*: $t_1 \ldots t_n$ is short for $(\cdots((t_1t_2)t_3) \cdots t_n)$. As constants we have for all $\sigma, \tau, \rho \in \mathcal{T}$:

$0 \in 0$ (zero), $S \in 00$ (successor),
$\mathbf{p}^{\sigma,\tau} \in \sigma\tau(\sigma \times \tau)$, (pairing)
$\mathbf{p}_0^{\sigma,\tau} \in (\sigma \times \tau)\sigma$, $\mathbf{p}_1^{\sigma,\tau} \in (\sigma \times \tau)\tau$, (unpairing)
$\mathbf{k}^{\sigma,\tau} \in \sigma\tau\sigma$, $\mathbf{s}^{\rho,\sigma,\tau} \in (\rho\sigma\tau)(\rho\sigma)\rho\tau$ (combinators),
$\mathbf{r}^\sigma \in \sigma(\sigma 0 \sigma)0\sigma$ (recursor).

Here again we use the same symbols $(\mathbf{k}, \mathbf{s}, \mathbf{p}, \mathbf{p}_0, \mathbf{p}_1)$ for operations closely analogous to the operations denoted by the same symbols in **APP**. We shall drop type suben superscripts wherever it is safe to do so; types are always assumed to be "fitting" (i.e. if tt' is written then $t \in \sigma\tau$, $t' \in \sigma$ for suitable σ, τ).

The logical basis of $\mathbf{HA}^\omega$ is many-sorted intuitionistic predicate logic with equality; the constants satisfy the following equations:

$$\mathbf{p}_0(\mathbf{p}xy) = x,\ \ \mathbf{p}_1(\mathbf{p}xy) = y,\ \ \mathbf{p}(\mathbf{p}_0 x)(\mathbf{p}_1 x) = x,$$
$$\mathbf{k}xy = x,\ \ \mathbf{s}xyz = xz(yz),\ \ \mathbf{r}xy0 = x,\ \ \mathbf{r}xy(Sz) = y(\mathbf{r}xyz)z.$$

Finally, we have $0 \neq Sx$, $Sx = Sy \to x = y$, and full induction. (Actually, $Sx = Sy \to x = y$ is redundant, since we can define a predecessor function P such that $P(St) = x$.)

There is defined λ-abstraction, as in for **APP**; we can use the second recipe mentioned in 2.2, with $\lambda x.t = \mathbf{k}t$ for all t not containing x. **HA** is embedded in $\mathbf{HA}^\omega$ in the obvious way.

3.2. The systems I-HA$^\omega$, E-HA$^\omega$

$\mathbf{I\text{-}HA}^\omega$, *intensional* finite-type arithmetic is a strengthening of $\mathbf{HA}^\omega$ obtained by including an equality functional $\mathbf{e}_\sigma \in \sigma\sigma 0$ for all $\sigma \in \mathcal{T}$, satisfying

$$\mathbf{e}x^\sigma y^\sigma \leq 1,\ \ \mathbf{e}x^\sigma y^\sigma = 0 \leftrightarrow x^\sigma = y^\sigma,$$

so equality is decidable at all types.

On the other hand, *extensional* finite-type arithmetic $\mathbf{E\text{-}HA}^\omega$ is obtained from $\mathbf{HA}^\omega$ by adding extensionality axioms for all types σ:

$$\forall x^\sigma (y^{\sigma\tau} x =_\tau z^{\sigma\tau} x) \leftrightarrow y^{\sigma\tau} =_{\sigma\tau} z^{\sigma\tau}.$$

This permits us to *define* equality of type σ in terms of $=_0$, via

$$y =_{\sigma\tau} z := \forall x^\sigma (yx =_\tau zx),$$
$$y =_{\sigma\times\tau} z := \mathbf{p}_0 y =_\sigma \mathbf{p}_0 z \wedge \mathbf{p}_1 y =_\tau \mathbf{p}_1 z.$$

Therefore we may assume $\mathbf{E\text{-}HA}^\omega$ to be formulated in a language which contains only $=_0$ as primitive equality, so that prime formulas are always decidable.

3.3. Models of HA$^\omega$

A model of $\mathbf{HA}^\omega$ is given by a type structure $\langle M_\sigma, \sim_\sigma \rangle_{\sigma\in\mathcal{T}}$, with M_σ a set, $\sim_\sigma$ an equivalence relation on M_σ, plus suitable interpretations of $\mathrm{App}_{\sigma,\tau}$ and the various constants.

(i) FTS, the *Full Type Structure.* Take $\mathbb{N}$ for M_0, for $M_{\sigma\tau}$ take the set of all functions from M_σ to M_τ, for $M_{\sigma\times\tau}$ take $M_\sigma \times M_\tau$; this is the full type structure; $\sim_\sigma$ at each type is set-theoretic equality, and it is obvious how to interpret App and the constants.

(ii) HRO, the *Hereditarily Recursive Operations.* Put

$$\mathrm{HRO}_0 := \mathbb{N},$$
$$\mathrm{HRO}_{\sigma\times\tau} := \{z : \mathbf{p}_0 z \in \mathrm{HRO}_\sigma \wedge \mathbf{p}_1 z \in \mathrm{HRO}_\tau\},$$
$$\mathrm{HRO}_{\sigma\tau} := \{z : \forall x \in \mathrm{HRO}_\sigma (z \bullet x \in \mathrm{HRO}_\tau)\}.$$

App is interpreted as partial recursive application (i.e. as $\bullet$), $=_\sigma$ as equality between numbers (as elements of HRO_σ),

$$[\![0]\!] := 0,\ [\![S]\!] := \Lambda x.Sx,\ [\![\mathbf{k}]\!] := \Lambda xy.x,\ [\![\mathbf{s}]\!] := \Lambda xyz.xz(yz),$$
$$[\![\mathbf{p}]\!] := \Lambda xy.\mathbf{p}(x,y),\ [\![\mathbf{p}_0]\!] := \Lambda x.\mathbf{p}_0 x,\ [\![\mathbf{p}_1]\!] := \Lambda x.\mathbf{p}_1 x,$$
$$[\![\mathbf{r}]\!] := \text{ a suitable code for a recursor, } [\![\mathbf{e}]\!] := \Lambda xy.\mathrm{sg}|x-y|.$$

The existence of a suitable code for a recursor either follows directly from the definition of recursive function, or by an application of the recursion theorem yielding a solution r to $r{\bullet}(x,y,0) \simeq 0$, $r{\bullet}(x,y,Sz) \simeq y{\bullet}(r{\bullet}(x,y,z),z)$, as in Troelstra and van Dalen [1988,3.7.5]. The result is a model of $\mathbf{I\text{-}HA}^\omega$.

(iii) HEO, the model of the *Hereditarily Effective Operations*. We define a partial equivalence relation $\sim_\sigma$ between natural numbers for each $\sigma \in \mathcal{T}$ by

$$\begin{aligned} x \sim_0 y \quad &:= x = y, \\ x \sim_{\sigma\times\tau} y &:= (\mathbf{p}_0 x \sim_\sigma \mathbf{p}_0 y) \wedge (\mathbf{p}_1 x \sim_\tau \mathbf{p}_1 y), \\ x \sim_{\sigma\tau} y \quad &:= \forall z z'(z \sim_\sigma z' \to x{\bullet}z \sim_\tau y{\bullet}z' \wedge x{\bullet}z \sim_\tau x{\bullet}z' \wedge y{\bullet}z \sim_\tau y{\bullet}z') \end{aligned}$$

where

$$z \in \mathrm{HEO}_\sigma := z \sim_\sigma z.$$

For the rest, the definition of interpretations of $0, S, \mathbf{k}, \mathbf{s}, \mathbf{p}, \mathbf{p}_0, \mathbf{p}_1, \mathbf{r}$ proceeds as before, we interpret $=_\sigma$ as $\sim_\sigma$, and we obtain a model of $\mathbf{E\text{-}HA}^\omega$.

3.4. Definition. (*Modified realizability*) We define $x^\sigma \, \underline{\mathtt{mr}} \, A$, for formulas of $\mathbf{HA}^\omega$, by induction on the complexity of A as follows. The type σ of x is determined by the structure of A.

$$\begin{aligned} x^0 \, \underline{\mathtt{mr}} \, (t = s) \; &:= (t = s), \\ x \, \underline{\mathtt{mr}} \, (A \wedge B) \; &:= \mathbf{p}_0 x \, \underline{\mathtt{mr}} \, A \wedge \mathbf{p}_1 x \, \underline{\mathtt{mr}} \, B, \\ x \, \underline{\mathtt{mr}} \, (A \to B) &:= \forall y (y \, \underline{\mathtt{mr}} \, A \to xy \, \underline{\mathtt{mr}} \, B), \\ x \, \underline{\mathtt{mr}} \, \forall x \, A \quad &:= \forall z (xz \, \underline{\mathtt{mr}} \, A), \\ x \, \underline{\mathtt{mr}} \, \exists z \, A \quad &:= \mathbf{p}_1 x \, \underline{\mathtt{mr}} \, A[z/\mathbf{p}_0 x]. \end{aligned}$$

We also consider $\underline{\mathtt{mrt}}$-realizability, which is similar to $\underline{\mathtt{rnt}}$-realizability. All clauses are the same as for $\underline{\mathtt{mr}}$, the implication clause excepted, which now reads

$$x \, \underline{\mathtt{mrt}} \, (A \to B) \; := \; \forall y (y \, \underline{\mathtt{mrt}} \, A \to xy \, \underline{\mathtt{mrt}} \, B) \wedge (A \to B). \quad \square$$

Remark. In the usual definition (cf. Troelstra [1973a,3.4.2]), one realizes with sequences of terms $\vec{t}$, of length and types depending on the structure of A. The attractive feature of this definition is that $\exists$-free formulas are literally self-realizing: for $\exists$-free A, $\vec{t} \, \underline{\mathtt{mr}} \, A \; := \; A$, so $\vec{t}$ is empty.

For our definition above, the choice of type 0 for the realizing objects of prime formulas is somewhat arbitrary; a more canonical choice might have been obtained by (conservatively) adding a singleton type to $\mathbf{HA}^\omega$ and letting the single element of this type realize $t = s$ iff true.

A concrete version of $\underline{\mathtt{mr}}$-realizability is obtained by interpreting $\mathbf{HA}^\omega$ in a model $\mathcal{M}$; this yields $\mathcal{M}$-$\underline{\mathtt{mr}}$-realizability. The difference between Kleene's realizability and $\underline{\mathtt{mr}}$-realizability becomes clear by comparing $\underline{\mathtt{rn}}$-realizability and HRO-$\underline{\mathtt{mr}}$-modified realizability of statements of the form

$$\forall y \neg \forall z \neg Txyz \to B$$

For $\underline{\mathtt{rn}}$, this requires a realizer t which must be applicable to the canonical realizer $\Lambda y.0$ of $\forall y \neg \forall z \neg Txyz$ *if* this is true. On the other hand, in the case of HRO-$\underline{\mathtt{mr}}$-realizability $t{\bullet}\Lambda y.0$ must be defined, whether $\forall y \neg \forall z \neg Txyz$ is true or not. In other words, in modified realizability, realizing objects for implications have a larger domain of definition than what is required by "pure" realizability.

Soundness now takes the form

3.5. Theorem. *(Soundness)*

$\mathbf{HA}^\omega \vdash A \Rightarrow \mathbf{HA}^\omega \vdash t\,\underline{\underline{\mathrm{mr}}}\,A \wedge t\,\underline{\underline{\mathrm{mrt}}}\,A$ *for some term* t *with* $\mathrm{FV}(t) \subset \mathrm{FV}(A)$.

Proof. By a straightforward induction on the length of derivations. □.

As noted above, for ∃-free formulas there are canonical realizers, and truth and realizability coincide for ∃-free formulas. Therefore the ∃-free formulas of $\mathbf{HA}^\omega$ play the same role w.r.t. $\underline{\underline{\mathrm{mr}}}$-realizability as the ∃-free formulas of $\mathbf{HA}^*$ w.r.t. $\underline{\underline{\mathrm{rn}}}$-realizability.

For an axiomatization we need the following

3.6. Lemma. *For each instance* F *of one of the following schemata*

$\mathrm{IP}_{\mathrm{ef}}$ $\quad (A \to \exists x^\sigma B) \to \exists y^\sigma (A \to B) \quad (y \notin \mathrm{FV}(A),\ A\ \exists\text{-free}),$

AC $\quad \forall x^\sigma \exists y^\tau A(x,y) \to \exists z^{\sigma\tau} \forall x^\sigma A(x, zx),$

there is a term t *such that* $\vdash t\,\underline{\underline{\mathrm{mr}}}\,F$, *with* $\mathrm{FV}(t) \subset \mathrm{FV}(A)$.

3.7. Theorem. *(Axiomatization of modified realizability)*

$$\mathbf{HA}^\omega + \mathrm{AC} + \mathrm{IP}_{\mathrm{ef}} \vdash A \leftrightarrow \exists x (x\,\underline{\underline{\mathrm{mr}}}\,A)$$

and for $\mathbf{H} \in \{\mathbf{HA}^\omega, \mathbf{I\text{-}HA}^\omega, \mathbf{E\text{-}HA}^\omega\}$

$$\mathbf{H} + \mathrm{AC} + \mathrm{IP}_{\mathrm{ef}} \vdash A \Leftrightarrow \mathbf{H} \vdash t\,\underline{\underline{\mathrm{mr}}}\,A$$

for some t *with* $\mathrm{FV}(t) \subset \mathrm{FV}(A) \setminus \{x\}$.

Remark. In a theory **T** with decidable prime formulas $\mathrm{IP}_{\mathrm{ef}}$ is implied by the schema IP defined in 1.13. To see this, note that in **T** ∃-free formulas are logically equivalent to negated formulas, since $\neg\neg B \leftrightarrow B$ (by induction on the construction of B). On the other hand, IP is $\underline{\underline{\mathrm{mr}}}$-realizable in $\mathbf{HA}^\omega$, so the preceding theorem also holds with IP replacing $\mathrm{IP}_{\mathrm{ef}}$, for **H** equal to $\mathbf{I\text{-}HA}^\omega$ or $\mathbf{E\text{-}HA}^\omega$.

3.8. Theorem. *(Applications of modified realizability) Let* $\mathbf{H} \in \{\mathbf{HA}^\omega, \mathbf{I\text{-}HA}^\omega, \mathbf{E\text{-}HA}^\omega\}$, *and let* $\mathbf{H}'$ *be* $\mathbf{H} \pm \mathrm{IP}_{\mathrm{ef}} \pm \mathrm{AC}$. *Then*

(i) $\mathbf{H}'$ *is consistent.*

(ii) $\mathbf{H}' \vdash A \vee B \Rightarrow \mathbf{H}' \vdash A$ *or* $\mathbf{H}' \vdash B$ *(for* $A \vee B$ *closed) (Disjunction Property DP).*

(iii) $\mathbf{H}' \vdash \exists x^\sigma A \Rightarrow \mathbf{H}' \vdash A[x/t^\sigma]$ *for a suitable term* t, $\mathrm{FV}(t) \subset \mathrm{FV}(A) \setminus \{x\}$ *(Explicit Definability ED).*

(iv) $\mathbf{H}' \vdash \forall x^\sigma \exists y^\tau A(x,y) \Rightarrow \mathbf{H}' \vdash \exists z^{\sigma\tau} \forall x^\sigma A(x, zx)$ *(Rule of choice* ACR*).*

(v) $\mathbf{H}' \vdash (A \to \exists x^\sigma B) \Rightarrow \mathbf{H}' \vdash \exists x^\sigma (A \to B)$ *where* A *is* ∃*-free (*$\mathrm{IPR}_{\mathrm{ef}}$*-rule).*

3.9. Concrete forms of modified realizability

The proof-theoretic applications of $\underline{\text{mr}}$-realizability obtained by specifying a model for $\mathbf{HA}^\omega$ have in fact two "levels of freedom": (a) the choice of a model $\mathcal{M}$, definable in a language $\mathcal{L}$ say, and (b) the theory formulated in $\mathcal{L}$ which is available for proving facts about $\mathcal{M}$, i.e. the metatheory for $\mathcal{M}$.

By "$\mathcal{M}$ definable in $\mathcal{L}$" we do not mean that $\mathcal{M}$ is globally definable in $\mathcal{L}$, but only that locally, for each A of $\mathbf{HA}^\omega$, we can express $[\![A]\!]_\mathcal{M}$ by a formula of $\mathcal{L}$. Thus choosing HRO for $\mathcal{M}$ is the first level of freedom, and choosing some theory Γ in the language of $\mathbf{HA}^*$ for proving facts about HRO is the second level of freedom.

An interesting example of this occurs in connection with two models of $\mathbf{HA}^\omega$ which are similar to HRO and HEO respectively, but based on partial continuous function application $|$ instead of partial recursive application $\bullet$.

The *Intensional Continuous Functionals* ICF are an analogue of HRO; we give the intuitively simplest definition (which does not mean the technically slickest) of the types:

$$\begin{aligned}
\mathrm{ICF}_0 &:= \mathbb{N},\\
\mathrm{ICF}_{00} &:= \mathbb{N} \to \mathbb{N},\\
\mathrm{ICF}_{\sigma 0} &:= \{\alpha : \forall \beta{\in}\mathrm{ICF}_\sigma(\alpha(\beta){\downarrow})\}\ (\sigma \neq 0),\\
\mathrm{ICF}_{0\sigma} &:= \{\alpha : \forall x(\lambda n.\alpha(\langle x\rangle * n) \in \mathrm{ICF}_\sigma)\}\ (\sigma \neq 0),\\
\mathrm{ICF}_{\sigma\tau} &:= \{\alpha : \forall \beta{\in}\mathrm{ICF}_\sigma(\alpha|\beta \in \mathrm{ICF}_\tau)\}\ (\sigma, \tau \neq 0).
\end{aligned}$$

Application is then defined in the obvious way: $\mathrm{App}_{\sigma,0}(\alpha,\beta) := \alpha(\beta)$, $\mathrm{App}_{0,\sigma}(\alpha, n) := \lambda m.\alpha(\langle n\rangle * m)$, $\mathrm{App}_{\sigma,\tau}(\alpha,\beta) := \alpha|\beta$, etc. Equality at type σ is interpreted by equality of numbers (for $\sigma = 0$) or functions (for $\sigma \neq 0$).

The *Extensional Continuous Functionals* ECF are related to ICF in the same way as HEO is related to HRO: one defines a hereditary equivalence relation based on $|$ instead of $\bullet$. ECF coincides with Kleene's countable functionals or Kreisel's continuous functionals.

Both ICF and ECF are locally definable in the language of $\mathbf{EL}^*$, and for soundness of ICF-$\underline{\text{mr}}$ and ECF-$\underline{\text{mr}}$ relative to $\mathbf{EL}^*$ nothing more is needed. But additional axioms added to $\mathbf{EL}^*$ may result in different properties of the models, and hence of $\mathcal{M}$-$\underline{\text{mr}}$-realizability. Two mutually incompatible additional axioms we can add to $\mathbf{EL}^*$ are $\mathrm{FAN_D}$ and a version of *Church's Thesis*

CT $\qquad \forall\alpha\exists x\forall y(\alpha x = x{\bullet}y)$.

CT states that the function variables in $\mathbf{EL}^*$ range over the total recursive functions; the incompatibility of CT with $\mathrm{FAN_D}$ follows from Kleene's well-known example of a primitive recursive tree well-founded w.r.t. all total recursive functions but not w.r.t. all functions, since the depth of the tree is unbounded (cf. Troelstra and van Dalen [1988,4.7.6]).

Assuming $\mathrm{FAN_D}$, we can show that ICF and ECF contain a *Fan Functional* ϕ_{uc} satisfying the axiom for a *Modulus of Uniform Continuity*

MUC $\qquad \forall z^2\forall\gamma\forall\alpha{\le}\gamma\forall\beta{\le}\gamma(\bar{\alpha}(\phi_{\mathrm{uc}}z\gamma) = \bar{\beta}(\phi_{\mathrm{uc}}z\gamma) \to z\alpha = z\beta)$.

If we add MUC to $\mathbf{HA}^\omega$, we can $\underline{\text{mr}}$-interpret $\mathrm{FAN_D}$. If, on the other hand, we use $\mathbf{EL}^* + \mathrm{CT}$ as our metatheory for ICF-$\underline{\text{mr}}$, we can realize a statement positively contradicting MUC. See Troelstra [1973a,2.6.4, 2.6.6, 3.4.16, 3.4.19].

As an example of an application of a concrete version of $\underline{\underline{\mathtt{mr}}}$-realizability we can show e.g. the consistency of $\mathbf{HA}^\omega$ + $\mathrm{IP_{ef}}$ + AC + WC-N + $\mathrm{FAN_D}$ + $\mathrm{EXT}_{1,0}$, where WC-N is the schema $\forall\alpha\exists n\, A(\alpha,n) \rightarrow \forall\alpha\exists n, m\forall\beta(\bar{\alpha}m = \bar{\beta}m \rightarrow A(\beta,n))$, and $\mathrm{EXT}_{1,0}$ is $\forall\alpha\beta z^2(\alpha = \beta \rightarrow z^2\alpha = z^2\beta)$. (Use ICF-$\underline{\underline{\mathtt{mr}}}$-realizability with $\mathbf{EL}^*$ + $\mathrm{FAN_D}$ as metatheory.)

Notation. Henceforth we write $\underline{\underline{\mathtt{mrn}}}$, $\underline{\underline{\mathtt{mrf}}}$ for HRO-$\underline{\underline{\mathtt{mr}}}$ and ICF-$\underline{\underline{\mathtt{mr}}}$-realizability respectively. $\Box$.

3.10. Notes

Modified realizability was first formulated by Kreisel [1962b]; a concrete version equivalent to our ICF-$\underline{\underline{\mathtt{mr}}}$-realizability was used in Kleene and Vesley [1965].

Cook and Urquhart [1993] and Harnik [1992] apply $\underline{\underline{\mathtt{mrt}}}$-realizability to bounded arithmetic and related systems, improving on earlier results obtained by Buss [1986] by means of numerical realizability.

Vesley [1970] used modified realizability to obtain consistency of intuitionistic analysis with a restricted form of IP (Vesley's principle). Moschovakis [1971] used a modified realizability interpretation to obtain consistency of a weak version of Church's thesis with Kleene's system for intuitionistic analysis (i.e **EL** with bar induction and GC for the case $A \equiv 0 = 0$), together with Vesley's Principle. The weak version of Church's thesis may be stated as: "each numerical function is not not recursive". In Troelstra [1973a,3.4.15] it is observed that the modified realizability of Moschovakis [1971] is essentially abstract modified realizability interpreted in the type structure consisting of the *recursive* elements of ICF, and that the consistency proof covers in fact full $\mathrm{IP_{ef}}$ (Troelstra [1973a,3.4.18]).

Some further examples of papers using or discussing modified realizability are Dragalin[1968], Diller[1980], Grayson[1981b,1982], van Oosten [1990], Scedrov and Vesley [1983]. See also 9.8 on Berger and Schwichtenberg [1995], Berger, Schwichtenberg and Seisenberger [1997].

4. Derivation of the Fan Rule

This section is devoted to an "indirect application" of modified realizability: it is shown how closure under the rule of choice ACR, obtained from $\underline{\underline{\mathtt{mrt}}}$-realizability, may be combined with the (intrinsically interesting) notion of "majorizable functional" to obtain closure under the Fan Rule.

We can define the so-called *majorizable* functionals relative to any finite-type structure. They are introduced via a relation of majorization, defined as follows.

4.1. Definition. $t^* \operatorname{maj}_\sigma t$, for $t^*, t \in \sigma$, is defined by induction on σ:

$$t^* \operatorname{maj}_0 t := t^* \geq t,$$
$$t^* \operatorname{maj}_{\sigma\times\tau} t := \mathbf{p}_0 t^* \operatorname{maj}_\sigma \mathbf{p}_0 t \wedge \mathbf{p}_1 t^* \operatorname{maj}_\tau \mathbf{p}_1 t,$$
$$t^* \operatorname{maj}_{\tau\sigma} t := \forall y^* y (y^* \operatorname{maj}_\tau y \to t^* y^* \operatorname{maj}_\sigma ty, t^* y).$$

Furthermore we put

$$t \in \mathrm{Maj} := \exists t^* \operatorname{maj}_\sigma t \quad (\text{“}t \text{ is majorizable”}).$$

Lemma. $t^* \operatorname{maj} t \Rightarrow t^* \operatorname{maj} t^*$.

Proof. Induction on the type of t.

4.2. Definition. For each $t \in 0\sigma$ we define $t^+ \in 0\sigma$ by induction on the structure of σ.

$$t^+0 = t0,\ t^+(Sz) = \max\{t^+z, t(Sz)\} \text{ for } \sigma = 0,$$
$$t^+ := \lambda n.[\lambda y((\lambda n.tny)^+ n)] \text{ for } \sigma \equiv \sigma_1\sigma_2,$$
$$t^+ = \lambda n.\mathbf{p}((\lambda n.\mathbf{p}_0(tn))^+ n)((\lambda n.\mathbf{p}_1(tn))^+ n) \text{ for } \sigma \equiv \sigma_1 \times \sigma_2.$$

Lemma. *If* $\forall n^0 (Fn \operatorname{maj} Gn)$, *then* $F^+ \operatorname{maj} G^+, G$.

Proof. We use induction on σ. Let $Fn, Gn \in \sigma$.
Case (i) $\sigma \equiv 0$. Almost immediate.
Case (ii) $\sigma \equiv \sigma_1\sigma_2$. The assumption yields

$$s^* \operatorname{maj} s \Rightarrow Fns^* \operatorname{maj} Fns, Gns$$

for all $n \in \mathbb{N}$. By the induction hypothesis we have

(1) $$(\lambda n.Fns^*)^+ \operatorname{maj} (\lambda n.Fns)^+, (\lambda n.Fns), (\lambda n.Gns)^+, (\lambda n.Gns),$$

Now by definition of F^+, G^+ and beta-conversion:

$$(\lambda n.Fns^*)^+ k = F^+ ks^*$$
$$(\lambda n.Fns)^+ k = F^+ ks$$
$$(\lambda n.Gns)^+ k = G^+ ks$$

If $n \geq m$, we obtain from (1)

$$F^+ ns^* \operatorname{maj} F^+ ns, F^+ ms, Fms, \quad F^+ ns^* \operatorname{maj} G^+ ms, Gms.$$

and from this $F^+ n \operatorname{maj} F^+ m, Fm, G^+ m, Gm$. Since $n \geq m$, it follows that $F^+ \operatorname{maj} G^+, G$.
Case (iii) $\sigma \equiv \sigma_1 \times \sigma_2$. We are given $\forall n (Fn \operatorname{maj} Gn)$, so

$$\forall n (\mathbf{p}_i(Fn) \operatorname{maj} \mathbf{p}_i(Gn)) \ (i \in \{0,1\}).$$

So we have

$$\forall n ((\lambda n.\mathbf{p}_i(Fn))n \operatorname{maj} (\lambda n.\mathbf{p}_i(Gn))n)$$

and hence by the induction hypothesis

$$(\lambda n.\mathbf{p}_i(Fn))^+ \operatorname{maj} (\lambda n.\mathbf{p}_i(Gn))^+, \lambda n.\mathbf{p}_i(Gn).$$

From this we obtain for $n \geq m$, $i \in \{0,1\}$

$$(\lambda n.\mathbf{p}_i(Fn))^+ n \operatorname{maj} (\lambda n.\mathbf{p}_i(Fn))^+ m, (\lambda n.\mathbf{p}_i(Gn))^+ m, (\lambda n.\mathbf{p}_i(Gn))m,$$
$$\mathbf{p}_i(F^+ n) \operatorname{maj} \mathbf{p}_i(F^+ m),\ \mathbf{p}_i(G^+ m), \mathbf{p}_i(Gm)$$

and therefore hence $F^+ \operatorname{maj} G^+, G$.

4.3. Proposition. *Let all free variables in $t \in \tau$ be of type 0 or 1; then there is a term $t^* \in \tau$ with $\mathrm{FV}(t^*) \subset \mathrm{FV}(t)$, such that $\mathbf{HA}^\omega \vdash t^* \,\mathrm{maj}\, t^*, t$.*

Proof. For each constant or variable of type 0 or 1 of $\mathbf{HA}^\omega$ (c^τ say) we show that there is a $c^* \in \tau$ with $c^* \,\mathrm{maj}_\tau\, c$.

(a) $0 \,\mathrm{maj}\, 0$, $S \,\mathrm{maj}\, S$ are immediate;

(b) $x^0 \,\mathrm{maj}\, x^0$; for y^1 define y^* by recursion as y^+;

(c) $\mathbf{k} \,\mathrm{maj}\, \mathbf{k}$, $\mathbf{s} \,\mathrm{maj}\, \mathbf{s}$, $\mathbf{p} \,\mathrm{maj}\, \mathbf{p}$, $\mathbf{p}_0 \,\mathrm{maj}\, \mathbf{p}_0$, $\mathbf{p}_1 \,\mathrm{maj}\, \mathbf{p}_1$;

(d) If $\mathbf{r}$ is the recursor with $\mathbf{r}0ts = t$ etc., take $\mathbf{r}^* := \mathbf{r}^+$.

4.4. Theorem. *(Fan Rule) Let A be a formula of $\mathbf{HA}^\omega$ containing only variables of types 0 or 1 free, then $\mathbf{HA}^\omega \vdash \forall \alpha{\leq}\beta \exists n\, A(\alpha, n) \Rightarrow \mathbf{HA}^\omega \vdash \exists m \forall \alpha{\leq}\beta \exists n{\leq}m\, A(\alpha, n)$, where $\alpha \leq \beta := \forall m(\alpha n \leq \beta n)$.*

Proof. Let $\mathbf{HA}^\omega \vdash \forall \alpha{\leq}\beta\, A(\alpha, F\alpha)$ for a suitable term $F \in (1)0$. F is majorizable, so there is an F^* such that $F^* \,\mathrm{maj}\, F^*, F$ which means in particular that $\forall \alpha\beta(\beta \geq \alpha \to F^*\beta^+ \geq F\alpha)$ and hence $\mathbf{HA}^\omega \vdash \forall \alpha{\leq}\beta \exists n{\leq}F^*\beta^+\, A(\alpha, n)$. □

Remarks. Switching from a recursor of type $\sigma(\sigma 0 \sigma)0\sigma$ to a recursor of type $0(\sigma 0 \sigma)\sigma\sigma$ is purely a matter of technical convenience; these recursors are interdefinable.

In Kohlenbach [1992] the following generalization is established for $\mathbf{E}\text{-}\mathbf{HA}^\omega$:

$$\vdash \forall \alpha \forall x {\leq_\rho} s\alpha \exists y^\tau A(\alpha, x, y) \Rightarrow \vdash \forall \alpha \forall x {\leq_\rho} s\alpha \exists y {\leq_\tau} t\alpha A(\alpha, x, y),$$

for some term t, where $\tau \in \{0, 1, 2\}$, $s \in 1\rho$, s closed, and where $\leq_\sigma$ is defined by induction on the type structure by $x^0 \leq_0 y_0 := x \leq y$, $x^{\sigma\tau} \leq_{\sigma\tau} y^{\sigma\tau} := \forall z^\sigma (xy \leq_\tau yz)$, $x^{\sigma\times\tau} \leq_{\sigma\times\tau} y^{\sigma\times\tau} := \mathbf{p}_0 x \leq_\sigma \mathbf{p}_0 y \wedge \mathbf{p}_1 x \leq_\tau \mathbf{p}_1 y$.

As observed above, the proof of closure under the Fan Rule given above depends on realizability only to the extent that we have used modified realizability to obtain closure under the fan rule. For other systems other interpretations, such as the Dialectica interpretation, yield closure under the rule of choice; cf. Kohlenbach [1992].

4.5. Notes

The notions of *majorization* and *majorizable functional* were introduced by Howard [1973]. The present version is a modification due to Bezem[1986,1989], called strong majorization by him; we have added a clause for product types.

Kohlenbach [1990] introduced a version of Bezem's definition with a special clause for types of the form $\sigma 0$; however, in the presence of product types we found it more convenient to stick to Bezem's definition.

The proof of the Fan Rule presented here is due to Kohlenbach [1992]. For other proofs, see e.g. Troelstra [1977c], Beeson [1985], Troelstra and van Dalen [1988,9.7.23].

5. Lifschitz realizability

This type of realizability was invented by Lifschitz [1979] to show that *Church's Thesis with Uniqueness*

$\mathrm{CT}_0!\qquad \forall x\exists! y\, A(x,y) \to \exists z\forall x(z{\bullet}x{\downarrow} \wedge A(x, z{\bullet}x))$

does not imply CT_0 in $\mathbf{HA}^*$. The idea to achieve this, is to use as realizer for an existential formula not a single instantiation for the quantifier, but a finite inhabited set of possible instantiations, such that in general there is no recursive procedure for selecting elements of such inhabited sets, although for singletons there is such a procedure. The sets we use are given by

$$V_x := \{y : y \leq \mathbf{p}_1 x \wedge \forall n \neg T(\mathbf{p}_0 x, y, n)\}.$$

If we know that V_x is a singleton, say $\{y : 0 = 0\}$, we can find y recursively in x as follows: we start computing $\mathbf{p}_0 x{\bullet}z$ for all values of $z \leq \mathbf{p}_1 x$; as soon as we have found terminating computations for $\mathbf{p}_1 x$ arguments, we know that the remaining argument $\leq \mathbf{p}_1 x$ is the required y.

5.1. Definition. The clauses for $\underline{\mathtt{rln}}$-realizability are identical to the clauses for $\underline{\mathtt{rn}}$-realizability, except for the existential quantifier:

$$x \,\underline{\mathtt{rln}}\, \exists y A := \mathrm{Inh}(V_x) \wedge \forall y \in V_x(\mathbf{p}_1 y \,\underline{\mathtt{rln}}\, A[y/\mathbf{p}_0 y])$$

where "$\mathrm{Inh}(W)$" means that $\exists z(z \in W)$. □

In this form the notion appears as a modification of numerical realizability. There is also a Lifschitz's analogue of function realizability. In that case the sets of realizers for the existential quantifiers take the form

$$V_\alpha := \{\gamma : \gamma \leq \mathbf{p}_1\alpha \wedge \forall n(\mathbf{p}_0\alpha(\bar{\gamma}n) = 0)\}.$$

The V_α are not finite, but compact. There is no general method for finding an element in inhabited V_α which is continuous in α, but there is a method for V_α's which are singletons. There is no interesting "abstract" version of Lifschitz realizability.

5.2. Definition. $\underline{\mathtt{rlf}}$-realizability is defined as $\underline{\mathtt{rf}}$-realizability, except for the clauses for the existential quantifiers, which become:

$$\alpha \,\underline{\mathtt{rlf}}\, \exists\beta A := \mathrm{Inh}(V_\alpha) \wedge \forall\gamma \in V_\alpha(\mathbf{p}_1\gamma \,\underline{\mathtt{rlf}}\, A[\beta/\mathbf{p}_0\gamma]),$$
$$\alpha \,\underline{\mathtt{rlf}}\, \exists x A := \mathrm{Inh}(V_\alpha) \wedge \forall\gamma \in V_\alpha(\mathbf{p}_1\gamma \,\underline{\mathtt{rlf}}\, A[x/(\mathbf{p}_0\gamma)0]). \quad \square$$

5.3. Summary of results for $\underline{\mathtt{rln}}$-realizability

Definition. In $\mathbf{HA}^*$ the *bounded Σ^0_2-formulas* ($\mathrm{B}\Sigma^0_2$-formulas) are formulas of the form $\exists x{<}t\neg(s = s')$; the *$\mathrm{B}\Sigma^0_2$-negative formulas* are the formulas constructed from prime formulas $s = s'$ and $\mathrm{B}\Sigma^0_2$-formulas by means of $\forall, \wedge, \to$. □

Corresponding classes in $\mathbf{HA}$ are defined as follows. A formula of the form $\exists x{\leq}y\forall z\, A$ with A primitive recursive is called a *bounded Σ^0_2-formula* ($\mathrm{B}\Sigma^0_2$-formula); the *$\mathrm{B}\Sigma^0_2$-negative formulas* are the formulas constructed from Σ^0_1-formulas and $\mathrm{B}\Sigma^0_2$-formulas by means of $\forall$, $\wedge$, $\to$.

N.B. Although the class of $\mathrm{B}\Sigma^0_2$-formulas in $\mathbf{HA}^*$ is somewhat wider than the corresponding class in $\mathbf{HA}$, the $\mathrm{B}\Sigma^0_2$-negative formulas for $\mathbf{HA}^*$ and $\mathbf{HA}$ are the same modulo logical equivalence. To see this, observe that (a) a Π^0_1-formula in **HA** can be written as $\neg s = s$ in $\mathbf{HA}^*$, and (b) $\exists x < t \neg(s = s')$ in $\mathbf{HA}^*$ is equivalent to a formula of the form $\exists y(t = y) \wedge \forall z(t = z \rightarrow \exists x{<}z.A(x))$, with A primitive recursive, which is $\mathrm{B}\Sigma^0_2$-negative in **HA** modulo logical equivalence.

In the case of numerical Lifschitz realizability, we cannot take as our basis theory $\mathbf{HA}^*$, but need instead an extension $\mathbf{HA}'$, which is $\mathbf{HA}^* + \mathrm{M} + \mathrm{CB}\Sigma^0_2$; here $\mathrm{CB}\Sigma^0_2$, the *Cancellation of double negations in Bounded* Σ^0_2*-formulas* is:

$\mathrm{CB}\Sigma^0_2$ $\quad \neg\neg A \rightarrow A$ (for A in $\mathrm{B}\Sigma^0_2$).

Van Oosten showed that in fact $\mathrm{CB}\Sigma^0_2$ is equivalent to the following principle

$$\forall nm(Pn \vee Qm) \rightarrow (\forall nPn \vee \forall mQm) \quad (P, Q \text{ primitive recursive}),$$

where $n \notin \mathrm{FV}(Q)$, $m \notin \mathrm{FV}(P)$. Soundness now holds w.r.t. $\mathbf{HA}'$, i.e. for all sentences A

$$\mathbf{HA}' \vdash A \;\Rightarrow\; \mathbf{HA}' \vdash \bar{n}\,\underline{\mathtt{rln}}\,A$$

for a suitable numeral $\bar{n}$. The following properties of the V_n are crucial in the proof of the soundness theorem:

(i) for some total recursive f_0, $\forall xy(y \in V_{f_0(x)} \leftrightarrow y = x)$, i.e. indices of singleton V_t's may be found recursively in their (unique) elements.

(ii) There is a partial recursive f_1 such that for any operation with code x, total on V_y, the image of V_y under x is $V_{f_1(x,y)}$.

(iii) There is a total recursive f_2 such that $V_{f_2(x)} = \bigcup\{V_z : z \in V_x\}$.

(iv) There is a partial recursive f_3 such that $\mathbf{HA}' \vdash \forall x(\mathrm{Inh}(V_x) \wedge \forall y{\in}V_x(y\,\underline{\mathtt{rln}}\,A) \rightarrow f_3(x)\,\underline{\mathtt{rln}}\,A)$.

With respect to the class of self-realizing formulas, we note an interesting deviation from the notions of realizability considered hitherto: these are not just the $\exists$-free formulas, but the wider class of $\mathrm{B}\Sigma^0_2$-negative formulas. Now we can axiomatize $\underline{\mathtt{rln}}$-realizability relative to $\mathbf{HA}'$ by means of the following scheme for $\mathrm{B}\Sigma^0_2$-negative A:

$\mathrm{ECT_L}$ $\quad \forall x(Ax \rightarrow \exists yBxy) \rightarrow \exists z\forall x(Ax \rightarrow z{\bullet}x{\downarrow} \wedge \mathrm{Inh}(V_{z\bullet x}) \wedge \forall u{\in}V_{z\bullet x}\, Bxu)$.

An interesting special case of $\mathrm{ECT_L}$ is $\mathrm{ECT_L}!$ which can be formulated as

$$\forall x(Ax{\rightarrow}\exists!y\, Bxy) \rightarrow \exists z\forall x(Ax \rightarrow z{\bullet}x{\downarrow}{\wedge}B(x, z{\bullet}x)) \quad (A\ \mathrm{B}\Sigma^0_2\text{-negative}),$$

with the help of the following

Lemma. *There is a partial recursive* f_5 *such that*

$$\mathbf{HA}' \vdash \forall z(\exists x\forall y(x = y \leftrightarrow y \in V_z) \rightarrow f_5(z) \in V_z).$$

5.4. Proposition. *(Applications)*

(i) $\mathbf{HA}' + \mathrm{ECT_L}$ *is consistent;*

(ii) $\mathbf{HA}^* + \mathrm{ECT_L}! \nvdash \mathrm{CT}_0$*;*

(iii) $\mathbf{HA}'$ *is closed under the rule* $\mathrm{ECR_L}$ *and a fortiori under the rule* $\mathrm{ECR_L}!$ *(as* $\mathrm{ECT_L}$ *and* $\mathrm{ECT_L}!$ *but with main* $\rightarrow$ *replaced by* $\Rightarrow$ *etc). Since* $\mathbf{HA}'$ *also satisfies the rules* DP *and* EDN*, we can formulate the rule* $\mathrm{ECR_L}!$ *even more strongly as: for* A *in* $\mathrm{B}\Sigma^0_2$,

$$\vdash \forall x(Ax \rightarrow \exists! yBxy) \Rightarrow \vdash \forall x(Ax \rightarrow \bar{n}\bullet x\downarrow \wedge B(x, \bar{n}\bullet x)$$

for a suitable numeral $\bar{n}$.

5.5. Summary of results for rlf- and rlft-realizability

The basis theory is now an extension of $\mathbf{EL}^*$, namely $\mathbf{EL}' \equiv \mathbf{EL}^* + \mathrm{M_{QF}} + \mathrm{KL_{QF}}$, where $\mathrm{M_{QF}}$ is Markov's principle for quantifier-free formulas, and *König's Lemma for quantifier-free formulas* is the schema for *quantifier-free* A.

$$\mathrm{KL_{QF}} \quad \begin{array}{l}\forall x \exists n(\mathrm{lth}(n) = x \wedge n{\le}\alpha \wedge An)\\ \qquad \wedge \forall nm(A(n * m) \rightarrow An) \rightarrow \exists \beta{\le}\alpha \forall n\, A(\bar{\beta} n)\end{array}$$

($n{\le}\alpha := \forall y{<}\mathrm{lth}(n)((n)_y \le \alpha y)$; $\exists \alpha{\le}\phi(\ldots) := \exists \alpha(\alpha \le \phi \wedge \ldots)$). The analogue in $\mathbf{EL}'$ of the $\mathrm{B}\Sigma^0_2$-negative formulas are the $\mathrm{B}\Sigma^1_2$-negative formulas (a $\mathrm{B}\Sigma^1_2$-formula is a formula of the form $\exists \alpha{\le}\phi \neg s = t$ (a *Bounded* Σ^1_2 *formula*); the class of $\mathrm{B}\Sigma^1_2$*-negative formulas* is obtained from formulas $\mathrm{B}\Sigma^1_2$-formulas and prime formulas by means of $\rightarrow, \wedge, \forall$). □

As a typical result one obtains that GC! (i.e.the special case of GC with uniqueness for the existential quantifier) does not imply GC, not even the special case of WC-N.

Remark. $\mathrm{KL_{QF}}$ for the language of $\mathbf{EL}^*$ follows from $\mathrm{KL_{QF}}$ for $\mathbf{EL}$ by observing that KL with $A \in \Sigma^0_1$ is derivable from $\mathrm{KL_{QF}}$ in $\mathbf{EL}$.

5.6. Notes

Khakhanyan [1980b] defined Lifschitz' realizability for certain set theories and uses it to obtain independence of CT_0 from $\mathrm{CT}_0!$ for these theories. Other relevant papers are van Oosten[1990,1991a,1991b]. As to van Oosten [1991b], see 8.31.

It is possible to combine Lifschitz realizability with modified realizability for HRO van Oosten [1991a].

It is not known whether for some or all results perhaps weaker theories than $\mathbf{HA}'$, $\mathbf{EL}'$ will suffice.

6. Extensional realizability

It is also possible to combine the idea of realizability with extensionality, by defining not just a notion of the form "x realizes A", but a relation between realizing objects: "x and y equally realize A". The definition below has been written out for **HA*** and partial recursive application, but also makes sense in the abstract setting of **APP**, if we read everywhere $\underline{\text{re}}$ for $\underline{\text{rne}}$.

6.1. Definition. We define "$x = x' \,\underline{\text{rne}}\, A$" ($x, x' \notin \text{FV}(A)$, $x \not\equiv y$), by induction on the complexity of A:

$$
\begin{aligned}
x = x' \,\underline{\text{rne}}\, P \quad &:= (x = x' \wedge P \wedge x\!\downarrow \wedge x'\!\downarrow) \;\; (P \text{ prime}),\\
x = x' \,\underline{\text{rne}}\, (A \wedge B) \quad &:= (\mathbf{p}_0 x = \mathbf{p}_0 x' \,\underline{\text{rne}}\, A) \wedge (\mathbf{p}_1 x = \mathbf{p}_1 x' \,\underline{\text{rne}}\, B),\\
x = x' \,\underline{\text{rne}}\, (A \to B) \quad &:= x\!\downarrow \wedge x'\!\downarrow \wedge \forall y y'(y = y' \,\underline{\text{rne}}\, A \to\\
&\quad x{\bullet}y = x{\bullet}y' \,\underline{\text{rne}}\, B \wedge x'{\bullet}y = x'{\bullet}y' \,\underline{\text{rne}}\, B \wedge x{\bullet}y = x'{\bullet}y \,\underline{\text{rne}}\, B),\\
x = x' \,\underline{\text{rne}}\, \forall y\, A \quad &:= \forall y(x{\bullet}y = x'{\bullet}y \,\underline{\text{rne}}\, A),\\
x = x' \,\underline{\text{rne}}\, \exists y\, A \quad &:= (\mathbf{p}_0 x = \mathbf{p}_0 x') \wedge (\mathbf{p}_1 x = \mathbf{p}_1 x' \,\underline{\text{rne}}\, A[y/\mathbf{p}_0 x]),
\end{aligned}
$$

and we put

$$x \,\underline{\text{rne}}\, A \;:=\; x = x \,\underline{\text{rne}}\, A.$$

As always, $\underline{\text{rnet}}$-realizability is obtained by adding "$\wedge\, (A \to B)$" in the implication clause. □

Remarks. Note that $x \,\underline{\text{rn}}\, A$ does not in general imply $x = x \,\underline{\text{rne}}\, A$; for if $x = x \,\underline{\text{rne}}\, [\forall y(t = 0) \to \forall z(s = 0)]$, then x must yield the same value when applied to extensionally equal realizers z, z' for $\forall y(t = 0)$; on the other hand, for an x such that $x \,\underline{\text{rn}}\, [\forall y(t = 0) \to \forall z(s = 0)]$ no such restriction applies.

The definition may also be formulated as a simultaneous inductive definition of "x extensionally realizes A" and "x and y are equivalent realizers for A", but this is more cumbersome.

It is straightforward to prove soundness.

The $\exists$-free formulas play the same role as in $\underline{\text{rn}}$-realizability. On the other hand, no simple axiomatization of the provably $\underline{\text{rne}}$-realizable formulas is known.

For proofs of the following facts we refer to van Oosten [1990].

6.2. The difference between ordinary realizability and extensional realizability is demonstrated by the fact that the following instance of ECT_0 is not $\underline{\text{rne}}$-realizable:

$$\forall z[\forall x \exists y(\neg\neg\exists u Tzxu \to Tzxy) \to \exists v \forall x(v{\bullet}x \wedge (\neg\neg\exists u Tzxu \to T(z, x, v{\bullet}x))]$$

On the other hand it is not hard to verify that the following *"Weak Extended Church's Thesis"* is provably $\underline{\text{rne}}$-realizable:

6.3. Proposition. *In* **HA** *we can* $\underline{\text{rne}}$*-realize:*

$\text{WECT}_0 \quad \forall x(A \to \exists y\, Bxy) \to \neg\neg\exists z \forall x(A \to z{\bullet}x\!\downarrow \wedge\, B(x, z{\bullet}x))$

for $\exists$*-free* A.

A nice application of $\underline{\text{rnet}}$-realizability is the following refinement of ECR.

6.4. Proposition. *Assume for* $\exists$*-free* B *that in* $\mathbf{HA}^*$

$$\vdash \forall z(\forall x \exists y\, Bzxy \rightarrow \exists u\, Czu)$$

then for some $\bar{n}$

$$\vdash \forall z(\bar{n}\bullet z\downarrow \wedge \forall v, v'(\forall x(v\bullet x = v'\bullet x \wedge B(z,x,v\bullet x)) \rightarrow \bar{n}\bullet z\bullet v = \bar{n}\bullet z\bullet v' \wedge C(z, \bar{n}\bullet z\bullet v)).$$

6.5. Notes

Extensional realizability appears for the first time explicitly in some unpublished notes by Grayson [1981c], and implicitly in Pitts [1981].

Renardel de Lavalette [1984] and Beeson[1979b,1985] use an abstract version of extensional realizability in combination with forcing, to prove that $\mathbf{ML}_0$ (the arithmetical fragment of the extensional version of Martin-Löf's type theory) is conservative over **HA**. $\mathbf{ML}_0$ includes $\mathbf{E\text{-}HA}^\omega + \mathrm{AC}$ as a subtheory. See also Eggerz [1987]. The proofs by Renardel and Beeson extend earlier work of Goodman [1978][3].

There is a close similarity between "$x = x'\ \underline{\mathrm{rne}}\ A$" and "$x = x' \in A$" in the type-theories of Martin-Löf [1982,1984], so it is not surprising that an interpretation akin to extensional realizability can be used to model (parts of) Martin-Löf's extensional type theories, cf. Beeson [1982]. See also 9.3.

$\underline{\mathrm{rne}}$ -realizability for the language of arithmetic does not lend itself to a straightforward axiomatization in the same manner as ECT_0 might be said to axiomatize $\underline{\mathrm{rn}}$-realizability relative to **HA**. But van Oosten [1993] showed that axiomatization is possible in a suitably chosen conservative extension of **HA** plus Markov's principle. The same paper discusses also $\underline{\mathrm{rne}}$-realizability for higher-order logic in the form of certain toposes.

7. Realizability for intuitionistic second-order arithmetic

7.1. The system HAS

HAS (*Heyting Arithmetic of Second order*) is a two-sorted extension of **HA** with quantifiers over $\mathcal{P}(\mathbb{N})$, the powerset of $\mathbb{N}$. So the language of **HA** is extended with set variables X, Y, Z, and corresponding (second-order) quantifiers $\forall X$, $\exists Y$; atomic formulas are now of the form $t = s$ or Xt (also written $t \in X$) for individual terms t, s and set variable X.

Instead of formally introducing set-terms $\lambda x.B$ (B any formula) we can formulate the axiom for second-order $\forall$ as

$$\forall X.A \rightarrow A[X/\lambda x.B]$$

where $A[X/\lambda x.B]$ is obtained from A by replacing every occurrence of Xt by $B[x/t]$. Alternatively, we restrict the $\forall^2$-axiom to

$$\forall X.A \rightarrow A[X/Y]$$

---

[3]We do not know whether the treatment in Beeson [1979b] is really equivalent to the one in Beeson [1985].

while adding the axiom schema of *full comprehension*

CA $\quad \exists X \forall x(Xx \leftrightarrow A) \quad (X \notin \mathrm{FV}(A))$.

Moreover, we require sets to respect equality

$$\forall Xxy(Xx \wedge x = y \rightarrow Xy).$$

$\mathbf{HAS}^*$ is related to $\mathbf{HAS}$ in the same way as $\mathbf{HA}^*$ to $\mathbf{HA}$. In particular, for the set variables we have strictness: $Xt \rightarrow t\downarrow$.

7.2. Realizability for HAS*

It is quite easy to extend $\underline{\mathrm{rn}}$-realizability from $\mathbf{HA}^*$ to $\mathbf{HAS}^*$ by "brute force"; we assign to each set variable X a new set variable X^*, representing the "realizability predicate" and then add the following clauses to $\underline{\mathrm{rn}}$-realizability for $\mathbf{HA}^*$:

$$x \,\underline{\mathrm{rn}}\, Xt \quad := X^*(x,t) \quad (x\downarrow \text{ is automatic by strictness})$$
$$x \,\underline{\mathrm{rn}}\, \forall X\, A := \forall X^*(x \,\underline{\mathrm{rn}}\, A),$$
$$x \,\underline{\mathrm{rn}}\, \exists X\, A := \exists X^*(x \,\underline{\mathrm{rn}}\, A).$$

Here $Y(t,t')$ for any set variable Y abbreviates $Y(\mathbf{p}(t,t'))$. (Nothing prevents us from taking $X^* \equiv X$, but in discussions this is sometimes inconvenient and confusing.)

7.3. Remark. In a second-order context, $\perp$, $\exists$ and $\wedge$ are definable in terms of $\rightarrow$ and $\forall$, in particular

$$\exists Y.A \ := \forall Z^0(\forall Y(A \rightarrow Z^0) \rightarrow Z^0,$$
$$A \wedge B := \forall Z^0(((A \rightarrow (B \rightarrow Z)) \rightarrow Z),$$
$$\perp \quad := \forall Z^0.Z,$$

where Z^0 ranges over propositions. (Strictly speaking, we do not have variables over propositions, only over sets, but the addition of proposition variables is conservative, since one may render $(\mathrm{Q}\,Z^0)A(Z^0)$ as $(\mathrm{Q}\,X)A(X0)$ for $\mathrm{Q} \in \{\forall, \exists\}$.) Using this definition of $\exists$, the clause for realizing $\exists X.A$ is in fact redundant, and we obtain an equivalent notion of realizability. A virtually immediate consequence of soundness for $\underline{\mathrm{rn}}$-realizability for $\mathbf{HAS}$ is the consistency of $\mathbf{HAS}$ with Church's thesis and the so-called *Uniformity principle*

UP $\quad \forall X \exists y\, A(X,y) \rightarrow \exists y \forall X\, A(X,y)$.

7.4. Proposition. $\mathbf{HAS}^* + \mathrm{ECT}_0 + \mathrm{UP} + \mathrm{M}$ *is consistent.*

Here ECT_0 is formulated as for $\mathbf{HA}^*$, except that A is restricted to $\exists$-free formulas of $\mathbf{HA}^*$, while B is arbitrary.

7.5. rnt-realizability for HAS*

Extension of $\underline{\mathrm{rnt}}$-realizability to $\mathbf{HAS}^*$ is similar to the extension of $\underline{\mathrm{rn}}$-realizability, but we have to be slightly more careful: we want to keep track of realizability *and* truth, so we want to associate with an arbitrary set X an arbitrary Y together with its realizability set Z. It is convenient to encode Y and Z into a single set X^*; we put

$$X^{*\mathrm{t}} := \{n : X^*(2n)\}, \quad X^{*\mathrm{r}} := \{n : X^*(2n+1)\}$$

representing the two components of truth and realizability respectively. The new clauses in the definition of $\underline{\mathtt{rnt}}$-realizability now become

$$\begin{aligned} x\,\underline{\mathtt{rnt}}\,Xt &:= X^{*\mathrm{t}}(t) \wedge X^{*\mathrm{r}}(x,t),\\ x\,\underline{\mathtt{rnt}}\,\forall X\,A &:= \forall X^*(x\,\underline{\mathtt{rnt}}\,A),\\ x\,\underline{\mathtt{rnt}}\,\exists X\,A &:= \exists X^*(x\,\underline{\mathtt{rnt}}\,A),\\ x\,\underline{\mathtt{rnt}}\,(A \to B) &:= \forall y(y\,\underline{\mathtt{rnt}}\,A \to x{\bullet}y\,\underline{\mathtt{rnt}}\,B) \wedge (A \to B)^*, \end{aligned}$$

where C^* is obtained from C by replacing all occurrences of Yt by $Y^{*\mathrm{t}}(t)$. It is readily verified that for all A with second-order variables contained in $\{X_1, X_2, \ldots X_n\}$

$$\vdash A[X_1,\ldots,X_n/X_1^{*\mathrm{t}},\ldots,X_n^{*\mathrm{t}}] \leftrightarrow A^*$$
$$\vdash x\,\underline{\mathtt{rnt}}\,A \to A^*$$

and we find that soundness holds. An interesting corollary is

7.6. Proposition. HAS* *is closed under the* Uniformity Rule

UR $\quad \vdash \forall X \exists y\, A(X,y) \;\Leftrightarrow \vdash \exists y \forall X\, A(X,y)$

and satisfies DP *and* EDN.

7.7. Second-order extensions of other types of realizability

The preceding two examples reveal something of a pattern for the extension to second-order languages. The pattern will become still clearer when we study the extension to higher-order logic in the next section, but let us already now indicate what has to be done to extend extensional and modified realizability.

In the case of extensional realizability, set variables should get assigned variables ranging over partial equivalence relations over $\mathbb{N}$. (A partial equivalence relation satisfies symmetry and transitivity, but not necessarily reflexivity). Second-order quantification is treated in the "uniform" way, just as for ordinary realizability.

In the case of modified realizability, there is no immediate generalization of the abstract version for $\mathbf{HA}^\omega$, but we can generalize HRO-$\underline{\mathtt{mr}}$-realizability; we shall abbreviate this as $\underline{\mathtt{mrn}}$-realizability ("modified realizability for numbers").

In this case we need to assign to each formula not only a set of realizers, but also a set of "potential realizers", which determine the domain of definition in the case of implication. (In the case of HRO-$\underline{\mathtt{mrn}}$-realizability restricted to $\mathbf{HA}^\omega$, the sets of potential realizers are always of the form HRO_σ.) In particular, we must assign to set variable X two variables X^{r} (representing the realizing numbers) and X^{d} (representing the set of potential realizers). We then define for each formula A the predicates $x\,\underline{\mathtt{mrn}}\,A$ ("x HRO-modified realizes A") and A^{d} (the set of potential realizers). Some typical clauses for the potential realizers: $(t = s)^{\mathrm{d}} := \mathbb{N}$, $(A \to B)^{\mathrm{d}} := \{x : \forall y{\in}A^{\mathrm{d}}(x{\bullet}y \in B^{\mathrm{d}})\}$, $(\forall X.A)^{\mathrm{d}} := \forall X^{\mathrm{d}} A^{\mathrm{d}}$, and for the realizability $x\,\underline{\mathtt{mrn}}\,Xt := x \in X^{\mathrm{d}} \wedge X^{\mathrm{r}}(x,t)$, $x\,\underline{\mathtt{mrn}}\,(A \to B) := x \in (A \to B)^{\mathrm{d}} \wedge \forall y(y\,\underline{\mathtt{mrn}}\,A \to x{\bullet}y\,\underline{\mathtt{mrn}}\,B)$, $x\,\underline{\mathtt{mrn}}\,\forall X.A := \forall X^{\mathrm{d}} X^{\mathrm{r}}(x\,\underline{\mathtt{mrn}}\,A)$.

The reader will have no difficulty in supplying the remaining ones, keeping in mind that this is to be an extension of HRO-$\underline{\mathtt{mr}}$-realizability. However, in verifying soundness, it turns out that there is on important extra property required of the A^{d}: there should always be a fixed number in the sets of potential realizers, so that operations defined over the A^{d} must be defined at least somewhere. If we let the variables X^{d} range over inhabited sets containing 0, and if we choose our gödelnumbering of partial recursive operations in such a way that $\mathbf{p}(0,0) = 0$ and $\Lambda x.0 = 0$, it follows that $0 \in A^{\mathrm{d}}$ for all A.

7.8. Realizability as a truth-value semantics

It is instructive to rewrite rn-realizability for $\mathbf{HA}^*$ in the form of a valuation in a set of truth-values. Let $X, Y \in \mathcal{P}(\mathbb{N})$; we define

Definition.

$$X \wedge Y := \{\mathbf{p}(x,y) : x \in X \wedge y \in Y\},$$
$$X \to Y := \{z : \forall x \in X(z{\bullet}x \in Y)\},$$
$$X \vee Y := \{\mathbf{p}(0,x) : x \in X\} \cup \{\mathbf{p}(Sz,y) : z \in \mathbb{N}, y \in Y\},$$
$$X \leftrightarrow Y := (X \to Y) \wedge (Y \to X).$$

We associate to each formula A of $\mathbf{HA}^*$ a set $[\![A]\!]$ of realizing numbers:

$$[\![t = s]\!] := \{x : t = s\},$$
$$[\![A \wedge B]\!] := [\![A]\!] \wedge [\![B]\!],$$
$$[\![A \to B]\!] := [\![A]\!] \to [\![B]\!],$$
$$[\![\forall x A]\!] := \{z : \forall x(z{\bullet}x \in [\![A]\!])\},$$
$$[\![\exists x A]\!] := \{\mathbf{p}(y,z) : z \in [\![A[x/y]]\!]\}.$$

The defined set contains the free variables of A as parameters. Furthermore we can put, in keeping with our definition of disjunction,

$$[\![A \vee B]\!] := [\![A]\!] \vee [\![B]\!]. \quad \square$$

The elements of $\mathcal{P}(\mathbb{N})$ act as truth-values; all inhabited elements represent "truth" in the sense of realizability.

If we now want to extend this to $\mathbf{HAS}^*$, we should put

$$[\![Xt]\!] := \{x : X^*(x,t)\},$$
$$[\![\forall X.A]\!] := \bigcap_X [\![A(X)]\!].$$

and now $[\![A]\!]$ contains for any X free in A a parameter X^*. (We may do without an explicit definition for the cases for $\wedge, \exists$ since these are definable in a second-order setting, cf. 7.3.)

Note that numerical and set quantifiers are treated in a completely different way. This can be remedied in this case in a more or less ad hoc manner: we associate with each domain D a set-valued function E_D on the elements, giving their "extent". In the case of $\mathbf{HAS}^*$ we take

$$E_{\mathbb{N}}(n) := \{n\}, \quad E_{\mathcal{P}(\mathbb{N})}(X) := \{0\},$$

and define for domains D

$$[\![\forall x \in D.A(x)]\!] := \bigcap_{x'} (Ex' \to [\![A(x)]\!])$$

where x' is the parameter in $[\![A(x)]\!]$ corresponding to x (i.e. $x \equiv x'$ for numerical x, $x' = X^*$ if x' is a set variable X). (To see that the resulting notion of realizability is equivalent in the sense of 1.2, take for the ϕ and ψ: $\phi_{\forall x A}(y) := \Lambda x.\phi_{A(x)}(y{\bullet}x)$,

$\psi_{\forall xA}(y) := \lambda x.\psi_{A(x)}(y \bullet x)$, $\phi_{\forall X.A}(y) := \lambda z.\phi_A(y)$ (z not free in $\phi_A(y)$), $\psi_{\forall X.A}(y) := \psi_A(y \bullet 0)$.) Such an ad hoc solution to enforce uniformity of definition will not be satisfactory in the case of higher-order logic, to be discussed in the next section.

7.9. Notes

Troelstra [1973b] extended mrn-realizability, and Friedman [1977a] extended q-realizability to **HAS**; here we have recast Friedman's definition as rnt-realizability.

The idea of realizability as a truth-value semantics occurred to several researchers independently, shortly before 1980. The first documented reference to "realizability treated as a truth-value semantics" I could find is Dragalin [1979], cf. also Dragalin [1988]. Other authors credit W. Powell, or D.S. Scott with the idea.

8. Realizability for higher-order logic and arithmetic

8.1. Formulation of HAH

Higher-order logic is based on a many-sorted language with a collection of *sorts* or *types*; we use $\sigma, \sigma', \dots, \tau, \tau', \dots$ for arbitrary types. There are variables $(x^\sigma, y^\sigma, z^\sigma, \dots)$ for each type, and an equality symbol $=_\sigma$ for each σ. Relation symbols and function symbols may take arguments of different types. For quantifiers ranging over objects of type σ we sometimes write $\forall x{\in}\sigma$, $\exists x{\in}\sigma$ instead of $\forall x^\sigma$, $\exists x^\sigma$.

For intuitionistic and classical *higher-order logic* there are certain type-forming operations generating new types with appropriate axioms connecting the types.

Definition. (*Axioms and language for higher-order logic*) In a many-sorted language for higher-order logic, the collection of types is closed under $\times, \mathrm{P}, \to$, i.e.

(i) with each type σ there is a *power type* $\mathrm{P}(\sigma)$;

(ii) with each pair of types σ, τ there is a *product type* $\sigma \times \tau$ and a *function type* $\sigma \to \tau$.

One often includes a type ω of truth-values; then $\mathrm{P}(\sigma)$ may be identified with $\sigma \to \omega$.

There is a binary relation $\in_\sigma$ with arguments of type $\sigma, \mathrm{P}(\sigma)$; instead of $\in_\sigma(x, y)$ we write $x \in_\sigma y$ and sometimes $y(x)$ (predicate applied to argument).

For types $\sigma \to \tau, \sigma$ there is an application operation $\mathrm{App}_{\sigma,\tau}$ such that for $t \in \sigma{\to}\tau$, $t' \in \sigma$, $\mathrm{App}_{\sigma,\tau}(t, t')$ is a term of type τ. Usually we write tt' for $\mathrm{App}(t, t')$.

For each pair σ, τ there are functional constants $\mathbf{p}^{\sigma,\tau}, \mathbf{p}_0^{\sigma,\tau}, \mathbf{p}_1^{\sigma,\tau}$ such that $\mathbf{p}$ takes arguments of type σ, τ and yields a value of type $\sigma \times \tau$, $\mathbf{p}_0, \mathbf{p}_1$ take arguments of type $\sigma \times \tau$ and yield values of type σ and τ respectively. The pairing axioms are assumed:

PAIR $\quad \forall x_0 x_1 (\mathbf{p}_i(\mathbf{p}(x_0, x_1) = x_i) \quad (i = 0, 1)$

SURJ $\quad \forall x^{\sigma\times\tau}(\mathbf{p}(\mathbf{p}_0 x, \mathbf{p}_1 x) = x)$.

For power-types we require replacement

REPL $\quad \forall X^{\mathrm{P}(\sigma)} \forall x^\sigma y^\sigma (x \in X \wedge x = y \to y \in X)$, as well as extensionality and comprehension:

EXT $\forall X^{\mathrm{P}(\sigma)} Y^{\mathrm{P}(\sigma)} (\forall x^\sigma (x \in X \leftrightarrow x \in Y) \to X = Y)),$

CA $\exists X^{\mathrm{P}(\sigma)} \forall x^\sigma (x \in X \leftrightarrow A(x)).$

For function types the corresponding requirements are

EXTF $\forall y^{\sigma \to \tau} z^{\sigma \to \tau} (\forall x^\sigma (yx = zx) \to y = z)$

CAF $\forall x^\sigma \exists! y^\tau A(x,y) \to \exists z^{\sigma \to \tau} \forall x^\sigma A(x, zx).$

If the type ω is present and $\mathrm{P}(\sigma)$ is identified with $\sigma \to \omega$, EXT and CA become special cases of EXTF and CAF, and REPL follows from the fact that functions respect equality.

8.2. Definition. **HAH**, *intuitionistic higher-order arithmetic* ("Heyting Arithmetic of Higher order") is a specialization of higher-order logic based on a single basic type 0 (or N) for the natural numbers; types are closed under power-type and function-type formation.

On the basis type 0 an injective function $S : 0 \to 0$ is given, with axioms $Sx = Sy \to x = y$, $0 \neq Sx$. Defining

$$x \in \mathbb{N} := \forall X (0 \in X \wedge \forall y (Xy \to X(Sy)) \to x \in X)$$

we add an axiom stating that all elements of type 0 are in $\mathbb{N}$: $\forall x^0 (x \in \mathbb{N})$. As a result, the induction axiom becomes valid. □

Remarks. (i) **E-HA**$^\omega$ is a fragment of **HAH** based on type 0 and function-type formation only.

(ii) It is well known, that if we consider in **HAH** any set X with a special element $x_0 \in X$ and a function $f : X \to X$, then there is a unique function $F : \mathbb{N} \to X$ such that $F0 = x_0$, $F(Sx) = f(Fx)$. In particular, if f is injective, then the image $f[X] \cup \{x_0\}$ is isomorphic to the type N.

8.3. Numerical realizability for many-sorted logic

Since our versions of intuitionistic higher-order logic, and the system **HAH** are based on intuitionistic many-sorted predicate logic, we first discuss realizability for many-sorted logic. Our definition of realizability will be motivated by the truth-functional reformulation of realizability for **HAS** in 7.8.

We start with realizability for many-sorted logic without function symbols. Below $\Omega \equiv \mathcal{P}(\mathbb{N})$, Ω^* is the collection of all inhabited subsets of $\mathbb{N}$. We first introduce Ω-sets, which will serve to interpret the types with their equalities.

8.4. Definition. An *Ω-set* $\mathcal{X} \equiv (X, =_\mathcal{X})$ is a set X together with a map $=_\mathcal{X}$: $X^2 \longrightarrow \Omega$ such that the following is true (writing $t =_\mathcal{X} t'$ for $=_\mathcal{X}(t,t')$):

$$\textstyle\bigcap_{x,y} (x =_\mathcal{X} y \to y =_\mathcal{X} x) \in \Omega^*,$$
$$\textstyle\bigcap_{x,y,z} (x =_\mathcal{X} y \wedge y =_\mathcal{X} z \to x =_\mathcal{X} z) \in \Omega^*.$$

Here $\wedge, \to$ on the left have to be understood as defined for elements of Ω, as in 7.8. We write $E_{\mathcal{X}}t$ for $t =_{\mathcal{X}} t$.5

The *Ω-product* of two Ω-sets $\mathcal{X} \equiv (X, \sim)$ and $\mathcal{Y} \equiv (Y, \sim')$ is the Ω-set $\mathcal{X} \times \mathcal{Y} \equiv (X \times Y, \sim'')$ where

$$(x, y) \sim'' (x', y') := (x \sim x') \wedge (y \sim' y').$$

A product of n factors $\mathcal{X}_1, \ldots, \mathcal{X}_n$ is defined as $(\mathcal{X}_1 \times \cdots \times X_{n-1}) \times \mathcal{X}_n$.

We use calligraphic capitals $\mathcal{X}, \mathcal{Y}, \ldots$ for Ω-sets. □

Examples. Ω itself may be viewed as an Ω-set $(\Omega, \leftrightarrow)$ where $X \leftrightarrow Y$ is defined as in 7.8. Another example is $\mathcal{N} := (\mathbb{N}, =_{\mathbb{N}})$, where $n =_{\mathbb{N}} m := \{n\} \cap \{m\} \equiv \{n : n = m\}$.

8.5. Definition. Let $\mathcal{X} \equiv (X, \sim)$ be an Ω-set and $F : X \longrightarrow \Omega$ a map. We put

$$\text{Strict}(F) := \bigcap_{x \in X} (Fx \to Ex),$$

$$\text{Repl}(F) := \bigcap_{x,y \in X} (Fx \wedge x \sim y \to Fy).$$

An *Ω-predicate* on $\mathcal{X}$ is an $F : X \to \Omega$ such that Strict(F) and Repl(F) are inhabited (belong to Ω^*). An *Ω-relation* on $\mathcal{X}_1, \ldots, \mathcal{X}_n$ is an Ω-predicate on $\mathcal{X}_1 \times \cdots \times \mathcal{X}_n$.

If $(X \times Y, \sim)$ is the product of the Ω-sets $(X, =_X)$ and $(Y, =_Y)$, and $F : X \times Y \longrightarrow \Omega$, we define

$$\text{Fun}(F) := \bigcap_{x,y,z} (F(x, y) \wedge F(x, z) \to y =_Y z)$$

$$\text{Total}(F) := \bigcap_{x} (Ex \to \bigcup_{y} F(x, y)).$$

An *Ω-function* from $\mathcal{X}$ to $\mathcal{Y}$ is an $F : X \times Y \to \Omega$ such that Strict(F), Repl(F), Fun(F), Total(F) are inhabited. The definition of Ω-function for more than one argument is reduced to this case via products of Ω-sets. □

8.6. Definition. An *interpretation* $[\![\;]\!]$ of a many-sorted relational language assigns

(i) to each type σ with equality $=_\sigma$ an Ω-set $[\![\sigma]\!] \equiv (\bar{\sigma}, [\![=_\sigma]\!])$; for $[\![=_\sigma]\!](x, x)$ we also write $E_\sigma x$, and we shall permit ourselves in the sequel a slight abuse of language, using $[\![\sigma]\!]$ also for the underlying set $\bar{\sigma}$.

(ii) to constants c of type σ an element $[\![c]\!]$ of $[\![\sigma]\!]$,

(iii) to each n-ary relation symbol R, taking arguments of sorts $\sigma_1, \ldots, \sigma_n$ respectively, an Ω-relation $[\![R]\!]$ on $[\![\sigma_1]\!], \ldots, [\![\sigma_n]\!]$. □

N.B. It is important to observe that for practical purposes the definition of Ex for an Ω-set $(X, \sim)$ may be liberalized, it suffices that $\bigcap_x (Ex \leftrightarrow x \sim x) \in \Omega^*$.

Remark. If in the definition above we take for Ω a complete Heyting algebra, and replace $\bigcap$ in the conditions above by the meet operator $\bigwedge$, and take $\Omega^* := \{\top\}$, $\top$ the top element of Ω, we obtain precisely the interpretation of many-sorted intuitionistic logic in Ω-sets, as described in Fourman and Scott [1979] or Troelstra and van Dalen [1988]. There Et measures the "degree of existence" of t.

8.7. Definition. Let $\mathcal{X} \equiv (X, \sim)$ be an Ω-set, and let $F : X \to \Omega$, then

$$\forall x{\in}\mathcal{X}F(x) := \bigcap_{x \in X}(E_{\mathcal{X}}x \to Fx), \quad \exists x{\in}\mathcal{X}F(x) := \bigcup_{x \in X}(E_{\mathcal{X}}x \wedge Fx). \ \square$$

N.B. "$\forall x \in \mathcal{X} \ldots$", "$\exists x \in \mathcal{X} \ldots$" indicate elements of Ω, but "$\forall x \in X \ldots$", "$\exists x \in X \ldots$" refer to ordinary quantification.

8.8. Definition. The *interpretation* of *formulas* of a many-sorted relational language may now be given modulo assignments ρ for the variables. Let ρ be an assignment of elements of $[\![\sigma]\!]$ to the variables of type σ, for all σ. For constants c the interpretation $[\![c]\!]_\rho$ is supposed to be given; for variables $[\![x]\!]_\rho := \rho(x)$, and for prime formulas

$$[\![t =_\sigma t']\!]_\rho := [\![=_\sigma]\!]([\![t]\!]_\rho, [\![t']\!]_\rho), \quad [\![R(t_1, \ldots, t_n)]\!]_\rho := [\![R]\!]([\![t_1]\!]_\rho, \ldots, [\![t_n]\!]_\rho),$$

and for compound formulas according to 7.8, i.e.

$$\begin{array}{ll} [\![A \wedge B]\!]_\rho & := [\![A]\!]_\rho \wedge [\![B]\!]_\rho, \\ [\![A \to B]\!]_\rho & := [\![A]\!]_\rho \to [\![B]\!]_\rho, \\ [\![\neg A]\!]_\rho & := [\![A]\!]_\rho \to \emptyset, \\ [\![\forall x{\in}\sigma.A]\!]_\rho & := (\forall d{\in}[\![\sigma]\!])[\![A]\!]_{\rho[x/d]}, \\ [\![\exists x{\in}\sigma.A]\!]_\rho & := (\exists d{\in}[\![\sigma]\!])[\![A]\!]_{\rho[x/d]}, \end{array}$$

where $\rho[x/d]$ is the assignment given by $\rho[x/d](y) = \rho(y)$ for $y \not\equiv x$, $\rho[x/d](x) = d$. Instead of using assignments, we may also use a language enriched with constants as names for each Ω-set used for the interpretation of the types, and define the interpretation only for sentences.

A sentence A is said to be *valid* if $[\![A]\!] \in \Omega^*$. $\square$

N.B. In the sequel we shall sometimes used "mixed" expressions: for an Ω-set $\mathcal{X} \equiv (X, \sim)$ $[\![\forall x \in \mathcal{X} A(x)]\!] := \bigcap_{x \in X}(Ex \to [\![A(x)]\!])$, $[\![\exists x \in \mathcal{X} A(x)]\!] := \bigcup_{x \in X}(Ex \wedge [\![A(x)]\!])$.

8.9. Proposition. *Intuitionistic many-sorted predicate logic is sound for realizability.*

Proof. The proof is routine. The definition of an interpretation says that for the Ω-relation $[\![R]\!]$ interpreting relation R of the language the following hold:

(1) $$\bigcap_{x,y}([\![x = y]\!] \to [\![y = x]\!]) \in \Omega^*,$$

(2) $$\bigcap_{x,y,z}([\![x = y]\!] \wedge [\![y = z]\!] \to [\![x = z]\!]) \in \Omega^*,$$

(3) $$\bigcap_{x_1,\ldots,x_n} ([\![R]\!](x_1,\ldots,x_n) \to Ex_1 \wedge \ldots \wedge Ex_n) \in \Omega^*,$$

(4) $$\bigcap_{\vec{x},\vec{y}}([\![R]\!](\vec{x}) \wedge [\![\vec{x}=\vec{y}]\!] \to [\![R]\!](\vec{y})) \in \Omega^*$$

where of course $[\![\vec{x}=\vec{y}]\!]$ abbreviates $[\![x_1=y_1]\!] \wedge \ \ldots\ \wedge [\![x_n=y_n]\!]$; similarly we may abbreviate $Ex_1 \wedge \ldots \wedge E_n$ as $E\vec{x}$.

(1) and (2) guarantee the validity of symmetry and transitivity of equality, and (3) and (4) the validity of strictness and replacement for R. Reflexivity translates into the trivial $\bigcap_x(Ex \to [\![x=x]\!]) \in \Omega^*$, so does not need an extra condition. □

The definition of the interpretation of a language with function symbols is reduced to the case of relational languages, by regarding functions as special relations (a partial function is a relation which is functional: $\forall\vec{x}yz(R(\vec{x},y) \wedge R(\vec{x},z) \to y=z)$, and a (total) function is a relation which is functional and total, i.e. satisfies $\forall\vec{x}\exists yR(\vec{x},y)$).

8.10. Definition. (*Interpretation of function symbols*) To each function symbol $F:\sigma_1\times\cdots\times\sigma_n\to\sigma$ we assign an Ω-function $[\![F]\!]$ from $[\![\sigma_1]\!]\times\cdots\times[\![\sigma_n]\!]$ to $[\![\sigma]\!]$. In full the conditions read:

$$\bigcap_{\vec{x}}(E\vec{x} \to \textstyle\bigcup_y[\![F]\!](\vec{x},y)) \in \Omega^*,$$

$$\bigcap_{\vec{x},y,z}([\![F]\!](\vec{x},y) \wedge [\![F]\!](\vec{x},z) \to [\![y=z]\!]) \in \Omega^*.$$

For partial functions the first condition may be omitted. □

As to the reason for using relations to interpret functions, see 8.13.

8.11. Definition. (*Interpretation of formulas for languages with function symbols*) We now assume a language with relation symbols and symbols for total functions.

We have to say how to interpret $t_1 = t_2$ and $Rt_1\ldots t_n$ for compound terms $t_1,\ldots,t_n$. This is done recursively: $t_1 = t_2$ for arbitrary t_1, t_2 is interpreted as $\exists x(t_1 = x \wedge t_2 = x)$; $Ft_1\ldots t_n = x$ is interpreted as $\exists x_1\ldots x_n(t_1 = x_1 \wedge \ldots \wedge t_n = x_n \wedge Fx_1\ldots x_n = x)$; the value of $Fx_1\ldots x_n = x$ is given by $[\![F]\!](\rho x_1,\ldots,\rho x_n,\rho x)$. $Rt_1\ldots t_n$ is interpreted as $\exists\vec{x}(\vec{t}=\vec{x} \wedge R(\vec{x}))$. □

8.12. Theorem. *The interpretation above is sound for many-sorted logic.*

Proof. Almost entirely routine. To see that e.g. all instances of $\forall xA \to A[x/t]$ are valid, one should note that the "unwinding" of $t_1 = t_2$ and $Rt_1\ldots t_n$ mentioned above is precisely what one does in showing syntactically that the addition of symbols for definable functions with the appropriate axiom is conservative; the standard proof, e.g. Kleene [1952], shows that the "unwinding" translation of $\forall xA \to A[x/t]$ is in fact derivable in the relational part of the language.

In our case this means that the soundness reduces to the soundness for a relational language with an extra relation symbol R_F for each function symbol F in the original language. □.

8.13. Remark. The reason that we have not imposed the stronger requirement that a function symbol $F : \sigma_1 \times \cdots \times \sigma_n \to \sigma$ is to be interpreted by a function $[\![F]\!] : [\![\sigma_1]\!] \times \cdots \times [\![\sigma_n]\!] \to [\![\sigma]\!]$ lies in the fact that this sometimes not sufficiently general: the interpretation of $\forall x \exists! y R(x, y)$ says that $\bigcap_x (Ex \to \bigcup_y [\![R]\!](x, y)) \in \Omega^*$, and $\bigcap_{x,y,z}([\![R]\!](x, y) \wedge [\![R]\!](x, z) \to [\![y = z]\!]) \in \Omega^*$, but there is no guarantee that we can find a function f such that $\bigcap_x [\![R]\!](x, fx) \in \Omega^*$. Cf. the similar situation for the interpretations where Ω is a complete Heyting algebra; only for sheaves the situation simplifies; see Troelstra and van Dalen [1988,1.3.16, chapter 14].

Example. Let f be a primitive recursive function with function symbol F in the language of **HA**. Over the domain $\mathcal{N} \equiv (\mathbb{N}, [\![=_{\mathbb{N}}]\!])$ as defined above we can introduce the interpretation of F by the relation

$$R_F(n, m) := En \wedge [\![m = fn]\!]$$

we add En to guarantee the strictness of R_F), so $R_F := \{\mathbf{p}(n, fn) : n \in \mathbb{N}\}$. It is now routine to see that this yields a realizability for **HA** (equivalent to) the one defined before.

8.14. Remark. The modelling of many-sorted logic described above is an interpretation in a certain category Eff, with as objects the Ω-sets, and as morphisms (equivalence classes of) Ω-functions. More precisely, the morphisms from $\mathcal{X}$ to $\mathcal{Y}$ are given by Ω-relations on $\mathcal{X} \times \mathcal{Y}$ such that (cf. definition 8.5)

$$\text{Strict}(F),\ \text{Repl}(F),\ \text{Fun}(F),\ \text{Total}(F) \in \Omega^*,$$

modulo an equivalence $\approx$ defined as

$$F \approx F' := [\![\forall xy(Fxy \leftrightarrow F'xy)]\!] = \bigcap_{x,y}(Ex \wedge Ey \to (Fxy \leftrightarrow F'xy)) \in \Omega^*.$$

Composition of morphisms $F : \mathcal{X} \to \mathcal{Y}$, $G : \mathcal{Y} \to \mathcal{Z}$ is given by the relational product: $G \circ F$ is the relation on $\mathcal{X} \times \mathcal{Z}$ given by

$$(G \circ F)(x, z) := \exists y \in \mathcal{Y}(F(x, y) \wedge G(y, z)).$$

The identity morphism $\mathrm{id}_{\mathcal{X}} : \mathcal{X} \to \mathcal{X}$ is simply (the equivalence class of) $\mathrm{id}_{\mathcal{X}}(x, y) := x =_{\mathcal{X}} y$.

Let Sets be the category of sets and set-theoretic mappings. There are functors Δ : Sets $\to$ Eff and Γ : Eff $\to$ Sets which may be described as follows.

Δ, the "*constant-objects functor*", maps a set X to the Ω-set $(X, \sim)$ with $x \sim y := \{n \in \mathbb{N} : x = y\}$, and if $f : X \to Y$, then Δf is represented by the Ω-relation R_f, $R_f := (fx \sim y)$.

$\Gamma(X, \sim) = \{x : Ex \text{ inhabited}\}/\simeq$, where $\simeq$ is the equivalence relation $x \simeq x' := (x \sim x' \text{ inhabited})$. Γ is usually called the *global-sections functor*, since it is naturally isomorphic tot the functor which assigns to $(X, \sim)$ the set of morphisms $\top \to (X, \sim)$; $\top$ is the terminal object $(\{*\}, =_*)$ with $(* =_* *) = \mathbb{N}$.

Δ preserves finite limits and is full and faithful; Γ also preserves finite limits, and Γ is left-adjoint to Δ. $\mathcal{N}$ is a natural-numbers object in **Eff**, and **Eff** is in fact a topos (see 8.17). For the theory of **Eff**, with proofs of the facts mentioned above, see Hyland [1982]; the general theory of realizability toposes is treated in Hyland, Johnstone and Pitts [1980]. Other sources of information on **Eff** are Robinson and Rosolini [1990] and Hyland, Robinson and Rosolini [1990].

The categorical view provided by **Eff** suggests the following

8.15. Definition. The Ω-sets $\mathcal{X} \equiv (X, \sim)$, $\mathcal{Y} \equiv (Y, \sim')$ are said to be *isomorphic* if there are Ω-functions $F : X \times Y \to \Omega$, $G : Y \times X \to \Omega$ such that $G \circ F \approx \mathrm{id}_{\mathcal{X}}$, $F \circ G \approx \mathrm{id}_{\mathcal{Y}}$, where $\approx$ is defined as above, i.e., $H \approx H' := [\![\forall xy(Hxy \leftrightarrow H'xy)]\!]$. □

It is easy to see that quantification over isomorphic sets yields equivalent results, in the following sense:

Lemma. *Let $\mathcal{X}$ and $\mathcal{Y}$ be isomorphic via F, G. Then*

$$\forall x{\in}\mathcal{X}.A(x) = \forall y{\in}\mathcal{Y}.A(Gy) = \forall y{\in}\mathcal{Y}\forall x{\in}\mathcal{X}.(G(y,x) \to A(x)),$$

and similarly for existential quantification.

The proof is routine, relying on the soundness of logic. Important special cases are (a) any Ω-set $\mathcal{X} \equiv (X, \sim)$ is isomorphic to $(X', \sim\restriction X' \times X')$ where $X' := \{x : x \in X \text{ and } Ex \in \Omega^*\}$, and (b) the following situation: let $\mathcal{X} \equiv (X, \sim)$ be an Ω-set, and let $X' \subset X$ such that

$$\forall x \in X(Ex \in \Omega^* \to \exists x' \in X'(x \sim x') \in \Omega^*).$$

8.16. Products, powersets and exponentials

Definition. The interpretation of a product $[\![\sigma_1 \times \sigma_2]\!]$ is $[\![\sigma_1]\!] \times [\![\sigma_2]\!]$. The functions $\mathbf{p}^{\sigma,\tau}, \mathbf{p}_0^{\sigma,\tau}, \mathbf{p}_1^{\sigma,\tau}$ are simply represented by the pairing and unpairing on the relevant Ω-sets. □

In order to interpret higher-order logic, the interpretation of type σ and types $P(\sigma)$ and $\sigma \to \tau$ relative to the interpretation of σ, τ, and the interpretation of the relation $\in_\sigma$ as well as the operator App must be such that extensionality and comprehension are valid.

Definition. Let $\mathcal{X} \equiv (X, \sim)$ be an Ω-set; the Ω-*powerset* of $\mathcal{X}$, $\mathrm{P}(\mathcal{X})$, is an Ω-set $(X \to \Omega, \simeq)$ where for F, G of $X \to \Omega$:

$$\begin{aligned} E(F) &:= \mathrm{Strict}(F) \wedge \mathrm{Repl}(F),\\ F \simeq G &:= E(F) \wedge E(G) \wedge \textstyle\bigcap_x (Fx \leftrightarrow Gx),\\ x \in_{\mathcal{X}} F &:= Fx \wedge EF. \end{aligned}$$

Let $\mathcal{X} \equiv (X, \sim), \mathcal{Y} \equiv (Y, \sim')$ be Ω-sets, then the Ω-*functionset* $\mathcal{X} \to \mathcal{Y}$ is $(X \times Y \to \Omega, \approx)$ such that for $F, G \in X \times Y \to \Omega$

$$\begin{aligned} E(F) &:= \mathrm{Str}(F) \wedge \mathrm{Repl}(F) \wedge \mathrm{Fun}(F) \wedge \mathrm{Total}(F)\\ F \approx G &:= \textstyle\bigcap_{x,y} (Ex \wedge Ey \to (Fxy \leftrightarrow Gxy)) \wedge EF \wedge EG,\\ \mathrm{App}_{\mathcal{X},\mathcal{Y}}(F, x, y) &:= EF \wedge Fxy. \end{aligned}$$

N.B. Here we have availed ourselves of the freedom to define $E(F)$ so as to be equivalent only to $F \approx F$ in the realizability sense, not literally identical. □

Remark. As noted above, Ω itself may be viewed as an Ω-set $(\Omega, \sim)$. It is then not hard to see that the Ω-powerset of a Ω-set $\mathcal{X}$ is in fact isomorphic to the Ω-functionset $\mathcal{X} \to \Omega$.

Definition. In an interpretation of intuitionistic higher-order logic powertypes and exponentials are interpreted such that $[\![\mathrm{P}(\sigma)]\!]$ is the $\mathrm{P}[\![\sigma]\!]$ and such that $[\![\sigma \to \tau]\!]$ is $[\![\sigma]\!] \to [\![\tau]\!]$. $[\![\in_\sigma]\!] :=\in_{[\![\sigma]\!]}$, $[\![\mathrm{App}_{\sigma,\tau}]\!] := \mathrm{App}_{[\![\sigma]\!],[\![\tau]\!]}$. □

Remark. We can easily introduce subtypes of a given type, as follows. Let $\mathcal{X} \equiv (X, \sim)$ be an Ω-set. Intuitively an Ω-predicate over $\mathcal{X}$ determines a subset of $\mathcal{X}$. We may again make this into an Ω-set:

Definition. Let $\mathcal{X} \equiv (X, \sim)$ be an Ω-set, and let $F : X \to \Omega$ be an Ω-predicate; then the Ω-subset of $\mathcal{X}$ determined by F is $(X', \sim')$ with $X' := \{x : Fx \in \Omega^*\}$, $x \sim' y := x \sim y$. □

N.B. An equivalent definition would have been obtained by taking $X' = X$, and $x \sim y := Fx \wedge Fy \wedge x \sim y$. The resulting Ω-set is isomorphic to the one defined above.

8.17. Proposition. *Extensionality and comprehension are valid.*
The proof is routine.

Remarks. (i) In categorical terms, the preceding facts mean that the category Eff has products and exponentials (i.e. is cartesian closed), and moreover has a classifying truth-value object, namely $(\Omega, \leftrightarrow)$, and hence is a topos.

The fact that the natural numbers are unique modulo isomorphism in higher-order logic (8.2) corresponds in categorical terms to the uniqueness of the natural number object in a topos.

(ii) Obviously, the notions needed for the realizability interpretation of **HAH** can be formalized in **HAH** itself. If we assign a *level* $\ell(\sigma)$ to types σ according to $\ell(0) = \ell(\omega) = 0$, $\ell(\mathrm{P}\sigma) = \ell(\sigma) + 1$, $\ell(\sigma{\to}\tau) = \max(\ell(\sigma) + 1, \ell(\tau)$, $\ell(\sigma{\times}\tau) = \max(\ell(\sigma), \ell(\tau))$, then the interpretation of a formula of level $\leq n$ (i.e. all variables are of level $\leq n$) is definable by a formula of level $\leq n$.

Our next aim will be to show that for **HAS** the resulting notion of realizability is in fact equivalent to realizability as defined in 7.2.

8.18. Definition. An Ω-set $\mathcal{X} \equiv (X, \sim)$ is called *canonically uniform* if $\bigcap_{x \in X} Ex$ is inhabited. $\mathcal{X}$ is *uniform* if it is isomorphic to a canonically uniform set.

8.19. Lemma. *For uniform Ω-sets $\mathcal{X} \equiv (X, \sim)$ interpretation of universal and existential quantifiers may be simplified to*

$$\forall x {\in} \mathcal{X}\, Fx := \bigcap_{x \in X} Fx, \quad \exists x {\in} \mathcal{X}\, Fx := \bigcup_{x \in X} Fx;$$

more precisely, $(\bigcap_x (Ex \to Fx) \leftrightarrow \bigcap_x Fx) \in \Omega^*$, $(\bigcup_x (Ex \wedge Fx) \leftrightarrow \bigcup_x Fx) \in \Omega^*$.
PROOF. It suffices to prove this for canonically uniform Ω-sets, and then it is easy: let $n \in \bigcap_x Ex$, and let $m \in \bigcap_x (Ex \to Fx)$, then $\forall x (m{\bullet}n \in Fx)$, i.e. $m{\bullet}n \in \bigcap_x Fx$, etc. □

8.20. Definition. Let $\mathcal{X} \equiv (X, \sim)$ be an Ω-set. $\mathcal{X}$ is *canonically separated* if

$$(x \sim y) \text{ inhabited } \Rightarrow x = y.$$

$\mathcal{X}$ is *canonically proto-effective* if

$$Ex \cap Ey \text{ inhabited } \Rightarrow x = y.$$

$\mathcal{X}$ is *separated* [*proto-effective*] if $\mathcal{X}$ is isomorphic to a canonically separated [canonically proto-effective] Ω-set. $\mathcal{X}$ is [*canonically*] *effective* if $\mathcal{X}$ is [canonically] separated and [canonically] effective. □

8.21. Proposition. *Let $\mathcal{X} \equiv (X, \sim)$ be a uniform and $\mathcal{Y} \equiv (Y, \sim')$ a proto-effective Ω-set. Then the uniformity principle*

$$\mathrm{UP}(\mathcal{X}, \mathcal{Y}) \qquad \forall x {\in} \mathcal{X}\, \exists y {\in} \mathcal{Y}\, A(x, y) \to \exists y {\in} \mathcal{Y}\, \forall x {\in} \mathcal{X}\, A(x, y)$$

is valid.

Proof. Without loss of generality we may assume $\mathcal{Y}$ to be canonically proto-effective. Let $n \in \bigcap_{x \in X} \exists y \in \mathcal{Y} A(x, y)$, so $n \in \bigcup_{y \in Y} (Ey \wedge A(x, y))$, then $\forall x {\in} X \exists y {\in} Y (\mathbf{p}_0 n \in Ey)$, i.e. $n \in \bigcup_{y \in Y} Ey \wedge \bigcap_{x \in X} A(x, y)$. □

The following proposition is not needed in what follows but describes the logical significance of separatedness:

8.22. Proposition. *An Ω-set $\mathcal{X} \equiv (X, \sim)$ is separated iff $\forall x, y {\in} \mathcal{X} (\neg\neg x {\sim} y \to x {\sim} y)$ is valid.*

Proof. It is easy to see that a canonically separated Ω-set $\mathcal{X}$ satisfies $\forall x, y {\in} \mathcal{X} (\neg\neg x {\sim} y \to x {\sim} y)$. Conversely, assume $\forall x, y {\in} \mathcal{X} (\neg\neg x {\sim} y \to x {\sim} y)$ to be valid. We define for $x, y \in X$:

$$\begin{aligned} [x] &:= \{x' \in X : x' \sim x \text{ inhabited}\}, \\ [x] \approx [y] &:= \{n \in \mathbb{N} : x \sim y \text{ inhabited}\}, \\ X' &:= \{[x] : x \in X\}. \end{aligned}$$

Then $\mathcal{X}' \equiv (X', \approx)$ is canonically separated and isomorphic to $\mathcal{X}$ via the Ω-relations F on $\mathcal{X} \times \mathcal{X}$ and G on $\mathcal{X} \times \mathcal{X}'$, defined by

$$F([x], y) \equiv G(x, [y]) := \{n \in \mathbb{N} : y \sim x \text{ inhabited}\}.$$

We have to show that F, G are strict, total, functional and that their composition is the identity. This is mostly routine. For example, to see that F is functional, observe that our hypothesis gives the existence of an n such that

(1) $$\forall m, m'(m \in Ex \wedge m' \in Ey \wedge (\exists m''(m'' \in x{\sim}y) \to n{\bullet}\mathbf{p}(m,m') \in (x{\sim}y)).$$

The functionality of F amounts to validity of $\forall [x], y, y'(F([x],y) \wedge F([x],y') \to y \sim y')$, i.e.

(2) $$\bigcap_{[x]\in X'} \bigcap_{y,y'\in X} (E[x] \wedge Ey \wedge Ey' \wedge \{n \in \mathbb{N} : \exists m(m \in x \sim y)\} \wedge \{n{\in}\mathbb{N} : \exists m(m \in x \sim y')\} \to y \sim y') \in \Omega^*.$$

Since $x \sim y$ and $x \sim y'$ inhabited implies $y \sim y'$ inhabited, (2) readily follows from (1). □

The following proposition, with a proof due to van Oosten, justifies the terminology "uniform".

8.23. Proposition. *An Ω-set $\mathcal{X} \equiv (X, \sim)$ is uniform iff $\mathcal{X}$ satisfies* UP$(\mathcal{X}, \mathcal{N})$.

Proof. One direction is a consequence of proposition 8.21. The other direction is proved as follows. Given $\mathcal{X}$, consider the Ω-set $\mathcal{Y} \equiv (Y, \simeq)$ defined by

$$Y := \{\mathbf{p}(x,n) : n \in E_{\mathcal{X}}x\},$$
$$(x,n) \simeq (y,m) := \{n : n \in x{\sim}y \wedge n = m\}.$$

We shall write (y,n) for $\mathbf{p}(y,n)$ in what follows. There is an Ω-function $G : \mathcal{Y} \to \mathcal{X}$ given by

$$G((x,n),x') := (x \sim x') \wedge \{n\}.$$

G is surjective as an Ω-function, i.e.

$$\forall y{\in}\mathcal{Y} \exists x{\in}\mathcal{X}. G(y,x),$$

i.e.

$$\bigcap_{(y,n)\in\mathcal{Y}} (E_{\mathcal{Y}}(y,n) \to \bigcup_{x\in X} (E_{\mathcal{X}}x \wedge G((y,n),x)) \in \Omega^*.$$

Let $H : \mathcal{Y} \to \mathcal{N}$ be the surjective Ω-function

$$H((x,n),m) := \{n : n = m\}.$$

Then

$$\forall x{\in}\mathcal{X}\, \exists y{\in}\mathcal{Y}\, \exists n{\in}\mathcal{N}(G(y,x) \wedge H(y,n))$$

is valid. If we assume UP$(\mathcal{X}, \mathcal{N})$, it follows that

$$\exists n{\in}\mathcal{N}\, \forall x{\in}\mathcal{X}\, \exists y{\in}\mathcal{Y}(G(y,x) \wedge H(y,n))$$

is valid, which means that for some $n \in \mathbb{N}$

(1) $$\bigcap_{x\in X}(Ex \to \bigcup_{x'\in X, n\in Ex'} (x \sim x'))$$

is inhabited. Let now $\mathcal{Z} := (Z, \sim')$, $Z := \{x{\in}X : n{\in}Ex\}$, $\sim'$ the restriction of $\sim$ to Z. Clearly F defined by $F(x,x') := (x \sim x')$ is an injection of $\mathcal{Z}$ into $\mathcal{X}$; and the formula (1) states that this injection is also a surjection, hence $\mathcal{Z}$ and $\mathcal{X}$ are isomorphic. □

8.24. Proposition. *The Ω-powerset of a separated Ω-set $\mathcal{X} = (X, \sim)$ is uniform.*

Proof. Let $\mathcal{X}$ be canonically separated, and let $\mathcal{Y} : X \to \Omega$ be an element of the Ω-powerset $\mathrm{P}(\mathcal{X})$ of $\mathcal{X}$, then $\Lambda k.\mathbf{p}_0 k$ realizes $\mathrm{Repl}(\mathcal{Y})$; $n \in \mathrm{Str}(\mathcal{Y})$ means $n \in \bigcap_{x \in X}(x \in \mathcal{Y} \to Ex)$.

By restricting attention to "normal" $\mathcal{Y}$ we can construct uniform realizers for Ey. Let us call $\mathcal{Y}$ *normal* if

$$m \in \mathcal{Y}(x) \Rightarrow \mathbf{p}_1 m \in Ex.$$

For normal $\mathcal{Y}$, $\Lambda m.\mathbf{p}_1 m \in \mathrm{Str}(\mathcal{Y})$, and so always

$$\mathbf{p}(\Lambda k.\mathbf{p}_0 k, \Lambda m.\mathbf{p}_1 m) \in E(\mathcal{Y}).$$

To show Ω-isomorphism of $\mathrm{P}(\mathcal{X})$ with the subset of normal elements, observe that if we map arbitrary $\mathcal{Y}$ to $\Phi(\mathcal{Y})$ with $\Phi(\mathcal{Y})(x) := \{\mathbf{p}(n,m) : n \in \mathcal{Y}(x) \wedge m \in Ex\}$, we have that $\mathcal{Y} =_{\mathrm{P}(\mathcal{X})} \Phi(\mathcal{Y})$ is inhabited:

$$\bigcap_{x \in X}(x \in \mathcal{Y} \leftrightarrow x \in \Phi(\mathcal{Y})) \in \Omega^*.$$

If $n \in \mathrm{Str}(\mathcal{Y})$, $k \in (x \in \mathcal{Y})$, then $n \bullet k \in Ex$, and $\mathbf{p}(k, n \bullet k) \in \Phi(\mathcal{Y})(x)$, etc. □

8.25. Proposition. *For* **HAS** *the realizability as defined above is equivalent to* $\underline{\mathrm{rn}}$ *as defined in 7.2.*

Proof. This result is now obtainable as a corollary to the preceding propositions. As to the second-order quantifiers, we may restrict attention to the normal elements of the Ω-powerset of $\mathcal{N}$. Let

$$\Phi'(X) := \{\mathbf{p}(\mathbf{p}(x,y),y) : \mathbf{p}(x,y) \in X\},$$
$$\Phi''(X) := \{\mathbf{p}(x,y) : \mathbf{p}(\mathbf{p}(x,y),y) \in X\}.$$

Φ' corresponds as operation on binary relations to Φ above, Φ'' is its inverse.

Let $\underline{\mathrm{rn}}$ be defined as for **HAS**-formulas as in 7.2 relative to an assignment $X \mapsto X^*$ for second-order variables; and let $\underline{\mathrm{rn}}'$ be the realizability notion as defined in this section, relative to an assignment $X \mapsto X^\circ$ of normal binary relations to the second-order variables (i.e. $x\,\underline{\mathrm{rn}}'\,Xt := x \in [\![Xt]\!] := X^\circ(x,t)$; $x\,\underline{\mathrm{rn}}'\,A := x \in [\![A]\!]$). Then for all formulas A of **HAS** there are ϕ_A, ψ_A such that

$$x\,\underline{\mathrm{rn}}\,A(X_1,\ldots,X_n) \to \psi_A\,\underline{\mathrm{rn}}'\,A(X_1,\ldots,X_n)[X_1^\circ,\ldots,X_n^\circ/\Phi' X_1^*,\ldots,\Phi' X_n^*],$$
$$x\,\underline{\mathrm{rn}}\,A(X_1,\ldots,X_n) \to \phi_A\,\underline{\mathrm{rn}}\,A(X_1,\ldots,X_n)[X_1^*,\ldots,X_n^*/\Phi'' X_1^\circ,\ldots,\Phi'' X_n^\circ],$$

where $X_1,\ldots,X_n$ is a complete list of the second-order variables free in A. □

8.26. Lemma. *For Ω-sets $\mathcal{X} \equiv (X,\sim)$, $\mathcal{Y} \equiv (Y,\sim')$, $\mathcal{Y}$ separated, the elements of the Ω-functionset $\mathcal{X} \to \mathcal{Y}$ may be represented by functions $f : X \to Y$.*

Proof. Let $F : \mathcal{X} \times \mathcal{Y} \to \Omega$ be an arbitrary element of the Ω-exponent, for which EF is inhabited. Then for certain k, k'

$$k \in \bigcap_{x\in X}(Ex \to \bigcup_{y\in Y} Fxy),$$
$$k' \in \bigcap_{x\in X,y,y'\in Y}(Fxy \wedge Fxy' \to y \sim' y').$$

By the second statement, it readily follows in combination with separatedness, that

$$Fxy \text{ inhabited, } Fxy' \text{ inhabited} \Rightarrow y = y',$$

and from the first statement that for all x with Ex inhabited, there is a y such that Fxy is inhabited; so let f be the function defined for x with Ex inhabited, such that $F(x, fx)$ is inhabited. □

8.27. Lemma. *For separated [effective] $\mathcal{Y} \equiv (Y, \sim')$ the Ω-function set $(X, \sim) \to (Y, \sim')$ is separated [effective].*

Proof. By the preceding lemma we may represent (there is a little checking to do, but we leave this to the reader) the elements of the exponent by the isomorphic $(X{\to}Y, \approx)$ with

$$(f \approx g) \text{ inhabited} \Rightarrow f = g,$$
$$f \approx f := \{m : \forall y \in Y \forall n \in Ey(m{\bullet}n \in E(fy))\},$$

or combined into a single definition:

$$f \approx g := \{m : f = g \wedge \forall y \in Y \forall n \in Ey(m{\bullet}n \in E(fy))\}.$$

The separatedness has been built into the definition; as to proto-effectiveness, suppose $Ef \cap Eg$ inhabited, then for some m

$$\forall y \in Y \forall n \in Ey(m{\bullet}n \in E(fy) \cap E(gy));$$

by the proto-effectivity of $\mathcal{Y}$ it follows that $\forall y \in Y(fy = gy)$, i.e. $f = g$. □

Remark. The fact that $\mathcal{X} \to \mathcal{Y}$ is separated for separated $\mathcal{Y}$ is also easily seen to hold for logical reasons (cf. 8.22): $\neg\neg f = g \leftrightarrow \neg\neg\forall x(fx = gx) \to \forall x(fx = gx) \leftrightarrow f = g$.

8.28. Proposition. *The structure of functional ω-sets generated from $\mathcal{N}$ is isomorphic to* HEO *as defined in 3.3.*

Proof. Induction on the type structure. □
The following is immediate:

8.29. Proposition. *For all function types σ generated from type 0 in* **HAH**, *the realizability interpretation validates a uniformity principle:*

$$\forall X^{\mathrm{P}[0]}\exists x^\sigma A(X, x) \to \exists x^\sigma \forall X^{\mathrm{P}[0]} A(X, x).$$

8.30. Generalization to other kinds of realizability

In the preceding section we have already indicated how the generalizations of realizabilities to second-order logic follow a pattern. If we combine this with the "truth-value semantics" idea introduced in the preceding section and used extensively above, we are led to consider other choices for Ω and Ω^*.

Examples. (a) If we want to generalize rnt-realizability, we take

$$\Omega^{\mathrm{rnt}} := \{(X,p) : X \subset \mathbb{N}, p \subset \{0\}\},$$
$$\Omega^{\mathrm{rnt}*} := \{(X,p) \in \Omega : X \text{ inhabited}, 0 \in p\}.$$

The crucial operations we have to define are $\wedge^{\mathrm{rnt}}$, $\rightarrow^{\mathrm{rnt}}$, $\bigcap^{\mathrm{rnt}}$:

$$(X,p) \wedge^{\mathrm{rnt}} (Y,q) := ((X \wedge Y, \{0 : 0 \in p \wedge 0 \in q\}),$$
$$(X,p) \rightarrow^{\mathrm{rnt}} (Y,q) := ((X \rightarrow Y), \{0 : 0 \in p \rightarrow 0 \in q\}),$$
$$\textstyle\bigcap^{\mathrm{rnt}}_{y \in Y}(Z_y, p_y) := (\bigcap_{y \in Y} Z_y, \bigcap_{y \in Y} p_y).$$

N.B. We do not really need to define an operation $\wedge^{\mathrm{rnt}}$, since in a second-order context we can define (7.3) operation $\wedge'$ in terms of the other operations, producing objects $(X,p) \wedge' (Y,q)$ isomorphic to $(X,p) \wedge^{\mathrm{rnt}} (Y,q)$.

(b) For modified realizability we can put

$$\Omega^{\mathrm{mrn}} := \{(X,Y) : X \subset Y \subset \mathbb{N}, 0 \in Y\},$$
$$\Omega^{\mathrm{mrn}*} := \{(X,Y) \in \Omega : X \text{ inhabited}\}.$$

$\bigcap^{\mathrm{mrn}}$ is defined component-wise, and for $\rightarrow^{\mathrm{mrn}}$ we take

$$(X,Y) \rightarrow^{\mathrm{mrn}} (X',Y') := ((X \rightarrow Y) \cap (X' \rightarrow Y'), X' \rightarrow Y')).$$

(In order to guarantee that 0 always occurs in the second component we must choose our gödelnumbering of the partial recursive functions such that $\Lambda x.0 = 0$.)

(c) For Lifschitz realizability another idea is needed, a reformulation of the original definition which makes Lifschitz realizability fit the general pattern van Oosten [1991b].

8.31. Notes

The proper definition of realizability for higher-order logic emerged from the study of special toposes (Hyland [1982], Hyland, Johnstone and Pitts [1980], Pitts [1981], Grayson[1981a,1981b,1981c]). Aczel[1980] described a less far-reaching common generalization of Heyting-valued and realizability semantics. The higher-order extension of rln-realizability is due to van Oosten [1991b]. By means of this extension he shows that the following principle RP (*Richman's Principle*)

$$\forall X^d(\forall Y^d(X \subset Y \vee X \cap Y = \emptyset \rightarrow \exists n \forall x(x \in X \rightarrow x = n))$$

(where $\forall X^d$, $\exists Y^d$ are quantifiers ranging over *decidable* subsets of $\mathbb{N}$) is false in the "Lifschitz topos" and true in **Eff**, that is to say false in the higher-order extension of Lifschitz realizability and true in the realizability interpretation for higher-order logic described in the preceding section. More information on the Lifschitz topos **Lif** is given in van Oosten [1996].

9. Further work

9.1. Realizability for set-theory

It is also possible to define $\underline{\mathrm{rn}}$-realizability, or the abstract version $\underline{\mathrm{r}}$-realizability for the language of set theory. The definition is straightforward except for the fact that we have to build in extensionality.

The problem becomes clear if we try to extend the definition of $\underline{\mathrm{rn}}$-realizability given in 7.2 to intuitionistic third-order arithmetic $\mathbf{HAS}^3$ (variables $X^2, Y^2, \ldots$) in which we can also quantify over $\mathcal{PP}(\mathbb{N})$, and with full impredicative comprehension and extensionality

EXT $\quad \forall X^2(Y^1 \in X^2 \wedge Y^1 = Z^1 \to Z^1 \in X^2)$

where $X^1 = Y^1 := \forall z(z \in X \leftrightarrow z \in Y)$. If we take as clauses $x \,\underline{\mathrm{rn}}\, X^2(Y^1) := X^{*2}(Y^{*1}, x)$, $x \,\underline{\mathrm{rn}}\, \forall X^2.A(X^2) := \forall X^{*2}(x \,\underline{\mathrm{rn}}\, A(X^2)$, etc., we discover that there is no problem in proving soundness except for the axiom EXT; this imposes a restriction on the sets over which the "starred variables" X^{*2} should range.

Some authors, e.g. Beeson [1985], solve the problem in the case of set theory by first giving a realizability interpretation for a set theory without the extensionality axiom, combined with an interpretation of the theory with extensionality into set theory without extensionality. Others such as McCarty [1984b] build the extensionality into the definition of realizability.

The earliest paper defining realizability for set theory is Tharp [1971]. Other papers using realizability for set theory are: Staples [1974], Friedman and Scedrov[1983,1984], McCarty[1984b,1986], Beeson [1985], and the series of papers by Khakhanyan.

9.2. Comparison with functional interpretations

Another type of interpretation which is in certain respects analogous to (modified) realizability, but in other respects quite different, is the so-called Dialectica interpretation devised by Gödel [1958]. There is also a modification due to Diller and Nahm [1974]. As we have seen, modified realizability associates to formulas A of $\mathbf{HA}^\omega$ $\exists$-free formulas of the form $A_{\mathrm{mr}}(x^\sigma)$ (x^σ a new variable not free in A), expressing "x^σ modified-realizes A". The Dialectica- and Diller–Nahm interpretation on the other hand associate with A formulas $\forall y^\tau A_D(x^\sigma, y^\tau)$ and $\forall y^\tau A_{DN}(x^\sigma, y^\tau)$ respectively, σ, τ depending on the logical structure of A alone, A_D, A_{DN} quantifier-free; we may read $\forall y^\tau A_D(x^\sigma, y^\tau)$, $\forall y^\tau A_{DN}(x^\sigma, y^\tau)$ as "x^σ D-interprets A"and "x^σ DN-interprets A" respectively.

For a soundness proof for the Dialectica interpretation, the prime formulas of the theory considered have to be *decidable* with a decision function of the appropriate type; for the Diller-Nahm interpretation this is not necessary. For theories with decidable prime formulas (e.g. $\mathbf{I\text{-}HA}^\omega$) the Diller-Nahm interpretation is equivalent to the Dialectica interpretation. For background information the reader may consult the commentary to Gödel [1958] in Gödel [1990], and the relevant chapter elsewhere in this volume.

Stein has constructed a whole sequence of interpretations intermediate between the DN-interpretation and modified realizability; see the papers by Stein, and Diller [1979].

9.3. Formulas-as-types realizability

In the formulas-as-types paradigm, formulas (representing propositions) are regarded as determined by (identified with) the set of their proofs. The idea is illustrated by taking a natural deduction formulation of intuitionistic predicate logic, and writing the deductions as terms in a typed lambda-calculus.

Normalization of the deductions suggests equations between the terms of such a calculus (in particular beta-conversion) and "t proves A" for compound A then behaves like an abstract realizability notion. Of particular interest is the realizability obtained by stripping the proof-terms of their types. With combinators instead of lambda-abstraction, such a realizability is already used in Staples[1973,1974]. Tait [1975] uses this concept for an elegant version of Girard's proof of the normalization theorem for second-order intuitionistic logic. For another version of the proof see Girard, Lafont and Taylor [1988].

Mints [1989], Barbanera and Martini [1994] study completeness questions for such realizabilities. More specifically, one is interested in completeness results of the following type: for all formulas A of a certain formal system $\mathbf{S}$, $\vdash_{\mathbf{S}} A$ iff $\vdash_{\mathbf{T}} \exists x(x\mathbf{r}A)$ where $\mathbf{T}$ is a suitable formal system (intuitionistic or classical logic), and $\mathbf{r}$ the abstract version of realizability studied.

"Formulas-as-types" has also been a leading idea in the formulation of various typed theories, such as the theories of Martin-Löf[1982,1984], permitting to absorb logical operations into type-forming operations (implication is subsumed under function-type formation, universal quantification under formation products of dependent types, etc.). In the proof-theoretic investigations of Martin-Löf's type theories by de Swaen[1989,1991,1992] realizability plays an important role.

9.4. Completeness questions for realizabilities

Rose [1953] gave an example of a classically valid, but not intuitionistically provable formula of propositional logic, such that all its arithmetical substitution examples are (classically) realizable; this result was improved by Kleene [1965b], who showed that the example also worked for $\underline{\mathtt{rf}}$- and $\underline{\mathtt{mrf}}$-realizability (in the latter case the substitution instances were provably realizable even intuitionistically). See also Tsejtin [1968]. On the other hand, Gavrilenko [1981] proved that the principle "Every formula for which all arithmetical substitution instances are realizable, is provable in intuitionistic propositional logic" is consistent with **HAS** (but not with **HAS** + M).

Kleene also showed that the class of formulas of *predicate* logic which are realizable under substitution is not recursive (for $\underline{\mathtt{rn}}$, $\underline{\mathtt{rf}}$, $\underline{\mathtt{mrn}}$). Similar questions have been studied at length in a series of papers by Plisko. A typical result of this kind is the following. Let $\mathcal{R}$ [$\mathcal{AR}$] be the class of all formulas $A(P_1, \ldots, P_n)$ of predicate logic such that all arithmetical substitution instances (i.e. formulas $A(P_1^*, \ldots, P_n^*)$ with $P_1^*, \ldots, P_n^*$ arithmetical) are $\underline{\mathrm{rn}}$-realizable [such that $\forall X_1 \ldots X_n\ A(X_1, \ldots, X_n)$ is $\underline{\mathrm{rn}}$-realizable as defined for **HAS**].

Plisko [1983] showed that $\mathcal{AR}$ is a complete Π^1_1-set, and that $\mathcal{AR} \subset \mathcal{R}$; $\mathcal{R}$ is also not arithmetical as shown in Plisko [1977].

Van Oosten[1991c], adapting a method originally due to de Jongh, gave a semantical proof of a result earlier established by proof-theoretic means by D. Leivant: if all arithmetical substitution instances of a formula of predicate logic are provable in **HA**, the formula itself is a theorem of intuitionistic predicate logic("maximality of intuitionistic arithmetic"). The method uses a realizability in which $\underline{\mathrm{rn}}$-realizability and Beth-semantics are combined. His proof also yields the following completeness result for realizability. Let $\mathbf{HA}^+$ be an extension of **HA** obtained by adding to the language primitive constants $\bullet$ (application), $\mathbf{k}$, $\mathbf{s}$ (combinators), with axioms saying that $(\mathbb{N}, \bullet, \mathbf{k}, \mathbf{s})$ is a partial combinatory algebra. Define $\underline{\mathrm{r}}$-realizability for $\mathbf{HA}^+$ relative to this combinatory algebra. Then a predicate formula A is provable in intuitionistic predicate logic iff all arithmetical instances of A are provably realizable in $\mathbf{HA}^+$.

A different sort of completeness result has been obtained by Läuchli [1970]. He defined a modified realizability for predicate logic with a set-theoretic hierarchy as models for the finite-type functionals. All formula of predicate logic is realizable by an element of this hierarchy iff it is classically provable; but if we require that the realizing functionals are invariant under permutations of the basic domains, we obtain precisely the intuitionistically provable formulas. Inspection shows that the "modified" aspect of Läuchli's construction is not really relevant. A modern recasting of Läuchli's result, linking it with the category-theoretic interpretation of logic, was given by Harnik and Makkai [1992]. See also Lipton and O'Donnell [1996].

9.5. Realizability for subsystems of intuitionistic arithmetic

Damnjanovic [1994] considers realizability for primitive recursive arithmetic, using primitive recursive functions instead of partial recursive functions. In particular the clauses for $\rightarrow$ and $\forall$ require modification. This is achieved using levels, which are provided by the Grzegorczyk hierarchy for the primitive recursive functions. In Damnjanovic [1995] the same idea is applied to obtain a realizability for **HA**, using so-called $< \varepsilon_0$–recursive functions instead of general recursive functions. (The $< \varepsilon_0$–recursive functions are precisely the functions provably recursive in **HA** and **PA**.) This permits reproving a number of metamathematical results for **HA** by realizability methods. The technique also applies to elementary analysis and finite–type arithmetic.

Wehmeier [1996] uses $\underline{\mathrm{rn}}$- and $\underline{\mathrm{rnt}}$-realizability in a study of intuitionistic arithmetic with Σ_1-induction, $\mathrm{I}\Sigma_1$. Wehmeier shows that whenever $\vdash \forall n \exists m A(n,m)$ then there is a primitive recursive $t(x)$ such that $\forall n A(n, t(n))$. Ferreira and Marques [1996] use a version of $\underline{\mathrm{rnt}}$-realizability for a language of arithmetic extended with infinite disjunctions in a study of the intuitionistic counterpart IS_2^1 of Buss's system S_2^1 (Buss [1986]), and of $\mathrm{I}\Sigma_1$ just mentioned. The authors obtain a new proof of the fact, established in Cook and Urquhart [1993] by mrt-realizability, that whenever $\vdash \forall n \exists m A(n,m)$ in IS_2^1, then $\vdash \forall n A(n, t(n))$ where $t(x)$ is polynomial-time computable in x. They also sketch another proof of Wehmeier's result.

9.6. Combining realizability with classical logic

Lifschitz[1982,1985] considered an extension of classical arithmetic with an additional predicate $K(x)$, "x is computable". The result is a combination of classical arithmetic and realizability. It is to be noted that in the category Eff we can obtain something similar by considering side by side $\mathcal{N}$ and $\Delta\mathbb{N}$.

9.7. Medvedev's calculus of finite problems

The calculus of finite problems as formulated by Medvedev, is somewhat reminiscent of, but actually diverges rather far from recursive realizability. See the papers by Medvedev, and by Maksimova, Shekhtman and Skvortsov [1979].

9.8. Applications to Computer Science

For some examples, see Scedrov [1990], Streicher [1991] (realizability modeling of the theory of constructions), Smith [1993] (slash relations for type theory) and the papers by Tatsuta (program synthesis by "realizability-cum-truth"). Within the effective topos one can find models for strong polymorphic type theories; see e.g. Hyland [1988].

In Berger and Schwichtenberg [1995] program-extraction from classical proofs of statements $\forall n \exists m A(n,m)$, A quantifier-free, is studied. Two methods are compared; one method is based on normalization of classical proofs formalized in a calculus of natural deduction, the other method uses modified realizability. The two methods yield terms $t_1(x), t_2(x)$ respectively such that $\forall n A(n, t_i(n))$ $(i = 1,2)$ and for all numerals $\bar{n}$, $t_1(\bar{n}) = t_2(\bar{n})$. An ingredient in the second method is the combination of the *Gödel–Gentzen* translation (Troelstra and van Dalen [1988,2.3]) with the Friedman-Dragalin *A-translation* (Troelstra and van Dalen [1988,3.5]); by this combination a classical proof of $\forall n \exists m\, A(n,m)$, A quantifier-free, can be transformed into an intuitionistic proof of the same statement.

Interesting applications of the second method are given in Berger, Schwichtenberg and Seisenberger [1997]. One of the examples concerns the following special case of Higman's lemma:

If x^1, y^1 are two number-theoretic functions, we can find $n < m$ such that $x^1 n \le x^1 m$ and $y^1 n \le y^1 m$.

This fact is easily proved classically. Applying the program extraction via modified realizability, one obtains an algorithm which is in suitable cases quadratically faster than "brute–force" search. The syntactic definition of modified realizability and of the Friedman-Dragalin translation are used in a variant which helps keeping the complexity of the extracted program (term) down.

Abbreviations in the references

AML = *Annals of Mathematical Logic.*
AMS Transl. = *American Mathematical Society Translations, Series 2.*
APAL = *Annals of Pure and Applied Logic.*
Archiv = *Archiv für mathematische Logik und Grundlagenforschung.*
Doklady = *Doklady Akademii Nauk SSSR.*
Izv.Akad.Nauk = *Izvestiya Akademii Nauk SSSR. Seriya Matematicheskaya.*
JSL = *The Journal of Symbolic Logic.*
LMPS = *Logic, Methodology and Philosophy of Science.*
Math. Izv. = *Mathematics of the USSR, Izvestiya.*
SM = *Soviet Mathematics. Doklady.*
ZLGM = *Zeitschrift für Logik und Grundlagen der Mathematik.*
Zapiski = *Zapiski Nauchnykh Seminarov Leningradskogo Otdeleniya Matematicheskogo Instituta imeni V. A. Steklova Akademii Nauk SSSR (LOMI).*

References

P. H. G. Aczel
[1968] Saturated intuitionistic theories, in: *Contributions to Mathematical Logic*, H. Schmidt, K. Schütte, and H. Thiele, eds., North-Holland, Amsterdam, pp. 1–11.
[1980] A note on interpreting intuitionistic higher-order logic. Handwritten note.

F. Barbanera and S. Martini
[1994] Proof–functional connectives and realizability, *Archive for Mathematical Logic*, 33, pp. 189–211.

H. P. Barendregt
[1973] Combinatory logic and the axiom of choice, *Indagationes Mathematicae*, 35, pp. 203–221.

J. Barwise, H. J. Keisler, and K. Kunen
[1980] eds., *The Kleene Symposium*, North-Holland, Amsterdam.

M. J. Beeson
[1975] The nonderivability in intuitionistic formal systems of theorems on the continuity of effective operations, *JSL*, 4l, pp. 321–346.
[1976a] Derived rules of inference related to the continuity of effective operations, *JSL*, 41, pp. 328–336.
[1976b] The unprovability in intuitionistic formal systems of the continuity of effective operations on the reals, *JSL*, 41, pp. 18–24.
[1977a] Continuity and comprehension in intuitionistic formal systems, *Pacific Journal of Mathematics*, 68, pp. 29–40.
[1977b] Principles of continuous choice and continuity of functions in formal systems for constructive mathematics, *AML*, 12, pp. 249–322.
[1979a] Continuity in intuitionistic set theories, in: *Logic Colloquium '78*, M. Boffa, D. van Dalen, and K. McAloon, eds., North-Holland, Amsterdam, pp. 1–52.
[1979b] Goodman's theorem and beyond, *Pacific Journal of Mathematics*, 84, pp. 1–28.

[1980] Extensionality and choice in constructive mathematics, *Pacific Journal of Mathematics*, 88, pp. 1–28.

[1981] Formalizing constructive mathematics: Why and how?, in: *Constructive mathematics*, F. Richman, ed., Springer Verlag, Berlin, Heidelberg, New York, pp. 146–190.

[1982] Recursive models for constructive set theories, *APAL*, 23, pp. 127–178.

[1985] *Foundations of constructive mathematics*, Springer Verlag, Berlin, Heidelberg, New York.

M. J. BEESON AND A. SCEDROV

[1984] Church's thesis, continuity and set theory, *JSL*, 49, pp. 630–643.

S. BERARDI, M. BEZEM, AND T. COQUAND

[1994] *On the computational content of the Axiom of Choice*, Tech. Rep. Logic Group Preprint Series, 116, Department of Philosophy, Utrecht University.

U. BERGER AND H. SCHWICHTENBERG

[1995] Program extraction from classical proofs, in: *Logic and Computaional Complexity. International Workshop LCC '94, Indianapolis, IN, USA october 1994*, D. Leivant, ed., Lecture Notes in Computer Science 960, Springer Verlag, Berlin, Heidelberg, New York, pp. 77–97.

U. BERGER, H. SCHWICHTENBERG, AND M. SEISENBERGER

[1997] From proofs to programs in the Minlog system. The Warshall algorithm and Higman's lemma. To appear in *Journal of Automated Reasoning.*

M. BEZEM

[1985] Strongly majorizable functionals of finite type: a model for bar recursion containing discontinuous functionals, *JSL*, 50, pp. 652–660.

[1986] *Bar Recursion and Functionals of finite Type*, PhD thesis, Rijksuniversiteit Utrecht.

[1989] Compact and majorizable functionals of finite type, *JSL*, 54, pp. 271–280.

W. BUCHHOLZ, S. FEFERMAN, W. POHLERS, AND W. SIEG

[1981] *Iterated Inductive Definitions and Subsystems of Analysis: Recent Proof-Theoretical Studies*, Springer Verlag, Berlin, Heidelberg, New York.

S. R. BUSS

[1986] The polynomial hierarchy and intuitionistic bounded arithmetic, in: *Structure in Complexity Theory*, Springer Verlag, Berlin, Heidelberg, New York, pp. 77–103. Springer Lecture Notes in Computer Science 223.

A. CARBONI, P. J. FREYD, AND A. SCEDROV

[1988] A categorical approach to realizability and polymorphic types, in: *Mathematical Foundations of Programming Language Semantics*, M. Main, M. Melton, A. Mislove, and D. Schmidt, eds., Springer Verlag, Berlin, Heidelberg, New York, pp. 23–42. Proceedings of the 3rd workshop, Tulane University, New Orleans, Louisiana, U.S.A., April 1987. Lecture Notes in Computer Science 298.

C. CELLUCCI

[1971] Operazioni di Brouwer e realizzabilita formalizzata (English summary), *Annali della Scuola Normale Superiore di Pisa. Classe di Science. Fisiche e Matiche, Seria III*, 25, pp. 649–682.

S. A. COOK AND A. URQUHART

[1993] Functional interpretations of feasibly constructive arithmetic, *APAL*, 63, pp. 103–200. The preprint dates from 1991.

Z. DAMNJANOVIC

[1994] Strictly primitive recursive realizability, I, *JSL*, 59, pp. 1210–1227.

[1995] Minimal realizability of intuitionistic arithmetic and elementary analysis, *JSL*, 60, pp. 1208–1241.

[1996] Elementary realizability. Submitted to *Journal of Philosophical Logic.*

J. Diller

[1979] Functional interpretations of Heyting's arithmetic in all finite types, *Nieuw Archief voor Wiskunde. Derde Serie*, 27, pp. 70–97.

[1980] Modified realization and the formulae–as–types notion, in: *To H.B. Curry: Essays on Combinatory Logic, Lambda Calculus and Formalism*, J. P. Seldin and J. R. Hindley, eds., Academic Press, New York, pp. 491–501.

J. Diller and W. Nahm

[1974] Eine Variante zur Dialectica–Interpretation der Heyting–Arithmetik endlicher Typen, *Archiv*, 16, pp. 49–66.

A. G. Dragalin

[1968] The computability of primitive recursive terms of finite type, and primitive recursive realization (Russian), *Zapiski*, 8, pp. 32–45. Translation *Seminars in Mathematics. V.A.Steklov Mathematical Institute Leningrad* 8(1970), pp. 13–18. This volume appeared as: A.O. Slisenko (ed.), *Studies in Constructive Mathematics and Mathematical Logic. Part II.* Consultants Bureau, New York, London.

[1969] Transfinite completions of constructive arithmetical calculus (Russian), *Doklady*, 189, pp. 458–460. Translation *SM* 10, pp. 1417–1420.

[1979] An algebraic approach to intuitionistic models of the realizability type (Russian), in: *Issledovaniya po Neklassicheskim Logikam i Teorii Mnozhestv (Investigations on Non-Classical Logics and Set Theory)*, A. I. Mikhajlov et al., eds., Nauka, Moskva, pp. 83–201.

[1980] New forms of realizability and Markov's rule (Russian), *Doklady*, 251, pp. 534–537. Translation *SM* 21, pp. 461–464.

[1988] *Mathematical Intuitionism*, American Mathematical Society, Providence, Rhode Island. Translation of the Russian original from 1979.

P. Eggerz

[1987] *Realisierbarkeitskalküle* $\mathbf{ML}_0$ *und vergleichbare Theorien im Verhältnis zur Heyting-Arithmetik*, PhD thesis, Ludwig-Maximilians-Universität, München.

S. Feferman

[1975] A language and axioms for explicit mathematics, in: *Algebra and Logic*, J. N. Crossley, ed., Springer Verlag, Berlin, Heidelberg, New York, pp. 87–139.

[1979] Constructive theories of functions and classes, in: *Logic Colloquium '78*, M. Boffa, D. van Dalen, and K. McAloon, eds., North-Holland, Amsterdam, pp. 159–224.

F. Ferreira and A. Marques

[1996] *Extracting algorithms from intuitionistic proofs*, Tech. Rep. Pré-publicações de Matemática 26/96, Universidade de Lisboa.

M. P. Fourman and D. S. Scott

[1979] Sheaves and logic, in: *Applications of Sheaves*, M. P. Fourman, C. J. Mulvey, and D. S. Scott, eds., Springer Verlag, Berlin, Heidelberg, New York, pp. 302–401.

H. M. Friedman

[1973] Some applications of Kleene's methods for intuitionistic systems, in: *Mathias and Rogers [1973]*, pp. 113–170.

[1975] The disjunction property implies the numerical existence property, *Proceedings of the National Academy of Sciences of the United States of America*, 72, pp. 2877–2878.

[1977a] On the derivability of instantiation properties, *JSL*, 42, pp. 506–514.

[1977b] Set theoretic foundations of constructive analysis, *Annals of Mathematics, Series 2*, 105, pp. 1–28.

H. M. FRIEDMAN AND A. SCEDROV

[1983] Set existence property for intuitionistic theories with dependent choice, *APAL*, 25, pp. 129–140. Corrigendum in *APAL* **26** (1984), p.101.

[1984] Large sets in intuitionistic set theory, *APAL*, 27, pp. 1–24.

[1986] Intuitionistically provable recursive well-orderings, *APAL*, 30, pp. 165–171.

J. H. GALLIER

[1993] *Proving properties of typed lambda-terms using realizability, covers, and sheaves*, Tech. Rep. MS-CIS-93-91, Computer and Information Science Department, School of Engineering and Applied Science, University of Pennsylvania, Philadelphia.

YU. V. GAVRILENKO

[1981] Recursive realizability from the intuitionistic point of view (Russian), *Doklady*, 256, pp. 18–22. Translation *SM* 23, pp. 9–14.

J. Y. GIRARD, Y. LAFONT, AND P. TAYLOR

[1988] *Proofs and Types*, Cambridge University Press, Cambridge U.K.

K. GÖDEL

[1958] Über eine bisher noch nicht benützte Erweiterung des finiten Standpunktes, *Dialectica*, 12, pp. 280–287.

[1990] *Collected Works, Volume 2*, Oxford University Press, Oxford.

N. D. GOODMAN

[1978] Relativized realizability in intuitionistic arithmetic of all finite types, *JSL*, 43, pp. 23–44.

R. J. GRAYSON

[1981a] Derived rules obtained by a model-theoretic approach to realisability. Handwritten notes from Münster University.

[1981b] Modified realisability toposes. Handwritten notes from Münster University.

[1981c] Note on extensional realizability. Handwritten notes from Münster University.

[1982] Appendix to modified realisability toposes. Handwritten notes from Münster University.

V. HARNIK

[1992] Provably total functions of intuitionistic bounded arithmetic, *JSL*, 57, pp. 466–477.

V. HARNIK AND M. MAKKAI

[1992] Lambek's categorical proof theory and Läuchli's abstract realizability, *JSL*, 57, pp. 200–230.

A. HEYTING

[1959] ed., *Constructivity in Mathematics*, North-Holland, Amsterdam.

W. A. HOWARD

[1973] Appendix: Hereditarily majorizable functionals of finite type, in: *Troelstra [1973a]*, Springer Verlag, Berlin, Heidelberg, New York, pp. 454–461.

J. M. E. HYLAND

[1982] The effective topos, in: *The L.E.J. Brouwer Centenary Symposium*, A. S. Troelstra and D. van Dalen, eds., North-Holland, Amsterdam, pp. 165–216.

[1988] A small complete category, *APAL*, 40, pp. 135–165.

J. M. E. HYLAND, P. T. JOHNSTONE, AND A. M. PITTS

[1980] Tripos theory, *Mathematical Proceedings of the Cambridge Philosophical Society*, 88, pp. 205–232.

J. M. E. HYLAND AND C.-H. L. ONG

[1993] Modified realizability toposes and strong normalization proofs, in: *Typed Lambda Calculi and Applications*, M. Bezem and J. F. Groote, eds., Springer Verlag, Berlin, Heidelberg, New York, pp. 179–194.

J. M. E. Hyland, E. Robinson, and G. Rosolini

[1990] The discrete objects in the effective topos, *Proceedings of the London Mathematical Society*, 60, pp. 1–36.

V. A. Jankov

[1963] Realizable formulas of propositional logic (Russian), *Doklady*, 151, pp. 1035–1037. Translation *SM* 4(, pp. 1146–1148.

D. H. J. de Jongh

[1968] *Investigations on the intuitionistic propositional calculus*, PhD thesis, University of Wisconsin, Madison.

[1970] A characterization of the intuitionistic propositional calculus, in: *Kino, Myhill and Vesley [1970]*, pp. 211–217.

[1980] Formulas of one propositional variable in intuitionistic arithmetic, in: *Troelstra and van Dalen [1980]*, pp. 51–64.

A. Joyal and I. Moerdijk

[1995] *Algebraic Set Theory*, London Mathematical Society Lecture Series 220, Cambridge University Press, Cambridge, U.K.

F. A. Kabakov

[1963] On the derivability of some realizable formulae of the sentential calculus (Russian), *ZMLG*, 9, pp. 97–104.

[1970a] Intuitionistic deducibility of some realizable formulae of propositional logic (Russian), *Doklady*, 192, pp. 269–271. Translation *SM* 11, pp. 612–614.

[1970b] On modelling of pseudo-boolean algebras by realizability (Russian), *Doklady*, 192, pp. 16–18. Translation *SM* 11, pp. 562–564.

V. Kh. Khakhanyan

[1979] Consistency of the intuitionistic set theory with Brouwer's principle, *Matematika*, 5. Not seen. Found in another bibliography and included for completeness.

[1980a] Comparative strength of variants of Church's thesis at the level of set theory (Russian), *Doklady*, 252, pp. 1070–1074. Translation *SM* 21, pp. 894–898.

[1980b] The consistency of intuitionistic set theory with Church's principle and the uniformization principle (Russian), *Vestnik Moskovskogo Universiteta. Seriya I. Matematika, Mekhanika*, 1980/5, pp. 3–7. Translation *Moscow University Mathematics Bulletin* 35/5, pp. 3–7.

[1980c] The consistency of intuitionistic set theory with formal mathematical analysis (Russian), *Doklady*, 253, pp. 48–52. Translation *SM* 22, pp. 46–50.

[1981] The consistency of some intuitionistic and constructive principles with a set theory, *Studia Logica*, 40, pp. 237–248.

[1983] Set theory and Church's thesis (Russian), in: *Issledovaniya po Neklascheskims Logikam i Formal'nym Sistemam (Studies in Nonclassical logics and Formal Systems)*, A. I. Mikhajlov, ed., Nauka, Moskva, pp. 198–208.

[1988] Nonderivability of the uniformization principle from Church's thesis (Russian), *Matematicheskie Zametki*, 43, pp. 685–691,703. Translation *Mathematical Notes* 43, pp. 394–398.

A. Kino, J. Myhill, and R. E. Vesley

[1970] eds., *Intuitionism and Proof Theory*, North-Holland, Amsterdam.

M. M. Kipnis

[1968] The constructive classification of arithmetic predicates and the semantic bases of arithmetic (Russian), *Zapiski*, 8, pp. 53–65. Translation *Seminars in Mathematics. V.A.Steklov Mathematical Institute Leningrad* 8(1970), pp. 22–27. This volume appeared as: A.O. Slisenko (ed.), *Studies in Constructive Mathematics and Mathematical Logic. Part II.* Consultants Bureau, New York, London.

[1971] On the realizations of predicate formulas (Russian) (English summary), *Zapiski*, 20, pp. 40–48. Translation *Journal of Soviet Mathematics* 1(1973), pp. 22–27.

S. C. Kleene

[1945] On the interpretation of intuitionistic number theory, *JSL*, 10, pp. 109–124.

[1952] *Introduction to metamathematics*, North-Holland, Amsterdam. Co-publisher: Wolters–Noordhoff; 8th revised ed.1980.

[1957] Realizability, in: *Summaries of Talks presented at the Summer Institute for Symbolic Logic*, Institute for Defense Analyses, Communications Research Division, Princeton, pp. 100–104. Also in *Heyting [1959]*, pp. 285–289. Errata in *Kleene and Vesley [1965]*, page 192.

[1960] Realizability and Shanin's algorithm for the constructive deciphering of mathematical sentences, *Logique et Analyse, Nouvelle Série*, 3, pp. 154–165.

[1962] Disjunction and existence under implication in elementary intuitionistic formalisms, *JSL*, 27, pp. 11–18. Addenda in *JSL* **28** (1963), pp. 154–156.

[1965a] Classical extensions of intuitionistic mathematics, in: *LMPS 2*, Y. Bar-Hillel, ed., North-Holland, Amsterdam, pp. 31–44.

[1965b] Logical calculus and realizability, *Acta Philosophica Fennica*, 18, pp. 71–80.

[1968] Constructive functions in "The Foundations of Intuitionistic Mathematics", in: *LMPS 3*, B. van Rootselaar and J. F. Staal, eds., North-Holland, Amsterdam, pp. 137–144.

[1969] *Formalized recursive functionals and formalized realizability*, vol. 89 of Memoirs of the American Mathematical Society, American Mathematical Society, Providence, Rhode Island.

[1973] Realizability: a retrospective survey, in: *Mathias and Rogers [1973]*, pp. 95–112.

S. C. Kleene and R. E. Vesley

[1965] *The foundations of intuitionistic mathematics, especially in relation to recursive functions*, North-Holland, Amsterdam.

S. Kobayashi and M. Tatsuta

[1994] Realizability interpretation of generalized inductive definitions, *Theoretical Computer Science*, 131, pp. 121–138.

U. W. Kohlenbach

[1990] *Theorie der majorisierbaren und stetigen Funktionale und ihre Anwendung bei der Extraktion von Schranken aus inkonstruktiven Beweisen: effektive Eindeutigkeitsmodule bei besten Approximationen aus ineffektiven Eindeutigkeitsbeweisen*, PhD thesis, J.W. Goethe-Universität, Frankfurt am Main.

[1992] Pointwise hereditary majorization and some applications, *Archive for Mathematical Logic*, 31, pp. 227–241.

G. Kreisel

[1953] A variant to Hilbert's theory of the foundations of arithmetic, *British Journal for the Philosophy of Science*, 4, pp. 107–127.

[1959] Interpretation of analysis by means of constructive functionals of finite types, in: *Heyting [1959]*, pp. 101–128.

[1962a] Foundations of intuitionistic logic, in: *LMPS*, E. Nagel, P. Suppes, and A. Tarski, eds., Stanford University Press, Stanford, pp. 198–210.

[1962b] On weak completeness of intuitionistic predicate logic, *JSL*, 27, pp. 139–158.

G. Kreisel and A. S. Troelstra

[1970] Formal systems for some branches of intuitionistic analysis, *APAL*, 1, pp. 229–387. Addendum in *APAL* 3, pp. 437–439.

M. D. Krol'

[1977] Disjunctive and existential properties of intuitionistic analysis with Kripke's scheme (Russian), *Doklady*, 234. Translation *SM* 18, pp. 755–758.

[1983] Various forms of the continuity principle (Russian), *Doklady*, 271, pp. 33–36. Translation *SM* 28, pp. 27–30.

[1992] On a version of realizability (Russian), in: *XI Interrepublican Conference on Mathematical Logic, University of Kazan, October 6–8, 1992*, p. 81.

S. A. Kurtz, J. C. Mitchell, and M. J. O'Donnell

[1992] *Connecting formal semantics to constructive intuitions*, Tech. Rep. CS 92–01, Department of Computer Science, University of Chicago.

J. Lambek and P. J. Scott

[1986] *Introduction to Higher Order Categorical Logic*, Cambridge University Press, Cambridge.

H. Läuchli

[1970] An abstract notion of realizability for which intuitionistic predicate calculus is complete, in: *Intuitionism and Proof Theory*, A. Kino, J. Myhill, and R. E. Vesley, eds., North-Holland, Amsterdam, pp. 227–234.

V. Lifschitz

[1979] CT_0 is stronger than $CT_0!$, *Proceedings of the American Mathematical Society*, 73, pp. 101–106.

[1982] Constructive assertions in an extension of classical mathematics, *JSL*, 47, pp. 359–387.

[1985] Calculable natural numbers, in: *Intensional Mathematics*, S. Shapiro, ed., North-Holland, Amsterdam, pp. 173–190.

J. Lipton

[1990] Constructive Kripke semantics and realizability, in: *Logic from Computer Science*, Y. N. Moschovakis, ed., Springer Verlag, Berlin, Heidelberg, New York. Also as a Technical Report, from Cornell University, nr.90–71.

J. Lipton and M. J. O'Donnell

[1996] Some intuitions behind realizability semantics for constructive logic: tableaux and Läuchli countermodels, *APAL*, 81, pp. 187–239.

L. L. Maksimova, V. B. Shekhtman, and D. P. Skvortsov

[1979] The impossibility of a finite axiomatization of Medvedev's logic of finitary problems (Russian), *Doklady*, 245, pp. 1051–1054. Translation *SM* 20, pp. 394–398.

P. Martin-Löf

[1982] Constructive mathematics and computer programming, in: *LMPS 6*, L. J. Cohen, J. Łoś, H. Pfeiffer, and K.-P. Podewski, eds., North-Holland, Amsterdam, pp. 153–175.

[1984] *Intuitionistic type theory. Notes by Giovanni Sambin of a series of lectures given in Padua, June 1980*, Bibliopolis, Napoli.

A. R. D. Mathias and H. Rogers

[1973] eds., *Cambridge Summer School in Mathematical Logic*, Springer Verlag, Berlin, Heidelberg, New York.

D. C. McCarty

[1984a] Information systems, continuity and realizability, in: *Logics of Programs*, E. Clarke and D. Kozen, eds., vol. 164 of Lecture Notes in Computer Science, Springer Verlag, Berlin, Heidelberg, New York, pp. 341–359.

[1984b] *Realizability and Recursive Mathematics*, Tech. Rep. CMU–CS–84–131, Department of Computer Science, Carnegie-Mellon University. Report version of the author's PhD thesis, Oxford University 1983.

[1986] Subcountability under realizability, *The Notre Dame Journal of Formal Logic*, 27, pp. 210–220.

[1988] Markov's principle, isols and Dedekind finite sets, *JSL*, 53, pp. 1042–1069.

[1991] Polymorphism and apartness, *The Notre Dame Journal of Formal Logic*, 32, pp. 513–532.

YU. T. MEDVEDEV

[1962] Finite problems (Russian), *Doklady*, 142, pp. 1015–1018. Translation *SM* 3, pp. 227–230.

[1963] Interpretation of logical formulae by means of finite problems and its relation to the realizability theory (Russian), *Doklady*, 148, pp. 771–774. Translation *SM* 4, pp. 180–183.

[1966] Interpretation of logical formulae by means of finite problems (Russian), *Doklady*, 169, pp. 20–23. Translation *SM* 7, pp. 857–860.

[1969] A method for proving the unsolvability of algorithmic problems (Russian), *Doklady*, 185, pp. 1232–1235. Translation *SM* 10, pp. 495–498.

[1972] Locally finitary algorithmic problems (Russian), *Doklady*, 203, pp. 285–288. Errata *Ibidem* 204, pp. 1286. Translation *SM* 13, pp. 382–386.

[1973] An interpretation of intuitionistic number theory, in: *LMPS 4*, P. Suppes, G. C. Moisil, and A. Joja, eds., North-Holland, Amsterdam, pp. 129–136.

G. E. MINTS

[1989] The completeness of provable realizability, *The Notre Dame Journal of Formal Logic*, 30, pp. 420–441.

J. R. MOSCHOVAKIS

[1967] Disjunction and existence in formalized intuitionistic analysis, in: *Sets, Models, and Recursion Theory*, J. N. Crossley, ed., North-Holland, Amsterdam, pp. 309–331.

[1971] Can there be no nonrecursive functions?, *JSL*, 36, pp. 309–315.

[1973] A topological interpretation of second-order intuitionistic arithmetic, *Compositio mathematica*, 26, pp. 261–275.

[1980] Kleene's realizability and "divides" notions for formalized intuitionistic mathematics, in: *Barwise, Keisler and Kunen [1980]*, pp. 167–179.

[1981] A disjunctive decomposition theorem for classical theories, in: *Constructive Mathematics*, F. Richman, ed., Springer Verlag, Berlin, Heidelberg, New York, pp. 250–259.

[1993] An intuitionistic theory of lawlike, choice and lawless sequences, in: *Logic Colloquium '90*, Springer Verlag, Berlin, Heidelberg, New York, pp. 191–209. (Lecture Notes in Logic 2).

[1994] More about relatively lawless sequences, *JSL*, 59, pp. 813–829.

[1996] A classical view of the intuitionistic continuum, *APAL*, 81, pp. 9–24.

J. MYHILL

[1973a] A note on indicator-functions, *Proceedings of the American Mathematical Society*, 29, pp. 181–183.

[1973b] Some properties of intuitionistic Zermelo-Fraenkel set theory, in: *Mathias and Rogers [1973]*, pp. 206–231.

[1975] Constructive set theory, *JSL*, 40, pp. 347–382.

D. NELSON

[1947] Recursive functions and intuitionistic number theory, *Transactions of the American Mathematical Society*, 61, pp. 307–368,556.

J. VAN OOSTEN

[1990] Lifschitz' realizability, *JSL*, 55, pp. 805–821.

[1991a] *Exercises in Realizability*, PhD thesis, Universiteit van Amsterdam.

[1991b] Extension of Lifschitz' realizability to higher order arithmetic, and a solution to a problem of F. Richman, *JSL*, 56, pp. 964–973.

[1991c] A semantical proof of De Jongh's theorem, *Archive for Mathematical Logic*, pp. 105–114.

[1993] *Extensional realizability*, Tech. Rep. ML-93-18, Institute for Logic, Language and Computation, University of Amsterdam. Submitted to *APAL*.

[1994] Axiomatizing higher-order Kleene realizability, *APAL*, 70, pp. 87–111.

[1996] Two remarks on the Lifschitz realizability topos, *JSL*, 61, pp. 70–79.

W. PHOA

[1989] Relative computability in the effective topos, *Mathematical Proceedings of the Cambridge Philosophical Society*, 106, pp. 419–420.

A. M. PITTS

[1981] *The Theory of Toposes*, PhD thesis, University of Cambridge.

V. E. PLISKO

[1973] On realizable predicate formulae (Russian), *Doklady*, 212, pp. 553–556. Translation *SM* 14, pp. 1420–1424.

[1974a] A certain formal system that is connected with realizability (Russian), in: *Teoriya Algorifmov i Matematicheskaya Logika: Sbornik Statej (Theory of Algorithms and Mathematical Logic. Collection of articles dedicated to Andrej Andrejevich Markov)*, B. Kushner and N. M. Nagornyi, eds., Vychislitel'nyj Tsentr Akademii Nauk SSSR, pp. 148–158,215.

[1974b] On interpretations of predicate formulae that are connected with constructive logic (Russian), in: *3 Vsesoyuznaya Konferentsiya po Matematicheskoj Logike. Tezitsy Doklady i Soobshcheniya (3rd All-Union Conference on Mathematical Logic)*, pp. 170–172.

[1974c] Recursive realizability and constructive predicate logic (Russian), *Doklady*, 214, pp. 520–523. Translation *SM* 15, pp. 193–197.

[1976] Some variants of the notion of realizability for predicate formulas (Russian), *Doklady*, 226, pp. 61–64. Translation *SM* 17, pp. 59–63.

[1977] The nonarithmeticity of the class of realizable predicate formulas (Russian), *Izv. Akad. Nauk.*, 41, pp. 483–502. Translation *Math. Izv.* 11, pp. 453–471.

[1978] Some variants of the notion of realizability for predicate formulas (Russian), *Izv. Akad. Nauk.*, 42, pp. 637–653. Translation *Math. Izv.* 12, pp. 588–604.

[1983] Absolute realizability of predicate formulas (Russian), *Izv. Akad. Nauk.*, 47, pp. 315–334. Translation *Math. Izv.* 22, pp. 291–308.

[1992] On the concept of relatively uniform realizability of propositional formulas (Russian), *Vestnik Moskovskogo Universiteta Seriya I. Mathematika, Mekhanika*, pp. 77–79. Translated in *Moscow University Mathematics Bulletin* 47 (1992), 41–42.

G. R. RENARDEL DE LAVALETTE

[1984] *Theories with type-free application and extended bar induction*, PhD thesis, Universiteit van Amsterdam.

[1990] Extended bar induction in applicative theories, *APAL*, 50, pp. 139–189.

E. ROBINSON AND G. ROSOLINI

[1990] Colimit completions and the effective topos, *JSL*, 55, pp. 678–699.

T. T. ROBINSON

[1965] Interpretations of Kleene's metamathematical predicate $\Gamma \mid A$ in intuitionistic arithmetic, *JSL*, 30, pp. 140–154.

G. F. ROSE

[1953] Propositional calculus and realizibility, *Transactions of the American Mathematical Society*, 75, pp. 1–19.

G. ROSOLINI

[1990] About modest sets, *International Journal of Foundations of Computer Science*, 1, pp. 341–353.

B. SCARPELLINI

[1970] A model for intuitionistic analysis, *Commentarii Mathematici Helvetici*, 45, pp. 440–471.

[1977] A new realizability notion for intuitionistic analysis, *ZLGM*, 23, pp. 137–167.

A. SCEDROV

[1984] Differential equations in constructive analysis and in the recursive realizability topos, *Journal of Pure and Applied Algebra*, 33, pp. 69–80.

[1986] On the impossibility of explicit upper bounds on lengths of some provably finite algorithms in computable analysis, *APAL*, 32, pp. 291–297.

[1990] Recursive realizability semantics for Calculus of Constructions. Preliminary report, in: *Logical Foundations of Functional Programming*, G. Huet, ed., Addison-Wesley Publishing Company, pp. 419–430.

A. SCEDROV AND P. J. SCOTT

[1982] A note on the Friedman slash and Freyd covers, in: *Troelstra and van Dalen [1980]*, pp. 443–452.

A. SCEDROV AND R. E. VESLEY

[1983] On a weakening of Markov's principle, *Archiv*, 23, pp. 153–160.

D. S. SCOTT

[1968] Extending the topological interpretation to intuitionistic analysis, *Compositio Mathematica*, 20, pp. 194–210.

P. SCOWCROFT

[1993] The disjunction and numerical existence properties for intuitionistic analysis, in: *Logical Methods*, J. N. Crossley, J. B. Remmel, R. A. Shore, and M. E. Sweedler, eds., Birkhaüser Boston, Inc., Boston MA, pp. 747–781.

N. A. SHANIN

[1958a] On the constructive interpretation of mathematical judgements (Russian), *Trudy Ordena Lenina Matematicheskogo Instituta imeni V. A. Steklova. Akademiya Nauk SSSR*, 52, pp. 226–311. Translation *AMS Transl.* 23, pp. 108–189.

[1958b] Über einen Algorithmus zur konstruktiven Dechiffrierung mathematischer Urteile (Russian) (German summary), *ZLGM*, 4, pp. 293–303.

[1964] Concerning the constructive interpretation of auxiliary formulas I (Russian), *Trudy Ordena Lenina Matematicheskogo Instituta imeni V. A. Steklova. Akademiya Nauk SSSR*, 72, pp. 348–379. Translation *AMS Transl.* 99, pp. 233–275.

J. M. SMITH

[1993] An interpretation of Kleene's slash in type theory, in: *Logical Environments*, G. Huet and G. Plotkin, eds., pp. 189–197.

J. STAPLES

[1973] Combinator realizability of constructive finite type analysis, in: *Mathias and Rogers [1973]*, pp. 253–273.

[1974] Combinator realizability of constructive Morse theory, *JSL*, 39, pp. 226–234.

M. STEIN

[1979] Interpretationen der Heyting–Arithmetik endlicher Typen, *Archiv*, 19, pp. 175–189.

[1980] Interpretations of Heyting's arithmetic – An analysis by means of a language with set symbols, *AML*, 19, pp. 1–31.

[1981] A general theorem on existence theorems, *ZLGM*, 27, pp. 435–452.

T. STRAHM

[1993] *Partial applicative theories and explicit substitutions*, Tech. Rep. IAM 93–008, Institut für Informatik und angewandte Mathematik, Universität Bern.

T. STREICHER

[1991] *Mathematical Independencies in the Pure Calculus of Constructions*, Tech. Rep. MIP–909, Fakultät für Mathematik und Informatik, Universität Passau.

M. D. G. SWAEN

[1989] *Weak and Strong Sum-Elimination in Intuitionistic Type Theory*, PhD thesis, Universiteit van Amsterdam.

[1991] The logic of first-order intuitionistic type theory with weak sigma-elimination, *JSL*, 56, pp. 467–483.

[1992] A characterization of **ML** in many-sorted arithmetic with conditional application, *JSL*, 57, pp. 924–953.

W. W. TAIT

[1975] A realizability interpretation of the theory of species, in: *Logic Colloquium*, R. Parikh, ed., Springer Verlag, Berlin, Heidelberg, New York, pp. 240–251.

M. TATSUTA

[1991a] Monotone recursive definition of predicates and its realizability interpretation, in: *Proceedings of Theoretical Aspects of Computer Software*, Springer Verlag, Berlin, Heidelberg, New York, pp. 38–52.

[1991b] Program synthesis using realizability, *Theoretical Computer Science*, 90, pp. 309–353.

[1992] Realizability interpretation of coinductive definitions and program synthesis with streams, in: *Proceedings of International Conference on Fifth Generation Computer Systems*, pp. 666–673.

L. H. THARP

[1971] A quasi–intuitionistic set theory, *JSL*, 36, pp. 456–460.

R. R. TOMPKINS

[1968] On Kleene's recursive realizability as an interpretation for intuitionistic elementary number theory, *Notre Dame Journal of Formal Logic*, 9, pp. 289–293.

A. S. TROELSTRA

[1971a] *Computability of terms and notions of realizability for intuitionistic analysis*, Tech. Rep. 71–02, Department of Mathematics, University of Amsterdam. Most, but not all, material reappears in *Troelstra [1973a]*.

[1971b] Notions of realizability for intuitionistic arithmetic and intuitionistic arithmetic in all finite types, in: *The Second Scandinavian Logic Symposium*, J. E. Fenstad, ed., North-Holland, Amsterdam, pp. 369–405.

[1973a] ed., *Metamathematical investigation of intuitionistic arithmetic and analysis*, Springer Verlag, Berlin, Heidelberg, New York. With contributions by A. S. Troelstra, C. A. Smoryński, J. I. Zucker and W. A. Howard.

[1973b] Notes on intuitionistic second order arithmetic, in: *Mathias and Rogers [1973]*, pp. 171–205.

[1977a] Axioms for intuitionistic mathematics incompatible with classical logic, in: *LMPS 5*, R. E. Butts and J. Hintikka, eds., vol. 1, D. Reidel, Dordrecht and Boston, pp. 59–84.

[1977b] A note on non–extensional operations in connection with continuity and recursiveness, *Indagationes Mathematicae*, 39, pp. 455–462.

[1977c] Some models for intuitionistic finite type arithmetic with fan functional, *JSL*, 42, pp. 194–202.

[1980] Extended bar induction of type zero, in: *Barwise, Keisler and Kunen [1980]*, pp. 277–316.

[1992] Comparing the theory of representations and constructive mathematics, in: *Computer Science Logic. 5th workshop, CSL '91*, E. Börger, G. Jäger, H. Kleine Büning, and M. M. Richter, eds., Springer Verlag, Berlin, Heidelberg, New York, pp. 382–395.

A. S. TROELSTRA AND D. VAN DALEN

[1980] eds., *The L. E. J. Brouwer Centenary Symposium*, North-Holland, Amsterdam.

[1988] *Constructivism in Mathematics*, North-Holland, Amsterdam. 2 volumes.

G. S. TSEJTIN

[1968] The disjunctive rank of the formulas of constructive arithmetic (Russian), *Zapiski*, 8, pp. 260–271. Translation *Seminars in Mathematics. V.A.Steklov Mathematical Institute Leningrad* 8(1970), pp. 126–132. This volume appeared as: A.O. Slisenko (ed.), *Studies in Constructive Mathematics and Mathematical Logic. Part II.* Consultants Bureau, New York, London.

F. L. VARPAHOVSKIJ

[1971] A class of realizable propositional formulas, *Zapiski*, 1, pp. 8–23. Translation *Journal of Soviet Mathematics* **1**,1–11 (1973).

R. E. VESLEY

[1970] A palatable substitute for Kripke's schema, in: *Kino, Myhill and Vesley [1970]*, pp. 197–207.

[1996] Realizing Brouwer's sequences, *APAL*, 81, pp. 25–74.

A. VORONKOV

[1991] *N-realizability: one more constructive semantics*, Tech. Rep. 71, Department of Mathematics, Monash University, Australia.

[1992] On completeness of program synthesis systems, in: *Computer Science Logic. 5th workshop, CSL '91*, E. Börger, G. Jäger, H. Kleine Büning, and M. M. Richter, eds., Springer Verlag, Berlin, Heidelberg, New York, pp. 411–418.

K. F. WEHMEIER

[1996] Fragments of HA based on Σ_1-induction. Preprint. Part of the author's dissertation project at he University of Münster, Westphalia, Germany.

CHAPTER VII

The Logic of Provability

Giorgi Japaridze

Department of Computer and Information Science, University of Pennsylvania
Philadelphia, Pennsylvania 19104-6389, USA

Dick de Jongh

Institute for Logic, Language and Computation, University of Amsterdam
NL-1018 TV Amsterdam, The Netherlands

This chapter is dedicated to the memory of George Boolos. From the start of the subject until his death on 27 May 1996 he was the prime inspirer of the work in the logic of provability.

Contents

HANDBOOK OF PROOF THEORY
Edited by S. R. Buss

1. Introduction, Solovay's theorems

Gödel's incompleteness theorems and Church's undecidability theorem for arithmetic showed that reasonably strong formal systems cannot be complete and decidable, and cannot prove their own consistency. Even at the time though these negative theorems were accompanied by positive results. Firstly, formal systems fare better in reasoning in restricted areas, and this reasoning can be formalized in the theories themselves. In Hilbert and Bernays [1939] one finds the formalization of the completeness theorem for the predicate calculus, i.e., reasoning in the predicate calculus is adequately described in strong enough theories. A fortiori, this is so for the propositional calculus in which reasoning is even (provably) decidable. Secondly, there is a positive component in the incompleteness theorems themselves. The formalized version of the second incompleteness theorem, i.e., if it is provable in **PA** that **PA** is consistent, then **PA** is inconsistent, is provable in **PA** itself. The area here called the logic of provability arose in the seventies when two developments took place almost simultaneously. The two facets mentioned above were, one might say, integrated by showing that propositional reasoning about the formalized provability predicate is decidable and can be adequately described in arithmetic itself (Solovay [1976]). And in the same period the de Jongh-Sambin fixed point theorem (see Sambin [1976], Smoryński [1978,1985]) was proved for modal-logical systems with the provability interpretation in mind. Since that time the main achievements have been to show that similar results mostly fail for predicate logic, to recognize reasoning about more complex notions like interpretability where arithmetic can be shown to reason adequately, and also to strengthen Solovay's results directly. Extensive overviews on the subject can be found in Boolos [1993b] and Smoryński [1985], a short history in Boolos and Sambin [1991].

Let us proceed somewhat farther in formulating Solovay's theorems, and call an *arithmetic realization* of the language of modal logic (see section 2) into the language of the arithmetic theory T (Σ_1-sound and extending $\mathbf{I\Sigma_1}$, sometimes $\mathbf{I\Delta_0}$) a mapping * that commutes with the propositional connectives and such that $(\Box A)^* = \mathrm{Pr}_T(\ulcorner A^* \urcorner)$ (where Pr_T is the formalized *provability* predicate for T, i.e., it is of the form $\exists y\, \mathrm{Proof}_T(x, y)$ where Proof_T is the formalized *proof* predicate of T). If we want to stress the dependency on T we write $(A)^*_T$ for $(A)^*$. More standard is the term "interpretation" for "realization" but that conflicts somewhat with our terminology with regard to interpretability. The term "realization" is used by Boolos [1993b].

1.1. Theorem. (Solovay's first arithmetic completeness theorem) *The modal formula A is provable in T under all arithmetic realizations iff A is provable in the modal logic* **L** (see sections 2, 3).

1.2. Theorem. (Solovay's second arithmetic completeness theorem) *The modal formula A is true under all arithmetic realizations iff A is provable in the modal logic* **S** (see sections 2, 3).

This chapter is to be thought of as divided into three parts: the first part consists of sections 2-10 and is devoted to propositional provability logic, i.e., the propositional logic of the provability predicate and its direct extensions, the second part consists of sections 11-15 and treats interpretability logic and related areas, the last part consists of section 16 and discusses predicate provability logic.

2. Modal logic preliminaries

The language of the modal propositional calculus consists of a set of propositional variables, connectives $\vee$, $\wedge$, $\rightarrow$, $\leftrightarrow$, $\neg$, $\top$, $\bot$ and a unary operator $\Box$. Furthermore, $\Diamond$ is an abbreviation of $\neg\Box\neg$. The modal logic **K** is axiomatized by the schemes 1 and 2:

1. All propositional tautologies in the modal language,

2. $\Box(A \rightarrow B) \rightarrow (\Box A \rightarrow \Box B)$,

together with the rules of modus ponens and *necessitation*, i.e., $A/\Box A$. The modal logic **L** is axiomatized by adding the scheme 3:

3. $\Box(\Box A \rightarrow A) \rightarrow \Box A$,

to **K** and keeping the rules of modus ponens and necessitation. The system is often called **GL**, e.g. in Boolos [1993b], and is named **PrL** in Smoryński [1985]. It is an exercise to show that $\Box A \rightarrow \Box\Box A$ is derivable in **L**, which makes **L** an extension of **K4**, the system axiomatized by the axioms of **K** together with $\Box A \rightarrow \Box\Box A$. Extensions of **K** such as **K4** and **L** that are closed under necessitation are called *normal* modal logics.

We will write $A_1, \ldots, A_n \vdash_{\mathbf{K}} B$ for: B is derivable in **K** from $A_1, \ldots, A_n$ without use of necessitation, or more precisely, B is derivable by modus ponens from theorems of **K** plus $A_1, \ldots, A_n$. Similarly for **K4**, **L**. To this notation the *deduction theorem* obviously applies: $A_1, \ldots, A_n, B \vdash_{\mathbf{K}} C$ iff $A_1, \ldots, A_n \vdash_{\mathbf{K}} B \rightarrow C$. We will write $\boxdot A$ for $A \wedge \Box A$. The results codified in the next proposition are readily proved.

2.1. Proposition.
(a) *If* $A_1, \ldots, A_n \vdash_{\mathbf{K}} B$, *then* $\Box A_1, \ldots, \Box A_n \vdash_{\mathbf{K}} \Box B$ *(also for* **K4**, **L***)*,
(b) *if* $\Box A_1, \ldots, \Box A_n \vdash_{\mathbf{K4}} B$, *then* $\Box A_1, \ldots, \Box A_n \vdash_{\mathbf{K4}} \Box B$ *(also for* **L***)*,
(c) *if* $\Box A_1, \ldots, \Box A_n, \Box B \vdash_{\mathbf{L}} B$, *then* $\Box A_1, \ldots, \Box A_n \vdash_{\mathbf{L}} \Box B$,
(d) $\vdash_{\mathbf{K}} \Box(A \wedge B) \leftrightarrow \Box A \wedge \Box B$,
(e) $\vdash_{\mathbf{K4}} \Box(A \leftrightarrow B) \rightarrow \Box(C(A) \leftrightarrow C(B))$,
(f) $\vdash_{\mathbf{K4}} \Box(A \leftrightarrow B) \rightarrow (\Box C(A) \leftrightarrow \Box C(B))$,
(g) $\vdash_{\mathbf{K4}} \boxdot(A \leftrightarrow B) \rightarrow (C(A) \leftrightarrow C(B))$,
(h) *if* $\vdash_{\mathbf{K},\mathbf{K4},\mathbf{L}} A \leftrightarrow B$, *then* $\vdash_{\mathbf{K},\mathbf{K4},\mathbf{L}} C(A) \leftrightarrow C(B)$,
(i) $\vdash_{\mathbf{K}} \Diamond\bot \rightarrow \bot$,

(j) $\vdash_{\mathbf{L}} \Diamond p \to \Diamond(p \wedge \Box\neg p)$ *and, hence,* $\vdash_{\mathbf{L}} \Diamond p \leftrightarrow \Diamond(p \wedge \Box\neg p)$ *and* $\vdash_{\mathbf{L}} p \to (p \wedge \Box\neg p) \vee \Diamond(p \wedge \Box\neg p)$.

The modal logic **S** is defined by:

$\vdash_{\mathbf{S}} A$ if and only if $\Box B_1 \to B_1, \ldots, \Box B_k \to B_k \vdash_{\mathbf{L}} A$ for some $B_1, \ldots, B_k$.

The logic **S** is not closed under necessitation, and is therefore a nonnormal modal logic.

2.2. Definition.
(a) A *Kripke-frame* for **K** is a pair $\langle W, R\rangle$ with W a nonempty set of so-called *worlds* or *nodes* and R a binary relation, the so-called *accessibility* relation.
(b) A *Kripke-frame* for **K4** is a Kripke-frame $\langle W, R\rangle$ with R a transitive relation.
(c) A *Kripke-frame* for **L** is a Kripke-frame $\langle W, R\rangle$ with R a transitive relation such that the converse of R is well-founded. (Of course, a finite transitive frame is conversely well-founded iff it is irreflexive.)
(d) A *root* of a Kripke-frame is a node w such that $w\,R\,w'$ for all $w' \neq w$ in the frame. (In the case of **K** put the transitive closure of R here in place of R.) The *depth* (also *height*) of a node w in a conversely well-ordered frame is the maximal m for which there exists a sequence $w = w_0 R\, w_1 \ldots R\, w_m$. The *height* of the model is the maximum of the height of its nodes.
(e) A *Kripke-model* for **K** (**K4**, **L**) is a tuple $\langle W, R, \Vdash\rangle$ with $\langle W, R\rangle$ a Kripke-frame for **K** (**K4**, **L**) together with a *forcing relation* $\Vdash$ between worlds and propositional variables. The relation $\Vdash$ is extended to a relation between worlds and all formulas by the stipulations

$w \Vdash \neg A$ iff $w \nVdash A$,
$w \Vdash A \wedge B$ iff $w \Vdash A$ and $w \Vdash B$,
and similarly for the other connectives,
$w \Vdash \Box A$ iff for all w' such that wRw', $w' \Vdash A$.

If $K = \langle W, R, \Vdash\rangle$, then $K \models A$ is defined as, $w \Vdash A$ for each $w \in W$, and we say that A is *valid* in M.

It is easy to check that Kripke-models are *sound* in the sense that each A derivable in **K** (**K4**, **L**) is valid in each Kripke-model for **K** (**K4**, **L**). In fact, the Kripke-models for **K4** (resp. **L**) are exactly the ones that validate the formulas derivable, respectively in **K4** and **L**. One says that **K4** and **L** *characterize* these classes of models. (For the main concepts of modal logic, see e.g., Chellas [1980], Hughes and Cresswell [1984].) Something stronger is true: in **K**, **K4** and **L** one can derive all the formulas that are valid in their respective model classes (*modal completeness*). The standard method in modal logic for proving completeness is to construct the necessary countermodels by taking maximal consistent sets of the logic as the worlds of the model and providing this set of worlds with an appropriate accessibility relation R. This method cannot be applied to **L**, since the logic is not compact: there exist infinite syntactically consistent sets of formulas that are semantically incoherent. We will apply to all three logics a method in which one restricts maximal consistent sets to a finite so-called adequate set of formulas. One obtains finite countermodels by this method and hence, immediately, decidability of the logics.

2.3. Definition. If A is not a negation, then $\sim A$ is $\neg A$, otherwise, if A is $\neg B$, then $\sim A$ is B.
An *adequate* set of formulas is a set Φ with the properties:
(i) Φ is closed under subformulas,
(ii) if $B \in \Phi$, then $\sim B \in \Phi$.

It is obvious that each formula is contained in a finite adequate set.

2.4. Theorem. (Modal completeness of **K**, **K4**, **L**.)
If A is not derivable in **K** (**K4**, **L**), *then there is a frame for* **K** (**K4**, **L**) *on which A is not valid.*

Proof. (**K**) Suppose $\nvdash_{\mathbf{K}} A$. Let Φ be a finite adequate set containing A. We consider the set W of all maximal **K**-consistent subsets of Φ. We define for $w, w' \in W$,

$$wRw' \iff \text{for all } \Box D \in w,\ D \in w'.$$

Furthermore, we define $w \Vdash p$ iff $p \in w$. It now follows that for each $B \in \Phi$, $w \Vdash B$ iff $B \in w$, by induction on the length of B. For propositional letters this is so by definition and the case of the connectives is standard, so let us consider the case that B is $\Box C$.
$\Longrightarrow$: Assume $\Box C \in w$. Then, for all w' such that wRw', $C \in w'$. By induction hypothesis, $w' \Vdash C$ for all such w'. So, $w \Vdash \Box C$.
$\Longleftarrow$: Assume $\Box C \notin w$. Consider the set $\{D | \Box D \in w\} \cup \{\sim C\}$. We will show this set to be **K**-consistent which means, by the conditions on adequate sets, that a maximal **K**-consistent superset w' of it exists inside Φ. By induction hypothesis, $w' \nVdash C$, and since wRw', this implies that $w \nVdash \Box C$.

To show that $\{D | \Box D \in w\} \cup \{\sim C\}$ is indeed **K**-consistent, suppose that it is not, i.e., $D_1, \ldots, D_k \vdash_{\mathbf{K}} C$ for some $\Box D_1, \ldots, \Box D_k \in w$. Then $\Box D_1, \ldots, \Box D_k \vdash_{\mathbf{K}} \Box C$ immediately follows, but that would make w inconsistent, contrary to what was assumed.

(**K4**) Suppose $\nvdash_{\mathbf{K4}} A$. Proceed just as in the case of **K**, except that now:

$$wRw' \iff \text{for all } \Box D \in w,\ \text{both } \Box D \in w' \text{ and } D \in w'.$$

The argument is as for **K**; only the case $B \equiv \Box C$ ($\Rightarrow$) needs special attention. Let $\Box C \notin w$. This time, consider the set $\{\Box D, D | \Box D \in w\} \cup \{\sim C\}$. The only additional fact needed in the argument to show that this set is **K4**-consistent is that $\Box D_i \vdash_{\mathbf{K4}} \Box\Box D_i$.

(**L**) Suppose $\nvdash_{\mathbf{L}} A$. Again proceed as before, except that now:

$$wRw' \iff \text{for all } \Box D \in w, \text{ both } \Box D \in w' \text{ and } D \in w' \text{ and, for some } \Box C \in w', \Box C \notin w.$$

The argument is as for **K4**; again, only the case $B \equiv \Box C$ ($\Longleftarrow$) merits some special attention. This time, consider $\{\Box D, D | \Box D \in w\} \cup \{\Box C, {\sim}C\}$. Note that the inclusion of $\Box C$ will insure that w' really is a successor of w. The argument that this set is consistent now rests on the fact that, if $\Box D_1, \ldots, \Box D_k \vdash_{\mathbf{L}} \Box(\Box C \to C)$, then $\Box D_1, \ldots, \Box D_k \vdash_{\mathbf{L}} \Box C$. ⊣

2.5. Definition. If Φ is an adequate set of formulas, then the rooted Kripke-model $\mathbf{M}$ is Φ-*sound* if in the root w of $\mathbf{M}$, $w \Vdash \Box B \to B$ for each $\Box B$ in Φ; $\mathbf{M}$ is A-*sound* if $\mathbf{M}$ is Φ-sound for the smallest adequate set Φ containing A.

If K is a Kripke-model with root w, then the *derived model* K' of K is constructed by adding a new root w' below w and giving w' exactly the same forcing relation as w for the atoms.

2.6. Lemma. *If K is a Φ-sound Kripke-model with root w, then w' forces in K' exactly the same formulas from Φ as w in K.*

Proof. Let K be a rooted Φ-sound Kripke-model with root w. We prove by induction on the length of A that $w' \Vdash A$ iff $w \Vdash A$. This is so by definition for the atomic formulas and otherwise obvious except for the $\Box$. If $w' \Vdash \Box A$, then $w \Vdash \Box A$, since $w R w'$. If $w \Vdash \Box A$, then not only for all w'' such that $w R w''$, $w'' \Vdash A$, but also, by the Φ-soundness of K, $w \Vdash A$. But then, for all w'' such that $w' R w''$, $w'' \Vdash A$, i.e., $w' \Vdash \Box A$. ⊣

2.7. Theorem. (Modal completeness of **S**) $\vdash_{\mathbf{S}} A$ *iff A is forced in the root of all A-sound $\mathbf{L}$-models.*

Proof. $\Longrightarrow$: Assume K is A-sound and root $w \nVdash A$. If we assume to get a contradiction that $\vdash_{\mathbf{S}} A$, then A is provable in $\mathbf{L}$ from k applications of the reflection scheme: $\Box B_1 \to B_1, \ldots, \Box B_k \to B_k$. Consider the model $K^{(k+1)}$ obtained from K by taking $k+1$ times the derived model starting with K. Each time that one takes the derived model, one or more of the $\Box B_i$ may change from being forced to not being forced (never the other way around). This implies, by the pigeon hole principle, that one of the times that one has taken the derived model $K^{(m)}$ $(0 \leqslant m \leqslant k+1)$ the forcing value of all the $\Box B_i$ remain the same. It is easy to check that, in the root of that model $K^{(m)}$ for that m, $\Box B_i \to B_i$ is forced for all $i \leqslant k$. By lemma 2.6, A is not forced however, and a contradiction has been reached.

$\Longleftarrow$: Assume $\nvdash_{\mathbf{S}} A$. Then a fortiori A is not derivable in $\mathbf{L}$ from the reflection principles for its boxed subformulas. The result then immediately follows by applying theorem 2.4. ⊣

An elegant formulation of the semantics of **S** in terms of infinite models, so-called *tail models* is given in Visser [1984].

The well-known normal modal system **S4** that is obtained by adding the scheme $\Box A \to A$ to **K4** plays a role in section 10. It can be shown that **S4** is modally complete with respect to the (finite) reflexive, transitive Kripke-models.

3. Proof of Solovay's theorems

We rely mostly on Buss's Chapter II of this Handbook. One can find there an *intensional* arithmetization of metamathematics worked out, the (Hilbert-Bernays)-Löb derivability conditions are given and proofs of the diagonalization lemma, Gödel's theorems and Löb's theorem are presented. An additional fact that we need is some formalization of the recursion theorem.

To repeat the statement of Solovay's first arithmetic completeness theorem (theorem 1.1), for Σ_1-sound r.e. theories T containing $\mathbf{I\Sigma_1}$:

$\vdash_{\mathbf{L}} A$ iff $T \vdash A^*$ for all arithmetic realizations *.

Proof of theorem 1.1. $\Rightarrow$: These are just the Hilbert-Bernays-Löb conditions and Löb's theorem (see Chapter II of this Handbook).
$\Leftarrow$: What we have to do is show that, if $\nvdash_{\mathbf{L}} A(p_1 \ldots, p_k)$, then there exist $\alpha_1, \ldots, \alpha_k$ such that $T \nvdash A^*$, where * denotes the realization generated by mapping $p_1, \ldots, p_k$ to $\alpha_1, \ldots, \alpha_k$.[1] Suppose $\nvdash_{\mathbf{L}} A$. Then, by theorem 2.4, there is a finite **L**-model $\langle W, R, \Vdash \rangle$ in which A is not valid. We may assume that $W = \{1, \ldots, l\}$, 1 is the root, and $1 \nVdash A$. We define a new frame $\langle W', R' \rangle$:

$$W' = W \cup \{0\},$$
$$R' = R \cup \{(0, w) \mid w \in W\}.$$

Observe that $\langle W', R' \rangle$ is a finite **L**-frame.

We are going to embed this frame into T by means of a function $h : \omega \to W'$ (with ω the nonnegative integers) and sentences Lim_w, for each $w \in W'$, which assert that w is the limit of h. This function will be defined in such a way that a basic lemma 3.2 holds about the statements that T can prove about the sentences Lim_w. These statements are tailored to prove the next lemma 3.3 that expresses that provability in T behaves for the relevant formulas on the Kripke-model in the same way as the modal operator $\Box$. This will allow us to conclude the proof.

3.2. Lemma.

(a) *T proves that h has a limit in W', i.e.,* $T \vdash \bigvee \{\text{Lim}_r \mid r \in W'\}$,
(b) *If $w \neq u$, then* $T \vdash \neg(\text{Lim}_w \wedge \text{Lim}_u)$,
(c) *If $w R' u$, then $T + \text{Lim}_w$ proves that* $T \nvdash \neg \text{Lim}_u$,
(d) *If $w \neq 0$ and not $w R' u$, then $T + \text{Lim}_w$ proves that* $T \vdash \neg \text{Lim}_u$,

[1] We will use italic capital letters for modal-logical formulas and Greek letters for arithmetic sentences and formulas, except that we will use Roman letters for descriptive names like "Proof".

(e) Lim_0 *is true,*
(f) *For each* $i \in W'$, Lim_i *is consistent with* T.

We now define a realization * by setting for each propositional letter p_i,

$$p_i^* = \bigvee\{\mathrm{Lim}_w \mid w \in W,\ w \Vdash p_i\}.$$

This p_i^* will then function as the above-mentioned α_i.

3.3. Lemma. *For any* $w \in W$ *and any* **L**-*formula* B,

(a) *if* $w \Vdash B$, *then* $T + \mathrm{Lim}_w \vdash B^*$,
(b) *if* $w \nVdash B$, *then* $T + \mathrm{Lim}_w \vdash \neg B^*$.

Proof. By induction on the complexity of B. If B is atomic, then clause (a) is evident, and clause (b) is also clear in view of lemma 3.2(b). The cases when B is a Boolean combination are straightforward. So, only the case that B is $\Box C$ will have to be considered.

(a) Assume that $w \Vdash \Box C$. Then, for each $w' \in W$ with $w R w'$, $w' \Vdash C$. By induction hypothesis, for each such w', $T + \mathrm{Lim}_{w'} \vdash C^*$, and this fact is then provable in T. On the other hand, by lemma 3.2(a) (proved in T itself) and (c), $T + \mathrm{Lim}_w$ proves that $T \vdash \bigvee\{\mathrm{Lim}_{w'} \mid w R' w'\}$. Together this implies that T proves that $T \vdash C^*$, i.e., $T \vdash (\Box C)^*$.

(b) Assume that $w \nVdash \Box C$. Then, for some $w' \in W$ with $w R' w'$, $w' \nVdash C$. By induction hypothesis, $T + \mathrm{Lim}_{w'} \vdash \neg C^*$, i.e., $T \vdash C^* \to \neg \mathrm{Lim}_{w'}$. By the second HBL-condition, $T \vdash (\Box C)^* \to \mathrm{Pr}_T(\neg \mathrm{Lim}_w)$. But lemma 3.2(c) implies that $T + \mathrm{Lim}_w \vdash \neg \mathrm{Pr}_T(\neg \mathrm{Lim}_w)$, i.e., $T + \mathrm{Lim}_w \vdash \neg(\Box C)^*$. ⊣

Observe by the way that lemma 3.3 expresses that $T + \mathrm{Lim}_w \vdash$ "$w \Vdash B$" $\leftrightarrow B^*$. Assuming lemma 3.2 we can now complete the proof of theorem 1.1. By the construction of the Kripke-model, $1 \Vdash \neg A$. By lemma 3.3, $T + \mathrm{Lim}_1 \vdash \neg A^*$. Since, by lemma 3.2(f), $T + \mathrm{Lim}_1$ is consistent, $T \nvdash A^*$. ⊣

Our remaining duties are to define the function h and to prove lemma 3.2. The recursion theorem enables us to define this function simultaneously with the sentences Lim_w (for each $w \in W'$), which, as we have mentioned already, assert that w is the limit of h.

3.4. Definition. (Solovay function h)
We define $h(0) = 0$.
If x is the code of a proof in T of $\neg \mathrm{Lim}_w$ for some w with $h(x) R w$, then $h(x+1) = w$.
Otherwise, $h(x+1) = h(x)$.

It is not hard to see that h is primitive recursive.

Proof of lemma 3.2. In each case below, except in (e) and (f), we reason in T.

(a) By induction on the nodes. For end nodes w (i.e., the ones with no R-successors), it can be proved that $T \vdash \forall x\,(h(x) = \bar{w} \to \forall y \geqslant x\, h(y) = \bar{w})$ by induction on x, and hence $T \vdash \exists x\, h(x) = \bar{w} \to \mathrm{Lim}_w$. Next, it is easily seen that, if for all successors w' of a node w, $T \vdash \exists x\, h(x) = \bar{w}' \to \bigvee\{\mathrm{Lim}_{w''} \mid w' = w'' \vee w'\, R\, w''\}$, then $T \vdash \exists x\, h(x) = \bar{w} \to \bigvee\{\mathrm{Lim}_{w'} \mid w = w' \vee w\, R\, w'\}$. Therefore, this will hold for $w = 0$, which implies (a).

(b) Clearly h cannot have two different limits w and u.

(c) Assume w is the limit of h and $w\, R'\, u$. Let n be such that for all $x \geqslant n$, $h(x) = w$. We need to show that $T \nvdash \neg\,\mathrm{Lim}_u$. Deny this. Then, since every provable formula has arbitrarily long proofs, there is $x \geqslant n$ such that x codes a proof of $\neg \mathrm{Lim}_u$; but then, according to definition 3.4, we must have $h(x+1) = u$, which, as $u \neq w$ (by irreflexivity of R'), is a contradiction.

(d) Assume $w \neq 0$, w is the limit of h and not $wR'u$. If $u = w$, then (since $w \neq 0$) there exists an x such that $h(x+1) = w$ and $h(x) \neq w$. Then x codes a proof of $\neg\,\mathrm{Lim}_w$ and $\neg\,\mathrm{Lim}_w$ is provable. Next suppose $u \neq w$. Let us fix a number z with $h(z) = w$. Since h is primitive recursive, T proves that $h(z) = w$. Now argue in $T + \mathrm{Lim}_u$: since u is the limit of h and $h(z) = w \neq u$, there is a number x with $x \geqslant z$ such that $h(x) \neq u$ and $h(x+1) = u$. This contradicts the fact that not $(w=)h(z) R'\, u$, Thus, $T + \mathrm{Lim}_u$ is inconsistent, i.e., $T \vdash \neg\,\mathrm{Lim}_u$.

(e) By (a), as T is sound, one of the Lim_w for $w \in W'$ is true. Since for no w do we have $wR'w$, (d) means that each Lim_w, except Lim_0, implies in T its own T-disprovability and therefore is false. Consequently, Lim_0 is true.

(f) By (e), (c) and the soundness of T. ⊣

To repeat the statement of Solovay's second arithmetic completeness theorem (theorem 1.2):

$\vdash_{\mathbf{S}} A$ iff $\mathbb{N} \models A^*$ for all arithmetic realizations *.

Proof of theorem 1.2. Since the $\Box A \to A$ are reflection principles and these are true for a sound theory, the soundness part is clear. So, assume $\nvdash_{\mathbf{S}} A$. Modal completeness of **S** then provides us with an A-sound (see definition 2.5) model in which A is not forced in the root. We can repeat the procedure of the proof of the first completeness theorem, but now directly to the model itself (which we assume to have a root 0) without adding a new root, and again prove lemmas 3.2 and 3.3. But this time we have forcing also for 0 and we can improve lemma 3.3 to apply it also to $w = 0$, at least for subformulas of A.

The proof of the (b)-part of that lemma can be copied. With respect to the (a)-part, restricting again to the $\Box$-case, assume that $0 \Vdash \Box C$. Then, for each $w \in W$ with $w \neq 0$, $w \Vdash C$. But now, by the A-soundness of the model, C is also forced in the root 0. By the induction hypothesis, for all $w \in W$, $T + \mathrm{Lim}_w \vdash C^*$. By lemma 3.2(a) then $T \vdash C^*$, so, $T \vdash (\Box C)^*$ and hence $T \vdash \mathrm{Lim}_0 \to (\Box C)^*$.

Applying the strengthened version of lemma 3.3 to $w = 0$ and A, we obtain $T \vdash \mathrm{Lim}_0 \to \neg\, A^*$, which suffices, since Lim_0 is true (lemma 3.2). ⊣

4. Fixed point theorems

For the provability logic **L** a fixed point theorem can be proved. One can view Gödel's diagonalization lemma as stating that in arithmetic theories the formula $\neg\Box p$ has a fixed point: the Gödel sentence. Gödel's proof of his second incompleteness theorem effectively consisted of the fact that the sentence expressing consistency, the arithmetic realization of $\neg\Box\bot$, is provably equivalent to this fixed point. Actually this fact is provable from the principles codified in the provability logic **L**, which means then that it can actually be presented as a fact about **L**. This leads to a rather general fixed point theorem, which splits into a uniqueness and an existence part. It concerns formulas A with a distinguished propositional variable p that only occurs *boxed* in A, i.e., each occurrence of p in A is part of a subformula $\Box B$ of A.

4.1. Theorem. (Uniqueness of fixed points) *If p occurs only boxed in $A(p)$ and q does not occur at all in $A(p)$, then* $\vdash_{\mathbf{L}} \boxdot((p \leftrightarrow A(p)) \wedge (q \leftrightarrow A(q)) \to (p \leftrightarrow q)$.

4.2. Corollary. *If p occurs only boxed in $A(p)$, and both $\vdash_{\mathbf{L}} B \leftrightarrow A(B)$ and $\vdash_{\mathbf{L}} C \leftrightarrow A(C)$, then $\vdash_{\mathbf{L}} B \leftrightarrow C$.*

4.3. Theorem. (Existence of fixed points) *If p occurs only boxed in $A(p)$, then there exists a formula B, not containing p and otherwise containing only variables of $A(p)$, such that $\vdash_{\mathbf{L}} B \leftrightarrow A(B)$.*

After the original proofs by de Jongh and Sambin (see Sambin [1976], Smoryński [1978,1985], and, for the first proof of uniqueness, Bernardi [1976]) many other, different, proofs have been given for the fixed point theorems, syntactical as well as semantical ones, the latter e.g., in Gleit and Goldfarb [1990]. It is also worthwhile to remark that theorem 4.3 follows from theorem 4.1 (which can be seen as a kind of implicit definability theorem) by way of Beth's definability theorem that holds for **L**. The latter can be proved from interpolation in the usual manner. Interpolation can be proved semantically in the standard manner via a kind of Robinson's consistency lemma (see Smoryński [1978]), and syntactically in the standard manner by cut-elimination in a sequent calculus formulation of **L** (Sambin and Valentini [1982,1983]).

In an important sense the meaning of the fixed point theorem is negative, namely in the sense that, if in arithmetic one attempts to obtain formulas with essentially new properties by diagonalization, one will not get them by using instantiations of purely propositional modal formulas (except once of course: the Gödel sentence, or the sentence Löb used to prove his theorem). That is one reason that interesting fixed points often use Rosser-orderings (see section 9).

5. Propositional theories and Magari-algebras

A *propositional theory* is a set of modal formulas (usually in a finite number of propositional variables) which is closed under modus ponens and necessitation, but not necessarily under substitution.

We say that such a theory is *faithfully interpretable* in **PA**, if there is a realization * such that $T = \{A \mid \mathbf{PA} \vdash A^*\}$. (This is an adaptation of definition 11.1 to the modal propositional language.) Each sentence α of **PA** generates a propositional theory which is faithfully interpretable in **PA**, namely $Th_\alpha = \{A(p) \mid \mathbf{PA} \vdash A^*(\ulcorner\alpha\urcorner)\}$. Of course, this theory is closed under **L**-derivability: it is an **L**-*propositional theory.* A question much wider than the one discussed in the previous sections is, which **L**-propositional theories are faithfully interpretable in **PA** and other theories. This question was essentially solved by Shavrukov [1993b]:

5.1. Theorem. *An r.e.* **L**-*propositional theory* T *is faithfully interpretable in* **PA** *iff* T *is consistent and satisfies* the strong disjunction property (*i.e.,* $\Box A \in T$ *implies* $A \in T$, *and* $\Box A \vee \Box B \in T$ *implies* $\Box A \in T$ *or* $\Box B \in T$).

Note that faithfully interpretable theories in a finite number of propositional variables are necessarily r.e. The theorem was given a more compact proof and at the same time generalized to all r.e. theories extending $\mathbf{I\Delta_0} + \mathrm{EXP}$ by Zambella [1994]. If one applies the theorem to the minimal **L**-propositional theory, an earlier proved strengthening of Solovay's theorem (Artëmov [1980], Avron [1984], Boolos [1982], Montagna [1979], Visser [1980]) rolls out.

5.2. Corollary. (Uniform arithmetic completeness theorem) *There exists a sequence of arithmetic sentences* $\alpha_0, \alpha_1, \ldots$ *such that, for any* n *and modal formula* $A(p_0, \ldots, p_n)$, $\vdash_{\mathbf{L}} A$ *iff, under the arithmetic realization* * *induced by setting* $p_0^* = \alpha_0, \ldots, p_n^* = \alpha_n$, A^* *is provable in* **PA**.

Sets of modal formulas that are the true sentences under some realization are closed under modus ponens, but not necessarily under necessitation; such sets are generally not propositional theories in the above sense. Let us call a set T of modal formulas *realistic* if there exists a realization * such that A^* is true, for every $A \in T$. Moreover, we say that T is *well-specified* if, whenever $A \in T$ and B is a subformula of A, we also have $B \in T$ or $\neg B \in T$. Strannegård [1997] proves a result that generalizes both theorem 5.1 and Solovay's second arithmetic completeness theorem. We give a weak but easy to state version of it.

5.3. Theorem. *Let* T *be a well-specified r.e. set of modal formulas. Then* T *is realistic iff* T *is consistent with* **S**.

An even more general point of view than propositional theories is to look at the Boolean algebras of arithmetic theories with one additional operator representing formalized provability and, more specifically, at the ones generated by a sequence of sentences in the algebras of arithmetic. The algebras can be axiomatized equationally and are called *Magari-algebras* (after the originator R. Magari) or *diagonalizable algebras.* Of course, theorem 5.1 can be restated in terms of Magari-algebras. Shavrukov proved two beautiful and essential additional results concerning the

Magari-algebras of formal theories that cannot naturally be formulated in terms of propositional theories.

5.4. Theorem. (Shavrukov [1993a]) *The Magari algebras of* **PA** *and* **ZF** *are not isomorphic, and, in fact not elementarily equivalent* (Shavrukov [1997]).

The proof only uses the fact that **ZF** proves the uniform Σ_1-reflection principle for **PA**. A corollary of the theorem is that there is a formula of the second order propositional calculus that is valid in the interpretation with respect to **PA**, but not in the one with respect to **ZF**. Beklemishev [1996b] gives a different kind of example of such a formula for the two theories **PA** and $\mathbf{I}\Delta_0 + \mathrm{EXP}$.

5.5. Theorem. (Shavrukov [1997]) *The first order theory of the Magari algebra of* **PA** *is undecidable.*

Japaridze [1993] contains some moderately positive results on the decidability of certain fragments of (a special version of) this theory.

6. The extent of Solovay's theorems

An important feature of Solovay's theorems is their remarkable stability: a wide class of arithmetic theories and their provability predicates enjoys one and the same provability logic **L**. Roughly, there are three conditions sufficient for the validity of Solovay's results: the theory has to be (a) sufficiently strong, (b) recursively enumerable (a provability predicate satisfying Löb's derivability conditions is naturally constructed from a recursive enumeration of the set of its axioms), and (c) sound. Let us see what happens if we try to do without these conditions. The situation is fully investigated only w.r.t. the soundness condition.

Consider an arbitrary arithmetic r.e. theory T containing **PA** and a Σ_1 provability predicate $\mathrm{Pr}_T(x)$ for T. *Iterated consistency assertions* for T are defined as follows:

$$\mathrm{Con}^0(T) := \top; \quad \mathrm{Con}^{n+1}(T) := \mathrm{Con}(T + \mathrm{Con}^n(T)),$$

where, as usual, $\mathrm{Con}(T + \varphi)$ stands for $\neg\,\mathrm{Pr}_T(\ulcorner \neg\,\varphi \urcorner)$. In other words, $\mathrm{Con}^n(T)$ is (up to provable equivalence) the unique arithmetic realization of the modal formula $\neg\,\Box^n\bot$. We say that T is of *height* n if $\mathrm{Con}^n(T)$ is true and $\mathrm{Con}^{n+1}(T)$ is false in the standard model. If no such n exists, we say that T has *infinite height.*

In a sense, theories of finite height are close to being inconsistent and therefore can be considered as a pathology. The inconsistent theory is the only one of height 0. All Σ_1-sound theories have infinite height, but there exist Σ_1-unsound theories of infinite height. The theory $T + \neg\,\mathrm{Con}^n(T)$ is of height n, if T has infinite height. Moreover, for each consistent but Σ_1-unsound theory T and each $n > 0$, one can construct a provability predicate for T such that T is precisely of height n with respect to this predicate (Beklemishev [1989a]).

Let us call *the provability logic of* T the set of all modal formulas A such that $T \vdash (A)^*_T$, for all arithmetic realizations * with respect to Pr_T. The *truth provability logic of* T is the set of all A such that $(A)^*_T$ is true in the standard model, for all realizations *.

6.1. Theorem. (Visser [1981]) *For an r.e. arithmetic theory* T *containing* **PA**, *the provability logic of* T *coincides with*

1. $\mathbf{L}$, *if* T *has infinite height,*
2. $\{A \mid \Box^n\bot \vdash_{\mathbf{L}} A\}$, *if* T *is of height* $0 \leqslant n < \infty$.

Proof. By Solovay's construction, using the fact that the formula $\Box^n\bot$ is valid in Kripke-models of height $< n$, and only in such models. ⊣

Generalization of Solovay's second theorem is more interesting. To formulate it, we first introduce a convenient notation. For a set of modal formulas X, let $\mathbf{L}X$ denote the closure under modus ponens and substitution of the set X together with all theorems of $\mathbf{L}$. In this notation, Solovay's logic $\mathbf{S}$ is the same as $\mathbf{L}\{\Box A \to A\}$. The following two logics have been introduced by respectively Artёmov [1980] and Japaridze [1986,1988b] with different provability interpretations in mind (see the next section):

$$\mathbf{A} := \mathbf{L}\{\neg\Box^n\bot \mid n \in \mathbb{N}\}$$
$$\mathbf{D} := \mathbf{L}\{\neg\Box\bot,\ \Box(\Box A \vee \Box B) \to (\Box A \vee \Box B)\}$$

Obviously, $\mathbf{A} \subset \mathbf{D} \subset \mathbf{S}$. The following theorem gives an exhaustive description of all truth provability logics.

6.2. Theorem. (Beklemishev [1989a]) *For an r.e. arithmetic theory* T *containing* **PA**, *the truth provability logic for* T *coincides with*

1. $\mathbf{S}$ *iff* T *is sound,*
2. $\mathbf{D}$ *iff* T *is* Σ_1*-sound but not sound,*
3. $\mathbf{A}$ *iff* T *has infinite height but is not* Σ_1*-sound,*
4. $\mathbf{L}\{\Box^{n+1}\bot, \neg\Box^n\bot\}$ *iff* T *is of height* $0 \leqslant n < \infty$.

It is known that, at least in some natural cases, the other two sufficient conditions can also be considerably weakened. Boolos [1979] shows that the non-r.e. predicate of ω*-provability* (dual to ω-consistency) over Peano arithmetic has precisely the same provability logic as Peano arithmetic itself, i.e., $\mathbf{L}$. The same holds for the natural formalization of the Σ_{n+1}-complete predicate *"to be provable in* **PA** *together with all true* Π_n*-sentences"* (Smoryński [1985]). Solovay (see Boolos [1993b]) showed that $\mathbf{L}$ is also the logic of the Π^1_1-complete predicate of provability in analysis together with the ω-rule. However, no results to the effect that Solovay's theorems hold for broad classes of non-r.e. predicates are known. On the other hand, Solovay found an axiomatization of the provability logic of the predicate "to be valid in all transitive models of **ZF**", which happens to be a proper extension of **L** (see Boolos [1993b]).

If one is somewhat careful, Solovay's construction can be adapted to show that Solovay's theorems hold for all (r.e., sound) extensions of $\mathbf{I\Delta_0}+\mathrm{EXP}$ (de Jongh, Jumelet and Montagna [1991]). The two theorems formulated earlier in this section can be similarly generalized. However, the most intriguing question, whether Solovay's theorems hold for essentially weaker theories, such as Buss's S^1_2 or even S_2, remains open. This problem was thoroughly investigated by Berarducci and Verbrugge [1993], where the authors, in a modification of the Solovay construction, only succeeded in embedding particular kinds of Kripke-models into bounded arithmetic. The main technical difficulty lies in the fact that Solovay's construction, in its known variations, presupposes (at least, sentential) provable $\exists\Pi^b_1$-completeness of the theory in question. This property happens to fail for bounded arithmetic under some reasonable complexity-theoretic assumption (Verbrugge [1993a]). As far as we know, it is not excluded that the answer to the question what is the provability logic of S^1_2 also depends on difficult open problems in complexity theory.

Solovay's theorem does not hold in its immediately transposed form for Heyting's arithmetic **HA** (the intuitionistic pendant of **PA**). The logic of the provability predicate of **HA** with regard to **HA** certainly contains additional principles beyond the obvious intuitionistic version of **L**: the intuitionistic propositional calculus **IPC** plus $\Box(\Box A\to A)\to\Box A$. The situation has been discussed by Visser in several papers (Visser [1985,1994], Visser et al. [1995]). It is unknown what the real logic is, for all we know it may even be complete Π^0_2. In any case it contains the additional principles:

- $\Box\neg\neg\Box A\to\Box\Box A$
- $\Box(\neg\neg\Box A\to\Box A)\to\Box\Box A$
- $\Box(A\vee B)\to\Box(A\vee\Box B)$ (Leivant's Principle[2])

But this is not an exhaustive list. It is well possible that the logic of the binary operator of Σ*-preservativity* over **HA** is better behaved than the logic of provability on its own: **PA**+A Σ-preserves **PA**+B, if from each Σ_1-sentence from which A follows in **PA**, B is also **PA**-derivable. In classical systems Σ-preservativity is definable in terms of Π_1-conservativity (see sections 12 to 14 for that concept) and vice versa, but constructively this is the proper version to study (see also Visser [1997]).

7. Classification of provability logics

One of the important methodological consequences of Gödel's second incompleteness theorem is the fact that, in general, it is necessary to distinguish between a *theory* T under study, and a *metatheory* U in which one reasons about the properties

[2]added as one of the "stellingen" (theses) to Leivant's Ph.D. thesis, Amsterdam, 1979

of T. Perhaps, the most natural choice of U is the full true arithmetic **TA**, the set of all formulas valid in the standard model, yet this is not the only possibility. Other meaningful choices could be T itself, or the reader's favorite minimal fragment of arithmetic, e.g., $\mathbf{I\Sigma_1}$. The separate role of the metatheory is emphasized in the definition of *provability logic of a theory T relative to a metatheory U* that was suggested independently by Artëmov [1980] and Visser [1981].

Let T and U be arithmetic theories extending $\mathbf{I\Delta_0} + \mathrm{EXP}$, T r.e. and U not necessarily r.e. The provability logic of T relative to, or simply *at*, U is the set of all modal formulas φ such that $U \vdash (\varphi)^*_T$, for all arithmetic realizations * (denoted $\mathsf{PRL}_T(U)$). Intuitively, $\mathsf{PRL}_T(U)$ expresses those principles of provability in T that can be verified my means of U. As a set of modal formulas, $\mathsf{PRL}_T(U)$ contains $\mathbf{L}$ and is closed under modus ponens and substitution, i.e., is a (not necessarily normal) modal logic extending $\mathbf{L}$.

Solovay's theorems can be restated as saying that, if T is a sound theory, then $\mathsf{PRL}_T(T) = \mathbf{L}$ and $\mathsf{PRL}_T(\mathbf{TA}) = \mathbf{S}$. A modal logic is called *arithmetically complete*, if it has the form $\mathsf{PRL}_T(U)$, for some T and U. The problem of obtaining a reasonable general characterization of arithmetically complete modal logics has become known as 'the classification problem', and was one of the early motivating problems for the Moscow school of provability logic founded by Artëmov.

The solution to this problem is the joint outcome of the work of several authors Artëmov [1980], Visser [1981,1984], Artëmov [1985b], Japaridze [1986,1988b], Beklemishev [1989a]. Artëmov [1980] (applying the so-called *uniform* version of Solovay's theorem, corollary 5.2) showed that all logics of the form $\mathbf{L}X$, for any set X of variable-free modal formulas, are arithmetically complete. In Artëmov [1985b], he showed that such extensions are exhausted by the following two specific families of logics:

$$\mathbf{L}_\alpha := \mathbf{L}\{F_n \mid n \in \alpha\},$$
$$\mathbf{L}^-_\beta := \mathbf{L}\{\textstyle\bigvee_{n \notin \beta} \neg F_n\},$$

where $\alpha, \beta \subseteq \mathbb{N}$, β is cofinite, and F_n denotes the formula $\Box^{n+1}\bot \to \Box^n\bot$. Some of the provability logics introduced above have this form: $\mathbf{L} = \mathbf{L}_\emptyset$, $\mathbf{A} = \mathbf{L}_{\mathbb{N}}$, $\mathbf{L}\{\Box^{n+1}\bot, \neg\Box^n\bot\} = \mathbf{L}^-_{\mathbb{N}\setminus\{n\}}$.

The families $\mathbf{L}_\alpha$ and $\mathbf{L}^-_\beta$ are ordered by inclusion precisely as their indices, and $\mathbf{L}_\alpha$ is included in $\mathbf{L}^-_\alpha$ for cofinite α. Note that the logics $\mathbf{L}^-_\beta$ are not contained in $\mathbf{S}$, and therefore correspond to unsound metatheories U if the theory T is sound. Visser [1984] showed that these are the only arithmetically complete logics not contained in $\mathbf{S}$. Artëmov [1985b] improved this by actually reducing the classification problem to the interval between $\mathbf{A}$ and $\mathbf{S}$. Any arithmetically complete logic ℓ from this interval generates a family of different arithmetically complete logics of the form $\ell \cap \mathbf{L}^-_\beta$, for cofinite β, and Artëmov showed that such logics, together with the families $\mathbf{L}_\alpha$ and $\mathbf{L}^-_\beta$, exhaust all arithmetically complete ones.

Japaridze [1986,1988b] found a new provability logic within the interesting interval by establishing that $\mathsf{PRL}_{\mathbf{PA}}(\mathbf{PA} + \omega\text{-Con}(\mathbf{PA})) = \mathbf{D}$, where $\omega\text{-Con}(\mathbf{PA})$ denotes the formalized ω-consistency of **PA**. The final step was made by Beklemishev [1989a], who showed that **D** is the only arithmetically complete modal logic within the interval between **A** and **S**, thus completing the classification. This result was also crucial for the proof of theorem 6.2 of the previous section. We denote $\mathbf{S}_\beta := \mathbf{S} \cap \mathbf{L}_\beta^-$, $\mathbf{D}_\beta := \mathbf{D} \cap \mathbf{L}_\beta^-$ and formulate the resulting theorem.

7.1. Theorem. (Classification Theorem, Beklemishev [1989a]) *The arithmetically complete modal logics are exhausted by the four families:* $\mathbf{L}_\alpha$, $\mathbf{L}_\beta^-$, $\mathbf{S}_\beta$, $\mathbf{D}_\beta$, *for* $\alpha, \beta \subseteq \mathrm{N}$, β *cofinite.*

From a purely modal-logical point of view, the meaning of the classification theorem is that only very few extensions of **L** are arithmetically complete. The word 'few' must not be understood here in terms of cardinality, because the family $\mathbf{L}_\alpha$ has already the cardinality of the continuum, but rather less formally. E.g., there is a continuum of different modal logics containing **A** (Artëmov [1985b]), but only four of them are arithmetically complete. Similar observations hold for other natural intervals in the lattice of extensions of **L**.

All arithmetically complete logics have nice axiomatizations, and are generally well-understood, although most of them are not normal. An adequate Kripke-type semantics is known for all arithmetically complete logics: for $\mathbf{L}_\alpha$ and $\mathbf{L}_\beta^-$ it can be formulated in terms of the height of the tree-like models for **L**; the so-called *tail models* for **S** were suggested independently by Boolos [1981] and Visser [1984]; a similar kind of semantics for **D** was produced by Beklemishev [1989b]. A corollary is that all logics of the families $\mathbf{S}_\beta$, $\mathbf{D}_\beta$, and $\mathbf{L}_\beta^-$ are decidable, and a logic of the form $\mathbf{L}_\alpha$ is decidable, iff its index α is a decidable subset of $\mathbb{N}$, i.e., iff it has a decidable axiomatization.

The fact that arithmetically complete logics are scarce tells us that inference 'by arithmetic interpretation' considerably strengthens the usual modal-logical consequence relation. In fact, the classification theorem can be understood as a classification of modally expressible arithmetic schemata. Familiar examples of such schemata are: the *local reflection principle* for T, that is the schema $\mathrm{Pr}_T(\ulcorner A \urcorner) \to A$ for all arithmetic sentences A, which is expressed by the modal formula $\Box p \to p$; ω *times iterated consistency* of T, which is expressed by $\{\neg \Box^n \bot \mid n \in \mathbb{N}\}$; the *local* Σ_1*-reflection principle*, which can be expressed by the axioms of **D**, etc. In general, a schema is *modally expressible* over a theory T, if it is deductively equivalent to the set of all arithmetic realizations with respect to T of a family of modal formulas. The classification theorem gives us a complete description of all modally expressible arithmetic schemata: they precisely correspond to axiomatizations of arithmetically complete modal logics.

It is very surprising that *all* such schemata are built up from instances of the local reflection principle, sometimes a little twisted by axioms of $\mathbf{L}_\beta^-$ type. This can be considered as a theoretical justification of the 'empirical' rule that in the study of provability all reasonable metatheories happen to be equivalent to some version of the reflection principle.

We round up the discussion of the classification theorem by giving some examples for natural pairs (theory, metatheory) of fragments of arithmetic.

$$
\begin{aligned}
\mathsf{PRL}_{\mathbf{I\Sigma_1}}(\mathbf{PA}) &= \mathbf{S}, \\
\mathsf{PRL}_{\mathbf{I\Sigma_1}}(\mathbf{I\Sigma_n}) &= \mathbf{D}, \quad \text{for } n>1, \\
\mathsf{PRL}_{\mathbf{I\Delta_0+}\mathrm{EXP}}(\mathbf{PRA}) &= \mathbf{D}, \\
\mathsf{PRL}_{\mathbf{I\Sigma_1}}(\mathbf{I\Sigma_1} + \mathrm{Con}(\mathbf{PA})) &= \mathbf{A}, \\
\mathsf{PRL}_{\mathbf{PRA}}(\mathbf{I\Sigma_1}) &= \mathbf{L}.
\end{aligned}
$$

All such results follow easily from the classification theorem and the usual proof-theoretic information about the provability of reflection principles for the theories in question. E.g., $\mathbf{I\Sigma_1} + \mathrm{Con}(\mathbf{PA})$ obviously contains ω-times-iterated consistency for $\mathbf{I\Sigma_1}$, but, being a finite Π_1-axiomatized extension of $\mathbf{I\Sigma_1}$, cannot contain the local Σ_1-reflection principle for $\mathbf{I\Sigma_1}$ (by Löb's theorem). Hence, $\mathsf{PRL}_{\mathbf{I\Sigma_1}}(\mathbf{I\Sigma_1} + \mathrm{Con}(\mathbf{PA}))$ contains $\mathbf{A}$ but does not contain $\mathbf{D}$. The classification theorem implies that, in this case, it must be $\mathbf{A}$.

8. Bimodal and polymodal provability logics

The fact that all reasonable theories have one and the same — Löb's — provability logic is, in a sense, a drawback: it means that the provability logic of a theory cannot distinguish between most of the interesting properties of theories, such as e.g., finite axiomatizability, reflexivity, etc. In fact, by Visser's theorem 6.1, the only recognizable characteristic of a theory is its height, and the situation does not become much better even if one considers truth provability logics.

One obvious way to increase the expressive power of the modal language is to consider provability operators in several different theories simultaneously, which naturally leads to *bi-* and *polymodal provability logic*. It turns out that the modal description of the joint behaviour of two or more provability operators is, in general, a considerably more difficult task than the calculation of unimodal provability logics. There is no single system that can justifiably be called *the* bimodal provability logic — rather, we know particular systems for different natural pairs of provability operators, and none of those systems occupies any privileged place among the others. Moreover, the numerous isolated results accumulated in this area, so far, give us no clue as to a possible general classification of bimodal provability logics for pairs of sound r.e. theories. This problem remains one of the most challenging open problems in provability logic. A short survey of the state of our knowledge in this field is given below.

The language $\mathcal{L}(\Box, \triangle)$ of bimodal provability logic is obtained from that of propositional calculus by adding two unary modal operators $\Box$ and $\triangle$. Let (T, U) be a pair of arithmetic r.e. theories, taken together with some fixed canonical Σ_1 provability predicates Pr_T and Pr_U. An *arithmetic realization* $(\cdot)^*_{T,U}$ *with respect to* (T, U) is a mapping of modal formulas to arithmetic sentences that commutes with

Boolean connectives and translates $\Box$ as provability in T and $\triangle$ as that in U:

$$(\Box A)^*_{T,U} = \mathrm{Pr}_T(\ulcorner (A)^*_{T,U} \urcorner), \qquad (\triangle A)^*_{T,U} = \mathrm{Pr}_U(\ulcorner (A)^*_{T,U} \urcorner).$$

The provability logic for (T, U), denoted $\mathsf{PRL}_{T,U}$, is the collection of all $\mathcal{L}(\Box, \triangle)$-formulas A such that $T \vdash (A)^*_{T,U}$ and $U \vdash (A)^*_{T,U}$, for every arithmetic realization *. In general, as in the unimodal case, one can consider bimodal provability logics for (T, U) relative to an arbitrary metatheory V (where $\mathsf{PRL}_{T,U}$ corresponds to $V = T \cap U$).

Not too much can a priori be said about $\mathsf{PRL}_{T,U}$, for arbitrary T and U. Clearly, $\mathsf{PRL}_{T,U}$ is closed under modus ponens, substitution and the $\Box$- and $\triangle$-necessitation rules. Moreover, $\mathsf{PRL}_{T,U}$ has to be a (normal) extension of the bimodal system **CS**, given by the axioms and rules of **L** formulated separately for $\Box$ and $\triangle$, and by the obvious mixed principles:

$$\Box A \to \triangle\Box A, \qquad \triangle A \to \Box\triangle A.$$

By Solovay's theorem we know that, whenever both T and U have infinite height, the fragment of $\mathsf{PRL}_{T,U}$ in the language $\mathcal{L}(\Box)$ of $\Box$ alone, as well as the one in the language of $\triangle$, actually coincides with **L**. Using the uniform version of Solovay's theorem, Smoryński [1985] showed that **CS** is the minimal bimodal provability logic, i.e., it coincides with $\mathsf{PRL}_{T,U}$ for a certain pair of finite extensions T, U of Peano arithmetic. Beklemishev [1992] showed that there is even a pair of provability predicates for Peano arithmetic itself for which the corresponding bimodal provability logic coincides with **CS**. Such predicates can be called *independent* in the sense that they 'know' as little about each other as is possible in principle. It should be noted however that, neither the theories in Smoryński's example, nor the independent provability predicates are natural — they are constructed by a tricky diagonalization. Thus, we are in the interesting situation that the bimodal logic **CS**, which structurally occupies a privileged place among the provability logics, does not correspond to any known *natural* pair of theories.

Deeper structural information on bimodal provability logics is provided by the Classification Theorem 7.1 for arithmetically complete modal logics. With every bimodal logic ℓ we can associate its *type*:

$$(\ell)^0 := \{A \in \mathcal{L}(\Box) \mid \ell \vdash \triangle A\}.$$

An easy analysis then shows that $(\cdot)^0$ surjectively maps normal extensions of **CS** onto the lattice of the unimodal logics containing **L**. Under the assumption of Σ_1-soundness of $T \cap U$ we obviously have:

$$\mathsf{PRL}_T(U) = (\mathsf{PRL}_{T,U})^0.$$

The classification theorem not only shows that not every type (of unimodal logic) is materialized as that of a bimodal provability logic, but also gives us a complete description of all such possible types.

Besides the general observations above, a number of particular bimodal provability logics for natural pairs of theories are known. These logics cover most of the examples of pairs of arithmetic theories that come to mind, but, unfortunately, are far from being an exhaustive list of all bimodal provability logics.

The best known system is the logic $\mathsf{PRL}_{\mathbf{PA},\mathbf{ZF}}$ discovered by Carlson [1986], and independently (with a different interpretation in mind) by Montagna [1987]. This logic can be axiomatized over **CS** by the principle of *essential reflexivity*

$$\triangle(\Box A \to A).$$

It is the *only* bimodal provability logic of type **S** and a maximal one among the bimodal logics for pairs of sound theories. In other words, $\mathsf{PRL}_{T,U} = \mathsf{PRL}_{\mathbf{PA},\mathbf{ZF}}$, whenever the theories T, U are sound and U contains the local reflection principle for T.

Furthermore, we know two natural bimodal provability logics of type **D**, introduced by Beklemishev [1996a]. The first one corresponds to pairs of theories (T, U) such that U is a finite extension of T that proves the local Σ_1-reflection principle for T. Typical examples are the pairs $(\mathbf{I\Delta_0} + \mathrm{EXP}, \mathbf{I\Delta_0} + \mathrm{SUPEXP})$, $(\mathbf{I\Sigma_m}, \mathbf{I\Sigma_n})$, for $n > m \geqslant 1$, etc. The logic can be axiomatized over **CS** by the *monotonicity* axiom $\Box A \to \triangle A$ and the schema

$$\triangle(\Box S \to S),$$

where S is an arbitrary (possibly empty) disjunction of formulas of the form $\Box B$ and $\triangle B$.[3]

The second one corresponds to Π_1-essentially reflexive (see definition 12.3) extensions of theories of bounded arithmetic complexity such as e.g., $(\mathbf{I\Delta_0} + \mathrm{EXP}, \mathbf{PRA})$, $(\mathbf{I\Sigma_n}, \mathbf{I\Sigma^R_{n+1}})$ for $n \geqslant 1$, where $\mathbf{I\Sigma^R_k}$ is defined like $\mathbf{I\Sigma_k}$ but with the induction for Σ_k-formulas formulated as a rule. The corresponding provability logic can be axiomatized over **CSM** by the Π_1*-essential reflexivity* schema

$$\triangle A \to \triangle(\Box(A \to S) \to S),$$

where S is as before.

We also know two natural provability logics of type **A** (Beklemishev [1994]). The first system corresponds to pairs of theories (T, U) such that U is an extension of T by finitely many Π_1-sentences and proves ω-times-iterated consistency of T, such as e.g., the pairs $(\mathbf{PA}, \mathbf{PA} + \mathrm{Con}(\mathbf{ZF}))$, $(\mathbf{I\Sigma_1}, \mathbf{I\Sigma_1} + \mathrm{Con}(\mathbf{I\Sigma_2}))$, etc. This logic can be axiomatized over **CSM** by the principle

$$\text{(P)} \qquad \triangle A \to \Box(\triangle\bot \vee A),$$

valid for all Π_1-axiomatizable extensions of theories, together with the schema

$$\triangle\neg\Box^n\bot, \qquad n \geqslant 1.$$

[3]In the following, **CS** together with the monotonicity axiom will be denoted **CSM**.

The second system corresponds to reflexive Π_1-axiomatizable extensions of theories, such as e.g., $(\mathbf{PA}, \mathbf{PA} + \{\mathrm{Con}^n(\mathbf{PA}) \mid n \geqslant 1\})$, $(\mathbf{I\Sigma_1}, \mathbf{I\Sigma_1} + \{\mathrm{Con}(\mathbf{I\Sigma_n}) \mid n \geqslant 1\})$. It can be axiomatized over **CSM** plus (P) by the *reflexivity* axiom

$$\triangle A \to \triangle \Diamond A.$$

Finally, we know by Beklemishev [1996a] a natural system of type **L** that corresponds to finite extensions of theories of the form $(T, T + A)$, where both $T + \varphi$ and $T + \neg\varphi$ are conservative over T with respect to Boolean combinations of Σ_1-sentences. Examples of such pairs are $(\mathbf{PRA}, \mathbf{I\Sigma_1})$, $(\mathbf{I\Sigma_n^R}, \mathbf{I\Sigma_n})$, for $n \geqslant 1$, and others. The logic is axiomatized over **CSM** by the $\mathcal{B}(\Sigma_1)$*-conservativity* schema

$$\triangle B \to \Box B,$$

where B denotes an arbitrary Boolean combination of formulas of the form $\Box C$ and $\triangle C$.

The six bimodal logics described above essentially exhaust all nontrivial cases for which natural provability logics have explicitly been characterized. It is worth mentioning that all these systems are decidable, and a suitable Kripke-style semantics is known for each of them. Smoryński [1985] contains an extensive treatment of $\mathsf{PRL}_{\mathbf{PA},\mathbf{ZF}}$ including proofs of three arithmetic completeness theorems due to Carlson. These theorems are extended by Strannegård [1997] to the setting of r.e. sets of bimodal formulas (as discussed in section 5). Visser [1995] presents a beautiful approach to Kripke semantics for bimodal provability logics. Beklemishev [1994, 1996a] gives a detailed survey of the current state of the field.

Apart from describing the joint behaviour of two 'usual' provability predicates, each of them being separately well enough understood, bimodal logic has been successfully used for the analysis of some nonstandard, not necessarily r.e., concepts of provability. The systems emerging from such an analysis often have not so much in common with **CS**, although different 'bimodal analyses' do share common technical ideas.

As early as 1986, Japaridze [1986,1988b] characterized the bimodal logic of provability and ω*-provability* (dual to ω-consistency) in Peano arithmetic. Later his study was simplified and further advanced by Ignatiev [1993a] and Boolos [1993b,1993a], who, among other things, showed that the same system corresponds to some other, so-called *strong*, concepts of provability (taken jointly with the usual one). Other examples of strong provability predicates are the Σ_{n+1}-complete *provability from all true arithmetic* Π_n*-sentences*, for $n \geqslant 1$, and the Π^1_1-complete *provability under the* ω*-rule in analysis*.

Japaridze's bimodal logic can be axiomatized by the axioms and rules of **L**, formulated separately for $\Box$ and for $\triangle$, the monotonicity principle $\Box A \to \triangle A$, and an additional Π_1*-completeness* principle

$$\Diamond A \to \triangle \Diamond A,$$

which reflects in so far as that is possible that $\triangle$ is strong enough to prove all true Π_1-sentences (if $\Box$ is the usual r.e. provability predicate and $\triangle$ a strong provability predicate). Japaridze's logic is decidable and has a reasonable Kripke semantics. An extensive treatment of Japaridze's logic is given in Boolos [1993b].

Bimodal analysis of other unusual provability concepts has been undertaken by Visser [1989,1995] and Shavrukov [1991,1994]. Using the work of Guaspari and Solovay [1979], Shavrukov [1991] found a complete axiomatization of the bimodal logic of the usual and *Rosser's provability predicate* for Peano arithmetic (see also section 9). It is worth noting that Rosser's provability predicate, although numerating (externally) the same theory as the usual one, has a very different modal behaviour; e.g., Rosser consistency of **PA** is a provable fact, but on the other hand, Rosser's provability predicate is not provably closed under modus ponens. Shavrukov [1994] characterizes the logic of the so-called *Feferman provability predicate.* This work was preceded by Visser [1989,1995], where the concept of *provability in* **PA** *from 'nonstandardly finitely many' axioms* and some other unusual provability concepts were bimodally characterized. These systems were motivated by their connections with interpretability logic, but another motivation originates with Jeroslow and Putnam who studied the Rosser and Feferman style systems as 'experimental' systems: their self-correcting behaviour is supposed to be closer to the way humans reason. Studying ordinary provability and self-correcting provability can provide a good heuristic for appreciating the differences between both kinds of systems.

A final example of such an analysis of an unusual proof predicate by the development of a bimodal logic was Lindström [1994]'s analysis of *Parikh provability,* i.e., the proof predicate that allows $\Box A/A$ as a rule of inference.

Additional early results in bimodal logic, e.g., a bimodal analysis of the so-called Mostowski operator, can be found in Smoryński [1985].

Many results in bimodal provability logic can be generalized to *polymodal logic.* Such a generalization is particularly natural in the modal-logical study of progressions of theories, a topic in proof theory that goes as far back as the work of Turing [1939]. From the modal-logical point of view, however, such a generalization, in all known cases, does not lead to any essentially new phenomena. Roughly, the resulting systems happen to be direct sums of their bimodal fragments; therefore we shall not go into the details.

Polymodal analogues are known for Japaridze's bimodal logic (modalities, indexed by natural numbers n, correspond to the operators *to be provable from all true* Π_n*-sentences*), and for natural provability logics due to Carlson and Beklemishev. Here, the modal operators correspond to the theories of the original Turing-Feferman progressions of transfinitely iterated reflection principles, and thus, are indexed by ordinals for some constructive system of ordinal notation, say, the natural one up to ϵ_0. Iterating full reflection leads to the polymodal analogue of $\mathsf{PRL}_{\mathbf{PA},\mathbf{ZF}}$, and transfinitely iterated consistency leads to a natural polymodal analogue of **A**-type provability logics (Beklemishev [1991,1994]).

9. Rosser orderings

To discuss Rosser sentences and more generally the so-called Rosser provability predicate in a modal context, Guaspari and Solovay [1979] enriched the modal language by adding, for each $\Box A$ and $\Box B$, the formulas $\Box A \prec \Box B$ and $\Box A \preccurlyeq \Box B$, with as their arithmetic realizations the Σ_1-sentences "A^* is provable by a proof that is smaller than any proof of B^*", and "A^* is provable by a proof that is smaller than or equal to any proof of B^*" (so-called *witness comparison formulas*). They axiomatized modal logics $\mathbf{R}^-$ and $\mathbf{R} = \mathbf{R}^- +$ the rule $\Box A/A$, and gave an arithmetic completeness result for $\mathbf{R}$. In this arithmetic completeness result they did have to allow arbitrary *standard* provability predicates in the arithmetic realizations however, i.e., arbitrary provability predicates satisfying the three Löb conditions. Shavrukov [1991] (see also the end of section 8) showed that this restriction can be dropped when one restricts the contexts for the new operator to $\Box A \prec \Box\neg A$ (the *Rosser provability predicate*, for short: $\Box^R A$), and de Jongh and Montagna [1991] showed that, allowing formulas with free variables as arithmetic substitutions leads to $\mathbf{R}^-$ as the arithmetically complete system. Guaspari and Solovay [1979] also showed that for some standard provability predicates all *Rosser sentences* (i.e., sentences α such that $\mathbf{PA} \vdash \alpha \leftrightarrow (\mathrm{Pr}_{\mathbf{PA}}(\ulcorner\neg\alpha\urcorner) \prec \mathrm{Pr}_{\mathbf{PA}}(\ulcorner\alpha\urcorner))$ are equivalent, and that for some other standard provability predicates this is not the case. This leaves open the question whether a reasonable notion of *usual* proof predicate can be defined for which the question "Is the Rosser sentence unique?" does have a definite answer. Hence also, uniqueness of fixed points is not provable in $\mathbf{R}$. Finally, they showed that also the existence part of the fixed point theorem fails for $\mathbf{R}$. Simpler proofs for the completeness theorems were given in de Jongh [1987] and Voorbraak [1988].

There are connections between this work in provability logic and speed up. First, de Jongh and Montagna [1988,1989] gave a new simpler proof of Parikh [1971]'s theorem that, for any provably recursive function g there is a sentence α provable in $\mathbf{PA}$ such that $\mathbf{PA}$ proves $\mathrm{Pr}_{\mathbf{PA}}(\ulcorner\alpha\urcorner)$ by a much shorter proof in the sense of g ($a <_g b$ iff $g(a) < b$) than it proves A itself. In de Jongh and Montagna [1988] this was done for g the identity function by showing that $\vdash_{\mathbf{R}} \boxdot(p \leftrightarrow (\Box\Box p \prec \Box p)) \to p$ ($\Box\Box p \prec \Box p$ *has only provable fixed points in* $\mathbf{R}$). The result shows that any fixed point α in $\mathbf{PA}$ of the arithmetic formula $\mathrm{Pr}_{\mathbf{PA}}[\mathrm{Pr}_{\mathbf{PA}}(x)] \prec \mathrm{Pr}_{\mathbf{PA}}(x)$[4] is provable in $\mathbf{PA}$, and the shortest proof of $\mathrm{Pr}_{\mathbf{PA}}(\ulcorner\alpha\urcorner)$ is shorter than the one of α. In the paper general conditions were given under which formulas have only provable fixed points in $\mathbf{R}$. In de Jongh and Montagna [1989] a Guaspari-Solovay theory of $\prec_g$ is developed for the notion "much shorter in the sense of g". Under reasonable conditions on g the resulting modal theory $\mathbf{R}^0_{\mathbf{g}}$ is not dependent on g. Parikh's theorem can then be proved in this setting. The theory of provable fixed points was extended to this setting. In his review of these and some consecutive papers Beklemishev [1993b] rightly remarked that the changing of the orders of the proofs in Guaspari-Solovay style interferes with the order induced by the function g and makes some of the

[4]for the meaning of [...] see notation 12.2.

results somewhat less clear than one might wish.

Montagna [1992] applied the results on provable fixed points in a study of *metamathematical rules*, i.e., rules like $\mathrm{Pr}_T(\ulcorner\alpha\urcorner)/\alpha$ that can be considered as realizations of modal-logical rules (in case: $\Box A/A$). He classified these rules into two types: rules giving only polynomial speed up in proofs in arithmetic, and rules giving a superexponential speed up. In Hájek, Montagna and Pudlak [1993] it was shown that the rule $\Box A/A$ is maximally powerful among these metamathematical rules in the sense that the use of any of them can be polynomially simulated by $\Box A/A$. Moreover, in that paper natural examples of statements of which the proof is superexponentially shortened by the above rule are given.

10. Logic of proofs

A provability reading of the modality $\Box$ as *"is (informally) provable"* was an intended semantics for the classical system **S4** of propositional modal logic (see end of section 2) since Gödel's paper (Gödel [1933]). However, as we have seen, the straightforward interpretation of $\Box F$ as $\mathrm{Pr}(\ulcorner F\urcorner)$ leads to the logics **L** and S which are incompatible with **S4**. The reflexivity principle $\Box F \to F$ fails in **L**, and the necessitation rule fails in **S**. Nevertheless, an interesting interpretation of the **S4**-modality as formal provability is possible. One can have the reflexivity principle as well as the necessitation rule if one incorporates into the modal language machinery to keep all proofs "real", i.e., given by actual natural numbers and not quantifying over them as in the provability predicate. Artëmov succeeded in doing this by replacing the quantifiers by a kind of Skolem functions in his logic of proofs **LP** (Artëmov [1994,1995]).

The language of **LP** contains besides the usual Boolean constants, connectives and sentence variables, *proof variables* $x_0, \ldots, x_n, \ldots$, *proof axiom constants* $a_0, \ldots, a_n, \ldots$, function symbols: monadic !, and binary $+$ and $\times$, and finally the modal operator symbol $[\![\ ]\!](\ )$.

Terms and formulas are defined in the natural way: proof variables and axiom constants are terms; sentence variables and Boolean constants are formulas; whenever s and t are terms $!t, (s+t), (s \times t)$ are again terms, Boolean connectives behave conventionally, and for t a term and F a formula, $[\![t]\!]\, F$ is a formula. We will write $s \cdot t$ or even st instead of $(s \times t)$ and skip parentheses when convenient. A term is *ground* if is does not contain variables. The system $\mathbf{LP}_{AS}$ has as its axioms all formulas of the of the forms below, and as its only rule *modus ponens*:

A0. The tautologies in the language of **LP**,

A1. $[\![t]\!]\, F \to F$ "reflexivity"

A2. $[\![t]\!]\, (F \to G) \to ([\![s]\!]\, F \to [\![ts]\!]\, G)$ "application"

A3. $[\![t]\!]\, F \to [\![!t]\!]\, [\![t]\!]\, F$ "proof checker"

A4. $[\![s]\!]\, F \to [\![s+t]\!]\, F, \quad [\![t]\!]\, F \to [\![s+t]\!]\, F$ "choice"

AS. A finite set of formulas of the form $[\![c]\!]\, A$, where c is an axiom constant, and A is an axiom A0-A4 "axiom specification"

The system **LP** is the generic name for the $\mathbf{LP}_{AS}$'s of the various axiom specifications *AS*. The intended understanding of **LP** is as a logic of operations on proofs, where $[\![t]\!]F$ stands for *"t is a code for a proof of F"*. For the usual Gödel proof predicate $\mathrm{Proof}(x,y)$ in **PA** there are provably recursive functions from codes of proofs to codes of proofs corresponding to $\times$ and !: $\times$ stands for an operation on proof sequences which realizes the *modus ponens* rule in arithmetic, and ! is a proof checker operation as it appears in the proof of the second Gödel Incompleteness theorem. The usual proof predicate has a natural nondeterministic version $\mathrm{PROOF}(x,y)$ here called *standard nondeterministic proof predicate*: *"x is a code of a derivation containing a formula with a code y"*. The predicate PROOF already has all three operations of the **LP**-language: the operation $s+t$ is in its case just the concatenation of the (nondeterministic) proofs s and t.

The system **LP** reminds one of propositional dynamic logic (see e.g., Harel [1984]), but is really quite different in character, since the modalities $[\![t]\!](\cdot)$ do not satisfy the property $[\![t]\!](p\to q)\to([\![t]\!]p\to[\![t]\!]q)$ in **LP**. This makes the logic **LP** nonnormal and not a polymodal logic in the sense of section 8. Nevertheless, the entire variety of labeled modalities in **LP** can simulate **S4**. For example, the necessitation rule $F/\Box F$ of normal modal logics has its constructive counterpart in **LP**: if $\mathbf{LP}\vdash F$, then $\mathbf{LP}\vdash[\![t]\!]F$ for some ground term t. In general, let F^o be the result of substituting $\Box$ for all occurrences of $[\![t]\!]$ in F, and $X^o=\{F^o\mid F\in X\}$ for any set X of **LP**-formulas. It is easy to see that **LP** is sound with respect to **S4**: $(\mathbf{LP})^o\subseteq\mathbf{S4}$. The converse inclusion $\mathbf{S4}\subseteq(\mathbf{LP})^o$ turns out to be valid as well: by an **LP**-*realization* $r=r(AS)$ of a modal formula F we mean

1. An assignment of **LP**-terms to all occurrences of $\Box$ in F,
2. a choice of an axiom specification *AS*.

Under F^r we denote the image of F under the realization r. Positive and negative occurrences of modality in a formula and a sequent are defined in the usual way. A realization r is *normal* if all negative occurrences of $\Box$ are realized by proof variables.

10.1. Theorem. (Artëmov [1995]) *If* $\vdash_{\mathbf{S4}}F$, *then* $\vdash_{\mathbf{LP}_{AS}}F^r$ *for some axiom specification AS and some normal realization* $r=r(AS)$.

The proof of the theorem provides an algorithm which, for a given cutfree derivation $\mathcal{T}$ in **S4**, assigns **LP**-terms to all appearances of the modality in $\mathcal{T}$.

Let us agree to use a new function symbol $\iota z\,\varphi(z)$ for any arithmetic formula $\varphi(z)$. A formula $\psi(\iota z\,\varphi(z))$ is now supposed to be decoded in the usual way (see van Dalen [1994]) as a pure arithmetic formula $\psi(\iota z\,\varphi(z))^-$: for the innermost occurrence of $\iota z\,\varphi(z)$ put $\psi(\iota z\,\varphi(z))^-$ to be $\exists z(\varphi(z)\wedge\psi(z))$, and then iterate this procedure when needed. Under $\mu z\,\varphi$ we understand the ι-term determined by the formula $\varphi(z)\wedge\forall v<z\neg\varphi(v)$. An arithmetic formula φ is *provably* Δ_1 iff both φ and $\neg\varphi$ are provably Σ_1. A term $\mu z\,\varphi$ is *provably recursive* iff φ is provably Σ_1 and *provably total* iff $\mathbf{PA}\vdash\exists z\,\varphi(z)$. A *closed recursive term* is a provably total and provably recursive term $\mu z\,\varphi$ such that φ contains no free variables other than z. Closed recursive terms represent all provably recursive names for natural numbers. We have to make

these distinctions, since some operations on proofs, e.g., the proof checker !, really depend on the name of the argument, not only on its value.

A *proof predicate* is a provably Δ_1-formula $\mathrm{Prf}(x,y)$ such that, for all φ, if $\mathbf{PA} \vdash \varphi$, then, for some $n \in \omega$, $\mathrm{Prf}(n, \ulcorner\varphi\urcorner)$ holds. A proof predicate $\mathrm{Prf}(x,y)$ is here called *normal* if

1. For every proof k, the set of corresponding theorems is finite and the function $T(k) =$ *the code of the set* $\{l \mid \mathrm{Prf}(k,l)\}$ is provably total recursive,

2. For any finite set X of codes of theorems of **PA** there exists a natural number n such that $X \subseteq T(n)$.

For each normal proof predicate Prf there are provably recursive terms $m(x,y)$, $a(x,y)$, $c(x)$ such that for all closed recursive terms s,t and for all arithmetic formulas φ, ψ the following formulas are valid:

$$\mathrm{Prf}(s, \ulcorner\varphi \to \psi\urcorner) \wedge \mathrm{Prf}(t, \ulcorner\varphi\urcorner) \to \mathrm{Prf}(m(s,t), \ulcorner\psi\urcorner)$$

$$\mathrm{Prf}(s, \ulcorner\varphi\urcorner) \to \mathrm{Prf}(a(s,t), \ulcorner\varphi\urcorner), \qquad \mathrm{Prf}(t, \ulcorner\varphi\urcorner) \to \mathrm{Prf}(a(s,t), \ulcorner\varphi\urcorner)$$

$$\mathrm{Prf}(t, \ulcorner\varphi\urcorner) \to \mathrm{Prf}(c(\ulcorner t\urcorner), \ulcorner\mathrm{Prf}(t, \ulcorner\varphi\urcorner)\urcorner).$$

As we have noted above, the nondeterministic Gödel proof predicate PROOF is a normal proof predicate.

Let AS be an axiom specification. An arithmetic *AS-realization* $*$ of the **LP**-language has the following parameters: AS, a normal proof predicate Prf, an evaluation of the sentence letters by sentences of arithmetic, and an evaluation of proof letters and axiom constants by closed recursive terms. We put $\top^* \equiv (0=0)$ and $\bot^* \equiv (0=1)$, $*$ commutes with Boolean connectives, $(t \cdot s)^* \equiv m(t^*, s^*)$, $(t+s)^* \equiv a(t^*, s^*)$, $(!t)^* \equiv c(\ulcorner t^*\urcorner)$, $([\![t]\!]F)^* \equiv \mathrm{Prf}(t^*, \ulcorner F^*\urcorner)$. We assume also that $\mathbf{PA} \vdash G^*$ for all $G \in AS$.

Under any *AS-interpretation* $*$ an **LP**-term t becomes a closed recursive term t^* (i.e., a recursive name of a natural number), and an **LP**-formula F becomes an arithmetic sentence F^*. Also note that the reflexivity principle is there, since $[\![t]\!]F \to F$ is provable in **PA** under any interpretation $*$. Indeed, let n be the value of t^*. If $\mathrm{Prf}(\overline{n}, \ulcorner F^*\urcorner)$ is true, then $\mathbf{PA} \vdash F^*$, thus $\mathbf{PA} \vdash \mathrm{Prf}(\overline{n}, \ulcorner F^*\urcorner) \to F^*$. If $\mathrm{Prf}(\overline{n}, \ulcorner F^*\urcorner)$ is false, then $\mathbf{PA} \vdash \neg\mathrm{Prf}(\overline{n}, \ulcorner F^*\urcorner)$, and again $\mathbf{PA} \vdash \mathrm{Prf}(\overline{n}, \ulcorner F^*\urcorner) \to F^*$.

10.2. Theorem. (Artëmov [1995], arithmetic completeness of **LP**) *If* $\vdash_{\mathbf{LP}_{AS}} F$, *then* $\mathbf{PA} \vdash F^*$ *and hence* $\mathbb{N} \models F^*$, *for any AS-interpretation* $*$.

Combining theorems 10.1 and 10.2 provides arithmetic completeness of **S4**:

10.3. Theorem. *If* $\vdash_{\mathbf{S4}} F$, *then* $\mathbf{PA} \vdash F^r$ *for some realization* r *and some axiom specification* AS.

By Gödel's translation of intuitionistic propositional logic into **S4**, which provides a faithful embedding of intuitionistic propositional logic in to **S4** (Gödel [1933],McKinsey and Tarski [1948]), this automatically includes an arithmetic completeness result for intuitionistic logic as well. If one considers this in the light of the Curry-Howard term interpretation of intuitionistic natural deduction (see e.g., Troelstra and Schwichtenberg [1996]), then one notes that many more terms are used in the **LP**-interpretation. It seems worthwhile to search in this light for a naturally restricted subsystem of **LP**.

The logic **LP** is a version of **S4** presented in a more rich operational language, with no information being lost, since **S4** is the exact term-forgetting projection of **LP**. A transliteration of an **S4**-theorem into **LP**-language may result in an exponential growth of its length, because the **S4**-derivations are included in the **LP**-formulas as proof terms. However, this increase looks much less dramatic if we calculate the complexity of the input **S4**-theorem F in an 'honest' way as the length of a proof of F in **S4**: the proof terms appearing in the realization algorithm have a size linear of the length of the proof, so, the total length of an **LP**-realization of an **S4**-theorem F is bounded by the quadratic function of the length of a given **S4**-derivation of F.

11. Notions of interpretability

In the part on interpretability and its logics (sections 11-15) we are going to investigate a family of concepts like interpretability and partial conservativity, which, in a sense, are generalizations of the notion of provability and for which we use the common name "interpretability". In the first two sections we will explain these concepts and relate them to each other. In the third section we develop an extension of provability logic to so-called interpretability logic with these concepts in mind. In the fourth section we will prove arithmetic completeness of the best-known interpretability logic **ILM** with regard to interpretability in as well as Π_1-conservativity over **PA**. In the fifth section we give a brief survey of the logics induced by some other concepts from the above family.

The concepts discussed in the first two sections are defined in terms of the comparison of the deductive strengths of theories like "one theory is included in another" or "one theory is consistent with another". To compare the strengths of two theories, these theories are not necessarily to be written in the same language, it is enough to organize a translation ("interpretation") of the language of one theory into the language of the other and just consider the translated variant of the first theory. While introducing the notions we will even assume that different theories always have different languages, even if the two languages coincide graphically.

For simplicity we restrict our considerations to theories formalized within the classical first order logic with identity; we suppose that the languages of the theories we consider contain finite or in the worst case countable sets of predicate constants and do not contain functional or individual constants. For a language K, Fm_K denotes the set of formulas of K and St_K the set of sentences, i.e., closed formulas of

K. If D is a nonempty set, St_K^D denotes the set of sentences of K with parameters in D. More precisely, the elements of St_K^D are pairs $\langle\varphi, f\rangle$, where $\varphi \in \mathrm{Fm}_K$ and f is a valuation of the free variables of φ in D; we usually write $\varphi(a_1, \ldots, a_n)$ instead of $\langle\varphi(x_1, \ldots, x_n), f\rangle$, if $a_1 = f(x_1), \ldots, a_n = f(x_n)$.

By a *theory* we mean a pair $T = \langle A, K\rangle$, where K is a language and $A \subseteq \mathrm{St}_K$. The set A contains the extra-logical axioms of T; *provability in* T means derivability from A in the classical predicate logic with identity. Thus, here we do not identify a theory with the set of its theorems, but rather with the set of its nonlogical axioms (in particular, we suppose that $\mathbf{I\Sigma_1}$ is finitely axiomatized). However we do say that a theory $T = \langle A, K\rangle$ is a *subtheory* of a theory $T' = \langle A', K'\rangle$ and write $T \subseteq T'$, if $K \subseteq K'$ and the set of theorems of T is a subset of that of T'; if at the same time A is finite, then T is said to be a *finite subtheory* of T'. If $K = K'$, we denote the theory $\langle A \cup A', K\rangle$ by $T + T'$; if $M \subseteq \mathrm{St}_K$ and $\varphi \in \mathrm{St}_K$, we may also use $T + M$ and $T + \varphi$ to denote the theories $\langle A \cup M, K\rangle$ and $\langle A \cup \{\varphi\}, K\rangle$, respectively.

Let us not be too lazy to define the well-known notion of *first order model:* for a language K, a *K-model* is a pair $M = \langle D, G\rangle$, where D is a nonempty set (of D- *"individuals"*) called the *domain* and G is a function that assigns to each n-place predicate constant P of K an n-ary relation G_P on D, such that $G_=$ is the identity relation. The *truth* of $\varphi \in \mathrm{St}_K^D$ in M, in symbols $M \models \varphi$, is defined in the standard way: an atom $P(a_1, \ldots, a_n)$ is true in M iff $G_P(a_1, \ldots, a_n)$ holds, truth commutes with the Boolean connectives and $M \models \forall x\varphi(x)$ iff for all $a \in D$, $M \models \varphi(a)$. *The theory T_M of a K-model M* is defined as $\langle\{\varphi \in \mathrm{St}_K | M \models \varphi\}, K\rangle$. And M is said to be *a model of a theory* $\langle A, K\rangle$, if $M \models \varphi$ for each $\varphi \in A$ (of course the latter also implies that $M \models \varphi$ for any closed theorem φ of T).

Let K and K' be languages. We may suppose that the set of individual variables of K is a subset of that of K' and that there are infinitely many variables of K' not belonging to K. Then a *relative translation* from K into K' is a pair $\langle \ell, \sigma(x)\rangle$, where:

- ℓ is a function which assigns to each n-place predicate constant P of K a formula $P^\ell(v_1, \ldots, v_n)$ of K' whose bounded variables do not belong to K and whose free variables are the first n variables of the alphabetical list of the variables of K',

- $\sigma(x)$ is a formula of K' (called the *relativizing formula*) with precisely x free whose bounded variables do not belong to K.

Henceforth we usually omit the word "relative(ly)" and we call translations from the language of **PA** into the same language *arithmetic translations*. Now, for each formula $\varphi \in \mathrm{Fm}_K$ we define $t\varphi$, the *t-translation* of φ into K', by the following induction on the complexity of φ:

- $t(x = y)$ is $x = y$,

- for any other atom $P(x_1, \ldots, x_n)$, $tP(x_1, \ldots, x_n)$ is $P^\ell(x_1, \ldots, x_n)$,

- t commutes with Boolean connectives: $t(\alpha \to \beta) = t\alpha \to t\beta$, etc.,
- $t(\forall x\, \alpha)$ is $\forall x(\sigma(x) \to t\alpha)$, and thus $t(\exists x\, \alpha)$ is $\exists x\,(\sigma(x) \wedge t\alpha)$.

If T and T' are theories in the languages K and K' and t is a translation from K into K', we define the theories

$$t(T) = \langle \{t\varphi \mid \varphi \in \mathrm{St}_K,\ T \vdash \varphi\}, K'\rangle \text{ and}$$
$$t^{-1}(T') = \langle \{\varphi \mid \varphi \in \mathrm{St}_K,\ T' \vdash t\varphi\}, K\rangle.$$

The notion of translation is a formal analog of that of model: a translation $t = \langle \ell, \sigma(x)\rangle$ from K into K' in fact defines a K-model in the language K', where $\sigma(x)$ plays the role of D and ℓ the role of G; as soon as we have a K'-model $M' = \langle D', G'\rangle$ such that $\{a \in D' \mid M' \vDash \sigma(a)\} \neq \emptyset$, a unique K-model $M = \langle D, G\rangle$ arises by taking $D = \{a \in D' \mid M' \vDash \sigma(a)\}$ and $G_P = \{\langle a_1, \dots, a_n\rangle \in D^n \mid M' \vDash P^\ell(a_1, \dots, a_n)\}$ for each n-place predicate letter P of K; we call this model the *K-model induced by* (t, M').

Suppose K and K' are languages and $M = \langle D, G\rangle$ and $M' = \langle D', G'\rangle$ are K- and K'-models, respectively. Then an *interpretation* of M in M' is a translation t from K into K' such that for all $\varphi \in \mathrm{St}_K$, $M \vDash \varphi \iff M' \vDash t\varphi$, i.e., the K-model induced by (t, M') is elementarily equivalent to M. And a *strong interpretation* of M in M' is a pair (t, f), where $t = \langle \ell, \sigma(x)\rangle$ is a translation from K into K' (in fact an interpretation of M in M') and f is an injection of D into $\{a \in D' \mid M' \vDash \sigma(a)\}$ such that for any n-place predicate letter P of K and any $a_1, \dots, a_n \in D$, we have $M \vDash P(a_1, \dots, a_n) \iff M' \vDash P^\ell(fa_1, \dots, fa_n)$; as is easily seen we have then also $M \vDash \varphi(a_1, \dots, a_n) \iff M' \vDash t\varphi(fa_1, \dots, fa_n)$ for all $a_1, \dots, a_n \in D$, $\varphi(x_1, \dots, x_n) \in \mathrm{Fm}_K$. If M and M' are models of theories T and T' respectively and t (or (t, f) for some f) is an interpretation (or a strong interpretation) of M in M', then M is also a model of $T + t^{-1}(T')$ and M' is a model of $T' + t(T)$, so we have a "truth-preserving" way of enriching both T and T'.

Gödel's method of arithmetization can be mentioned here as an impressive example of a strong interpretation (t, f) of the "standard model" of meta-arithmetic (though the latter is not formal) in the standard model of arithmetic: f is just the Gödel numbering function which injects the "domain" of the "standard model" of meta-arithmetic, the set of finite strings of arithmetic symbols, into the domain ω of the standard model of arithmetic, and t is the function which assigns to each meta-predicate its what we call "arithmetic formalization".

The above-defined notion of interpretability of models can also be considered as a relation between complete and consistent theories (of these models). It is easily seen that, if $T = \langle A, K\rangle$ and $T' = \langle A', K'\rangle$ are two such theories and t is a translation from K into K', then the assertions $t(T) \subseteq T'$, $t^{-1}(T') \subseteq T$, $T' \subseteq t(T)$, $T \subseteq t^{-1}(T')$ are equivalent. But in the general case that is not so, and we get at least the following four natural binary relations between theories (T and T' are arbitrary theories and t ranges over translations from the language of T into the language of T'):

11.1. Definition.

- T is *interpretable* in T', if there exists t, called an *interpretation* of T in T' such that $t(T) \subseteq T'$,
- T' is *cointerpretable* in T, if there is t, called a *cointerpretation* of T' in T, such that $t^{-1}(T') \subseteq T$,
- T is *faithfully interpretable* in T', if there is t, called a *faithful interpretation* of T in T', which is both an interpretation of T in T' and a cointerpretation of T' in T,
- T is *weakly interpretable* in T' if there exists t, called a *weak interpretation* of T in T' such that $T' + t(T)$ is consistent (which is also equivalent to the assertion that $T + t^{-1}(T')$ is consistent).

The binary relation of weak interpretability has a natural many-place generalization. Observe that T is weakly interpretable in T' if and only if T is interpretable in some consistent extension of T' which has the same language as T'. Instead of pairs we can consider arbitrary nonempty finite sequences of theories and say that such a sequence $T_1, \ldots, T_n$ is (linearly) *tolerant*, if there are consistent extensions $T_1^+, \ldots, T_n^+$ of these theories such that for each $1 \leqslant i < n$, T_{i+1}^+ is interpretable in T_i^+. Thus, consistency is the unary case of linear tolerance and weak interpretability the binary case. Further generalization consists in removing linearity and passing from sequences of theories to trees: a finite tree of theories is *tolerant*, if there are consistent extensions of these theories, of which each one is interpretable in its predecessors in the tree. The intuition here is that in a tolerant tree of theories we can add to each theory the translated information contained in its children (which already have been augmented in the similar manner), obtaining this way a consistent "avalanche" of information. Changing in the above definition the word "interpretable" for "cointerpretable", we obtain the notion of *cotolerance* of a tree of theories.

The notions of interpretability and weak interpretability between theories were introduced by Tarski, Mostowski and Robinson [1953]; faithful interpretability was first considered by Feferman, Kreisel and Orey [1960], and cointerpretability, tolerance and cotolerance by Japaridze [1992,1993]. These relations between axiomatic theories can be used, and actually have been used many times, to prove relative consistency results, different kinds of conservativity results, decidability and undecidability of theories.

12. Interpretability and partial conservativity.

In many cases the interpretability relations can be characterized in terms of partial conservativity or consistency. We are going to study these characterizations only for theories in the language of **PA** with primitive recursive sets of axioms which contain the axioms of **PA**. Let us call such theories *superarithmetic theories* (again we consider the variant of **PA** without functional symbols; however, below we speak, without any confusion, about terms for primitive recursive functions in superarithmetic theories). As usual, the theorems proved below for this special class of theories are of a much more general character; actually they hold for all reasonable so-called (locally) essentially reflexive theories (see definition 12.3). The main theorems that we are going to prove in this section establish that for such theories interpretability and cointerpretability are nothing but Π_1- and Σ_1-conservativity, respectively (theorems 12.7 and 12.13); weak interpretability corresponds to what we call Π_1-consistency (theorem 12.8), and faithful interpretability of T in S takes place exactly when we have interpretability of T in S and cointerpretability of S in T (theorem 12.14). For finitely axiomatizable theories the situation is considerably different. We will make some remarks and give references on this at the end of this section.

12.1. Definition. Let R be an n-ary relation on ω, $\alpha(x_1, \ldots, x_n)$ an arithmetic formula, and T a superarithmetic theory. We say that:

- α *defines* R,

  if for all $k_1, \ldots, k_n \in \omega$, we have $R(k_1, \ldots, k_n) \iff \mathbb{N} \models \alpha(\bar{k}_1, \ldots, \bar{k}_n)$, $\mathbb{N}$ the standard model of arithmetic,

- α *numerates* R *in* T,

  if for all $k_1, \ldots, k_n \in \omega$, $R(k_1, \ldots, k_n) \Longrightarrow T \vdash \alpha(\bar{k}_1, \ldots, \bar{k}_n)$,

- α *binumerates* R *in* T,

  if α numerates R and $\neg\alpha$ numerates the complement of R in T.

We need some more terminology and notation. The formula class $\Sigma_1!$ is the set of arithmetic formulas which have an explicit Σ_1 form, i.e., $\exists x\, \varphi$ for some primitive recursive formula φ. Similarly for $\Pi_1!$. Simply Σ_1 (resp. Π_1) denotes the class of formulas which are $\mathbf{I\Sigma_1}$-equivalent to some $\Sigma_1!$- (resp. $\Pi_1!$-) formula.

It is known (see Smoryński [1977]) that the predicate "x codes a true $\Sigma_1!$-sentence" can be formalized by a $\Sigma_1!$-formula, which we will denote by True(x). This formula is such that ($\mathbf{I\Sigma_1}$ proves that)

$$\textit{for each } \Sigma_1!\textit{-sentence } \varphi, \quad \mathbf{I\Sigma_1} \vdash \varphi \leftrightarrow \mathrm{True}(\ulcorner \varphi \urcorner).$$

Next, we denote by Regwit(y, x) the very primitive recursive formula for which

$$\mathrm{True}(x) \;\equiv\; \exists y\, \mathrm{Regwit}(y, x)$$

and say that k is a *regular witness* of a $\Sigma_1!$-sentence φ, iff Regwit$(\bar{k}, \ulcorner \varphi \urcorner)$ is true. And k is said to be a *regular counterwitness* of a $\Pi_1!$-sentence $\forall z\, \varphi$, iff k is a regular witness of $\exists z \neg \varphi$.

We write $T \vdash_k \varphi$ to express that k is the code of a T-proof of φ, and denote by **PC** the pure predicate calculus (with identity); theorems of **PC** will be referred to as *tautologies* and $\mathrm{Pr}(x)$ will denote in this section an intensional formalization of the predicate "x codes a tautology"; dually, $\mathrm{Con}(x)$ expresses that x codes a formula φ such that $\mathbf{PC} \nvdash \neg\varphi$. For a theory T and a natural number m, $T{\downarrow}m$ denotes the finite subtheory of T obtained by restricting the set of axioms of T to those whose codes are $\leqslant m$.

Suppose T is a theory in the arithmetic language. Given an arithmetic formula α defining the set of (codes of) axioms of T, we can build in a uniform way (see Feferman [1960]) a formula $\mathrm{Prf}_\alpha(z,x)$ (resp. $\mathrm{Prf}_{\alpha\downarrow y}(z,x)$) expressing that x codes some sentence φ and $T \vdash_z \varphi$ (resp. $T{\downarrow}y \vdash_z \varphi$). The formulas $\mathrm{Pr}_\alpha(x)$ and $\mathrm{Pr}_{\alpha\downarrow y}(x)$ will abbreviate $\exists z\, \mathrm{Prf}_\alpha(z,x)$ and $\exists z\, \mathrm{Prf}_{\alpha\downarrow y}(z,x)$, respectively. If T is a finite theory, there is a canonical formula defining T, namely the formula $x=\bar{n}$, where n is the code of the conjunction of all axioms of T; in this case we will write Pr_T instead of $\mathrm{Pr}_{x=\bar{n}}$. The sentences Con_α, $\mathrm{Con}_{\alpha\downarrow y}$ and Con_T will abbreviate $\neg\mathrm{Pr}_\alpha(\ulcorner x \neq x \urcorner)$, $\neg\mathrm{Pr}_{\alpha\downarrow y}(\ulcorner x \neq x \urcorner)$ and $\neg\mathrm{Pr}_T(\ulcorner x \neq x \urcorner)$, respectively. Using the formula Pr_α, we can also construct in a standard way the sentence Compl_α expressing that T is syntactically complete, i.e., that for every $\varphi \in \mathrm{St}_T$ we have $T \vdash \varphi$ or $T \vdash \neg\varphi$.

12.2. Notation. For any arithmetic formula φ, let $[\varphi]$ denote the term (with exactly the same free variables as φ) for the primitive recursive function which, if the free variables of φ and $[\varphi]$ are $x_1,\ldots,x_n$, assigns to each n-tuple $k_1,\ldots,k_n$ of numbers the code of the formula $\varphi(\bar{k}_1,\ldots,\bar{k}_n)$.

We prefer this notation to the more common dot notation, because it avoids the need to specify the free variables.

12.3. Definition. A theory T, the language of which contains that of **PA**, is said to be *locally essentially reflexive*, if for any sentence φ of the language of T, $T \vdash \mathrm{Pr}_{T\downarrow n}(\ulcorner\varphi\urcorner) \to \varphi$ for all n. The theory is said to be *globally essentially reflexive* if for any formula φ of the language of T, $T \vdash \mathrm{Pr}_{T\downarrow n}[\varphi] \to \varphi$ for all n.

It is known that superarithmetic theories are globally essentially reflexive. At the same time no consistent finite(ly axiomatized) theory which satisfies the conditions of Gödel's second incompleteness theorem can even be locally essentially reflexive, for otherwise such a theory would prove its own consistency. In fact, it is shown in Visser [1990] that essential reflexivity is equivalent to full induction. (The idea of the proof, by the way, is already present in Kreisel and Lévy [1968].) That local essential reflexivity is much weaker than global essential reflexivity follows from the following observation. For any reasonable theory T, T plus local reflection for T is easily seen to satisfy local essential reflexivity. However, by a result from Feferman [1962], T plus local reflection for T is contained in T plus all true Π_1-sentences, which for weaker theories certainly does not entail full induction. It turns out that for our results we just need local essential reflexivity. This is the reason that, in the following we will with "essential reflexivity", perhaps nonstandardly, refer to its local version.

12.4. Definition. Let $T = \langle A, K\rangle$ and $T' = \langle A', K'\rangle$ be theories and suppose that $\Gamma \subseteq \mathrm{Fm}_K \cap \mathrm{Fm}_{K'}$. Then

- T is Γ-*conservative* over T', if for any $\varphi \in \Gamma \cap \mathrm{St}_K$, we have that $T \vdash \varphi$ implies $T' \vdash \varphi$,

- T is Γ-*consistent* with T', if for any $\varphi \in \Gamma \cap \mathrm{St}_K$, we have that $T \vdash \varphi$ implies $T' \nvdash \neg\varphi$; in other words, if T is Γ-conservative over some consistent extension of T' in the same language.

Note that for sufficiently strong theories the notions of $\Sigma_1!$- and Σ_1-conservativity (as well as $\Pi_1!$- and Π_1-conservativity) are equivalent.

12.5. Lemma. (**PA** $\vdash$:) *Suppose $t = \langle \ell, \sigma(x)\rangle$ is a translation from a language K into a language K' and $\varphi \in \mathrm{St}_K$. Then* $\mathbf{PA} \vdash \mathrm{Pr}(\ulcorner\varphi\urcorner) \to \mathrm{Pr}(\ulcorner \exists x\, \sigma(x) \to t\varphi\urcorner)$.

Proof. Argue in **PA**. Suppose P is a proof of φ in **PC**, and let $x_1, \ldots, x_n$ be all variables occurring freely in P. Let then $\Delta = \sigma(x_1) \wedge \ldots \wedge \sigma(x_n)$. By induction on the length of P, one can easily verify that $\vdash_{\mathbf{PC}} \Delta \to t\varphi$ and hence (as φ is closed) $\mathbf{PC} \vdash \exists\Delta \to t\varphi$, where $\exists\Delta$ is the existential closure of Δ. On the other hand, $\vdash_{\mathbf{PC}} \exists x\, \sigma(x) \to \exists\Delta$. Consequently, $\vdash_{\mathbf{PC}} \exists x\, \sigma(x) \to t\varphi$. ⊣

12.6. Lemma. (**PA** $\vdash$:) *For any formula $\alpha(x)$ defining a set of arithmetic sentences, there is an arithmetic translation t such that for any sentence φ,*

(a) $\mathbf{PA} + \mathrm{Con}_\alpha \vdash \mathrm{Pr}_\alpha(\ulcorner\varphi\urcorner) \to t\varphi$,
(b) $\mathbf{PA} + \mathrm{Con}_\alpha + \mathrm{Compl}_\alpha \vdash \mathrm{Pr}_\alpha(\ulcorner\varphi\urcorner) \leftrightarrow t\varphi$.

It is easier to explain the idea of the proof of this lemma than to give a strict proof. Gödel's completeness theorem for the classical predicate calculus (with identity) says that every consistent theory has a model. An analysis of Henkin's proof of this theorem shows how to construct for a consistent arithmeticly definable (say, superarithmetic) theory $T = \langle A, K\rangle$ a model $M = \langle D, G\rangle$, where both D and each relation G_P are arithmetically defined; the whole proof can be formalized in **PA** (Hilbert and Bernays [1939]). As we noted above, to define a K-model in some language (in our case in the language of **PA**) means to give a translation from K into this language; for each concrete sentence φ, **PA** plus the assumption that T is consistent then proves that as soon as φ is a theorem of T, φ is true in M, and the clause (a) of the lemma expresses just this fact; as for clause (b), it is an immediate consequence of (a), for $\mathbf{PA} + \mathrm{Compl}_\alpha \vdash \neg\mathrm{Pr}_\alpha(\ulcorner\varphi\urcorner) \to \mathrm{Pr}_\alpha(\ulcorner\neg\varphi\urcorner)$ and $\mathbf{PA} \vdash t\neg\varphi \leftrightarrow \neg t\varphi$.

12.7. Theorem. (Orey [1961], Hájek [1971,1972]) (**PA** $\vdash$:) *For superarithmetic theories T and S the following are equivalent:*

(i) *T is interpretable in S,*
(ii) *for all m, $S \vdash \mathrm{Con}_{T\downarrow m}$,*
(iii) *T is Π_1-conservative over S.*

Proof. (i)$\Rightarrow$(ii): Suppose $t = \langle \ell, \sigma \rangle$ is an interpretation of T in S. Let φ be the conjunction of all axioms of T with codes $\leqslant m$. Then $S \vdash t\varphi$, i.e., $S \vdash \neg t \neg \varphi$; as $T \vdash \exists x\,(x = x)$, we also have $S \vdash \exists x\, \sigma(x)$. Then, by lemma 12.5 and since S is essentially reflexive, $S \vdash \neg \mathrm{Pr}(\ulcorner \neg \varphi \urcorner)$, i.e., $S \vdash \mathrm{Con}_{T\downarrow m}$.

(ii)$\Rightarrow$(i): Let $\tau(x)$ be a primitive recursive formula defining the set of axioms of T, and $\alpha(x)$ the formula $\tau(x) \wedge \mathrm{Con}_{T\downarrow x}$. Then, as soon as condition (ii) holds, $\alpha(x)$ binumerates the set of axioms of T in S and, thus, $\mathrm{Pr}_\alpha(x)$ numerates the set of theorems of T in S. According to lemma 12.6(a), there is a translation t such that $\mathbf{PA} + \mathrm{Con}_\alpha \vdash \mathrm{Pr}_\alpha(\ulcorner \varphi \urcorner) \rightarrow t\varphi$ for all φ, whence, taking into account the obvious fact that $S \vdash \mathrm{Con}_\alpha$, we have $S \vdash \mathrm{Pr}_\alpha(\ulcorner \varphi \urcorner) \rightarrow t\varphi$ for each φ. This, together with the fact that $\mathrm{Pr}_\alpha(x)$, numerates the set of theorems of T in S, implies that t is an interpretation of T in S.

(ii)$\Rightarrow$(iii): Suppose φ is a Π_1!-sentence and $T \vdash \varphi$. This means that $T{\downarrow}m \vdash \varphi$ for some m. We can suppose that m is large enough for $T{\downarrow}m$ to prove all axioms of **Q** (Robinson's arithmetic). It is known that the latter disproves all false Π_1!-sentences. This fact is provable in S and, as $S \vdash \mathrm{Pr}_{T\downarrow m}(\ulcorner \varphi \urcorner)$, S proves that the consistency of $T{\downarrow}m$ implies φ. Then, if (ii) is satisfied, we have $S \vdash \varphi$.

(iii)$\Rightarrow$(ii): Suppose T is Π_1-conservative over S and let m be an arbitrary natural number. It is obvious that T being essentially reflexive, proves $\mathrm{Con}_{T\downarrow m}$. And as $\mathrm{Con}_{T\downarrow m}$ is a Π_1!-sentence, $S \vdash \mathrm{Con}_{T\downarrow m}$. $\dashv$

Taking into account that weak interpretability of T in S is nothing but interpretability of T in some consistent extension of S, we get:

12.8. Corollary. (**PA** $\vdash$:) *For superarithmetic theories T and S the following are equivalent:*

(i) *T is weakly interpretable in S,*
(ii) *for all m, $S \nvdash \neg \mathrm{Con}_{T\downarrow m}$,*
(iii) *T is Π_1-consistent with S.*

Our next goal is to find a similar characterization for cointerpretability. We need some preparatory lemmas.

12.9. Lemma. (Guaspari [1979]) (**PA** $\vdash$:) *Suppose S is a superarithmetic theory and Γ is a recursively enumerable set of natural numbers. Then there is a Σ_1-formula $\gamma(x)$ such that:*

(i) *$\gamma(x)$ numerates Γ in S;*
(ii) *for any $k \in \omega$, if $k \notin \Gamma$, then $S + \neg \gamma(\bar{k})$ is Σ_1-conservative over S.*

Proof. Let $\theta(x)$ be a Σ_1-formula which defines Γ. Let $\Sigma_1!(x)$ be a primitive recursive formula expressing that x is (the code of) a $\Sigma_1!$-sentence and let $\dot{\rightarrow}$ be a term for the primitive recursive function which assigns to each pair m_1, m_2 of numbers, as soon as they code some formulas ϵ_1 and ϵ_2, the code of the formula $\epsilon_1 \rightarrow \epsilon_2$. Finally, let $\sigma(x)$ be a primitive recursive formula defining the set of axioms of S. Applying self-reference, we can construct a $\Sigma_1!$-formula $\gamma(x)$ such that

$$(1)\quad \mathbf{PA} \vdash \gamma(x) \leftrightarrow \exists y \Big(\mathrm{Regwit}(y, [\theta(x)]) \wedge \forall z, t \leq y$$
$$(\Sigma_1!(z) \wedge \mathrm{Prf}_\sigma(t, [\neg\gamma(x)] \dot{\rightarrow} z) \rightarrow$$
$$\exists r(\mathrm{Regwit}(r, z) \wedge \forall r' \leqslant r \neg \mathrm{Regwit}(r', [\gamma(x)]))))\Big).$$

The formula $\gamma(x)$ expresses that there is a regular witness y of $\theta(\bar{x})$ and any $\Sigma_1!$-sentence λ with $S \vdash_{\leq y} \neg\gamma(\bar{x}) \rightarrow \lambda$ has a regular witness less than any regular witness of $\gamma(\bar{x})$.

(i). Suppose $n \in \Gamma$. Then $\theta(\bar{n})$ is true; let k be the smallest regular witness of $\theta(\bar{n})$. Let $\lambda_1, \ldots, \lambda_m$ be all the $\Sigma_1!$-sentences with

$$(2)\quad S \vdash_{\leq k} \neg\gamma(\bar{n}) \rightarrow \lambda_i.$$

Then

$$(3)\quad S \vdash \gamma(\bar{n}) \leftrightarrow \bigwedge\{\exists r(\mathrm{Regwit}(r, \ulcorner\lambda_i\urcorner) \wedge$$
$$\forall r' \leqslant r \neg \mathrm{Regwit}(r', [\gamma(\bar{n})])) \mid 1 \leqslant i \leqslant m\}.$$

Argue in S. Suppose $\neg\gamma(\bar{n})$. Then, by (3), there is i $(1 \leqslant i \leqslant m)$ such that for any regular witness r of λ_i there is an r' with $r' \leqslant r$ which is a regular witness of $\gamma(\bar{n})$. By (2), λ_i is true and has a regular witness. But $\gamma(\bar{n})$ has no regular witness, because (as we have assumed) it is false, which is a contradiction. Thus $S \vdash \gamma(\bar{n})$, and this proves that $\gamma(x)$ numerates Γ in S.

(ii). Now suppose $n \notin \Gamma$ (i.e., $\theta(\bar{n})$ is false), λ is a $\Sigma_1!$-sentence and $S + \neg\gamma(\bar{n}) \vdash \lambda$. Then $S \vdash_e \neg\gamma(\bar{n}) \rightarrow \lambda$ for some e (under the standard Gödel numbering, $e > \ulcorner\lambda\urcorner$) and S proves that if $\theta(\bar{n})$ has a regular witness, the latter is larger than e. Argue in S, and suppose $\gamma(\bar{n})$. Then, by (1) and the above remark, $\theta(\bar{n})$ has a regular witness and the smallest such witness is larger than e. Then, again by (1), λ has a regular witness (smaller than any regular witness of $\gamma(\bar{n})$), so λ is the case. Thus $S + \gamma(\bar{n}) \vdash \lambda$ and, since we have assumed that $S + \neg\gamma(\bar{n}) \vdash \lambda$, we have $S \vdash \lambda$. This proves the desired conservativity. ⊣

12.10. Lemma. (Lindström [1984]) ($\mathbf{PA} \vdash$:) *For superarithmetic theories T and S, there is a formula $\beta(x)$ such that:*

(i) *for all λ, if $S \vdash \mathrm{Pr}_\beta(\ulcorner\lambda\urcorner)$, then, for some m, $S \vdash \mathrm{Pr}_{T\downarrow m}(\ulcorner\lambda\urcorner)$,*

(ii) *$\beta(x)$ binumerates the set of axioms of T in S.*

Proof. Let X be the set of all the sentences ϵ such that $S \vdash \mathrm{Pr}_{T\downarrow m}(\ulcorner\epsilon\urcorner)$ for some m. By lemma 12.9, there is a $\Sigma_1!$-formula $\gamma(x)$ such that for all sentences λ,

(4) *if* $\lambda \in X$, *then* $S \vdash \gamma(\ulcorner\lambda\urcorner)$,

(5) *if* $\lambda \notin X$, *then* $S + \neg\gamma(\ulcorner\lambda\urcorner)$ *is* Σ_1*-conservative over* S.

Let $\tau(x)$ and $\sigma(x)$ be primitive recursive formulas defining the sets of axioms of T and S, respectively. Applying self-reference, we define $\beta(x)$ by

$$\beta(x) \;\equiv\; \tau(x) \wedge \forall y, z \leqslant x(\mathrm{Prf}_\sigma(y, [\mathrm{Pr}_\beta(z)]) \rightarrow \gamma(z)).$$

To prove (i), suppose $S \vdash_m \mathrm{Pr}_\beta(\ulcorner\lambda\urcorner)$. Clearly $S \vdash \ulcorner\lambda\urcorner \leqslant \bar{m}$ (unless we have some pathological Gödel numbering) and thus $S + \neg\gamma(\ulcorner\lambda\urcorner) \vdash \forall x(\beta(x) \rightarrow \tau{\downarrow}\bar{m}(x))$, where $\tau{\downarrow}\bar{m}(x)$ denotes $\tau(x) \wedge x \leqslant \bar{m}$. Hence $S + \neg\gamma(\ulcorner\lambda\urcorner) \vdash \mathrm{Pr}_\beta(\ulcorner\lambda\urcorner) \rightarrow \mathrm{Pr}_{\tau\downarrow\bar{m}}(\ulcorner\lambda\urcorner)$ and, since $S \vdash_m \mathrm{Pr}_\beta(\ulcorner\lambda\urcorner)$ and τ is primitive recursive,

(6) $S + \neg\gamma(\ulcorner\lambda\urcorner) \vdash \mathrm{Pr}_{T\downarrow m}(\ulcorner\lambda\urcorner)$.

Suppose $\lambda \notin X$. Then, by (5) and (6) (as $Pr_{T\downarrow m}(\ulcorner\lambda\urcorner) \in \Sigma_1!$), $S \vdash \mathrm{Pr}_{T\downarrow m}(\ulcorner\lambda\urcorner)$, and (i) is proved.

If x is not the code of an axiom of T, then $S \vdash \neg\tau(\bar{x})$ and thus $S \vdash \neg\beta(\bar{x})$. If x is a code an axiom of T, then $S \vdash \tau(\bar{x})$, and to show that $s \vdash \beta(\bar{x})$ it is enough to show that for all $y, z\ (\leqslant x)$, $S \vdash \mathrm{Prf}_\sigma(\bar{y}, [\mathrm{Pr}_\beta(\bar{z})]) \rightarrow \gamma(\bar{z})$.

This is obvious if y is not the code of an S-proof of $\mathrm{Pr}_\beta(\bar{z})$. And if $S \vdash_y \mathrm{Pr}_\beta(\bar{z})$, then, by (i), z codes an element of X, whence, by (4), $\gamma(\bar{z})$. Hence (ii). ⊣

In the sequel we will use the following convention: λ^i is λ if $i = 0$, and is $\neg\lambda$ if $i = 1$.

12.11. Lemma. (Scott [1962]) (**PA** $\vdash$:) *Suppose S is a superarithmetic theory and $\nu(x)$ is a $\Sigma_1!$-formula. There is then a formula $\zeta(x)$ such that for all (arithmetically defined) functions $g, h : \omega \rightarrow \{0,1\}$, if the set $S_g = S + \{\nu(\bar{n})^{g(n)} \mid n \in \omega\}$ is consistent, then so is the set $S_{g,h} = S_g + \{\zeta(\bar{n})^{h(n)} \mid n \in \omega\}$.*

Proof. Let $\sigma(x)$ be a primitive recursive formula defining the set of axioms of S. We define the formula $\alpha(x)$ by:

$$\sigma(x) \vee \exists y \leqslant x\Big((x = [\nu(y)] \wedge \mathrm{True}([\nu(y)])) \vee (x = [\neg\nu(y)] \wedge \neg\mathrm{True}([\nu(y)]))\Big).$$

Explanation: Let $t : \omega \rightarrow \{0,1\}$ be such that $t(n) = 0$ iff $\nu(\bar{n})$ is true, and let $S^+ = S + \{\nu(\bar{n})^{t(n)} \mid n \in \omega\}$. Then the formula $\alpha(x)$ expresses that x is the code of an axiom of S^+. It is easy to see that

(7) *for each function $g : \omega \rightarrow \{0,1\}$, $\alpha(x)$ binumerates the set of axioms $S_g = S + \{\nu(\bar{n})^{g(n)} \mid n \in \omega\}$ in S_g. Consequently,* $\mathrm{Prf}_\alpha(x, y)$ *binumerates the relation "... is an S_g-proof of ..." in S_g.*

(for, roughly speaking, S_g thinks that $S^+ = S_g$).

Let $\mathrm{Seq}(s,l)$ be a primitive recursive formula expressing that s is the code of a $\{0,1\}$-valued sequence of length l (i.e., of a function: $M \to \{0,1\}$, where $M=\{0,\ldots,l-1\}$ if $l>0$, and $M=\emptyset$ if $l=0$). Let $\mathrm{Conj}(s,u)$ be a term for the primitive recursive function that assigns to each pair (s,u), of which s codes a finite (possibly empty) $\{0,1\}$-valued sequence $f=\langle f(0),\ldots,f(m)\rangle$ and u codes a formula λ which contains exactly one free variable, the code of the conjunction $\lambda(\bar{0})^{f(0)} \wedge \ldots \wedge \lambda(\bar{m})^{f(m)}$. Now, applying self-reference, we construct a formula $\zeta(x)$ such that

$$(8)\quad \mathbf{PA} \vdash \ \zeta(x) \leftrightarrow \ \forall y \forall s\Big(\mathrm{Seq}(s,x) \wedge \mathrm{Prf}_\alpha(y, \mathrm{Conj}(s,\ulcorner\zeta\urcorner) \dot{\to} [\zeta(x)]) \to$$
$$\exists t<y\, \exists s'(\mathrm{Seq}(s',x) \wedge \mathrm{Prf}_\alpha(t, \mathrm{Conj}(s',\ulcorner\zeta\urcorner) \dot{\to} [\neg\zeta(x)]))\Big).$$

The formula $\zeta(x)$ asserts that, if $\zeta(\bar{x})$ is proved by some extension of S^+ of the type $S^+ + \pm\zeta(\bar{0}) \wedge \ldots \wedge \pm\zeta(\overline{x-1})$ (where $\pm$ means the presence or absence of $\neg$), then there is a shorter proof of $\neg\zeta(\bar{x})$ in some extension of S^+ of the same type.

Assume that the set $S_g = S + \{\nu(\bar{n})^{g(n)} \mid n \in \omega\}$ is consistent. Let $U_0 = \{S_g\}$ and $U_{l+1} = \{R+\zeta(\bar{l}), R+\neg\zeta(\bar{l}) \mid R \in U_l\}$. To prove the lemma it suffices to show by induction on l that each $R \in U_l$ is consistent. The only element S_g of U_0 is consistent by our assumption. Suppose there is a theory in U_{l+1} which is inconsistent. Then there is a theory in U_l which proves $\zeta(\bar{l})$ or $\neg\zeta(\bar{l})$. Let then k be the smallest number such that, for some $R \in U_l$ and $i \in \{0,1\}$, we have $R \vdash_k \zeta(\bar{l})^i$. More precisely, k is the smallest number such that, for some $\{0,1\}$-valued sequence f of length l and some $i \in \{0,1\}$, we have $S_g \vdash_k \bigwedge\{\zeta(\bar{n})^{f(n)} \mid 0 \leqslant n < l\} \to \zeta(\bar{l})^i$, and R is the theory $S_g + \bigwedge\{\zeta(\bar{n})^{f(n)} \mid 0 \leqslant n < l\}$.

Below we employ, without explicit mention, proposition (7), the primitive recursiveness of $\mathrm{Seq}(.,.)$, $\mathrm{Conj}(.,.)$, and the fact that the number of $\{0,1\}$-valued sequences of length l is finite.

Case 1: $i=0$. Then

$$(9)\quad S_g \vdash \mathrm{Seq}(\ulcorner f\urcorner, \bar{l}) \wedge \mathrm{Prf}_\alpha(\bar{k}, \mathrm{Conj}(\ulcorner f\urcorner, \ulcorner\zeta\urcorner) \dot{\to} [\zeta(\bar{l})]).$$

By our choice of k, there is no number $t<k$ and no $\{0,1\}$-valued sequence f' of length l such that $S_g \vdash_k \bigwedge\{\zeta(\bar{n})^{f'(n)} \mid 0 \leqslant n < l\} \to \neg\zeta(\bar{l})$. Therefore we also have:

$$(10)\quad S_g \vdash \neg\exists t<\bar{k}\, \exists s'(\mathrm{Seq}(s',\bar{l}) \wedge \mathrm{Prf}_\alpha(t, \mathrm{Conj}(s',\ulcorner\zeta\urcorner) \dot{\to} [\neg\zeta(\bar{l})])).$$

Now, (9) and (10) imply by (8) that $S_g \vdash \neg\zeta(\bar{l})$, whence the theory R is inconsistent, which is in contradiction with the induction hypothesis.

Case 2: $i=1$. Then $S_g \vdash \mathrm{Seq}(\ulcorner f\urcorner, \bar{l}) \wedge \mathrm{Prf}_\alpha(\bar{k}, \mathrm{Conj}(\ulcorner f\urcorner, \ulcorner\zeta\urcorner) \dot{\to} [\neg\zeta(\bar{l})])$, whence

$$(11)\quad S_g \vdash \forall y>\bar{k}\, \forall s\, \exists t<y\, \exists s'(\mathrm{Seq}(s',\bar{l}) \wedge \mathrm{Prf}_\alpha(t, \mathrm{Conj}(s',\ulcorner\zeta\urcorner) \dot{\to} [\neg\zeta(\bar{l})])).$$

By our choice of k, for each $y \leqslant k$ and any $\{0,1\}$-valued sequence f' of length l, $S_g \nvdash_y \bigwedge\{\zeta(\bar{n})^{f'(n)} \mid 0 \leqslant n < l\} \to \zeta(\bar{l})$, whence

(12) $S_g \vdash \forall y \leqslant \bar{k} \forall s \neg(\mathrm{Seq}(s, \bar{l}) \wedge \mathrm{Prf}_\alpha(y, \mathrm{Conj}(s, \ulcorner\zeta\urcorner) \dot{\to} [\zeta(\bar{l})]))$.

Now, (11) and (12) immediately imply by (8) that R is inconsistent, which is in contradiction with the induction hypothesis. ⊣

12.12. Lemma. (Lindström [1984]) ($\mathbf{PA} \vdash$:) *Suppose T and S are super-arithmetic theories, $\alpha(x)$ binumerates the set of axioms of T in S, and $S \vdash \mathrm{Con}_\alpha$. There is then an interpretation t of T in S such that, for any sentence λ, $S \vdash t\lambda \Rightarrow S \vdash \mathrm{Pr}_\alpha(\ulcorner\lambda\urcorner)$.*

Proof. Assume the conditions of the lemma and let us fix an enumeration $\{\Psi_n\}_{n\in\omega}$ of all arithmetic sentences. Consider the following recursive definition, where X is any subset of ω:

$$(13)\quad \Phi_n = \begin{cases} \Psi_n, \text{ if} & \text{(a) } T \vdash \bigwedge\{\Phi_m \mid m<n\} \to \Psi_n, \text{ or} \\ & \text{(b) } T \nvdash \bigwedge\{\Phi_m \mid m<n\} \to \neg\Psi_n \text{ and } n \in X, \\ \neg\Psi_n, \text{ otherwise.} & \end{cases}$$

(If $n=0$, $\bigwedge\{\Phi_m \mid m<n\}$ is identified with $\bar{0}=\bar{0}$.)

Let $\xi(x)$ be the formula given by lemma 12.11 for $\nu(x) = \mathrm{Pr}_\alpha(x)$. Next, let $\chi(x,y)$ be a formalization of the result of converting (13) into an explicit definition in the usual way using $\mathrm{Pr}_\alpha(x)$ and $\xi(x)$ to represent the predicates "$T \vdash \ldots$" and "$\ldots \in X$", respectively, and let $\beta(x)$ be $\exists y\, \chi(x,y)$. Obviously, $S \vdash \mathrm{Con}_\beta$ and $S \vdash \mathrm{Compl}_\beta$, whence by lemma 12.6(b), there is a translation t such that for each sentence λ, $S \vdash t\lambda \leftrightarrow \mathrm{Pr}_\beta(\ulcorner\lambda\urcorner)$; clearly we also have $\mathbf{PA} \vdash \mathrm{Pr}_\beta(\ulcorner\lambda\urcorner) \leftrightarrow \beta(\ulcorner\lambda\urcorner)$ and thus

(14) $S \vdash t\lambda \leftrightarrow \beta(\ulcorner\lambda\urcorner)$.

Suppose now $S \nvdash \mathrm{Pr}_\alpha(\ulcorner\lambda\urcorner)$. To complete the proof we must show that $S \nvdash t\lambda$. Let $g : \omega \to \{0,1\}$ be such that $g(\ulcorner\lambda\urcorner) = 1$ and

(15) $S + \mathrm{Y}_g$ is consistent,

where $\mathrm{Y}_g = \{\mathrm{Pr}_\alpha(\bar{n})^{g(n)} \mid n \in \omega\}$. Next we define Φ'_n as follows:

$$\Phi'_n = \begin{cases} \Psi_n, \text{ if} & \text{(a) } \mathrm{Pr}_\alpha(\ulcorner \bigwedge\{\Phi'_m \mid m<n\} \to \Psi_n\urcorner) \in \mathrm{Y}_g, \text{ or} \\ & \text{(b) } \mathrm{Pr}_\alpha(\ulcorner \bigwedge\{\Phi'_m \mid m<n\} \to \neg\Psi_n\urcorner) \notin \mathrm{Y}_g \text{ and} \\ & \quad\; \mathrm{Pr}_\alpha(\ulcorner \bigwedge\{\Phi'_m \mid m<n\} \wedge \Psi_n \to \lambda\urcorner) \notin \mathrm{Y}_g, \\ \neg\Psi_n, \text{ otherwise.} & \end{cases}$$

Let $h : \omega \to \{0,1\}$ be such that

$$h(n) = 0 \text{ iff } Pr_\alpha(\ulcorner \bigwedge\{\Phi'_m \mid m<n\} \wedge \Psi_n \to \lambda\urcorner) \notin \mathrm{Y}_g,$$

and let $\mathrm{Y}_{g,h} = \mathrm{Y}_g \cup \{\xi(\bar{n})^{h(n)} \mid n \in \omega\}$. Then, by lemma 12.11 and the choice of ξ, (15) implies that

(16) $S + \mathrm{Y}_{g,h}$ is consistent.

By induction on n we can easily check that $S + \mathrm{Y}_{g,h} \vdash \chi(\ulcorner \Phi'_n \urcorner, \bar{n})$, whence

(17) $S + \mathrm{Y}_{g,h} \vdash \beta(\ulcorner \Phi'_n \urcorner)$.

We now show by induction on n that for every n,

(18) $\mathrm{Pr}_\alpha(\ulcorner \bigwedge\{\Phi'_m \mid m < n\} \to \lambda \urcorner) \notin \mathrm{Y}_g$.

Observe that $\{\epsilon \mid \mathrm{Pr}_\alpha(\ulcorner \epsilon \urcorner) \in \mathrm{Y}_g\}$ is closed under logical deduction. If $n = 0$, (18) holds by our choice of Y_g. Suppose (18) holds for $n = k$.

Case 1: $\Phi'_k = \Psi_k$. Then either (a) $\mathrm{Pr}_\alpha(\ulcorner \bigwedge\{\Phi'_m \mid m < k\} \to \Psi_k \urcorner) \in \mathrm{Y}_g$, or (b) $Pr_\alpha(\ulcorner \bigwedge\{\Phi'_m \mid m < k+1\} \to \lambda \urcorner) \notin \mathrm{Y}_g$. The subcase (b) just means that (18) holds for $n = k+1$, and the subcase (a) together with the induction hypothesis also implies (18) for $n = k+1$.

Case 2: $\Phi'_k = \neg\Psi_k$. Then $\mathrm{Pr}_\alpha(\ulcorner \bigwedge\{\Phi'_m \mid m < k\} \wedge \Psi_k \to \lambda \urcorner) \in \mathrm{Y}_g$, for otherwise $\mathrm{Pr}_\alpha(\ulcorner \bigwedge\{\Phi'_m \mid m < k\} \to \neg\Psi_k \urcorner) \notin \mathrm{Y}_g$ and so $\Phi'_k = \Psi_k$. But then (18) for $n = k+1$ easily follows from the induction hypothesis. This proves (18).

Finally, in view of (18), it follows that for some k, $\Phi'_k = \neg\lambda$. Hence, by (17), $S + \mathrm{Y}_{g,h} \vdash \beta(\ulcorner \neg\lambda \urcorner)$; clearly the latter implies $S + \mathrm{Y}_{g,h} \vdash \mathrm{Pr}_\beta(\ulcorner \neg\lambda \urcorner)$ and, (as $S \vdash \mathrm{Con}_\beta$) $S + \mathrm{Y}_{g,h} \vdash \neg\mathrm{Pr}_\beta(\ulcorner \lambda \urcorner)$, whence, by (16), $S \nvdash \mathrm{Pr}_\beta(\ulcorner \lambda \urcorner)$, and by (14), $S \nvdash t\lambda$. The proof of lemma 12.12 is complete. ⊣

12.13. Theorem. (Japaridze [1993]) (**PA** $\vdash$:) *For superarithmetic theories T and S the following are equivalent:*

(i) *S is cointerpretable in T,*
(ii) *for all λ and m, if $S \vdash \mathrm{Pr}_{T\downarrow m}(\ulcorner \lambda \urcorner)$, then $T \vdash \lambda$,*
(iii) *S is Σ_1-conservative over T.*

Proof. (i)$\Rightarrow$(ii): Suppose $t = \langle \tau, \delta \rangle$ is a cointerpretation of S in T, and also $S \vdash \mathrm{Pr}_{T\downarrow m}(\ulcorner \lambda \urcorner)$, i.e., $S \vdash \mathrm{Pr}(\ulcorner \bigwedge T\downarrow m \to \lambda \urcorner)$, where $\bigwedge T\downarrow m$ is the conjunction of all axioms of $T\downarrow m$. Then, by lemma 12.5, $S \vdash \mathrm{Pr}(\ulcorner \exists x\, \delta(x) \to t(\bigwedge T\downarrow m \to \lambda) \urcorner)$, which implies, since S is essentially reflexive, $S \vdash \exists x\, \delta(x) \to t(\bigwedge T\downarrow m \to \lambda)$. Therefore, $S \vdash t\,(\exists x(x = x) \to (\bigwedge T\downarrow m \to \lambda))$, whence (as t is a cointerpretation of S in T) $T \vdash \exists x(x = x) \to (\bigwedge T\downarrow m \to \lambda)$, and $T \vdash \lambda$.

(ii)$\Rightarrow$(i): Let us fix the formula β from lemma 12.10. By clause (ii) of that lemma, β binumerates the set of axioms of T in S. Then, by lemma 12.11, there is a translation t such that for all sentences λ,

(19) if $S + \mathrm{Con}_\beta \vdash t\lambda$, then $S + \mathrm{Con}_\beta \vdash \mathrm{Pr}_\beta(\ulcorner \lambda \urcorner)$.

We claim that if condition (ii) of theorem 12.13 holds, then t is a cointerpretation of S in T. Indeed, suppose $S \vdash t\lambda$. Then, by (19), $S + \mathrm{Con}_\beta \vdash \mathrm{Pr}_\beta(\ulcorner\lambda\urcorner)$; on the other hand, we clearly have $S + \neg\mathrm{Con}_\beta \vdash \mathrm{Pr}_\beta(\ulcorner\lambda\urcorner)$. Consequently, $S \vdash \mathrm{Pr}_\beta(\ulcorner\lambda\urcorner)$. Then, by lemma 12.10(i), $S \vdash \mathrm{Pr}_{T\downarrow m}(\ulcorner\lambda\urcorner)$ for some m, which together with the condition (ii) of our theorem, implies that $T \vdash \lambda$.

(ii)$\Rightarrow$(iii): Assume (ii). Suppose λ is a Σ_1-sentence and $S \vdash \lambda$. Let m be such that $T\!\downarrow m$ contains Robinson's arithmetic. Then $S \vdash \mathrm{Pr}_{T\downarrow m}(\ulcorner\lambda\urcorner)$, whence, by (ii), $T \vdash \lambda$.

(iii)$\Rightarrow$(ii): Suppose S is Σ_1-conservative over T and $S \vdash \mathrm{Pr}_{T\downarrow m}(\ulcorner\lambda\urcorner)$. Since $\mathrm{Pr}_{T\downarrow m}(\ulcorner\lambda\urcorner)$ is a Σ_1-sentence, it follows that $T \vdash \mathrm{Pr}_{T\downarrow m}(\ulcorner\lambda\urcorner)$ and T, being essentially reflexive, proves λ. ⊣

12.14. Theorem. (Lindström [1984])(**PA** $\vdash$:) *A superarithmetic theory T is faithfully interpretable in a superarithmetic theory S iff T is Π_1-conservative over S and S is Σ_1-conservative over T.*

Proof. In view of theorems 12.7 and 12.13, the direction ($\Rightarrow$) is straightforward. To prove ($\Leftarrow$), suppose T is Π_1-conservative over S and S is Σ_1-conservative over T. Then by theorems 12.7 and 12.13, we have:

(20) for all m, $S \vdash \mathrm{Con}_{t\downarrow m}$,

(21) for all m and λ, if $S \vdash \mathrm{Pr}_{t\downarrow m}(\ulcorner\lambda\urcorner)$, then $T \vdash \lambda$.

Let β be the formula from lemma 12.10, and let $\alpha(x)$ be the formula $\beta(x) \wedge \mathrm{Con}_{\beta\downarrow x}$. Then (arguing as in the proof of theorem 12.7(ii)$\Rightarrow$(i)), (20) implies that $\alpha(x)$ binumerates the set of axioms of T in S and $S \vdash \mathrm{Con}_\alpha$. Consequently, by lemma 12.10, there is an interpretation t of T in S such that

(22) for all λ, $S \vdash t\lambda \Longrightarrow S \vdash \mathrm{Pr}_\alpha(\ulcorner\lambda\urcorner)$.

To show that t is also a cointerpretation of S in T, suppose $S \vdash t\lambda$. Then, by (22), $S \vdash \mathrm{Pr}_\alpha(\ulcorner\lambda\urcorner)$. It is obvious that $\mathbf{PA} \vdash \mathrm{Pr}_\alpha(\ulcorner\lambda\urcorner) \rightarrow \mathrm{Pr}_\beta(\ulcorner\lambda\urcorner)$. Then, by lemma 12.10(i), $S \vdash \mathrm{Pr}_{T\downarrow m}(\ulcorner\lambda\urcorner)$ for some m, whence, by (21), $T \vdash \lambda$. ⊣

12.15. Finitely axiomatized theories

In the case of finitely axiomatized theories the interpretability relations have other interesting characterizations. E.g., a theorem due to Harvey Friedman (improved by Visser [1990]) establishes that for finitely axiomatized sequential theories T and S, T is interpretable in S if and only if the weak theory $\mathbf{I\Delta_0} + \mathrm{EXP}$ proves that the consistency of S (with respect to cutfree proofs) implies the consistency of T (with respect to cutfree proofs).

12.16. Feasible interpretability

Visser introduced the notion of feasible interpretability. A theory T is *feasibly interpretable* in a theory T' iff there is a translation t from the language of T into the language of T' and a polynomial function $P(x)$ such that for any λ and x, if $T \vdash_x \lambda$, then $T' \vdash_{\leq P(x)} t\lambda$. In a similar manner we can define the notion of feasible Π_1-conservativity: T is feasibly Π_1-conservative over S iff there is a polynomial $P(x)$ such that for any x and Π_1-sentence λ, if $T \vdash_x \lambda$, then $S \vdash_{\leq P(x)} \lambda$. Verbrugge [1993b] showed that theorem 12.7(i)$\Leftrightarrow$(iii) continues to hold when "interpretable" and "Π_1-conservative" are replaced by "feasibly interpretable" and "feasibly Π_1-conservative".

The main difference between interpretability and feasible interpretability appears when one estimates the arithmetic complexity of the two relations: the relation of interpretability between extensions of arithmetic by finite sets of additional axioms is Π_2-complete (this is originally due to Solovay), whereas the relation of feasible interpretability between such theories turns out to be Σ_2-complete (Verbrugge [1993b]).

13. Axiomatization, semantics, modal completeness of ILM

The idea of interpretability logics arose in Visser [1990] in which they were already developed to a large extent. The modal completeness with respect to the Kripke-semantics due to Veltman was, for the most important systems, proved in de Jongh and Veltman [1990]. Realizing that one cannot cover the concept as well as provability, since interpretability has a more infinitary character, one has to choose primitives of course, and, somewhat surprisingly, it turns out that choosing a binary connective is much more rewarding than choosing a unary connective. The arithmetic realization of $A \rhd B$ in a theory T will be that T plus the realization of B is interpretable in T plus the realization of A (T plus A *interprets* T plus B), or, alternatively (and, as we have seen, in the case of **PA** equivalently), that T plus the realization of B is Π_1-conservative over T plus the interpretation of A. The unary pendant "T interprets T plus A" is much less expressive and was studied in de Rijke [1992]. For a recent complete overview, see Visser [1997].

We first introduce a basic interpretability logic **IL**: it contains, besides the usual axiom $\Box(\Box A \to A) \to \Box A$ for the provability logic **L** and its rules, modus ponens and necessitation, the axioms:

(1) $\Box(A \to B) \to (A \rhd B)$,
(2) $(A \rhd B) \wedge (B \rhd C) \to (A \rhd C)$,
(3) $(A \rhd C) \wedge (B \rhd C) \to (A \vee B \rhd C)$,
(4) $(A \rhd B) \to (\Diamond A \to \Diamond B)$,
(5) $\Diamond A \rhd A$.

With respect to priority of parentheses $\rhd$ is treated as $\to$. Furthermore, in this section, we will consider the extension **ILM** = **IL** + M of **IL** where M is the axiom $(A \rhd B) \to (A \wedge \Box C \rhd B \wedge \Box C)$. We will write $\vdash_{\mathbf{IL}}$ and $\vdash_{\mathbf{ILM}}$ for derivability in **IL** and **ILM**, but sometimes we may leave off the subscript. As will be proved further on, the logic **ILM** is the logic of Π_1-conservativity of **PA**, and therefore also, as shown in the previous section, its interpretability logic. We will not treat here the logic **ILP** which arises by extending **IL** by the scheme $(A \rhd B) \to \Box(A \rhd B)$ that axiomatizes the interpretability logic of the most common finitely axiomatizable theories (Visser [1990], using a modal completeness result of de Jongh and Veltman [1990]).

13.1. Lemma.
(a) $\vdash_{\mathbf{IL}} \Box\neg A \to (A \rhd B)$,
(b) $\vdash_{\mathbf{IL}} A \vee \Diamond A \rhd A$,
(c) $\vdash_{\mathbf{IL}} A \rhd A \wedge \Box\neg A$.

Proof. The parts (a) and (b) are easy. For part (c) use lemma 2.1(j) to obtain $\vdash_{\mathbf{L}} A \to (A \wedge \Box\neg A) \vee \Diamond(A \wedge \Box\neg A)$. Then use the necessitation rule, axiom (1), part (b) and axiom (2). ⊣

13.2. Corollary.
(a) *The formulas* $A \rhd B$, $A \wedge \Box\neg A \rhd B$ *and* $A \rhd B \wedge \Box\neg B$ *are* **IL**-*equivalent.*
(b) *The formulas* $A \rhd \bot$ *and* $\Box\neg A$ *are* **IL**-*equivalent.*

Proof. (a) By lemma 13.1(c) and its converse, which is derivable from axiom (1), and transitivity of $\rhd$ (axiom (2)).

(b) The direction from right to left follows from lemma 13.1(a). The other direction is obtained by using axiom (4) with $\bot$ for B, lemma 2.1(i) and transitivity of $\rhd$. ⊣

13.3. Definition. An **IL**-*frame* (also *Veltman-frame*) is an **L**-frame $\langle W, R\rangle$ with, for each $w \in W$, an additional relation S_w, which has the following properties:
(i) S_w is a relation on $w\!\uparrow = \{w' \in W \mid w R w'\}$,
(ii) S_w is reflexive and transitive,
(iii) if $w', w'' \in w\!\uparrow$ and $w' R w''$, then $w' S_w w''$.

We may write S for $\{S_w | w \in W\}$.

13.4. Definition. An **IL**-*model* is given by an **IL**-frame $\langle W, R, S\rangle$ combined with a forcing relation $\Vdash$ with the clauses:

$$u \Vdash \Box A \iff \forall v(uRv \Rightarrow v \Vdash A),$$
$$u \Vdash A \rhd B \iff \forall v(uRv \text{ and } v \Vdash A \Rightarrow \exists w(vS_u w \text{ and } w \Vdash B)).$$

13.5. Definition.
1. If F is a frame, then we write $F \models A$ iff $F = \langle W, R, S\rangle$ and $w \Vdash A$ for every $w \in W$ and every $\Vdash$ on F.
2. If $\mathcal{K}$ is a class of frames, we write $\mathcal{K} \models A$ iff $F \models A$ for each $F \in \mathcal{K}$.
3. $\mathcal{K}_{\mathrm{M}}$, the class of **ILM**-frames, is the class of **IL**-frames satisfying
   (iv) if $u S_w v R z$, then $u R z$.
4. An **ILM**-model is an **IL**-model on an **ILM**-frame.

The scheme M characterizes (see section 2) the class of frames $\mathcal{K}_M$; that is the content of part (b) of the next soundness lemma.

13.6. Lemma. *For all* **IL**-*frames* F,

(a) *For each* A, *if* $\vdash_{\mathbf{IL}} A$, *then* $F \models A$.

(b) $F \models$ **ILM** *iff* $F \in \mathcal{K}_M$.

(c) *For each* A, *if* $\vdash_{\mathbf{ILM}} A$, *then* $\mathcal{K}_M \models A$.

As before, in the case of **L**, we work inside a so-called adequate set. It is convenient to use the fact that $\Box$ is definable in **IL** in terms of $\rhd$: $\Box A$ is **IL**-equivalent to $\neg A \rhd \bot$ (corollary 13.2(b)). This means that we can, in constructing countermodels, restrict our attention to formulas that do not contain $\Box$. The entire following discussion will be based on the presumption the formulas discussed do not contain $\Box$.

The other side of the coin is that this will allow us to use $\Box$ as a defined symbol. The most convenient way to this turns out to be the following: $\Diamond A$ will be an abbreviation of $\neg(A \rhd \bot)$ and $\Box A$ will then abbreviate the formula $\sim\Diamond\sim A$ (i.e., $\sim A \rhd \bot$). We need to adapt the concept of adequate set to the new situation.

13.7. Definition. An *adequate* set of formulas is a set Φ that satisfies the following conditions:

1. Φ is closed under taking subformulas,
2. if $A \in \Phi$, then $\sim A \in \Phi$,
3. $\bot \rhd \bot \in \Phi$,
4. $A \rhd B \in \Phi$ if A is an antecedent or succedent of some $\rhd$-formula in Φ, and so is B.

13.8. Lemma. *If* Φ *is an adequate set, then* $A \rhd B \in \Phi$ *iff both* $\Diamond A$ *and* $\Diamond B$ *are in* Φ (*and in case* Φ *contains no doubly negated formulas*) *iff both* $\Box\sim A$ *and* $\Box\sim B$ *are in* Φ.

It is obvious that each formula is contained in a finite adequate set. In proving completeness we can of course restrict our attention to formulas without double negations, and will therefore be able to use adequate sets with formulas without double negations, so that we can apply the last part of lemma 13.8. We will write **ILS** if our remarks apply to both **IL** and **ILM**.

13.9. Definition. Let Γ and Δ be maximal **ILS**-consistent subsets of some finite adequate Φ. Then $\Gamma < \Delta \iff$ for each $\Box A \in \Gamma$, $\Box A, A \in \Delta$, and, for some $\Box A \notin \Gamma$, $\Box A \in \Delta$. In this case we say that Δ is a *successor* of Γ (see the proof for **L** of theorem 2.4).

13.10. Definition. Let Γ and Δ be maximal **ILS**-consistent subsets of some given adequate Φ. Then Δ is a C-*critical successor* of Γ ($\Gamma <_C \Delta$) iff

(i) $\Gamma < \Delta$,

(ii) $\sim A, \Box\sim A \in \Delta$ for each A such that $A \rhd C \in \Gamma$.

13.11. Lemma. *If* $\Gamma <_C \Delta$ *and* $\Delta < \Theta$, *then* $\Gamma <_C \Theta$.

13.12. Lemma. *Suppose* Γ *is maximal* **ILS**-*consistent in* Φ *and* $\neg(B \rhd C) \in \Gamma$. *Then there exists a* C-*critical successor* Δ *of* Γ, *maximal* **ILS**-*consistent in* Φ, *such that* $B \in \Delta$.

Proof. Let Γ, Φ, B and C satisfy the conditions of the lemma. Take Δ to be a maximal **ILS**-consistent extension of

$$\{D, \Box D | \Box D \in \Gamma\} \cup \{\Box{\sim}A, {\sim}A | A \rhd C \in \Gamma\} \cup \{B, \Box{\sim}B\}.$$

Note first that the adequacy of Φ ensures that all the formulas in Δ are indeed available, and second that such a Δ, if it exists, is a C-critical successor of Γ. (It is a successor, because $\Box{\sim}B$ **IL**-implies $B \rhd C$ and, hence, cannot be a member of Γ.) To prove that such a Δ exists it is sufficient to prove that the above set is **IL**-consistent. Suppose not. Then there exist $A_1, \ldots, A_m$ and $D_1, \ldots, D_k$ with

$$D_1, \ldots, D_k, \Box D_1, \ldots, \Box D_k, \neg A_1, \ldots \neg A_m, \Box\neg A_1, \ldots, \Box\neg A_m, B, \Box\neg B \vdash \bot$$

or equivalently

$$D_1, \ldots, D_k, \Box D_1, \ldots, \Box D_k \vdash B \wedge \Box\neg B \to A_1 \vee \ldots \vee A_m \vee \Diamond(A_1 \vee \ldots \vee A_m).$$

Applying what we know of **L** gives

$$\Box D_1, \ldots, \Box D_k \vdash \Box(B \wedge \Box\neg B \to A_1 \vee \ldots \vee A_m \vee \Diamond(A_1 \vee \ldots \vee A_m)).$$

Axiom (1) then implies

$$\Box D_1, \ldots, \Box D_k \vdash B \wedge \Box\neg B \rhd A_1 \vee \ldots \vee A_m \vee \Diamond(A_1 \vee \ldots \vee A_m).$$

From lemmas and axiom (2) it follows then that

$$\Box D_1, \ldots, \Box D_k \vdash B \rhd A_1 \vee \ldots \vee A_m.$$

Given that $A_1 \rhd C, \ldots, A_m \rhd C \in \Gamma$, we also have, by using axiom (3) that $\Gamma \vdash A_1 \vee \ldots \vee A_m \rhd C$. So, finally, we obtain $\Gamma \vdash B \rhd C$ which contradicts the consistency of Γ. ⊣

13.13. Lemma. *Let* $B \rhd C \in \Gamma$. *Then, if there exists an* E-*critical successor* Δ *of* Γ *with* $B \in \Delta$, *there also exists an* E-*critical successor* Δ' *of* Γ *with* $C, \Box\neg C \in \Delta'$.

Proof. Suppose B, C, E, Γ and Δ satisfy the assumptions of the lemma and there is no such Δ'. Then there would be $\Box D_1, \ldots, \Box D_n \in \Gamma$, and $F_1 \rhd E, \ldots, F_k \rhd E \in \Gamma$ such that

$$D_1, \ldots, D_n, \Box D_1, \ldots, \Box D_n, \neg F_1, \ldots, \neg F_k, \Box\neg F_1, \ldots, \Box\neg F_k, C, \Box\neg C \vdash \bot$$

and, therefore,

$$D_1, \ldots, D_n, \Box D_1, \ldots, \Box D_n \vdash C \wedge \Box\neg C \to F_1 \vee \ldots \vee F_k \vee \Diamond(F_1 \vee \ldots \vee F_k),$$

which as before implies $\Gamma \vdash B \rhd E$. Since B and E are respectively an antecedent and a succedent of some $\rhd$-formula in Φ, the adequacy conditions imply then that this can be strengthened to $B \rhd E \in \Gamma$. As Δ is supposed to be an E-critical successor of Γ, this implies ${\sim}B \in \Delta$ and we have arrived at a contradiction. ⊣

13.14. Theorem. (Completeness and decidability of **IL**) *If* $\nvdash_{\mathbf{IL}} A$, *then there is a finite* **IL**-*model* K *such that* $K \not\models A$.

Proof. Take some finite adequate set Φ containing A, and let Γ be a maximal **IL**-consistent subset of Φ containing $\sim A$. The intuitive idea of the construction of the model is to divide the set of successors of each constructed world w, starting with Γ, into different parts, each part containing the E-critical successors w for some $\rhd$-succedent E in the adequate set. For occurrences of the same maximal consistent set in different parts we use distinct copies. The S_w are defined to be the universal relation inside each part consisting of the E-critical successors for some E, but to be such as to make no other connections between worlds. Then lemmas 13.12 and 13.13 give the theorem rather straightforwardly. With some care this program can be executed, but we take a slightly more complicated road that points the way to the completeness proof for **ILM** where the straightforward manner does not work.

Set W_Γ to be the smallest set of pairs $\langle \Delta, \tau \rangle$ with Δ a maximal consistent subset of Φ and τ a finite sequence of formulas from Φ that satisfy the following requirements:

(i) $\Gamma < \Delta$ or $\Gamma = \Delta$,

(ii) τ is a finite sequence of formulas from Φ, the length of which does not exceed the depth of Γ minus the depth of Δ. (So, e.g. Γ is only paired off with the empty sequence.)

It is clear that W_Γ is finite, since, for any Δ, if Δ' is a successor of Δ, then Δ' has fewer successors than Δ. We define R on W_Γ by $w\,R w'$ iff $(w)_0 < (w')_0$ and $(w)_1 \subseteq (w')_1$. The required properties check out easily. Let $u\,S_w v$ apply if (I) and (II) hold (writing $*$ for concatenation):

(I) $u, v \in w{\uparrow}$,

(II) either $(w)_1 = (u)_1 \subseteq (v)_1$, or $(u)_1 = (w)_1 * \langle C \rangle * \tau$ and $(v)_1 = (w)_1 * \langle C \rangle * \tau'$ for some C, τ, τ', and if in the latter case $(u)_0$ is a C-critical successor of $(w)_0$, then so is $(v)_0$.

Let us check that under this definition the S_w will have the properties (i)–(iii) required by definition 13.3:

(i) That S_w is a relation on $w{\uparrow}$ is instantaneous.

(ii) Reflexivity and transistivity of S_w are also easy to check.

(iii) If $w', w'' \in w{\uparrow}$ and $w' R w''$, then (I) is immediate. For (II) it suffices to recall that successors of C-critical successors are C-critical (lemma 13.11).

Finally, we define $w \Vdash p$ iff $p \in (w)_0$. We will now prove that, for each $B \in \Phi$ and $w \in W_\Gamma$, $w \Vdash B$ iff $B \in (w)_0$, by induction on the length of B. Of course, the connectives are trivial, so it suffices to prove that

$$B \rhd C \in (w)_0 \Longleftrightarrow \forall u (w\,R u \wedge B \in (u)_0 \rightarrow \exists v (u\,S_w\,v \wedge C \in (v)_0)).$$

$\Longleftarrow$: Suppose $B \rhd C \notin (w)_0$. Then $\neg (B \rhd C) \in (w)_0$. We have to show that, for some u with $w\,R u$, $B \in (u)_0$ and $\forall v\,(u\,S_w v \rightarrow \sim C \in (v)_0)$. Let Δ with $(w)_0 <_C \Delta$ be as given by lemma 13.12, and take u to be $\langle \Delta, (w)_1 * \langle C \rangle \rangle$. It is clear that u fulfills the requirements.

$\Longrightarrow$: Suppose $B \rhd C \in (w)_0$. Consider any u such that $B \in (u)_0$ and $w R u$, and first assume $(u)_1 = (w)_1 * \langle E \rangle * \tau$ and $(u)_0$ is an E-critical successor of $(w)_0$. By lemma 13.13 we can find an E-critical successor Δ' of $(w)_0$ with $C \in \Delta'$. It is clear that $v = \langle \Delta', (w)_1 * \langle E \rangle \rangle$ is a member of W_Γ and fulfills all the requirements to make $u S_w v$.

If $(u)_1 = (w)_1 * \langle E \rangle * \tau$ but $(u)_0$ is not an E-critical successor of $(w)_0$, then we find a successor Δ' of $(w)_0$ with $C \in \Delta'$ by using axiom (4) instead of lemma 13.13. Again it is clear that $v = \langle \Delta', (w)_1 * \langle E \rangle \rangle$ is a member of W_Γ and fulfills all the requirements to make $u S_w v$. The final case is that $(u)_1 = (w)_1$. In that case also we apply axiom (4) to obtain Δ' with $C \in \Delta'$ and take $v = \langle \Delta', (w)_1 \rangle$. ⊣

13.15. Theorem. (Completeness and decidability of **ILM**) *If* $\nvdash_{\mathbf{ILM}} A$, *then there is a finite* **ILM**-*model* K *such that* $K \nVdash A$.

The main problem in the proof of this theorem is the following. To apply the characteristic axiom $(A \rhd B) \to (A \wedge \Box C \rhd B \wedge \Box C)$ we seem to be forced to add the succedent of this formula to the adequate set whenever we have the antecedent. A straightforward definition of adequate set for the case of **ILM** would therefore lead adequate sets to be always infinite, which is of course unacceptable. After some searching we are lead to the following definition.

13.16. Definition. An **ILM**-*adequate* set Φ is an adequate set that satisfies the additional condition:

if $B \rhd C, \Box D \in \Phi$, then there is in Φ a formula $B' \rhd C'$ such that
B' is **ILM**-equivalent to $B \wedge \Box D$ and C' to $C \wedge \Box D$.

Even though we require only equivalents to be present in Φ it is of course no longer evident that each finite set of formulas is contained in a finite **ILM**-adequate set, since each newly constructed $B \wedge \Box D$ gives rise to a new $\Box$-formula: $B \wedge \Box D \rhd \bot$. But we will show that this is nevertheless true. To make it easier on ourselves we assume that in our formula A all antecedents and succedents of $\rhd$-formulas have the form $B \wedge \Box \sim B$, except for $\bot$. In view of corollary 13.2(a) this is not an essential restriction. (The restriction is not really necessary, see Berarducci [1990].)

13.17. Lemma. *Each formula* A *is contained in an* **ILM**-*adequate set* Φ *that contains only a finite number of* **ILM**-*equivalence classes.*

Proof. Let Φ be the smallest **IL**-adequate set containing A. Let Ψ be the set of antecedents and succedents of $\rhd$-formulas in Φ including $\bot$. We obtain Ψ^* by closing Ψ off under the operation that forms $D \wedge E$ from each formula D in the class and each formula E that, either is a $\Box$-formula in Φ, or is of the form $\Box \sim F$ for some F in the class. The claim is that Ψ^* contains only a finite number of equivalence classes. Given that claim we can construct a finite **ILM**-adequate set by joining to Φ

the subformulas of a finite set of representatives of all equivalence classes in Ψ^*, and finally adding all the interpretability formulas combining two members of this finite set of representatives.

It remains to prove the claim. This will be done by induction on the cardinality of Ψ. If that cardinality is 1 (i.e., $\Psi = \{\bot\}$), the result is obvious. So, we can assume that the cardinality is larger than 1. We note that each element of Ψ^* is of the form $B \wedge \Box \sim B \wedge \Box C_1 \wedge \ldots \wedge \Box C_k$, with $B \wedge \Box \sim B$ from Ψ. That $\Box \sim B$ is a member of this conjunction means that in the C_i's all occurrences of $B \wedge \Box \sim B$ can be replaced by $\bot$. Also one will recognize that $B \wedge \Box \sim B$ will only be thrown in by the operation into the C_i in conjuncts of the form $\neg(B \wedge \Box \sim B \wedge ...)$. Replacing those occurrences of $B \wedge \Box \sim B$ by $\bot$ means that one can drop the whole conjunct and keep an equivalent formula. If one drops all those conjuncts containing $B \wedge \Box \sim B$, then the resulting formula is of the form $B \wedge \Box \sim B \wedge \Box D_1 \wedge \ldots \wedge \Box D_m$ with $B \wedge \Box \sim B$ not (relevantly) occurring in the D_i. This means that the D_i have been constructed from the $\Box$-formulas in Φ and the other elements of Ψ. Thus, by the induction hypothesis, there are only a finite number of such D_i (up to equivalence) and hence only a finite number of equivalence classes of elements of Ψ^* that start with $B \wedge \Box \sim B$. The same holds for each of the other elements of Ψ, so that the resulting set is finite. ⊣

Proof of theorem 13.15. Take some finite **ILM**-adequate set Φ containing A and some maximal consistent subset Γ of Φ containing $\sim A$. We define both W_Γ and R as in the previous proof. This time, however, we let $u\, S_w v$ apply if (I) holds as well as (II′) and (III),

(II′) $(u)_1 \subseteq (v)_1$, and if $(u)_1 = (w)_1 * \langle C \rangle * \tau$ and $(v)_1 = (w)_1 * \langle C \rangle * \tau'$ for some C, τ, τ', and $(u)_0$ is a C-critical successor of $(w)_0$, then so is $(v)_0$.

(III) each $\Box A \in (u)_0$ is also a member of $(v)_0$,

That under this definition the S_w will have the properties (i)–(iii) is shown in almost the same manner as before; that the S_w has the property (iv) required by definition 13.5 is shown as follows:

Suppose that $\langle \Delta', \tau' \rangle S_w \langle \Delta'', \tau'' \rangle R \langle \Gamma', \sigma \rangle$. We must show $\langle \Delta', \tau' \rangle R \langle \Gamma', \sigma \rangle$. That $\tau' \subseteq \sigma$, is immediate. That $\Delta' < \Gamma'$, follows from $\Delta'' < \Gamma'$ combined with the fact that, by (III), $\Box$-formulas are preserved from Δ' to Δ''.

Naturally, we again define $w \Vdash p$ iff $p \in (w)_0$, and it will be sufficient to prove that, for each $D \in \Phi$, $w \Vdash D$ iff $D \in (w)_0$. The only interesting case is the one that D is $B \rhd C$, i.e., we have to show that

$$B \rhd C \in (w)_0 \Longleftrightarrow \forall u(w\, R\, u \wedge B \in (u)_0 \rightarrow \exists v(u\, S_w v \wedge C \in (v)_0).$$

$\Longleftarrow$: Basically as in the proof for **IL**.

$\Longrightarrow$: Assume that $B \rhd C \in (w)_0$, and that u is such that $w\, R\, u$ and $B \in (u)_0$. Let $\{\Box D_1, \ldots, \Box D_n\}$ be the set of $\Box$-formulas in $(u)_0$. By axiom M (see proposition 2.1(d)) and the adequacy of Φ, $(w)_0$ will contain a formula $B' \rhd C'$ with B' and C' respectively **ILM**-equivalent to $B \wedge \Box D_1 \wedge \ldots \wedge \Box D_n$ and $C \wedge \Box D_1 \wedge \ldots \wedge \Box D_n$.

Let us just treat the case that $(u)_1 = (w)_1 * \langle E \rangle * \tau$ and $(u)_0$ is an E-critical successor of $(w)_0$. (The other cases are easy, given our experience with **IL**.) We can find, by lemma 13.13, with $(w)_0$, $(u)_0$ and $B' \rhd C'$ as input, an E-critical successor Δ' of $(w)_0$ with both C and $\Box D \in \Delta'$ for each $\Box D \in (u)_0$. It suffices to take $v = \langle \Delta', (u)_1 \rangle$. Given that each $\Box$-formula in $(u)_0$ appears also in Δ', the depth of Δ' cannot be larger than the depth of $(u)_0$. Therefore, $v \in W_\Gamma$ and v fulfills all requirements. ⊣

Visser (see Berarducci [1990]) showed that, from the models constructed in the above proof, one can construct models with an S relation that is independent of the world w (see also definition 15.4). These models may have to be infinite however. The first arithmetic completeness proofs used these models instead of the finite models constructed in the above proof, but we will not introduce them in this section, since our arithmetic completeness proof (section 14) uses the finite models directly.

The fixed point theorem of **L** can be extended to **IL** and hence to **ILM** and **ILP** (de Jongh and Visser [1991]).

14. Arithmetic completeness of ILM

We fix a theory T containing $\mathbf{I\Sigma_1}$. For safety we assume that T is in the language of arithmetic and T is sound, i.e., all its axioms are true (in the standard model of arithmetic), although in fact it is easy to adjust our proof of the completeness theorem to the weaker condition of Σ_1-soundness of T.

14.1. Definition. The definition of a *realization* given in section 1 is extended to the language of **ILM** by stipulating that $(A \rhd B)^* = \mathrm{Conserv}(\ulcorner A^* \urcorner, \ulcorner B^* \urcorner)$, where $\mathrm{Conserv}(\ulcorner A^* \urcorner, \ulcorner B^* \urcorner)$ is an intensional formalization (see Chapter II of this Handbook) of "$T + B^*$ is Π_1-conservative over $T + A^*$".

If $T = \mathbf{PA}$, then, in view of theorem 12.7, the interpretability and Π_1-conservativity relations over its finite extensions are the "same" in all reasonable senses, so we can take $\mathrm{Conserv}(\ulcorner A^* \urcorner, \ulcorner B^* \urcorner)$ to be a formalization of "$T + B^*$ is interpretable in $T + A^*$". Below we prove the completeness of **ILM** as the logic of Π_1-conservativity over T and thus at the same time the completeness of **ILM** as the logic of interpretability over $T = \mathbf{PA}$. The fact that **ILM** is the logic of interpretability over **PA** was proven more or less simultaneously and independently by Berarducci [1990] and Shavrukov [1988]. Later, Hájek and Montagna [1990,1992] proved that **ILM** is the logic of Π_1-conservativity over $T = \mathbf{I\Sigma_1}$ and stronger theories.

14.2. Theorem. $\vdash_{\mathbf{ILM}} A$ *iff for every realization* *, $T \vdash A^*$.

Proof. The ($\Longrightarrow$) part can be verified by induction on **ILM** proofs. Since the soundness of **L** is already known, we only need to verify that if D is an instance of one of the additional 6 axiom schemata of **ILM**, then, for any realization *, $T \vdash D^*$. All the arguments below are easily formalizable in T:

Axiom (1): $\Box(A \to B) \to (A \rhd B)$. If $T \vdash A \to B$, then clearly $T + B^*$ is conservative over $T + A^*$.

Axiom (2): $(A \rhd B) \wedge (B \rhd C) \to (A \rhd C)$. Evidently, the relation of conservativity is transitive.

Axiom (3): $(A \rhd C) \wedge (B \rhd C) \to A \vee B \rhd C$. It is easy to see that if $T + C^*$ is (Π_1-) conservative over $T + A^*$ and $T + B^*$, then so is it over $T + A^* \vee B^*$.

Axiom (4): $(A \rhd B) \to (\Diamond A \to \Diamond B)$. Clearly, if $T + B^*$ is Π_1-conservative over $T + A^*$ and $T + A^*$ is consistent, then so is $T + B^*$.

Axiom (5): $\Diamond A \rhd A$. Suppose λ is a Π_1!-sentence provable in $T + A^*$. We need to show, arguing in $T + (\Diamond A)^*$, that then λ is true. Indeed, suppose $T + A^*$ is consistent. Then it cannot prove a false Π_1!-sentence (by Σ_1!-completeness), and hence λ must be true.

Axiom (M): $(A \rhd B) \to (A \wedge \Box C \rhd B \wedge \Box C)$. Suppose $T + B^*$ is Π_1-conservative over $T + A^*$ and λ is a Π_1!-sentence provable in $T + B^* \wedge (\Box C)^*$. Then $T + B^*$ proves $(\Box C)^* \to \lambda$. But the latter is a Π_1-sentence and therefore it is also proved by $T + A^*$. Hence, $T + A^* \wedge (\Box C)^* \vdash \lambda$.

The following proof of the ($\Longleftarrow$) part of the theorem is taken from Japaridze [1994b] and has considerable similarity to proofs given in Japaridze [1992,1993] and Zambella [1992]. Just as in Japaridze [1992,1993], the Solovay function is defined in terms of regular witnesses rather than provability in finite subtheories (as in Berarducci [1990], Shavrukov [1988], Zambella [1992]). Disregarding this difference, the function is almost the same as the one given in Zambella [1992], for both proofs, unlike the ones in Berarducci [1990] and Shavrukov [1988], employ finite **ILM**-models rather than infinite Visser-models.

Suppose $\nvdash_{\mathbf{ILM}} A$. Then, by theorem 13.15, there is a finite **ILM**-model $\langle W, R, \{S_w\}_{w \in W}, \Vdash \rangle$ in which A is not valid. We may assume that $W = \{1, \ldots, l\}$, 1 is the root of the model in the sense that $1Rw$ for all $1 \neq w \in W$, and $1 \nVdash A$. We define a new frame $\langle W', R', \{S'_w\}_{w \in W'} \rangle$:

$$W' = W \cup \{0\},$$

$$R' = R \cup \{(0, w) \mid w \in W\}.$$

$$S'_0 = S_1 \cup \{(1, w) \mid w \in W\} \text{ and for each } w \in W,\ S'_w = S_w.$$

Observe that $\langle W', R', \{S'_w\}_{w \in W'} \rangle$ is a finite **ILM**-frame.

Just as in section 3, we are going to embed this frame into T by means of a Solovay style function $g : \omega \to W'$ and sentences Lim_w for $w \in W'$ which assert that w is the limit of g. This function will be defined in such a way that the following basic lemma holds:

14.3. Lemma.

(a) *T proves that g has a limit in W', i.e.,* $T \vdash \bigvee \{\mathrm{Lim}_r \mid r \in W'\}$,

(b) *If* $w \neq u$, *then* $T \vdash \neg(\mathrm{Lim}_w \wedge \mathrm{Lim}_u)$,

(c) *If* $w R' u$, *then* $T + \mathrm{Lim}_w$ *proves that* $T \nvdash \neg \mathrm{Lim}_u$,

(d) *If* $w \neq 0$ *and not* $w\,R'\,u$, *then* $T + \mathrm{Lim}_w$ *proves that* $T \vdash \neg\,\mathrm{Lim}_u$,
(e) *If* $u\,S'_w v$, *then* $T + \mathrm{Lim}_w$ *proves that* $T + \mathrm{Lim}_v$ *is* Π_1*-conservative over* $T + \mathrm{Lim}_u$,
(f) *Suppose* $w\,R'\,u$ *and* V *is a subset of* W' *such that for no* $v \in V$, $u\,S_w v$;

then $T + \mathrm{Lim}_w$ *proves that* $T + \bigvee\{\mathrm{Lim}_v \mid v \in V\}$ *is not* Π_1*-conservative over* $T + \mathrm{Lim}_u$,
(g) Lim_0 *is true,*
(h) *For each* $i \in W'$, Lim_i *is consistent with* T.

To deduce the main thesis from this lemma, we define a realization * by setting for each propositional letter p,

$$p^* = \bigvee\{\mathrm{Lim}_r \mid r \in W,\ r \Vdash p\}.$$

14.4. Lemma. *For any* $w \in W$ *and any* **ILM***-formula* B,
(a) *if* $w \Vdash B$, *then* $T + \mathrm{Lim}_w \vdash B^*$;
(b) *if* $w \nVdash B$, *then* $T + \mathrm{Lim}_w \vdash \neg\,B^*$.

Proof. By induction on the complexity of B. The cases when B is atomic or has the form $\Box C$ are handled just as in the proof of lemma 3.3, so we consider only the case when $B = C_1 \rhd C_2$.

Assume $w \in W$. Then we can always write $w\,R\,x$ and $x\,S_w\,y$ instead of $w\,R'\,x$ and $x\,S'_w\,y$. Let $\alpha_i = \{r \mid w\,R\,r,\ r \Vdash C_i\}$ $(i = 1, 2)$. First we establish that for both $i = 1, 2$,

$$(*) \qquad T + \mathrm{Lim}_w \text{ proves that } T \vdash C_i^* \leftrightarrow \bigvee\{\mathrm{Lim}_r \mid r \in \alpha_i\}.$$

Indeed, argue in $T + \mathrm{Lim}_w$.

Since each $r \in \alpha_i$ forces C_i, we have by the induction hypothesis for clause (a) that for each such r, $T \vdash \mathrm{Lim}_r \to C_i^*$, whence $T \vdash \bigvee\{\mathrm{Lim}_r \mid r \in \alpha_i\} \to C_i^*$. Next, according to lemma 14.3(a), $T \vdash \bigvee\{\mathrm{Lim}_r \mid r \in W'\}$ and, according to lemma 14.3(d), T disproves every Lim_r with *not* wRr; consequently, $T \vdash \bigvee\{\mathrm{Lim}_r \mid wRr\}$; at the same time, by the induction hypothesis for clause (b), C_i^* implies in T the negation of each Lim_r with $r \nVdash C_i$. We conclude that $T \vdash C_i^* \to \bigvee\{\mathrm{Lim}_r \mid wRr,\ r \Vdash C_i\}$, i.e., $T \vdash C_i^* \to \bigvee\{\mathrm{Lim}_r \mid r \in \alpha_i\}$. Thus, (*) is proved. Now continue:

(a) Suppose $w \Vdash C_1 \rhd C_2$. Argue in $T + \mathrm{Lim}_w$. By (*), to prove that $T + C_2^*$ is Π_1-conservative over $T + C_1^*$, it is enough to show that $T + \bigvee\{\mathrm{Lim}_r \mid r \in \alpha_2\}$ is Π_1-conservative over $T + \bigvee\{\mathrm{Lim}_r \mid r \in \alpha_1\}$. Consider an arbitrary $u \in \alpha_1$ (the case with empty α_1 is trivial, for any theory is conservative over $T + \bot$). Since $w \Vdash C_1 \rhd C_2$, there is $v \in \alpha_2$ such that $u\,S_w\,v$. Then, by lemma 14.3(e), $T + \mathrm{Lim}_v$ is Π_1-conservative over $T + \mathrm{Lim}_u$. Then so is $T + \bigvee\{\mathrm{Lim}_r \mid r \in \alpha_2\}$ (which is weaker than $T+\mathrm{Lim}_v$). Thus, for each $u \in \alpha_1$, $T+\bigvee\{\mathrm{Lim}_r \mid r \in \alpha_2\}$ is Π_1-conservative over $T+\mathrm{Lim}_u$. Clearly this implies that $T+\bigvee\{\mathrm{Lim}_r \mid r \in \alpha_2\}$ is Π_1-conservative over $T+\bigvee\{\mathrm{Lim}_r \mid r \in \alpha_1\}$.

(b) Suppose $w \nVdash C_1 \rhd C_2$. Let us then fix an element u of α_1 such that $u\, S_w\, v$ for no $v \in \alpha_2$. Argue in $T+\mathrm{Lim}_w$.

By lemma 14.3(f), $T+\bigvee\{\mathrm{Lim}_r \mid r \in \alpha_2\}$ is not Π_1-conservative over $T+\mathrm{Lim}_u$. Then, neither is it Π_1-conservative over $T+\bigvee\{\mathrm{Lim}_r \mid r \in \alpha_1\}$ (which is weaker than $T+\mathrm{Lim}_u$). This means by (*) that $T+C_2^*$ is not Π_1-conservative over $T+C_1^*$. ⊣

Now we can pass to the desired conclusion: since $1 \nVdash A$, lemma 14.4 gives $T \vdash \mathrm{Lim}_1 \to \neg A^*$, whence $T \nvdash \neg\mathrm{Lim}_1 \Rightarrow T \nvdash A^*$. But we do have $T \nvdash \neg\mathrm{Lim}_1$ according to lemma 14.3(h). This ends the proof of theorem 14.2. ⊣

Our remaining duty is to define the function g and to prove lemma 14.3. The recursion theorem enables us to define this function simultaneously with the sentences Lim_w (for each $w \in W'$), which, as we have mentioned already, assert that w is the limit of g, and the formulas $\Delta_{wu}(y)$ (for each pair (w,u) with $wR'u$), which we define by

$$\Delta_{wu}(y) \equiv \exists t > y\,(g(t) = \bar{u} \wedge \forall z (y \leqslant z < t \to g(z) = \bar{w})).$$

14.5. Definition. (function g)
We define $g(0)=0$.

Assume that $g(y)$ has already been defined for every $y \leqslant x$, and let $g(x)=w$. Then $g(x+1)$ is defined as follows:

(1) Suppose $wR'u$, $n \leqslant x$ and for all z with $n \leqslant z \leqslant x$ we have $g(z)=w$. Then, if $\vdash_x \mathrm{Lim}_u \to \neg\Delta_{wu}(\bar{n})$, we define $g(x+1)=u$.

(2) Else, suppose $m \leqslant x$, λ is a $\Pi_1!$-sentence and the following holds:
(a) λ has a regular counterwitness which is $\leqslant x$,
(b) $\vdash_m \mathrm{Lim}_u \to \lambda$,
(c) $w\, S_{g(m)}\, u$,
(d) m is the least number for which such λ and u exist, i.e., there are no $m' < m$, world u' and $\Pi_1!$-sentence λ' satisfying the conditions (a)–(c) with m', u' and λ' substituted for m, u and λ.
Then we define $g(x+1)=u$.

(3) In all remaining cases $g(x+1)=g(x)$.

It is not hard to see that g is primitive recursive. Before we start proving lemma 14.3, let us agree on some jargon and prove two auxiliary lemmas. When the transfer from $w=g(x)$ to $u=g(x+1)$ is determined by definition 14.5(1), we say that at the moment $x+1$ the function g makes (or we make) an *R'-move* from the world w to the world u. If this transfer is determined by definition 14.5(2), then we say that an S'*-transfer* takes place and call the number m from definition 14.5(2) the *rank* of this S'-transfer. Sometimes the S'-transfer leads to a new world, but 'mostly' it does not, i.e., $(u=)g(x+1)=g(x)(=w)$, and then it is not a move in the proper sense. Those S'-transfers which lead to a new world we call S'*-moves.* As for R'-transfers, they (by irreflexivity of R') always lead to a new world, so we always say "R'-move" instead of "R'-transfer".

In these terms, the formula $\Delta_{wu}(n)$ asserts that starting at or before the moment n and until some moment t, we stay at the world w without moving and then, at the moment t, we move directly to u.

Intuitively, we make an R'-move from w to some u with $wR'u$ in the following situation: since some moment n and up to the present we have been staying at world w, and just now we have reached evidence that $T+\mathrm{Lim}_u$ thinks that the first (proper) move which happens after passing moment n (and thus our next move) cannot lead directly to the world u; then, to spite this belief of $T+\mathrm{Lim}_u$, we immediately move to u.

And the conditions for an S'-transfer from w to u can be described as follows: we are staying at the world w and by the present moment we have reached evidence that $T+\mathrm{Lim}_u$ proves a false $\Pi_1!$-sentence λ. This evidence consists of two components: (1) a regular counterwitness, which indicates that λ is false, and (2) the rank m of the transfer, which indicates that $T+\mathrm{Lim}_u \vdash \lambda$. Then, as soon as $w\, S_{g(m)}\, u$, the next moment we must be at u (move to u, if $u \neq w$, and remain at w, if $u=w$); if there are several possibilities for such a transfer, we choose the one with the least rank. An additional necessary condition for an S'-transfer is that in the given situation an R'-move is impossible; R'-moves have priority over S'-moves.

Note that the condition for an R'-move here is weaker than for the function h defined in section 3: T only needs to prove $\mathrm{Lim}_u \to \neg\Delta_{wu}(\bar{n})$. This feature will play a crucial role in the verification of 14.3(f).

14.6. Lemma. $(T\vdash:)$ *For each natural number m and each $w \in W'$, $T+\mathrm{Lim}_w$ proves that no S'-transfer to w can have rank less than m.*

Proof. Indeed, "the rank of an S'-transfer is $<m$" means that $T+\mathrm{Lim}_w$ proves a false (i.e., one with a regular counterwitness) $\Pi_1!$-sentence λ and the code of this proof (i.e., of the T-proof of $\mathrm{Lim}_w \to \lambda$) is smaller than m. But the number of all $\Pi_1!$-sentences with such short proofs is finite, and as $T+\mathrm{Lim}_w$ proves each of them, it also proves that none of these sentences has a regular counterwitness (recall our assumptions about the formula $\mathrm{Regwit}(x,y)$ from section 12). ⊣

14.7. Lemma. $(T\vdash:)$ *If $g(x)R'w$, then for all $y \leqslant x$, $g(y)R'w$.*

Proof. Suppose $g(x)R'w$ and $y \leqslant x$. We proceed by induction on $n=x-y$. If $y=x$, we are done. Suppose now $g(y+1)R'w$. If $g(y)=g(y+1)$, we are done. If not, then at the moment $y+1$ the function makes either an R'-move or an S'-move. In the first case we have $g(y)R'g(y+1)$ and, by transitivity of R', $g(y)R'w$; in the second case we have $g(y)S'_v\,g(y+1)$ for some v, and the desired thesis then follows from property (iv) of **ILM**-frames (definition 13.4).

Proof of lemma 14.3. In each case below, except in (g) and (h), we reason in T.

(a): First observe that there exists some z such that for all $z' \geqslant z$, not $g(z')\,R'\,g(z'+1)$. Indeed, suppose this is not the case. Then, by lemma 14.7, for all z, there is z' with $g(z)R'\,g(z')$. This means that there is an infinite (or "sufficiently long") chain $w_1 R'\, w_2 R' \cdots$, which is impossible because W' is finite and R' is transitive and irreflexive.

So, let us fix this number z. Then we never make an R'-move after the moment z. We claim that S'-moves can also take place at most a finite number of times (whence it follows that g has a limit and this limit is, of course, one of the elements of W').

Indeed, let $x+1$ be an arbitrary moment after z at which we make an S'-move, and let m be the rank of this move. That is, for some Π_1!-sentence λ with a $\leqslant x$ regular counterwitness, we have $\vdash_m \mathrm{Lim}_u \to \lambda$ and $wS_{g(m)}u$, where $w=g(x)$ and $u=g(x+1)$. Suppose we make the next S'-move, with rank m', at some moment $x'+1$, $x'>x$, from the world u to a world v, $v \neq u$. Since $S_{g(m)}$ is reflexive, conditions (a)-(c) of definition 14.5(2) hold for x', u, u, λ, m in the roles of x, w, u, λ, m, respectively, and then, according to condition (d) of definition 14.5(2), the only reason for moving to v instead of u — instead of remaining at u, that is — could be that $m > m'$ (the case $m = m'$ is ruled out because $\mathrm{Lim}_u \neq \mathrm{Lim}_v$). Similarly, the rank m'' of the following S'-move will be less than m', etc. Thus, consecutive S'-moves without an R'-move between them have decreasing ranks. Therefore, S'-moves can take place at most m times after passing x.

(b): Clearly g cannot have two different limits w and u.

(c): Assume w is the limit of g and $w\,R'\,u$. Let n be such that for all $x \geqslant n$, $g(x)=w$. We need to show that $T \nvdash \neg\,\mathrm{Lim}_u$. Deny this. Then $T \vdash \mathrm{Lim}_u \to \neg\,\Delta_{wu}(\bar{n})$ and, since every provable formula has arbitrary long proofs, there is $x \geqslant n$ such that $\vdash_x \mathrm{Lim}_u \to \neg\,\Delta_{wu}(\bar{n})$; but then, according to definition 14.5(1), we must have $g(x+1)=u$, which, as $u \neq w$ (by irreflexivity of R'), is a contradiction.

(d): Assume $w \neq 0$, w is the limit of g and not $wR'\,u$. If $u=w$, then (since $w \neq 0$) there is x such that $g(x) = v \neq u$ and $g(x+1)=u$. This means that at the moment $x+1$ we make either an R'-move or an S'-move. In the first case we have $T \vdash \mathrm{Lim}_u \to \neg\,\Delta_{vu}(\bar{n})$ for some n for which, as it is easy to see, the Σ_1!-sentence $\Delta_{vu}(\bar{n})$ is true, whence, by Σ_1!-completeness, $T \vdash \neg\,\mathrm{Lim}_u$. And if an S'-move is taken, then again $T \vdash \neg\,\mathrm{Lim}_u$ because $T + \mathrm{Lim}_u$ proves a false (with a $\leqslant x$ regular counterwitness) Π_1!-sentence.

Next, suppose $u \neq w$. Let us fix a number z with $g(z)=w$. Since g is primitive recursive, T proves that $g(z)=w$. Now argue in $T+\mathrm{Lim}_u$: since u is the limit of g and $g(z)=w \neq u$, there is a number x with $x \geqslant z$ such that $g(x) \neq u$ and $g(x+1)=u$. Since not $(w=)g(z)R'\,u$, we have by lemma 14.7 that

$$(*) \qquad \text{for each } y \text{ with } z \leqslant y \leqslant x, \text{ not } g(y)R'\,u.$$

In particular, not $g(x)R'\,u$ and the transfer from $g(x)$ to $g(x+1)(=u)$ can have been determined only by definition 14.5(2). Then $(*)$ together with the property (i) of **IL**-frames and definition14.5(2c), implies that the rank of this S'-move is less than z, which, by lemma 14.6, is a contradiction. Thus, $T+\mathrm{Lim}_u$ is inconsistent, i.e., $T \vdash \neg\,\mathrm{Lim}_u$.

(e): Assume $u\, S'_w\, v \neq u$ (the case $v=u$ is trivial). Suppose w is the limit of g, λ is a Π_1-sentence and $T \vdash_z \mathrm{Lim}_v \to \lambda$. We may suppose that $\lambda \in \Pi_1!$ and that z is sufficiently large, namely, $g(z)=w$. Fix this z. We need to show that $T+\mathrm{Lim}_u \vdash \lambda$.

Argue in $T+\mathrm{Lim}_u$. Suppose not λ. Then there is a regular counterwitness c for λ. Let us fix a number $x>z,c$ such that $g(x)=g(x+1)=u$ (as u is the limit of g, such a number exists). Then, according to definition 14.5, the only reason for $g(x+1)=u\neq v$ can be that we make an S'-transfer from u to u and the rank of this transfer is less than z, which, by lemma 14.6, is not the case. Conclusion: λ (is true).

(f): Assume w is the limit of g, $wR'u$, $V \subseteq W'$ and for each $v \in V$, not $u\, S'_w\, v$. Let n be such that for all $z \geqslant n$, $g(z)=w$. By primitive recursiveness of g, T proves that $g(n)=w$. By definition 14.5(1), $T+\mathrm{Lim}_u \nvdash \neg\,\Delta_{wu}(\bar{n})$. So, as $\neg\,\Delta_{wu}(\bar{n})$ is a Π_1-sentence, in order to prove that $T+\bigvee\{\mathrm{Lim}_v \mid v \in V\}$ is not Π_1-conservative over $T+\mathrm{Lim}_u$, it is enough to show that for each $v\in V$, $T+\mathrm{Lim}_v \vdash \neg\,\Delta_{wu}(\bar{n})$. Let us fix any $v \in V$. According to our assumption, not $u\, S'_w\, v$ and, by reflexivity of S'_w, $u \neq v$.

Argue in $T+\mathrm{Lim}_v$. Suppose, for a contradiction, that $\Delta_{wu}(n)$ holds, i.e., there is $t>n$ such that $g(t)=u$ and for all z with $n\leqslant z<t$, $g(z)=w$. As v is the limit of g and $v\neq u$, there is $t'>t$ such that $g(t'-1)\neq v$ and at the moment t' we arrive at v to stay there for ever. Let then $x_0<\ldots<x_k$ be all the moments in the interval $[t,t']$ at which R'- or S'-moves take place, and let $u_0=g(x_0),\ldots,u_k=g(x_k)$. Thus $t=x_0$, $t'=x_k$, $u=u_0$, $v=u_k$ and $u_0,\ldots,u_k$ is the route of g after departing from w (at the moment t).

Now let j be the least number among $1,\ldots,k$ such that for all $j\leqslant i\leqslant k$, not $u_0\, R'\, u_i$. Note that such a j does exist because at least $j=k$ satisfies the condition (otherwise, if $(u=)\ u_0\, R'\, u_k\ (=v)$, property (iv) of **ILM**-frames would imply $u\, S'_w\, v$). Note also that, for each i with $j\leqslant i\leqslant k$, the move to u_i cannot be an R'-move. Otherwise, we must have $u_{i-1}R'\, u_i$, whence, by lemma 14.7, $u_0\, R'\, u_i$, which is impossible for $i\geqslant j$.

Thus, from the moment x_j onwards, each move is an S'-move. Moreover, for each i with $j\leqslant i\leqslant k$, the rank of the S'-move to u_i is less than x_0. Indeed, suppose, for such an i, the rank of the S'-move to u_i is m for $m\geqslant x_0$. We have $g(m)=u_e$ for some e with $0\leqslant e\leqslant k$ and, by definition 14.5(2c), we should have $u_{i-1}S'_{u_e}u_i$, and by property (i) of **IL**-frames, $u_eR'u_i$, whence, as above, lemma 14.7 gives that $u_0\, R'\, u_i$, which is impossible for $i\geqslant j$. On the other hand, since consecutive S'-moves decrease the rank (as we noted in the proof of (a) above), and since the rank of the S'-move to u_k cannot be less than n (lemma 14.6), we conclude: for each i with $j\leqslant i\leqslant k$, the rank of the S'-move to u_i is in the interval $[n,x_0-1]$. But the value of g in this interval is w, and by definition 14.5(2c) this means that $u_{j-1}\, S'_w\, u_j\, S'_w \ldots S'_w\, u_k$. At the same time, we have either $u_0=u_{j-1}$ or $u_0\, R'\, u_{j-1}$. In both cases we then have $u_0\, S'_w\, u_{j-1}$ (in the first case by reflexivity of S'_w and in the second case by property (iii) of **IL**-frames), whence, by transitivity of S'_w, $u_0\, S'_w\, u_k$, i.e., $u\, S'_w\, v$, which is a contradiction. Conclusion: $T + \mathrm{Lim}_v \vdash \neg \Delta_{wu}(\bar{n})$.

(g): By (a), as T is sound, one of the Lim_w for $w \in W'$ is true. Since for no w do we have $wR'w$, (d) means that each Lim_w, except Lim_0, implies in T its own T-disprovability and therefore is false. Consequently, Lim_0 is true.

(h): As 3.2(f). ⊣

The proof of theorem 14.2 is complete. In de Jongh and Pianigiani [1998] this theorem and its extension to an interpretability logic with witness comparison formulas (Hájek and Montagna [1992]) was applied to solve a conjecture of Guaspari [1983]. This conjecture stated that those formulas of modal logic that under each arithmetic realization are interpreted as Σ_1-sentences are **L**-equivalent to disjunction of $\Box$-sentences (already proved in Visser [1995]), and those of modal logic extended with witness comparison formulas are **R**-equivalent to disjunctions which contain as their members conjunctions of witness comparison formulas and $\Box$-formulas. A companion paper is Beklemishev [1993a], in which it is shown that the realization of other formulas, i.e., the ones that are not always realized as Σ_1-sentences, cannot be restricted to any particular class in the arithmetic hierarchy, thereby improving Guaspari [1983]'s results as well.

Visser [1990] showed that **ILP** is the interpretability logic for all reasonable finitely axiomatizable theories that contain $\mathbf{I}\Delta_0 + \mathrm{SUPEXP}$. An open problem is the axiomatization of the logic of the principles valid for interpretability in all reasonable r.e. theories. Visser [1991] showed that this logic is not just the intersection of **ILM** and **ILP**.

15. Tolerance logic and other interpretability logics

15.1. The logics of cointerpretability and faithful interpretability

Unlike interpretability, no modal axiomatization for the logic of cointerpretability or faithful interpretability (over **PA** or any other reasonable theory) has been found so far. Even the question of decidability of these logics remains open.

However, the logics of weak interpretability and the more general relations of tolerance and cotolerance (see section 11) have been studied thoroughly. Here is a brief history of research in this field, which starts from some digression from the subject.

15.2. The logic of the arithmetic hierarchy

Japaridze [1990b,1994a] introduced a decidable propositional logic **HGL** with infinitely many unary modal operators: $\Box, \Sigma_1, \Sigma_1^+, \Sigma_2, \Sigma_2^+, \ldots$ and proved its soundness and completeness with respect of the arithmetic interpretation where $\Box A$ is understood as a formalization of "A^* is provable (in **PA**)", $\Sigma_n A$ as "A^* is (**PA**-equivalent to) a Σ_n-sentence" and $\Sigma_n^+ A$ as "A^* is (**PA**-equivalent to) a Boolean combination of Σ_n-sentences". The logic has a reasonable axiomatization and Kripke semantics.

15.3. The logic of tolerance and its fragments

Ignatiev [1990] (see Ignatiev [1993b]) strengthened the $(\Box, \Sigma_1)$-fragment of the logic of the arithmetic hierarchy by switching from the unary modal operator Σ_1 to the more general binary operator $\gg$, where $A \gg B$ is interpreted as "there is a Σ_1-sentence φ such that $\mathbf{PA} \vdash (A^* \to \varphi) \wedge (\varphi \to B^*)$" (for comparison: the interpretation of $\Sigma_1 A$ is nothing but "there is a Σ_1-sentence φ such that $\mathbf{PA} \vdash (A^* \to \varphi) \wedge (\varphi \to A^*)$"). He constructed a logic **ELH** in this language, called "the logic of Σ_1-interpolability", and proved its arithmetic completeness. Although the author of the logic of Σ_1-interpolability did not suspect this, he actually had found the logic of weak interpretability over **PA**, because, as it is now easy to see in view of corollary 12.8, the formula $\neg(A \gg \neg B)$ expresses that $\mathbf{PA} + B^*$ is weakly interpretable in $\mathbf{PA} + A^*$.

We know that weak interpretability is a special (binary) case of linear tolerance, and the latter is a special (linear) case of tolerance of a tree of theories. Japaridze [1992] gave an axiomatization of the logic **TOL** of linear tolerance over **PA**, and Japaridze [1993] did the same for the logic **TLR** of the most general relation of tolerance for trees.

All three logics **ELH**, **TOL** and **TLR** are decidable. Among them **TOL** has the most elegant language, axiomatization and Kripke semantics, and although **TOL** is just a fragment of **TLR**, here we are going to have a look only at this intermediate logic.

The language of **TOL** contains the single variable-arity modal operator $\Diamond$: for any n, if $A_1, \dots, A_n$ are formulas, then so is $\Diamond(A_1, \dots, A_n)$. This logic is defined as classical logic plus the rule $\neg A / \neg \Diamond(A)$ plus the following axiom schemata:

1. $\Diamond(\vec{C}, A, \vec{D}) \to \Diamond(\vec{C}, A \wedge \neg B, \vec{D}) \vee \Diamond(\vec{C}, B, \vec{D})$,
2. $\Diamond(A) \to \Diamond(A \wedge \neg \Diamond(A))$,
3. $\Diamond(\vec{C}, A, \vec{D}) \to \Diamond(\vec{C}, \vec{D})$,
4. $\Diamond(\vec{C}, A, \vec{D}) \to \Diamond(\vec{C}, A, A, \vec{D})$,
5. $\Diamond(A, \Diamond(\vec{C})) \to \Diamond(A \wedge \Diamond(\vec{C}))$,
6. $\Diamond(\vec{C}, \Diamond(\vec{D})) \to \Diamond(\vec{C}, \vec{D})$.

(Here $\vec{A}$ stands for $A_1, \dots, A_n$ for an arbitrary $n \geqslant 0$, $\Diamond(\langle\rangle)$ is identified with $\top$.)

15.4. Definition. A *Visser-frame* (see Berarducci [1990]) is a triple $\langle W, R, S\rangle$, where $\langle W, R\rangle$ is a Kripke-frame for **L** and S is a transitive, reflexive relation on W such that $R \subseteq S$ and, for all $w, u, v \in W$, we have $wSuRv \Longrightarrow wRv$.

A **TOL**-*model* is a quadruple $\langle W, R, S, \Vdash\rangle$ with $\langle W, R, S\rangle$ a Visser-frame combined with a forcing relation $\Vdash$ with the clause

$w \Vdash \Diamond(A_1 \ldots, A_n)$ iff there are $u_1, \ldots, u_n$ with $u_1 S \ldots S u_n$ such that, for all i, wRu_i and $u_i \Vdash A_i$.

Such a model is said to be *finite*, if W is finite.

15.5. Theorem. (Japaridze [1992]) *For any* **TOL**-*formula* A, $\vdash_{\mathbf{TOL}} A$ *iff* A *is valid in every* **TOL**-*model; the same is true if we consider only finite* **TOL**-*models.*

15.6. Theorem. (Japaridze [1992]) *Let* T *be a sound superarithmetic theory, and let, for* * *an arithmetic realization,* $(\Diamond(A_1, \ldots, A_n))^*$ *be interpreted as a natural formalization of "the sequence* $T + A_1^*, \ldots, T + A_n^*$ *is tolerant". Then, for any* **TOL**-*formula* A, $\vdash_{\mathbf{TOL}} A$ *iff for every realization* *, $T \vdash A^*$.

With the arithmetic interpretation in mind, note that **L** is the fragment of **TOL** in which the arity of $\Diamond$ is restricted to 1. This is because consistency of A^* with T, expressed in **L** by $\Diamond A$, means nothing but tolerance of the one-element sequence $\langle T + A^*\rangle$ of theories, expressed in **TOL** by $\Diamond(A)$.

As for cotolerance, one can easily show, using theorems 12.7 and 12.13 ((i)$\Longleftrightarrow$(iii)), that a sequence of superarithmetic theories is cotolerant iff the sequence where the order of these theories is reversed is tolerant. Moreover, it was shown in Japaridze [1993] that cotolerance — though not tolerance — for trees can also be expressed in terms of linear tolerance. In particular, a tree of superarithmetic theories is cotolerant iff one of its topological sortings is. Hence, given a tree Tr of modal formulas, cotolerance of the corresponding tree of theories can be expressed in **TOL** by $\Diamond(\vec{A_1}) \vee \ldots \vee \Diamond(\vec{A_n})$, where $\vec{A_1}, \ldots, \vec{A_n}$ are all the reverse-order topological sortings of Tr. Thus **TOL**, being the logic of linear tolerance, can, at the same time, be viewed as the logic of (unrestricted) cotolerance over **PA**.

Just like tolerance, the notion of Γ-consistency (see definition 12.4) can be generalized to finite trees, including sequences as special cases of trees: a tree Tr of theories is Γ-*consistent* iff there are consistent extensions of these theories, of which each one is Γ-conservative over its predecessors in the tree.

Then the corollaries of theorems 12.7 and 12.13 generalize to the following:

15.7. Theorem. (Japaridze [1993], **PA** $\vdash$) *For any finite tree* Tr *of superarithmetic theories,*

(a) Tr *is tolerant iff* Tr *is* Π_1-*consistent;*
(b) Tr *is cotolerant iff* Tr *is* Σ_1-*consistent.*

Just as in the case of **ILM**, in the arithmetic completeness theorems for **TOL** and **TLR**, the requirement of superarithmeticity (essential reflexivity) of T can be weakened to $\mathbf{I\Sigma_1} \subseteq T$ if we view these logics as logics of Π_1-consistency rather than tolerance.

15.8. Truth interpretability logics

We want to finish our discussion of propositional interpretability logics by noting that the closure under modus ponens of the set of theorems of **ILM**, or any other of the logics mentioned in this section, supplemented with the axiom $\Box A \rightarrow A$ or its equivalent, yields the logic (in case of **ILM** called $\mathbf{ILM}^\omega$) that describes all *true* principles expressible in the corresponding modal language, just as this was shown to be the case for **L** in section 3. The original sources usually contain proofs of both versions of the arithmetic completeness theorems for these logics.

Strannegård [1997] considers infinite r.e. sets of modal formulas of interpretability logic. He generalizes his theorem 5.3 for the specific case of interpretability over **PA** to the following theorem.

15.9. Theorem. *Let T be a well-specified r.e. set of formulas of interpretability logic. Then T is realistic iff it is consistent with* $\mathbf{ILM}^\omega$.

As in the case of **L** (corollary 5.2), a stronger version of this theorem implies as a corollary a uniform version of the arithmetic completeness of **ILM** with regard to **PA**. For a further consequence, let us first note that the existence of *Orey-sentences* in **PA**, i.e., arithmetic sentences A such that both $\mathbf{PA} + A$ and $\mathbf{PA} + \neg A$ are interpretable in **PA** (first obtained by Orey [1961]), follows immediately from the arithmetic completeness of **ILM** with regard to **PA**. In Strannegård's terminology this can be phrased as: Orey [1961] showed that the set $\{\top \rhd p, \top \rhd \neg p\}$ is realistic. Orey continued by asking what similar sets (such as $\{\top \rhd p,\ \top \rhd q,\ \top \rhd \neg(p \wedge q),$ $\neg(\top \rhd \neg p),\ \neg(\top \rhd \neg q),\ \neg(\top \rhd p \wedge q)\}$) are realistic. Let an *Orey set* be a set of modal formulas of the form $(\neg)(B \rhd C)$, where B and C are Boolean formulas. Strannegård can then give the following answer to Orey's question.

15.10. Theorem. *Let T be an r.e. Orey set. Then T is realistic iff it is consistent with* $\mathbf{ILM}^\omega$.

16. Predicate provability logics

16.1. The predicate modal language and its arithmetic interpretation

The language of predicate provability logic is that of first order logic (without identity or function symbols) together with the operator $\Box$. We assume that this language uses the same individual variables as the arithmetic language. Throughout this section T denotes a sound theory in the language of arithmetic containing **PA**. We also assume that T satisfies the Löb derivability conditions.

As in the previous sections, we want to regard each modal formula $A(P_1, ..., P_n)$ as a schema of arithmetic formulas arising from $A(P_1, ..., P_n)$ by substitution of arithmetic predicates $P_1^*, ..., P_n^*$ for the predicate letters $P_1, ..., P_n$ and replacing $\Box$ by $\mathrm{Pr}_T()$. However, some caution is necessary when we try to make this approach precise. In particular, we need to forbid for P_i^* to contain quantifiers that bind variables occurring in A.

16.2. Definition. A *realization* for a predicate modal formula A is a function * which assigns to each predicate symbol P of A an arithmetic formula $P^*(v_1, \ldots, v_n)$, whose bound variables do not occur in A and whose free variables are just the first n variables of the alphabetical list of the variables of the arithmetic language if n is the arity of P. For any realization * for A, we define A^* by the following induction on the complexity of A:

- in the atomic cases, $(P(x_1, \ldots, x_n))^* = P^*(x_1, \ldots, x_n)$,
- * commutes with quantifiers and Boolean connectives: $(\forall x B)^* = \forall x(B^*)$, $(B \to C)^* = B^* \to C^*$, etc.,
- $(\Box B)^* = \mathrm{Pr}_T[B^*]$.

For an explanation of the notation "[ ]" see notation 12.2. Observe from this that A^* always contains the same free variables as A. We say that an arithmetic formula φ is a *realizational instance* of a predicate modal formula A, if $\varphi = A^*$ for some realization * for A.

The main task is to investigate the set of predicate modal formulas which express valid principles of provability, i.e., all of whose realizational instances are provable, or true in the standard model.

16.3. The situation here is not as smooth as in the propositional case, ...

Having been encouraged by the impressive theorems of Solovay on the decidability of propositional provability logic, one might expect that the valid principles captured by the predicate modal language are also axiomatizable (decidability is ruled out of course). However, the situation here is not as smooth as in the propositional case. The first doubts about this were raised by Montagna [1984]. In fact, it turned out afterwards that we have very strong negative results, one of which is the following theorem on nonarithmeticity of truth predicate logics of provability.

16.4. Theorem. (Artëmov [1985a]) *Suppose T is recursively enumerable. Then (for any choice of the provability predicate* Pr_T*) the set* Tr *of predicate modal formulas all of whose realizational instances are true, is not arithmetic.*

It was later shown by Vardanyan [1986], and also by Boolos and McGee [1987] that Tr is in fact Π_1-complete in the truth set of arithmetic.

Proof of theorem 16.4. We assume here that the arithmetic language contains one two-place predicate letter E and two three-place predicate letters A and M, and does not contain any other predicate, functional or individual letters. Thus, this language is a fragment of our predicate modal language. In the standard model $E(x,y)$, $A(x,y,z)$ and $M(x,y,z)$ are interpreted as the predicates $x=y$, $x+y=z$ and $x \times y = z$, respectively.

One variant of a well-known theorem of Tennenbaum (see e.g., Chapter 29 of Boolos and Jeffrey [1989]) asserts the existence of an arithmetic sentence β such that:

(1) β is true ("true" here always means "true in the standard model"),

(2) any model of β, with domain ω, E interpreted as the identity relation, and A and M as recursive predicates, is isomorphic to the standard model.

We assume that β conjunctively contains the axioms of Robinson's arithmetic **Q**, including the identity axioms. Therefore, using standard factorization, we can pass from any model D of β with domain ω and such that E, A and M are interpreted as recursive predicates, to a model D' which satisfies the conditions of (2) and which is elementarily equivalent to D. Thus, (2) can be changed to the following:

(2$'$) any model D of β, with domain ω and E, A and M interpreted as recursive predicates, is elementarily equivalent to the standard model (i.e., $D \models \gamma$ iff γ is true, for all sentences γ).

Let C be the formula

$$\begin{aligned}&\forall x, y\,(\Box E(x,y) \vee \Box\neg E(x,y)) \wedge\\ &\quad \forall x,y,z\,(\Box A(x,y,z) \vee \Box\neg A(x,y,z)) \wedge\\ &\quad \forall x,y,z\,(\Box M(x,y,z) \vee \Box\neg M(x,y,z)).\end{aligned}$$

The following lemma yields the algorithmic reducibility of the set of all true arithmetic formulas (which, by Tarski's theorem, is nonarithmetic) to the set Tr, and this proves the theorem.

16.5. Lemma. *For any arithmetic formula φ, φ is true if and only if every realizational instance of $\beta \wedge C \to \varphi$ is true.*

Proof. $\Longrightarrow$: Suppose φ is true, * is a realization for $\beta \wedge C \to \varphi$ and $\beta^* \wedge C^*$ is true. We want to show that φ^* is also true. It is not hard to see that, since T is consistent and recursively enumerable (this condition is essential!), the truth of C^* means that the relations defined on ω in the standard model by the formulas E^*, A^* and M^* are recursive. Let us define a model D with domain ω such that, for all $k, m, n \in \omega$,

$$\begin{aligned}&D \models E(k,m) \text{ iff } E^*(k,m) \text{ is true,}\\ &D \models A(k,m,n) \text{ iff } A^*(k,m,n) \text{ is true,}\\ &D \models M(k,m,n) \text{ iff } M^*(k,m,n) \text{ is true.}\end{aligned}$$

Observe that for every arithmetic formula γ (for which the realization * is legal), we have $D \models \gamma$ iff γ^* is true. In particular $D \models \beta$, and thus D satisfies the conditions of (2′), i.e., D is elementarily equivalent to the standard model, whence (as φ is true) $D \models \varphi$, whence φ^* is true.

$\Longleftarrow$: Suppose φ is false. Let * be the trivial realization, i.e., such that $E^*(x,y) = E(x,y)$, $A^*(x,y,z) = A(x,y,z)$, $M^*(x,y,z) = M(x,y,z)$. Then $\beta^* = \beta$, $\varphi^* = \varphi$ and therefore it suffices to show that $\beta \wedge C^* \to \varphi$ is false, i.e., that $\beta \wedge C^*$ is true. But β is true by (1), and from the decidability in T of the relations $x = y$, $x + y = z$ and $x \times y = z$, it follows that C^* is also true. ⊣

Formalizing in arithmetic the idea employed in the above proof, Vardanyan [1986] also proved that if T is recursively enumerable, then the set of predicate modal formulas whose realizational instances are provable in T (or in **PA**) is not recursively enumerable and is in fact Π_2-complete.

There is one perhaps even more unpleasant result which should also be mentioned here. For recursively enumerable T, the answer to the question whether a predicate modal formula expresses a valid provability principle, turns out to be dependent on the choice of the formula Pr_T, that is, on the concrete way of formalization of the predicate "x is the code of an axiom of T", even if a set of axioms is fixed (Artëmov [1986]). Note that the proofs of Solovay's theorems for propositional provability logic are insensitive in this respect and actually the only requirement is that the three Löb-conditions must be satisfied.

16.6. ... but still not completely desperate

Against this gloomy background one still can succeed in obtaining positive results in two directions. Firstly, although the predicate logic of provability in full generality is not (recursively) axiomatizable, some natural fragments of it can be so and may be stable with respect to the choice of the formula Pr_T.

And secondly, all the above-mentioned negative facts exclusively concern recursively enumerable theories, and the proofs hopelessly fail as soon as this condition is removed. There are however many examples of interesting and natural theories which are not recursively enumerable (e.g., the theories induced by ω-provability or the other strong concepts of provability mentioned in section 8), and it well might be that the situation with their predicate provability logics is as nice as in the propositional case.

The main positive result we are going to consider is the following: the "arithmetic part" of Solovay's theorems, according to which the existence of a Kripke countermodel (with a transitive and converse well-founded accessibility relation) implies arithmetic nonvalidity of the formula, can be extended to the predicate level. This gives us a method of establishing nonvalidity for a quite considerable class of predicate modal formulas.

16.7. Kripke-models for the predicate modal language

A *Kripke-frame* for the predicate modal language is a system

$$M = \langle W, R, \{D_w\}_{w \in W} \rangle,$$

where $\langle W, R \rangle$ is a Kripke-frame in the sense of section 2, $\{D_w\}_{w \in W}$ are nonempty sets ("domains of individuals") indexed by elements of W such that if $w\,R\,u$, then $D_w \subseteq D_u$, and a *Kripke-model* is a Kripke-frame together with a forcing relation $\Vdash$, which is now a relation between worlds $w \in W$ and closed formulas with parameters in D_w; for the Boolean connectives and $\Box$, $\Vdash$ behaves as described in section 2, and we have only the following additional condition for the universal quantifier:

- $w \Vdash \forall x A(x)$ iff $w \Vdash A(a)$ for all $a \in D_w$,

and a similar one for the existential quantifier. A formula is said to be *valid* in a Kripke-model $\langle W, R, \{D_w\}_{w \in W}, \Vdash \rangle$, if A is forced at every world $w \in W$. Such a model is said to be finite if W as well as all D_w are finite.

16.8. The predicate version of Solovay's theorems

For every predicate modal formula A, let REFL(A) denote the universal closure of $\bigwedge\{\Box B \to B \mid \Box B \in \mathrm{Sb}\}$, where Sb is the set of the subformulas of A.

16.9. Theorem. (Artëmov and Japaridze [1987,1990]). *For any closed predicate modal formula A,*
(a) *if A is not valid in some finite Kripke-model with a transitive and converse well-founded accessibility relation, then there exists a realization * for A such that $T \nvdash A^*$,*
(b) *if* REFL(A) $\to A$ *is not valid in such a model, then there exists a realization * for A such that A^* is false.*

Proof. We prove here only clause (a), leaving (b) as an exercise for the reader. Some details in this proof are in fact redundant if we want to prove only (a), but they are of assistance in passing to a proof of (b).

Assume that $\langle W, R, \{D_w\}_{w \in W}, \Vdash \rangle$ is a model with the above-mentioned properties in which A is not valid. As before, without loss of generality we may suppose that $W = \{1, \ldots, l\}$, 1 is the root and $1 \nVdash A$. We suppose also that $D_w \subseteq \omega$ and $0 \in D_w$ for each $w \in W$. Let us define a model $\langle W', R', \{D'_w\}_{w \in W'}, \Vdash' \rangle$ by setting

- $W' = W \cup \{0\}$,
- $R' = R \cup \{(0, w) \mid w \in W\}$,
- $D'_0 = D_1$ and, for all $w \in W$, $D'_w = D_w$,
- for any atomic formula P, $0 \Vdash' P$ iff $1 \Vdash P$ and, if $w \in W$, $w \Vdash' P$ iff $w \Vdash P$.

We accept the definitions of the Solovay function h and the sentences Lim_w from section 3 without any changes; the only additional step is the following:

For each a from $D = \bigcup\{D_w \mid x \in W\}$ we define an arithmetic formula $\gamma_a(x)$ with only x free by setting

$$\gamma_a(x) = \bigvee\{\exists t \leqslant x(\, h(t) = h(x) = w \wedge \neg \exists z < t(h(z) = w) \wedge x = t + a) \mid a \in D_w\}.$$

Thus, using the jargon from section 14, $\gamma_a(x)$ says that we have reached some world w such that $a \in D_w$, at the moment x we are still at w, and exactly a moments have passed since we moved to this world (we assume that the first "move", to the world 0, happened at the initial moment 0). We define the predicates γ'_a by

- for each $0 \neq a \in D$, $\gamma'_a(x) = \gamma_a(x)$, and
- $\gamma'_0(x) = \gamma_0(x) \vee \neg \bigvee\{\gamma_a(x) \mid a \in D\backslash\{0\}\}$.

(It is easy to check that the left disjunct of $\gamma'_0(x)$ is redundant; it implies the right disjunct.) Since we employ the same Solovay function h as in section 3, lemma 3.2 continues to hold. In addition, we need the following lemma:

16.10. Lemma.

(i) $T \vdash \neg(\gamma'_a(x) \wedge \gamma'_b(x))$ *for all* $a \neq b$,

(ii) $T \vdash \mathrm{Lim}_w \to \bigwedge\{\exists x\, \gamma'_a(x) \mid a \in D_w\}$ *for all* $w \in W'$,

(iii) $T \vdash \mathrm{Lim}_w \to \forall x(\bigvee\{\gamma'_a(x) \mid a \in D_w\})$ *for all* $w \in W'$.

Proof. (i): The formulas $\gamma_a(x)$ and $\gamma_b(x)$ for $a \neq b$ are defined so that each disjunct of $\gamma_a(x)$ is inconsistent with each disjunct of $\gamma_b(x)$. And the right disjunct of $\gamma'_0(x)$, by definition, is inconsistent with each $\gamma_a(x)$, $a \neq 0$.

(ii): Suppose $a \in D_w$ and argue in $T + \mathrm{Lim}_w$. Since w is the limit of h, there is a moment t at which we arrive in w, and stay there for ever (more formally: we have $\neg \exists y < t\,(h(y) = w)$ and $\forall y \geqslant t\,(h(y) = w)$). Then, by definition, $\gamma_a(t + a)$ holds, whence $\gamma'_a(t + a)$ holds, whence $\exists x\, \gamma'_a(x)$. And so for each $a \in D_w$.

(iii): Argue in $T + \mathrm{Lim}_w$. Consider an arbitrary number x. We must show that $\gamma'_a(x)$ holds for some $a \in D_w$. The definition of h implies that, either $h(x) R' w$, or $h(x) = w$; in both cases we then have $D_{h(x)} \subseteq D_w$. Let t be the least number such that $h(t) = h(x)$, and let $a = x - t$. By definition, if $a \in D_{h(x)}$ (and thus $a \in D_w$), then $\gamma_a(x)$ holds, whence $\gamma'_a(x)$ holds and we are done; and if $a \notin D_{h(x)}$, then (the right disjunct of) $\gamma'_0(x)$ holds and we are also done, because $0 \in D_w$. ⊣

We now define a realization *. For each n-place predicate letter P, let P^* be

$$\bigvee\{\mathrm{Lim}_w \wedge \gamma'_{a_1}(v_1) \wedge \ldots \wedge \gamma'_{a_n}(v_n) \mid a_1, \ldots, a_n \in D_w,\ w \Vdash' P(a_1, \ldots, a_n)\}.$$

16.11. Lemma. *Let B be a predicate modal formula with precisely $x_1, \ldots, x_n$ free. Then, for each $w \in W$ and for all $a_1, \ldots, a_n \in D_w$,*

(a) *if $w \Vdash' (B(a_1, \ldots, a_n)$, then $T \vdash \mathrm{Lim}_w \wedge \gamma'_{a_1}(x_1) \wedge \ldots \wedge \gamma'_{a_n}(x_n) \to B^*$;*

(b) *if $w \nVdash' (B(a_1, \ldots, a_n)$, then $T \vdash \mathrm{Lim}_w \wedge \gamma'_{a_1}(x_1) \wedge \ldots \wedge \gamma'_{a_n}(x_n) \to \neg B^*$.*

Proof. We proceed by induction on the complexity of B. Suppose $B(x_1, \ldots, x_n)$ is atomic. If $w \Vdash' B(a_1, \ldots, a_n)$, then $\mathrm{Lim}_w \wedge \gamma'_{a_1}(x_1) \wedge \ldots \wedge \gamma'_{a_n}(x_n)$ is one of the disjuncts of B^* and the desired result is obvious. If $w \nVdash' B(a_1, \ldots, a_n)$, then that formula is not a disjunct of B^* and, according to lemma 3.2(b) and 16.10(i), it implies in T the negations of all the disjuncts of B^*.

Next suppose that $B(x_1, \ldots, x_n)$ is $\forall y\, C(y, x_1, \ldots, x_n)$.
If $w \Vdash \forall y\, C(y, a_1, \ldots, a_n)$, then $w \Vdash C(b, a_1, \ldots, a_n)$ for all $b \in D_w$. Then, by the induction hypothesis, for all $b \in D_w$,
$T \vdash \mathrm{Lim}_w \wedge \gamma'_b(y) \wedge \gamma'_{a_1}(x_1) \wedge \ldots \wedge \gamma'_{a_n}(x_n) \to (C(y, x_1, \ldots, x_n))^*$.

Therefore,

$$T \vdash \mathrm{Lim}_w \wedge \left(\bigvee\{\gamma'_b(y) \mid b \in D_w\}\right) \wedge \gamma'_{a_1}(x_1) \wedge \ldots \wedge \gamma'_{a_n}(x_n) \to (C(y, x_1, \ldots, x_n))^*.$$

Note that there is no free occurrence of y in either Lim_w or $\gamma'_{a_1}(x_1) \wedge \ldots \wedge \gamma'_{a_n}(x_n)$. Universal quantification over y gives

$$T \vdash \mathrm{Lim}_w \wedge \forall y\left(\bigvee\{\gamma'_b(y) \mid b \in D_w\}\right) \wedge \gamma'_{a_1}(x_1) \wedge \ldots \wedge \gamma'_{a_n}(x_n) \to \\ \forall y(C(y, x_1, \ldots, x_n))^*.$$

By lemma 16.10(iii), we can eliminate the conjunct $\forall y\left(\bigvee\{\gamma'_b(y) \mid b \in D_w\}\right)$ in the antecedent of the above formula and conclude that

$$T \vdash \mathrm{Lim}_w \wedge \gamma'_{a_1}(x_1) \wedge \ldots \wedge \gamma'_{a_n}(x_n) \to \forall y\, (C(y, x_1, \ldots, x_n))^*.$$

If on the other hand $w \nVdash \forall y\, C(y, a_1, \ldots, a_n)$, then there is $b \in D_w$ such that $w \nVdash C(b, a_1, \ldots, a_n)$. By the induction hypothesis,

$$T \vdash \mathrm{Lim}_w \wedge \gamma'_b(y) \wedge \gamma'_{a_1}(x_1) \wedge \ldots \wedge \gamma'_{a_n}(x_n) \to \neg\, (C(y, x_1, \ldots, x_n))^*.$$

Again, neither Lim_w nor $\gamma'_{a_1}(x_1) \wedge \ldots \wedge \gamma'_{a_n}(x_n)$ contains y free, and existential quantification over y gives

$$T \vdash \mathrm{Lim}_w \wedge \exists y\, \gamma'_b(y) \wedge \gamma'_{a_1}(x_1) \wedge \ldots \wedge \gamma'_{a_n}(x_n) \to \exists y\, \neg(C(y, x_1, \ldots, x_n))^*.$$

According to lemma 16.10(ii), $T \vdash \mathrm{Lim}_w \to \exists y\, \gamma'_b(y)$. Therefore,

$$T \vdash \mathrm{Lim}_w \wedge \gamma'_{a_1}(x_1) \wedge \ldots \wedge \gamma'_{a_n}(x_n) \to \neg(\forall y\, C(y, x_1, \ldots, x_n))^*.$$

Finally, suppose that B is $\Box C$. If $w \Vdash \Box C(a_1, \ldots, a_n)$, then for each u such that $wR'u$, we have $u \Vdash C(a_1, \ldots, a_n)$ and, by the induction hypothesis,

$$T \vdash \mathrm{Lim}_u \wedge \gamma'_{a_1}(x_1) \wedge \ldots \wedge \gamma'_{a_n}(x_n) \to (C(x_1, \ldots, x_n))^*.$$

Therefore,

$$T \vdash \left(\bigvee\{\mathrm{Lim}_u \mid wR'u\}\right) \wedge \gamma'_{a_1}(x_1) \wedge \ldots \wedge \gamma'_{a_n}(x_n) \to (C(x_1,\ldots,x_n))^*,$$

and, by the first two Löb conditions,

$$T \vdash \mathrm{Pr}_T[\left(\bigvee\{\mathrm{Lim}_u \mid wR'u\}\right) \wedge \gamma'_{a_1}(x_1) \wedge \ldots \wedge \gamma'_{a_n}(x_n)] \to (\Box C(x_1,\ldots,x_n))^*.$$

Observe that the formulas $\gamma_a(x)$ are primitive recursive and we have that $T \vdash \gamma_a(x) \to \mathrm{Pr}_T[\gamma_a(x)]$; together with lemma 3.2(d) this means that

$$\begin{aligned} T \vdash\ & \mathrm{Lim}_w \wedge \gamma'_{a_1}(x_1) \wedge \ldots \wedge \gamma'_{a_n}(x_n) \to \\ & \mathrm{Pr}_T[\left(\bigvee\{\mathrm{Lim}_u \mid wR'u\}\right) \wedge \gamma'_{a_1}(x_1) \wedge \ldots \wedge \gamma'_{a_n}(x_n)]. \end{aligned}$$

Thus, we get $T \vdash \mathrm{Lim}_w \wedge \gamma'_{a_1}(x_1) \wedge \ldots \wedge \gamma'_{a_n}(x_n) \to (\Box C(x_1,\ldots,x_n))^*$.

If $w \nVdash \Box C(a_1,\ldots,a_n)$, then there is u such that $wR'u$ and $u \nVdash C(a_1,\ldots,a_n)$. By the induction hypothesis,

$$T \vdash \mathrm{Lim}_u \wedge \gamma'_{a_1}(x_1) \wedge \ldots \wedge \gamma'_{a_n}(x_n) \to \neg\,(C(x_1,\ldots,x_n))^*.$$

Therefore,

$$T \vdash (C(x_1,\ldots,x_n))^* \to \neg\,(\mathrm{Lim}_u \wedge \gamma'_{a_1}(x_1) \wedge \ldots \wedge \gamma'_{a_n}(x_n)),$$

$$T \vdash \mathrm{Pr}_T[(C(x_1,\ldots,x_n))^*] \to \mathrm{Pr}_T[\neg\,(\mathrm{Lim}_u \wedge \gamma'_{a_1}(x_1) \wedge \ldots \wedge \gamma'_{a_n}(x_n))],$$

$$T \vdash \neg\,\mathrm{Pr}_T[\neg\,(\mathrm{Lim}_u \wedge \gamma'_{a_1}(x_1) \wedge \ldots \wedge \gamma'_{a_n}(x_n))] \to \neg\,(\Box C(x_1,\ldots,x_n))^*.$$

On the other hand, we have

$$\begin{aligned} T \vdash\ & \neg\,\mathrm{Pr}_T[\neg\,\mathrm{Lim}_u] \wedge \mathrm{Pr}_T[\gamma'_{a_1}(x_1) \wedge \ldots \wedge \gamma'_{a_n}(x_n)] \to \\ & \neg\,\mathrm{Pr}_T[\neg\,(\mathrm{Lim}_u \wedge \gamma'_{a_1}(x_1) \wedge \ldots \wedge \gamma'_{a_n}(x_n))] \end{aligned}$$

(this is a realizational instance of the principle $\Diamond p \wedge \Box q \to \Diamond(p \wedge q)$ which is provable in **K**). According to lemma 3.2(c), and since $T \vdash \gamma'_a(x) \to \mathrm{Pr}_T[\gamma'_a(x)]$, we have

$$\begin{aligned} T \vdash\ & \mathrm{Lim}_w \wedge \gamma'_{a_1}(x_1) \wedge \ldots \wedge \gamma'_{a_n}(x_n) \to \\ & \neg\,\mathrm{Pr}_T[\neg\,\mathrm{Lim}_u] \wedge \mathrm{Pr}_T[\gamma'_{a_1}(x_1) \wedge \ldots \wedge \gamma'_{a_n}(x_n)]. \end{aligned}$$

Therefore, $T \vdash \mathrm{Lim}_w \wedge \gamma'_{a_1}(x_1) \wedge \ldots \wedge \gamma'_{a_n}(x_n) \to \neg\,(\Box C(x_1,\ldots,x_n))^*$. ⊣

To finish the proof of theorem 16.9: since A is closed and $1 \nVdash A$, we have by lemma 16.11, $T \vdash \mathrm{Lim}_1 \to \neg\,A^*$. By lemma 3.2(f), Lim_1 is consistent with T, and consequently $T \nvdash A^*$. ⊣

16.12. Further positive results

One of the applications of theorem 16.9 is the following. Consider the fragment of our predicate modal language which arises by restricting the set of variables to one single variable x. In this case, without loss of generality, we may assume that every predicate letter is one-place. Since the variable x is fixed, it is convenient to omit it in the expressions $\forall x$, $P(x)$, $Q(x)$, ... and simply write $\forall$, p, q, ... In fact, we then have a bimodal propositional language with the modal operators $\Box$ and $\forall$. The modal logic **Lq**, introduced by Esakia [1988], is axiomatized by the following schemata:

1. all propositional tautologies in the bimodal language,
2. the axioms of **L** for $\Box$,
3. the axioms of **S5** for $\forall$, i.e.,
   - $\forall(A \to B) \to (\forall A \to \forall B)$,
   - $\forall A \to A$,
   - $\exists A \to \forall \exists A$ ($\exists$ abbreviates $\neg\forall\neg$),
4. $\Box \forall A \to \forall \Box A$,

together with the rules modus ponens, $A/\Box A$ and $A/\forall A$. For this logic (the language of which is understood as a fragment of the predicate modal language) we have the following modal completeness theorem:

16.13. Theorem. (Japaridze [1988a,1990a]) *For any* **Lq**-*formula* A, $\vdash_{\mathbf{Lq}} A$ *iff* A *is valid in all finite predicate Kripke-models with a transitive and converse well-founded accessibility relation.*

In view of the evident arithmetic soundness of **Lq**, this modal completeness theorem together with the above predicate version of Solovay's first theorem implies the arithmetic completeness of **Lq**:

16.14. Corollary. *For any* **Lq**-*formula* A, $\vdash_{\mathbf{Lq}} A$ *iff every realizational instance of* A *is provable in* T.

Japaridze [1988a,1990a] also introduced the bimodal version **Sq** of **S** and proved that $\vdash_{\mathbf{Sq}} A$ iff every realizational instance of A is true. The axioms of **Sq** are all theorems of **Lq** plus $\Box A \to A$, and the rules of inference are Modus Ponens and $A/\forall A$.

Taking into account that we deal with a predicate language, the requirement of finiteness of the models in theorem 16.9 is a very undesirable restriction however. In Japaridze [1990a] a stronger variant of this theorem was given with the condition of finiteness replaced by a weaker one. What we need instead of finiteness, is roughly the following:

(1) The relations $w \in W$, $w R u$, $a \in D_w$ must be binumerable in T (see definition 12.1), and this fact must be provable in T.

(2) The relation $\Vdash$ must be numerable in T and T must prove that fact. To be more precise, $\Vdash$ need not be defined for all worlds and all formulas, but only for those which are needed to falsify the formula A in the root of the model (i.e., in some cases we may have neither $w \Vdash B$ nor $w \Vdash \neg B$); T should just prove that $\Vdash$ behaves "properly", e.g., $w \Vdash B \implies w \nVdash \neg B$, $w \Vdash B \vee C \implies (w \Vdash B$ or $w \Vdash C)$, $w \Vdash \neg (B \vee C) \implies (w \Vdash \neg B$ and $w \Vdash \neg C)$, ...

(3) T also must "prove" that the relation R is transitive and converse well-founded. Of course, well-foundedness is not expressible in the first order language, and T should somehow simulate a proof of this property of R. This is the case if, e.g., T proves the scheme of R-induction for the elements of W, i.e.,

$$T \vdash \forall w \in W \Big(\forall u (w R u \to \varphi(u)) \to \varphi(w) \Big) \to \forall w \in W\, \varphi(w).$$

We want to end this section by mentioning one last positive result. Let **QL** be the logic which arises by adding to **L** (written in the predicate modal language) the axioms and rules of the classical predicate calculus. Similarly, let **QS** be the closure of **S** with respect to classical predicate logic.

16.15. Theorem. (Japaridze [1990a,1991]). *Suppose T is strong enough to prove all true Π_1-sentences, and A is a closed predicate modal formula which satisfies one of the following conditions:*

(i) *no occurrence of a quantifier is in the scope of some occurrence of $\Box$ in A, or*

(ii) *no occurrence of $\Box$ is in the scope of some other occurrence of $\Box$ in A, or*

(iii) *A has the form $\Box^n \bot \to B$.*

Then we have:

(a) $\vdash_{\mathbf{QL}} A$ *iff all realizational instances of A are provable in T,*

(b) $\vdash_{\mathbf{QS}} A$ *iff all realizational instances of A are true.*

(Of course, clause (b) is trivial in case (iii).) The proof for the (ii) and (iii)-fragments in Japaridze [1990a] is based on the above-mentioned strong variant of the predicate version of Solovay's theorems. Both Vardanyan's and Artëmov's theorems on nonenumerability and nonarithmeticity hold for the (i) and (ii)-fragments as well, but this is not in contradiction with theorem 16.15. The point is that the use of Tennenbaum's theorem in the proofs of these negative results is possible only on assumption of the recursive enumerability of T, whereas no consistent theory which proves all the true Π_1-sentences can be recursively enumerable. Thus there are no immediate objections against the optimistic conjecture that **QL** and **QS** are complete for such strong theories without any restriction on the language.

17. Acknowledgements

In the first place we are very grateful to Lev Beklemishev for providing us with a draft for the sections 6, 7 and 8 in a near perfect state. He also gave extensive comments on other sections. Sergei Artëmov supported us with section 10, and in answering some questions for us. Albert Visser was very helpful with comments and discussions, answering questions, and pointing out mistakes. Giovanni Sambin provided valuable comments. Claes Strannegård shared his expertise with us. Joost Joosten, Rosalie Iemhoff and Eva Hoogland found quite a number of inaccuracies in the manuscript. Anne Troelstra stood by us with advice. Sam Buss was a very helpful editor and careful reader.

The first author was supported by N.W.O., the Dutch Foundation for Scientific Research, while working on this chapter in 1992-1993, and by the National Science Foundation (grant CCR-9403447) while working on its final version in 1997.

References

S. N. Artëmov

[1980] Arithmetically complete modal theories, *Semiotika i Informatika, VINITI, Moscow*, 14, pp. 115–133. In Russian, English translation in: *Amer. Math. Soc. Transl.* (2), 135: 39–54, 1987.

[1985a] Nonarithmeticity of truth predicate logics of provability, *Doklady Akademii Nauk SSSR*, 284, pp. 270–271. In Russian, English translation in *Soviet Math. Dokl.* 32 (1985), pp. 403-405.

[1985b] On modal logics axiomatizing provability, *Izvestiya Akad. Nauk SSSR, ser. mat.*, 49, pp. 1123–1154. In Russian, English translation in: *Math. USSR Izvestiya* 27(3).

[1986] Numerically correct logics of provability, *Doklady Akademii Nauk SSSR*, 290, pp. 1289–1292. In Russian.

[1994] Logic of proofs, *Annals of Pure and Applied Logic*, 67, pp. 29–59.

[1995] *Operational Modal Logic*, Tech. Rep. MSI 95-29, Cornell University.

S. N. Artëmov and G. K. Japaridze (Dzhaparidze)

[1987] On effective predicate logics of provability, *Doklady Akademii Nauk SSSR*, 297, pp. 521–523. In Russian, English translation in *Soviet Math. Dokl.* 36 (1987), pp. 478-480.

[1990] Finite Kripke models and predicate logics of provability, *Journal of Symbolic Logic*, 55, pp. 1090–1098.

A. Avron

[1984] On modal systems having arithmetical interpretations, *Journal of Symbolic Logic*, 49, pp. 935–942.

L. D. Beklemishev

[1989a] On the classification of propositional provability logics, *Izvestiya Akademii Nauk SSSR, ser. mat. D*, 53, pp. 915–943. In Russian, English translation in *Math. USSR Izvestiya* 35 (1990) 247–275.

[1989b] A provability logic without Craig'sprotect interpolation property, *Matematicheskie Zametkie*, 45, pp. 12–22. In Russian, English translation in *Math. Notes* 45 (1989).

[1991] Provability logics for natural Turing progressions of arithmetical theories, *Studia Logica*, pp. 107–128.

[1992] Independent numerations of theories and recursive progressions, *Sibirskii Matematichskii Zhurnal*, 33, pp. 22–46. In Russian, English translation in *Siberian Math. Journal*, 33 (1992).

[1993a] On the complexity of arithmetic applications of modal formulae, *Archive for Mathematical Logic*, 32, pp. 229–238.
[1993b] Review of de Jongh and Montagna [1988,1989], Carbone and Montagna [1989,1990], *Journal of Symbolic Logic*, 58, pp. 715–717.
[1994] On bimodal logics of provability, *Annals of Pure and Applied Logic*, 68, pp. 115–160.
[1996a] Bimodal logics for extensions of arithmetical theories, *Journal of Symbolic Logic*, 61, pp. 91–124.
[1996b] Remarks on Magari-algebras of PA and $I\Delta_0 + EXP$, in: *Logic and Algebra*, A. Ursini and P. Aglianò, eds., Marcel Dekker, Inc., New York, pp. 317–326.

A. Berarducci
[1990] The interpretability logic of Peano arithmetic, *Journal of Symbolic Logic*, 55, pp. 1059–1089.

A. Berarducci and R. Verbrugge
[1993] On the provability logic of bounded arithmetic, *Annals of Pure and Applied Logic*, 61, pp. 75–93.

C. Bernardi
[1976] The uniqueness of the fixed-point in every diagonalizable algebra, *Studia Logica*, 35, pp. 335–343.

G. Boolos
[1979] *The Unprovability of Consistency*, Cambridge University Press.
[1981] Provability, truth and modal logic, *Journal of Philosophic Logic*, 9, pp. 1–7.
[1982] Extremely undecidable sentences, *Journal of Symbolic Logic*, 47, pp. 191–196.
[1993a] The analytical completeness of Dzhaparidze's polymodal logics, *Annals of Pure and Applied Logic*, 61, pp. 95–111.
[1993b] *The Logic of Provability*, Cambridge University Press.

G. Boolos and R. C. Jeffrey
[1989] *Computability and Logic, 3rd ed.*, Cambridge University Press.

G. Boolos and V. McGee
[1987] The degree of the set of sentences of predicate provability logic that are true under every interpretation, *Journal of Symbolic Logic*, 52, pp. 165–171.

G. Boolos and G. Sambin
[1991] Provability: the emergence of a mathematical modality, *Studia Logica*, 50, pp. 1–23.

A. Carbone and F. Montagna
[1989] Rosser orderings in bimodal logics, *Zeitschrift für Mathematische Logik und Grundlagen der Mathematik*, 35, pp. 343–358.
[1990] Much shorter proofs: a bimodal investigation, *Zeitschrift für Mathematische Logik und Grundlagen der Mathematik*, 36, pp. 47–66.

T. Carlson
[1986] Modal logics with several operators and provability interpretations, *Israel Journal of Mathematics*, 54, pp. 14–24.

B. F. Chellas
[1980] *Modal Logic: An Introduction*, Cambridge University Press.

P. Clote and J. Krajíček
[1993] eds., *Arithmetic, Proof Theory and Computational Complexity*, Oxford University Press.

D. van Dalen
[1994] *Logic and Structure*, Springer Verlag, Berlin, Amsterdam.

L. L. Esakia
[1988] Provability logic with quantifier modalities, in: *Intensional Logics and the Logical Structure of Theories: Material from the fourth Soviet-Finnish Symposium on Logic, Telavi, May 20-24, 1985*, Metsniereba, Tbilisi, pp. 4–9. In Russian.

S. FEFERMAN
[1960] Arithmetization of metamathematics in a general setting, *Archive for Mathematical Logic*, 6, pp. 52–63.
[1962] Transfinite recursive progressions of axiomatic theories, *Journal of Symbolic Logic*, 27, pp. 259–316.

S. FEFERMAN, G. KREISEL, AND S. OREY
[1960] 1-consistency and faithful interpretations, *Fundamenta Mathematicae*, 49, pp. 35–92.

Z. GLEIT AND W. GOLDFARB
[1990] Characters and fixed points in provability logic, *Notre Dame Journal of Formal Logic*, 31, pp. 26–551.

K. GÖDEL
[1933] Eine Interpretation des intuitionistischen Aussagenkalkuls, *Ergebnisse Math. Colloq.*, Bd. 4, pp. 39–40.

D. GUASPARI
[1979] Partially conservative extensions of arithmetic, *Transactions of the American Mathematical Society*, 254, pp. 47–68.
[1983] Sentences implying their own provability, *Journal of Symbolic Logic*, 48, pp. 777–789.

D. GUASPARI AND R. M. SOLOVAY
[1979] Rosser sentences, *Annals of Mathematical Logic*, 16, pp. 81–99.

P. HÁJEK
[1971] On interpretability in set theories I, *Comm. Math. Univ. Carolinae*, 12, pp. 73–79.
[1972] On interpretability in set theories II, *Comm. Math. Univ. Carolinae*, 13, pp. 445–455.

P. HÁJEK AND F. MONTAGNA
[1990] The logic of Π_1-conservativity, *Archiv für Mathematische Logik und Grundlagenforschung*, 30, pp. 113–123.
[1992] The logic of Π_1-conservativity continued, *Archiv für Mathematische Logik und Grundlagenforschung*, 32, pp. 57–63.

P. HÁJEK, F. MONTAGNA, AND P. PUDLAK
[1993] Abbreviating proofs using metamathematical rules, in: *Clote and Krajíček [1993]*, pp. 387–428.

D. HAREL
[1984] Dynamic logic, in: *Handbook of Philosophic Logic, Volume II, Extensions of Classical Logic*, D. Gabbay and F. Guenthener, eds., Kluwer Academic Publishers, Dordrecht, Boston, pp. 497–604.

D. HILBERT AND P. BERNAYS
[1939] *Grundlagen der Mathematik II*, Springer, Berlin.

G. E. HUGHES AND M. J. CRESSWELL
[1984] *A Companion to MODAL LOGIC*, Methuen, London, New York.

K. N. IGNATIEV
[1990] *The logic of Σ_1-interpolability over Peano arithmetic.* Manuscript. In Russian.
[1993a] On strong provability predicates and the associated modal logics, *Journal of Symbolic Logic*, 58, pp. 249–290.
[1993b] The provability logic of Σ_1-interpolability, *Annals of Pure and Applied Logic*, 64, pp. 1–25.

G. K. JAPARIDZE (DZHAPARIDZE)
[1986] *The Modal Logical Means of Investigation of Provability*, PhD thesis, Moscow State University. In Russian.
[1988a] The arithmetical completeness of the logic of provability with quantifier modalities, *Bull. Acad. Sci. Georgian SSR*, 132, pp. 265–268. In Russian.

[1988b] The polymodal logic of provability, in: *Intensional Logics and the Logical Structure of Theories: Material from the fourth Soviet-Finnish Symposium on Logic, Telavi, May 20-24, 1985*, Metsniereba, Tbilisi, pp. 16–48. In Russian.

[1990a] Decidable and enumerable predicate logics of provability, *Studia Logica*, 49, pp. 7–21.

[1990b] Provability logic with modalities for arithmetical complexities, *Bull. Acad. Sci. Georgian SSR*, 138, pp. 481–484.

[1991] Predicate provability logic with non-modalized quantifiers, *Studia Logica*, 50, pp. 149–160.

[1992] The logic of linear tolerance, *Studia Logica*, 51, pp. 249–277.

[1993] A generalized notion of weak interpretability and the corresponding logic, *Annals of Pure and Applied Logic*, 61, pp. 113–160.

[1994a] The logic of arithmetical hierarchy, *Annals of Pure and Applied Logic*, 66, pp. 89–112.

[1994b] A simple proof of arithmetical completeness for Π_1-conservativity logic, *Notre Dame Journal of Formal Logic*, 35, pp. 346–354.

D. H. J. de Jongh

[1987] A simplification of a completeness proof of Guaspari and Solovay, *Studia Logica*, 46, pp. 187–192.

D. H. J. de Jongh, M. Jumelet, and F. Montagna

[1991] On the proof of Solovay's theorem, *Studia Logica*, 50, pp. 51–70.

D. H. J. de Jongh and F. Montagna

[1988] Provable fixed points, *Zeitschrift für Mathematische Logik und Grundlagen der Mathematik*, 34, pp. 229–250.

[1989] Much shorter proofs, *Zeitschrift für Mathematische Logik und Grundlagen der Mathematik*, 35, pp. 247–260.

[1991] Rosser-orderings and free variables, *Studia Logica*, 50, pp. 71–80.

D. H. J. de Jongh and D. Pianigiani

[1998] Solution of a problem of David Guaspari, *Studia Logica*, 57. To appear.

D. H. J. de Jongh and F. Veltman

[1990] Provability logics for relative interpretability, in: *Petkov [1990]*, pp. 31–42.

D. H. J. de Jongh and A. Visser

[1991] Explicit fixed points in interpretability logic, *Studia Logica*, 50, pp. 39–50.

G. Kreisel and A. Lévy

[1968] Reflection principles and their use for establishing the complexity of axiomatic systems, *Zeitschrift für Mathematische Logik und Grundlagen der Mathematik*, 14, pp. 97–142.

P. Lindström

[1984] On faithful interpretability, in: *Computation and Proof Theory*, M. M. Richter, E. Börger, W. Oberschelp, B. Schinzel, and W. Thomas, eds., Lecture Notes in Mathematics #1104, Springer Verlag, Berlin, Berlin, pp. 279–288.

[1994] *The Modal Logic of Parikh Provability*, Tech. Rep. Filosofiska Meddelanden, Gröna serien, No. 5, University of Göteborg.

J. C. C. McKinsey and A. Tarski

[1948] Some theorems about the calculi of Lewis and Heyting, *Journal of Symbolic Logic*, 13, pp. 1–15.

F. Montagna

[1979] On the diagonalizable algebra of Peano arithmetic, *Bulletino della Unione Matematica Italiana*, 5, 16B, pp. 795–812.

[1984] The predicate modal logic of provability, *Notre Dame Journal of Formal Logic*, 25, pp. 179–189.

[1987] Provability in finite subtheories of PA, *Journal of Symbolic Logic*, 52, pp. 494–511.

[1992] Polynomially and superexponentially shorter proofs in fragments of arithmetic, *Journal of Symbolic Logic*, 57, pp. 844–863.

S. OREY
[1961] Relative interpretations, *Zeitschrift für Mathematische Logik und Grundlagen der Mathematik*, 7, pp. 146–153.

R. PARIKH
[1971] Existence and feasibility, *Journal of Symbolic Logic*, 36, pp. 494–508.

P. P. PETKOV
[1990] ed., *Mathematical Logic, Proceedings of the Heyting 1988 Summer School*, New York, Plenum Press.

M. DE RIJKE
[1992] Unary interpretability logic, *Notre Dame Journal of Formal Logic*, 33, pp. 249–272.

G. SAMBIN
[1976] An effective fixed-point theorem in intuitionistic diagonalizable algebras, *Studia Logica*, 35, pp. 345–361.

G. SAMBIN AND S. VALENTINI
[1982] The modal logic of provability. The sequential approach., *Journal of Philosophical Logic*, 11, pp. 311–342.
[1983] The modal logic of provability: cut elimination., *Journal of Philosophical Logic*, 12, pp. 471–476.

D. S. SCOTT
[1962] Algebras of sets binumerable in complete extensions of arithmetic, in: *Recursive Function Theory*, American Mathematical Society, Providence, R.I., pp. 117–121.

V. Y. SHAVRUKOV
[1988] *The Logic of Relative Interpretability over Peano Arithmetic*, Tech. Rep. Report No.5, Stekhlov Mathematical Institute, Moscow. (in Russian).
[1991] On Rosser's provability predicate, *Zeitschrift für Mathematische Logik und Grundlagen der Mathematik*, 37, pp. 317–330.
[1993a] A note on the diagonalizable algebras of PA and ZF, *Annals of Pure and Applied Logic*, 61, pp. 161–173.
[1993b] Subalgebras of diagonalizable algebras of theories containing arithmetic, *Dissertationes mathematicae (Rozprawy matematycne)*, 323. Instytut Matematyczny, Polska Akademia Nauk, Warsaw.
[1994] A smart child of Peano's, *Notre Dame Journal of Formal Logic*, 35, pp. 161–185.
[1997] Undecidability in diagonalizable algebras, *Journal of Symbolic Logic*, 62, pp. 79–116.

C. SMORYŃSKI
[1977] The incompleteness theorems, in: *Handbook of Mathematical Logic*, J. Barwise, ed., vol. 4, North–Holland, Amsterdam, Amsterdam, pp. 821–865.
[1978] Beth's theorem and self–referential statements, in: *Computation and Proof Theory*, A. Macintyre, L. Pacholski, and J. B. Paris, eds., North–Holland, Amsterdam, Amsterdam, pp. 17–36.
[1985] *Self-reference and modal logic*, Springer-Verlag, Berlin.

R. M. SOLOVAY
[1976] Provability interpretations of modal logic, *Israel Journal of Mathematics*, 25, pp. 287–304.

C. STRANNEGÅRD
[1997] *Arithmetical Realizations of Modal Formulas*, PhD thesis, University of Göteborg, Acta Philosophica Gothoburgensia 5.

A. TARSKI, A. MOSTOWSKI, AND R. M. ROBINSON
[1953] *Undecidable Theories*, North–Holland, Amsterdam, Amsterdam.

A. S. TROELSTRA AND H. SCHWICHTENBERG

[1996] *Basic Proof Theory*, Cambridge University Press.

A. TURING

[1939] System of logics based on ordinals, *Proceedings of the London Mathematical Society, Ser. 2*, 45, pp. 161–228.

V. A. VARDANYAN

[1986] Arithmetic complexity of predicate logics of provability and their fragments, *Doklady Akademii Nauk SSSR*, 288, pp. 11–14. In Russian, English translation in *Soviet Math. Dokl.* 33 (1986), pp. 569-572.

R. VERBRUGGE

[1993a] *Efficient Metamathematics*, PhD thesis, Universiteit van Amsterdam, ILLC-disseration series 1993-3.

[1993b] Feasible interpretability, in: *Clote and Krajíček [1993]*, pp. 197–221.

A. VISSER

[1980] Numerations, λ-calculus and arithmetic, in: *To H.B. Curry: Essays on Combinatory logic, lambda calculus and formalism*, J. P. Seldin and J. R. Hindley, eds., Academic Press, Inc., London, pp. 259–284.

[1981] *Aspects of Diagonalization and Provability*, PhD thesis, University of Utrecht, Utrecht, The Netherlands.

[1984] The provability logics of recursively enumerable theories extending Peano arithmetic at arbitrary theories extending Peano arithmetic, *Journal of Philosophical Logic*, 13, pp. 97–113.

[1985] *Evaluation, Provably Deductive Equivalence in Heyting's arithmetic of Substitution Instances of Propositional Formulas*, Tech. Rep. LGPS 4, Department of Philosophy, Utrecht University.

[1989] Peano's smart children. a provability logical study of systems with built-in consistency, *Notre Dame Journal of Formal Logic*, 30, pp. 161–196.

[1990] Interpretability logic, in: *Petkov [1990]*, pp. 175–209.

[1991] The formalization of interpretability, *Studia Logica*, 50, pp. 81–106.

[1994] *Propositional Combinations of Σ-Sentences in Heyting's Arithmetic*, Tech. Rep. LGPS 117, Department of Philosophy, Utrecht University. To appear in the Annals of Pure and Applied Logic.

[1995] A course in bimodal provability logic, *Annals of Pure and Applied Logic*, 73, pp. 109–142.

[1997] An overview of interpretability logic, in: *Advances in Modal Logic '96*, M. Kracht, M. de Rijke, and H. Wansing, eds., CSLI Publications, Stanford.

A. VISSER, J. VAN BENTHEM, D. H. J. DE JONGH, AND G. R. RENARDEL DE LAVALETTE

[1995] *NILL*, a study in intuitionistic propositional logic, in: *Modal Logic and Process Algebra, a Bisimulation Perspective*, A. Ponse, M. de Rijke, and Y. Venema, eds., CSLI Lecture Notes #53, CSLI Publications, Stanford, pp. 289–326.

F. VOORBRAAK

[1988] A simplification of the completeness proofs for Guaspari and Solovay's R, *Notre Dame Journal of Formal Logic*, 31, pp. 44–63.

D. ZAMBELLA

[1992] On the proofs of arithmetical completeness of interpretability logic, *Notre Dame Journal of Formal Logic*, 35, pp. 542–551.

[1994] Shavrukov's theorem on the subalgebras of diagonalizable algebras for theories containing $I\Delta_0 + \mathrm{EXP}$, *Notre Dame Journal of Formal Logic*, 35, pp. 147–157.

CHAPTER VIII

The Lengths of Proofs

Pavel Pudlák
Mathematical Institute, Academy of Sciences of the Czech Republic
115 67 Prague 1, The Czech Republic

Contents

HANDBOOK OF PROOF THEORY
Edited by S. R. Buss

1. Introduction

In this chapter we shall consider the problem of determining the minimal complexity of a proof of a theorem in a given proof system. We shall deal with propositional logic and first order logic. There are several measures of complexity of a proof and there are many different proof systems. Let us give some reasons for this research, before we discuss particular instances of the problem.

1.1. Our subject could be called *the quantitative study of the proofs.* In contrast with the classical proof theory we want to know not only whether a theorem has a proof but also whether the proof is feasible, i.e., can be actually written down or checked by a computer. An ideal justification for such research would be a proof that a particular theorem for which we have only long proofs (such as the four color theorem), or a conjecture for which we do not have any proof (such as $\mathcal{P} \neq \mathcal{NP}$), does not have a short proof in some reasonable theory (such as **ZF**). Presently this seems to be a very distant goal; we are only able to prove lower bounds on the lengths of proofs for artificial statements, or for natural statements, but in very weak proof systems. The situation here is similar to the situation in the study of (weak) fragments of arithmetic and complexity theory. In fragments of arithmetic we can prove unprovability of Π_1^0 sentences only for sentences obtained by diagonalization, and in complexity theory we can separate complexity classes also only when diagonalization is possible. These three areas are very much connected and it is not possible to advance very much in one of them without making progress in the others.

Nevertheless there are already now some practical consequences of this research. For instance in first order logic we know quite precisely how much cut-elimination increases the size of proofs. In propositional logic we have simple tautologies which have only exponentially long resolution proofs. This is very important information for designers of automated theorem provers.

Another reason for studying the lengths of proofs is that information about the size of proofs is very important in the study of weak fragments of arithmetic, namely when metamathematics of fragments is considered. For instance, in bounded arithmetic the exponentiation function is not provably total. Therefore the cut-elimination theorem is not provable in bounded arithmetic (in fact first order cut-elimination requires more than elementary increase in the size of proofs). Consequently we have (at least) two different concepts of consistency in bounded arithmetic: the usual one and cut-free consistency.

Furthermore there is a relation between provability in bounded arithmetic and the lengths of proofs in certain proof systems for propositional logic. This seems to be the most promising way of proving concrete independence results for bounded arithmetic.

Finally this area is important because of its tight relation to complexity theory. Actually, research into the lengths of proofs should be considered as a part of complexity theory. There are two kinds of connections with computational complexity.

On the one hand there are explicit connections such as the fact that a proof system for propositional logic is a nondeterministic algorithm for the ($co\mathcal{NP}$ complete) set of tautologies. On the other hand there are intuitive connections which are not supported by theorems. For example the relation between Frege systems and extension Frege systems (see below for definitions) for propositional logic is very much like the relation between the complexity measures of boolean functions based on formula size and circuit size, respectively. It is an open problem whether Frege systems are as powerful as extension Frege systems and also it is an open problem whether formulas are as powerful as circuits; but we are not able to prove any of two implications between these apparently related problems. Some people think that the difficult problems in complexity theory such as $\mathcal{P} = \mathcal{NP}$? are essentially logical (not combinatorial) problems. If it is so, then proof theory, and in particular the lengths of proofs, should play an important role in their solution.

1.2. Now we shall briefly outline the contents of this chapter. Section 2 introduces some basic concepts. In section 3 we describe a technique of constructing short formulas for inductively defined concepts. This technique has various applications. Section 4 contains results about dependence of different measures of complexity of proofs and a remark on the popular Kreisel Conjecture. In section 5 we shall consider the cut-elimination theorem from the point of view of the lengths of proofs; namely, we shall show a lower bound on the increase of the length. In section 6 we shall prove a version of the second incompleteness theorem for finite consistencies. This enables us to prove some concrete lower bounds and speed-up. In section 7 we survey speed-up theorems, namely results about shortening of proofs when a stronger theory is used instead of a weaker one and related results. Section 8 is a survey of the most important propositional proof systems. In section 9 we give a nontrivial example of a lower bound on the lengths of propositional proofs in the resolution system. In section 10 we present important relations between the lengths of proofs in propositional logic and provability in fragments of arithmetic. The final section 11 surveys especially those results which have not been treated in the main text.

2. Types of proofs and measures of complexity

In this section we introduce notation and some basic concepts used in both propositional logic and first order logic.

2.1. One can consider many different formalizations and it is difficult to find a classification schema which would cover all. There is however one basic property which all formalizations of the concept of a proof must satisfy: it must be computable in polynomial time whether a given sequence is a proof of a given formula. Here we assume, as usual, that proofs and formulas are encoded as strings in a finite alphabet and we identify feasible computations with polynomial time computations. This trivial observation gives us important link to computational complexity. The proof systems in such a general setting are just nondeterministic decision procedures for the set of tautologies or the set of theorems of a theory in question. More specifically, an upper bound on the size of proofs for a particular proof system gives a nondeterministic decision procedure with the bound on the running time and, conversely, a lower bound on the nondeterministic time complexity is a lower bound for any proof system.

In particular, let $TAUT$ be the set of propositional tautologies in some fixed complete basis of connectives. A *propositional proof system* is a binary relation $P(x, y)$ which is *computable in polynomial time* and

$$\varphi \in TAUT \;\equiv\; \exists y\, P(\varphi, y).$$

Since the set of propositional tautologies is $\mathcal{NP}$-complete, we get the following immediate corollary.

2.1.1. Theorem. (Cook and Reckhow [1979]) *There exists a proof system for propositional logic in which all tautologies have proofs of polynomial length if and only if* $\mathcal{NP} = co\mathcal{NP}$. □

This general concept of a proof system can be generalized even further. Firstly, we can allow randomized computations; secondly, we can assume that the proof is not given to us, but we can access parts of the proof via an oracle. Usually such an *interactive proof system* is presented as a two player game, where we are the *Verifier* and the oracle is the *Prover*. It turns out that the Verifier can check with high probability that a proof for a given formula exists without learning almost anything about the proof. The most striking example is the so-called PCP theorem by Arora et al. [1992]. Roughly speaking, they showed, that there there are interactive proof systems, where the Verifier needs to check only a constant number of randomly selected bits of the sequence in order to check with high probability that the proof is correct.

Note, however, that these results concern only the question *how can be proofs checked* but do not give new information about *the lengths of proofs.*

2.2. We turn now to more structured proofs, which are typical for logic, while the above concepts rather belong to complexity theory. Such proof systems are usually defined using a finite list of *deduction rules.* The basic element of a proof, called *a proof step,* or *a proof line*, is a formula, a set of formulas, a sequence of formulas or a sequent (pair of sequences of formulas). A *proof* is either a *sequence* or a *tree* of proof steps such that each step is an axiom or follows from previous ones by a deduction rule. The complete information about the intended way of proving a given theorem should also contain the information for each step of which rule is applied and to which previous steps it is applied. However in most cases this does not influence the complexity of the proofs essentially.

It is important to realize that when proof lines and deduction rules are determined, there are two possible forms of proofs: *the tree form* and *the sequence form.* In the tree form, a proof line may be a premise of an application of a rule only once, while in the sequence form it can be used again and again. The trivial transformation from the sequence form to tree form results in exponential increase of size.

The most important measure of complexity of proofs is the *size* of a proof. We take a finite alphabet and a natural encoding of proofs as sequences (words) in a finite alphabet. Then the size of a proof is the length of its code.

The next one is the *number of proof lines.* Trivially, the number of proof lines is at most the size, however, a proof may contain very large formulas, thus there is an essential difference between the two measures.

Quite often it is important to bound the maximal complexity of formulas in the proof. Usually we consider the quantifier complexity or the number of logical symbols. Thus we get other measures.

Comparing the above measures with the complexity measures in computational complexity we see that the size corresponds clearly to time. At first glance it may seem that the maximal size of a formula (or proof line) should correspond to space, but this is not correct. In order to present a proof in a lecture, or to check it on a computer we cannot show a single formula (proof line) at a time, we have to keep the formulas (lemmas) on the blackboard until they are used for the last time as premises. The minimal size of a blackboard on which the proof can be presented is the right concept corresponding to space. Note that a suitable choice of the concept of a proof line and rules leads to *linear proofs*, where each rule has at most one premise (Craig [1957a]). In such proofs the maximal size of a proof line is the measure corresponding to space.

In first order logic we consider also the proofs in a *theory* T. This means that we can use axioms of T in proofs. Talking about theories is not quite precise here; different axiomatizations give clearly different concepts of proofs and hence the smallest size proofs of a given formula may be different. Therefore we shall use preferably the term *axiomatization.*

2.3. We shall use the following notation. The size of a formula φ resp. a proof d will be denoted by $|\varphi|$ resp. $|d|$. Let A be a proof system or a proof system plus an axiomatization of a theory. Then we write $d : A \vdash \varphi$, if d is a proof of φ in A; $A \vdash \varphi$, if $\varphi$is provable in A; and $A \vdash^n \varphi$, if φ has a proof of size $\leq n$ in A. Note that the same notation is often used for the number of proof steps. We shall distinguish it by writing $A \vdash^n_{\text{steps}} \varphi$. Often it is more convenient to use the alternative notation:

$$||\varphi||_A = \begin{cases} \text{minimal } n \text{ such that } A \vdash^n \varphi \text{ if } A \vdash \varphi \\ \infty \text{ otherwise.} \end{cases}$$

This enables us to write inequalities such as

$$||\psi||_A \leq ||\varphi||_A + ||\varphi \rightarrow \psi||_A + |\psi| + O(1),$$

which holds in the presence of modus ponens in A.

2.4. Suppose that we consider a particular logical calculus. In propositional logic, this simply means that we fix a set of connectives; in first order logic, this means that we fix a language and, possibly consider some theory. Then we can compare the power of different proof systems with respect to the complexity of proofs. If we consider the *size* of proofs, then it is quite natural to disregard polynomial differences in proofs. In particular we define $P_1 \preceq P_2$, if there exists a polynomial $p(x)$ such that for each tautology (resp. theorem) φ, if $d_1 : P_1 \vdash \varphi$, then for some $d_2, |d_2| \leq p(|d_1|), d_2 : P_2 \vdash \varphi$, (using the norm notation: $\|\varphi\|_{P_1} \leq p(\|\varphi\|_{P_2})$).

Usually, if $P_1 \preceq P_2$, then there exists a polynomial time algorithm to construct d_2 from d_1; in such a situation we say that P_2 *polynomially simulates* P_1, (see Cook and Reckhow [1979]). We say that P_1 is *polynomially equivalent* to P_2, if P_1 and P_2 polynomially simulate each other.

A well-known theorem of Craig states that a theory has a recursive axiomatization, if it is recursively enumerable. It is an easy exercise to prove the following modification of the theorem.

2.4.1. Proposition. *Let P_1 be an arbitrary proof system for a calculus with the connective of implication. Then there exists a polynomially equivalent calculus P_2 based on a polynomial time decidable set of axioms and the single rule of* modus ponens. □

Consequently one has to consider stronger assumptions in order to restrict the class of proof systems. The usual approach is to work with the *schematic theories* of Parikh [1973], where we have a *finite* set of rules and *axiom schemas.*

2.5. We shall conclude this section by presenting the most often used proof systems for first order logic; we consider those used in mathematical logic, there are several others used in artificial intelligence, see Chang and Lee [1973] and Eder [1992].

2.5.1. Gentzen [1935] attributes the following system to Hilbert and Glivenko:

Rules

$$(6.1) \quad \frac{A, A \to B}{B}$$

$$(6.2) \quad \frac{A \to \Phi(x)}{A \to \forall y \Phi(y)}, \text{ where } x \text{ does not occur in } A,$$

$$(6.3) \quad \frac{\Phi(x) \to A}{\exists x \Phi(x) \to A}, \text{ where } x \text{ does not occur in } A.$$

Axiom Schemas

(1.1) $A \to A$
(1.2) $A \to (B \to A)$
(1.3) $(A \to (A \to B)) \to (A \to B)$
(1.4) $(A \to (B \to C)) \to (B \to (A \to C))$
(1.5) $(A \to B) \to ((B \to C) \to (A \to C))$
(2.1) $(A \wedge B) \to A$
(2.2) $(A \wedge B) \to B$
(2.3) $(A \to B) \to ((A \to C) \to (A \to (B \wedge C)$
(3.1) $A \to (A \vee B)$
(3.2) $B \to (A \vee B)$
(3.3) $(A \to C) \to ((B \to C) \to ((A \vee B) \to C))$
(4.1) $(A \to B) \to ((A \to \neg B) \to \neg A)$
(4.2) $\neg A \to (A \to B)$
(5.1) $\forall x \Phi(x) \to \Phi(t)$
(5.1) $\Phi(t) \to \exists x \Phi(x)$

(t stands for a term, x, y are variables).

We shall refer to this calculus as the *Hilbert style* calculus.Note that in a system such as above we can either say that we have *axiom schemas* or that we have *axioms and allow the substitution rule to be applied only to axioms.* We shall consider the power of various proof systems for propositional logic in section 8. The propositional part of the Hilbert style system is a special case of calculi called Frege systems. Contrary to the history, the general substitution rule is not permitted in Frege systems.

There are more compact Hilbert style systems, e.g. the one considered by Hilbert and Ackermann [1928], use only the connectives $\vee$ and $\neg$. As we shall see, the propositional parts simulate each other and (unless we use some strange quantifier rules) this can be extended to the whole systems.

Let us note that there are natural proof systems for first order logic which have only modus ponens as a rule and the quantifier rules are replaced by a finite number of simple axiom schemas, see e.g. Grzegorczyk [1974].

2.5.2. Another important system has been introduced by Gentzen [1935]. The basic elements of the proof are *sequents* which are sequences $\varphi_1, \ldots, \varphi_n \longrightarrow \psi_1, \ldots, \psi_m$. Here $\longrightarrow$ is a syntactical symbol, a different symbol $\to$ is used for implication. The interpretation of such a sequent is $\varphi_1 \wedge \ldots \wedge \varphi_n \to \psi_1 \vee \ldots \vee \psi_m$ (with $\to$ standing now for implication). The system has a single axiom scheme $A \longrightarrow A$, where A is a formula, and several rules which have one or two sequents as assumptions and one sequent as a conclusion. A proof is a tree of sequents where leaves are instance of the axiom and every other sequent follows from its predecessors by a rule. The tree structure is very convenient for analyzing proofs, but one can also consider sequences of sequents as a proof. The most important rule as the *cut rule*

$$\frac{\alpha_1, \ldots, \alpha_k \longrightarrow \beta_1, \ldots, \beta_l, \varphi \qquad \varphi, \gamma_1, \ldots, \gamma_m \longrightarrow \delta_1, \ldots, \delta_n}{\alpha_1, \ldots, \alpha_k, \gamma_1, \ldots, \gamma_m \longrightarrow \beta_1, \ldots, \beta_l, \delta_1, \ldots, \delta_n}$$

Observe that for $k = l = m = 0, n = 1$ we get essentially modus ponens. The whole system is described in Chapter I. Gentzen presented transformations of proofs from the Hilbert style calculus to his sequent calculus and vice versa. In Eder [1992] it is shown that this in fact gives polynomial simulations of the systems if

1. in both we take tree-proofs or
2. in both we take sequence-proofs.

In section 4 we shall show that there is also polynomial simulation of sequence-proofs by tree-proofs in the Hilbert style calculus. Thus the most commonly used systems are polynomially equivalent.

The systems above are prototypes of what is called *a schematic theory*. This concept is a natural extension of the concept of the Frege system used in propositional logic. In first order logic, however, it is not easy to define precisely such a concept especially because restrictions on occurrences of variables in quantifier rules (or axioms) are needed. For possible definitions of schematic theories see Vaught [1967], Parikh [1973], Krajíček [1989a], Farmer [1984,1988] and Buss [1994].

Hilbert's *ε-calculus* is based on a different language. Instead of quantifiers it uses ε-terms $\varepsilon_{\varphi(x)}$ whose meaning is an element which satisfies the formula $\varphi(x)$ if there is any, otherwise $\varepsilon_{\varphi(x)}$ is an arbitrary element. This system is described in the famous book of Hilbert and Bernays [1934,1939]. Other popular systems are Beth's system of *semantic tableaux*, described in Beth [1959] and Smullyan [1968], and Prawitz's *natural deduction* system, described in Prawitz [1970] and Girard [1989]. Brief descriptions of these systems can be found in Chapter I of this volume.

For most systems polynomial simulations have been found and it seems very likely that mutual polynomial simulations with other systems can be found. Thus I do not expect that interesting results on the length of proof can be obtained here. Nevertheless various systems may be useful in some other situations. E.g., as Matthias Baaz pointed out, the ε-calculus is useful in situations where we study the structure of the terms in the proof; the prominent example is Kreisel's Conjecture (see section 4).

2.6. The main theorem of Gentzen [1935] asserts that the sequent calculus *without the cut rule* is still complete. This is a very strong statement, since the cut rule is the only rule where some structure present in the assumptions is missing in the conclusion (if we disregard the terms). Some of the numerous application of the *cut-elimination theorem* and its proof can be found in Chapters I and II of this handbook. Of course we have to pay something for it and the price is high: the increase of the size cannot be bounded by an elementary recursive function, i.e., cannot be bounded by a constant number of iterations of the exponential function. We shall prove such a lower bound in section 5.

2.6.1. There are several theorems which are in a sense equivalent to the cut-elimination theorem: Herbrand's theorem, Hilbert's ε-theorem (see Hilbert and Bernays [1934,1939]), semantic tableaux. Each of them can be used to define a concept of a proof and the resulting measures are closely related. Namely, the known transformations give mutual simulations in time bounded by iterated exponential functions, see 5.1. Thus we have two main classes of proof systems for first order logic: (1) the unrestricted ones, and (2) cut-free (and the equivalent ones).

Considering the undecidability of first order logic, which means that we cannot bound the size of a proof of a formula by *any computable function*, it is quite surprising that the spectrum of natural complexity measures consists essentially of two elements. Can this empirical evidence be supported by a mathematical theorem?

3. Some short formulas and short proofs

In this section we discuss basic concepts used in the study of the length of proofs in first order logic and prove some bounds on the length of proofs. The upper bounds have two applications: firstly they enable us to show big differences in lengths between different types of proofs, the so-called speed-up; secondly, they are needed for reductions of the lower bounds on the length of proofs of one set of formulas to another one.

3.1. We shall use the Hilbert style proof system described in the previous section with the following axioms of equality:

$$x = x$$
$$x = y \rightarrow y = x,$$
$$x = y \wedge y = z \rightarrow x = z,$$
$$x_1 = y_1 \wedge \ldots \wedge x_n = y_n \rightarrow (R(x_1, \ldots, x_n) \rightarrow R(y_1, \ldots, y_n))$$

for each predicate symbol R, and

$$x_1 = y_1 \wedge \ldots \wedge x_n = y_n \rightarrow F(x_1, \ldots, x_n) = F(y_1, \ldots, y_n)$$

for each function symbol F.

3.2. The first question that we consider is the length of formulas defined by iterating a certain construction (some examples will be considered below). Let $\Phi(R, a_1, \ldots, a_k, b_1, \ldots, b_l)$ be a formula with a specified k-ary predicate symbol and where $a_1, \ldots, a_k, b_1, \ldots, b_l$ are all free variables of Φ. Let us abbreviate the strings of variables by $\bar{a}$ and $\bar{b}$. Now suppose a formula $\varphi_0(\bar{a}, \bar{b})$ is given and we need a sequence of formulas $\varphi_1, \varphi_2, \ldots$ such that

$$\varphi_{n+1}(\bar{a}, \bar{b}) \equiv \Phi(\varphi_n, \bar{a}, \bar{b}) \tag{1}$$

is provable in first order logic. Here $\equiv$ denotes the biconditional.

In order to understand better what is going on, let us write Φ as

$$\Phi(R(\bar{x}_1), \ldots, R(\bar{x}_t), \bar{a}, \bar{b}), \tag{2}$$

where $R(\bar{x}_i)$ denote particular occurrences of R in Φ and $\bar{x}_i$ is a string of k bound, not necessarily distinct, variables of Φ. Thus (1) is better represented by:

$$\varphi_{n+1}(\bar{a}, \bar{b}) \equiv \Phi(\varphi_n(\bar{x}_1, \bar{b}), \ldots, \varphi_n(\bar{x}_t, \bar{b}), \bar{a}, \bar{b}). \tag{3}$$

The variables $\bar{b}$ do not change, they are "parameters", thus we shall omit them from now on.

A trivial solution is to take $\varphi_{n+1}(\bar{a})$ to be *equal* to $\Phi(\varphi_n, \bar{a})$. However often we need φ_n to be of polynomial size and, in fact, we need a polynomial (in n) size proof of (3). If $t > 1$, which is usually the case, then mere substitutions lead to exponentially large formulas. The solution is to replace $\Phi(R, \bar{a})$ by an equivalent formula, in which R occurs only once. This is always possible if $\equiv$ as a connective is present in our language.

3.2.1. Theorem. (Ferrante and Rackoff [1979]) *Suppose $\equiv$ is present in the language. Then, for every formula $\Phi(R, \bar{a})$, there exists an equivalent formula $\Psi(R, \bar{a})$, in which R occurs only once.* □

We shall not prove this theorem here, because we want to construct polynomial size formulas not using biconditional. Several people observed that the assumption about biconditional is essential for Theorem 3.2.1, (of course the negation of $\equiv$ is sufficient too). If we consider, say, all binary connectives without biconditional and its negation, then one can define *positive* and *negative* occurrences of R and it is not possible to replace one by the other. Therefore the theorem fails to hold in this case.

Let us consider the construction of formulas satisfying the inductive condition (1) using Theorem 3.2.1. If we disregard the size of variables, i.e., we assign a unit cost to each variable, we clearly get formulas $\varphi_n(\bar{a})$ of *linear* size by iterating $\varphi_{n+1}(\bar{a}) =_{def} \Phi(\varphi_n(\bar{a}), \bar{a})$.

In order to obtain a polynomial size proof of (1) we prove, for every n,

$$\Psi(\varphi_n, \bar{a}) \equiv \Phi(\varphi_n, \bar{a}). \tag{4}$$

The proof of this formula is obtained from the proof d of $\Psi(R, \bar{a}) \equiv \Phi(R, \bar{a})$ by substituting φ_n for each occurrence of R in d. Hence the proof is also of linear size in n.

In a more precise computation of proof size, we have to take into account the size of variables. After the reduction to one occurrence the inductive condition is

$$\varphi_{n+1}(\bar{a}) = \Psi(\varphi_n(\bar{x}), \bar{a}), \tag{5}$$

where $\bar{x}$ is a string of variables bound in Ψ. Clearly, we cannot use the same string $\bar{x}$ for all n (except in trivial cases) because of the possible clashes. If we use different strings $\bar{x}$ for each n, we get formulas of size of the order $n \cdot \log n$, since the n-th variable can be coded by a word of length $O(\log n)$. Alternatively we can use just two strings: one for odd n's and one for even n's. The resulting formulas are of linear size but a little unnatural, since one variable occurs in the scope of several quantifiers bounding it. Though unnatural it is usually permitted.

3.2.2. Now we prove the existence of polynomial size formulas defined by iteration in the case when $\equiv$ is not present in the language.

3.2.3. Theorem. (Solovay [unpublished]) *Suppose $\neg$ and at least one of the connectives $\to$, $\vee$, $\wedge$ are present in the language. Let $\varphi_0(\bar{a})$ and $\Phi(R, \bar{a})$ be given. Then it is possible to construct formulas $\varphi_1(\bar{a}), \varphi_2(\bar{a}), \ldots$ such that*

$$\varphi_{n+1}(\bar{a}) \equiv \Phi(\varphi_n, \bar{a}) \tag{6}$$

have polynomial size proofs.

Proof. We shall use only the fact that $p \to q$ has an equivalent formula in the language where q occurs once. We use $p \equiv q$ in (6) and below as an abbreviation of an equivalent formula in the language, e.g. $(p \to q) \wedge (q \to p)$, if both $\to$ and $\wedge$ are present.

The idea of the proof is the same as for the case with $\equiv$ plus an additional trick. The trick is to first define *the graph of the truth value function* for φ_n's. If $f_n(\bar{a})$ is the truth value function, then both $\varphi_n(\bar{a})$ and $\neg\varphi_n(\bar{a})$ can be expressed as positive statements $f_n(\bar{a}) = 1$ and $f_n(\bar{a}) = 0$ respectively.

In order to get simpler formulas we shall use inessential assumptions that a constant 0 is in the language and $\exists x \exists y (x \neq y)$ is a logical axiom. Consider the formula

$$\Phi(R, \bar{a}) \equiv y = 0.$$

Take a prenex normal form of it

$$\overline{Q}\, \Theta(R(\bar{x}_1), \ldots, R(\bar{x}_t), \bar{a}, y),$$

where $\overline{Q}$ denotes the quantifier prefix which bounds, among others, the variables $\bar{x}_1, \ldots, \bar{x}_t$, and all occurrences of R in Θ are displayed. Now we define formulas for the graphs of $f_n(\bar{a})$'s. In order to simplify the formulas, truth will be represented by 0 and falsehood by anything different from 0. Define $\Psi_0(\bar{a}, y)$ to be the formula

$$\varphi_0(\bar{a}) \equiv y = 0 \tag{7}$$

and define $\Psi_{n+1}(\bar{a}, y)$ to be the formula

$$\begin{aligned} \overline{Q} \forall y_1 \cdots \forall y_z (\forall \bar{z} \forall u (((\bar{z} = \bar{x}_1 \wedge u = y_1) \vee \cdots \vee (\bar{z} = \bar{x}_t \wedge u = y_t)) \to \Psi_n(\bar{z}, u)) \\ \to \Theta(y_1 = 0, \ldots, y_t = 0, \bar{a}, y)), \end{aligned} \tag{8}$$

where $\overline{Q}$ is as above, $\bar{z} = \bar{x}_i$ is an abbreviation for $z_1 = x_{i1} \wedge \ldots \wedge z_k = x_{ik}$ and $y_i = 0$ are substituted for $R(\bar{x}_i)$ in Θ. Note that the meaning of the antecedent in the definition is that y_i codes the truth value of $\varphi_n(\bar{x}_i)$; below we give a formal proof of this.

Since Ψ_n occurs only once in the recurrence relation, we get Ψ_n of polynomial size ($O(n \log n)$ if we use different variables and linear if we "recycle" variables).

Define formulas

$$\begin{aligned}
\varphi_n(\bar{a}) &=_{df} \Psi_n(\bar{a}, 0), \\
\alpha_{n+1}(\bar{a}, y) &=_{df} \Psi_{n+1}(\bar{a}, y) \equiv (\Phi(\varphi_n, \bar{a}) \equiv y = 0), \\
\beta_n(\bar{a}, y) &=_{df} \Psi_n(\bar{a}, y) \equiv (\varphi_n(\bar{a}) \equiv y = 0).
\end{aligned}$$

3.2.4. Lemma. *Let $\bar{x}$ be the string of all free variables of formulas μ and ν; let $\Gamma(\mu)$ and $\Gamma(\nu)$ be obtained by substituting μ and ν in $\Gamma(R)$ for R. Then*

$$\forall \bar{x}(\mu(\bar{x}) \equiv \nu(\bar{x})) \to \Gamma(\mu) \equiv \Gamma(\nu)$$

has a polynomial size proof in the size of μ, ν and Γ.

The idea of the proof is to use induction on the depth of Γ. □

3.2.5. Lemma.
(i) $\exists y \Psi_0(\bar{a}, y)$ *is provable.*

The following formulas have polynomial size proofs.
(ii) $\forall y \alpha_{n+1}(\bar{a}, y) \to \exists y \Psi_{n+1}(\bar{a}, y)$;
(iii) $\forall y \alpha_{n+1}(\bar{a}, y) \to \varphi_{n+1}(\bar{a}) \equiv \Phi(\varphi_n, \bar{a})$;
(iv) $\forall y \alpha_{n+1}(\bar{a}, y) \to \forall y \beta_{n+1}(\bar{a}, y)$;
(v) $\forall \ldots \alpha_n(\bar{a}, y) \to \forall \ldots \alpha_{n+1}(\bar{a}, y)$; *where $\forall \ldots$ denotes the universal closure.*

Proof. (i) Use (7): if $\varphi_0(\bar{a})$, then take $y = 0$, if $\neg\varphi(\bar{a})$, take an arbitrary $y \neq 0$.

(ii) Similar as in (i): to find y such that $\Psi_{n+1}(\bar{a}, y)$ holds distinguish the cases $\Phi(\varphi_n, \bar{a})$ and $\neg\Phi(\varphi_n, \bar{a})$. In the first case take $y = 0$, in the second any $y \neq 0$. The formulas involved are of polynomial size, the number of steps is constant, thus the whole proof is polynomial.

(iii) Assume $\forall y \alpha_{n+1}(\bar{a}, y)$; in particular we have $\alpha_{n+1}(\bar{a}, 0)$ which is

$$\Psi_{n+1}(\bar{a}, 0) \equiv (\Phi(\varphi_n, \bar{a}) \equiv 0 = 0).$$

Using the definition of $\varphi_{n+1}(\bar{a})$ this reduces to the statement

$$\varphi_{n+1}(\bar{a}) \equiv \Phi(\varphi_n, \bar{a}).$$

(iv) Assume $\forall y \alpha_{n+1}(\bar{a}, y)$. By (iii) we can substitute $\varphi_{n+1}(\bar{a})$ for $\Phi(\varphi_n, \bar{a})$ in $\forall y\ \alpha_{n+1}(a, y)$, which we assume. Thus we get $\forall y\ \beta_n(\bar{a}, y)$. As we do not have the substitution rule, we must use Lemma 3.2.4 to estimate the length of the proof.

(v) Assume $\forall \ldots \alpha_n(\bar{a}, y)$. By the definition of Ψ_0 and (iv) we have also $\forall \ldots \beta_n(\bar{a}, y)$. From definition (8) we immediately get

$$\begin{aligned}
\Psi_{n+1}(\bar{a}, y) \equiv\ & \overline{Q} \forall y_1 \ldots \forall y_t (\Psi_n(\bar{x}_1, y_1) \wedge \ldots \wedge \Psi_n(\bar{x}_t, \bar{y}_t) \to \\
& \to \Theta(y_1 = 0, \ldots, y_t = 0, \bar{a}, y)),
\end{aligned}$$

using a polynomial size proof. By $\forall \ldots \beta_n(\bar{a}, y)$ we can substitute $\varphi_n(\bar{x}_i)$ for $y_i = 0$.
Thus we get

$$\Psi_{n+1}(\bar{a}, y) \equiv \overline{Q}\forall y_t(\Psi_n(\bar{x}_1, y_1) \wedge \ldots \wedge \Psi_n(\bar{x}_t, y_t) \rightarrow$$

$$\rightarrow \Theta(\varphi_n(\bar{x}_1), \ldots, \varphi_n(\bar{x}_t), \bar{a}, y)).$$

Pushing the universal quantifies inside, we get

$$\Psi_{n+1}(\bar{a}, y) \equiv \overline{Q}(\exists y_1 \Psi_n(\bar{x}_1, y_1) \wedge \ldots \wedge \exists y_t \Psi_n(\bar{x}_t, y_t) \rightarrow$$

$$\rightarrow \Theta(\varphi_n(\bar{x}_1), \ldots, \varphi_n(\bar{x}_t)\bar{a}, y)).$$

Now, by (ii), $\exists y_i \Psi_n(\bar{x}_i, y_i)$ have polynomial size proofs, thus we get

$$\Psi_{n+1}(\bar{a}, y) \equiv \overline{Q}\ \Theta(\varphi_n(\bar{x}_1), \ldots, \varphi_n(\bar{x}_t), \bar{a}, y).$$

By definition of Θ, this is equivalent to

$$\Psi_{n+1}(\bar{a}, y) \equiv (\Phi(\varphi_n, \bar{a}) \equiv y = 0).$$

The calculation that the proofs are of polynomial size use Lemma 3.2.4 and the same ideas as we have already used before. For instance, the last equivalence is obtained by taking the constant size proof of

$$(\Phi(R, \bar{a}) \equiv y = 0) \equiv \overline{Q}\ \Theta(R(\bar{x}_1), \ldots, R(\bar{x}_t)\bar{a}, y)$$

and substituting φ_n for each occurrence of R in the proof.

□

To finish the proof of the theorem we first prove $\forall \ldots \alpha_1(\bar{a}, y)$. The proof is identical with (v) above, except that we get $\forall \ldots \beta_0(\bar{a}, y)$ directly from the defining equation (7). Now we combine the polynomial size proofs of $\forall \ldots \alpha_1(\bar{a}, y) \rightarrow \forall \ldots \alpha_2(\bar{a}, y), \ldots, \forall \ldots \alpha_n(\bar{a}, y) \rightarrow \forall \ldots \alpha_{n+1}(\bar{a}, y)$ to obtain a polynomial size proof of $\forall \ldots \alpha_{n+1}(\bar{a}, y)$. Then, by (iii), we get a polynomial size proof of

$$\varphi_{n+1}(\bar{a}) \equiv \Phi(\varphi_n, \bar{a}).$$

□

3.2.6. Suppose we allow repeated use of the same variables, hence φ_n's are of linear size. Then one can easily check that the sentences in Lemma 3.2.5 have linear size proofs hence (6) has quadratic size proofs.

3.3. We consider two applications of Theorem 3.2.3. The first application is to construct a partial truth definition.

We shall consider T, a sufficiently strong fragment of arithmetic or set theory; namely, we need to be able to formalize syntax in T. A natural assumption is that the theory T is *sequential* which means that T contains Robinson arithmetic **Q** (see Chapter II) and a formula formalizing the relation *"x is the i-th element of y"*; we only require that there exists an empty sequence and each sequence can be prolonged by adding an arbitrary element. E.g. in the Gödel-Bernays set theory **GB** we can define the i-th element of the sequence coded by a class X by

$$(X)_i =_{def} \{x \; ; \; (x,i) \in X\}.$$

Let us stress that it is important to code all elements, it would not suffice to code, say, only sets in **GB**.

Let $\ulcorner\alpha\urcorner$ denote the Gödel number (the code) of a formula α. By a well-known theorem of Tarski [1936], there is no formula $\varphi(x)$ such that

$$T \vdash \varphi(\ulcorner\alpha\urcorner) \equiv \alpha$$

for all sentences α (it is a simple application of the diagonalization lemma). However it is possible to construct such a formula for some classes of sentences α, in particular for α with bounded quantifier complexity. We shall need the following particular case. We would like to define *satisfaction* for formulas $\alpha(\bar{x})$ of bounded size and a string $\bar{x}$ of elements. Let $(x)_i$ denote some coding function in T, i.e., $(x)_i$ is the i-th element of the sequence x (we may assume that every element is a code of some sequence). We want to construct formulas $\varphi_n(x,y), n = 1,2,\ldots,$ such that for every $\alpha(y_1,\ldots,y_n)$ of depth $\leq n$,

$$T \vdash \varphi_n(\ulcorner\alpha\urcorner, x) \equiv \alpha((x)_1,\ldots,(x)_k), \tag{9}$$

using a polynomial size proof, (depth 0 are atomic formulas etc.). In fact we need more: we want to have polynomial size proofs of *Tarski's conditions* for φ_n. Tarski's conditions are conditions which define satisfaction by induction on the depth of formulas. For each connective and each quantifier there is one condition. E.g., for implication Tarski's condition is

$$\varphi_n(\ulcorner\beta \to \gamma\urcorner, x) \equiv (\varphi_n(\ulcorner\beta\urcorner, x) \to \varphi_n(\ulcorner\gamma\urcorner, x)).$$

It is assumed that satisfaction for open formulas is easily definable. This is true in our case, since we assume that T is sufficiently strong. Let $R(x,y)$ be a new binary predicate. Let φ_0 be a formula defining satisfaction for open formulas and let $\Phi(R,x,y)$ be a formula expressing the following:

1. if x is atomic then $\varphi_0(x,y)$,
2. if x is $\neg u$ then $\neg R(u,y)$
3. if x is $u \to v$ then $R(u,y) \to R(v,y)$,

4. if x is $\forall z_i\, u$ and $R(u, y')$ for every sequence y' identical with y on all coordinates $j \neq i$, then $R(x, y)$,

etc. for the other connectives and for quantifiers.

By Theorem 3.2.3 we have polynomial size formulas $\varphi_n(x, y)$ and polynomial size proofs of

$$\varphi_{n+1}(x, y) \equiv \Phi(\varphi_n, x, y). \tag{10}$$

What we need is a little different; namely, we need polynomial size proofs in T of

$$dpt_n(x) \rightarrow \varphi_n(x, y) \equiv \Phi(\varphi_n, x, y), \tag{11}$$

where $dpt_n(x)$ is a formula saying that x is a formula of depth $\leq n$. To prove this, it suffices to prove, using polynomial size proofs in T,

$$dpt_n(x) \rightarrow \varphi_{n+1}(x, y) \equiv \varphi_n(x, y). \tag{12}$$

To prove (12) we observe that for $n = 0$ it follows from the definition of Φ and for $n > 0$

$$\begin{aligned} &\forall x, y(dpt_n(x) \rightarrow \varphi_{n+1}(x, y) \equiv \varphi_n(x, y)) \rightarrow \\ &\rightarrow \forall x, y(dpt_{n+1}(x) \rightarrow \varphi_{n+2}(x, y) \equiv \varphi_{n+1}(x, y)) \end{aligned} \tag{13}$$

have polynomial size proofs. Let us prove the implication. Assume the antecedent and $dpt_{n+1}(x)$. We distinguish the cases: x is atomic, x is a negation, x is an implication etc. If x is atomic, then, by the definition of Φ and (10), both $\varphi_{n+2}(x, y)$ and $\varphi_{n+1}(x, y)$ are equivalent to $\varphi_0(x, y)$. If x is the negation of x' then we have (by the definition of Φ)

$$\varphi_{n+2}(x, y) \equiv \neg\varphi_{n+1}(x', y)$$

and

$$\varphi_{n+1}(x, y) \equiv \neg\varphi_n(x'y).$$

By our assumption we have

$$\varphi_{n+1}(x', y) \equiv \varphi_n(x', y),$$

since we have also $dpt_n(x')$. Thus $\varphi_{n+2}(x, y) \equiv \varphi_{n+1}(x, y)$. The other cases are proved in the same way. In order to see that the resulting proof has polynomial size, observe that it does not use the structure of formulas $\varphi_n, \varphi_{n+1}, \varphi_{n+2}$. Thus these proofs can be constructed from a *finite* proof schema by substituting the formulas $\varphi_n, \varphi_{n+1}, \varphi_{n+2}$. The resulting proof has size linear in the size of $\varphi_n, \varphi_{n+1}, \varphi_{n+2}$. This finishes the proof of (11).

Now we can show easily that (9) have also polynomial size proofs provided α is of depth $\leq n$ (and of polynomial size, i.e., does not use variables with long codes). This is done by induction on the depth of α; we leave out the details.

The following theorem summarizes what we have proved above.

3.3.1. Theorem. (Pudlák [1986]) *Let T be a sequential theory. There exists a sequence of formulas $\varphi_n(x,y)$ (of polynomial size) and such that there are polynomial size proofs in T of Tarski's condition for $\varphi_n(x,y)$ where x is of depth $\leq n$ and polynomial size proofs of*

$$\varphi_n(\ulcorner\alpha\urcorner, x) \equiv \alpha((x)_1, \dots, (x)_n)$$

for α of depth $\leq n$. □

3.4. Now we consider another application of Theorem 3.2.3. Let T be a fragment of arithmetic, let S be a function symbol for the successor. Now T can be much weaker; we shall specify the condition that we need later. Let $\varphi(x)$ be a formula. We say that $\varphi(x)$ is a *cut* in T of T proves

$$\varphi(0), \tag{14}$$
$$\forall x(\varphi(x) \to \varphi(S(x))), \tag{15}$$
$$\forall x, y(x \leq y \wedge \varphi(y) \to \varphi(x)). \tag{16}$$

If $\varphi(x)$ satisfies only (14) and (15), then $\varphi(x)$ is called *inductive.* Let $\varphi(x)$ be inductive and assume that T proves $x + 0 = 0$, $x + S(y) = S(x + y)$ and the associative law for $+$.

Define $\psi(x)$ by

$$\psi(x) \equiv_{df} \forall z(\varphi(z) \to \varphi(z + x)). \tag{17}$$

Then one can easily show that $\psi(x)$ is also inductive in T and

$$T \vdash \psi(x) \wedge \psi(y) \to \psi(x + y); \tag{18}$$

(this construction is due to Solovay, unpublished). If (18) is satisfied, we say that ψ is closed under addition. Assuming a little bit more about T and that $\varphi(x)$ is a cut, we get that $\psi(x)$ is also a cut. Suppose that T contains exponentiation 2^x along with axioms

$$2^0 = S(0),\ 2^{S(x)} = 2^x + 2^x. \tag{19}$$

Then we can continue by first taking

$$\varphi_1(x) \equiv_{df} \psi(2^x). \tag{20}$$

We get that $\varphi_1(x)$ is inductive (resp., is a cut) and

$$T \vdash \forall x\ \varphi_1(x) \to \varphi(2^x), \tag{21}$$

since $T \vdash \psi(x) \to \varphi(x)$. Then we can repeat the construction, since $\varphi_1(x)$ is inductive, and we obtain some $\varphi_2(x)$ with

$$T \vdash \forall x\ \varphi_2(x) \to \varphi(2^{2^x})$$

etc. Observe that the above construction is *schematic*, we could have assumed that φ is just a second order variables and derive (21) from (14) - (16). More precisely it means the following: let $R(x)$ be a unary predicate, let $Ind(R(x))$ (resp. $Cut(R(x))$ denote the conjunction of (14) and (15) (and (16) resp.). Then there exists a formula $\Psi(R,x)$ and a finite fragment of arithmetic T_0 such that

$$T_0 \vdash Ind(R(x)) \rightarrow Ind(\Psi(R,x)) \wedge \forall y(\Psi(R,y) \rightarrow R(2^y)).$$

Applying Theorem 3.2.3 to $\Phi(R,x)$ defined by

$$\Phi(R,x) \equiv_{df} \bigwedge T_0 \rightarrow \Psi(R,x)$$

we obtain the following theorem.

3.4.1. Theorem. *Let T be a sufficiently strong fragment of arithmetic; suppose $\varphi_0(x)$ is inductive (resp. is a cut) in T. Then there exists a sequence of formulas $\varphi_1(x), \varphi_2(x), \ldots$, such that for each n*

$$Ind(\varphi_{n+1}(x)) \wedge \forall x(\varphi_{n+1}(x) \rightarrow (\varphi_n(x) \wedge \varphi_n(2^x)))$$

(resp. the formula with Cut instead of Ind) has a proof in T of size polynomial in n.
□

3.5. Since cuts are quite important in the study of theories containing some part of arithmetic, we shall mention a few basic facts about them, though they are not needed in this chapter. More can be found e.g. in Hájek and Pudlák [1993].

In order to obtain a cut closed under multiplication and contained in $\varphi(x)$ one can apply the trick of (17) to $\psi(x)$. It is possible to go on and get cuts closed under more rapidly growing functions, but not for 2^x (unless $\varphi(x)$ has some special properties). There is another way to get such cuts, using which we can better see what these functions are. Let 2^x_n denote n-times iterated exponential function. Let $\omega_n(x)$ be nondecreasing functions such that

$$2_n^{S(x)} = \omega_n(2_n^x); \tag{22}$$

we assume that these properties are provable in T.

Let $\psi_n(x)$ be defined by

$$\psi_n(x) \equiv_{df} \exists y(\varphi_n(y) \wedge x \leq 2_n^y). \tag{23}$$

By construction

$$T \vdash \varphi_n(y) \rightarrow \varphi(2_n^y), \tag{24}$$

hence $\psi_n(x)$ is contained in $\varphi(x)$. Also it is easy to check that $\varphi_n(x)$ is a cut in T. To see that $\psi_n(x)$ is closed under $\omega_n(x)$ just observe that

$$T \vdash x \leq 2_n^y \rightarrow \omega_n(x) \leq \omega_n(2_n^y) = 2_n^{S(y)}.$$

For instance take $\omega_2(x) = x^2$, then we obtain $\psi_2(x)$ closed under x^2, hence closed under multiplication.

4. More on the structure of proofs

In this section we shall prove two basic results. First we prove that for the usual calculi for predicate logic proofs as sequences can be replaced by tree-proofs with only a polynomial increase. Thus the size measures based on proofs as sequence and proofs as trees are polynomially related. (We shall sketch a different proof of the same result for some propositional proof systems in section 8.) The second result says that the depth of a proof can be bounded by a square root of its size, provided the proved sentence has negligible size. The proof is based on the theory of unification of terms. We shall survey a few other results which use unification, in particular Kreisel's Conjecture on generalizations of proofs in arithmetic.

4.1. Theorem. (Krajíček [1994a]) *Let $\|\varphi\|^{\text{sequence}}$ resp. $\|\varphi\|^{\text{tree}}$ be the size of the smallest sequence-proof resp. tree-proof of a provable sentence φ in the Hilbert style calculus. Then there exists a polynomial $p(x)$ such that*

$$\|\varphi\|^{\text{tree}} \leq p(\|\varphi\|^{\text{sequence}})$$

for every provable sentence φ.

We consider here only the Hilbert style calculus, but the result can be extended to the Gentzen sequent calculus, as there are polynomial simulations for both versions - tree and sequence, cf. Eder [1992].

This result is quite surprising, since there is an obvious similarity between sequence-proofs and circuits on the one hand, and tree-proofs and formulas on the other hand. In circuits the output of a gate can be connected with several other gates, thus we can use the boolean function computed at this gate several times. While a formula is represented by a tree, thus each node has at most one successor. Similarly in a sequence-proof we can use a formula several times as a premise of a rule, while in a tree proof it is allowed only once. It is generally accepted, through still a difficult open problem, that circuits are exponentially more powerful for computations of boolean formulas than formulas. Still the corresponding statement for proofs is false as we shall see below.

Proof. We shall first prove the theorem for the propositional calculus. The idea is quite simple. Let $(\varphi_1, \ldots, \varphi_n)$ be a proof. We shall replace this sequence by $\varphi_1, \varphi_1 \wedge \varphi_2, \ldots, (\ldots(\varphi_1 \wedge \varphi_2) \wedge \ldots) \wedge \varphi_n$. In this sequence each formula follows from the previous one. This sequence is, however, not a proof, thus we have to insert some proof trees in it such that a leaf of a tree is $(\cdots(\varphi_1 \wedge \varphi_2) \wedge \cdots) \wedge \varphi_i$ and the root is $(\cdots(\varphi_1 \wedge \varphi_2) \wedge \cdots) \wedge \varphi_{i+1}$.

In order to simplify notation we agree to omit parenthesis in expressions like $(\cdots((\varphi_1 \wedge \varphi_2) \wedge \varphi_3) \cdots) \wedge \varphi_n$. Furthermore let us say that a class of sentences has *polynomial size tree proofs*, abbreviated by *pst-proofs*, if there is a polynomial upper bound on the size of tree proofs in terms of the size of a formula. We shall use this also for proofs from assumptions.

The proof now reduces to the two statements in the following lemma.

4.1.1. Lemma.
(a) $\alpha \to \alpha \wedge \beta$ *has a pst-proof provided* β *is an instance of an axiom;*
(b) $\alpha_1 \wedge \cdots \wedge \alpha_n \wedge \gamma$ *has a pst-proof from* $\alpha_1 \wedge \cdots \wedge \alpha_n$ *provided* $\alpha_i \to \gamma$ *is* α_j *for some* $1 \le i, j \le n$.

Proof. (a) The proof of (a) is trivial, but since we shall use the same argument several times below we shall spell it out at least once. Suppose β is an instance of an axiom $\psi(p_1, \ldots, p_k)$, thus β is $\psi(\beta_1, \ldots, \beta_k)$ for some $\beta_1, \ldots, \beta_k$. Consider the following formula $p_{k+1} \to p_{k+1} \wedge \psi(p_1, \ldots, p_k)$. It is a tautology, thus it has a tree-proof d_0. Let arbitrary α and $\beta_1, \ldots, \beta_n$ be given. Then we obtain a tree-proof of $\alpha \to \alpha \wedge \psi(\beta_1, \ldots, \beta_n)$ simply by substituting $\beta_1, \ldots, \beta_k, \alpha$ for $p_1, \ldots, p_{k+1}$ in d_0. Thus the size of this tree-proof will be bounded by $c \cdot (|\alpha| + |\beta_1| + \cdots + |\beta_k|) \le c \cdot (|\alpha| + |\beta|) = O(|\alpha \to \alpha \wedge \beta|)$, where the constant c is determined by d_0.

(b) To prove the second statement we derive another lemma. (We shall not need it in its full strength.)

4.1.2. Lemma. *Let* π *be a permutation on* $\{1, \ldots, n\}, \alpha_1, \ldots, \alpha_n$ *formulas. Then*

$$\alpha_1 \wedge \alpha_2 \wedge \cdots \wedge \alpha_n \to \alpha_{\pi(1)} \wedge \alpha_{\pi(2)} \wedge \cdots \wedge \alpha_{\pi(n)} \tag{25}$$

has a pst-proof.

Proof. First we prove that

$$\delta \wedge \beta \wedge \alpha_1 \wedge \alpha_2 \cdots \wedge \alpha_n \wedge \gamma \to \delta \wedge \gamma \wedge \alpha_1 \wedge \alpha_2 \cdots \wedge \alpha_n \wedge \beta \tag{26}$$

has a *pst*-proof. Clearly

$$(\xi \wedge \gamma \to \zeta \wedge \beta) \to (\xi \wedge \eta \wedge \gamma \to \zeta \wedge \eta \wedge \beta) \tag{27}$$

have a *pst*-proofs for any $\beta, \gamma, \xi, \zeta, \eta$. Thus start with a *pst*-proof of

$$\beta \wedge \gamma \to \gamma \wedge \beta,$$

take the *pst*-proofs of

$$(\xi_i \wedge \gamma \to \zeta_i \wedge \beta) \to (\xi_{i+1} \wedge \beta \to \zeta_{i+1} \wedge \gamma)$$

given by (27) for ξ_j equal to $\delta \wedge \beta \wedge \alpha_1 \wedge \cdots \wedge \alpha_j$ and ζ_j equal to $\delta \wedge \gamma \wedge \alpha_1 \wedge \cdots \wedge \alpha_i$, and then apply modus ponens inferences to get (26).

Notice that (26) shows that also

$$\alpha_1 \wedge \cdots \wedge \alpha_{i-1} \wedge \beta \wedge \alpha_{i+1} \wedge \cdots \wedge \alpha_{n-1} \wedge \gamma \to \alpha_1 \wedge \cdots \wedge \alpha_{i-1} \wedge \gamma \wedge \alpha_{i+1} \wedge \cdots \wedge \alpha_{n-1} \wedge \beta \tag{28}$$

has *pst*-proofs. Since

$$(\xi \to \zeta) \to (\xi \wedge \eta \to \zeta \wedge \eta)$$

has a *pst*-proof, we get from (28) a *pst*-proof of (25) for any transposition π. In order to get it for a general π just recall the well-known fact that each π can be decomposed into a polynomial number of transpositions. □

Using the same argument as in (a) one can show that

$$\xi \wedge (\beta \to \gamma) \wedge \beta \to \xi \wedge (\beta \to \gamma) \wedge \beta \wedge \gamma \tag{29}$$

has a *pst*-proof for any ξ, β, γ. Now we can finish the proof of (b). For a given $\alpha_1 \wedge \cdots \wedge \alpha_n$, first move $\alpha_i \to \gamma$ and α_j to the end of the conjunction using Lemma 4.1.2, then apply (29) to add γ at the end, and finally, again using Lemma 4.1.2 move $\alpha_i \to \gamma$ and α_j back.

□

This finishes the proof of the theorem for the case of propositional logic.

4.1.3. Now we sketch how the above argument should be modified in order to get the result for the predicate calculus. We cannot simply take the conjunction of formulas in the proof, since clashes of variables may occur and it is no longer true that $\varphi_1 \wedge \cdots \wedge \varphi_{i+1}$ follows from $\varphi_1 \wedge \cdots \wedge \varphi_i$. Therefore we work with universal closures of the formulas $\varphi_1, \ldots, \varphi_n$. Let $\forall \ldots \varphi$ denote a universal closure of a formula φ. Instead of modus ponens and the two quantifier rules we need now, for some formulas $\psi, \varphi, \alpha, \beta$,

(1) to derive $\forall \ldots \psi$ from $\forall \ldots (\varphi \to \psi)$ and $\forall \ldots \varphi$;

(2) to derive $\forall \ldots (\alpha \to \forall y \beta(y))$ from $\forall \ldots (\alpha \to \beta(x))$, where x does not occur in α;

(3) to derive $\forall \ldots (\exists y \alpha(y) \to \beta)$ from $\forall \ldots (\alpha(x) \to \beta)$, where x does not occur in β.

This can be done as follows. For each particular case of (1)–(3) prove the corresponding implication, i.e.,

$$\begin{array}{c} \forall \ldots \varphi \to (\forall \ldots (\varphi \to \psi) \to \forall \ldots \psi); \\ \forall \ldots (\alpha \to \beta(x)) \to \forall \ldots (\alpha \to \forall y \beta(y)); \\ \forall \ldots (\alpha(x) \to \beta) \to \forall \ldots (\exists y \alpha(y) \to \beta). \end{array} \tag{30}$$

Insert these subproofs in the sequence $\forall \ldots \varphi_1, \forall \ldots \varphi_2, \ldots \forall \ldots \varphi_n$ obtained from the original proof and then use the proof for the propositional calculus. Thus the proof reduces now to following lemma whose proof we omit.

4.1.4. Lemma. *The sentences (30) have pst-proofs.* □

Let us note that the above formulas (30) are essentially the schemas used to formalize first order logic by a finite number of schemas and the single rule modus ponens. Thus what we actually did above was replacing the quantifier rules by quantifier axiom schemas and applying the result for the propositional logic.

4.2. We shall prove the next result, Theorem 4.2.5, using an estimate on unification. It is also possible to prove it directly, but unification is a very useful tool and it is natural to express the combinatorial statement that we need using it.

Consider terms in a language consisting of variables constants and function symbols. A substitution σ is a mapping from the set of variables into the set of terms. We shall write $t\sigma$ for the result of substitution σ applied to t which means that we replace each variable x in t by $\sigma(x)$. A *unification problem* is a set of pairs of terms $\{(t_1, t_2), (t_3, t_4), \dots, (t_{2k-1}t_{2k})\}$. A substitution σ is a *unifier* if

$$t_1\sigma = t_2\sigma, \dots, t_{2k-1}\sigma = t_{2k}\sigma.$$

We think of a unification problem as a system of equations with variables being unknown terms; however variables may occur in a solution (=unifier) too. A unifier Σ is a *most general unifier* if for every unifier σ there exists a substitution δ such that $\Sigma\delta = \sigma$. The following result is easy but has very important applications, see Chapter I and Chang and Lee [1973].

4.2.1. Proposition. *If there exists a unifier then there exists a most general unifier.* □

We shall use this proposition later. Now we only need to observe that the most general unifier gives the smallest possible solution.

As usual, we shall think of terms as rooted trees. We say that the root is in depth 0, its sons in depth 1 etc. The *depth of a term t*, denoted by $d(t)$ is the maximal depth that occurs in it; the *size* of t, denoted by $|t|$, is the number of subterms (i.e., the nodes in the tree). We say that a *subterm s of a term t is in depth d* if the root of s in depth d in t.

The following lemma is the combinatorial substance of the bound on the depth of formulas which we are going to prove.

4.2.2. Lemma. *Let Σ be a most general unifier of a unification problem*

$$\{(t_1, t_2), (t_3, t_4), \dots, (t_{2k-1}, t_{2k})\}.$$

Let $d = \max_i d(t_i)$, $S = \sum_i |t_i\Sigma|$ and $D = \max_i d(t_i\Sigma)$. Then

$$D \leq \sqrt{(2 + o(1))(d+1)S}.$$

Proof. Let w be a term $t_{i_0}\Sigma$ with the maximal depth; thus $d(w) = D$. Consider a branch B in w of length D. Let $B_1, \dots, B_r$, $r = \lfloor \frac{D}{d+1} \rfloor$, be the end segments of B of lengths $d+1, 2(d+1), \dots, r(d+1)$ respectively.

4.2.3. Claim. *For each subterm u of any $t_i\Sigma$ with $d(u) > 0$, there exists j such that u occurs in $t_j\Sigma$ in depth $< d(t_j)$.*

To prove the Claim, suppose it is false. Then u can occur only in the part of the terms $t_j\Sigma$ which belongs to σ. Thus we can obtain a smaller unifier by replacing all occurrences of u by a variable. □

We shall show that $B_1, \ldots, B_r$ have disjoint occurrences. For each B_i take the term w_i corresponding to the first vertex of B_i and take an occurrence of w_i in the depth $\leq d(t_j)$ in some $t_j\Sigma$. Then the occurrences of D_i's in these occurrences of w_i's must be disjoint. Thus we have

$$\begin{aligned} S &\geq |B_1| + \ldots + |B_r| \\ &> d+1+2(d+1)+\ldots+\left\lfloor \frac{D}{d+1} \right\rfloor (d+1) \\ &= (d+1)\cdot\frac{1}{2}\cdot\left\lfloor \frac{D}{d+1} \right\rfloor \cdot \left(\left\lfloor \frac{D}{d+1} \right\rfloor + 1\right) \\ &\geq \frac{1}{2}\left\lfloor \frac{D}{d+1} \right\rfloor \cdot D \\ &\geq \frac{D^2}{2(d+1)} - \frac{D}{2} = \frac{D^2}{2(d+1)} \cdot (1 - o(1)). \end{aligned}$$

□

4.2.4. Suppose that $\Delta = (\varphi_1, \ldots, \varphi_n)$ is a proof of φ, i.e., $\varphi_n = \varphi$. The *skeleton* of Δ is a sequence of the same length where each φ_i is replaced by an axiom schemas or a rule used in Δ at this step; moreover, if a rule was used, then there is also information about the proof lines to which the rule was applied. E.g. a formula obtained by modus ponens from formulas φ_j and φ_k will be replaced by (MP, j, k).

We shall show that for a given formula φ and a skeleton Σ there exists in a sense a most general proof. This proof will be constructed from a most general unifier for a unification problem obtained from Σ. In defining the unification problem assigned to the proof Δ we shall follow Baaz and Pudlák [1993], the idea goes back to Parikh [1973].

Replace all atomic formulas in Δ by a single constant c; let $\Delta' = (\varphi'_1, \ldots, \varphi'_n)$ be the resulting sequence. The language for the terms in the unification problem will consists of the constant c, distinct variables v_β for every subformula β of Δ' and a function symbol for each connective and quantifier, i.e., $f_\rightarrow, f_\neg, f_\exists$ etc. We shall write the pairs of the unification problem as equations:

(1) For each propositional axiom schema used in the proof we add an equation which represents it; e.g., if φ'_i is $\alpha \rightarrow (\beta \rightarrow \alpha)$, then we add equation

$$v_{\varphi'_i} = f_\rightarrow(v_\alpha, f_\rightarrow(v_\beta, v_\alpha));$$

(2) if φ_i is derived from φ_j and φ_k via modus ponens, where φ_k is $\varphi_j \rightarrow \varphi_i$, we add

$$v_{\varphi'_k} = f_\rightarrow(v_{\varphi'_j}, v_{\varphi'_i});$$

(3) if φ_i is an instance of a quantifier axiom, say φ_i is $\Phi(t) \to \exists x\Phi(x)$, then we add the equation

$$v_{\varphi'_i} = f_{\to}(v_\alpha, f_{\exists}(v_\alpha));$$

here α is the formula obtained from $\Phi(t)$ by substituting c for atomic formulas, this is the same formula which we thus obtain from $\Phi(x)$;

(4) in the same way we add equations for quantifier rules: e.g., suppose φ'_j is $\alpha \to \beta, \varphi'_i$ is $\exists x\alpha \to \beta$ and φ_i is derived from φ_j by the quantifier rule (6.3), then we add equations

$$v_{\varphi'_j} = f_{\to}(v_\alpha, v_\beta),$$
$$v_{\varphi'_i} = f_{\to}(f_{\exists}(v_\alpha), v_\beta);$$

(5) finally we add

$$v_{\varphi'_n} = \tau,$$

where τ is obtained from φ'_n by replacing connectives and quantifiers by the corresponding function symbols $f_{\to}, f_{\neg}, f_{\exists}, \ldots$.

Now we are ready to prove the result. Let $dp(\varphi)$ denote the *depth* of φ a formula, where we consider φ as a term but we treat atomic formulas as atoms. Let $dp(\Delta)$, for a proof Δ, denote the maximal depth of a formula in Δ.

4.2.5. Theorem. (Krajíček [1989a], Pudlák [1987]) *Let Δ be a proof of φ and suppose Δ has smallest possible size. Then*

$$dp(\Delta) = O\left(\sqrt{|\Delta| \cdot (dp(\varphi) + 1)}\right).$$

Proof. Consider a proof Δ of φ of minimum size. Let $\mathcal{U}$ be the unification problem assigned to Δ. Clearly Δ determines a unifier σ for $\mathcal{U}$ in the natural way. Let Σ be a most general unifier of $\mathcal{U}$. We shall construct a proof $\Gamma = (\psi_1, \ldots, \psi_n)$ from Σ. Let δ be the substitution such that $\sigma = \Sigma\delta$. Choose a small formula ξ which does not contain any variable which occurs in Δ, e.g., $0 = 0$ if it is in the language. Consider the i-th formula in the proof Δ, i.e., φ_i, and terms $v_{\varphi'_i}\sigma$ and $v_{\varphi'_i}\Sigma$. We have $v_{\varphi'_i}\Sigma\delta = v_{\varphi'_i}\sigma$; this means that $v_{\varphi'_i}\Sigma$ is $v_{\varphi'_i}\sigma$ with some subformulas replaced by variables. Thus we define ψ_i to be φ_i with the subformulas corresponding to variables in $v_{\varphi'_i}\Sigma$ replaced by ξ.

4.2.6. Let us consider an example. Suppose φ_i has been obtained from φ_j by the quantifier rule (6.3). Suppose φ_i is

$$\exists x(P(x) \to (Q(x) \to R(y))) \to R(y),$$

Then $v_{\varphi'_i}\sigma$ is

$$f_{\to}(f_{\exists}f_{\to}(c, f_{\to}(c, c)), c).$$

By case (4) of the definition of $\mathcal{U}$, $v_{\varphi_i}\Sigma$ has form

$$f_{\to}(f_{\exists}(t), s),$$

for some terms t and s. Because Σ is most general, s is either c or a variable and t is either as in $v_{\varphi'_i}\sigma$ or $f_{\rightarrow}(c, v_\alpha)$, or a variable. Let us suppose that $v_{\varphi_i}\Sigma$ is

$$f_{\rightarrow}(f_{\exists}f_{\rightarrow}(c, v_\alpha), c).$$

Then ψ_i is

$$\exists x(P(x) \rightarrow \xi) \rightarrow R(y).$$

Furthermore ψ_j must be

$$(P(x) \rightarrow \xi) \rightarrow R(y).$$

We see that the structure of formulas needed in axiom schemas and rules is preserved. Note that also the restrictions on variables in quantifier rules are satisfied, since ξ does not contain any variable which should be bounded. Finally we have also $\psi_n = \varphi_n = \varphi$.

4.2.7. Now we can apply Lemma 4.2.2. The terms in $\mathcal{U}$ have constant depth (where the constant is determined by our choice of the proof system) except for the last equation where we have a term whose depth is equal to $dp(\varphi)$; thus the maximal depth is $O(dp(\varphi))$. Hence the maximal depth of a term $v_{\varphi'_i}\Sigma$ is $O(\sqrt{dp(\varphi)S})$, where $S = \sum_i |v_{\varphi'_i}\Sigma|$. Clearly also $S = O(\sum_i |\psi_i|)$. Furthermore $|\Gamma| = O(|\Delta|)$, since we have replaced some subformulas in Δ by a constant size formula ξ in Γ. This finishes the proof of Theorem 4.2.5. □

4.2.8. Remarks. (1) Clearly the theorem holds for a variety of other systems. In particular it holds for every Frege system, (see section 8 for the definition).

(2) In the proof we have actually constructed "a most general" proof Γ with the same skeleton Δ. To make it more precise, we should allow propositional variables in our first order formulas and then keep the variables v_α in Γ and treat them as propositional variables.

4.3. Now we consider the relation of the number of steps to the size and depth of a proof. A relation to the depth is easy to obtain, since the depth does not include information about terms. For instance we can also bound the depth of a most general unifier as follows (see Krajíček and Pudlák [1988]).

4.3.1. Lemma. *Let Σ be a most general unifier of a unification problem $\{(t_1, t_2), \ldots, (t_{2n-1}, t_{2n})\}$. Then*

$$\max_i d(t_i\Sigma) \leq \Sigma_i |t_i|.$$

□

Then using a similar proof as above derive:

4.3.2. Theorem. (Parikh [1973], Farmer [1984], Krajíček [1989a]) *If φ has a proof with n steps, then φ has a proof with n steps and depth bounded above by*

$$O(n + |\varphi|).$$

□

This result gives a bound on the size of a proof in terms of the number of steps, if we disregard terms or use a language without function symbols.

It is more difficult to bound the size of a proof using the number of steps and the size of the formula, if we use the usual definition of the size which includes terms. The technique based on unification works only in cut-free Gentzen sequent calculi. An ordinary proof must be first replaced by a cut-free proof, which results in a big increase. Again we state the result without a proof; see Krajíček and Pudlák [1988] for a more precise bound and a proof (the idea will be also sketched in the proof of Theorem 4.4.1).

4.3.3. Theorem. *There exists a primitive recursive function F such that for every sentence φ and number n, if φ has a proof with n steps, then it has a proof with size bounded by $F(\varphi, n)$.* □

4.3.4. Problem. (Krajíček and Pudlák [1988], Clote and Krajíček [1993]) Can F be elementary, i.e., bounded by a constant time iterated exponential function (in $|\varphi| + n$)?

The following interesting result of S. Buss shows very nicely that it is hard to determine the structure of terms in first order proofs. He proved this theorem for a particular version of a sequent calculus.

4.3.5. Theorem. (Buss [1991b]) *Given a number n and a sequent $\Gamma \to \Delta$, it is not decidable whether $\Gamma \to \Delta$ has a proof with $\leq n$ steps.* □

At first it may seem that this contradicts Theorem 4.3.3, however notice, that Theorem 4.3.3 does not claim that given a proof of φ with n steps, there must exist a proof of φ with size $\leq F(|\varphi|, n)$ *and n steps.* Consequently it is not possible to minimize the size and the number of steps at the same time. For some solvable cases see Farmer [1988].

4.4. Finally we mention a related topic which is very popular in this field and also demonstrates that the structure of terms in first order proofs is rather complex. Kreisel stated the following conjecture, see Friedman [1975] and Takeuti [1987]:

> **Kreisel's Conjecture** *Suppose for a formula $\varphi(x)$ and a number k, one can prove $\varphi(S^n(0))$ in Peano Arithmetic using $\leq k$ steps for every n. Then $\forall x \varphi(x)$ is provable in Peano arithmetic.*

Here $S^n(0)$ stands for the term obtained by applying the successor function S n-times to 0. The statement seems to be also quite sensitive on particular formalization of Peano Arithmetic. We shall sketch the idea of a proof for the case where Peano Arithmetic is replaced by a finite fragment of arithmetic. The validity of Kreisel's Conjecture for finite fragments was first proved by Miyatake [1980] using a different proof.

4.4.1. Theorem. *There exists a primitive recursive function G such that for every formula $\varphi(x)$ and numbers k, n, if $\varphi(S^n(0))$ has a proof with k steps and $n > G(\varphi, k)$ then $\forall x \varphi(S^n(x))$ is provable.*

In the theorem we use the provability in pure logic; note that this implies that the theorem is true also for any finitely axiomatized theory T as we can incorporate finitely many axioms in φ. We need to add only a very weak assumption about T in order to deduce Kreisel's conjecture.

4.4.2. Corollary. *Let T be a finite fragment of arithmetic such that*

$$T \vdash \forall x (x = 0 \lor x = S(0) \lor \ldots \lor x = S^{n-1}(0) \lor \exists y (x = S^n(y)))$$

for every n. Then Kreisel's Conjecture holds for T.

Proof-hint. By the assumption on T we have

$$T \vdash \varphi(0) \land \varphi(S(0)) \land \ldots \land \varphi(S^{n-1}(0)) \land \forall x \varphi(S_n(x)) \to \forall x \varphi(x),$$

for every formula $\varphi(x)$. □

Proof-idea of Theorem 4.4.1. Let $\varphi(x)$ and k, n be given such that $\varphi(S^n(0))$ has a proof with k steps. We shall see how large n must be.

First we transform the proof into a cut-free proof in the Gentzen system. By Corollary 5.2.2 below, the number of steps in a cut-free proof can be bounded by a constant which depends only k and $|\varphi(x)|$.

Then we apply the technique of unification. This time, however, we consider also the terms in the proof. This is done in two stages. First we consider all proof-skeletons of length K, (there are finitely many). For each of them we find a most general proof (with respect to the propositional and quantifier structure) as in the proof of Theorem 4.2.5. Then for each of these proofs we find most general terms which can be used in them. This can also be done using the theorem about a most general unifier. However, now we treat terms $S^n(0)$ in the sentence $\varphi(S^n(0))$ as unknown, which means that it is represented by a variable in the unification problem. If in terms in the most general solution remain variables for terms, we replace them by first order variables. Thus we obtain a proof whose size is bounded by a primitive recursive function in K and $\varphi(x)$, thus also in k and $\varphi(x)$. Let us denote the bound by L. (This was essentially the idea of the proof of Theorem 4.3.3, except for the treatment of the term $S^n(0)$.)

Let us have a look on what happens with $\varphi(S^n(0))$ in the most general proof. This formula is replaced by $\varphi(t)$ for some term t which has two properties

(1) $|t| \leq L$;

(2) $t\sigma = S^n(0)$, for some substitution σ.

Thus t is either $S^m(y)$ for $m \leq L$ and some variable y, or $S^n(0)$ and $n \leq L$. Hence, if we choose $n > L$, we get a proof of $\varphi(S^m(y))$, with $m < n$. Then, applying generalization, we get a proof of $\forall y \varphi(S^m(y))$ with $m < n$, which in turn implies $\forall x \varphi(S^n(x))$. □

4.4.3. If we now consider full Peano Arithmetic, we can also perform the first part of the proof. But in the second part, where we want to bound the size of terms, the proof fails. It is not possible to write the conditions on terms in the form of a unification problem. Some time ago Baaz proposed a program for proving Kreisel's Conjecture. Among the most important ideas of his are the use of Hilbert's ε-calculus and semiunification (a generalization of unification). This program has been so far realized only for a subtheory of existential induction Baaz and Pudlák [1993]; the proof uses Herbrand's theorem instead of the ε-calculus.

5. Bounds on cut-elimination and Herbrand's theorem

The undecidability of first order logic is caused by the fact that we cannot bound the size of a proof in terms of the size of the proved sentence. Nevertheless it is still possible to deduce something about the proof from the structure of the formula. (Fortunately proof theoretical studies in this direction started before the undecidability was discovered and therefore they were not hindered by this negative fact.) The theorems of this type are Herbrand's theorem, Hilbert's ε-theorem and Gentzen's cut-elimination theorem. The important consequence for all natural systems is that one can bound the quantifier complexity of the proof in terms of the quantifier complexity of the formula. This is achieved on the expense of lengthening the proof, however the lengthening can be bounded by a primitive recursive function. This raises an interesting question which we are going to deal with in this section: determine the growth rate of this function.

These theorems give more information about the structure of proofs. The most important is the cut-elimination theorem, which states that a general proof can always be replaced by a cut-free proof. Cut-free proofs have the so-called *subformula property*, which means that all formulas in the proof are subformulas of the proved formula φ. Here the concept of being a subformula is slightly weaker: the terms in the subformula may be different from those in φ. Hence there are infinitely many subformulas of φ, (even if we do not use function symbols, since there are infinitely many variables).

The three theorems are equivalent in the sense that there are easy proofs of one from another one. More important the simulations are polynomial, or at most exponential (depending on particular proof systems). Hence, if we are satisfied with a precision up to an exponential function, it is sufficient to give bounds only to one of them.

5.1. Let us consider the important specific case of the relation of the Herbrand theorem and the cut-elimination theorem. An easy extension of the cut-elimination theorem is the *Midsequent Theorem.* It states that each proof of a formula in the prenex form can be transformed into a proof where there is a sequent above which no quantifier rule is used and below which only quantifier rules are used. This can, in fact, be easily constructed from a cut-free proof. An easy analysis of the midsequent shows that it is essentially a Herbrand disjunction (see Hájek and Pudlák [1993,Chapter V]). Recall that a *Herbrand disjunction* is a disjunction of term instances of a *Herbrand variant* of a formula, where the Herbrand variant is obtained by systematically omitting the quantifiers, starting from the outermost, and replacing each universally bounded variable x by $F(y_1, \ldots, y_k)$, where F is a new function symbol and $y_1, \ldots, y_k$ are the free variables of the current formula. A midsequent does not contain these new function symbols, but the dependencies among the occurrences of variables allow us to replace variables by such terms while preserving the propositional validity of the disjunction.

Now suppose we are given a Herbrand disjunction. First replace the maximal terms whose outermost function symbol is a Herbrand function symbol by distinct variables. Then omit disjunctions and interpret it as a sequent. It has a propositional proof in the sequent calculus. Now each sequent provable in the propositional sequent calculus has a proof of at most exponential size. Thus we get the upper part of the sequential proof. The lower part is obtained by applying quantifier rules in a suitable order. This is possible due to the structure of the Herbrand disjunction. The number of the proof lines with quantifier rules is, of course, bounded by the number of variables. For more details see Takeuti [1987] and Hájek and Pudlák [1993,Chapter V, section 5].

5.2. We shall use the Hilbert style system of Chapter 1. Note however that when no restrictions are posed on the complexity of formulas in the proof the Hilbert style and Gentzen's sequent calculi are equivalent up to a polynomial increase of size. By Theorem 4.1 it is true even if we take proofs in a tree form in one of them and in a sequence form in the other one.

We shall start with an upper bound to cut-elimination.

5.2.1. Theorem. *Suppose a sentence φ has a proof of size n and depth d (i.e., each formula in the proof has logical depth at most d). Then φ has a cut-free proof of size $2^n_{O(d)}$.* □

The proof can be found in Chapter I.

5.2.2. Corollary. *If φ has a proof of size n, then φ has a cut-free proof of size $2^n_{O(|\varphi|\cdot\sqrt{n})}$.*

Proof. This follows from Theorems 5.2.1 and 4.2.5. □

Now we consider a lower bound. The proof will be easy since we have already developed the theory of definable cuts.

5.2.3. Theorem. *There exists a sequence of sentences $\psi_1, \psi_2, \ldots$ such that ψ_n has a proof of size $p(n)$, $n = 1, 2, \ldots$, where p is a fixed polynomial, and there is no cut-free proof of ψ_n with less than 2^0_n proof-lines for $n = 1, 2 \ldots$.*

Proof. Consider the following very weak fragment of arithmetic. It has the constant 0, the successor function $S(x)$, addition $+$ and exponentiation 2^x. It has axioms of equality, say those considered in section 3, and the following mathematical axioms:

$$0 + x = x$$

$$x + (y + z) = (x + y) + z,$$

$$x + S(x) = S(x + y),$$

$$2^0 = S(0),$$

$$2^{S(x)} = 2^x + 2^x.$$

Furthermore the theory contains a unary predicate symbol $I(x)$ with interpretation "an initial segment of integers without the last element". Thus we also include the axioms saying that I is inductive:

$$\begin{array}{c} I(0) \\ I(x) \rightarrow I(S(x)). \end{array} \tag{31}$$

Let us call this theory A. For a natural number n and a term t we denote by $E^n(t)$ the term defined inductively by

$$E^0(t) = t,$$

$$E^{n+1}(t) = 2^{(E^n(t))}.$$

In particular the value of $E^n(0)$ is 2^0_n. Now we define ψ_n by

$$\exists \ldots (\bigwedge A \rightarrow I(E^n(0))),$$

where $\bigwedge A$ denotes the conjunction of the axioms of A and $\exists \ldots$ denotes the existential closure.

5.2.4. Claim. *ψ_n's have polynomial size proofs.*

Proof. We shall use Theorem 3.4.1. By this theorem there exists a sequence of formulas $\varphi_0(x), \varphi_1(x), \varphi_2(x), \ldots$ with $\varphi_0(x)$ equal to $I(x)$ and

$$\varphi_{i+1}(0) \wedge \forall x(\varphi_{i+1}(x) \rightarrow \varphi_{i+1}(S(x))), \tag{32}$$

$$\forall x(\varphi_{i+1}(x) \rightarrow \varphi_i(2^x)), \tag{33}$$

having polynomial size proofs in A for $i = 0, 1, \ldots$. Combining (33) for $i = 0, \ldots, n-1$ we get a polynomial size proof of

$$\forall x(\varphi_n(x) \rightarrow I(E^n(x)))$$

in A. The first half of (32) for $i + 1 = n$ together with the last sentence give a polynomial size proof of $I(E^n(0))$ in A, hence a polynomial size proof of ψ_n in first order logic. □

5.2.5. Claim. *Let t be a closed term of A with value in $\mathbb{N}$ equal to m. Let β be a conjunction of term instances of axioms of A such that*

$$\beta \rightarrow I(t)$$

is provable in first order logic. Then β contains at least m term instances of the axiom $I(x) \rightarrow I(S(x))$.

Proof. W.l.o.g. we may assume that all the terms in β are closed (otherwise substitute 0). Suppose β contains fewer m occurrences of the axiom. Consider the values of terms t such that $I(t) \rightarrow I(S(t))$ occurs in β. By the pigeonhole principle there is an $i_0 < m$ which is not the value of any such a term. Assign truth values to the atomic subformulas of $\beta \rightarrow I(t)$ as follows: assign an identity a truth value according to its interpretation in natural numbers, and assign $I(t)$ the value TRUE if the value of t is less then or equal to i_0 and FALSE if it is bigger. Thus all the instances of axioms of A get the value TRUE giving this value also to β, while $I(t)$ gets FALSE. Thus $\beta \rightarrow I(t)$ cannot be provable. □

Now we derive the lower bound. Let a cut-free proof d of ψ_n be given. Let γ denote $\exists \ldots (\bigwedge A \rightarrow I(E^n(0)))$. All the quantifier rules of d are the rules of $\exists$-introduction applied to a term instance of γ or a term instance of a formula obtained from γ in this way. Let d' be the proof obtained by applying the same rules to initial segments of d but omitting the quantifier rules. We have to omit also the contractions applied to formulas with $\exists$, since such formulas will not appear in the new proof d'. Thus the end sequent of d' is a sequent $\longrightarrow \gamma_1, \ldots, \gamma_k$ where γ_i's are term instances of γ. Let γ_i be $\alpha_i \rightarrow I(E^n(0))$, (where α_i is a term instance of $\bigwedge A$). Then

$$\alpha_1 \wedge \cdots \wedge \alpha_k \rightarrow I(E^n(0))$$

is a tautology. By Claim 5.2.5, $k \geq 2^0_n$. Since in the original proof d all $\gamma, \ldots, \gamma_k$ must eventually merge into one formula, d must contain at least $k \geq 2^0_n$ proof-lines.
□

Let us note that the above proof can be applied directly to Herbrand theorem too. Namely, the above argument also shows that any Herbrand disjunction for ψ_n must have at least 2^0_n disjuncts.

5.3. The question whether mathematical reasoning as represented by Zermelo-Fraenkel set theory is consistent has intrigued a lot of mathematicians and philosophers. The approach of finitists is to discard it as meaningless and ask instead whether there is a *feasible* proof of contradiction from our axioms of set theory. We shall say more about this modified question in the next section. Now we only want to show that there are theories, not quite unnatural, which are inconsistent but in which no feasible proof of contradiction exists. Such theories have been considered by several researchers including Parikh [1971], Dragalin [1985], Gavrilenko [1984] and Orevkov [1990]; the first and the most influential was the paper of Parikh.

Let T be any fragment of arithmetic (it can be even the set of all true sentences in the standard model). Let t be a closed term whose value m is so large that no proof of size m can be ever constructed. Note that t can be quite simple, say 2^{100}. Extend T to T' by adding axioms

$$I(0),$$
$$I(x) \to I(S(x)),$$
$$\neg I(2^0_t).$$

Clearly T' is not consistent. We shall show, however, that there is no feasible contradiction in T'.

Suppose we can derive a contradiction in T' of size less than n. Then, by the bound on cut-elimination, there is a cut-free proof of contradiction of size less than 2^0_n. This means that we have such a proof of $\longrightarrow \neg \bigwedge T_0$, for a finite fragment T_0 of T'. Let T_1 be a Skolemization of T_0. Then the proof of $\longrightarrow \neg \bigwedge T_1$ is at most polynomially larger than 2^0_n (since each sentence has a polynomial size proof from its Skolemization). Thus by taking m, hence also t, only slightly larger than n, we get an upper bound 2^0_m to the *open* theory T_1. Then we use the same "interpretation" argument as in the lower bound proof above to show that such a proof cannot exist.

Let us note that we can add also other closure properties such as $I(x) \wedge I(y) \to I(x+y)$ and the same for multiplication, if we take t a little larger, since we can interpret such a theory in T' using small formulas and short proofs (see 3.5).

6. Finite consistency statements – concrete bounds

We have already remarked that there are almost no concrete examples of sentences for which one can prove nontrivial bounds on the length of proofs. There is, however, one exception; namely, the sentences expressing that a theory T does not prove a contradiction using a proof of length $\leq n$; (we shall say that the theory T is consistent up to n).

These are not real mathematical theorems, which would be interesting for an ordinary mathematician, but they are very interesting for people who study foundations of mathematics. We shall prove bounds on the length of such a statement in the theory T itself. This could be called a *finite* (or, if you prefer the word, *feasible*) version of the second Gödel Theorem. Furthermore, these bounds (especially the lower bounds) have interesting applications.

6.1. Formalization of syntax. We shall derive a strengthening of the second Gödel Incompleteness Theorem and some speed-up results. We shall try to avoid the boring subject of the formalization of syntax as much as possible. However we have to say something about it, since the classical way of formalizing syntax cannot be used here.

6.1.1. First we need a more efficient way of representing numbers by terms. The classical numerals $S^n(0)$ cannot be used, since their length is already greater than n, while we want to bound the lengths of proofs by a polynomial in $|n|$ – the length of the binary representation of n. Thus we define the n-th numeral $\underline{n}$ as follows. If $n = \sum_{i=0}^{k} 2^i a_i,\ a_i \in \{0,1\}$, then $\underline{n}$ is the closed term

$$\underline{a_0} + \underline{2} \cdot (\underline{a_1} + 2 \cdot (\underline{a_2} + \cdots (\underline{a_{k-1}} + 2 \cdot \underline{a_k}) \cdots))),$$

where $\underline{1} = S(0)$, $\underline{2} = \underline{1} + \underline{1}$.

We need also to represent sequences by numbers. A suitable one-to-one mapping from $\{0,1\}^*$ onto $\mathbb{N}$ is given by

$$(a_0, \ldots, a_n) \mapsto \sum 2^i (a_i + 1).$$

A formula φ is first represented as a $0-1$ sequence a, then we take the number m which codes a as the Gödel number of φ. We shall use the symbol $\ulcorner \varphi \urcorner$ for such a Gödel number of φ.

6.1.2. Suppose that we want to formalize a concept which can be represented as a subset $R \subseteq \mathbb{N}^k$. If R is formalized by a formula $\rho(x_1, \ldots, x_k)$ in a theory of T, then we clearly need that

$$(n_1, \ldots, n_k) \in R \Leftrightarrow T \vdash \rho(\underline{n_1}, \ldots, \underline{n_k}).$$

This alone is usually not sufficient. The key property for our proof is that the above formula has a proof of polynomial length. As it is an important concept, we shall define it precisely.

6.1.3. Definition. Let an axiomatization of a theory T be fixed, let $R \subseteq \mathbb{N}^k$ and let $\rho(x_1, \ldots, x_k)$ be a formula. We say that ρ *polynomially numerates* R in T, if for some polynomial p and every $n_1, \ldots, n_k \in \mathbb{N}$, the following holds: $R(n_1, \ldots, n_k)$ iff $T \vdash \rho(\underline{n}, \ldots, \underline{n_k})$ by a proof of length $\leq p(|n_1|, \ldots, |n_k|)$.

It turns out that, for a sufficiently strong theory T, the polynomially numerable relations are just the $\mathcal{NP}$ relations.

6.1.4. Theorem. *The following are equivalent*
(1) R is $\mathcal{NP}$*;*
(2) R is polynomially numerable in Robinson arithmetic $\mathbf{Q}$.

Since (2) $\Rightarrow$ (1) is trivial for any finitely axiomatized theory T, the same theorem holds for any finite consistent extension of $\mathbf{Q}$.

Before we sketch the proof of the converse implication, we state a lemma whose proof we defer to section 6.3.4.

6.1.5. Lemma. *For every bounded formula* $\varphi(x)$*, with* x *the only free variable, there exists a polynomial* p *such that*

$$I\Delta_0 + \text{Exp} \vdash \forall x \varphi(x)$$

implies that for every $n \in \mathbb{N}$,

$$\mathbf{Q} \vdash \varphi(\underline{n})$$

by a proof of length $\leq p(\log n)$.

This lemma allows us to replace $\mathbf{Q}$ by $I\Delta_0 + \text{Exp}$ in the proof of the implication (1) $\Rightarrow$ (2). If we are proving some property of a concept formalized by a Δ_0 formula in $I\Delta_0 + \text{Exp}$, then this statement may not be provable in $\mathbf{Q}$, but each numeric instance has a polynomial proof. Thus for instance we are free to use commutative and associative laws.

Proof-sketch of Theorem 6.1.4. Let an $R \in \mathcal{NP}$ be given. We formalize computations of a Turing machine defining R. Thus $R(n_1, \ldots, n_k)$ is equivalent to the existence of a $0-1$ string s whose length is bounded by a polynomial in $|n_1|, \ldots, |n_k|$ and which satisfies a certain property (namely, s codes an accepting computation). This property states that each c particular bits of s have one of some particular forms, where c is some constant. For a given s, there are polynomially many such conditions. Denote by $\sigma(x_1, \ldots, x_k, y)$ such a formula, where $x_1, \ldots, x_n$ stand for $n_1, \ldots, n_k$ and y for the string s. If $R(n_1, \ldots, n_k)$ is true, then $\sigma(\underline{n}_1, \ldots, \underline{n}_k, \underline{m})$ holds for some number m, whose length is bounded by a polynomial in $|n_1|, \ldots, |n_k|$. To prove $\sigma(\underline{n}_1, \ldots \underline{n}_k, \underline{m})$ by a polynomial proof in $\mathbf{Q}$, transform it into statements about single bits of the string encoded by m. Since the string really witnesses $R(n_1, \ldots, n_k)$, these elementary statements are true, hence provable. Finally derive $\exists y \sigma(\underline{n}_1, \ldots, \underline{n}_k, y)$ from $\sigma(\underline{n}_1, \ldots, \underline{n}_k, \underline{m})$. Thus $\exists y \sigma(x_1, \ldots, x_n)$ polynomially numerates R. □

Now we apply Theorem 6.1.4 to the provability predicate. Suppose a theory T is given by an $\mathcal{NP}$, resp. $\mathcal{P}$, set of axioms. Let $R(x, y)$ denote that x is a proof of y in T. Then R is in $\mathcal{NP}$, resp. $\mathcal{P}$, also. By Theorem 6.1.4 there is a formalization Proof_T of this relation, such that every true numeric instance has a polynomial proof in $\mathbf{Q}$. Since the relation "$|m| < n$" can also be polynomially numerated, we get the following corollary:

6.1.6. Corollary. *There exists formalization* $\mathrm{Pr}_T(\underline{n}, \ulcorner\varphi\urcorner)$ *of the relation* $\|\varphi\|_T \leq n$ *such that whenever it is true that* $\|\varphi\|_T \leq n$*, then* $\mathrm{Pr}_T(\underline{n}, \ulcorner\varphi\urcorner)$ *has a proof of polynomial length in* n *in* **Q**. □

Recall that $\|\varphi\| \leq n$ is a convenient notation for the statement that there exists a proof d of φ in T whose length is $\leq n$. Furthermore we shall denote by

$$\mathrm{Con}_T(x) \equiv_{df} \neg \mathrm{Pr}_T(x, \ulcorner \underline{0} = \underline{1} \urcorner),$$

the consistency of T up to the length x.

6.2. Now we are ready to prove the main lemma of the lower bound.

6.2.1. Lemma. *Let* T *be a sufficiently strong fragment of arithmetic. Let* $f : \mathbb{N} \to \mathbb{N}$ *be a polynomial time computable, increasing function. Suppose that*

> *for every* n *and every sentence* φ *of length* $\leq \log n$*, if* $\|\varphi\|_T \leq n$ *then* $\|\mathrm{Pr}_T(n, \varphi)\|_T \leq f(n)$*;*

and moreover the formalization of this sentence is provable in T*. Then*

$$n = O\left(f\left(\|\mathrm{Con}_T(\underline{n})\|_T\right)\right)$$

Hence if f can be extended to an increasing function defined on positive real numbers, then we can write the conclusion as

$$\|\mathrm{Con}_T(\underline{n})\|_T = \Omega(f^{-1}(n)).$$

Proof. We shall denote by $\ulcorner\varphi(\dot{x})\urcorner$ a formalization of the function $n \mapsto$ "the Gödel number of $\varphi(\underline{n})$". (E.g. the statement *"for all* n *the formula* $\varphi(\underline{n})$ *has property* P*"* is formalized by $\forall x\ P(\ulcorner\varphi(\dot{x})\urcorner)$.) Strictly speaking, in most cases this function cannot be formalized by a term and one has to use a formula with two variables which defines the graph of this function. This would however make the notation awkward.

Using the diagonalization lemma (see Chapter II) define a formula δ such that

$$T \vdash \delta(x) \equiv \neg \mathrm{Pr}(x, \ulcorner\delta(\dot{x})\urcorner). \tag{34}$$

6.2.2. Claim. $\|\delta(\underline{n})\|_T \leq n \to \|\underline{0} = \underline{1}\|_T \leq g(n)$*, where* $g(n) = O(f(n))$*.*

To prove the claim assume

$$\|\delta(\underline{n})\|_T \leq n. \tag{35}$$

Substituting $\underline{n}$ in (34) we get

$$\|\delta(\underline{n}) \equiv \neg Pr(\underline{n}, \ulcorner\delta(\underline{n})\urcorner)\|_T = (\log n)^{O(1)}. \tag{36}$$

Here we implicitly use the assumption that $\ulcorner\varphi(\dot{x})\urcorner$ has the polynomial numerability property. The assumption of the lemma and $\|\delta(\underline{n})\|_T \leq n$ implies that

$$\|\mathrm{Pr}(\underline{n}, \ulcorner\delta(\underline{n})\urcorner)\|_T \leq f(n). \tag{37}$$

Thus we get from (35), (36) and (37)

$$\|\underline{0} = \underline{1}\|_T = O(n + (\log n)^{O(1)} + f(n)) = O(f(n)),$$

which proves the claim. □

Since the assumption of the lemma is formalized in T, the claim can also be proved in T, thus we get

$$T \vdash \mathrm{Pr}_T(x, \ulcorner\delta(\dot{x})\urcorner) \to \neg\,\mathrm{Con}_T(\ulcorner g(x)\urcorner). \tag{38}$$

Since T is consistent, the claim implies in particular that $\|\delta(\underline{n})\|_T > n$. Thus it suffices to upper-bound $\|\delta(\underline{n})\|_T$ using $\|\,\mathrm{Con}_T(\underline{n})\|_T$. (Observe that the proof follows very much the structure of the proof of the second Gödel Incompleteness Theorem.) First substitute $\underline{n}$ in (38), thus we obtain

$$\|\,\mathrm{Con}_T(\ulcorner g(\underline{n})\urcorner) \to \neg\,\mathrm{Pr}(\underline{n}, \ulcorner\delta(\underline{n})\urcorner)\|_T = (\log n)^{O(1)}.$$

Combining it with (36) we get

$$\|\,\mathrm{Con}_T(\ulcorner g(\underline{n})\urcorner) \to \delta(\underline{n})\|_T = (\log n)^{O(1)},$$

hence

$$\|\,\mathrm{Con}_T(\ulcorner g(\underline{n})\urcorner)\|_T \geq \|\delta(\underline{n})\|_T - (\log n)^{O(1)} \geq n - (\log n)^{O(1)}.$$

Since $g(n) = O(f(n))$ and $\|\ulcorner g(\underline{n})\urcorner = \underline{g(n)}\|_T = (\log n)^{O(1)}$, the conclusion of the lemma follows. □

6.2.3. Theorem. (Friedman [1979], Pudlák [1986]) *Let T be a sufficiently strong fragment of arithmetic axiomatized by an $\mathcal{NP}$ set of axioms. Then there exists $\varepsilon > 0$ such that for all n,*

$$\|\,\mathrm{Con}_T(\underline{n})\| > n^{\varepsilon}.$$

Proof. by Corollary 6.1.6 and Lemma 6.2.1. □

With a little more additional work one can reduce the assumption about the strength to the condition $T \supseteq \mathbf{Q}$. Also it is possible to give a more precise lower bound by improving the bound in Corollary 6.1.6. The best lower bound has been proved in Pudlák [1987]. In that paper we considered first order logic augmented with Rosser's C-rule, which allows to introduce names for objects whose existence has been proved. Formally it means that we can derive $\varphi(c)$ from $\exists x \varphi(x)$ for a new constant c. (This *apparently* enables to shorten some proofs, but we are not able *to prove* a speed-up of this calculus versus the ordinary one.) For such a calculus we obtained a lower bound $\Omega(n/(\log n)^2)$.

6.3. Now we turn to the upper bound. Recall that in section 3 we proved that for a sequential theory T, there exists a sequence of formulas φ_n which define satisfaction for formulas of depth $n = 1, 2, \ldots$. Moreover

$$\varphi_n(\ulcorner\alpha\urcorner, x) \equiv \alpha((x)_1, \ldots, (x)_n) \tag{39}$$

and Tarski's conditions have polynomial size proofs. The following is an immediate consequence.

6.3.1. Lemma. *(1) For every axiom α of T, $dp(\alpha) \leq n$, T proves $\forall x \varphi_n(\ulcorner \alpha \urcorner, x)$ using a polynomially long proof.*

(2) For every n, T proves that any axiom of depth $\leq n$ is true and the truth of formulas of depth $\leq n$ is preserved by every rule, furthermore these proof are bounded by a polynomial in n.

Proof. The first part follows directly from (39). For part (2), let us consider only modus ponens. Thus we need a proof of

$$\forall x, y\,(\ulcorner dp(x \to y) \leq \underline{n} \urcorner \wedge \forall z \varphi_n(x, z) \wedge \forall z \varphi_n(x \to y, z) \to \forall z \varphi_n(y, z))\,. \tag{40}$$

We know that Tarski's condition

$$\ulcorner dp(x \to y) \leq \underline{n} \urcorner \to (\varphi_n(x \to y, z) \equiv (\varphi_n(x, z) \to \varphi_n(y, z)))$$

has a polynomial proof, thus also (40) has a polynomial proof. □

6.3.2. Theorem. (Pudlák [1986]) *Let T be a sequential theory axiomatized by a finite set of axioms. Then*

$$\mathrm{Con}_T(\underline{n}) = n^{O(1)}.$$

Proof. Let n be given. Let $\alpha(x)$ be the following formula

$$\begin{gathered}\forall y, z(\text{“}y \text{ is a proof of depth } \leq \underline{n} \text{ and size } \leq x\text{”} \wedge \\ \wedge \text{“}z \text{ is a formula of } y\text{”} \to \forall v \varphi_n(y, v)).\end{gathered}$$

Lemma 6.3.1 implies that $\alpha(\underline{0})$ and $\forall x(\alpha(x) \to \alpha(S(x))$ have polynomial size proofs. Thus by proving $\alpha(\underline{0}), \alpha(\underline{1}), \ldots, \alpha(\underline{n})$ one by one we get a polynomial proof of $\alpha(\underline{n})$. On the other hand, by (39), we have $\forall v \neg \varphi_n(\ulcorner \underline{0} = \underline{1} \urcorner, v)$, also by a polynomial proof. Thus we have a polynomially long proof that a proof of length $\leq n$ does not contain the formula $\underline{0} = \underline{1}$. □

This theorem has been proved also for some theories which are not finitely axiomatized, namely for theories axiomatized by a certain kind of axiom schemas. These results include the theories Peano Arithmetic and Zermelo-Fraenkel set theory.

Furthermore for finitely axiomatized sequential theories it is possible to improve the bound to $O(n)$. This improvement is based on the following ideas.

Firstly, by counting more precisely, it is possible to prove that (39) and Tarski's conditions for the truth definition for formulas of depth d have proofs of size $O(d^2)$.

Secondly, by Theorem 4.2.5, a proof of contradiction of length n can be transformed into a proof of depth $O(\sqrt{n})$. Thus we need the truth definition only for such a depth and hence the auxiliary formulas have linear size proofs.

Finally, one can use a shorter way to prove $\alpha(\underline{n})$. This is because of the following lemma, which gives us a proof even much shorter than $O(n)$.

6.3.3. Lemma. *Suppose $T \supseteq \mathbf{Q}$ and $\alpha(x)$ is a formula such that T proves*

$$\alpha(0) \wedge \forall x\,(\alpha(x) \rightarrow \alpha(S(x)))\,.$$

Then

$$\|\alpha(\underline{n})\|_T = O\left((\log n)^2\right)$$

(thus $\alpha(\underline{n})$ have proofs polynomial in $\log n$*).*

Proof. First define a subcut α' of α by

$$\alpha'(x) \equiv_{df} \forall y \leq x \alpha(y).$$

Then take a subcut α'' of α which is closed under addition and multiplication. This is easy, if the integers in T satisfy the laws of a ring; in 3.5 we have sketched a possible definition of such an α''. This is more technical for $\mathbf{Q}$ alone, so we refer the reader to Nelson [1986]. Then, in order to prove $\alpha''(\underline{n})$, prove $\alpha''(t)$ inductively for the subterms of $\underline{n}$. There are $O(\log n)$ such subterms and they have length $O(\log n)$. Finally use the fact that $T \vdash \alpha''(x) \rightarrow \alpha(x)$. □

6.3.4. Proof-sketch of Lemma 6.1.5. We shall use a similar idea in the proof that we still owe to the reader.

First we consider the length of proof of $\varphi(\underline{n})$ in $I\Delta_0$. An easy model-theoretical argument shows that the assumption of the lemma implies

$$I\Delta_0 \vdash \forall x(\exists y = 2^x_k \rightarrow \varphi(x))$$

for some $k \in \mathbb{N}$. Take a cut $\psi(x)$ in $I\Delta_0$ such that

$$I\Delta_0 \vdash \forall x(\psi(x) \rightarrow \exists y = 2^x_k).$$

Then we have

$$I\Delta_0 \vdash \forall x(\psi(x) \rightarrow \varphi(x)).$$

Thus we only need to construct a short proof of $\psi(\underline{n})$ in $I\Delta_0$, which is done as in the lemma above.

To get the theorem for $\mathbf{Q}$, use the well-known fact that $I\Delta_0$ has an interpretation in $\mathbf{Q}$ (this is an unpublished result of Wilkie, for a proof see Nelson [1986]). □

6.4. Some applications of the bounds. The lower bound can be used to show some strengthenings of the second Gödel Incompleteness Theorem. While the original theorem says only that T is consistent with its formal inconsistency (cf. Chapter II), we shall show that T is consistent with a statement saying that there is a *short* proof of contradiction. We have two such results.

6.4.1. Corollary. (Pudlák [1985]) *Let $T \supseteq \mathbf{Q}$ be consistent and axiomatized by an $\mathcal{NP}$ set of axioms. Then the following hold:*

(1) if I is a cut in T, then

$$T + \exists x(I(x) \& \neg \operatorname{Con}_T(x))$$

is consistent,

(2) if $\beta(x)$ is a bounded arithmetical formula such that $\beta(\underline{n})$ is true for every $n \in \mathbb{N}$, then there exists $k \in \mathbb{N}$ such that

$$T + \exists x(\beta(x) \& \neg \operatorname{Con}(x^k)).$$

is consistent, (x^k is $\underbrace{x \cdot x \cdot \ldots \cdot x}_{k-\text{times}}$).

Proof. (1) Suppose that the statement is false. Then

$$T \vdash I(x) \rightarrow \operatorname{Con}_T(x).$$

But, by Lemma 6.3.3, we have that $\|I(\underline{n})\|_T = O((\log n)^2)$, whence also $\|\operatorname{Con}_T(\underline{n})\|_T = O((\log n)^2)$ which is a contradiction with $\|\operatorname{Con}_T(\underline{n})\| \geq n^{\varepsilon}$, $\varepsilon > 0$.

(2) We need the following lemma.

6.4.2. Lemma.

$$s(\underline{n}) = \underline{s(n)},\ \underline{m} + \underline{n} = \underline{m+n},\ \underline{m} \cdot \underline{n} = \underline{m \cdot n}$$

have proofs of size polynomial in $\log n$ and $\log m$.

The proof is easy, if we assume ring operations, otherwise we have to work in a suitable cut. □

One consequence is that the same holds for arbitrary arithmetical terms. This can be used to show that, for a bounded formula $\beta(x)$ in the language of $\mathbf{Q}$, there exists a polynomial p_1 such that

$$\|\beta(\underline{n})\|_T \leq p_1(n), \tag{41}$$

whenever $\beta(\underline{n})$ is true.

The second consequence is that there is a polynomial p_2 such that

$$\|\underline{n}^k = \underline{n^k}\|_T \leq p_2(k, \log n). \tag{42}$$

(Hint: prove $\underline{n} \cdot \underline{n} = \underline{n^2}, \underline{n} \cdot \underline{n^2} = \underline{n^3}, \ldots, \underline{n} \cdot \underline{n^{k-1}} = \underline{n^k}$.)

We continue with the proof of (2). Assume that $\beta(\underline{n})$ is true for every $n \in \mathbb{N}$. Then we have (41) for all n. Clearly

$$\|\operatorname{Con}_T(\underline{n^k})\|_T \leq O(\|\beta(\underline{n})\|_T + \|\forall x(\beta(x) \rightarrow \operatorname{Con}_T(x^k))\|_T + \|\underline{n}^k = \underline{n^k}\|_T),$$

thus for some polynomial p_3

$$\| \operatorname{Con}_T(\underline{n}^k)\|_T \leq p_3(n, k, \|\forall x(\beta(x) \rightarrow \operatorname{Con}_T(x^k)\|_T). \tag{43}$$

By Theorem 6.2.3 there exists an $\varepsilon > 0$ be such that for every r

$$\| \operatorname{Con}_T(\underline{r})\|_T \geq r^{\varepsilon}. \tag{44}$$

Let d be the degree of n in $p_3(n, k, m)$. Take k so that $k\varepsilon > d$. Now suppose (2) fails for k, thus

$$T \vdash \forall x(\beta(x) \rightarrow \operatorname{Con}_T(x^k)).$$

Let m be the length of this proof. Take n so large that

$$p_3(n, k, m) < n^{k\varepsilon}.$$

Then, by (43),

$$\| \operatorname{Con}(\underline{n}^k)\|_T \leq p_3(n, k, m) < n^{k\varepsilon},$$

which is a contradiction with (44). □

6.5. Let us observe that also the upper bound on $\| \operatorname{Con}_T(\underline{n})\|_T$ can be used to obtain interesting corollaries. This is because the upper and the lower bounds are quite close, especially in the case of the calculus with the C-rule, hence the results that we used in the proof of the upper bound cannot be substantially improved. There are two such results. One is Theorem 3.2.3, where more precise calculations give a bound $O(n^2)$. The second one is the bound $O(\sqrt{n})$ on the depth of the shortest proof of a fixed size formula. Due to our bounds, the first result cannot be proved for a function which is $o(n^2/(\log n)^2)$, while the second for $o(\sqrt{n}/\log n)$. (The first statement is true also for first order logic without the C-rule, see Pudlák [1987].) However we feel that it should be possible to find direct arguments showing even the sharp bounds $\Omega(n^2)$ and $\Omega(\sqrt{n})$.

Further applications of the lower bounds will be shown in the next section.

7. Speed-up theorems in first order logic

The *speed-up* phenomenon is a situation, where we have two systems (proof systems, theories) such that some theorems have much shorter proofs in one of them. After the problem of proving lower bounds on proofs of concrete statements, this is the second most interesting problem. Note that in the intuitive relation between complexity of computations and complexity of proofs, speed-up theorems should correspond to separations of complexity classes.

We have already encountered a speed-up theorem in section 5, where we showed that cut-free proofs can be much longer than the proofs with cuts or proofs in a Hilbert style calculus. We shall consider such questions about propositional logic in sections 8 and 9. In this section we shall talk about two most important speed-up phenomena in first order logic. The one is the speed-up caused by having a stronger theory. The second one appears when we prove $\mathrm{Pr}_T(\varphi)$ in T instead of φ itself. It turns out that a dramatic speed-up is obtained in the first case almost by any extension of the theory. The second question, as we shall see, is related to the first.

From the point of view of the principal goal of proving concrete lower bounds these results are rather disturbing. Suppose that with a lot of effort we succeeded in proving that a statement, such as $\mathcal{P} \neq \mathcal{NP}$, does not have a feasible proof, say, in **ZF**, (which is extremely unlikely to happen in the near future) which would explain why we are not able to prove it. Then a simple and intuitively correct additional axiom, say $\mathrm{Con}_{\mathbf{ZF}}$, could change the situation completely, because in $\mathbf{ZF} + \mathrm{Con}_{\mathbf{ZF}}$ some proofs can be much shorter and we still would not know, why we cannot decide $\mathcal{P} = \mathcal{NP}$ in such a theory.

This suggests that a more reasonable goal is not to look for proofs of lower bounds in as strong as possible theories, but rather to consider also weak theories and to try to find out which ones are adequate for which problems.

Finally we shall consider an interesting situation where we can get a large speed-up by a *conservative* extension of a theory. A typical case is extending **ZF** (Zermelo-Fraenkel set theory with the axiom of choice) to **GB** (Gödel-Bernays set theory). Such extensions are called *predicative.*

7.1. We shall start with a very strong result of Ehrenfeucht and Mycielski which has a very simple proof.

7.1.1. Theorem. (Ehrenfeucht and Mycielski [1971]) *If the theory $T + \neg\alpha$ is undecidable, then there is no recursive function f such that*

$$\|\varphi\|_T \leq f(\|\varphi\|_{T+\alpha}),$$

for every sentence φ provable in T.

Proof. Suppose there is such an f. Suppose $T + \neg\alpha \vdash \varphi$, hence $T \vdash \alpha \vee \varphi$. Then, ignoring an additive constant, we have

$$\|\alpha \vee \varphi\|_T \leq f(\|\alpha \vee \varphi\|_{T+\alpha}) \leq f(|\alpha| + |\alpha \vee \varphi|).$$

(To prove $\alpha \vee \varphi$ from axioms $T + \alpha$, we first derive α, then $\alpha \vee \varphi$). Thus we can decide whether $T + \neg\alpha$ proves φ by checking all proofs of length $\leq f(|\alpha| + |\alpha \vee \varphi|)$ – a contradiction. □

7.1.2. Corollary. *Let T be a recursively axiomatized theory containing Robinson Arithmetic* **Q**. *Then* any *proper extension of T has arbitrary recursive speed-up over T.*

Proof. This follows from the fact that **Q** is essentially undecidable. □

The following result of Statman, improved by Buss, concerns the number of steps. We state it without a proof.

7.1.3. Theorem. (Statman [1981], Buss [1994]) *Let T be a theory axiomatized by a finite number of axiom schemas. Suppose α is a sentence undecided by T and such that $T + \neg\alpha$ is consistent. Then $T + \alpha$ has an infinite speed-up with respect to the number of steps over T, i.e., there exists an infinite set Φ of sentences and a k such that*

$$\forall \varphi \in \Phi \quad T + \alpha \vdash^k_{\text{steps}} \varphi,$$

but there is no m such that

$$\forall \varphi \in \Phi \quad T \vdash^m_{\text{steps}} \varphi,$$

□

We shall give some intuition about this theorem by an example due to Baaz. Suppose Kreisel's Conjecture holds for T. Let $\varphi(x)$ be a formula such that

1. $T \nvdash \forall x \varphi(x)$;
2. $\forall n \quad T \vdash \varphi(\underline{n})$.

A typical example of such a formula is $\mathrm{Con}_T(x)$. Then, clearly, for some constant c

$$T + \forall x \varphi(x) \vdash^c_{\text{steps}} \varphi(\underline{n}),$$

for every n, since we only need to substitute the term $\underline{n}$ into $\varphi(x)$. If, however, for some m

$$T \vdash^m_{\text{steps}} \varphi(\underline{n}),$$

for every n, we would get $T \vdash \forall x \varphi(x)$ by Kreisel's Conjecture. Thus $T + \forall x \varphi(x)$ has infinite speed-up over T.

7.2. Next we shall show that the lower bound on the length of proof of $\mathrm{Con}_T(\underline{n})$ in T (Theorem 6.2.3) can be used to obtain speed-up when T is extended to $T + \mathrm{Con}_T$ or when $Pr(\varphi)$ is used instead of φ.

Let T be a sufficiently strong fragment of arithmetic. We want to use sentences $\mathrm{Con}_T(f(\underline{n}))$ for fast-growing functions f. Such functions needn't be representable by terms in T. So we take the sentence

$$\exists y\, (\varphi(\underline{n}, y) \vee \mathrm{Con}_T(\varphi)) \tag{45}$$

instead, where φ defines the graph of f. Then we need two conditions to be satisfied

1. f is a *provably total* function in T, i.e., $T \vdash \forall x \exists! y \varphi(x, y)$;
2. φ polynomially numerates the graph of f.

Let us make a simple observation.

7.2.1. Lemma. *For every recursive function f, there exists a recursive function g such that*

1. *$\forall n \in \mathbb{N} \qquad f(n) \leq g(n)$,*
2. *the graph of g is a polynomial time computable relation.*

Proof. Let M be a Turing machine for f. Then one can construct a Turing machine M' which on input n prints 2^m in binary, where m is the number of steps of M on n. We take g to be the function computed by M'. □

If T is sufficiently strong, Lemma 7.2.1 can be formalized in T. Thus polynomial numerability is not an essential restriction. We shall abbreviate (45) by $\mathrm{Con}_T(f(\underline{n}))$.

7.2.2. Theorem. *Let T be a sufficiently strong theory. Let f be a provably total increasing recursive function in T whose graph has a polynomial numeration in T. Then there exists a $\delta > 0$ such that*

$$\| \mathrm{Con}_T(f(\underline{n}))\|_T \geq f(n)^\delta,$$

while

1. $\| \mathrm{Con}_T(f(\underline{n}))\|_{T+\mathrm{Con}_T} = O(\log n)$;
2. $\| \mathrm{Pr}_T(\ulcorner \mathrm{Con}_T(f(\underline{n}))\urcorner)\|_T = O(\log n)$.

Thus in both cases we get a speed-up by any provably total recursive function of T.

Proof. *Lower bound.* Let $\varphi(x, y)$ be the formula which polynomially numerates $f(x) = y$. Thus we want to bound

$$\|\exists y(\varphi(\underline{n}, y) \wedge \mathrm{Con}_T(y))\|_T.$$

Let $m = f(n)$. Clearly

$$\exists! y \varphi(\underline{n}, y) \wedge \varphi(\underline{n}, \underline{m}) \rightarrow \mathrm{Con}_T(\underline{m}). \tag{46}$$

Thus

$$\| \mathrm{Con}_T(\underline{m})\|_T \leq$$
$$\|\exists y(\varphi(\underline{n}, y) \wedge \mathrm{Con}_T(y))\|_T + \|\exists! y\varphi(\underline{n}, y)\|_T + \|\varphi(\underline{n}, \underline{m})\|_T + K,$$

where K is the length of the proof of (46). The proof of (46) depends only linearly on the lengths of n and m, thus $K = O(\log m)$. Similarly

$$\|\exists! y\varphi(\underline{n}, y)\|_T = O(\log n),$$

since we assume $T \vdash \forall x \exists! y \varphi(x, y)$. Finally we have a bound $(\log m)^{O(1)}$ on $\|\varphi(\underline{n}, \underline{m})\|_T$ by polynomial numerability. Thus, using Theorem 6.2.3 we have

$$m^\varepsilon \leq \| \mathrm{Con}_T(\underline{m})\|_T \leq \|\exists y(\varphi(\underline{n}, y) \wedge \mathrm{Con}_T(y))\|_T + (\log m)^{O(1)},$$

which gives the lower bound.

Upper bound (1). Recall that Con_T denotes $\forall x \,\mathrm{Con}_T(x)$ and that we assume $T \vdash \forall x \exists! y \varphi(x, y)$. Again, the proof of

$$\forall x \,\mathrm{Con}_T(x) \wedge \forall x \exists! y \varphi(x, y) \rightarrow \exists y(\varphi(\underline{n}, y) \wedge \mathrm{Con}_T(y))$$

depends only linearly on the length of n, thus

$$\|\exists y(\varphi(\underline{n}, y) \wedge \mathrm{Con}_T(y)\|_{T+\mathrm{Con}_T} = O(\log n).$$

Upper bound (2). We want to bound

$$\| \Pr_T(\ulcorner \exists y(\varphi(\underline{n}, y) \wedge \mathrm{Con}_T(y))\urcorner)\|_T \tag{47}$$

The argument will be similar to the one above. We have

$$\forall y \Pr_T(\ulcorner \mathrm{Con}_T(\dot{y})\urcorner) \wedge \Pr_T(\ulcorner \forall x \exists y \varphi(x, y)\urcorner) \to \Pr_T(\ulcorner \exists y(\varphi(\underline{n}, y) \wedge \mathrm{Con}_T(y))\urcorner),$$

by a proof of linear size in $\log n$. So it remains only to show that the conjunction in the antecedent is provable in T.

The provability of the second term follows from the assumptions.

The proof of $\forall y \Pr_T(\ulcorner \mathrm{Con}_T(\dot{y})\urcorner)$ can be constructed by formalizing the following argument (we assume that T is sufficiently strong): "Either T is inconsistent, and then it proves everything, or T is consistent, and then we can prove $\mathrm{Con}_T(\underline{n})$ by checking all proofs of length $\leq n$."

Hence (47) has a proof of length $O(\log n)$. □

For further improvements of these theorems, see Buss [1994].

7.2.3. Theorem 7.2.2 gives a worse speed-up for the length of proofs in proper extensions of T, moreover it requires that the extension proves Con_T. On the other hand it gives more explicit formulas on which the speed-up is attained. In particular it enables us to study the trade-off between the speed-up and the complexity of formulas. For instance consider statements $\mathrm{Con}_T(f(\underline{n}))$ for a primitive recursive f. Then $\mathrm{Con}_T(f(\underline{n}))$ are numeric instances of a (formalization of a) primitive recursive predicate. Thus we get *primitive recursive speed-up* on *primitive recursive formulas* etc.

We mention without a proof a related result where a speed-up is obtained for a simple formula in the fragment $T = I\Delta_0 + \Omega_1$. We shall denote 2^1_n by 2_n (the stack of n 2's).

7.2.4. Theorem. (Hájek, Montagna and Pudlák [1993]) *Let* $T = I\Delta_0 + \Omega$. *Then we have*

(1) $\quad \| \Pr_T(\ulcorner \exists y(y = 2_{2_{\underline{n}}})\urcorner)\|_T = n^{O(1)}$;

(2) $\quad \|\exists y(y = 2_{2_{\underline{n}}})\|_T = \Omega(2_n)$. □

7.3. Speed-up of GB over ZF. It is well-known that **GB** proves the same set of formulas as **ZF**. (We consider **ZF** with the axiom of choice.) Therefore it is very interesting to find out if the set formulas have proofs of approximately the same length in both theories. The answer is no; in fact there is a nonelementary speed-up as for cut-elimination. This seems to be typical for results obtained from cut-elimination or Herbrand theorem (and where a direct proof is not known).

This result is based on the following lemma due to Solovay (unpublished); a similar construction was considered by Vopěnka (unpublished).

7.3.1. Lemma. *There is a cut $I(x)$ in* **GB** *such that*

$$\mathbf{GB} \vdash \forall x(I(x) \rightarrow \mathrm{Con}_{\mathbf{ZF}}(x)).$$

It is outside of the scope of this chapter to give a proof of this lemma. Let us only very briefly describe the main idea. One can construct a sort of inner model of **ZF** in **GB** where the universe of sets is some cut. This model is constructed along with a satisfaction relation for it. Since the satisfaction relation is defined by a formula with class quantifiers, we cannot use induction to show the consistency of **ZF**. Instead we only show that the segment of numbers x such that there is no contradiction of length $\leq x$ is closed under successor. But this is exactly what we need. □

7.3.2. Theorem. (Pudlák [1986])
(1) $\| \mathrm{Con}_{\mathbf{ZF}}(2_{\underline{n}})\|_{\mathbf{GB}} = n^{O(1)}$;
(2) $\| \mathrm{Con}_{\mathbf{ZF}}(2_{\underline{n}})\|_{\mathbf{ZF}} = (2_n)^{\varepsilon}$, *for some constant* $\varepsilon > 0$.

Proof. To prove (1) we need, by Lemma 7.3.1, only to have a short proof of $I(2_{\underline{n}})$ in **GB**. The bound

$$\|I(2_n)\|_{\mathbf{GB}} = n^{O(1)}$$

follows from Theorem 3.4.1. The second part is contained in Theorem 7.2.2. □

A more precise computation gives a bound $\| \mathrm{Con}_{\mathbf{ZF}}(2_{\underline{n}})\|_{\mathbf{GB}} = O(n^2)$, which implies a lower bound on the speed-up of **GB** over **ZF** of the form $2_{\Omega(\sqrt{n})}$. An upper bound $2_{O(\sqrt{n})}$ complementing the above result was proved by Solovay [1990]. Let us observe that such bounds can be used to show that the estimate on the depth of formulas in a proof, Theorem 4.2.5, is asymptotically optimal.

8. Propositional proof systems

In this section we consider some concrete propositional proof systems. There are several reasons for studying these systems. Firstly they are natural systems which are used for formalization of the concept of a proof, in fact, they are good approximations of human reasoning. Some systems, especially resolution, are also used in automated theorem proving. Therefore it is important to know how efficient they are. Secondly they are suitable benchmarks for testing our lower bound techniques. Presently we are able to prove superpolynomial lower bounds only for the weakest systems; we shall give an example of a lower bound in section 9. Thirdly there are important connections between provability in some important theories of bounded arithmetic and the lengths of proofs in these propositional proof systems. This will be the topic of section 10.

8.1. Frege systems and its extensions. Frege systems are the most natural calculi for propositional logic and they are also used to axiomatize the propositional part of the first order logic in the Hilbert style formalizations. We have used a particular special case of a Frege system for presenting results on the lengths of proofs in first order logic.

8.1.1. To define a general Frege system, we need the concept of a Frege rule. A *Frege rule* is a pair $(\{\varphi_1(p_1,\ldots,p_n),\ldots,\varphi_k(p_1,\ldots,p_n)\},\varphi(p_1,\ldots,p_n))$, such that the implication $\varphi_1\wedge\ldots\wedge\varphi_k\rightarrow\varphi$ is a tautology. We use $p_1,\ldots,p_n$ to denote propositional variables. Usually we write the rule as

$$\frac{\varphi_1,\ldots,\varphi_k}{\varphi}.$$

When using the rule, we use actually its *instances* which are obtained by substituting arbitrary formulas for the variables $p_1,\ldots,p_n$. A Frege rule can have zero assumptions, in which case it is an *axiom schema.* A *Frege proof* is a sequence of formulas such that each formula follows from previous ones by an application of a Frege rule from a given set.

8.1.2. Definition. A *Frege system* F is determined by a finite complete set of connectives B and a finite set of Frege rules. We require that F be implicationally complete for the set of formulas in the basis B.

Recall that *implicationally complete* means that whenever an implication $\psi_1\wedge\ldots\wedge\psi_k\rightarrow\psi$ is a tautology, then ψ is derivable from $\psi_1,\ldots,\psi_k$. Rules such as modus ponens and cut ensure that the system is implicationally complete whenever it is complete.

An example of a Frege system is the propositional part of the proof system considered in section 2; it has 14 axiom schemas and one rule with two assumptions (modus ponens).

8.1.3. Note that in an application of a Frege rule (in particular also in axioms) we substitute arbitrary formulas for the variables in the rule, however we are not allowed to substitute in an arbitrary derived formula. It is natural to add such a rule. The rule is called *the substitution rule* and allows to derive from $\varphi(p_1,\ldots,p_k)$, with propositional variables $p_1,\ldots,p_k$, any formula of the form $\varphi(\psi_1,\ldots,\psi_k)$.[1]

8.1.4. Definition. A *substitution Frege system* SF is a Frege system augmented with the substitution rule.

8.1.5. The *extension rule* is the rule which allows to introduce the formula

$$p\equiv\varphi,$$

where p is a propositional variable and φ is any formula and the following conditions hold:

1. when introducing $p\equiv\varphi$, p must not occur in the preceding part of the proof or in φ;
2. such a p must not be present in the proved formula.

[1]In fact Frege used this rule originally and the idea of axiom schemas was introduced by von Neumann later.

If $\equiv$ is not in the basis, we can use an equivalent formula instead, e.g. $(p \to \varphi) \wedge (\varphi \to p)$. This rule is not used to derive a new tautology quickly, as it is the case of Frege and substitution rules, but its purpose is to abbreviate long formulas.

8.1.6. Definition. An *extension Frege system EF* is a Frege system augmented with the extension rule.[2]

8.1.7. The first question that we shall address is: how does the lengths of proofs depend on a particular choice of the basis of connectives and the Frege rules. If the basis is the same for two Frege systems F_1 and F_2, it is fairly easy to prove that they are polynomially equivalent. We have used this argument already in the previous sections. Let for instance

$$\frac{\varphi_1, \ldots, \varphi_k}{\varphi}$$

be a rule in F_1. Since F_2 is implicationally complete, there exists a proof π of φ from $\varphi_1, \ldots, \varphi_k$ in F_2. To simulate an instance of this rule obtained by substituting some formulas into it, we simply substitute the same formulas in π. Thus we get only linear increase of the size.

If the two bases are different, the proof is not so easy, but the basic idea is simple. One uses a well-known fact from boolean complexity theory that a formula in one complete basis can be transformed into an equivalent formula in another complete basis with at most polynomial increase in size, in fact, using a polynomial algorithm. This, of course, does not produce a proof from a proof, but one can show that it suffices to add pieces of proofs of at most polynomial size between the formulas to get one. Details are tedious, so we leave them out.

The same holds for substitution Frege and extension Frege systems. Thus we have:

8.1.8. Theorem. (Cook and Reckhow [1979], Reckhow [1976]) *Every two Frege systems are polynomially equivalent, every two substitution Frege systems are polynomially equivalent, and every two extension Frege systems are polynomially equivalent.* □

8.1.9. This still leaves three classes, moreover each can be considered also in the tree form and we can count the number of steps instead of the size, which gives altogether twelve possibilities. We shall show that these cases reduce to only three, if we identify polynomially equivalent ones (namely, Frege, extension Frege and the number of steps in substitution Frege).

The question about a speed up of sequence versus tree proofs has been solved in Theorem 4.1 for Frege systems which contain modus ponens. The same holds for extensions of such Frege systems by extension and substitution rules. We shall return to this question below. Now we shall consider the remaining ones. Let us first consider the relation of substitution Frege systems and extension Frege systems.

[2]Here I deviate slightly from the literature where the name *extended Frege system* is used, which, I think, is rather ambiguous.

8.1.10. Theorem. (Dowd [1985], Krajíček and Pudlák [1989]) *Every substitution Frege system is polynomially equivalent to every extension Frege system.*

Proof. By the theorem above, we can assume that both systems have the same language.

1. First we show a polynomial simulation of an extension Frege system by a substitution Frege system. Let an extension Frege proof of a tautology ψ be given, let $p_1 \equiv \varphi_1, \ldots, p_m \equiv \varphi_m$ be all formulas introduced by the extension rule listed in the order in which they were introduced. By Theorem 8.1.8 we can assume w.l.o.g. that our systems contain suitable connectives and suitable Frege rules. Using an effective version of the deduction theorem (whose easy proof we leave to the reader) we get, by a polynomial transformation, a proof of

$$p_1 \equiv \varphi_1 \wedge \ldots \wedge p_m \equiv \varphi_m \to \psi \tag{48}$$

which does not use the extension rule (i.e., a Frege proof). Now apply the substitution rule to (48) with the substitution $p_m \mapsto \varphi_m$. Thus we get

$$p_1 \equiv \varphi_1 \wedge \ldots \wedge p_{m-1} \equiv \varphi_{m-1} \wedge \varphi_m \equiv \varphi_m \to \psi. \tag{49}$$

From (49) we get by a polynomial size Frege proof

$$p_1 \equiv \varphi_1 \wedge \ldots \wedge p_{m-1} \equiv \varphi_{m-1} \to \psi.$$

We repeat the same until we get a proof of ψ.

2. The polynomial simulation of substitution Frege systems by extension Frege systems is not so simple. The idea of the proof is the following. Suppose we have simulated a substitution Frege proof until φ_j which is derived by a substitution from φ_i, say $\varphi_j = \varphi_i(\bar{p}/\bar{\alpha})$. Then we can derive φ_j by repeating the previous part of the proof with variables $\bar{p}$ replaced by $\bar{\alpha}$. However, repeating this, the new *Frege* proof would grow exponentially. The trick is to prove the formulas of the substitution Frege proof not for a particular substitution, but for the substitution, where the proof fails for the first time. In reality the proof does not fail (we start with a real proof), but it enables us to argue: "if it failed, then we could go on, hence we can go on in any case". Now the problem is for which propositional variables should the proof fail. Therefore we introduce an extra set of variables for each formula of the substitution Frege proof. The variables at some step will be defined using variables at the following steps of the proof. As this nesting may result in an exponential growth, we introduce them using the extension rule.

Now we shall argue formally. W.l.o.g. we can assume that the substitution Frege system has only modus ponens and axiom schemas as Frege rules. Let $(\varphi_1, \ldots, \varphi_m)$ be a substitution Frege proof. Let $\bar{p} = (p_1, \ldots, p_n)$ be all propositional variables of the proof. Take sequences $\bar{q}_i$ of length n consisting of new distinct variables, for $i = 1, \ldots, m-1$, and denote by $\bar{q_m} = \bar{p}$. Let ψ_i be $\varphi_i(\bar{p}/\bar{q}_i)$, for $i = 1, \ldots, m$; thus $\psi_m = \varphi_m$. Let $\bar{\beta}_j$ be a sequence of n formulas defined as follows:

1. if φ_j is an axiom or is derived by modus ponens then $\bar{\beta}_j = \bar{q}_j$;
2. if φ_j is derived by substitution from φ_i, namely $\varphi_j = \varphi_i(\bar{p}/\bar{\alpha})$, then $\bar{\beta}_j = \bar{\alpha}(\bar{p}/\bar{q}_j)$.

The extension Frege proof will start by introducing

$$q_{i,l} \equiv (\Psi_i \wedge \neg\psi_{i+1} \wedge \beta_{i+1,l}) \vee \ldots \vee (\Psi_{m-1} \wedge \neg\psi_m \wedge \beta_{m,l}),$$

where Ψ_j is $\psi_1 \wedge \ldots \wedge \psi_j$. We have to introduce these formulas in the order $i = m-1, \ldots, 1$.

Then we add polynomial size proofs of

$$\Psi_{j-1} \wedge \neg\psi_j \rightarrow \psi_i \equiv \psi_i(\bar{q}_i/\bar{\beta}_j), \tag{50}$$

for $i < j$. To prove (50) we first derive

$$\Psi_{j-1} \wedge \neg\psi_j \rightarrow q_{i,l} \equiv \beta_{j,l}$$

from the axioms introducing $q_{i,l}$ and then successively construct proofs of the corresponding statements for subformulas of ψ_i.

Now we derive $\psi_1, \ldots, \psi_m$. Suppose we have proved $\psi_1, \ldots, \psi_{j-1}$. Consider three cases.

1. φ_j is an axiom. Then ψ_j is also an axiom.

2. φ_j was derived from φ_u, φ_v, $u, v < j$, $\varphi_u = \varphi_v \rightarrow \varphi_j$ by modus ponens. First derive Ψ_{j-1}. Then, using (50) with $i = u, v$, we get

$$\neg\psi_j \rightarrow \psi_u(\bar{\beta}_j) \wedge \psi_v(\bar{\beta}_j).$$

Since

$$\psi_u(\bar{\beta}_j) = \psi_v(\bar{\beta}_j) \rightarrow \psi_j(\bar{\beta}_j),$$

we get

$$\neg\psi_j \rightarrow \psi_j(\bar{\beta}_j).$$

As we are considering the case of modus ponens, $\bar{\beta}_j = \bar{q}_j$, hence we have derived $\neg\psi_j \rightarrow \psi_j$, whence we get ψ_j immediately.

3. φ_j was derived from φ_i, $i < j$ by substitution. Then ψ_j is just $\psi_i(\bar{q}_i/\bar{\beta}_j)$, thus (50) gives

$$\Psi_{j-1} \wedge \neg\psi_j \rightarrow \psi_i \equiv \psi_j$$

and we get ψ_j easily from $\psi_1, \ldots, \psi_{j-1}$.

Finally recall that ψ_m is the conclusion of the proof φ_m, thus we have the simulation. □

The simulation of extension Frege systems by substitution Frege systems was shown already in Cook and Reckhow [1979]. The other simulation has a simple "higher order" proof based on a relation to bounded arithmetic. Namely by Theorem 10.3.6 below, it suffices to prove the reflection principle for substitution Frege system in S^1_2, which is easy. This was observed independently by Dowd [1985] and Krajíček and Pudlák [1988].

8.1.11. It is an open problem whether Frege systems can simulate extension and substitution Frege systems. We conjecture that the answer is *no*. It seems, though it is not supported by any mathematical result, that the relation of Frege systems to extension Frege systems is the same as the relation of boolean formulas to boolean circuits in complexity theory. It is generally accepted that it is unlikely that formulas can simulate circuits with only polynomial increase in size; the uniform version of this conjecture is $\mathcal{NC}^1 \neq \mathcal{P}$; both conjectures are also open.

8.1.12. We shall now consider the number of steps in these systems. It turns out that the number of steps in Frege and extension Frege systems is, up to a polynomial, the size of extension Frege proofs.

8.1.13. Lemma. *If φ can be proved by a proof with n steps in an extension Frege system, then it can be proved by a proof with n steps in a Frege system; namely, we can omit the extension rule from the given system.*

Proof-sketch. Omit every instance of the extension rule $p \equiv \varphi$ and at the same time replace all occurrences of the variable p by φ. □

8.1.14. Lemma. *There exists a polynomial $f(x)$ such that for every tautology φ and every extension Frege proof of φ with n steps, there exists an extension Frege proof of φ whose size is $\leq f(|\varphi| + n)$.*

The idea of the proof is to introduce propositional variables for each *relevant subformula* of the proof and work with these variables instead of formulas. A subformula is relevant, if it is in constant depth from the root of some formula in the proof, where the constant is determined by the Frege system. We leave out further details. □

From these two lemmas we get immediately:

8.1.15. Theorem. (Statman [1977], Cook and Reckhow [1979]) *For every tautology φ, the minimal number of steps of a proof of φ in a Frege system, the minimal number of steps of a proof of φ in an extension Frege system and the minimal size of a proof of φ in an extension Frege system are polynomially related.* □

8.1.16. It can be shown that the minimal number of steps in a proof of a tautology in an extension Frege system can be exponentially larger than in a substitution Frege system Tsejtin and Čubarjan [1975], Krajíček [1989b]. Consider the tautology $\neg^{2^n}(p \vee \neg p)$, where $\neg^{2^n}$ denotes 2^n-times $\neg$. It is not very hard to show that the number of steps needed to prove it in a Frege system is $\Omega(2^n)$ (it is also possible to prove it by defining a winning strategy for Adversary in the game below). Thus the minimal size of an extension Frege proof for $\neg^{2^n}(p \vee \neg p)$ must be $2^{\Omega(n)}$. On the other hand it can be proved using only $O(n)$ steps in a substitution Frege system. This is based on the fact that it is possible to derive $q \to \neg^{2^{k+1}} q$ from $q \to \neg^{2^k} q$ in constant number of steps: first derive $\neg^{2^k} q \to \neg^{2^{k+1}} q$ by substitution $q \mapsto \neg^{2^k} q$, and then use the transitivity of implication.

8.2. A game. The following game was devised as an approach to proving lower bounds on the lengths of propositional proofs. So far we were not able to get new lower bounds result using it; in fact it is even not so easy to interpret the known lower bounds using this game. However the game can be used at least to prove something about the structure of propositional proofs.

8.2.1. We shall call the game *Prover-Adversary game.* The game is determined by a complete set of propositional connectives B. There are two players *Prover* and *Adversary.* The aim of Prover is to prove a proposition φ and the aim of Adversary is to pretend that, for some assignment, the formula φ can have value 0 (=false). The game starts with Prover's asking φ and Adversary answering 0, and then Prover asks other propositions and Adversary assigns values to them. The game ends when there is a simple contradiction in the statements of the Adversary which means the following. Suppose we consider propositions in a basis of connectives B. Then a *simple contradiction* means that for some connective $\circ \in B$, and propositions $\varphi_1, \ldots, \varphi_k$, Adversary has assigned values to $\varphi_1, \ldots, \varphi_k$, $\circ(\varphi_1, \ldots, \varphi_k)$ and they do not satisfy the truth table of $\circ$; e.g., he assigned 0 to φ, 1 to ψ and 1 to $\varphi \wedge \psi$.

We define that a proposition φ is *provable in this game,* if Prover has a winning strategy. A natural measure of complexity of such proofs is *the minimal number of rounds needed to convict any Adversary.*

It is easy to prove that the Prover-Adversary game as a proof system is sound and complete, (however it does not satisfy the definition of a propositional proof system 2.5). To prove the soundness, suppose φ is not a tautology. Then Adversary can simply evaluate the propositions on an input a for which $\varphi[a] = 0$. To prove the completeness, let Prover ask all subformulas of φ, including the variables.

The most interesting fact about the Prover-Adversary game is the relation of the number of rounds in the game to the number of steps in a Frege proof.

8.2.2. Proposition. *The minimal number of rounds in the Prover-Adversary game needed to prove φ is proportional to the logarithm of the minimal number of steps in a Frege proof of φ.*

More precisely, for every basis B and every Frege system F, there are constants c_1, c_2 such that for every tautology φ,

(i) if it has a proof with k steps in F, then it can be proved in $\leq c_1 \log k$ rounds and

(ii) if it can be proved in r rounds, then it can be proved in F in k steps with $\log k \leq c_2 r$.

Proof. 1. Let a Frege proof of φ be given, say $\varphi_1, \ldots, \varphi_k$, with $\varphi_k = \varphi$. Consider conjunctions

$$\psi_i = (\cdots(\varphi_1 \wedge \varphi_2) \wedge \cdots) \wedge \varphi_i.$$

If Adversary tries to be consistent as long as possible, Prover needs only a constant numbers of questions to force him to assign 1 to an axiom. Thus he can force value 1 for ψ_1. Also he needs only a constant number of questions to get 0 for ψ_k, since $\varphi_k \mapsto 0$. Then he uses binary search to find an i such that $\psi_i \mapsto 1$ and $\psi_{i+1} \mapsto 0$. This takes $O(\log k)$ rounds. A constant number of rounds is needed to get $\varphi_{i+1} \mapsto 0$. Suppose φ_{i+1} was derived from $\varphi_{i_1}, \ldots, \varphi_{i_l}$, $i_1, \ldots, i_l \leq i$. For each of these premises it takes only $\leq \log i$ rounds to force 1 (or to get an elementary contradiction), since $\psi_i \mapsto 1$, – use binary search again. Once the premises got 1's and the conclusion 0, Prover needs only a constant number of questions to force an elementary contradiction.

2. Let a winning strategy for Prover be given, suppose it has r rounds in the worst case. We construct a sequent calculus proof of φ of size $2^{O(r)}$, which, as we know, can be transformed into a Frege proof with at most polynomial increase; we shall consider this transformation in more details below.

Consider a particular play P, let $\alpha_1, \ldots, \alpha_t$, $t \leq r$ be the questions asked by Prover, where we have added (or removed) negations, if Adversary answered 0 (in particular α_1 is $\neg\varphi$). Thus $\alpha_1 \wedge \cdots \wedge \alpha_t$ is false, hence $\rightarrow \neg\alpha_1, \ldots, \neg\alpha_t$ is a true sequent. Moreover, as easily seen, it has a proof with constant number of lines, since there is a *simple contradiction* in the statements $\alpha_1, \ldots, \alpha_t$. The proof of φ is constructed by taking proofs of all such sequents and then using cuts eliminating successively all formulas except of φ. This is possible due to the structure of the possible plays. Namely,

1. for each play P there is another play P' in which all the questions and answers are the same except for the answer which corresponds to the last question of P; P' may be longer than P;
2. for every two plays P, P', if they have the same questions up to the i-th one, say $\alpha_1, \ldots, \alpha_i$, then they have the same answers up to the $i-1$-st one and different i-th answer.

Finally observe that the number of such sequents is at most 2^r, which gives the bound. □

Let us note that the proof constructed from the game has a very special structure. Firstly it is in a *tree form*; secondly, it is like a dual to cut-free proofs, since it uses everywhere *only the cut rule*, except for the leaves of the proof tree.

Let us note also that we can characterize the *size* of proofs in Frege systems in a similar way, we have only to add the logarithm of the maximal size of a query to the cost of the play.

8.2.3. We return to the problem about the relation of the lengths of proofs as sequences and lengths of proofs as trees. We would like to use the proof obtained by transforming a general proof into the Prover-Adversary game and then back to a proof. The resulting proof has the number of steps polynomial in the original number of steps k and it is in a tree form, but it is a sequent proof. We shall analyze its transformation into a Frege proof in order to see that the tree structure can be preserved also in the Frege form.

First we shall assume that we have a Frege system F_0 with suitable rules. Let $\rightarrow \neg\alpha_1, \ldots, \neg\alpha_t$ be a sequent on the leaf of the sequent proof. We shall replace it by

$$(\cdots(\neg\alpha_1 \vee \neg\alpha_2)\cdots) \vee \neg\alpha_t. \tag{51}$$

We shall start with proofs of such sequents. We know that some $(\cdots(\neg\alpha_{i_1} \vee \neg\alpha_{i_2})\cdots) \vee \neg\alpha_{i_c}$, where c is a constant determined by the basis that we use, is a tautology. Moreover this tautology has a constant size proof, as it comes from a simple contradiction. Hence it also has a constant size tree proof. To get (51) we have to add the remaining disjuncts using a tree proof. It is quite easy, if we use the ideas shown in section 4.

Let us note that $t = O(\log k)$, thus also the proof of (51) is $O(\log k)$.

The rest of the proof is the same as in the sequent case, provided that we have a cut rule in the form

$$\frac{A \vee B,\ C \vee \neg B}{A \vee C}.$$

If we have a general Frege system F, we have to simulate each application of a rule of F_0 by several, however a constant number, of rules of F. Structurally it means that we replace each node with its in-going edges of the original tree by a constant size tree. Thus in general we get a larger tree; but the point is that the new tree has depth also $O(\log k)$, thus it has size also polynomial in k. Hence have proved:

8.2.4. Theorem. (Krajíček [1994a]) *For* every *Frege system there exists a polynomial $p(x)$ such that for every tautology φ*

$$\|\varphi\|^{tree}_{steps} \leq p(\|\varphi\|^{sequence}_{steps}).$$

□

8.3. Resolution. The most important propositional calculus for automated theorem proving is the resolution system. It is fairly easy to implement and there is a variety of heuristics there that one can try in the proof search.

The idea can be simply explained as follows. Suppose that we want to prove a tautology which is a DNF. Thus it suffices to derive a contradiction from its negation, which is a CNF, say $\bigwedge_{i\in I} \delta_i$. This is the same as to derive a contradiction from the set $\{\delta_i\}_{i\in I}$. If we think of disjunctions as obtained by applying the *set* operator of disjunction to a *set* of variables and its negations, then we need only a single rule – the cut. The contradiction then would be the disjunction of an empty set.

In the usual terminology we call variables and negated variables *literals*; the disjunctions are represented simply as sets of literals and they are called *clauses*, the cut rule is called *resolution*. As we are proving a contradiction from assumptions, we rather talk about a *refutation* than a proof. Thus a *resolution refutation* of a set of clauses $\mathcal{C}$ is a sequence starting with the clauses of $\mathcal{C}$, the following clauses are derived by resolution and the last clause should be $\emptyset$.

8.4. Extended resolution. Though a lot of interesting tautologies are DNF's, we would like to be able to prove also others. There is a natural way, in which we can extend the resolution system to be able to talk about arbitrary formulas; namely, we introduce variables for formulas and add the defining clauses.

Formally *extended resolution* for the basis $\{\wedge, \vee, \neg\}$ and variables $p_1, \ldots, p_n$ is resolution augmented with the clauses obtained from the CNF's of

$$q_{p_i} \equiv p_i,\ q_{\neg\varphi} \equiv \neg q_\varphi,\ q_{\varphi_1 \wedge \varphi_2} \equiv q_{\varphi_1} \wedge q_{\varphi_2},\ q_{\varphi_1 \vee \varphi_2} \equiv q_{\varphi_1} \vee q_{\varphi_2},$$

for all formulas in the language $\{p_1, \ldots, p_n, \wedge, \vee, \neg\}$, where q's are some new distinct variables. E.g., $q_{\neg\varphi} \equiv \neg q_\varphi$ is replaced by the two clauses $\{q_{\neg\varphi}, q_\varphi\}$ and $\{\neg q_{\neg\varphi}, \neg q_\varphi\}$. We define it for other bases similarly.

While resolution is much weaker than Frege systems, the extended resolution system is polynomially equivalent to extension Frege systems. The simulation of extension Frege system by extended resolution is based on essentially the same idea as Lemma 8.1.14.

8.5. Bounded depth Frege systems. Intermediate between the resolution system and the Frege systems are bounded depth Frege systems. They are very important for bounded arithmetic, see section 10. Also they are the strongest systems for which we are able to prove exponential lower bounds.

Consider formulas in basis $\{\wedge, \vee, \neg\}$. We define inductively classes Σ_i and Π_i of such formulas. Σ_0 and Π_0 are just literals. A formula φ is in Σ_{i+1}, if it is the disjunction of formulas from Σ_{i+1} or the conjunction of formulas from Σ_i or a negation of a formula from Σ_i. The classes Π_{i+1} are defined dually. A formula has *depth d*, if it is in $\Sigma_d \cup \Pi_d$.

A *depth d Frege proof* is a Frege proof, where all formulas are depth d. If a suitable set of rules is chosen such a system is complete for depth d tautologies.

Krajíček [1994a] has shown that there are depth d tautologies which have polynomial size tree-like proofs in a depth $d+1$ Frege system, but only exponential size tree-like proofs in a depth d Frege system, and, conversely, there are depth d tautologies which have polynomial size (general) proofs in a depth d Frege system, but only exponential size tree-like proofs in a depth d Frege system. (More precisely, one has to use refutations instead of proofs.) It is not known, if there is such a speed up for sequence-like proofs. Also it is an open problem, if there is a d_0 such that for every $d \geq d_0$, depth d and $d+1$ systems can be separated in such a way using tautologies of depth $\leq d_0$. On the other hand there is a sequence of tautologies of depth 3 which have polynomial size (unbounded depth) Frege proofs, but only exponentially large depth d Frege proofs for every constant d (see Buss [1987], Krajíček, Pudlák and Woods [1995], Pitassi, Beame and Impagliazzo [1993] and Beame et al. [1992]). The tautologies express a very simple theorem – the pigeonhole principle. We shall prove a lower bound for resolution refutations of sets of clauses expressing the pigeonhole principle in the next section.

8.6. Propositional sequent calculus. We have already mentioned that sequent proof systems are polynomially equivalent to the Frege system that we considered in the first part of the chapter, hence to all Frege systems. Thus it remains to mention the cut-free propositional sequent calculus. Since we know about nonelementary speed-up in the case of first order logic, it is not surprising that there is a speed-up also for propositional logic. The speed-up is exponential Takeuti [1990], and, trivially, cannot be larger. An exponential speed-up (slightly worse) follows also from the speed-up of unbounded depth Frege system versus bounded depth Frege system (using the fact that a cut-free proof of a bounded depth tautology is also bounded depth).

8.7. Propositional natural deduction. The natural deduction system is essentially a Frege system with an additional rule which allows to prove an implication $\varphi \to \psi$ by taking φ as an assumption and deriving ψ. The fact that this rule can be simulated in a Frege system is called *the deduction theorem* and the rule is called *the deduction rule.* Mutual simulations of the sequent calculus and natural deduction were shown by Gentzen [1935] and they are actually polynomial simulations, see Eder [1992]. The power of the deduction rule has been investigated in more detail by Bonet and Buss [1993].

8.8. Quantified propositional proof systems. It seems unlikely that there is a proof system for propositional logic which can polynomially simulate all other proof systems (see Krajíček and Pudlák [1989] for the relation of this question to problems in computational complexity). Thus it is interesting to look for stronger and stronger proof systems. How can one construct a system stronger than extension Frege systems? One possible way is to extend the expressive power of the language used in the proofs and the most natural extension is to take quantified propositional formulas.[3]

The language of quantified propositional logic consists of *quantified propositional formulas* which are usual propositional formulas with quantifiers binding some propositional variables. The semantics of such formulas is clear. E.g., the following is a quantified propositional tautology

$$\forall p, q \exists r ((p \to q) \to (p \to r) \wedge (r \to q)).$$

As a logical calculus we simply modify either a Hilbert style or Gentzen sequent first order calculus. Again, we give only an example. Consider the axiom schema (5.1) of section 2

$$\Phi(t) \to \exists x \Phi(x).$$

We use this schema in quantified propositional logic as it stands, the only point is that now there is no distinction between terms and subformulas. So precisely stated it is as follows. Let $\varphi(p)$ be a quantified propositional formula with a free variable p

[3] It is interesting that a quantified propositional calculus was introduced by Russell [1906] as a *"theory of implication"*.

and let ψ be any quantified propositional formula, then the following formula is an axiom

$$\varphi(p/\psi) \to \exists x \varphi(x).$$

It is interesting to investigate the proof systems for all of quantified propositional logic, but we would also like to know, if such systems enable us to prove ordinary propositions faster. This seems plausible, as quantified propositional formulas can define functions in $\mathcal{PSPACE}$, thus very likely they have stronger expressive power than ordinary propositional formulas. But even if this were true, it would not necessarily imply that, say, the quantified propositional sequent calculus has shorter proofs for some propositional tautologies. Also we cannot exclude that the quantified propositional sequent calculus is stronger, but at the same time quantified propositional formulas of polynomial size define the same functions as ordinary propositional formulas of polynomial size. The only relation that we know for sure is that it polynomially simulates substitution (hence also extension) Frege systems, see Krajíček and Pudlák [1990].

Important applications in bounded arithmetic were found by Dowd [1979] and Krajíček and Takeuti [1990]; for further applications in bounded arithmetic, see section 10 and Krajíček and Pudlák [1990].

A related question is, how strong is the propositional part of the first order calculus. Let us consider the Hilbert style calculus of section 2. If we had only propositional variables, then it is just a Frege system. However, if we have some predicate, say $P(x)$, then we can code the propositional variable p_i using the first order variable x_i as $P(x_i)$. This enables us to code all *quantified* propositional formulas, thus we get at least the power of the quantified propositional calculus. However, using a suitable representation we can simulate arbitrarily strong propositional proof system, see 10.4.1 below.

8.9. "Mathematical" proof systems. Let us have look at the problem about the length of proofs in propositional logic from the point of view of complexity theory. The set of propositional tautologies is a $co\mathcal{NP}$-complete sets, say L. A proof system is a relation $R(x, y)$ computable in polynomial time such that

$$x \in L \equiv \exists y R(x, y).$$

A proof of x is a y such that $R(x, y)$. Thus we can take an arbitrary $co\mathcal{NP}$-complete set and an arbitrary R for it and ask what are the lengths of such proofs. We shall consider three examples of such calculi.

8.9.1. The Hajós calculus. In the first example the set L consists of graphs which cannot be colored by three colors. Hajós [1961] has proved that every such graph can be obtained as follows.

1. Start with K_4, the complete graph on four vertices, and apply the following operations:

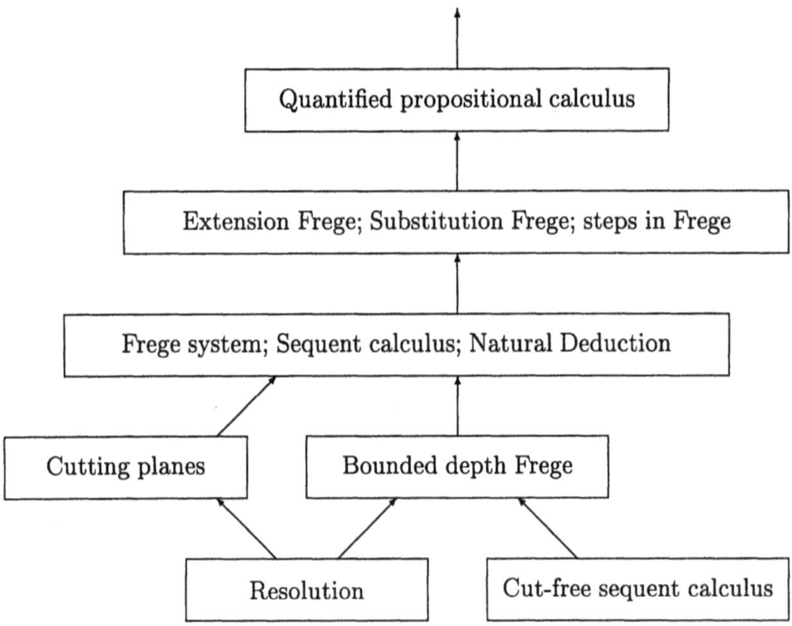

Figure 1: The hierarchy of propositional proof systems

2. **edge/vertex introduction:** add any new vertices and any new edges to a constructed graph;
3. **join:** if G_1 and G_2 has already been constructed, G_1 and G_2 with disjoint sets of vertices, (a_1, b_1) an edge in G_1, (a_2, b_2) an edge in G_2, then construct a new graph by contracting a_1 with a_2, deleting the edges (a_1, b_1), (a_2, b_2) and adding the edge (b_1, b_2);
4. **contraction:** contract any two non-adjacent vertices in a constructed graph.

On the other hand it is quite easy to prove that no graph obtained in this way is 3-colorable. A *proof* of the fact that G is not 3-colorable in the Hajós calculus is a sequence where K_4 is used as an axiom, the three rules above are used to construct new graphs and where the last graph is G. Hajós' theorem asserts that this calculus is complete for graphs which are not 3-colorable.

Surprisingly Pitassi and Urquhart [1992] have shown that the Hajós calculus is polynomially equivalent to extension Frege systems. This means that

1. there is a polynomial time computable function which to each tautology φ and its extension Frege proof d assigns a graph G and a proof h in the Hajós calculus that G is not 3-colorable;
2. and *vice versa,* there is a polynomial time computable function which to each graph G and a proof h in the Hajós calculus that G is not 3-colorable assigns a tautology φ and its extension Frege proof d .

This shows that the concept of extension Frege systems is quite robust and that it will be very hard to prove that there is no polynomial bound on shortest proofs in the Hajós calculus.

8.9.2. Nullstellensatz. The second example are systems of algebraic equations over finite fields. Let

$$\begin{aligned} f_1(x_1, \ldots, x_n) &= 0 \\ &\vdots \\ f_m(x_1, \ldots, x_n) &= 0 \end{aligned} \tag{52}$$

be a system of algebraic equations over a field F. The famous Hilbert's Nullstellensatz says that (52) does *not* have a solution in $\overline{F}$ (the algebraic closure of F) iff there exist polynomials $g_1(x_1, \ldots, x_n), \ldots, g_m(x_1, \ldots, x_n)$ such that

$$\sum_{i=1}^{m} g_i(x_1, \ldots, x_n) f_i(x_1, \ldots, x_n) = 1 \tag{53}$$

in the ring of polynomials over F.

We shall make some additional assumptions. We shall assume that F is finite and that equations (52) can have solutions only in F. The last condition can be ensured by adding equations $\prod_{a \in F}(x_i - a) = 0$ for $i = 1, \ldots, n$. Then such sets of unsolvable systems of equations are $co\mathcal{NP}$-complete. Furthermore we shall assume that polynomials are given as sums of monomials. Then (53) can be decided in polynomial time, since we can expand the sum of products into a sum of monomials where the number of monomials is polynomial in the number of monomials of polynomials g_i and f_i, $i = 1, \ldots, n$. Thus we can think of the system of polynomials $g_1, \ldots, g_m$ as *a proof that (52) is unsolvable.* (Let us remark that in this special case the proof of the Nullstellensatz is easy, so this proof system is not based on a deep result.)

This system is not known to be equivalent to another proof system, it is weaker than Frege systems and there are superpolynomial lower bounds for a sequence of unsolvable systems. The main application of this approach is in the works of Beame et al. [1996] and Buss et al. [1996/1997], proving independence of counting principles in bounded depth Frege systems and in bounded arithmetic (this was first proved by Ajtai [1994b] using a different, and very deep proof).

A related system, called the *polynomial calculus*,[4] was introduced in Clegg, Edmonds and Impagliazzo [1996]. In this system we derive equations sequentially using additions and multiplications by arbitrary polynomials. Alternatively, it is just equational calculus with no variables allowed. For a given bound d on degree of polynomials occurring in the proof, the system is stronger than the Nullstellensatz system. If d is a constant, it is still decidable in polynomial time, if there is a proof of a given polynomial from a given set of polynomials.

Finally we consider a proof system which uses ideas of *linear programming* which was introduced in W. Cook, Coullard and Turán [1987].

8.9.3. Cutting plane proof system. This system is, in a sense, an extension of resolution; in particular it is also a refutation system for a set of clauses. However, instead of clauses we use linear inequalities which adds power to the system.

A *proof line* is an expression

$$a_1 p_1 + \cdots + a_n p_n \geq B, \tag{54}$$

where $a_1, \ldots, a_n, B$ are integers. We allow also expressions of the form $0 \geq B$. For a given clause C we represent literals p_i identically and $\neg p_i$ by $1 - p_i$. Let $f_1, \ldots, f_k$ be the linear terms expressing the literals of C. Then we represent C by the expression

$$f_1 + \cdots + f_k \geq 1.$$

(Of course, to get an expression of the form (54), we have to collect the constant terms on the right hand side; also we collect constant and other terms after each application of a rule.) The axioms and derivation rules are

1. axioms are all translations of the clauses in question and the expressions $p_i \geq 0$, $-p_i \geq -1$;
2. **addition:** add two lines;
3. **multiplication:** multiply a line by a positive integer;

[4]Another name proposed for this calculus is the *Groebner proof system*.

4. **division:** divide a line (54) by a positive integer c which divides evenly $a_1, \ldots, a_k$ and *round-up the constant term on the right hand side*, i.e., we get

$$\frac{a_1}{c}p_1 + \cdots + \frac{a_n}{c}p_n \geq \left\lceil \frac{B}{c} \right\rceil .$$

(Note that on the left hand side we have integers, thus rounding up is sound.)

A contradiction is obtained, when we prove $0 \geq 1$.

We suggest to the reader, as an easy exercise, to check that this system simulates resolution. Goerdt [1991] proved that Frege systems polynomially simulate the cutting plane proof system. Furthermore, Buss and Clote [1996] proved that the cutting plane system with the division rule restricted to the division by 2 (or any other constant > 1) polynomially simulates the general system. Recent success in proving exponential lower bounds on the lengths of cutting plane proofs (see section 9.3) gives us also interesting separations. The cutting plane proof system cannot be simulated by bounded depth Frege systems as it proves the pigeonhole principle (see Cook, Coullard and Turán [1987]) using polynomial size proofs. The cutting plane proof system does not polynomially simulate bounded depth Frege systems Bonet, Pitassi and Raz [1997a], Krajíček [1997a], Pudlák [1997].

9. Lower bounds on propositional proofs

In this section we give an example of a lower bound proof in propositional logic. Our lower bound will be an exponential lower bound on the size of resolution proofs of the *pigeonhole principle*. The first such bound for unrestricted resolution was proved by Haken [1985]. Unfortunately his proof cannot be generalized to stronger systems, (at least nobody has succeeded in doing it). Therefore we shall apply a technique of Ajtai [1994a], which he used for bounded depth Frege systems. The case of resolution, which can be considered as a depth one Frege system, is simpler than for larger depths and thus can serve as a good introduction to more advanced results.

9.1. A general method. Before we consider the concrete example, we shall present a general framework for lower bound proofs, which can be applied to some existing proofs and, maybe, can be also used for some new proofs. A general description of what is going on in lower bound proofs is always useful, since, when proving a lower bound, we are working with nonexisting things (the short proofs whose existence we are disproving) and therefore it is difficult to give any intuition about them.

The basic idea of our approach is as follows. Suppose that we want to show that $(\alpha_1, \alpha_2, \ldots, \alpha_m)$ is not a proof of α. Let L be the set of subformulas of $\alpha_1, \alpha_2, \ldots, \alpha_m$ and α. L is a partial algebra with operations given by the connectives. Suppose that we have a boolean algebra B and a homomorphism $\lambda : L \to B$ such that $\lambda(\alpha) \neq 1_B$. Then α cannot be among $\alpha_1, \ldots, \alpha_m$, since $\lambda(\varphi) = 1_B$ for every axiom and this is preserved by Frege rules. In this form the method cannot work: if α is a tautology (and we are interested only in tautologies), then $\lambda(\alpha) = 1_B$. Therefore we have to modify it. We take only some subsets $L_i \subseteq L$ and $\lambda_i : L_i \to B_i$ for different boolean algebras B_i.

Now we shall describe this method in details. Let

$$\frac{\varphi_1(p_1,\ldots,p_\ell),\ldots,\varphi_k(p_1,\ldots,p_\ell)}{\varphi(p_1,\ldots,p_\ell)}$$

be a Frege rule R. We shall associate with it the set L_R of all subformulas of $\varphi_1,\ldots,\varphi_k$ and φ. If

$$\frac{\varphi_1(\psi_1,\ldots,\psi_\ell),\ldots,\varphi_k(\psi_1,\ldots,\psi_\ell)}{\varphi(\psi_1,\ldots,\psi_\ell)}$$

is an instance of R, we associate with it the set

$$L_{R(\vec{\psi})} = L_{R(\psi_1,\ldots,\psi_\ell)} = \{\alpha(\psi_1,\ldots,\psi_\ell); \alpha(p_1,\ldots,p_\ell) \in L_R\}.$$

Let B be a boolean algebra. A *homomorphism* $\lambda : L_{R(\vec{\psi})} \to B$ is a mapping which maps connectives onto corresponding operations in B, i.e.,

$$\lambda(\neg\varphi) = \neg_B \lambda(\varphi)$$

$$\lambda(\varphi \vee \psi) = \lambda(\varphi) \vee_B \lambda(\psi)$$

etc.

The following lemma formalizes our method.

9.1.1. Lemma. *Let $(\alpha_1, \alpha_2, \ldots, \alpha_m)$ be a Frege proof using a set of assumptions S. Suppose the following conditions are satisfied:*

1. *For every formula α_i of the proof we have a boolean algebra B_i and an element $b_i \in B_i$. Furthermore, if $\alpha_i \in S$, then $b_i = 1_{B_i}$.*
2. *For every instance of a rule $R(\vec{\psi})$ of the proof we have a boolean algebra $B_{R(\vec{\psi})}$ and a homomorphism $\lambda_{R(\vec{\psi})} : L_{R(\vec{\psi})} \to B_{R(\vec{\psi})}$.*
3. *For every formula α_i of the proof and and every instance of a rule $R(\vec{\psi})$ where $\alpha_i \in L_{R(\vec{\psi})}$, we have an embedding $\kappa_{i,R(\vec{\psi})} : B_i \to B_{R(\vec{\psi})}$ so that $\kappa_{i,R(\vec{\psi})}(b_i) = \lambda_{R(\vec{\psi})}(\alpha_i)$.*

Then

$$b_1 = 1_{B_1}, \ldots, b_m = 1_{B_m}.$$

The proof of this lemma is based on the following observation:

9.1.2. Lemma. *A Frege rule is sound in any boolean algebra.*

Proof. Suppose for some assignment of values from B we get the value 1_B for the assumptions but a value $b < 1_B$ for the conclusion. Take a homomorphism $\kappa : B \to \{0,1\}$ such that $\kappa(b) = 0$. Then we get a contradiction with the soundness of the rule for the algebra $\{0,1\}$. □

Proof of Lemma 9.1.1. We shall use induction. If $\alpha_1 \in S$, then $b_1 = 1_{B_1}$, otherwise α_1 is an instance of a logical axiom, say, $R(\vec{\psi})$ (a rule without assumptions). Thus, by Lemma 9.1.2, $\lambda_{R(\vec{\psi})}(\alpha_1) = 1_{B_{R(\vec{\psi})}}$. Hence

$$\kappa_{1,R(\vec{\psi})}(b_1) = \lambda_{R(\vec{\psi})}(\alpha_1) = 1_{B_R(\vec{\psi})}.$$

Since $\kappa_{1,R(\vec{\psi})}$ is an embedding, $b_1 = 1_{B_1}$. The induction step is similar. □

It may seem at the first glance that it does not make sense to talk about boolean algebras B_i, since we can simply take all of them to be 4-element algebras, and thus isomorphic. It turns out, however, that in applications nontrivial boolean algebras B_i appear quite naturally. They have to reflect the properties of the tautologies that we consider.

Let us remark that this is only one possible interpretation of some lower bound proofs and there are other interpretations. In particular other interpretations are based on the idea of forcing, due to Ajtai [1994a], and partial boolean algebras, due to Krajíček [1994b]; let us note that using partial boolean algebras one can characterize (up to a polynomial) the length of proofs in Frege systems.

Another tool which we shall use are *random restrictions.* They have been successfully applied for proving lower bounds on bounded depth boolean circuits and later also for lower bounds on proofs in bounded depth Frege systems by Ajtai [1994a,1990,1994b], Beame et al. [1992], Beame and Pitassi [1996], Bellantoni, Pitassi and Urquhart [1992], Krajíček [1994a], Krajíček, Pudlák and Woods [1995] and Pitassi, Beame and Impagliazzo [1993]. The idea is to assign more or less randomly 0's and 1's to some variables. Then many conjunctions and disjunctions became constant and thus the circuits, or the formulas, can be simplified. For the reduced formulas it is then much easier to construct boolean algebras with the required properties.

9.2. An exponential lower bound on the pigeonhole principle in resolution.

Let D and R be disjoint sets with cardinalities $|D| = n+1$ and $|R| = n$. We shall consider the proposition PHP_n stating that there is no $1-1$ mapping from D onto R. (This is a weaker proposition than just: "there is no $1-1$ mapping of D *into* R", thus the lower bound is a stronger statement.) We denote by $p_{ij}, i \in D, i \in R$ propositional variables (meaning that i maps onto j in the alleged mapping). We shall consider clauses made of p_{ij} and $\neg p_{ij}$, furthermore we shall use the true clause $\top$ and the false, or empty, clause $\bot$. PHP_n is the negation of the following set of clauses

$$\begin{array}{lll} \bigvee_{j \in R} p_{ij}, & \text{for} & i \in D; \\ \bigvee_{i \in R} p_{ij}, & \text{for} & j \in R; \\ \neg p_{ij} \vee \neg p_{ik}, & \text{for} & j \in D, j, k \in R, j \neq k; \\ \neg p_{ji} \vee \neg p_{ki}, & \text{for} & j, k \in D, j \neq k, i \in R. \end{array}$$

Let M be the set of partial one-to-one mappings $D \to R$. We shall consider boolean algebras determined by subsets $T \subseteq D \cup R, |T| < n$, as follows. Let V_T be a subset of partial matchings g such that

1. $T \subseteq \mathrm{dom}(g) \cup \mathrm{rng}(g)$;
2. $\forall (i,j) \in g(i \in T \vee j \in T)$.

The boolean algebra associated with T is $P(V_T)$, the boolean algebra of subsets of V_T.

In order to be able to assign a value to a clause Δ in $P(V_T)$, a certain relation of T to Δ must be satisfied. We define that Δ is *covered* by T if

1. $p_{ij} \in \Delta \Rightarrow i \in T \vee j \in T$;
2. $\neg p_{ij} \in \Delta \Rightarrow i \in T \wedge j \in T$.

The clauses $\top$ and $\bot$ are covered by any set T. Suppose Δ is covered by T, then the *value* of Δ in $P(V_T)$ is the set

$$b^T_\Delta = \{g \in V_T;\quad g(i) = j \text{ for some } p_{ij} \in \Delta, \quad \text{or} \quad g(i) \neq j \text{ for some } \neg p_{ij} \in \Delta$$
$$\text{or} \quad g^{-1}(j) \neq i \text{ for some } \neg p_{ij} \in \Delta\}.$$

The following can be easily checked:

9.2.1. Lemma. *If Δ is one of the clauses of PHP_n and T covers Δ, $|T| < n$, then $b^T_\Delta = 1_{P(V_T)}$.* □

Suppose $T \subseteq T'$, $|T'| < n$, then there is a natural mapping $\lambda_{T,T'} : P(V_T) \to P(V_{T'})$ defined by

$$\lambda_{T,T'}(b) = \{g' \in V_{T'} : \exists g \in b(g \subseteq g')\}.$$

9.2.2. Lemma. *Let $T \subseteq T'$, $|T'| < n$. Then $\lambda_{T,T'}$ is an embedding of the boolean algebra $P(V_T)$ into $P(V_{T'})$.*

Proof. All properties are trivial except for the following one: $\lambda_{T,T'}$ is injective. This property follows from the fact that each $g \in V_T$ can be extended to a $g' \in V_T$, which holds due to the fact that $|T'| < n$. □

Consider an instance of the cut rule

$$\frac{\Gamma \vee p_{ij} \qquad \Delta \vee \neg p_{ij}}{\Gamma \vee \Delta}$$

Suppose we have chosen $P(V_{T_i}), i = 1,2,3$ as the boolean algebra for $\Gamma \vee p_{ij}$, $\Delta \vee \neg p_{ij}$ and $\Gamma \vee \Delta$ respectively. Then we choose $P(V_T)$ with $T = T_1 \cup T_2 \cup T_3$ for this rule. We only have to ensure that $|T| < n$. Since T covers all subformulas involved, we can define their values in $P(V_T)$. The condition that this is a homomorphism of $\neg$ and $\vee$ is easy to check. By Lemma 9.2.2 we have also the necessary embeddings.

The simplest way to ensure $|T| < n$ for the rules is to choose the covering sets of the formulas of size $< n/3$. This is not always possible (take e.g. $p_{11} \vee p_{22} \vee \ldots p_{nn}$), therefore we apply random restrictions.

Suppose we assign 0's and 1's to some variables p_{ij} and leave the other as they are. If we do it for all formulas in a proof, the resulting sequence will be a proof again. However some initial clauses may reduce to $\bot$, so we cannot argue that $\bot$

cannot be derived from them by a short proof. Therefore the restrictions must reflect the nature of the tautology in question.

Let $g \in M$ be a partial one-to-one mapping. We shall associate with g the partial assignment defined by

$$\begin{array}{ll} p_{ij} \to 1 & \text{if} \quad (i,j) \in g; \\ p_{ij} \to 0 & \text{if} \quad i \in \mathrm{dom}(g) \text{ or } j \in \mathrm{rng}(g), \quad \text{but} \quad (i,j) \notin g; \\ p_{ij} \to p_{ij} & \text{otherwise.} \end{array}$$

Given a clause Δ, we define Δ^g to be

1. $\top$ if some $p_{ij} \in \Delta$ is mapped to 1 or some p_{ij} such that $\neg p_{ij} \in \Delta$ is mapped to 0,
2. otherwise it is the clause consisting of all literals which are not 0.

Let us denote by $D' = D - \mathrm{dom}(g), R' = R - \mathrm{rng}(g), n' = |R'|$. Clearly Δ^g is a clause with variables $p_{ij}, i \in D', j \in R'$. If Δ is a clause of PHP_n then Δ^g is either $\top$ or becomes a clause of $PHP_{n'}$ (on D' and R'). Denote by $M_{n,n'}$, the set of all partial one-to-one mappings of size $n - n'$. The following is the key combinatorial lemma for the proof of the lower bound.

9.2.3. Lemma. *Let $n' = \lfloor n^{1/3} \rfloor$, let Δ be an arbitrary clause. Then for a $g \in M_{n,n'}$, chosen with uniform probability, the probability that Δ^g can be covered by a set of size $< \frac{1}{3}n'$ is at least*

$$1 - 2^{\varepsilon n^{1/3}},$$

where $\varepsilon > 0$ is a constant.

We shall use the following simple estimate.

9.2.4. Lemma. *Let $a, b, l \leq n, A \subseteq \{1, \ldots, n\}, |A| = a$. Take a random $B \subseteq \{1, \ldots, n\}, |B| = b$, with uniform probability. Then*

$$\mathrm{Prob}(|A \cap B| \geq l) \leq \left(\frac{eab}{nl} \right)^l .$$

Proof.

$$\begin{aligned} \mathrm{Prob}(|A \cap B| \geq l) &\leq \sum_{\{a_1,\ldots,a_\ell\} \subseteq A} \mathrm{Prob}(a_1 \in B, \ldots, a_\ell \in B) \\ &= \binom{a}{l} \cdot \frac{b}{n} \cdot \frac{b-1}{n-1} \cdot \ldots \cdot \frac{b-l+1}{n-l+1} \\ &\leq \left(\frac{ea}{l}\right)^l \cdot \left(\frac{b}{n}\right)^l \leq \left(\frac{eab}{nl}\right)^l . \end{aligned}$$

□

Proof of Lemma 9.2.3. Let us denote by $l = \lfloor \frac{1}{3} n' \rfloor$. Let Δ be given. We shall simplify the situation by replacing each $\neg p_{ij} \in \Delta$ by

$$\bigvee_{i' \neq i} p_{i'j} \vee \bigvee_{j' \neq j} p_{ij'}.$$

This operation commutes with the restriction and the new clause is covered by $T, |T| \leq l$, iff the old one is, since $\ell < n' - 2$. Thus we can assume that Δ contains only positive literals. Such a Δ is determined by the graph

$$E = \{(i,j); p_{ij} \in \Delta\}.$$

Let

$$a = \frac{n^{2/3}}{40}.$$

From now on we shall omit the integer part function and assume that all numbers are integers. This introduces only inessential errors. Furthermore denote by

$$A = \{j \in R; \deg_E(j) \geq 2a\}.$$

We shall consider two cases.

Case 1: $|A| \geq 2a$. We shall show that in this case $\Delta^g = \top$ with high probability. First we estimate $|A \cap \operatorname{rng}(g)|$. Note that $\operatorname{rng}(g)$ is a random subset of R of size $n - n'$, thus also $R' = R \setminus \operatorname{rng}(g)$ is a random subset of size n'. Hence we can apply Lemma 9.2.4.

$$\begin{aligned} \operatorname{Prob}(|A \cap \operatorname{rng}(g)| < a) &= \operatorname{Prob}(|A \cap R'| \geq |A| - a) \\ &\leq \left(\frac{e|A|n^{1/3}}{n(|A| - a)} \right)^{|A|-a} \leq \left(\frac{2e}{n^{2/3}} \right)^a . \end{aligned} \tag{55}$$

The probability that Δ^g is not $\top$ is bounded by

$$\begin{aligned} &\operatorname{Prob}\left(\forall j \in A \cap \operatorname{rng}(g)((g^{-1}(j), j) \notin E)\right) \leq \\ &\quad \operatorname{Prob}(|A \cap \operatorname{rng}(g)| < a) + \\ &\quad \operatorname{Prob}\left(\forall j \in A \cap \operatorname{rng}(g)((g^{-1}(j), j) \notin E) \;\mid\; |A \cap \operatorname{rng}(g)| \geq a\right). \end{aligned} \tag{56}$$

The second term can be estimated by

$$\max_{C \subseteq A,\ |C| \geq a} \operatorname{Prob}\left(\forall j \in A \cap \operatorname{rng}(g)((g^{-1}(j), j) \notin E) \mid A \cap \operatorname{rng}(g) = C\right),$$

thus it suffices to consider a fixed such C and bound the probability. Let $C = \{j_1, j_2, \ldots, j_{|C|}\}$; think of the vertices $g^{-1}(j_1), g^{-1}(j_2), \ldots, g^{-1}(j_{|C|})$ as chosen one by one independently, except that they must be different.

$$\begin{aligned} &\operatorname{Prob}\left((g^{-1}(j_{t+1}), j_{t+1}) \notin E \;\mid\; g(i_1) = j_1, \ldots, g(i_t) = j_t\right) = \\ &= 1 - \frac{|E^{-1}(j_{t+1}) - \{i_1, \ldots, i_t\}|}{n + 1 - t} \leq 1 - \frac{\deg_E(j_{t+1}) - t}{n+1} \leq 1 - \frac{2a - t}{n+1}. \end{aligned}$$

Thus the probability that $(g^{-1}(j_t), j_t) \notin E$ for all $t = 1, \dots, |C|$ is

$$\leq \left(1 - \frac{2a}{n+1}\right)\left(1 - \frac{2a-1}{n+1}\right)\dots\left(1 - \frac{2a-|C|+1}{n+1}\right) \leq \left(1 - \frac{a}{n+1}\right)^a .$$

Since $\frac{a}{n+1} \sim \frac{1}{n^{1/3}}$ and $a \sim n^{2/3}$, this expression is $e^{-\Omega(n^{1/3})}$. The first term of (56) is estimated in (55) and is even smaller. Thus in Case 1 the probability is $1 - e^{-\Omega(n^{1/3})}$ as required.

Case 2: $|A| < 2a$. In this case we cover Δ^g by the set

$$(A \cap R') \cup (E^{-1}(R' \backslash A) \cap D').$$

We need only to estimate the probability that the size of the two sets in the union is small. We shall use Lemma 9.2.4 again.

$$\text{Prob}\left(|A \cap R'| > \frac{\ell}{2}\right) \leq \left(\frac{e2an^{1/3}}{n \cdot n^{1/3}/6}\right)^{n^{1/3}/6} = \left(\frac{12e}{40n^{1/3}}\right)^{n^{1/3}/6} = e^{-\Omega(n^{1/3})}. \qquad (57)$$

To estimate the second set, first observe that

$$|E^{-1}(R' \backslash A)| \leq |R'| \cdot 2a = n^{1/3} \cdot 2\frac{n^{2/3}}{40} = \frac{n}{20}.$$

Thus

$$\begin{aligned} \text{Prob}(|E^{-1}(R' \backslash A) \cap D'| > \frac{\ell}{2}) &\leq \left(\frac{e \cdot \frac{n}{20} \cdot (n^{1/3}+1)}{(n+1)n^{1/3}/6}\right)^{n^{1/3}/6} \qquad (58) \\ &= \left(\frac{3e}{10} \cdot \frac{n(n^{1/3}+1)}{(n+1)n^{1/3}}\right)^{n^{1/3}/6} = e^{-\Omega(n^{1/3})}, \end{aligned}$$

since the term in the parentheses converges to $\frac{3e}{10} < 1$. By (57) and (58) we get the required bound in Case 2. □

Now we are ready to prove the lower bound which was originally proved by Haken [1985] with a better exponent than we give here.

9.2.5. Theorem. (Haken [1985]) *Every resolution proof of PHP_n has size at least $2^{\varepsilon n^{1/3}}$, where $\varepsilon > 0$ is a constant.*

Proof. Suppose a proof of size $< 2^{\varepsilon n^{1/3}}$ is given. Take a random $g \in M_{n'}, n' = \lfloor n^{1/3} \rfloor$. Then, by Lemma 9.2.2, for every formula Δ of the proof the probability that Δ^g is not covered by a set of size $< \frac{n'}{3}$ is at most $2^{\varepsilon n^{1/3}}$. Thus we have positive probability that, for some $g \in M_{n,n'}$, all formulas are covered by sets $< \frac{n'}{3}$. Hence there is at least one such a g.

Consider the proof restricted using such a g; it is a derivation of $\perp$ from clauses of $PHP_{n'}$. Choose a covering set of size $< \frac{n'}{3}$ for each clause in this proof. Then take boolean algebras $P(V_T)$ for clauses and for each application of the rule as described above. As we have observed, the clauses of $PHP_{n'}$ get value 1 in their boolean algebras. Now we can apply Lemma 9.1.1. The conclusion should be that $\bot$ gets also 1. But $\bot$ gets the value 0 by the definition of the boolean algebras.

Hence the proof must have size $\geq 2^{\varepsilon n^{1/3}}$. □

9.3. Lower bounds based on effective interpolation theorems. We are going to discuss an approach which is not based on such *ad hoc* proofs, but instead it uses some general theorems interesting in their own right. These theorems are versions of the *interpolation theorem,* a classical result of Craig [1957a,1957b], see Chapter I. The interpolation theorem has a first order logic version and a propositional version. Recently some strengthenings of the propositional interpolation theorem have been successfully applied to prove lower bounds on the length of propositional proofs.

The propositional interpolation theorem states that for a given propositional tautology $\Phi(\bar{p},\bar{q}) \to \Psi(\bar{p},\bar{r})$, where $\bar{p},\bar{q},\bar{r}$ are disjoint strings of propositional variables, there exists a formula $I(\bar{p})$, which contains only the common variables $\bar{p}$, such that both $\Phi(\bar{p},\bar{q}) \to I(\bar{p})$ and $I(\bar{p}) \to \Psi(\bar{p},\bar{r})$ are also tautologies. Such a formula $I(\bar{p})$ is called an *interpolant* of $\Phi(\bar{p},\bar{q}) \to \Psi(\bar{p},\bar{r})$. The proof of this statement is trivial: Take the quantified boolean formula $\exists\bar{x}\ \Phi(\bar{p},\bar{x})$ (or $\forall\bar{x}\ \Psi(\bar{p},\bar{x})$); clearly, it interpolates $\Phi(\bar{p},\bar{q}) \to \Psi(\bar{p},\bar{r})$. As any boolean function can be defined by an ordinary propositional formula, there is a propositional formula $I(\bar{p})$ equivalent to $\exists\bar{x}\ \Phi(\bar{p},\bar{x})$.

Craig gave constructive proofs of his theorems, i.e., he showed how to construct an interpolant $I(\bar{p})$ from a proof d of $\Phi(\bar{p},\bar{q}) \to \Psi(\bar{p},\bar{r})$. Thus the complexity of $I(\bar{p})$ depends on the complexity of the proof d. This led Krajíček [1994a] to propose a method of lower bounds proofs whose idea can be stated as follows: suppose we can show that $\Phi(\bar{p},\bar{q}) \to \Psi(\bar{p},\bar{r})$ does not have a simple interpolant, then it cannot have a simple proof.

Another relationship of interpolation theorems to questions in complexity theory had earlier been considered by Mundici [1984], but he did not consider the lengths of proofs.

9.3.1. The original proof of Craig was based on cut-elimination, so the constructed interpolant can be exponentially large. His proof can be used to get a good bound on interpolants for cut-free sequent propositional proofs, but we have to consider a different measure of the complexity of interpolants. The new idea is that we can look at an interpolant as a boolean function and then we can apply any of the measures of complexity of boolean functions. Here the right measure is the size of the smallest *circuit* computing the boolean function.

9.3.2. Theorem. (Krajíček [1997a]) *Let d be a cut-free proof with k lines of a sequent*

$$\Phi_1(\bar{p},\bar{q}),\ldots,\Phi_m(\bar{p},\bar{q}) \longrightarrow \Psi_1(\bar{p},\bar{r}),\ldots,\Psi_l(\bar{p},\bar{r})$$

where $\bar{p}, \bar{q}, \bar{r}$ are disjoint sets of propositional variables (i.e. no $\bar{q}$ occurs in the consequent and no $\bar{r}$ occurs in the antecedent). Then it is possible to construct a an interpolant $I(\bar{p})$ of $\bigwedge_i \Phi_i(\bar{p}, \bar{q}) \rightarrow \bigvee_j \Psi_j(\bar{p}, \bar{r})$ which is a boolean circuit of size $k^{O(1)}$.

The proof is essentially the original one of Craig [1957a,1957b]. The idea is to construct interpolants for each sequent in the proof successively starting with the initial sequents and going down to the end sequent. As the proof is cut-free, each sequent contains only formulas containing either only variables $\bar{p}, \bar{q}$, or only variables $\bar{p}, \bar{r}$, so it makes sense to talk about an interpolant for it. □

The reason for using circuit size is because we consider proofs in the sequence form. For *tree-like proofs* we actually get a polynomial size *formula* as an interpolant.

9.3.3. Suppose we have an interpolant $I(\bar{p})$ for $\Phi(\bar{p}, \bar{q}) \rightarrow \Psi(\bar{p}, \bar{r})$ i.e., the implications $\Phi(\bar{p}, \bar{q}) \rightarrow I(\bar{p})$ and $I(\bar{p}) \rightarrow \Psi(\bar{p}, \bar{r})$ are true. Let a truth assignment $\bar{a}$ to the variables $\bar{p}$ be given. Then either $\neg\Phi(\bar{p}, \bar{a})$ or $\Psi(\bar{a}, \bar{r})$ is true. The interpolant can be used to decide which of the two possibilities holds, namely, if $I(\bar{a})$ is true, then $\Psi(\bar{a}, \bar{r})$ is true, otherwise $\neg\Phi(\bar{a}, \bar{r})$ is true. (It is possible that both $\neg\Phi(\bar{p}, \bar{a})$ and $\Psi(\bar{a}, \bar{r})$ are true, in which case $I(\bar{a})$ could be true or false.) Thus there is an alternative way of looking at interpolant: Let $\alpha(\bar{p}, \bar{q}) \vee \beta(\bar{p}, \bar{r})$ be a valid disjunction; an interpolant is a procedure which produces one of the two disjuncts which becomes a tautology after assigning given truth values to $\bar{p}$.

We can look at the interpolation theorem even more abstractly (see Razborov [1994]). Let A and B be disjoint $\mathcal{NP}$ sets. Then we can define the set of input strings $\bar{a}$ of length n which are not in A, resp. not in B, by a polynomial size formula $\alpha_n(\bar{p}, \bar{q})$, resp. $\beta_n(\bar{p}, \bar{r})$ ($\bar{a}$ is not in A if $\alpha_n(\bar{a}, \bar{q})$ is a tautology, similarly for B, see next section). Since A and B are disjoint i.e., the complements cover all inputs, the disjunction $\alpha_n(\bar{p}, \bar{q}) \vee \beta_n(\bar{p}, \bar{r})$ is a tautology. If we have a polynomial time computable set C which separates A from B i.e., $A \subseteq C$, $C \cap B = \emptyset$, then we have a polynomial time decision algorithm for finding a true disjunct from $\alpha_n(\bar{a}, \bar{q}) \vee \beta_n(\bar{a}, \bar{r})$. Cook's theorem implies that then there exists also a polynomial size circuit $C_n(\bar{p})$ for this problem. Clearly, $C_n(\bar{p})$ is an interpolant for $\alpha_n(\bar{p}, \bar{q}) \vee \beta_n(\bar{p}, \bar{r})$.

9.3.4. The most interesting application of the effective interpolation is in the case of resolution.

9.3.5. Theorem. (Krajíček [1997a]) *Let d be a resolution proof of the empty clause from clauses $A_i(\bar{p}, \bar{q}), i \in I$, $B_j(\bar{p}, \bar{r}), j \in J$ where $\bar{p}, \bar{q}, \bar{r}$ are disjoint sets of propositional variables. Then it is possible to construct a circuit $C(\bar{p})$ such that for every 0-1 assignment $\bar{a}$ for $\bar{p}$*

$$C(\bar{a}) = 0 \quad \Rightarrow \quad A_i(\bar{a}, \bar{q}), i \in I \text{ are unsatisfiable, and}$$

$$C(\bar{a}) = 1 \quad \Rightarrow \quad B_j(\bar{a}, \bar{r}), j \in J \text{ are unsatisfiable;}$$

the size of the circuit C is bounded by $O(|d|)$.

Moreover, one can construct a resolution proof of the empty clause from clauses $A_i(\bar{a}, \bar{q}), i \in I$ *if* $C(\bar{a}) = 0$, *respectively from* $B_j(\bar{a}, \bar{r}), j \in J$ *if* $C(\bar{a}) = 1$, *whose size is at most the size of* d.

We shall sketch two proofs of this theorem. The idea of the first one, due to Krajíček [1997a], is to reduce it to Theorem 9.3.2. This looks strange, as resolution proofs consist only of cuts and we know that cut-elimination does not work. The trick is to eliminate cuts by replacing them by conjunctions.

For each initial clause $s_1 \vee \ldots \vee s_k$, where s_i are literals, first prove the sequent $\longrightarrow \bigwedge_{i=1}^k \overline{s_i}, s_1, \ldots, s_k$; we denote by $\overline{s_i}$ the literal complementary to s_i. Our goal is to derive a sequent consisting only of such conjunctions obtained from initial clauses $\longrightarrow \ldots, \bigwedge_{i=1}^k \overline{s_i}, \ldots$. Thus we want to replace the refutation proof by a proof of the corresponding tautology (DNF). We shall not quite succeed, we have to add also conjunctions of the form $s_i \wedge \overline{s_i}$, which are, however, false, hence do not influence interpolants at all.

In transforming the resolution proof into a cut-free sequent proof we follow the given resolution proof, but instead of applying cut with some cut literal s_i, we introduce $s_i \wedge \overline{s_i}$. Thus a general sequent in the proof will consist of conjunctions of negated literals of $A_i(\bar{p}, \bar{q})$'s and $B_j(\bar{p}, \bar{r})$'s, conjunctions of complementary literals $s_i \wedge \overline{s_i}$ and single literals. The single literals of the sequent are just the literals of the corresponding clause in the resolution proof. In the last sequent, as in the resolution proof, the single literals will be eliminated and we are left only with conjunctions of negated literals of the initial clauses $A_i(\bar{p}, \bar{q})$'s and $B_j(\bar{p}, \bar{r})$'s and conjunctions of complementary literals. Since conjunctions of complementary literals are false, an interpolant for this sequent is also an interpolant for the sequent without them. So we can apply Theorem 9.3.2 to get an interpolant for this sequent which is an interpolant for $A_i(\bar{p}, \bar{q}), i \in I$, $B_j(\bar{p}, \bar{r}), j \in J$ in the sense of the theorem. □

The idea of the second proof (Pudlák [1997]) is to construct a refutation proof either from $A_i(\bar{a}, \bar{q}), i \in I$, or from $B_j(\bar{a}, \bar{r}), j \in J$ for every given truth assignment. If there is a polynomial time algorithm for constructing such a proof, then there is one also for deciding which of the two sets is unsatisfiable, hence also a polynomial size circuit. Substitute the truth assignment $\bar{a}$ into the initial clauses and discard those which contain a literal which is true under the truth assignment $\bar{a}$ and delete the $\perp$ produced by the substitution from the others. Then we follow the proof. What we want is to never mix variables $\bar{q}$ with variables $\bar{r}$. So when we should resolve along a variable q_i or r_i we do it, since this will not produce a mixed clause. However, if we should resolve along some p_i, we must do something else. Now we simply take the clause which corresponds to an original clause where the literal p_i, resp. $\overline{p_i}$, is false under the truth assignment $\bar{a}$. This clause will be a subclause of the next original clause, hence we can continue and eventually obtain an empty clause. Since variables $\bar{q}$ and $\bar{r}$ are never mixed, the new proof will split into at least two disconnected parts. We can backtrack which initial clauses are actually needed to get the empty clause (they must be of the same kind) and take only that component as the new proof. □

A closer analysis of this proof shows that we can use the directed graph of the proof as the graph for the circuit, provided we take suitable connectives. So the relation between the proof and the circuit is very close.

One can also easily show that we have to use circuits instead of formulas, unless formulas are as powerful as circuits (which most researcher doubt) Krajíček [1994a]. Namely, for every circuit we can write a tautology stating that the computation is unique. We use variables $\bar{p}$ for the input values of the circuit, variables $\bar{q}$ for the values at the gates in the first computation and variables $\bar{r}$ for the values of at the gates in the second computation. The tautology asserts that if the output value, say q_k, in the first computation is 1, then the output value r_k in the second computation is also 1. Clearly, any interpolant of this tautology computes the same function as the circuit. On the other hand, the tautology has a resolution proof of linear size.

9.3.6. In order to apply Theorem 9.3.5 we need to have good lower bounds on the size of circuits computing some explicitly defined boolean functions. Presently all the known lower bounds for explicitly defined functions are only linear. Fortunately there is a version of the theorem which can be combined with currently known lower bounds. Quite surprisingly a very mild condition on the clauses implies that the interpolating circuits can be constructed monotone. A *monotone boolean circuit* is a circuit in the basis $\{\wedge, \vee, 0, 1\}$, i.e., a circuit whose gates are monotone boolean functions.

9.3.7. Theorem. (Krajíček [1997a]) *Assume that clauses as in Theorem 9.3.5 are given. Suppose moreover that either all variables $\bar{p}$ occur in $A_i(\bar{p},\bar{q}), i \in I$ only positively or all variables $\bar{p}$ occur in $B_j(\bar{p},\bar{r}), j \in J$ only negatively, then there exists a circuit C satisfying the conclusion of Theorem 9.3.5 which is moreover monotone.*

The proof of this theorem is obtained by inspection of either of the proofs of Theorem 9.3.5. □

There are well-known exponential lower bounds for the monotone circuit complexity of explicit boolean functions. This alone would not suffice to get an exponential lower bound on resolution proofs. By another lucky coincidence the lower bounds on monotone circuits give more: they actually show that some pairs of disjoint $\mathcal{NP}$ sets cannot be separated by monotone circuits.

In particular such a lower bound can be derived for tautologies related to the clique problem. Let $Clique_{n,k}(\bar{p},\bar{q})$ denote a set of clauses expressing that the graph with n vertices coded by $\bar{p}$ has a clique of size at least k coded by $\bar{q}$. The variables $\bar{p}$ represent edges of the graph and the variables $\bar{q}$ represent the graph of a one-to-one function from a k-element set into the set of vertices of the graph. Formally, we take variables $p_{i,j}$, $1 \le i < j \le n$, $q_{i,r}$, $1 \le i \le n$, $1 \le r \le k$ and clauses

$$\begin{array}{ll} \bigvee_i q_{i,r} & \text{for all } 1 \le r \le k; \\ \neg q_{i,r} \vee \neg q_{i,r'} & \text{for all } 1 \le i \le n,\ 1 \le r < r' \le k; \\ \neg q_{i,r} \vee \neg q_{i',r'} \vee p_{i,i'} & \text{for all } 1 \le i < i' \le n,\ 1 \le r < r' \le k. \end{array}$$

Let $Color_{n,l}(\bar{p}, \bar{r})$ denote a set of clauses expressing that the graph with n vertices coded by $\bar{p}$ is l-colorable. The variables $\bar{r}$ code a mapping from the set of vertices of the graph into a set of size l such that no edge is mapped on a single point. This can be expressed by a similar set of clauses as above.

If $k > l$, the set of graphs containing a k-clique is disjoint with the set of l-colorable graphs (a clique needs at least k colors), hence the two sets of clauses $Clique_{n,k}(\bar{p}, \bar{q})$ and $Color_{n,l}(\bar{p}, \bar{r})$ cannot be satisfied simultaneously. For suitable parameters it has been shown that these sets of graphs cannot be separated by small monotone circuits.

9.3.8. Theorem. (Razborov [1985], Alon and Boppana [1987]) *Let $l < k$ and $\sqrt{kl} \le \frac{n}{8 \log n}$. Then every monotone circuit which outputs* 1 *on graphs with a k-clique and* 0 *on l-colorable graphs has size $2^{\Omega(\sqrt{l})}$.* □

9.3.9. Corollary. (Krajíček [1997a]) *Any resolution refutation of the set of clauses $Clique_{n,k}(\bar{p}, \bar{q}) \cup Color_{n,l}(\bar{p}, \bar{r})$ has size $2^{\Omega(\sqrt{l})}$.* □

Using this approach we do not avoid combinatorial technicalities, since the proof of Theorem 9.3.8 is nontrivial. Its advantage is that an exponential lower bound on the length of resolution proofs is easily accessible to those who already know lower bounds on the size of monotone circuits.

9.3.10. Another advantage of this approach is that it can be applied to cutting plane proofs, where the random restriction method does not seem to work.

The version of Theorem 9.3.5 for cutting plane proofs is almost identical. There are two versions of the monotone case, Theorem 9.3.7, for cutting plane proofs. The first one (Bonet, Pitassi and Raz [1997a], Krajíček [1997a]) gives monotone boolean circuits, but requires that the coefficients in the proof are polynomially bounded by its size (put otherwise, the size of the monotone circuit is bounded not only by the number of lines but also by the size of the coefficients). The second version (Pudlák [1997]) works without any restriction on the coefficients, but the interpolating circuit is not an ordinary monotone boolean circuit. We have to consider circuits which are monotone and compute with *arbitrary real numbers.* Again fortunately, the known proofs of the lower bounds for monotone boolean circuits can be easily extended to the more general model (Pudlák [1997], Haken and Cook [n.d.]). In particular, an exponential lower bound can be proved for the clauses $Clique_{n,k}(\bar{p}, \bar{q}) \cup Color_{n,l}(\bar{p}, \bar{r})$ presented as inequalities in the cutting plane proof system.

9.3.11. At first this approach to lower bounds looked very promising. Unfortunately, it became clear very soon that it cannot be extended much further beyond resolution. We do not know, if an effective interpolation theorem in the style of Theorem 9.3.5 holds for bounded depth Frege systems and we rather think it does not hold even for such weak proof systems (cf. Krajíček [1997a] for some arguments). For Frege systems we have strong evidence that it does not hold. Namely one can prove that such a theorem does not hold for Frege systems using the widely accepted conjecture that factoring of integers is not in polynomial time.

9.3.12. Theorem. (Bonet, Pitassi and Raz [1997b]) *There exists a sequence of tautologies of the form $\alpha_n(\bar{p}, \bar{q}) \vee \beta_n(\bar{p}, \bar{r})$ which have polynomial size Frege proofs, but for which there is no sequence of polynomial size interpolation circuits, provided that factoring of integers is not in polynomial time.* □

Instead of proving this theorem we shall explain in general terms the rather surprising connection between propositional calculus and cryptography. The basic concept of cryptography is the *one-way function*, which is, roughly speaking, a function which can be easily computed (in polynomial time) but whose inverse function is hard.[5] It is not known if such functions exist; in fact, we even do not know how to prove their existence assuming $\mathcal{P} \neq \mathcal{NP}$. We do know, however, that a one-way function exists iff there exist disjoint $\mathcal{NP}$ sets which cannot be separated by a set in $\mathcal{P}$. We shall see in the next section (see 10.3) that arithmetical theorems of certain logical complexity can be translated into a sequence of propositional tautologies. Furthermore for each first order theory we can construct a propositional proof system where the translations of such theorems have polynomial size proofs. Now, if we have a pair of disjoint $\mathcal{NP}$ sets A, B which cannot be separated by a set in $\mathcal{P}$, we can take a theory T in which this fact is provable (just include this statement as an axiom). Hence in the propositional proof system P derived from T we can prove the tautologies derived from A, B. On the other hand, polynomial time interpolation for P would give us a separating set for A, B as noted above. (For sake of simplicity we are talking about polynomial time algorithms instead of polynomial size circuits; the distinction between the two concepts is not essential for our argument.)

A weaker version of Theorem 9.3.12, which gave the result only for extension Frege systems, was originally proved by taking the conjectured one-way function $x \mapsto g^x \bmod n$ and proving in S^1_2, which is a theory associated with extension Frege proof systems, that the corresponding pair of $\mathcal{NP}$ sets is disjoint (see Krajíček and Pudlák [1998] for a full proof).

9.4. Other lower bounds. The method of random restrictions (exemplified in section 9.2) has been extended by Ajtai [1994a] to any fixed depth Frege system. It gives, however, only slightly superpolynomial lower bounds. In order to get

[5]For practical cryptography one needs *hard in the average;* here we consider only the worst case complexity.

exponential lower bounds one needs a more substantial change in which the concept of covering sets is replaced by certain decision trees and a Switching Lemma, of the type used by Yao [1985] and Håstad [1986], is applied to reduce the depth of formulas; see Beame et al. [1992], Krajíček, Pudlák and Woods [1995] and Pitassi, Beame and Impagliazzo [1993].

Let us define at least the concept of the *decision tree* which is used in these bounds for PHP_n. We use the same notation as above. Such a tree is a labelled rooted tree, where the vertices are labelled by elements of $D \cup R$, except for the leaves, which are labelled by 0 – reject, and 1 – accept; the edges are labelled by pairs (i, j), $i \in D$, $j \in R$. We require that for a nonleaf vertex v with a label $i \in D$, resp. $i \in R$, the outgoing edges are labelled by (i, j), resp (j, i), one edge for every j which does not occur on the path leading to v. Consequently, the edge labels on every branch are independent, i.e., they form a partial one-to-one mapping.

In the lower bound proof we assign to each formula the boolean algebra of all subsets of leaves of such a tree and the value of the formula $\lambda(\varphi)$ is the subset of leaves labelled by 1.

The intuitive meaning of this concept is the following. We think of truth values of the propositional variables $p_{i,j}$ as given by some imaginary one-to-one mappings from D onto R. In fact, in a nonstandard model with n infinite, there are such external mappings. The decision tree enables us to decide in a natural way if such a mapping is accepted or not. Then all the boolean algebras defined by trees can be embedded into a single one which is the boolean algebra of subsets of one-to-one mappings from D onto R. Put otherwise our logic is a logic of one-to-one mappings from D onto R.

PHP_n is not the only sequence for which one can prove exponential lower bounds on bounded depth Frege proofs. Another such sequence is PAR_n — the parity principle — where PAR_n, n odd, expresses that a set of cardinality n cannot be partitioned into pairs. Similarly, one can consider the counting principle $COUNT_{p,n}$ which expresses that a set of size n, n not divisible by p, cannot be partitioned into blocks of size p. Ajtai [1990] has shown that PAR_n does not have polynomial size bounded depth proofs, even if we use instances of PHP_m as premises, and similar independence results have been proved for the counting principles by Ajtai [1994b], Beame et al. [1996], Buss et al. [1996/1997].

Together with exponential lower bounds for cutting plane proof systems and degree lower bounds for the polynomial calculus, these are the strongest results so far. For unrestricted Frege system we have only an $\Omega(n^2)$ lower bound for tautologies such as $\neg^{2n}(p \vee \neg p)$. The proof is based on the claim that all subformulas of this tautology must occur essentially (i.e., in a constant depth) in the proof. This is essentially the same idea as in Claim 4.2.3, see also 8.1.16. Apart from this rather simple proof we do not have anything for Frege and stronger systems.

10. Bounded arithmetic and propositional logic

In this section we shall show an important relation between the lengths of proofs of propositional tautologies and provability in fragments of arithmetic. By this connection certain arithmetical formulas can be translated to a sequence of propositions, and if the formula is provable in some theory, then the propositions have small (e.g., polynomial size) proofs in some propositional proof system associated with the theory. Surprisingly, there are pairs of such a theory T and a propositional proof system P where both the theory T and the proof system T are quite natural. In this situation we can think of T and P to be just two facets of a single concept, where T is a *uniform* version of the *nonuniform* P. This is just another parallel to boolean circuit complexity, where the uniform model is the Turing machine and the nonuniform model is a sequence of boolean circuits.

The main application of this relation is in showing independence results. If we could prove superpolynomial lower bounds on strong propositional proof systems, then we could show interesting independence results in bounded arithmetic such as unprovability of $\mathcal{NP} = co\mathcal{NP}$.

There is also practical use of this relation which is necessary to take into account even if you are not interested in first order theories. It might be fairly difficult to find and describe short proofs of some tautologies directly, while in a bounded arithmetic we can often see easily that the corresponding first order formula is provable. This was used, e.g. in Pudlák [1991], to disprove a conjecture saying that formulas expressing Ramsey's theorem in propositional logic do not have polynomial size proofs in Frege systems. Similarly, it is possible to prove the existence of a polynomial simulation of a proof system P by a proof system Q by proving the reflection principle (see below) for P in a theory associated with Q. In such a way the polynomial simulation of substitution Frege by extension Frege system was discovered by Dowd [1985] and Krajíček and Pudlák [1989].

This subject requires some familiarity with fragments of arithmetic considered in bounded arithmetic. The reader, who does not know that subject should consult Chapter II Buss [1986], Hájek and Pudlák [1993] or Krajíček [1995].

10.1. There are basically two translations of bounded formulas into propositions. They are determined by the particular way in which we represent truth assignments. A truth assignment is a finite sequence $\bar{a}$ of 0's and 1's. We can code it either by a subset of a finite segment of integers or by a number whose binary representation is $1\bar{a}$. We start with the simpler one.

10.2. First translation. Let $L_0(\alpha)$ be the language of arithmetic with nonlogical symbols $\underline{0}, S, +, \cdot, \leq$ augmented with a second order variable α for l-ary relations. We consider the class $\Delta_0(\alpha)$ of bounded formulas in the language $L_0(\alpha)$. Assume that we use the same connectives in the first order language and the propositional calculus and they include $\wedge, \vee$; moreover we shall assume that we have propositional constants $\bot, \top$ in propositional logic and that the propositional variables are indexed by l-tuples of nonnegative integers.

Let $\theta \in \Delta_0(\alpha)$ be a formula with k free variables. Then for each sequence $n_1, \dots, n_k$ of nonnegative integers we define *a propositional formula*

$$\langle\theta\rangle_{n_1,\dots,n_k}$$

inductively as follows.

1. for terms $s(n_1,\dots,n_k), t(n_1,\dots,n_k)$, we define

$$\langle s(n_1,\dots,n_k) = t(n_1,\dots,n_k)\rangle_{n_1,\dots,n_k} =_{df} \bot \text{ (resp. } = \top),$$

   if $s(n_1,\dots,n_k) = t(n_1,\dots,n_k)$ is false, (resp. true); we use the same definition for $\leq$ in place of $=$;
2. for terms $t_1(x_1,\dots,x_k),\dots,t_l(x_1,\dots,x_k)$, we define

$$\langle\alpha(t_1(x_1,\dots,x_k),\dots,t_l(x_1,\dots,x_k))\rangle_{n_1,\dots,n_k} =_{df} p_{i_1,\dots,i_l},$$

   where $i_1,\dots,i_l$ are the values of $t_1(n_1,\dots,n_k),\dots,t_l(n_1,\dots,n_k)$;
3. propositional connectives are translated identically, e.g.,

$$\langle\theta_1 \wedge \theta_2\rangle_{n_1,\dots,n_k} =_{df} \langle\theta_1\rangle_{n_1,\dots,n_k} \wedge \langle\theta_2\rangle_{n_1,\dots,n_k};$$

4. bounded quantifiers are translated to long disjunctions and conjunctions, thus

$$\langle\exists y \leq s(x_1,\dots,x_k)\, \theta(x_1,\dots,x_k,y)\rangle_{n_1,\dots,n_k} =_{df}$$

$$\langle\theta(x_1,\dots,x_k,y)\rangle_{n_1,\dots,n_k,0} \vee \dots \vee \langle\theta(x_1,\dots,x_k,y)\rangle_{n_1,\dots,n_k,m},$$

   where m is the value of $s(n_1,\dots,n_k)$; in the case of bounded universal quantifier the propositional formula is defined dually.

10.2.1. Example. Let $\theta(x)$ be the formula expressing the pigeonhole principle for the binary relation α (for sake of simplicity we use a little stronger form than in section 9):

$$\exists u \leq S(x)\forall v \leq x(\neg\alpha(u,v)) \vee \exists u_1, u_2 \leq S(x)\exists v \leq x(u_1 \neq u_2 \wedge \alpha(u_1,v) \wedge \alpha(u_2,v)).$$

For a given n, the translation $\langle\theta(x)\rangle_n$ has form:

$$\bigvee_{i\leq n+1}\bigwedge_{j\leq n}\neg p_{ij} \vee \bigvee_{i_1,i_2\leq n+1}\bigvee_{j\leq n}\bar{\delta}_{i_1,i_2} \wedge p_{i_1,j} \wedge p_{i_2,j},$$

where $\bar{\delta}_{i_1,i_2}$ denotes $\bot$ if $i_1 = i_2$ and denotes $\top$ otherwise. The constants $\top$ and $\bot$ can be easily eliminated; namely, the formula is equivalent, using a polynomial size bounded depth Frege proof, to

$$\bigvee_{i\leq n+1}\bigwedge_{j\leq n}\neg p_{ij} \vee \bigvee_{i_1,i_2\leq n+1, i_1\neq i_2}\bigvee_{j\leq n} p_{i_1,j} \wedge p_{i_2,j}.$$

We have obtained the usual form of the propositional formula expressing the pigeonhole principle.

Let us observe, which is quite clear from the example, that the translation is a formula of polynomial size in the indices $n_1, \dots, n_k$ and, moreover, the depth is bounded by a constant, namely by the depth of the first order formula.

Let $I\Delta_0(\alpha)$ denote $I\Delta_0$ with the induction schema extended to all $\Delta_0(\alpha)$ formulas.

10.2.2. Theorem. (implicit in Paris and Wilkie [1985]) *If $I\Delta_0(\alpha)$ proves $\forall x_1 \dots \forall x_k \theta(x_1, \dots, x_k)$, where $\theta(x_1, \dots, x_k) \in \Delta_0(\alpha)$, then there exists a polynomial p and a constant d such that the propositions $\langle \theta(x_1, \dots, x_k) \rangle_{n_1, \dots, n_k}$ have Frege proofs of size $\leq p(n_1, \dots, n_k)$ and depth $\leq d$.*

Proof-sketch. Suppose $\forall x_1 \dots \forall x_k \theta(x_1, \dots, x_k)$ is provable in $I\Delta_0(\alpha)$, let $n_1, \dots, n_k$ be given. By cut elimination in the sequent calculus formalization of $I\Delta_0(\alpha)$, we have a free-cut-free proof of this sentence. From this proof we get a proof of the sequent $\to \theta(a_1, \dots, a_k)$ which contains only $\Delta_0(\alpha)$ formulas. Starting at the bottom, i.e., with $\to \theta(a_1, \dots, a_k)$, we shall gradually translate the first order proof into a propositional proof. The structural and propositional rules are, of course, translated identically.

Consider an instance of the induction rule

$$\frac{A(b), \Gamma \to \Delta, A(S(b))}{A(\underline{0}), \Gamma \to \Delta, A(t)},$$

where we have already translated the part of the proof from the lower sequent on. Suppose that in the course of translation we have assigned numbers $m_1, \dots, m_r$ to the free variables of the lower sequent. Observe that $\langle A(b_1, \dots, b_r, t(b_1, \dots, b_r)) \rangle_{m_1, \dots, m_r}$ is equal to $\langle A(b_1, \dots, b_r, b_{r+1}) \rangle_{m_1, \dots, m_r, m_{r+1}}$, where m_{r+1} is the value of $t(m_1, \dots, m_r)$. We take m_r translations of the upper sequent with indices $m_1, \dots, m_r, m$, $m = 0, \dots, m_r - 1$ (m stands for the free variable b). The translation of the lower sequent follows from them by applying $r - 1$ cuts.

The quantifier rules for bounded quantifiers are treated similarly. Eventually we reach initial sequents which are translated to initial sequents in propositional logic. □

This theorem can be used, as mentioned above, to construct short bounded depth Frege proofs, but, what is more interesting, also to prove, for instance that the pigeonhole principle for a free second order variable α is not provable in $I\Delta_0(\alpha)$. The first proof of this independence, by Ajtai [1994a], was based on model theory and a lower bound on the length of proofs of the propositional pigeonhole principle was derived as a corollary. Nowadays it is clear that the right and simpler way is to prove the lower bound for propositional logic first, see Beame et al. [1992], Pitassi, Beame and Impagliazzo [1993] and Krajíček, Pudlák and Woods [1995].

Let us mention by passing another parallel with computational complexity. The results for theories augmented with an extra free second order variable are alike to the oracle results in complexity theory. The "absolute" results, e.g. unprovability of the pigeonhole principle for Δ_0-formulas in $I\Delta_0$, are beyond present means, as well as unrelativized separation results in computational complexity theory.

10.2.3. The same translation can be applied to second order theories where we have also true second order axioms. For U_1^1 we get a bound $2^{(\log n)^{O(1)}}$ on the size of Frege proofs (with a $(\log n)^{O(1)}$ bound on the depth); for V_1^1, Krajíček [1994b] gives a polynomial bound on the size of extension Frege proofs.

10.3. Second translation. For the second translation we consider the language L_2 of the theories S_2 and T_2 introduced by Buss [1986], see Chapter II. This language extends L_0 by $\lfloor x/2 \rfloor, x\#y, |x|$. The interpretation of these function symbols is $|x| = \lceil \log_2(x+1) \rceil$, $x\#y = 2^{|x|\cdot|y|}$. The # function is used to obtain faster growth rate of terms, namely $2^{p(\log x)}$, p a polynomial. This means that the *lengths* of the numbers increase polynomially, which renders formalization of polynomial time computations possible. The $|x|$ function is used to define *sharply bounded* quantifiers

$$\forall x \leq |t|, \qquad \exists x \leq |t|,$$

where t is a term. The basic property is that there are only polynomially many elements x less than or equal than $|t|$, since the outermost function in this term is, essentially, the logarithm.

The class Π_1^b consists of formulas of L_2 which contain only sharply bounded quantifiers and strong bounded quantifiers (positive occurrences of universal bounded and negative occurrences of existential bounded); the other classes Π_i^b, Σ_i^b are defined similarly.

We want to define propositional translations of a Π_1^b formula $\varphi(x_1, \ldots, x_k)$. The translation will be denoted by $[\varphi(x_1, \ldots, x_k)]_{n_1,\ldots,n_k}$. Now we index the translation with strings of integers again, but the meaning is that we express propositionally that the sentence $\varphi(x_1, \ldots, x_k)$ holds for all $x_1, \ldots, x_k$ with $|x_1| \leq n_1, \ldots, |x_k| \leq n_k$. The intuition behind the translation is the following. We identify truth assignments with (binary representations of) numbers. Since the terms are polynomial time computable functions, we can express atomic first order formulas by polynomial size propositions. Sharply bounded quantifiers are translated to polynomial size disjunctions and conjunctions. The strong bounded quantifiers are represented by sequences of propositional variables; this is a correct interpretation, since, by definition, a propositional tautology must be satisfied *for all* truth assignments.

A formal definition is fairly involved, thus most authors do not give a full definition, and we shall also only sketch how to resolve some technical problems of the definition.

First consider an atomic formula, say, with only one free variable, $s(x) = t(x)$. Let n be given for which we want to express propositionally that the sentence $s(x) = t(x)$ holds for all x with $|x| \leq n$. Ideally we would take propositional variables $\bar{p} = (p_1, \ldots, p_n)$ representing such numbers and formulas $\sigma_i(\bar{p}), \tau_i(\bar{p})$, $i = 1, \ldots, m$,

$(m = n^{O(1)})$, representing the bits of $s(x)$ resp. $t(x)$, and define

$$[s(x) = t(x)]_n =_{df} \bigwedge_{i=1,\dots,m} \sigma_i(\bar{p}) \equiv \tau_i(\bar{p}).$$

There are such formulas of polynomial size for each of the basic functions, hence by composing them we get polynomial size formulas for all terms. Probably one can use these formulas, but it would require to find short extension Frege proofs of basic properties of these functions, which is by no means obvious for such a formalization. Therefore, instead of it, we take the natural circuits for the functions and introduce propositional variables for the functions computed at the vertices of the circuits. Then the translation will be an implication with the antecedent being the conjunction of simple clauses relating the values of the vertices of the circuits and consequent being

$$\bigwedge_{i=1,\dots,n} q_i \equiv r_i,$$

where q_i and r_i are the propositional variables for the outputs of the circuits for $s(x)$ and $t(x)$ respectively. Thus the translation will have a polynomial number of extra variables which do not code bits of the numbers representing the free variables of the first order formula. For such a formalization it is much easier to prove the basic properties of the translation

As explained above, the strong bounded quantifiers are simply omitted, (except that the bounds on the variables are left as a part of the formula) and the sharply bounded quantifiers are translated using disjunctions and conjunctions. Consider for instance a formula Φ starting with a sharply bounded quantifier followed by a universal bounded quantifier, say

$$\exists y \leq |t(x)| \forall z \leq s(x,y)\ \varphi(x,y,z),$$

where φ is an open formula. We want to define the translation $[\Phi]_n$. We first replace the quantified variable y by the numerical instances, and then translate

$$\bigvee_{i=0,\dots,|t(n)|} (\underline{i} \leq |t(x)| \rightarrow \varphi(x, \underline{i}, z_i))\,,$$

where $z_0, \dots, z_{|t(n)|}$ are new distinct variables.

After this example it should not be difficult for the reader to go on and handle more complex cases.

10.3.1. S_2 is a theory based on a finite number of basic open axioms with induction for bounded formulas of the form

$$\varphi(\underline{0}) \wedge \forall x(\varphi(\lfloor x/2 \rfloor) \rightarrow \varphi(x)) \ \rightarrow \forall x \varphi(x).$$

The most important fragment of bounded arithmetic S_2 is the theory S^1_2 where the induction schema is restricted to Σ^b_1 formulas. This theory is adequate for formalization of polynomial time computations, see Buss [1986]. Furthermore it is related to extension Frege proof systems:

10.3.2. Theorem. (Cook [1975], Buss [1986]) *If* S_2^1 *proves* $\forall x_1 \ldots \forall x_k \varphi(x_1, \ldots, x_k)$, *where* $\varphi(x_1, \ldots, x_k) \in \Pi_1^b$, *then there exists a polynomial* p *such that the propositions* $[\varphi(x_1, \ldots, x_k)]_{n_1,\ldots,n_k}$ *have extension Frege proofs of size* $\leq p(n_1, \ldots, n_k)$. □

The proof of this theorem is similar to the proof of Theorem 10.2.2, but much more involved due to the difficulties with the basic axioms.

10.3.3. This theorem naturally rises the question: is extension Frege proof system the weakest system for which we can prove this theorem? We do not know; it is possible that one can construct some pathological counterexample, but there is another reason for associating extension Frege systems with S_2^1, which we shall consider next. Following Krajíček and Pudlák [1990], we shall define a natural relation between theories and propositional proof systems.

10.3.4. Definition. (1) For a propositional proof system P we denote by $RFN(P)$ (the reflection principle for P) the $\forall\Pi_1^b$ sentence

$$\forall d, u((d : P \vdash u) \to Taut(u)),$$

where $Taut(x)$ is a Π_1^b formula defining the set of propositional tautologies. Note that $d : P \vdash u$ (d is a P proof of a proposition u) can be written as a Σ_1^b formula, since it is a polynomial time computable predicate.

(2) A propositional proof system P *simulates a theory* T, if for every $\varphi(x) \in \Pi_1^b$

$$T \vdash \forall x \varphi(x) \quad \Rightarrow \quad S_2^1 \vdash \forall y \exists d \, (d : P \vdash [\varphi(x)]_{|y|}).$$

(3) A propositional proof system P is *associated to* a theory T, if P simulates T and $T \vdash RFN(P)$.

Probably in (2) you expected rather a statement like in Theorems 10.2.2 and 10.3.2. In fact the condition (2) is stronger: by Buss's Theorem II.3.2, the provability of such a Π_2 statement in S_2^1 implies that it can be witnessed by a polynomial time computable function. Thus, in particular, the P proofs of $[\varphi(x)]_n$'s must be of polynomial size. So (2) means that there is a polynomial bound on the lengths of P proofs of $[\varphi(x)]_n$'s *provably in a weak theory.*

Let us also note that $RFN(P)$ is equivalent to the consistency of P assuming some "mild conditions" on P.

We shall denote by G the quantified propositional proof system based on the sequent calculus, see 8.8. Let G_i denote the subsystem of G obtained by imposing the restriction of at most i alternations of quantifiers in each formula of a proof. Let G_i^* denote G_i where we allow only tree-like proofs.

The following theorem gives some known pairs of a proof system associated to a theory (for definitions of the theories see Chapter II).

10.3.5. Theorem. (Cook [1975], Krajíček and Takeuti [1990], Krajíček and Pudlák [1990]) *The following are pairs of a theory and a proof system associated to it:* (S^1_2, *extension Frege*), (S^i_2, G^*_i) *for* $i \geq 1$, (T^i_2, G_i) *for* $i \geq 1$, (U^1_2, G). □

Note that for U^1_2 we have two related systems, depending on which translation we take. Further results of this type were proved in Clote [1992].

Next theorem shows that under reasonable conditions the associated propositional proof system is determined up to polynomial simulation.

10.3.6. Theorem. (Krajíček and Pudlák [1990]) *Let P be a propositional proof system associated to a theory T. Suppose T contains S^1_2 and the following is provable in S^1_2: P simulates extension Frege systems and it is closed under modus ponens. Then P polynomially simulates any propositional proof system for which T proves the reflection principle.*

Thus, e.g. by Theorem 10.3.5, extension Frege systems and G^*_1 are polynomially equivalent.

Proof. Suppose $T \vdash RFN(Q)$. Let $\rho_Q(x, y)$ be the Π^b_1 formula which defines the reflection principle, i.e.,

$$\rho_Q(d, u) \equiv d : Q \vdash u \rightarrow Taut(u).$$

By the assumptions $S^1_2 \vdash \forall z(P \vdash [\rho_Q(x, y)]_z)$. We now we argue in the theory S^1_2. Thus we have

$$P \vdash [x : Q \vdash y \rightarrow Taut(y)]_z.$$

Since $[x : Q \vdash y \rightarrow Taut(y)]_z$ is $[x : Q \vdash y]_z \rightarrow [Taut(y)]_z$ and P is closed under modus ponens, we get

$$P \vdash [x : Q \vdash y]_z \quad \rightarrow \quad P \vdash [Taut(y)]_z.$$

We have also

$$P \vdash [Taut(y)]_z \quad \rightarrow \quad P \vdash y,$$

since it is true already for extension Frege systems (we leave this claim without a proof). Thus we have obtained in S^1_2

$$P \vdash [x : Q \vdash y]_z \quad \rightarrow \quad P \vdash y.$$

Back in the real world, by Buss's witnessing theorem it means that one can construct in polynomial time a proof of φ in P from a proof of $[d : Q \vdash \varphi]_n$ in P.

Now suppose that we are given a proof d of φ in Q. Substituting the numbers which encode d and φ we get *a true variable-free propositional formula* $[\underline{d} : Q \vdash \underline{\varphi}]_n$. Such formulas always have polynomial size proofs even in a Frege system. Thus we get a P proof of φ in polynomial time. □

The meaning of this theorem is that the proof system associated to a theory T is, from the point of view of T, the strongest proof system, i.e., stronger systems may be inconsistent. Let us state it formally:

10.3.7. Corollary. *Under the same assumptions as in Theorem 10.3.6, if* $T \vdash \mathcal{NP} = co\mathcal{NP}$*, then* P *is polynomially bounded.*

Proof. Since the set of propositional tautologies is $co\mathcal{NP}$-complete, the assumption $T \vdash \mathcal{NP} = co\mathcal{NP}$ means that

$$T \vdash \forall x(\sigma(x) \equiv Taut(x)), \tag{59}$$

for some $\sigma(x) \in \Sigma_1^b$. So σ defines a polynomially bounded propositional proof system Q (proofs are the witnesses for the existential bounded quantifiers). The sentence (59) implies $T \vdash RFN(Q)$. Hence, by Theorem 10.3.6, P polynomially simulates Q. But if Q is polynomially bounded, then also P must be. □

As we believe that $\mathcal{NP} \neq co\mathcal{NP}$, we expect that the corollary will be used in the contrapositive form. Let us state the nicest special case of it (proved directly by Wilkie in 1987, unpublished; as observed in Krajíček and Pudlák [1989] it also follows from results of Cook [1975] and Buss [1986]).

10.3.8. Corollary. *If extension Frege proofs are not polynomially bounded, then* S_2^1 *does not prove* $\mathcal{NP} = co\mathcal{NP}$. □

10.4. Optimal proof systems and consistency statements. The second translation can be used to show a link between a fundamental problem about the lengths of proofs of finite consistency statements and the existence of an optimal propositional proof system. Furthermore there is a statement from structural complexity theory which is equivalent to these problems. A set $Y \subseteq \{0,1\}^*$ is called *sparse*, if for every n, the size of $Y \cap \{0,1\}^n$ is bounded by a polynomial.

10.4.1. Theorem. (Krajíček and Pudlák [1989]) *The following are equivalent:*

1. *There exists a consistent finitely axiomatized theory* $T \supseteq S_2^1$ *such that for every consistent finitely axiomatized theory* S

$$\|Cons_S(\underline{n})\|_T = n^{O(1)}.$$

2. *There exists an* optimal propositional proof system, *i.e., a propositional proof system* P *such that for every propositional proof system* Q

$$\|\varphi\|_Q = \|\varphi\|_P^{O(1)},$$

for every tautology φ.

3. *For every* $co\mathcal{NP}$*-set* X *there exists a nondeterministic Turing machine which accepts* X *and uses only polynomial time on every sparse subset* $Y \subseteq X$, $Y \in \mathcal{P}$.

The proof of the equivalence of 1. and 2. is based on the following two constructions. If T is an optimal theory in the sense of 1., we take a propositional proof system P defined by:

$$d : P \vdash \varphi \equiv_{df} d : T \vdash Taut(\underline{\varphi}).$$

If, on the other hand, P is an optimal propositional proof system, we take the theory T defined by:

$$T =_{df} S^1_2 + RFN(P).$$

We omit the rest of the proof. □

Given a propositional proof system P which is not polynomially bounded, we can produce, using this theorem, a sequence of tautologies which surely do not have polynomial size proofs in P. Unfortunately, the tautologies will be rather complex artificial statements, thus not amenable to a combinatorial analysis. However, as noted by Krajíček [1995], one can use the polynomial reductions, by which $\mathcal{NP}$ completeness results are proved, to turn these tautologies into simple combinatorial statements. For instance one can construct a sequence of nonhamiltonian graphs, such that there are no polynomial size proofs in P of the tautologies expressing that the graphs are nonhamiltonian. Thus the problem reduces to finding a class of nonhamiltonian graphs for which it is difficult to prove in P that they are nonhamiltonian.

11. Bibliographical remarks for further reading

In this section we shall give a few more references which have not been mentioned in the main text. This should serve to the reader who is interested in the history of the subject or who wants to learn more about it. Our aim is not to complete the list of references about *results* on the lengths of proofs, rather we want to partially complement the above presentation which concentrated on *methods* used in this research area. Thus, in particular, we shall not repeat results described above.

Probably the oldest recorded paper on the subject is Gödel [1936]. In this two-page abstract he stated the result that there is a speed-up between the lengths of proofs of formulas provable in i-th order and $i+1$-st order arithmetics. To quote him: *The transition to the logic of the next higher type not only results in certain previously unprovable propositions becoming provable, but also in it becoming possible to shorten extraordinarily infinitely many of the proofs already available.* The length of proofs is considered to be the number of steps and the speed-up is $\phi(n)$ for any function ϕ "computable" in the lower system. There was no proof given in the paper. For a full proof of this statement see Buss [1994].

Another important writing of Gödel which was discovered only a few years ago, is the letter by Gödel [1993]. In that letter he posed the question whether one can decide in linear, quadratic, etc. time in n whether a given formula has a proof of length (= number of symbols) n. Now we know that this problem is $\mathcal{NP}$-complete. See Buss [1995a] for a discussion and a proof of an unproven claim of Gödel.

Looking at the literature it seems that the subject lay dormant for several decades. I think that many people thought about problems on the lengths of proofs, but the things that they actually could prove did not look interesting enough, especially when compared with other fancy topics like set theory. Furthermore some basic concepts were missing (one of such crucial things was the distinction between polynomial size and exponential size). This can be documented by a remark of Kreisel [1967,page 241], who mentions a conversation with Gödel where Gödel asked the question of what are the lengths of proofs of finite consistency statements. No paper had been written about it until Friedman [1979], but he did not consider it to be worth publishing.

At the early stages, Georg Kreisel was one of the main proponents of this field. His student Statman [1978] determined the increase of the lengths of proofs in cut-elimination and Herbrand's Theorem. Another of his students, Baaz (see Baaz and Pudlák [1993], Baaz and Zach [1995]), made significant progress in Kreisel's Conjecture. As seen on Kreisel's Conjecture, Kreisel was more interested in *positive* results in the sense of deriving more information from the proofs than just the mere fact that the statement is true. Logic should help mathematicians to get more or better results, rather than only to show impossibilities of certain proofs, see e.g. Kreisel [1990]. From this point of view, one of the greatest successes in proof theory was the result of Luckhardt [1989], deriving explicit bounds on approximation of algebraic numbers by rational numbers (Roth's theorem), using Herbrand's theorem.

Originally the interest in the lengths of proofs was based mainly on philosophical and methodological considerations. With the advent of computers a new practical reason appeared: *automated theorem proving.* The main tool in automated theorem proving is the resolution system for first order logic, see e.g. Chang and Lee [1973]. For us, theoreticians, most of the papers are too much applied, however there are several results which are important also for theory. Such a notable result is the exponential lower bound for propositional *regular* resolution of Tsejtin [1968]. The question about the efficiency of proof-search strategies are often nontrivial mathematical problems, let us mention at least some results of this type Baaz and Leitsch [1992,1994]. There are several books about the complexity of logical calculi, e.g. Eder [1992]; they deal mainly with the first order logic.

The next important stimulus was the rise of complexity theory. The lengths of proofs is just one of several research areas which combine logic and complexity theory. Another one, which is closely related to it, is the complexity of logical theories. The problem is how efficiently can we decide if a sentence is provable in a given decidable theory T (e.g., Presburger arithmetic). Note that an upper bound on the lengths of proofs in T gives an upper bound on a *nondeterministic* procedure for decidability. Often this bound is not very far from the best. We refer the reader to the surveys Rabin [1977] and Compton and Henson [1990].

We can say that the research into complexity of proofs really started with the seminal paper of Parikh [1971] which introduced several important concepts and proved basic results about them: *speed-up for first order theories, theories which are inconsistent but are consistent for practical purposes, and bounded arithmetic.* Soon after it, he published a basic result on Kreisel's Conjecture in Parikh [1973]. He proved that the conjecture is true, if we take Peano arithmetic with $+$ and $\times$ as ternary relations instead of function symbols. That proof has been a paradigm for all subsequent proofs of instances of Kreisel's Conjecture.

After that several people started to work on these subjects. One of the most influential researchers in this field has been Orevkov. We shall mention only the most important papers of the many that he published. Orevkov [1982] gave a different proof of the lower bounds on the lengthening of proofs in cut-elimination and Orevkov [1986] gave more precise upper bounds. Orevkov [1987b] introduced explicitly the concept of the skeleton and Orevkov [1987a] proved several results related to Kreisel's Conjecture. All these results, and many more, are covered in Orevkov [1993].

There are more results on the complexity of first order proofs. Of those that we have not presented yet, let us mention the dissertation of Ignjatović [1990]. He proved a nonelementary speed up between Primitive Recursive Arithmetic and $I\Sigma_0$.

Currently the most active area is propositional logic and bounded arithmetic. The fundamental paper is Cook [1975], where a relation of the lengths of proofs in propositional logic and provability in arithmetic was considered for the first time. The most influential papers in bounded arithmetic after Parikh [1971] were written by Paris and Wilkie; let us mentioned at least the Paris and Wilkie [1985] paper on counting problems which influenced very much research on the complexity of propositional logic. The basic book on bounded arithmetic is due to Buss [1986]. Another fundamental paper is by Ajtai [1994a], where he introduced the method of random restrictions into propositional logic, which had already been used in complexity theory. This development has been partially described in this chapter and also in Chapter II; much more can be found in the monograph by Krajíček [1995], which covers the whole area in detail except for the most recent results. As this manuscript is being finalized, new exciting results are being obtained on the polynomial calculus by Razborov [n.d.], Krajíček [1997b] and Riis and Sitharam [1997].

Acknowledgments

I would like to thank Sam Buss for helping me with the preparation of the manuscript and suggesting several improvements and Jan Krajíček for checking the manuscript. The preparation of the article was supported by grant #A1019602 of the Academy of Sciences of the Czech Republic and the cooperative research grant INT-9600919/ME-103 of the U.S. National Science Foundation and the Czech Republic Ministry of Education.

References

M. Ajtai

[1990] Parity and the pigeonhole principle, in: *Feasible Mathematics: A Mathematical Sciences Institute Workshop held in Ithaca, New York, June 1989*, S. R. Buss and P. J. Scott, eds., Birkhäuser, Boston, pp. 1–24.

[1994a] The complexity of the pigeonhole principle, *Combinatorica*, 14, pp. 417–433. Extended abstract in *Proc. 29th Annual IEEE Symposium on Foundations of Computer Science*, 1988, pp. 346-355.

[1994b] The independence of the modulo p counting principles, in: *Proceedings of the 26th Annual ACM Symposium on the Theory of Computing*, Association for Computing Machinery, New York, pp. 402–411.

[1995] On the existence of modulo p cardinality functions, in: *Feasible Mathematics II*, P. Clote and J. B. Remmel, eds., Birkhäauser, Boston, pp. 1–14.

N. ALON AND R. BOPPANA

[1987] The monotone circuit complexity of boolean functions, *Combinatorica*, 7, pp. 1–22.

S. ARORA, C. LUND, R. MOTWANI, M. SUDAN, AND M. SZEGEDY

[1992] Proof verification and hardness of approximation problems, in: *Proceedings of the 33rd Annual Symposium on Foundations of Computer Science*, IEEE Computer Society, Piscataway, New Jersey, pp. 14–23.

M. BAAZ AND A. LEITSCH

[1992] Complexity of resolution proofs and function introduction, *Annals of Pure and Applied Logic*, 20, pp. 181–215.

[1994] On Skolemization and proof complexity, *Fundamenta Mathematicae*, 20.

M. BAAZ AND P. PUDLÁK

[1993] Kreisel's conjecture for $L\exists_1$, in: *Arithmetic Proof Theory and Computational Complexity*, P. Clote and J. Krajíček, eds., Oxford University Press, pp. 30–39.

M. BAAZ AND R. ZACH

[1995] Generalizing theorems in real closed fields, *Annals of Pure and Applied Logic*, 75, pp. 2–23.

P. BEAME, R. IMPAGLIAZZO, J. KRAJÍČEK, T. PITASSI, AND P. PUDLÁK

[1996] Lower bounds on Hilbert's Nullstellensatz and propositional proofs, *Proceedings of the London Mathematical Society*, 73, pp. 1–26.

P. BEAME, R. IMPAGLIAZZO, J. KRAJÍČEK, T. PITASSI, P. PUDLÁK, AND A. WOODS

[1992] Exponential lower bounds for the pigeonhole principle, in: *Proceedings of the 24th Annual ACM Symposium on the Theory of Computing*, Association for Computing Machinery, New York, pp. 200–221.

P. BEAME AND T. PITASSI

[1996] Exponential separation between the matching principles and the pigeonhole principle, *Annals of Pure and Applied Logic*, 80, pp. 195–228.

S. BELLANTONI, T. PITASSI, AND A. URQUHART

[1992] Approximation and small-depth Frege proofs, *SIAM Journal on Computing*, 21, pp. 1161–1179.

E. W. BETH

[1959] *The Foundations of Mathematics*, North-Holland, Amsterdam.

M. L. BONET AND S. R. BUSS

[1993] The deduction rule and linear and near-linear proof simulations, *Journal of Symbolic Logic*, 58, pp. 688–709.

M. L. BONET, T. PITASSI, AND R. RAZ

[1997a] Lower bounds for cutting planes proofs with small coefficients, *Journal of Symbolic Logic*, 62, pp. 708–728. An earlier version appeared in *Proc. Twenty-Seventh Annual ACM Symposium on the Theory of Computing*, 1995, pp. 575-584.

[1997b] No feasible interpolation for TC^0-Frege proofs, in: *Proceedings of the 38th Annual Symposium on Foundations of Computer Science*, IEEE Computer Society, Piscataway, New Jersey, pp. 254–263.

S. R. Buss

[1986] *Bounded Arithmetic*, Bibliopolis, Napoli. Revision of 1985 Princeton University Ph.D. thesis.

[1987] Polynomial size proofs of the propositional pigeonhole principle, *Journal of Symbolic Logic*, 52, pp. 916–927.

[1991a] Propositional consistency proofs, *Annals of Pure and Applied Logic*, 52, pp. 3–29.

[1991b] The undecidability of k-provability, *Annals of Pure and Applied Logic*, 53, pp. 75–102.

[1994] On Gödel's theorems on lengths of proofs I: Number of lines and speedup for arithmetics, *Journal of Symbolic Logic*, 59, pp. 737–756.

[1995a] On Gödel's theorems on lengths of proofs II: Lower bounds for recognizing k-symbol provability, in: *Feasible Mathematics II*, P. Clote and J. B. Remmel, eds., Birkhäauser, Boston, pp. 57–90.

[1995b] Some remarks on lengths of propositional proofs, *Archive for Mathematical Logic*, 34, pp. 377–394.

S. R. Buss and P. Clote

[1996] Cutting planes, connectivity and threshold logic, *Archive for Mathematical Logic*, 35, pp. 33–62.

S. R. Buss, R. Impagliazzo, J. Krajíček, P. Pudlák, A. A. Razborov, and J. Sgall

[1996/1997] Proof complexity in algebraic systems and constant depth Frege systems with modular counting, *Computational Complexity*, 6, pp. 256–298.

S. R. Buss and T. Pitassi

[1997] *Resolution and the Weak Pigeonhole Principle.* Typeset manuscript, to appear in *CSL'97*.

S. R. Buss and Gy. Turán

[1988] Resolution proofs of generalized pigeonhole principles, *Theoretical Computer Science*, 62, pp. 311–317.

C.-L. Chang and R. C.-T. Lee

[1973] *Symbolic Logic and Mechanical Theorem Proving*, Academic Press, New York.

M. Clegg, J. Edmonds, and R. Impagliazzo

[1996] Using the Groebner basis algorithm to find proofs of unsatisfiablility, in: *Proceedings of the Twenty-Eighth Annual ACM Symposium on the Theory of Computing*, Association for Computing Machinery, New York, pp. 174–183.

P. Clote

[1992] ALOGTIME and a conjecture of S. A. Cook, *Annals of Mathematics and Artificial Inteligence*, 6, pp. 57–106.

P. Clote and J. Krajíček

[1993] *Arithmetic, Proof Theory and Computational Complexity*, Oxford University Press.

K. J. Compton and C. W. Henson

[1990] A uniform method for proving lower bounds on the computational complexity of logical theories, *Annals of Pure and Applied Logic*, 48, pp. 1–79.

S. A. Cook

[1975] Feasibly constructive proofs and the propositional calculus, in: *Proceedings of the Seventh Annual ACM Symposium on the Theory of Computing*, Association for Computing Machinery, New York, pp. 83–97.

S. A. Cook and R. A. Reckhow

[1979] The relative efficiency of propositional proof systems, *Journal of Symbolic Logic*, 44, pp. 36–50.

W. Cook, C. R. Coullard, and Gy. Turán

[1987] On the complexity of cutting plane proofs, *Discrete Applied Mathematics*, 18, pp. 25–38.

W. Craig

[1957a] Linear reasoning. A new form of the Herbrand-Gentzen theorem, *Journal of Symbolic Logic*, 22, pp. 250–268.

[1957b] Three uses of the Herbrand-Gentzen theorem in relating model theory and proof theory, *Journal of Symbolic Logic*, 22, pp. 269–285.

M. Dowd

[1979] *Propositional Representation of Arithmetic Proofs*, PhD thesis, University of Toronto.

[1985] *Model-Theoretic Aspects of $P \neq NP$*. Typewritten manuscript.

A. G. Dragalin

[1985] Correctness of inconsistent theories with notions of feasibility, in: *Computation Theory, Fifth Symposium Proceedings*, A. Skowron, ed., vol. 108 of Lecture Notes in Computer Science #208, Springer-Verlag, Berlin, pp. 58–79.

E. Eder

[1992] *Relative Complexities of First Order Calculi*, Verlag Vieweg.

A. Ehrenfeucht and J. Mycielski

[1971] Abbreviating proofs by adding new axioms, *Bulletin of the American Mathematical Society*, 77, pp. 366–367.

W. M. Farmer

[1984] *Length of Proofs and Unification Theory*, PhD thesis, University of Wisconsin, Madison.

[1988] A unification algorithm for second order monadic terms, *Annals of Pure and Applied Logic*, 39, pp. 131–174.

J. Ferrante and C. W. Rackoff

[1979] *The Computational Complexity of Logical Theories*, Lecture Notes in Mathematics #718, Springer-Verlag, Berlin.

H. M. Friedman

[1975] One hundred and two problems in mathematical logic, *Journal of Symbolic Logic*, 40, pp. 113–129.

[1979] *On the consistency, completeness, and correctness problems.* Ohio State University, unpublished.

Yu. V. Gavrilenko

[1984] Monotone theories of feasible numbers, *Doklady Akademii Nauk SSSR*, 276, pp. 18–22.

G. Gentzen

[1935] Untersuchungen über das Logische Schliessen, *Mathematische Zeitschrift*, 39, pp. 176–210 and 405–431.

J.-Y. Girard

[1989] *Proofs and Types*, Cambridge University Press.

K. Gödel

[1936] Über die Länge von Beweisen, *Ergebnisse eines Mathematischen Kolloquiums*, pp. 23–24. English translation in *Kurt Gödel: Collected Works, Volume 1*, pages 396-399, Oxford University Press, 1986.

[1993] A letter to von Neumann, March 20, 1956, in: *Arithmetic Proof Theory and Computational Complexity*, P. Clote and J. Krajíček, eds., Oxford University Press, pp. vii–ix.

A. GOERDT

[1991] Cutting plane versus Frege proof systems, in: *Computer Science Logic: 4th workshop, CSL '90*, E. Börger and et al., eds., Lecture Notes in Computer Science #533, Springer-Verlag, Berlin, pp. 174–194.

A. GRZEGORCZYK

[1974] *An Outline of Mathematical Logic*, D. Reidel Publishing Co., Dordrecht-Boston, Mass., PWN-Polish Scientific Publishers, Warsaw. Translation of *Zarys logiki matematicznej*, Państwowe Wyadwnictwo Naukowe, 1969.

P. HÁJEK, F. MONTAGNA, AND P. PUDLÁK

[1993] Abbreviating proofs using metamathematical rules, in: *Arithmetic Proof Theory and Computational Complexity*, P. Clote and J. Krajíček, eds., Oxford University Press, pp. 197–221.

P. HÁJEK AND P. PUDLÁK

[1993] *Metamathematics of First-order Arithmetic*, Perspectives in Mathematical Logic, Springer-Verlag, Berlin.

G. HAJÓS

[1961] Über eine Konstruktion nicht n-färberer Graphen, *Wiss. Zeitschr. M. Luther Univ. Halle-Wittenberg*, A 10, pp. 116–117.

A. HAKEN

[1985] The intractability of resolution, *Theoretical Computer Science*, 39, pp. 297–308.

A. HAKEN AND S. A. COOK

[n.d.] *An Exponential Lower Bound for the Size of Monotone Real Circuits.* To appear in *J. of Computer and System Science.*

J. HÅSTAD

[1986] *Computation Limits of Small Depth Circuits*, MIT Press.

D. HILBERT AND W. ACKERMANN

[1928] *Grundzüge der theoretischen Logik*, Springer-Verlag, Berlin.

D. HILBERT AND P. BERNAYS

[1934] *Grundlagen der Mathematik I*, Springer-Verlag, Berlin.

[1939] *Grundlagen der Mathematik II*, Springer-Verlag, Berlin.

A. IGNJATOVIĆ

[1990] *Fragments of First and Second Order Arithmetic and Length of Proofs*, PhD thesis, University of California, Berkeley.

R. IMPAGLIAZZO, P. PUDLÁK, AND J. SGALL

[1997] *Lower Bounds for the Polynomial Calculus and the Groebner Basis Algorithm*, Tech. Rep. TR97-042, Electronic Colloquium on Computational Complexity (ECCC).

J. JOHANNSEN

[1997] *Lower Bounds for Monotone Real Circuit Depth and Formula Size and Tree-like Cutting Planes*, Tech. Rep. TR97-032, Electronic Colloquium on Computational Complexity, http://www.eccc.uni-trier.de/eccc/.

J. KRAJÍČEK

[n.d.] *Discretely Ordered Modules as a First-Order Extension of the Cutting Planes Proof System.* To appear in the *J. of Symbolic Logic.*

[1989a] On the number of steps in proofs, *Annals of Pure and Applied Logic*, 41, pp. 153–178.

[1989b] Speed-up for propositional Frege systems via generalizations of proofs, *Commentationes Mathematicae Universitatis Carolinae*, 30, pp. 137–140.

[1994a] Lower bounds to the size of constant-depth propositional proofs, *Journal of Symbolic Logic*, 59, pp. 73–86.

[1994b] On Frege and extended Frege proof systems, in: *Feasible Mathematics II*, J. Krajíček and J. B. Remmel, eds., Birkhäuser, Boston, pp. 284–319.

[1995] *Bounded Arithmetic, Propositional Logic and Complexity Theory*, Cambridge University Press.

[1997a] Interpolation theorems, lower bounds for proof systems and independence results for bounded arithmetic, *Journal of Symbolic Logic*, 62, pp. 457–486.

[1997b] *On the Degree of Ideal Membership Proofs from Uniform Families of Polynomials over a Finite Field.* Typeset manuscript.

J. KRAJÍČEK AND P. PUDLÁK

[1988] The number of proof lines and the size of proofs in first-order logic, *Archive for Mathematical Logic*, 27, pp. 69–84.

[1989] Propositional proof systems, the consistency of first-order theories and the complexity of computations, *Journal of Symbolic Logic*, 54, pp. 1063–1079.

[1990] Quantified propositional calculi and fragments of bounded arithmetic, *Zeitschrift für Mathematische Logik und Grundlagen der Mathematik*, 36, pp. 29–46.

[1998] Some consequences of cryptographical conjectures for S^1_2 and EF, *Information and Computation*, 140, pp. 82–94.

J. KRAJÍČEK, P. PUDLÁK, AND A. WOODS

[1995] An exponential lower bound to the size of bounded depth Frege proofs of the pigeonhole principle, *Random Structures and Algorithms*, 7, pp. 15–39.

J. KRAJÍČEK AND G. TAKEUTI

[1990] On bounded Σ^1_1-polynomial induction, in: *Feasible Mathematics*, S. R. Buss and P. J. Scott, eds., Birkhäuser, Boston, pp. 259–280.

G. KREISEL

[1967] Mathematical logic: What has it done for the philosophy of mathematics, in: *Bertrand Russell: Philosopher of the Century, Essays in his Honour*, R. Shoenemann, ed., George Allen and Unwin, pp. 201–272.

[1990] Logical aspects of computation: Contributions and distractions, in: *Logic and Computer Science*, Academic Press, New York, pp. 205–278.

H. LUCKHARDT

[1989] Herbrand-Analysen zweier Beweise des Satzes von Roth: polynomiale Anzahlschranken, *Journal of Symbolic Logic*, 54, pp. 234–263.

T. MIYATAKE

[1980] On the length of proofs in formal systems, *Tsukuba Journal of Mathematics*, 4, pp. 115–125.

D. MUNDICI

[1984] NP and Craig's interpolation theorem, in: *Logic Colloquium '82*, G. Lolli, G. Longo, and A. Marcja, eds., North-Holland, Amsterdam, pp. 345–358.

E. NELSON

[1986] *Predicative Arithmetic*, Princeton University Press.

V. P. OREVKOV

[1982] Lower bounds on the increase in complexity of deductions in cut elimination, *Journal of Soviet Mathematics*, 20. Original Russian version in *Zap. Nauchn. Sem. L.O.M.I. 88* (1979), pp.137-162.

[1986] Upper bound on the lengthening of proofs by cut elimination, *Journal of Soviet Mathematics*, 34, pp. 1810–1819. Original Russian version in *Zap. Nauchn. Sem. L.O.M.I. 137* (1984), pp.87-98.

[1987a] Lower bounds on the lengths of derivations in arithmetic in terms of the complexity of terms involved in the derivations, *Soviet Mathematics Doklady*, 35, pp. 579–582. Original Russian version in *Dokl. Akad. Nauk. 294/4* (1987).

[1987b] Reconstruction of a proof from its scheme, *Soviet Mathematics Doklady*, 35, pp. 326–329. Original Russian version in Dokl. Akad. Nauk. **293** (1987) 313-316.

[1990] Correctness of short proofs in theory with notions of feasibility, in: *COLOG-88: International Conference on Computer Logic, Tallinn, USSR, Dec. 1988, Proceedings*, P. Martin-Löf and G. E. Mints, eds., Lecture Notes in Computer Science #417, Springer-Verlag, Berlin, pp. 242–245.

[1993] *Complexity of Proofs and Their Transformations in Axiomatic theories*, vol. 128 of Translations of Mathematical Monographs, American Mathematical Society, Providence, Rhode Island.

R. Parikh

[1971] Existence and feasibility in arithmetic, *Journal of Symbolic Logic*, 36, pp. 494–508.

[1973] Some results on the lengths of proofs, *Transactions of the American Mathematical Society*, 177, pp. 29–36.

J. B. Paris and A. J. Wilkie

[1985] Counting problems in bounded arithmetic, in: *Methods in Mathematical Logic, Proceedings of the 6-th Latin American Symposium, Caracas, Venezuella*, C. A. Di Prisco, ed., Lecture Notes in Mathematics #1130, Springer-Verlag, Berlin, pp. 317–340.

T. Pitassi, P. Beame, and R. Impagliazzo

[1993] Exponential lower bounds for the pigeonhole principle, *Computational Complexity*, 3, pp. 97–140.

T. Pitassi and A. Urquhart

[1992] The complexity of the HajósnameindexHajós, G. calculus, in: *Proceedings of the 33th Annual Symposium on Foundations of Computer Science*, IEEE Computer Society, Piscataway, New Jersey, pp. 187–196.

D. Prawitz

[1970] Ideas and results in proof theory, in: *Proceedings of the Second Scandinavian Logic Symposium*, J. E. Fenstad, ed., North-Holland, Amsterdam.

P. Pudlák

[1985] Cuts, consistency statements and interpretation, *Journal of Symbolic Logic*, 50, pp. 423–441.

[1986] On the lengths of proofs of finitistic consistency statements in first order theories, in: *Logic Colloquium '84*, J. B. Paris, A. J. Wilkie, and G. M. Wilmers, eds., North-Holland, Amsterdam, pp. 165–196.

[1987] Improved bounds to the lengths of proofs of finitistic consistency statements, in: *Logic and Combinatorics*, S. G. Simpson, ed., vol. 65 of Contemporary Mathematics, American Mathematical Society, Providence, Rhode Island, pp. 309–331.

[1991] Ramsey's theorem in bounded arithmetic, in: *Computer Science Logic '90*, E. Börger and et al., eds., Lecture Notes in Computer Science #533, Springer-Verlag, Berlin, pp. 308–312.

[1997] Lower bounds for resolution and cutting planes proofs and monotone computations, *Journal of Symbolic Logic*, 62, pp. 981–998.

M. O. Rabin

[1977] Decidable theories, in: *Handbook of Mathematical Logic*, J. Barwise, ed., North-Holland, Amsterdam, pp. 595–629.

A. A. Razborov

[n.d.] *Lower Bounds for the Polynomial Calculus*. To appear in *Computational Complexity*.

[1985] Lower bounds on the monotone complexity of some boolean functions, *Doklady Akademii Nauk SSSR*, 282, pp. 1033–1037. English translation in: Soviet Mathem. Doklady, 31, pp. 354-357.

[1994] *On provably disjoint NP-pairs*, Tech. Rep. RS-94-36, Basic Research in Computer Science Center, Aarhus, Denmark, November. http://www.brics.dk/index.html.

[1996] Lower bounds for propositional proofs and independence results in Bounded Arithmetic, in: *Automata, languages and programming: 23rd international colloquium, ICALP '96*, F. Meyer auf der Heide and B. Monien, eds., Lecture Notes in Computer Science #1099, Springer-Verlag, Berlin, pp. 48–62.

A. A. Razborov, A. Widgerson, and A. C.-C. Yao

[1997] Read-once branching programs, rectangular proofs of the pigeonhole principle and the transversal calculus, in: *Proceedings of the 29th Annual ACM Symposium on the Theory of Computing*, Association for Computing Machinery, New York, pp. 739–748.

R. A. Reckhow

[1976] *On the Lengths of Proofs in the Propositional Calculus*, PhD thesis, Department of Computer Science, University of Toronto. Technical Report #87.

S. Riis and M. Sitharam

[1997] *Non-constant Degree Lower Bounds imply Linear Degree Lower Bounds*, Tech. Rep. TR97-048, Colloquium on Computation Complexity, ECCC, http://www.eccc.uni-trier.de/eccc/.

B. Russell

[1906] The theory of implication, *American Journal of Mathematics*, 28, pp. 159–202.

R. M. Smullyan

[1968] *First-Order Logic*, Springer-Verlag, Berlin.

R. M. Solovay

[1990] *Upper Bounds on the Speedup of GB over ZF.* preprint.

R. Statman

[1977] Complexity of derivations from quantifier-free Horn formulae, mechanical introduction of explicit definitions, and refinement of completeness theorems, in: *Logic Colloquium '76*, R. O. Gandy and J. M. E. Hyland, eds., North-Holland, Amsterdam, pp. 505–517.

[1978] Proof search and speed-up in the predicate calculus, *Annals of Mathematical Logic*, 15, pp. 225–287.

[1981] Speed-up by theories with infinite models, *Proceedings of the American Mathematical Society*, 81, pp. 465–469.

G. Takeuti

[1987] *Proof Theory*, North-Holland, Amsterdam, 2nd ed.

[1990] Some relations among systems for bounded arithmetic, in: *Mathematical Logic, Proceedings of the Heyting 1988 Summer School*, P. P. Petkov, ed., Plenum Press, New York, pp. 139–154.

A. Tarski

[1936] Der Warheitsbegriff in den formalisierten Sprachen, *Studia Philosophica, Commentarii Societatis Philosophicae Polonorum*, 1, pp. 261–405.

G. S. Tsejtin

[1968] On the complexity of derivations in propositional calculus, in: *Studies in mathematics and mathematical logic, Part II*, A. O. Slisenko, ed., pp. 115–125. in Russian.

G. S. Tsejtin and A. A. Čubarjan

[1975] On some bounds to the lengths of logical proofs in classical propositional calculus, *Trudy Vyčisl. Centra AN ArmSSR i Erevan. Univ.*, 8, pp. 57–64. In Russian.

R. L. VAUGHT

[1967] On axiomatizability by a schema, *Journal of Symbolic Logic*, 32, pp. 473–479.

A. C.-C. YAO

[1985] Separating the polynomial time hierarchy by oracles, in: *Proceedings of the 26th Annual Symposium on Foundations of Computer Science*, IEEE Computer Society, Piscataway, New Jersey, pp. 1–10.

CHAPTER IX

A Proof-Theoretic Framework for Logic Programming

Gerhard Jäger
Institut für Informatik und angewandte Mathematik, Universität Bern
Neubrückstrasse 10, CH-3012 Bern, Switzerland
jaeger@iam.unibe.ch

Robert F. Stärk
Institut für Informatik, Universität Freiburg
Rue Faucigny 2, CH-1700 Fribourg, Switzerland
robert.staerk@unifr.ch

Contents

HANDBOOK OF PROOF THEORY
Edited by S. R. Buss

1. Introduction

The purpose of this article is to present one specific proof-theoretic framework for first order logic programming, but of course it is not claimed that our approach is the only possible one. However, we hope to succeed in providing a perspicuous and satisfactory explanation of the most central concepts in this area, where our emphasis is put on a deductive and procedural point of view.

The basic principles of logic programming, its history, and its relationship to the programming language Prolog are well presented in many other publications (cf. e.g. Apt [1990], Doets [1994] and Lloyd [1987]) so that we can omit details. A first important distinction is between definite logic programs which are based on so called definite Horn clauses, and extensions thereof which provide means for treating negative information.

For definite logic programs the situation is quite simple. We have the straightforward observation that a closed atomic formula A is valid in the least Herbrand model M_P of a definite logic program P if and only if A is a logical consequence of P. Moreover, M_P is the least fixed point of the immediate consequence operator T_P introduced in van Emden and Kowalski [1976]. T_P also provides the link to the so called SLD-resolution, which is the standard proof procedure for definite logic programs and equivalent to direct proofs in suitable sequent calculi.

Although formulated in a very restricted language, definite logic programs are computationally complete in the sense that they can represent all recursively enumerable relations. Nevertheless definite programs do not adequately reflect the paradigm of programming in logic since there is no way to express negative information.

The standard method to introduce negative information into logic programming environments is by Clark's famous negation as failure rule. The result of adding this rule to SLD-resolution is called SLDNF-resolution and will be described below in detail. Negation as failure has a strong procedural character and is easy to implement. On the other hand, negation as failure must not be identified with, for example, classical or intuitionistic negation, and its exact logical meaning is quite intricate. The survey articles Apt and Bol [1994] and Shepherdson [1992] are dedicated to the logical environment of negation in logic programming and are a good guide to the relevant literature in this field. Jäger [1989] is concerned with the treatment of negative information by means of so called default operators and axiomatic extensions.

Our article focuses on the interpretation of logic programs (with negation) as deductive systems and provides a natural reconstruction of logic programming in terms of traditional proof theory. By following this line we can exploit the close interplay between proof search and computation and can profit from the fact that proof theory gives more insight into the procedural behavior of logic programs than most model-theoretic approaches.

The paper consists of five major parts. We begin in Section 2 with introducing the basic syntactic and semantic notions. This is essentially a repetition of standard terminology including three- and four-valued structures.

The backbone of Section 3 is formed by the general theory of inductive definability for three- and four-valued structures plus the notions of *adequate structure* and *envelope generated by a logic program*. This machinery is used to introduce in the fastest possible way that part of model theory which will be needed later.

The aim of Section 4 is to set up deductive systems for logic programs. We introduce sequent calculi with additional program rules, consider their identity-free subsystems and prove cut-elimination for them. In addition they are shown to be sound and complete with respect to the semantics introduced before.

In Section 5 we study SLDNF-resolution. Starting point is the negation as failure rule which is carefully integrated into the resolution process. Modes and mode assignments are then introduced in order to specify the input/output behavior of logic programs. They provide a powerful tool for setting up large and natural syntactically definable classes of logic programs for which SLDNF-resolution is shown to be a sound and complete proof procedure.

Partiality is considered in the last section. We show how a simple syntactic transformation makes it possible to regard logic programs (with negation) as a system of closure conditions of simultaneous positive inductive definitions so that the proof theory of inductive definitions becomes immediately applicable to logic programming. This section concludes with a brief indication of the importance of induction principles for proving properties about logic programs and with presenting an adequate formal basis for such activities.

2. Basic notions

2.1. Syntactic framework

In the following we will deal with countable first order languages $\mathcal{L}$ with equality which consist of the following basic symbols:

1. Countably many free variables $(u, v, w, u_1, v_1, w_1, \ldots)$ and countably many bound variables $(x, y, z, x_1, y_1, z_1, \ldots)$;
2. one or more 0-ary function symbols (= constants) and an arbitrary countable number of function symbols of finite arities greater than 0;
3. the symbols $=$ for equality and $\neq$ for inequality;
4. countably many relation symbols $(R, S, T, R_1, S_1, T_1, \ldots)$ of every finite arity greater than 0;
5. the symbol $\sim$ for the formation of complementary relations;
6. the propositional constants $\top$ and $\bot$, the propositional connectives $\vee$ and $\wedge$ and the quantifiers $\exists$ and $\forall$.

As auxiliary symbols we have parentheses and commas. To simplify the notation we do not denote the equality and inequality symbols as relation symbols. Apart from the function and relation symbols, the basic vocabulary of all languages which we will consider is the same. Each of our first order languages is thus determined by its function and relation symbols.

The *terms* $(a, b, c, d, a_1, b_1, c_1, d_1, \ldots)$ of the language $\mathcal{L}$ are defined as usual. The *literals* $(L, M, L_1, M_1 \ldots)$ of $\mathcal{L}$ are all expressions $R(a_1, \ldots, a_n)$ and $\widetilde{R}(a_1, \ldots, a_n)$ so that R is an n-ary relation symbol of $\mathcal{L}$; the literals $R(a_1, \ldots, a_n)$ are called *positive*, and the literals $\widetilde{R}(a_1, \ldots, a_n)$ are called *negative*; the positive literals are sometimes also denoted as *atoms*. The *atomic formulas* of $\mathcal{L}$ are the literals plus $\top$, $\bot$, $(a = b)$ and $(a \neq b)$. The *formulas* $(A, B, C, A_1, B_1, C_1, \ldots)$ of $\mathcal{L}$ are generated as follows:

1. If A is an atomic formula of $\mathcal{L}$, then A is an $\mathcal{L}$ formula.
2. If A and B are $\mathcal{L}$ formulas, then $(A \vee B)$ and $(A \wedge B)$ are $\mathcal{L}$ formulas.
3. If $A(u)$ is an $\mathcal{L}$ formula, then $\exists x A(x)$ and $\forall x A(x)$ are $\mathcal{L}$ formulas.

The vector notation $\vec{V}$ is used as shorthand for a finite string $V_1, \ldots, V_n$ whose length will be specified by the context. We write $A[\vec{u}]$ to indicate that all free variables of A come from the list $\vec{u}$; analogously, $a[\vec{u}]$ stands for a term with no variables different from $\vec{u}$. The formulas $A(\vec{u})$ and the terms $a(\vec{u})$ may contain other free variables besides $\vec{u}$. We denote the set of all free variables of the formula A by $var(A)$. The universal closure of a formula A is denoted by $\forall(A)$ and its existential closure by $\exists(A)$.

So far we have no negation for $\mathcal{L}$ formulas. However, it can be easily introduced by means of the complementary relations, the law of double negation and de Morgan's laws. The negation $\neg A$ of an $\mathcal{L}$ formula A is inductively defined as follows:

1. If R is an n-ary relation symbol of $\mathcal{L}$, then we set

$$\neg R(\vec{a}) := \widetilde{R}(\vec{a}) \qquad \text{and} \qquad \neg \widetilde{R}(\vec{a}) := R(\vec{a}).$$

2. For the other formulas we have

$$\begin{aligned} \neg\top &:= \bot, & \neg\bot &:= \top, \\ \neg(a = b) &:= (a \neq b), & \neg(a \neq b) &:= (a = b), \\ \neg(A \vee B) &:= (\neg A \wedge \neg B), & \neg(A \wedge B) &:= (\neg A \vee \neg B), \\ \neg\exists x A(x) &:= \forall x \neg A(x), & \neg\forall x A(x) &:= \exists x \neg A(x). \end{aligned}$$

Logical implication $(A \to B)$ and logical equivalence $(A \leftrightarrow B)$ are defined as usual. In the following we shall omit parentheses whenever the meaning is evident from the context. The complexity of formulas will often be measured in terms of their rank.

2.1.1. Definition. The *rank* $rn(A)$ of an $\mathcal{L}$ formula A is inductively defined as follows:

1. If A is an atomic formula, then $rn(A) := 0$.
2. If A is a formula $(B \vee C)$ or $(B \wedge C)$ so that $rn(B) = m$ and $rn(C) = n$, then $rn(A) := \max(m, n) + 1$.
3. If A is a formula $\exists x B(x)$ or $\forall x B(x)$ so that $rn(B(u)) = n$, then $rn(A) := n+1$.

Terms and formulas without free variables are called *closed*. The *equality formulas* $(E, E_1, \ldots)$ of $\mathcal{L}$ are the $\mathcal{L}$ formulas which do not contain relation symbols; the

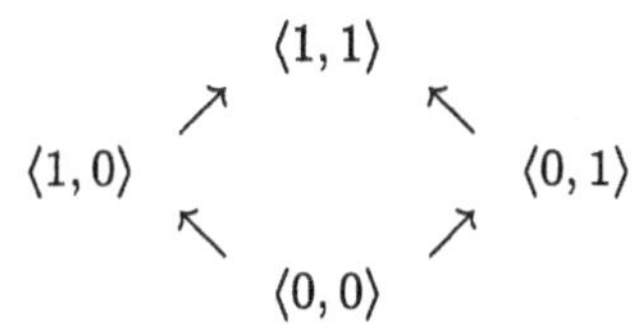

Figure 1: The information ordering on FOUR.

positive formulas of $\mathcal{L}$ are the $\mathcal{L}$ formulas which do not contain negative literals. Observe, however, that equations $(a = b)$ and inequations $(a \neq b)$ are not considered as literals in our terminology so that positive formulas may contain equations and inequations.

Following Shepherdson [1988], a language $\mathcal{L}$ is called *finite* if its set of function symbols is finite, otherwise it is called *infinite*. Thus finite languages with at least one function symbol of positive arity have an infinite number of closed terms.[1] The *Herbrand universe* $\mathsf{U}_{\mathcal{L}}$ of $\mathcal{L}$ is the collection of all closed terms of $\mathcal{L}$. By our assumptions on $\mathcal{L}$ we know that $\mathsf{U}_{\mathcal{L}}$ contains at least one element.

2.2. Two-valued, three-valued and four-valued structures

Classical logic just employs two truth values **t** (true) and **f** (false). On the other hand, recent research in logic programming indicates that a third truth value **u** (undefined) and a fourth truth value **c** (contradictory) have their natural place (cf. e.g. Lassez and Maher [1985], Mycroft [1984], Fitting [1985] and Kunen [1987, 1989]).

We follow the presentation of Fitting [1991]; similar approaches are due to Belnap [1977] and Ginsberg [1987]. The set of truth values is the set FOUR $:= \{0,1\} \times \{0,1\}$. If $\langle \mathsf{x}, \mathsf{y} \rangle \in$ FOUR is assigned to some statement A, then x represents the degree of evidence for A and y the degree of evidence against A. The usual truth values can be embedded into this framework by setting $\mathbf{t} := \langle 1, 0 \rangle$ and $\mathbf{f} := \langle 0, 1 \rangle$; the third truth value $\mathbf{u}$ is represented by $\langle 0, 0 \rangle$ and $\mathbf{c}$ is $\langle 1, 1 \rangle$. On FOUR a binary relation is defined by

$$\langle \mathsf{x}_1, \mathsf{y}_1 \rangle \sqsubseteq \langle \mathsf{x}_2, \mathsf{y}_2 \rangle \quad :\Longleftrightarrow \quad \mathsf{x}_1 \leq \mathsf{x}_2 \quad \text{and} \quad \mathsf{y}_1 \leq \mathsf{y}_2$$

where the relation $\leq$ on the right hand side is the usual ordering relation of the natural numbers (see Fig. 1). This partial ordering is sometimes denoted as *information-ordering*.

On FOUR one defines the following operations which will be used below to

[1]Hence "finite" refers to the number of function symbols and not to the number of closed terms.

interpret the logical connectives:

$$
\begin{aligned}
-\langle \mathsf{x}, \mathsf{y} \rangle &:= \langle \mathsf{y}, \mathsf{x} \rangle, \\
\langle \mathsf{x}_1, \mathsf{y}_1 \rangle + \langle \mathsf{x}_2, \mathsf{y}_2 \rangle &:= \langle \max(\mathsf{x}_1, \mathsf{x}_2), \min(\mathsf{y}_1, \mathsf{y}_2) \rangle, \\
\langle \mathsf{x}_1, \mathsf{y}_1 \rangle \cdot \langle \mathsf{x}_2, \mathsf{y}_2 \rangle &:= \langle \min(\mathsf{x}_1, \mathsf{x}_2), \max(\mathsf{y}_1, \mathsf{y}_2) \rangle, \\
\textstyle\sum_{i \in I} \langle \mathsf{x}_i, \mathsf{y}_i \rangle &:= \langle \max\{\mathsf{x}_i : i \in I\}, \min\{\mathsf{y}_i : i \in I\} \rangle, \\
\textstyle\prod_{i \in I} \langle \mathsf{x}_i, \mathsf{y}_i \rangle &:= \langle \min\{\mathsf{x}_i : i \in I\}, \max\{\mathsf{y}_i : i \in I\} \rangle.
\end{aligned}
$$

Observe that the sets $\{\mathbf{t}, \mathbf{f}\}$, $\{\mathbf{t}, \mathbf{f}, \mathbf{c}\}$ and $\{\mathbf{t}, \mathbf{f}, \mathbf{u}\}$ are closed under these operations. This is not the case for the limit of elements of FOUR which is defined by taking the pointwise maxima:

$$
\lim_{i \in I} \langle \mathsf{x}_i, \mathsf{y}_i \rangle := \langle \max\{\mathsf{x}_i : i \in I\}, \max\{\mathsf{y}_i : i \in I\} \rangle.
$$

It is clear, however, that all these operations are monotone on FOUR with respect to the relation $\sqsubseteq$.

2.2.1. Definition.

1. A *four-valued structure* $\mathfrak{M}$ for $\mathcal{L}$ consists of a non-empty domain $|\mathfrak{M}|$ together with assignments $\mathfrak{M}(f)$ and $\mathfrak{M}(R)$ to all function symbols f and relation symbols R of $\mathcal{L}$ so that
   (a) $\mathfrak{M}(f)$ is an n-ary function from $|\mathfrak{M}|$ to $|\mathfrak{M}|$ if f is n-ary,
   (b) $\mathfrak{M}(R)$ is an n-ary function from $|\mathfrak{M}|$ to FOUR if R is n-ary.
2. An *upper three-valued structure* for $\mathcal{L}$ is a four-valued structure for $\mathcal{L}$ so that the functions $\mathfrak{M}(R)$ do not take the value $\mathbf{u}$ for any relation symbol R of $\mathcal{L}$.
3. A *lower three-valued structure* for $\mathcal{L}$ is a four-valued structure for $\mathcal{L}$ so that the functions $\mathfrak{M}(R)$ do not take the value $\mathbf{c}$ for any relation symbol R of $\mathcal{L}$.
4. A *two-valued structure* for $\mathcal{L}$ is a four-valued structure for $\mathcal{L}$ so that the functions $\mathfrak{M}(R)$ do not take the values $\mathbf{c}$ and $\mathbf{u}$ for any relation symbol R of $\mathcal{L}$.

For a four-valued $\mathcal{L}$ structure $\mathfrak{M}$ one introduces the language $\mathcal{L}[\mathfrak{M}]$ by adding to $\mathcal{L}$ new constants $\overline{m}$ for all $m \in |\mathfrak{M}|$. Yet in order to simplify notation we often write $A[m_1, \ldots, m_n]$ instead of $A[\overline{m}_1, \ldots, \overline{m}_n]$. The value of each closed expression of $\mathcal{L}[\mathfrak{M}]$ is now inductively defined as follows:

2.2.2. Definition. Let $\mathfrak{M}$ be a four-valued $\mathcal{L}$ structure. We assign to each closed term a and closed formula A of $\mathcal{L}[\mathfrak{M}]$ a value $\mathfrak{M}(a) \in |\mathfrak{M}|$ and a value $\mathfrak{M}(A) \in$ FOUR.

1. If a is the term $\overline{m}$ for some element m of $|\mathfrak{M}|$, then $\mathfrak{M}(a) := m$.
2. If a is the term $f(a_1, \ldots, a_n)$ for some n-ary function symbol f and terms $a_1, \ldots, a_n$, then $\mathfrak{M}(a) := \mathfrak{M}(f)(\mathfrak{M}(a_1), \ldots \mathfrak{M}(a_n))$.

3. If A is the formula $\top$, then $\mathfrak{M}(A) := \mathbf{t}$; if A is the formula $(a = b)$, then

$$\mathfrak{M}(A) := \begin{cases} \mathbf{t}, & \text{if } \mathfrak{M}(a) = \mathfrak{M}(b), \\ \mathbf{f}, & \text{if } \mathfrak{M}(a) \neq \mathfrak{M}(b). \end{cases}$$

4. If A is the formula $R(a_1, \ldots, a_n)$ for some n-ary relation symbol R and terms $a_1, \ldots, a_n$, then $\mathfrak{M}(A) := \mathfrak{M}(R)(\mathfrak{M}(a_1), \ldots, \mathfrak{M}(a_n))$.
5. If A is the formula $\neg B$ and B the formula $\top$, an equation or a positive literal, then $\mathfrak{M}(A) := -\mathfrak{M}(B)$.
6. If A is a formula $(B \vee C)$, then $\mathfrak{M}(A) := \mathfrak{M}(B) + \mathfrak{M}(C)$.
7. If A is a formula $(B \wedge C)$, then $\mathfrak{M}(A) := \mathfrak{M}(B) \cdot \mathfrak{M}(C)$.
8. If A is a formula $\exists x B(x)$, then $\mathfrak{M}(A) := \sum_{m \in |\mathfrak{M}|} \mathfrak{M}(B(m))$.
9. If A is a formula $\forall x B(x)$, then $\mathfrak{M}(A) := \prod_{m \in |\mathfrak{M}|} \mathfrak{M}(B(m))$.

Obviously one has $\mathfrak{M}(A) \in \{\mathbf{t}, \mathbf{f}, \mathbf{c}\}$ for all upper three-valued $\mathfrak{M}$, $\mathfrak{M}(A) \in \{\mathbf{t}, \mathbf{f}, \mathbf{u}\}$ for all lower three-valued $\mathfrak{M}$ and $\mathfrak{M}(A) \in \{\mathbf{t}, \mathbf{f}\}$ for all two-valued $\mathfrak{M}$. Hence these three-valued and four-valued structures are natural generalizations of the two-valued case. Observe that equality is always handled as the usual two-valued identity. If A is an equality formula of $\mathcal{L}[\mathfrak{M}]$, then $\mathfrak{M}(A) = \mathbf{t}$ or $\mathfrak{M}(A) = \mathbf{f}$, also for three-valued and four-valued structures $\mathfrak{M}$.

If $\mathfrak{M}$ is a four-valued structure for $\mathcal{L}$ and A is a closed $\mathcal{L}[\mathfrak{M}]$ formula so that $\mathfrak{M}(A) = \langle \mathsf{x}, \mathsf{y} \rangle$, then we often write $\mathfrak{M}_{\mathrm{1st}}(A)$ for x and $\mathfrak{M}_{\mathrm{2nd}}(A)$ for y; hence we have $\mathfrak{M}(A) = \langle \mathfrak{M}_{\mathrm{1st}}(A), \mathfrak{M}_{\mathrm{2nd}}(A) \rangle$.

The *Herbrand structures* for $\mathcal{L}$ are the $\mathcal{L}$ structures so that the domain of these structures is the set $\mathsf{U}_{\mathcal{L}}$ and the function symbols have their obvious interpretations over $\mathsf{U}_{\mathcal{L}}$. Hence every Herbrand structure $\mathfrak{M}$ is characterized by the interpretation of its relation symbols. In the following we write $\mathfrak{U}_{\mathcal{L}}$ for the two-valued Herbrand structure for $\mathcal{L}$ which interprets each relation symbol as identically $\mathbf{f}$, and $3\mathfrak{U}_{\mathcal{L}}$ to denote the lower three-valued Herbrand structure for $\mathcal{L}$ which interprets each relation symbol as identically $\mathbf{u}$.

There is a natural notion of extension on the four-valued $\mathcal{L}$ structures which is obtained by lifting the above defined relation $\sqsubseteq$ on FOUR pointwise to the four-valued $\mathcal{L}$ structures:

2.2.3. Definition. Let $\mathfrak{M}$ and $\mathfrak{N}$ be four-valued structures for $\mathcal{L}$ which have the same universe and the same interpretations of the function symbols. $\mathfrak{N}$ is called an *extension* of $\mathfrak{M}$ if we have $\mathfrak{M}(R(\vec{m})) \sqsubseteq \mathfrak{N}(R(\vec{m}))$ for all relation symbols R of $\mathcal{L}$ and $\vec{m} \in |\mathfrak{M}|$. In this case we write $\mathfrak{M} \sqsubseteq \mathfrak{N}$.

The relation $\sqsubseteq$ is a partial ordering on the four-valued structures for $\mathcal{L}$, and $3\mathfrak{U}_{\mathcal{L}}$ is the least Herbrand structure with respect to this ordering. In addition it is easy to see that the $\mathcal{L}$ formulas are monotone with respect to $\sqsubseteq$ in the sense of the following remark.

2.2.4. Remark. Let $\mathfrak{M}$ and $\mathfrak{N}$ be four-valued $\mathcal{L}$ structures. Then we have for all closed $\mathcal{L}[\mathfrak{M}]$ formulas A:

$$\mathfrak{M} \sqsubseteq \mathfrak{N} \quad \Longrightarrow \quad \mathfrak{M}(A) \sqsubseteq \mathfrak{N}(A).$$

As a special case this means that as soon as an $\mathcal{L}[\mathfrak{M}]$ formula has obtained a value **t** or **f** in a lower three-valued structure $\mathfrak{M}$, it will keep this value in all lower three-valued extensions of $\mathfrak{M}$.

2.3. Four-valued versus two-valued structures

In this section we introduce the extension $\mathcal{L}^\sharp$ of a first order language $\mathcal{L}$ and show that the four-valued structures for $\mathcal{L}$ can be identified with two-valued structures for $\mathcal{L}^\sharp$.

2.3.1. Definition. Let $\mathcal{L}^\sharp$ be the first order language which results from $\mathcal{L}$ by replacing each relation symbol R of $\mathcal{L}$ by two new independent relation symbols R^+ and R^-, which are of the same arity as R.

Hence each relation symbol R of $\mathcal{L}$ corresponds to a pair (R^+, R^-) of relation symbols of $\mathcal{L}^\sharp$. Four-valued interpretations for R can therefore be associated to two-valued interpretations for R^+ and R^-.

2.3.2. Definition. Let $\mathfrak{M}$ be a four-valued structure for $\mathcal{L}$ and $\mathfrak{N}$ a two-valued structure for $\mathcal{L}^\sharp$.

1. $\mathfrak{M}^\sharp$ is the two-valued $\mathcal{L}^\sharp$ structure which has the same universe as $\mathfrak{M}$ and agrees with $\mathfrak{M}$ on the interpretation of the function symbols; for relation symbols $R^\pm$ of $\mathcal{L}^\sharp$ and $\vec{m} \in |\mathfrak{M}|$ we set

$$\mathfrak{M}^\sharp(R^+)(\vec{m}) := \langle \mathfrak{M}_{1\text{st}}(R(\vec{m})), 1 - \mathfrak{M}_{1\text{st}}(R(\vec{m}))\rangle,$$
$$\mathfrak{M}^\sharp(R^-)(\vec{m}) := \langle \mathfrak{M}_{2\text{nd}}(R(\vec{m})), 1 - \mathfrak{M}_{2\text{nd}}(R(\vec{m}))\rangle.$$

2. $\mathfrak{N}^\diamond$ is the four-valued $\mathcal{L}$ structure which has the same universe as $\mathfrak{N}$ and agrees with $\mathfrak{N}$ on the interpretation of the function symbols; for relation symbols R of $\mathcal{L}$ and $\vec{m} \in |\mathfrak{N}|$ we set

$$\mathfrak{N}^\diamond(R)(\vec{m}) \;:=\; \langle \mathfrak{N}_{1\text{st}}(R^+(\vec{m})), \mathfrak{N}_{1\text{st}}(R^-(\vec{m}))\rangle.$$

The previous two constructions are inverse to each other in the strongest possible sense. We have for all four-valued $\mathcal{L}$ structures $\mathfrak{M}$ and all two-valued $\mathcal{L}^\sharp$ structures $\mathfrak{N}$ that $(\mathfrak{M}^\sharp)^\diamond = \mathfrak{M}$ and $(\mathfrak{N}^\diamond)^\sharp = \mathfrak{N}$. Hence it is perfectly legitimate to identify the four-valued structures for the language $\mathcal{L}$ with the two-valued structures for the extension $\mathcal{L}^\sharp$ of $\mathcal{L}$. In view of the following remark it is possible to identify the lower three-valued $\mathcal{L}$ structures with the two-valued $\mathcal{L}^\sharp$ structures which satisfy the uniqueness condition that all R^+ and R^- are interpreted as disjoint relations.

2.3.3. Remark. Let $\mathfrak{M}$ be a four-valued $\mathcal{L}$ structure. Then $\mathfrak{M}$ is upper three-valued if and only if $\mathfrak{M}^\sharp(R^+(\vec{m}) \vee R^-(\vec{m})) = \mathbf{t}$ for all relation symbols R of $\mathcal{L}$ and all $\vec{m} \in |\mathfrak{M}|$; analogously, $\mathfrak{M}$ is lower three-valued if and only if $\mathfrak{M}^\sharp(R^+(\vec{m}) \wedge R^-(\vec{m})) = \mathbf{f}$ for all relation symbols R of $\mathcal{L}$ and all $\vec{m} \in |\mathfrak{M}|$.

Based on the extension of the language $\mathcal{L}$ to the language $\mathcal{L}^\sharp$ we now translate every $\mathcal{L}$ formula A into $\mathcal{L}^\sharp$ formulas A^+ and A^- as follows:

1. If A is an atomic equality formula, then $A^+ := A$ and $A^- := \neg A$.
2. If A is of the form $R(\vec{a})$ for some n-ary relation symbol of $\mathcal{L}$, then $A^+ := R^+(\vec{a})$ and $A^- := R^-(\vec{a})$.
3. If A is of the form $\widetilde{R}(\vec{a})$ for some n-ary relation symbol of $\mathcal{L}$, then $A^+ := R^-(\vec{a})$ and $A^- := R^+(\vec{a})$.
4. If A is of the form $(B \vee C)$, then $A^+ := (B^+ \vee C^+)$ and $A^- := (B^- \wedge C^-)$.
5. If A is of the form $(B \wedge C)$, then $A^+ := (B^+ \wedge C^+)$ and $A^- := (B^- \vee C^-)$.
6. If A is of the form $\exists x B(x)$, then $A^+ := \exists x B^+(x)$ and $A^- := \forall x B^-(x)$.
7. If A is of the form $\forall x B(x)$, then $A^+ := \forall x B^+(x)$ and $A^- := \exists x B^-(x)$.

This means that the $\mathcal{L}^\sharp$ formula A^+ is obtained from the $\mathcal{L}$ formula A by changing all positive literals $R(\vec{a})$ in A into $R^+(\vec{a})$ and all negative literals $\widetilde{R}(\vec{a})$ in A into $R^-(\vec{a})$; A^- is obtained from A by replacing all positive literals $R(\vec{a})$ in $\neg A$ by $R^+(\vec{a})$ and all negative literals $\widetilde{R}(\vec{a})$ in $\neg A$ by $R^-(\vec{a})$. If there are axioms available which express that the formulas $R^-(\vec{a})$ are the negations of the formulas $R^+(\vec{a})$, then one may identify A^+ with A and A^- with $\neg A$.

2.3.4. Remark. If A is an $\mathcal{L}$ formula, then A^+ and A^- are positive $\mathcal{L}^\sharp$ formulas.

The following remark shows how the close connection between four-valued $\mathcal{L}$ structures and two-valued $\mathcal{L}^\sharp$ structures extends to arbitrary $\mathcal{L}$ formulas.

2.3.5. Remark. We have for all two-valued $\mathcal{L}^\sharp$ structures $\mathfrak{N}$ and all closed $\mathcal{L}[\mathfrak{N}^\diamond]$ formulas A that $\mathfrak{N}^\diamond(A) = \langle \mathfrak{N}_{1st}(A^+), \mathfrak{N}_{1st}(A^-) \rangle$.

2.4. Logical consequences

An $\mathcal{L}$ theory is a (possibly infinite) set of $\mathcal{L}$ formulas. By $\mathsf{Th} \vdash A$ we express that the formula A can be deduced from the theory Th by the usual axioms and rules of first order predicate logic.

If $\mathfrak{M}$ is a two-valued structure, $A[\vec{u}]$ is an $\mathcal{L}$ formula and Th an $\mathcal{L}$ theory, then we define as usual: $A[\vec{u}]$ is *valid* in $\mathfrak{M}$ if $\mathfrak{M}(A[\vec{m}]) = \mathbf{t}$ for all $\vec{m} \in |\mathfrak{M}|$. Then we call $\mathfrak{M}$ a *model* of $A[\vec{u}]$ and write $\mathfrak{M} \models A[\vec{u}]$. Th is *valid* in $\mathfrak{M}$ if all elements of Th are valid in $\mathfrak{M}$. Then we call $\mathfrak{M}$ a *model* of Th and write $\mathfrak{M} \models \mathsf{Th}$. $A[\vec{u}]$ is a *logical consequence* of Th if $A[\vec{u}]$ is valid in all models of Th. Then we write $\mathsf{Th} \models A[\vec{u}]$.

The usual completeness result for first order logic states that derivability is equivalent to logical consequence, i.e., that $\mathsf{Th} \vdash A$ if and only if $\mathsf{Th} \models A$ for all $\mathcal{L}$ theories Th and $\mathcal{L}$ formulas A.

This form of logical consequence is based on two-valued structures. Special forms of consequences of logic programs with respect to certain three-valued and four-valued structures will be introduced in Section 3.1.

2.5. Clark's equational theory

Unification plays a major role in practically all implementations of logic programming environments. In general the most simple form of unification is employed which treats two closed expressions as equal if and only if they are syntactically identical. The corresponding unification theorem goes back to Robinson [1965] and states the existence of an algorithm which for any two expressions produces an idempotent most general unifier if they are unifiable and otherwise reports the nonexistence of a unifier.

Space does not permit to go into details, and only the basic terminology can be repeated. An $\mathcal{L}$ *substitution* θ is a finite set $\{u_1/a_1, \ldots, u_n/a_n\}$ of bindings so that the $\mathcal{L}$ terms a_i are different from the variables u_i for $1 \leq i \leq n$ and u_i is different from u_j for $1 \leq i < j \leq n$. We shall use $\theta, \sigma, \tau, \theta_1, \sigma_1, \tau_1, \ldots$ for substitutions; the empty substitution is denoted by ε.

The *instance* $\mathcal{Z}\theta$ of an expression $\mathcal{Z}$ and a substitution $\theta = \{u_1/a_1, \ldots, u_n/a_n\}$ is the expression obtained from $\mathcal{Z}$ by simultaneously replacing each occurrence of the variable u_i in $\mathcal{Z}$ by the term a_i $(i = 1, \ldots, n)$. In addition, an expression $\mathcal{Z}_1$ is called a *variant* of the expression $\mathcal{Z}_2$ if there exist substitutions σ and τ so that $\mathcal{Z}_1\sigma = \mathcal{Z}_2$ and $\mathcal{Z}_2\tau = \mathcal{Z}_1$.

Let $\mathcal{S}$ be the set of equations $\{a_1 = b_1, \ldots, a_n = b_n\}$. A *unifier* of $\mathcal{S}$ is an $\mathcal{L}$ substitution θ with the property that $a_i\theta$ and $b_i\theta$ are identical for $1 \leq i \leq n$. This unifier is *most general* if for any other unifier σ of $\mathcal{S}$ there exists a substitution τ so that σ is the composition of θ and τ, i.e., $\sigma = \theta\tau$. A (most general) unifier of the two atoms $R(a_1, \ldots, a_n)$ and $R(b_1, \ldots, b_n)$ is a (most general) unifier of the set of equations $\{a_1 = b_1, \ldots, a_n = b_n\}$. For further unexplained notions we refer to Apt [1990], Doets [1994] and Lloyd [1987].

Clark's equational theory $CET_{\mathcal{L}}$ (cf. Clark [1978]) may be understood as the axiomatic counterpart of this form of unification. The theory $CET_{\mathcal{L}}$ depends on the language $\mathcal{L}$ and comprises the following equality axioms (E1) and (E2). First we have

(E1) $\neg(a_1 = b_1 \wedge \ldots \wedge a_n = b_n)$

for all $\mathcal{L}$ terms $a_1, \ldots, a_n, b_1, \ldots, b_n$ so that $\{a_1 = b_1, \ldots, a_n = b_n\}$ is not unifiable. The second group of axioms states

(E2) $(a_1 = b_1 \wedge \ldots \wedge a_n = b_n) \rightarrow c = d$

provided that $\{a_1 = b_1, \ldots, a_n = b_n\}$ is unifiable with a most general unifier θ and $c\theta$ and $d\theta$ are syntactically identical.

A four-valued structure $\mathfrak{M}$ is called an *equational structure* if the universal closures of the equality axioms (E1) and (E2) are true in $\mathfrak{M}$. $\mathfrak{U}_{\mathcal{L}}$ is an equational structure and is sometimes called the *standard model* of $CET_{\mathcal{L}}$. This is justified by the obvious fact that every model of $CET_{\mathcal{L}}$ contains an isomorphic copy of the standard model. $CET_{\mathcal{L}}$ does not contain equality assertions for relation symbols. Hence, for example, a formula of the form $a = b \wedge R(a) \to R(b)$ for some unary relation symbol R is not derivable from $CET_{\mathcal{L}}$.

A result of Mal'cev [1971] states that $CET_{\mathcal{L}}$ is complete if $\mathcal{L}$ is an infinite language. Observe, however, that $CET_{\mathcal{L}}$ is in general not complete for finite languages. Let $\mathcal{L}$, for example, be a language with a constant a and no other function symbols. Then $\mathfrak{U}_{\mathcal{L}}$ is a model of $\forall x(x = a)$ but $CET_{\mathcal{L}}$ does not prove this equality formula. In order to obtain completeness also for finite languages $\mathcal{L}$ one has to strengthen $CET_{\mathcal{L}}$ by the so called *domain closure axiom* $DCA_{\mathcal{L}}$,

$$DCA_{\mathcal{L}} := \forall x \bigvee_{f \in \mathcal{L}} \exists \vec{y}(x = f(\vec{y})),$$

which says that every element of the universe belongs to the range of some function symbol of $\mathcal{L}$. Then the following theorem follows for example from Mal'cev [1971] or Shepherdson [1988].

2.5.1. Theorem. *Let E be a closed equality formula of the language $\mathcal{L}$. Then we have the equivalence*

$$CET_{\mathcal{L}} \vdash E \iff \mathfrak{U}_{\mathcal{L}} \models E,$$

provided that $\mathcal{L}$ is infinite. On the other hand, if $\mathcal{L}$ is finite, then one can only show that

$$CET_{\mathcal{L}} + DCA_{\mathcal{L}} \vdash E \iff \mathfrak{U}_{\mathcal{L}} \models E.$$

2.6. Logic programs and their completions

Finally the stage is set for introducing the central object of this article: logic programs. What we simply call a logic program here is sometimes denoted as a normal or general logic program, in contrast to definite logic programs (cf. e.g. Apt [1990], Doets [1994] and Lloyd [1987]).

Goals $(G, H, G_1, H_1, \ldots)$ in the language $\mathcal{L}$ are finite (possibly empty) sequences of $\mathcal{L}$ literals. The empty goal is denoted by $\varnothing$. A *program clause* in $\mathcal{L}$ is an expression of the form

$$A :- G$$

so that A is a positive literal of $\mathcal{L}$. The atom A is the *head* and the sequence of literals G the *body* of the clause. If the body of a program clause is empty, we simply write A instead of $A :- \varnothing$. A *logic program* in $\mathcal{L}$ is a pair $(\mathcal{L}, \mathsf{PC})$ which consists of a first order language $\mathcal{L}$ and a finite set PC of program clauses in $\mathcal{L}$.

Suppose that P is a logic program, R is an n-ary relation symbol and that there are m clauses in P whose heads are of the form $R(\ldots)$ so that the ith clause is of the form

$$R(a_{i,1}[\vec{v}], \ldots, a_{i,n}[\vec{v}]) :- L_{i,1}[\vec{v}], \ldots, L_{i,k(i)}[\vec{v}]$$

and has $k(i)$ literals in its body. Then the *definition form* of R with respect to P is defined to be the formula

$$D_R[u_1, \ldots, u_n] \;:=\; \bigvee_{i=1}^{m} \exists \vec{x} \left(\bigwedge_{j=1}^{n} (u_j = a_{i,j}[\vec{x}]) \wedge \bigwedge_{j=1}^{k(i)} L_{i,j}[\vec{x}] \right).$$

The special cases $m = 0$ and $k(i) = 0$ are included by interpreting empty disjunctions as $\bot$ and empty conjunctions as $\top$.

In Section 4 we will introduce deductive systems for logic programs, and in these systems so called program rules are associated to the program clauses. However, from a declarative point of view a logic program P is often identified with the theory consisting of all formulas $\forall \vec{x}(D_R[\vec{x}] \to R(\vec{x}))$ so that each $D_R[\vec{u}]$ is the definition form of R with respect to P. Other schools in the model-theoretic approach to logic programming argue that the intended meaning of a logic program P is better reflected by the so called *Clark completion* of P, in which the implications of the previous formulas are replaced by equivalences (cf. Clark [1978]).

More formally, let P be a logic program in $\mathcal{L}$ and assume that the definition form of each relation symbol R of $\mathcal{L}$ is the formula $D_R[\vec{u}]$. Then we call

$$\forall \vec{x}(D_R[\vec{x}] \leftrightarrow R(\vec{x}))$$

the *completed definition* of R with respect to P. The *completion* of P is the $\mathcal{L}$ theory $comp(P)$ which consists of $CET_{\mathcal{L}}$ plus the completed definitions of all relation symbols of $\mathcal{L}$.

3. Some model-theoretic properties of logic programs

There are some central model-theoretic properties of logic programs which are crucial for our proof-theoretic approach, in particular from the point of view of providing a semantic platform and motivation of the following steps. We will now recall these results and present them in a form tailored for our later applications.

3.1. Adequate structures

We start with structures which are adequate to logic programs. Informally, adequate structures are structures which reflect the meaning of a logic program in the sense that the information content of the definition form is inherited to the corresponding relation.

3.1.1. Definition. Let P be a logic program in $\mathcal{L}$. A four-valued equational structure $\mathfrak{M}$ for $\mathcal{L}$ is called *adequate to* P if

$$\mathfrak{M}(D_R[\vec{m}]) \sqsubseteq \mathfrak{M}(R(\vec{m}))$$

for all $\vec{m} \in |\mathfrak{M}|$ and all relation symbols R of $\mathcal{L}$ plus their definition form $D_R[\vec{u}]$ with respect to P.

This definition implies that a two-valued equational structure $\mathfrak{M}$ for $\mathcal{L}$ is adequate to a logic program P if and only if $\mathfrak{M}$ is a model of $comp(P)$. Moreover, Remark 3.2.6 below describes the relationship between three-valued models of $comp(P)$ in the sense of Fitting [1985] and Kunen [1987] and three-valued structures which are adequate to P.

3.1.2. Definition. Let P be a logic program in $\mathcal{L}$ and $A[\vec{u}]$ an $\mathcal{L}$ formula.

1. $A[\vec{u}]$ is called a *4-adequate consequence* of P if $\mathfrak{M}_{1st}(A[\vec{m}]) = 1$ for all four-valued equational $\mathcal{L}$ structures $\mathfrak{M}$ which are adequate to P and all $\vec{m} \in |\mathfrak{M}|$. Then we write $P \models_4 A[\vec{u}]$.
2. $A[\vec{u}]$ is called an *upper consequence* of P if $\mathfrak{M}_{1st}(A[\vec{m}]) = 1$ for all upper three-valued equational $\mathcal{L}$ structures $\mathfrak{M}$ which are adequate to P and all $\vec{m} \in |\mathfrak{M}|$. Then we write $P \models_\triangle A[\vec{u}]$.
3. $A[\vec{u}]$ is called a *lower consequence* of P if $\mathfrak{M}_{1st}(A[\vec{m}]) = 1$ for all lower three-valued equational $\mathcal{L}$ structures $\mathfrak{M}$ which are adequate to P and all $\vec{m} \in |\mathfrak{M}|$. Then we write $P \models_\nabla A[\vec{u}]$.
4. $A[\vec{u}]$ is called a *2-adequate consequence* of P if $\mathfrak{M}_{1st}(A[\vec{m}]) = 1$ for all two-valued equational $\mathcal{L}$ structures $\mathfrak{M}$ which are adequate to P and all $\vec{m} \in |\mathfrak{M}|$. Then we write $P \models_2 A[\vec{u}]$.

Since the two-valued structures which are adequate to a logic program P agree with the models of $comp(P)$, it is obvious that a formula A is a 2-adequate consequence of P if and only if it is a logical consequence of the completion of P.

3.2. Envelopes generated by logic programs

A four-valued structure $\mathfrak{M}$ can be viewed as providing some partial information about the intended scope of interest, and a logic program P as a means of modifying this information $\mathfrak{M}$ to a new structure $\mathfrak{M}[P]$, which we call the P-envelope of $\mathfrak{M}$.

3.2.1. Definition. Let $\mathfrak{M}$ be a four-valued $\mathcal{L}$ structure and P a logic program in $\mathcal{L}$. Then the *P-envelope* $\mathfrak{M}[P]$ *of* $\mathfrak{M}$ is the $\mathcal{L}$ structure which has the same universe as $\mathfrak{M}$ and agrees with $\mathfrak{M}$ on the interpretation of the function symbols; if R is a relation symbol of $\mathcal{L}$ and $D_R[\vec{u}]$ its definition form with respect to P, then we set $\mathfrak{M}[P](R)(\vec{m}) := \mathfrak{M}(D_R[\vec{m}])$ for all $\vec{m} \in |\mathfrak{M}|$.

It follows from this definition that the P-envelope of a two-valued structure is two-valued and that of an upper or lower three-valued structure is upper or lower three-valued, respectively. In general it is not the case that $\mathfrak{M}[P]$ is an extension of $\mathfrak{M}$, but at least the following property is given.

3.2.2. Remark. Let $\mathfrak{M}$ and $\mathfrak{N}$ be four-valued $\mathcal{L}$ structures and assume that P is a logic program in $\mathcal{L}$. Then we have:

$$\mathfrak{M} \sqsubseteq \mathfrak{N} \quad \Longrightarrow \quad \mathfrak{M}[P] \sqsubseteq \mathfrak{N}[P].$$

Hence the formation of envelopes is monotone. Making use of envelopes, it is now an easy task to characterize those structures which are adequate to a logic program P.

3.2.3. Remark. Let $\mathfrak{M}$ be a four-valued equational structure for $\mathcal{L}$ and assume that P is a logic program in $\mathcal{L}$. Then $\mathfrak{M}$ is adequate to P if and only if $\mathfrak{M}[P] \sqsubseteq \mathfrak{M}$.

There is a close relationship between four-valued and lower three-valued adequate structures. Every four-valued structure which is adequate to a logic program P extends a lower three-valued structure which is invariant under the formation of its P-envelope and thus adequate to P by the previous remark:

3.2.4. Proposition. *Let P be a logic program in $\mathcal{L}$ and $\mathfrak{M}$ a four-valued $\mathcal{L}$ structure which is adequate to P. Then there exists a lower three-valued structure $\mathfrak{N}$ for $\mathcal{L}$ so that $\mathfrak{N} \sqsubseteq \mathfrak{M}$ and $\mathfrak{N}[P] = \mathfrak{N}$.*

Proof. Let K be the nonempty set of all lower three-valued structures for $\mathcal{L}$ which are extended by $\mathfrak{M}$. Then $(\mathsf{K}, \sqsubseteq)$ is a complete partial ordering. In addition, the operation which maps an element of K to its P-envelope, which belongs to K as well, is monotone according to Remark 3.2.2. Therefore there exists a structure $\mathfrak{N}$ as claimed in the assertion. □

3.2.5. Corollary. *We have for all logic programs P in $\mathcal{L}$ and for all $\mathcal{L}$ formulas that A is a 4-adequate consequence of P if and only if A is a lower consequence of P, i.e.,*

$$P \models_4 A \iff P \models_\nabla A.$$

Fitting [1985] and Kunen [1987] use slightly different definitions and introduce the notion of a three-valued model of the completion $comp(P)$ of a logic program P. Then it is obvious that one has the following correspondence:

3.2.6. Remark. Let P be a logic program in $\mathcal{L}$ and $\mathfrak{M}$ a lower three-valued structure for $\mathcal{L}$. Then $\mathfrak{M}$ is a three-valued model of the completion $comp(P)$ if and only if $\mathfrak{M}[P] = \mathfrak{M}$.

3.3. Least adequate structures

Standard techniques of the theory of inductive definitions, as presented, for example, in Moschovakis [1974], provide the means to show that all logic programs P have least (lower three-valued) structures which are adequate to P. These structures are generated by iterating the formation of P-envelopes through a sufficiently large initial segment of the ordinals.

If $\mathfrak{M}$ is a four-valued structure for $\mathcal{L}$ then $3\mathfrak{M}$ is the lower three-valued structure for $\mathcal{L}$ which has the same universe and the same interpretation of all function symbols as $\mathfrak{M}$ and interprets each relation symbol as identically $\mathbf{u}$. A family $(\mathfrak{M}_i : i \in I)$ of four-valued structures for $\mathcal{L}$ is *based on* a four-valued structure $\mathfrak{M}$ if $3\mathfrak{M}_i = 3\mathfrak{M}$ for all $i \in I$.

Now let $(\mathfrak{M}_i : i \in I)$ be a non-empty family of four-valued structures for $\mathcal{L}$ which are based on a four-valued $\mathcal{L}$ structure $\mathfrak{M}$. Then the *limit* $\bigsqcup_{i\in I} \mathfrak{M}_i$ of this family is the four-valued structure $\mathfrak{N}$ for $\mathcal{L}$ so that $3\mathfrak{N} = 3\mathfrak{M}$ and

$$\mathfrak{N}(R)(\vec{m}) := \lim_{i\in I} \mathfrak{M}_i(R)(\vec{m})$$

for all relation symbols R of $\mathcal{L}$ and $\vec{m} \in |\mathfrak{M}|$. This implies that the degree of evidence for (against) a positive literal $R(\vec{a})$ is 1 in $\bigsqcup_{i\in I} \mathfrak{M}_i$, if it is 1 in some $\mathfrak{M}_i$, and 0 otherwise.

3.3.1. Definition. Let P be a logic program in $\mathcal{L}$ and $\mathfrak{M}$ a four-valued $\mathcal{L}$ structure. Then we define by recursion on the ordinals the following four-valued structures for $\mathcal{L}$:

$$\mathfrak{IM}_P^0 := 3\mathfrak{M}, \qquad \mathfrak{IM}_P^{\alpha+1} := \mathfrak{IM}_P^{\alpha}[P], \qquad \mathfrak{IM}_P^{\lambda} := \bigsqcup_{\xi<\lambda} \mathfrak{IM}_P^{\xi}$$

for all limit ordinals λ. Then $\mathfrak{IM}_P$ is the limit of all $\mathfrak{IM}_P^{\xi}$ for $\xi \in On$; i.e.,

$$\mathfrak{IM}_P := \bigsqcup_{\xi\in On} \mathfrak{IM}_P^{\xi}.$$

If $\mathfrak{M}$ is the lower three-valued Herbrand structure $3\mathfrak{U}_{\mathcal{L}}$, then we simply write $\mathfrak{I}_P^{\alpha}$ and $\mathfrak{I}_P$ instead of $\mathfrak{IM}_P^{\alpha}$ and $\mathfrak{IM}_P$, respectively.

It follows that $\mathfrak{IM}_P^{\alpha} \sqsubseteq \mathfrak{IM}_P^{\beta}$ for $\alpha \leq \beta$ and that there exists an ordinal γ so that $\mathfrak{IM}_P^{\gamma} = \mathfrak{IM}_P$. The least such ordinal is called the *three-valued closure ordinal of P with respect to* $\mathfrak{M}$.

Further, standard results about inductive definability (cf. e.g. Blair [1982] and Moschovakis [1974]) show: (i) The three-valued closure ordinal of P with respect to $3\mathfrak{U}_{\mathcal{L}}$ is less than or equal to the first non-recursive ordinal ω_1^{CK} for all logic programs P; (ii) for every ordinal $\alpha \leq \omega_1^{CK}$ there exists a (definite) logic program P whose three-valued closure ordinal with respect to $3\mathfrak{U}_{\mathcal{L}}$ is α.

3.3.2. Example. This example shows that the three-valued closure ordinal of a logic program with respect to $3\mathfrak{U}_{\mathcal{L}}$ can be greater than ω. Let $\mathcal{L}$ be a language with a constant 0, a unary function symbol f and two relation symbols R and S. Let P be the logic program in $\mathcal{L}$ which consists of the following clauses:

$R(f(u)) :- R(u)$
$S(f(u)) :- S(u)$
$S(u) :- R(v)$

It is easy to see that the three-valued closure ordinal of P with respect to $3\mathfrak{U}_{\mathcal{L}}$ is $\omega + \omega$, i.e., $\mathfrak{I}_P^{\omega+\omega} = \mathfrak{I}_P$. One simply has to consider the definition forms $D_R[u]$ and $D_S[u]$ of R and S with respect to P:

$$\exists x\,(u = f(x) \wedge R(x)) \quad \text{and} \quad \exists x\,(u = f(x) \wedge S(x)) \vee \exists x\, R(x).$$

Trivial induction on the ordinals shows that the structures $\mathfrak{IM}_P^\alpha$ and the structure $\mathfrak{IM}_P$ are lower three-valued. Furthermore, if a four-valued $\mathcal{L}$ structure $\mathfrak{N}$ is adequate to P and $3\mathfrak{M} = 3\mathfrak{N}$, then $\mathfrak{IM}_P^\alpha \sqsubseteq \mathfrak{N}$ for all $\alpha \in On$. In view of Remark 3.2.3 we therefore obtain the following theorem.

3.3.3. Theorem. *Let P be a logic program in $\mathcal{L}$ and $\mathfrak{M}$ a four-valued structure for $\mathcal{L}$. Then $\mathfrak{IM}_P$ is a lower three-valued structure which is adequate to P. In addition, if $\mathfrak{N}$ is a four-valued structure for $\mathcal{L}$ which is adequate to P and satisfies $3\mathfrak{M} = 3\mathfrak{N}$, then we have $\mathfrak{IM}_P \sqsubseteq \mathfrak{N}$.*

This theorem implies, in particular, that the structures $\mathfrak{I}_P$ are the least four-valued Herbrand structures which are adequate to the logic programs P. Some simple reflections on the high logical complexity of the structures $\mathfrak{I}_P$, which is reflected by the fact that the three-valued closure ordinals may be as large as ω_1^{CK}, make it clear that the least adequate structures can only be of limited use for a procedural approach to logic programming. It follows, for example, that in general the collection of all closed formulas true in $\mathfrak{I}_P$ is not even first order definable. We agree with Kunen [1987] that the "procedural content" of a logic program P is better approached by the finite stages $(\mathfrak{I}_P^n : n < \omega)$ of $\mathfrak{I}_P$.

3.4. The finite stages of least adequate structures

This thesis is also supported by the following observations. Taking up an idea of Kunen [1987] and Shepherdson [1988] one assigns to each $\mathcal{L}$ formula A and natural number n an equality formula $E_P^n(A)$, which depends on the given logic program P.

3.4.1. Definition. Let P be a logic program in $\mathcal{L}$. For every $n < \omega$ and every $\mathcal{L}$ formula A we define the equality formula $E_P^n(A)$ by induction on n.

1. If A is an atomic equality formula, then $E_P^n(A) := A$.
2. If R is a relation symbol of $\mathcal{L}$ and $D_R[\vec{u}]$ the definition form of R with respect to P, then we set

$$E_P^0(R(\vec{a})) := \bot, \qquad E_P^0(\neg R(\vec{a})) := \bot,$$
$$E_P^{n+1}(R(\vec{a})) := E_P^n(D_R[\vec{a}]), \quad E_P^{n+1}(\neg R(\vec{a})) := E_P^n(\neg D_R[\vec{a}]).$$

3. The propositional connectives and quantifiers are dealt with in the obvious way:

$$E_P^n(A \vee B) := E_P^n(A) \vee E_P^n(B), \quad E_P^n(A \wedge B) := E_P^n(A) \wedge E_P^n(B),$$
$$E_P^n(\exists x A(x)) := \exists x E_P^n(A)(x), \qquad E_P^n(\forall x A(x)) := \forall x E_P^n(A)(x).$$

Now the lemma below reduces truth of A in the finite stages $\mathfrak{IM}_P^n$ of the lower three-valued structures $\mathfrak{IM}_P$ to two-valued validity of the formulas $E_P^n(A)$. It is essentially proved by main induction on n and side induction on the rank of the formulas involved.

3.4.2. Lemma. *Let P be a logic program in $\mathcal{L}$ and $\mathfrak{M}$ a four-valued $\mathcal{L}$ structure. Then we have for all closed $\mathcal{L}$ formulas A and all $n < \omega$:*

$$\mathfrak{IM}^n_P(A) = \mathbf{t} \iff \mathfrak{IM}^0_P \models E^n_P(A).$$

This lemma may be combined with Theorem 2.5.1. As a result we have a reduction of truth in the finite stages $\mathfrak{I}^n_P$ to purely equational reasoning.

3.4.3. Theorem. *Let P be a logic program in $\mathcal{L}$, A an $\mathcal{L}$ formula and $n < \omega$. Then we have the equivalence*

$$\mathfrak{I}^n_P(A) = \mathbf{t} \iff CET_{\mathcal{L}} \vdash E^n_P(A),$$

provided that $\mathcal{L}$ is infinite. On the other hand, if $\mathcal{L}$ is finite, then one only obtains that

$$\mathfrak{I}^n_P(A) = \mathbf{t} \iff CET_{\mathcal{L}} + DCA_{\mathcal{L}} \vdash E^n_P(A).$$

4. Deductive systems for logic programs

After the preceding semantic considerations we will now approach logic programming in a more deductive and procedural way. Traditionally, a logic program is often regarded as a set of axioms. Alternatively, however, it is also possible to replace this *programs-as-theories* interpretation of logic programs by a *programs-as-deductive-systems* paradigm (cf. e.g. Hallnäs and Schroeder-Heister [1990], Jäger [1994], Schroeder-Heister [1991], Stärk [1991,1994a]) so that one is conceptually closer to a procedural understanding of logic programming.

We begin with introducing a calculus $\mathcal{R}(P)$ for each logic program P, which will provide the widest framework for the following considerations. The systems $\mathcal{R}(P)$ are designed for a proof-theoretic treatment and form a link between the model theory of logic programming and very specific proof procedures (like SLDNF-resolution) suited for implementations. Later we will also study subsystems of the $\mathcal{R}(P)$ which arise naturally in the context of logic programming.

4.1. The calculi $\mathcal{R}(P)$

The following deduction systems are presented in a Tait-style manner. Accordingly, the axioms and derivation rules are formulated for finite sets of $\mathcal{L}$ formulas which are interpreted disjunctively. The capital Greek letters $\Gamma, \Delta, \Pi, \Sigma, \ldots$ (possibly with subscripts) denote finite sets of $\mathcal{L}$ formulas, and we write (for example) Γ, Δ, A, B for the union of Γ, Δ and $\{A, B\}$. Given a four-valued structure $\mathfrak{M}$ for $\mathcal{L}$, we sometimes simply write $\mathfrak{M}(\Gamma)$ for the truth value of the universal closure of the disjunction of the formulas in Γ according to $\mathfrak{M}$. We say that Γ is *valid* in $\mathfrak{M}$ if $\mathfrak{M}(\Gamma) = \mathbf{t}$.

The systems $\mathcal{R}(P)$ for logic programs P are extensions of the usual Tait calculus for predicate logic (cf. e.g. Tait [1968]) by adding equality axioms and so called *program rules* which take care of the program clauses in P. Altogether we have the following five classes of axioms and rules.

I. Logical axioms. For all atomic $\mathcal{L}$ formulas A:

(L1) $\Gamma,\ \neg A,\ A,$

(L2) $\Gamma,\ \top.$

The axioms (L1) are often called *identity axioms* . They will play an important role later.

II. Equality axioms. For all $\mathcal{L}$ terms $a_1, \ldots, a_n, b_1, \ldots, b_n$ so that the set of equations $\{a_1 = b_1, \ldots, a_n = b_n\}$ is not unifiable:

$$\text{(E1)} \qquad \Gamma,\ a_1 \neq b_1, \ldots,\ a_n \neq b_n.$$

For all $\mathcal{L}$ terms $a_1, \ldots, a_n, b_1, \ldots, b_n, c, d$ so that $\{a_1 = b_1, \ldots, a_n = b_n\}$ is unifiable with a most general unifier θ and $c\theta$ is syntactically identical to $d\theta$:

$$\text{(E2)} \qquad \Gamma,\ a_1 \neq b_1, \ldots,\ a_n \neq b_n,\ c = d.$$

III. Logical rules. For all $\mathcal{L}$ formulas A and B, all $\mathcal{L}$ terms a and all free variables u which do not occur in $\Gamma, \forall x A(x)$:

$$\frac{\Gamma,\ A}{\Gamma,\ A \vee B}\ (\vee 1), \qquad \frac{\Gamma,\ B}{\Gamma,\ A \vee B}\ (\vee 2), \qquad \frac{\Gamma,\ A \qquad \Gamma,\ B}{\Gamma,\ A \wedge B}\ (\wedge),$$

$$\frac{\Gamma,\ A(a)}{\Gamma,\ \exists x A(x)}\ (\exists), \qquad \frac{\Gamma,\ A(u)}{\Gamma,\ \forall x A(x)}\ (\forall).$$

IV. Cut rules. For all $\mathcal{L}$ formulas A:

$$\frac{\Gamma,\ A \qquad \Gamma,\ \neg A}{\Gamma}\ (cut).$$

The formulas A and $\neg A$ are called the *cut formulas* of this cut; the rank of a cut is the rank of its cut formulas.

V. Program rules for P. For every relation symbol R of $\mathcal{L}$ and its definition form $D_R[\vec{u}]$ with respect to the logic program P and all $\mathcal{L}$ terms $\vec{a}$ we have the following positive and negative program rules:

$$\frac{\Gamma,\ D_R[\vec{a}]}{\Gamma,\ R(\vec{a})}\ (+R), \qquad \frac{\Gamma,\ \neg D_R[\vec{a}]}{\Gamma,\ \neg R(\vec{a})}\ (-R).$$

One should emphasize that the program rules are impredicative in the sense that the rank of the main formula of the premise of such a rule is in general greater than the rank of the main formula of the corresponding conclusion.

Based on these axioms and rules of inference derivability in $\mathcal{R}(P)$ is introduced in the standard way. The notation $\mathcal{R}(P) \vdash_r^n \Gamma$ expresses that Γ is provable in $\mathcal{R}(P)$ by a proof whose length and cut complexity are bounded by n and r, respectively.

4.1.1. Definition. Let P be a logic program in $\mathcal{L}$ and Γ a finite set of $\mathcal{L}$ formulas. Then we define $\mathcal{R}(P) \vdash_r^n \Gamma$ for all $n, r < \omega$ by induction on n.

1. If Γ is an axiom of $\mathcal{R}(P)$, then we have $\mathcal{R}(P) \vdash_r^n \Gamma$ for all $n, r < \omega$.
2. If $\mathcal{R}(P) \vdash_r^{n_i} \Gamma_i$ and $n_i < n$ for every premise Γ_i of a logical rule, a program rule or a cut of $\mathcal{R}(P)$ whose rank is less than r, then we have $\mathcal{R}(P) \vdash_r^n \Gamma$ for the conclusion Γ of that rule.

We write $\mathcal{R}(P) \vdash_r \Gamma$ if there exists an $n < \omega$ so that $\mathcal{R}(P) \vdash_r^n \Gamma$. Similarly, $\mathcal{R}(P) \vdash \Gamma$ is written if there exists an $r < \omega$ so that $\mathcal{R}(P) \vdash_r \Gamma$.

4.1.2. Remark. Each $\mathcal{R}(P)$ is a deductive system which corresponds to the completion $comp(P)$ of the logic program P; i.e., we have for all logic programs P in $\mathcal{L}$ and all $\mathcal{L}$ formulas A:

$$\mathcal{R}(P) \vdash A \quad \Longleftrightarrow \quad comp(P) \models A.$$

The notation $\mathcal{R}(P) \vdash_0 \Gamma$ means that there exists a cut-free proof of Γ in $\mathcal{R}(P)$. Later we will say more about cut elimination in $\mathcal{R}(P)$ and related systems. For the moment we only mention a *partial cut elimination theorem* for $\mathcal{R}(P)$. Since the ranks of the main formulas of the logical and equality axioms and the program rules of $\mathcal{R}(P)$ are 0, it is proved by standard techniques, as presented for example in Girard [1987b], Schütte [1977] or Takeuti [1987].

4.1.3. Theorem. *We have for all logic programs P in $\mathcal{L}$ and all finite sets Γ of $\mathcal{L}$ formulas:*

$$\mathcal{R}(P) \vdash \Gamma \quad \Longrightarrow \quad \mathcal{R}(P) \vdash_1 \Gamma.$$

4.2. Identity-free derivations in $\mathcal{R}(P)$

Now we turn to the subsystems of the calculi $\mathcal{R}(P)$ which deal with identity-free derivations only. They are of great importance in connection with lower three-valued structures which are adequate to logic programs P and are also related to SLDNF-resolution.

If we consider the system $\mathcal{R}(P)$ under the aspect of lower three-valued logic, we make the following observation: (i) The identity axioms of $\mathcal{R}(P)$ are not valid in the sense of lower three-valued structures; (ii) the equality axioms of $\mathcal{R}(P)$ are valid in all lower three-valued equational structures; (iii) if every premise of a propositional rule, quantifier rule or cut rule of $\mathcal{R}(P)$ is valid in a lower three-valued structure $\mathfrak{M}$, then the conclusion of that rule is also valid in $\mathfrak{M}$. This means that — besides the identity axioms and the program rules — all axioms and rules of $\mathcal{R}(P)$ are correct in the lower three-valued sense.

Moreover, if $\mathfrak{M}$ is a lower three-valued structure which is adequate to a logic program P, then also the program rules are valid in $\mathfrak{M}$. Hence from the point of view of lower three-valued logic only the identity axioms create some problems. For this reason it is very pleasing that just deleting the identity axioms yields an interesting subsystem of $\mathcal{R}(P)$.

4.2.1. Definition. Let P be a logic program in $\mathcal{L}$ and Γ a finite set of $\mathcal{L}$ formulas. Then we define $\mathcal{R}(P) \Vdash_r^n \Gamma$ for all $n, r < \omega$ by induction on n.

1. If Γ is an axiom of $\mathcal{R}(P)$ but not an identity axiom, then we have $\mathcal{R}(P) \Vdash_r^n \Gamma$ for all $n, r < \omega$.
2. If $\mathcal{R}(P) \Vdash_r^{n_i} \Gamma_i$ and $n_i < n$ for every premise Γ_i of a logical rule, a program rule or a cut of $\mathcal{R}(P)$ whose rank is less than r, then we have $\mathcal{R}(P) \Vdash_r^n \Gamma$ for the conclusion Γ of that rule.

The shorthand notations $\mathcal{R}(P) \Vdash_r \Gamma$ and $\mathcal{R}(P) \Vdash \Gamma$ will be used in the obvious sense. Hence $\mathcal{R}(P) \Vdash \Gamma$ means that Γ is derivable in $\mathcal{R}(P)$ by a proof which does not make use of the identity-axioms.

4.2.2. Remark. The omission of the identity axioms does not affect the equality formulas. We have for all logic programs in $\mathcal{L}$ and all equality formulas A:

$$CET_{\mathcal{L}} \vdash A \quad \Longrightarrow \quad \mathcal{R}(P) \Vdash A.$$

One can immediately conclude that the identity-free derivations in $\mathcal{R}(P)$ are correct with respect to the lower three-valued structures which are adequate to P. In Stärk [1991] also the corresponding completeness result is proved. The approach presented there makes use of techniques similar to Schütte's deductive chains (see Schütte [1977]) and is dual to Girard's proof of the completeness of the cut-free rules of the sequent calculus in Girard [1987b]. Semantic cut elimination can be derived from the proof of the following completeness result.

4.2.3. Theorem. *Let P be a logic program in $\mathcal{L}$. Then we have for all $\mathcal{L}$ formulas A:*

$$\mathcal{R}(P) \Vdash A \quad \Longleftrightarrow \quad P \models_{\nabla} A.$$

It is possible to tie identity-free derivations in $\mathcal{R}(P)$ very closely to finitely iterated program envelopes of three-valued structures. Given a lower three-valued $\mathcal{L}$ structure $\mathfrak{M}$ and a logic program P in $\mathcal{L}$, we define for all $n < \omega$:

$$\mathfrak{M}_P^0 := \mathfrak{M} \qquad \text{and} \qquad \mathfrak{M}_P^{n+1} := \mathfrak{M}_P^n[P].$$

This (iterated) formation of envelopes is the semantic counterpart of the program rules. Therefore a formula proved in n steps is true in the n-times iterated envelope.

4.2.4. Theorem. *Let P be a logic program in $\mathcal{L}$ and $\mathfrak{M}$ a lower three-valued equational structure for $\mathcal{L}$ so that $\mathfrak{M} \sqsubseteq \mathfrak{M}[P]$. Then we have for all finite sets of $\mathcal{L}$ formulas and all $n, r < \omega$:*

$$\mathcal{R}(P) \Vdash_r^n \Gamma \quad \Longrightarrow \quad \mathfrak{M}_P^n(\Gamma) = \mathbf{t}.$$

This assertion is verified by simple induction on n. Together with Theorem 4.2.3 it implies the following corollary, which was first obtained by Kunen [1987] by a purely model-theoretic proof based on ultrapower constructions. The above theorem and its corollary are quite remarkable since the structures $\mathfrak{M}_P^n$ are in general not adequate to P (cf. Example 3.3.2).

4.2.5. Corollary. *Let P be a logic program in $\mathcal{L}$, A a closed $\mathcal{L}$ formula and $\mathfrak{M}$ a lower three-valued equational structure for $\mathcal{L}$ so that $\mathfrak{M} \sqsubseteq \mathfrak{M}[P]$. Then $P \models_\nabla A$ implies that there exists an $n < \omega$ so that $\mathfrak{M}^n_P(A) = \mathbf{t}$.*

In a next step we compare identity-free derivability in $\mathcal{R}(P)$ with truth in the finite stages of $\mathfrak{I}_P$. For this purpose we make use of some notions and results which have been presented at the end of Section 3.4.

4.2.6. Lemma. *Let P be a logic program in $\mathcal{L}$. Then we have for all $\mathcal{L}$ formulas A and all $n < \omega$:*

$$\mathcal{R}(P) \Vdash \neg E^n_P(A),\ A.$$

The proof of this lemma by main induction on n and side induction on the rank of A is a matter of routine. This relationship between the formulas A and $E^n_P(A)$ permits the integration of Theorem 3.4.3 into the context of $\mathcal{R}(P)$ and provides soundness and completeness with respect to the family $(\mathfrak{I}^n_P : n < \omega)$.

4.2.7. Theorem. *Let P be a logic program in $\mathcal{L}$ and A a closed $\mathcal{L}$ formula. Then we have the equivalence*

$$\sum_{n<\omega} \mathfrak{I}^n_P(A) = \mathbf{t} \quad \Longleftrightarrow \quad \mathcal{R}(P) \Vdash A,$$

provided that $\mathcal{L}$ is infinite. On the other hand, if $\mathcal{L}$ is finite, then one only obtains that

$$\sum_{n<\omega} \mathfrak{I}^n_P(A) = \mathbf{t} \quad \Longleftrightarrow \quad \mathcal{R}(P) \Vdash (DCA_{\mathcal{L}} \to A).$$

Proof. We assume first that $\mathcal{L}$ is infinite. If $\mathfrak{I}^n_P(A) = \mathbf{t}$ for some $n < \omega$, then we can apply Theorem 3.4.3 and Remark 4.2.2 and conclude that $\mathcal{R}(P) \Vdash E^n_P(A)$. The previous lemma and a cut yield the assertion. On the other hand, if we have $\mathcal{R}(P) \Vdash A$, then Theorem 4.2.4 implies $\mathfrak{I}^n_P(A) = \mathbf{t}$ for some $n < \omega$. In the second case $\mathcal{L}$ is finite. Then we have only to observe that the domain closure axiom $DCA_{\mathcal{L}}$ is valid in all Herbrand structures and can proceed as before. □

Theorem 4.2.3 and Theorem 4.2.7 provide a proof-theoretic approach to an interesting result which exhibits the close connections between statements with respect to the finite stages of $\mathfrak{I}_P$ and lower three-valued consequences of P. Its first part is originally due to Kunen and proved in Kunen [1987].

4.2.8. Corollary. *Let P be a logic program in $\mathcal{L}$ and A an $\mathcal{L}$ formula. Then we have the equivalence*

$$\sum_{n<\omega} \mathfrak{I}^n_P(A) = \mathbf{t} \quad \Longleftrightarrow \quad P \models_\nabla A,$$

provided that $\mathcal{L}$ is infinite. On the other hand, if $\mathcal{L}$ is finite, then one only obtains that

$$\sum_{n<\omega} \mathfrak{I}^n_P(A) = \mathbf{t} \quad \Longleftrightarrow \quad P \models_\nabla (DCA_{\mathcal{L}} \to A).$$

4.3. Cut elimination for identity-free derivations

We end this section with a theorem which states that all identity-free derivations in $\mathcal{R}(P)$ can be transformed in identity-free derivations without cuts.

4.3.1. Theorem. *We have for all logic programs P and finite sets Γ of $\mathcal{L}$ formulas:*

$$\mathcal{R}(P) \Vdash \Gamma \quad \Longrightarrow \quad \mathcal{R}(P) \Vdash_0 \Gamma.$$

Proof. This proof is based on the following idea: We interpret $\mathcal{R}(P)$ with respect to its identity-free derivations into a ramified system $\mathcal{R}^\star(P)$, eliminate the cuts in $\mathcal{R}^\star(P)$ and then go back from $\mathcal{R}^\star(P)$ to $\mathcal{R}(P)$.

Each relation symbol R of $\mathcal{L}$ is replaced in $\mathcal{R}^\star(P)$ by an infinite family $R^0, R^1, \ldots$ of stratified relation symbols. Given a natural number n, we obtain the n-translation of the $\mathcal{L}$ formula A into the formula A^n of $\mathcal{R}^\star(P)$ by replacing all relation symbols R in A by R^n. Then it is easy to see that the definition of rank can be modified so that

$$rn(D_R^n(\vec{a})) < rn(R^{n+1}(\vec{a})) \tag{1}$$

for all relation symbols R, all terms $\vec{a}$ and all natural numbers n; the rank of disjunctions, conjunctions and quantified formulas is defined from its subformulas as before.

The ramified system $\mathcal{R}^\star(P)$ contains the logical axiom (L2), the equality axioms (E1) and (E2), the logical rules and the cut rules of $\mathcal{R}(P)$, all formulated for the language of $\mathcal{R}^\star(P)$. The identity axioms of $\mathcal{R}(P)$ are omitted, and instead of the program rules we have for all natural numbers m and n so that $m < n$ the following *stratified program rules*:

$$\frac{\Gamma,\ D_R^m[\vec{a}]}{\Gamma,\ R^n(\vec{a})} \qquad\qquad \frac{\Gamma,\ \neg D_R^m[\vec{a}]}{\Gamma,\ \neg R^n(\vec{a})}.$$

If Γ is a finite set of $\mathcal{L}$ formulas, then Δ is called n-suited for Γ if it results from Γ by replacing each occurrence of a relation symbols R in Γ by R^m for some m so that $n \leq m$. Then the following is proved by straightforward induction on n:

$$\mathcal{R}(P) \vdash^n \Gamma \text{ and } \Delta \text{ is } n\text{-suited for } \Gamma \quad \Longrightarrow \quad \mathcal{R}^\star(P) \vdash \Delta. \tag{2}$$

Because of (1) it is further possible by standard techniques to eliminate all cuts, i.e.,

$$\mathcal{R}^\star(P) \vdash \Gamma \quad \Longrightarrow \quad \mathcal{R}^\star(P) \vdash_0 \Gamma \tag{3}$$

for all finite sets Γ of formulas of $\mathcal{R}^\star(P)$. Note, that the equality axioms (E1) and (E2) are closed under cuts. Finally, if Γ is a finite set of formulas of $\mathcal{R}^\star(P)$, then we write Γ° for the set of $\mathcal{L}$ formulas which results from Γ by replacing all relation symbols R^m by R. Then an easy induction on the length of the derivation shows:

$$\mathcal{R}^\star(P) \vdash_0 \Gamma \quad \Longrightarrow \quad \mathcal{R}(P) \vdash_0 \Gamma^\circ. \tag{4}$$

From (2)–(4) we can conclude that all cuts can be eliminated from the identity-free derivations in $\mathcal{R}(P)$. □

A corresponding result for derivations which make use of identity axioms is in general not possible. However, there are syntactically definable classes of logic programs P (for example call-consistent and stratified programs) so that the systems $\mathcal{R}(P)$ permit the elimination of all cuts. Unfortunately space does not permit to say more about this interesting direction of research.

5. SLDNF-resolution

The well-known SLD-resolution[2] is a sound and complete proof procedure for definite logic programs (cf. e.g. Lloyd [1987]). Prolog as well as many other relational programming languages are based on some form of SLD-resolution. However, SLD-resolution does not permit to derive negative information. To overcome this restriction, the negation as failure rule NF has been introduced in Clark [1978].

5.1. Negation as failure

Intuitively, one wants a (closed) negative literal $\neg A$ to be derivable from a logic program P by means of NF if and only if all possible attempts to derive A from P fail after finitely many steps. This is often reflected as follows:

1. If the atom A finitely fails, then $\neg A$ succeeds.
2. If A succeeds, then $\neg A$ fails.

There exist various versions of SLDNF-resolution, many of which are discussed in the survey article Apt and Bol [1994] and in Shepherdson [1989]. For notational simplicity we restrict ourselves in the following to one particular (and rather general) form of SLDNF-resolution.

In order to have a compact form of notation let $*$ be a new symbol which does not belong to the language $\mathcal{L}$. *Goal forms* of $\mathcal{L}$ are all expressions of the form $G_1, *, G_2$ so that G_1 and G_2 are arbitrary goals in $\mathcal{L}$. Thus, goal forms are finite sequences of literals and the symbol $*$ in which $*$ occurs exactly once. If $G[*]$ is a goal form and H a goal, then $G[H]$ is the goal which results from $G[*]$ by replacing $*$ by H; $G[]$ is the goal which we obtain if we delete the symbol $*$ from $G[*]$.

The following definition is based on Kunen [1989]. For a given logic program P in $\mathcal{L}$, we define a set $\mathbf{R}(P)$ consisting of pairs formed from goals and substitutions in $\mathcal{L}$, and a set $\mathbf{F}(P)$ consisting of $\mathcal{L}$ goals. We write $G\ \mathbf{R}(P)\ \theta$ for $\langle G, \theta \rangle \in \mathbf{R}(P)$.

$G\ \mathbf{R}(P)\ \theta$ means that the goal G succeeds from the program P using SLDNF-resolution and returns answer θ. $G \in \mathbf{F}(P)$, on the other hand, means that G finitely fails from P using SLDNF-resolution.

5.1.1. Definition. Let P be a logic program in $\mathcal{L}$. Then the sets $\mathbf{R}(P)$ and $\mathbf{F}(P)$ are defined to be the least sets which are closed under the following conditions for all positive literals A.

(R0) $\varnothing\ \mathbf{R}(P)\ \varepsilon$.

[2]SLD stands for 'linear resolution for definite clauses with selection function'

(R1) Let $B :- H$ be a variant of a program clause of P which contains no variables from the goal $G[A]$, and let σ be a most general unifier of A and B. If $G[H]\sigma\ \mathbf{R}(P)\ \tau$ and $(var(A\sigma) \setminus var(G[H]\sigma)) \cap var(G[H]\sigma\tau) = \emptyset$, then $G[A]\ \mathbf{R}(P)\ \theta$ for the restriction θ of $\sigma\tau$ to the variables of $G[A]$.

(R2) If $A \in \mathbf{F}(P)$, A is closed and $G[]\ \mathbf{R}(P)\ \theta$, then $G[\neg A]\ \mathbf{R}(P)\ \theta$.

(F1) Let $\hat{P}(A)$ be the collection of all clauses of P which have a variant so that its head is unifiable with A. If for each element of $\hat{P}(A)$ there exists a variant $B :- H$ containing no variables from $G[A]$ so that $G[H]\sigma \in \mathbf{F}(P)$ for a most general unifier σ of A and B, then $G[A] \in \mathbf{F}(P)$.

(F2) If $A\ \mathbf{R}(P)\ \theta$ and there exists a substitution σ so that $A\theta\sigma$ is identical to A, then $G[\neg A] \in \mathbf{F}(P)$.

Condition (R0) says that the empty goal returns the identity substitution. In rule (R1), an atom A is selected from the goal G and a program clause of P is chosen. The clause is renamed so that it does not contain any variables from G. If the head of the clause is unifiable with A, then a new goal is derived from G by replacing the selected atom A with the body of the clause and applying the most general unifier σ to the new goal. If the derived goal returns an answer substitution τ, then the composition of σ and τ, restricted to the variables of the original goal G, is returned. Rule (F1) is dual to rule (R1). An atom A is selected from the goal G and if every derived goal from G using A fails, then G fails. The rules (R2) and (F2) switch from success to failure and vice versa. Usually, it is required in rule (F2) that the atom A is closed. In this article we work with an asymmetric version of SLDNF-resolution. The reason for choosing this variant of SLDNF-resolution is that it is exactly what is needed for the Completeness Theorem 5.3.5.

5.1.2. Example. This example explains why the condition on variables in rule (R1) of the previous definition is necessary. Let P be the following logic program:

$R(f(u))$
$S(u,v) :- R(u)$

Obviously, we have $R(u)\ \mathbf{R}(P)\ \{u/f(v)\}$. If we omit the variable condition, then we obtain $S(u,v)\ \mathbf{R}(P)\ \{u/f(v)\}$ as well. But the answer $\{u/f(v)\}$ is not most general for $S(u,v)$. A most general answer is $\{u/f(w)\}$.

The definition of the sets $\mathbf{R}(P)$ and $\mathbf{F}(P)$ is very much directed towards an actual implementation of SLDNF-resolution, and as a consequence we are willing to accept a rather unperspicuous formulation, which is partly caused by the fact that one has to deal with most general unifiers and variants of clauses. Later we will prove soundness and completeness of SLDNF-resolution. In order to simplify the proofs of these results, we are now going to present a theoretically more transparent approach to SLDNF-resolution by means of sets $\mathbf{Y}(P)$ and $\mathbf{N}(P)$. They correspond to the sequent calculus $\mathcal{S}_P$ of Buchholz [1992].

5.1.3. Definition. Let P be a logic program in $\mathcal{L}$. Then the sets $\mathbf{Y}(P)$ and $\mathbf{N}(P)$ are defined to be the least sets of $\mathcal{L}$ goals which are closed under the following conditions for all positive literals A and goal forms $G[*]$:

(Y0) $\varnothing \in \mathbf{Y}(P)$.

(Y1) Let $B :- H$ be a program clause from P and let σ be an $\mathcal{L}$ substitution so that $A = B\sigma$. If $G[H\sigma] \in \mathbf{Y}(P)$, then $G[A] \in \mathbf{Y}(P)$.

(Y2) Let G be the goal $\neg A_1, \ldots, \neg A_k$ which consists of negative literals only. If $A_i \in \mathbf{N}(P)$ for all $1 \leq i \leq k$, then $G \in \mathbf{Y}(P)$.

(N1) If $G\sigma[H\tau] \in \mathbf{N}(P)$ for all program clauses $B :- H$ of P and all $\mathcal{L}$ substitutions σ and τ so that $A\sigma = B\tau$, then $G[A] \in \mathbf{N}(P)$.

(N2) If $A \in \mathbf{Y}(P)$, then $G[\neg A] \in \mathbf{N}(P)$.

The sets $\mathbf{Y}(P)$ and $\mathbf{N}(P)$ have several natural closure properties. In the following remark we list three of those which will be needed later.

5.1.4. Remark. Let P be a logic program in $\mathcal{L}$ and σ an $\mathcal{L}$ substitution.

1. If $G \in \mathbf{Y}(P)$, then $G\sigma \in \mathbf{Y}(P)$.
2. If $G \in \mathbf{N}(P)$, then $G\sigma \in \mathbf{N}(P)$.
3. $G[H] \in \mathbf{Y}(P)$ if and only if $G[] \in \mathbf{Y}(P)$ and $H \in \mathbf{Y}(P)$.

The relationship between the sets $\mathbf{Y}(P)$ and $\mathbf{N}(P)$ and $\mathbf{R}(P)$ and $\mathbf{F}(P)$ is rather intricate. A first and easy observation shows that $\mathbf{Y}(P)$ and $\mathbf{N}(P)$ comprise the previously defined $\mathbf{R}(P)$ and $\mathbf{F}(P)$ in the sense of the following lemma, which is proved by induction on the definition of $\mathbf{R}(P)$ and $\mathbf{F}(P)$.

5.1.5. Lemma. *Let P be a logic program in $\mathcal{L}$ and G be a goal in $\mathcal{L}$.*

1. *If $G\ \mathbf{R}(P)\ \theta$, then $G\theta \in \mathbf{Y}(P)$.*
2. *If $G \in \mathbf{F}(P)$, then $G \in \mathbf{N}(P)$.*

The sets $\mathbf{Y}(P)$ and $\mathbf{N}(P)$, however, are not "equivalent" to the sets $\mathbf{R}(P)$ and $\mathbf{F}(P)$. The following example shows that the converse of the previous lemma is not true in general.

5.1.6. Example. Let $\mathcal{L}$ be a language with two constants a and b and two relation symbols R and S. Consider the logic program P in $\mathcal{L}$ which consists of the following clauses:

$R(a)$
$S(b) :- \neg R(u)$

Then we have $R(b) \in \mathbf{N}(P)$, $\neg R(b) \in \mathbf{Y}(P)$ and $S(b) \in \mathbf{Y}(P)$. But there is no substitution θ so that $S(v)\ \mathbf{R}(P)\ \theta$.

The sets $\mathbf{Y}(P)$ and $\mathbf{N}(P)$ are more general than the sets $\mathbf{R}(P)$ and $\mathbf{F}(P)$. Therefore we define for every logic program P two sets of goals, $\mathbf{S}^+(P)$ and $\mathbf{S}^-(P)$, so that SLDNF-resolution in the sense of $\mathbf{R}(P)$ and $\mathbf{F}(P)$ and "derivability" in the sense of $\mathbf{Y}(P)$ and $\mathbf{N}(P)$ do agree for goals from $\mathbf{S}^+(P)$ and $\mathbf{S}^-(P)$.

5.1.7. Definition. Let P be a logic program in $\mathcal{L}$. A pair $\langle \mathsf{S}^+, \mathsf{S}^- \rangle$ of sets of goals is called *safe* for P if the following conditions are satisfied:

(A1) If σ is an $\mathcal{L}$ substitution and $G \in \mathsf{S}^+$, then $G\sigma \in \mathsf{S}^+$.

(A2) Let $B :- H$ be a program clause of P and σ be an $\mathcal{L}$ substitution so that $A = B\sigma$. If $G[A] \in \mathsf{S}^+$, then $G[H\sigma] \in \mathsf{S}^+$.

(A3) Let G be the goal $\neg A_1, \ldots, \neg A_k$ which consists of negative literals only. If $G \in \mathsf{S}^+$, then A_i is closed and $A_i \in \mathsf{S}^-$ for all $1 \le i \le k$.

(B1) If σ is an $\mathcal{L}$ substitution and $G \in \mathsf{S}^-$, then $G\sigma \in \mathsf{S}^-$.

(B2) Let $B :- H$ be a program clause of P and σ be an $\mathcal{L}$ substitution so that $A = B\sigma$. If $G[A] \in \mathsf{S}^-$, then $G[H\sigma] \in \mathsf{S}^-$.

(B3) If $G[\neg A] \in \mathsf{S}^-$, then $A \in \mathsf{S}^+$.

Obviously, $\langle \emptyset, \emptyset \rangle$ is safe for P, and if $\langle \mathsf{S}_i^+, \mathsf{S}_i^- \rangle_{i \in I}$ is a family of safe pairs for P, then also the union $\langle \bigcup_{i \in I} \mathsf{S}_i^+, \bigcup_{i \in I} \mathsf{S}_i^- \rangle$ is safe for P. Therefore every program P has a largest safe pair, called $\langle \mathbf{S}^+(P), \mathbf{S}^-(P) \rangle$.

The complements of the sets $\mathbf{S}^+(P)$ and $\mathbf{S}^-(P)$ consist of goals which always flounder. They have the following inductive definition. We say that a goal of the form $G[H]\theta$ is a *resolvent* of the goal $G[A]$ with an input clause of P if there exists a variant $B :- H$ of a clause of P so that θ is a most general unifier of A and B.

5.1.8. Definition. Let P be a logic program in $\mathcal{L}$. Then the sets $\mathbf{F}\boldsymbol{\ell}^+(P)$ and $\mathbf{F}\boldsymbol{\ell}^-(P)$ are defined to be the least sets of $\mathcal{L}$ goals which are closed under the following conditions for all positive literals A:

1. Let G be the goal $\neg A_1, \ldots, \neg A_k$ which consists of negative literals only. If A_i is non-ground or $A_i \in \mathbf{F}\boldsymbol{\ell}^-(P)$ for at least one $1 \le i \le k$, then $G \in \mathbf{F}\boldsymbol{\ell}^+(P)$.
2. If $H \in \mathbf{F}\boldsymbol{\ell}^+(P)$ and H is a resolvent of G with an input clause from P, then $G \in \mathbf{F}\boldsymbol{\ell}^+(P)$.
3. If $A \in \mathbf{F}\boldsymbol{\ell}^+(P)$, then $G[\neg A] \in \mathbf{F}\boldsymbol{\ell}^-(P)$.
4. If $H \in \mathbf{F}\boldsymbol{\ell}^-(P)$ and H is a resolvent of G with an input clause from P, then $G \in \mathbf{F}\boldsymbol{\ell}^-(P)$.

It is easy to see that $\mathbf{S}^+(P)$ is the complement of $\mathbf{F}\boldsymbol{\ell}^+(P)$ and that $\mathbf{S}^-(P)$ is the complement of $\mathbf{F}\boldsymbol{\ell}^-(P)$.

5.1.9. Lemma. *Let P be a logic program in $\mathcal{L}$ and G be an $\mathcal{L}$ goal. Then we have:*

1. $G \in \mathbf{S}^+(P) \iff G \notin \mathbf{F}\boldsymbol{\ell}^+(P)$,
2. $G \in \mathbf{S}^-(P) \iff G \notin \mathbf{F}\boldsymbol{\ell}^-(P)$.

The next lemma is a lifting lemma. A "derivation" of a goal $G\theta$ in $\mathbf{Y}(P)$ is lifted down to an SLDNF-resolution proof of the goal G which returns a substitution σ so that $G\sigma$ is more general than $G\theta$. The lemma is proved by induction on the definition of $\mathbf{Y}(P)$ and $\mathbf{N}(P)$; cf. Stärk [1994b].

5.1.10. Lemma. *Let P be a logic program in $\mathcal{L}$ and G be an $\mathcal{L}$ goal.*

1. *If $G\theta \in \mathbf{Y}(P)$ and $G \in \mathbf{S}^+(P)$, then there exist $\mathcal{L}$ substitutions σ and τ so that $G\, \mathbf{R}(P)\, \sigma$ and $G\sigma\tau = G\theta$.*
2. *If $G \in \mathbf{N}(P)$ and $G \in \mathbf{S}^-(P)$, then $G \in \mathbf{F}(P)$.*

Is is important to observe that this lemma does not remain true if we restrict rule (F2) of Definition 5.1.1 to closed atoms A.

5.1.11. Example. Let $\mathcal{L}$ be the language with one unary function symbol f and one unary relation symbol R, and let P be the logic program which consists of the bodyless clause $R(f(u))$ only. Then we have $R(f(u)) \in \mathbf{Y}(P)$ and thus $\neg R(f(u)) \in \mathbf{N}(P)$. Since the goal $\neg R(f(u))$ belongs to $\mathbf{S}^-(P)$, the goal $\neg R(f(u))$ must be in $\mathbf{F}(P)$ according to the previous lemma. Indeed this is true, because $R(f(u))\, \mathbf{R}(P)\, \varepsilon$. If rule (F2) of Definition 5.1.1 were restricted to closed atoms, then we would not have $\neg R(f(u)) \in \mathbf{F}(P)$.

5.2. Mode assignments

Some results concerning the completeness of SLDNF-resolution are formulated and proved with reference to the sets $\mathbf{S}^+(P)$ and $\mathbf{S}^-(P)$. In general, however, it is not decidable whether a goal belongs to one of these sets. In fact, the set $\{\langle G, P\rangle : G \in \mathbf{S}^+(P)\}$ is Π^0_1-complete. In this section we will therefore introduce something like a syntactically decidable "approximation" of $\mathbf{S}^+(P)$ and $\mathbf{S}^-(P)$ which covers most practically relevant cases.

Modes $(\alpha, \beta, \gamma, \alpha_0, \beta_0, \gamma_0, \ldots)$ for n-ary relation symbols R are n-tuples $\langle \mathsf{x}_1, \ldots, \mathsf{x}_n\rangle$ so that $\mathsf{x}_i \in \{\mathsf{out}, \mathsf{log}, \mathsf{in}\}$. If $\mathsf{x}_i = \mathsf{in}$, then the ith argument of R is called an *input* argument. If $\mathsf{x}_i = \mathsf{out}$, then the ith argument of R is called an *output* argument. Otherwise, if $\mathsf{x}_i = \mathsf{log}$, then the ith argument of R is called a *logical* argument.

Let A be the atom $R(a_1, \ldots, a_n)$ and α be the mode $\langle \mathsf{x}_1, \ldots, \mathsf{x}_n\rangle$. Then the set of *input variables* of A with respect to α is the set of variables which occur in input arguments. The set of *output variables* is defined analogously:

$$in(A,\alpha) := in(\neg A, \alpha) := \bigcup\{\, var(a_i) : 1 \le i \le n,\ \mathsf{x}_i = \mathsf{in}\,\},$$
$$out(A,\alpha) := out(\neg A, \alpha) := \bigcup\{\, var(a_i) : 1 \le i \le n,\ \mathsf{x}_i = \mathsf{out}\,\}.$$

A *mode assignment for $\mathcal{L}$* is a function μ which assigns to every n-ary relation symbol R of $\mathcal{L}$ a set $\mu^+(R) \subseteq \{\mathsf{out}, \mathsf{log}, \mathsf{in}\}^n$ of so called positive modes for R and a set $\mu^-(R) \subseteq \{\mathsf{log}, \mathsf{in}\}^n$ of so called negative modes for R. In negative modes output arguments are not used since negative calls do not compute any output. Negative calls should be considered as tests. Both $\mu^+(R)$ or $\mu^-(R)$ can be empty. Positive modes are used in positive calls and negative modes are used in negative calls.

A mode assignment μ is extended in a natural way from the set of relation symbols to the set of all literals. If A is the atom $R(\vec{a})$, we set

$$\mu^+(A) := \mu^+(R), \qquad \mu^+(\neg A) := \mu^-(R),$$
$$\mu^-(A) := \mu^-(R), \qquad \mu^-(\neg A) := \mu^+(R).$$

A program must be written in such a way that in a computed answer the output variables are contained in the input variables. Thus in a certain way a mode assignment reflects the data flow of a logic program. Consider a clause

$$A :- B_1, \dots, B_m, \neg C_1, \dots, \neg C_n$$

and suppose that A is called with some input values. Then these values are passed to B_1, and B_1 computes some output values which are passed to B_2. This value passing is continued until one reaches $\neg C_1$. Finally, if the tests $C_1, \dots, C_n$ fail, the output values of B_m are returned to A and are the output values of A. Formally, this can be expressed by the following three definitions.

5.2.1. Definition. Let μ be a mode assignment for $\mathcal{L}$. We call a program clause $A :- L_1, \dots, L_n$ *μ-correct* if the following conditions are satisfied:

(C1) For all modes $\alpha \in \mu^+(A)$ there exist $\beta_1 \in \mu^+(L_1), \dots, \beta_n \in \mu^+(L_n)$ so that for all $1 \le i \le n$:

- (a) $out(A, \alpha) \subseteq in(A, \alpha) \cup \bigcup\{ out(L_j, \beta_j) : 1 \le j \le n, L_j \text{ positive} \}$
- (b) $in(L_i, \beta_i) \subseteq in(A, \alpha) \cup \bigcup\{ out(L_j, \beta_j) : 1 \le j < i, L_j \text{ positive} \}$ if L_i is positive,
- (c) $var(L_i) \subseteq in(A, \alpha) \cup \bigcup\{ out(L_j, \beta_j) : 1 \le j \le n, L_j \text{ positive} \}$ if L_i is negative,

(C2) For all modes $\alpha \in \mu^-(A)$ there exist $\beta_1 \in \mu^-(L_1), \dots, \beta_n \in \mu^-(L_n)$ so that $in(L_i, \beta_i) \subseteq in(A, \alpha)$ for all $1 \le i \le n$.

In this definition condition (C1) means that for all positive modes α of the head A there exist positive modes β_i of the literals L_i in the body so that we have: (a) Every output variable of A (with respect to α) is an input variable of A (with respect to α) or an output variable of some positive L_i (with respect to β_i); (b) every input variable of a positive L_i (with respect to β_i) is an input variable of A (with respect to α) or an output variable of some positive L_j (with respect to β_j) which is left of L_i; (c) every variable of a negative L_i is an input variable of A (with respect to α) or an output variable of some positive L_j (with respect to β_j). In particular, if the body of the clause is empty, then the clause is μ-correct if and only if $out(A, \alpha)$ is a subset of $in(A, \alpha)$ for all modes $\alpha \in \mu^+(A)$.

5.2.2. Definition. Let μ be a mode assignment for $\mathcal{L}$. Then a logic program P in $\mathcal{L}$ is called *μ-correct* if all clauses of P are μ-correct.

For clauses and programs the notion of correctness will be sufficient for our purposes. In the case of goals we need more and introduce the sets of μ-correct and μ-closed goals, thus obtaining an analogue to the sets $\mathbf{S}^+(P)$ and $\mathbf{S}^-(P)$; see Proposition 5.2.5 below.

5.2.3. Definition. Let μ be a mode assignment for $\mathcal{L}$. A goal $L_1, \ldots, L_n$ is called *μ-correct* if there exist modes $\beta_1 \in \mu^+(L_1), \ldots, \beta_n \in \mu^+(L_n)$ so that for all $1 \leq i \leq n$:

(G1) $in(L_i, \beta_i) \subseteq \bigcup\{\, out(L_j, \beta_j) : 1 \leq j < i, L_j \text{ positive}\,\}$ if L_i is positive,

(G2) $var(L_i) \subseteq \bigcup\{\, out(L_j, \beta_j) : 1 \leq j \leq n, L_j \text{ positive}\,\}$ if L_i is negative.

5.2.4. Definition. Let μ be a mode assignment for $\mathcal{L}$. A goal $L_1, \ldots, L_n$ is called *μ-closed* if there exist modes $\beta_1 \in \mu^-(L_1), \ldots, \beta_n \in \mu^-(L_n)$ so that $in(L_i, \beta_i) = \emptyset$ for all $1 \leq i \leq n$.

A goal which consists of a single atom A is μ-correct if and only if there exists a mode $\beta \in \mu^+(A)$ so that $in(A, \beta) = \emptyset$. If a goal G is μ-closed and contains the negative literal $\neg A$, then the goal A is μ-correct.

5.2.5. Proposition. *Let μ be a mode assignment and P be a μ-correct program in $\mathcal{L}$. Then we have:*

1. *If the goal G is μ-correct, then $G \in \mathbf{S}^+(P)$.*
2. *If the goal G is μ-closed, then $G \in \mathbf{S}^-(P)$.*

Proof. Let D^+ be the set of all μ-correct goals and D^- be the set of all μ-closed goals. Then it is easy to see that $\langle \mathsf{D}^+, \mathsf{D}^- \rangle$ is safe for P. Therefore we have $\mathsf{D}^+ \subseteq \mathbf{S}^+(P)$ and $\mathsf{D}^- \subseteq \mathbf{S}^-(P)$ by the definition of $\mathbf{S}^+(P)$ and $\mathbf{S}^-(P)$. □

It is decidable whether logic programs are correct and whether goals are correct or closed with respect to a given mode assignment μ. Hence we have gained a lot since the previous proposition enables us to make use of results based on the sets $\mathbf{S}^+(P)$ and $\mathbf{S}^-(P)$ provided that the syntactic criteria of correctness and closedness with respect to suitable mode assignments are satisfied. Moreover, there are well-known classes of logic programs which are special cases of programs that are correct with respect to some canonical mode assignment.

5.2.6. Example (Definite logic programs). Definite logic programs are programs which contain no negative literals. If we set

$$\mu^+(R) := \{\langle \mathsf{log}, \ldots, \mathsf{log} \rangle\} \quad \text{and} \quad \mu^-(R) := \{\langle \mathsf{log}, \ldots, \mathsf{log} \rangle\}$$

for every relation symbol R of $\mathcal{L}$, then it is easy to see that definite programs and definite goals are μ-correct.

5.2.7. Example (Allowed logic programs). Often a program clause is called allowed if every variable of the clause occurs also in a positive literal of its body; a program is called allowed if it consists of allowed clauses only. This class of programs is of interest since Cavedon and Lloyd [1989] and Kunen [1989] could show that SLDNF-resolution is complete for allowed programs. If we set

$$\mu^+(R) := \{\langle \mathsf{out}, \dots, \mathsf{out} \rangle\} \quad \text{and} \quad \mu^-(R) := \{\langle \mathsf{log}, \dots, \mathsf{log} \rangle\}$$

for every relation symbol R of $\mathcal{L}$, then it is easy to see that allowed programs and allowed clauses are μ-correct. Hence the completeness results due to Cavedon, Lloyd and Kunen also follow from our more general completeness results below.

In the following we present several examples which show the usefulness of mode assignments and indicate that the programs used in practice are always correct with respect to some mode assignment. We use the Prolog notation for lists. The constant `nil` is denoted by [] and the term $\mathtt{cons}(u, v)$ is written as $[u|v]$. A term of the form $\mathtt{cons}(u_1, \mathtt{cons}(u_2, \dots \mathtt{cons}(u_n, \mathtt{nil}) \dots))$ is written as $[u_1, u_2, \dots, u_n]$.

5.2.8. Example. The concatenation of two lists u_1 and u_2 to a list u_3 is described by the following definite program:

$\mathsf{append}([], u, u)$
$\mathsf{append}([v|u_1], u_2, [v|u_3]) :- \mathsf{append}(u_1, u_2, u_3)$

According to Example 5.2.6 these program clauses are correct with respect to the following trivial mode assignment μ:

$$\mu^+(\mathsf{append}) = \{\langle \mathsf{log}, \mathsf{log}, \mathsf{log} \rangle\}, \quad \mu^-(\mathsf{append}) = \{\langle \mathsf{log}, \mathsf{log}, \mathsf{log} \rangle\}.$$

For applications, however, this trivial mode-assignment is not sufficient. One is interested in having modes with as many output arguments as possible. The clauses of the append relation are also correct with respect to the following mode assignment μ:

$$\mu^+(\mathsf{append}) = \{\langle \mathsf{in}, \mathsf{in}, \mathsf{out} \rangle,\ \langle \mathsf{out}, \mathsf{out}, \mathsf{in} \rangle\}, \quad \mu^-(\mathsf{append}) = \{\langle \mathsf{log}, \mathsf{log}, \mathsf{log} \rangle\}.$$

Examples of μ-correct goals are (`a,b,c,d` are constants):

$$\mathsf{append}([\mathsf{a}, \mathsf{b}], [\mathsf{c}, \mathsf{d}], u) \quad \text{and} \quad \mathsf{append}(u_1, u_2, [\mathsf{a}, \mathsf{b}, \mathsf{c}, \mathsf{d}]).$$

The first goal is used to concatenate two lists, the second one to decompose a list.

5.2.9. Example. The actual reason to introduce mode assignments are programs with nested negation. The subset relation between lists u_1 and u_2 is described by the following program:

$\mathtt{subset}(u_1, u_2) :- \neg\, \mathtt{notsubset}(u_1, u_2)$
$\mathtt{notsubset}(u_1, u_2) :- \mathtt{member}(v, u_1),\ \neg\, \mathtt{member}(v, u_2)$
$\mathtt{member}(u, [u|v])$
$\mathtt{member}(u, [v|w]) :- \mathtt{member}(u, w)$

These clauses are correct with respect to the following mode assignment μ:

$$\begin{array}{lll} \mu^+(\texttt{subset}) = \{\langle \mathsf{in}, \mathsf{in}\rangle\}, & \mu^-(\texttt{subset}) = \{\langle \mathsf{in}, \mathsf{in}\rangle\}, \\ \mu^+(\texttt{notsubset}) = \{\langle \mathsf{in}, \mathsf{in}\rangle\}, & \mu^-(\texttt{notsubset}) = \{\langle \mathsf{log}, \mathsf{log}\rangle\}, \\ \mu^+(\texttt{member}) = \{\langle \mathsf{out}, \mathsf{in}\rangle\}, & \mu^-(\texttt{member}) = \{\langle \mathsf{log}, \mathsf{log}\rangle\}. \end{array}$$

Although the two clauses for the `member` relation are definite, one cannot take the canonical mode assignment of Example 5.2.6 for `member`.

5.3. Soundness and completeness of SLDNF-resolution

SLDNF-resolution can be justified in the following sense: If a goal $L_1, \ldots, L_n$ returns some answer θ using SLDNF-resolution, then the formula $L_1\theta \wedge \ldots \wedge L_n\theta$ is a lower three-valued consequence of the logic program; if a goal $L_1, \ldots, L_n$ fails finitely using SLDNF-resolution, then the formula $\neg\exists(L_1 \wedge \ldots \wedge L_n)$ is a lower three-valued consequence of the logic program. The proof of these two facts follows from Clark [1978].

5.3.1. Theorem (Soundness of SLDNF-resolution). *Let P be a logic program and G be the $\mathcal{L}$ goal $L_1, \ldots, L_n$. Then we have:*

1. *If $G\ \mathbf{R}(P)\ \theta$, then $P \models_\nabla \forall(L_1\theta \wedge \ldots \wedge L_n\theta)$.*
2. *If $G \in \mathbf{F}(P)$, then $P \models_\nabla \neg\exists(L_1 \wedge \ldots \wedge L_n)$.*

The converse of this theorem, the completeness of SLDNF-resolution, does not hold for arbitrary programs and goals. In the next example we present a logic program and a goal for which the converse of the first assertion of the previous theorem is not true.

5.3.2. Example. Let $\mathcal{L}$ be the language with a constant c and two unary relation symbols R and S. Let P be the logic program in $\mathcal{L}$ which consists of the following clauses:

$R(c)$
$S(u) :- R(u)$
$S(u) :- \neg R(u)$

Then $P \models_\nabla \forall x\, S(x)$ since $P \models_\nabla \forall x(x = c \vee x \neq c)$. But we do not have $S(u)\ \mathbf{R}(P)\ \varepsilon$; we only have $S(u)\ \mathbf{R}(P)\ \{u/c\}$. Observe that the goal $S(u)$ does not belong to $\mathbf{S}^+(P)$.

One can prove completeness of SLDNF-resolution, however, for arbitrary logic programs P and all goals which belong to $\mathbf{S}^+(P)$ or $\mathbf{S}^-(P)$. Below we will generalize a theorem of Kunen [1989] which he proved for allowed programs and allowed goals only. Instead of working in a language which contains infinitely many constants (as Kunen does), we consider structures which are generated from *all* terms of the language.

A four-valued structure $\mathfrak{M}$ for $\mathcal{L}$ is called a *free term structure for* $\mathcal{L}$ if it has the following properties: (i) The domain $|\mathfrak{M}|$ of $\mathfrak{M}$ is the set of all $\mathcal{L}$ terms, including those which contain variables; (ii) for all n-ary function symbols f of $\mathcal{L}$ and all $\mathcal{L}$ terms $a_1, \ldots, a_n$ we have $\mathfrak{M}(f)(a_1, \ldots, a_n) = f(a_1, \ldots, a_n)$.

The proof of the completeness of SLDNF-resolution is based on the following lemma which relates the finite stages $\mathfrak{IM}_P^n$ to SLDNF-resolution.

5.3.3. Lemma. *Let P be a logic program in $\mathcal{L}$ and $\mathfrak{M}$ a free term structure for $\mathcal{L}$. If G is the $\mathcal{L}$ goal $L_1, \ldots, L_k$, then we have for all natural numbers $n_1, \ldots, n_k$:*

1. *If $G \in \mathbf{S}^+(P)$ and $\mathfrak{IM}_P^{n_i}(L_i) = \mathbf{t}$ for all i so that $1 \le i \le k$, then $G \in \mathbf{Y}(P)$.*
2. *If $G \in \mathbf{S}^-(P)$ and if for each $\mathcal{L}$ substitution θ there exists an i so that $1 \le i \le k$ and $\mathfrak{IM}_P^{n_i}(L_i\theta) = \mathbf{f}$, then $G \in \mathbf{N}(P)$.*

Proof. For the sake of this proof we define the body-length of a program clause to be the number of literals in its body. Now let r be a natural number which is larger than the body-length of every clause from P. Such an upper bound always exists since logic programs are finite. Assertions *1* and *2* are proved by simultaneous induction on the natural number $r^{n_1} + \cdots + r^{n_k}$. There are four cases.

Case 1. $G \in \mathbf{S}^+(P)$; $\mathfrak{IM}_P^{n_i}(L_i) = \mathbf{t}$ for all $1 \le i \le k$; L_j is positive for some $1 \le j \le k$: We can assume that L_1 is positive. Since $\mathfrak{IM}_P^{n_1}(L_1) = \mathbf{t}$, n_1 is greater than 0 and by the definition of $\mathfrak{IM}_P^{n_1}$ there exists a clause $B :- M_1, \ldots, M_\ell$ from P and a substitution σ so that $L_1 = B\sigma$ and $\mathfrak{IM}_P^{n_1-1}(M_1\sigma \wedge \ldots \wedge M_\ell\sigma) = \mathbf{t}$. By (A2) of Definition 5.1.7 the goal $M_1\sigma, \ldots, M_\ell\sigma, L_2, \ldots, L_k$ belongs to $\mathbf{S}^+(P)$. Since

$$\ell \cdot r^{n_1-1} + r^{n_2} + \cdots + r^{n_k} < r^{n_1} + r^{n_2} + \cdots + r^{n_k},$$

we obtain by the induction hypothesis that the goal $M_1\sigma, \ldots, M_\ell\sigma, L_2, \ldots, L_k$ is in $\mathbf{Y}(P)$. By (Y1) we obtain that $G \in \mathbf{Y}(P)$.

Case 2. $G \in \mathbf{S}^-(P)$; for every $\mathcal{L}$ substitution θ it is $\mathfrak{IM}_P^{n_i}(L_i\theta) = \mathbf{f}$ for some $1 \le i \le k$; L_j is positive and $n_j > 0$ for some $1 \le j \le k$: We can assume that L_1 is positive and $0 < n_1$. Let $B :- M_1, \ldots, M_\ell$ be a clause from P and σ and τ be substitutions so that $L_1\sigma = B\tau$. We have to show that the goal $M_1\tau, \ldots, M_\ell\tau, L_2\sigma, \ldots, L_k\sigma$ is in $\mathbf{N}(P)$. Note that by (B1) and (B2) the goal $M_1\tau, \ldots, M_\ell\tau, L_2\sigma, \ldots, L_k\sigma$ belongs to $\mathbf{S}^-(P)$. Let θ be a substitution. By assumption, there exists an $1 \le i \le k$ so that $\mathfrak{IM}_P^{n_i}(L_i\sigma\theta) = \mathbf{f}$. If $i = 1$, then $\mathfrak{IM}_P^{n_1}(L_1\sigma\theta) = \mathbf{f}$ and, by the definition of the finite stages $\mathfrak{IM}_P^n$, there exists a $1 \le j \le \ell$ so that $\mathfrak{IM}_P^{n_1-1}(M_j\tau\theta) = \mathbf{f}$. Since

$$\ell \cdot r^{n_1-1} + r^{n_2} + \cdots + r^{n_k} < r^{n_1} + r^{n_2} + \cdots + r^{n_k},$$

we can apply the induction hypothesis and obtain that $M_1\tau, \ldots, M_\ell\tau, L_2\sigma, \ldots, L_k\sigma$ is in $\mathbf{N}(P)$. Since the clause $B :- M_1, \ldots, M_\ell$ and the substitutions σ and τ are chosen arbitrarily, we obtain that $G \in \mathbf{N}(P)$ by (N1).

Case 3. $G \in \mathbf{S}^+(P)$; $\mathfrak{IM}_P^{n_i}(L_i) = \mathbf{t}$ for all $1 \le i \le k$; L_i is negative for all $1 \le i \le k$: Then L_i is of the form $\neg A_i$ for all $1 \le i \le k$ and therefore $\mathfrak{IM}_P^{n_i}(A_i) = \mathbf{f}$.

Because of condition (A3) we know that A_i is closed and $A_i \in \mathbf{S}^-(P)$. Hence *Case 2* implies $A_i \in \mathbf{N}(P)$, and by (Y2) we obtain that $G \in \mathbf{Y}(P)$.

Case 4. $G \in \mathbf{S}^-(P)$; for every $\mathcal{L}$ substitution θ it is $\mathfrak{IM}_P^{n_i}(L_i\theta) = \mathbf{f}$ for some $1 \le i \le k$; $n_i = 0$ for all $1 \le i \le k$ so that L_i is positive: We apply the assumption to the identity substitution ε and obtain an $1 \le i \le k$ so that $\mathfrak{IM}_P^{n_i}(L_i) = \mathbf{f}$. The literal L_i must be negative, i.e., of the form $\neg A$. By (B3) we have $A \in \mathbf{S}^+(P)$ and $\mathfrak{IM}_P^{n_i}(A) = \mathbf{t}$. By *Case 1* we obtain that $A \in \mathbf{Y}(P)$. Therefore we have $G \in \mathbf{N}(P)$ because of (N2). □

Theorem 4.2.3, Theorem 4.2.4, Lemma 5.1.10 and the previous lemma yield the completeness of SLDNF-resolution for goals from $\mathbf{S}^+(P)$ and $\mathbf{S}^-(P)$.

5.3.4. Theorem (Completeness of SLDNF-resolution). *Let P be a logic program in $\mathcal{L}$, and let G be the $\mathcal{L}$ goal $L_1, \ldots, L_k$. Then we have:*

1. *If $G \in \mathbf{S}^+(P)$ and $P \models_\nabla \forall(L_1\theta \wedge \ldots \wedge L_k\theta)$, then there exist $\mathcal{L}$ substitutions σ and τ so that $G\,\mathbf{R}(P)\,\sigma$ and $G\sigma\tau = G\theta$.*
2. *If $G \in \mathbf{S}^-(P)$ and $P \models_\nabla \neg\exists(L_1 \wedge \ldots \wedge L_k)$, then $G \in \mathbf{F}(P)$.*
3. *If the atom A is in $\mathbf{S}^+(P)$ and $P \models_\nabla \exists x\, A(x)$, then there exists a term t so that $A(u)\,\mathbf{R}(P)\,\{u/t\}$.*
4. *If the atom A is in $\mathbf{S}^+(P)$ and in $\mathbf{S}^-(P)$ and $P \models_\nabla \exists(A) \vee \forall(\neg A)$, then either there exists an $\mathcal{L}$ substitution σ so that $A\,\mathbf{R}(P)\,\sigma$ or $A \in \mathbf{F}(P)$.*

Proof. Let $\mathfrak{M}$ be a free term structure for $\mathcal{L}$. To prove the first assertion we assume that $G \in \mathbf{S}^+(P)$ and $P \models_\nabla \forall(L_1\theta \wedge \ldots \wedge L_k\theta)$. By Theorem 4.2.3 there exists a natural number n so that $\mathcal{R}(P) \Vdash^n \forall(L_1\theta \wedge \ldots \wedge L_k\theta)$. Since $\mathfrak{IM}_P^0 \sqsubseteq \mathfrak{IM}_P^1$ and $\mathfrak{IM}_P^1 = \mathfrak{IM}_P^0[P]$, we obtain $\mathfrak{IM}_P^n(\forall(L_1\theta \wedge \ldots \wedge L_k\theta)) = \mathbf{t}$ by Theorem 4.2.4. From the previous lemma we conclude that $G\theta \in \mathbf{Y}(P)$. By Lemma 5.1.10 there exist substitutions σ and τ so that $G\,\mathbf{R}(P)\,\sigma$ and $G\sigma\tau = G\theta$.

Now we turn to the second assertion and assume that G is an element of $\mathbf{S}^-(P)$ and $P \models_\nabla \neg\exists(L_1 \wedge \ldots \wedge L_k)$. As in the first case there exists a natural number n so that $\mathfrak{IM}_P^n(\neg\exists(L_1 \wedge \ldots \wedge L_k)) = \mathbf{t}$. This means that $\mathfrak{IM}_P^n(\forall(\neg L_1 \vee \ldots \vee \neg L_k)) = \mathbf{f}$. Hence, for every substitution θ there exists an $1 \le i \le k$ so that $\mathfrak{IM}_P^n(L_i\theta) = \mathbf{f}$. By the previous lemma it follows that $G \in \mathbf{N}(P)$. By Lemma 5.1.10, we obtain that $G \in \mathbf{F}(P)$.

The third and the fourth assertion are proved in a similar way. □

Proposition 5.2.5 yields in addition the completeness of SLDNF-resolution for mode assignments μ.

5.3.5. Theorem. *Let P be a logic program in $\mathcal{L}$ which is correct with respect to a mode assignment μ and let G be the $\mathcal{L}$ goal $L_1, \ldots, L_k$. Then we have:*

1. *If the goal G is μ-correct and $P \models_\nabla \forall(L_1\theta \wedge \ldots \wedge L_k\theta)$, then there exist $\mathcal{L}$ substitutions σ and τ so that $G\,\mathbf{R}(P)\,\sigma$ and $G\sigma\tau = G\theta$.*
2. *If the goal G is μ-closed and $P \models_\nabla \neg\exists(L_1 \wedge \ldots \wedge L_k)$, then $G \in \mathbf{F}(P)$.*

3. *If* $P \models_{\nabla} \forall x \exists y\, R(x,y)$ *and* $\langle \mathsf{in}, \mathsf{out} \rangle \in \mu^{+}(R)$, *then for every closed term a there exists a closed term b so that* $R(a,v)\ \mathbf{R}(P)\ \{v/b\}$.
4. *If the atom A is* μ*-correct and* μ*-closed and* $P \models_{\nabla} \exists(A) \vee \forall(\neg A)$, *then either there exists an* $\mathcal{L}$ *substitution* σ *so that* $A\ \mathbf{R}(P)\ \sigma$ *or* $A \in \mathbf{F}(P)$.

6. Partiality in logic programming

The omission of the identity axioms has the effect of disconnecting a relation symbol R from its complement $\widetilde{R}$, and thus an adequate framework for discussing the procedural aspects of logic programs and for SLDNF-resolution is provided. However, the identity-free derivation in the calculi $\mathcal{R}(P)$ are extremely weak and sometimes considered unnatural. In addition, on the semantical side we have to deal with additional truth values in order to obtain a decent model theory for identity-free derivations.

Now we want to further the conceptual clarity and present an alternative approach to logic programming which is based on two-valued logic only. To this end we introduce a form of partiality into logic programming: We present the partial completion $comp^{\sharp}(P)$, the corresponding deductive system $\partial(P)$ and the inductive extension $ind^{\sharp}(P)$ of logic programs P.

These formalizations are also discussed in Jäger [1994] and Stärk [1996]. Furthermore, similar concepts are studied in Drabent and Martelli [1991] and Van Gelder and Schlipf [1993].

6.1. The partial completion of logic programs

The syntactic framework for defining the partial completion of logic programs in $\mathcal{L}$ is provided by the extension $\mathcal{L}^{\sharp}$ of $\mathcal{L}$ introduced in Section 2.3. Given this rich language, the partial completion of a logic program P just consists of closure conditions for the relations R^{+} and R^{-} plus some basic equality and freeness axioms.

6.1.1. Definition. Let P be a logic program in $\mathcal{L}$. Then the *partial completion* $comp^{\sharp}(P)$ of P is the $\mathcal{L}^{\sharp}$ theory which consists of the following axioms.

1. $CET_{\mathcal{L}}$ plus equality axioms for all relation symbols $R^{\pm}$ of $\mathcal{L}^{\sharp}$:
$$(a_1 = b_1 \wedge \ldots \wedge a_n = b_n \wedge R^{\pm}(a_1,\ldots,a_n)) \rightarrow R^{\pm}(b_1,\ldots,b_n).$$
2. For all relation symbols R of $\mathcal{L}$ and their definition forms $D_R[\vec{u}]$ with respect to P:
$$\forall \vec{x}(D_R^{+}[\vec{x}] \rightarrow R^{+}(\vec{x})) \quad \text{and} \quad \forall \vec{x}(D_R^{-}[\vec{x}] \rightarrow R^{-}(\vec{x})).$$

These formulas express that the relations R^{+} and R^{-} are closed with respect to the partial definition forms D_R^{+} and D_R^{-} of P. If $\mathfrak{M}$ is a model of the partial completion of a program P and if we know, in addition, that the relations $\mathfrak{M}(R^{-})$ are the (set-theoretic) complements of the relations $\mathfrak{M}(R^{+})$, then the $\mathfrak{M}(R^{+})$ describe fixed points of the system of inductive definitions which is associated to P. But in general we do not know whether $\mathfrak{M}(R^+)$ and $\mathfrak{M}(R^-)$ are complementary so that the prefix "partial" is in place. This is similar to the distinction between truth definitions and partial truth definitions as for example in Feferman [1991].

The equality axioms for the relation symbols of $\mathcal{L}^\sharp$ have to be added in the partial case since it is not possible to derive them from the other axioms. This is different from the completions $comp(P)$ of logic programs P and the calculi $\mathcal{R}(P)$ since there the equality axioms for relations can always be proved.

Because of the close relationship between the four-valued structures for $\mathcal{L}$ and the two-valued structures for $\mathcal{L}^\sharp$ and Remark 2.3.5 in particular, the following is obvious.

6.1.2. Remark. Let P be a logic program in $\mathcal{L}$ and $\mathfrak{M}$ a four-valued equational structure for $\mathcal{L}$. Then $\mathfrak{M}$ is adequate to P if and only if $\mathfrak{M}^\sharp$ is a (two-valued) model of $comp^\sharp(P)$.

Hence we obtain a theorem which reduces the 4-adequate consequences of a logic program P to logical consequences of the partial completion $comp^\sharp(P)$ of P; in view of Corollary 3.2.5 this is also true for lower consequences of P.

6.1.3. Theorem. *Let P be a logic program in $\mathcal{L}$ and A an $\mathcal{L}$ formula. Then we have the following two equivalences:*

$$P \models_4 A \iff comp^\sharp(P) \models A^+ \iff P \models_\nabla A.$$

The partial completion reduces to the Clark completion if we add the two further axioms $\forall\vec{x}(R^+(\vec{x}) \lor R^-(\vec{x}))$ and $\forall\vec{x}\neg(R^+(\vec{x}) \land R^-(\vec{x}))$ for all relation symbols R of $\mathcal{L}$. They express that each $R^-(\vec{x})$ is equivalent to the negation of $R^+(\vec{x})$.

The first of these axioms is a kind of totality assertion, stating that at least one of $R^+(\vec{a})$ or $R^-(\vec{a})$ has to be true. From the point of view of SLDNF-resolution this means that $R(\vec{a})$ succeeds or fails. However, this is generally not the case, so that this totality assertion is rejected.

The second axiom is a uniqueness condition which means that $R^+(\vec{a})$ and $R^-(\vec{a})$ must not both be true. This corresponds, in the context of SLDNF-resolution, to the statement that it is not possible that $R(\vec{a})$ succeeds and fails. This is correct, but the previous theorem makes it clear that the logical power of the partial completion is not increased by adding this axiom.

6.2. The calculi $\partial(P)$

It is now easy to set up deductive systems which correspond to the partial completion of programs. As in Section 4 we work in extensions of the Tait calculus for predicate logic; the only significant difference between $\mathcal{R}(P)$ and $\partial(P)$ is the way in which the programs are incorporated into the calculi.

The systems $\partial(P)$ for logic programs P in $\mathcal{L}$ comprise the logical axioms, the equality axioms, the logical rules and the cut rules of $\mathcal{R}(P)$, but all formulated for $\mathcal{L}^\sharp$

instead of $\mathcal{L}$. In addition we have the following equality axioms for the relation symbols of $\mathcal{L}^\sharp$ and the following partial program rules.

Equality axioms for the relation symbols of $\mathcal{L}^\sharp$. If $a_1, \ldots, a_m$, $b_1, \ldots, b_m$, $c_1, \ldots, c_n$ and $d_1, \ldots, d_n$ are $\mathcal{L}^\sharp$ terms, $R^\pm$ is a n-ary relation symbol of $\mathcal{L}^\sharp$ so that the set $\{a_1 = b_1, \ldots, a_m = b_m\}$ is unifiable and, for the most general unifier θ, $R^\pm(c_1, \ldots, c_n)\theta$ and $R^\pm(d_1, \ldots, d_n)\theta$ are identical, then we have:

$$\Gamma,\ a_1 \neq b_1, \ldots,\ a_m \neq b_m,\ \widetilde{R^\pm}(c_1, \ldots, c_n),\ R^\pm(d_1, \ldots, d_n).$$

Partial program rules for P. We have for every relation symbol R of $\mathcal{L}$ and its definition form $D_R[\vec{u}]$ with respect to the logic program P and for all $\mathcal{L}$ terms $\vec{a}$:

$$\frac{\Gamma,\ D_R^+[\vec{a}]}{\Gamma,\ R^+(\vec{a})}\ (R^+), \qquad \frac{\Gamma,\ D_R^-[\vec{a}]}{\Gamma,\ R^-(\vec{a})}\ (R^-).$$

There is an important difference between the program rules of $\mathcal{R}(P)$ and the partial program rules of $\partial(P)$: The formulas $D_R^+[\vec{u}]$ and $D_R^-[\vec{u}]$ in the premises of the partial program rules of $\partial(P)$ are positive in all relation symbols whereas the formulas $D_R[\vec{u}]$ and $\neg D_R[\vec{u}]$ in the premises of the program rule of $\mathcal{R}(P)$ may contain positive and negative occurrences of R and other relation symbols.

Let P be a logic program in $\mathcal{L}$ and Γ a finite set of $\mathcal{L}^\sharp$ formulas. Then $\partial(P) \vdash_r^n \Gamma$ and $\partial(P) \Vdash_r^n \Gamma$ is defined for all $n, r < \omega$ in analogy to $\mathcal{R}(P) \vdash_r^n \Gamma$ and $\mathcal{R}(P) \Vdash_r^n \Gamma$ (cf. Section 4). We will also make use of the abbreviated forms $\partial(P) \vdash_r \Gamma$, $\partial(P) \vdash \Gamma$, $\partial(P) \Vdash_r \Gamma$ and $\partial(P) \Vdash \Gamma$, in the same sense as before.

6.2.1. Remark. It is obvious that $\partial(P)$ is a Tait-style formalization of the partial completion of P; i.e., we have for all logic programs P in $\mathcal{L}$ and all $\mathcal{L}^\sharp$ formulas A:

$$\partial(P) \vdash A \iff comp^\sharp(P) \models A.$$

Because of the positivity of the partial program rules it is possible by standard proof-theoretic techniques to eliminate all cuts. Further, the identity axioms can be removed as well if we restrict ourselves to the positive fragment of the system $\partial(P)$.

6.2.2. Theorem. *Let P be a logic program in $\mathcal{L}$. Then we have for all finite sets Γ of $\mathcal{L}^\sharp$ formulas and all finite sets Δ of positive $\mathcal{L}^\sharp$ formulas:*

1. $\partial(P) \vdash \Gamma \implies \partial(P) \vdash_0 \Gamma$,
2. $\partial(P) \vdash_0 \Delta \implies \partial(P) \Vdash_0 \Delta$.

The following shorthand notation will be used from now on: If Γ is the set $\{A_1, \ldots, A_n\}$ of $\mathcal{L}$ formulas, then Γ^+ stands for the corresponding set $\{A_1^+, \ldots, A_n^+\}$ of $\mathcal{L}^\sharp$ formulas. Then the relationship between the identity-free and cut-free derivations in $\mathcal{R}(P)$ and $\partial(P)$ is obvious: they are identical in the sense of the following lemma.

6.2.3. Lemma. *We have for all logic programs P in $\mathcal{L}$ and finite sets Γ of $\mathcal{L}$ formulas:*

$$\mathcal{R}(P) \Vdash_0 \Gamma \iff \partial(P) \Vdash_0 \Gamma^+.$$

Together with Theorem 4.3.1 which states cut elimination for identity-free derivations in $\mathcal{R}(P)$ we therefore obtain the following result about the relationship between $\mathcal{R}(P)$ and $\partial(P)$.

6.2.4. Corollary. *Let P be a logic program in $\mathcal{L}$ and Γ be a finite set of $\mathcal{L}$ formulas. Then we have the following equivalences:*

$$\mathcal{R}(P) \Vdash_0 \Gamma \iff \partial(P) \vdash \Gamma^+ \iff \mathcal{R}(P) \Vdash \Gamma.$$

This means that the identity-free and the identity- and cut-free derivations in $\mathcal{R}(P)$ correspond exactly to the positive fragment of $\partial(P)$. The following side remark refers to cut-free derivations in $\mathcal{R}(P)$ which permit identity-axioms.

6.2.5. Remark. Let (Tot) consist of the following sets of $\mathcal{L}^\sharp$ formulas which express that all pairs (R^+, R^-) are total in the sense that at least one of the two formulas $R^-(\vec{a})$ or $R^+(\vec{a})$ is true:

$$\Gamma,\ R^-(\vec{a}),\ R^+(\vec{a}). \qquad \text{(Tot)}$$

Then one immediately has for all logic programs P in $\mathcal{L}$ and all finite sets Γ of $\mathcal{L}$-formulas:

$$\mathcal{R}(P) \vdash_0 \Gamma \iff \partial(P) + \text{(Tot)} \Vdash_0 \Gamma^+.$$

The general role of cuts and cut-free derivations in the sequent calculus is, for example, analyzed in Girard [1987b] and Girard, Lafont and Taylor [1989]. Similar results about the identity-free derivations in the sequent calculus are contained in Hösli and Jäger [1994]. This article also studies the close dualities between cut-free and identity-free derivations.

6.3. The inductive extension of logic programs

The partial completions of logic programs are comparatively weak theories. They are not powerful enough to prove many interesting properties of logic programs and, for instance, the equivalence of logic programs. To make this point clearer, consider the following two examples.

6.3.1. Example (Termination). We use the same notions as in Example 5.2.8 and let P_1 be the logic program which consists of the following clauses:

```
list([])
list([u|v]) :- list(v)
member(u, [u|v])
member(u, [v|w]) :- member(u, w)
```

Suppose, that we want to prove that for every term a and every list b the goal $\texttt{member}(a,b)$ either succeeds or fails using SLDNF-resolution. By our previous results we know that this is equivalent to the statement that the partial completion of P_1 proves the formula

$$\texttt{list}^+(v) \to (\texttt{member}^+(u,v) \lor \texttt{member}^-(u,v)).$$

However, it is easy to see that this is not possible without making use of some form of induction.

6.3.2. Example (Equivalence). Now we define the addition of natural numbers in two different ways: by recursion on the first argument and by recursion on the second argument. Let P_2 be the following logic program:

$\texttt{nat}(0)$
$\texttt{nat}(s(u)) :- \texttt{nat}(u)$
$\texttt{add1}(0,u,u)$
$\texttt{add1}(s(u),v,s(w)) :- \texttt{add1}(u,v,w)$
$\texttt{add2}(u,0,u)$
$\texttt{add2}(u,s(v),s(w)) :- \texttt{add2}(u,v,w)$

It would be nice if one could show that both definitions have the same input/output behavior. Unfortunately, this is not possible in $comp^\sharp(P_2)$; for example, the following formula is not provable there:

$$\texttt{nat}^+(u) \land \texttt{nat}^+(v) \to (\texttt{add1}^+(u,v,w) \leftrightarrow \texttt{add2}^+(u,v,w)).$$

In order to overcome these deficiencies we add to the partial completion suited forms of induction. The basic idea is the following. Suppose we are given a logic program P in $\mathcal{L}$ which contains the relation symbol $R_0,\ldots,R_n$. Then we collect all positive formulas $D^+_{R_i}[\vec{x}]$ and $D^-_{R_i}[\vec{x}]$ and consider them as the definition clauses of a simultaneous inductive definition *SID* of the relations $R^+_0, R^-_0, \ldots, R^+_n, R^-_n$ in the sense of, for example, Moschovakis [1974].

The partial completion expresses that these relations are closed under this simultaneous inductive definition. However, $comp^\sharp(P)$ does not say that the relations R^+_i and R^-_i are fixed points, let alone least fixed points of *SID*. The next step is therefore to add further induction principles which enforce the relations $R^+_0, R^-_0, \ldots, R^+_n, R^-_n$ to be least fixed points.

For notational convenience we have to introduce some shorthand notations: Let P be a logic program in $\mathcal{L}$ which contains the relation symbols $R_0,\ldots,R_n$. Then we write *closed*(P) for the formula

$$\bigwedge_{i=0}^{n} (\forall\vec{x}\,(D^+_{R_i}[\vec{x}] \to R^+_i(\vec{x})) \land \forall\vec{x}\,(D^-_{R_i}[\vec{x}] \to R^-_i(\vec{x}))). \qquad \text{(CLOSURE)}$$

Obviously *closed*(P) is provable in $comp^\sharp(P)$. Now suppose further that we have for each relation symbol R_i two $\mathcal{L}^\sharp$ formulas $A_i(\vec{u})$ and $B_i(\vec{u})$ with distinguished free variables $\vec{u} = u_1, \dots, u_m$, provided that R_i is m-ary. Then $closed(P, \vec{R}^+/\vec{A}, \vec{R}^-/\vec{B})$ is the formula which results from $closed(P)$ by simultaneously replacing each occurrence of $R_i^+(\vec{a})$ by $A_i(\vec{a})$ and $R_i^-(\vec{a})$ by $B_i(\vec{a})$ for $0 \leq i \leq n$. As additional abbreviation we write $sub(\vec{R}^+, \vec{A}, \vec{R}^-, \vec{B})$ for

$$\bigwedge_{i=0}^{n} (\forall \vec{x}\,(R_i^+(\vec{x}) \to A_i(\vec{x})) \land \forall \vec{x}\,(R_i^-(\vec{x}) \to B_i(\vec{x}))).$$

6.3.3. Definition. Let P be a logic program in $\mathcal{L}$. Then the *inductive extension* $ind^\sharp(P)$ of P is the $\mathcal{L}^\sharp$ theory which consists of $comp^\sharp(P)$ and comprises the following additional axioms

$$closed(P, \vec{R}^+/\vec{A}, \vec{R}^-/\vec{B}) \to sub(\vec{R}^+, \vec{A}, \vec{R}^-, \vec{B}) \qquad \text{(MINIMALITY)}$$

for all $\mathcal{L}^\sharp$ formulas $\vec{A}$ and $\vec{B}$ with a suitable number of distinguished free variables.

From the point of view of inductive definitions $ind^\sharp(P)$ is an extremely natural theory. We can show in $ind^\sharp(P)$ that for each relation symbol R of P the relation symbols R^+ and R^- are least fixed points of the simultaneous inductive definition which corresponds to P in the sense described above. This means, in particular, that induction on all R^+ and R^- is available in $ind^\sharp(P)$. Further one can prove in the inductive extension of P that R^+ and R^- have no elements in common.

Although strong induction principles are added to $comp^\sharp(P)$, the theory $ind^\sharp(P)$ is a conservative extension of $comp^\sharp(P)$ with respect to positive $\mathcal{L}^\sharp$ formulas. The proof of the following theorem is obvious from the elementary theory of inductive definitions.

6.3.4. Theorem. *Let P be a logic program in $\mathcal{L}$. Then we have for all positive $\mathcal{L}^\sharp$ formulas A:*

$$ind^\sharp(P) \vdash A \iff comp^\sharp(P) \vdash A.$$

The combination of the previous result, Theorem 6.1.3, Theorem 5.3.1 (soundness of SLDNF-resolution) and Theorem 5.3.5 (completeness of SLDNF-resolution) provides a powerful framework for the analysis of logic programs. We show this by continuing the discussion of the programs P_1 and P_2 introduced above.

First we turn to the question of termination in Example 6.3.1. Since induction is available, it is easy to see that we have

$$ind^\sharp(P_1) \vdash \forall x \forall y (\texttt{list}^+(y) \to (\texttt{member}^+(x,y) \lor \texttt{member}^-(x,y))). \qquad (1)$$

Now choose arbitrary terms a and b so that $\texttt{list}(b)$ succeeds using SLDNF-resolution, i.e., $\texttt{list}(b)\ \mathbf{R}(P)\ \varepsilon$. Then we can conclude that $\texttt{member}(a,b)$ either succeeds or fails using SLDNF-resolution by the following argument: In view of Theorem 5.3.1 and Theorem 6.1.3 we have

$$comp^\sharp(P_1) \vdash \texttt{list}^+(b). \qquad (2)$$

Since $ind^{\sharp}(P_1)$ is an extension of $comp^{\sharp}(P_1)$, we obtain from (1) and (2) that $ind^{\sharp}(P_1)$ proves $\texttt{member}^+(a,b) \vee \texttt{member}^-(a,b)$. Hence the previous theorem implies

$$comp^{\sharp}(P_1) \vdash \texttt{member}^+(a,b) \vee \texttt{member}^-(a,b). \tag{3}$$

Applying Theorem 5.3.5 and Theorem 6.1.3 with the mode assignment of Example 5.2.6 yields that either $\texttt{member}(a,b)\ \mathbf{R}(P)\ \sigma$ for some substitution σ or $\texttt{member}(a,b) \in \mathbf{F}(P)$. In other words, the goal $\texttt{member}(a,b)$ succeeds or fails using SLDNF-resolution.

After the treatment of termination of logic programs we come back to the problem of the equivalence of the logic programs in Example 6.3.2. It is easy to verify that we have

$$ind^{\sharp}(P_2) \vdash \forall x \forall y \forall z(\texttt{nat}^+(x) \wedge \texttt{nat}^+(y) \rightarrow (\texttt{add1}^+(x,y,z) \leftrightarrow \texttt{add2}^+(x,y,z))). \tag{4}$$

Let a, b and c be terms so that the goals $\texttt{nat}(a)$ and $\texttt{nat}(b)$ succeed using SLDNF-resolution. Since the program P_2 is definite, we can use the trivial mode assignment of Example 5.2.6, and by the same argument as above we can conclude that the goal $\texttt{add1}(a,b,c)$ succeeds using SLDNF-resolution just in case that the goal $\texttt{add2}(a,b,c)$ succeeds.

The two relations add1 and add2 do not only have the same behavior with respect to success but also share the same behavior with respect to failure. We first observe that

$$ind^{\sharp}(P_2) \vdash \forall x \forall y \forall z(\texttt{nat}^+(x) \rightarrow (\texttt{add1}^+(x,y,z) \vee \texttt{add1}^-(x,y,z))) \tag{5}$$

and

$$ind^{\sharp}(P_2) \vdash \forall x \forall y \forall z(\texttt{nat}^+(y) \rightarrow (\texttt{add2}^+(x,y,z) \vee \texttt{add2}^-(x,y,z))). \tag{6}$$

Lines (4), (5), (6) and the fact that $ind^{\sharp}(P_2)$ proves $\forall \vec{x} \neg(R^+(\vec{x}) \wedge R^-(\vec{x}))$ for any relation R of $\mathcal{L}$ yield that

$$ind^{\sharp}(P_2) \vdash \forall x \forall y \forall z(\texttt{nat}^+(x) \wedge \texttt{nat}^+(y) \rightarrow (\texttt{add1}^-(x,y,z) \leftrightarrow \texttt{add2}^-(x,y,z))).$$

From this we can conclude that the goal $\texttt{add1}(a,b,c)$ fails using SLDNF-resolution if and only if the goal $\texttt{add2}(a,b,c)$ fails if a, b and c are terms so that $\texttt{nat}(a)$ and $\texttt{nat}(b)$ succeed. Thus, the relations add1 and add2 have the same behavior with respect to success and failure.

7. Concluding remark

In discussing the foundations of logic programming it is often possible to distinguish between three levels of abstraction:

I. Declarative semantics. Semantical considerations about logic programming are often guided by the attempt of constructing suitable minimal models of logic programs. However, in general the logical complexity of minimal models of logic programs is very high and the corresponding semantics is noneffective.

II. Proof theory. It deals with the development and analysis of deductive systems for logic programs and is often directed to the *proofs as computations* paradigm. In the ideal case there is a close connection between the proof theory and the procedural aspects of logic programming in the sense that query-answering mechanisms can be interpreted as formal proofs and suitable formal proofs can be transformed into successful computations.

III. Procedural semantics. It is concerned with the general principles behind the implementations of logic programming. Since today most procedural approaches to first order logic programming are based on some form of SLDNF-resolution, the distinguished role of this concept is evident.

In our article we followed this general pattern. The results we presented can be roughly summarized as follows. If P is a "decent" logic program and A a closed atom, then the following assertions are equivalent:

(i) A is true in the ω-segment of the least adequate Herbrand structure $\mathfrak{I}_P$ of P.

(ii) A is true in all structures which are adequate to P.

(iii) A is identity and cut-free provable in the deductive system $\mathcal{R}(P)$.

(iv) A is derivable by SLDNF-resolution.

Furthermore, by a simple syntactic transformation it is possible to associate to each logic program P a system of positive inductive definitions $ind^\sharp(P)$ so that for closed atoms A derivability from $ind^\sharp(P)$ is equivalent to each of the four assertions above. In addition induction principles are available in $ind^\sharp(P)$ which make it possible to prove properties about logic programs. In this sense we hope that we could provide a proof-theoretic framework for logic programming.

Of course there exist other proof-theoretic approaches to logic programming which we did not mention at all, and we conclude this article with mentioning two of them.

Some interesting activities in this area start off from Girard's linear logic (cf. e.g. Girard [1987a]) and study the connections between logic programming and linear logic. Another important area in the general field of logic programming deals with higher order logic programming, and we refer the reader for example to Miller [1991] and Pfenning [1992] for further reading.

References

K. R. Apt

[1990] Logic programming, in: *Handbook of Theoretical Computer Science, Volume B*, J. van Leeuwen, ed., Elsevier, ch. 10, pp. 495–574.

K. R. Apt and R. Bol

[1994] Logic programming and negation: A survey, *J. of Logic Programming*, 19/20, pp. 9–72.

N. D. Belnap

[1977] A useful four-valued logic, in: *Modern Uses of Multiple-Valued Logic*, J. M. Dunn and G. Epstein, eds., D. Reidel, Dordrecht, pp. 8–37.

H. A. Blair

[1982] The recursion-theoretic complexity of the semantics of predicate logic as a programming language, *Information and Control*, 54, pp. 25–47.

W. Buchholz

[1992] *A negation as failure calculus*, tech. rep., University of Munich.

L. Cavedon and J. W. Lloyd

[1989] A completeness theorem for SLDNF-resolution, *J. of Logic Programming*, 7, pp. 177–191.

K. L. Clark

[1978] Negation as failure, in: *Logic and Data Bases*, H. Gallaire and J. Minker, eds., Plenum Press, New York, pp. 293–322.

K. Doets

[1994] *From Logic to Logic Programming*, MIT Press.

W. Drabent and M. Martelli

[1991] Strict completion of logic programs, *New Generation Computing*, 9, pp. 69–69.

M. H. van Emden and R. A. Kowalski

[1976] The semantics of predicate logic as a programming language, *J. of the Association for Computing Machinery*, 4, pp. 733–742.

S. Feferman

[1991] Reflecting on incompleteness, *J. of Symbolic Logic*, 56, pp. 1–49.

M. Fitting

[1985] A Kripke-Kleene semantics for logic programs, *J. of Logic Programming*, 2, pp. 295–312.

[1991] Bilattices and the semantics of logic programming, *J. of Logic Programming*, 11, pp. 91–116.

M. L. Ginsberg

[1987] Multi-valued logics, in: *Readings in Nonmonotonic Reasoning*, M. L. Ginsberg, ed., Morgan Kaufmann, pp. 251–255.

J.-Y. Girard

[1987a] Linear logic, *Theoretical Computer Science*, 50, pp. 1–102.

[1987b] *Proof Theory and Logical Complexity*, Bibliopolis, Napoli.

J.-Y. Girard, Y. Lafont, and P. Taylor

[1989] *Proofs and Types*, Cambridge University Press.

L. Hallnäs and P. Schroeder-Heister

[1990] A proof-theoretic approach to logic programming. I. Clauses as rules, *J. of Logic and Computation*, 1, pp. 261–283.

B. Hösli and G. Jäger

[1994] About some symmetries of negation, *J. of Symbolic Logic*, 59, pp. 473–485.

G. Jäger

[1989] Non-monotonic reasoning by axiomatic extensions, in: *Logic, Methodology and Philosophy of Science VIII*, J. E. Fenstad, I. T. Frolov, and R. Hilpinen, eds., North-Holland, Amsterdam, pp. 93–110.

[1994] A deductive approach to logic programming, in: *Proof and Computation*, H. Schwichtenberg, ed., Series F: Computer and Systems Sciences, NATO Advanced Study Institute, International Summer School held in Marktoberdorf, Germany, 1993, Springer-Verlag, Berlin, pp. 133–172.

K. Kunen

[1987] Negation in logic programming, *J. of Logic Programming*, 4, pp. 289–308.

[1989] Signed data dependencies in logic programs, *J. of Logic Programming*, 7, pp. 231–245.

J.-L. LASSEZ AND M. J. MAHER

[1985] Optimal fixed points of logic programs, *Theoretical Computer Science*, 39, pp. 15–25.

J. W. LLOYD

[1987] *Foundations of Logic Programming*, Springer-Verlag, Berlin, second ed.

A. I. MAL'CEV

[1971] Axiomatizable classes of locally free algebras of various types, in: *The Metamathematics of Algebraic Systems, Collected Papers*, North-Holland, Amsterdam, ch. 23, pp. 262–281.

D. MILLER

[1991] A logic programming language with lambda-abstraction, function variables and simple unification, *J. of Logic and Computation*, 1, pp. 497–536.

Y. N. MOSCHOVAKIS

[1974] *Elementary Induction on Abstract Structures*, North-Holland, Amsterdam.

A. MYCROFT

[1984] Logic programs and many-valued logic, in: *STACS 84: Symposium on Theoretical Aspects of Computer Science*, M. Fontet and K. Mehlhorn, eds., Lecture Notes in Computer Science #166, Springer-Verlag, Berlin, pp. 274–286.

F. PFENNING

[1992] ed., *Types in Logic Programming*, MIT Press.

J. A. ROBINSON

[1965] A machine-oriented logic based on the resolution principle, *J. Ass. Comp. Mach.*, 12, pp. 23–41.

P. SCHROEDER-HEISTER

[1991] Hypothetical reasoning and definitional reflection in logic programming, in: *Extensions of Logic Programming*, P. Schroeder-Heister, ed., Lecture Notes in Computer Science #475 (Lecture Notes in Artificial Intelligence), Springer-Verlag, Berlin, pp. 327–339.

K. SCHÜTTE

[1977] *Proof Theory*, Springer-Verlag, Berlin.

J. C. SHEPHERDSON

[1988] *Language and Equality Theory in Logic Programming*, Tech. Rep. PM-88-08, University of Bristol.

[1989] A sound and complete semantics for a version of negation as failure, *Theoretical Computer Science*, 65, pp. 343–371.

[1992] Logics for negation as failure, in: *Logic from Computer Science*, Y. N. Moschovakis, ed., Springer-Verlag, Berlin, pp. 521–583.

R. F. STÄRK

[1991] A complete axiomatization of the three-valued completion of logic programs, *J. of Logic and Computation*, 1, pp. 811–834.

[1994a] Cut-property and negation as failure, *International Journal of Foundations of Computer Science*, 5, pp. 129–164.

[1994b] Input/output dependencies of normal logic programs, *J. of Logic and Computation*, 4, pp. 249–262.

[1996] From logic programs to inductive definitions, in: *Logic: From Foundations to Applications. Proceedings of Logic Colloquium '93*, W. Hodges, ed., Oxford University Press, pp. 453–481.

W. W. Tait

[1968] Normal derivability in classical logic, in: *The Syntax and Semantics of Infinitary Languages*, J. Barwise, ed., Lecture Notes in Mathematics #72, Springer-Verlag, Berlin, pp. 204–236.

G. Takeuti

[1987] *Proof Theory*, North-Holland, Amsterdam.

A. Van Gelder and J. S. Schlipf

[1993] Commonsense axiomatizations for logic programs, *J. of Logic Programming*, 17, pp. 161–195.

CHAPTER X

Types in Logic, Mathematics and Programming

Robert L. Constable
Computer Science Department, Cornell University
Ithaca, New York 14853, USA

Contents

HANDBOOK OF PROOF THEORY
Edited by S. R. Buss

1. Introduction

Proof theory and computer science are jointly engaged in a remarkable enterprise. Together they provide the practical means to formalize vast amounts of mathematical knowledge. They have created the subject of *automated reasoning* and a digital computer based *proof technology*; these enable a diverse community of mathematicians, computer scientists, and educators to build a new artifact — a globally distributed digital library of formalized mathematics. I think that this artifact signals the emergence of a new branch of mathematics, perhaps to be called *Formal Mathematics.*

The theorems of this mathematics are completely formal and are processed digitally. They can be displayed as beautifully and legibly as journal quality mathematical text. At the heart of this library are completely formal proofs — created with computer assistance. Their correctness is based on the axioms and rules of various foundational theories; this formal accounting of correctness supports the highest known standards of rigor and truth. The need to formally relate results in different foundational theories opens a new topic in proof theory and foundations of mathematics.

Formal proofs of interesting theorems in current foundational theories are very large rigid objects. Creating them requires the speed and memory capacities of modern computer hardware and the expressiveness of modern software. Programs called *theorem provers* fill in tedious detail; they recognize many kinds of "obvious inference," and they automatically find long chains of inferences and even complete subproofs or proofs. The study of these theorem provers and the symbolic algorithms that make them work is part of the subject of automated reasoning. This science and the proof technology built on it are advancing all the time, and the new branch of mathematics that they enable will have its own standards, methods, surprises and triumphs.

This article is about the potent mixture of proof theory and computer science behind automated reasoning and proof technology. The emphasis is on proof theory topics while stressing connections to computer science.

Computer science is concerned with automating computation. Doing this well has made it possible to formalize real proofs. Computing well requires fast and robust hardware as well as expressive high level programming languages. High level languages are partially characterized by their *type systems*; i.e., the organization of *data types* expressible in the language. The evolution of these languages has led to type systems that resemble mathematical type theories or even computationally effective set theories. (This development underlines the fact that high level programming is an aspect of computational mathematics.) *This article will focus mainly on relating data types and mathematical types.*

The connection between data types and mathematical types in the case of formal mathematics and automated reasoning is even tighter than the general connection. Here is why. To preserve the highest standards of rigor in formalized mathematics built with computer assistance (the only way to produce it), it is necessary to reason about programs and computations. This is what intuitionists and constructivists do at a very high level of abstraction. So as the programming languages for automating reasoning become more abstract and expressive, constructive mathematics becomes directly relevant to Formal Mathematics and to the "grand enterprise" of building it using theorem provers. We will see that connections are quite deep.

It turns out that proof technology is relevant to other technologies of economic and strategic importance. For instance, the *type checkers* in commercial programming languages like ML are actually small theorem provers. They check that arguments to a function match the type of the function (see section 3). Industrial *model checkers* systematically search for errors in the design of finite state systems, such as hardware circuits or software protocols. More general tools are *program verification systems*. These combine type checkers, model checkers, decision procedures, and theorem provers that use formalized mathematics. They are employed to prove that programs have certain formally specified properties. Such proofs provide the highest levels of assurance that can be given that programs operate according to specifications. There are also software systems based on proof technology which synthesize correct programs from proofs that specifications are realizable. We will examine the proof theory underlying some of these systems.

My approach to the subject comes from the experience of designing, studying, and using some of the earliest and then some of the most modern of these theorem provers. Currently my colleagues and I at Cornell are working with the system we call Nuprl ("new pearl").[1] We call it a *proof development system* in Constable et al. [1986], but some call it a *problem solving environment* (PSE) or a *logical framework* (LF). From another point of view it is a *collaborative mathematics environment*, c.f., Chew et al. [1996]. Whatever Nuprl is called, I am concerned with systems like it and their evolution. We will examine the logical features common to a variety of current systems of a similar kind, such as ACL2, Alf, Coq, HOL, IMPS, Isabelle, Kiv, LA, Lego, Mizar, NqThm and Otter. So while I will refer to Nuprl from time to time, most of the ideas are very general and will apply to the systems of the 21st century as well. Before saying more about the article, let me put the work into historical perspective. Doing this will allow me to state my goals more exactly (especially after each topic of Section 1.1).

1.1. Historical perspective on a grand enterprise 1875-1995. From *Begriffsschrift* [1879] onwards until *Grundgesetze* [1903], logic was re-surveyed by Gottlob Frege, and the ground was cleared to provide a firm foundation for mathematics.[2] In *Principia Mathematica*, Whitehead and Russell [1925-27] revised Frege's flawed architectural plans, and then using these plans, Hilbert [1926] laid out a formalist

[1] We have released Version 4.2, see http://www.cs.cornell.edu/Info/Projects/NuPrl/nuprl.html. Version 5 and "Nuprl Light" will be available at this World Wide Web site in 1999.

[2] *Begriffsschrift* ("concept script") analyzed the notion of a proposition into function and argument, introduced the quantifiers, binding, and a theory of identity. This created the entire predicate calculus. *Grundgesetze* presented a theory of classes based on the comprehension principle and defined the natural numbers in terms of them.

program to build the completely formal theories which would be used to explain and justify the results and methods of mathematics. His program would defend mathematical practice against critics like Brouwer who saw the need to place the foundation pilings squarely on the natural numbers and build with constructive methods.[3]

Hilbert called for workers, training some himself, and began with them the task which proved to be so compelling and attractive to many talented mathematicians like Church, von Neumann, Herbrand, Gentzen, Skolem, Turing, Tarski, Gödel, and many more. Boring deep into the bedrock to explore the foundation site, Kurt Gödel [1931] unexpected limitations to the planned activity. It could never be completed as envisioned by Hilbert.[4] His surprising discovery changed expectations, but the tools Gödel created transformed the field and stimulated enormous interest in the enterprise. More remarkable discoveries followed.

Within two decades, computer science was providing new "power tools" to realize in software the formal structures needed to support mathematics. By 1960 computer hardware could execute programming languages like Lisp, c.f. McCarthy [1963], designed for the symbolic processing needed to build formal structures. Up in the scaffolding computer scientists began to encounter their own problems with "wiring and communications," control of resource expenditure, design of better tools, etc. But already even in the 1970's poised over the ground like a giant drilling rig, the formal structures supported still deeper penetration into the bedrock designed to support mathematics (and with it the mathematical sciences and much of our technical knowledge). The theory of computational complexity, arising from Hartmanis and Stearns [1965], led to further beautiful discoveries like Cook's $P = NP$ problem, and to a theory of algorithms needed for sophisticated constructions, and to a theory of *feasible mathematics* (see Buss [1986], Leivant [1994b,1994a,1995]), and to ideas for the *foundations of computational mathematics.*

By 1970 the value of the small formal structure already assembled put to rest the nagging questions of earlier times about why mathematics should be formalized. The existing structure provided economic benefit to engineering, just as Leibniz dreamed, Frege foresaw, McCarthy planned [1962], and many are realizing.

Even without the accumulating evidence of economic value, and without counting the immediate utility of the software artifacts, scientists in all fields recognized that the discoveries attendant on this "grand enterprise" illuminate the very nature of knowledge while providing better means to create and manage it. The results of this enterprise have profound consequences because all scholars and scientists are in the business of processing information and contributing to the accumulation and dissemination of knowledge.

The construction of the foundational structure goes on; it is forming a new kind

[3]I refer to Hilbert's *formalist program* founded on a finitistic analysis of formal systems to prove their consistency and to justify non-constructive reasoning as a (possibly meaningless) *detour* justified by the *consistency* of a formal system.

[4]Gödel showed that *consistency* is not sufficient to justify the detour because there are formulas of number theory such that both P and $\neg P$ can be consistently added (P an unprovable formula).

of place, like a biosphere made out of bits. We might call it a "cybersphere" since it encloses the space we call "cyberspace." Many people now live in this space... which supports commerce and recreation as well as scholarship and science.

It is in the context of this "grand enterprise" that I have framed the article. I see it concerned with two of the major modes of work in assembling the formal structures — logical analysis and algorithmic construction. I will briefly mention the aspects of these activities that I treat here.

Logical analysis. When looking back over the period from 1879 to now, we see that the formal analysis of mathematical practice started with logical language. Frege [1879] said:

> "To prevent anything intuitive from penetrating [into an argument] unnoticed, I had to bend every effort to keep the chain of inferences free of gaps. In attempting to comply with this requirement in the strictest possible way I found the inadequacy of language to be an obstacle. This deficiency led me to the present ideography...
> Leibniz, too, recognized — and perhaps overrated the advantages of adequate notation. His idea of a... *calculus philosophicus*... was so gigantic that the attempt to realize it could not go beyond the bare preliminaries. The enthusiasm that seized its originator when he contemplated the immense increase in intellectual prover of mankind that [the calculus would bring] caused him to underestimate the difficulties.... But even if this worthy goal cannot be reached in one leap, we need not despair of a slow step by step approximation."

So Frege began with very limited goals and took what he characterized as "small steps" (like creating all of predicate logic!). He did not include a study of computation and its language; he limited his study of notation to logical operators, and he ruled out creating a natural expression of proofs or classifying them based on how "obvious" they are.

In addition, Frege focused on understanding the most fundamental types, natural numbers, sequences, functions, and classes. He adopted a very simple approach to the domain of functions, forcing them all to be *total*. He said (in his *Collected Papers*) "the sign $a + b$ should always have a reference, whatever signs for definite objects may be inserted in place of 'a' and 'b'." *Principia* took a different approach to functions, introducing types, but also it excluded from consideration an analysis of computation or natural proofs or the notational practices of working mathematics. It too developed only basic mathematics with no attempt to treat abstract algebra or computational parts of analysis.

Principia Mathematica, the monumental work of Whitehead and Russell [1925-27], was indeed the first comprehensive rendering of mathematics in symbolic logic. Gödel's celebrated 1931 paper "On Formally Undecidable Propositions of *Principia Mathematica* and Related Systems" begins:

> "The development of mathematics toward greater precision has led, as is well known, to the formalization of large tracts of it, so that one

> can prove any theorems using nothing but a few mechanical rules. The most comprehensive formal systems that have been set up hither to are the system of *Principia Mathematica* (PM) on the one hand and the Zermelo-Fraenkel axiom system of set theory... on the other."

Principia presents a logic based on types and derives in it a theory of classes, while ZF set theory provided informal *axioms* and, logic was incidental.[5] *Principia* deals with the topics that I find fundamental in my own work of "implementing mathematics" in the Nuprl system, Constable et al. [1986]. Thus much of what I say here is related to PM. Indeed, in a sense Nuprl is a modern style *Principia* suitable for computational mathematics.

Hilbert introduced a greater degree of formalization, essentially banishing semantics, and he began as a result to deal with computation in his metalanguage. But he took a step backwards from *Principia* in terms of analyzing when expressions are meaningful. He reduced this to an issue of parsing and "decidable type checking" of formulas as opposed to the semantic judgments of *Principia.*

It wasn't until Gentzen that the notion of proofs as they occur in practice was analyzed, first in *natural deduction* and then in *sequent calculi.* It wasn't until Herbrand, Gödel, Church, Markov, and Turing that computation was analyzed and not until de Bruijn that the organization of knowledge (into a "tree of knowledge" with explicit contexts) was considered. De Bruijn's Automath project also established links to computing, like those simultaneously being forged from computer science.

Recently Martin-Löf has widened the logical investigation to reintroduce a semantic approach to logic, to include computation as part of the language, and to make manifest the connections to knowledge. Martin-Löf [1983,p.30] says:

> "to have proved = to know = to have understood, comprehended, grasped, or seen. It is now manifest, from these equations, that proof and knowledge are the same. Thus, if proof theory is construed, not in Hilbert's sense, as metamathematics but simply as the study of proofs in the original sense of the word, then proof theory is the same as theory of knowledge..."

We are now in a position where mathematical logic can consider all of these elements: an analysis of the basic judgments of typing, truth, and computational equality; an analysis of natural proofs and their semantics; the integration of computational concepts into the basic language; an analysis of the structure of knowledge and its role in practical inference; and classification of inference according to its computational complexity.

We will attempt a consideration of logic that takes all this into account and is linked to computing practice, and yet is accessible. I begin the article with an account of typed logic that relates naturally to the Automath conception. The connection is discussed explicitly in Section 2.12.

---

[5] *Principia* is not *formal* in the modern sense. There are semantic elements in the account which Wittgenstein [1953,1922] objected to. Hilbert made a point of formalizing logic, and we follow in that purely formal tradition.

The article stresses the nature of the underlying logical language because that is so basic — everything else is built upon it. The structures built are so high that a small change in the foundation can cause a large movement at the top of the structure. So any discoveries that improve the foundation for formal mathe atics are among the most profound in their effect. As it is, we are standing on the shoulders of giants. I take the time in section 2 to review this heritage that is so crucial to everything else.

Algorithmic construction. Computer science completely transformed the "grand enterprise." First it introduced computational procedure and *procedural knowledge*, and it gradually widened the scope of its successes. It could check formulas and synthesize them; later it could check proofs and synthesize them. In all this, the precision that Gödel referred to reached new levels of *mechanical precision*. The vast change of scale from processing a few hundred lines by hand to tens of thousands by machine (now hundreds of millions) caused a qualitative change and created new fields like Automated Deduction and Formal Mathematics in which formalisms became usable.

The success of procedural knowledge created the questions of relating it to declarative knowledge, a question at the heart of computer science, studied extensively in the database area, also in "logic programming" and in AI. It is a question at the heart of AD, McAllester [1989], as is clearly seen in Bundy's work on *proof plans* [1991]. From the AI perspective, one can see this impact as reintroducing "mind" and "thought" into the enterprise McAllester [1989]. From the logical perspective one can see this as reintroducing the study of *intension* and Frege's notion of *sense* into logic. As Jean-Yves Girard put it in Girard, Taylor and Lafont [1989,p.4]:

> "In recent years, during which the algebraic tradition has flourished, the syntactic tradition was not of note and would without a doubt have disappeared in one or two more decades, for want of any issue or methodology. The disaster was averted because of computer science — that great manipulator or syntax — which posed some very important theoretical problems."

Computer science produced new high level languages for expressing algorithms. These have evolved to modern programming languages such as ML (for Meta Language) designed to help automate reasoning. ML and its proposed extensions have such a rich system of data types that the type system resembles a constructive theory of mathematical types. We discuss this observation in section 3. Our concern for the relationship between data types and mathematical types is a reason that I will talk so much about typed logic in section 2.

Computer science also created a new *medium* for doing mathematics — the digital electronic medium now most visible through the World Wide Web. This affects every aspect of the enterprise. For example, the "surface" or concrete syntax can be disconnected from the abstract syntax, and we can *display* the underlying terms in a large variety of forms. To take a trivial point, the typed universal quantifier, "for all x of type A" can be displayed as $\forall x : A.$ or as $\bigwedge x : A.$ or as "For all x in A" or $\forall x \in A.$ or $\forall x^A$: . The key parts of the abstraction are the operator name, *all*, a binding variable, x, and a type, A.

The new medium provides *hypertext* and *hyperproofs*, e.g. Nuprl's proof editor and Hyperproof (Barwise and Etchemendy [1991]).[6] It is difficult to render these proofs on paper as we will see in the appendix; one must "live in the medium" to experience it. This medium also creates a new approach to communication and doing mathematics. It is routine to embed computation in documents and proofs, e.g. *Mathematica* notebooks, Wolfram [1988]. Databases of definitions, conjectures, theorems, and proofs will be available on line as *digital mathematics libraries.* There will be new tools for collaborating remotely on proofs as we can now do on documents (with OLE, OpenDoc, and similar tools).

These are the overt changes brought by computer science, but as Girard says, the theoretical questions raised are very important to logic. We can see some of these by looking at the stages of work in automated reasoning.

Stage 1. From the late 50's to the late 60's was an *algorithmic phase* during which the basic symbolic procedures and low level data structures were discovered and improved. The basic matching, unification, and rewriting algorithms were coded and tested in the resolution method and various decision procedures. We learned the extent and value of these basic symbolic procedures, and they were made very efficient in systems like Otter (Wos et al. [1984]) and Prolog. Now there are links being formed with other non-numerical computing methods from symbolic algebra. A deep branch of computational mathematics has been formed, and communities of scientists are using the tools.

Stage 2. The 70's saw the creation of several systems for use in writing more reliable software. These were called *program verifiers*, e.g. the Stanford Pascal Verifier (Igarashi, London and Luckham [1975]) and Gypsy (Good [1985]) were targeted to Pascal and NqThm (Boyer and Moore [1979]) for pure Lisp, PL/CV (Constable, Johnson and Eichenlaub [1982]) for a subset of PL/I, and LCF (Gordon, Milner and Wadsworth [1979]) for a higher order functional programming language. These systems could also be seen as implementing *logics of computable functions* (hence LCF) or *programming logics.* The motivation (and funding) for this work came from computer science. The goal was to "prove that programs met their specifications." This technology drew on the algorithms from the earlier period, but also contributed new techniques such as *congruence closure* (Kozen [1977], Nelson and Oppen [1979], Constable, Johnson and Eichenlaub [1982]) and new decision procedure such as Arith (Chan [1982]) and SupInf (Bledsoe [1975], Shostak [1979]) and the notion of theorem proving *tactics* (from LCF). During this period there also appeared systems for *checking* formalizing mathematics such as Automath (de Bruijn [1970], Nederpelt, Geuvers and Vrijer [1994]) and FOL (Weyrauch [1980]).

[6]This capability can be explored at the Nuprl project home page www.cs.cornell.edu/Info/Projects/NuPrl.

Stage 3. In the 80's and 90's the so-called program verification idea was refined a great deal. The Larch (Guttag, Horning and Wing [1985]) system was multi-lingual, and it represents the change in emphasis from "verification" to *property checking.* The LCLint system is a result of this evaluation (see www.larch.lcs.mit.edu). The prover is seen as a *bug detector* or "falsification system." The LCF system spawned HOL (Gordon and Melham [1993]), Nuprl (Constable et al. [1986]), Isabelle (Paulson [1994]) and many others.[7] Nuprl, in turn, spawned others like Coq (Coquand and Huet [1988]), Lego (Pollack [1995]), PVS (Owre et al. [1996]), and Nuprl-Light (Hickey [1997]).

This period saw also the second generation efforts to build computer systems to help create formalized mathematics. The Mizar checking effort (Trybulec [1983]) included starting the *Journal of Formalized Mathematics* (see www.mcs.anl.gov/qed). The Nuprl system focused on tools for synthesizing proofs and will be discussed in this article, but numerous other systems were created, e.g. Alf, Coq, HOL, IMPS, Isabelle, PVS, STeP by Constable et al. [1986], Paulin-Mohring and Werner [1993], Farmer [1990], Farmer, Guttman and Thayer [1991], Owre, Rushby and Shankar [1992], Owre et al. [1996], Bjørner et al. [1996], and others.[8] Some of these were integrated systems with special editors, a library of theories and various "logic engines." The 90's is seeing a move toward more modular and open systems, and in the 21st century we will see cooperative systems such as Howe's HOL-Nuprl effort [1996b].

This article will introduce some of the theory behind tactic-oriented theorem proving and will show examples of modern synthesized proofs. It will relate this to developments in typed programming languages.

1.2. Outline. Section 2 covers Typed Logic. The development moves from a *pure* logic of typed propositions to a calculus with specific types—**N** the natural numbers, cartesian products of types, *lists* over any types, functions from one type to another, sets built over a type, and types defined by imposing a new equality on our existing type... so called quotient or congruence types.

The exposition is very similar to the way logic is presented in Nuprl, but instead of deriving logic from type theory, we start with logic in the spirit of *Principia.* We also present logic in the manner of Automath, so one might call this logic "*AutoMathematica.*" It contrasts with the *polymorphic predicate calculus* of LCF (Gordon, Milner and Wadsworth [1979]) and HOL (Gordon and Melham [1993]).

Section 3 covers Type Theory. This is essentially an introduction to *Martin-Löf semantics* using an axiomatic theory close to his Intuitionistic Type Theory circa 1982-88 and its Nuprl variants (Martin-Löf [1982,1984,1983]). My approach is very "expository" since there are many accessible accounts in the books of Nordstrom, Petersson and Smith [1990], Martin-Löf [1984], Thompson [1991], the articles of Backhouse [1989], Martin-Löf [1982], Allen [1987a], and the theses of Allen [1987b],

---

[7]A comprehensive history of theorem proving up to 1970 can be found in Siekmann and Wrightson [1983]; see MacKenzie [1995] for recent developments.

[8]The web site www-formal.stanford.edu/clt/ARS/systems.html at Stanford lists these systems.

Palmgren [1991,1995b], Poll [1994], Setzer [1993], Rezus [1985], Helmink [1992] which provide technical detail. I use a small fragment to illustrate the theory and present Stuart Allen's non-type theoretic account [1987a] as well as a *semantics of proofs as objects*. Many of the ideas of type theory are discussed for the *impredicative theories* based on Girard's system F (see Girard, Taylor and Lafont [1989]). There is a large literature here as well Coquand and Huet [1988], Reynolds [1974], Luo [1994], Poll [1994]. Dependent types and universes are in this section. These types can then be added to the Typed Logic. So in a sense, this section extends Typed Logic.

Section 4 covers Typed Programming. Here I explain the notion of "partial types" common in programming and relate them to type theory. This account is expository, designed to make connections. But I discuss the recursive types that Constable and Mendler [1985] introduced which are closely related to the subsequent accounts for Coq and Alf of Coquand and Paulin-Mohring [1990], Dybjer [1994]. I then discuss a new type constructor due to Jason Hickey [1996a] — the *very dependent type*. These recursive and very dependent types can be added to the type theory and hence, to the Typed Logic. So this section too can be viewed as extending the logic of Section 2. This section provides the theoretical basis for understanding tactic-oriented proving, but there is no space to treat the subject further.

1.3. Highlights. Section 2 stresses a typed presentation of the predicate calculus because we deal with a general mechanism for making the judgment that a formula is *meaningful*. This is done first in the style of Martin-Löf as expressed in Nuprl, but I also mention the essential novelty in Nuprl's use of *direct computation* rules which go beyond Martin-Löf's inference rules by allowing reasoning about general recursive functions (say defined by the Y combinator). This is a powerful mechanism that is not widely known.

Section 2 also stresses the notion that *proof expressions* are sensible for classical logic (see Constable [1989]). This separates an important technique from its origins in constructive logic. So the account of Typed Logic covers both constructive as well as classical logic. It also lays the ground work for tactics.

Section 3 features a very simple fragment of type theory which illustrates the major design issues. The fragment has a simple (predicative) inductive semantics. We keep extending it until the semantics requires the insights from Allen [1987a].

The treatment of recursive types in Section 4 is quite simple because it deals only with partial types. It suggests that just as the notion of program "correctness" can be nicely factored into partial and total correctness, the rules for data types can also be factored this way.

Table 1 shows how similar concepts change their appearance as they are situated in the three different contexts.

2. Typed logic

Ordinary mathematical statements are usually expressed in a *typed language*. Consider the trivially true proposition: "if there is a rational number q whose

Typed Logic	Type Theory	Typed Programming
terms with binding	terms with binding	terms with binding
definitional equality	computation rules	observational equivalence & bisimulation
proofs as objects	content of proofs	programs
$Prop_i$	$Type_i$ or U_i	*Type* (partial types)
equality propositions	equality judgments	equality functions
consistency	nontriviality	representation integrity (internal & external)
standard models	inductive definitions	termination conditions
proof synthesis	element synthesis	program synthesis
tactic-tree proof	derivation	proof data type & tactics

Table 1: Type concepts

absolute value is less than the reciprocal of any natural number n, then for every natural number n, there is a real number r whose absolute value is also less than $1/n$." Symbolically we express this as follows for $\mathbb{N}$ the natural numbers, $\mathbb{Q}$ the rationals, and $\mathbb{R}$ the reals:

$$* \quad \exists q{:}\mathbb{Q}.\ \forall n{:}\mathbb{N}.\ (|q| < 1/n) \Rightarrow \forall n{:}\mathbb{N}.\ \exists r{:}\mathbb{R}.\ (|r| < 1/n)\,.$$

We can abstract on the relation $|r| < 1/n$, and speak of any relation L on $\mathbb{N} \times \mathbb{R}$ and recognize that $\mathbb{Q}$ is a *subtype* of $\mathbb{R}$ to obtain

$$\exists q{:}\mathbb{Q}.\ \forall n{:}\mathbb{N}.\ L\,(n,q) \Rightarrow \forall n{:}\mathbb{N}.\ \exists r{:}\mathbb{R}.\ L\,(n,r)\,.$$

Abstracting further, we know that for any types A, A' and B where A' is a *subtype* of A, say $A' \sqsubseteq A$ the following is true

$$\exists a{:}A'.\ \forall x{:}B.\ L(x,a) \Rightarrow \forall x{:}B.\ \exists a{:}A.\ L\,(a,x)\,.$$

This last statement is an *abstraction* of $*$ with respect to $\mathbb{Q}, \mathbb{N}, \mathbb{R}$ and the relation $|r| < 1/n$. It is these purely *abstract typed propositions* that we want to study in the beginning. We want to know what statements are true regardless of the types and the exact propositions. We are looking for those properties of mathematical propositions that are invariant under arbitrary replacement of types and propositions.

Here logic is presented in a way that relates it closely to the type theory of section 3 and the programming language of section 4. Essentially, we will be able to lay one presentation on top of the others and see a striking correspondence as Table 1 suggests. This goal leads to a novel presentation of logic because of the role of explicit typing judgments. We begin now to gradually make these ideas more precise.

2.1. Propositions

Relation to sentences. Use of the word *proposition* in logic refers to an idea that is new in this century. The English usage is from Russell's reading of Frege. To explain propositions it is customary to talk first about *declarative sentences* in some natural language such as English. A sentence is an aggregation of words which expresses a complete thought, and when that thought is an assertion, e. g. $0 < 1$, then the sentence is declarative. The thought expressed is the *sense* of the sentence. We are interested in the conditions under which we can *assert a sentence* or *judge it to be true.*

Logic is not concerned directly with the nature of natural language and sentences. It is a more abstract subject. The abstract object corresponding to a sentence is a *proposition.* As Church [1960] says "...a proposition as we use the term, is an abstract object of the same general category as class, number or function." He says that any concept of *truth-value* is a proposition whether or not it is expressed in a natural language.[9]

This definition from Frege [1903] as explained by Church [1956] will suffice even for the varieties of constructive logic we will consider. We can regard truth-values themselves as abstractions from a more concrete relationship, namely that we know evidence for the truth of a formula; by forgetting the details of the evidence, we come to the notion of a *truth-value.* We say that the asserted sentence is *true.* Thus, when we judge that a sentence or expression is a proposition, we are saying that we know what counts as evidence for its truth, that is, we understand what counts as a proof of it.

It is useful to single out two special propositions, say $\top$ for a proposition agreed to be true, accepted as true without further analysis. We can say it is a *canonical true proposition*, a generalization of the concrete proposition $0 = 0$ in $\mathbb{N}$. We say that $\top$ is atomically true. Likewise, let $\bot$ be a canonically false proposition, a generalization of the idea $0 = 1$ in $\mathbb{N}$; it has no proof.

The category of propositions, *Prop*. In order to relate this account of typed logic to the type theory of section 3 and to programming section 4, I would like to consider the collection of all propositions and refer to it as *Prop.*[10] But we know already from *Principia* that the concept of proposition is *indefinitely extensible.* Here is how Whitehead and Russell put the matter.

> "...vicious circles...[arise] from supposing that a collection of objects may contain members which can only be defined by means of the collection as a whole. Thus, for example, the collection of *propositions* will be supposed to contain a proposition stating that 'all propositions are either true or false.' It would seem, however, that such a statement

[9]There are, of course, opposing views; Wittgenstein [1953,1922] was concerned with sentences or *formulas* as is Quine [1960], and we access propositions through specific formulas.

[10]In topos theory *Prop* and the true propositions form the *subobject classifier*, Ω, $\top$ see Bell [1988], MacLane and Moerdijk [1992], Lambek and Scott [1986].

> could not be legitimate unless 'all propositions' referred to some already definite collection, which it cannot do if new propositions are created by statements about 'all propositions.' We shall, therefore, have to say that statements about 'all propositions' are meaningless. By saying that a set has 'no total,' we mean, primarily, that no significant statement can be made about 'all its members.' In such cases, it is necessary to break up our set into smaller sets, each of which is capable of a total. This is what the theory of types aims at effecting."

So we must be very careful about introducing this notion. There are *predicative* approaches to this which lead to *level restrictions* as in *Principia* and allow writing $Prop_i$ as the "smaller collections" into which *Prop* is broken (Martin-Löf [1982], Constable et al. [1986]).[11] There are *impredicative* approaches that allow *Prop* but restrict the logic in other ways (Girard, Taylor and Lafont [1989]). Ultimately we will impose level restrictions on *Prop* and relate it to another indefinitely extensible concept which we denote as *Type*. These restrictions and relationships will occupy us in section 3.

From the beginning I want to recognize that we intend to treat collections of propositions as mathematical objects — some collections will be types, some will be too "comprehensive" to be types. We will call these large collections *categories* or *classes* or *kinds* or *large types* to use names suggestive of "large collections."

The use of *Prop* as a concept in this account of logic, while not common in logic texts, provides an orientation to the subject which will gradually reveal problems that I think are both philosophically and practically important. So I adopt it here. We recognize at the onset that *Prop* will not denote *all* propositions. We will be developing an understanding of how to talk about such open-ended notions in a meaningful way. This development will confirm that at the very least we can *name them.* The issue is what else can we say?

The category of propositional functions. Frege's analysis of propositions into functions and arguments is central to modern logic. It requires us to consider the notion of a *propositional function.* Frege starts with concrete propositions such as $0 < 1$, then abstracts with respect to a term to obtain a form like $0 < x$ which denotes a *function in x.* In *Principia* notation, given a proposition ϕa, ϕx is the ambiguous value, and the propositional function is $\phi\hat{x}$; so $0 < 1$ can be factored as $\phi 1$ and abstracted as $0 < \hat{x}$. Also, in *Principia* a type is defined to be the range of significance of a propositional function; we might write the type as $type(x.\phi x)$. So the function maps this type to *Prop*. For example, $\frac{1}{3} < \frac{1}{2}$ might be abstracted to $\frac{1}{x} < \frac{1}{2}$; it is not meaningful for $x = 0$.

In the logic presented here, propositional functions also map types T to *Prop*. Given a type T we denote the category of propositional functions over T as $T \to Prop$. Instead of using $0 < x$, we denote the abstractions in the manner of Church

[11] In topos theory this leads to the Grothendieck topos which is definable in our predicative type theory of section 3.

with lambda notation, $\lambda(x.\ 0 < x)$. The details of the function notation will not concern us until section 2.9. It suffices now to say that given a specific proposition such as $0 < 1$, we require that the individuals such as $0, 1$ be elements of types, here N. The function maps N to *Prop*, thus an element of the category $(\mathrm{N} \to \mathit{Prop})$. Given a function P in $(T \to \mathit{Prop})$ and t in the type T, then $P(t)$ denotes the application of P to t just as in ordinary mathematics.

Types. Types are structured collections of objects such as natural numbers, N, or pairs of numbers $\mathrm{N} \times \mathrm{N}$ or lists of numbers, etc. In the section on type theory we will present specific concrete types, here we treat the notion abstractly. We think of the elements of the types as possibly given first without specifying the type; they might be built from sets for example or be the raw data in a computer (the bytes) or be physical objects or be given by constructions. Even when we are thinking of constructing objects according to a method specified by the type (as for the natural numbers based on zero and successor), still we imagine the object as existing without type information attached to it, and thus objects can be in more than one type.

The critical point about judging that T is a type is knowing what it means for an expression to be an element of T. This is what we know when we show that T is a type.

Assertions about objects require their classification into types. Moreover, we need relative or hypothetical classification because an object t occurring in a proposition may be built contingently from other objects x_i of type T_i, and will be in some type T. We understand these relative type membership judgments as follows. To judge that $t[x]$ is a member of T_2 provided x is a member of T_1 means that for any object t_1 of type T_1, $t[t_1]$ is an object of T_2. We write this as $x : T_1 \vdash t[x] \in T_2$. We extend the notation to n assumptions by writing $x_1 : T_1, \ldots, x_n : T_n \vdash t \in T$ for x_i distinct identifiers. (We write $t[x_1, \ldots, x_n]$ only when we need to be explicit about the variables; this notation is a *second order variable.*) We give concrete examples of this judgment below and discuss them at length in section 3.

We don't treat this judgment as an assertion. It summarizes what must hold for a statement about t to make sense. It is not the kind of expression that is true or false. For example, when we say $0 \in \mathrm{N}$, we are thinking that 0 is a natural number. This is a *fact.* We are thinking "0 *is_a* N," and it does not make sense to deny this, thinking that 0 is something else.

In order to consider propositional functions of more than one argument, say $P(x, y)$, we explicitly construct pairs of elements, $\langle x, y\rangle$ and express $P(x, y)$ as $P(\langle x, y\rangle)$. The pair $\langle x, y\rangle$ belongs to the Cartesian product type. So our types have at least this structure.

Definition. If A, B are types, then so is their *Cartesian product*, $A \times B$. If $a \in A$ and $b \in B$ then the *ordered pair* $\langle a, b\rangle$ belongs to $A \times B$. We write $\langle a, b\rangle \in A \times B$. We write $1of\ (\langle a, b\rangle) = a$ and $2of\ (\langle a, b\rangle) = b$. Let A^n denote $A \times \cdots \times A$ taken n times.

Propositions are elements of the large type (or category) *Prop*. We use membership judgements

$$x_1 : T_1, ..., x_n : T_n \vdash P \in Prop$$

or an abbreviation $\mathcal{T} \vdash P \in Prop$ to define them. Atomic propositions are constants: $\top$ for a canonically true one and $\perp$ for a canonically false one, or propositional variables or applications of propositional function variables in the large type $(T \to Prop)$ for some type T. Compound propositions are built from the logical connectives $\&, \vee, \Rightarrow$ or the logical operators $\exists$ and $\forall$.

The membership rules are that if $\mathcal{T} \vdash F \in Prop$ and $\mathcal{T} \vdash G \in Prop$, then $\mathcal{T} \vdash (F\ op\ G) \in Prop$ for *op* a connective; and if $x_1 : T_1, ..., x_n : T_n \vdash F \in Prop$, then $\mathcal{T}' \vdash (Qx_i : T_i . F) \in Prop$ where Q is a quantification operator and where $\mathcal{T}'$ is obtained by removing the typing assumption $x_i : T_i$.

The usual names for the compound propositions are:

proposition		English equivalent	operator name
$(F \& G)$	is	F and G	conjunction
$(F \vee G)$	is	F or G	disjunction
$(F \Rightarrow G)$	is	F implies G	implication
$F(t)$	is	F at t	instance
$\forall x{:}A.F(x)$	is	for all x of type A, $F(x)$	universal quantification
$\exists x{:}A.F(x)$	is	for some x of type A, $F(x)$	existential quantification

With these definitions in place we can write examples of general typed propositions. We use *definiendum* == *definiens* to write definitions.

Definition. For any propositions F and G define

$$\begin{aligned} \neg F &== (F \Rightarrow \perp) \\ (F \Leftarrow G) &== (G \Rightarrow F) \\ (F \Leftrightarrow G) &== ((F \Leftarrow G) \& (F \Rightarrow G)). \end{aligned}$$

For $P : ((A \times B) \to Prop)$ let $P(x, y) == P(\langle x, y \rangle)$.

Examples. Let $P \in (A \to Prop), Q_i \in (B \to Prop), R \in ((A \times B) \to Prop)$ and $C \in Prop$.

1. $\forall x{:}A.\ \forall y{:}B.\ (P(x) \& Q(y)) \Leftrightarrow \forall x{:}A.\ P(x) \& \forall y{:}B.\ Q(y)$
2. $\exists x{:}A.P(x) \& \exists y{:}B.\ Q(y) \Rightarrow \exists z{:}A \times B.\ R(z)$
3. $\neg \forall x{:}A.\ P(x) \Leftrightarrow \exists x{:}A.\ \neg P(x)$
4. $\forall x{:}A.\ (C \Rightarrow P(x)) \Leftrightarrow (C \Rightarrow \forall x{:}A.\ P(x))$
5. $\exists y{:}B.\ \forall x{:}A.\ R(x, y) \Rightarrow \forall x{:}A.\ \exists y{:}B.\ R(x, y)$

2.2. Judgments and proofs

Knowledge arises when we judge a proposition to be true. A proposition P becomes an *assertion* when we judge it to be true or assert it. In *Principia* this judgment is called an *assertion* and is written $\vdash P$.

Normally we assert a proposition when we *know it to be true*, but people also make so-called "blind assertions" which are made without this knowledge but happen to be true because someone else knows this or the person "speaking blindly" discovers it later. (These "blind assertions" normally make a person, especially students in exams, anxious.)

Assertions are not the only form of knowledge. We follow Per Martin-Löf [1982] and speak also of judgments of the form $P \in Prop$ that we discussed above. These are *typing judgments.* They also convey knowledge and need to be made evident, but we consider them as a different category of knowledge from assertions. Indeed, we see that knowing these judgments is part of knowing truth judgments because we must know that the object P we are asserting is a proposition before (or as) we judge it to be true.[12]

In section 2.1 we treated this notion in the background with the notation $P \in Prop$, $P \in A \to Prop$, without explaining judgments in general. In most logic textbooks this judgment is reduced to a question of correct parsing of formulas, i.e., is made *syntactic* and is thus prior to truth. But we follow Russell and Martin-Löf in believing it to be a semantic notion that can be treated in other ways when we formalize a theory.

For the predicate calculus, we leave most of these typing judgments implicit and adopt the usual convention that all the formulas we examine represent well-formed propositions. In the full typed logic this condition cannot be checked syntactically, and explicit typing judgments must be made.

In general, to make a judgment is to know it or for it to be *evident.* It does not make sense for a judgment to be evident without a person knowing it. However, we recognize that there are propositions which are true but were not known at a previous time. So at any time there are propositions which are not known, but will be asserted in the future.

True propositions and proofs. One of the most interesting properties of a proposition is whether the thought it expresses is *true.* Here is a more concrete way to say this. To grasp the sense of a proposition is to understand what counts as evidence for its truth. To know whether a proposition is true, we must find evidence for it. Trying to find evidence is a *mathematical task*, and it is an abstract activity. [13]

The evidence for an assertion is called a *proof.* Proving a proposition is the way

[12]In Martin-Löf's theorem these judgments are *prior* to assertion; in Nuprl they can be simultaneous with assertion.

[13]But the consequences of finding a proof or a disproof can be very concrete and very significant. For instance, a purported proof might cause someone to test a device in the belief that it is safe to do so. If the proof is flawed, the device might destroy something of great value.

of knowing it.[14] When we judge that an expression is a proposition, we specify what counts as a proof of it. For the propositions $\top, \bot$ $(P\&Q), (P \vee Q), \exists x : A.\ P(x)$ it is easy to specify this. To prove $\top$ we could cite an axiom, that is, we say it is *self-evident.* As we said before, $\bot$ has no proofs. To prove $(P\&Q)$ we prove P and prove Q. To prove $(P \vee Q)$ we either prove P or prove Q. To prove $\exists x : A.\ P(x)$ we exhibit an element a of A and prove $P(a)$.

To say what it means to prove $(P \Rightarrow Q)$ we need to understand logical consequence or what is the same in this article, *hypothetical proof.* We write this as *hypothetical judgment* $P \vdash Q$ which reads, assuming P is true, we can prove that Q is. What this means is that if we have a proof p of P, then we can build a proof q of Q using p.

To discuss nested implications such as $P \Rightarrow (Q \Rightarrow P)$ we need to understand hypothetical judgments of the form $P_1, \ldots, P_n \vdash Q$ which means that assuming we have proofs p_i of P_i, we can find a proof q of Q.

To prove $\forall x{:}A.P(x)$ we need the hypothetical judgment $x{:}A \vdash P(x)$ which says that given any a of type A, we can find a proof of $P(a)$. Combining these two forms of hypothetical judgments we will need to consider $H_1, \ldots, H_n \vdash P$ where H_i can be either a proposition or a type declaration such as $x{:}A$.

This account of provability gives what we call a *semantics of evidence.* Depending on the interpretation of the functionality, judgments like $P \vdash Q$ or $x{:}A \vdash P(x)$ we can explain both classical and constructive logic in this way. These ideas will become clearer as we proceed.

In general, there is no systematic way to search for a proof. Indeed, the notion of proof we have in mind is not for any *fixed* formal system of mathematics. We are interested mainly in open-ended notions. Like the concept of a proposition, the concept of a proof is inexhaustible or *open* or *creative.* By Gödel [1933] and Tarski [1956] we know that for any consistent *closed non-trivial formal system,* we can systematically enlarge it, namely by adding a rule asserting its consistency or giving a definition of truth for that system.

For the collection of *pure* abstract propositions, there is a systematic way to search for a proof. If we want to describe this procedure, then it is necessary to have a *representation* of propositions as data that lets us access their structure. We can do this with the inductive definition of propositions given next, but it is more natural to build a representation that directly expresses the *linguistic structures* that we use when writing and manipulating formulas. This is the traditional approach to the study of logic and provability. We take it up in the subsections on Formulas and Formal Proofs.

[14] When we need to present evidence for a typing judgment, we will incorporate that into our proofs as well and speak of proving a typing. One might want to give this a special name, such as a *derivation,* but in Nuprl we use the term proof. These typing "proofs" never have interesting computational content.

2.3. Pure propositions

The traditional approach to studying logic is to first isolate the propositional calculus and then the first-order predicate calculus, and to study completeness results for these calculi, since completeness results tie together the semantic and proof-theoretic concepts. These topics are covered by defining restricted formal languages. We will take this approach to typed logic starting in section 2.4 on formulas. But many of the concepts can also be presented by a mathematical analysis of propositional functions without resort to linguistic mechanisms and formulas.

There is one outstanding technical problem about the propositional calculus which benefits from this direct analysis, namely a functional approach to completeness of the intuitionistic propositional calculus. The most widely known completeness theorem for this calculus is based on Kripke semantics (Kripke [1965]); this account of propositional calculus semantics, while illuminating and technically elegant, and even constructively meaningful, is not *faithful* to constructive semantics. In particular, it is not based on the type-theoretic semantics we offer in part 3. Providing a completeness result for a *constructively faithful* semantics is an open area, and it seems to me that Martin-Löf's inductive semantics has created new opportunities here to produce definitive results.

The key to a constructive semantics might be a careful study of a functional approach to propositions that allows us to express the functional *uniformity of proofs* that is central to completeness.[15] As a start, we can try to understand the basic concept of a pure propositional function without resort to formal languages.

Consider a propositional function of one argument, $P \Rightarrow P$. This can be understood as a function from *Prop* to *Prop*.[16] The function $P \Rightarrow (Q \Rightarrow P)$ in variables P, Q is a two-argument function, most naturally from *Prop* to ($Prop \rightarrow Prop$) as would be clear from writing the function as

$$\lambda(P.\ \lambda(Q.\ P \Rightarrow (Q \Rightarrow P))).$$

We could also think of this as a mapping from $Prop \times Prop$ to *Prop* if we took pairs as arguments (writing $\lambda(z.\ 1of(z) \Rightarrow (2of(z) \Rightarrow 1of(z)))$ in $Prop^2 \rightarrow Prop$). For ease of analysis, we will consider propositional functions from the Cartesian power, $Prop^n$, into *Prop*. The constants $\top$ and $\bot$ are regarded as zero-ary functions, and for convenience define $Prop^0 = \mathbf{1}$ for $\mathbf{1}$ the *unit type*. Then $f(x) = \top$ and $f(x) = \bot$ are in $Prop^0 \rightarrow Prop$.

The idea is to define the pure propositional functions inductively as a subtype of $Prop^n \rightarrow Prop$ constructed using only constant functions, simple projections like $proj_i^n(P_1, \ldots, P_n) = P_i$ and the operations $\&, \vee, \Rightarrow$ lifted up to the level of functions.

Each connective $\&, \vee, \Rightarrow$ can be lifted to the functions $Prop^n \rightarrow Prop$, namely given f and g, define $(f\ op\ g)(\bar{P}) = f(\bar{P})\ op\ g(\bar{P})$ where $\bar{P} \in Prop^n$. For example,

[15] Läuchli [1970] tries to express this uniformity using permutations.

[16] We will deal later with the issue of equality on *Prop*, which seems necessary to talk about *functions*.

if $f(P,Q) = P$ and $g(P,Q) = (Q \Rightarrow P)$ then $f \Rightarrow g$ is a function h such that $h(P,Q) = (P \Rightarrow (Q \Rightarrow P))$.

We can now define the general abstract propositional functions of n variables call the class $\mathcal{P}_n$ as the inductive subset of $Prop^n \to Prop$ whose base elements are the constant and projection functions,

$$C_\top(\bar{P}) = \top \qquad C_\bot(\bar{P}) = \bot$$
$$proj_i^n(\bar{P}) = P_i \quad \text{where } \bar{P} = \langle P_1, \ldots, P_n \rangle \text{ and } 1 \le i \le n.$$ [17]

Then given $f, g \in \mathcal{P}_n$ and given any lifted connective op, we have $(f\ op\ g) \in \mathcal{P}_n$. Nothing else belongs to $\mathcal{P}_n$. When we want to mention the underlying type, we write $\mathcal{P}_n$ as $\mathcal{P}(Prop^n \to Prop)$. Let $\mathcal{P} = \bigcup_{n=0}^{\infty} \mathcal{P}_n$; these are the *pure* propositions. Note that $\mathcal{P} = \bigcup_{n=0}^{\infty} \mathcal{P}(Prop^n \to Prop)$ is inductively defined. The *valid* elements of $\mathcal{P}$ are those functions $f \in \mathcal{P}$ such that for $f \in \mathcal{P}_n$ and $\bar{\mathcal{P}}$ *any* element of $Prop^n$, $f(\bar{\mathcal{P}})$ is true. Call these $True(\mathcal{P})$.

Using these concepts we can express the idea of a uniform functional proof. The simplest approach is probably to use a Hilbert style axiomatic base. If we take Heyting's or Kleene's axioms for the intuitionistic propositional calculus, then we can define $Provable_H(\mathcal{P})$ inductively. The completeness theorem we want is then $True(\mathcal{P}) = Provable_H(\mathcal{P})$.

We can use the same technique to define the pure typed propositional functions. First we need to define *pure type functions* $\mathcal{T}$ as a subset of $\text{Type}^n \to \text{Type}$ for $n = 1, 2, \ldots$. We take $n \ge 1$ since there are as yet no constant types to include. An example is $t(A, B) = A \times B$. Next we define the typed propositional functions $p : t(\bar{T}) \to Prop$.

In general we need to consider functions whose inputs are n-tuples of the type

$$(t_1(\bar{T}) \to Prop) \times \ldots \times (t_n(\bar{T}) \to Prop)$$

and whose output is a $Prop$. We do not pursue this topic further here, but when we examine the proof system for typed propositions we will see that it offers a simple way to provide abstract proofs for pure typed propositions that use only rules for the connectives and quantifiers — say a *pure proof.* There are various results suggesting that if there is any proof of these pure propositions, then there is a *pure proof.* These are *completeness results* for this typed version of the predicate calculus. We will not prove them here.

2.4. Formulas

Propositional calculus. Consider first the case of formulas to represent the pure propositions. The standard way to do this is to inductively define a class of *propositional formulas, PropFormula.* The base case includes

[17] Since we do not study any mapping of formulas to pure propositions, I have not worried about relating elements of P_n and P_m, $n < m$, by coherence conditions.

the propositional constants, *Constants* $= \{\top, \bot\}$, and propositional variables, *Variables* $= \{P, Q, R, P_1, Q_1, R_1, \ldots\}$. These are propositional formulas. The inductive case is

If F, G are *PropFormulas*, then so are $(F \,\&\, G)$, $(F \vee G)$, and $(F \Rightarrow G)$. Nothing else is a *PropFormula*.

We can assign to every formula a mathematical meaning as a pure proposition. Given a formula F, let $P_1, \ldots, P_n$ be the propositional variables occurring in it (say ordered from left to right); let $\bar{P}$ be the vector of them. Define a map from n variable propositional formulas, $PropFormulas_n$, into $(Prop^n \to Prop)$ inductively by

$$\begin{aligned}
[\![P_i]\!] &= proj_i^n \\
[\![(F \,\&\, G)]\!] &= [\![F]\!] \,\&\, [\![G]\!] \\
[\![(F \vee G)]\!] &= [\![F]\!] \vee [\![G]\!] \\
[\![(F \Rightarrow G)]\!] &= [\![F]\!] \Rightarrow [\![G]\!].
\end{aligned}$$

For each variable P_i, corresponds to the projection function $proj_i^n(\bar{P}) = P_i$. Say that F is *valid* iff $[\![F]\!]$ is a valid pure proposition.

Boolean valued formulas. If we consider a single-valued relation from propositions to their truth values, taken as Booleans, then we get an especially simple semantics. Let $\mathbb{B} = \{tt, ff\}$ and let $\boldsymbol{B} : Prop \times \mathbb{B} \to Prop$ such that $P \Leftrightarrow \boldsymbol{B}(P, tt)$.

In classical mathematics one usually assumes the existence of a *function* like $\boldsymbol{b}$, say $\boldsymbol{b} : Prop \to \mathbb{B}$ where $P \Leftrightarrow \boldsymbol{b}(P) = tt$ in $\mathbb{B}$. But since $\boldsymbol{b}$ is not a computable function, this way of describing the situation would not be used in constructive mathematics. Instead we could talk about "decidable propositions" or "boolean propositions."

$$BoolProp == \{\langle P, v\rangle : Prop \times \mathbb{B} | P \Leftrightarrow (v = tt \text{ } in \text{ } \mathbb{B})\}$$

Then there is a function $\boldsymbol{b} \in BoolProp \to \mathbb{B}$ such that $P \Leftrightarrow (\boldsymbol{b}(P) = tt \text{ in } \mathbb{B})$.

If we interpret formulas as representing elements of pure *boolean propositions*, then each variable P_i denotes an element of $\mathbb{B}$. An *assignment* α is a mapping of variables into $\mathbb{B}$, that is, an element of $Variables \to \mathbb{B}$. Given an assignment α we can compute a boolean value for any formula F. Namely

$$\begin{aligned}
Value(F, \alpha) = \; & \text{if } F \text{ is a variable, then } \alpha(F) \\
& \text{if } F \text{ is } (F_1 \; op \; F_2) \text{ then } Value(F_1, \alpha) \; bop \; Value(F_2, \alpha)
\end{aligned}$$

where *bop* is the boolean operation corresponding to the propositional operator *op* in the usual way, e. g.

P	Q	P $\&_b$ Q	P $\vee_b$ Q	P $\Rightarrow_b$ Q
tt	tt	tt	tt	tt
ff	tt	ff	tt	tt
tt	ff	ff	tt	ff
ff	ff	ff	ff	tt

Typed propositional formulas. To define typed propositional formulas, we need the notion of a type expression, a term, and a *type context* because formulas are built in a type context. Then we define propositional variables and propositional-function variables which are used along with terms to make atomic propositions in a context. From these we build compound formulas using the binary connectives $\&, \vee, \Rightarrow$, and the *typed quantifiers* $\forall x : A, \exists x : A$. We let *op* denote any binary connective and $Qx\!:\!A$ denote either of the quantifiers.

Type expressions. Let $A_1, A_2, \ldots$ be type variables, then A_i are type expressions.

If T_1, T_2 are type expressions, then so is $(T_1 \times T_2)$.
Nothing else is a type expression for now.

Terms. Let $x_1, x_2, \ldots$ be individual variables (or element variables); they are *terms.* If s, t are terms, then so is the ordered pair $\langle s, t\rangle$. Nothing else is a term for now.

Typing contexts. If $T_1, \ldots, T_n$ are type expressions and x_i, $i = 1, \ldots, n$ are *distinct* individual variables, then $x_i : T_i$ is a *type assumption* and the list $x_1\!:\!T_1, \ldots, x_n\!:\!T_n$ is a *typing context.* We let $\mathcal{T}, \mathcal{T}', \mathcal{T}_j$ denote typing contexts.

Typing judgments. Given a typing context, $\mathcal{T}$, we can assign types to terms built from the variables in the context. The judgment that term t has type T in context $\mathcal{T}$ is expressed by writing

$$\mathcal{T} \vdash t \in T.$$

If we need to be explicit about the variables of $\mathcal{T}$ and t, we use a *second-order variable* $t[x_1, \ldots, x_n]$ and write

$$x_1\!:\!T_1, \ldots, x_n\!:\!T_n \vdash t[x_1, \ldots, x_n] \in T$$

When using a second-order variable we know that the only variables occurring in t are x_i. We call these variables of t *free variables.*

Later, we give rules for knowing these judgments. Just as we said in section 2.2, it should be noted that $t \in T$ is *not a proposition*; it is *not* an expression that has a truth value. We are saying what an ordered pair *is* rather than giving a property of it. So the judgment $t \in T$ is giving the meaning of t and telling us that the expression t is *well-formed* or meaningful. In other presentations of predicate logic these judgments are incorporated into the syntax of terms, and there is an algorithm to check that terms are meaningful before one considers their truth. We want a more flexible approach so that typing judgments need not be decidable.

We let $P_1, P_2, \ldots$ denote *propositional variables,* writing $P_i \in Prop$, for propositional function variables, writing $P_i \in (T \rightarrow Prop)$ for T a type expression.

If $\mathcal{T} \vdash t \in T$ and $P \in (T \rightarrow Prop)$, then $P(t)$ is an *atomic formula in the context* $\mathcal{T}$ with the variables occurring in t *free*; it is an *instance* of P. Note, we abbreviate $P(\langle t_1, \ldots, t_n\rangle)$ by $P(t_1, \ldots, t_n)$. If t is a variable, say x, then $P(x)$ is an *arbitrary instance* or *arbitrary value* of P with free variable x. A propositional variable, $P_i \in Prop$, is also an atomic formula.

If F and G are formulas with free variables $\bar{x}, \bar{y}$ respectively in contexts $\mathcal{T}$, and if *op* is a connective and $Qx{:}A$ a quantifier, then

$$(F \; op \; G)$$

is a formula with free variables $\{\bar{x}\} \cup \{\bar{y}\}$ in context $\mathcal{T}$ and with *immediate subformulas* F and G;

$$Qv{:}T.F$$

is a formula in context $\mathcal{T}'$ where $\mathcal{T}'$ is $\mathcal{T}$ with $v{:}A$ removed; this formula has leading *binding operator* $Qv:A$ with *binding occurrence* of v whose *scope* is F, and its free variables are $\{\bar{x}\}$ with v removed, and all free occurrences of v in F become *bound* by $Qv{:}A$; its immediate subformula is F.

A formula is *closed* iff it has no free variables; such a formula is well-formed in an empty context, but its subformulas might only be well-formed in a context. A *subformula* G of a formula F is either an immediate subformula or a subformula of a subformula.

Examples. Here are examples, for $P_1 : A \to Prop,\ P_2 : B \to Prop,\ P_3 : A \times B \to Prop$.

$$(\forall x{:}A.\ \exists y{:}B.\ P_3(x,y) \Rightarrow (\exists x{:}A.\ P_1(x)\ \&\ \exists y : B.\ P_2(x)))$$

is a closed formula. $\forall x{:}A.\ \exists y{:}B.\ P_3(x,y)$ is an immediate subformula which is also closed, but $\exists y{:}B.\ P_3(x,y)$ is not closed since it has the free variable $x{:}A$; this latter formula is well-formed in the context $x{:}A$.

The atomic subformulas are $P_1(x), P_2(y)$, and $P_3(\langle x,y\rangle)$ which are formulas in the context $x{:}A,\ y{:}B$, and the typing judgment $x{:}A,\ y{:}B \vdash \langle x,y\rangle \in A \times B$ is used to understand the formation of $P_3(x,y)$ (which is an abbreviation of $P_3(\langle x,y\rangle)$).

2.5. Formal proofs

There are many ways to organize formal proofs of typed formulas, e. g. natural deduction, the sequent calculus, or its dual, tableaux, or Hilbert style systems to name a few. We choose a sequent calculus presented in a top-down fashion (as with tableaux). We call this a *refinement logic* (RL). The choice is motivated by the advantages sequents provide for automation and display.[18] Here is what a simple proof looks like for $A \in Type$, $P \in A \to Prop$; only the relevant hypotheses are mentioned and only the first time they are generated.

[18] This is the mechanism used in Nuprl and HOL; PVS uses multiple conclusion sequents.

	$\vdash \forall x : A.\ (\forall y : A.\ P(y) \Rightarrow \exists x : A.\ P(x))$	by $\forall R$
1.	$x : A \vdash \forall y : A.\ P(y) \Rightarrow \exists x : A.\ P(x)$	by $\Rightarrow R$
1.1	$f : (\forall y : A.\ P(y)) \vdash \exists x : A.\ P(x)$	by $\forall L$ on f with x
1.1.1	$l : P(x) \vdash \exists x : A.\ P(x)$	by $\exists R$ with x
1.1.1.1	$\vdash P(x)$	by *hyp* l
1.1.1.2	$\vdash x \in A$	by *hyp* x
1.1.2	$\vdash x \in A$	by *hyp* x

The schematic tree structure with path names is

```
                    ⊢ G
                     |
               1.  H1 ⊢ G1
                     |
             1.1   H2 ⊢ G2
                     /\
     1.1.1  H3 ⊢ G2      1.1.2  H2 ⊢ G3
              /\
1.1.1.1  H3 ⊢ G4   1.1.1.2  H3 ⊢ G3
```

Sequents. The nodes of a proof tree are called *sequents*. They are a list of *hypotheses* separated by the assertion sign, $\vdash$ (called turnstile or proof sign) followed by the *conclusion*. A hypothesis can be a typing assumption such as $x\!:\!A$ for A a type or a *labeled assertion*, such as $l\!:\!P(x)$. The label l is used to refer to the hypothesis in the rules. The occurrence of x in $x\!:\!A$ is an *individual variable*, and we are assuming that it is an object of type A. So it is an assumption that A is inhabited. Here is a sequent,

$$\boxed{x_1 : H_1, \ldots, x_n : H_n \vdash G}$$

where H_i is an assertion or a type and x_i is either a label or a variable respectively. The x_i are all distinct. G is always an unlabeled formula. We can also refer to the hypothesis by number, $1 \ldots n$, and we refer to G as the 0-th component of the sequent. We abbreviate a sequent by $\bar{H} \vdash G$ for $\bar{H} = (x_1 : H_1, \ldots, x_n : H_n)$; sometimes we write $\bar{x} : \bar{H} \vdash G$.

Rules. Proof rules are organized in the usual format of the single-conclusion sequent calculus. They appear in a table shortly. We explain now some entries of this table. There are two rules for each logical operator (connective or quantifier). The *right* *rule* for an operator tells how to decompose a conclusion formula built with that operator, and the *left rule* for an operator tells how to decompose such a formula when it is on the left, that is when it is a hypothesis. There are also trivial rules for the constants $\top$ and $\bot$ and a rule for hypotheses. So the rules fit this pattern and are named as shown.

	Left	Right
&	&L	&R
$\vee$	$\vee$L	$\vee$Rl $\vee$Rr
$\Rightarrow$	$\Rightarrow$L	$\Rightarrow$R
$\forall$	$\forall$L	$\forall$R
$\exists$	$\exists$L	$\exists$R
$\top$	-	$\top$R
$\bot$	$\bot$L	-

$$x_1 : H_1, \ldots, x_n : H_n \vdash H_i \text{ by } hyp\ x_i$$
$$1 \leq i \leq n$$

The $\vee$R rule splits into two, $\vee$Rl and $\vee$Rr. To prove $\bar{H} \vdash P \vee Q$ we can prove $\bar{H} \vdash P$ or we can prove $\bar{H} \vdash Q$. The first rule is $\vee$Rl because we try to prove the *left* disjunct. We pronounce these somewhat awkwardly as "or right left" and "or right right."

The other rules all have obvious pronunciations, e. g. "and left" for &L, "and right" for &R, etc. Some of them have ancient names, such as *ex falso libet* for "false left" (meaning "anything follows from false"). In Nuprl we use *any* for $\bot$L. Sometimes $\vee$L is called "cases", and instead of "by $\vee$L l" we might say "by cases on l" for l a label. For $\forall$L on l with a term t we might say "instantiating l with t."

The rule $\Rightarrow L$ is similar to the famous *modus ponens* rule usually written as

$$\frac{A \quad A \Rightarrow B}{B}.$$

In top down form it would be

$$\begin{array}{ll} A, A \Rightarrow B \vdash B & \text{by} \Rightarrow L \\ \quad A, B \vdash B & \\ \quad\quad A \vdash A & \end{array}$$

Some of the rules such as $\forall$L and $\exists$R require *parameters*. For example, to decompose $\forall x : T.P(x)$ as a hypothesis, we need a term $t \in T$. So the rule is $\forall$L on t. For $\exists x{:}T.\, P(x)$ as a goal, to decompose it, we also need a term $t \in T$; the decomposition generates the subgoal $P(t)$.

Magic rule. These rules do not allow us to prove the formula $P \vee \neg P$ nor $\neg\neg P \Rightarrow P$ nor any equivalent formula. If we add one of these formulas as an axiom scheme then we can prove the others. We can also prove them by adopting the *proof by contradiction* rule

$$H \vdash P \text{ by } \mathit{contradiction}$$
$$H, \neg P \vdash \bot$$

My preference is to base arguments for these formulas on the axiom scheme $P \vee \neg P$ called the *law of excluded middle* because these arguments have a special status in relating logic to computation and because this law is so important in philosophical and foundational discussions. In the organization I adopt, this is the only rule which does not fit the sequent pattern and it is the only rule not constructively justifiable as we will see later. I sometimes call the rule "magic" based on the discussion of justification to follow.

Justifications. The rule names and parameters to them make up a very important part of the proof called the *justification* of the inference step. We can think of the justification as an *operator on sequents* which decomposes the goal sequent into a subgoal sequents. This format for the justification reveals that role graphically.

$$\begin{array}{ll} \bar{x} : \bar{H} \vdash G & \text{by } r(\bar{x}; \bar{t}) \\ \quad 1.\bar{H}_1 \vdash G_1 & \\ \quad \vdots & \\ \quad k.\bar{H}_k \vdash G_k & \end{array}$$

For example

$$\begin{array}{l} \bar{H} \vdash (P \vee Q) \text{ by } \vee Rl \\ \quad 1.\bar{H} \vdash P \end{array}$$

The justification takes the variables and labels of $\bar{x}$ plus some parameters $\bar{t}$ and *generates* the k subgoals $\bar{H}_i \vdash G_i$. The hypothesis rule generates no subgoals and so terminates a branch of the proof tree. Such rules are thus found at the leaves.

By putting into the justifications still more information, we can reveal all the links between a goal and its subgoals. To illustrate this information, consider the $\Rightarrow$L rule and the &L rule.

$$\begin{array}{l} \bar{H}, f : (P \Rightarrow Q), \bar{J} \vdash G \text{ by } \Rightarrow L \text{ on } f \\ 1.\ \bar{H}, f : (P \Rightarrow Q), \bar{J} \qquad \vdash P \\ 2.\ \bar{H}, f : (P \Rightarrow Q), \bar{J}, y{:}Q \vdash G \end{array}$$

$$\begin{array}{l} \bar{H}, pq{:}P \,\&\, Q \vdash G \text{ by } \&L \\ \bar{H}, pq{:}P \,\&\, Q, p{:}P, q{:}Q, \bar{J} \vdash G \end{array}$$

If the $\Rightarrow$R justification provided the label y, then all the information for generating the subgoal would be present. If the &L rule provided the labels p, q then the data is present for generating its subgoals as well. So we will add this information to form a *complete justification.*

Notice that these labels behave like the variable names x_i in the sense that we can systematically *rename* them without changing the meaning of a sequent or a justification. They act like *bound variables* in the sequent. The phrase new u, v in a justification allows us to explicitly name these bound variables.

Structural rules. Sequents as defined here are based on lists of formulas, so the rules for decomposing on the left must refer to the position of the formula. This is indicated by supplying a context around the formula, typically of the form $\bar{H}, x : F, \bar{J} \vdash G$. The *cut* rule specifies the exact location at which the formula is to be introduced into a hypothesis list, and *thin* does the same.

By combining applications of *cut* and *thin*, hypotheses can be moved (exchanged) or contracted. The so-called *structural rules* are included among these rules.

2.6. Proof expressions and tactics

Complete justifications. If there is enough information in a justification to generate the subgoals, then the tree of justifications and the top goal can *generate* the whole proof. Moreover, the tree of justifications can be combined into a single "algebraic expression" describing the whole proof. Indeed, the proof tree stripped of the sequents is just a parse tree for this expression.

If we present the justifications in the right way we can read the rules annotated by them as an *attribute grammar* (c.f. Reps and Teitelbaum [1988], Reps [1982], Griffin [1988a]) for generating an expression describing the proof called a *proof expression.* Consider the case of the $\Rightarrow L$ and $\& L$ rules again. Suppose we let p and g denote proof expressions for the subgoals, then

$$\begin{array}{ll} \bar{x} : \bar{H}, f : (P \Rightarrow Q), \bar{J} \vdash G & \text{by } \Rightarrow L \text{ on } f \\ \quad - \quad \vdash P & \text{by } p(\bar{x}) \\ \quad -, y : Q \vdash G & \text{by } g(\bar{x}, y) \end{array}$$

If we think of the proof expressions $p(\bar{x})$ and $g(\bar{x}, y)$ as being *synthesized* up from the subtrees, then the complete proof information for the goal sequent is

$$\boxed{\Rightarrow L \text{ on } f \text{ from } p(\bar{x}) \text{ and from } g(\bar{x}, y) \text{ with new } y}$$

Organizing this into a more compact expression and recognizing that y is a new bound variable, a suggestive expression is

$$\boxed{\Rightarrow L(f; p(\bar{x}); y . g(\bar{x}, y))}$$

Here we use the "dot notation" $y.g(\bar{x}, y)$ to indicate that y is a new *bound label* in the proof expression $g(\bar{x}, y)$. The dot notation is used with quantifiers as in $\forall x{:}A.\,F$ to separate the binding operator $\forall x{:}A$ from the formula F. Likewise, in the lambda notation, $\lambda(x.b)$, the dot is used to indicate the beginning of the scope of the binding of x.

In the case of $\&L$, the rule with proof expressions looks like

$$\bar{x}{:}\bar{H}, z{:}P\&Q \vdash G \text{ by } \&L \text{ in } z \text{ with new } u, v$$
$$\bar{x}{:}\bar{H}, u{:}P, v{:}Q \vdash G \text{ by } g(\bar{x}, u, v)$$

A compact notation is

$$\&L(z; u, v.\ g(\bar{x}, u, v))$$

Here u, v are new labels which again behave like bound variables in the proof expression.

The justification for $P \vee \neg P$ will be the term $magic(P)$. This is the only justification term that requires the formula as a subterm.

With this basic typed predicate logic as a basis, we will now proceed to add a number of specific types, namely natural numbers, lists, functions, sets over a type, and so-called quotient types. Each of these shows an aspect of typed logic. Note, in these rules we are presupposing that P, Q, and the formulas in $\bar{H}$ are well-formed according to the definition of a formula and that the type expressions are also well-formed in accordance with the typing rules. As we introduce more types, it will be necessary to incorporate typing judgments as subgoals. The Nuprl logic of Constable et al. [1986] relies on such subgoals from the beginning so that the caveat just stated for this table of rules is unnecessary there.

Tactics. Complete justifications will generate the entire proof given the goal formula because the rule name, and labeling formation and parameters are enough data to generate subgoals from the goals. So the subgoals are computable from the part of the justification that does not include the proof expression for the subproofs (the synthesized expressions). This fact suggests a way to automate interactive proof generation. Namely, a program called a *refiner*, takes a goal and a complete justification and produces the subgoals. Nuprl works this way.

Nuprl also adapts tactics from LCF (Gordon, Milner and Wadsworth [1979]) into the proof tree setting to get a notion of *tactic-tree proof* (Allen et al. [1990], Basin and Constable [1993], Griffin [1988b]). In this setting the justifications are called primitive refinement tactics. These can be combined using procedures called *tacticals*. For example, if a refinement r_o generates subgoals $G_1, \ldots, G_n$ when applied to sequent G_o, then the compound refinement tactic written r_o THENL$[r_1; \ldots; r_n]$ executes r_o, then applies r_i to subgoal G_i generated by r_o.

There are many tacticals (c.f. Jackson [1994a], Constable et al. [1986]); two basic ones are ORELSE and REPEAT. The ORELSE tactical relies on the idea that a refinement might fail to apply, as in trying to use $\&R$ on an implication. In r_o ORELSE r_1, if r_o fails to decompose the goal, then r_1 is applied.

Table of justification operators

	Left(L)	Right(R)
&	$\bar{H},\ x{:}P\&Q,\ \bar{J} \vdash G$ by $\&L(x;u,\ v.\ g(u,v))$ 1. $\bar{H},\ x{:}P\&Q,\ u{:}P,\ v{:}Q,\ \bar{J} \vdash G$ by $g(u,v)$	$\bar{H} \vdash P\&Q$ by $\&R(p;q)$ 1. $\bar{H} \vdash P$ by p 2. $\bar{H} \vdash Q$ by q
$\vee$	$\bar{H},\ x{:}P \vee Q,\ \bar{J} \vdash G$ by $\vee L(x;u.\ g_l(u);v.\ g_r(v))$ 1. $\bar{H},\ x{:}P \vee Q, u{:}P,\ \bar{J} \vdash G$ by $g_l(u)$ 2. $\bar{H},\ x{:}P \vee Q, v{:}Q,\ \bar{J} \vdash G$ by $g_r(v)$	$\bar{H} \vdash P \vee Q$ by $\vee Rl(p)$ 1. $\bar{H} \vdash P$ by p $\bar{H} \vdash P \vee Q$ by $\vee Rr(q)$ 1. $\bar{H} \vdash Q$ by q
$\Rightarrow$	$\bar{H},\ x{:}P \Rightarrow Q,\ \bar{J} \vdash G$ by $\Rightarrow L(x;p;y.\ g(y))$ 1. $\bar{H},\ x{:}P \Rightarrow Q,\ \bar{J} \vdash P$ by p 2. $\bar{H},\ x{:}P \Rightarrow Q,\ \bar{J},\ y{:}Q \vdash G$ by $g(y)$	$\bar{H} \vdash P \Rightarrow Q$ by $\Rightarrow R(x.\ q(x))$ $\bar{H},\ x{:}P \vdash Q$ by $q(x)$
$\forall$	$\bar{H},\ x{:}\ \forall z{:}A.\ P(z),\ \bar{J} \vdash G$ by $\forall L(x;a;y.\ g(y))$ 1. $\bar{H},\ x{:}\ \forall z{:}A.\ P(z),\ \bar{J} \vdash a \in A$ 2. $\bar{H},\ x{:}\ \forall z{:}A.\ P(z),\ \bar{J},\ y{:}P(a) \vdash G$ by $g(y)$	$\bar{H} \vdash \forall z{:}A.P(z)$ by $\forall R(z.\ p(z))$ new w $\bar{H},\ w{:}A \vdash P(w)$ by $p(w)$
$\exists$	$\bar{H},\ x{:}\ \exists z{:}A.\ P(z),\ \bar{J} \vdash G$ by $\exists L(x;u,v.\ g(u,v))$ new u,v $\bar{H},\ x{:}\ \exists z{:}A.\ P(z),\ u{:}A, v{:}P(u),\ \bar{J} \vdash G$ by $g(u,v)$	$\bar{H} \vdash \exists z{:}A.\ P(z)$ by $\exists R(a;p)$ 1. $\bar{H} \vdash a \in A$ 2. $\bar{H} \vdash P(a)$ by p
$\bot$ $\top$	$\bar{H},\ x{:}\ \bot,\ \bar{J} \vdash G$ by $any(x)$	$\bar{H} \vdash \top$ by true
H_i	$x_1{:}H_1, \ldots, x_n{:}H_n \vdash H_i$	by $hyp\, x_i$ $i = 1, \ldots, n$ (recall x_i are *distinct*)
cut	$\bar{H},\ \bar{J} \vdash G$ by cut$(x.g(x);c)$ $\bar{H},\ x{:}C,\ \bar{J} \vdash G$ by $g(x)$ $\bar{H},\ \bar{J} \vdash C$ by c	Assert C @ i where i locates C in $\bar{H}, \bar{J}$.
thin	$\bar{H},\ x{:}P,\ \bar{J} \vdash G$ by g $\bar{H},\ \bar{J} \vdash G$ by g	Thin @ i where i locates $x{:}P$ in $\bar{H},\ \bar{J}$.

Magic: $H \vdash P \vee \neg P$ by $magic(P)$

We will use tacticals to put together compound justifications when the notation seems clear enough.

2.7. Natural numbers

One of the most basic mathematical types is $\mathbf{N}$, the natural numbers. This type is formed by the rule $H \vdash \mathbf{N} \in \text{Type}$. The type is inductively defined by the rules which say that $0 \in \mathbf{N}$, and if $n \in \mathbf{N}$ then $suc(n) \in \mathbf{N}$. The typing judgments we need are

$H \vdash 0 \in \mathbf{N}$	type_of_zero
$H \vdash suc(n) \in \mathbf{N}$ $\quad H \vdash n \in \mathbf{N}$	type_of_successor

To express the fact that $\mathbf{N}$ is inductively defined we use the rule of mathematical induction. In its unrestricted form, this essentially says that nothing else is a member of $\mathbf{N}$ except what can be generated from 0 using *suc*. But the form of the rule given here does not quantify over all propositional functions on $\mathbf{N}$, so it is not a full statement of the principle.

Suppose $P\colon(\mathbf{N} \times A) \to Prop$, then

$$\begin{array}{l} \bar{x}\colon\bar{H},\ n\colon\mathbf{N} \vdash P(n,\bar{x}) \text{ by } ind(n; p_o; u, i,\ p_s(u,i,\bar{x})) \\ \quad \bar{x}\colon\bar{H}, n\colon\mathbf{N} \vdash P(0) \text{ by } p_o \\ \quad \bar{x}\colon\bar{H}, n\colon\mathbf{N}, u\colon\mathbf{N}, i\colon P(u,\bar{x}) \vdash P(suc(u),\bar{x}) \text{ by } p_s(u,i,\bar{x}) \end{array}$$

Arithmetic. When we display proofs of arithmetical propositions, we will assume that there is an automatic proof procedure which will prove any true quantifier free conclusion in a sequent involving 0, $suc(n)$, $+$, $-$, $*$, $=$ and $<$. So for example, here are some arithmetic facts in this category

$$0 < x, y < suc(z) \vdash y * x < suc(z * x).$$

Although there is no proof procedure with this power (the problem is undecidable), there are good arithmetical proof procedures for *restricted arithmetic* (Arith, see Boyer and Moore [1988], Church [1960]) and *linear arithmetic* (SupInf, see Chan [1982], Shostak [1979], Bledsoe [1975]). We refer the interested reader to the citations for details. The use of Arith allows us to present proofs in a form close to that of Nuprl (Constable et al. [1986], Jackson [1994c]). Here are two proofs of the same trivial theorem, one inductive, one not. We write Arith$(l_1, \ldots, l_n)$ to show which labeled hypotheses are used by the proof procedure. For readability we intentionally elided parts of the justification, using —.

$$\begin{array}{ll} \vdash \forall x\colon\mathbf{N}.\ \exists y\colon\mathbf{N}.\ (x < y) & \text{by } \forall R(x, \_\_) \\ \quad x\colon\mathbf{N} \vdash \exists y\colon\mathbf{N}.\ (x < y) & \text{by } \exists R(s(x); \_\_) \\ \quad\quad \vdash x < suc(x) & \text{by Arith} \end{array}$$

The complete proof expression is $\forall R(x.\ \exists R(suc(x); \text{Arith}))$.

$$\begin{array}{lr} \vdash \forall x{:}\mathbb{N}.\ \exists y{:}\mathbb{N}.\ (x < y) & \text{by } \forall R(x.\ ---) \\ x{:}\mathbb{N} \vdash \exists y{:}\mathbb{N}.\ (x < y) & \text{by } ind(x; ---;\ ) \\ \vdash \exists y{:}\mathbb{N}(0 < y) & \text{by } \exists R(suc(0); \text{Arith}) \\ x{:}\mathbb{N}, u{:}\mathbb{N}, i{:}\ \exists y{:}\mathbb{N}.\ (u < y) \vdash \exists y{:}\mathbb{N}.\ (suc(u) < y) & \text{by } \exists L(i; y_o, ---) \\ x{:}\mathbb{N}, u{:}\mathbb{N}, y_o{:}\mathbb{N}, l{:}(u < y_o) \vdash \exists y{:}\mathbb{N}(suc(u) < y) & \text{by } \exists R(suc(y_o); ---) \\ \vdash (suc(u) < suc(y_o)) & \text{by Arith}(l) \end{array}$$

The complete proof expression is

$$\forall R(x.\ ind(x; \exists R(suc(0);\ \text{Arith}); u, i.\ \exists L(i; y_0, l.\ \exists R(suc(y_0);\ \text{Arith}(l))))).$$

The following example will provide another compact proof expression. It shows that integer square roots exist without using Magic (O'Leary et al. [1995]). First we specify these roots. Let $Root(r,n) == r^2 \le n < (r+1)^2$.

Theorem. $\vdash \forall n{:}\mathbb{N}.\ \exists r{:}\mathbb{N}.\ Root(r,n)$.

by $\forall R$ THEN *ind new u*
base case
1. $n{:}\mathbb{N}$
   $\vdash \exists r{:}\mathbb{N}.\ Root(r,0)$
   by $\exists R\, 0$ THEN Arith

induction case
1. $n{:}\mathbb{N}$
2. $u{:}\exists r{:}\mathbb{N}.\ Root(r,n)$
   $\vdash \exists r{:}\mathbb{N}.\ Root(r, suc(n))$
   by $\exists L$ *on u new* r_o, v
3. $r_o{:}\mathbb{N}$
4. $v{:}Root(r_o, n)$
   $\vdash \exists r{:}\mathbb{N}.\ Root(r, suc(n))$
   by *cut* $(r_o+1)^2 \le suc(n)\ \vee\ suc(n) < (r_o+1)^2$ with label d THENA Arith.
   (This rule generates two subgoals. The "auxiliary one" is to prove the cut formula. That subgoal can be proved by Arith, so we say THENA Arith to indicate this.)

(The "main" subgoal is this one.)
5. $d{:}(r_o+1)^2 \le suc(n)\ \vee\ suc(n) < (r_o+1)^2$
   $\vdash \exists r{:}\mathbb{N}.\ Root(r,n)$
   by $\vee L$ *on d*
   (This is case analysis on the cases in hypothesis 5.)
6. $(r_o+1)^2 \le suc(n)$
   $\vdash \exists r{:}\mathbb{N}.\ Root(r,n)$
   by $\exists R(r_o+1)$ THEN *SupInf*
   (Since $r_o^2 \le n < (r_o+1)^2$, from $(r_o+1)^2 \le suc(n)$ we know $(r_o+1)^2 \le suc(n) < ((r_o+1)+1)^2$ since $n+1 < (r_o+1)^2+1$. The SupInf procedure can find this proof.)

6. $suc(n) < (r_o + 1)^2$
   $\vdash \exists r : \mathrm{N}.\ Root(r, n)$
   by $\exists R r_o$ THEN Arith
   (Since $r_o^2 \leq n$ we know immediately that $r_o^2 \leq suc(n)$.)

The proof expression corresponding to this is

$$\forall R(n.\ ind(n; \exists R(0; Arith);\ n, u.\ \exists L(u; r_o, v.\ cut(d.\ VL(d; \exists R(r_o + 1; SupInf); \exists R(r_o; Arith)); Arith)))).$$

In the appendix, section 6.2, we consider another simple arithmetic example and show a complete Nuprl proof.

2.8. Lists

Sequences are basic forms of construction in mathematics, often written $(a_1, \ldots, a_n)$. With the widespread use of programming languages we have come to distinguish several data types associated with sequences as distinct types. There are *lists*, *arrays*, and *sequences*, and they have different mathematical properties. We first look at lists.

If A is a type, then so is A list. We could write an informal rule like this

$$\frac{A \in Type}{A\ list \in Type}.$$

The elements of an A list are nil and $(a.l)$ where $a \in A$ and $l \in A$ list. If A is N, these are lists:

$$nil \quad (2.\ nil) \quad (1.\ (2.\ nil))$$

All lists are built up from nil and the operation of pairing an element of A with a list. The typing rules are

$\bar{H} \vdash nil \in A\ list$ *type_of_nil*

$\bar{H} \vdash (a.\ l) \in A\ list$ *type_of_cons*
  $\bar{H} \vdash a \in A$
  $\bar{H} \vdash l \in A\ list$

Equality on lists is given by these rules.

$\bar{H} \vdash (h.\ t) = (h'.\ t')$ in $A\ list$ by $list - eq$
  $\bar{H} \vdash h = h'$ in A
  $\bar{H} \vdash t = t'$ in $A\ list$

$\bar{H} \vdash nil = nil$ in $A\ list.$ by $nil - eq$

For every type A, A list is an inductive type with base case of nil. The inductive character of the type is given by its induction rule stated below for $P \in (A \text{ list} \times T \Rightarrow Prop)$

$$\bar{x}\!:\!\bar{H}, l\!:\!A\ list \vdash P(l,\bar{x}) \text{ by } list\_ind\,(l;p_o;h,t,i.\,p(h,t,i,\bar{x}))$$
$$\bar{x}\!:\!\bar{H} \qquad\qquad \vdash P(nil,\bar{x}) \text{ by } p_o$$
$$\bar{x}\!:\!\bar{H}, l\!:\!A\ list, h\!:\!A, t\!:\!A\ list, i\!:\!P(t,\bar{x}) \vdash P((h.\ t),\bar{x}) \text{ by } p(h,t,i,\bar{x})$$

We can define the usual head-and-tail functions by induction. Define $Compound(x) == \exists h\!:\!A.\ \exists t\!:\!A\ list.\ x = (h.\ t) \text{ in } A\ list.$

$\vdash \forall x\!:\!A\ list.\ (x = nil \text{ in } A\ list\ \vee\ Compound(x))$
by $\forall R$
$x\!:\!A\ list \vdash (x = nil \text{ in } A\ list\ \vee\ Compound(x))$
by $list\_ind$
$\vdash nil = nil \text{ in } A\ list\ \vee\ Compound(x)$
by $\vee Rl$
$x\!:\!A\ list, h\!:\!A, t\!:\!A\ list \vdash (h.\ t) = nil \text{ in } A\ list \vee Compound(h.\ t))$
by $\vee Rr$
$\vdash \exists h'\!:\!A.\ \exists t'\!:\!A\ list.\ (h.\ t) = (h'.\ t') \text{ in } A\ list.$
by $\exists R\, h$ THEN $\exists R\, t$

We can also prove

$$\vdash \forall x\!:\!A\ list.\ \exists! h\!:\!A.\ \exists! t\!:\!A\ list.\ (\neg(x = nil) \Rightarrow x = (h.\ t) \text{ in } A\ list)$$

where $\exists!$ expresses unique existence (see the end of section 2.9).

2.9. Functions

The function type is one of the most important in modern mathematics. As we have noted, Frege patterned his treatment of logic which we are following on the concept of a function. In some ways this type represents the divide between abstract and concrete mathematics. By quantifying over functions we enter the realm of abstract mathematics. Indeed, the very notion of obtaining a function from an expression is called *abstraction.*

The day-to-day notation for functions at the beginning of the century was that one wrote phrases like "the function $sin(x)$ in x or e^x in x". Russell's notation, $\phi\hat{x}$, and Church's lambda notation, $\lambda x.e^x$, brought flexibility to the notation, creating a single name for the function with a binding operator (λ) to indicate the arguments. The modern *working notation* in mathematical articles and books (used in Bourbaki for example) is $x \mapsto b$ for a function with argument x and value given by the expression b in x; for example $x \mapsto x$ for the identity, $x \mapsto e^x$ for the exponential.

As we did for propositional functions, we will adopt the lambda notation in the form $\lambda(x.\ b)$ for $x \mapsto b$. In Nuprl one can *display* this in a variety of ways, including $x \mapsto b$ or $b\hat{x}$ or *fun* $x \Rightarrow b$. The important points are:

- There is an operator name, *lambda* that distinguishes functions. Their canonical value is $\lambda(x.\ b)$.
- A *binding phrase*, $x.\ b$ is used to identify the name of the *argument* (or formula parameter), x, and the body of the function.

- The usual rules about binding phrases apply concerning bound variables, scope, and α-equality.

Essentially the only way to use a function is to apply it to an argument.[19] Informal *notation* for applying a function f to an argument a is to write $f(a)$ or fa or even to show the substitution of "actual" argument for the "formal" one as in $sin(a)$ or e^a. We adopt an operator name to remind ourselves that application is a distinct *operation.* So we write $ap(f;a)$. But again, Nuprl can *display* this anyway the user pleases, e.g. as $f(a)$ or fa or even $f.\ a$ or $f@a$.

One of the major discoveries from a systematic study of function notations, especially the lambda calculus and combinatory calculus and later programming languages, is that rules for *formally calculating* with functions can be given independently of their meaning, especially independently of types.

The rules for calculation or for "definitional equality" can be expressed nicely as evaluation rules. Here is the so called "call_by_name" evaluation rule.

$$\frac{f \downarrow \lambda(x.\ b) \quad b[z/x] \downarrow c}{ap(f;a) \downarrow c}$$

The "call_by_value" rule is this

$$\frac{f \downarrow \lambda(x.\ b) \quad a \downarrow a' \quad b[a'/x] \downarrow c}{ap(f;a) \downarrow c}$$

Closed expression functions like $I == \lambda(x.\ x)$ or $K == \lambda(x.\ \lambda(y.\ x))$ are called *combinators*; these two are "polymorphic" in that we can compute their values regardless of the form of the input. Thus $ap(\lambda(x.\ x);K) \downarrow K$ and $ap(\lambda(x.\ x);0) \downarrow 0$, and $ap(K;I) \downarrow \lambda(x.\ I)$.

Other functions like $\lambda(z.1of(z))$ or $\lambda(z.add(1of(z);2of(z)))$ can only be reduced to values on inputs of a specific form, and others like $\lambda(x.suc(x))$ or $\lambda(x.\ 4/x)$ reduce to meaningful values (typed values) only on specific inputs. For example, $ap(\lambda(z.1of(z);0) \downarrow 1of(0)$ but $1of(0)$ is not a canonical value let alone a sensible value. In the case $ap(\lambda(x.\ suc(x));pair(0;0))$ the result of evaluation is the value $suc(pair(0;0))$, but this value has no type.

Typing functions. The space of functions from type A to type B is denoted $A \rightarrow B$. The *domain* type is A, the *range* (or *co-domain*) is B. The typing rule for functions is intuitively simple. We say that $\lambda(x.\ b) \in A \rightarrow B$ provided that on each input $a \in A$, $ap(\lambda(x.\ b);a) \in B$. This judgment is usually made symbolically by assuming $x \in A$ and judging by typing rules that $b \in B$. This is the form of typing judgment we adopt. So the typing rule has the form

$$\begin{array}{l} \bar{H} \vdash \lambda(x.\ b) \in A \rightarrow B \text{ by fun\_type} \\ \bar{H}, x{:}A \vdash b \in B \end{array}$$

[19] Although if functional equality is defined intensionally, then it is also possible to analyze their structure. Of course, function can also be passed as data.

More generally, given an expression f we allow

$$\bar{H} \vdash f \in A \to B \text{ by fun\_type}$$
$$\bar{H}, x{:}A \vdash ap(f;x) \in B$$

In the course of judging that an expression t has a type T, we allow replacing t by any term t' that is definitionally equal or by a term t' that t evaluates to. So if t in T and $t \downarrow t'$, then $t \in T$. In the logic over $(A \to B)$ we add the rule for function equality

$$\bar{H} \vdash f = g \text{ in } A \to B \text{ by extensional\_equalityR}$$
$$\bar{H}, x{:}A \vdash ap(f;x) = ap(g;x) \text{ in } B$$

$$\bar{H}, f = g \text{ in } A \to B \vdash ap(f;a) = ap(g;b) \text{ in } B \text{ by extensional\_equalityL}$$
$$\vdash a \in A$$

Here is Cantor's interesting argument about functions based on the method of *diagonalization.* It illustrates the rules for functions. (See the appendix for a Nuprl proof.)

Definition. Call f in $(A \to B)$ *onto* iff $\exists g{:}(B \to A)$ such that $\forall y{:}B.\ f(g(y)) = y$ in B.

Cantor shows that for inhabited types A with two distinct elements there is no function from A onto $(A \to A)$—essentially because $(A \to A)$ is "too big" to be enumerated by A. We state the condition on A using functions. We require that there is a function $\mathit{diff} \in A \to A$ such that $\mathit{diff}(x) \neq x$ for all x in A. The theorem is

Cantor's Theorem.
$(\exists\ \mathit{diff} : (A \to A).\ \forall x{:}A.\ \mathit{diff}(x) \neq x \ \mathit{in}\ A) \Rightarrow (\neg \exists e{:}A \to (A \to A).\ e \ \mathit{is\ onto})$

Proof. by $\Rightarrow R$ THEN $\Rightarrow R$

1. $\exists\ \mathit{diff} : (A \to A).\ \forall x{:}A.\ \mathit{diff}(x) \neq x$ in A
2. $\exists e{:}A \to (A \to A).\ e$ is onto

$\vdash \bot$

Next use $\exists L$ on 2 THEN unfold "onto" THEN $\exists L$

2. $e{:}A \to (A \to A)$
3. $g{:}(A \to A) \to A$
4. $\forall h{:}(A \to A).\ e(g(h)) = h$ in $(A \to A)$

Next $\exists L$ on 1 to replace 1 by 1.1 $\mathit{diff} : A \to A$, 1.2 $\forall x{:}A.\ \mathit{diff}(x) \neq x$ in A

Let $h_0 == \lambda(x.\ \mathit{diff}(e(x)(x)))$

Now $\forall L$ on 4 with h_0

5. $e(g(h_0)) = h_0$ in $A \to A$

Let $d == g(h_0)$, by *extensional_equalityL*

6. $e(d)(d) = h_0(d)$ in A

Now evaluate $h_0(d)$ to rewrite 6 as

6. $e(d)(d) = \mathit{diff}(e(d)(d))$

Now by $\forall L$ on 1.2 with $e(d)(d)$

7. $\mathit{diff}(e(d)(d)) \neq e(d)(d)$ (which is $(\mathit{diff}(e(d)(d)) = e(d)(d)) \to \perp$)

$\vdash \perp$

Finish by $\Rightarrow L$ on 7. and 6. □

Implicit functions from relations. A common way to define functions is implicitly in terms of relations. Suppose R is a relation on $A \times B$ and we know that for every $x \in A$ there is a unique y in B such that $R(x, y)$. Then we expect to have a function $f \in A \to B$ such that $R(x, f(x))$. How do we specify this function?

To facilitate consideration of this matter, let us define $\exists! y{:}A.\ P(y)$ to mean there is a y satisfying P, and any z that satisfies it is y. Thus

Definition. $\exists! y{:}A.\ P(y) == \exists y{:}A.\ P(y)\ \&\ \forall z{:}A.\ (P(z) \Rightarrow y = z \text{ in } A)$.

We expect the following formula to be true.

Function Comprehension. $\forall x : A.\ \exists! y : B.\ R(x, y) \Rightarrow \exists f : A \to B.\ \forall x : A.\ R(x, f(x))$.

For many instances of types A, B and relation R we can prove this formula by exhibiting a specific function. For example, if we define $\mathit{Root}(n, r)$ for n, r in N as $r^2 \leq n\ \&\ n < (r+1)^2$ then not only can we prove $\forall x{:}N.\ \exists! r{:}N.\ \mathit{Root}(n, r)$ but we can also define a function root by primitive recursion, namely

$$\mathit{root}(0) = 0$$
$$\mathit{root}(\mathit{suc}(n)) = \text{if } (\mathit{root}(n) + 1)^2 \leq n \text{ then } \mathit{root}(n) + 1 \ \mathit{else}\ \mathit{root}(n).$$

We know that $\lambda\,(x.\ \mathit{root}(x)) \in N \to N$ and $\mathit{Root}\,(n, \mathit{root}(n))$ is true. So perhaps if there are enough expressions for defining functions, we can prove the conjecture.

In set theory, functions are usually *defined* as single-valued total relations, i.e., a relation R on $A \times B$ is a function iff for all x in A there is a unique y in B such that $R(x, y)$. The relation R is a subset of $A \times B$, and this R is taken to *be* the function.

If the underlying logic has a choice function (or Hilbert $\in$-operator) as in Bourbaki [1968b] or HOL (Gordon and Melham [1993]), then the value of the function defined by R on input x is $\mathit{choice}(y.\ R(x, y))$ and a λ form for the function is $\lambda(x.\ \mathit{choice}(y.\ R(x, y)))$.

The choice operator would not only prove the implicit function conjecture, but it would prove the closely related axiom of choice as well. That axiom is

Axiom of Choice. $\forall x{:}A.\ \exists y{:}B.\ R(x, y) \Rightarrow \exists f{:}(A \to B).\ \forall x{:}A.\ R(x, f(x))$.

We will see in section 3 that in constructive type theory this axiom is provable because the theory has enough expressions for functions.

2.10. Set types and local set theories

Another of the most fundamental concepts of modern mathematics is the notion of *set* or class. Class theory arose out of Frege's foundation for mathematics in *Grundgesetze* and in *Principia* along similar lines. Even before 1900 Cantor was creating a rich *naive set theory* which was axiomatized in 1908 by Zermelo and improved by Skolem and Fraenkel into modern day axiomatic set theories such as ZF (Bernays [1958]) and BG (Gödel [1931]) and Bourbaki's set theory ([1968b]).

We could formulate a full blown axiomatic set theory based on the type Set. But the type theory of section 3 is an alternative into which ZF can be encoded (Aczel [1986]). So instead we pursue a much more modest treatment of sets along the lines of *Principia's* classes. In *Principia*, given a propositional function $\phi\hat{x}$ whose range of significance is the type A, we can form the class $\hat{x}(\phi x)$ of those elements of A satisfying ϕ. We write this as $\{x : A \mid \phi(x)\}$. We call this a *set type* or a *class.* Given two classes α, β we can form the usual combinations of union, $\alpha \cup \beta$, intersection, $\alpha \cap \beta$, complement, $\bar{\alpha}$, universal class, A, and empty class, ϕ.

The typing judgment associated with a set type is what one would expect. Suppose A is a type and $P \in A \to Prop$, then

$$\begin{array}{l} \bar{H} \vdash a \in \{x{:}A \mid P(x)\} \ \text{ by } \ setR \\ \quad \bar{H} \vdash a \in A \\ \quad \bar{H} \vdash P(a) \end{array}$$

The rule for using an assumption about set membership is

$$\begin{array}{l} \bar{H}, y : \{x{:}A \mid P(x)\} \vdash G \quad by\ setL \\ \bar{H}, y{:}A,\ P(y) \vdash G \end{array}$$

As with the other rules, we can choose to name the assumption $P(y)$ by using the justification by *setL new u*. In Nuprl there is the option to "hide" the proof of $P(y)$. This hidden version is the default in Nuprl. A hypothesis is hidden to prevent the proof object from being used in computations. This is necessary because the set membership rule, *setR*, does not keep track of the proof $P(a)$; so the constructive elimination rule is

$$\begin{array}{l} \bar{H}, y{:}\{x{:}A \mid P(x)\}, \bar{J} \vdash G \ \text{ by } \ IsetL, new\ u \\ \bar{H}, y{:}A, [u{:}P(y)], \bar{J} \vdash G. \end{array}$$

In local set theories, the concept of the power set, $\mathcal{P}(A)$ is introduced (c.f. Bell [1988], MacLane and Moerdijk [1992]). This type collects all sets built over A and $Prop$. If A is a type, then $\mathcal{P}(A)$ is a type.

In order to express rules about this type, we need to treat the judgments $A \in Type$ and $P \in A \to Prop$ in the rules. Thus far we have expressed these judgments only implicitly, not as explicit goals, in part because $Type$ and $A \to Prop$ are not types themselves, but "large types." However, it makes sense to write a rule such as

$$\begin{array}{l} \bar{H} \vdash \{x{:}A \mid P(x)\} \in P(A) \\ \quad \bar{H} \vdash A \in Type \\ \quad \bar{H} \vdash P \in A \to Prop \end{array}$$

We can also imagine the rule

$$\bar{H}, X{:}\mathcal{P}(A) \vdash \exists P{:}A \to Prop.\,(X = \{x{:}A \mid P(x)\} \text{ in } \mathcal{P}(A))\,.$$

This introduces the large type, $(A \to Prop)$ into the type position. Treating this concept precisely requires that we consider explicit rules for *Type* and *Prop*, especially their stratification as $Type_i$ and $Prop_i$. We defer these ideas until section 3.7.

Let us note at this point that the notion of *Prop* and set types be at the heart of *topos theory* as explained in Bell [1988]. Essentially, the *subobject classifier*, Ω and $T : 1 \to \Omega$, of topos theory is an (impredicative) notion of *Prop* and the subtype of true propositions. The notion of a *pullback* is used to define subtypes of a type A by "pulling back" a characteristic function $P : A \to Prop$ and the truth arrow $T : 1 \to Prop$ to get the domain of P, $\{x : A \mid P(x)\}$. A topos is essentially a category with Cartesian products (n-ary) a subobject classifier and power objects. In other words, it is an abstraction of a type theory which has *Prop*, a collection of true propositions, subtypes and a power type, $\mathcal{P}(A)$ for each type. The notion of a *Grothendieck topos* (c.f. Bell [1988], MacLane and Moerdijk [1992]) is essentially a *predicative* version of this concept. It can be defined in Martin-Löf type theory and in Nuprl, but that is beyond the scope of these notes. (However, see section 5.)

2.11. Quotient types

The equality relation on a type, written $s = t$ in T or $s =_T t$, defines the element's referential nature. The semantic models we use in section 3.9 take a type to be a partial equivalence relation (per) on a collection of terms.

Given a type T, other types can be defined from it by specifying new equality relations on the elements of T. For example, given the integers $\mathbb{Z}$, we can define the *congruence integers* $\mathbb{Z}//mod\, n$ to be the type whose elements are those of $\mathbb{Z}$ related by

$$x = y \; mod\; n \text{ iff } n \; divides \; (x - y).$$

More symbolically, let $n \mid m$ mean that n divides m, i.e., $\exists k{:}\mathbf{N}.\ m = k \star n$. Then $x = y\; mod\; n$ iff $n \mid (x - y)$. If $rm(x, n)$ is the remainder when x is divided by n, then clearly $x = y\; mod\; n$ iff $rm(x, n) = rm(y, n)$ in $\mathbb{Z}$. It is easy to see that $x = y\; mod\; n$ is an equivalence relation on $\mathbb{Z}$. In general, this is all we require to form a quotient type. If A is a type and E is an equivalence relation on A, then $A//E$ is a new type, the quotient of A by E. The equality rule is $x = y$ in $A//E$ iff $E(x, y)$ for x, y in A. Here are the new rules.

$A//E$ is a type iff A is a type and E is an equivalence relation on A

$\bar{H} \vdash a$ in $A//E$ by *quotient_member*
  $\bar{H} \vdash a$ in A

$\bar{H}, x{:}A//E, \bar{J} \vdash b[x]$ in B by *quotientL*
  $\bar{H}, x{:}A, \bar{J} \vdash b[x]$ in B
  $\bar{H}, x{:}A, x'{:}A, E(x, x'), \bar{J} \vdash b[x] = b[x']$ in B

For P to be a propositional function on a type A, we require that when $a = a'$ in A then $P(a)$ and $P(a')$ are the same proposition. If we consider atomic propositions $P(x)$ iff $x = t$ in $A//E$, then $a = t$ in $A//E$. The rules for equality of expressions built from elements of $A//E$ will guarantee the functional nature of propositions over $A//E$. We discuss the topic in detail in section 3.9 and in the literature on Nuprl Constable et al. [1986], Allen [1987b].

The quotient type is very important in many subjects. We have found it especially natural in automata theory (Constable et al. [1998]), rational arithmetic and of course, for congruences. For congruence integers we have proved Fermat's little theorem in this form:

Theorem. $\forall p{:}\{x{:}\mathrm{N} \mid prime(p)\}\,.\,\forall x{:}\mathbb{Z}//mod\, p.\ (x^p = x)$

Here the display mechanism suppresses the type on equality when it can be immediately inferred from the type of the equands.

Equivalence classes. It is noteworthy that quotient types offer a *computationally tractable* way of treating topics normally expressed in terms of equivalence classes. For example, if we want to study the algebraic properties of $\mathbb{Z}//mod\, n$ it is customary to form the set of equivalence classes of $\mathbb{Z}$ where the equivalence class of an element z is $[z] = \{i{:}\mathbb{Z} \mid i = z\ mod\ n\}$. The set of these classes is denoted $\mathbb{Z}/mod\ n$. The algebraic operations are extended to classes by

$$[z_1] + [z_2] = [z_1 + z_2],$$
$$[z_1] \star [z_2] = [z_1 \star z_2], \quad \text{etc.}$$

All of this development can be rephrased in terms quotient types. We show that $+$ and $\star$ are well-defined on $\mathbb{Z}//mod\ n$, and the elements are ordinary integers instead of equivalence classes. What changes is the equality on elements.

2.12. Theory structure

So far we have introduced a typed mathematical language and a few examples of specific types and then rules—for N, lists, Cartesian products, functions, subsets, and quotients. The possibilities for new types are endless, and we shall see more of them in sections 3 and 4. For example, we could introduce the type *Set* and explore classical and computational set theories. We can introduce partial objects via the *bar types* that Constable and Smith [1993] developed. As we have seen, we can use the Magic rule or not or various weaker forms of it.

Some choices of rules are inconsistent, e.g. bar types and Magic or the impredicative $\triangle$ type of Mendler [1988] and dependent products on the fixed point rule with all types. How are we to keep track of the consistent possibilities?

One method is to postulate *fixed theories* in the typed logic such as Heyting Arithmetic (HA) (c.f. Troelstra [1973]) or Peano Arithmetic (HA + Magic) or IZF (c.f. Beeson [1985], Friedman and Scedrov [1983], Joyal and Moerdijk [1995], Moerdijk and Reyes [1991]) or Intuitionistic Type Theory (ITT) or Higher Order Logic (HOL). We rely on a community of scholars to establish the consistency of various collections of axioms. Books like Troelstra [1973] study relationships between dozens of these theories. The space of them is very large.

Another possibility is to explore the "tree of knowledge" formed by doing mathematics in various *contexts* determined by the definitions and axioms used for any result. We can think of definitions and axioms as establishing contexts. N.G. de Bruijn [1980] has proposed a way to organize this knowledge, including derivation of inconsistency on certain paths.

Essentially de Bruijn defined typed mathematical languages, PAL, Aut-68, Aut-QE, Aut-Π, which were used for writing definitions and axioms.[20] He proposed a *logical framework* for organizing definitions, axioms and theorems into *books*. We will explore these typed languages in the next section. They are more primitive than our typed logic.

The apparatus of Automath is completely *formal*; it is a mechanism whose meaning is to be found completely in its ability to organize information and classify it without regard for content. Extending this attitude to the mathematics being expressed leads to the formalist philosophy of mathematics espoused by Hilbert [1926]. This is de Bruijn's view in fact, and it surely contrasts with *Principia* which found its meaning in the logical truths written into a fixed foundational theory. It will contrast as well to the Martin-Löf view, Girard's [1987] view, the views of Coquand and Huet in Coq and my own view (as expressed to a large extent in Nuprl) in which the logical framework is organized to express computational meaning. It is noteworthy that the three influential philosophical schools—Formalism, Logicism, and Intuitionism, can be characterized rather sharply in this setting (and coexist!).

An Automath book is a sequence of *lines*. A line has four parts as indicated in Table 2. Each line introduces a unique identifier which is either a *primitive notion*, PN, or a *block opener* or is defined. The category part provides the grammatical category; type is a built-in category, defined types like *nat* are another.

The lines form two structures, one the *linear order* and the other a *rooted tree.*

[20] "Automath is a language which we claim to be suitable for expressing very large parts of mathematics, in such a way that the correctness of the mathematical contents is guaranteed as long as the rules of the grammar are obeyed." de Bruijn [1980].

indicator	identifier	definition	category
0	nat	PN	type
0	n	—	nat
0	real	PN	type
n	x	—	real

Table 2: Sequence of lines

The nodes of the tree are identifiers, x, and the edges are from x to the indicator of the line having x as its identifier part. The *complete context* of x is the list of indicators from x back to the root. So each line uses as its indicator the last block opener in its context. When the definition and category components are included with x, the result is what de Bruijn calls the *tree of knowledge.*

Nuprl has a similar structure to its knowledge base, called a *library.* A library consists of lines. Each one is uniquely named by an identifier. These can include the equivalent of block openers, called *theory delimiters* (begin_thyname, end_thyname). The library is organized by a *dependency graph* which indicates the logical order among theories (the lines between delimiters). Unlike in Automath, the theory structure is a *directed acyclic graph* (dag). Theories can also be linked to a file system or a database which provides additional "nonlogical" structuring.

The Nuprl 5 system also provides a structured library with mechanisms to control access to theories. There are two modes of accessing information. One is by collecting axioms, definitions, and theorems into *controlled access theories.* These theories can only use the specific rules and axioms assembled at its root. Each type such as N or $S \times T$ is organized into a small theory consisting of its rules.[21] More complex theories are built by collecting axioms.[22] We will be specifying certain important theories later. One of them is Nuprl 4, the fixed logic in the Nuprl 4.2 release. Another theory could be Nuprl 4_bar, the theory with partial objects (Constable and Smith [1993]) or NuIZF, the formulation of IZF in type theory.

Another way to use the library we might call *free access.* A user can prove theorems using any rules whatsoever, even inconsistent collections. Once a theorem is proved, the system can define its *root_system,* the collection of all rules and definitions used to state and prove it. The root_system determines the class of theories into which the result can be "planted."

2.13. Proofs as objects

The notion of proof plays a fundamental role in logic as we have seen here. Hilbert's *proof theory* is a study of proofs, and for philosophical reasons he conceived

[21]The associated tactics are attached as well, see Hickey [1996b,1997].

[22]The associated tactics can also enforce global constraints on the theory such as "decidable type checking."

of it as a constructive theory, and a *metatheory*.[23] Given the central role of proofs in all of mathematics, it is not a great leap to begin thinking about proofs as mathematical objects with the same "reality" as numbers. This viewpoint is central to intuitionistic and constructive mathematics, and it seems to be coherent classically as well. De Bruijn designed the Automath formalisms around notion of *formal* proofs as objects, and ordinary objects such as functions could depend on proofs. In order to treat what was called classical mathematics he had to add a *principle of irrelevance of proofs*.[24] However, to bring proof expressions fully into the mathematics as objects means more than allowing them into the language. As the proof irrelevance principle shows, they can be regarded as part of the underlying linguistic apparatus.[25] To make proofs *explicit objects* with a referential character, we must define equality on them (the kind of equality called *book equality* in Automath as opposed to definitional equality which holds for all terms whether referential or not).

There are two sources to guide the discovery of equality rules for proof objects. We can turn to intuitionistic mathematics and its semantics for the logical operators or we can look to proof theory and the reduction (or normalization rules). Neither account is definitive for classically conceived mathematics. In the case of using intuitionistic reasoning as a guide, we must handle classical rules, such as contradiction, or classical axioms like the law of excluded middle "magic". There are various ways to approach this with promising results (Allen et al. [1990], Murthy [1991], Girard [1991]). The subject is still very active.

Another approach is suggested by the normalization theorems for classical and constructive logics natural deduction systems, or N-systems (due to Prawitz [1965]), and the body of results on cut elimination in the sequent calculi, or L-systems (arising from Gentzen [1935]). Unfortunately, the results give somewhat conflicting notions of proof equality (c.f. Zucker [1974,1977], Ungar [1992]). It is perhaps premature to suggest the appropriate classical theory, so instead we will sketch the constructive ideas and leave the technical details to section 3 where we will explore carefully Martin-Löf's interpretation in which the *computational content* of a proof is taken as the object.

Another prerequisite to treating proofs as objects is that we understand the domain of significance, the type of assertions about proofs. This is another point that is not entirely clear. For instance, the views of Kreisel [1981], Scott [1976], and Tait [1967,1983] differ sharply from those of Martin-Löf [1982,1983] and Girard [1987].

One of the key points is whether we understand a proof p as a proof of a proposition P, *p proves P*, or whether provability is a relation on proofs so that $Proves(p, P))$ is the appropriate relationship. In the latter case there arises the

---

[23]That it had to be so was part of *Hilbert's Program* for a formal foundation of mathematics. Classical parts of mathematics were to be considered as ideal elements ultimately justified by constructive means.

[24]"... we extend the language by proclaiming that proofs of one and the same proposition are always definitionally equal. This extra rule was called 'proof irrelevance'...."

[25]This is quite different from taking them to be *metamathematical* objects as is done in proof theory... a theory that could be formalized in Automath.

danger of an infinite regress since we will require a proof p' of $(p\ Proves\ P)$. At some level it seems that provability must be a basic judgment, like the typing judgment $t \in T$.

If we start with the view of the relationship $p\ proves\ P$ as a typing judgment, then we are led to the view that the type of a proof is the proposition that it proves. Thus propositions play the role of types according to the *propositions_as_types principle.*

This principle is designed into Automath (but can be regarded as "linguistic"), and it is the core of both Martin-Löf type theory (Martin-Löf [1982,1984,1983], Nordstrom, Petersson and Smith [1990]) and Girard type theory (Stenlund [1972], Constable et al. [1986], Girard, Taylor and Lafont [1989]). According to this principle, a proposition P is provable (constructivists would say *true*) iff there is a proof p whose type is P, that is

$$\vdash P \quad \text{iff} \quad \text{for some } p,\ \vdash p \in P$$

Indeed, on this interpretation and recognizing that proof expressions p denote proofs, we can see the sequent notation $\bar{H} \vdash P$ by p as just another way of writing $\bar{H} \vdash p \in P$.

The $\bar{H} \vdash P$ by p judgment form can be considered *implicit.* Attention is focused on P, and the main concern is that there is some inhabitant. The $\bar{H} \vdash p \in P$ form is *explicit*, and attention is focused on the actual proof. The rules could all be presented in either implicit (logical) form or explicit (type theoretic) form. Consider the $\forall L$ and $\forall R$ rules, for example. Here is an implicit form.

$$\begin{array}{r}
H, f{:}\forall x{:}A.\ P(x,)\bar{J} \vdash G \ \text{ by } \ \forall L(f; a; y.\ g[y]) \\
H, f{:}\forall x{:}A.\ P(x), \bar{J}, y{:}P(a) \vdash G \ \text{ by } \ g[y] \\
\bar{H} \vdash A \ \text{ by } \ a
\end{array}$$

$$\begin{array}{l}
H \vdash \forall x{:}A.\ P(x) \quad \text{by} \quad \forall R(x.\ p[x]) \\
\quad H, x{:}A \vdash P(x) \quad \text{by} \quad p[x]
\end{array}$$

Here is the explicit form of the $\forall L$ rule.

$$\begin{array}{r}
\bar{H}, f{:}\forall x{:}A.\ P(x), \bar{J} \vdash \forall L(f; a; y.\ g[y]) \in G \\
\bar{H}, f{:}\forall x{:}A.\ P(x), \bar{J}, y{:}P(a) \vdash g[y] \in G \\
\bar{H} \ \vdash \ a \in A
\end{array}$$

We will discover in section 3.11 that there is a reasonable notion of reduction on proof expressions (which can either be considered as computation or definitional equality) and that this gives rise to a minimal concept of equality on proofs that is sufficient to give them the status of mathematical objects.

2.14. Heyting's semantics

Here is Heyting's interpretation of the judgment $p\ proves\ P$.

1. For atomic P we cannot base the explanation on propositional components of P because there aren't any. But it might depend on an analysis of terms and

their type which could be compound.

We recognize certain atomic propositions, such as $0 = 0$ *in* N as "atomically true." That is, the proofs are themselves atomic, so the proposition is an axiom. In the case when the terms are atomic and the type is as well, there is little left to analyze. But other atomic propositions can be reduced to these axioms by computation on terms, say $5 * 0 = 1 * 0$ *in* N.
Some atomic propositions are proved by *computation on terms and proofs.* For example, $suc(suc(suc(0))) = suc(suc(suc(0)))$ *in* N is proved by thrice iterating the inference rule *suc_eq*

$$\frac{n = m}{suc(n) = suc(m)}$$

We might take the object $suc\_eq(suc\_eq(suc\_eq(zero\_eq)))$ as a proof expression for this equality. On the other hand, in such a case we can just as well consider the proof to be a computation procedure on the terms whose result is some token indicating success of the procedure.
In general, the proofs of atomic propositions depends on an analysis of the terms involved and the underlying type and its components. For example, $a = b$ *in* $A//E$ might involve a proof the proposition $E(a, b)$.

So we cannot say in advance what all the forms of proof are in these cases. As a general guide, in the case of completely atomic propositions such as $0 = 0$ *in* N in which the terms and type are atomic, we speculate that the proof is atomic as well. For these atomic proofs we might have a special symbol such as *axiom.*[26]

2. A proof of $P \,\&\, Q$ is a pair $\langle p, q\rangle$ where p proves P and q proves Q.
3. A proof of $P \vee Q$ is either p or q where p proves P and q proves Q. To be more explicit we say it is a pair $\langle tag, e\rangle$ where if the tag designates P then e is p and if it designates Q, then e is q.
4. A proof of $P \Rightarrow Q$ is a procedure f which maps any proof p of P to $f(p)$, a proof of Q.
5. A proof of $\exists x\!:\!A.\ P[x]$ is a pair $\langle a, p\rangle$ where $a \in A$ and p proves $P[a]$.
6. A proof of $\forall x\!:\!A.\ P[x]$ is a procedure f taking any element a of A to a proof $f(a)$ of $P[a]$.

Note, we treat $\neg P$ as $P \Rightarrow \perp$, so these definitions give an account of negation, but there are other approaches, such as Bishop [1967].

We will see a finer analysis of this definition in the section on type theory; there following Martin-Löf [1982] and Tait [1967,1983], we will distinguish between *canonical proof expressions* and non-canonical ones such as $add(suc(0); suc(suc(0)))$ (which reduces to a canonical one $suc(suc(suc(0)))$). In this more refined analysis

[26] In Martin-Löf type theory and in Nuprl all proofs of atomic formulas are reduced to a token (*axiom* in Nuprl). Information that might be needed from the proof is kept only at the metalevel.

we say that the above clauses define the *canonical proofs*, e.g. a canonical proof of $P \,\&\, Q$ is a pair $\langle p, q\rangle$, but $\Rightarrow L(\Rightarrow R(x.\langle x, q\rangle); p)$ is a noncanonical proof of $P \,\&\, Q$ which reduces to $\langle p, q\rangle$ when we "normalize" the proof.

Although this is a suggestive semantics of both proofs and propositions, several questions remain. Given a proposition P, can we be sure that all proofs have the structure suggested by this semantics? Suppose $P \,\&\, Q$ is not proved by proving P and proving Q but instead by a case analysis or by decomposing an implication and then decomposing an existential statement, etc.; so if t proves $P \,\&\, Q$, do we know t is a pair?

If proofs are going to be objects, then what is the right equality relation on them? If t proves $P\&Q$ then is t at least *equal* to a pair $\langle p, q\rangle$? What is the right equality on propositions? If $P = Q$ and p proves P does p prove Q? How can we make sense of Magic as a proof object? It is a proof of $P \vee \neg P$ yet it has no structure of the kind Heyting suggests. We will see that the type theories of the next section provide just the right tools for answering these questions.

3. Type theory

3.1. Introduction

Essential features. In this section I want to give a nontechnical overview of the subject I am calling *type theory*. I will discuss these points:

- It is a *foundational theory* in the sense of providing definitions of the basic notions in logic, mathematics, and computer science in terms of a few primitive concepts.
- It is a *computational theory* in the sense that among the primitive built-in concepts are notions of algorithm, data type, and computation. Moreover these notions are so interwoven into the fabric of the theory that we can discuss the computational aspects of every other idea in the theory. (The theory also provides a foundation for *noncomputational* mathematics, as we explain later.)
- It is *referential* in the sense that the terms denote *mathematical objects*. The referential nature of a term in a type T is determined by the *equality relation* associated with T, written $s = t$ in T. The equality relation is basic to the meaning of the type. All terms of the theory are *functional* over these equalities.
- When properly formalized and *implemented*, the theory provides practical tools for expressing, performing, and reasoning about computation in all areas of mathematics.

A detailed account of these three features will serve to explain the theory. Understanding them is essential to seeing its dynamics. In a sense, the axioms of the theory serve to provide a very abstract account of mathematical data, its transformation by effective procedures, and its assembly into useful knowledge. I summarized my ideas on this topic in Constable [1991].

Language and logic. In a sense, the theory is *logic free*. Unlike our account of typed logic, we do not start with propositions and truth. Instead we begin with more elementary parts of language, in particular, with a theory of computational equality of terms (or expressions). In *Principia* these elementary ideas are considered as part of the meaning of propositions. We separate them more clearly. We examine the mechanism of naming and definition as the most fundamental and later build upon this an account of propositions and truth.

This analysis of language draws on the insights of Frege, Russell, Brouwer, Wittgenstein, Church, Curry, Markov, de Bruijn, Kolmogorov, and Martin-Löf, and it draws on technical advances made by numerous computer scientists and logicians. We can summarize the insights in this way. The notion of *computability* is grounded in rules for processing language (Church [1940], Curry and Feys [1958], Markov [1949]). In particular, they can be organized as rules for a basic (type free) *equality on expressions* closely related to Frege's theory of identity in [1903]. The rules explain when two expressions will have the same reference if they have any reference. (We call these *computation rules*, but they could also be considered simply as general rules of *definitional equality* as in Automath.) De Bruijn showed that to fully understand the definitional rules, we need to understand how expressions are organized into *contexts* in a *tree of knowledge* as we discussed in section 2.12.

Frege not only realized the nature of identity rules, but he explained that the very notion of an *object* (or mathematical object) depends on rules for equality of expressions which are intended to denote objects. The equality rules of a theory serve to define the objects and prepare the ground for a *referential language*, one in which the expressions can be said to denote objects.

Frege also believed that the equality rules were not arbitrary but expressed the primitive truths about *abstract objects* such as numbers and classes. We build on Brouwer's theme that an understanding of the *natural numbers* **N** is an especially clear place to begin, and we try to build as much as possible with them. Here the insights of Brouwer [1975] (see van Stigt [1990]) show how to connect intuitions about number to the rules for equality of expressions. Brouwer shows that the idea of natural number and of pairing numbers are meaningful because they arise from mental operations. Moreover, these are the same abilities needed to manipulate the language of expressions (see Chomsky [1988]).[27]

So like Frege and Brouwer (and unlike formalists), we understand type theory to be *referential*, that is, the theory is about mathematical objects, and the meaningful expressions denote them.

Following Russell, we believe that a referential theory is created by classifying expressions into types. Not every expression is meaningful, for example, school children know that 0/0 is not. We sometimes say that the meaningful expressions are those that refer to mathematical objects, but this seems to presuppose that we

[27]For Brouwer this language is required by an individual only because of the limits and flaws in his or her mental powers. But for our theory, language is essential to the communication among agents (human and artificial or otherwise) needed to establish public knowledge.

know what such objects are. So we prefer to say that the task of type theory is *to provide the means to say when an expression is meaningful.* This is done by classifying expressions into types. Indeed to define a type is to say what *expressions* are of that type. This process also serves to define mathematical objects.[28]

Martin-Löf suggested a particular way of specifying types based on ideas developed by W. W. Tait [1967,1983]. First designate the standard irreducible names for elements of a type, say $t_1, t_2, \ldots$ belong to T. Call these *canonical values.* Then based on the definition of evaluation, extend the membership relation to all t' such that t' evaluates to a canonical value of T; we say that membership is extended by *pre-evaluation.*

Level restrictions. Russell [1908] observed that it is not possible to regard the collection of all types as a type itself. Let Type be this collection of all types. So Type is not an element of Type. Russell suggested schemes for layering or stratifying these "inexhaustible concepts" like Type or Proposition or Set. The idea is to introduce notions of types of various *levels.* In our theory these levels are indicated by *level indexes* such as Type_i. They will be defined later.

Architecture of type theory. What we have said so far lays out a basic structure for the theory. We start with a class of *terms.* This is the linguistic material needed for communication. We use variables and substitution of terms for variables to express relations between terms. Let x, y, z be variables and s, t be terms. We denote the substitution of term s for all free occurrences of variable x in t by $t[s/x]$. The details of specifying this mechanism vary from theory to theory. Our account is conventional and general.

Substitution introduces a primitive *linguistic relationship* among terms which is used to define certain basic *computational equalities* such as $ap(\lambda(x.b); a) = b[a/x]$.

There are other relations expressed on terms which serve to define computation. We write these as evaluation relations

$$t \; evals\_to \; t' \quad \text{also written} \quad t \downarrow t'.$$

Some terms denote types, e. g. $\mathbb{N}$ denotes the type of natural numbers. There are type forming operations that build new types from others, e. g. the Cartesian product $T_1 \times T_2$ of T_1 and T_2. Corresponding to a type constructor like $\times$ there is usually a constructor on elements, e. g. if $t_1 \in T_1$, $t_2 \in T_2$ then $pair(t_1; t_2) \in T_1 \times T_2$. By the Tait pre-evaluation condition above

$$\frac{t' \; evals\_to \; pair(t_1; t_2)}{t' \in T_1 \times T_2}$$

[28] The interplay between expressions and objects has seemed confusing to readers of constructive type theory. In my opinion this arises mainly from the fact that computability considerations cause us to say more about the underlying language than is typical, but the same relationship exists in any formal account of mathematics.

Part of defining a type is defining equality among its numbers. This is written as $s = t$ in T. The idea of defining an equality with a type produces a concept like Bishop's sets (see Bishop [1967], Bishop and Bridges [1985]), that is Bishop [1967,p.63] said "... a *set* is defined by describing what must be done to construct an element of the set, and what must be done to show that two elements are *equal.*"

The basic forms of judgment in this type theory are

- t is a term
  This is a simple context-free condition on strings of symbols that can be checked by a parser. We stress this by calling these *readable expressions.*
- T *is_a type*
  We also write $T \in \mathsf{Type}$ and prefer to write capital letters, S, T, A, B for types. This relationship is not decidable in general and cannot be checked by a parser. There are rules for inferring typehood.
- $t \in T$ (type membership or elementhood)
  This judgement is undecidable in general.
- $s = t$ in T (equality on T)
  This judgement is also undecidable generally.

Inference mechanism. Since Post it has been the accepted practice to define the class of formulas and the notion of proof inductively. Notice our definition of formula in section 2.4, also, for example, a *Hilbert style proof* is a sequence of closed formulas $F_1, \ldots, F_n$ such that F_i is an axiom or follows by a rule of inference from F_j, F_k for $j < i, k < i$. A typical inference rule is expressed in the form of hypotheses above a horizontal line with the conclusion below as in *modus poneus.*

$$\frac{A,\ A \Rightarrow B}{B}$$

This definition of a proof includes a specific presentation of evidence that an element is in the class of all proofs.

The above form of a rule can be used to present any inductive definition. For example, the natural numbers are often defined inductively by one rule with no premise and another rule with one.

$$0 \in \mathbf{N} \qquad \frac{n \in \mathbf{N}}{suc(n) \in \mathbf{N}}$$

This definition of $\mathbf{N}$ is one of the most basic inductive definitions. It is a pattern for all others, and indeed, it is the clarity of this style of definition that recommends it for foundational work.

Inductive definitions are also prominent in set theory. The article of Aczel [1986] "An Introduction to Inductive Definitions" surveys the methods and results. He bases his account on sets Φ of rule instances of the form $\frac{X}{x}$ where X are the *premises* and x the *conclusions.* A set Y is called Φ-closed iff $X \subseteq Y$ implies $x \in Y$. The set inductively defined by Φ is the intersection of all subsets Y of A which are Φ-closed.

3.2. Small fragment — arithmetic

We build a small fragment of a type theory to illustrate the points we have just made. The explanations are all inductive. We let S and T be metavariables for types and let, s, t, s_i, t_i, also s', t', s'_i, t'_i denote terms.

We arrange the theory around a *single judgment*, the equality $s = t$ in T. We avoid membership and typehood judgments by "folding them into equality" just to make the fragment more compact. First we look at an informal account of this theory.

The intended meaning of $s = t$ in T is that T is a type and s and t are equal elements of it. Thus a premise such as $s = t$ in T implies that T is a type and that s and t are elements of T (thus subsuming membership judgment).[29]

The only atomic type is N. If S and T are types, then so is $(S \times T)$; these are the only compound types.

The canonical elements of N are 0 and $suc(n)$ where n is an element of N, canonical or not. The canonical elements of $(S \times T)$ are $pair(s;t)$ where s is of type S and t of type T. The expressions $1of(p)$ and $2of(p)$ are noncanonical. The evaluation of $1of(pair(s;t))$ is s and of $2of(pair(s;t))$ is t.

The inference mechanism must generate the evident judgments of the form $s = t$ in T according to the above semantics. This is easily done as an inductive definition. The rules are all given as clauses in this definition of the usual style (recall Aczel [1977] for example).

We start with *terms* and their evaluation. The only atomic terms are 0 and N. If s and t are terms, then so are $suc(t), (s \times t), pair(s;t), 1of(t), 2of(t)$. Of course, not all terms will be given meaning, e.g. $(0 \times \mathrm{N})$, $suc(\mathrm{N})$, $1of(\mathrm{N})$ will *not* be.

Evaluation. Let s and t be terms.

$$0\ evals\_to\ 0 \quad \mathrm{N}\ evals\_to\ \mathrm{N} \quad suc(t)\ evals\_to\ suc(t) \quad pair(s;t)\ evals\_to\ pair(s;t)$$

$$1of(pair(s;t))\ evals\_to\ s \qquad 2of(pair(s;t))\ evals\_to\ t$$

Remark: $s(\mathrm{N})\ evals\_to\ s(\mathrm{N})$, $1of(pair(\mathrm{N};0))\ evals\_to\ \mathrm{N}$. So evaluation applies to meaningless terms. It is a purely formal relation, an *effective calculation.* Thus the base of this theory includes a formal notion of *effective computability* (c.f. Rogers [1967]) compatible with various formalizations of that notion, but not restricted necessarily to them (e.g. Church's thesis is not assumed). Also note that *evals_to* is idempotent; if $t\ evals\_to\ t'$ then $t'\ evals\_to\ t'$ and t' is a value.

general equality

$$\frac{t_1 = t_2 \text{ in } T}{t_2 = t_1 \text{ in } T} \qquad \frac{t_1 = t_2 \text{ in } T \quad t_2 = t_3 \text{ in } T}{t_1 = t_3 \text{ in } T} \qquad \frac{t_1 = t_2 \text{ in } T \quad t_1\ evals\_to\ t'_1}{t'_1 = t_2 \text{ in } T}$$

[29] In the type theory of Martin-Löf [1982], a premise such as $s = t$ in T presupposes that T is a type and that $s \in T, t \in T$. This must be known *before* the judgment makes sense.

typehood and equality

$$0 = 0 \text{ in } \mathbf{N} \qquad \frac{t = t' \text{ in } \mathbf{N}}{suc(t) = suc(t') \text{ in } \mathbf{N}} \qquad \frac{s = s' \text{ in } S \quad t = t' \text{ in } T}{pair(s;t) = pair(s';t') \text{ in } (S \times T)}$$

The inductive nature of the type N and of the theory in general is apparent from its presentation. That is, from *outside* the theory we can see this structure. We can use induction principles from the informal mathematics (the *metamathematics*) to say, for example, every canonical expression for a number is either 0 or $suc(n)$. But so far there is no construct *inside* the theory which expresses this fact. We will eventually add one in section 3.3.

Examples. Here are examples of true judgments that we can make: $suc(0) = suc(0)$ in N. This tells us that N is a type and $suc(0)$ an element of it. Also $pair(0; suc(0)) = pair(0; suc(0))$ in $(\mathbf{N} \times \mathbf{N})$ which tells us that $(\mathbf{N} \times \mathbf{N})$ is a type with $pair(0; suc(0))$ a member. Also $1of(pair(0; a))$ belongs to N and $suc(1of(pair(0; a)))$ does as well for arbitrary a.

Here is a derivation that $suc(1of(pair(0; suc(0)))) = 2of(pair(0; suc(0)))$ in N.[30]

$$0 = 0 \text{ in } \mathbf{N}$$

$$\begin{aligned} 0 = 0 \text{ in } \mathbf{N} suc(0) &= suc(0) = suc(o) \text{ in } \mathbf{N} \\ pair(0; suc(0)) &= pair(0; suc(0)) \text{ in } \mathbf{N} \times \mathbf{N} \\ 1of(pair(0; suc(0))) &= 1of(pair(0; suc(0))) \text{ in } \mathbf{N} \quad 1of(pair(0; suc(0)))\ evals\_to\ 0 \end{aligned}$$

$$2of(pair(0; suc(0))) = 2of(pair(0; suc(0))) \text{ in } \mathbf{N} \quad 2of(pair(0; suc(0)))\ evals\_to\ suc(0)$$

$$1of(pair(0; suc(0))) = 0 \text{ in } \mathbf{N} \qquad 2of(pair(0; suc(0))) = suc(0) \text{ in } \mathbf{N}$$

$$suc(1of(pair(0; suc(0)))) = suc(0) \text{ in } \mathbf{N} \qquad suc(0) = 2of(pair(0; suc(0))) \text{ in } \mathbf{N}$$

$$\overline{suc(1of(pair(0; suc(0)))) = 2of(pair(0; suc(0))) \text{ in } \mathbf{N}}$$

Analyzing the fragment. This little fragment illustrates several features of the theory.

First, *evaluation is defined prior to typing.* The *evals_to* relation is purely formal and is grounded in language which is a prerequisite for communicating mathematics. Computation does not take into account the meaning of terms. This definition of computability might be limiting since we can imagine a notion that relies on the information in typehood, and it is possible that a "semantic notion" of computation must be explored in addition, once the types are laid down.[31] Our approach to

[30] In type theory, we will write the derivations in the usual bottom-up style with the conclusion at the bottom, leaves at the top.

[31] In IZF this is precisely the way computation is done, based on the information provided by a membership proof.

computation is compatible with the view taken in computation theory (c.f. Rogers [1967]).

Second, the semantics of even this simple theory fragment shows that the concept of a proposition involves the notion of its meaningfulness (or well-formedness). For example, what appears to be a simple proposition, $t = t$ in T, expresses the judgments that T is a type and that t belongs to this type. These judgments are part of understanding the judgment of truth.

To stress this point, notice that by postulating $0 = 0$ in **N** we are saying that **N** is a type, that 0 belongs to **N** and that it equals itself. The truth judgment is entirely trivial; so the significance of $t = t$ in T lies in the well-formedness judgments implicit in it. These judgments are normally left *implicit* in accounts of logic.

Notice that the well-formedness judgments cannot be *false*. They are a different category of judgment from those about truth. To say that $0 \in \mathbf{N}$ is to *define* zero, and to say **N** is a type is to define **N**. We see this from the rules since there are no separate rules of the form "**N** is_a type" or 0 *is_a* **N**." Note, because $t = t$ whenever t is in a type, *the judgment $t = t$ in T happens to be true exactly when it is well-formed.*

Finally the points about $t = t$ in T might be clarified by contrasting it with $suc = suc$ in 0. This judgment is meaningless in our semantics because 0 is not a type. Likewise $suc = suc$ in **N** is meaningless because although **N** is a type, *suc* is not a member of it. Similarly, $0 = suc$ in **N** is meaningless since *suc* is not a member of **N** according to our semantics. None of these expressions, which read like propositions, is false; they are just *senseless.* So we cannot understand, with respect to our semantics, what it would mean for them to be false.

Third, notice that the *semantics of the theory were given inductively* (although informally), and the proof rules were designed to directly express this inductive definition. This feature will be true for the full theory as well, although the basic judgments will involve variables and will be more complex both semantically and proof theoretically.

Fourth, the semantic explanations are *rooted in the use of informal language.* We speak of terms, substitution and evaluation. The use of language is critical to expressing computation. We do not treat terms as mathematical objects nor evaluation as a mathematical relation. To do this would be to conduct metamathematics *about* the system, and that metamathematics would then be based on some prior informal language. When we consider implementing the theory, it is the informal language which we implement, translating it to a programming notation lying necessarily outside of the theory.

Fifth, although the theory is grounded in language, it *refers to abstract objects.* This abstraction is provided by the equality rules. So while $1of(pair(0; suc(0)))$ is not a canonical integer in the term language, we cannot observe this linguistic fact in the theory. This term denotes the number 0. The theory is referential in this sense.

Sixth, the theory is *defined by rules.* Although these rules reflect concepts that we have mastered in language, so are meaningful, and although all of the judgments we assert are evident, it is the rules that define the theory. Since the rules reflect a semantic philosophy, we can see in them answers to basic questions about the objects of the theory. We can say what a number is, what 0 is, what successor is. Since the fragment is so small, the answers are a bit weak, but we will strengthen it later.

Seventh, the theory is *open-ended*. We expect to extend this theory to formalize ever larger fragments of our intuitions about numbers, types, and propositions. As Gödel showed, this process is never complete. So at any point the theory can be extended. By later specifying how evaluation and typing work, we provide a framework for future extensions and provide the guarantees that extensions will preserve the truths already expressed.

3.3. First extensions

We could extend the theory by adding further forms of computation such as a term, *prd*, for predecessor along with the evaluation

$$prd(suc(n))\ evals\_to\ n.$$

We can also include a term for addition, $add(s; t)$ along with the evaluation rules

$$add(0; t)\ evals\_to\ t \qquad \frac{add(n; t)\ evals\_to\ s'}{add(suc(n); t)\ evals\_to\ suc(s')}$$

We include, as well, a term for multiplication, $mult(s; t)$ along with the evaluation rule

$$mult(0; t)\ evals\_to\ 0 \qquad \frac{mult(n; t)\ evals\_to\ m \quad add(m; t)\ evals\_to\ a}{mult(suc(n); t)\ evals\_to\ a}$$

These rules enable us to type more terms and assert more equalities. We can easily prove, for instance, that

$$add(suc(0); suc(0)) = mult(suc(0); add(suc(0); suc(0)))\ \text{in}\ \mathbf{N}.$$

But this "theory" is woefully weak. It cannot

- internally express general statements such as $prd(suc(x)) = x$ in $\mathbf{N}$ or $add(suc(x); y) = suc(add(x; y))$ for any x because there is no notion of variable, but these are true in the metalanguage.
- express function definition patterns such as the *primitive recursions* which were used to define add, multiply and for which we know general truths.
- express the inductive nature of $\mathbf{N}$ and its consequences for the uniqueness of functions defined by primitive recursion.

Adding capability to define new functions and state their "functionality" takes us from a concrete theory to an abstract one; from specific equality judgments to *functional judgments*. These functional judgments are the essence of the theory, and they provide the basis for connecting to the propositional functions of typed logic. So we add them next.

The simplest new construct to incorporate is one for constructing any object by following the pattern for the construction of a number. We call it a (*primitive*) *recursion combinator*, R. It captures the pattern of definition of *prd*, *add*, *mult* given above. It will later be used to explain induction as well.

The defining property of R is its rule of computation and its respect for equality. We present the computation rule using substitution.[32] The simplest way to to this as to use the standard mechanism of *bound variables* (as in the lambda calculus or in quantifier notation). To this end we let u, v, w, x, y, z be variables, and given an expression *exp* of the theory, we let $u.exp$ or $u, v.exp$ or $u, v, x.exp$ or generally $u_1, \ldots, u_n.exp$ (also written $\bar{u}.exp$) be a *binding phrase*. We say that the u_i are *binding occurrences* of variables whose *scope* is *exp*. The occurrences of u_i in *exp* are *bound* (by the smallest binding phrase containing them). The unbound variables of *exp* are called *free*, and if x is a free variable of $\bar{u}.exp$, then $\bar{u}.exp[t/x]$ denotes the substitution of t for every free occurrence of x in *exp*. If any of the u_i occur free in t, then as usual $\bar{u}.exp[t/x]$ produces a new binding phrase $\bar{u}'.exp'$ where the binding variables are renamed to prevent *capture* of free variables of t.[33]

$$\frac{b[t/v]\ evals\_to\ c}{R(0; t; v.b; u, v, i.h)\ evals\_to\ c}$$

$$\frac{R(n; t; v.b; u, v, i.h)\ evals\_to\ a \quad h[n/u, t/v, a/i]\ evals\_to\ c}{R(suc(n); t; v.b; u, v, i.h)\ evals\_to\ c}$$

Here is a typical example of R used to define addition in the usual primitive recursive way.

$$R(n; m; v.v; u, v, a.suc(a))$$

We see that

$$\begin{aligned} &R(0; m; --)\ evals\_to\ m\ ,\ i.e.\ \ 0 + m = m \\ &R(suc(n); m; --)\ evals\_to\ suc(R(n; m; --)), \\ &\qquad \text{i.e.}\ \ suc(n) + m\ evals\_to\ suc(n + m) \end{aligned}$$

Once we have introduced binding phrases into terms, the format for equality and consequent typing rules must change. Consider typing R. We want to say that if $v.b$ and $u, v, i.h$ have certain types, then R has a certain type. But the type of b and h will depend on the types of u, v and i. For example, the type of $v.v$ will be T in a context in which the variable v is assumed to have type T. Let us agree to use the judgment $t \in T$ to discuss typing issues, but for this theory fragment (as for Nuprl) this notation is just an abbreviation for $t = t$ in T. We will use it when we intend to focus on typing issues. We might write a rule like

[32] R can also be defined as a combinator without variables. In this case the primitive notion is application rather than substitution.

[33] If u_i is a free variable of t then it is *captured* in $\bar{u}.exp[t/x]$ by the binding occurrence u_i.

$$\frac{n \in \mathbf{N} \quad t \in A_1 \quad \dfrac{v \in A_1}{b \in B_2} \quad \dfrac{u \in \mathbf{N} \quad v \in A_1 \quad i \in B_2}{h \in B_2}}{R(n; t; v.b; u, v, i.h) \in B_2}$$

The premises

$$\frac{u \in N \; v \in A_1 \; i \in B_2}{h \in B_2}$$

reads "h has type B_2 under the assumption that u has type $\mathbf{N}$, u has type A_1 and i has type B_2."

For ease of writing we render this *hypothetical typing judgment* as $u : N,\ v : A_1,\ i : B_2 \vdash h \in B_2$. The syntax $u : N$ is a variant of $u \in N$ which stresses that u is a variable. Now the typing of R can be written

$$\frac{n \in N \quad t \in N \quad v{:}A_1 \vdash b \in B_2 \quad u{:}N,\ v{:}A_1,\ i{:}B_2 \vdash h \in B_2}{R(n; t; v.b; u, v, i.\ h) \in B_2}$$

This format tells us that n, t, b and h are possibly compound expressions of the indicated types with v, u, i as variables assumed to be of the indicated types.

Following our practice of subsuming the typing judgment in the equality one, we introduce the following rule.

First let

$$\begin{aligned}
Principle_argument &== n = n' \text{ in } \mathbf{N} \\
Aux_argument &== t = t' \text{ in } \mathbf{N} \\
Base_equality &== v = v' \text{ in } A_1 \vdash b = b' \text{ in } B_2 \\
Induction_equality &== u = u' \text{ in } \mathbf{N}, v = v' \text{ in } A_1, i = i' \text{ in } B_2 \vdash h = h' \text{ in } B_2
\end{aligned}$$

Then the rule is

$$\frac{Principle\_argument \quad Aux\_argument \quad Base\_equality \quad Induction\_equality}{R(n; t; v.\ b; u, v, e.\ h) = R(n'; t'; v'.\ b'; u', v', e'.\ h') \text{ in } B_2}$$

Unit and empty types. We have already seen a need for a type with exactly one element, called a unit type. We take **1** as the type name and $\bullet$ as the element, and adopt the rules:

$$\bullet = \bullet \text{ in } \mathbf{1}$$

We adopt the convention that such a rule automatically adds the new terms $\bullet$ and **1** to the collection of terms. We also automatically add

$$\bullet \ \textit{evals\_to} \ \bullet \qquad \mathbf{1} \ \textit{evals\_to} \ \mathbf{1}$$

to indicate that the new terms are canonical unless we stipulate otherwise with a different evaluation rule.

We will have reasons later for wanting the "dual" of the unit type. This is the empty type, $\mathbf{0}$, with no elements. There is no rule for elements, but we postulate $\mathbf{0}$ *is_a type* from which we have that we $\mathbf{0}$ as a term and $\mathbf{0}$ *evals_to* $\mathbf{0}$

An interesting point about handling $\mathbf{0}$ is to decide what we mean by *assuming* $x \in \mathbf{0}$. Does

$$x : \mathbf{0} \vdash x \in \mathbf{0}$$

make sense? Is this a sensible judgment? We seem to be saying that if we assume x belongs to $\mathbf{0}$ and that $\mathbf{0}$ is type, then x indeed belongs to $\mathbf{0}$. We clearly know functionality vacuously since there are no closed terms t, t' with $t = t'$ in $\mathbf{0}$. It is more interesting to ask about such anomalies as

$$x : \mathbf{0} \vdash x \in \mathbb{N} \quad \text{or} \quad x : \mathbf{0} \vdash x \in \mathbf{1}$$

or even the possible nonsense

$$x : \mathbf{0} \vdash \mathbb{N} \in \mathbb{N}.$$

What are we to make of these "boundary conditions" in the design of the theory?

According to our semantics and Martin-Löf's typing judgments, even $x : \mathbf{0} \vdash (suc = \mathbb{N} \text{ in } \mathbb{N})$ is a true judgment because we require that $\mathbf{0}$ is a type and for t, t' in $\mathbf{0}$, if $t = t'$ in $\mathbf{0}$, then $suc \in \mathbb{N}, \mathbb{N} \in \mathbb{N}$ and $suc = \mathbb{N}$ in $\mathbb{N}$. Since anything is true for all t, t' in $\mathbf{0}$, the judgment is true.

This conclusion is somewhat bizarre, but we will see later that there will be other types, of the form $\{x : A \mid P(x)\}$ whose emptiness is unknown. So our recourse is to treat types uniformly and not attempt to make a special judgment in the case of assumptions of the form $x : T$ for which T might be empty.

List types. The list data type is almost as central to computing as the natural numbers. We presented this type in the logic as well, and we follow that example even though we can see lists as a special case of the recursive types to be discussed later (section 4). The rules are more compact and pleasing to examine if we omit the typing context $\mathcal{T}$ and use the typing abbreviation of $t \in T$ for $t = t$ in T. So although we will write a rule like[34]

$$\boxed{\frac{a \in A, \; l \in list(A)}{cons(a; l) \in list(A)}}$$

Without its typing context, we intend the full rule

$$\frac{\mathcal{T} \vdash a = a' \text{ in } A \quad \mathcal{T} \vdash l = l' \text{ in } list(A)}{\mathcal{T} \vdash cons(a; l) = cons(a'; l') \text{ in } list(A).}$$

[34] In this section we use *list*(A) instead of A *list* to stress that we are developing a different theory than in section 2.

We also introduce a form of primitive recursion on lists, the combinator L whose evaluation rule and typing rules are:

$$\frac{b[t/v] \ evals\_to \ c}{L(nil; p; v.b; h, t, v, i.g) \ evals\_to \ c}$$

$$\frac{L(l, s, v.b, h, t, v, i.g) \ evals\_to \ c_1 \quad g[a/h, l/t, s/v, c_1/i] \ evals\_to \ c_2}{L(cons(a; l); s; v.b; h, t, v, i.g) \ evals\_to \ c_2}$$

Let $L[x; b, g] = L(x; v.\ b; h, t, v, e.\ g)$, and

$H_B == v = v'$ in $S \vdash b = b' \in B$,

$H_S == h = h'$ in A, $t = t'$ in $list(A)$, $v = v'$ in S, $i = i'$ in $B \vdash g = g'$ in B,

$C_A == \vdash a = a'$ in A,

$C_S == \vdash s = s'$ in S, and

$C_{Alist} == \vdash l = l'$ in $list(A)$, then

$$\frac{H_B \qquad H_S \qquad C_A \qquad C_S \qquad C_{Alist}}{L[cons(a; l), b; g] = L[cons(a'; l'), b', g'] \text{ in } list(A)}$$

$$L(nil; v.b; h, t, v, i.g) = L(nil; v.b'; h', t, v, i.g') \text{ in } list(A)$$

Here are typical generalizations of the functions add, mult, exp to N list to illustrate the use of L. For the list (3, 8, 5, 7, 2) the operations behave as follows. Add *addL* is $(3 + (8 + (5 + (7 + (2 + 0)))))$, *multL* is $3 * 8 * 5 * 7 * 2 * 1$, $expL_2$ is $(((((2)^2)^7)^5)^8)^3$.

$$\begin{aligned} addL(l) &== L(l; 0; h, t, a.add(h, a)) \\ multL(l) &== L(l; 1; h, t, m.mult(h, m)) \\ expL(l)_k &== L(l; k; h, t, e.exp(h, e)). \end{aligned}$$

The induction rule for lists is expressed using L as follows. Let $H_S ==$

$$x \in list(A), y \in S, \ v \in S \vdash f[nil/x, v/y] = b \text{ in } B$$

and let $H_{list} ==$

$$x \in list(A), y \in S, h \in A, t \in list(A), v \in S, i \in B \vdash f[cons(h; t)/x, v/y] = g \text{ in } B,$$

then

$$\frac{H_S \qquad H_{list}}{x \in list(A), y \in S \vdash f = L(x; y; v.b; h, t, v, i.g) \text{ in } B}$$

This says that L defines a unique functional expression over $list(A)$ and S because the values as inductively determined by the evaluation rule completely determine functions over $list(A)$.

3.4. Functions

The judgment $x = x$ in $A \vdash b = b$ in B defines a function from A to B whose rule is given by the expression b. We know this from the functionality constraint implicit in the judgment, i.e. if $a = a'$ in the type A, then $b[a/x] = b[a'/x]$ in the type B. Likewise if b_1 is an expression in x and b' is an expression in x' then $x = x'$ in $A \vdash b = b'$ in B defines such a function. The two rules b, b' are considered equal on equal a, a' in A. Also it is part of the judgment that $b[a/x] = b'[a'/x']$. To this extent at least the notion of equality on these functions is *extensional*.

Let us look at patterns of functionality that involve functions as arguments. The addition function on N is represented by

$$z \in N \times N \vdash add(1of(z); 2of(z)) \in N$$

We also know that

$$l \in list(N) \vdash addL(l) \in N.$$

We know that the pattern of definition used to form $addL, multL, expL$ can be extended to any binary function f from $N \times N$ to N using $fL_k(l) = L(l; k; h, t, a.f(h, a))$. For any specific f we can write this function $fL_k(l)$, but we would like to express the general fact as a function of f, saying: for any function from $N \times N$ to N and any k in N, $L(l; k; h, t, a.f(h, a))$ is a functional expression in l, k and f.

In order to say this, we need a type for f. The notation $(N \times N) \to N$ is the type used in section 2. We can add $(A \to B)$ as a type expression for A and B types. But we also need canonical values for the type, what should they be? Can we use $(x \in A \vdash b \in B)$ as a notation for a function in $(A \to B)$?

It would be acceptable to use just that notation; it is even similar to the Bourbaki notation $x \mapsto b(x \in A, b \in A)$ (see Bourbaki [1968a]). But in fact we do not need the type information to define the evaluation relation nor to describe the typing rule. So we could simply use $(x \mapsto b)$. Instead we adopt the lambda notation $\lambda(x.b)$ more familiar in computer science as we did in sections 1 and 2.

We also need notation for function application. We write $ap(f; a)$ for the application of function f to argument a, but often display this as $f(a)$. The new evaluation rules are:

$$\lambda(x.b) \ evals\_to \ \lambda(x.b)$$

$$\frac{b[a/x] \ evals\_to \ c}{ap(\lambda(x.b); a) \ evals\_to \ c}$$

The typing rule is

$$\frac{x = x' \text{ in } A \ \vdash \ b = b' \text{ in } B}{\lambda(x.\, b) = \lambda(x'.\, b') \text{ in } (A \to B)}$$

This rule generates the type $(A \to B)$ as a term.[35]

3.5. Duality and disjoint unions

The types **0** and **1** are called *duals* of each other in a category theory. Here is what this means. The object **1** is called *terminal* (or final) because for every type A, there is a unique map in $A \to \mathbf{1}$, i.e. a map terminating in **1**, namely $\lambda(x.\ \bullet)$. The object **0** is *initial* since for every type A, there is a unique map initiating in **0**, i.e. $\mathbf{0} \to A$, namely $\lambda(x.x)$.[36]

The duality concept is that the arrows of the types are reversed in the definition.

1 is final iff for all A there is a unique element in $A \to \mathbf{1}$.

0 is initial iff for all A there is a unique element in $\mathbf{0} \to A$.

We will examine another useful duality next.

The type $A \times B$ can be characterized in terms of functions. In category theory this is done with a diagram

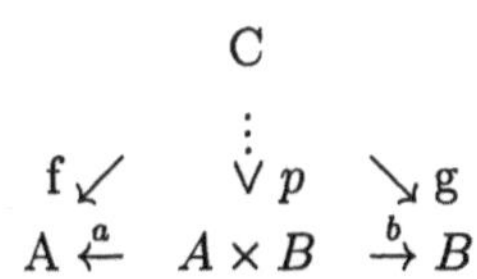

which says that given the projection functions $a == \lambda(x.1of(x))$, $b == \lambda(x.2of(x))$ and any functions $f : C \to A$, $g : C \to B$, there is exactly one map p denoted $\langle f, g\rangle \in C \to A \times B$ such that $f = a \circ p$ and $g = b \circ p$; that is, for $z \in C$

$$f(z) = a(\langle f, g\rangle(z))$$
$$g(z) = b(\langle f, g\rangle(z)).$$

We can show that $\lambda(z.pair(f(z); g(z)))$ is the unique map $\langle f, g\rangle$.

In category theory there is a construction that is dual to the product, called *co-product.* Duals are created by reversing the arrows in the diagram, so for a dual we claim this.

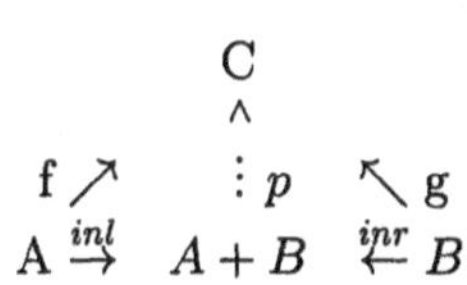

Given A, B with maps $inl \in A \to A + B$, $inr \in B \to A + B$ and maps $f \in A \to C$, $g \in (B \to C)$ there is a unique map $[f, g] \in A + B \to C$ such that

$$[f, g] \circ inl = f \quad \text{and} \quad [f, b] \circ inr = g.$$

In type theory we take $inl(a)$, $inr(b)$ to be canonical values with evaluation

[35] Martin-Löf would only need the premise $x{:}A \vdash b \in B$ since this means that A is a type. But in his system to prove $x{:}A \vdash b \in B$ requires proving A *is_a type*.

[36] We could also use $\lambda(x.\ a)$ for any $a \in A$ if there is one since under the assumption that $x \in 0$, $x = a$ for any a, thus $\lambda(x.\ x) = \lambda(x.\ a)$ in $\mathbf{0} \to A$.

$$inl(a)\ evals\_to\ inl(a) \qquad inr(b)\ evals\_to\ inr(b).$$

For A and B types, $A+B$ is a new type called the *disjoint union* of A and B. But the typing rules present a difficulty. If we simply write

$$\frac{a = a' \text{ in } A}{inl(a) = inl(a') \text{ in } A+B} \qquad \frac{b = b' \text{ in } B}{inr(b) = inr(b') \text{ in } A+B}$$

then we can deduce a judgment like $inl(0) = inl(0)$ in $\mathbf{N} + suc(0)$ which does not make sense because $\mathbf{N} + suc(0)$ is not a type. That is, the rules would no longer propagate the invariant that if $t = t$ in T then T is a type.

We could solve this problem by including a new judgment, T *is_a type*, into the theory. The rules would be quite clear for the types already built, namely:

$$\mathbf{N}\ is\_a\ type \quad \mathbf{1}\ is\_a\ type \quad \mathbf{0}\ is\_a\ type$$

$$\frac{A\ is\_a\ type \quad B\ is\_a\ type}{\begin{array}{ll}(A \times B) & is\_a\ type\\ list(A) & is\_a\ type\\ (A \to B) & is\_a\ type\\ (A+B) & is\_a\ type\end{array}}$$

We can then use the rules

$$\frac{a = a' \text{ in } A \quad B\ is\_a\ type}{inl(a) = inl(a') \text{ in } A+B} \qquad \frac{b = b' \text{ in } B \quad A\ is\_a\ type}{inr(b) = inr(b') \text{ in } A+B}$$

We will see in section 3.7 how to avoid adding this new judgment T *is_ type*.

The map $[f,g]$ is built from a new form called $decide(d; u.f(u); v.g(v))$ whose evaluation rules are

$$\frac{f(a)\ evals\_to\ c}{decide(inl(a); u.f(u); v.g(v))\ evals\_to\ c}$$

$$\frac{g(b)\ evals\_to\ c}{decide(inr(b); u.f(u); v.g(v))\ evals\_to\ c}$$

The function $[f,g]$ is $\lambda(x.decide(x; u.f(u); v.g(v)))$. It is easy to see that

$[f,g](inl(a)) = f(a)$ and $[f,g](inr(b)) = g(b)$.

3.6. Metamathematical properties of the type theory fragment

The theory with base types **0**, **1**, N and type constructors $\times, list, \rightarrow$ and $+$ is sufficiently complex that it is worthwhile analyzing its properties.

First, it is based on a simple inductive model of computability and typing that is *intuitively clear*. So we could accept it based on self-evidence. Indeed it is like PRA Church [1960] in that regard—a manifestly correct theory baring mistakes of formalization of the intuitive ideas. Discussing this type evidence for the theory leads us into philosophy and Formal Methods studies of formalization which are beyond the scope of the work.

Second, we can prove various properties of the formalism by syntactic means. For instance:

Termination of Evaluation: If $\vdash t = t$ in T then there is a term t' such that t *evals_to* t' and t' *evals_to* t'.

Subject Reduction: If $\vdash t = t$ in T and t *evals_to* t' then $\vdash t' = t'$ in T.

Typehood: If $\vdash t_1 = t_2$ in T then T *is_a type*, and $\vdash t_1 = t_1$ in T and $\vdash t_2 = t_2$ in T.

Nontriviality: There is no term t such that $\vdash t = t$ in **0**.

Consistency: It is not possible to derive $0 = suc(0)$ in N.

Third, we can translate this theory into various well-known mathematical theories including Heyting Arithmetic of ω order, HA^{ω}, IZF set theory and ZF set theory, and the theories of Feferman [1970,1975]. There are also categorical models of this simple fragment using topoi (Bell [1988]).

3.7. Inductive type classes and large types

The types defined so far belong to an inductively defined collection according to the scheme for T *is_a type* in the last section. Let U_1 denote this inductively defined collection of types; it has the characteristic of a type in that it has elements and is structured. Evaluation is defined on the elements, e.g. N *evals_to* N, $(\mathrm{N} \times \mathrm{N})$ *evals_to* $(\mathrm{N} \times \mathrm{N})$, etc. So all of the elements are canonical and are built up inductively themselves. In this regard U_1 resembles N. It has all the properties of a type.

We want to make U_1 a type. So we add rules for its elements in terms of equalities. For example, there are rules $\mathbf{0} = \mathbf{0}$ in U_1 and

$$\frac{A = A' \text{ in } U_1 \quad B = B' \text{ in } U_1}{A \times B = A' \times B' \text{ in } U_1}$$

The equality rules we have in mind are these

$$\mathrm{N} = \mathrm{N} \text{ in } U_1 \quad \mathbf{0} = \mathbf{0} \text{ in } U_1 \quad \mathbf{1} = \mathbf{1} \text{ in } U_1$$

$$\frac{A = A' \text{ in } U_1 \quad B = B' \quad \text{in } U_1}{\begin{array}{ll} (A \times B) = (A' \times B') & \text{in } U_1 \\ list(A) = list(A') & \text{in } U_1 \\ (A \to B) = (A' \to B') & \text{in } U_1 \\ (A + B) = (A' + B') & \text{in } U_1 \end{array}}$$

This is a structural or *intensional equality* (used in both Nuprl and Martin-Löf [1982]). It turns out that this equality is also extensional since $A = B$ in U_1 iff $a \in A$ implies $a \in B$ and conversely. This is the only type so far whose *elements are types*, but it does not include *all* types, in particular U_1 is not in U_1 according to our semantics.

We have no way to prove that U_1 is not in U_1. We don't even have a way to say this. But it would be possible to add a recursion combinator on U_1 that expressed the idea that U_1 is the least type closed under these operations. The combinator would have the form of a primitive recursive definition

$$\begin{array}{l} f(\mathbf{0}, x) = b_0(x) \\ f(\mathbf{1}, x) = b_1(x) \\ f(N, x) = b_2(x) \\ f((A \times B), x) = h_1(A, B, f(A, x), f(B, x)) \\ \vdots \\ f((A + B), x) = h_4(A, B, f(A, x), f(B, x)) \end{array}$$

With this form of recursion and the corresponding induction rule we could prove that every element of U_1 was either $\mathbf{0}$, $\mathbf{1}$, N, a product, a union, etc.

Once we can regard types as elements of a type like U_1, then we can extend our methods for building objects, say over $\mathbb{N}$ or by case analysis over a type of Booleans, say $\mathbb{B}$ etc. to building types. Here are two examples, taking $\mathbb{B}$ as an abbreviation of $\mathbf{1} + \mathbf{1}$ and abbreviating $inl(\bullet)$ as tt and $inr(\bullet)$ as ff.

Let $T(tt) = A, T(ff) = B$, then $\lambda(x.T(x))$ is a function $\mathbb{B} \to U_1$. If we build a generalization of $\mathbb{B}$ to n distinct values, say $\mathbb{B}_n = ((\mathbf{1} + \mathbf{1}) + \cdots + \mathbf{1})$ n times defined by $\mathbb{B} = \mathbf{1}, \mathbb{B}(suc(n)) = \mathbb{B}(n) + \mathbf{1}$ with elements $1_b, \ldots, n_b$, then we can build a function $T(x)$ selecting n types, $T(i_b)$.

It is worth thinking harder about functions like $T : \mathbb{B}_n \to U_1$. This is an indexed collection of types, $\{T(1_b), \cdots, T(n_b)\}$. We can imagine putting them together to form types in various ways, for instance by products or unions or functions

$$\begin{array}{l} T(1_b) \times \cdots \times T(n_b) \text{ or} \\ T(1_b) + \cdots + T(n_b) \text{ or} \\ T(1_b) \to \cdots \to T(n_b). \end{array}$$

We could define these types recursively, say by functions Π, Σ and Θ if we could have inputs like this: m in $\mathbf{N}$, T in $\mathbb{B}_m \to U_1$,

$$\begin{array}{l} \Pi_m(0)(T) = T(i_m(1)) \\ \Pi_m(n)(T) = \Pi_m(n-1)(T) \times T(i_m(suc(n))) \end{array}$$

where $i_m(k)$ selects the k-th constant of $\mathbb{B}_n$, k_b.[37] Likewise for Σ and Θ. However, we are unable to type these functions Π, Σ, Θ with the current type constructors. We could type them with the new ones we are trying to define!

In the case of Π and Σ the operations make sense even for infinite families of types, say indexed by $T \in A \to U_1$ for any type A. We can think of Π over $T \in A \to U_1$ as functions f such that on input $a \in A$, we have $f(a) \in T(a)$. For Σ over $T \in A \to U_1$ we can use the elements a as "tags" so that elements are pairs $\langle a, t\rangle$ where $t \in T(a)$.

These ideas give rise to two new type constructors, Π and Σ over an indexed family of types $T \in A \to U_1$. We write the new constructors as $\Pi(A;T)$ and $\Sigma(A;T)$. We could use typing rules like these

$$\frac{A \in U_1 \quad T \in A \to U_1}{\begin{array}{c}\Pi(A;T) \in U_1 \\ \Sigma(A;T) \in U_1\end{array}}$$

$$\frac{x \in A \vdash f \in T(x)}{\lambda(x.f) \in \Pi(A;T)} \qquad \frac{\vdash a \in A \quad \vdash b \in T(a)}{pair(a;b) \in \Sigma(A;T)}$$

The dotted lines forming the box indicate that this is an exploratory rule which will be supplanted later. We treat $\lambda(x.f)$ and $pair(a;b)$ just as before, so we are not adding new elements to the theory, just new ways to type existing ones.

With Π and Σ and using induction over $\mathbf{N}$ we can build types that are not in this U_1. For example, let $f(0) = A, f(suc(n)) = A \times f(n)$. Then f is a function $\mathbf{N} \to U_1$ where $f(n) = A \times \cdots \times A$ taken n times. The actual function is $\lambda(n.R(n;A;u,t.A \times t))$. Now we can build types like $\Sigma(N;\lambda(n.R(n;A;u,t.A \times t)))$ and $\Pi(A;\lambda(n.R(n;A;u,t.A \times t)))$ which are not in U_1. We could imagine trying to enlarge the inductive type class U_1 by adding these operators to the inductive definition. We will take up this topic in the next section.

Dependent types. The construction of Π and Σ types over U_1 suggests something more expressive. Instead of limiting the dependent constructions to functions from $T \in A \to U_1$, we could allow dependency whenever we can form a type expression $B[x]$ that is meaningful for all x of type A. We are led to consider a rule of the form

$$\frac{\vdash A \in U_1 \quad x : A \vdash B[x] \in U_1}{\begin{array}{c}fun(A;x.B) \in U_1 \\ prod(A;x.B) \in U_1\end{array}}$$

[37] $i_n(0) = inl^{m-1}(inl(\bullet))$ and $i_m(n) = inl^{m-n}(inr(\bullet))$.

We call *fun* a *dependent function* constructor and *prod* a *dependent product.*[38] We adopt a different notation from Π and Σ to suggest the more fundamental character of the construction. If we have $T \in A \to U_1$, then $\Pi(A;T)$ is the same as $fun(A;x.T(x))$ and $\Sigma(A;T)$ is the same as $prod(A;x.T(x))$. But now we can iterate the construction without going beyond U_1. That is, we postulate that U_1 is closed under dependent functions and products.

This conception of Π and Σ is reminiscent of the collection axiom in set theory. For example, in ZF if $R(x,y)$ is a single-valued relation on sets, then we can form $\{y \mid \exists x \in A.R(x,y)\}$. Another way to think of collection is to have a function $f : A \to Set$ where $A \in Set$ and postulate the existence of the set $\{f(x) \mid x \in A\}$.

The similarity between collection and these rules is that we can consider B in $fun(A;x.B)$ to define a function $\lambda(x.B)$ from A into U_1. With the addition of dependent types, the intuitive model becomes more complex. What assurance can we offer that the theory is still consistent, e.g. that we can't derive $0 = 1$ in N or that we derive $t \in T$ but evaluation of t fails to terminate? Can we continue to understand the model inductively? If we can build an inductive model of U_1 then we can be assured of not only consistency but of a constructive explanation. We answer these questions next.

3.8. Universes

We can consider U_1 and the rules for it in the last section as partial axiomatization of the concept of *Type*. On this view, we think of U_1 as open-ended, and we do not adapt an axiom capturing its closed inductive character, such as the recursion combinator for U_1 discussed above.

On the other hand, we can also think of U_1 as a large type belonging to *Type*. On this view the axioms for U_1 *reflect* the rules of type construction on *Type* into the collection of types. The axioms postulate a certain enrichment of the concept *Type* in the same way that the axiom of inaccessible cardinals postulates an enrichment of Set. Similarly, from the foundations of category theory (Kreisel [1959]), Grothendieck's concept of a *universe* is a way of modeling large categories (and is equivalent to inaccessible cardinals).

If we take the view that U_1 is a universe (rather than *Type*), then it makes sense to form larger universes, say U_2, then U_3, etc. To form U_2 we extend U_1 by adding the type U_1 itself, like this: $U_1 = U_1$ in U_2.

Martin-Löf and Nuprl axiomatize a universe hierarchy indexed by natural numbers, U_i. The method of doing this is to add $U_i = U_i$ to U_{i+1} and to postulate *cumulativity*, that any type A in U_i belongs to all U_j for $i < j$. So the *universe rules* are:

$$U_i = U_i \text{ in } U_{i+1} \qquad \frac{A = A \text{ in } U_i}{A = A \text{ in } U_j \quad \text{for } i < j.}$$

[38]Martin-Löf calls this a *dependent sum* and writes $\Sigma(x \in A)B$. We think of it as a generalization of $A \times B$ and display the constructor in Nuprl as $x : A \times B$.

It is possible to extend the universe hierarchy further, say indexed by ordinal numbers *ord*. It is possible to postulate closure of *Type* under various schemes for generating larger universes; Palmgren [1991] considers such matters.

Nuprl has been designed to facilitate index free, or "polymorphic", treatment of U_i. Generally, the user simply writes a universe as U_i and the system keeps track of providing relative level numbers among them in terms of expressions called *level expressions* which allow forming $i + 1$ and $max(i, j)$. The theoretical basis for this is in Allen [1987b] and was implemented by Howe and Jackson (see Jackson [1994c]).

3.9. Semantics: PER models

The principal mathematical method that we have used to prove the soundness of Nuprl (and Martin-Löf type theory) has been to interpret equality relations on a type as partial equivalence relations ("pers") over terms — thereby building a variety of *term model* (see Stenlund [1972]). We use a method pioneered by Stuart Allen [1987a,1987b] to define the model inductively.[39] In his thesis Allen compares his models to those of Aczel [1986], Beeson, and Smith [1984]. The modeling techniques also borrow from Tait [1967,1983] in that the membership relation is extended from values to all terms by the pre-evaluation relation; in that regard it follows closely Martin-Löf's informal semantics.[40]

Allen's method has been remarkably potent in all of our work. Mendler [1988] used the technique to model the recursive types defined in Mendler, Constable and Panangaden [1986]), and Smith [1984] used it to model our bar types for partial objects Constable and Smith [1993]. Harper [1992] gave one of the most accessible accounts; I draw heavily on the accounts of Allen, Mendler, and Harper to explain the method.

The first step is to fix the collection of terms. The next step is to equip the terms with an evaluation relation, written now as $t \downarrow t'$. Allen [1987b] gives an abstract account of the syntax of terms and the properties of evaluation. We follow Mendler and Harper in supplying less detail.

Assume that on *closed* terms t evaluation satisfies E1 and E2:

E1. if $t \downarrow t'$ and $t \downarrow t''$ then $t' = t''$, so $\downarrow$ is *deterministic*, and
E2. if $t \downarrow t'$ then $t' \downarrow t'$, so $\downarrow$ is *idempotent.*

If $t \downarrow t'$ then we call t' a (*canonical*) *value.*

Our task now is to specify those terms which are intended to be expressions for mathematical objects and to specify those terms which are expressions for types. We carry this out for the types built from $\mathbf{N}$ using products and dependent functions. We *distinguish* these as two tasks. The first one is to consider *membership*, and the

[39]In his introduction Allen says "The principal content of this thesis is a careful development of ... a semantic reinterpretation [of type theory] with the intention of making the bulk of type-theoretic practice ... independent of its original type-theoretic and constructive basis. ... Moreover, in the unfamiliar domain of intuitionistic type theory, the reinterpretation can serve as a *staff made of familiar mathematical material.*"

[40]We say that if $t \downarrow t'$ and $t' \in T$, then $t \in T$. This is the *preevaluation* relation of t' to t.

second is to determine *type expressions.* We look at membership first since it is more basic.

According to Martin-Löf, to specify a type is to say what its members are, i.e., which terms are members, and to say what equality means on those terms which are members. Equality will be an equivalence relation, E, on some collection of terms. Considered over the entire collection, the relation need only be partial. The *field* of the relation (those elements in the relation to themselves, xEx) are the members of the type. These relations are called *partial equivalence relations* or per for short.

The built-in notion of computation places an additional requirement on the pers, namely, they must respect evaluation. That is, if $t \downarrow t'$ and tEr, then $t'Er$. We can say this succinctly by defining *Kleene equality* on terms, $t \simeq t'$ means that if $t \downarrow s$ or $t' \downarrow s$, then $t \downarrow s$ and $t' \downarrow s$, i. e. if either term has a value, then both have that same value. So we require that $t \simeq t'$ and tEr implies $t'Er$. We next introduce notation for this notion of type, starting with the idea that types themselves are mathematical objects with an equality defined on them.

Let us see how this notion of type membership looks for the natural numbers, i. e. for the type $\mathbb{N}$. Define the relation *Neq* on terms inductively by

$$0 \; Neq \; 0$$
$$a \; Neq \; b \text{ implies } suc(a) \; Neq \; suc(b)$$
$$a' \; Neq \; b' \text{ and } a \downarrow a' \text{ and } b \downarrow b' \text{ implies } a \; Neq \; b.$$

Neq is a partial equivalence relation which determines a minimal notation for numbers on which we can compute by primitive recursion (lazily). That is, we know what elements are zero, and for nonzero numbers, we can find the predecessor.

Next, we define a membership per for the Cartesian product of two types A and B with α and β as the membership relations. Let α^* denote the pre-evaluation relation of a relation α, that is $a\alpha b$ iff there are a', b' such that $a'\alpha b'$ and $a{\downarrow}a', b{\downarrow}b'$. Define $\alpha \otimes \beta$ as $\{\langle pair(a;b), pair(a';b')\rangle \mid a\alpha a' \;\&\; b\beta b'\}^*$.

We can see that if α and β are value respecting pers, then so is $\alpha \otimes \beta$. It clearly defines membership in a Cartesian product according to our account of products.

Finally, we need a membership condition for the dependent function space constructor, $fun(A; x.\, B)$. This is a bit more complex because for each element a of A, $B[a/x]$ is a type. So we need to consider a family of membership relations indexed by a type. The members of the function type will be lambda terms, $\lambda(x.\, b)$. Let α be a value respecting per and for each a such that $a\alpha a$, let $\Phi(a)$ be a value respecting per. Define $\Pi\alpha\Phi$ as the following partial equivalence relation:

$$\{\langle \lambda(x.\, b), \lambda(x'.\, b')\rangle \mid \forall a, a'.\; a\alpha a' \Rightarrow b[a/x] \; \Phi(a) \; b'[a'/x']\}^*.$$

In order for this per to define type membership for the function space, we require that whenever $a\alpha a'$, then $\Phi(a) = \Phi(a')$. We have in mind that these membership conditions are put together inductively. This is made explicit by the following inductive definition of a relation $\boldsymbol{K}$ on pers.

$\boldsymbol{K}(Neq)$
$\boldsymbol{K}(\alpha \otimes \beta)$ if $\boldsymbol{K}(\alpha)$ and $\boldsymbol{K}(\beta)$
$\boldsymbol{K}(\Pi\alpha\Phi)$ if $\boldsymbol{K}(\alpha)$ and $\forall a, a'.\, a\alpha a' \Rightarrow \Phi(a) = \Phi(a')\ \&\ \boldsymbol{K}(\Phi(a))$.

We can prove inductively that all the pers in $\boldsymbol{K}$ are value respecting and all define type membership. $\boldsymbol{K}$ provides a per semantics for the small type theory based on N, products and dependent functions. Notice equality on pers is *extensional.*

Type expressions. The inductively defined set $\boldsymbol{K}$ determines a collection of membership pers which represent types, but it does not relate these to the terms used to name types, e. g. terms such as $N, N \times N$, $fun\,(N; x.\ decide(s; u.\ N; v.\ N \times N))$ and so forth. We establish this relationship next by modifying the definition of $\boldsymbol{K}$ to include names for types. Let $\boldsymbol{M}$ be the following inductively defined binary relation.

$$N\ \ \boldsymbol{M}\ \ Neq$$
$$A \times B\ \ \boldsymbol{M}\ \ \alpha \otimes \beta \text{ if } A\boldsymbol{M}\alpha \text{ and } B\boldsymbol{M}\beta$$
$$fun(A; x.\ B)\ \ \boldsymbol{M}\ \ \Pi\alpha\Phi \text{ if } A\boldsymbol{M}\alpha \text{ and } \forall a,a'.\, a\alpha a' \Rightarrow \Phi(a) = \Phi(a') \text{ and } B[a/x]\boldsymbol{M}\Phi(a)$$

This is an ordinary inductive definition of a binary relation. Also, it is easy to see that $A\boldsymbol{M}\alpha$ implies $\boldsymbol{K}\alpha$. The only membership pers described by $\boldsymbol{M}$ are those whose constituents are also described by $\boldsymbol{M}$. Moreover, all the membership pers are represented by terms, i. e. are related to terms by $\boldsymbol{M}$. This is critical for the $\Pi\alpha\Phi$ pers because it guarantees that Φ is represented by a term. Here are three critical facts about $\boldsymbol{M}$.

Fact 1 $A\boldsymbol{M}\alpha \Rightarrow \boldsymbol{K}(\alpha)$
Fact 2 $A\boldsymbol{M}\alpha$ and $A\boldsymbol{M}\alpha' \Rightarrow \alpha = \alpha'$
Fact 3 $A\boldsymbol{M}\alpha$ and $A \simeq A' \Rightarrow A'\boldsymbol{M}\alpha$.

These facts can be proved by $\boldsymbol{M}$ induction. Fact 1 means that all member pers are value respecting, and Fact 3 means that the type names are value respecting as well.

Pers for intensional type equality. We now want to define a per on type expressions which represents type equality and is value respecting. There is already a sensible equality that arises from $\boldsymbol{M}$, namely, $A = A'$ if $A\boldsymbol{M}\alpha, A'\boldsymbol{M}\alpha'$ and $\alpha = \alpha'$. This is an extensional equality. We want to model the structural equality of section 3.7, thus $A \times B = A' \times B'$ iff $A = A'$ and $B = B'$. Here is the appropriate definition of a binary relation $\boldsymbol{E}$ on terms.

$N\boldsymbol{E}N$
$A \times B\boldsymbol{E}A' \times B'$ if $A\boldsymbol{E}A'$ and $B\boldsymbol{E}B'$
$fun(A; xB)\boldsymbol{E}\ fun(A'; x'.\ B')$ if $A\boldsymbol{E}A'$ and $\exists\alpha\ A\boldsymbol{M}\alpha$ and $A'\boldsymbol{M}\alpha$ and $\forall a, a'.\, a\alpha a' \Rightarrow B[a/x]\boldsymbol{E}\ B'[a'/x']$
$A\boldsymbol{E}A'$ if $B\boldsymbol{E}B'$ and $A \downarrow B$ and $A' \downarrow B$.

We say that A is an intensional type expressions of model $\boldsymbol{M}$ iff $\exists\alpha(A\boldsymbol{M}\alpha$ and $A\boldsymbol{E}A)$. Clearly, the per representing type equality is value respecting. The relations $\boldsymbol{E}$ and $\boldsymbol{M}$ provide a model of our type theory fragment. The methods extend to Martin-Löf's '82 theory and to Nuprl 3 as Allen has shown.

We now summarize the approach described above, starting from $\boldsymbol{E}$ and following Harper's method of using least fixed points to present the inductive relations.

Summary of per semantics. Here is a summary of the per semantics along the lines developed by Harper [1992]. Let $\mathcal{T}$ be the collection of terms. We define on $\mathcal{T}$ a *partial equivalence relation* E intended to denote type equality. If aEa then we say that a is a type. If aEa' then a and a' are equal types. Let $|E| = \{t : \mathcal{T} \mid tEt\}$, called the *field* of E. Let $\mathcal{T}/E$ be the set of equivalence classes of terms; say $[t]_E = \{x : \mathcal{T} \mid xEt\}$. Let PER denote the set of all partial equivalence relations on $\mathcal{T}$.

Associated with each type is a membership equality, corresponding to $a = b$ in A; thus for each $a \in |E|$, there is a partial equivalence relation, $\boldsymbol{L}(a)$. So $\boldsymbol{L} \in \mathcal{T}/E \to$ PER.

We require of E and each $\boldsymbol{L}(a)$ that they respect evaluation, i.e. if $a_1 E a_2$ and $a_1 \downarrow a_1'$ and $a_2 \downarrow a_2'$, then $a_1' E a_2'$. Likewise for $\boldsymbol{L}(a)$ in place of E.

Now consider how we might define E and $\boldsymbol{L}$ mutually recursively to build a model of the type theory. For the sake of simplicity, we start with an extensional notion of type equality, as above. Call it Ext. We define Ext and $\boldsymbol{L}$ mutually recursively.

$$a\ Ext\ b \quad \text{iff} \quad \forall x, y.\ ((x\boldsymbol{L}(a)y) \Leftrightarrow (x\boldsymbol{L}(b)y))$$

$$s\boldsymbol{L}(a_1 \times a_2)t \quad \text{iff} \quad \exists s_1, s_2, t_1, t_2.\ s \downarrow pair(s_1; s_2) \ \&\ t \downarrow pair(t_1; t_2) \ \&\ s_1\boldsymbol{L}(a_1)t_1 \ \&\ s_2\boldsymbol{L}(a_2)t_2.$$

$$f\boldsymbol{L}(fun(a_1; x.\ a_2))f' \quad \text{iff} \quad \exists x, b, x', b'.\ f \downarrow \lambda(x.\ b) \ \&\ f' \downarrow \lambda(x'.\ b') \ \&\ \forall y, y' \colon |Ext(a_1)|.\ (y\boldsymbol{L}(a_1)y' \Rightarrow b[y/x]\boldsymbol{L}(a_2[y/x])b'[y'/x'].$$

This is a mutually recursive definition of Ext and $\boldsymbol{L}$, and it reflects our intuitive understanding, but the definition is not a standard positive (hence monotone) inductive definition because of the negative occurrence of $y\boldsymbol{L}(a_1)y'$ in the clause defining equal functions.

Allen calls these "half-positive" definitions; his method of using $\boldsymbol{K}$ and $\boldsymbol{M}$ as above shows how to replace this nonstandard definition with a standard positive induction which can be interpreted in either classical or constructive settings (for example, in ZF or IZF or in a theory of inductive definitions, see Troelstra [1973] and Feferman [1970]).

Definition. A *type system* τ is a pair $\langle E, \boldsymbol{L}\rangle$ where E is a value respecting per on $\mathcal{T}$ and for each $a \in |E|$, $\boldsymbol{L}(a)$ is a value respecting per. Given type systems $\tau = \langle E, \boldsymbol{L}\rangle$ and $\tau' = \langle E', \boldsymbol{L}'\rangle$, define $\tau \sqsubseteq \tau'$ iff $E \sqsubseteq E'$ and $\forall a \colon |E|.\ \boldsymbol{L}(a) = \boldsymbol{L}'(a)$; that is $\tau \sqsubseteq \tau'$ iff τ' has possibly more types, and on the types in τ it has the same equality.

Let TS be the collection of all type systems over $\mathcal{T}$ with evaluation $\downarrow$. It is easy to see that TS under $\sqsubseteq$ is a complete partially ordered set, a *cpo*. The relation

$\tau \sqsubseteq \tau'$ is a partial order on TS, and there is a least type system in this ordering, namely $\langle \phi, \phi \rangle$ where ϕ is the empty set. A non-empty subset D of TS is *directed* iff every pair of elements in D, say τ, τ', has an upper bound in D. Given any directed set D of type systems, say τ_i for $i \in I$, it has a least upper bound $\hat{\tau}$ where $\hat{E} = \cup E_i$ and $\boldsymbol{L}_\omega(a) = \boldsymbol{L}_i(a)$ if any $\boldsymbol{L}_i(a)$ is defined. If $\boldsymbol{L}_i(a)$ is defined, then since τ_i, τ_j for $i \neq j$ has an upper bound in D, say τ_k, we know that $\boldsymbol{L}_j(a) = \boldsymbol{L}_i(a)$ for $a \in E_i \cup E_j$, so the type system τ_a is well defined. Also τ_ω is least since if $\tau_i \sqsubseteq \tau'$ for all i, then $E_i \sqsubseteq E'$ for all i, so $\cup E_i \sqsubseteq E'$.

Theorem. *For any cpo D with order $\sqsubseteq$, if $F \in D \to D$ and F is monotone, i.e. $x \sqsubseteq y \Rightarrow F(x) \sqsubseteq F(y)$ for all x, y in D, then there exists a least fixed point of F in D, i.e. an element x_0 such that (i) $F(x_0) = x_0$, (ii) for all z such that $F(z) = z$ and $x_0 \sqsubseteq z$.*

We now define an operation $T \in TS \to TS$ which is monotone and whose least fixed point, $\hat{\tau}$, is a type system which models our rules.

Definition. Let $T(\langle E, \boldsymbol{L} \rangle) = \langle F^*, \boldsymbol{M} \rangle$ where

$$\begin{aligned} F = \{\langle N, N \rangle\} & \\ & \cup \{\langle a \times b, a' \times b' \rangle \mid aEa' \;\&\; bEb'\} \\ & \cup \{\langle fun(a; x.\, b), fun(a'; x'.\, b') \rangle \mid aEa' \;\& \\ & \qquad \forall y, y'.\, y\boldsymbol{L}(a)y' \Rightarrow b[y/x]Eb'[y'/x']\} \end{aligned}$$

$$\boldsymbol{M}(a) = \begin{cases} Neq \text{ if } a = N \\ \boldsymbol{L}(a_1) \otimes \boldsymbol{L}(a_2) \text{ if } a = a_1 \times a_2 \\ \Pi(\boldsymbol{L}(a_1), \lambda(x.\, \boldsymbol{L}(b))) \text{ if } a = fun(a_1; x.\, b) \;\& \\ a_1 \in |E| \;\&\; \forall y\colon |\, \boldsymbol{L}(a_1)|.\, b[y/x] \in |E| \end{cases}$$

Theorem. *T is monotone in $\sqsubseteq$ on TS.*

3.10. Using type systems to model type theories

Allen's techniques enable us to model a variety of type theories. Let us designate some models for the theories discussed earlier. We'll fix the terms and evaluation relation to include those of the richest theory; so the terms are: **0**, **1**, $\bullet$, N, 0, $suc(t)$, $prd(s)$, $add(s;t)$, $mult(s;t)$, $exp(s;t)$, $R(n;t;v.b;u,v,i.h)$, $s \times t$, $pair(s;t)$, $prod(s;x.t)$, $fun(s;x.t)$, $\lambda(x.t)$, $list(t)$, $(s.t)$, $L(s;a;v.b;h,t,v,i.g)$, $s+t$, $inl(s)$, $inr(t)$, and $decide(p;u.s;v.t)$.

There is also the evaluation relation s *evals_to* t which we abbreviate as $s \downarrow t$. We consider various mappings $\mathbf{T}_l : TS \to TS$ where l is a label such as N, G, ML, Nu, etc. The most elementary "theory" we will examine is a subtheory of arithmetic involving only equalities over **N** built from $0, suc(t), prd(s)$, and $add(s;t)$. This is modeled by $\mathbf{T}_N$. The input to $\mathbf{T}_N$ is any pair $\langle E, \boldsymbol{L} \rangle$ and the output is $\langle F, \boldsymbol{M} \rangle$ where the only type name is N, so NFN, thus $N \in |F|$, and the only type equality is $\boldsymbol{M}(N)$ which is defined inductively. We have that $s\boldsymbol{M}(N)t$ is the least relation *Neq* such that

$$s\ Neq\ t\ iff\ (s \downarrow 0\ \&\ t \downarrow 0\ v\ \exists s', t'.\ s \downarrow suc(s')\ \&.\ t \downarrow suc(t') s' Neq\ t').$$

The map T_N takes any $\langle E, \boldsymbol{L}\rangle$ to this $\langle F, \boldsymbol{M}\rangle$, so its least fixed point, $\mu(\boldsymbol{T}_N)$ is just $\langle F, \boldsymbol{M}\rangle$. This is the model for *successor arithmetic.* We see that in this model, N is a type, that $s = t$ in N iff s evaluates to a canonical natural number and t evaluates to the same canonical number.

The rules can be confirmed as follows. First, notice that evaluation is deterministic and idempotent on the terms. As we observed, the general equality rules hold in any type system (because $\boldsymbol{M}(N)$ is in equivalence relation on canonical numbers). This follows by showing inductively that $0, suc(0), suc(suc(0)), \ldots$ are in the relation $\boldsymbol{M}(N)$, i.e. in the field of the relation. The fact that $\boldsymbol{M}(N)$ respects evaluation validates the last equality rule.

$$\begin{array}{l} \mu(T_N) \models 0 = 0 \text{ in } N \\ \mu(T_N) \models s = t \text{ in } N \text{ implies } \mu(T_N) \models suc(s) = suc(t) \text{ in } N. \end{array}$$

The typing rule for successor is also confirmed by induction on *Neq*; namely, if $s'\boldsymbol{L}(N)t'$, then since $suc(s') \downarrow suc(s')$ and $suc(t') \downarrow suc(t')$, then we have $suc(s)\boldsymbol{M}(N)suc(t)$ as required for the typing rule.

In the case of N, the model $\mu(T_N)$, and the informal semantics are essentially the same. So the theory fragment for N can *stand on its own* with respect to the model. Even a set theoretic semantics for N will have the same essential ingredient of an inductive characterization. For instance, Frege's definition was that

$$N == \{x \mid \forall X.\ (0 \in X\ \&\ \forall y.\ (y \in X \Rightarrow s(y) \in X)) \Rightarrow x \in X\}$$

where $s(x)$ is $\{z \mid \exists u.\ (u \in z\ \&\ z - \{u\} = x)\}$. In ZF we can use the postulated infinite set, *inf*, and form $\omega == \{i : inf \mid \forall x.\ nat\_like(x) \Rightarrow i \in x\}$ where $nat\_like(x)$ *iff* $(0 \in x\ \&\ \forall y.\ (y \in x \Rightarrow suc(y) \in x))$ for $Suc(y) = y \cup \{y\}$. In both of these definitions, the inductive nature of **N** is expressed. But Frege's theory and ZF allow very general ways of using this inductive character. So far we have only used it for specifying the canonical values.

The same approach can be used to define a model for the type theory with cartesian products. In this case we denote the operator on type systems as ${T_N}^2$. Given ${T_N}^2(\langle E, \boldsymbol{L}\rangle) = \langle F, \boldsymbol{M}\rangle$, if $S \in |E|$ and $T \in |E|$ then $S \times T \in |F|$, and $\boldsymbol{M}(S \times T)$ is $\boldsymbol{L}(S) \otimes \boldsymbol{L}(T)$. For this system, ${T_N}^2$ is continuous, i. e. if $\tau_0 = \langle \phi, \phi\rangle$ and ${T_N}^2(\tau_i) = \tau_{i+1}$, then $\mu({T_N}^2) = \tau_\omega$.

In $\mu({T_N}^2)$ all the rules for the fragment of section 3.2 are true. Again the theory is so close to the semantics that it stands on its own. Notice that in confirming the rule for typing pairs, we rely on the fact that $\mu({T_N}^2)$ is a fixed point.

$$\begin{array}{ll} \mu({T_N}^2) & \models s = s' \text{ in } S \text{ and } \mu({T_N}^2) \models (t = t' \text{ in } T) \text{ imply} \\ \mu({T_N}^2) & \models pair(s;t) = pair(s';t') \text{ in } S \times T. \end{array}$$

Note, this fact would not be true in any fixed τ_i since $S \times T$ might be defined only in τ_{i+1}.

To provide a semantics for $fun(A; x. B)$ and $prod(A; x. B)$ we use the map T_{ML} defined in section 3.9. The model is $\mu(T_{ML})$. To prove the rules correct, we recall the meaning of sequents such as $x \in A \vdash B$ type and $x \in A \vdash s = t$ in T.

$$\mu(T_{ML}) \models (x \in A \vdash B\ type) \text{ implies } \mu(T_{ML}) \models fun(A; x.B)\ type$$
$$\mu(T_{ML}) \models (x \in A \vdash b \in B) \text{ implies } \mu(T_{ML}) \models \lambda(x.b) \text{ in } fun(A; x.B)$$

Modeling hypothetical judgments. The meaning of $x \in A \vdash b \in B$ is that A is a type and for any two elements, a, a' of A, $B[a/x]$ is a type and $B[a/x] = B[a'/x]$ (i.e. B is *type functional* in A), and moreover, $b[a/x] \in B[a/x]$ and $b[a/x] = b[a'/x]$ in $B[a/x]$. We have extended this notion to multiple hypotheses inductively to define $x_1 \in A_1, \ldots, x_n \in A_n \vdash b \in B$. This definition can be carried over to type systems.

3.11. A semantics of proofs

The discussion of proofs as objects and Heyting semantics in section 2 suggested treating proofs as objects and propositions as the types they inhabit. True propositions are those inhabited by proofs. But there were several questions left open in section 2.14 about the details of carrying out this idea.

The type theory of this section can answer these questions, and in so doing it provides a *semantics of proofs.* The basic idea is to consider a proposition as the type of all of its proofs and to take proof expressions to denote objects of these types. Based on Heyting's semantics we have a good idea of how to assign a type to compound propositions in terms of types assigned to the components. For atomic propositions there are several possibilities, but the simple one will turn out to provide a good semantics. The idea is to consider only those atomic propositions which can plausibly have *atomic proofs* and to denote the canonical atomic proofs by the term *axiom.* We will assign types to the compound propositions in such a way that the canonical elements will represent what we will call *canonical proofs.* Moreover, the reduction relation on the objects assigned to proof expressions will correspond to meaningful reductions on proofs. Proofs corresponding to noncanonical objects will be called noncanonical proofs. The correspondence will guarantee that noncanonical proofs p' of a proposition P will reduce to canonical proofs of P.

We now define the correspondence between propositions and types and between proofs and objects. Sometimes this correspondence is called the Curry-Howard isomorphism.

Curry-Howard isomorphism. For the sake of this definition, if P is a proposition, we let $[\![P]\!]$ be the corresponding type, and if p is a proof expression, we let $[\![p]\!]$ be the corresponding element of $[\![P]\!]$. We proceed to define $[\![\ ]\!]$ inductively on the structure of proposition P from section 2.5.

1. We consider only atomic propositions of the form $a = b$ in A. The type $[\![a = b \text{ in } A]\!]$ will have the atomic proof object *axiom* if the proposition is

axiomatically true.

If the proof expression e for $a = b$ in A evaluates to a canonical proof built only from equality rules, then we arrange that $e \downarrow axiom$. This is a simple form of correspondence that *ignores equality information*. For instance

$$[\![symmetry(3)]\!] \downarrow axiom \quad [\![transitivity(e_1, e_2)]\!] \downarrow axiom.$$

$$\frac{[\![e]\!] \downarrow e'}{[\![equality\_intro(e)]\!] \downarrow e'}$$

We also need these evaluation rules for the proof expressions for substitution and type equality.

$$\frac{[\![p]\!] \downarrow p'}{[\![subst(p; e)]\!] \downarrow p'} \quad \frac{[\![p]\!] \downarrow p'}{[\![eq(p; e)]\!] \downarrow p'}$$

2. $[\![P \mathbin{\&} Q]\!] = [\![P]\!] \times [\![Q]\!]$ and
   $[\![\&R(e_1, e_2)]\!] = pair([\![e_1]\!]; [\![e_2]\!])$, and
   $[\![\&L(e_1; u, v.\ e_2)]\!] = [\![e_2]\!](1of([\![e_1]\!])/u, 2of([\![e_1]\!]/v))$.
3. $[\![P \vee Q]\!] = [\![P]\!] + [\![Q]\!]$,
   $[\![\vee Rl(a)]\!] = inl([\![a]\!])$,
   $[\![\vee Rr(b)]\!] = inr([\![b]\!])$,
   $[\![\vee L(d; u.\ e_1; v.\ e_2)]\!] = decide([\![d]\!]; u.[\![e_1]\!]; v.[\![e_2]\!])$.
4. $[\![P \Rightarrow Q]\!] = [\![P]\!] \rightarrow [\![Q]\!]$,
   $[\![\Rightarrow R(x.\ e)]\!] = \lambda(x.[\![e]\!])$,
   $[\![\Rightarrow L(f; p; y.\ q)]\!] = [\![g]\!][ap([\![f]\!]; [\![p]\!]/y]$.
5. $[\![\exists x{:}A.\ P[x]]\!] = prod(A; x.\ [\![P[x]]\!])$,
   $[\![\exists R(a; p)]\!] = pair(a; [\![p]\!])$,
   $[\![\exists L(p; u, v.\ g)]\!] = [\![g]\!](1of([\![p]\!])/u, 2of([\![p]\!]/v))$.
6. $[\![\forall x{:}A.\ P[x]]\!] = fun(A; x.\ [\![P[x]]\!])$,
   $[\![\forall R(x.\ e)]\!] = \lambda(x.[\![e]\!])$,
   $[\![\forall L(f; a; y.\ e)]\!] = [\![g]\!][ap([\![f]\!]; a)/y]$.

Sequents to typing judgments. We can now translate *deductions* of sequents $\bar{H} \vdash P$ *by* p to *derivations* of $[\![\bar{H}]\!] \vdash [\![p]\!] \in [\![P]\!]$. Given $\bar{H} = x_1 : H_1, \ldots, x_n : H_n$ we take $[\![\bar{H}]\!]$ to be $x_1 \in H_1', \ldots, x_n \in H_n'$ where if H_i is a type then $H_i' = H_i$ and if H_i is a formula then $H_i' = [\![H_i]\!]$. In this case we treat the label x_i as a variable.

Now to translate a deduction tree to a derivation tree we work up from the leaves translating sequents as prescribed and changing the rule names. The proof system was designed in that we need not change the variable names.

Expressing well-formedness of formulas. The introduction of U_1 combined with the propositions-as-types interpretation allows us to express the pure proposition of typed logic more generally, and we can solve the small difficulty of insuring that $A + B$ is a type discussed at the end of section 3.5.

According to the propositions-as-types principle, U_1 represents the type of (small) propositions, and a function $P \in A \to U_1$ can be interpreted as a propositional function. When we want to stress this logical interpretation, we use the display form $Prop_1$ for U_1 and generally $Prop_i$ for U_i, and we call $Prop_i$ the proposition of $level_i$.

We can express general propositions in typed logic by quantifying over $Prop_i$ and U_i. Here are some examples from section 2.

1. $\forall A, B\!:\!U_1.\ \forall P\!:\!A \to Prop_1 \forall Q\!:\!B \to Prop_1.$
   $\forall x\!:\!A.\ \forall y\!:\!B.\ (P(x) \& Q(y)) \Leftrightarrow \forall x\!:\!A.\ P(x) \& \forall y\!:\!B.\ Q(y).$
2. $\forall A, B\!:\!U_1. \forall R\!:\!A \times B \to Prop_1.\ (\exists y\!:\!B.\ \forall x\!:\!A.\ R(x,y) \Rightarrow \forall x\!:\!A.\ \exists y\!:\!B.\ R(x,y))$.

At this level of generality, we need to express the well-formedness of typed formulas in the logic rather than as preconditions on the formulas as we did in section 2. This can be accomplished easily using U_i and $Prop_i$. We incorporate into the rules the conditions necessary for well formedness. For example, in the rule

$$\bar{H} \vdash P \Rightarrow Q \text{ by } \Rightarrow\!R$$
$$\bar{H}, P \vdash Q$$

We need to know that P and Q are propositions. We express this by additional well-formedness subgoals. A complete rule might be

$$\bar{H} \vdash P \Rightarrow Q \text{ by } \Rightarrow\!R \text{ at } i$$
$$\bar{H}, P \vdash Q$$
$$\bar{H} \vdash P \in Prop_i$$
$$\bar{H} \vdash Q \in Prop_i$$

If we maintain the invariant that whenever we can prove $\bar{H} \vdash a \in A$ then we know A is in a U_i, and whenever we prove $\bar{H} \vdash P$ then we know P is in $Prop_i$, then we can simplify the rule to this

$$\bar{H} \vdash P \Rightarrow Q \text{ by } \Rightarrow\!R \text{ at } i$$
$$\bar{H}, P \vdash Q$$
$$\bar{H} \vdash P \in Prop_i$$

We need to add well-formedness conditions to the following rules, $\vee R$, $\Rightarrow\!R$, $\forall R$, *Magic*. We already presented $\Rightarrow\!R$; here are the others.

VR

$$\bar{H} \vdash P \vee Q \text{ by } \vee R_l \text{ at } i$$
$$\bar{H} \vdash P$$
$$\bar{H} \vdash Q \in Prop_i$$

The $\vee R_r$ case is similar.

$\forall R$

$$\bar{H} \vdash \forall x\!:\!A.\ P(x) \text{ by } \forall R \text{ at } i$$
$$\bar{H}, x\!:\!A \vdash P(x)$$
$$\bar{H} \vdash A \in U_i$$

Magic

$$\bar{H} \vdash P \vee \neg P \text{ by } Magic \text{ at } i$$
$$\bar{H} \vdash P \in Prop_i$$

3.12. Proofs as programs

The type corresponding to a proposition of the form $(\forall x{:}A.\ \exists y{:}B.\ S[x,y])$ is the function space $x : A \to y : B \times [\![S[x,y]]\!]$. The proof expressions, say p, for this object denotes a canonical element of the type. That element is a function $\lambda(x.\ b)$ where for each $a \in A, b[a/x] \in y{:}B \times [\![S[a,y]]\!]$ and if $1of(b[a/x]) \in B$ and $2of(b[a/x]) \in [\![S[a, 1of(b[a/x])]]\!]$. So the function $\lambda(x.\ 1of(b)) \in A \to B$ and let $f = \lambda(x.\ 1of(b))$, then $f \in A \to B$ and $\lambda(x.\ 2of(b))$ proves $\forall x{:}A.\ S[x, f(x)]$.

So we can see that the process of proving the "specification" $\forall x{:}A.\ \exists y{:}B.\ S[x,y]$ constructively creates a program f for solving the *programming task* given by the specification, and it simultaneously produces the verification $\lambda(x.\ 2of(b))$ that the program meets its specification (c.f. Constable [1972], Bates and Constable [1985] and Kreitz [n.d.]).

Refinement style programming. This style of programming provides a way to build the program and its justification hand-in-hand. It is possible to gradually refine these two objects, filling only as much detail as necessary for clarity. So for example, proof detail can be omitted for programming steps that are obvious. The extreme case of "unbridled" programming arises when we omit all proof steps except those that come automatically as part of the programming, e.g. certain "type checking steps" and the over all logical structure of the proof.

Explicit programming style. We can *program* a solution to $\forall x{:}A.\ \exists y{:}B.\ S[x,y]$ directly by writing a function $f \in A \to B$ and then proving $\forall x : A.\ S[x, f(x)]$. Christine Paulin-Mohring [1989] is studying how to use the program information to help drive the derivation of the proof.

4. Typed programming languages

4.1. Background

Programming at its "lowest level" involves communicating with specific digital hardware in "machine language," sequences of *bits* (0's and 1's). The particular machine model will classify sequences of bits into a fixed number of "types," say instructions, signals, addresses, and data; the data might be further classified as *floating point* or *integer* or *audio* or *video*, etc. Programming at this machine level or just above at *assembly language* level is generally regarded as "untyped" in part because everything is ultimately bits.

We are mainly concerned with so-called *higher-level programming* languages, and for the purpose of this discussion, higher-level languages will be classified into two groups as *typed* or essentially *untyped*. Two high level languages from the earliest period are still "alive," Fortran and Lisp. Fortran is considered typed (though minimally) as are more modern languages like Pascal, C^{++}, ML, and Java. Two of the most historically significant typed languages were Algol 68 and Simula 67. Lisp is considered *untyped* as is its modern descendent Scheme. These languages have a notion of *run-time typing* in which data is tagged with type information during execution. Whereas Algol 68, ML, and Java, for example, are *statically typed* in that data and expressions are typed before execution (at "compile time").[41]

One of the major design debates in the computer science community over the years has been about the value of rich static typing, represented by Algol 68 and Simula, and "untyped" programming represented by Lisp and Scheme. There are formal languages that capture the essence of this distinction. Lisp and Scheme are represented by the *untyped lambda calculus* of Church [1960] (see Barendregt [1981], Stenlund [1972], Hindley, Lercher and Seldin [1972]) on which they were modeled, and ML by the *typed lambda calculus* (see Barendregt [1977], de Bruijn [1972]).

We have seen the untyped lambda calculus in section 3.4. Its terms are variables, abstractions, and applications denoted respectively $x_i, \lambda(x.\ t)$, and $ap(s;t)$ for s and t terms. The typed calculus introduces some system of types T and requires that the variables are typed, x^T. Usually the types include the individuals, ι, and if α, β are types, then so is $(\alpha \rightarrow \beta)$. The untyped lambda calculus can express the full range of sequential control structures and hence the class of general recursive functions. For example, the Y combinator $\lambda(f.\ ap(\lambda(x.\ ap(f;ap(x;x)))); \lambda(x.\ ap(f;ap(x;x))))$ more commonly written $\lambda(f.\ f\lambda(x.\ xx))\ \lambda(x.\ xx)$ is used to define recursive functions. We have that $Y(\lambda(f.\ F[f])) = F[Y(\lambda(f.\ F[f]))]$ so that Y "solves" the recursive definition $f = F[f]$.

In the typed lambda calculus, Y is not typeable because the self-application $\lambda(x.\ ap(x;x))$ cannot be typed. This situation summarizes for "typeless programming devotees" the inherent limitations of typed programming; for them types "get in the way."

The debate about typed or untyped languages illustrates one of the many design issues that have been studied and debated over the years. Other topics include: functional *versus* imperative, lazy *versus* eager evaluation, manual *versus* automatic storage allocation, reflection or not, and so forth.

Many of these issues have been explored with theoretical models, and much is known about the design consequences. Indeed many programming language constructs arose first in the setting of formal logical theories, e.g. the lambda calculus, type systems, binding mechanisms, block structure, abstract data types (as algebraic structures) and modules. Just as assembling a good formal theory is high art, so is assembling a good programming language. Both are formal systems which can be processed by computers. But there is at least one major difference.

Good programming languages are widely used, perhaps by tens of thousands of people over their life times. Most logical theories are never implemented, and the best of those that are might be used by less than one hundred people over a lifetime.[42]

---

[41] A compiler translates high-level language programs into another language, typically a lower-level language such as assembly code or native code (machine language).

[42] We hope that the fact that Nuprl contains a programming language and that proofs are executable will attract a significant audience.

I believe that this fact has a major consequence for "theory designers," namely they must learn about programming language evolution.

We see from a history of programming languages what ideas "work", what combinations of features are most expressive, what constructs are heavily used. As with the evolution of natural languages, the speakers exert a force to mold the language to its purpose. One of the lessons of programming language history is that types are critical. A language's type system is its *most important component.* We also know that modularity mechanisms are critical, but this too is defined by the type system.

The evolutionary trend is toward ever richer type systems—from the fixed types of Fortran to the polymorphic recursive types of ML and the classes of Java. One might argue that *this development must eventually subsume the type systems of the mathematical theories.* I believe this is true, and our discussion of type systems will reveal why.

Role of types in programming. Let us examine the role of types in programming (see the excellent article by Hoare [1972] as well). Fortran used variable names beginning with i, j, k, l, m, n to denote integers (fixed point numbers), the other letters indicated reals (floating point numbers). This type distinction facilitated connection to mathematical practice where the same conventions were used, and it provided information to the compiler about how to translate expressions into assembly language which also made the distinction between fixed and floating numbers.

Another important type in Fortran and Algol was the *array.* Arrays represent sequences, matrices, tensors, etc. A typical specification (or declaration) of this type might be *real array*$[n, m]$, a two dimensional array (matrix) of reals. The declaration provides a link to important mathematical types such as sequences or matrices, and provides information to the compiler on how much memory needs to be allocated for this data.

The *record type* (or Algol structure) also provides links to mathematical types and provides information for the compiler. A typical record syntax is $record(a_1 : T_1; \ldots ; a_n : T_n)$ where T_i are types and a_i are identifiers called *field selectors.* This type corresponds to a *cartesian product* $T_1 \times \ldots \times T_n$, and if t is an expression of this record type, then $t.a_i$ indicates the i-th component, which has type T_i. We discuss the field selectors in Section 4.4.

In this case the type declaration also introduces new identifiers (or names) into the language. This was a convenience not systematically used in mathematics. But it also led to some confusion about the status of these names a_i; are they bound variables or free? And if bound, what is their scope? Here a small "convenience" leads to interesting new questions about scope and naming in formal languages.

Algol 68 introduced a union type, $union(T_1, \ldots, T_n)$. This was an obvious attempt to link to mathematical types, but it created problems for efficient language translation since the compiler might have to reserve storage based on the type T_i needing the most memory. This type also brought language designers face to face with the problems of a "computable set theory." A programmer given data t in the type $union(A, B, C)$ will need to know which type it is in. So there must be an operation, like $decide(t)$ which will decide what type t belongs to. This operation is not available as a computable operation in set theory, so new mathematics had to be worked out. Algol 68 was rich in a "computable mathematics of types," and its reference manual is a type theory which inspired both logician and computer scientist alike.

In Pascal the union type was considered to be a *variant* of the record type. The simplest such structure is essentially $record(x_1 : A_1; x_2 : A_2[x_1])$ which is thought of as a union indexed by the (necessarily finite) type A_1. This is a restricted version of our dependent product type $prod(A_1; x.\ A_2[x])$ from Section 3.7. The Pascal conception reveals both the computational way to treat unions, namely use *disjoint unions*, and reveals the implementation strategy (borrowed from set theory)—use elements from a type A_1 as tags on the data to keep track of the disjunct. So if the tag types are the booleans, $\mathbb{B}$, and $A_1(i) = \textit{if } i \textit{ then } S \textit{ else } T \textit{ fi}$ then $prod(\mathbb{B}; i.\ A_1(i))$ is the Algol 68 $union(S, T)$ and the Pascal variant record $record(i{:}\mathbb{B}; x{:}A_1[i])$.[43]

Algol 60 and Algol 68 considered the notion of higher order functions. Algol 68 essentially had the idea of the type $\mathsf{fun}(x : A)B$ as the type of function from A to B. But the implementation technology was not up to the task of returning functions as values. This type challenged the community to implement it correctly as done in Scheme and ML with *closures*.

The function space concept $\mathsf{fun}(x : A)B$ does not mean the same thing as the corresponding mathematical notion, $A \to B$ even in the constructive case. In computational mathematics the elements f of $A \to B$ are *total* functions; that is, on every element a of $A, f(a)$ converges to a value b in B. Whereas, the elements ψ of $\mathsf{fun}(x{:}A)B$ are *partial functions*, that is, $\psi(a)$ might diverge or abort without returning a value. This is a major difference between programming types and mathematical types.

There are two reactions to the difference. It is possible to give total function semantics to $\mathsf{fun}(x{:}A)B$ and claim that current implementations are just approximations to the idea. The full concept emerges in a programming logic with termination rules (Dijkstra [1968]). On the other hand, one can regard the partial function space as a new mathematical construct and try to work out axioms and models for it (Scott [1976], Plotkin [1977]). Both approaches have been pursued.

Notice that it is a simple manner to extend $\mathsf{fun}(x : A)B$ to dependent function types by allowing B to depend on x. This type is then closely related to $\mathsf{fun}(A; x.\ B)$ of Section 3.

A more modern addition to the type structure of programming languages is the *module* or *object* (or ADT or package or unit). This concept can be traced to Simula 67 and is well developed in Modula and SML. Among the interesting experimental languages for modules were Russell at Cornell (Demers and Donahue [1980], Boehm et al. [1986]), CLU at MIT (Liskov and Guttag [1986]), and Modula at DEC. The basic idea is that a module is a type, say D, and a collection of operations f_i

[43]The actual Pascal syntax is very baroque, and the so called free unions are a well known place for "breaking" the type discipline since the user must keep track of the dependency. Note the *if_then_else_fi* notation is the Algol 68 way of "bracketing" the conditional with delimiters *if*, *fi*.

on D and auxiliary types. This is the type of a *structure* in algebra (Bourbaki [1968a]) and model theory (Chang and Keisler [1990]). For example, we might have $\langle D, f, g, e\rangle$ where the *signature* of the module is list of types of the components, e.g. $D \in Type,\ f : D \times D \to D, g : D \to \mathbb{B}, e \in D$. A *group* would have signature $G \in Type, op : G \times G \to G, inv : G \to G, e \in G$, and then there would be axioms saying that *op* is associative, *inv* is an inverse and e an identity.

The module concept corresponds exactly to dependent types over Type. In Section 2 we would denote the type of groups (signature) as

$$G : Type\ \times op : (G \times G \to G) \times\ inv : (G \to G) \times e : G$$

Except for the fact that the function types in the programming type are partial and Type has less mathematical structure, the algebraic concept and the programming one are similar.

We will see that the notion of subtype and inheritance that is so critical to modern programming practice can be nicely captured in our type theory. This leads to a mathematical treatment of the central concepts in *object-oriented programming* (c.f. Meyer [1988]).

Looking over the types described above we discern these uses.

1. Types relate data in the machine to standard mathematical concepts.
2. Types express the domain of significance of a programming problem and impose constraints on the data for it to be "meaningful" in the sense that the computer will not "crash" (attempt to execute a meaningless instruction) and the data will not fail to represent mathematical objects.[44] Usually these constraints can be rapidly checked to provide some level of assurance that a program is sensible.
3. Types provide a notation for structuring a solution by decomposing a task into components (modules) and levels of abstraction.
4. Types provide an interface language for analyzing ("debugging") a computation.
5. Type information can be used to increase the performance of the compiled code.

There is a direct historical link from Russell and Church to languages like Algol and Lisp. Also we are seeing a close correspondence between mathematical types and data types: Cartesian products correspond to record types, unions to disjoint unions (or variant record types), function spaces to procedure types, inductive types to recursive data types, algebraic structures to modules (and superstructures correspond to subtypes). The integers are included in some programming languages as the data type "bignums", and real numbers are (badly) approximated by "floating point numbers". In a sense the system of data types provides a computational type theory capable of organizing and unifying programming problems and solutions in

[44] Crashing can mean a complete failure to respond or an unwanted response from the operating system ("bus error") or from the hardware ("segmentation fault").

the same way that type theory organizes and unifies computational (also constructive and intuitionistic) mathemataical problems and solutions. The continuing (rapid) evolution of programming languages will probably lead to data type theories that subsume mathematical type theories. There may be new data types appropriate for expressing the problems of *interaction* as well as those of "functional action" which now dominate.

Although the similarities between types and data types just enumerated is compelling and interesting, I think it is also important to understand the differences. These differences challenge us to find logical foundations for new types.

4.2. Type ∈ type and domain theory

Given that programming types are not the same as mathematical ones, might it be sensible to allow a type of all types, precisely the notion that type theory was created to disallow in accordance with the vicious circle principle? One fact we know from the work of Meyer and Reinhold [1986] and Howe [1991,1989,1987,1996b] is that adding the typing rule *Type* ∈ *Type* to the simply typed lambda calculus allows new terms to be typed among which are applications that fail to terminate. No such terms can be typed without this new rule. On the other hand, this rule would not cause the type system to "collapse" in the sense that every term could be typed or every term belongs to every type (as would happen if we added the rule $T_1 = T_2$ for any two types T_1 and T_2). Indeed, we know that such a type system has a nontrivial mathematical model (Cardelli [1994], Meyer [1988]).

The discovery of interesting mathematical models for programming language types is a flourishing topic in the field of programming language semantics. It has led directly to the rich subject of *domain theory* pioneered by Dana Scott [1970a,1970b,1972,1976] led early on by Gordon Plotkin [1975]. (The results of Plotkin [1981], Abramsky [1993], Reynolds [1981], Cardelli [1994], Mitchell [1996], Gunter [1994], Egli and Constable [1976], and Abadi and Cardelli [1996] are quite relevant to the work discussed here.)

One of the major early discoveries of domain theory is that there are referential or "denotational" mathematical models of partial function spaces, in particular, of the untyped lambda calculus in which function equality is extensional (see Scott [1976]). The challenge for domain theory has been to relate these models to the standard mathematical types and type theories. This remains an active area of research with especially promising recent results in analysis (Edalat [1994]).

Let us call types which allow diverging elements *partial types*. Given that there is a consistent theory of partial types allowing *Type* ∈ *Type* and that this rule drastically simplifies the theory, we proceed to explore it.

One view of this theory is that it speaks about a domain. Another is that it is a "partial type theory" which will require refinement as more constraints are added, such as totality restrictions. But until we require totality, the vicious circle principle has no force since its consequence is merely a nonwell founded concept (nonterminating term). This approach to type theory permits a great deal of freedom—partial objects are allowed, *illogical comprehension* is possible, e.g. $\{x : \mathit{Type} \mid x \in x\}$, negative recursive definitions are allowed (see Section 4.3), and concepts need not be referential since equality relations are not required. It will be left to the programming logics to impose more logical order on these "unruly" types.

One of the first benefits of this theory is that dependent products taken over Type provide a notion of *module*. The signature (or type) of a module is

$$M : \mathit{Type} \times F(M)$$

where $F(M)$ is a type built from M such as $M \times M \to M$. By iterating this construct we get the general structure of a module

$$x_0 : \mathit{Type} \times x_1 : T_1(x_0) \times \cdots \times x_n : T_n(x_0, \ldots, x_{n-1}).$$

4.3. Recursive types

As we have seen, inductive definitions and principles of inductive reasoning lie at the heart of computational mathematics and logic. The inductive definition of the natural numbers, lists, and formulas come immediately to mind. The elements introduced inductively can be represented in computer memory by *linked data* structures constructed from *pointers*. For example, a list of elements of type A, say $(a_1, \ldots, a_n)$, would be represented by

where the arrows are pointers (data of type *address* or in Algol 68 terminology, *references* to A objects, thus of type $\mathit{ref}(A)$). A seminal discussion of these methods can be found in C.A.R. Hoare's article *Notes on Data Structuring* [1972].[45]

One of the most decisive uses of types in programming languages is in defining recursive data types at the same level of abstraction used in mathematics. This innovation was pioneered by Lisp and its treatment of lists without explicit mention of pointers. The pointer representation is managed by the run-time system of programming language, and a program called a *garbage collector* is used to dynamically manage the allocation and deallocation of memory for lists and other inductive structures.

In programming these inductive types are called *recursive types* or recursive data structures by analogy with recursive programs. They include circular data structures, unfounded lists (or streams) and other "nonwell-founded" recursive data that would not be considered as properly "inductive."[46] The definition of such a

---

[45]The small book *Structured Programming*, Dahl, Dijkstra and Hoare [1972], is one of the gems of computer science. All three articles are closely related to the subject of this section.

[46]Perhaps the reason for the popularity of the term "recursive data type" comes from Hoare's evocative analogy: "There are certain close analogies between the methods used for structuring data and the methods for structuring a program which processes that data ... a discriminated union corresponds to a conditional ... arrays to *for* statements ... sequence structure ... to unfounded looping .... The question naturally arises whether the analogy can be to a data structure corresponding to recursive procedures."

type is disarmingly simple to paraphrase Hoare: "write the name of the type being defined inside its own definition." In his notation we write

$$\mathit{type}\ T = F[T]$$

where $F[X]$ is a type definition in X. If we use $+$ for disjoint union and $\mathbf{1}$ for the unit type and $\times$ for cartesian product, then here are the definitions for natural numbers and lists over a type A.

$$\begin{aligned}&\mathit{type}\ N = \mathbf{1} + N\\ &\mathit{list}\ L = \mathbf{1} + (A \times L).\end{aligned}$$

We will use a more compact notation, writing a single term with a binding construct. Our notations for these types are $\mu(N.\ \mathbf{1} + N), \mu(L.\ \mathbf{1} + A \times L)$ where N and L are bound variables. In general, if $F[T]$ is a type expression in T, then $\mu(T.\ F[T])$ denotes the *recursive type* used above to illustrate Hoare's notation. In giving the rules for recursive types, we will use $A \to B$ and $x{:}A \to B[x]$ for the programming type $fun(x{:}A)B$; so the elements are *partial functions.*

1. $H \vdash \mu\,(x.\ F[x]) \in \mathit{Type}$ *rec_type_def*
   $H, x{:}\mathit{Type} \vdash F[x] \in \mathit{Type}$
2. $H \vdash t \in \mu\,(x.\ F[x])$ *rec_type_member*
   $H \vdash t \in F\,[\mu(x.\ F[x])]$
3. $\bar{H} \vdash \mu\,(t; f, y.\ g[f, y]) \in G$ *rec_type_elim*
   $\bar{H}, x{:}\mathit{Type}, f{:}x \to G, y{:}F[x] \vdash g[f, y] \in G$
   $\bar{H} \vdash t \in \mu(x.\ F[x])$

The term $\mu\,(t; f, y.\ g[f, y])$ is called a *recursion combinator.* It is the recursive program associated with the recursive definition. The evaluation rule is

$$\frac{g\,[\lambda\,(z.\ \mu\,(z; f, y.\ g[f, y]))\,/f, t/y] \downarrow a}{\mu\,(t; f, y.\ g[f, y]) \downarrow a}$$

The operational intuition behind these rules is this. A recursive type $\mathit{type}\ T = F[T]$ is well formed exactly when its "body" $F[T]$ is a type under the assumption that T is a type. This is "writing the name of the type being defined in its own definition." To construct a member of the type, build a member of $F[T]$, and if this construction requires an element of T, then apply the construction recursively (in the implementation, use a pointer to T and build recursively). The process may not terminate unless there is a "base case" which does not mention T, as in the left disjunct of $\mathbf{1} + T$ or of $\mathbf{1} + A \times T$. A definition like $\mu(X.\ X)$ is empty because no element can be created, likewise for $\mu(X.\ X + X)$ or $\mu(T.\ T \times T)$. Note however that $\mu(T.\ T \to T)$ will contain the element $\lambda(x.\ x)$ by this application of rules

$$\begin{aligned}&\vdash \lambda(x.\ x) \in \mu(T.\ T \to T)\\ T{:}\mathit{Type} &\vdash \lambda(x.\ x) \in T \to T\\ T.\ \mathit{Type},\ x{:}T &\vdash x \in T.\end{aligned}$$

Associated with $\mu(x.\ F[x])$ is a method of *recursive computation* (as Hoare suggested and as we know from inductive definitions in mathematics). If the recursive type is "well-founded" then this procedure will terminate, otherwise it might not. The recursive procedure is the following. Given $t \in \mu(X.\ F[x])$, to compute an element of type G, use a program g that computes on elements of $F[x]$. This procedure g may decompose t into components t' of type $\mu(x.\ F[x])$. In this case, call the procedure recursively. To specify this we note that if we consider that t belongs to $F[x]$, then component t' will belong to X. The recursive call of the procedure is represented in the rule by the function variable f from X to G. We see from the evaluation rule that this is used exactly as a recursive call.

This method of organizing the rules comes from Constable and Mendler [1985] and Mendler [1988]; it can be made more expressive using the subtyping relation $S \sqsubseteq T$ and dependent function types and parameterized recursions. First, with dependent types we get

$$\bar{H}, u{:}\mu(X.\ F[X]) \vdash \mu(u; f, y.\ g) \in G[u]$$
$$\bar{H}, X{:}Type, f{:}\ (x{:}X \to G[x]), y{:}F[X] \vdash g \in G[y]$$

The parameterized form of recursive type allows the defined type to depend on a parameter of type A. The syntax is $\mu(X.\ F[x])\,@a$

1p. $\bar{H} \vdash \mu(X.\ F[X])\,@a \in Type$
$\quad \bar{H}, X{:}A \to Type \vdash F[x] \in (A \to Type)$
$\quad \bar{H} \vdash a \in A$

2p. $\bar{H}t \in \mu(X.\ F[x])\,@a$
$\quad \bar{H} \vdash t \in F\,[\lambda(y.\ \mu(X.\ F[X]))\,@y]\,(a)$

3p. $\bar{H} \vdash \mu(a; t; f, u, y.\ g) \in G$
$\quad \bar{H}, X{:}(A \to Type), \forall u{:}A.\ (X(u) \sqsubseteq \mu(X.\ F[X])@u)$
$\quad \vdash g[f, u, y] \in G$
$\quad \vdash a \in A$
$\quad \vdash t \in \mu(X.\ F[x])@a$

4p. $$\frac{g\,[\lambda(u.\ \lambda(r.\ \mu(u; r; f, u, y.\ g)))\,/f, a/u, t/y] \downarrow c}{\mu(a; t; f, u, y.\ g) \downarrow c}$$

We can combine the parameterized form and the dependent form; such rules are given in Constable et al. [1986] and Mendler [1988], but we won't use this level of complexity here.

The parameterized recursive types can be used to define mutually recursive types since we can think of $\mu(X.\ F[x])@u$ as a family of simultaneously recursively defined types. With the propositions-as-types principle and restricting the recursive types to be well-founded, we get recursively defined relations. These have been exploited well in the Coq theorem prover (Coquand and Paulin-Mohring [1990], Coquand [1990], Paulin-Mohring and Werner [1993]).

With recursive types and disjoint unions and a unit type we can define natural numbers and lists as we have shown. Using record types we can define pairs of numbers which gives us integers and rational numbers. (Using function types we can define the *computable reals*; see Bishop [1967], Chirimar and Howe [1991], Forester

[1993].) Booleans can be defined as $\mathbf{1} + \mathbf{1}$. So the number of primitives for a rich type theory can be reduced to a very small set. We will examine some especially interesting reductions in the next section.

Example defining primitive recursion on N. To illustrate the workings of the recursion combinator $\mu()$, we use it to define primitive recursive functions from $\mathbf{N}$ to G. Suppose f is defined primitive recursively on $\mu(X.\ \mathbf{1}+X)$ to G by

$$\begin{aligned} &f(0) = b \\ &f(suc(u)) = h(n, f(n)). \end{aligned}$$

Then the corresponding combinator is $\mu\,(u; f, u.\ decide\,(u; v.\ b; v.\ h(v, f(v))))$ whose typing is seen from the judgment.

$$X{:}Type, f{:}X \to G, u{:}\mathbf{1}+X \vdash decide(u; v.\ b; v.\ h(v, f(v))) \in G.$$

Typing a fixed point combinator. While the recursion combinators are essential for inductive types, indeed they characterize them, in a rich enough partial type theory they can be defined. The idea is to use the richness of the recursive types to assign a type to a fixed point combinator, like Y. Recall that the Y combinator is abbreviated $\lambda\,(g.\ \lambda(x.\ g(xx))\lambda(x.\ g(xx)))$ or still further by letting $w = \lambda(x.\ g(xx))$ and writing Y as $\lambda(g.\ ww)$. We show that Y has type $(T \to T) \to T$ for any type T by using the auxiliary recursive type $S == \mu(X.\ X \to T)$. Here is the derivation.

The type of g will be $T \to T$, the type of w is $S \to T$. The "trick" is to type $ap(x; x)$ to be of type T. We examine the typing derivation for w.

$$\begin{array}{lll} g{:}T \to T & \vdash \lambda(x.\ g(xx)) \in \mu(X.\ X \to T) & \text{by } \mu\text{\_membership} \\ & \vdash \lambda(x.\ g(xx)) \in S \to T & \text{by} \to R \\ g{:}T \to T, x{:}S & \vdash g(xx) \in T & \text{by} \to L \\ & \vdash xx \in T & \text{by } ap \\ & \vdash x \in S \to T & \text{by } unroll\ x \\ & \vdash x \in S & \text{by } hyp\ x \end{array}$$

Once we know that $w \in S \to T$ and $w \in S$, then $ww \in T$ and $g(ww) \in T$.

One corollary of this typing is that $Y(\lambda(x.\ x))$ belongs to the empty type $\mu(X.\ X)$ called *void*, since $\lambda(x.\ x) \in void \to void$. But $Y(\lambda(x.\ x))$ is a diverging term, so it is not a *value* belonging to *void*. Indeed, we can easily show that there are no values of type *void*.

Now we can use Y to define any partial recursive function, including the recursion combinators of type $\mu(X.\ F) \to G$. In general, $\lambda\,(x.\ \mu(x; f, u.\ g[f, u]))$ is just $Y\,(\lambda\,(f.\ \lambda(u.\ g[f, u])))$. The type of f is $(\mu\,(X.\ F) \to G) \to (\mu(X.\ F) \to G)$, and we observed that $g[f, u] \in G$ can be derived from this typing of f.

Applying this general construction to primitive recursion we get the term $Y(\lambda(f.\ \lambda(n.\ decide(u; v.\ b; v.\ h(v, f(v))))))$, which is R, the primitive recursion combinator, (with b and h as parameters).

Inductive types. Constable and Mendler [1985] and Mendler [1988] gave conditions needed to guarantee that recursive types $\mu(X.\ F)$ define only total objects. One such condition is that F be a monotone operation on types in the sense that $X \sqsubseteq Y \Rightarrow F[X] \sqsubseteq F[Y]$. We also studied conditions to guarantee that elements of these types are functional. The result is a set of rules used in Nuprl for inductive types (c.f. Constable et al. [1986], Hickey [1996a]).

When F is required to be monotone, then we cannot define the type $\mu(X.X \to T)$ used in typing Y. Indeed, it is not possible to type Y nor divergent elements. For this reason the $\mu(x; f, u.\ g)$ recursion forms are needed. They provide the structural induction rules for inductive types. In Nuprl these induction rules for recursive types can be used to prove that certain applications of the Y combinator, $Y(\lambda(f.\ b))$ are indeed total objects (see Constable et al. [1986]). So we get the advantages of general recursive programs without losing the logical structure of type theory.

4.4. Dependent records and very dependent types

We are aiming to exhibit a small core type system that will generate all of the types we have studied. The step in this direction that we take here is of considerable practical value—it builds record types from dependent function spaces.

Consider the record type $record(x_1 : A_1, \ldots, x_n : A_n)$. Let $N_n == \{1, \ldots, n\}$ be an n element enumeration type—it can simply be $\mathbf{1} + \cdots + \mathbf{1}$ taken n times. Define a function $B(i) = A_i$ from N_m to Type. Then the essential structure of the record is given by the dependent function space $e : N_n \to B(i)$. Given f in this type, $f(i)$ is the i-th component. We obtain a nice display form for record selection if we define $f.\ x_i == f(i)$.

This definition of records has nice subtyping properties. In a standard record calculus a record type, r_1, is a subytpe of record type r_2, written $r_1 \sqsubseteq r_2$, iff r_1 has additional fields. So a colored point is a subtype of a point or a group type is a subtype of monoid type, etc. Our definition provides this subtyping directly from the subtyping relation on function spaces. Recall that if $A_1 \sqsubseteq A_2, B_1 \sqsubseteq B_2$ then $A_2 \to B_1 \sqsubseteq A_1 \to B_2$. Also if $N_n \sqsubseteq N_m$, and $n \leq m$, and $B_1(i) = B_2(i)$ for $i \in N_n$ then

$$i : N_m \to B_1(i) \sqsubseteq i : N_n \to B_2(i).$$

Notice that $f \in (i : N_m \to B_1(i))$ is an element of $i : N_n \to B_2(i)$ simply by the polymorphic nature of functions (i.e. they are rules given by λ terms).

Encoding dependent records. The dependent product types, $x : A_1 \times A_2[x]$ offer a form of dependent record as mentioned above. The general form is $record(x_1 : A_1; x_2 : A_2[x_1]; \ldots ; x_n : A_n[x_1, \ldots , x_{n-1}])$. Can we also define these records as dependent functions?

The existing dependent function space is not adequate for this task, but Jason Hickey [1996a] has discovered an extension that he calls *very dependent* function spaces. The basic notation is $fun(A; f, x.\ B[f,x])$ as opposed to $fun(A; x.\ B[x])$. The idea is that the type B can depend not only on the argument to the function so that $g(a) \in B[a]$, but now the type of B can depend on "previous values" of g, so $g(a) \in B[g, a]$. To see how the idea works, let's use it to define the elements of $x_1 : A_1 \times x_2 : A_2(x_1)$. Note $A_2 : A_1 \to Type$, and an element is $\langle a_1, a_2 \rangle$ where $a_1 \in A_1, a_2 \in A_2(a_1)$. The encoding is based on $N_2 = \{1, 2\}$. Imagine that $B(1) = A_1$, and we want $B(2) = A_2(a_1)$ where $a_1 \in A_1$. We could say this if we had the element g such that $g(1) \in A_1$. So if we add g as a parameter to B we can say

$$B(g, 1) = A_1$$
$$B(g, 2) = A_2(g(1)).$$

This particular definition makes sense because at $B(g, 2)$, g is referenced only at previous arguments. Hickey takes this as the basis for defining the simplest *very dependent function space.* He requires a well-ordering on g as prerequisite to forming the type (see Hickey [1996a]). In a partial type theory we can get away with less. A particular computable function g will generate an ordering on values via its computation. So we can allow arbitrary expressions $B[g, x]$ in forming the type, but it will be empty unless there is a function satisfying the constraints of B. The (viciously circular) rules are:

1. $\bar{H} \vdash fun(A; f, x.\ B) \in Type$
   $\bar{H} \vdash A \in Type$
   $\bar{H}, x{:}A, f{:}fun(A; f, x.\ B) \vdash B \in Type$
2. $\bar{H} \vdash \lambda(x.\ b) \in fun(A; f, x.\ B)$
   $\bar{H}, x{:}A \vdash b \in B[\lambda(x.\ b)/f]$
3. $\bar{H} \vdash g(a) \in B[g/f, a/x]$ by *ap over* $fun(A; f, x.\ B)$
   $\bar{H} \vdash g \in fun(A; f, x.\ B)$
   $\bar{H} \vdash a \in A$

With this type we can define dependent products as

$$prod(A; x.\ B[x]) == fun(N_2; f, x.\ \text{if } x = 1 \text{ then } A \text{ else } B[f(1)]).$$

4.5. A very small type theory

The previous reductions show that we can define a very rich type theory using only three primitive type constructors and one primitive type, namely *Type*.

types:	$Type \quad A + B \quad fun(A; f, x.\ B) \quad \mu(X.\ B)$
values:	$inl(a),\ inr(b),\ \lambda(x.\ b)$
forms:	$decide(t; u.\ a; v.\ b) \quad ap(t; a)$

This language can be seen as a combination of the ideas from Constable and Mendler [1985], Mendler [1988], Hickey [1996a]; it is in the style of Mendler's thesis using Hickey's key reduction. The language FPC in Gunter's textbook [1992] considers the nondependent recursive types in a similar spirit.

5. Conclusion

In the main, this article is a snapshot of three subjects recently come into alignment. This conclusion addresses research dynamics driving these subjects.

Typed logic. Many standard topics in logic must be reworked for typed logic. We have already seen that its *deductive machinery* is different, so we need to ask about normalization results for natural deduction (as in Prawitz [1965]) or cut elimination for numerous variants of the sequent calculus (with structural rules or without, tableau style or bottom up, etc.) What properties of the normal syntax of proofs reflect their deeper semantic content? What symmetries of the sequent calculus reveal properties of evidence?

The emergence of automated deduction systems has introduced new issues and questions. For example, the notion of a *tactic-tree proof* (Allen et al. [1990]) illustrated here is a novel structure, and its use in *refinement logics* (Bates [1979], Bates and Constable [1985]) raises questions, such as, how is soundness and type correctness of the metalevel programming language for tactics related to the soundness of the logic?

The traditional questions about the relative "power" of logical theories can be posed for typed logics, and the various translation results such as the Kolmogorov and Gödel translations are being studied (Troelstra and Schwichtenberg [1996]). Chet Murthy [1990,1992] discovered remarkable results relating these translations to Plotkin's CPS translations, and he proved Friedman's [1978] theorem for a fragment of Nuprl as part of this work (see also Palmgren [1995a]). These results have been applied in interesting ways in program extraction by Murthy [1992] and Berger and Schwichtenberg [1996]. Friedman's program of "reverse mathematics" can be elaborated in this context as well, and now *programming logics* can be considered in a more uniform manner (Kozen [1977], Kozen and Tiuryn [1990]).

The subject of *applied logic* has emerged in the intersection of logic and computer science. This includes the study of specification languages such as Z (Spivey [1989]), a main topic in *formal methods.* The languages of typed logic (say in Coq, HOL, Nuprl, and PVS) provide alternative specification languages which seem to have advantages over Z in automation. These rich typed logics can accommodate special languages such as those needed in temporal logic and for hybrid systems (Nerode and Shore [1994], Henzinger and Ho [1994]).

The field of *automated deduction* is a flourishing part of applied logic. Presently, specialized tools such as *model checkers* (c.f. Clarke, Long and McMillan [1989], Burch et al. [1991], Henzinger and Ho [1994]), *type checkers* (c.f. Milner, Tofte and Harper [1991]), and arithmetic *decision procedures* are already used by industry in production. Integrated systems like Coq, HOL, Nuprl, and PVS are also valuable to industry.[47] The deployment of logic-based industrial systems has led to a wealth of research problems and challenges (Kreitz, Hayden and Hickey [n.d.]). For example,

[47]The late IBM Fellow, Harlam Mills, said in December 1984, "It is the kind of research that can change the course of industrial history."

it is becoming imperative to share libraries of mathematics between provers. Howe's work [1996a] with HOL libraries in Nuprl is one of the first examples of how this can be done. Practical deployment relies on several years of investigating the underlying semantic issues involved in translating between theories (Howe [1996b,1991]).

The need to share results between provers is only one example of a more general need to build more *open* theorem proving systems. These systems should be able to interface with several text and proof editors, with other provers, with programming languages to evaluate computable terms, and with metalanguages for managing proof planning and generation. Nuprl Version 5 is one such system. We discuss these problems in a wider context in Collaborative Mathematics Environments (Chew et al. [1996]).

Type theory. The research agenda in type theory is strongly tied to logic as this article illustrates, providing a new semantics. In addition, there are strong ties to pure and applied mathematics (Gallier [1993]). Indeed, Martin-Löf type theory arose as an attempt to find a foundational account of the practice of constructive mathematics, especially in the style of Bishop (Bishop [1967], Bishop and Bridges [1985], Mines, Richman and Ruitenburg [1988]). This constructive mathematics is more similar to the practice of computational mathematics than to Intuitionistic mathematics in that its results are consistent classically. Indeed, Bishop's book can be read as a piece of classical analysis or as computational or Intuitionistic mathematics. Nuprl, in fact, arose as an attempt to provide a foundation for computer science — numerical analysis, computer algebra, the theory of algorithms and computability. It was based on programming concepts (Constable [1972], Constable and Zlatin [1984]) and influenced by Algol68 and Simula, but we recognized in 1978 the power of Martin-Löf semantics to organize this activity, and in Constable and Zlatin [1984] used his semantics to improve our earlier design.

As computational mathematics has gained importance, more work has been done to systematize it. For example, the algebra underlying a computer algebra system such as AXIOM (Jenks and Sutor [1992]) is constructive: consider the definition of an integral domain; it provides a function, *div*, which will divide $a \star c$ by $c \neq o$. In general, in computer algebra, to claim that an object "exists" is to give an algorithm to construct it. A current active area of research is expressing the concepts of computer algebra in constructive type theory. It is especially promising that the work provides an orderly account of the types and domains used in algebra systems — for example, compare AXIOM (Jenks and Sutor [1992]) or Weyl (Zippel [1993]) to Jackson's account in Nuprl [1994b,1994a]. Peter Aczel is considering Galois theory in LEGO (Pollack [1995]), and more work of this sort will be done.

Another important topic in the same vein is the use of type theory to organize the foundations of numerical mathematics by Boehm et al. [1986], Chirimar and Howe [1991]. It will be interesting to see whether floating point numbers could be incorporated into a rigorous theory, perhaps even arranging that the notion of a constructive real number as a sequence of approximations each of which was a "floating point" number. It is intriguing to imagine that this work might extend to a computational treatment of nonstandard analysis (see Nelson [1968], Wattenberg [1988]). This is potentially interesting because it is now realized since the work of Loeb that nonstandard accounts of probability applications can be significantly more intuitive than their classical counterparts.

Category theory can be seen as an abstract organization of type theory, and just as type theory provides an alternative and more general foundation for mathematics than set theory, so too, category theory provides such a foundation. The category called an elementary topos generalizes set theory.[48] It is possible to develop a predicative version of topos theory (the *Grothendieck topos*) in Martin-Löf type theory (Palmgren [1995a]). Likewise, category theory can provide models of type theory (Crole [1993], Seely [1987]). The categorical models allow new kinds of constructive completeness theorems for the Intuitionistic predicate calculus Palmgren [1995a], and from these it is possible to give a uniform computational interpretation to nonstandard analysis (Palmgren [1995a]).

Typed programming languages. The research agenda in programming languages is the most fast-paced of the three; like everything in computer science it is driven by curiosity, by technology, and by market forces. Research is put to use before the "ink is dry." Each small result seems to explode into an industry. Needs for secure mobile code will now be a major influence as code reuse and modularity were before.

Language research depends on a deeper understanding of the *design space* and on a range of semantic tools to rapidly validate experimental designs. Our approach of "partial types" is one of many attempts to provide this knowledge, domain theory, and theories of operational semantics (c.f. Plotkin [1981]) are others (see also Crary [1998]).

Acknowledgments. I want to thank Kate Ricks for preparing this manuscript and Stuart Allen for helpful comments on earlier drafts and for helping with a new account of his 1987 thesis work.

6. Appendix

6.1. Cantor's Theorem. Here is a Nuprl proof of Cantor's theorem from Section 2.9.

```
*T cantor
⊢ ∀A:U. (∃diff:A → A. ∀x:A. ¬(diff x = x))
            ⇒ (∀e:A → A → A. ∃d:A → A. ∀x:A. ¬(e x = d))
|
BY UnivCD THENW Auto
|
1. A: U
```

[48] "The startling aspect of topos theory is that it unifies two seemingly wholly distinct mathematical subjects: on the one hand, topology and algebraic geometry, and on the other hand, logic and set theory." MacLane and Moerdijk [1992,p.1]

```
2. ∃diff:A → A. ∀x:A. ¬(diff x = x)
3. e: A → A → A
⊢ ∃d:A → A. ∀x:A. ¬(e x = d)
|
BY D 2
|
2. diff: A → A
3. ∀x:A. ¬(diff x = x)
4. e: A → A → A
|
BY With ⌜λa.diff (e a a)⌝ (D 0) THENW Auto
|
⊢ ∀x:A. ¬(e x = (λa.diff (e a a)))
|
BY D 0 THENW Auto
|
5. x: A
⊢ ¬(e x = (λa.diff (e a a)))
|
BY D 0 THENW Auto
|
6. e x = (λa.diff (e a a))
⊢ False
|
BY With ⌜e x x⌝ (D 3) THENW Auto
|
3. e: A → A → A
4. x: A
5. e x = (λa.diff (e a a))
6. ¬(diff (e x x) = e x x)
|
BY D 6
|
⊢ diff (e x x) = e x x
|
BY RW (AddrC [3;1] (HypC 5)) 0 THENW Auto
|
⊢ diff (e x x) = (λa.diff (e a a)) x
|
BY Reduce 0 THEN Auto

*C cantor_end              ********************************
```

6.2. Stamps problem. Here is a complete Nuprl proof for a simple arithmetic problem. We show that any number greater than or equal to 8 can be written as a sum of 3's and 5's. We call this the "stamps" problem. When Sam Buss saw this theorem we discussed a generalization which is included in Section 6.3. Christoph

Kreitz proved the generalization following Sam's handwritten notes. It is interesting that Nuprl caught a missing case in this proof. The arguments seem sufficiently self-contained that we present them without further comment.

```
⊢ ∀i:{8...}. ∃m,n:N . 3 * m + 5 * n = i
|
BY D 0 THENA Auto·
|
1. i: {8...}
⊢ ∃m,n:N . 3 * m + 5 * n = i
|
BY NSubsetInd 1
| THEN Auto·
|\
| 1. i: Z
| 2. 0 < i
| 3. 8 = i
| |
1 BY DTerm ⌜1⌝ 0 THENM DTerm ⌜1⌝ 0 THEN Auto·
 \
  1. i: Z
  2. 8 < i
  3. ∃m,n:N . 3 * m + 5 * n = i - 1
  |
  BY D 3 THEN D 4·
  |
  3. m: N
  4. n: N
  5. 3 * m + 5 * n = i - 1
  |
  BY Decide ⌜n > 0⌝ THENA Auto·
  |\
  | 6. n > 0
  | |
  1 BY DTerm ⌜m + 2⌝ 0 THENM DTerm ⌜n - 1⌝ 0 THEN Auto·
   \
    6. ¬(n > 0)
    |
    BY DTerm ⌜m - 3⌝ 0 THENM DTerm ⌜n + 2⌝ 0 THEN Auto·
    |
    ⊢ 0 ≤ m - 3
    |
    BY SupInf THEN Auto
```

6.3. Generalized stamps problem

Lemmata from the Standard Nurpl Library.

```
*T mul_bounds_1a      ⊢ ∀a,b:N. 0 ≤ a*b
*T mul_bounds_1b      ⊢ ∀a,b:N+. 0 < a*b
*T mul_preserves_lt  ⊢ ∀a,b:Z. ∀n:N+. a < b ⇒ n*a < n*b
*T mul_preserves_le  ⊢ ∀a,b:Z. ∀n:N. a ≤ b ⇒ n*a ≤ n*b
*T multiply_functionality_wrt_le
           ⊢ ∀i1,i2,j1,j2:N. i1 ≤ j1 ⇒ i2 ≤ j2 ⇒ i1*i2 ≤ j1*j2

*T rem_bounds_1       ⊢ ∀a:N. ∀n:N+. 0 ≤ a rem n ∧ a rem n < n
*T div_rem_sum        ⊢ ∀a:Z. ∀n:Z-0. a = (a ÷ n)*n + a rem n
*A pm_equal           i = ± j
           == i = j ∨ i = -j

*A divides            b | a
           == ∃c:Z. a = b*c
*T divisor_bound      ⊢ ∀a:N. ∀b:N+. a | b ⇒ a ≤ b
```

Newly Introduced Notions and Lemmata.

```
STAMPS
*T pm_equal_nat       ⊢ ∀a:N+. a = ± 1 ⇒ a = 1
*T divisor_of_sub     ⊢ ∀a,b,c:Z. a | b ⇒ a | c ⇒ a | b - c
*T divisor_of_sub_self ⊢ ∀a,b:Z. a | b ⇒ a | b - a
*A even               a is even == 2 | a
*A odd                a is odd == 2 | a + 1
*T odd_mul            ⊢ ∀b,m:Z. m*b is odd ⇒ b is odd
*T odd_or_even        ⊢ ∀z:Z. z is even ∨ z is odd
*A stampproperty      a and b are useful stamp values
           == ∀i:{a + b...}. ∃n,m:N. i = n*a + m*b
```

Proof of the 'Induction' Step.

```
*T stamp_pre ⊢ ∀a,b:N+. a < b
        ⇒ (∀i:{(a + b)..(2*a + b)⁻}. ∃n,m:N. i = n*a + m*b)
         ⇒ a and b are useful stamp values
BY    simple_prover
|
1. a: N+
2. b: N+
3. a < b
4. ∀i:{(a + b)..(2*a + b)⁻}. ∃n,m:N. i = n*a + m*b
5. i: {a + b...}
```

```
⊢ ∃n,m: N . i = n*a + m*b
BY Cases [⌈i < 2*a + b⌉; ⌈i ≥ 2*a + b ⌉]....
|\
| 6. i < 2*a + b
1 BY allE 4 ⌈i⌉....
|
 \
  6. i ≥ 2*a + b
  BY      Assert ⌈(a + b) + (i - b) rem a ∈ {(a + b)..(2*a + b)⁻}⌉
  |  THENL [Id; allE 4 ⌈(a + b) + (i - b) rem a⌉]
  |\
  | ⊢ (a + b) + (i - b) rem a ∈ {(a + b)..(2*a + b)⁻}
  1 BY  FLemmaOn 'rem_bounds_1' [⌈i - b⌉; ⌈a⌉] THEN SupInf....
  |
   \
    7. x: (a + b) + (i - b) rem a ∈ {(a + b)..(2*a + b)⁻}
    8. ∃n,m: N . (a + b) + (i - b) rem a = n*a + m*b
    BY    thin 7
    | THEN Repeat existentialE
    | THEN exI ⌈((i - b) ÷ a - 1) + n⌉
    |\
    | 4. i:{(a + b)..(2*a + b)⁻} → (n: N  × m: N  × (i = n*a + m*b))
    | 6. 2*a + b ≤ i
    | 7. n: N
    | 8. m: N
    | 9. (a + b) + (i - b) rem a = n*a + m*b
    | ⊢ 0 ≤ ((i - b) ÷ a - 1) + n
    1 BY Assert ⌈2*a ≤ i - b⌉
    |   THENL [ SupInfWf; Assert ⌈2 ≤ (i - b) ÷ a⌉ THEN SupInfWf ]
     \
     7. n: N
     8. m: N
     9. (a + b) + (i - b) rem a = n*a + m*b
     ⊢ ∃m: N . i = (((i - b) ÷ a - 1) + n) *a + m*b
     BY exI ⌈m⌉
     |
     ⊢ i = (((i - b) ÷ a - 1) + n) *a + m*b
     BY    SubstInConcl ⌈(((i - b) ÷ a - 1) + n) *a + m*b
                      = (((i - b) ÷ a) *a - a) + n*a + m*b⌉....
       THEN RevHypSubst 9 0 ...
       THEN FLemmaOn 'div_rem_sum' [⌈i - b⌉; ⌈a⌉]
       THEN SupInf....
```

Main Theorem.

```
*T StampThm ⊢ ∀a,b: N⁺. a < b
          ⇒ (a and b are useful stamp values
            ⟺ a = 1
               ∨ (a = 2 ∧ b is odd)
               ∨ (a = 3 ∧ b = 4)
               ∨ (a = 3 ∧ b = 5))
```

References

M. ABADI AND L. CARDELLI

[1996] *A Theory of Objects*, Springer-Verlag, Berlin.

S. ABRAMSKY

[1993] Computational interpretations of linear logic, *Theoretical Computer Science*, 111, pp. 3–57.

P. H. G. ACZEL

[1977] An introduction to inductive definitions, in: *Handbook of Mathematical Logic*, J. Barwise, ed., North-Holland, Amsterdam, pp. 739–782.

[1986] The type theoretic interpretation of constructive set theory, in: *Logic, Methodology and Philosophy of Science VII*, R. B. Marcus, G. J. W. Dorn, and P. Weingartner, eds., Elsevier Science Publishers, Amsterdam, pp. 17–49.

S. F. ALLEN

[1987a] A non-type-theoretic definition of Martin-Löf's types, in: *Proceedings of the Second Symposium on Logic in Computer Science*, IEEE Computer Society, Washington, D.C., pp. 215–224.

[1987b] *A non-type-theoretic semantics for type-theoretic language*, PhD thesis, Cornell University.

S. F. ALLEN, R. L. CONSTABLE, D. J. HOWE, AND W. AITKEN

[1990] The semantics of reflected proof, in: *Proceedings of the Fifth Symposium on Logic in Computer Science*, IEEE Computer Society, Washington, D.C., pp. 95–197.

R. C. BACKHOUSE

[1989] Constructive type theory–an introduction, in: *Constructive Methods in Computer Science, NATO ASI Series, Vol. F55: Computer & System Sciences*, M. Broy, ed., Springer-Verlag, Berlin, pp. 6–92.

H. P. BARENDREGT

[1977] The typed lambda calculus, in: *Handbook of Mathematical Logic*, J. Barwise, ed., North-Holland, Amsterdam, pp. 1091–1132.

[1981] *The Lambda Calculus: Its Syntax and Symantics*, North-Holland, Amsterdam.

J. BARWISE AND J. ETCHEMENDY

[1991] *The Language of First-Order Logic*, Lecture Notes Number 23, Center for the Study of Language and Information, Stanford University, second ed.

D. BASIN AND R. L. CONSTABLE

[1993] Metalogical frameworks, in: *Logical Environments*, G. Huet and G. Plotkin, eds., Cambridge University Press, ch. 1, pp. 1–29.

D. BASIN AND M. KAUFMANN

[1991] The Boyer-Moore prover and Nuprl: An experimental comparison, in: *Logical Frameworks*, G. Huet and G. Plotkin, eds., Cambridge University Press, pp. 89–119.

J. L. BATES

[1979] *A Logic for Correct Program Development*, PhD thesis, Cornell University.

J. L. BATES AND R. L. CONSTABLE

[1985] Proofs as programs, *ACM Transactions on Programming Languages and Systems*, 7, pp. 53–71.

M. J. BEESON
[1985] *Foundations of Constructive Mathematics*, Springer-Verlag, Berlin.

J. L. BELL
[1988] *Toposes and Local Set Theories*, Oxford Logic Guides #14, Oxford University Press.

U. BERGER AND H. SCHWICHTENBERG
[1996] The greatest common divisor: a case study for program extraction from classical proofs, in: *Types for Proofs and Programs: International Workshop TYPES'95*, S. Berardi and M. Coppo, eds., Lecture Notes in Computer Science #1158, Springer Verlag, Berlin, pp. 36–46.

P. BERNAYS
[1958] *Axiomatic Set Theory*, North-Holland, Amsterdam. With an introduction by A.A. Fraenkel

E. BISHOP
[1967] *Foundations of Constructive Analysis*, McGraw Hill, New York.

E. BISHOP AND D. BRIDGES
[1985] *Constructive Analysis*, Springer-Verlag, Berlin.

N. BJØRNER, A. BROWN, E. CHANG, M. COLÓN, A. KAPUR, Z. MANNA, H. B. SIPMA, AND T. E. URIBE
[1996] STep: Deductive-algorithmic verification of reactive and real-time systems, in: *CAV'96: Proceedings of the Eighth International Conference on Computer Aided Verification*, R. Alur and T. A. Henzinger, eds., Lecture Notes in Computer Science #1102, Springer-Verlag, Berlin, pp. 415–418.

W. W. BLEDSOE
[1975] A new method for proving certain Presburger formulas, in: *Proceedings of the Fourth International Joint Conference on Artificial Intelligence, held in Tbilisi, Georgia, USSR.*

H.-J. BOEHM, R. CARTWRIGHT, M. RIGGLE, AND M. J. O'DONNELL
[1986] Exact real arithmetic: A case study in higher order programming, in: *ACM Symposium on LISP and Functional Programming*, Association for Computing Machinery, New York.

N. BOURBAKI
[1968a] *Elements of Mathematics, Algebra, Volume 1*, Addison-Wesley, Reading, Massachusetts.
[1968b] *Elements of Mathematics, Theory of Sets*, Addison-Wesley, Reading, Massachusetts.

R. S. BOYER AND J. S. MOORE
[1979] *A Computational Logic*, Academic Press, New York.
[1988] Integrating recursive procedures into hueristic theorem provers: A case study in linear arithmetic, in: *Machine Intelligence II*, Oxford University Press.

L. E. J. BROUWER
[1975] *Collected Works*, vol. 1, North-Holland, Amsterdam. Edited by A. Heyting.

N. G. DE BRUIJN
[1970] The mathematical language Automath, its usage and some of its extensions, in: *Symposium on Automatic Demonstration*, M. Laudet, D. Lacombe, L. Nolin, and M. Schützenberger, eds., Lecture Notes in Mathematics #125, Springer-Verlag, Berlin, pp. 29–61.
[1972] Lambda calculus notation with nameless dummies, a tool for automatic formula manipulation, with application to the Church-Rosser theorem, *Indagationes Mathematicae*, 34, pp. 381–392.
[1980] A survey of the project Automath., in: *To H.B. Curry: Essays in Combinatory Logic, Lambda Calculus, and Formalism*, J. P. Seldin and J. R. Hindley, eds., Academic Press, New York, pp. 589–606.

A. BUNDY

[1991] The use of proof plans for normalization, in: *Essays in Honor of Woody Bledsoe*, R. S. Boyer, ed., Kluwer Academic Publishers, Dordrecht, Boston, pp. 149–166.

J. R. BURCH, E. CLARKE, K. L. MCMILLAN, D. L. DILL, AND L. J. HWANG

[1991] Symbolic model checking: 10^{20} states and beyond, in: *Proceedings of the Fifth Symposium on Logic in Computer Science*, IEEE Computer Society, Los Alamitos, California, pp. 428–439.

S. R. BUSS

[1986] The polynomial hierarchy and intuitionistic bounded arithmetic, in: *Structure in Complexity Theory*, A. L. Selman, ed., Lecture Notes in Computer Science #223, Springer-Verlag, Berlin, pp. 77–103.

L. CARDELLI

[1994] Extensible records in a pure calculus of subtyping, in: *Theoretical Aspects of object-oriented Programming: Types, Semantics and Language Design*, C. A. Gunter and J. C. Mitchell, eds., MIT Press, Cambridge, Massachusetts.

T. CHAN

[1982] An algorithm for checking PL/CV arithmetic inferences, in: *An Introduction to the PL/CV2 Programming Logic*, R. L. Constable, S. D. Johnson, and C. D. Eichenlaub, eds., Lecture Notes in Computer Science #135, Springer-Verlag, Berlin, pp. 227–264.

C. C. CHANG AND H. J. KEISLER

[1990] *Model Theory*, vol. 73 of Studies in Logic and the Foundations of Mathematics, North-Holland, Amsterdam, 3rd ed.

L. P. CHEW, R. L. CONSTABLE, S. VAVASIS, K. PINGALI, AND R. ZIPPEL

[1996] *Collaborative Mathematics Environments.* www@cs.cornell.edu/Info/Projects/Nuprl.

J. CHIRIMAR AND D. J. HOWE

[1991] Implementing constructive real analysis: a preliminary report, in: *Symposium on Constructivity in Computer Science*, J. P. Myers and M. J. O'Donnell, eds., Lecture Notes in Computer Science #613, Springer-Verlag, Berlin, pp. 165–178.

N. CHOMSKY

[1988] *Language and Problems of Knowledge: Managua Lectures*, MIT Press, Cambridge, Massachusetts.

A. CHURCH

[1940] A formulation of the simple theory of types, *Journal of Symbolic Logic*, 5, pp. 55–68.

[1956] *Introduction to Mathematical Logic, Vol I*, Princeton University Press.

[1960] Application of recursive arithmetic to the problem of circuit synthesis,, in: *Summaries of talks presented at the Summer Institute for Symbolic Logic, Cornell University 1957*, Institute for Defense Analyses, Princeton, pp. 3–50.

E. CLARKE, D. E. LONG, AND K. L. MCMILLAN

[1989] Compositional model checking, in: *Proceedings of the Fourth Symposium on Logic in Computer Science*, IEEE Computer Society, Washington, D.C.

R. L. CONSTABLE

[1972] Constructive mathematics and automatic program writers, in: *Proceedings of Information Processing 71 (IFIP)*, C. V. Freiman, J. E. Griffith, and J. L. Rosenfeld, eds., North-Holland, Amsterdam, pp. 229–233.

[1989] Assigning meaning to proofs: a semantic basis for problem solving environments, in: *Constructive Methods in Computer Science, NATO ASI Series, Vol. F55: Computer & System Sciences*, M. Broy, ed., Springer-Verlag, Berlin, pp. 63–91.

[1991] Type theory as a foundation for computer science, in: *Theoretical Aspects of Computer Software, International Conference TACS '91*, T. Ito and A. R. Meyer, eds., Lecture Notes in Computer Science #526, Springer-Verlag, Berlin, pp. 226–243.

R. L. Constable, S. F. Allen, H. M. Bromley, W. R. Cleaveland, J. F. Cremer, R. Harper, D. J. Howe, T. B. Knoblock, N. P. Mendler, P. Panangaden, J. T. Sasaki, and S. F. Smith

[1986] *Implementing Mathematics with the Nuprl Development System*, Prentice-Hall, Englewood Cliffs, New Jersey.

R. L. Constable, P. B. Jackson, P. Naumov, and J. Uribe

[1998] Constructively formalizing automata theory I: Finite automata, in: *Robin Milner Festschrift*, Springer-Verlag, Berlin. To appear.

R. L. Constable, S. D. Johnson, and C. D. Eichenlaub

[1982] Introduction to the PL/CV2 progamming logic, in: *An Introduction to the PL/CV2 Programming Logic*, R. L. Constable, S. D. Johnson, and C. D. Eichenlaub, eds., Lecture Notes in Computer Science #135, Springer-Verlag, Berlin, pp. 165–178.

R. L. Constable and N. P. Mendler

[1985] Recursive Definitions in Type Theory, in: *Logics of Programs*, R. Parikh, ed., Lecture Notes in Computer Science #193, Springer-Verlag, Berlin, pp. 61–78.

R. L. Constable and S. F. Smith

[1993] Computational foundations of basic recursive function theory, *Theoretical Computer Science*, 121, pp. 89–112.

R. L. Constable and D. R. Zlatin

[1984] The type theory of PL/CV3, *ACM Transactions on Programming Languages and Systems*, 6, pp. 94–117.

T. Coquand

[1990] Metamathematical investigations of a calculus of constructions, in: *Logic and Computer Science*, P. Odifreddi, ed., Academic Press, London, pp. 91–122.

T. Coquand and G. Huet

[1988] The Calculus of Constructions, *Information and Computation*, 76, pp. 95–120.

T. Coquand and C. Paulin-Mohring

[1990] Inductively defined types, preliminary version, in: *COLOG '88, International Conference on Computer Logic*, P. Martin-Löf and G. E. Mints, eds., Lecture Notes in Computer Science #417, Springer-Verlag, Berlin, pp. 50–66.

K. Crary

[1998] *Programming Language Semantics in Foundational Type Theory.* To appear in *IFIP Working Conference on Programming Concepts and Methods, New York, June 1998.*

R. L. Crole

[1993] *Categories for Types*, Cambridge University Press, Cambridge.

H. B. Curry and R. Feys

[1958] *Combinatory Logic, Volume I*, North-Holland, Amsterdam.

O.-J. Dahl, E. W. Dijkstra, and C. A. R. Hoare

[1972] *Structured Programming*, Academic Press, London, New York.

A. J. Demers and J. Donahue

[1980] Type-completeness as a language principle., in: *Proceedings of the Seventh Annual ACM Symposium on Principles of Programming Languages*, Association for Computing Machinery, New York, pp. 234–244.

E. W. DIJKSTRA

[1968] A constructive approach to the problem of software correctness, *BIT*, 8, pp. 174–186.

M. A. E. DUMMETT

[1981] *Frege: Philosophy of Mathematics*, Harvard University Press, Cambridge and Duckworth, London, 2nd ed.

P. DYBJER

[1994] Inductive families, *Formal Aspects of Computing*, 6, pp. 1–26.

A. EDALAT

[1994] Domain theory and integration, in: *Proceedings of the Ninth Annual Symposium on Logic in Computer Science*, IEEE Computer Society, Los Alamitos, California, pp. 115–124.

H. EGLI AND R. L. CONSTABLE

[1976] Computability concepts for programming language semantics, *Theoretical Computer Science*, 2, pp. 133–145.

W. M. FARMER

[1990] A partial functions version of Church's simple theory of types, *Journal of Symbolic Logic*, 55, pp. 1269–1291.

W. M. FARMER, J. D. GUTTMAN, AND F. J. THAYER

[1991] *IMPS: an interactive mathematical proof system*, Tech. Rep. M90-19, The MITRE Corp., Bedford, Massachusetts.

S. FEFERMAN

[1970] Formal theories for transfinite iterations of generalized inductive definitions and some subsystems of analysis, in: *Intuitionism and Proof Theory: Proceedings of the Summer Conference at Buffalo N.Y. 1968*, A. Kino, J. Myhill, and R. E. Vesley, eds., North-Holland, Amsterdam, pp. 303–326.

[1975] A language and axioms for explicit mathematics, in: *Algebra and Logic*, J. N. Crossley, ed., Lecture Notes in Mathematics #450, Springer-Verlag, Berlin, pp. 87–139.

M. B. FORESTER

[1993] *Formalizing Constructive Real Analysis*, Tech. Rep. TR93-1382, Computer Science Dept., Cornell University, Ithaca, New York.

G. FREGE

[1879] *Begriffsschrift, eine der arithmetischen nachgebildete Formelsprache des reinen Denkens*, Halle. English translation in van Heijenoort [1967], pp.1-82.

[1903] *Grundgesetze der Arithmetik, Begriffsschriftlich Abgeleitet*, Pohle, Jena. Reprinted 1962, Olms, Hildesheim.

H. M. FRIEDMAN

[1978] Classically and intuitionistically provably recursive functions, in: *Higher Set Theory*, D. S. Scott and G. H. Muller, eds., Lecture Notes in Mathematics #669, Springer-Verlag, Berlin, pp. 21–28.

H. M. FRIEDMAN AND A. SCEDROV

[1983] Set existence property for intuitionistic theories with countable choice, *Annals of Pure and Applied Logic*, 25, pp. 129–140.

J. H. GALLIER

[1993] Constructive logics. Part I: A tutorial on proof systems and typed λ-calculi, *Theoretical Computer Science*, 110, pp. 249–339.

G. GENTZEN

[1935] Untersuchungen über das logische Schliessen, *Mathematische Zeitschrift*, 39, pp. 176–210, 405–431. English translation in: Gentzen [1969], pp.68-131.

[1969] *Collected Papers of Gerhard Gentzen*, North-Holland, Amsterdam. Edited by M. E. Szabo.

J.-Y. GIRARD

[1987] *Proof Theory and Logical Complexity, Volume 1*, Bibliopolis, Napoli.

[1991] A new constructive logic: classical logic, *Mathematical Structures in Computer Science*, 1, pp. 255–296.

J.-Y. GIRARD, P. TAYLOR, AND Y. LAFONT

[1989] *Proofs and Types*, Cambridge Tracts in Computer Science, Vol. 7, Cambridge University Press.

K. GÖDEL

[1931] Über formal unentscheidbar Sätze der *principa mathematica* und verwandter Systeme i, *Monatshefte für Mathematik und Physik*, 38, pp. 173–198. Reprinted and translated in Gödel [1986,pp.144-195].

[1933] Zur intuitionistisched Arithmetic und Zahlentheorie, *Ergebnisse eines Mathematischen Kolloquiums*, 4, pp. 34–38. Reprinted and translated in Gödel [1986,pp.282-295].

[1986] *Collected Works, Volume 1*, Oxford University Press. Edited by S. Feferman, J. W. Dawson., S. C. Kleene, G. H. Moore, R. M. Solovay, J. van Heijenoort.

D. GOOD

[1985] Mechanical proofs about computer programs, in: *Mathematical Logic and Programming Languages*, C. A. R. Hoare and J. C. Shepherdson, eds., Prentice-Hall, Englewood Cliffs, New Jersey, pp. 55–75.

M. GORDON AND T. MELHAM

[1993] *Introduction to HOL*, University Press, Cambridge.

M. GORDON, R. MILNER, AND C. WADSWORTH

[1979] *Edinburgh LCF: a mechanized logic of computation*, Lecture Notes in Computer Science #78, Springer-Verlag, Berlin.

T. G. GRIFFIN

[1988a] Notational definition - a formal account, in: *Proceedings of the Third Annual Symposium on Logic in Computer Science*, IEEE Computer Society, Los Alamitos, California, pp. 372–383.

[1988b] *Notational Definition and Top-Down Refinement for Interactive Proof Development Systems*, PhD thesis, Cornell University.

C. A. GUNTER

[1992] *Semantics of Programming Languages: Structures and Techniques*, Foundations of Computing Series, MIT Press, Cambridge, Massachusetts.

E. GUNTER

[1994] *Studying the ML Module System in HOL90.* Presented at Logic & Computation Seminar, Univ. of Pennsylvania, December.

J. GUTTAG, J. HORNING, AND J. M. WING

[1985] *Larch in five easy pieces*, Tech. Rep. 5, Digital Equipment Corporation, Systems Research Center, July.

R. HARPER

[1992] Constructing type systems over an operational semantics, *Journal of Symbolic Computing*, 14, pp. 71–84.

R. HARPER, F. HONSELL, AND G. PLOTKIN

[1993] A framework for defining logics, *Journal of the Association for Computing Machinery*, 40, pp. 143–184.

J. HARTMANIS AND R. STEARNS

[1965] On the computational complexity of algorithms, *Transactions of the American Mathematics Society*, 117, pp. 285–306.

J. VAN HEIJENOORT

[1967] *From Frege to Gödel: A sourcebook in mathematical logic, 1879-1931*, Harvard University Press.

L. HELMINK

[1992] *Tools for Proofs and Programs*, PhD thesis, Universiteit van Amsterdam, The Netherlands.

T. A. HENZINGER AND P.-H. HO

[1994] Model checking strategies for hybrid systems, in: *Seventh International Conference on Industrial and Engineering Applications of Artificial Intelligence and Expert Systems*, F. D. Anger, R. V. Rodriguez, and M. Ali, eds., Gordon and Breach, Langhorne, Pennsylvania.

J. J. HICKEY

[1996a] *Formal Abstract Data Types.* unpublished manuscript.

[1996b] Objects and theories as very dependent types, in: *Proceedings of FOOL 3*, July.

[1997] Nuprl-Light: An implementation framework for higher-order logics, in: *CADE-14: 14th International Conference on Automated Deduction*, W. McCune, ed., Lecture Notes in Computer Science #1249, Springer-Verlag, Berlin, pp. 395–399.

D. HILBERT

[1926] Über das Unendliche, *Mathematische Annalen*, 95, pp. 161–190.

J. R. HINDLEY, B. LERCHER, AND J. P. SELDIN

[1972] *Introduction to Combinatory Logic*, Cambridge University Press, London.

C. A. R. HOARE

[1972] Notes on data structuring, in: *Structured Programming*, Academic Press, New York.

J. E. HOPCROFT AND J. D. ULLMAN

[1969] *Formal Languages and Their Relation to Automata*, Addison-Wesley, Reading, Massachusetts.

D. J. HOWE

[1987] The computational behaviour of Girard's paradox, in: *Proceedings of the Second Annual Symposium on Logic in Computer Science*, IEEE Computer Society, Washington, D.C., pp. 205–214.

[1989] Equality in lazy computation systems, in: *Proceedings of the Fourth Annual Symposium on Logic in Computer Science*, IEEE Computer Society, Washington, D.C., pp. 198–203.

[1991] On computational open-endedness in Martin-Löf's type theory, in: *Proceedings of the Sixth Symposium on Logic in Computer Sciience*, IEEE Computer Society, Los Alamitos, California, pp. 162–172.

[1996a] Importing mathematics from HOL into Nuprl, in: *Proceedings of the Ninth International Conference on Theorem Proving in Higher Order Logics*, J. von Wright, J. Grundy, and J. Harrison, eds., Lecture Notes in Computer Science #1125, Springer-Verlag, Berlin, pp. 267–281.

[1996b] Semantic foundations for embedding HOL in Nuprl, in: *Proceedings the Fifth International Conference on Algebraic Methodology and Software Technology (AMAST'96)*, M. Wirsing and M. Nivat, eds., Lecture Notes in Computer Science #1101, Springer-Verlag, Berlin, pp. 85–101.

G. HUET AND B. LANG

[1978] Proving and applying program transformations expressed with second-order patterns, *Acta Informatica*, 11, pp. 31–55.

J. M. E. HYLAND

[1982] The effective topos, in: *The Brouwer Centenary Symposium*, A. S. Troelstra and D. van Dalen, eds., North-Holland, Amsterdam, pp. 165–216.

S. IGARASHI, R. LONDON, AND D. LUCKHAM

[1975] Automatic program verification I: a logical basis and its implementation., *Acta Informatica*, 4, pp. 145–82.

P. B. JACKSON

[1994a] *Enhancing the Nuprl Proof Development System and Applying it to Computational Abstract Algebra*, PhD thesis, Cornell University, August. Forthcoming.

[1994b] Exploring abstract algebra in constructive type theory, in: *CADE-12: International Conference on Automated Deduction*, A. Bundy, ed., Lecture Notes in Computer Science #814 (Lecture Notes in Artificial Intelligence), Springer-Verlag, Berlin, June, pp. 590–604.

[1994c] *The Nuprl Proof Development System, Version 4.1 Reference Manual and User's Guide*, Cornell University, Ithaca, New York, February.

R. D. JENKS AND R. S. SUTOR

[1992] *Axiom: The Scientific Computation System*, Springer-Verlag, Berlin.

A. JOYAL AND I. MOERDIJK

[1995] *Algebraic Set Theory*, Cambridge University Press, Cambridge.

A. KENNY

[1995] *Frege*, Penguin Books, London.

D. E. KNUTH

[1984] *The TEXbook*, Addison-Wesley, Reading, Massachusetts.

D. KOZEN

[1977] *Complexity of Finitely Presented Algebras*, PhD thesis, Computer Science Department, Cornell University, Ithaca, New York.

D. KOZEN AND J. TIURYN

[1990] Logics of programs, in: *Handbook of Theoretical Computer Science, Volume B*, J. van Leeuwen, ed., North Holland, Amsterdam, pp. 789–840.

G. KREISEL

[1959] Interpretation of analysis by means of constructive functionals of finite type, in: *Constructivity in Mathematics: Proceedings of the Colloquium held at Amsterdam, 1957*, A. Heyting, ed., North-Holland, Amsterdam, pp. 101–128.

[1981] Neglected possibilities of processing assertions and proofs mechanically: choice of problems and data, in: *University-Level Computer-Assisted Instruction at Stanford: 1968-1980*, P. Suppes, ed., Institute for Mathematical Studies in the Social Sciences, Stanford University, pp. 131–147.

C. KREITZ

[n.d.] Program synthesis, in: *Automated Deduction – A Basis for Applications*, W. Bibel and P. Schmitt, eds., Kluwer Academic Publishers, Dordrecht, Boston, ch. III.2.10. To appear.

C. KREITZ, M. HAYDEN, AND J. J. HICKEY

[n.d.] *A Proof Environment for the Development of Group Communication Systems.* to appear.

S. KRIPKE

[1965] Semantical analysis of intuitionistic logic I, in: *Formal Systems and Recursive Functions*, J. N. Crossley and M. A. E. Dummett, eds., North-Holland, Amsterdam, pp. 92–130.

J. LAMBEK AND P. J. SCOTT

[1986] *Introduction to Higher-Order Categorical Logic*, vol. 7 of Cambridge Studies in Advanced Mathematics, Cambridge University Press, Cambridge, UK.

H. LÄUCHLI

[1970] An abstract notion of realizability for which intuitionistic predicate calculus is complete, in: *Intuitionism and Proof Theory: Proceedings of the Summer Conference at Buffalo N.Y. 1968*, A. Kino, J. Myhill, and R. E. Vesley, eds., North-Holland, Amsterdam, pp. 227–34.

D. LEIVANT

[1994a] A foundational delineation of poly-time, *Information and Computation*, 110, pp. 391–420.

[1994b] Predicative recurrence in finite type, in: *Logical Foundations of Computer Science*, A. Nerode and Yu. V. Matijacevič, eds., Lecture Notes in Computer Science #813, Springer-Verlag, Berlin, pp. 227–239.

[1995] Ramified recurrence and computational complexity I: Word recurrence and polynomial time, in: *Feasible Mathematics II, Perspectives in Computer Science*, P. Clote and J. B. Remmel, eds., Birkhäuser, Boston.

B. LISKOV AND J. GUTTAG

[1986] *Abstraction and Specification in Program Development*, MIT Press, Cambridge, Massachusetts.

C. L. LIU

[1985] *Elements of Discrete Mathematics*, McGraw-Hill, New York, 2nd ed.

Z. LUO

[1994] *Computation and Reasoning, A Type Theory for Computer Science*, Oxford University Press.

D. MACKENZIE

[1995] The automation of proof: A historical and sociological exploration, *IEEE Annals of the History of Computing*, 17, pp. 7–29.

S. MACLANE AND I. MOERDIJK

[1992] *Sheaves in Geometry and Logic, a First Introduction to Topos Theory*, Springer-Verlag, Berlin.

A. A. MARKOV

[1949] On the representation of recursive functions (Russian), *Izvestiya Akad. Nauk SSSR. Ser. Mat.*, 13, pp. 417–424. English translation: American Mathematical Society Translation 54 (1950) 13 pp.

P. MARTIN-LÖF

[1982] Constructive mathematics and computer programming., in: *Proceedings of the Sixth International Congress on Logic, Methodology, and Philosophy of Science*, L. J. Cohen, J. Łoś, H. Pfeiffer, and K.-P. Podewski, eds., North-Holland, Amsterdam, pp. 153–75.

[1983] *On the Meaning of the Logical Constants and the Justification of the Logical Laws.* Lectures in Siena.

P. MARTIN-LÖF

[1984] *Intuitionistic Type Theory, Studies in Proof Theory, Lecture Notes*, Bibliopolis, Napoli.

D. A. MCALLESTER

[1989] *ONTIC: A Knowledge Representation System for Mathematics*, MIT Press, Cambridge, Massachusetts.

J. McCarthy

[1963] A basis for a mathematical theory of computation, in: *Computer Programming and Formal Systems*, P. Braffort and D. Hirschberg, eds., North-Holland, Amsterdam, pp. 33–70.

J. McCarthy et al.

[1962] *Lisp 1.5 Users Manual*, Mit Press, Cambridge, Massachusetts.

N. P. Mendler, R. L. Constable, and P. Panangaden

[1986] Infinite Objects in Type Theory, in: *Proceedings of the First Annual Symposium on Logic in Computer Science*, IEEE Computer Society, Washington, D.C., pp. 249–257.

P. F. Mendler

[1988] *Inductive Definition in Type Theory*, PhD thesis, Cornell University, Ithaca, New York.

A. R. Meyer and M. B. Reinhold

[1986] Type is not a type, in: *Proceedings of the 13th Annual ACM Symposium on Principles of Programming Languages*, Association for Computing Machinery, New York, pp. 287–295.

B. Meyer

[1988] *Object-oriented software construction*, Prentice-Hall, Englewood Cliffs, New Jersey.

R. Milner, M. Tofte, and R. Harper

[1991] *The Definition of Standard ML*, MIT Press, Cambridge, Massachusetts.

R. Mines, F. Richman, and W. Ruitenburg

[1988] *A Course in Constructive Algebra*, Springer-Verlag, Berlin.

J. C. Mitchell

[1996] *Foundations for Programming Languages*, MIT Press, Cambridge, Massachusetts.

I. Moerdijk and G. E. Reyes

[1991] *Models for Smooth Infinitesimal Analysis*, Springer-Verlag, Berlin.

C. Murthy

[1990] *Extracting Constructive Content for Classical Proofs*, PhD thesis, Cornell University, Dept. of Computer Science. (TR 89-1151).

[1991] An evaluation semantics for classical proofs, in: *Proceedings of the Sixth Symposium on Logic in Computer Science*, IEEE Computer Society, Los Alamitos, California, pp. 96–109.

[1992] A computational analysis of Girard's translation and LC, in: *Proceedings of the Seventh Symposium on Logic in Computer Science*, IEEE Computer Society, Los Alamitos, California, pp. 90–101.

R. P. Nederpelt, J. H. Geuvers, and R. C. D. Vrijer

[1994] eds., *Selected Papers in Automath*, Studies in Logic and the Foundations of Mathematics #133, Elsevier Science Publishers, Amsterdam.

G. Nelson and D. Oppen

[1979] Simplification by cooperating decision procedures, *ACM Transactions on Programming Languages and Systems*, 1, pp. 245–257.

R. J. Nelson

[1968] *Introduction to Automata*, John Wiley & Sons, Inc., New York.

A. Nerode and R. A. Shore

[1994] *Logic for Applications*, Springer-Verlag, Berlin.

B. Nordstrom, K. Petersson, and J. M. Smith

[1990] *Programming in Martin-Löf's Type Theory*, Clarendon Press, Oxford University Press.

J. O'LEARY, M. LEESER, J. J. HICKEY, AND M. AAGAARD
[1995] Non-restoring integer square root: A case study in design by principled optimization, in: *Theorem Provers in Circuit Design: Second International Conference (TPCD '94)*, T. Kropf and R. Kumar, eds., Lecture Notes in Computer Science #901, Springer-Verlag, Berlin, pp. 52–71.

S. OWRE, S. RAJAN, J. M. RUSHBY, N. SHANKAR, AND M. SRIVAS
[1996] PVS: Combining specification, proof checking, and model checking, in: *CAV'96: Proceedings of the Eighth International Conference on Computer Aided Verification*, R. Alur and T. A. Henzinger, eds., Lecture Notes in Computer Science #1102, Springer-Verlag, Berlin, pp. 411–414.

S. OWRE, J. M. RUSHBY, AND N. SHANKAR
[1992] PVS: A prototype verification system, in: *CADE-11: International Conference on Automated Deduction*, D. Kapur, ed., Lecture Notes in Computer Science #607 (Lecture Notes in Artificial Intelligence), Springer-Verlag, Berlin, pp. 748–752.

E. PALMGREN
[1991] *On Fixed Point Operators, Inductive Definitions and Universes in Martin-Löf's Type Theory*, PhD thesis, Uppsala University, March.
[1995a] The Friedman translation for Martin-Löf's type theory, *Mathematical Logic Quarterly*, 41, pp. 314–326.
[1995b] *A sheaf-theoretic foundation for nonstandard analysis*, Tech. Rep. 1995:43, Department of Mathematics, Uppsala University.

C. PAULIN-MOHRING
[1989] Extracting F_ω'*s* programs from proofs in the calculus of constructions, in: *Proceedings of the Sixteenth Annual ACM Symposium on Principles of Programming Languages*, Association for Computing Machinery, New York, pp. 89–104.

C. PAULIN-MOHRING AND B. WERNER
[1993] Synthesis of ML programs in the system Coq, *Journal of Symbolic Computation*, 15, pp. 607–640.

L. C. PAULSON
[1993] Set theory for verification: I from foundations to functions, *Journal of Automated Reasoning*, 11, pp. 353–389.
[1994] *Isabelle: A Generic Theorem Prover*, Lecture Notes in Computer Science #828, Springer-Verlag, Berlin.

A. M. PITTS
[1987] Polymorphism is set-theoretic, constructively, in: *Category Theory and Computer Science*, D. H. Pitt, A. Poigne, and D. E. Rydeheard, eds., Lecture Notes in Computer Science #283, Springer-Verlag, Berlin, pp. 12–39.

G. PLOTKIN
[1975] Call-by-name, call-by-value, and the λ-calculus, *Theoretical Computer Science*, pp. 125–59.
[1977] LCF considered as a programming language, *Theoretical Computer Science*, 5, pp. 223–255.
[1981] *A Structural Approach to Operational Semantics*, Tech. Rep. DAIMI-FN-19, Computer Science Department, Aarhus University.

E. POLL
[1994] *A Programming Logic Based on Type Theory*, PhD thesis, Technische Universiteit Eindhoven.

R. POLLACK
[1995] *The Theory of LEGO: A Proof Checker for the Extended Calculus of Constructions*, PhD thesis, Department of Computer Science, University of Edinburgh, April.

D. Prawitz
[1965] *Natural Deduction*, Almquist and Wiksell, Stockholm.

W. Quine
[1960] *Word and Object*, MIT Press, Cambridge, Massachusetts.

T. Reps
[1982] *Generating Language-Based Environments*, PhD thesis, Cornell University, Ithaca, New York.

T. Reps and T. Teitelbaum
[1988] *The Synthesizer Generator Reference Manual*, Springer-Verlag, Berlin, third ed.

J. C. Reynolds
[1974] Towards a theory of type structure., in: *Programming Symposium*, B. Robinet, ed., Lecture Notes in Computer Science #19, Springer-Verlag, Berlin, pp. 408–425.
[1981] The essence of Algol, in: *International Symposium on Algorithmic Languages*, J. de Bakker and J. van Vliet, eds., North-Holland, Amsterdam, pp. 345–372.

A. Rezus
[1985] *Semantics of Constructive Type Theory*, Tech. Rep. 70, Informatics Dept., Nijmegen University, September.

H. Rogers
[1967] *Theory of Recursive Functions and Effective Computability*, McGraw-Hill, New York.

B. Russell
[1908] *The Principles of Mathematics*, Cambridge University Press, Cambridge.

D. S. Scott
[1970a] Constructive validity, in: *Symposium on Automatic Demonstration*, M. Laudet, D. Lacombe, L. Nolin, and M. Schützenberger, eds., Lecture Notes in Mathematics #125, Springer-Verlag, Berlin, pp. 237–275.
[1970b] Outline of a mathematical theory of computation, in: *Proc. 4th Annual Princeton Conf. on Information Sciences & Systems*, Princeton, New Jersey, pp. 169–176.
[1972] Lattice theoretic models for various type-free calculi., in: *Proceedings of the Fourth International Congress on Logic and Methodology and Philosophy of Science*, R. J. Bogdan and I. Niiniluoto, eds., D. Reidel, Dordrecht, pp. 157–87.
[1976] Data types as lattices., *SIAM Journal on Computing*, 5, pp. 522–87.

R. A. G. Seely
[1987] Categorical semantics for higher order polymorphic lambda calculus, *Journal of Symbolic Logic*, 52, pp. 969–989.

A. Setzer
[1993] *Proof theoretical strength of Martin-Löf Type Theory with W-type and one universe*, PhD thesis, Ludwig-Maximilians-Universität, München, September.

N. Shankar
[1994] *Metamathematics, Machines, and Gödel's Proof*, Cambridge University Press, Cambridge, Massachusetts.

R. E. Shostak
[1979] A practical decision procedure for arithmetic with function symbols, *Journal of the Association for Computing Machinery*, 26, pp. 351–360.

J. Siekmann and G. Wrightson
[1983] *Automation of Reasoning*, vol. 1, Springer-Verlag, Berlin.

B. C. SMITH

[1984] Reflection and semantics in LISP, in: *Proceedings of the Eleventh Annual ACM Symposium on Principles of Programming Languages*, Association for Computing Machinery, Washington, D.C., pp. 23–35.

J. M. SPIVEY

[1989] *The Z Notation*, Prentice-Hall, Englewood Cliffs, New Jersey.

S. STENLUND

[1972] *Combinators, λ-Terms, and Proof Theory.*, D. Reidel, Dordrechte.

W. P. VAN STIGT

[1990] *Brouwer's Intuitionism*, North-Holland, Amsterdam.

W. W. TAIT

[1967] Intensional interpretation of functionals of finite type, *Journal of Symbolic Logic*, 32, pp. 189–212.

[1983] Against intuitionism: Constructive mathematics is part of classical mathematics, *Journal of Philosophical Logic*, 12, pp. 173–195.

A. TARSKI

[1956] The concept of truth in formalized languages, Clarendon Press, Oxford, pp. 152–278. Translation of 1933 paper in Polish.

S. THOMPSON

[1991] *Type Theory and Functional Programming*, Addison-Wesley, Reading, Massachusetts.

A. S. TROELSTRA

[1973] *Metamathematical Investigation of Intuitionistic Mathematics*, Lecture Notes in Mathematics #344, Springer-Verlag, Berlin.

A. S. TROELSTRA AND D. VAN DALEN

[1988] *Constructivism in Mathematics, An Introduction, Vol. I,II*, North-Holland, Amsterdam.

A. S. TROELSTRA AND H. SCHWICHTENBERG

[1996] *Basic Proof Theory*, Cambridge University Press.

A. TRYBULEC

[1983] On a system of computer-aided instruction of logic, *Bulletin of the Section of Logic*, 12, pp. 214–220. Polish Academy of Science.

A. M. UNGAR

[1992] *Normalization, Cut-elimination and the theory of proofs,* Lecture Notes 28, CLSI, Stanford.

F. WATTENBERG

[1988] Nonstandard analysis and constructivism, *Studia Logica*, 47, pp. 303–309.

R. WEYRAUCH

[1980] Prolegomena to a theory of formal reasoning, *Artificial Intelligence*, 13, pp. 133–170.

A. N. WHITEHEAD AND B. RUSSELL

[1925-27] *Principia Mathematica*, vol. 1, 2, 3, Cambridge University Press, 2nd ed.

L. WITTGENSTEIN

[1922] *Tractatus Logico-Philosophicus*, Routledge & Kegan Paul, London.

[1953] *Philosophical Investigations*, Basil Blackwell, Oxford.

S. WOLFRAM

[1988] *Mathematica: A System for Doing Mathematics by Computer*, Addison-Wesley, Reading, Massachusetts.

L. Wos, R. Overbeek, E. Lusk, and J. Boyle
[1984] *Automated Reasoning*, Prentice–Hall, Englewood Cliffs, New Jersey.

R. Zippel
[1993] *Effective Polynomial Computation*, Kluwer Academic Publishers, Dordrecht, Boston.

J. I. Zucker
[1974] The correspondence between cut-elimination and normalization, *Annals of Mathematical Logic*, 7, pp. 1–112. Errata *ibid* 7, p.156.
[1977] Formalization of classical mathematics in Automath, in: *Colloque International de Logique*, Colloques Internationaux du Centre National de la Recherche Scientifique, CNRS, Paris, pp. 135–145.

Name Index

Subject Index

Kolibri 360 GmbH
Ferdinand-Jühlke-Straße 7
99095 Erfurt, Deutschland
produktsicherheit@kolibri360.de